ISC

2022 EXAMINATION

10 YEARS SOLVED PAPERS

SCIENCE STREAM

CLASS XII

- ENGLISH (PAPER - I)
- ENGLISH (PAPER - II)
- HINDI
- PHYSICAL EDUCATION
- MATHEMATICS
- COMPUTER SCIENCE
- PHYSICS
- CHEMISTRY
- BIOLOGY
- ENVIRONMENTAL SCIENCE

DISCLAIMER

With the ambition of providing standard academic resources, we have exercised extreme care in publishing the content. In case of any discrepancies in the matter, we request readers to excuse the unintentional lapse and not hold us liable for the same. Suggestions are always welcome.

EDITION : YEAR 2021

ISBN : 978-93-91184-44-5

PUBLISHED BY

GURUKUL BOOKS & PACKAGING

(Unit of Oswal Printers & Publishers Pvt. Ltd.)

 1/12, Sahitya Kunj, M.G. Road, Agra - 282002

 (0562) 2527771-4, +91 7534077222

 info@oswalpublishers.in

 www.oswalpublishers.com

The cover of this book has been designed using resources from Freepik.com

In accordance with the latest syllabus prescribed by the Council for the Indian School Certificate Examinations, New Delhi

PREFACE

- The present volume titled 'ISC 10 YEARS SOLVED PAPERS' is an effort to cater to the needs of the students preparing for the board examinations of ISC class 12th.
- This volume has been structured in a manner so that students get a fair idea of the type and style of questions asked in the examinations. It will acquaint them with all the facets and dimensions of the subjects. Despite the dynamic pattern of the ISC exam papers, extensive practice of last years papers will instill a sense of confidence amongst the students.
- A team of experienced academicians have made sincere efforts to provide the best possible answers in a lucid way so that students may overcome their doubts. Time-bound paper solving, acts like a stimulator in preparing for the exams. It also gives an idea about the important topics of the syllabus that are often asked in the exams. Since all the answers have been written in accordance with the Board Marking Scheme, the students get to learn the keywords that the Board is looking for in their answers by going through these solved papers.
- Lastly, we hope that the present volume, patterned on the questions set by the Council, will go a long way in preparing the students for the examinations. This learning aid will definitely help them to be in a positive frame of mind to face the examinations confidently.

FINDING THE RIGHT BIN FOR MASKS?

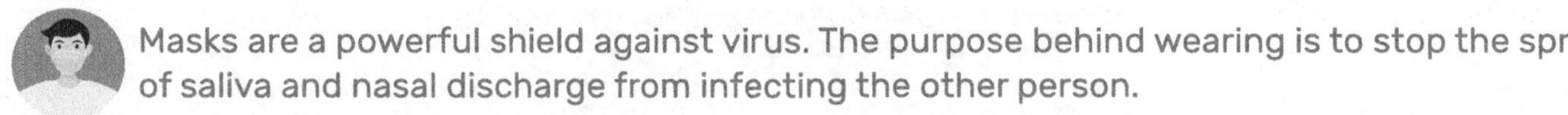

Masks are a powerful shield against virus. The purpose behind wearing is to stop the spread of saliva and nasal discharge from infecting the other person.

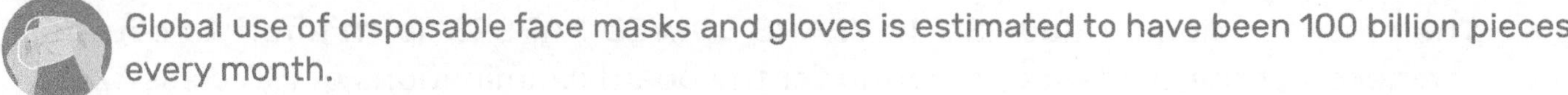

Global use of disposable face masks and gloves is estimated to have been 100 billion pieces every month.

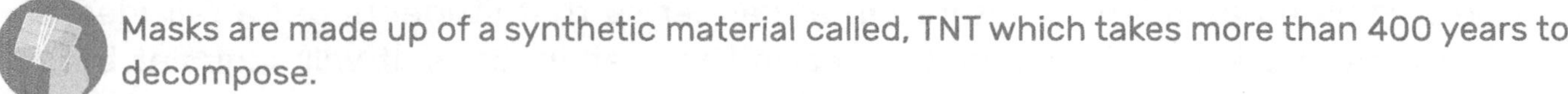

Masks are made up of a synthetic material called, TNT which takes more than 400 years to decompose.

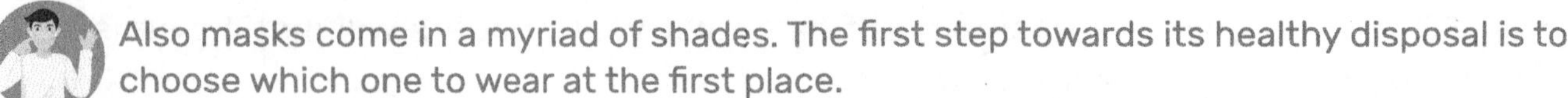

Also masks come in a myriad of shades. The first step towards its healthy disposal is to choose which one to wear at the first place.

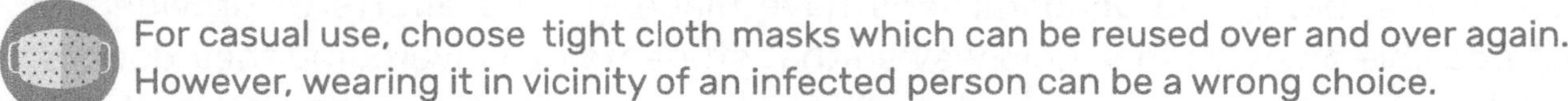

For casual use, choose tight cloth masks which can be reused over and over again. However, wearing it in vicinity of an infected person can be a wrong choice.

When visiting hospitals and other contaminated zones, choose loose fitting surgical masks which blocks droplets and splashes of saliva. However, they are ineffective against very small particles that come off sneezing and coughing in close vicinity.

N95 Masks are not for routine use and must be spared for medical experts who are in direct contact with the patient.

When it comes to disposing a mask, the very first step is to wash your hands before removing it from your face.

Wash your hands for atleast 20 seconds, using either an alcohol based sanitizer (having 60% alcohol) or using soap and water.

Remove your mask either using the mask strings or downside up. Don't touch the center of mask.

If you are wearing a cloth mask, put it separately into wash without mixing it with other clothes.

In case of a surgical mask, fold it inwards such that the droplets are not exposed to the air. Make it into a roll and wrap it up in two polythene bags or tissue bag, one inside the other.

Prefer cloth masks for casual use and avoid one time used masks.

CONTENTS

STUDY LOG

TOPICS I'VE MASTERED

TOPICS I'VE TO WORK ON

PRACTICE STRATEGY

TIPS

1. Divide lengthy topics into smaller chunks and work on understanding one sub-topic at a time.
2. At the end of each study session, make short notes for quick revision.
3. Revisit previously learnt topics regularly from self-made notes.
4. Try to understand the topic than just mugging it up. It will stay in your memory for longer.

YOGA, MEDITATION AND MINDFULNESS

Yoga has ailing nature and as a result, it brings numerous benefits to life. It helps us grow into a delightful personality. Yoga is essential for Teenagers as they come across many hormonal and physical changes in their mind & body during this time. These subsequent changes need powerful physical exercises and intellectual moral inputs to ensure healthy growth.

• **Tadasana :** Many times while studying we forget to sit upright by leaning forward or backward. This can cause serious issues like neck or back strains. Tadasana will help you improve your body posture by strengthening the knees & thighs, helps by increasing height, and regulates the digestive, nervous, and respiratory systems.

• **Uttanasana :** Stress is something everyone has to encounter during their school days. There are an uncountable amount of worries that one confronts during this time period. Uttanasana helps the students by calming their minds and by relieving their anxiety and stress. It relieves headaches, cures insomnia, improves digestion, and activates the kidneys.

• **Adho Mukha Savasana:** Working mentally and physically our bodies sometimes crave rejuvenation and energy. Just like our brain, our body also needs to be re-charged and re-energized to work effectively. Adho Mukha Svanasana energizes and rejuvenates the body by relieving mild depression and cures back pain and fatigue. This pose strengthens the bones and also increases lung capacity by strengthening the chest muscles.

Meditation : There are three types of meditation practices that can prove to be useful in your student's life. So, let's find out which one works best for you.

• **Focused Meditation :** This meditation helps you increase focus by paying all your attention to a point, a flickering flame, or your breaths. As its name suggests focussed meditation helps you improve your focus and learning.

• **Mindfulness Meditation :** Do you like to practice alone? Or don't have an access to a teacher. In this practice You simply allow your mind to flow with the environment around you. This exercise helps you in becoming keenly aware of the sights, smells, and sounds around you. Hence making you more attentive, positive, and full of life.

• **Movement Meditation :** Are you action-oriented? Or have trouble sitting at a place? If yes, movement meditation can prove to be very useful as it helps you to focus on yourself and your body. You move your body in simple exercising postures by focusing on your breath and by stretching your muscles. Which distracts you from external objects or sounds.

Lastly, we need to know that Great things come to those who know "Who they are and What They Want". The courage to take the path less traveled comes from within and so, meditation and mindfulness together prepare you into a great warrior in the field of life.

So, Work hard and let your results hustle. Good luck.

ENGLISH-I

2020

(Maximum Marks : 100)
(Time allowed : Three hours)
*(Candidates are allowed **additional 15 minutes** for **only** reading the paper.*
They must NOT start writing during this time.)
*Attempt all **four** questions*
The intended marks for questions or parts of questions are given in brackets [ ].
*(You are advised to spend not more than **45** minutes on Question **1**, **55** minutes on Question **2**, **30** minutes on Question **3** and **50** minutes on **Question 4**).*
(You should begin each answer on a fresh page.)

QUESTIONS

Question 1.

*Write a composition (in approximately 400–450 words) on **any one** of the following subjects :* **[25]**

(You are reminded that you will be rewarded for orderly and coherent presentation of material, use of appropriate style and general accuracy of spelling, punctuation and grammar.)

(a) *Recently, you travelled to a place you found peaceful and rewarding. Describe the place, the weather, the people and the local activities that make this destination particularly attractive.*

(b) *Narrate an experience when you took on a responsibility in your school that you usually would not, and you were pleasantly surprised at the outcome.*

(c) *"If winter comes, can spring be far behind?" Express your views on this statement.*

(d) *Solitude.*

(e) *Students should be allowed to use electronic gadgets in the classroom. Argue for or against this proposition.*

(f) *Write an original short story begins with the line :*
The lone young man who sat hunched on the park bench had an unusual gleam in his eyes.

Question 2.

(a) *As the Secretary of the Creative Arts Club of XYZ School, you organised an inter-school drama competition. Write a report for the school magazine in not more than 300 words, detailing the success of the event based on the following points :* **[20]**
Date, time and place—objective of the event—inauguration—schools involved—judges invited—response of participants—reaction of audience—plan for similar events in future.

(b) *As a member of the Student Council, you have been given the responsibility of arranging for a career guidance workshop. Write a proposal in not more than 150 words, stating the steps you would take to make the workshop a success.* **[10]**

Question 3.

*Answer sections **(a)**, **(b)** and **(c)**.*

(a) *In each of the following items, sentence **A** is complete, while sentence **B** is not. Complete sentence **B**, making it as similar as possible to sentence A. Write sentence B in each case.* **[10]**

Example : (0) (A) As soon as the bell rings, the children run out to play.
(B) No sooner ....
Answer : (0) No sooner does the bell ring than the children run out to play.

(1) *(A) When he returned, we asked him many questions.*
(B) On his ....

(2) *(A) He arrived and the crowd went hysterical with joy.*
(B) Scarcely ....

(3) *(A) The bus service has been interrupted by floods.*
(B) Floods ....

(4) *(A) This scenery is breath-taking.*
(B) How ....

(5) *(A) You are entitled to a discount as long as the offer lasts.*
(B) So long ....

(6) *(A) He is so tall that he cannot enter through this door.*
(B) He is too ....

(7) *(A) The doctor said to him, "Drink at least three litres of water every day."*
(B) The doctor advised ....

(8) *(A) Sunita is the best debater in our team.*
(B) No other ....

(9) *(A) Not only did he score a century but also took three important wickets.*
(B) Besides ....

(10) (A) *I am sorry I am unable to accompany you.*

(B) *I regret my ....*

(b) *Fill in each blank with a suitable word. (Do not write the sentence.)* **[5]**

(1) *The antique furniture was sold ________ an auction.*

(2) *The tickets were sold ________ within an hour.*

(3) *I realised I had walked ________ a trap, but it was too late.*

(4) *She had walked ________ two hours before she realised she was lost.*

(5) *Can he take ________ the additional responsibility ?*

(6) *She takes ________ her mother in all her mannerisms.*

(7) *Neha turned ________ a new leaf after the incident.*

(8) *I did not expect so many people to turn ________ for the conference.*

(9) *The students brought ________ the first issue of their magazine.*

(10) *Mowgli was brought ________ by wolves in the jungle.*

(c) *Fill in the blanks in the passage given below with the appropriate form of the verb given in brackets. Do not write the passage but write the verbs in the correct order.* **[5]**

When Albert Einstein ________ (1) (be) three years old, his parents ________ (2) (be) very worried. He ________ (3) (seem) intelligent, but had not spoken a single word. They ________ (4) (take) him to many doctors who ________ (5) (say) that they ________ (6) (find) nothing wrong with the child. One day, at the dinner table, Albert suddenly said, "The soup ________ (7) (be) too hot." The shocked parents ________ (8) (ask) him why he ________ (9) (remain) silent for so long. The young child solemnly replied, "Everything ________ (10) (be) in order until now."

Question 4 .

Read carefully the passage given below and answer the questions (a), (b) and (c) that follow :

(1) My job as the District Veterinarian in Darrowby, Yorkshire has thrown up some interesting moments. I was driving up to Pawson's sheep farm one morning, when I saw this rare sight.

(2) I suppose it isn't unusual to see a man pushing a pram in a town, but on a lonely moorland road this scene merits a second glance. Especially when the pram contains a large gray dog. I drew up beside them and stopped.

(3) I looked at the pram, ancient and rusty, and at the big animal sitting upright inside it. He was a cross-bred greyhound, and he gazed back at me with unruffled dignity. I couldn't help but admire his healthy and happy disposition.

(4) "Nice dog," I said.

(5) "Aye, that's Jake." The man smiled, introducing himself as Roddy.

(6) That same evening, about eight o'clock, the doorbell rang. I answered it and found Roddy on the front doorstep. Behind him, stood the ubiquitous pram. One look at the prone dog warned me that something was terribly wrong.

(7) I threw the door wide open. "Bring him in."

(8) I grabbed the animal round the **<u>middle</u>** and we lifted him onto the table. I watched in disbelief as the huge form lay there. There was no fight for breath, he was unconscious. His pulse was rapid and feeble, yet he didn't breathe.

(9) Understanding the symptoms, I said, "Roddy, he is choking. I'm going to have a look at his throat."

(10) I pushed Jake's jaws apart, depressed his tongue with a forefinger, and shone my torch into the depths. He was the kind of good-natured dog who offered no resistance as I prodded around. I shall always be thankful that at that very instant the dog coughed, opening up the cartilages of the larynx and giving me a glimpse of the cause of all the trouble. There, beyond the drooping epiglottis, I saw for a fleeting moment a smooth round object no bigger than a pea.

(11) "I think it's a pebble," I gasped. "Right inside his larynx. I'll have to get it out."

(12) I seized a pair of scissors and clipped away the hair from the ventral surface of the larynx. I dared not use a general anesthetic, and therefore I infiltrated the area with local before swabbing with antiseptic.

(13) "Hold his head steady," I said hoarsely, and gripped a scalpel. I cut down through skin, fascia, and the thin layers of the muscle until the ventral surface of the larynx was revealed.

(14) And there it was. A pebble right enough—gray and glistening and tiny, but big enough to kill.

(15) I had to fish it out quickly and cleanly without pushing it into the trachea. I leaned back and rummaged in the tray until I found some broad-bladed forceps, then I poised them over the **wound**. Great surgeons' hands, I felt sure, didn't shake like this.

(16) I clenched my teeth, **introduced** the forceps, and my hand magically steadied as I clamped them over the pebble.

(17) I didn't breathe at all as I bore the shining little object slowly and tenderly through the opening and dropped it with a gentle rat-tat on the table.

(18) "Is that it ?" asked Roddy, almost in a whisper.

(19) "That's it." I reached for a needle and suture silk. "All is **well** now."

(20) The stitching took only a few minutes and by the end of it, Jake was bright-eyed and alert, paws shifting impatiently, ready for anything.

(21) He seemed to know his troubles were over.

Adapted from : *James Herriot's Favorite Dog Stories*

(a) (i) *Given below are four words and phrases. Find the words which have a similar meaning in the passage :* [4]

(1) *deserves*
(2) *undisturbed*
(3) *being unable to breathe*
(4) *wet and shining*

(ii) *For each of the words given below, write a sentence of at least ten words using the same word **unchanged in form, but with a different meaning** from that which it carries in the passage :* [4]

(1) *middle (line 16)* (2) *wound (line 38)*
(3) *introduced (line 40)* (4) *well (line 45)*

(b) *Answer the following questions in your own words as briefly as possible :*

(i) *What was unusual about the pram on the moorland road ?* [2]

(ii) *How did the narrator know that the dog was choking ?* [2]

(iii) *Describe how the narrator found the cause of the choking ?* [3]

(iv) *What preparations did the narrator make before the surgery ?* [2]

(c) *Summarise how the narrator removed the pebble and saved Jake's life. (Paragraphs 11 to 19). You are required to write the summary in the form of a connected passage in about 100 words. Failure to keep within the word limit will be penalised.* [8]

ANSWERS

Answer 1.

(a) A TRIP TO LANSDOWNE

Recently my mother and I took a trip to Lansdowne. It is a small hill station nestled in the lap of Uttarakhand in the northern part of the country. This holiday was all the more special because it was the first time when me and my mother had gone together for a holiday. Before this it had always been the four of us – my father, mother, my brother and me. But my brother and father were busy this time, so we left on our own.

We hired a cab and left for our little holiday. It took us about six hours to reach Lansdowne from New Delhi. Our hotel was called Windsor Cottages. It was a beautiful and an old property with Uttarakhand's traditional interiors and architecture. It had a certain old-world charm and feel to it which attracts tourists from all parts of the world. It was surrounded on all sides by huge and mighty grey oak trees which dance to the quiet Himalayan breeze. I liked it all the more because it was located on the outskirts of the city; a place to relax and rejuvenate.

The weather was cool and there were cold winds blowing constantly which made us yearn for the sun and sit near the fireplace in the evenings. The staff of the hotel was hospitable and the local people generally were very warm to talk to.

On the first day, early in the morning, we set out for Tip N Top which is the sunrise point of the hill station. Then we had lunch at a famous café in the hill station. In the evening, we went to the Santoshi Maa temple which is also a famous sunset and sunrise point in the area. Then, we walked down to the Mall road. The local market is usually very less crowded. We then walked around the town in nature and our last destination in Lansdowne was the Garhwali museum which traces down the history of the Garhwali Rifles regiment. It was a treat to get to know so much about Indian Army and to see weapons and war artefacts.

All in all, it was a very relaxing trip and helped me get over the stress of my studies and daily life.

(b) RESPONSIBILITY

To be chosen to become the class monitor was one thing I had always hated and dreaded. It involves a lot of hard work and discipline which I thought I did not have. It also involves acting in a mature and responsible manner, being on your feet at all times and taking wise decisions for the entire class.

With the title of the monitor comes a long list of responsibilities. Some of these duties which are to be shouldered by a monitor include maintaining discipline and silence in the class in the teacher's absence, checking the uniform of all the students, distributing corrected notebooks and test papers. It involves running to the principal's office if a particular teacher does not turn up for a class, switching off the lights and fans if the students are not in class, taking care of the cupboard keys, and managing the stationery. It also requires you to be vigilant and help keep the classroom clean, update the notice boards and make all the relevant announcements.

With this long list of duties, it was always my opinion that I wouldn't make a good monitor as I might not be able to control the class or do multi-tasking. But my opinion changed last month when my class teacher made me the monitor! I couldn't say no to her, and all my friends encouraged me by saying that I would make a rather efficient and a good monitor whom everyone would listen to.

On the first day of my duty, I was scared and nervous. I collected everyone's homework, checked their uniforms and helped maintain silence when the teachers were not in class. Everything was more or less in order. This was when I thought to myself that everything was under control and being a monitor wasn't really as bad as I had imagined it to be. My class teacher was really proud of me by the way I was handling things. I was learning and improving each day.

At the end of the month, I thought about the entire experience. How scared I was at the beginning and how I always tried to escape this responsibility and where I am now. All my classmates respected and listened to me. I was in the good books of the teachers and also of my parents and what more I had learnt is to be more patient, responsible and wise. This really was a life-changing experience for me.

(c) "IF WINTER COMES CAN SPRING BE FAR BEHIND"

This is a classic line taken from the poem—*Ode to the West Wind* written by Percy Bysshe Shelley. It is not only a great classical metaphor but also a rhetorical question that the poet is asking all his readers.

Throughout the poem, the poet evokes the West wind and explains its trajectory throughout the year. This line symbolically represents the ever-changing nature of seasons in a cyclical way - from spring to winter and then back to spring. While on the one hand, winter stands for death, decay, and stillness in the atmosphere. Spring on the other hand stands for regeneration, vibrancy, and growth. Most plants and flowers, blossom in the spring season and shed their leaves during the onset of winter. Also, winter is always described by coldness in the air, hibernation of animals and an icy cover over the landscape. Spring on the contrary is represented by colourful flowers, lushness of plants, and a plethora of wildlife all around.

Metaphorically, the line has deeper connotations when connected to human life. It is the epitome of positive thinking. And a metaphor to explain that sadness is transitory and momentary and is always followed by happier times. This means that one should never lose hope during tough times and remember that happier days are sure to follow. Though it may seem like the days of despair are never-ending, but a life full of joy and comforts is always waiting at the end of the dark tunnel. We must always have a firm belief in the goodness of things and self-confidence to overcome the sad times and let ourselves in our happy place. Like the cycle of seasons continues unhindered, we must continue our lives unhindered by agony and misery.

In the poem, Shelley further connects this idea to his art of writing poetry in general. He hopes that his poems are like the leaves and when the wind blows, he hopes that these poems will land on people's minds and help to nurture new growth in terms of new ideas and fresh thoughts and drive away the negative and stale ones. He opines that even if it seems like his poems will die and be ignored or be left unread during a symbolic winter in literature, there will still be a 'spring' to follow soon where his poems will help create new intellectual or spiritual ideas in the future and among generations to come.

(d) SOLITUDE

Solitude refers to the state of being alone. Some people like being in this state while others dislike being in the same. Each person has his or her own opinion about solitude.

As we are aware, man is a social animal. Therefore, being in complete and total solitude at all times is not possible. We all have our family members, friends, colleagues, and neighbours that we talk to so that we can get through our day. Whether we like it or not some amount of human interaction takes place every day.

Some people enjoy other people's company and like to talk and mix up with others to share their feelings be it joys or sorrows. Solitude becomes a burden to such people as they don't know what to do in their own company. This is not always a good situation to be in. People like teachers, doctors, lawyers, actors, and sportsmen need people for their professional work as they cannot function in solitude.

On the other hand, some people loathe company and prefer being alone and on their own. These people do not like to talk to others about their feelings. They enjoy being with themselves and feel pressurised or stressed when too many people are around them. Scientists, philosophers, writers, sages, and monks like to live in solitude. They like to think and do their work in peace.

It is often argued that some amount of solitude is necessary for our brain to re-wire itself and to think clearly. It also helps to know ourselves better. Many people misinterpret solitude. A lot of people believe that they have to necessarily be with other people at all times. If you are always spending time with others, you usually do what they want to do or you do something that interests you. You may learn things that you enjoy doing—playing a sport, singing, cooking—but you never really learn about what your inner thoughts are.

As a matter of fact, solitude helps you believe in yourself and become independent. If you are always dependent on other people, you will not be able to deal with life's changes and problems. Solitude helps you rely on your own self and to utilise your independent capabilities.

(e) STUDENTS SHOULD BE ALLOWED TO USE ELECTRONIC GADGETS IN THE CLASSROOM

(I) FOR THE MOTION

Electronic gadgets allow instant access to knowledge and multitude of information. If students are allowed to use electronic gadgets in classrooms, they can find a lot of new ideas by expanding their capability in educational technology, in applying 21st century skills, building interpersonal communication skills, working more effectively with teams, and creating and innovating new ideas, concepts, and products.

According to Willie Mata of Centre Technologies, 'Readiness for the future mobility is currently the next great movement in the workplace, and students who use technology in the classroom today will be more adapted to using it in the future.' The National Science Board in the USA says that in 2013, President Obama announced the Connect ED initiative, pledging to connect 99% of American students to next-generation broadband and high-speed wireless in their schools and libraries by 2018. And with the help of the new ideas, the youth or students today can use their knowledge and skills for their future careers. The world will have a higher technology use in the future compared to today.

There is proven research on the fact that blended learning styles keep students focused for longer and makes them more excited to learn more, especially for STEM (Science, Technology, Engineering, and Math) subjects. Also, laptops, tablets, mobile phones and other digital devices can help the students enhance their interaction with their classmates and instructors by encouraging collaborative learning. It positively affects how students live, work, play, and most importantly, learn. With the integration of technology in education, students have the ability to learn at their own pace. They have the option to slow down and go back over lessons and concepts, and, if they are fast learners they can go ahead. It also encourages the teacher to help their students on a more one-on-one level.

So, schools should allow students to use their gadgets in the classroom because it contributes a lot to the students' ability, understanding, knowledge and skills, by having new features or a multitude of resources on the internet or apps which can be used in individual or group work. It makes students more advanced in their thinking and enhances what they can learn in the classrooms. Also, schools should let students have access to Wi-Fi connections in doing physical activities and group work. Technology is very significant, so it is crucial for the students' benefit from this opportunity to have an advanced knowledge and be better prepared for the future.

(II) AGAINST THE MOTION

Electronic gadgets such as cell phones or tablets should not be allowed in schools as they pose to be a major distraction for both students and teachers. For students, it will prove to be a bigger hindrance as they will not be able to keep up with the class if they are busy with their gadgets. For the teachers, it will pose to be a problem in the process of their teaching as the entire flow and focus of the class will get disturbed if any student is found fiddling with their phone. It has been found that on an average, students spend up to

20% of their time in their classrooms texting, emailing and checking their social media accounts. This leaves lesser time for the real studies.

Moreover, students might start browsing irrelevant content, inappropriate websites, and watch distracting videos and in the process waste their precious time which can be utilised for studies or extra-curricular activities.

There have also been cases where cell phones were used for cheating during tests as many such cases have been reported. Even if the student is not caught, it can lead to poor understanding of the concepts and lead it to becoming a bad habit.

In certain scenarios, phone cameras have been used for nasty and malicious purposes like for clicking inappropriate pictures or making videos that are inappropriate. This could hamper the overall growth, mindset, and personality of the students and have adverse effects on their physical and mental health.

Also, allowing gadgets in schools will give rise to an unhealthy, unwanted competition among students. This would not be healthy for the students and would encourage wrong practices such as bullying, teasing, and harassing other students. More so, cyberbullying can be harder to see than other forms of bullying, making it difficult for teachers to identify and stop when it is happening.

Not only would there be social and economic disparity among friends, but the poor parents would also suffer as they would be under constant pressure to get new and expensive phones for their children. It is for the same reason that each school has its separate and assigned uniforms which students are required to wear every day to school so that all students look the same without any biases.

(f) THE DISTRACTION

The lone young man who sat hunched on the park bench had an unusual gleam in his eyes. It was a rather chilly evening. The mighty grey oaks were casting soft shadows on the isolated house at the centre of the park. The sky was in a tinge of bluish grey which made the atmosphere dull, gloomy and morose.

But the man on the bench or William was rather fond of such evenings. He loved his long walks in solitude where he spent long hours thinking about his life which was like the evening itself. Suddenly he saw from the bench, an object which caused a distraction in his train of thought; a rather unusual looking creature peeking through one of the windows of the house. "Fancy seeing anybody in this part of the park and at this hour of the day," he muttered to himself.

At first, he tried to dismiss the sight for the shadow of a long creeper or that of old furniture kept inside the house. But the image from behind the window was getting clearer. It seemed like a short old man wearing a pointed cap. In a matter of no time, the old man started crying in pain and screaming in a strange language. William had no idea of what he was talking about. Fed up of being disturbed thus, William wore his woollen cap, adjusted his scarf, wore his gloves and began walking towards the house.

On entering the house, he knocked firstly. The screams had now stopped. He was confused and wondered if everything that had just happened was a part of his imagination. Nevertheless, he decided to enter the house. The door opened with a loud creak. William hated creaking doors. He made a mental note to check the oiling of the hinges of his doors when he got back home. Slowly he started finding his way through the dark interiors of the house. "Who's there?" he called out multiple times but there was no answer. Finally, he reached the very window that he had thought the sounds were coming from.

Surprisingly, there was no one. He looked around but all in vain. Then he tripped on a nail of the floor and landed on the ground with a thud. Finally, he woke up, "Oh! No! I have been sleep walking again!"

Answer 2.

(a) IT'S SHOWTIME!

By Reena Mishra

19th January 20XX

New Delhi

Recently, XYZ School, New Delhi, organised their annual inter-school drama competition titled – It's Showtime! The event was spread over 3 days and saw a humungous participation by over 50 schools across the city. Some of which are : St. Josephs Academy, St. Thomas, Narmada Public School, etc.

The objective of the event was to make students understand the importance of getting involved in the performing arts alongside paying attention to their mainstream curriculum. The theme of the festival was 'Unity in diversity' which helped sensitise students to the diverse cultures, customs and laws of our country. Esteemed theatre artists Saurabh Shukla and Kamala Das were the judges of the competition. They even lectured the students on the importance of 'Expressions on the stage' and kindly consented to enact a few scenes of their upcoming play 'The Window' for the students.

The event began with a traditional lighting of the lamp by the Principal and revered guests.

This was followed by a welcome note by the Headmistress. She welcomed and introduced the judges of the event and invited them to grace the occasion with their words. After this, there were the performances which were painstakingly prepared by students of grades 6-10 across all the participating schools.

The audience was cheering throughout the acts and seemed to have a gala time. Finally, the winners were announced by the judges. Each winning team was given a certificate and a cash prize worth rupees two thousand.

The event was a huge success and we expect to see a bigger and better show next year because it's always show time in here folks!

(b) **PROPOSAL FOR ARRANGING A CAREER GUIDANCE WORKSHOP**

To help students take better informed decisions about their career paths in life, and to help them make the right choices in their life we propose to arrange a career guidance workshop for them.

A career guidance workshop will help students choose the correct streams for themselves in their high school options. It will be instrumental in shaping the course of their further studies and their lives ahead.

- The auditorium would be used for conducting the sessions.
- The workshops will only take place on Saturdays for two months and would be held between 12–2 p.m.
- The workshop will be open for students and parents of classes IX and X.
- The fee would be INR 200 per session.
- The workshop would include activities such as SWOT analysis, aptitude tests and talks by industry experts.

We hope that the proposal will be accepted so that the career guidance workshops become a reality in the life of the school.

Answer 3.

(a) **(1)** On his return, we asked him many questions.

(2) Scarcely did he arrive when the crowd went hysterical with joy

(3) Floods have disrupted the bus service.

(4) How breathtaking this scenery is!

(5) So long as the offer lasts, you are entitled to a discount.

(6) He is too tall to enter this door.

(7) The doctor advised him to drink at least three litres of water every day.

(8) No other debater in our team is as good as Sunita.

(9) Besides scoring a century, he also took three important wickets.

(10) I regret my inability to accompany you.

(b) **(1)** at **(2)** out **(3)** into
(4) for **(5)** on **(6)** after
(7) over **(8)** up **(9)** out
(10) up

(c) **(1)** was **(2)** were **(3)** seemed
(4) took **(5)** said **(6)** found/could find
(7) is **(8)** asked **(9)** had remained
(10) was

Answer 4.

(a) **(i)** **(1)** merits **(2)** unruffled
(3) choking **(4)** glistening

(ii) **(1)** Rahul was standing in the **middle** of the field.

(2) They **wound** up the clock.

(3) Later they were taken to a hall where they were **introduced** to their newest colleagues.

(4) **Well**, I will give three pieces of silver for that fish.

(b) **(i)** The pram on moorland road was carrying a large greyhound instead of a baby. Secondly, since the road itself was pretty lonely, the scene was classified as an unusual one.

(ii) The narrator knew that the dog was choking because its pulse was feeble and rapid but he was not breathing. Moreover, the dog was not fighting for breath and was unconscious.

(iii) The narrator understood the cause of choking only after prodding the dog. The narrator first opened the dog's mouth wide, pushed his tongue with his forefinger and saw into the depths of his throat. There, beyond the dog's epiglottis, he saw a pea-sized pebble stuck in his throat.

(iv) The narrator seized a pair of scissors and clipped the hair from the ventral surface, used local anaesthesia and swabbed the area with antiseptic.

(c) The narrator saved Jake's life in a scientific and logical manner. First, he did a clinical observation to discover that he was choking. He prodded to understand the cause. He first opened Jake's mouth wide and saw that there was a pea-sized pebble stuck in his throat. Then, he clipped the dog's hair from the ventral surface of the larynx and used local anaesthetic on the area. Using a scalpel, he cut down the skin therein until the ventral surface of the larynx was revealed. Finally, he removed the pebble using a pair of broad-bladed forceps and saved Jake's life.

(100 words)

●●

ENGLISH-I 2019

QUESTIONS

Question 1.

*Write a composition (in approximately 450–500 words) on **any one** of the following subjects :* **[25]**

(You are reminded that you will be rewarded for orderly and coherent presentation of material, use of appropriate style and general accuracy of spelling, punctuation and grammar.)

(a) *You had gone on a trekking expedition with a group of friends. As night closed in, you realised that you were separated from the group. Describe your initial feelings and the place where you got lost. How did you spend the night alone before you were reunited with your group ?*

(b) *It is easy to advise others not to worry, but difficult for oneself to follow the same. Narrate an experience which made you realise the truth of this statement.*

(c) *'The best things in life come free'. Write for or against the statement.*

(d) *Photographs.*

(e) *In order to be successful, one should rely on oneself and not expect others to help. Express your views on the given statement.*

(f) *Write an original short story beginning with the following words :*

She was delighted to find the sealed envelope lying on her table. As she reached for it ....

Question 2.

(a) *You are Nitin/Nita from XYZ School. You wish to start applying to various colleges for admission after your ISC examination. In about 300 words, draft your personal profile using the guidelines given below:* **[20]**

Your name, years spent in the school you are passing out from—academic interests and performance—cultural activities you participated in—sports positions of leadership held during your stay in the school—your future plans and goals—two characteristics that you consider to be your strengths—two characteristics that set you apart from the others.

(b) *As a member of the Student Council of your school, you wish to start an old clothes collection drive. These clothes are to be donated to an orphanage nearby. Write a proposal in not more than 150 words, stating the steps you would take to make it a success.* **[10]**

Question 3.

*Answer sections **(a)**, **(b)** and **(c)**.*

(a) *In each of the following items, sentence **A** is complete, while sentence **B** is not. Complete sentence **B**, making it as similar as possible to sentence **A**. Write sentence **B** in each case.* **[10]**

Example : (0) (A) All nooks and corners were checked.

(B) No ....

Answer : (0) No nook or corner was left unchecked.

(1) *(A) He is so old that he cannot remember events of the past.*

(B) He is too......................

(2) *(A) As soon as the bell rings, the children run out to play.*

(B) No sooner......................

(3) *(A) No other planet is as big as Jupiter.*

(B) Jupiter......................

(4) *(A) He is forgetful as well as careless with his work.*

(B) Besides......................

(5) *(A) It is probable that the ship will reach the shore.*

(B) In all......................

(6) *(A) She said that she would return the book to the library the next day.*

(B) She said, "......................"

(7) *(A) Only trained graduates need to apply for the post.*

(B) None......................

(8) *(A) His attitude towards his parents has always puzzled me.*

(B) I have......................

(9) *(A) He loved his little daughter, but he was in mortal fear of her fierce temper.*

(B) Although......................

(10) *(A) When she heard the news of her daughter's result; she was very happy.*

(B) On......................

(b) *Fill in each blank with a suitable word. (Do not write the sentence.)* **[5]**

(1) *She turned _________ the offer to work at the factory.*

(2) *To everyone's surprise, he turned _________ at the party.*

(3) *He broke _________ in the middle of his speech at the function.*

(4) *The burglars broke _________ the house when the family was away.*

(5) *He agreed to carry _________ the manager's orders.*

(6) *Sita's daughter carried _________ the family business when Sita went abroad.*

(7) *He ran _________ from home at the age of ten.*

(8) *I ran _________ my old friend in the market place.*

(9) *Her aunt looks _________ her aging grand-parents.*

(10) *The Principal looks _________ matters of complaint from teachers and students.*

(c) *Fill in the blanks in the passage given below with the appropriate form of the verb given in brackets. Do not write the passage but write the verbs in the correct order.* **[5]**

*One day I _________ **(1)** (be) in the lunch line, and there I saw a pile of apples. The teacher-in-charge _________ **(2)** (stare) at me and said, "Just _________ **(3)** (take) one. God _________ **(4)** (watch)."*

*So, I _________ **(5)** (take) an apple, and the line _________ **(6)** (move) along. At the next table there _________ **(7)** (be) a pile of chocolate chip cookies. I _________ **(8)** (not know) what to do.*

*"Psst", the kid behind me _________ **(9)** (whisper). "_________ **(10)** (take) all you want. God's watching the apples."*

Question 4 .

Read carefully the passage given below and answer the questions (a), (b) and (c) that follow :

(1) At the Literary Society's meeting, Isola read out the letters written to her Granny Pheen, when she was but a little girl. They were from a very **kind** man—a complete stranger. Isola told us how these letters came to be written.

(2) When Granny Pheen was nine years old, her cat died. Heartbroken, sitting in the middle of the road, she was sobbing her heart out.

(3) A carriage, driving far too fast, came within a whisker of running her down. A very big man in a dark coat with a fur collar, jumped out, leaned over Pheen, and asked if he could help her. Granny Pheen said she was beyond help. Muffin, her cat, was dead.

(4) The man said, 'Of course, Muffin's not dead. You do know cats have nine lives, don't you ?' When Pheen said yes, the man said, 'Well, I happen to know your Muffin was only on her third life, so she has six lives left.' Pheen asked how he knew. He said he always knew– cats would often appear in his **mind** and chat with him. Well, not in words, of course, but in pictures.

(5) He sat down on the road beside her and told her to keep **still**—very still. He would see if Muffin wanted to visit him. They sat in silence for several minutes, when suddenly the main grabbed Pheen's hand.

(6) 'Ah – yes! There she is! She's being born this minute! In a mansion—in France. There's a little boy petting her, he's going to call her Solange. This Solange has great spirit, great verve — I can tell already! She is going to have a long, venturesome life.'

(7) Granny Pheen was so rapt by Muffin's new fate that she stopped crying. The man said he would visit Solange every so often and find out how she was faring.

(8) He asked for Granny Pheen's name and the name of the farm where she lived, got back into the carriage, and left.

(9) Absurd as all this **sounds**, Granny Pheen did receive eight long letters. Isola then read them out. They were all about Muffin's life as the French cat—Solange. She was, apparently, something of a feline musketeer. She was no idle cat, lolling about on cushions, lapping up cream—she lived through one wild adventure after another—the only cat ever to be awarded the red rosette of the Legion of Honour.

(10) What a story this man had made up for Pheen — lively, witty, full of drama and suspense. We were enchanted, speechless at the reading. When it was over (and much applauded), I asked Isola if I could see the letters, and she handed them to me.

(11) The writer had signed his letters with a grand flourish
VERY TRULY YOURS,
O.F. O'F. W.W.

It was highly possible that Isola had inherited eight letters written by Oscar Wilde, for who else could have had such a preposterous name as Oscar Fingal O'Flahertie Willis Wilde.

Adapted from : *The Guernsey Literary & Potato Peel Pie Society—* by Mary Ann Shaffer & Annie Barrows

(a) (i) *Given below are four words and phrases. Find the words which have a similar meaning in the passage :* **[4]**

(1) *adventurous*

(2) *cat-like*

(3) *appreciated*

(4) *received something on someone's death*

(ii) *For each of the words given below, write a sentence of at least ten words using the same word* ***unchanged in form, but with a different meaning*** *from that which it carries in the passage :* **[4]**

(1) *kind (line 2)* (2) *mind (line 13)*

(3) *still (line 15)* (4) *sounds (line 26)*

(b) *Answer the following questions in your own words as briefly as possible :*

(i) *Where did Isola get the letters from to read at the Literary Society's meeting ?* **[2]**

(ii) *Who consoled Granny Pheen when she was heart-broken ? What did he say about Muffin's lives ?* **[2]**

(iii) *What did the man say when Granny Pheen asked him how he knew about cats' lives ?* **[3]**

(iv) *According to the man, what was Muffin's new fate ?* **[2]**

(c) *In not more than 100 words, summarise why the eight letters were a treasure to Granny Pheen. (Paragraphs 2 to 10). Failure to keep within the word limit will be penalised. You will be required to write the summary in the form of a connected passage in about 100 words.* **[8]**

ANSWERS

Answer 1.

(a) A TREKKING EXPEDITION WITH FRIENDS

This happened almost a week ago. We had just finished with our gruelling board exams. My friends and I had gone out to celebrate having survived the exams when Amit, my closest friend suggested we do something new, something that we will remember for the rest of our lives for we had realized that after getting our results, we would be going our separate ways to the colleges of our choices.

This thought made the decision for us. We decided to go to Ladakh for a trekking expedition. Our parents agreed on the provision that we take along a guide with us. All seven of us packed our bags and booked tickets to Ladakh. Upon reaching, the tour instructor told us about the importance of being fit and prepared us for the trek ahead. For a couple of days, we exercised with him and built up a bit of stamina.

Finally, three days later, we started our trek up a craggy mountain. Having climbed quite a considerable height of the mountain, we made camps to spend the night. Upon completing all our chores, we bunked down in our tents and decided to sleep the exhaustion off. In the middle of the night however, I had to leave the camp to answer nature's call.

I tried not to make any sound as everyone was pretty tired from the day's adventure. Taking my flashlight with me, I left the clearing and walked to find a secluded spot. It was when I was trying to come back that I realized, I had gone farther than I should have !

I was lost! I sunk down on my knees in shock. The first thing I felt was panic. What if they left me behind ? What if I could never find my way out ? Trying to move past these sad feelings, I tried to make myself comfortable as it was very cold and dark.

Thankfully, I had layered up completely before leaving my tent. I spent some time walking forward in the direction that I hoped was the campsite, but then I decided to wait for the sunrise. It would be better to let them start searching for me. I was feeling pretty hungry and extremely exhausted. I should have woken up Amit and brought him along. At least we could have found our way back together.

I prayed to God to help me reunite with my friends, all night. I don't know when I fell asleep, but I remember waking up to my friends' voices, the next morning. I had only strayed a few meters away from the camp, and they had found me easily. I could not help thanking God!

(b) IT IS EASY TO ADVISE OTHERS NOT TO WORRY, DIFFICULT FOR ONESELF TO FOLLOW THE SAME

In the year 2013, I was in Toronto, Canada, completing my graduation in English Literature. I had worked very hard for the scholarship through which I was able to get admission in the University of Toronto. I had wanted to study and finally settle down there ever since I had visited the city with my family, some ten years ago. It was the happiest moment of my life when I finally got through.

My parents and siblings, though extremely proud and happy for me, had reservations regarding my going abroad. This was so because I was the youngest in the house. Nobody thought I would be able to manage living in a new city all by myself. This made them hesitant about the final preparations. Yet, I was adamant! It was my dream, and I was not going to let anyone take it away from me.

I understood their worries, and I was not annoyed at their warnings and advices. I heard them all. I also reassured them that it would turn out just fine. I was a smart woman, an 18-year-old adult who knew what she was doing. I kept telling my parents that I had saved the numbers of all the professors, help-cells and the head of the English department.

I also reassured them that I had been in touch with the hostel management and I was on a first-name basis with the warden. I had done it all. I even went to the extent of checking out the local police station which was located close to where University campus was, and I had also saved the contact details of the police station besides the emergency number.

I was all set, and it took a week to make my family come around the idea of me living alone in Toronto. I wanted them to be happy and not worry about me. The day I was to leave for Toronto, my mother cried as if I was getting married and she was bidding me farewell for the final time. I was sad too, but the thrill and excitement reigned supreme in my heart. I was ready to embark on this new adventure.

However, once I landed in the new city, the panic set in. My luggage had been lost in transit, but thankfully I had all my bank cards, cash and important documents with me. I was told that the luggage was back in India, and I would have to wait for three days for it to reach Toronto. The moment I stepped out of the airport, I realized that the cab I had booked for myself had already left due to the delay regarding the baggage fiasco.

I started hyper-ventilating now. I sat down and nearly broke into tears. I realized that it was easy to allay the fears of family and friends and to tell them that everything would be smoothly carried out, but once it came to ourselves, it was difficult to control the anxiety. However, I was a grown up and had to make my own decisions and face the repercussions. I tried to pull myself together and hired a cab for the hostel. Once I reached there, I contacted the college authorities, and they assured me of prompt assistance. My roommate turned out to be a fantastic person, and she took me shopping where I bought some clothes and toiletries.

Six years later, today, I am working in Toronto and call this city my home. Sometimes I do recall the panic-stricken girl who nearly cried at the airport and laugh for hours on my stupidity.

(c) THE BEST THINGS IN LIFE COME FOR FREE

(I) FOR THE MOTION

The proverb, "The best things in life come for free", in every sense of the word, is true. It is an archaic notion, and this kind of philosophy is usually mocked at in the current way of life yet the truth of the matter remains, the best things that make our lives worthwhile are free.

Relationships are free in life. They have no market price or selling price. It only requires our consistent efforts to be true to our family and friends, return the love we get from them and share all our highs and lows with them. One does not have to go to the mall to buy the love or the warmth of a mother's hand when one is ill. One can never buy a father's proud pat on the back when one achieves something or his hand on one's head, his unconditional support, when one loses. A handshake of a friend or a hug from a daughter is all free and in abundance, if we try and work on our relationships with sincerity.

Secondly, the experiences we undergo in life are free of cost too. Yes, education comes at a price. We pay our way through school and then college, yet the time we spend there and the memories and experiences we take with us are free. The support and camaraderie we receive from our teachers through the years is free and frankly, invaluable. The pride that our principal shows in us when we participate in inter or intra-school competitions is real and cannot be paid for. We also get constant attention by everyone involved in the well-being of students. All this is not a part of the fees we pay to the administration. The experiences we receive, when we organise a fest or school function for the first time, the experience of fighting and then later coaxing our friends to forgive us, the experience of working under our professors and then getting appreciated or scolded later, the experience of getting stuck in some serious situation and then coming out of it later on our own, are all priceless, all too valuable to even think about buying through paper currency.

Furthermore, we tend to forget that the friendships that make our lives more worthy to live, are for free. They are gifts of time and love we get from people we have chosen to remain in our lives. Friends are unique because we choose them, they are not born with us. Their companionship is for life and all for free. Then, the people we fall in love with and decide to spend the rest of our lives with, are also free. We may spend a lot of money buying gifts, taking them for treats or spend lavishly even on a wedding, yet, the bond is strong, not because it was purchased with money but because it was built on trust.

The air we breathe and the beauty of nature we find solace in, is all free. The sunshine after days of rain, the rain after months of heat, one cannot put a price on the feeling we get when we observe the changes in seasons. The beaches, or the forests, the deserts or the mountains, whatever gives you happiness, is free for you to enjoy.

(II) AGAINST THE MOTION

I do not agree with those who believe that all the best things in life are free. I have always come across people who wax eloquent about how the air, water and nature are free for all, how all the emotions and relationships are free. I agree that all these things are free, but one still has to pay the price for almost all of them. There is no such thing as free air or free water or free nature scenes.

I am sorry, but the air we breathe in, is poison. We cannot deny that the amount of particles, dust and harmful gases in the air we breathe, is frightening. Take the case of stubble burning and what has it done to the national capital. It was reported that children being born in December 2017, were breathing air as poisonous as smoking more than 20 cigarettes. This air starts weakening our lungs and affecting our whole system since we take birth. Air purifiers have become a necessity in the current scenario. We have to invest in purifying the air or allow it to slowly suffocate us.

The water which is free is regretfully not drinkable anymore ! It is two hundred years after the industrial phase. The water is contaminated and will seriously harm us if we consume it for either bathing or drinking. If we want to drink water, we must buy the bottled one or get a water purifier installed to be able to drink water and not get sick.

Our loved ones love us for free, I agree. Our relationships give us the most deeply felt satisfaction. Due to the kind of lifestyle we have started living, the pollution that gets into our system even before we take birth renders most of us ill. We have health-related complications. All of us and if you want yourself and your loved ones to live a long and happy life, you need to spend a lot of money on their health. The medical services have expensive price tags. And we have to pay to make sure that our family and even friends remain healthy for a long time.

Furthermore, if we talk about nature, then sorry to burst someone's bubble but nobody lives in a place where you get to see the mountains, deserts and beaches all in the same area. If you want to enjoy any of these sites, then you have to pay for the trip—travel tickets, accommodation and for food. The sight is free and certainly beautiful, yet to see that view, money has to be paid! Thus, nothing in the world, whether education, love, meditation, health, relationship is free! All of it comes with a price tag.

(d) PHOTOGRAPHS

Susan was a hard working girl. She lived in a tiny studio apartment and studied in a school. The only thing she wanted ever was to study as hard as she could and make her parents proud. With this goal in mind, she did odd jobs in the city after returning from school.

There were nights when she had to sleep on an empty stomach, yet she would be happy. "I will make you both proud", she would tell her parents. She would look at them and would be filled with a determination that one day, she would buy a huge house and take them with her. One day, she received a call. The Trinity College in London had read her admission essays and came across the profile that she had sent. She had qualified for the scholarship for English honours. The girl was thrilled. She told her parents that she had taken a step towards their collective dream of seeing her make a name for herself.

While moving into college hostel, she took her parents along with her. After all, she could not do without them. They were the source of her strength. She talked to them non-stop through her journey to London. Susan worked continuously and sincerely until she had received her degree. The college recruitment process was easy for she sailed through all the interviews and finally landed with a job at a publishing house as an assistant to the copy editor. There, she remained for two years till she got a contract to write a book.

Susan knew this was the opportunity she had been waiting for. The book not only sold millions of copy but also went on to win an award. In the press conference where she was to be honoured, she was requested to bring along a guest. She told them that she was going to bring her parents.

When she arrived, everyone was surprised to see her carrying a large photograph. She kept the picture on the table for all to see. It was a picture of a man and a woman, smiling happily.

"My parents died, trying to save me from drowning. We were all in a car which had skidded off a bridge and fell in the lake. Though they could have escaped, they kept trying to unbuckle my seat belt which had jammed. They kept at it, till they had rescued me. I was a kid at that time, but I knew that it was too much for my parents. My mother had asthma, and my father had used up all his strength while trying to save us."

"Both of them died that night. But they stayed in my heart. This photograph is my life. If not for this photograph, I would have lost memory of my parents moving from one foster home to another."

"This is not just a photograph but the symbol of my love for them."

(e) SELF-RELIANCE IS IMPORTANT

To be successful, one should rely on oneself and not expect others to help. It means developing the prowess and abilities to do a job without outside help. Its core is to create an opportunity for ourselves by working hard and learning the capabilities that are vital for moving forward in life. It may also mean to wait for the right opportunity and being prepared. For students both in school and college, it means regular studies to do well in the examination and later interviews.

Self-reliance makes and breaks destiny. If one does not rely on others, then they do not have anyone else but themselves to blame. They do not even blame fate, or situations or society. They create their tools and utilise them with high concentration. Self-reliant people focus on their goals. They depend upon themselves for their success and failure. People who do not wait for others to begin work but are proactive, tend to be creative in their ideas.

With single-minded concentration, they can attain their goals because they are determined and self-disciplined. They are aware of their strengths and weaknesses. They do not allow their flaws to affect others. Since they accomplish their plans by themselves, they have the benefit to mould things in their way.

People who depend on others lose their abilities to do even the smallest of acts. This makes them want to have someone on their back at all times. They become complacent and lazy. Such people ask too many things and tend to annoy others. They keep asking questions and lose their sense of initiative.

Self-reliance and success have a direct connection. Dhirubhai Ambani managed his work by himself and went on to create one of the biggest conglomerates—Reliance Group of Industries. Lal Bahadur Shastri was determined to study and swam through a lake everyday, to get to his school. Such stories inspire us and make us realise that self-reliance is one of the most important qualities in the world.

A.P.J. Abdul Kalam is the perfect example of self-dependence. He collected tamarind seeds and sold them to earn his living. He also collected newspapers and distributed them. Though he was not born in a low-income family, yet such acts of self-reliance helped him achieve what he did in his lifetime.

Mahatma Gandhi was born to a wealthy family. He left the comforts of rich home and well-paying job and started working for the country. He was confident that even if he was alone, he was strong enough to win freedom for the country. He may have begun alone, yet millions joined him later. This was the power of self-reliance! Thus, depending on others only makes one lazy and a procrastinator, while self-reliance and independence makes one faster, brighter and more successful.

(f) ANNA AND THE MAGIC SCHOOL

She was delighted to find the envelope lying on her table. As she reached for it, the envelope flew and hovered in front of her in the air. Anna had been expecting a reply from her favourite aunt for weeks now! Shocked and scared out of her wits, Anna jumped back. The hovering envelope split at its edges and formed a mouth.

"Dear Miss Anna Zachary, You have been accepted in the Exetram School of Witchcraft and Wizardry. Please follow the instruction as mentioned in the letter enclosed and board the Exetram boat from Plymouth, on June 30, 2019. Hope you and your family are doing well. Maggie Smith, Head Mistress, Exetram School of Witchcraft and Wizardry."

Bless the world! The whole Harry Potter world was real! Anna screamed with joy. She knew it! She had always known that magic existed. She had felt it when she was angry or extremely happy. She had dreamt of flying through the clouds. She had seen people vanishing and re-appearing in public places a number of times. During those times, she knew only she could see them but had chalked it all up to her active imagination. She loved the wizarding world of Harry Potter after all! She had a feeling that the author, JK Rowling herself was a witch and thus had written about the world of magic with pertinent details.

Anna ran to her mother, Jessy, and told her about it. Her mother apparently did not believe her. "You must have been dreaming honey. Envelopes do not fly in the air nor is the world of Mr. Potter real," said Jessy.

"But it is real mom. I saw the envelope fly up and hover in front of me. I swear mom!" Anna was adamant to make her mother believe.

"Seriously, Anna. Stop it now. I am frankly glad that there is no such thing as a wizard school. Imagine that even Lord Voldemort must be real! What horror!" Jessy turned away, and switched on the television to take her daughter's mind off the visions she was imagining.

Finally, Anna felt it. The magic moving through her. She raised her hand and pointed at the television. It switched off. Her mother blinked. She looked at Anna who again pointed at the T.V. and on it came back! Jessy finally understood. Her daughter was special. She had magic.

"Oh, Anna! I have been waiting for this day since you turned ten. I have magic too my love! Your father and I went to Exetram, and that is where we met. We show magic normally at the age of ten but you are twelve, and we had lost all hope," Jessy was beaming.

Anna was shocked again, but a blast of happiness was running through her joy. She was going to a magic school!

Answer 2.

(a) My name is Nitin Kumar. I studied in Buzzword Public School, Delhi from Kindergarten onwards and passed my class 12th with 75 per cent from the ISC board this year. Having always an aptitude for English Literature, I want to pursue my graduation in the same subject. My plan is to explore my capabilities and contribute as a writer to the literary world.

I have been participating in speech, essay writing and poetry recital competitions ever since I was in class 1st. From class 5th onwards, my interest in football saw me become a part of the junior football team of our school. With time, my interest increased and I led the senior school football team as the captain. I was an active member of the drama club all my school life and kept winning accolades and awards along with my team in several inter and intra-school competitions.

I have always been a team player and have performed well in roles of leadership. Under my captaincy, our school team won the 2018 ISC Inter-School Football tournament. I have also been an active member of the eco-club and performed a lot of activities like road-side skits, donation camps, quizzes, competitions and cleanliness drives for the same.

I feel that my strength lies in the fact that I devote my entire focus and attention to the task entrusted to me. If it is something connected to my interests like performing arts or football then my concentration levels are the strongest. It does not mean that my enthusiasm or contribution wanes if the task given is different or out of my comfort zone. I make it a point to give my best in everything I do.

I feel that what sets me apart from others is the kind of devotion and determination I can bring to the work I do. I think that I will prove to be an active student of your college who will contribute, both academically and through co-curricular activities, in taking your college to greater heights.

(b) **PROPOSAL FOR AN OLD CLOTHES COLLECTION DRIVE**

To start an old clothes collection drive in the school to be donated to the nearby orphanage.

This drive will be aimed at making the students realize the plight of the children living in the orphanage. This shall foster a feeling of compassion and empathy among the students towards the less fortunate.

List of measures :

- The drive shall begin from 10th of August 20XX, and continue till the 15th of August which would be a great day to spread some joy among the needy.
- A group of 5 students will be selected from each class from 8th till 11th and they will be responsible to prepare a short performance for the children of the orphanage.
- The donated clothes will be handed over to the caretakers of the orphanage.
- The students will be instructed to donate only those clothes that are good and clean and not torn.
- The children will spend some time teaching the kids something enriching and interesting and this shall be monitored by the teacher's representative.

We hope that this proposal would be accepted so that we would be able to instill empathy, compassion and brotherhood among our students.

Answer 3.

(a) (1) He is too old to remember events of the past.

(2) No sooner does the bell ring, than the children run out to play.

(3) Jupiter is the biggest planet.

(4) Besides being forgetful, he is careless with his work.

(5) In all probability, the ship will reach the shore.

(6) She said, "I will return the book to the library tomorrow."

(7) None other than trained graduates should apply for the post.

(8) I have always been puzzled by his attitude towards his parents.

(9) Although he loved his little daughter, he was in mortal fear of her fierce temper.

(10) On hearing the news of her daughter's result, she became very happy.

(b) (1) down (2) up (3) down
(4) into (5) out (6) on
(7) away (8) into (9) after
(10) into

(c) (1) was (2) stared (3) take
(4) is watching (5) took (6) moved
(7) was (8) did not know
(9) whispered (10) take

Answer 4.

(a) **(i)** **(1)** venturesome **(2)** feline
(3) applauded **(4)** inherited

(ii) **(1)** The *kind* of work little children are made to do in factories is inhumane.

(2) When I broke my sister's watch, she did not really *mind*.

(3) Ram had failed to crack the college entrance exams though he *still* hopes that he might get the acceptance letter because of his achievements in cricket.

(4) The *sounds* that the cat was making made us feel that she had a terrible pain in her foot.

(b) **(i)** Isola got the letters from Granny Pheen. They had been written to her granny by a kind stranger who may have been Oscar Wilde.

(ii) A very big man in a dark coat with a fur collar, consoled Granny Pheen, that muffin, her cat, was not dead. He said that cats had nine lives and that muffin had only been on her third life and still had six lives left.

(iii) The man said that cats often appeared in his mind and chatted with him in form of pictures.

(iv) The man told her that Muffin had been reborn in France in a mansion. A little boy named her Solange and that she was going to have a long, venturesome life.

(c)

A	complete	stranger
told	Granny	Pheen
that	Muffin	was
only	on	her
third	life	and
had	six	more
lives	left.	Muffin
had	been	reborn
as	Solange,	a
little	boy's	cat,
in	a	mansion
in	France.	She
lived	like	a
feline	musketeer	and
was	not	at
all	idle.	She
lived	through	one
wild	adventure	after
adventure.	She	was
the	only	cat
to	have	been
awarded	with	the
Red	Rosette	of
the	Legion	of
Honour.	These	letters
were	written	to
her	by	a
stranger	probably	to
comfort	her	innocent
little	self.	Pheen
loved	her	cat
a	lot.	

(95 words)

ENGLISH-I

2018

QUESTIONS

Question 1.

*Write a composition (in approximately 450–500 words) on **any one** of the following subjects :* **[30]**

(You are reminded that you will be rewarded for orderly and coherent presentation of material, use of appropriate style and general accuracy of spelling, punctuation and grammar.)

(a) *You have recently moved to a new neighbourhood in your city. Describe the new neighbourhood, comparing it to the one you have just left.*

(b) *Relate something unpleasant that happened to you during your childhood that nevertheless helped you to grow up and mature.*

(c) *'The end justifies the means.' Argue for or against the statement.*

(d) *'Appearances can be deceptive.' Give your views on the topic.*

(e) *Dreams*

(f) *Write a short story that ends with the words :*

'... I really doubt if things could have turned out any better.'

Question 2.

You are the editor of your school magazine. You have recently attended a cultural programme in your city. Write an account of this programme (in approximately 300 words) using the points given below : **[20]**

Date and venue — occasion — Chief Guest — other invitees — inauguration — events — other important features — highlights — reaction of audience—conclusion.

Question 3.

*Answer sections **(a)**, **(b)** and **(c)**.*

(a) *In each of the following items, sentence **A** is complete, while sentence **B** is not. Complete sentence **B**, making it as similar as possible to sentence **A**. Write sentence **B** in each case.* **[10]**

Example : (0) (A) The heavy showers of rain revived the plants.

(B) The plants ....

Answer : (0) The plants were revived by the heavy showers of rain.

(1) *(A) Ashita is the most beautiful girl in the class.*

(B) No ....

(2) *(A) As soon as she reached home, she learnt that she had to leave for London.*

(B) Hardly ....

(3) *(A) Ramesh said, "Rajiv, please bring your physics book to school tomorrow."*

(B) Ramesh ....

(4) *(A) Some of the cakes had been eaten by the boys before the party began.*

(B) The boys ....

(5) *(A) If you are not ready to come with me, I will not go.*

(B) Unless ....

(6) *(A) It is such a wonderful opportunity that we must not miss it.*

(B) It is too ....

(7) *(A) We did not know that Mr. Francis was retiring and leaving for Ooty.*

(B) Little ....

(8) *(A) He knew the culprit, but refused to admit it.*

(B) Although ....

(9) *(A) I am afraid you cannot get admission to the college without clearing the admission test.*

(B) I am afraid you cannot be ....

(10) *(A) If Rita had finished her work earlier, she would have gone to the market.*

(B) Had ....

(b) *Fill in each blank with a suitable word. (Do not write the sentence.)* **[5]**

(1) *After his parents died in a car accident, his uncle looked ________ him.*

(2) *The Chairman says that he will look ________ their complaints.*

(3) *The soldiers laid ________ their lives defending their country.*

(4) *I was laid ________ for three weeks with a broken leg.*

(5) *This book runs ________ a hundred and fifty pages.*

(6) *The programme ran ________ six hours.*

(7) *The ailing company was taken ________ by the government.*

(8) *He takes ________ his grandfather.*

(9) *I have an appointment ________ the dentist.*

(10) *The best candidate should be appointed ________ the post.*

(c) *Fill in the blanks in the passage given below with the appropriate form of the verb given in brackets. Do not write the passage but write the verbs in the correct order.* **[5]**

One day a millionaire went to a hotel in New York and ______ **(1)** *(ask) for the cheapest room they ______* **(2)** *(have).*

"What ______ **(3)** *(be) the price of the room?" he ______* **(4)** *(ask). The manager ______* **(5)** *(tell) him. "Is that the cheapest room you ______* **(6)** *(have) ? I ______* **(7)** *(stay) by myself and only ______* **(8)** *(need) a small room." The manager said, "Why do you choose a poor room like that? Your son always ______* **(9)** *(stay) in our most expensive room." "Yes," said the millionaire, "but his father ________* **(10)** *(be) a wealthy man; mine is not."*

Question 4.

Read carefully the passage given below and answer the questions (a), (b) and (c) that follow :

(1) I could hear the squeaking that heralded the evening arrival of the **bats**. I listened to the noises of the approaching night. Every day my hearing grew sharper. I was learning to filter out whatever I did not need to listen to, and giving no **sign** that I could hear everything that went on in the house.

(2) I could not sleep. The air was heavy and still, the moon hidden behind thick **banks** of cloud. Lord Otori was sound asleep. I did not want to leave the house I'd come to love so much, but I seemed to be bringing nothing but trouble to it. Perhaps it would be better for everyone if I just vanished in the night

(3) Now I heard the hiss of hot water as the bath was prepared, the clatter of dishes from the kitchen, the sliding sigh of the cook's knife, a dog barking two streets away, and the sounds of feet on the wooden bridges on the canals. I knew the sounds of the house, day and night, in sunshine and under the rain. This evening I realized I was always listening for something more. I was waiting too. For what ?

(4) I began to wonder if I could get out of the house without setting the dogs barking and arousing the guards. I started consciously listening for the dogs. Usually I heard them bark on and off throughout the night, but I'd learned to distinguish their barks and to ignore them. I set my ears for them but heard nothing. Then I started listening for the guards; the sound of a foot on stone or a whispered conversation. Nothing. Sounds that should have been there were missing from the night's familiar web.

(5) Now I was wide-awake, straining my ears to hear. There came the slightest of sounds, hardly more than a tremor, between the window and the ground.

(6) For a moment I thought it was the earth shaking, as it so often did. Another tiny tremble followed, then another. Someone was climbing up the side of the house.

(7) My first instinct was to yell out, but cunning took over. I rose from the mattress and crept silently to Lord Otori's side. I knelt beside him and whispered in his ear, "Lord Otori, someone is outside."

(8) He woke instantly, and then reached for the sword and knife that lay beside him. I gestured to the window. The faint tremor came again.

(9) Lord Otori passed the knife to me and stepped to the wall. I moved to the other side of the window. We waited for the assassin to climb in.

(10) Step by step he came up the wall, stealthy and unhurried, as if he had all the time in the world. We waited for him with the same patience.

(11) He paused on the sill to take out the knife he planned to use on us, and then stepped inside. Lord Otori took him in a stranglehold. The intruder wriggled backwards. I leaped at him, and the three of us fell into the garden like a flurry of fighting cats.

(12) The man fell first, across the stream, striking his head on a boulder. Lord Otori landed on his feet. My fall was broken by one of the shrubs. The intruder groaned, tried to rise, but slipped **back** into the water.

(13) "Get a light", Lord Otori said.

(14) I ran to the house, took a light that still burned in one of the candle stands and carried it back to the garden.

(15) The assassin had died without regaining consciousness. It turned out he had a poison pellet in his mouth and had crushed it as he fell. He was dressed in black, with no marking on his clothes. I held the light over him. There was nothing to tell us who he was.

Adapted from : *Tales of the Otori* by Lian Hearn

(a) (i) *Given below are four words and phrases. Find the words which have a similar meaning in the passage :* **[4]**

(1) *Coming near*

(2) *Disappeared suddenly*

(3) *Awakening from sleep*

(4) *Moved slowly and gradually*

(ii) *For each of the words given below, write a sentence of at least ten words using the same word* ***unchanged in form, but with a different meaning*** *from that which it carries in the passage:* **[4]**

(1) *Bats (line 1)* **(2)** *Sign (line 4)*

(3) *Banks (line 6)* **(4)** *Back (line 43)*

(b) *Answer the following questions in your own words as briefly as possible :*

(i) *What could the narrator hear as he was lying down ?* **[3]**

(ii) *Why couldn't the narrator sleep ?* **[2]**

(iii) *When did he realise that there was something wrong ?* **[2]**

(iv) *How did the narrator and Lord Otori overpower the intruder ?* **[3]**

(c) *Describe the incident of the assassination attempt that took place during night, in not more than 100 words (Paragraphs 4 to 15). Failure to keep within the word limit will be penalized. You will be required to :*

(i) *List your ideas clearly in point form.* **[6]**

(ii) *In about 100 words, write your points in the form of a connected passage.* **[6]**

ANSWERS

Answer 1.

(a) NEW NEIGHBOURHOOD IN COMPARING TO THE OLD WE LEFT

Having recently moved to Marathahalli, Bangalore, I often reminisce the things I cherished in Hennur and all that I now miss in Marathahalli.

I still remember the days I learnt to ride my bicycle, enjoying the days riding around the colony. The colony where we lived was a very beautiful and idyllic place, away from the city with lots of blossoming trees. I remember that alternate streets had various coloured flowering trees. They were bright yellow and lavender shaded flowers. One could literally walk on the carpet of flowers. I enjoyed picking flowers, climbing trees, chasing friends around the park, playing ball, chatting with friends and playing with puppies that were plenty in number as many of our neighbours had dogs as pets. So very often there were new set of puppies to play with.

That was a time when I was not worried about anything. I enjoyed school and my life to the fullest.

The one thing, I enjoyed the most was the space around the colony that allowed us to gallivant and relax without getting our parents worried.

In comparison, the place we have moved to this year is an apartment, where there is not much space around our buildings, let alone the presence of nature. The well-built closed community surely has better roads and well-trimmed gardens, but it is nothing like the previous neighbourhood where we were surrounded by untamed nature.

The most important thing that I miss are my friends. As we have moved recently, I have only made a few friends and I am yet to know the others. They are nice, but I still remember the crazy things I would do with my friends in Hennur such as shake the tamarind tree, eat the fresh and sour tamarind. We would run away laughing when the old uncle from our neighbourhood would come after us, yelling at us to stop the ruckus we were creating. Nothing ever stopped us from having fun.

On the other hand, I like our house here as we live on the 10th floor of the building which gives a great view of the city around us and the crisp cold air in winters makes me want to snuggle a little longer in bed. I love having my own room.

I am enjoying the summer holidays now as much as I can and I hope I get better acquainted with my new neighbourhood, make new friends as I await joining my new school and look forward to have great experiences.

(b) UNPLEASANT EXPERIENCE OF CHILDHOOD

This happened when I was in school, and when buses were the major way of commuting from place to place. It reminds me of days where I had to wait at the crowded and dirty bus stands. I still remember the day when as a family we had to travel from Dindigul to Oddanchatram, both fairly small towns in Tamil Nadu.

It was a hot afternoon, when we were standing at the crowded Dindigul bus stand. My father got us some cold drinks to cool us down during that hot summer day. While my father was moving around to find the right bus, he took us near a bus that we thought was the correct bus. I climbed the bus, thinking the rest of the family had got in with me. When I suddenly realized I was alone, I started to panic.

On the other hand, my father started looking around for me, climbing every bus standing at the stand. It took him about twenty minutes to find me. Both my parents were in distress and were frantically looking around for me.

At the same time, I got off the bus looking for them. It seemed like an eternity for all of us as we kept searching for each other. Finally, we found each other and were relieved and scared at the same time, thinking of what could have happened.

My parents were terrified and gave me a good scolding for not being conscious enough to be with them and stick together.

We then travelled to Oddanchatram and later took another bus to reach home. I felt guilty for a very long time thinking how I had been careless earlier and for troubling my parents. I learnt the importance of staying together when one is in a crowded place and of never letting go of one's family.

It scares me when I think of it even now and wonder that it was a miracle that all turned out well in the end. So, now whenever I travel in a group with friends or family, I always look out for everyone and keep a check if they are around.

Life has taught me many lessons but most importantly to be careful and stay safe. I wish for every child that he/she stays together whenever they travel with family and friends and be safe.

(c) THE END JUSTIFIES THE MEANS

(I) FOR THE MOTION

The phrase "The end justifies the means" connotes that a good outcome excuses wrongs committed to attain it. The phrase is attributed to one of the major political thinkers, Niccolo Machiavelli.

The phrase makes it clear that one can engage in wrong things if in the end it is justifiable and noble. For example, in Tamil, there is a saying, one can tell a thousand lies to make a wedding happen. People for long have followed this maxim, when families lie about various details pertaining to the character and capabilities of the boy or girl to the other party. The girl's family might state that their girl is excellent in singing or playing musical instruments and she is a wonderful cook when in reality these things might not be entirely true. The parents and relatives often justify saying it is important that good alliances are formed for a good family life. In this case, it becomes necessary as the happiness of a generation depends on few lies or half-truths spoken about the boy or girl or about the families as a whole. It becomes imperative in order to be married into the right family so as to avoid future issues.

Another simple example which many might identify with are braces for teeth. Braces are messy, uncomfortable and difficult to maintain with food getting stuck in them and anyone who has worn them would vouch for this. But the end result is all worth it with perfectly straight teeth. The straight teeth definitely justify the whole ordeal.

It would also be appropriate to add at this juncture that one might fight and also harm another person, if that person intended to harm their family. Anyone fierce enough to protect their loved ones will take a step to harm anyone in their way. In such a case, the end justifies the means.

In conclusion, it would be right to say that in many situations and events, the end justifies the means.

(II) AGAINST THE MOTION

Brutus justified that Caesar had to be assassinated for the greater good of Rome and all that he did was to root out over-reaching ambition. But that did not end well for Brutus as Mark Anthony aroused the anger of the mob against Brutus and other conspirators, and it set the ball rolling as all the leaders involved in the killing of Caesar ended being killed after the battle.

Here the end, that is the future of Rome, did not bode well for either the leader or the citizens as there was more trouble than ever before. The means to attain greatness for Rome could not be justified either.

The end cannot justify the means as killing or assassination do not solve any critical situations in a country.

Often in history we find that people have done various things such as assassinations because they believed that they are doing the country a greater good by killing certain men or even the thousands of people.

For example, Joseph Stalin justified the death of millions of his people as he felt that the population was too large to be effective in the building of the nation and support communism. Does this act bring about growth in the country? It seems to be an apparent yes, but in reality killing innocent lives cannot bring about greater good as it brings so much bloodshed and turmoil. Mass killing or ethnic cleansing has happened in many countries, all with a similar aim to bring about a purer race or better community with superior qualities. The sad part is that in spite of such drastic actions, they cannot be justified.

Another example can be building wealth based on someone else's hardwork who deserves it more. Exploitation of wealth and resources in the name of growth cannot be justified when lives of millions are affected in a country. There can be no true growth but just a sense of growth when many in a country might be in utter poverty. Many examples can be taken from developing countries where policies benefit the rich and undermine the necessities of the common man.

When ethical stands are broken to achieve a goal that might seem 'rosy', it cannot be justified because many actually are scratched by 'thorns' while others enjoy the 'roses.'

In conclusion, the thought that we need to pay attention to, is that a goal however glorious it might seem if it causes hunger, poverty and illness, emotional torment, physical pain or displacement due to wars or human rights violation, it cannot be justified.

(d) APPEARANCES CAN BE DECEPTIVE

Appearances can be deceptive is an old adage that most of us can relate to or might have experienced at some point in our lives. Appearance does play a very important role in many aspects of life.

People are often enamoured by physical appearances, social standings, ways of dressing and even footwear. Appearances hold a clear advantage for businesses, For example, in real estate business, the façade and external looks of a building might fetch great customers. In truth, there might be many hidden faults in various aspects of the building's structure, plumbing or electrical lines. This can be true for any said product or service in today's world. Packaging has gained greater priority as appearance can attract customers. With online shopping becoming the in-thing, customers are attracted to products because of the manner it is presented in, but when the product arrives it is far from what we had imagined or seen on the website.

Even higher credibility of appearances can prove faulty in some fields like medicine, finance or education. Promises are squandered and people fall into misfortune because they are gullible.

This can be true for human beings also. People give so much significance to the appearance that we fail to see beyond the mask. The mask might hide flaws, bad character or even suffering and pain.

In some cases, it might hide bravery, for example, the legend of Fa Mulan, popularized by the Disney movie Mulan; talks about the story of a girl named Mulan of the Chinese Han dynasty who impersonates a man. The appearance of a man held within a fierce and brave girl who fought along her men for the country.

In other instances, we meet people who seem sophisticated and refined but are proud and selfish, seeking their own end, never bothering about the good of others. We might find such people in the family, community, state or country.

There are many examples, in our day to day lives when appearances have been deceptive and it has been true in history across all nations too where people followed great leaders for their charisma but were brutal and heartless in their pursuits against humanity.

In contrast, at times some people who had been declared as outcast have been found to be wonderful people. Similarly, a person who might appear apprehensive, might turn out to be a 'good Samaritan' when the need occurs.

To conclude, one can boldly say that appearances can be deceptive and the question is whether we are able to see what lies beyond the veil or not.

(e) DREAMS

The very word 'dream' evokes a million thoughts bustling in our mind. Dreams at night, day dreams and nightmares are all a part of the psyche.

People often dream of a better world. A world where there is less hatred, lesser strife, brighter days, a world where there is kindness and no greed, a world where every man is free, every child is safe, a world where one can wake up to a better day and a million small things that would make the world a better place to live.

Dreams on the other hand, have also been studied and explained as a succession of images, ideas, emotions and sensations that occur while we sleep. Though this aspect of dreams has been researched and studied, no one is sure as to why they occur. Dreams are attributed to rapid eye movements, that occur when the brain is active, triggering a sequence of images that tend to be vivid or memorable.

The explanation of dreams does not stop there, as many have tried to work towards the interpretation of dreams. It is a whole new world out there in the arena of interpretation of dreams. Dreams are analysed and help is offered to those who find trouble living with such dreams. Characters are studied based on the dreams that people have.

Apart from the serious stuff, people tend to spend most of their day day-dreaming which is a wonderful thing. Dreams become the base for many success stories. When dreams are worked upon with dedication and determination, there are positive results.

People do rely on dreams a lot and feel that dreams are sometimes signboards that indicate some direction in their lives. For example, Calpurnia believed in her dream that Caesar was in danger and invariably Caesar and his 'friends' dismissed the belief as foolish and goaded Caesar to go ahead and attend the council. We know that all did not bode well with either Caesar or for Rome. It is said that the Romans believe a lot in the dreams. And it can be said for everyone across the globe, though each community might interpret them differently.

(f) I REALLY DOUBT IF THINGS COULD HAVE TURNED OUT ANY BETTER

It was the end of July and we were on our way to visit our uncle's newborn baby in Coimbatore. It was a beautiful day and we had started the journey pretty early in the morning. We started off in our car and had driven for about two hours when we stopped for breakfast at a hotel.

After about half an hour as we started on our way again we suddenly saw a dog crossing the road from the other side of the highway. Before we could realise, the dog ran under our car and got injured and died. As we stopped the car we realized that the car was badly hit.

The car wouldn't move and when we checked we saw that the chassis had been damaged by the impact. We were stranded by the road. The place where we were, was in between towns and it was a village side with nothing in sight.

As my father called for roadside assistance for the car, we began to explore the area as there were plenty of trees around with plenty of shade. We found huge stones where we could sit.

Our dad told us that it would take at least two hours for the roadside assistance to reach us from the nearest car workshop.

We sighed and started walking around only to find a bund near the spot where we were stranded. We climbed up that bund and saw a lake on the other side. We were happy to see such a place and as there was not much water we walked to the lake skipping the water puddles. My little sister plucked wild flowers and my dad and I skipped stones on the water and began to enjoy ourselves.

Before we realised the tow truck had come and we waited as the workshop personnel maneuvered our car on to the tow truck. Finally, our car was 'loaded' on the tow truck, while we waited for our uncle to come and pick us up. We had some more time to play in the lake and we had finished all the snacks we had brought for our journey and were terribly hungry.

Soon our uncle came by and we went back to Chennai. We were safe and that's all that mattered and I really doubt if things could have turned out any better.

Answer 2.

I enjoy cultural programmes more than anything and I was extremely glad when I got the tickets to attend the most popular cultural extra-vaganza at the Museum theatre in Chennai, which held a week-long cultural programme between February 5th to 9th, 20XX.

I got the tickets to attend the folk dances of India on the 9th February which was also a finale for all the programmes that included music, puppetry, storytelling and regional short films from all across the country.

The evening at Museum Theatre resonated with excitement and expectation. It was wonderful to see families dressed for the part as people came in traditional clothes adding colour to the evening. The enthusiasm to watch was wonderful as people are usually glued to screens at home. This seemed like a welcoming change as people were out to experience their heritage and the passing of their own traditions to the next generation.

Soon it was time to be seated and the programme started exactly at 6:30 as announced. The programme was inaugurated by the Chief Guest who was none other than Shobana, the veteran actress and dancer of Bharatnatyam. Her presence added charm and grace to the entire programme and her students performed a traditional 'Kutrala kuravanji' dance depicting the life of tribes of 'kuravas', the nomadic hunters and fortune tellers from Andhra Pradesh. It also included the changes that have happened to these tribes over the years due to modernisation. The compelling story was a blend of both classical and modern interpretation of the legend of yore.

The other dances included the Rasleela, Bhangra, Garba, Bihu, Mayilattam, Karagattam and Puli kali. Every dance heightened the delight among the audience and the vibrancy could not be missed.

On the whole, I celebrated every minute of the programme and it was a joy to witness some of the best dances of India. I was proud to be a part of it all. (321 words)

Answer 3.

(a) **(1)** No other girl in the class is as beautiful as Ashita.

(2) Hardly did she reach home, when she learnt that she had to leave for London.

(3) Ramesh requested Rajiv to bring his physics book to school the next day.

(4) The boys had eaten some of the cakes before the party began.

(5) Unless you are ready to come with me, I will not go.

(6) It is too wonderful an opportunity for us to miss.

(7) Little did we know that Mr. Francis was retiring and leaving for Ooty.

(8) Although he knew the culprit, he refused to admit it.

(9) I am afraid you cannot be admitted to the college without clearing the admission test.

(10) Had Rita finished her work earlier, she would have gone to the market.

(b) **(1)** after **(2)** into **(3)** down
(4) up **(5)** into **(6)** for
(7) over **(8)** after **(9)** with
(10) to

(c) **(1)** asked **(2)** had **(3)** is
(4) asked **(5)** told **(6)** have
(7) am staying **(8)** need **(9)** stays
(10) is

Answer 4.

(a) **(i)** **(1)** approaching **(2)** vanished
(3) arousing **(4)** crept

(ii) **(1)** When mobile phones first came out they were something of a novelty. These days, nobody *bats* an eyelid.

(2) They *sign* to the waiter that they want more drinks.

(3) The exam aims to encourage the recruitment and placement of young graduates in public sector *banks* in India

(4) The aggravation associated with *back* pain can range from a dull, constant ache to sudden sharp pains.

(b) **(i)** As he was lying down, the narrator would hear the squeaking of the bats, the hiss of hot water for the bath, a clatter of dishes in the kitchen, sliding sigh of the cook's knife, barking of a dog and the sound of the guards walking on stone.

(ii) The narrator was restless because the air was heavy and still. He was upset because he felt he was bringing trouble to the house he loved. He wished that he could vanish in the night without causing trouble.

(iii) He realized something was wrong when he heard none of the usual sounds which he hears at night. He could not hear the barking of the dogs or the footsteps of the guards nor the hush whispers.

(iv) The narrator and Lord Otori waited patiently for the murder. Lord Otori took the intruder in a stranglehold and when he wriggled, the narrator leaped on to him and the three fell into the garden where they fought like a flurry of cats.

(c) **(i)** **(1)** The narrator could not hear the usual sounds of the guards walking on stone as all was very quiet.

(2) He was wide awake, when he heard a mild tremor between the window and the ground. He realised that someone was climbing up the side of the house.

(3) He whispered to Lord Otori that someone was outside and both of them waited patiently for the assassin on either side of the window.

(4) The assassin reached the window and readied himself for the kill.

(5) Lord Otori took him in a strangehold and the narrator attacked him.

They fell into the garden.

The man struck his head on a boulder but Lord Otori and the narrator were unhurt.

(6) The assassin fell into the stream and died. They could not make out who he was.

(ii) The narrator finds it strange when there are no sounds that night. He heared a mild tremor between the window and ground. Though alert, he woke up Lord Otori and both of them waited patiently for the assassin. When the assassin reached the window and readied himself for the kill, Lord Otori took him in a strange hold and the narrator attacked him. They fell into the garden and fight like cats. Thereafter, the assassin struck his head on a boulder across the stream and died. There was nothing on his dress to indicate who he was. (96 words)

●●

ENGLISH-I 2017

QUESTIONS

Question 1.

Write a composition (in approximately 450–500 words) on ***any one*** *of the following subjects :* **[30]**

(You are reminded that you will be rewarded for orderly and coherent presentation of material, use of appropriate style and general accuracy of spelling, punctuation and grammar.)

(a) *Imagine that you visited a remote village untouched by the influence of technology. Describe the village, the people, their routine and at least one person you met there. How did you feel at first ? What were your feelings when you had to return ?*

(b) *You were waiting at the ATM booth of Focus Bank, when a white car drew up and four men with masks rushed into the bank. Narrate how an attempted bank robbery was foiled by an alert security guard.*

(c) *'Today's youth is more practical and less swayed by emotions.' Argue for or against the given statement.*

(d) *Sacrifice.*

(e) *'It is easier to preach than to practise.' Express your views on the given statement.*

(f) *Write an original short story beginning with the following words :*

'Everything went horribly wrong that evening ......'

Question 2.

As a reporter you have visited a number of hospitals in your city. Write a newspaper report entitled 'City Hospitals', based on the following points : **[20]**

Buildings and infrastructure—maintenance of the hospitals—quality of services—doctors—nursing staff—waiting period for treatment—emergency facilities—other facilities available—medical equipment—overall opinion.

Question 3.

Answer sections ***(a), (b)*** *and* ***(c).***

(a) *In each of the following items, sentence* ***A*** *is complete, while sentence* ***B*** *is not. Complete sentence* ***B****, making it as similar as possible to sentence* ***A****. Write sentence* ***B*** *in each case.* **[10]**

Example : (0) (A) He always obeys his teachers.

(B) His teachers ....

Answer : (0) His teachers are always obeyed by him.

(1) *(A) Not only did he buy a desktop but also a laptop.*

(B) Besides ....

(2) *(A) The heavy showers of rain revived the plants.*

(B) The plants ....

(3) *(A) The men were forced to work although the light was poor.*

(B) In spite of ....

(4) *(A) No sooner did we see a flash of light in the sky than we heard a loud explosion near us.*

(B) As soon as ....

(5) *(A) As she has not brought a pen, she is writing with a pencil.*

(B) Not ....

(6) *(A) Mumbai is the most vibrant city in India.*

(B) No ....

(7) *(A) Their dog is too friendly to be an effective guard dog.*

(B) Their dog is so ....

(8) *(A) Sameer said, "Ramesh, can you help me with my homework ?"*

(B) Sameer ....

(9) *(A) All nooks and corners were checked.*

(B) No nook or corner ....

(10) *(A) How long I shall stay is doubtful ?*

(B) The ....

(b) *Fill in each blank with a suitable word. (Do not write the sentence.)* **[5]**

(1) *If we set ______ early, we can return before sunset.*

(2) *He left the company for good and set ______ his own business.*

(3) *The teacher jumped ________ a conclusion and punished the child.*

(4) *Ramesh jumped ________ my offer of a job.*

(5) *I am a simple man and have no use ________ much money.*

(6) *My father lost the use ________ his right arm in an accident.*

(7) *Rahul decided to carry ________ with tennis and give up volleyball.*

(8) *Dinanath is a good worker and is always ready to carry ________ his master's orders.*

(9) *Reena agreed _______ me when I proposed a meeting.*

(10) *The four sisters could not agree ________ themselves.*

(c) *Fill in the blanks in the passage given below with the appropriate form of the verb given in brackets. Do not write the passage but write the verbs in the correct order.* **[5]**

*The other day we ________ **(1)** (discuss) the setting up of a Debating Club in our school. We ________ **(2)** (talk) about the club and how many students ________ **(3)** (be) interested in it. Then we ________ **(4)** (begin) to talk about the rules and regulations________ **(5)** (govern) the club. None of us could ________**(6)** (agree) and we began ________ **(7)** (argue). Finally, our English teacher________ **(8)** (have) to be informed and she ________**(9)** (come) and ________**(10)** (break) up the meeting.*

Question 4 .

Read carefully the passage given below and answer the questions (a), (b) and (c) that follow :

(1) My father and I left the hall and made our way out to walk across to the headquarters. There were street lights round the square, but underfoot the cobblestoned ground was dark and lumpy. My father tripped on the uneven surface and went down on one knee, trying not to fall entirely.

(2) At exactly the same moment there was a loud bang and sharp zinging sound and a scrunch of glass breaking.

(3) I bent over my father and saw that his eyes were stretched wide and his mouth grim with pain.

(4) "Run", he said. "Run for cover. For God's sake, Ben! That was a gunshot!"

(5) We were halfway across the square, easy immobile targets. He struggled to get to his **<u>feet</u>** and told me again to run and for once I disobeyed him.

(6) "Stay down," I told him.

(7) "You don't understand..." His voice was anguished.

(8) "Are you bleeding ?"

(9) "What ? I don't think so. I twisted my ankle."

(10) People ran out of the hotel, drawn by the bang that re-echoed around the **<u>square</u>**. There was confusion and people saying, "What happened, what happened ?" and hands stretching down to my father to help him up. When he was well surrounded he finally took my arm. Putting his left foot down caused him much discomfort.

(11) "That noise..." a woman said.

(12) Heads nodded. "It sounded like... Was it ... a gun ?"

(13) "But where ? There's no one here with a gun."

(14) Everyone looked round, but it was far too late to see the rifle, let alone the person shooting. My father put his arm round my shoulders for support and indicated that we should set off.

(15) We walked in a procession to the headquarters and my father called the police.

(16) I went to the open door, looking across the square to the hotel.

(17) I remembered the zing of the bullet. If the bullet had been aimed at my father, and if he'd stumbled at the exact **<u>second</u>** that the trigger was squeezed and if the bullet had smashed some glass so that I heard the tinkle, then why was every pane of the window in the headquarters intact ?

(18) I told myself that the whole thing had been a coincidence and turned to go back inside, and saw for an instant a flash of light on broken glass down on the ground.

(19) It was a window of the shop next door that had been hit.

(20) Zing. Ricochet. Smash. The straight line could have been deflected by the curve of a cobblestone.

(21) The police arrived at headquarters, and it was nearly two o'clock when I finally closed and bolted the doors and switched off the lights.

(22) My father moved himself into one of the single beds in the bedroom. I lay on the second, not at all sleepy.

(23) My father said, "Ben, why didn't you run ?"

(24) I answered, "I didn't want you to get shot."

(25) "So you stood in the way..."

(26) After a while. I said, "I'd say it was a .22 rifle with a high-velocity bullet. Hitting you in the body, it quite likely won't kill you. You need to hit the **head** or the neck to be lethal. All I did was shield your head."

(27) There was a silence. Then he said, "I'd forgotten you could shoot."

(28) "I was in the school team. We were taught by one of the country's best marksmen." I smiled in the dark. "You paid for it, you know."

Adapted from : *Penalty* by Dick Francis

(a) (i) *Given below are four words and phrases. Find the words which have a similar meaning in the passage :* **[4]**

(1) *unmoving*

(2) *full of severe pain and agony*

(3) *complete, not damaged*

(4) *protect*

(ii) *For each of the words given below, write a sentence of at least ten words using the same word* ***unchanged in form, but with a different meaning*** *from that which it carries in the passage :* **[4]**

(1) *feet (line 11)* **(2)** *square (line 16)*

(3) *second (line 29)* **(4)** *head (line 45)*

(b) *Answer the following questions in your own words as briefly as possible :*

(i) *What two things happened simultaneously when Ben and his father were walking to the headquarters ?* **[2]**

(ii) *What did Ben see when he looked at his father on the ground ?* **[3]**

(iii) *Why did Ben's father tell him to run and why did Ben disobey him ?* **[2]**

(iv) *What was Ben's logical answer to his father explaining why he did not obey him ?* **[3]**

(c) *Describe the incident of the shooting in not more than 100 words (Paragraphs 1 to 10). Failure to keep within the word limit will be penalized. You will be required to :*

(i) *List your ideas clearly in point form.* **[6]**

(ii) *In about 100 words, write your points in the form of a connected passage.* **[6]**

ANSWERS

Answer 1.

(a) A VISIT TO A REMOTE VILLAGE

I was too excited ! The journey to Kambaska village, a remote area in the Garo Khasi hills, was soon to be my destination. The survey on influence of technology, taken up by us, the team of journalists was the reason behind it.

I was all set for this enthralling journey. With hardly any rail routes to reach this village, I underwent a lot of hitch-hiking, bus-travelling and cartrides. It was tiring but fun too. The summer rains made walking into woods difficult, but it was thrilling all the way. It was my first job as a journalist.

The cart-rider dropped me close to the nuclear yet small village in the afternoon hours. The canopied tall trees above me and wet leaves on the ground, the smell of wet earth, the sun playing truant amongst the leaves, was a beautiful sight. It felt ethereal.

Far amongst the trees, I saw smoke and was happy to find a few huts build close to each other. A few children almost naked were playing in the puddles of rainwater. I reminisced my childhood days.

I was welcomed by an old man who was the head of the village clan. He asked me the reason for being there and then there was no stopping. Since I knew Garo language, communication with them became very easy. I was offered lunch, given a charpai to sleep under a canopied hut, and I dozed off soon. I woke up to the chirpings of birds and the cock's crowing. A new day had started. I saw the women ready with their wooden stoves to prepare food. The kids were up, playing here and there and the men had left for work. The old men sat around the big Banyan tree, sharing their experiences of life.

There was no mobile, radio, iPad or T.V. in their lives. They were living a life without electricity. Technology had not influenced their lives, yet they were happy and satiated.

They spent time with each other. Soon I saw a group of children running towards a young man, who was carrying something with him. He laughed and then sat down with the kids happily, opening a book and reading out to them. I watched the children's faces and heard the narration quietly.

He was Sohana, an educated boy of 22 years, living there with his family.

As he saw me watching him, Sohana came forward and that's when I knew I had a friend in this wilderness. He showed me around the village, the forest and what made him stay back here. Sohana replied, "The life of technology filled in cities gave me everything but with no parents around and no way to stay connected with them, brought me back". He had studied in Shillong for 8 years at his maternal place and decided that knowledge must reach his home, family and friends too.

I was surprised to see what changes this young innovative mind had brought in the village. Children were making wood carvings, which were sold back in the city. The youngsters were busy making compost from dead leaves, plants and decaying food. The men were involved in organic food and vegetable plantations, while the women collected herbs and dried them. A huge pit was dug at one side of the huts where water was collected. It acted as a water hole for all.

Sohana had changed the life of so many. Without the influence of technology this village and its members were much more educated and innovative in their way of life.

The four days' stay came to an end soon. The trips with Sohana unraveled what selfish city goers had become. There was no community in cities but wonderful spirit of camaraderie here. Google had not reached here but the villagers of Kambaska had googled out their life and aims in detail.

I left the place and all members with a heavy heart, not because I was feeling bad but with the unselfish love and gifts that I was loaded with. I had learnt my first lesson of life and shall remember it always.

(b) SECURITY GUARD SAVES THE DAY

Demonetisation wretched havoc on all of us. Scared of losing out of any cash, next day, in the morning I rushed to a close by ATM booth. It was empty with the security guard dozing away. I knocked at the door and waited for him to come out.

Before I could reach the door, a white car came to a screeching halt and four masked men rushed towards the booth. They threw me out of the way uncaringly. The security guard who had walked out a few minutes before, to attend to his nature's call, hearing the noise, chose to stay away. All I could say 'a coward' he was!

I gathered myself up and was too a coward to fight against these four men with guns.

I knew what their intention was. It was a day light looting of the ATM. The money scare had just started, I tried to run here and there looking for help but the morning joggers just whisked away, ignoring my request. I decided to be brave, I picked up the log next to me, and attacked one of them, who stood at the door. His strong arms pushed me so hard in retaliation to the blow.

I lost balance and fell with a thud. Feeling weak with the sudden blow, through the corner of my eyes, I saw some movement behind the booth. The men busy with uprooting the ATM machine were unaware of something brewing up. Hating to be a game spoiler, I kept quiet feigning that I was hurt, I lied down.

Suddenly, a group of men with sticks and bats pounced on to the door and the men at work caught them by surprise, and got beaten with some strong blows. Pulling away their weapons they made the thieves handcuffed them with ropes. The police van pulled up and the men were carried away. Our hero appeared, the security guard Ramu, who intelligently spoiled the masked men's plans. When questioned how it all happened, his story revealed his presence of mind and plan of action. He had seen the men coming but moved away quietly ignoring it. He was aware that fighting with them single-handedly would be a foolish act, so he quietly went to every house, made everyone aware of things but requested all to be patient. He called up the police from the PCO and asked all to get bats and sticks for attacking the men. And the rest was good. All ended well.

As a gift for his bravery, all of us decided to contribute some money and the policeman on duty promised to get him a gun when on duty to safeguard himself and the bank property henceforth.

I finally withdrew some money and left with a thousand rupee note to the valiant hero Ramu.

I knew it was not enough, but he was indeed braver and wiser than me.

(c) TODAY'S YOUTH IS MORE PRACTICAL AND LESS SWAYED BY EMOTIONS

(I) FOR THE MOTION

A recent incident in my family forced me to pen down my thoughts. My maternal uncle and aunt, live by themselves in Khandwa. Their kids are well settled in the USA, as famous doctors. An urgent call to the USA one day by my aunt made the difference.

Uncle, a diabetic fell in the park while he was taking his morning walk. The worried aunt called up her son. The son quietened her down and said he will do something. His mother was crying helplessly over the phone. With the help of neighbours, uncle was taken to the nearest city hospital. Frequent calls to the son seemed to come on deaf ears. The mother was broken that her son didn't care about them anymore.

I reached the hospital, within an hour of the incident. I took charge of things, money, doctors and everything.

Everyone who came to meet uncle and aunt, sympathesised with them and also questioned them about their son and they had no answers. That brings us to the discussion of– 'Is today's youth more practical and less emotional'. Didn't the son feel any pain for his father? Was he so selfish that he did not come to help his father in this situation ? These queries crossed my mind. But let's pause and think why was I there ! Who had sent me ?

It was an early morning call from the US that had forced me to be at uncle's side. The RTGS transfers, constant call to the doctors and their instant availability at the hospital, the constant messages on whatsapp tell us a different story. It was all lived, planned and executed by Bhaiya from US. He didn't loose his presence of mind, he maintained balance, kept himself calm and diligently handled all immediate help to get the first medical aid to uncle. I here by say he was not emotionless but yes he was practical. Covering under emotion would have made him weak and incapable of making such decisions that helped to save uncle's broken leg and slip disc situation.

With this I agree vociferously, that today's youth is more practical and less emotional. If they need to help their family in times of need, it is not their physical presence but their presence of mind which they use to save the situation. Making a long journey and not providing instant help could have led to disastrous results.

So I firmly believe that it is better to be practical than to be emotional. It is a real world with real problems and to save our dear ones, one needs to be a source of help, even though from distance.

(II) AGAINST THE MOTION

We are the most coddled generation in history and it is apparent from both our work ethic and our outlook of life. We expect things to be hand delivered or spoon-fed to us. Everything is too convenient for our generation. We understand that our parents want to give us the life they never had, but that may just be the reason of making this generation of ours softer and emotional, away from the ingenious ability of practically handling things out.

By receiving everything at one place we spent our time in useless things which deviates us from our goals. A big misconception in this new generation is the idea that getting rich quick is actually a feasible feat. But in reality, success takes time and hard work, nothing comes easily in this world. When someone criticizes us with the intention of helping us and improving us as human beings, we either get angry or upset. This spoon-feeding has led to blocking of our imaginations and we just see what the world shows us, we become weak and we often panic. We turn help and support instead of facing the obstacles. Sometimes we overreact and take sudden decisions without thinking practically.

Since elementary school, our teachers and parents reward us for every minor thing that we do—did well on a test, here's a sticker; good report card, parents buy us a present. This has made us attention seekers and now no task can be done without seeking some sort of reward for it.

We are one of the laziest generations in history. Although we are smart and extremely skilled, our downfall starts at the very point where we do not seem to work well in environments that don't keep us interested. We always look for the easy way out. We constantly wait until the last minute to get things done and we always find something more interesting to do than the duties we are obligated to get done.

We have all heard stories that our grandparents and parents have told us of how they used to be physically rebuked for their mischiefs and mis-deeds. Nowadays, it's considered child abuse if you lay a hand on your child and also beating the kids for their betterment, make the kids misunderstand their parents or elders and the results are bad.

We live in an era where we have to be extremely cautious with what we say because people have become way too sensitive. More terms have been labeled as offensive in this generation than at any time before. We are easily one of the softest and very emotional generations in history and it's time we highlight this issue, before it gets too out of hand.

(d) SACRIFICE

'Sacrifice' is a word with many connotations both negative and positive. Sacrifice can be giving up something away or taking away something for the sake of happiness. The end is one-happiness for oneself.

A sacrifice will always give us happiness, according to the holy scriptures. But does it do so? Do we find happiness by offering/sacrificing a poor animal to appease the Gods and shower their blessings on us? Should we consider this practice a sacrifice or a slaughter? Sacrificing someone else in the name of God is easy for a coward, but the real sacrifice to me lies in sanity, humility and dignity, giving away something dear to a person that may give happiness to someone else. Someone rightly stated- "People who sacrifice and give up their life and happiness for others are martyrs". Such people do exist. Our freedom is what we owe to these selfless sacrifices. The soldiers are also like us, they also feel fear, they also want to live long, but still, out of love for the country, for the people in it, they sacrifice themselves so that we can live peacefully.

Happiness is a way of life; it comes back to you if you give it. Sacrifices can never beget the happiness for yourself, but happiness of others whom you love, will definitely give you happiness. I would like to put 'the mothers' as example, in fact mothers are the best example of this aspect. They sacrifice for their kids, for their whole family, for God. They are the most beautiful and incredible things that God has ever provided us with, as they live for others but not for their own, they try to spread happiness by giving their best and sacrificing their independence, for others that is why they are considered next to God.

The sages sacrifice worldly pleasures to attain unity with the Almighty. But today such acts of sacrifice are acts of cowardice, running away from responsibility and attaining eternity in all wrong. The word sacrifice loses its meaning here.

To conclude, sacrifices can be made out of love, due to greed, under fear or because of other reasons and in every case one loses something for some gain. If made with a good intention, the good would surely return to the person.

(e) IT IS EASIER TO PREACH THAN TO PRACTISE

Well, the above statement is a fact, that is very prevalent in today's world but was not before (a few decades ago), when the world's population and the hosting countries were evolving.

I firmly believe that years ago when man was fighting for his rights to live happily, respectfully and healthily, preaching such facts was very much practiced by all. Those who preached the goodness of healthy living, practiced the same and made others realize that problems could be resolved. Progress and well being could only be achieved when all is practised what was preached by others or themselves. It was this that led to changes in life socially and politically, problems were resolved and solutions were widely accepted. Goodness that prevailed then made faith, in preaching good things effectively.

People who preached and practiced the same, became leaders, role models and harbingers of change. The change around their lives reinforced their belief that we must practice what we preach and the result indeed will be profitable/generous. But sadly today with increasing corruption, malpractice and dearth of good and genuine people, this statement has lost its significance. With changing times and the chameleon trait being dominant in us humans false promises and statements are given away. Leaders and role models are no longer looked with respect because what they preach now is not practiced in spirit by them. It's just words that are spoken and that are not meant to be believed and acted upon.

Only a man/woman who understands his/her own views or opinions exhibits grit and determination, and thus will practice what he/she preaches. It is indeed easier to preach than to practice. People in today's world are manipulative and convincing. Very few are genuine preachers, who practice what they state. As a grown up individual, I myself realize that the statement is true today. Men/women are masters of their own selfish needs, wants and ambitions which keep changing according to time. Sigmund Freud a great psychologist once rightly stated, 'Human beings are great deceivers of others and can camouflage things with utmost clarity'.

(f) EVERYTHING WENT HORRIBLY WRONG THAT EVENING...

Everything went horribly wrong that evening of the College Reunion, 1992 batch at my farmhouse. We woke up to a wonderful, breezy yet bright morning. The warmth made me feel warm too, it warmed my heart. I was eagerly waiting to relive the past after so many years. I thought to myself that nothing had changed amongst us all, only the surname did!

Everyone was busy. My hired men, me and my growing daughters with their wonderful dad. Every detail was planned well and I knew my husband, a manager to perfection would see to it. The constant ringing of the phone, incoming messages and noises kept us all on our toes. Arranging the sprawling garden was a mammoth task. Working out in the garden under the sun was nice but a dark cloud hovered above my head. I crossed my fingers.

With everything ready and the countdown starting we got ready. The first bell excited us all. My guests had arrived; the drinks ready to be served; the barbeque burning; the coal and wood all laid out and the Elvis Presley song loud and clear. It was such a beautiful atmosphere with all my old schoolmates back together with me. We were cracking jokes, singing songs and laughing a lot, while remembering our old days. Amidst the fun and noise, I decided to look at the barbeque. A droplet fell on my head and it did not stop, to my dismay it had started to rain. I stood outside watching the wood, coal and tables with linen getting soaked. I felt broken. 'Hell had struck the part!'

I rushed indoors. Seeing me wet all reacted with a loud sigh. All the hard work was in rain. The barbeque, the tables and chairs, the snacks and drinks all laid out in the garden, getting wet in the inopportuned rain. All went horribly wrong but in a few seconds all changed. The crowd of young hearted oldies ran out to play in the rain. It felt wonderful seeing them not bothered by the spoils. They said that they wanted fun and that they shall get it in plenty.

Food was fried and warmed in the kitchen, the music was played louder, steps because faster and so was the laughter with hugs and hurrahs! The warm hug of yesteryears made me realize, all was not lost. Though everything went horribly wrong, but the evening party that ventured into late night, was loaded with love, happiness and wonderful memories to be released and reminisced forever.

Answer 2.

CITY HOSPITALS

Most city hospitals, government and private, have inadequate infrastructure and the buildings do not meet the minimum criteria in terms of space and build quality. The rooms are small and crowded with modifications made to accommodate extra beds even if they do not have the capacity for such additions. There is hardly any maintenance because the rooms are damp and there are lumps of paint and cement peeling off from the walls and ceilings. Cleaning is done only once a day and the stench from the toilets is unbearable because they are used both by the patients and their visitors.

Even the food served to the patients is of poor quality because it is either too spicy or too watery. No care is taken to provide nutrition to the patients. The utensils are hardly cleaned and one can easily find oil stains in them even after washing.

The quality of services is very poor because the nursing and support staff do not respond on time. In fact, at times they even refuse to look at the patients during night duty. Most of them are even reluctant to push injections and never bother about accuracy of readings for the different tests that are done on the patients.

The doctors attached to such hospitals are more interested in getting these patients referred to their private clinics and fleece their relatives like anything. They tend to visit the hospitals only once a day, or even a week, and devote time only in summarily checking them. In case of only major surgeries, they refer these patients to bigger hospitals with better facilities. Sometimes, there are allegations of doctors making guinea pigs of patients by experimenting on them. This is done to test the effects of newer and more powerful medicines of pharmaceutical companies provided to them with hefty commissions.

There are also unending queues in getting patients admitted to hospitals and securing beds as there is a racket being run by all the service providers so that the relatives of patients can be overcharged and fleeced. Even in case of emergencies, relatives have to fill the forms and deposit advances before their near and dear ones can be admitted. The emergency facilities are also inadequate in most hospitals either because of shortage of staff or poor condition of equipments.

The overall opinion is that right from the ambulance driver to the doctor and to the owners/ directors of hospitals, everybody is corrupt and uncaring. Their only concern seems to be in minting money by taking advantage of the helplessness of the patients and their relatives, as well as by accepting huge commissions from pharma companies.

However, all is not bleak because there are good doctors and good hospitals where treatments are done either for free or for nominal costs. Such hospitals are generally run by philanthropic societies or trusts endowed by the rich and famous.

Answer 3.

(a) **(1)** Besides buying a desktop, he also bought a laptop.

(2) The plants were revived by the heavy showers of rain

(3) In spite of poor light, the men were forced to work.

(4) As soon as we saw a flash of light in the sky, we heard a loud explosion near us.

(5) Not having brought a pen, she is writing with a pencil.

(6) No city in India is as vibrant as Mumbai.

(7) Their dog is so friendly that it cannot be an effective guard dog.

(8) Sameer asked Ramesh whether he could help him with the homework.

(9) No nook or corner was left unchecked.

(10) The length (duration) of my stay is doubtful.

(b) **(1)** out **(2)** up **(3)** to
(4) at **(5)** for **(6)** of
(7) on **(8)** out **(9)** with
(10) amongst

(c) **(1)** were discussing **(2)** were talking
(3) would be **(4)** began **(5)** governing
(6) agree **(7)** arguing **(8)** had
(9) came **(10)** broke

Answer 4.

(a) **(i)** **(1)** immobile **(2)** anguished
(3) intact **(4)** shield

(ii) **(1)** All the metrical feet used in the English poetry have exactly one stressed and one or two unstressed syllables.

(2) Avinash was blamed four-square for the fiasco at the Annual Function of the school.

(3) The second day of the test match between India and Australia saw the fall of 19 wickets as both the teams witnessed unprecedented batting collapses.

(4) Sayani's selection as the Head Girl of the school was absolutely surprising.

(b) (i) When Ben and his father were walking towards the headquarters, the two things that happened simultaneously were that the latter tripped on the uneven surface of the cobblestoned ground and at the same instant there was also a loud bang and a sharp zinging sound with a scrunch of glass breaking.

(ii) When Ben looked at his father on the ground, the former's eyes were stretched wide and his mouth was grim with pain.

(iii) Ben's father asked Ben to run because the former was aware that a gun had been fired and both father and son were easy targets because of their immobility. Ben chose to disobey his father because he did not want to leave his father alone in that painful and critical condition.

(iv) Ben's logic for disobeying his father was that the shooter had used a .22 rifle with a high-velocity bullet, which was quite likely not to kill his father if the bullet hit the latter's body. For a lethal shot, the bullet needed to hit the neck or the head and Ben was just shielding the head.

(c) (i) (1) Father and son left the hall to walk across to the headquarters. In spite of street lights round the square, the cobbled road was dark and lumpy. The father tripped and went down on his knee.

(2) Simultaneously, there was a loud bang, a sharp zinging sound and a scrunch of glass breaking. Father's eyes were wide opened and mouth was grim with pain.

(3) He asked Ben to run for cover because it was the sound of a gunshot. Being halfway across the square, they were easy immobile targets, so Ben disobeyed his father.

(4) He told his father to stay down and asked whether he was bleeding. His father told him that he twisted his ankle.

(5) People ran out of the hotel. Amidst the confusion, his father was helped to his feet.

(6) When he was well surrounded, he finally took his son's help to walk.

(ii) They walked through the cobbled, dark and lumpy road, though, there were street lights. As the father tripped, there was a bang, a sharp sound and a scrunch of glass breaking. His father grimaced with pain but asked Ben to run, for it was a gunshot. Ben disobeyed and told his father to stay down and asked whether he was bleeding. His father told him that he had twisted his ankle. People from the hotel helped the father to his feet. They helped Ben's father to get up. When he was properly surrounded he took his son's help to walk. (100 words)

●●

ENGLISH-I 2016

QUESTIONS

Question 1.

Write a composition (in approximately 450–500 words) on ***any one*** *of the following subjects :* **[30]**

(You are reminded that you will be rewarded for orderly and coherent presentation of material, use of appropriate style and general accuracy of spelling, punctuation and grammar.)

(a) *Describe any three important lessons that you have learnt through the course of your life. How did you learn these lessons and why do you think they are so important ?*

(b) *Relate how the reading of literature has enriched and refined you.*

(c) *'Every cloud has a silver lining'. Express your views on the above statement.*

(d) *Deception.*

(e) *Argue for or against the statement : 'We live in deeds and not in years'.*

(f) *Write an original short story beginning with the following words :*

'The voice was familiar but I could not recognize the face......'

Question 2.

Imagine that you are the Head Boy/Head Girl of ABC Public School. You have to deliver a speech at the Farewell Party organised by your juniors. Prepare a suitable speech for the occasion using the notes given below : **[20]**

Years spent—early memories—association with juniors—skills acquired—values learnt—memorable moment—gratitude to teachers—message for juniors—future plans—thanks giving.

Question 3.

Answer sections ***(a), (b)*** *and* ***(c).***

(a) *In each of the following items, sentence A is complete, while sentence B is not. Complete sentence B, making it as similar as possible to sentence A. Write sentence B in each case.* **[10]**

Example : (0) (A) No sooner had the match finished than it began to rain.

(B) Hardly ....

Answer : (0) Hardly had the match finished when it began to rain.

(1) (A) *If he does not run fast he will lose the race.*

(B) *Unless ....*

(2) (A) *Nobody knew much about his plans for the future.*

(B) *Little ....*

(3) (A) *On seeing that rat, she shrieked.*

(B) *As soon as ....*

(4) (A) *You must never tell a lie.*

(B) *Under no circumstances ....*

(5) (A) *The boss said, "What an extraordinary success !"*

(B) *The boss exclaimed ....*

(6) (A) *It is too cold for me to step out without a shawl.*

(B) *It is so ....*

(7) (A) *I play both tennis and squash.*

(B) *Not only ....*

(8) (A) *The school team lost the cricket match although the captain scored a century.*

(B) *Despite the fact ....*

(9) (A) *Charles Dickens is the most interesting writer.*

(B) *No other ....*

(10) (A) *They had to shut down the computers before leaving the office.*

(B) *The computers ....*

(b) *Fill in each blank with a suitable word. (Do not write the sentence.)* **[5]**

(1) *After the argument, his friends turned ______ him.*

(2) *Ramesh turned ______ his father for help.*

(3) *The Headmaster will look ________ the cheating case.*

(4) *I look ________ to my grandfather for what he has achieved.*

(5) *She takes ________ her mother, she has her lovely green eyes.*

(6) *Ramesh has taken ________ computers like a duck to water.*

(7) *Will you entrust him ________ that message ?*

(8) *He is dishonest; you must not entrust any important job ________ him.*

(9) *Do not put ________ till tomorrow what you can do today.*

(10) *I put ________ at my friend's house for the night.*

(c) *Fill in the blanks in the passage given below with the appropriate form of the verb given in brackets. Do not write the passage, but write the verbs in the correct order.* **[5]**

I ______ (1) (realize) that I ______ (2) (know) one of the two men by sight, and I ______ (3) (spend) a few seconds thinking why he______ (4) (seek) me out on a Sunday afternoon.

During this pause, three small boys______ (5) (walk) up the passage from the house behind me, ______ (6) (thread) a way around me and the two men outside, and silently ______ (7) (climb) like cats up into a tree in the middle of the lawn outside. There, the three figures______ (8) (rest), ______ (9) (become) immobile, ______ (10) (lie) on their stomachs, deep in a secret game.

Question 4 .

Read the passage carefully given below and answer the questions (a), (b) and (c) that follow :

(1) Inside the cockpit of the World War II Mosquito plane, I could make out, against the light of the moon, the muffled head of its pilot and the twin circles of his goggles as he looked out of the side window towards me. Carefully he raised his right hand till I could see it in the window, fingers straight, **palm** downwards. He jabbed the fingers forward and down, meaning that we were going to descend and that I should follow him.

(2) I nodded and quickly brought up my own left hand so he could see it, first pointing forwards to my own control panel with one forefinger, and then holding up my five spread fingers. Finally I drew my hand across my throat. By common agreement this sign meant that I had only five minutes fuel left, and then my engine would cut out. I saw the muffled, goggled, oxygen-masked head nod in understanding, and then we were heading downwards towards the sheet of fog

(3) My plane stopped trembling and plunged ahead of the Mosquito. I pulled back on the throttle, hearing the engine die to a low whistle, and the other pilot was back beside me. We were diving straight towards the shrouded **land**. I glanced at my altimeter. two thousand feet, still diving.

(4) He pulled out at three hundred feet. The fog was still below us. Probably the fog bank was only from the ground to two hundred feet up, but that was more than enough to prevent a plane from landing without guidance. I could imagine the **stream** of instructions coming from the radar hut into the earphones of the man flying beside me, eighty feet away. I kept my eyes on him, following as closely as possible, afraid of losing sight for an instant, watching for his every hand-signal. Two minutes later he held up his clenched left fist in the window, then opened the fist to splay all five fingers against the glass, indicating that I should lower the undercarriage. I moved the lever downwards and felt the dull thunk as all three wheels went down, happily powered by hydraulic pressure and not dependent on the failed electrical system.

(5) The pilot of the shepherd aircraft pointed down again, for another descent. I managed to flick a glance at my fuel gauge : it was on zero, flickering feebly. For God's sake, hurry up, I prayed, for if my fuel failed me now there would be no time to climb to the minimum five hundred feet needed for bailing out. A jet fighter at one hundred feet without an engine is a death-trap with no chances for survival.

(6) For two or three minutes he seemed content to hold his position, while the sweat seat broke out behind my neck and began to run in streams down my back, sticking the light nylon flying suit to my skin.

(7) Quite suddenly he straightened out, so fast I almost lost him. I caught him a second later and saw his left hand flash the dive signal to me. Then he dipped towards the fog **bank**, I followed, and we were in it, a shallow, flat descent, but a descent nevertheless, and from a mere hundred feet, towards nothing.

Adapted from : ***The Shepherd*** **by Fredrick Forsyth**

(a) **(i)** *Given below are four words and phrases. Find the words which have a similar meaning in the passage :* **[4]**

(1) *go down* **(2)** *covered*

(3) *instrument* **(4)** *satisfied*

(ii) *For each of the words given below, write a sentence of at least ten words using the same word* ***unchanged in form, but with a different meaning*** *from that which it carries in the passage :* **[4]**

(1) *palm (line 4)* **(2)** *land (line 13)*

(3) *stream (line 18)* **(4)** *bank (line 35)*

(b) *Answer the following questions in your own words as briefly as possible :*

(i) *What was the first instruction of the pilot of the Mosquito plane ?* **[2]**

(ii) *What was the narrator's reply ?* **[3]**

(iii) *How did the other pilot let the narrator know that the wheels had to be lowered?* **[2]**

(iv) *Why did the narrator begin to pray towards the end ?* **[3]**

(c) *Describe how the pilot of the Mosquito aircraft guided the narrator to land, in not more than 100 words (Paragraphs 3 to 7). Failure to keep within the word limit will be penalised. You will be required to :*

(i) *List your ideas clearly in point form.* **[6]**

(ii) *In about 100 words, write your points in the form of a connected passage.* **[6]**

ANSWERS

Answer 1.

(a) THREE IMPORTANT LESSONS OF LIFE

Although not very old, I am old enough to clearly perceive, think and judge. These attributes have contributed in helping me to learn three very important lessons which will last with me a life-time: honesty, punctuality and respect.

The age old adage says that 'Honesty is the best policy' and on numerous occasions it has been proved true. In fact, being honest with oneself is more important than being honest to others because that gives us a clear conscience and enables us to endure fearlessly. The most significant incident that helped me to learn this lesson was when I admitted to have dropped the catch during the final match of an interschool tournament that we ultimately lost. However, because of the frank confession, I won the admiration of one and all.

Punctuality, as mentioned earlier, is another virtue I have cultivated. Consequently, I have never been reprimanded ever during my entire school life. This has also allowed me to value the essence of time and it has saved me in more than one occasion. One of my friends, who has never been punctual, actually missed out an hour of a paper, during the final board exams because he failed to reach the exam venue on time. That was again an important lesson learnt, not only by my friend, but also by me.

Respect is another important thing that lasts a lifetime for every individual and if our respect is taken away from us, then our lives become meaningless. In order to gain the respect of others, it is important to show respect to others as well. If somebody insults or humiliates us then it is quite natural that we try to pay back in the same coin. The same theory applies to others also, if we humiliate or insult others then they are bound to retort in the same manner. So it is imperative to show respect in order to gain respect. The best example of this that I have experienced is in the form of our driver, Uncle Raju. My father admonished me severely whenever I addressed him as 'Driver' or 'Raju'—it was always 'Uncle Raju'. It is due to this reason that he has stayed around with our family for close to two decades. His loyalty and devotion has been unwavering to us and were revealed at the time when my grandfather lost consciousness during a car accident. In spite of being severely injured and bleeding profusely himself, uncle Raju managed to reach the hospital and saved my grandfather's life.

Also, it is important to respect our own selves because that prevents us from falling low and leading immoral lives. Self-respect is the most crucial aspect of one's life. When everything else in the world fails you, you will always have your self-respect to fall back on.

(b) ENRICHMENT AND REFINEMENT THROUGH LITERATURE

In today's world of cut-throat competition and skill-based learning, parents, particularly in India, encourage their children to pursue higher education in the science field, especially in engineering and medicine as the number of job opportunities in these fields are far more than in any other field. Consequently, there is a predominant idea among large sections of the society in India that those who pursue subjects related to Humanities and Social Sciences are less intelligent and that these subjects are only of academic importance, having no real value in the practical world as they offer limited employment opportunities.

However, such people are mostly ignorant or not highly educated as they tend to make injudicious comparisons between fields of studies. Most of them do not realize that the Arts and the Sciences are interconnected. As there have been scientific movements in the course of history, whereby earlier pseudo-scientific notions were replaced with modern inventions and discoveries, so there also have been artistic and literary movements that have displaced the age-old concepts in a similar manner, transposing them with newer approaches. Accordingly even the study of literature has become more practical in approach and has adopted the experiential method of learning. Few people are aware of the fact that the inspiration behind Einstein's theory of Relativity was the writings of the sixteenth century philosopher and visionary, Giordano Bruno.

The study of literature has actually had a humanizing effect on people. It has opened the eyes of people to the problems of the world and has enabled them to empathize with the victims of an unequal social order. Knowledge of these has induced many industrialists and business people to involve themselves in philanthropic activities. Also, literature has provided us insights into the cultures and practices of different countries across time. Through the study of literature, we have come to know about the social orders of ancient Greece, India and Mesopotamia as well as of Medieval Europe and Early Modern America and Europe.

Literature also provides us with a welcome respite from the hectic day-to-day lives because it is also a source of entertainment in the form of theatrical productions and films. Good films and plays provide families the opportunity to be together, even if it is for a short while. This is important in maintaining the familial bond because every individual is always busy in chasing his or her dreams.

Lastly, one cannot help but admit that good stories are always welcomed and they are an integral part of human existence, and all good stories are nothing but literature. Thus, we cannot deny the fact that literature helps in enhancing the whole personality of a person. It has the power to lead the whole society towards the right path.

(c) EVERY CLOUD HAS A SILVER LINING

Ups and Downs are part and parcel of life. In fact, if these did not exist, life would have been boring and meaningless. All these experiences enrich us in some way or the other and help us to become determined individuals. In fact, all the hardships and difficulties we face in life make us richer in experience and enlighten us about tackling adverse situations. Therefore, it can be aptly said that 'Every Cloud has a Silver Lining'.

To explain the above, let us examine the difficulties of my friend Rajiv. He was an average person, coming from a middle-class family. His parents, who were well-educated themselves, wanted him to have the best education and so he was admitted to the best school in the city. As a student, he excelled neither in academics nor in sports and was always placed in the middle of his class in both. Even in extra co-curricular activities, Rajiv did not leave behind any distinctive mark. Being an obedient child, he was always guided by parental advice, which prevented the development of his own personality and as a result, he lacked self-confidence.

However, Rajiv's life went topsy-turvy with the sudden demise of his father in an accident. The family was staring at imminent poverty because his father was the sole bread-winner and the pension that they were to get was not enough to fund the education of both Rajiv and his younger sister. All of a sudden the entire burden of the family descended on Rajiv's young shoulders, but that was a blessing in disguise because the misfortune made him resolute. He developed into a very strong-willed individual who was ready to take on the world legitimately. Rajiv took stock of the situation and started providing private tuitions to supplement the family income. In doing so, he realized that his actual calling in life was to be an academician. Gradually, he cleared all the examinations with flying colours and at a very young age joined an elite institution as a teacher. There has been no looking back since then. Had the mishap never occurred, then Rajiv might have followed and done what his parents deemed fit for him instead of realizing his own potential and calling in life and society would have been deprived of a very good teacher.

Like Rajiv, there are millions of such individuals who suffer major setbacks in their lives but still manage to bounce back. So, one should never feel despondent because of problems, as no individual in this world leads a trouble-free existence. Clouds of difficulties may appear, but they always carry an indication of something better that might happen.

(d) DECEPTION

Lexically, the word 'deception' implies 'make-believe', that is, convincing a person that something exists whereas, in reality, it does not exist. Deception is a simple tool for enticing people and trapping them to believe in something unbelievable.

This is an evil that has subsisted among human beings since time immemorial. In literature and history we come across several uses of deceptive methods; even in our contemporary world, deception is rampant. In fact, such is the enormity of deception that it is often difficult to distinguish the real from the fake. Consequently, even today we come across reports of how many large corporations, especially related to chit funds, dupe their investors through unrealistic schemes. As a result of their 'get-rich-quickly' schemes lakhs of innocent investors lose their lifelong savings because they are shown the dream of coming out of their miseries but are pushed to the brink of perishing.

Deception is also a major tool used in forging currency notes and other negotiable instruments like cheques and drafts. Governments across the world have introduced many new techniques to ensure that their markets are not flooded by fake currency notes or negotiable instruments but, so far, there has been no foolproof methodology to prevent the menace. Master criminals always manage to deceive by preparing exact copies of the originals. The same applies for important documents like passports and visas.

With the dawn of online shopping, fraudsters have had a free-run as it is practically easy to dupe people and escape without leaving a trace. People end up paying money for products bought online but never actually receive the goods on many occasions, or even if they do, such goods are either defective or elaborately wrapped empty boxes are provided. Consequently, tracking cybercriminals is very difficult and the number of solved cases is low.

However, like everything else, even deception is a double-edged weapon: it can be used against a deceitful enemy and so it serves as an important military tactic, especially in guerrilla warfare. We have read how Babur, the first Mughal Emperor enticed Ibrahim Lodhi to charge headlong and then fall into the covered trenches on the battlefield or how Chhatrapati Shivaji, the great Maratha warrior, used deception to kill Afzal Khan and harass Aurangzeb. Nobody can ever forget the deception carried out by Mark Antony in inciting the Romans from the Forum to rebel against the conspirators who killed Julius Caesar.

In present day military tactics, elaborate plans are drawn out to combat the enemy. The best known methodology is to plant spies among them so that delicate information can be easily retrieved and the opponent's actions can be monitored.

(e) WE LIVE IN DEEDS AND NOT IN YEARS FOR THE MOTION

A typical Indian tradition is to touch the feet of seniors and respectable persons on any festive occasion and seek their blessings and the typical blessings that are given are — 'may you live long' or 'may you live a hundred years' and so on. The idea is to wish for someone's prolonged and healthy life. Very rarely do we come across people wishing us for our deeds. In fact, all over the world, the emphasis is to have a long and healthy life, not a fruitful one.

However, in reality, we actually live by our deeds and not our years. There are many people who live more than hundred years and their names get published in print and digital media, but that fame is too short-lived, not even lasting an hour. That is because such individuals only live long, without making any significant contribution to the world.

On the other hand, there are people who have had shorter life-spans and still have managed to leave behind a mark on people's minds, either for their good or bad deeds, but definitely for their deeds. They include kings, rulers, scientists, teachers, philosophers, researchers, builders, artists, men of letters, sports persons, warriors, explorers and so forth. Their actions touched and influenced the lives of people who were not only their contemporaries but who were their progeny or successors.

Also, not all of these great men were either completely good or completely bad. While some people loved them unconditionally, others hated them in equal measure. This is because, if a person cannot be dear to everybody, he or she will always have an antagonist. The great Sir Isaac Newton, who not only discovered gravitation but also many scientific phenomena, was a lonely individual and had an unhappy childhood. Born as posthumous child, he was neither welcomed in the household of his step-father nor at his grandparents'. At school too, he did not have many friends. However, the world does not remember him for the tragic loner that he was but for the great scientist. The same can be said about Nikola Tesla.

From another point of view, posterity can never forget the actions of such notorious rulers like Napoleon, Hitler, Mussolini, Stalin and Pol Pot. They represent some of the most horrific chapters in human history because of the bloodshed and the genocide that followed their reigns. But, from their life histories we also learn a great deal about the military leadership, their administrative skills and the strength of their personalities.

(f) THE VOICE WAS FAMILIAR......

The voice was familiar, but I could not recognize the face. It was difficult to recognize him from that angle, especially at that time of the day. He was already getting down from the bus and I could barely manage a glimpse of his face while he walked past my window seat. During the evening rush hour, everybody is in a hurry to get back home at the earliest and so they are almost always on the run after alighting from trains and buses.

However, I am sure it was Pankaj's voice, except the face was not his. We were in the same school but three years ago his father got transferred to Mumbai and the family moved out. Since then, Pankaj had been completely out of the radar; nobody knew where he was because of his absence from social media also.

Although quite average in studies, Pankaj was very good in Mathematics. In almost every exam he used to top the class in Mathematics. Maybe he was too lazy to study other subjects with equal fervor or maybe he was plainly dull in them. Pankaj was a brilliant footballer also. He got selected for the senior school team while he was studying only in Class 5. Our sports teacher used to call him the 'Midfield Wizard'.

As long as he played for the school, we never lost a match in any of the inter-school tournaments. In fact, on two occasions he was also adjudged the Man of the Tournament. The first time was in Shimla during the All-India ICSE Schools Football Championship and the other was in Kalimpong during the Independence Interschool Football Tournament. When the team returned with the trophies, the Principal had specially honoured him with a medal and a certificate.

A week later, I heard the same voice once again, asking for his change from the bus-conductor. At that very moment, I turned round and our eyes met. The voice was the same but the face was different! But, he recognized me! He came to meet me and we started talking. I was absolutely dumb-founded to at what I heard next.

Pankaj's family was holidaying in Goa and they had rented a private taxi to explore Hilly Goa. On the way, a lorry hit them from behind and as he was not wearing a seat-belt, his face was completely smashed by the shards of the broken windscreen. When he regained consciousness after some days, he was in hospital. Doctors had to do a series of plastic surgeries to reconstruct his face. But, he never got back his original face.

Answer 2.

It is always an emotionally difficult moment to bid farewell to something or someone with whom you are associated with all your heart and soul. Like our parents, who have nurtured us from our very infancy, our school also has a great role to play in defining who we are. For fourteen long years, it has nurtured us carefully so that we become responsible citizens of tomorrow. And to be invited to share the memories on the occasion of the farewell is a special honour that shall be cherished forever.

I joined ABC Public School in Nursery and since then it has been a long journey. I still remember some special memories; for example, the long lunch break that was spent in playing games and the eagles swooping down to snatch our food, the kind teachers with an occasional severity, when we were mischievous. I shall never forget the enthusiasm during the interschool tournaments, hockey, cricket, football, basketball and athletics, when me and my classmates were pumped up with energy, or the long hours of practice in preparation for the Annual Sports, Quiz and Debate Competitions.

None of us can ever forget the fun we used to have, when teachers went to attend meetings and the senior students used to come over and manage the classes and also the time when we did the same. All of us looked forward to these opportunities because they helped us to bond with our seniors and juniors and has immensely contributed to the learning and sharing process. This is a wonderful tradition that has been kept alive by generations of students who have been part of this hallowed institution.

Our school has not only helped us to learn from books but has also taken care to enhance our skills, based on the individual aptitudes of each student and through a scientific approach.

Thus, all the students who go out into the larger world after the initial years, are equipped in every sense of the term—be it with superior language and communication skills, computers or the lab-based subjects—through continuous field-trips, laboratory and library work as well as hands-on training programmes. All these have combined together to foster a spirit of experiential learning. Consequently, a lot of values have also been instilled in us as a result. We have all realised the importance of punctuality, of team spirit, of learning and the importance of showing respect in order to gain respect. In fact, one of my most memorable moments was when I was in class 8.

One of my seniors got punished for not being able to answer a simple question and I happened to laugh at him for it. The very next moment, the teacher gave me the same punishment and then I was laughed at. At that moment I learnt, albeit the hard way, how important it is to show respect in order to gain respect. And I thank not only that particular teacher but all the teachers throughout my school life for rectifying me whenever I made mistakes. They were like second parents to me. Their admonishments were as important as the love they showered. It is also an important message that I would like to pass on to my juniors before leaving my second home.

As for our future, all of us have different individual plans. Some will prepare for engineering, some for medical, some will try to join the defense forces and others will try to pursue professional and academic careers as per their liking. We don't know where we will reach in the end but we will never be able to deny the influence of our school in shaping us and our lives.

Finally, I would like to thank our Principal, administrators, all the teachers, staff members and all my juniors for this wonderful opportunity to express not only my views but also the views of the entire batch and for listening to me patiently.

Answer 3.

(a) **(1)** Unless he runs fast, he will lose the race.

(2) Little did anybody know about his plans for the future.

(3) As soon as she saw the rat, she shrieked.

(4) Under no circumstances must you tell a lie.

(5) The boss exclaimed that it was an extra-ordinary success.

(6) It is so cold that I cannot step out without a shawl.

(7) Not only do I play tennis, but also squash.

(8) Despite the fact that the captain scored a century, the school team lost the cricket match.

(9) No other writer is as interesting as Charles Dickens.

(10) The computers had to be shut down by them before leaving the office.

(b) (1) against (2) to (3) into
(4) up (5) after (6) to
(7) with (8) to (9) off
(10) up

(c) (1) realised (2) knew (3) spent
(4) had sought/was seeking (5) walked
(6) threading/threaded (7) climbed
(8) rested (9) became/becoming
(10) lying

Answer 4.

(a) (i) (1) descend (2) shrouded
(3) gauge (4) content

(ii) (1) The only cover that was available for the platoon was a series of *palm* trees.

(2) To *land* such a heavy plane, in such weather conditions, is always a difficult proposition for even the most experienced pilots

(3) It was a wondrous sight: the clear blue sky, the crystal-clear sparkling waters of the *stream* flowing with a musical gurgle and the outlines of the blue-white mountains in the distance.

(4) The *bank* reported huge losses in the last quarter of the 2014-15 financial year because of bad debts.

(b) (i) The pilot of the Mosquito plane raised his right hand with the fingers straight and the palm downwards. He then jabbed his fingers forward and downwards indicating that they were to descend and that the narrator was to follow him.

(ii) In response, the narrator pointed at the control panel of his craft with the forefinger of his left hand and then held up five spread fingers. Finally, he drew his hand across the throat. All this meant to communicate that he had only five minutes of fuel left in the craft.

(iii) The other pilot signaled him to lower the rear wheels by first clenching his left fist in the window and then opening it to spread all five fingers against the glass.

(iv) The narrator began to pray towards the end because he knew that if the fuel got emptied at that point of time, he would not be able to climb to the minimum five hundred feet required for bailing out. A jet at hundred feet, and without an engine, was a veritable death-trap.

(c) (i) (1) The narrator plunged head of the Mosquito plane only to pull back the throttle so as to reduce the engine speed almost to a whistle, to help in the descent.

(2) At three hundred feet, the other pilot pulled out ahead so as to guide the narrator.

(3) From a distance of eighty feet, the other pilot indicated to the narrator to lower the under-carriage.

(4) Then the other pilot indicated to the narrator about another descent.

(5) For the next two or three minutes, the other pilot held on to his position, before straightening out suddenly and then indicating the narrator to dive.

(6) Lastly, the other pilot dipped into the fog bank ahead of the narrator to guide the latter to a successful landing.

(ii) Plunging ahead of the Mosquito plane, the narrator pulled the throttle to reduce engine speed almost to a whistle, enabling descent. To guide the narrator, the other pilot pulled ahead at three hundred feet. From eighty feet away, he indicated to the narrator to lower the wheels and once again, another descent. For the next few minutes, the other pilot held on to his position before straightening out suddenly for another dive. Finally, he dipped into the fog bank, guiding the narrator to a successful landing.

(86 words)

●●

ENGLISH-I 2015

QUESTIONS

Question 1.

*Write a composition (in approximately 450–500 words) on **any one** of the following subjects :* **[30]**

(You are reminded that you will be rewarded for orderly and coherent presentation of material, use of appropriate style and general accuracy of spelling, punctuation and grammar.)

(a) *Courage does not mean the absence of fear, but implies an attempt to conquer fear. Describe some of your worst fears and how you managed to overcome them.*

(b) *Grandparents are a source of joy to us. Narrate the most memorable experience you have had with your grandparents.*

(c) *'Films should be made to escape from reality, not to remind us of how grim life is.' Argue for or against this statement.*

(d) *How far do you agree that nice people are seldom successful and thoroughly successful people are seldom nice ?*

(e) *Eyes.*

(f) *Write an original story beginning with the following words :*

'The news came as a pleasant surprise ....'

Question 2.

Write an article for your School magazine on a competition that was recently held in your school. Write the article in about 300 words using the points given below : **[20]**

Name of the competition—nature of event—organizers—number of participants—chief guest—judges—quality of the competition—criteria for judgment—winners—overall experience.

Question 3.

*Answer sections **(a)**, **(b)** and **(c)**.*

(a) *In each of the following items, sentence A is complete, while sentence **B** is not. Complete sentence **B**, making it as similar as possible to sentence **A**. Write sentence **B** in each case.* **[10]**

Example : (0) (A) If you want to earn well, you must work hard.

(B) To earn ....

Answer : (0) To earn well, you must work hard.

(1) (A) *Sachin is an outstanding sportsman and a good artist.*

(B) *Besides ....*

(2) (A) *Mother said to Paul, "A courier came for you yesterday."*

(B) *Mother told Paul that ....*

(3) (A) *As soon as the minister took the oath, the spectators started applauding.*

(B) *No sooner ....*

(4) (A) *He is so short that he cannot be a soldier.*

(B) *He is too ....*

(5) (A) *The lawyer examined the documents with utmost care.*

(B) *The documents ....*

(6) (A) *I will buy the horse if it is quite sound.*

(B) *Unless ....*

(7) (A) *Only Shakespeare could write such a tragedy.*

(B) *No one ....*

(8) (A) *Rita is a journalist and writer.*

(B) *Not only ....*

(9) (A) *Although Lakshmi worked very hard, she failed in the examination.*

(B) *Notwithstanding Lakshmi's ....*

(10) (A) *No one would deny that he was totally fit before his death.*

(B) *Everyone ....*

(b) *Fill in each blank with a suitable word. (Do not write the sentence.)* **[5]**

(1) *Do you want to take ________ every single object from the box ?*

(2) *The plane must take ________ before dark.*

(3) *The children spoke ________ a whisper as they were scared.*

(4) *We must not speak ill about a person ________ his back.*

(5) *He disliked the play so much that he walked ________ in the middle of the first act.*

(6) *Walking barefoot ________ the grass is very good for health.*

(7) *This rule does not apply ________ you as you are less than eighteen years old.*

(8) *You must apply ______ the job immediately.*

(9) *They have been living in Delhi ________ ten years.*

(10) *Mrs. Kapur has been living in Chennai _____ 2000.*

(c) *Fill in the blanks in the passage given below with the appropriate form of the verb given in brackets. Do not write the passage but write the verbs in the correct order.* **[5]**

Two carpenters ________ (1) (work) on Mr. Sharma's roof. When they ________ (2) (stop) work at 6 p.m., they________ (3) (leave) their ladder________ (4) (lean) against the house. At 7 p.m., Raju, a thief, passed by the house and ________ (5) (see) the ladder. The house ________ (6) (be) now empty as Mr. and Mrs. Sharma ________ (7) (go) to the market. Raju ________ (8) (climb) up the ladder, ________ (9) (get) in through a first-floor window, and ________ (10) (go) straight to the main bed room where he stole all of Mrs. Sharma's jewelry.

Question 4 .

Read carefully the passage given below and answer the questions (a), (b) and (c) that follow :

(1) To be encumbered with a corpse is to be in a difficult position. True, any doctor, even one just out of medical school, would have been able to diagnose the cause of death. The man had died of heart failure or what the doctors call cardiac **arrest**. The cause of his heart having stopped pumping blood was that someone had slid a sharp sliver of steel between his ribs just far enough to penetrate the great muscle of the heart and to cause a serious and irreversible leakage of blood so that it stopped beating. Cardiac arrest, as I said.

(2) I wasn't too anxious to find a doctor because the knife was mine and the hilt had been in my hand when he died. I stood on the open road with the body at my feet and I was scared, so scared that the nausea rose in my throat to choke me. This particular body had been a stranger—I had never seen him before in my life.

(3) I was unarmed, except the 'sgian dubh'—the black knife—which I always carry. The 'sgian dubh' is a much underrated weapon. Mine is at least a hundred and fifty years old. The ebony handle is ribbed on one side to give a good grip, but smooth on the other side so it **draws** clear without catching; the blade is less than four inches long; the stone **set** in the handle balances the knife so that it makes a superlative throwing weapon. I carry it in a flat sheath in my left sock.

(4) This is how it had happened.

(5) A little after I had driven out of the city, I saw a car ahead, pulled off the road, and a man waving the universally recognized distress signal of the stranded motorist. It turned out, quite naturally, that there was something wrong with his car and he couldn't get it to move. I got out, walked over to his car and peered at the exposed engine.

(6) He didn't use the gun straight away. He first tried to take a swipe at me with a well-designed little club. I turned my head and saw his upraised arm and dodged sideways. If the **club** had connected with my skull it would have jarred my brain loose; instead it hit my shoulder and my whole arm went numb

(7) I hopped back and groped for the 'sgian dubh' as I went. Fortunately it's a left-handed weapon which was just as well because my right arm wasn't going to be of any use.

(8) He came for me again but when he saw the knife he hesitated. He dropped the club and dipped his hand beneath his jacket and it was my turn to hesitate. But his club has a leather wrist loop and the dangling weapon spoilt his draw and I jumped him just as the pistol came out.

(9) I didn't stab him. He swung around and ran straight into the blade. He sagged against me with a look of surprise on his face. Then he went down at my feet and the knife came free.

(10) So there I was on a lonely road with a newly created corpse at my feet and a knife in my hand, a bad taste in my mouth and a frozen brain.

(11) From the time I had got out of my car to the moment of death had been less than two minutes.

Adapted from : *Running Blind* by Desmond Bagley

(a) **(i)** *Given below are four words and phrases. Find the words which have a similar meaning in the passage :* **[4]**

(1) *burdened*
(2) *enter*
(3) *not given much importance*
(4) *most effective*

(ii) *For each of the words given below, write a sentence of at least ten words using the same word* ***unchanged in form, but with a different meaning*** *from that which it carries in the passage :* **[4]**

(1) *arrest (line 3)* **(2)** *draws (line 15)*
(3) *set (line 15)* **(4)** *club (line 25)*

(b) *Answer the following questions in your own words as briefly as possible :*

(i) *How did the stranger die ?* **[3]**

(ii) *Why was the narrator scared ?* **[2]**

(iii) *Describe the narrator's weapon.* **[3]**

(iv) *Why did the narrator stop his car and get out ?* **[2]**

(c) *Describe the incident of the killing of the stranger in not more than 100 words (paragraphs 5 to 9). Failure to keep within the word limit will be penalized. You will be required to :*

(i) *List your ideas clearly in point form.* **[6]**

(ii) *In about 100 words, write your points in the form of a connected passage.* **[6]**

ANSWERS

Answer 1.

(a) ON COURAGE

Fear is natural to all humans. It doesn't spare anyone—babies, young ones or adults—all come under its grip. Some fears develop due to bitter experiences; some are the figments of imagination; some develop due to insecurities. Whatever may be the reason behind it, it is one of the most negative emotions that always comes in the way of success. The only antidote for it, is courage which is definitely not absence of fear; but an indomitable will to master it.

Can we say, a soldier stepping onto the battlefield does not have fear? Doesn't the one who dare to climb the Everest experience fear? Do not people who address a public gathering have stage fright? A child who learns to balance on a bicycle, a young person who ventures out on the road to test his driving skills don't they undergo fear? Yes, fear is a part and parcel of our life but we do conquer it, don't we?

A fear that literally brought me nightmares was that of swimming. Just the image of a swimming pool brought the fear of drowning. It was mainly due to the memory of a tragic incident that once happened in front of my eyes. I decided that I had to blot out that memory or I would never learn to swim. I was still brooding on the side of the pool when an instructor laid a friendly hand on my shoulder; the next thing I knew was that I was deep in the water trying to get my moorings by frantically flapping my limbs. Plunging into the waters never posed a problem for me after that. Sometimes, taking the plunge is the only way to beat fear.

Another phantom that haunted me was my inability to address a gathering. The very presence of the mike was an intimidating proposition. The moment I looked at the gathering, my mind would go blank. Then I read about George Bernard Shaw, the famous writer, who was shy of speaking. He couldn't even muster up enough courage to call on a friend at home. Then he realized if he wanted to be somebody in the world, he had to get rid of his timidity. He straight away joined a debating society and started to speak on anything and everything. In a matter of few days he overcame his fear. As days went by, he came to be known as one of the best speakers of the Twentieth Century. This was exactly what I did! I took inspiration from his example and enrolled myself in the next public speaking competition in my school. I gave myself a lot of auto-suggestion and visualized myself doing so well that it actually happened!

So I would say, fear is common to all of us. Courage is to identify its source, seek remedies, cultivate belief in yourself, inculcate a never say die attitude and the world would lie before you, studded with opportunities!

(b) GRANDPARENTS, A GIFT OF JOY

It's my strong belief that since angels cannot live in the world, God blessed children with their grandparents. My grandparents mean the world to me. When I am hurt by the world, when I feel lost, when I am riddled with doubts, I run to them for solace. They treat me with infinite patience. They lavish their unconditional love on me. A few moments with them make me feel substantially happy.

I still remember the day, when I reached home with tears of humiliation pouring down my cheeks, and planking down on the bed, sobbing my heart out, my Grandma came and inquired but I didn't reply. Grandpa cleared his throat in obvious discomfort at this blatant show of weakness. I didn't lift my head.

A few minutes later when the avalanche of emotions was over, grandpa sat beside me offering me a bar of chocolate and a glass of milk. He softly asked me why I was upset. Something in his quizzical smile and his gentle tone, prompted me to speak to him. I had gone out to try my new bicycle, which was a wee bit high for me. I had just climbed on it when I lost balance and fell. The local bully, Rajiv shouted, "Jyotika can't ride a bicycle even at this age". His friend Manoj crooned, "Such a big girl, and she can't get on to the cycle". Grandpa listened with patience to all my outpourings, smiled and said, "If you give up due to some taunts, you are proving that you are a loser. Fight back, instead of wallowing in self-pity. Show them you have the pluck to prove them wrong."

Grandma added,"It's all in the mind, baby. If you think you can, you can. It's all about having belief in yourself. You have to turn failure into success." She took my hands and continued, "Use all the pent up anger to your advantage. If you succumb to circumstances, they will break your back . If you stand up and fight, there's nothing in this world that you can't accomplish."

Slowly, sense dawned on me. I went out with my head held up. There was a new zest in me, I looked around. My tormentors were playing cricket. Without looking at them, I took a deep breath and heaved myself on to the cycle. To my delight I found myself whizzing past them, waving to them with a new found confidence.

When I came back, after an exhilarating ride, I found the boys waiting for me. They had a bunch of flowers in their hands; they clapped and shouted, "Here's a bouquet for the brave lady." My grand-parents were proudly watching from a distance. I was sure they had persuaded the bullies to congratulate me. My cup of happiness overflowed when I stepped inside. On that day I realized how much they cared for me.

Today they are old. Their eyes have become dim. Time has drawn many a wrinkle on their faces. Their steps are faltering; but one thing that has not changed is their love for their children and grand children. I singularly feel blessed with their presence. Just as they have made my childhood wonderful, I wish to make their old age wonderful so that when they say the inevitable goodbye, there will be lingering smile lighting up their faces.

(c) FILMS SHOULD BE MADE TO ESCAPE FROM REALITY NOT TO REMIND US OF HOW GRIM LIFE IS

FOR THE MOTION

You open the newspaper in the morning, filled with stories of theft, rape, or murder stare at your face. You switch on the television, stories of the same kind, or of scheming, plotting characters in the daily soaps, assail your sense and sensibilities. Do we also want movies to portray such grim realities ? O, lord save me from them ! I want to keep depression at bay.

I'm a youngster, stressed with the daily dose of homework, project submission, periodical tests and examinations. I want to see movies that take me to a world of laughter and entertainment. In short, I'm all for movies that give me a feel-good experience, lighten me and make me forget about the stress and difficulties around me.

You may call me an escapist—running away from realities. You may also argue that there are many columns in papers where you can read about beautiful things. There are many channels that give you ample opportunities to watch dance shows, music shows and laughter shows. So, what's the sense in watching mindless movies where actors run around trees, dance on streets and mountain tops, and indulge in fistic-fights? Well, I may jump into your wagon a few years later, when I'm a little more matured to realize the cathartic effects of watching tragedies; but not now.

Let's take, for example, out and out entertainers that have been recently released. It's worth the money you spend. You laugh, feel thrilled, marvel at the technical finesse of the scenes and come back with your mind refreshed. In contrast, you spend a lot of money on tickets, spend three hours shedding tears or feeling awful, watching all those atrocities unfolding before your eyes, only to come home with a heavy heart. Well, the choice is yours.

Movies must be a source of entertainment, presented in such a manner that it certainly leaves a positive impression on its audiences. They must teach you how to deal with difficult situations; they must be a source of inspiration. For instance, the main character revolves around drama, romance, comedy, all emotional aspects are dealt with, in order to give a complete package, which certainly entertains the audience, but the central idea revolves around the basic ethical values, which are indirectly budded in us.

What I strongly feel is that, the type of movies one likes or dislikes depends on one's age, disposition and mood. Take for example, small children. They like fairy tales, cartoon films and other animated movies. The intellectuals seem to like art movies in which social themes are dealt with. The college going youngsters have a penchant for thrillers and action movies. You cannot have a blanket rule that only movies that show reality are good. To each his own; but my vote goes for clean entertainers with a good story line. I don't mind even if they are pure fantasies!

(d) SUCCESS AND ITS CONSEQUENCES

There is a general belief that successful people are selfish, self-centered and ruthless. In contrast, many feel that the not so successful people are gentle, caring, honest and kind-hearted. What I feel is that good and bad are everywhere; success definitely is not a measuring stick for determining character. Another factor that we have to keep in mind is what we actually consider as success. Is it money, status or prestige? Rich people, especially business people, are deemed to be money-minded. They have that killer instinct, which makes them seem rather unkind. They may not give so much consideration for emotions. If they do that, they can't survive in business but this doesn't mean they are intrinsically bad. Take for example, Bill Gates, one of the richest man who is yet so charitable. Coming to our own neighborhood, there are many business tycoons, who have a heart of gold. They build hospitals, orphanages and donate money for noble causes.

As said earlier, success does not always mean money. Anyone who accomplishes something for self, others or society in general, are successful. What about the Father of our Nation, or Mother Teresa? Aren't they successful people? Can you think of anybody nicer than them? Pandit Nehru, with his infinite love for children, and his countrymen– wasn't he rich, nice and successful?

When we think of ordinary people who have not scaled the heights of worldly success, there are plenty of unsung heroes. They fulfill the duties of good parents, sons and daughters, wives and husbands; doing good in their own spheres, contributing to the welfare of near and dear ones; and contributing to the well being of society in general. Pitch in along with them the social workers, the do-gooders who work without fan fare, doing their might/deed for others: they are the real people and ever so nice!

We should also not forget that sometimes the lack of money and success brings out the worst in humans. Inferiority complex leads to anger, jealousy and frustration. This mentality paves way for many undesirable traits. If such people retain their niceness, we should consider them and ourselves blessed. Neither should we forget that sometimes money and position make people arrogant and cunning. The basic nature of an individual cannot be determined with their rate of success or their bank balance. These are not rules but it just happens, it is a matter of chance and not a matter of choice.

So, ultimately, everything boils down to the fact that good and bad does not discriminate success or lack of it. It's all a matter of a human being, being human.

(e) EYES

'You are the apple of my eye' says the mother lovingly to her child. "I shut my eyes and all the world drops dead; I lift my eyes and all is born again' says the poetess, Sylvia Plath. All these tell us how important eyes are in our life. We realize their importance when our eyesight is impaired. Imagine a world without colours, without the sight of your near and dear ones: unable to read directly and watch television or a movie! 'Of all the senses, sight must be the most delightful'—these words seem to be wrenched out of the heart of the visually impaired Helen Keller, summing up the feelings of being blind.

Eyes are our windows to the world. We see and understand our surroundings with their help. The good, the bad, the ugly—all seem to flow into our consciousness through these important visual organs. And there is also the mind's eye through which we perceive matters of importance.

Eyes do speak volumes. They are the mirrors of our souls. Our innermost feelings are reflected through our eyes. It is said, one can smile and smile and be a villain; but eyes are a true give away. According to a saying, "Eyes shout what the lips fear to say." They reflect the truth. That's why babies have honest, serene eyes, a reflection of absolute innocence. Our innermost feelings of happiness, sorrow, love, tenderness, anger, jealousy all come out in undiluted form through our eyes. When in sorrow, it is the eyes that lighten our hearts by shedding tears. The very same organs sparkle in laughter to express our happiness.

Eyes are also indicators of our health. Tired people are blear-eyed. Energetic people are bright eyed. Doctors examine our eyes to diagnose problems like anemia, jaundice, etc. As old age creeps in, our eyes become weak and we have to rely on glasses.

Our eyes are the most beautiful of all our features. Literature is full of lotus eyed, doe eyed, or almond eyed beauties, though their colors differ according to races. While the occidental beauties have blue, green or topaz eyes, the oriental ones are famed for their jet black or dark brown hues.

Lastly, we can say, life may be bleak without eyes; but that is not the end of life. Many, who are not blessed with sight, overcome their affliction with indomitable courage. They can maneuver through crowded areas, cross roads and do innumerable things by making use of other sharpened senses. People like Helen Keller educate themselves by mastering Braille and other educative programs. Long live their spirit!

(f) A PLEASANT SURPRISE

The news came as a pleasant surprise. It was uncanny to think, that the mail that came a couple of minutes back had the power to change my life. I have been selected to be a part of the mission to Antarctic. It was a dream come true.

The icy beauty of the polar region had always fascinated me. I had often wondered what life would be there! The bone freezing winds, the vast expanse of ice-sheets, and the life in an igloo, how do people manage to live there! And my favorite penguins!

Ever since I saw a picture of an Antarctic penguin in a magazine, I had fallen in love with the cute, fluffy, cuddly-looking creatures. That prompted me to know about the penguin communities, their way of life, and their adaptations to survive in the coldest region of the world. Just the imagination of the way they huddle like our cricketers do, brings a smile on my face. I'm sure the way they shift on their feet to take turns, to get a patch of sunlight looks like a well coordinated dance movement. It seems, they do have 'high temperature' days! They keep their flippers away from their bodies to cool themselves on such days. Definitely, they will make a pretty picture in black and white. I'm also told that the sight of rows of penguins walking out of the sea in rhythmic steps is a magnificent one.

Another thing which fascinates me is the life in an igloo. I have always wondered how could a roof made of thick ice slabs be habitable, and that too an extent that it provides shelter and warmth in such harsh climatic conditions. The glowing lights from the igloos situated between the widely spread serene ice sheets, are truly a panoramic view.

Now, I'll be able to see all these with my own eyes! The journey so far has not been an easy one. It required a lot of paper work, answering quizzes over the phone, seeking help from an uncle who is involved in expeditions, cajoling my parents to permit me to go on this risky journey and at last—an idea that loomed so far away, suddenly looks so near at hand!

I can very well understand my parent's anxiety. I'll be so far away from civilisation, on the other end of the world. I'll have to survive the inclemency of nature. I, myself, am aware of the dangers involved—I know I'll have to undergo rigorous training: difficult times, difficult tasks—but as said, 'nothing ventured, nothing gained'. So, here I am, all geared up, for the chance of a life time. A chance to live my imaginations. Here I go, my dear friends, to say 'Hi' to my dear penguins.

Answer 2.

'SATTVA' THE DRAMA FESTIVAL

The Interact Club of Children's Academy organised an interschool drama competition called 'Sattva' on 26th January 2015 at the school's open air theatre. Fourteen prestigious schools from Jaipur participated in this mega event. Keeping with the importance of the day, the theme was based on 'Patriotism' and 'Nation-Building'. The participants surprised all with their dramatic skills. The sensitivity with which each of the characters was portrayed was indeed commendable. Each performance provided entertainment and food for thought.

The President of Rotary Club, Jaipur presided over the function. Quite a few Rotary members graced the occasion. The Chief Guest unfurled the National Flag after which the competition started.

In the keenly fought competition, it was MRC School, Jaipur, that lifted the trophy for the Best Performance. Sushant City School came second, while there was a tie between Pink City School and Vivekanada Baal Bhavan for the third place. The host school was not a part of the competition, although they presented a very thought- provoking drama on youth and nation building. The presentations were adjudged by well known theatre personalities like Sharat Chandra Mohan, the veteran theatre artist, Ms. Priyadarshini Sharma, actor-director and Kapil Mishra, Director of Jaipur Natya Manch. The judges opined that the standard of the competition was so good that it was difficult to choose the winners.

The entries were adjudicated on the basis of theme, dialogue delivery, expression and over all presentation. A special prize was presented to Vinti Sonkya of Children's Academy for putting up the most memorable performance. There was also a prize for the Best Comic Role which went to Abhinav Nirwan of Cambridge School.

The Chief Guest lauded the initiative taken by the host school in conducting such a fabulous competition. He gave away the prizes to the winners and motivated them to participate in many more competitions like this. He stated that participation matters more than winning. The programme ended with a zestful rendering of the National Anthem.

Answer 3.

(a) **(1)** Besides being an outstanding sportsman, Sachin is a good artist.

(2) Mother told Paul that a courier had come for him the previous day.

(3) No sooner did the minister take the oath than the spectators started applauding.

(4) He is too short to be a soldier.

(5) The documents were examined by the lawyer with utmost care.

(6) Unless the horse is quite sound, I will not buy it.

(7) No one but Shakespeare could write such a tragedy.

(8) Not only is Rita a journalist but also a writer.

(9) Notwithstanding Lakshmi's hard work, she failed in the examination.

(10) Everyone would agree that he was totally fit before his death.

(b) **(1)** out **(2)** off **(3)** in

(4) behind **(5)** out **(6)** on

(7) to **(8)** for **(9)** for

(10) since

(c) (1) were working (2) stopped (3) left
(4) leaning (5) saw (6) was
(7) had gone (8) climbed (9) got
(10) went

Answer 4.

(a) (i) (1) encumbered (2) penetrate
(3) underrated (4) superlative

(ii) (1) The judge ordered the police to *arrest* the culprit on the basis of clear evidence against him.

(2) He *draws* such beautiful pictures in sand that the onlookers are astounded at his skill.

(3) Leander Paes made India proud by winning the *set* and the match in a short period of time.

(4) The Literary *club* in our school conducts a number of activities to hone the skills of its members.

(b) (1) The man had died of cardiac arrest or heart failure. Somebody had stabbed him with a sharp knife which penetrated his heart muscle, causing irreparable damage. The unstoppable blood flow stopped his heart beat.

(2) The narrator was scared as the murder weapon belonged to him. Although, it was in self-defense that he killed the other person, the fact was that he was a murderer and he could be booked by law.

(3) The narrator's weapon was a hundred and fifty year old 'sgian dubh', or the black knife. It was a 150 years old. Its blade was only four inches long but sharp and deadly. One side of its ebony handle was ribbed enabling a strong grip but the other side was so smooth that it would slide effortlessly. The stone set in the handle gave perfect balance, making it an excellent throwing weapon.

(4) The narrator was driving along a road, a little distance away from the city. Seeing a car parked on the side of the road and a man signing for him to stop in the universally recognized 'distress signal'. The narrator stopped his car and stepped out to offer help.

(c) (i) (1) While driving, the narrator saw a car parked off the road. A man was waving distress signal.

(2) The narrator got out of the car and the driver attacked him with a club. The narrator dodged the weapon but got hit on his shoulder. His arm became numb.

(3) The narrator groped for his own knife. The knife was a left handed weapon so he could use it with his good hand.

(4) The stranger tried to take out his gun but could not.

(5) The narrator jumped on him just as he took his gun out.

(6) The stranger swung around and ran into the narrator's blade. He fell at the narrator's feet and the knife came free.

(ii) The narrator was driving when he saw a man waving a distress signal, his car parked off the road. He got out of his car to help. He was attacked with a club. Sensing the blow, the narrator swerved and was hit on his right shoulder. He took out his knife in self-defense. Seeing it, the man decided to attack with his pistol kept inside his jacket. The leather strap of his club prevented smooth movement. In panic he ran into the knife. He fell at the narrator's feet and the knife came free. (94 words)

●●

ENGLISH-I

2014

QUESTIONS

Question 1.

*Write a composition (in approximately 450–500 words) on **any one** of the following subjects :* **[30]**

(You are reminded that you will be rewarded for orderly and coherent presentation of material, use of appropriate style and general accuracy of spelling, punctuation and grammar.)

(a) *The small things that we own are the most precious to us. Give your views on the statement.*

(b) *Write about an incident in your life when you experienced the pain of defeat. It had seemed to you then that life would come to an end. What lessons did you learn from the experience and how did you move ahead after that incident?*

(c) *Describe the career you have decided to pursue and the factors that have influenced you to make that decision.*

(d) *Capital punishment should be abolished. Argue for or against the proposition.*

(e) *Luck.*

(f) *Write an original story beginning with the following words :*

"The silence of the evening was broken when I heard my sister's shrill cry…"

Question 2.

You have recently visited a tourist destination. Write a description of it for a travel magazine in about 300 words using the points given below: **[20]**

Name of the place—location—means of traveling to the destination—climate—best season to visit—picturesque landscape—lodging and food—recreational facilities—places of interest in the area—local language/dress—handicrafts/ products —overall experience.

Question 3.

*Answer sections **(a)**, **(b)** and **(c)**.*

(a) *In each of the following items, sentence **A** is complete, while sentence **B** is not. Complete sentence **B**, making it as similar as possible to sentence **A**. Write sentence **B** in each case.* **[10]**

Example : (0) (A) I lost the book .

(B) The book ……………………

Answer : (0) The book was lost by me.

(1) (A) *Raju plays both cricket and football.*

(B) *Not only ……………………*

(2) (A) *She was too full for another meal.*

(B) *She was so ……………………*

(3) (A) *Ranjeet said, 'Sheela, why don't you take the advice of your parents in this matter ?'*

(B) *Ranjeet asked Sheela ……………………*

(4) (A) *Rahul has not been to school for over two months.*

(B) *It has ……………………*

(5) (A) *As soon as he entered the room, he slipped and fell.*

(B) *Hardly ……………………*

(6) (A) *Candidates may not bring textbooks into the examination hall.*

(B) *Candidates are ……………………*

(7) (A) *Although it was a sunny day, it was very cold.*

(B) *Despite ……………………*

(8) (A) *The book I had read earlier was better than this book.*

(B) *This book is ……………………*

(9) (A) *They said that he had broken the chair.*

(B) *They accused ……………………*

(10) (A) *All the girls have brought their books with them.*

(B) *Each of the girls ……………………*

(b) *Fill in each blank with a suitable word. (Do not write the sentence.)* **[5]**

(1) *We failed to agree _______ a common plan of action.*

(2) *When I explained my plan of action to him, he did not agree _______ me.*

(3) *The Principal does not approve _______ indiscipline.*

(4) *I need his approval _______ I can start the work.*

(5) *An explanation about the evolution of species is given _______ Chapter 2.*

(6) *The answers to these questions are given _______ page 44.*

(7) *One fine day, he set _______ on his adventurous trip.*

(8) *As soon as he entered the room, he set _______ his heavy bag.*

(9) *It has been a long time _______ I saw her.*

(10) *I have not seen Ravi _______ ten years.*

(c) *Fill in the blanks in the passage given below with the appropriate form of the verb given in brackets. Do not write the passage but write the verbs in the correct order.* **[5]**

*We decided to travel by car and ________ **(1)** (leave) the house early. We _______ **(2)** (be) on the road for two hours, when our car_______ **(3)** (hit) a stone that _______ **(4)** (lie) in the middle of the road. The car _______ **(5)** (go) off the road but we _______ **(6)** (escape) with minor injuries. Had we______**(7)** (travel) faster, we _______ **(8)** (involve) in a serious accident. The mishap _______ **(9)** (delay) us but did not _______ **(10)** (damp) our enthusiasm.*

Question 4 .

Read carefully the passage given below and answer the questions (a), (b) and (c) that follow :

THE SNOW GOOSE

(1) In the late spring of 1930, Philip Rhayader came to the abandoned lighthouse at the mouth of the river Elder on the Essex coast. He was a painter of birds and of nature and had withdrawn from all human society. He was afflicted with a hunched back and a deformed, twisted hand.

(2) Although physical deformity often breeds hatred for humanity in people, Rhayader did not hate any one. His heart was filled with pity and understanding. He had mastered his handicap, but he could not **master** the rebuffs he suffered because of his appearance. The thing that drove him into seclusion was his failure to find anybody who loved him as much as he loved nature and humanity.

(3) One November afternoon, three years after Rhayader had come to the Great Marsh, a child approached his lighthouse studio. In her arms, she carried a burden. She was no more than twelve, slender, dirty, nervous and timid as a bird, but beneath the dirt, as beautiful as a fairy. She was desperately frightened of the ugly man she had come to see, but greater than her fear was the need of that which she carried. For locked in her child's heart was the knowledge picked up somewhere in swamp-land, that this ogre who lived in the lighthouse had magic that could heal injured things.

(4) She had never seen Rhayader before and was close to fleeing in panic at the dark apparition that appeared at the studio door, drawn by her footsteps-the black head and beard, the sinister hump and the crooked hand, bent at the wrist. She stood there staring, poised like a disturbed marsh bird for instant **flight**. But his voice was deep and kind when he spoke to her.

"What is it, child ?"

(5) She stood her ground, and then edged timidly forward. The things she carried in her arms was a large white bird, and it was quite **still**. There were stains of blood on its whiteness and on her dress where she had held it to her.

(6) The girl placed it in his arms. "I found it, sir. It is hurt. Is it still alive ?" "Yes Yes, I think so. Come in, child, come in." Rhayader went inside bearing the bird, which he placed upon a table, where it moved feebly. Curiosity overcame fear. The little girl followed and found herself in a room warmed by a coal fire, shining with many colored pictures that covered the walls, and full of a strange but pleasant smell.

(7) The bird fluttered. With his good right hand Rhayader spread one of its. Immense white pinions. The end was beautifully tipped with black.

Rhayader looked and marveled, and said, "Child, where did you find it ?"

(8) "In the marsh, sir where fowlers had been. What — what is it, sir ?"

"It is a snow goose from Canada. But how in heaven did it come here ?"

(9) The name seemed to mean nothing to the little girl. Her deep, violet eyes, shining out of the dirt on her thin face, were fixed with concern on the injured bird.

She said, "Can you heal it, sir ?"

"Yes, yes," said Rhayader. "We will try. Come, you shall help me."

(10) There were scissors and bandages and splints on a shelf and he was marvelously deft, even with the crooked hand that managed to hold things. He said, "Ah, she has been shot, poor thing. Her leg is broken, and the wing tip, but not badly. We will bandage the wing closer to her body, so that she cannot move it until it has set, and then make a splint for the poor leg."

(11) Her fears forgotten, the child watched, fascinated, as he worked, and fixed a fine splint to the shattered leg. "A **bitter** reception for a visiting princess," concluded Rhayader. "We will call her the Lost Princess. And in a few days, she will be feeling much better. See ?"

(12) He reached into his pocket and produced a handful of grains. The snow goose opened its round yellow eyes and nibbled at it. The child laughed with delight.

Adapted from : *The Snow Goose* by Paul Gallico

(a) (i) *Given below are four words and phrases. Find the words which have a similar meaning in the passage :* **[4]**

(1) *Cruel and frights person*

(2) *Seemingly evil and dangerous*

(3) *Wings*

(4) *Skillful*

(ii) *For each of the words given below, write a sentence of at least ten words using the same word* ***unchanged in form, but with a different meaning*** *from that which it carries in the passage :* **[4]**

(1) *master (line 6)* (2) *flight (line 19)*

(3) *still (line 23)* (4) *bitter (line 44)*

(b) *Answer the following questions in your own words as briefly as possible :*

(i) *Why did Rhayader live alone ?* **[3]**

(ii) *What was the child's reaction on first seeing Rhayader ?* **[3]**

(iii) *What was the child's burden ?* **[2]**

(iv) *How did Rhayader manage to fascinate the child and make her happy ?* **[3]**

(c) *Describe how Rhayader attended to the bird (paragraphs 6 to 12) in not more than 100 words. Failure to keep within the word limit will be penalized. You will be required to :*

(i) *List your ideas clearly in point form.* **[6]**

(ii) *In about 100 words, write your points in the form of a connected passage.* **[6]**

ANSWERS

Answer 1.

(a) THE SMALL THINGS THAT WE OWN ARE THE MOST PRECIOUS TO US

We hanker after success, money, social position, big bungalows, luxury cars, and expensive gadgets. We feel we have arrived with these paraphernalia that surround us. But do they provide us actual happiness and satisfaction ? No. The more you have, the more you yearn for.

Take the example of a wrist watch given to you by your father when you were a child, a camera presented to you by your uncle, or a painting gifted to you by your sister. Aren't they more valuable sources of true happiness than all these electronic gadgets you possess ? Mobiles get outdated in a jiffy and you throw them in a jiffy! Do we do that in case of the lovely possessions cherished from childhood ? Not at all because they are a part of our childhood memories, of unconditional love, of a trust that we grew up with.

What about the certificates of merit we have accumulated from Nursery classes ? We cannot easily throw them away because they are a part of our growing up. They are the bricks of our character formation. They give us courage, confidence and a sense of worth ! Obviously, these 'small' things are more precious to us than any of the things money can buy !

And what about that album whose snapshots taken with simple cameras take us on a nostalgic trip down the memory lane ? The wedding photos of your parents, the images of your grandparents, you in your short pants or frocks, your first step captured so thoughtfully by your parents, your first birthday, thousand images flash across! Can anything else give you so much pleasure ?

One may relish a Five Star lunch once in a while but ultimately you yearn for simple food made by your mother. Small things we own like our books, letters of near and dear ones, our simple comfortable dresses, soulful songs ultimately prove more valuable to us than any other ostentatious object.

None can deny the fact, that we enjoy the most while reading a book or give expression to our creativity through pen or paints. You escape into a new world where you indulge in your fantasies, where you get acquainted with yourself, where you achieve your full potential. The fish in the pond, the flowers in the garden, the flute that you own, the talents you possess, the love of your family, the easy camaraderie with your friends, your character, your learning, your upbringing, your values, these are your real possessions and these are the most precious ones. Other things are transitory and should be given only that much attention that they deserve.

(b) INCIDENT OF LIFE WHEN YOU EXPERIENCE DEFEAT

I was touted to be the best contestant of 'Indian Idol'. Votes of the public poured in. I was supremely confident, I cruised along, practising with my mentors. Yes, I was poised to win. It was not only the fame that was most important, I wanted the winner's money very badly. My father was seriously ill, he had a rare heart problem. He needed surgery. He had been slogging for the comfort of the family for years. With the result, he didn't have money for his own treatment. I saw the advertisement in the TV for the auditions of the music show. I considered it to be an opportunity to use my talent. I gave in my name and the next thing I knew was that I was called for an audition. I prayed to God like never before to get selected. God was kind to me. The judges gave me the ticket to Heaven!

The day finally arrived. There were butterflies in my stomach. No one in my family could attend the function as my father's health had taken a serious turn.

My performance would decide my fate; my father's fate. I started perspiring at the collar. Rivulets of sweat were pouring down my cheeks. I felt like crying. It was a sad song that I selected to sing. I thought it suited my mood; but it suited too much! I broke down in the middle. Everyone wanted me to win. I withdrew, my head bent. It felt like a total defeat written large over my face.

There was desperation on the faces of my friends. The judges looked disappointed too. I didn't have to wait for the declaration. I came third. I stood there dumbfounded, empty-handed. Future seemed bleak. My dreams were shattered, crumbled to the ground ! Then came the sound of a loud applause. All were looking at me. The anchor came towards me and lifted my hand. Sanjay Leela Bhansali, one of the judges promised to give me all the songs in his upcoming film. The TV channel announced a special prize for me !

My cup of happiness was full again. Luck comes in different guises. One should never lose hope. Today, I am a happy man. I could get my father treated with the money I got and he is perfectly all right. Proper post operative treatment could be given by the rest of the money. Later, I felt proud of my country; not only for the expert treatment given, but for all the love and care of my countrymen. The experience made me realize, that victory need not come in the form of prize. My defeat enabled me to earn much more than what I could have earned by winning the title. Today I trust God, trust humanity and trust myself. I am a confident youngster full of optimism. I will never lose my positive outlook.

(c) CAREER YOU WANT TO PURSUE

Very often, I have thought about the career I would like to pursue. Ultimately, I have decided to become an actor. There is no guarantee that you'll succeed but I am enamored by the prospects of living many roles ! The only field where you can live all your dreams is through the career of an actor.

An actor lives in the memory of all. Even when the years pass, an actor remains a star. Take for example, Mr. Amitabh Bachchan. He is a living legend. The Star of the Millennium ! His charisma had brushed on to many young men of the seventies and eighties in such a way that they started identifying with the characters he gave life to. 'Sholey', 'Coolie', 'Deewar', 'Lawaris', and many more such films have won accolade of the viewers. The adulation he gets wherever he goes, the affection with which he is greeted by the young and the old, the global recognition he has got by the dint of his work, it is a phenomenal achievement. I would want to be like him.

Amir Khan, Shahrukh Khan, Rajnikant, Kamal Haasan, Mammooty, Mohan Lal, all are super stars in their own right. They have done meaningful cinema and have left their impact in the minds of their audience. Like Amir Khan, who got to scratch the conscience of a nation with his television programme 'Satyamev Jayate'. Like Salman Khan, the founder of 'Being Human' or the number of charity works done by many of our stars go a long way in improving the condition of fellow beings. A popular star can make crores of rupees. If I can earn that much, I'll be able to do a lot of good work, like establishing hospitals, schools and social service organisations.

Cinema is a powerful medium that can be successfully used to change the mindset of people. This is the reason why the popular stars are roped in when there is a need to convey a message. Awareness is created through their charismatic personae. Whether it is Aids, eye donation, kidney donation, physical fitness, respect for girl child, child labor, when these iconic personalities advertise them, people follow. Patriotic feelings can be aroused, psychological insight can be given into the mindset of a wheel chair ridden victim as in 'Guzarish' convincingly enacted by Hritik Roshan, or the dyslexic child in 'Taare Zamin Par' or the victim of Progeria, so beautifully brought out by Amitabh Bachchan in 'Pa'.

An actor can educate, convince, be an emotional anchor, and touch the hearts of the people. This is the reason why I want to become an actor. So, watch out all.

Here I am, the next super star on screen !

(d) CAPITAL PUNISHMENT SHOULD BE ABOLISHED

FOR THE MOTION

I think capital punishment works great. 'Every killer you kill never kills again' goes one argument. 'When you can't give life, you have no right to take life' goes the other argument. The crux of the matter is purely ethical. We are not Gods and are not in a position to sentence anybody to death. But the flip side is that you cannot allow a cold blooded murderer to go free. He deserves punishment, otherwise there will be total anarchy in society. People will have license to kill. Only a drastic punishment will prove as a deterrent to prevent further crimes.

Now, the question arises that is death penalty the severest form of punishment ? What about Life Imprisonment ? This will make the life of the criminal sheer hell. You get only a small piece of sky to see, bland food to eat, no company of the close ones, no friends to banter with, an apology for food and endless time to brood over what one has done ! Isn't this life like a living death? Will it not act as a deterrent for crimes in society ?

As for Capital Punishment, once the victim is dead, they're dead, nothing will bring them back. Killing the criminal is simply state sanctioned murder. How can one say murder is wrong but then commit it to serve as an example !

The second point is that a criminal should be forced to confront the consequences of his actions. The death penalty prevents this. It's undoubtedly frightening for the condemned, but soon they're dead, therefore, they have nothing to worry about once that conclusion is reached. With life imprisonment, there's at least a chance they'll be influenced to at least think about what they've done.

Another salient point is that as long as you have capital punishment, there is no guarantee that innocent people won't be put to death. Such possibilities become impossible in case of death penalty.

Again there is a possibility that money or power can buy justice ! How many millionaires get capital punishment? Worst crimes remain unproved due to the skill of the lawyers. And the same skills can prove an innocent person guilty ! Imposing death on innocent ones, becomes the most unjust deed.

Data showing crime rate in many countries where capital punishment occurs, prove that it doesn't necessarily deter crime. Heinous crimes like murders and rapes occur despite the provision of death penalty. Whereas many states where the capital punishment is not allowed, show lesser crime rate. Long term imprisonment serves the purpose of punishment better than the savage method of killing.

Death Penalty should be abolished as it is like fighting the symptom rather than the root cause. Just give a thought to the effect of death penalty on public's psychology ? Our society is already too violence-saturated as it is, without adding this horrendous relic from the past.

(e) LUCK

'He worked by day, He worked by night
He burnt a lot of midnight oil,
He spent all day in study and toil
Then he reached a point of success, And people called it Luck !'

Everyone wants to be lucky; lucky in job, lucky in marriage, lucky in a game, lucky in getting preferred subject in college ! But what is luck exactly? It is difficult to define. Is it being blessed by God? Is it something co-incidental? Is it something that happens out of the world ?

Most of us mistake luck as something that drops from the sky but most of the time it is the result of our own action, attitude and mind set. Luck comes in the garb of hard work, persistence and perseverance. When people succeed, we see only their success, not the preparation and hard work behind it.

We envy the rich. We consider them lucky. They can get admissions anywhere in the world. They travel in luxury cars. They have palatial houses, hundreds of servants, a big bank balance ! But do you actually think they can maintain such high standard of living if they don't plan well, manage their finances well, run offices, manage people and other things ? One cannot remain 'lucky' forever. One has to sweat for it. It is truly said that there are no free meals in this world. If not physical toil, one has to pitch in mentally, emotionally and socially for remaining lucky. Yes, things can happen by fluke any time, out of the blue. One cannot dismiss this phenomenon as altogether impossible; but one can also maneuver luck by being present in the right place in the right time.

Pandit Nehru was born with a silver spoon in his mouth. But he attained global stature on the ground of his intelligence, vision, scientific thinking, education, talent in writing, speaking and interacting with people. His achievement was not mere luck. He moulded his life in such a way that he remained a lucky personality in the eyes of the world.

In sports, people become famous due to their efforts, coaching, training, maintaining stamina and what not. You can't score a six or a goal out of sheer luck. That requires skill, talent and will-power! In movies, there are many first time Lucky ones. But to maintain that luck, they have to persevere. Take the Khans, the Kapoors and Kumars. They are lucky; they are prosperous, they seem to be happy and they are always in the lime light. The common factor underlying all the lucky ones is progress. Once they achieve something, they don't rest on their haunches. Wide, wider and still wider. That should be the policy to remain lucky. That is the secret everybody should be aware of. 'Once lucky, always lucky' is the mantra everybody should practice !

(f) THE SILENCE OF THE EVENING WAS BROKEN WHEN I HEARD MY SISTER'S SHRILL CRY......

The silence of the evening was broken when I heard my sister's shrill cry. I was sitting in the garden of our ancestral house, enjoying the sheer calm and beauty of the typical country side.

I was happy I could make it to the village in the midst of my busy life. Grandma had been calling the whole family for ages. I remember her telling me on the phone 'Beta, if you don't come fast, you'll regret that you couldn't meet your dadi, before she breathed her last'. I decided then and there that I have to meet her.

She was always busy making laddoos and jalebis. Her 'sambar', 'rasam' and buttermilk curry were just out of the world. Her love extended to the animals and birds. She had a brood of chicks, ducks, parrots in her farm house. She would talk to the myriad plants and they seemed to flourish under her care. If there was a great nature lover, that was grandma.

Years passed and here I was with my whole family. Grandma's face was radiant when we arrived at the farm house and when I touched her feet. She looked much older than I had imagined. She flashed a toothless smile. It was heartbreaking to see the once energetic woman, wobbling around with the help of a stick. But the eyes remained the same-smiling eyes, now shimmering with tears of happiness. That night she joked with us, told us the stories of the village, how the flood had destroyed the crops in the previous monsoon, how the villagers had sought her help in times of difficulty, how she adopted a boy of ten who was orphaned in the floods, how she helped the parents to get their daughter married. She was not boasting; she was not even looking for praise. It was a simple narration of events. I just couldn't help but compare the self-centered, ego centric city dwellers and this calm, serene woman whose heart went out to all those who needed help. Definitely, in my eyes she was an unsung hero.

The next morning we sat on the rope cots, basking in the glory of the rising sun. She listened to our experiences in the city. She was happy to know that we were doing well in our studies; she was fascinated with my laptop. I showed the photos of the family, showed her old movie songs, and clicked her picture in my mobile.

I felt so happy sitting there watching the evening sky, thinking of the indefatigable woman, when I heard the shrill cry of my sister. I rushed in. I saw my grandmother lying stretched on her bed. Her eyes were closed. Her face still had that radiant smile. She was at peace with herself. Tears blurred my vision. With a lump in my throat, I muttered to console myself. "People become memories when they leave and memories become treasures for their dear ones."

Answer 2.

Jaipur, the capital city of Rajasthan, has been a hub of tourism for so many years. Very few cities in India can match the Pink City for its sheer opulence and rich heritage. Located about 15 kilometers from the Sanganer Airport, 8 kilometers from the railway station and a bus terminus within the city, its easy accessibility is the main factor in making it a much favored tourism centre. As any North Indian city, Jaipur sports extreme climate, the temperature goes up to 45 degrees in peak summer and below 10 degrees in winters. The best season to visit the place is October - November when the weather is comfortably pleasant.

Encircled by the rugged Aravali Hills, the city offers many a scope for picturesque viewing. The beautiful 'Jal Mahal', the summer palace of erstwhile Maharajas, placed in the centre of a fabulous lake, shimmering in the morning sunlight, offers a fairy land experience.

As you go further into the city, you are greeted with the world famous Hawa Mahal. Its rose pink color, lattice work windows and mandna work make it a unique work of architecture. At the heart of the city, there is a sprawling garden called Ram Nivas Bagh. Albert Museum shines as a jewel right in the centre of the garden. Jaipur zoo and Aviary are situated in the garden itself, offering a delightful time for the children as well as adults.

The city is dotted with many tourist attractions. City Palace is the one to begin with, where the rich heritage of Rajasthan is displayed in all its glory. The royal apparels, the finest carpets, the ornaments, marble figurines and statues, the brightly lit royal chambers, thrones and well planned garden, all give you a glimpse of the regal background of the city. Amber Palace, the old capital city is a sprawling structure made in sandstone, decorated with marbles and gems. The Sheesh Mahal in the palace is an enchanting sight.

Jaipur is famous for its Bandhej and Laharia work, block printing, precious stones, Meenakari work. The bright shops overflowing with such items including puppets and Blue Pottery prove irresistible to tourists.

While there are several big, Seven and Five Star hotels like Rajvilas, Ram Bagh Palaces, Sheraton and other ITC hotels for the elite, Rajasthan Tourism hotels offer comfort in fair price.

People speak Hindi in general, though, Marwari is the local dialect. The bright turbans worn by the gents and ethnic lehengas sparkling with 'gota' work worn by the women reflect the mood of the local people who are approachable, generous and friendly.

Daal Baati Churma, gatte ki sabji, red and white meat preparation give a taste of typical Rajasthani food.

All in all, if you are planning for a trip that can give you tradition and modernity on a platter, there is no better place than Jaipur.

Answer 3.

(a) (1) **(B)** Not only does Raju play cricket but also football.

(2) **(B)** She was so full that she could not eat another meal.

(3) **(B)** Ranjeet asked Sheela why she didn't take the advice of her parents in that matter.

(4) **(B)** It has been two months since Rahul has been to school.

(5) **(B)** Hardly did he enter the room when he slipped and fell.

(6) **(B)** Candidates are not permitted to bring textbooks into the examination hall.

(7) **(B)** Despite it being a sunny day, it was very cold.

(8) **(B)** This book is not as good as the book I had read earlier.

(9) **(B)** They accused him of having broken the chair.

(10) **(B)** Each of the girls has brought their books with her.

(b) (1) on (2) with (3) of
(4) before (5) in (6) on
(7) out (8) down (9) since
(10) for

(c) (1) left (2) had been (3) hit
(4) was lying (5) went (6) escaped
(7) been traveling
(8) would have been involved
(9) delayed
(10) dampen

Answer 4.

(a) (i) (1) ogre (2) sinister
(3) pinions (4) deft

(ii) (1) His *master* gave him a lot of hard tasks to do.

(2) The *flight* was canceled on account of bad weather and persistent rainfall.

(3) My father knew that something tragic happened; *still* he managed to remain calm.

(4) The medicine tasted *bitter* but I managed to drink it.

(b) **(i)** Rhayader lived alone as he had withdrawn from the human society. He could not find anybody who loved him and could not master the rebuff of the society, as his back was hunched and he had a deformed, twisted hand. This made him look ugly in the eyes of the people as deformity often makes them hate humanity.

(ii) The girl was frightened or rather panic stricken at the sight of the hunchbacked figure with crooked hand. At first she wanted to flee away like a disturbed bird.

(iii) The burden the girl carried was a big, wounded, bloody bird. It was a Snow Goose from Canada.

(iv) Rhayader invited her into his room brightly lit with a fire and decorated with pictures. In a deep, kind, reassuring voice, he gently inquired the details of how and where she found it and deftly attended to the wound of the bird. She was fascinated by his kind approach and expertise in dealing with the emergency. After reassuring the girl that the bird will be all right, he fed it with some grains. As the bird nibbled at them, the girl laughed in happiness.

(c) (i) (1) The girl brought Rhayader a wounded migratory bird.

(2) He took the bleeding bird from her hands and gently put it on the table.

(3) He then spread one of the bird's big black-tipped white wings. He did this to check what was wrong with the bird.

(4) There were scissors and bandages and splints on a shelf in his house which Rhayader used defty.

(5) He bandaged its wing closely to the body to immobilize it.

(6) He made a splint for its broken leg and then fed it with grains. The girl laughed delightfully at this site

(ii) The girl brought Rhayader a wounded migratory bird. He took the bleeding bird and gently put it on the table. He deftly took scissors, bandages and splints which he managed expertly in spite of his deformity. He bandaged its wing closely to the body to immobilize it. He made a splint for its broken leg. He reassured the girl that the 'Lost Princess' will be soon much better. He fed it with grains, and the girl laughed delightfully at this site. (81 words)

ENGLISH-I

2013

QUESTIONS

Question 1.

Write a composition (in approximately 450–500 words) on ***any one*** *of the following subjects :* **[30]**

(You are reminded that you will be rewarded for orderly and coherent presentation of material, use of appropriate style and general accuracy of spelling, punctuation and grammar.)

(a) *One should judge a man not by what he has, but by what he is. Express your views on this statement.*

(b) *Write a description of the most interesting person you have ever met.*

(c) *It is right for a person to interfere in the affairs of another. Write for or against this statement.*

(d) *Memories.*

(e) *Relate how travel has been of educational value to you.*

(f) *Write an original story with the following ending : "...... It was exactly as he had said it would be."*

Question 2.

Write a review of a book you have recently read (in approximately 300 words) for your school magazine. Make use of the given points. **[20]**

Name of the book–author–publisher–readership– cover and illustrations–plot–characters–insight into some interesting incidents–message if any– appeal the book holds for you.

Question 3.

Answer sections ***(a), (b)*** *and* ***(c).***

(a) *In each of the following items, sentence* ***A*** *is complete, while sentence* ***B*** *is not. Complete sentence* ***B****, making it as similar as possible to sentence* ***A****. Write sentence* ***B*** *in each case.* **[10]**

Example : (0) (A)) Though it was a cloudy day, the temperature was 45°C.

(B) Despite ....

Answer : (0) Despite it being a cloudy day, the temperature was 45°C.

(1) *(A) They are demolishing the entire block of flats.*

(B) The entire ....

(2) *(A) Anil thought the examination would be held later.*

(B) Anil was ....

(3) *(A) I do not like this food as much as you do.*

(B) You ....

(4) *(A) Raju failed in the examination, yet he was not upset.*

(B) In spite of ....

(5) *(A) "Don't lend Ramesh any money," I said to Rita.*

(B) I advised ....

(6) *(A) As soon as the arrangements were made to release extra water from the reservoirs, the rain came in a downpour.*

(B) No sooner ....

(7) *(A) He had to sign or be arrested.*

(B) If he ....

(8) *(A) Meena complained of not being able to see the performance from the back row.*

(B) Meena complained that ....

(9) *(A) Many ambitious people see politics as a stepping stone to personal power.*

(B) Politics ....

(10) *(A) Rahul has wasted his wealth and ruined his health by his drinking habit.*

(B) Rahul has not only ....

(b) *Fill in each blank with a suitable word. (Do not write the sentence.)* **[5]**

(1) *Let us talk _______ them tomorrow about their homework.*

(2) *The Principal gave us a talk _______ bullying.*

(3) *Manoj was not at home when his grandfather passed _______ .*

(4) *When we visited Dalhousie, we passed _______ the house in which we had lived twenty years ago.*

(5) *This novel has been written _______ Jeffrey Archer.*

(6) *In the Middle Ages, books were written _______ a variety of subjects.*

(7) *He had promised her that he would come to the party but he didn't turn _______ .*

(8) *I was forced to turn _________ Maya's invitation as I was going to be out of town the following week.*

(9) *Rashi has applied _________ a scholarship in an American university.*

(10) *Arjun applied _________ the university for higher studies but was denied admission.*

(c) *Fill in the blanks in the passage given below with the appropriate form of the verb given in brackets. Do not write the passage but write the verbs in the correct order.* **[5]**

Breathing is good, on this we ______ ***(1)*** *(can) agree. That is why sleep apnoea, a condition in which* *sufferers repeatedly _______ (2) (stop) breathing and _______ (3) (awake) during the night, _______ (4) (be) so distressing and dangerous. Doctors _______ (5) (have) long known that the majority of people with sleep apnoea also _______ (6) (have) gas and reflux, so they often _______ (7) (treat) the reflux with heartburn medication. To prove the connection between reflux and sleep apnoea, scientists at an army hospital _______ (8) (observe) apnoea patients as they _______ (9) (sleep). They _______ (10) (find) that reflux could, in fact, be the cause of many night-time awakenings.*

Question 4 .

Read carefully the passage given below and answer the questions (a), (b) and (c) that follow :

BOYHOOD AND BALANCE

(1) My eldest boy had just turned nine. One day not long after, he was on his way to the pool on his bike when he slammed on his brakes in front of me and did a really neat skid. He looked back and gave me his cheesy smile of satisfaction.

(2) Normally that little skid would have provoked a stern warning not to "wear out your tyres." I was always taught to take care of my things and keep them in good **shape**. I've come to learn, though, that there is a difference between the natural course of wearing things out and trying to make things last at any price. The difference became acute to me one bright sunny day four years ago.

(3) My then five-year-old boy had spent a normal summer afternoon playing in the **park** and swimming at the neighbourhood pool. At his bath time that night, I noticed he had small red spots all over his body. Some kind of measles, I thought, and called my wife to take a look. By pure chance, the next day he had an appointment with his pediatrician.

(4) Several tests (and long hours) later, the doctors figured my son had an illness, called ITP (idiopathic thrombocytopenic purpura), which causes your spleen to kell off your body's own blood-clotting platelets. If it got worse, he could bleed to death internally. Only time would tell whether it would get better on its own. You can imagine the fear that crept into our hearts.

(5) Three days into his stay at the hospital, I went to pick out a **present** for him. I reached for a yellow toy convertible (he loves cars). My hand hesitated. I knew that it wouldn't last long with a five-year-old boy. Then I thought to myself : So what if the doors break off and the wheels fall off. If those things happen, it will mean he is alive and well. I bought it in the hope that he would be **well** enough to play with it. He was thrilled to receive it, and it helped him pass the week he spent at the hospital.

(6) I saw the car the other day, sitting on a shelf in his room. The wheels have fallen off, the doors are broken and all of the chrome has worn off. I see it and I smile to myself. My boy has been perfectly well these past four years, and is full of vitality. His mysterious illness came and went.

(7) And I learned that things are things and can be replaced if necessary. If any one of my three boys happens to break something or wear it out playing with it, instead of chastising him for carelessness, I now prefer to celebrate his boyhood. The empty shell of what once was a nice car, the balding tyre of his new bike, the lost pieces of his Monopoly game, all bear witness to the fact that there lives a healthy, happy boy.

(8) Prudence and preservation have their place, as do experimentation and curiosity. In the end, it is my relationship with my sons and my wife that have real lasting power. I choose to celebrate our lives and live with the losses.

(9) Besides, I used to skid my tyres the same way—just not in front of my dad.

Adapted from : The Globe and Mail by Elliot Van Egmond

(a) **(i)** *Given below are four words and phrases. Find the words which have a similar meaning in the passage :* **[4]**

(1) *strict* **(2)** *energy and vigour*

(3) *scolding* **(4)** *strong desire to know*

(ii) *For each of the words given below, write a sentence of at least ten words using the same word* ***unchanged in form, but with a different meaning*** *from that which it carries in the passage :* **[4]**

(1) *shape (line 5)*

(2) *park (line 9)*

(3) *present (line 17)*

(4) *well (line 22)*

(b) *Answer the following questions in your own words as briefly as possible :*

(i) *Why is the disease ITP so serious ?* **[2]**

(ii) *What does the yellow car look like at present?* **[2]**

(iii) *What did the writer learn from his experience ?* **[3]**

(iv) *What is most important to the writer now?* **[3]**

(c) *Summarize the given passage in not more than 100 words. Failure to keep within the word limit will be penalized. You will be required to:*

(i) *List your ideas clearly in point form.* **[6]**

(ii) *In about 100 words, write your points in the form of a connected passage.* **[6]**

ANSWERS

Answer 1.

(a) ONE SHOULD JUDGE A MAN NOT BY WHAT HE HAS, BUT BY WHAT HE IS

In any society it is a general behavior to be materialistic. All of us tend to be very materialistic in our thinking, whether we like it or not. So people tend to give more importance to a person's bank balance than other aspects. Money is one of the most important aspects of life. Everyone needs money to survive and lead a comfortable standard of living. Money is needed for education, to build homes, to live life comfortably and also in times of emergency. It is an established fact that along with money, what you own because of it is also very important. It is also important to give credit to a person who has worked hard to earn money.

It is not right to judge a man by what he has and not by what kind of person he is. Often we have seen people praising someone they know and being extra sweet, almost subservient to people who have a lot of wealth. They tend to look at people superficially and ignore the qualities that the person may or may not possess.

For example, if you do not talk to the person and do not take time to understand his positive qualities and just try to please the person because of his wealth, you may never know what kind of person he or she is. They may be really interesting and good people, but you may never find out. On the contrary it is also not right if a person is not so good and one has made a positive opinion about that person just because of the large bank balance the person possesses.

Another side to the story is not behaving properly and treating people very badly just because they do not own money or wealth of any sort. It is often seen that rich people are looked at differently as compared to the poor people. When rich people are around, there are some individuals who tend to ignore the less fortunate ones and also treat them badly. This often happens which is not right. Poor people may also have negative attributes, but you must respect and treat a person according to what he or she is.

Human psychology tends to be like this even if we like it or not. So, individually we need to make more effort to understand people before forming an opinion about them. We need to look beyond the wealth and see for ourselves what sort of people we are engaging with. Also, if we come to know what the person is like, we can avoid such people that we personally feel have a lot of negative attributes, irrespective of their bank balance. Therefore, we need to judge a man by what he is and not by what he has.

(b) MOST INTERESTING PERSON I HAVE EVER MET

You meet so many people in your life. There can be new people who walk into your lives every day. You meet new friends, new people at work and at various social occasions. When one interacts with people they get to know what type of a person they are dealing with. I have also met a lot of people. Some I liked, some not so much. But there are only a few people that have left significant impression on me. These are the people who you feel awestruck about and want to meet often and find extremely interesting in life.

There is one person who I feel is the most interesting person I have ever met. Here are a few of the qualities about that person which made him extremely interesting to me.

- This person takes each day with positivity and lot of joy. The enthusiasm of this person is effortless and when you are with this person you feel enthusiastic about life as well.
- Another reason why I found this person interesting is because of music and sports. This person is interested in similar music as I like. Similar sports interests also helped.
- The sense of humor of this person is amazing. He does not crack stupid jokes all the time but well timed effortless jokes makes everyone around this person laugh.
- Another reason I find this person very interesting is the ability to take decisions fast and efficiently in life. This person is very intelligent. Intelligence is something that makes a person look smart and stand out.

- This person I know always seems to understand what is happening and reads between the lines. I will not say cunning but alert is a better word to describe this person.
- Also apart from being so smart and interesting, this person is humble and you can make out that the person is ultimately a nice human being and that you can trust this person.

Sure the person I am talking about is not completely perfect and does have some flaws but just overall it is a lot of fun to talk and interact with this person. I always appreciate the way this person looks at most situations and also has an opinion that I value. This person, I believe is the most interesting person I have ever met.

(c) IT IS RIGHT FOR A PERSON TO INTERFERE IN THE AFFAIRS OF ANOTHER

(I) FOR THE MOTION

We all are surrounded by many people related to us as friends, family, colleagues, competitors and even enemies. So it is not wrong to say that people linked to us play a major role. With this thought, most of the people do not like others interfering in their personal affairs.

Generally, a whole family likes to handle their issues personally without involving any outsider. They believe that an outsider's interference may deteriorate the situation. Sharing your problems and predicaments with any third person is not at all wrong, instead, there is someone who you can look up to.

Most of the time when anyone tries to interfere in one's personal issues, he or she is thought to be ill-mannered, but such persons genuinely try to help others.

Also, while snooping in the affairs of others, one must keep in mind his own limitations. People trust only those outsiders who are capable enough to give them right suggestion. Therefore, a person who likes to interfere in others matters must be an 'agony aunty' only then he or she will be allowed to enter the world of issues.

So, if you ask me is it right to peep inside other's issues, I would say it's not wrong if you are ready to help others with a clear heart and mind rather than making fun of their issues by spreading it among others.

(II) AGAINST THE MOTION

Human beings live in communities. We form social groups with our friends and people of similar interests. But there is always a line that we need to maintain. We have no right to interfere in other people's lives. Every individual has his own privacy. It is very important to maintain privacy and also respect the privacy of one another. Curiosity about what might be going on in other people's lives tends to get the better of us many times. Human beings are very curious, it is a natural fact. Some people can be very nosy and it's a need for them to find out what is happening in other's lives. To develop a habit of it and try to find out as much as you can about other people is wrong.

Also, there are people who tend to give unnecessary and unwanted advice to others. Most people do not appreciate such kind of interference. Even if they may keep the respect of the person and not openly tell them to stop meddling. Also it can badly affect a situation if one gives wrong advice and people follow it just to keep that person's respect.

One must realize if he or she is interfering too much in other's affairs. People must value each other's privacy. If advice is asked, be sure to give the best advice that one possibly can give but otherwise interfering in other's lives is not a good habit. Interference can cause problems for others and also for the person who interferes can lose respect. So, it is not right to interfere in other's affairs.

(d) MEMORIES

A memory is something that you remember which happened to you at some point of time in your life. Generally, every life event that happens in a person's life is not remembered by that person. Some people have a very strong memory and remember almost everything that happens in their life. Such memory is called photographic memory, eidetic memory or identic memory. These people tend to remember pretty much everything that has happened to them in their lives. Then there are others who remember very little and have the tendency to move on easily in life and forget about their past, this is called a weak memory. Mostly normal people remember some of the things that have happened in their past, both negative and positive in nature.

If you come in contact with a person from the past, a situation you remember or you see a thing that reminds you of something of the past, it generates a trigger in your memory and you remember that particular incident. Often, people who have suffered amnesia (where a person tends to forget parts of their lives) are introduced to people they knew or places they have visited in the past to help them remember their lives which they have forgotten.

Memories are very powerful. Because of memories, people avoid making the same mistakes in their lives again and again. Memories remind you of people that you love and have loved in the past. Thinking of positive memories from the past is always a pleasant experience. People tend to look at old photos and videos from the past to remember occasions from their past.

Bad memories can also haunt people and people tend to suffer trauma because of them. They often need help and counseling to help them leave bad memories behind and look towards the brighter future. It is always advisable not to dwell in the past and to look at the future in such situations, but it is more easily said than done.

We sometimes have short term memories of the events that take place in our lives in a short period of time. Long term memories are those memories which we remember, of incidents that have happened in our lives many years ago. Our work is also based on what experience we have and our education. Past work experience is nothing but memories of how work is to be done from the past experience. Our education is also in our past and we can use this as experience in our work which is based on the memories of that education.

Our memories are very important to help us in the future course of our lives. We all have both positive and negative memories in our lives and those memories are a part of who we are today. It is human nature to remember memories. One must not dwell too much on negative memories but must learn from them and cherish the positive experiences and memories of the past.

(e) HOW TRAVEL HAS BEEN OF EDUCATIONAL VALUE TO YOU

Travel means moving from one place to another. Many people love to travel. Earth is a very large planet and there is so much to see on this planet. Traveling to different places of the world and seeing different aspects is very educational in nature. Even if the place is very close to my home, it still teaches me something new. They say traveling is not just about reaching from one place to another but it is about the journey as well as the destinations.

Nowadays, traveling has become more convenient and easy as compared to older days due to great innovations and new technology in transport facilities. I can travel from one part of the world to another in less than a day. Airways, railways, roadways and seaways brought the world together. Also means of communication help me learn more about different parts of the world.

When I travel to a new place, be it a new city or a new country I get to learn about the culture of the people there and also learn their way of life. I also get to experience new cuisines, dressing styles, clothes of the people of that region and also get to know more about people from that part of the world. Travelling is always educational in nature. For example, when a person travels from a developing country to a developed country he or she gets to learn about the ways and innovation adopted by people in the developed countries to make their lives better.

When I visit another country with a culture completely different from my own it broadens the perspective of life for me. Travel teaches you so many things about life. It makes you more humble and also helps you appreciate your own life once more.

Travel is also used for educational purposes. I have been on college trips that have taught me a lot. Many college courses especially in the fields of engineering and architecture have study tours. For example, the students on an architecture study tour go to a city with plenty of historical architecture or modern architecture where they can get firsthand knowledge of the buildings and the great architectural work done in the past.

I love to find out more geographical locations and also love to travel to these locations. We can learn about the natural beauty of our planet by visiting places that have abundant natural beauty. I have been on trips designed as animal safaris where I saw and learned about the life of animals and birds which I saw on the trip.

Whether the trip is recreational, a study tour or a business tour, traveling has always been educational in nature. I love to travel and plan trips every year. Thus, traveling for me has been very educational and satisfying in nature.

(f) THE SACRIFICE OF A SOLDIER

I have been writing stories for quite some time now, but never got the chance of talking about my friend who was a soldier in the Indian Army. Although he is no more in this world, his sacrifice is talked about across the nation. We had played together through our childhood and along with that we even went to the same school. It was his dream to join the Indian Army and serve the nation. He always told me that he wanted to do something for the nation, so that every single person of this nation will remember him. In addition, he even told me that he would surely do something in his life that will force me to write something about him.

Years passed glibly and both of us went on our chosen paths. As I had flair of writing, I was soaring through it and my friend passed the examination of NDA (National Defense Academy) and went to his training. On his passing out, he was a Commissioned Lieutenant. On his return, we both had a great party and talked about the ways of workings. He informed me about the wars going on between India and Pakistan and the intense complications between both the countries.

Soon after some days, my friend was called back to the unit. He called me after few days and gave me a good news that he had been selected in the most coveted Black Cat Commando unit and had been promoted to the rank of Major. Things were going smooth until the fateful day of 26/11 came into the lives of every Indian. Terror had Mumbai in its grip as some Pakistani terrorists had attacked several parts of Mumbai including Hotel Taj. I was in a state of shock to hear this news and when I was watching details of the attack, I came to know that Black Cat Commandos of the Indian army had come to Hotel Taj and my best friend was there in the team.

I was just praying to God that everything should go fine and my friend should come back in one piece. However, fate was about to strike back on me as I heard the news that my friend was gunned down by the terrorists. The world broke down on me and even his parents, as it was a devastating news, which I had heard. After things were under control, the body of my friend, Major Sandeep Unnikrishnan, the soldier and a son, was brought back and everybody was chanting his name all across India because of his heroic deed. As I was looking at the coffin, his body covered with the Indian flag, his words resonated in my mind : "I will do something that will make the entire world remember me". It was exactly as he had said it would be.

Answer 2.

THE WHITE TIGER BY ARAVIND ADIGA

Son of a rickshaw puller, the neo conventional protagonist from the book, The White Tiger, Balram Halwai, narrates the actual story connected with his lifestyle and his movement through darkness in order to attain light. In a tongue-in-cheek manner he unravels how the ruling capitalist ideology never ever allows the actual poverty stricken to come up into the limelight. The publisher of the book is Harper Collins India Limited.

- **Summary of the book :** In essentially the most impoverished along with destitute villages in the Indian subcontinent, a young boy is hungry for knowledge. However, the circumstances plus the cultural disorders prevalent causes it to become almost unattainable for him to do so. Prevailing among the limited opportunities, the son of a rickshaw-puller, finds a means to escape the vice-grip associated with his family and he becomes the driver for the son of a wealthy landowner. Brilliant and quick-witted, he quickly relates to terms while using the divide concerning the rich and the poor. He realizes that he should now break the centuries-old shackles and should flee intended for his existence and vie his old life to satisfy his fate.
- **Detailed review :** White Tiger is the story of Balram, the son of a rickshaw puller, who lives within a small Indian village. He detects the destitution of his family members, repulses and decides to break clear of it. He is searching for opportunities that can alleviate his poverty. He learns to drive and manages to obtain a driver's job with the property owner of his village. Lady Luck smiles upon him as Balram was asked to accompany the property owner's son to Delhi as a driver. In Delhi, Balram understands the ways of the city society. A keen observer and a fast learner, Balram realizes fastly that slight dishonesty should bring him sufficient money for any security in near future. He kills his master along with which he runs off to Bengaluru and years later, Balram is viewed as a good influential member of the Bengaluru power circle productively steering his career derived from one height completely to another.
- **About Aravind Adiga :** Born on October 23, 1974, Aravind Adiga is working as an Indian journalist and writer. In the year 2008, his debut novel The White Tiger won the Man Booker Prize.

Answer 3.

(a) **(1)** (B) The entire block of flats is being demolished by them.

(2) (B) Anil was under the impression that the examination would be held later.

(3) (B) You like this food more than I do.

(4) (B) In spite of failing in the examination, Raju was not upset.

(5) (B) I advised Rita not to lend any money to Ramesh.

(6) (B) No sooner were the arrangements made to release extra water from reservoirs than the rain came in a downpour.

(7) (B) If he hadn't signed, he would have been arrested.

(8) (B) Meena complained that she was not able to see the performance from the back row.

(9) (B) Politics is seen as a stepping stone to personal power by many ambitious people.

(10) (B) Rahul has not only wasted his wealth, but also ruined his health due to his drinking habit.

(b) (1) to (2) on (3) away
(4) by (5) by (6) on
(7) up (8) down (9) for
(10) in

(c) (1) can (2) stop (3) awake
(4) is (5) have (6) have
(7) treat (8) observed (9) slept
(10) found

Answer 4.

(a) (i) (1) stern (2) vitality
(3) chastising (4) curiosity

(ii) (1) The *shape* of the cake we ordered was supposed to be round and not rectangular.

(2) Please *park* your car at the other side of the road as this is a No-Parking area.

(3) The Managing Director of the company was *present* at the meeting yesterday.

(4) The villagers went to the nearby *well* to fetch water.

(b) (i) The disease ITP is serious because if it gets worsened, one could internally bleed to death.

(ii) The wheels of the yellow car have fallen off, the doors have broken and all the chrome has worn off.

(iii) The writer learned that things are things and can be replaced if necessary. He has also learned to enjoy the boyhood of his son.

(iv) It is most important for the writer to celebrate the boyhood of his sons.

(c) (i) (1) The passage is about a man who has always been taught to take good care of his things.

(2) Four years ago his son was diagnosed of ITP, a fatal disease, because of which he could die.

(3) During hospitalisation, he presented a yellow car to his son. Though his son was too young for it but he still bought it.

(4) His son's disease came and went but the writer learned a lesson. Things are things and can be replaced, but life is more precious.

(ii) The passage is about a man who was taught to take care of his things. Later he realized that natural course of wearing things differs from making things last at any point. Four years ago his son was diagnosed of ITP, a fatal disease. While his son was hospitalised, he bought a yellow car for him. He knew the car would break but he still bought it because if the car broke, it would mean his son is in good health to play with it. His son's disease came and went but the writer learned a lesson that things are things and can be replaced if required, but life is invaluable. (111 words)

●●

ENGLISH-I 2012

QUESTIONS

Question 1.

Write a composition (in approximately 450–500 words) on ***any one*** *of the following subjects :* **[30]**

(You are reminded that you will be rewarded for orderly and coherent presentation of material, use of appropriate style and general accuracy of spelling, punctuation and grammar.)

(a) *"We are happier than our forefathers." Express your views on the given statement.*

(b) *The sounds of laughter and loud voices attract your attention to a room in your house. The door is closed. Describe what you hear, as well as what you see, as you gently open the door.*

(c) *"Today, in our global society, if young people go abroad to study or to seek jobs, they cannot be termed unpatriotic." Write for or against this topic.*

(d) *Diversity.*

(e) *"Nowadays, there are careers that cater to every interest and aptitude." Discuss this in the light of your own interests.*

(f) *Write an original story with the following beginning: "I was not more than fourteen at that time....."*

Question 2.

A student, at the end of Class XII, requests the Principal of his school for a certificate or testimonial which he needs for his further studies. Imagine that you are the Principal and draft a certificate, keeping the following points in mind : **[20]**

*Duration of acquaintance—student's academic record—student's conduct—involvement in Co-curricular and extra-curricular activities—leadership qualities—feed-back obtained from teachers and any other relevant information.***

Question 3.

Answer sections (a), (b) and (c).

(a) *In each of the following items, sentence A is complete, while sentence B is not. Complete sentence B, making it as similar as possible to sentence A. Write sentence B in each case.* **[10]**

Example : (0) (A) We must endure what we cannot cure.

(B) What cannot be ....

Answer : (0) What cannot be cured must be endured.

** Answer is not given due to change in the present syllabus.

(1) (A) *He knows everything about the business.*

(B) *There is ....*

(2) (A) *A bridge would have been built if the funds had not run out.*

(B) *Had the ....*

(3) (A) *As soon as winter comes, birds migrate to the south.*

(B) *No sooner ....*

(4) (A) *Man is exploring the sea in order to increase his resources.*

(B) *The sea is ....*

(5) (A) *They were the only ones who completed the task.*

(B) *Except for ....*

(6) (A) *He's the richest man I've ever met.*

(B) *I've never ....*

(7) (A) *I have never seen a film as bad as this.*

(B) *This is ....*

(8) (A) *He said, "I will definitely visit you tomorrow."*

(B) *He said that ....*

(9) (A) *We should have sympathy for the less fortunate.*

(B) *We should sympathize ....*

(10) (A) *My praise pleased her.*

(B) *What gave ....*

(b) *Fill in each blank with a suitable word. (Do not write the sentence.)* **[5]**

(1) *The soccer player was shown a yellow card for arguing _______ the referee.*

(2) *Mr. Donald will call _______ him tomorrow.*

(3) *The exhibition was a complete sell _______.*

(4) *How can she put _______ with such indisciplined students ?*

(5) *The man broke _______ the cordon and rushed to the dais.*

(6) *We must cut _______ on wasteful expenditure.*

(7) *Though I was young, my father had great confidence _______ me.*

(8) *The novelty of being a celebrity wears _______ ultimately.*

(9) *Anna wasn't looking where she was going and she walked_______ a lamp post.*

(10) *She is the nicest person I have come_______.*

(c) *Fill in the blanks in the passage given below with the appropriate form of the verb given in brackets. Do not write the passage but write the verbs in the correct order.* **[5]**

After they________(1) (walk) for a while, they________ (2) (see) an isolated cottage. They________ (3) (peep) through the broken window,________(4) (look) for someone who ________(5) (guide) them back to the main road. They________(6) (find) that the cottage ________ (7) (be) dirty, cobwebs ________ (8) (hang) across the roof and it________ (9) (seem) to be unoccupied. They________ (10) (be) disappointed.

Question 4 .

Read carefully the passage given below and answer the questions (a), (b) and (c) that follow :

THE HIDDEN BOX

(1) I opened the iron gate and walked as softly as I could up the weed-tufted gravel drive. My mind was strangely empty. I felt no glow of pleasure and was unexcited at the prospect of becoming rich. I was occupied only with the mechanical task of **finding** a black box.

(2) The front door was closed and set far back in a very deep porch. The wind and rain had whipped a coating of gritty dust against the panels, showing that it had been shut for years. Standing on a derelict flower-bed. I tried to push open the first window on the left. It yielded to my strength stubbornly. I clambered through the opening and found myself crawling along the deepest window-ledge I had ever seen. After I had jumped noisily down upon the floor, I looked up.

(3) The room was thick with dust, musty and empty of all furniture. Spiders had erected great stretching of their webs about the fireplace. I made my way quickly to the room where the box was and paused on the threshold. It was a dark morning and the weather had stained the windows with blears of gray wash which kept the brightest part of the weak light from coming in. The far corner of the room was a blur of shadow. I had a sudden impulse to have done with my task and be out of this house forever. I walked across the bare boards, knelt down in the corner and passed my hands about the floor in search of the lose board. To my surprise, I found it easily. I lifted it up, laid it aside and struck a match. I saw a black metal cash-box nestling dimly in the hole. I put my hand down and crooked a finger into the loose reclining **handle**, but the match suddenly flickered and went out and the handle of the box, which I had lifted up about an inch, slid heavily off my finger. Without stopping to light another match, I **thrust** my hand into the opening and, just when it should be closing about the box, something happened.

(4) I cannot hope to describe what it was but it had frightened me very much. It was some change which came upon me or upon the room, indescribably subtle, yet momentous. It was as if the daylight had changed with unnatural suddenness, as if the temperature had altered greatly in an instant or as if the air had become twice as rare or twice as dense as it had been in the twinkling of an eye. Perhaps all of these or other things, happened together, for all my senses were bewildered all at once and could give me no explanation. The fingers of my right hand, thrust in the opening in the floor, had closed mechanically, found nothing at all, and came up again empty. The box had gone.

(5) I heard a cough behind me, soft and natural, yet more disturbing than any **sound** that could ever come upon the human ear. That I did not die of fright was due, I think, to two things : the fact that my senses were already disarranged and able to interpret to me only gradually what they had perceived, and also the fact that the utterance of the cough seemed to bring with it some more awful alteration in everything. I collapsed weakly into a limp sitting-down position upon the floor. Sweat broke out on my brow and my eyes remained open for a long time without a wink.

(6) In the darkest corner of the room near the window, a man was sitting in a chair, eyeing me with a mild but unwavering interest.

(a) **(i)** *Given below are four words and phrases. Find the words which have a similar meaning in the passage :* **[4]**

(1) *neglected* **(2)** *urge*
(3) *significant* **(4)** *explain*

(ii) *For each of the words given below, write a sentence of at least ten words using the same word* **unchanged in form, but with a different meaning** *from that which it carries in the passage:* **[4]**

(1) *finding (line 3)* (2) *handle (line 19)*
(3) *thrust (line 21)* (4) *sound (line 31)*

(b) *Answer the following questions in your own words as briefly as possible :*

(i) *What was the narrator's state of mind when he came to the house in search of a black box ?* **[2]**

(ii) *What details show that the house had been shut for years ?* **[2]**

(iii) *Why did the narrator have a feeling of uneasiness and fear when he entered the room ?* **[3]**

(iv) *What was the strange thing that happened about the box ? What impact did it have on the narrator ?* **[3]**

(c) *Describe the narrator's purpose behind entering the abandoned house and the strange atmosphere he encountered there. You will be required to :*

(i) *List your ideas clearly in point form.* **[6]**

(ii) *In about 100 words, write your points in the form of a connected passage.* **[6]**

ANSWERS

Answer 1.

(a) WE ARE HAPPIER THAN OUR FOREFATHERS

Happiness is a state of mind. If there's something money can't buy, it's happiness. Compared to these days, I strongly feel, that our forefathers were happier. Their life was simple, their needs minimal. They had less expectations. Their days started with the morning sun and ended with the setting sun. They listened to the chirping of birds as they tilled their fields, breathing the fresh air around. The pristine beauty of Nature, with its greens, the golden harvests, the flowers, the fragrance–added colors and substance to their lives.

In short, they had time and the sensibilities to appreciate and marvel at the wonderful things God has blessed us with. After a day's hard work, the simple meal served on the home platter was relished by our forefathers. Organic food with unpolluted water–fresh and healthy ! What would we not give for such a luxury !

The simple lifestyle our forefathers led, kept them away from diseases. The sheer routine of their life with oodles of exercise never allowed them to become obese. Diabetes and heart burns were unheard of. Cooling devices and heaters never sucked up the oxygen around. A big fire in the center kept their huge family warm; the fronds above fanned them in summer to keep them cool and fit. The big joint families of the olden times shared their joy and sorrows, helped each other in hours of need, and shared responsibilities.

Children grew up in the lap of Nature, under the protective supervision of their elders. How unlike the children of today's nuclear families where parents go to work and children are left to the care of domestic help or in the company of modern gadgets! Festivals and weddings used to be occasions for the huge extended family to come together and enjoy. They were extended holidays; unlike the hurried two day affairs in five star hotels amidst filmy settings! Festivals like Holi were played with natural ingredients made from fragrant flowers and leaves. It kept away skin rashes and eye sores. Spurious drinks did not take away lives on Diwali nights. What if our forefathers had to walk miles to reach their destinations, what if they had to till their land in scorching sun, what if they were not able to travel around the world?

The truth is that they led a contented life. Contentment, the chief ingredient for happiness! Do we have it? Can we ever hope to get it? Coming to think of it, I feel, our forefathers were stress free, guilt free, more innocent, led uncomplicated life and consequently were much, much happier than us.

(b) THE SOUNDS OF LAUGHTER AND LOUD VOICES ATTRACT YOUR ATTENTION TO A ROOM IN YOUR HOUSE. THE DOOR IS CLOSED

I stretched my arms after a long journey and looked around. It'd been a long time–too long for me to remember the lanes and by-lanes of this small village. In a remote corner was my ancestral house. My childhood memories were attached with it.

Far away in the foreign land, I had been gripped with nostalgia. India calling! Oh! How desperately I wanted to get in touch with my roots, my village, my home.

A motley of buildings had come up changing the landscape. I could see a factory of some sort at a distance. I asked a group of youngsters for 'Manuhaar Haveli'. They silently pointed beyond the chimney of the factory. The fear on their faces perturbed me. It was a bit of a walk but I was thrilled to see the silhouette of the palatial house I was heading for. As I stopped to admire the picturesque view, a voice from behind startled me. An old lady asked me, in her crackling voice. "Beta, where are you going." I pointed at my house. All of a sudden, her expression changed. "That house? That accursed one? Don't go there beta, you'll land in trouble."

I nodded in confusion but continued my trek. I walked around the house and saw the back door ajar. I discreetly peeped inside and saw a group of youngsters smoking and joking with half filled glasses in their hands. Lots of wooden cases were put as though they were going to be dispatched. Some equipments for distillation were also visible at the far end. Two country-made pistols were lying on a stool. Then I heard a burly fellow with a deep scar on his face saying 'What a lark to find this house as our hide out. No one, not even the police, will suspect us'. It was quite frightening. "Grandfather Uday Nath Singh must be turning in his grave What about his son and the grandson? If they land here some day ?," a young voice asked. "Oh! They are busy with their life. The son is bedridden and the grandson, he's in the US. He won't be interested in this Godforsaken place," the other voice replied.

To my horror, I realised that my dear house has been transformed into an 'adda' for making spurious liquor. My blood boiled. I made a bee line towards the nearest police station. When I showed my credentials, the police were convinced of my identity. We rushed back in the police vehicle. By the time we reached, most of the gang were in drunken stupor. The less intoxicated ones made their escape. The remaining ones were nabbed and hurdled into the van.

I heaved a sigh of relief. I went in front of the huge portrait of my grandfather and took an oath– "Never, never I will leave this house uncared for. Sorry Grandpa, for what happened but your grandson will now put the things right. Promise!" Did I see a smile on the lips of the portrait? I wonder.

(c) TODAY, IN OUR GLOBAL SOCIETY, IF YOUNG PEOPLE GO ABROAD TO STUDY OR TO SEEK JOBS, THEY CANNOT BE TERMED UNPATRIOTIC.

AGAINST THE MOTION

"I am neither an Athenian nor a Greek, I am a citizen of the world," said the sage of Athens. These words echoing from the past, captures the mood of the citizens of the modern world. Today, as the world is shrinking in size and is more like a global village, we have to examine the connotation of the word 'patriotism' in a new light.

Is it unpatriotic to go abroad to study or to work? The question seems rather absurd! Patriotism, dear friends, is love for one's own country. Distance does not matter if one is devoted to the welfare of one's country and accepts the responsibility for its growth.

We cry hoarse over the brain drain; saying that the best brains in our country after getting quality education, leave the country to lavish their skills for the development of other nations. Just as every coin has two sides, can't there be a positive side to it ? Many an Indians, settled abroad have brought name and fame to our country by exhibiting super talents that vie with the best in the world.

The value of this for India is very real. This process took a quantum jump with our success in the Information Technology sector. Today, multinational companies feel privileged if they get an opportunity to do business in partnership with Indian companies any where in the world, thereby contributing to our economic growth.

After a promising youngman has completed his training, he usually expects work which should not only bring in enough money and other emoluments but also give him sufficient professional satisfaction. An overwhelming majority of such repatriates go abroad as students seeking academic, scientific or technological excellence which is no where available in our country. And what's wrong with that.

It caught the international headlines when Dr. Jayant Narlikar discovered his Hoyle-Narlikar Theory, when Dr. Hargobind Khurana was awarded the Nobel Prize for Medicine in 1968, and also when the famous Indian-born U.S. astrophysicist Dr. S. Chandrashekhara made a name for himself or Kalpana Chawla got the precious chance to explore the space ! Moreover, why do we forget the very important fact in these cases, the loss of one single country is a colossal gain to the world at large.

So, lets think about the world in general and consider ourselves as the citizens of one world society envisaged by our great philosopher, Dr. S. Radhakrishnan. In this technologically advanced age, there are no nationalistic barriers. On the other hand there is the need to build up a world brain, a world mind and a world culture. So let's be broad minded to say that if young people go abroad to study or to seek jobs, they cannot be termed unpatriotic.

(d) DIVERSITY

The other day someone presented me with a lovely bouquet. Sprigs of blue orchids, deep pink gladioli, a big bright orange rose and lots of ferns– seemingly clashing in color, texture and fragrance; but the total effect–just stunning! Each quality offsetting the individuality of the other; and yet merging to form a unified whole !

Such is diversity. It adds zing to life. In fact, God intended the world to be diverse. He has painted the vast canvas of the universe with diverse creatures and objects that are unique in character.

Look at the rich biodiversity that exists among our fauna and flora. The ones that exist on deserts are different from that of the seas and the skies and the mountains. Some swim, some crawl and some fly. There are the beautiful peacocks proudly strutting with their gold speckled feathers, the eagles that scale the dizzying heights, the flightless ostriches, kiwis and our own home grown hens; each serving their own functions in life. There are the majestic lions, the fierce tigers, the lovable jumbos and the cute cats and dogs. The thorny but the most beautiful roses, the stately lilies, the pink lotus, the carnations, the cacti, the fir trees, the maple trees, the coconut trees, the poplars and the casuarinas–there is infinite variety that makes this world a veritable paradise.

There are the Whites, the Blacks and the Browns, the blondes, the brunettes, the raven–haired ones, the blue eyed, the emerald eyed, the brown and the black eyed people inhabiting the different lands. They live in continents that have geographical and climatic variations. People think in different ways; have different aspirations, ways of life, attitude and behavioral patterns. This diversity is what that makes life interesting.

Diverse ideas have led to the progress in the world. If all the people had thought alike, worked alike, there would not have been so many inventions and discoveries in this world. There would not have been such rich choice in art, literature, dances and songs.

This is the unique quality that gives our land the pride of place. People of many religions, caste and creed performing their rituals, celebrating their own festivals, speaking different languages, attired in dresses unique to their culture where else in the world can one see such diversity ? Yet here is this amazing co-existence of diversity in this largest democracy of the world. An invisible thread of harmony binds the different people. There is mutual tolerance and respect for each other that enable our land to progress through centuries. Yes, India is a perfect example of Unity in Diversity. It is the same factor that will lead to our vision of the world as a global village.

In order to survive, humanity also has to sing a harmonious song. The secret lies in maintaining diversity and individuality but, at the same time attain unity in diversity.

(e) NOWADAYS, THERE ARE CAREERS THAT CATER TO EVERY INTEREST AND APTITUDE

Gone are the days when one was expected to be either a doctor, or an engineer, or a lawyer; the only jobs considered worth following. They were like prestige symbols that added to the family's status in society. Today, as I stand on the brink of choosing a profession, I stand spoilt for choice. My interests are many-my aptitude? I still have to unveil it ! It is time for me to mull over the possibilities. After all, I'm special and I want to do something unconventional. I think I'll go for anything that requires oodles of creativity like advertising, writing, music, painting, etc. But, my talents in these fields are still nascent. Will I be able to turn them into career prospective?

I've always been interested in travelling. I've travelled widely in India and aspire to go abroad. I know quite a lot about the countries of the world. A travel consultant; that's what I want to be. My dream job ! Having all the inside dope on exotic destinations, planning people's itineraries and organising their trip - well, maybe I can do this some day ! My interests don't stop there. I'm good at managing events; whether they are home parties, community get–together, or school functions . I can give a try in Event Management. One can do it even without a degree in the subject. A lot of zest and energy and oodles of confidence; I've got them all !

I'm interested in cooking. I dole out exotic food stuffin a jiffy! Healthy, tasty, eye-catching and mouth-watering! May be I can start a restaurant ! A speciality, up- market restaurant. But alas ! It requires oodles of money.

Taking into consideration my computer skills' I can become a Video game designer. Designers can also move up to be creative directors as well. However, one needs to have strong skills in computer programming, computer engineering, software development, computer animation, graphic design and computer graphics–or all of them ! Will I be able to master the tricks?

But, what about my inherent desire to do something for the country, or the world, or the environment at large ! I'm really pained at the way people treat Mother Nature. If I can become an Environmental Engineer, I can contribute my mite for the well being of the humanity. It is the application of science and engineering principles to improve the natural environment. Environmental engineers also address local and worldwide environmental issues such as the effects of acid rain, global warming, ozone depletion, water pollution and air pollution from automobile exhausts and industrial sources.

How deeply satisfying a job it would be for me! Yes, I've made up my mind. This is going to be the aim of my life. It'll be exciting, and at the same time give me a sense of worth as I'll be doing a service to the humanity in general and specifically for the future generation.

(f) I WAS NOT MORE THAN FOURTEEN AT THAT TIME......

I was not more than fourteen at that time. As I was born in an aristocratic family, I thought the world lay at my feet. My parents doted on me as I happened to be their only son and I loved all the attention I got at school. Somehow all these pampering went to my head. I was intoxicated with self importance.

It was at that time, Mohan, a Twelve year old orphan boy came to live in our house. My father took pity on the orphan and kept him for his help.

Mohan turned out to be a bright boy who spent all his free time reading my old, discarded, text books. Some times he used to accompany me for my horse riding sessions and there also he stole the show. I don't know when I started disliking the boy.

Then one day, all hell broke loose. My mother's favorite pearls went missing. In fact I had taken them from her dressing table and placed them in a tiny box in the servant's quarters. A thorough search was made. At the suitable time, I announced that I had seen Mohan passing the door of her room. And Lo! The beautiful string lay shining among his clothes. All this time he kept on whimpering, "No sir, I haven't taken it. I swear by God."

But no one took any heed. The next day the boy went missing. No one bothered about him. But there was guilt niggling at the back of my mind.

Years later, on my way back from one of our factories, I stopped at a small but neat restaurant. When I asked for the bill, the waiter told me that my meal was on the house. Bewildered, I looked towards the manager. I found a pleasant young man moving towards me. He greeted me cheerfully and said, "You won't remember me sir but, I'm Mohan, your servant boy." All the memories came rushing back. I felt awkward but I knew this was the moment to wash off all the guilt. I started apologising. "Mohan I'm sorry...." He waived off my apology, and said "Oh it's a simple rags to riches story. I was picked up by a gentleman, who wanted help in his dhaba. The business improved so much, that he made me his manager and now I own this restaurant!.

Again, I started blubbering, "Mohan, it was bad of me...." He said with a wide grin. "Oh sir, you were not more than fourteen then. Such things happen, and you see what happened to me. If I had not left the Haveli at that time, do you think I would be in this place now ? I thank you Sir, for what you did to me," I'm happy for Mohan, but it was a cold consolation for the nasty thing I did to him.

Answer 3.

(a) **(1)** (B) There is nothing he doesn't know about the business.

(2) (B) Had the funds not run out, a bridge would have been built.

(3) (B) No sooner does the winter come, birds migrate to the south.

(4) (B) The sea is being explored by man to increase his resources.

(5) (B) Except for them no one completed the task.

(6) (B) I've never met such a rich man

(7) (B) This is the worst film I have ever seen.

(8) (B) He said that he would definitely visit her the next day.

(9) (B) We should sympathise with the less fortunate.

(10) (B) What gave her pleasure was my praise for her.

(b) **(1)** with **(2)** on **(3)** out
(4) up **(5)** through **(6)** down
(7) in **(8)** off/out **(9)** into
(10) across

(c) **(1)** walked **(2)** saw **(3)** peeped
(4) looked **(5)** could guide/guide
(6) found **(7)** was **(8)** hung
(9) seemed **(10)** were

Answer 4.

(a) **(i)** **(1)** Neglected – derelict
(2) urge – impulse
(3) significant – momentous
(4) explain – interpret

(ii) **(1)** The accidental *finding* of fire by our forefathers was a revolutionary one.

(2) Teachers, over the years, learn how to *handle* difficult situations in class rooms.

(3) Motivational words from his mentor gave him the necessary *thrust* to make a name for himself.

(4) The *sound* of the traffic didn't allow me to study.

(b) **(i)** The narrator's mind was empty. He was totally unexcited at the thought of becoming rich. It gave him no pleasure. He was mechanically focusing on the task of finding the black box.

(ii) There was a coating of dust on the panels. A neglected flower bed added to the desolate look of the house. The rooms were full of cobwebs; dusty, musty, without furniture, giving an impression that the house was shut for years.

(iii) The room was dark and the far corner of the room was a blur of shadow. The whole atmosphere was eerie and this caused a feeling of uneasiness and fear.

(iv) When the writer put his hand down on the handle, the match went off and the atmosphere of the room changed with unnatural suddenness. The box suddenly vanished. The writer almost died of fright at this strange happening.

(c) **(i)** **(1)** Intention to find the black box.

(2) The house looked abandoned.

(3) Room musty, dusty, cobwebs all around.

(4) Dark morning; shadows made it look eerie.

(5) Box disappearing at his touch.

(6) A cough behind, frightening him.

(7) A man, sitting on a chair, staring at him.

(ii) The writer's intention was to find a black box that would make him rich. He walked silently to a deserted house. Without excitement, he focused on his job. The house looked shut for years. It was dusty, devoid of furniture. The room where the box was kept looked dark with frightening shadows. He lifted the loose board to touch the black box. The light went off. The handle slid from his hand. When he tried again to lift it, there was no box. Hearing a cough, he turned around in fear. He saw a man at the far end of the room staring at him. (105 words)

●●

ENGLISH-I

2011

Question 1.

Write a composition (in approximately 450-500 words) on any one of the following subjects : **[30]**

(You are reminded that you will be rewarded for orderly and coherent presentation of material, use of appropriate style and general accuracy of spelling, punctuation and grammar.)

(a) *'Advancement in Science and Technology has made life more convenient and comfortable' Describe 'one day' in the life of a common man, highlighting the significance of the above statement.*

(b) *'The joys and challenges of adolescence'. Explore the theme by referring to your own experiences.*

(c) *Inspiration.*

(d) *'The end justifies the means'. Argue for or against the proposition.*

(e) *'We fight for our rights but forget our duties'. Bring out the implication of this statement giving your own views.*

(f) **(i)** *Write an original story which has for its ending : ... I said to myself, "Truth indeed is stranger than fiction".*

OR

(ii) *Write an original story entitled : 'The Secret'.*

Question 2.

The Interact Club of your school held a week long literacy campaign recently. As a member of the club, write a report for the school magazine in about 300 words on the activities undertaken by the club members during the campaign. Use the notes given below. You may include other evant details. **[20]**

Announcement about the campaign in the morning assembly—purpose of the campaign—banners and posters made—visit to a slum—interaction between members of the club and the children of the slum—books, notebooks and stationery distributed—children taught in groups—food packets distributed —response from learners.

Question 3.

*Answer sections **(a)**, **(b)** and **(c)**.*

(a) *In each of the following items, sentence **A** is complete, while sentence **B** is not. Complete sentence, **B**, making it as similar in meaning as possible to sentence **A**. Write down sentence **B** in each case.* **[10]**

Example : (0) (A) Everyone knows that Ram is brave.

(B) There is no one.................

Answer : *(0) There is no one who does not know that Ram is brave.*

(1) (A) *Those who are honest are trusted by all.*
(B) *The................................*

(2) (A) *Kusum is not as intelligent as her brother Kabir.*
(B) *Kabir.............................*

(3) (A) *There were many problems in the beginning, all the same it has been a successful event.*
(B) *Notwithstanding the.......................*

(4) (A) *She was in a hurry and I was in hurry too.*
(B) *She was in a hurry and so.................*

(5) (A) *Mr. Kapoor taught us western music in school.*
(B) *We were.............................*

(6) (A) *I am sorry, I am unable to accompany you.*
(B) *I regret my.............................*

(7) (A) *Smoking can cause damage to your health.*
(B) *Your health.............................*

(8) (A) *The prices of food grains have increased due to hoarding.*
(B) *Hoarding has led.......................*

(9) (A) *Not many people listen to the radio these days.*
(B) *Few..................................*

(10) (A) *Arun is fat, yet he runs fast.*
(B) *In spite ....*

(b) *Fill in each blank with a suitable word. (Do not write the sentence.)* **[5]**

(1) *This is ________ all means the better book.*

(2) *The robbers broke ________the house in the middle of the night.*

(3) *The kitchen was infested ________ rats.*

(4) *During the vacations, the students will be busy preparing________the examinations.*

(5) *The workers toiled ________dawn to dusk.*

(6) *I take ________all that I said about him as I realize I had been wrong.*

(7) *He returned to his country ________ten long years.*

(8) *Ravi distributed the chocolates ________his cousins.*

(9) *The workers said they would call________the strike if their demands were met.*

(10) *The old man passed________in his sleep.*

(c) *Fill in each blank with the appropriate form of the word given in brackets. (Do not write the sentence) :* **[5]**

(1) *We________ (be) all, the caretakers of this earth.*

(2) *He need not go to the doctor as his fever ________ (subside).*

(3) *The plane landed at Bangalore only two hours after________ (take) off from Delhi.*

(4) *As soon as the teacher entered the class, the students ________ (stand) up to wish her.*

(5) *The river Ganga________ (has) its origin in the Himalayas.*

(6) *The patient ________ (die) before the doctor arrived.*

(7) *The joys and sorrows of life________ (weave) together.*

(8) *David ________ (exercise) regularly when he was in college.*

(9) *Avoid________ (eat) junk food in order to remain healthy.*

(10) *Let me know as soon as they________ (finish).*

Question 4.

Read the passage given below and answer the questions that follow :

(1) The boy found the kites from Japan hidden in a dark passage of the house. Snehamoy helped him take out the kites from the box, lifting each delicately by its **staff**, and spread them out on the courtyard.

(2) They fixed the day of celebration for Biswakarma the god of machines to be their kite-flying day. But first, they'd have to be ready with rolls *line* to fly. The kites, **lines** sharpened with a coating of finely **ground** glass to cut those of their rivals. A suitable **field** must be found, one with a good view of the **horizon** and free of trees to prevent the lines getting tangled up in the branches. Their hearts pounded at the thought of a kite-fight. Like a sensible teacher, Snehamoy sat the boy down with his exercise book. 'If each kite takes five hundred feet of line to fly, how many feet would twenty kites take ?' he asked the boy, The boy frowned. 'Only five hundred feet for each kite ! What if they get tangled with others during a fight and keep circling away needing more and more ?'

(3) The two spent a whole afternoon asking neighbors for empty glass bottles. Snehamoy's aunt smashed them in her kitchen, then ground the shards in a pestle, pricking her thumb in the process. Glue was added to the glass granules,and a touch of vermilion dye to have the lines stand out against the blue sky and the clouds. After returning from school, Snehamoy joined the boy in stringing lines from one end of their courtyard *to* another. Then, like workers, they wrapped their palms in soft cotton cloth, took lavish dips in the pail of abrasive, and went from one end of a line to another, carefully applying a **uniform** layer over it. In the end, they sat back in the kitchen and admired the brilliant maze turning their courtyard into a field of red pepper.

(4) For days, pedaling to and fro from school, Snehamoy heard the village buzzing with the impending kite-fight. **Rival** groups of boys formed teams and were spurred on by a shopkeeper who was offering his regular kites free to anyone who was willing to brave the foreign invasion. Yet on the day of the grand **contest,** as the first of the Japanese kites came out of the box and heaved by its flyers, went up over the school building, there was a pall of silence.

It was a giant Baromon, crushing the helmet of a samurai warrior between its teeth. Scarlet and ochre, it hung still, face up, refused to be cowed by the stiff breeze.

(5) Standing in a circle with the boys, Snehamoy strung the kites,gaze fixed on the sky. Like characters in a play, they made their appearance one by one. Tsugaru whirred like an airplane through a clever loop fixed on its back. The crowd went into a roar over the kabuki faces, which floated in the sky like stained glass windows lit by the afternoon sun. Soon the lines were crossing and tangling up the kites with a steady stream of taunts spurring on the flyers. The boy reached inside the kite box and **scooped** out more. Instantly the field buzzed with humming and whistling kites shaped as cicadas, gnats, dragonflies and dancing carps. It felt like a carnival and drew the crowd's applause.

Adapted from **The Japanese Wife** by Kunal Basu

(a) (i) *Use each of the following words as used in the passage in a sentence of your own construction so as to bring out its meaning very clearly. Using the word in a context very similar to the passage will be penalized.* **[5]**

(1) *field (line 6)* (2) *horizon (line 7)*
(3) *rival (line 23)* (4) *contest (line 25)*
(5) *scooped (line 35)*

(ii) *For each of the words given below, write a sentence of at least 10 words using the word* ***unchanged in form, but with a different meaning*** *from that it carries in the passage :* **[4]**

(1) *staff (line 2)* (2) *line (line 5)*
(3) *ground (line 6)* (4) *uniform (line 19)*

(iii) *Explain, in the context of the passage in not more than two sentences of your own, the meaning of each of the following expressions taken from the passage (mey using phrases will not do).* **[4]**

(1) *... have the lines stand out against the blue sky and the clouds. (lines 15 - 16)*

(2) *... the brilliant maze turning their court-yard into a field of red pepper. (line 20)*

(3) *...were spurred on by a shopkeeper who was offering his regular kites free to anyone who was willing to brave the foreign invasion. (lines 23 - 24).*

(4) *It felt like a carnival and drew the crowd's applause. (line 36 - 37)*

(b) *Answer the following questions in your own words as briefly as possible :*

(i) *Where did the boy find the kites ?* **[1]**

(ii) *How did Snehamoy and the boy decide how much line each kite would need ?* **[2]**

(iii) *What effect did the impending kite-fight have on the villagers ?* **[2]**

(iv) *Describe the first kite which was heaved into the sky.* **[1]**

(v) *How did the kites flown by Snehamoy and the boy add to the atmosphere of joy and brightness ?* **[1]**

(c) *In not more than 100 words of your own, describe the arrangements made by Snehamoy and his companion before the kite-fight. (Failure to keep within the word limit will be penalized.)* **[10]**

ANSWERS

Answer 1.

(a) ADVANCEMENT IN SCIENCE AND TECHNOLOGY HAS MADE LIFE MORE CONVENIENT AND COMFORTABLE

Post liberalization, there has been a tremendous impetus to science and technology in India. All those gizmos which were earlier affordable and available to only a few selected people are now within the reach of common man also. Information technology and BPOs have transformed the living standards of an average citizen. Media exposure, advertising in a big way and more disposable income has led to this change in almost every field and profession. New discoveries and inventions in Transportation and Telecommunication have brought a sea-change in a common man's life. It has definitely made life not only more comfortable and convenient but has also put it on a fast track.

In present scenario, an ordinary man might not be having a dwelling of his own or a car, but if you go to his house he will definitely be having a colored television, a music system, a gas stove and not to forget the mobile phone. These have almost became a 'must have' for any common person, whether he is living in a village or a small town or even a big city. The lady of the house has cleaner mode of cooking—the L.P.G. For entertainment and news from all over the world, there is television with many channels, the operation of which is known to everyone in the family, as they bat their eyelids so do the channels, they flip them over fast from one to another.

India perhaps is a country which has maximum mobile phone users. All kinds of models are available to suit almost every pocket. This exchange and sharing of information is the major cause of common people's awareness about science and technology. A plumber, an electrician or a person working in an organization, are all dependent in some or the other way on mobiles and computers, which undoubtedly has made their life more organized and comfortable.

The invention of computers is another land-mark in the field of science and technology. A person is almost considered an illiterate if he does not have the working knowledge of computers. So a person who is working in the office or a bank is totally dependent on this wonder machine and why not ? It can do so many things on his behalf. One can talk, sort, find or enter any information that one needs by just one click. One can do shopping that too at one's convenience. The list is endless and so are the uses of science and technology to mankind.

In spite of all the innovations and progress made in this field, still man remains supreme. After all, it is the human brain that has made all these inventions and innovations.

(b) THE JOYS AND CHALLENGES OF ADOLESCENCE

'A lad of fourteen is neither ornamental nor useful' That is how Rabindranath Tagore described an adolescent in his story 'The Home Coming', which really touched my heart when I read it. This sentence can be held true partly in today's scenario as it is very much valid in that transition period from a child to an adult. The metamorphoses is odd, strange and at times painful. Their character and personality is highly vulnerable and fragile. The thinking of young girls and boys at this stage can be overpowered and swayed easily. Very often, we have observed such traits in them, which makes us anxious and concerned at the same time. Careful handling by the parents at home and teachers in school is very much needed at this stage. They are unable to cope up at times with sudden spurts of changes in them.

At the same time, there is no need to get panicky or paranoid on this issue as it is very much normal.

Adolescence is synonynous with adventure. It is that phase in life when the thrill of the 'New' is overwhelming. When the excitement of discovery soars we must not forget that our biggest asset is our future. This is a generation that we should not take for granted and it is a generation about which we should not be judgmental either. We should try to understand them by giving them our quality time. What they need is our unconditional love and support at all times.

Media explosion and advancement in technology has opened new doors for them also. The adolescent of today spends about ten hours a day on media ranging from radio to 'YouTube', on social networking sites like Facebook, quite sometime on mobile phones talking and messaging, not to forget playing games on computers. Television and mobiles have become almost an addiction to them. There statistical findings open a window to the mind of a generation. Like any other, this generation of today is shaped by the spirits of time.

It is a generation born in affluence where shortages are unknown and there is a plethora of choices in almost everything. Technology dominates life of this generation. The world of today's teenager has enough danger zones in it to make an adult anxious. Take drugs or drinking alcohol for instance, most of them get hooked on to such vices at this age only. This is a generation with a million possibilities. Because of technology and freedom, they enjoy, they are able to keep large parts of their lives away from the prying eyes of adults. Some will inevitably fall by the wayside but most will take it in their stride and grab the great opportunities which will come in their way. They also need to be focused in achieving their goals of life professionally, along with all the fun and enjoyments they are having.

(c) INSPIRATION

The word 'Inspiration' rings out many bells in one's mind. It evokes many feelings such as commitment and confidence which further charges up the other emotions to move ahead in whatever sphere one want to. Inspiration is a highly potent dose to bring out all the hidden traits and characters in one's personality, which helps him or her to realize the potential and achieve the set goals.

All the well known and famous people who have made a mark in their life have been inspired and motivated at some or the other point of their life. We can take examples of people in the field of Art, Music, Medicine or in the field of Technology. Even politicians and great saints can not be ruled out. One can get inspired in umpteen ways, it can be verbal, written or may be through some of the work done or an example set by the person who kind of became role models. When you are totally enamored by a person, you start believing and have a blind faith in his thinking and ideology and just start following the footprints without even being sure of that.

The person you are impressed with or whom you are following becomes your mentor and you have to consider him like your 'guru'. Only then, benefits can be reaped and dreams can be realized. If we turn the pages of history Arjuna got so inspired from Lord Krishna's sermons and he realized that if he did not listen to Lord Krishna at that time, then he will not be able to fight and protect 'Dharma'.

There are many saints who have earned name and fame in people's heart by simply following their 'guru' for an inspiration in life. So an inspiration coming from the right person helps us to internalize our goals and we attain our mission. Sometimes when we are depressed and are literally down in the dumps, a soothing sentence or a gentle hand on the head gives us much wanted and needed confidence and assurance. Such motivation goes a long way.

Therefore, one can work hard only if there is moral support and inspiration. As the saying goes that there can be perspiration only if there is an inspiration.

(d) THE END JUSTIFIES THE MEANS

(I) FOR THE MOTION

Life is never satiating but ever demanding. Right from the time we are born till the time we live on this earth, every stage and every phase of life has its own aspirations and ambitions. To attain these goals we have to adopt certain paths or means.

If we quit the race midway, one is branded as a failure or a non-achiever. Living with that label on us is again very difficult and at times unbearable too. If we are in that race with full vigour or force and are persuing our goals energetically, then at times we might have to take a road which has not been taken by many. Spiritually also, the same theory was advocated by Lord Krishna too, that in the end every thing comes out to be straight and simple, then it is immaterial to know as to what kind of means have been used, it may be 'Saam, daam dand bhed'. People in general have got in the habit of taking shortcuts and simple routes as they are in a hurry to reach fast and also be the first. There is cut-throat competition everywhere, if they do not show agility and aggressive behavior then they are going to be lagged behind in whatever sphere they have chosen.

Therefore, I feel, 'the end justifies the means'.

(II) AGAINST THE MOTION

It is true that life is short and in that short span we have to achieve whatever we can. Strategically moving forward is the call of the hour and those who do not adopt this approach remain at the lower rung of the ladder. They are being mocked at by their peers and other close people.

Now, the question arises, if all five fingers are not of the same size then how can there be the same socio-economic pattern for every one. The unfair means used in everyday life and corruption is like a worm which is slowly eating our roots. That means if everybody is adopting same way of living, the world will be full of rogues, wicked and cunning people. There will be no trust left for any one.

Are we thinking of building that kind of society where nobody has any value for relationships but is only worried about the status and the money that one has. If we let our children tread this path we are responsible for the consequences of their downfall. We can neither fall prey ourselves nor encourage others to propagate that 'The end justifies the means'. History has shown that 'only balanced societies exist and flourish, the rest just perish.

(e) WE FIGHT FOR OUR RIGHTS BUT FORGET OUR DUTIES

India is a free, secular and democratic country. All the citizens living here have all the rights and freedom to choose the way they want to live and lead their life. Our constitution has provided us with many rights and over the years some more have been added to the list. We are all aware of our rights but our duty towards other is often ignored by us. We open the newspaper in the morning and we read sensational headlines murders, thefts and burglaries are common things that we keep on reading or hearing from time to time.

Let us review different scenes—We are driving on the road, it is true that it is our right to use the road whenever we feel like but we are also responsible to be considerate to other road users. There are many who do not think like that. They drive on the road thinking it is meant for them only and rest of the people are just 'extras' on the road. They will also disobey and violate traffic rules frequently, thus, forgetting their duty.

We all have parks and open spaces in our communities, it is certainly our right to use them and enjoy the natural beauty. People throwing plastic bottles and wrappers indiscriminately is a common sight, even if the dustbin might be a few steps away. Next thought that comes to the mind is practicing religion. As citizens of India we are all free to preach and practice which ever religion we want, but at the same time we have the responsibility to respect other people's religion and the sentiments attached to it. Their beliefs, their places of worship are as sacred as your own, so why such callous and insensitive attitude towards one another, when we also believe that God is one.

To make us aware of our duties, we do not have to read the pages of our constitution or some strict policing, for that purpose but we just need to open our eyes and be aware and tolerant of other people's rights as well. We cannot be selfish and self-centered in this respect. Such awareness has to come from our heart and mind both. The general civic habits which we appreciate so much when we travel abroad are not cultivated by us in our own country and just criticise India.. Until we shed such feelings from our heart and mind we will remain immune to our duties but will keep on making loud noises about our rights.

(f) (i) TRUTH INDEED IS STRANGER THAN FICTION

I was sitting in my bed reading an interesting article about 'True Stories' that was printed in the magazine 'Reader's Digest'. I was fully engrossed because it was about a family who gets entrapped in the aftermath of an earthquake and how they, with the help of neighbors and friends come out of it. It was indeed a nightmarish experience for them which gave me goose bumps as I was reading about them. Outside my bedroom window I could hear the pitter-patter sound of rain drops; the wind was cool that prompted me to have another cup of coffee, though I had already consumed two since evening, but "Never mind, another cup will do no harm," I thought to myself.

As I got out of the bed and went to the kitchen, suddenly my mobile phone rang. I skipped my heart beat and wondered who could that be at that unearthly hour. Millions of thoughts flashed my mind. Anxiously, I picked up the phone and on the other end was my dear friend, Anil, who has gone to Kullu for a holiday. He was sounding very worried and I asked him what the matter was ? His voice was echoing so much, may be, because of network problem on the hills. What I could hear was that, there has been a serious cloud burst as it was raining incessantly and there was a lot of gushing of water. The water, on the hills generally becomes ferocious and turbulent, on it's down hill journey. In brief he said that, here people and their homes along with animals have just been washed away, and he needed my help.

Being in a good position, I always had a car and a driver at my disposal. So I immediately set off for Kullu along with lot of foodstuff, clothes and blankets dumped in the boot of the car. As I reached there it was a heart wrenching scene. There were dead bodies of human beings, animals and small children. Those who survived were crying and wailing and were looking out for their near and dear ones. Their houses and animals were washed away with the strong currents of water. It was one of the biggest natural catastrophe, there was total devastation. I was appalled to see so much destruction due to nature's fury in such a short time.

The most disturbing was that there was hardly any rescue team working. Only locals who survived, were trying to render a helping hand to those in need. People were hungry and feeling cold but there was neither enough food nor warm clothing for them. With great difficulty, I was able to locate my friend and his family. Fortunately they were safe, but imagine the condition of those who are exposed to nature's misery and have no one to turn to. In that need of the hour, I said to myself, "Truth indeed is stranger than fiction".

(f) (ii) THE SECRET

Two brothers Naveen and Pramod belonged to a modest family and lived in a mediocre house in a small township of Uttar Pradesh. Their father was working with government as a superintendent. The family was happy with their life. The father would go to the office on his old Bajaj Scooter, mother, like a traditional woman would look after the house, and the sons as usual were busy with their school, home-work and tests. Life was going on smoothly, with no high ambitions or aspirations. They did not feel bad or jealous as to what their neighbors were eating, what vehicle they were driving. Naveen and Pramod's father, Ram Lal was quite satisfied and contented with whatever he had.

Out of the two brothers, Naveen the elder one was a highly unassuming child, very quiet but intelligent and hard working, whereas, Pramod was just the opposite, very playful, a big prankster and care-free type. He did not take his studies or for that matter any work seriously. Since they were studying in the same school, so all teachers and students grew very fond of Naveen because of his good behavior and excellent academic performance. He became the darling of all and sundry. Pramod too was very fond of his brother and affectionate towards him but somehow temperamentally he could not imbibe good qualities that Naveen had. Since there was not much age difference between them, they were more like good friends than brothers.

Exams were drawing near and both had started preparing for that. After a month, the result was declared. Naveen stood first whereas Pramod had failed in all the major subjects. He was very depressed and started crying. He was feeling bad, thinking that they have both studied together but how come his brother has done so well, but his own performance is dismal. Their father Ram Lal summoned him and told him the secret of Naveen's success. He explained that consistent hard work and not loosing sight of your goal is the mantra of success. Those who can internalize their goals can actually achieve them also. The father continued and told Pramod that nothing much has been lost as yet. You still have many more years to accomplish what you have lost now, but you also must remember you can not waste or lose time again and again and then keep hoping that things will be alright. *'THE SECRET'* is—set your targets and then strive hard to achieve those.

Answer 2.

THE INTERACT CLUB OF SCHOOL

The last week in our school was full of fun. The Interact Club of our school had a literacy campaign from 7th March till 12th March. Since the time it was announced in the morning assembly, everybody was excited and was looking forward to not only fun-filled activities, but enriching experience throughout the week. All the members were assigned duties, since it was going to be a team effort, so everyone was involved. This whole campaign was aimed for children belonging to economically weaker section and those living in slum areas. It was to make them aware of the value and importance of being literate, so that they are able to live their lives comfortably, without being exploited by others.

To advertise about the event, we all made posters and banners with catchy slogans written on them. All the members had a meeting and planned their visit from 9:00 a.m. to 12:00 p.m. Six of our members had prepared lectures on different topics and all of them, were received very enthusiastically by the children as it had many visuals to keep them engrossed. A questionnaire with cartoons and pictures regarding the importance of education was also prepared and response sheets were given to them. The children participated happily and took interest in answering them.

They were also given books, notebooks and stationary and were quite amused and excited by all the knick-knacks provided to them. Then the members of the club divided them into small groups, so that there could be one to one interaction. The children asked many questions as they were curious to know about our school and what we do over there. We also had a small entertainment programme on the fifth day. They presented a small skit which laid emphasis on the importance of Education. Later we distributed sweets, chips and cake to them along with cold drinks. At the time when it was slightly informal the members of the Interact Club also had the chance to know as to what had they learnt and some of their other related experiences.

They all shared enthusiastically and in fact they wanted such activities more often in their area. At the end of the 6th day we were also slightly fatigued but the satisfaction of having achieved something was definitely there.

Answer 3.

(a) (1) (B) : The honest people are trusted by all.

(2) (B) : Kabir is more intelligent than his sister Kusum.

(3) (B) : Notwithstanding the problems in the beginning it had been a successful event.

(4) (B) : She was in a hurry and so was I.

(5) (B) : We were taught western music by Mr. Kapoor in school.

(6) (B) : I regret my inability to accompany you.

(7) (B) : Your health can be damaged due to smoking.

(8) (B) : Hoarding has led to an increase in the prices of food grains.

(9) (B) : Few people listen to the radio these days.

(10) (B) : In spite of being fat, Arun runs fast.

(b) (1) by **(2)** into **(3)** with

(4) for **(5)** from **(6)** in

(7) after **(8)** among **(9)** off

(10) away

(c) (1) being **(2)** has subsided **(3)** taking

(4) stood **(5)** has

(6) had died **(7)** are woven

(8) exercised **(9)** eating

(10) have finished.

Answer 4.

(a) (i) (1) A school is going to be build in 400 acres.

(2) Scientific attitude broadens our *horizon* of thinking.

(3) Both the brothers, Anil and Mukesh Ambani have not become business *rivals*.

(4) The *contest* was tough as there were many participants in the musical reality show.

(5) My mother *scooped* out a big chunk of ice cream and served me on a plate.

(ii) (1) I saw a short man roaming in the streets with a big staff in his hand.

(2) The sergeant ordered the soldiers to fall in a straight line.

(3) As I was walking, I felt the ground was wet.

(4) My elder brother looks very smart in his army uniform.

(iii) (1) Snehamoy and the boy spent time in breaking and grinding the glass, mixed with the glue and just a bit of vermilion dye so that the line can be seen clearly against blue sky and clouds.

(2) When both of them had applied the abrasive mixture on to the line, it was looking nice in bright red, and their courtyard appeared to be a field of red pepper.

(3) The rival groups had already formed their teams and were being coaxed by the local shopkeepers to fight bravely against the foreign Japanese kites and for that they were also offering free kites to the boys.

(4) As many heads make a lively carnival which portrays lot of excitement and joy. Similarly, the sky was looking beautiful with different shapes, sizes and colors of the kites that made the sky look like a beautiful carnival and people looking at them were enjoying and having good times.

(b) (i) The boy found the kites imported from Japan hidden in a box that was placed in a dark passage of the house.

(ii) Snehamoy and the boy sat down with an exercise book and calculated mathematically as if each kite takes five hundred feet of line to fly, then how many feet would twenty kites take ?

(iii) Due to the impending kite-fight, the rival groups cropped up and the shopkeepers approached them and offered them free kites and accessories in order to defeat foreign invasion.

(iv) On the day of the contest, as the first kite heaved into sky by its flyers, it went up over the school building and everybody was watching with abated breath. It was moving like a giant baromon, crushing the helmet of a samurai warrior between its teeth. Its color was a mix of scarlet and ochre and in the sky it hung still, its face up and was indifferent to the still breeze.

(v) Snehamoy took out his kites and flew them one by one, soon the whole sky was full with kites of different lines and sizes. The sky was looking like a stage and all the kites were looking like different characters. The crowd was watching in excitement. Tsugarh whirred like an aeroplane and the crowd went in roar over the Kabuki faces. All the kites seem to be floating in the sky.

(c) Before the start of the event, the two spent the whole afternoon asking neighbours for empty glass bottles. With the help of his aunt, the bottles were smashed in the kitchen in mortar and pestle, in the process, she pricked her thumb also. Glue was added to the glass with a touch of vermilion dye, so that color of line should stand out against the sky and the clouds, then both the boys after tying their hands with cotton cloth dipped it in abrasive and the whole line was dried up in the courtyard. (94 words)

ENGLISH-II

2020

(Maximum Marks : 100)
(Time allowed : Three hours)
(Candidates are allowed additional 15 minutes for **only** reading the paper.
They must NOT start writing during this time.)
Answer **one** question from Section A and **four** questions from Section B.
In Section B choose questions on at least **three** textbooks which may include **EITHER**
Shakespeare's **The Tempest OR** Bernard Shaw's **Candida.**
Note : You are required to select questions on **one play only,**
EITHER The Tempest **OR** Candida in Sections A and B.
The intended marks for questions or parts of questions are given in brackets [ ].

QUESTIONS

Section – A

*(Answer **one** question)*

THE TEMPEST—*Shakespeare*

Question 1.

*Choose any **two** of the passages from **(a)** to **(c)** and answer briefly the questions that follow :*

(a) ***Prospero*** : *This blue-eyed hag was hither brought with child,*
And here was left by th' sailors. Thou, my slave,
As thou report'st thyself, was then her servant,
And for thou wast a spirit too delicate
To act her earthy and abhorr'd commands,
Refusing her grand hests, she did confine thee,
By help of her more potent ministers
And in her most unmitigable rage,
Into a cloven pine ......

(i) *What leads Prospero to talk about the 'blue-eyed hag'?* **[1]**

(ii) *Why was she brought to this island ?* **[2]**

(iii) *What account do we get of Ariel's suffering in the cloven pine ? Who had saved him and how ?* **[2]**

(iv) *How does Prospero describe the hag's son ?* **[2]**

(v) *What does Prospero threaten to do to Ariel now ? What effect does his threat have on Ariel ?* **[2]**

(vi) *Give the meanings of the following words in the context of the passage :* **[1]**

abhorr'd, *cloven*

(b) ***Caliban*** : *I prithee let me bring thee where crabs grow,*
And I with my long nails will dig thee pig-nuts,
Show thee a jay's nest, and instruct thee how
To snare the nimble marmoset. I'll bring thee
To clust'ring filberts, and sometimes I'll get thee
Young scamels from the rock. Wilt thou go with me ?

(i) *To whom is Caliban speaking these lines ?* **[1]**

(ii) *Briefly describe the circumstance under which he had met them.* **[2]**

(iii) *Who does he think they are ? Why ?* **[2]**

(iv) *To which other person had Caliban shown such a slavish attitude ? What had the result been ?* **[2]**

(v) *How do his present listeners respond to his words ?* **[2]**

(vi) *Give the meanings of the following words as they are used in the context of the passage :* **[1]**

snare, *marmoset*

(c) ***Ariel*** : *Thy thoughts I cleave to. What's thy pleasure ?*

Prospero : *Spirit, we must prepare to meet with Caliban.*

Ariel : *Ay, my commander. When I presented Ceres*
I thought to have told thee of it, but I fear'd
Lest I might anger thee.

Prospero : Say again, where didst thou leave these varlets ?

(i) *For whom is the term varlets being used ?* **[1]**

(ii) *What comical behaviour of theirs does Ariel narrate now ?* **[2]**

(iii) *Where had these people been going ? Why ?* **[2]**

(iv) *Where did Ariel leave them eventually ? What did they do then ?* **[2]**

(v) *What does Prospero tell Ariel to do ? Why ?* **[2]**

(vi) *Give the meanings of the following words as they are used in the context of the passage :* **[1]**

cleave, varlets

Section – B

*(Answer **four** questions on at least **three** textbooks which may include **EITHER** The Tempest **OR** Candida)*

THE TEMPEST—*Shakespeare*

Question 3.

Referring closely to Act 1V Scene i :

(a) *Describe the blessings showered by Ceres and Iris on Miranda and Ferdinand. What is the significance of these blessings ?* **[8]**

(b) *What did Prospero tell Ferdinand just before the masque ? What did he emphasize and how did Ferdinand answer him ?* **[6]**

(c) *What do you conclude about Prospero from the events that occur in the act ?* **[6]**

Question 4.

Prospero refers to Caliban as "A devil, a born devil." In this context, evaluate the character and role of Caliban in the play. **[20]**

ECHOES

Question 13.

*Referring closely to the short story **B. Wordsworth:***

(a) *Describe the interaction between B. Wordsworth and the young narrator.* **[8]**

(b) *What did the narrator learn from his friend ?* **[6]**

(c) *Do you feel that B. Wordsworth was an escapist ? Give reasons for your answer.* **[6]**

Question 14.

*How does Roald Dahl depict the theme of obsession in his short story, **The Sound Machine** ?* **[20]**

Question 15.

*How successful was the writer in bringing out the protagonist's innate goodness in the short story, **Salvatore?*** **[20]**

REVERIE

Question 16.

*Referring closely to the poem, **Dover Beach** :*

(a) *Describe the moonlit scene described by the poet at the beginning of the poem.* **[8]**

(b) *How does the poet introduce the note of melancholy and disturbance ? In this context, what does he say about Faith ?* **[6]**

(c) *What does the poet say about the world in which he finds himself ? What is the only hope left for mankind?* **[6]**

Question 17.

*How is the power of the Music Makers brought out in **We are the Music Makers** ?* **[20]**

Question 18.

*The poem **John Brown** presents two attitudes to war. Describe the two attitudes. Which one do you sympathise with ? Why ?* **[20]**

ANSWERS

Section – A

Answer 1.

(a) (i) After Ariel narrates how he accomplished the task of the tempest, Prospero tells him to do more work. When Ariel reminds him he was promised freedom, Prospero reminds Ariel, how he should be grateful to him for his present freedom from the blue-eyed Sycorax.

(ii) Sycorax was brought to the island because she was practicing sorcery. She was pregnant with the child of an evil spirit and was banished from Algiers for her wicked acts.

(iii) Ariel, the spirit of air, valued freedom but he was imprisoned in a cloven pine for twelve years by Sycorax and then she died without freeing him. Ariel cried in agony which resounded in the island. Even the wolves howl on his groans, and it also penetrated the breasts of the ever angry bears. It was Prospero with his superior magic who freed him from his agony.

(iv) Prospero describes Caliban, the son of Sycorax as, "A freckle whelp hag born" which means he was hag-born,that is, he was not born in a human shape. He also says that the whelp is a dull thing.

(v) Prospero warns Ariel that if he complained again, he will tear open an oak tree and confine him as a prisoner within its knotty roots, for another twelve winters Ariel apologises and promises to follow his master's instructions like an obedient spirit.

(vi) The meaning of 'abhorr'd' is most hated. 'Cloven' means split into two; this is in reference to the pine tree that pinned Ariel so tightly that he could not escape from it for twelve years.

(b) **(i)** Caliban is speaking these lines to Trinculo and Stephano.

(ii) Caliban was gathering wood when Trinculo appeared. Caliban thought that he was a spirit sent by Prospero and hid under his gabardine. Trinculo, fearing storm, also hid inside it. At this time, Stephano came with a bottle in his hand and was surprised to see a creature with two heads and four legs.

(iii) Stephano thinks they are some strange creatures, may be a monster in the island with four legs and two voices. He thinks so because he is unaware of the fact that Trinculo has crept into the gabardine in which Caliban had taken refuge. Caliban thinks they are divine creatures sent by Prospero to torment him as he has never seen many human beings in his life. He thinks Stephano is a god with a celestial drink in his hand.

(iv) Earlier Caliban had shown a slavish attitude to Prospero. He showed him all the places on the island that provided clear water and delicious fruits. He brought him berries, fetched water and did all the menial jobs. However, in the end, he starts resenting being dominated and hates Prospero for his high handed ways. Prospero snatched the island from him and confined him to a rock.

(v) The listeners did not pay much heed to Caliban and Stephano tells Caliban to hold his bottle and lead the way.

(vi) 'Snare' is trap and 'marmoset' is a monkey.

(c) **(i)** By varlets, Prospero is referring to Caliban, Stephano and Trinculo who have no scruples.

(ii) Ariel talks about the drunken behaviour of the rogues. With faces red from drinking, they smote at the air for blowing against their faces and stomped the ground for touching their feet.

(iii) They had been going to Prospero's cell to kill him.

(iv) Ariel with his drumbeats cast magic on them and led them through thorny bushes to a filthy pond covered in scum located beyond Prospero's cell. They started dancing in it and were immersed in the dirty water up to their chins.

(v) Prospero told Ariel to retain his invisible shape and bring him the "trumpery" or the fine garments from the house. He wanted to use those gaudy clothes as a bait to catch those rogues.

(vi) 'Cleave' in the context means ready to obey. 'Varlets' means rogues or unprincipled people.

Section – B

Answer 3.

(a) In order to celebrate the betrothal of Miranda and Ferdinand, and to demonstrate his prowess as a magician. Prospero entrusts Ariel to organise a masque. Goddesses are invited to celebrate "a contract of true love" and grant blessings to the couple.

Juno, the goddess of all goddesses confers honour, riches, all happiness in married life, and the continuance of all these throughout their life. She wishes that their happiness will double and multiply day by day.

Iris, the goddess of the rainbow summons Ceres, the goddess of agriculture to bestow her blessings on the young couple. She promises Earth's increase and foison plenty and blesses that their barns will never be empty, thick bunches of grapes will always grow on their vines, and trees will be overloaded with fruits throughout the year. She wishes "Spring come to you at the farthest/In the very end of the harvest", so that there cannot be winter in their lives. In other words, scarcity or lack of anything will never trouble them, and that is assured by the goddess of plenty.

Iris is the rainbow which unites the earth and the sky. She blesses their wedded lie and summons the nymphs of the stream and the reapers to the dance on the green. The blessings emphasize fertility- a desired quality in any marriage. The significance of this most "majestic vision/ and harmonious charmingly" as Ferdinand describes, is that this is a father's gift to her daughter and her betrothed, a most befitting way to shower his blessing through the words of the divine beings. These are the divine gifts every parent wants their children to have. Prospero only presents the masque after a frank discussion of sexuality as he feels Ferdinand may feel tempted by his attraction towards Miranda. It is pertinent to note, together, these two deities symbolise family and prosperity, giving the couple a blessing that moves away from any suggestion of physicality. The role of marriage in society is emphasised.

(b) Prospero told Ferdinand that if he thought his punishments had been severe, then he was also compensated amply by winning his daughter's hand. He further explained that all the arduous tasks given to Ferdinand were to test his love for his beloved daughter. He was happy because the young man had passed all the tests admirably. He said, "Here afore Heaven/I ratify this my rich gift". He also told the young man not to laugh at his pride as very soon he would realise that Miranda is worthy of all the praises showered upon her. He also warned Ferdinand that if he violated her virginity before the marriage ceremonies, no sweet blessings will descend upon them. Moreover, "barren hate and sour-eyed discord" would disturb the peace of their marriage. Angry words would be exchanged leading to mutual disagreement and frictions. He told Ferdinand to pay heed to these warnings and follow the guidance which Hymen, the God of Marriage offers to those in love.

Ferdinand answered that he hoped for peaceful days, beautiful children and a long life full of love. His love for Miranda was pure and no temptation could convert it into lust. He said that he would reject even the strongest evil urge offered by a suitable place. He would await his wedding night with such eagerness and anxiety that he wondered whether the evening would ever come.

(c) Prospero comes across as a caring father who takes immense pride in her upbringing. He loves his daughter deeply and calls her a rich gift which he is giving to Ferdinand and he also does not want her to be hurt. He is extra cautious that Ferdinand does not take undue advantage of her and strictly warns him of the dire consequences if that happens.

Prospero is a master magician, which is proved by the fabulous masque he conjured up with the help of Ariel. It requires high magical skills to incorporate supernatural elements to put up such an extravagant show. He is also proud of his skills and wants to show off.

The philosophical side of the master magician comes to the fore when he explains the sudden disappearance of the masque. He tells Ferdinand, real life is also similar as all things, "the cloud-capped towers, the gorgeous palaces the solemn temples, the globe itself" are transient and "we are such stuff as dreams are made on".

Prospero can get extremely angry which can be seen in his treatment of the three rogues towards the end of the scene.

Answer 4.

Caliban is one of the most interesting characters in the play. He is an anomaly of nature, the offspring of the witch Sycorax and the devil. He is half-monster and half-human. Prospero has made Caliban his servant or, more accurately, his slave. Caliban showed Prospero all the places on the island. He used to get fruits and pure water for Prospero and Miranda. However, in most of the play, Caliban is insolent and rebellious for all the menial work he does for the father and daughter. Whenever he rebels, he is punished by Prospero with pinches and stitches and is only controlled through the use of magic. Caliban claims the island as his own and maintains that Prospero has tricked him in the past.

Caliban represents the evil of his mother and initially appears bad, especially when judged by conventional civilized standards. Modern day analysts interpret the relationship between Prospero and Caliban as that of the colonist and the colonised. He curses Prospero all the time, not showing any gratitude for training him and teaching him the language. However, he has learned only to curse. He has even tried to molest Miranda, and that makes the daughter and father hate him. This is the reason why Prospero refers to him as "born devil". According to Prospero, "as with age, his body uglier grows; so his mind cankers."

However, one can see another side of him, when he becomes poetic describing the natural beauty of the island. At this time, he does not seem to be mean or without sensibilities. His resentment is that of a colonised subject, over the domination of their rights and Prospero, like any other coloniser, feels he is being ungrateful. However, Caliban has a better set of values than Stephano and Trinculo. They are distracted from their plan by their greed for Prospero's rich garments. Only Caliban realizes that such finery is unimportant: "Leave it alone, thou fool, it is but trash." Caliban is not a good judge of character. He decides, for example, that Stephano is a god because he dispenses 'celestial liquor' but then it must be remembered that he has only known his mother, Prospero, Miranda and the spirits that torture him. In many ways, Caliban is an innocent, reacting to emotional and physical needs without the ability to think through and fully understand the events and people who surround him. He is truly a child of nature, uneducated and reacting to his surroundings in much the same way that an animal does. Perhaps Prospero's words reflect the European attitudes to the natives of the regions they conquered.

Answer 13.

(a) Naipaul's B. Wordsworth is a story that deals with the friendship between a little boy from ordinary circumstances and an educated man with a poetic mind and sensibilities.

A tidily dressed man knocks on a house of a little boy who lives with his mother. The man asks whether he can see the bees in their compound. His mother reluctantly agrees to let him in the yard. The boy and the man watch the bees together. The man says that he likes watching the bees and asks the child if he likes it as well. He replies that he never has the time to. The man shakes his head. The boy, intrigued by the stranger, asks various questions.

The man introduces himself as Black Wordsworth. He then says that he can watch a flower like a morning glory and cry. The kid asks what he cries for and B. Wordsworth replies "When you're a poet you can cry for everything". After this. Wordsworth pulls out a printed paper from his pocket. He tells the kid that on that paper, the greatest poem about mothers is written and he'd sell it to him for 4 cents if he likes. The kid rushes up to his mother to ask if she'd buy it but his mother declines. The kid goes back to the man and tells him that his mother doesn't have four cents. Wordsworth says that this way he expects to meet new poets. But not a single copy has been sold till then. Later the young narrator B. Wordsworth become friends quite unknown to the narrator's mother. The boy meets B. Wordsworth at the corner of Miguel Street as he was coming back from school one day. B. Wordsworth declared that he had been waiting for this young friend for a long time. He invited the young boy to come to his yard and eat mangoes. The boy noticed that the yard seemed all green with mango, coconut and plum trees. The place looked wild as though it wasn't in the city at all. One couldn't see the big concrete houses in the street. The boy enjoyed the sweet, juicy mangoes. Of course, his mother was very angry with him and beat him rather badly for his escapade. In anger, the boy goes to B. Wordsworth's house. Instead of commenting on what had happened, B. Wordsworth took him for a walk to the race-course. There, both of them lay on the grass and looked up at the sky. B. Wordsworth also told the young boy an oblique story of a relationship between a boy poet and girl poet who died with a young poet inside her. They went to the Botanical Garden and the Rock Gardens for walks. They climbed Chancellor hill in the late afternoon and watched the darkness fall on port of Spain and the lights on in the city and on the ships in the harbour. B. Wordsworth did everything as though he were doing it for the first time in his life. He did enjoy as though he were doing some Church rite.

(b) The child in Miguel Street learned to understand many things from B. Wordsworth who created in him a sensitivity and understanding about nature. Watching bees, crying at the beauty of the flowers were all new to the child. He also revealed the mystery of stars. Lying on the sands, looking at the sky, the poet told him to think how far they are from them. The kid does so. "Never in his life had he felt so great yet nothing". Thus, the poet taught the young boy about the joys of being alive and present at the moment; he helped the young boy to realise his poetic sensibilities.

B. Wordsworth, one day, told him a story about a poet couple who fell in love and got married. Another day, he told the boy that he was writing the greatest poem of all time, which will take twenty-two years to complete for he only wrote one line every month. After that, he revealed that everything he had ever told him about even his poetry and the greatest poem in the world, had all been made up. In spite of this disappointment, the boy realised the value of friendship.

(c) The theme of alienation and escapism is well brought out by the writer. It can be said that B. Wordsworth is living a make-belief existence– a made-up story in which he is the main character. Throughout the story, a lot of insights are given by B. Wordsworth on what it is like to be a poet. Watching the bees and stars for hours together, finding new meanings in whatever he sees, reacting to them in a deeply sensitive way are all signs of an alien life far away from realities of his life. The young boy provides the poet a medium to live the dream of whoever he wishes to be. The stories of his love and marriage, of being a writer are all figments of imagination. It is thoroughly, a classic case of escapism.

Answer 14.

'Sound Machine' written by Roald Dahl deals with the theme of obsession. Klausner, the protagonist is obsessed with sound. He spends his entire time with his sound machine. It first starts out as an innocent experiment that excites him but becomes something that he cannot keep away from. At no stage in the story do we get mention of any other activity that Klausner is involved in. This is clearly unhealthy and abnormal to be so focused only on one thing and not exploring anything else. It affects the balance of life.

Klausner believes that there are numerous sounds in the world that humans are unable to hear due to their high frequencies. He has invented a machine that will allow him to listen to those frequencies and convert them into audible sound. The first time he hears shrieking in his headphones was when his neighbour cuts roses from her garden. Each time a flower is cut, he hears a shriek. The next day, he takes an axe and aims at a large beech tree. He is horrified to hear the deep and pathetic moan that the tree makes in response. Klausner rushes back to the house and calls his doctor. "Please come. Come quickly. I want someone to hear it. It's driving me mad!" he says. The doctor comes over and listens, but just as Klausner takes a second swing at the tree a large branch crashes down and destroys the machine. Klausner is deeply shaken and asks the doctor to paint the tree's cuts with iodine.

Klausner is attempting to treat the tree as a human would be treated. He is empathising with the tree based on its cry. He really cares for the tree but this heightened sensitivity is not realistic. It is as though Klausner's involvement with the sound machine has played havoc with his imagination. He may not be thinking straight because of the excitement he feels over his discovery.

Dr. Scott cannot confirm that he heard the sound that Klausner heard. This leaves the reader suspecting that Klausner has worked too hard on the sound machine and as a result has a breakdown. Thus, Roald Dahl depicts the theme of obsession in the most apt way.

Answer 15.

Somerset Maugham expresses his doubt whether he will succeed in whatever he plans to do at the beginning of the story 'Salvatore'. At the end of the story, we come to know that he has been really successful in his intention to capture the essence of the character of a simple, humble human being.

"Goodness, just goodness" is what Salvatore is as described by the writer. He is a gem among humans, not because of exemplary courage, nor because of any noble sacrifice; but just because of being a human. A true human who does not have any complaint with anything or anyone. The author gives certain glimpses of his life and makes us realise how sheer goodness can make a person content and happy.

Salvatore's life was beset with challenges and upheavals. However, his manner did not reveal any of his troubles. He was cheerful and loving to others; not even a trace of bitterness appeared in his words or actions. He learned to live with his incapacities and limitations and he bore them with "the mute and uncomprehending patience of a dog".

The first challenge in his life came when he had to join military force for the sake of marrying the girl he loved. It was agony for him to leave his beloved island, but he did it as it had to be done. He noiselessly adjusted with the hard life.

The true test came when he contacted a debilitating disease which forced him to leave service. He saw the brighter side of it by thinking that it gave him the opportunity to go back to the girl he missed so much. But the trusting, simple-hearted boy was in for a shock. The girl whom he rushed to see rejected him saying that he was "not man enough to support her". He cried his heart out, but never blamed the girl instead empathized with her.

As days passed, Salvatore came to terms with his loss. He worked in the vineyard, got married to a woman of his mother's choice, and proved an exemplary father to his children. It was a hard life, fishing in the sea, working in the vineyard, fighting with his rheumatism but he forgot his troubles when he gleefully played with his children. A strong man, he treated his children with utmost tenderness as though they were flowers. The comparison of large hands with flowers reveals the gentleness of his nature.

It was his innate goodness, his simple heartedness, his tendency to see good in others that enabled Salvatore to live a contented life. An ordinary fisherman who possessed nothing else in the world except a quality which was "the rarest, the most precious and the loveliest that anyone can have"—the quality of goodness. Yes, Maugham with infinite ease has portrayed the picture of goodness.

Answer 16.

(a) Mathew Arnold portrays the beauty of the sea which is calm at night. He captures an idyllic scene. "The tide is full, the moon lies fair". He can see the French coast gleaming faraway. Then he focuses on the cliffs of England, "Glimmering and vast, out in the tranquil bay"; "Sweet is the night-air!" exclaims the poet and talks about the long line of spray seen where the sea meets "the moon-blanched land,"

However, the tranquility is broken by the grating roar of pebbles which the waves fling, "Begin, and cease, and then again begin, with tremulous cadence slow, and bring the eternal note of sadness in". The cliffs that the speaker mentions are largely chalk, meaning, that they are white, that is what makes them glimmer in the moonlight. They also have a sheer drop, with the coastline ending abruptly and giving way to the sea.

The sea that the speaker looks out upon is the English channel, which divides England from France. The scene is outwardly beautiful—the cliffs are very impressive.

"Which seems
To lie before us like a land of dreams,
So various, so beautiful, so new."

But very soon, the poet notices the pebbles and shingles instead of smooth sand. The sea cries a melancholic song as though it is the tragic song of human reality.

(b) The unpleasant roar of the waves brings 'a sense of melancholy' to the speaker's mind. The poet is reminded of the ancient Greek playwright Sophocles who also heard the sounds of the Aegean Sea and then wrote tragedies on human misery. Going further, the speaker laments the lack of faith in the modern society. Here, he refers to the frustration and hopelessness of the modern chaotic world. He also expresses his view that this kind of situation where there is "neither joy, nor love, nor light, nor certitude, nor peace, nor help for pain" has been created by the decline of 'faith'.

The poet says that human faith, religious faith, and faith in fellow people once covered the earth like sea water. It was at its fullest just as the tide then. And this strong faith made everything easy and solved many problems. It united mankind and brought meaning to life. "But now I only hear melancholy's long withdrawing roar", the speaker feels sorry to think those days are now of the past. He compares the human situation to the "naked shingles of the world" prone to disaster and unhappiness.

(c) "Ah, love, let us be true
To one another!"

It is human love and compassion that must get them through this new, ever changing world, awarding to the poet. They must comfort each other, be faithful to one another. Only the loyalty and comfort of personal relationships can fill the void produced by the disappearing faith in God. One cannot prevent the march of science; the pebbles and shingles of wave swept coast is symbolic of a land swept by new scientific changes. The world can offer none of the promises it makes. Joy, love light, certitude, peace, help for pain. What the world is really like is a battlefield at night where soldiers clash, fighting with shadows, unable to tell friend from foe. It is a dark plain "where ignorant armies clash by night". It is filling human minds with confusion. It is creating chaos in thoughts and actions. Religious beliefs, one by one are questioned and at times ridiculed. The poet feels this will only bring sadness to all of humanity. A deep faith, in oneself and in fellow beings, is the only hope left for mankind. Amidst the hopelessness, the poet can only find a glimmer of hope in personal bonds and relationships.

Answer 17.

'We are the Music Makers' is a thought provoking poem. It is an ode to celebrate the power of all artists. The underlying theme of the poem is the appreciation of art. The poet encourages the readers to dream like artists who shape the world through their imaginative faculties. The transformative power of art, the ability of the artists to cause changes are highlighted in the poem.

The poet describes the artists and their soul-searching as they wander alone by sea-breakers, and sit by the desolate streams, to shape their creativity worthy of admiration. He highlights the difficult life of the artists seen as the world losers and world forsakers. With the sacrifices comes the meagre sustenance they manage to carve out for themselves through their art. The power of music makers is brought out by the poet as he says that the music makers are dreamers of dreams and they have a role to fulfill in the world. They are the 'movers and shakers'. They are needed for rebellion against stereotypes, for changing the status quo; "With wonderful deathless ditties/We build up the world's great cities". "Built Nineveh with our sighing, And Babel itself with our mirth." In other words, artistic fantasy, and stories have the power to push society forward. Artists have been the cause of society's march towards progress. It implies that the myth-makers of the modern world are to be considered as the leaders of that world. "And out of a fabulous story/We fashion an empire's glory", the lasting effect of poetry outlives most societies, and it is this that must be celebrated and enjoyed. This is the true wonder of poetry: Art is man's greatest achievement as much as it is a powerful tool of transformation.

Such is the power of an artist, "One man with a dream, can conquer a crown; And three with a new song's measure can trample an empire down." This means, creation and destruction are in their hands and judicious use of both can change the destiny of any land. God gives the artists a divine power. They are master creators that can overthrow the outdated and create a new order. "For each age is a dream that is dying, or one that is coming to birth". Thus, art is eternal and required for the dynamic movement of civilisations.

Answer 18.

John Brown is a touching song composed by Nobel laureate Bob Dylan. It presents two attitudes to war—the illusion nurtured by the mother and the reality as portrayed by the son.

An American mother sent her son John Brown to the war on some foreign land. The mother was so proud of her son who "stood so straight and tall in his uniform and all". She told him "Oh son, you look so fine, I'm glad you are a son of mine". Then she instructed him to obey whatever his captain said so that he would win a number of medals with which they could decorate the wall when he came home. She was reflecting the belief of the times that serving in armed forces was considered as the most dignified and noble professions of the world. Bravery, victory, and laurels won, were the ideals of the times and mothers were proud when their sons joined the forces. Totally ignorant of the grim realities of war, this mother also nurtured a romantic illusion and dreamt of a heroic welcome for her son, coming unhurt and victorious from far.

However, when the son came back, it was a shock for her. He was totally unrecognizable, his face shot up, hand blown off and he wore a metal brace around his waist. When he spoke, his voice was barely audible. Such was the change that his own mother could not recognize him. "O Lord, not even recognize his face!" It is then that her son tells her about the reality of war that he had faced. He asked her whether this was what she had wanted. It was on the battlefield that the young boy learnt how terrible and useless a battle was.

John Brown recalled his mother teaching him that it was the noblest thing to fight for the nation. But he soon realised it was not noble at all. He had to fight people of his own kind. "I'm tryin' to kill somebody or die trying". "He said with horror that, "when my enemy came close/And I saw that his face looked like mine". For him, all the illusions were shattered. He was just a puppet in a horrible play, instructed to kill others without remorse, and when finally this realisation struck, a cannonball came through the smoke and blew his eyes away. The bitter reality of the son's words devastated the mother, her romantic illusion about war was completely shattered in a complete rejection of what she thought war stood for. It also shatters the notions that society has of war and becomes a kind of reality check. The son brought his medals and gave them into her hands, before walking away. This becomes a bitter indictment of war. The medals were symbolic of the empty glory that he had won in the war and the high price he had to pay to do so.

I, most vehemently, support the view that war is horrible. It only serves the vanity and selfish interests of the powerful. It is just a ploy to instigate young people to join the fight in the name of a cause. There is nothing noble about a war. It does not solve any problem. It is a waste of innocent lives. This is proved by the poet's words on the medal which seem to ask derisively, "Was it worth all the suffering he went through ?" It is a hard-hitting question on the reality of war.

●●

ENGLISH-II

2019

QUESTIONS

Section – A

*(Answer **one** question)*

THE TEMPEST—*Shakespeare*

Question 1.

*Choose any **two** of the passages from **(a)** to **(c)** and answer briefly the questions that follow :*

(a) ***Prospero*** : *Ariel, thy charge*
Exactly is perform'd; but there's more work.
What is the time o'th day ?
Ariel : *Past the mid-season*
Prospero : *At least two glasses. The time 'twixt six and now*
Must by us both be spent most preciously.
Ariel : *Is there more toil ? Since thou dost give me pains,*
Let me remember thee what thou hast promised
Which is not yet perform'd me.

(i) *Where are Ariel and Prospero ? Where had Prospero sent Ariel ?* **[1]**

(ii) *What account of his task does Ariel give Prospero ?* **[2]**

(iii) *What had Ariel done with Alonso and his companions?* **[2]**

(iv) *Which promise does Ariel remind Prospero about ? What does this tell you about Ariel ?* **[2]**

(v) *Why does Ariel think that he deserves what has been promised to him ? What effect do his words have on Prospero ?* **[2]**

(vi) *Give the meanings of the following words in the context of the passage :* **[1]**

twixt, *remember*

(b) ***Antonio*** : *Then let us both be sudden.*
Gonzalo : *(Waking) Now, good angels*
Preserve the king !
Alonso : *Why, how now ? Ho! Awake ?*
Why are you drawn ?
Wherefore this ghastly looking ?
Gonzalo : *What's the matter ?*

(i) *Where are the speakers ? Briefly explain how they were at this place.* **[1]**

(ii) *What does Antonio mean by "Then let us both be sudden ?" What does he convince Sebastian to do ?* **[2]**

(iii) *What makes Gonzalo and Alonso wake up suddenly?* **[2]**

(iv) *What reason do Sebastian and Antonio give for their strange behaviour ?* **[2]**

(v) *Which sounds had Gonzalo heard ? What did the king's party decide to do at the end of the scene ?* **[2]**

(vi) *Give the meanings of the following words as they are used in the context of the passage :* **[1]**

sudden, *drawn*

(c) ***Stephano*** : *Put off that gown, Trinculo [Reaches for it]. By this hand, I'll have that gown.*
Trinculo : *Thy grace shall have it.*
Caliban : *The dropsy drown this fool! What do you mean*
To dote thus on such luggage ? Let't alone,
And do the murder first. If he awake,
From toe to crown he'll fill our skins with pinches
Make us strange stuff.

(i) *Where are Trinculo and Stephano ? What has distracted Trinculo ?* **[1]**

(ii) *Which murder is Caliban referring to ? Why does he want this ?* **[2]**

(iii) *What temptations does he offer to make them carry out the plan ?* **[2]**

(iv) *What would the two have to do first to ensure the success of the plan ?* **[2]**

(v) *What does Caliban fear would happen if they fail ?* **[2]**

(vi) *Give the meanings of the following words as they are used in the context of the passage :* **[1]**

dropsy, *crown*

Section – B

*(Answer **four** questions on at least **three** textbooks which may include **EITHER** The Tempest **OR** Candida)*

THE TEMPEST—*Shakespeare*

Question 3.

Referring closely to the conversation between Prospero and Miranda in Act 1 Scene II, answer the following :

(a) *Narrate what Prospero tells Miranda about his past and the circumstances which brought him to the island.* **[8]**

(b) *How does Miranda react to this tale ? What is Prospero's purpose in telling her this tale ?* **[6]**

(c) *What do you conclude about Prospero from this conversation ?* **[6]**

Question 4.

Miranda is a perfect blend of innocence and determinations. Discuss her role in the play with close reference to the text. **[20]**

ECHOES

Question 13.

Referring closely to the short story **To Build a Fire**, *answer the following :*

(a) *Describe the man's repeated attempts to keep himself warm.* **[8]**

(b) *Contrast the attitudes of the Man and his Dog towards the weather.* **[6]**

(c) *What does the man finally learn from his experience?* **[6]**

Question 14.

With close reference to the short story, **The Singing Lesson;** *describe how the lesson as it progresses, becomes an echo of Miss Meadows' feelings.* **[20]**

Question 15.

Referring closely to the short story, **The Story of the Hour**, *give an account of the thoughts and conflicts which go through Mrs. Mallard's mind when she hears about the untimely and sudden death of her husband. Comment on the ending of the story.* **[20]**

REVERIE

Question 16.

Referring closely to the poem, **The Darkling Thrush**, *answer the following :*

(a) *Describe the bleak picture of the day as described by the poet.* **[8]**

(b) *How does the atmosphere suddenly change ?* **[6]**

(c) *What are the poet's feelings at the end of the poem ?* **[6]**

Question 17.

Describe the interaction between the Spider and the Fly in the poem **The Spider and the Fly.** **[20]**

Question 18.

The poem **Crossing the Bar** *is a poem of affirmation and faith. Give your views.* **[20]**

ANSWERS

Section – A

Answer 1.

(a) (i) Ariel and Prospero are on the enchanted island where Prospero lived with his daughter. Prospero had sent Ariel to create mayhem on the ship carrying Sebastian, Antonio, Alonso and Ferdinand.

(ii) While Prospero had created the storm, Ariel had created fear and amazement on the ship. He appeared on the ship as a 'flickering flame'. Sometimes he would burn in many places and then meet and burn as a single flame. Simultaneously frightening the passengers and crew with thunder and lightening. When the mariners jumped overboard because of the fear, Ariel made sure that they were all separated but all safely ashore.

(iii) Ariel had made sure that all Alonso and his companions come to the island unhurt. He had dispersed all of them in different parts of the island and Ferdinand had been left alone in one spot.

(iv) Prospero had promised Ariel his freedom if he served him dutifully a full year. This tells us that Ariel was a free-spirit. He disliked serfdom and wanted his freedom at all costs.

(v) Ariel feels that he deserves freedom and says that he had done worthy service, told Prospero no lies, made no mistakes and served without grudge or grumbling. Prospero gets angry and tells Ariel that he kept complaining, then Prospero would bind him to a tree again using his magic.

(vi) i. between **ii.** remind

(b) (i) The speakers are on Prospero's island. After the shipwreck, Ariel had dispersed the king and his companions in different parts of the island.

(ii) Antonio means that they should be quick with the act they have decided. Antonio has convinced Sebastian to kill both Alonso and Gonzalo. Both Antonio and Sebastian believe Ferdinand has died in the ship wreck, thus by killing Alonso and Gonzalo, Sebastian shall be entitled to the throne of Naples.

(iii) Ariel comes and sings a warning song in Gonzalo's ears causing him to wake with a start, Gonzalo then also wakes up Alonso.

(iv) Both Sebastian and Antonio had hatched a plan to kill Alonso, but were rendered unsuccessful by Ariel. Alonso and Gonzalo woke up to see them with their swords drawn. To cover up they claimed to have heard the roar of lions or bellowing of bulls. They said that they unsheathed their swords to save the king.

(v) Gonzalo claimed to have heard a strange humming sound. They all decided to be vigilant and leave the area immediately.

(vi) i. Quick

ii. Standing with the swords unsheathed

(c) (i) Caliban and his fellow conspirators were near Prospero's cell as they had planned to murder him. Ariel had placed fancy apparel in their way to distract them. Trinculo and Stephano were fascinated by the apparel and were in danger of forgetting their purpose.

(ii) Caliban is referring to Prospero's murder. He wants to kill Prospero because he thinks that Prospero has deprived him the ownership of the island which belongs to him. He hates Prospero and wishes to be free from his tyranny.

(iii) Caliban offers Stephano ownership of the island, and have Caliban and Trinculo as his lieutenants, and all the household items of Prospero, and marry Miranda in return of killing Prospero.

(iv) They would have to kill Prospero while he was sleeping and seize the magic books of Prospero without which he would be rendered powerless.

(v) Caliban fears that if they failed then Prospero would torture them relentlessly and very cruelly.

(vi) i. a disease in which body retains fluids

ii. head

Answer 3.

(a) Prospero informs her that he was the Duke of Milan and was renowned for his incredible intelligence.

He narrates to her, how he slowly grew disinterested in politics, turned to his 'secret studies' and thus began neglecting his obligations as a duke. This made his brother Antonio conspire against him. He began hatching plans to overthrow him. The King of Naples Alonso and Antonio, working together, overthrew Prospero. The pact between Antonio and the King of Naples was that he would pay Antonio, a yearly tribute and some homage. On the appointed night, Antonio opened the gates of Milan and the officials appointed to perform the task conducted Prospero and Miranda out of the city. They were taken out to sea and put aboard a "rotten carcass" of a "butt" which had no tackle sail or mast and left them at the mercy of the sea and wind. But Gonzalo, an honest and truthful Neapolitan, supplied them with clothing and food, and he also gave him books from Prospero's collection. By divine providence Prospero and his daughter were brought to the enchanted island.

(b) Miranda listens to Prospero's tale with rapt attention. At first, she is full of wonder and disbelief that her uncle Antonio could be so full of treachery. She also feels sorry for her father and the travails that he had to bear. She feels that she must have been a great burden to her father in his hour of trial. As Prospero's tale is a long one, he has to check from time to time whether she is actually listening to him. She sympathises with her father and also asks many pertinent questions ("Wherefore did they not/That hour destroy us") which reveals that she has been listening to the tale very closely. Finally, she feels grateful to Gonzalo for aiding them in their hours of trial. The reason why Prospero reveals the truth to Miranda is because he is preparing her for the subsequent events and his plans involving her and Ferdinand.

(c) Through this conversation we can conclude. Prospero, we can deduce that he is an intelligent man who is also a bit selfish. He could have worked for his people but chose to pursue his own interests. He also trusted people around him blindly due to which he was betrayed. We notice his pride in Milan which he declares was the foremost state in all of Italy. We also note his love and concern for Miranda. Prospero is also a man who likes to keeps things in control as not only does he raise a storm to ensure that his enemies are brought to shore but he also uses his magic to ensure that Miranda goes to sleep while he attends to his other tasks.

Answer 4.

Miranda appears as a symbol of beauty, innocence and modesty in the play. Upon seeing her, Ferdinand exclaims, "What is this maid / Is she the goddess who hath severed us. And brought us thus together."

She is deeply concerned for those involved in the shipwreck and also asks her father to justify his actions. "If by our Art my dearest father, you have put the wild waters in this roar, ally them. The sky, it seems, would pour down stinking pitch. But that the sea, mounting to the welkin's cheek, dashes the fire out. O, I have suffered with those that I saw suffer."

She shows sympathy for everyone in pain except Caliban, whom she dislikes because he attempted to violate her modesty. She says, "Tis a villain, sir I do not love to look on".

She berates him by saying that she had tried to school him but in vain. Even during the telling of Prospero's tale she is more worried about the agony and suffering endured by her father. Prospero admits that it was Miranda his cherubin who helped him survive when in the seas in the rotten carcass of a boat.

When they reached the island, Prospero ensured that Miranda is well educated and takes good care of her.

Miranda falls in love at first sight and she is simple and forthright enough to express her love openly. It is here that we see her strong determination. In spite of her father's command, when she falls in love with Ferdinand and visits him as often as possible. "I am your wife if you marry me; if not, I'll die your maid."

Her innocence is again seen when she sees the king and his companions as she had never seen so many humans in her life. She innocently refers to them as "goodly creatures" and also says, "How beauteous mankind is!" Not knowing that those very "goodly creatures" were responsible for making her and her father landed upon the island.

Miranda shows great strength when she pleads with her father asking him to be merciful to Ferdinand. She disregards her father's orders and comes out of her house to see Ferdinand and bravely stands for her love. Her tender heart is touched when she sees him bearing logs and she even offers to do the job for him. "If you'll sit down,I'll bear your logs the while: pray give me that; I'll carry it to the pile." Once in love she shows great enterprise in making the transition to womanhood. She is even willing to go against her father to marry Ferdinand.

Answer 13.

(a) The man, a newcomer to the region, sets out alone to a lumber camp. He had been warned by an old timer that this was not the weather to be out alone. However, he was a man with little imagination, and therefore could not imagine the extent of the cold. Once he begins walking however, the truth of the old timer's statement strikes home. He feels he has equipped himself well for the weather. He is adequately covered and carries his food well wrapped next to his skin so that it does not freeze. The first fire was built easily and the man was able to eat his biscuits. He was forced to make another fire when his feet got wet, when he broke through the surface ice while walking. He pulled twigs off the brush piled around trees to light the fire, but it was extinguished. He builds another fire with the help of his matches but it all gets pulled apart when he tried to take a piece of moss out. He tries to put the dog to the fire, to feed the flames yet he could not do so as he had no life left in his hands. Finally he attempts at running around the area to regulate his circulation but to no avail. He then decides he should meet death in a more dignified manner. He imagines his friends finding his body the next day. The man slowly drifts into the sleep of death, unable to fight the elements any longer.

(b) It was instinct that helped the dog to survive. The dog has an inbuilt respect for nature and her power. The dog is very anxious throughout the incidents of the story, he only forges ahead as he is frightened of the man. The dog is cautious and stays afar from the man, as he does not trust him. The dog only thinks of the man in two ways, one who provides fire and food. The dog is naturally insulated and thus prepared. Unlike the man, the dog recognizes the rhythms and signs that nature gives. Described as a "big native husky, the proper wolf dog, gray-coated and without any visible or temperamental difference from its brother the wild wolf", the dog acts as a foil to the unnamed man.

The man lacks imagination and cannot imagine what such cold could mean to a human being. Being a newcomer to the region, he feels that he is adequately equipped to combat the cold which is not the case. The man being emotional is conceited enough to feel he can defeat anything including what nature sends to his way. The old timer at Sulphur Creek had cautioned the man to listen to the warnings of nature. His cardinal mistake was of course, being out in such cold, alone and without a human companion. The fact that he is a newcomer to the region is repeatedly mentioned. He cannot foresee the dangers that lay in his path. But he did not do any such thing and moves out at his expedition and eventually gets frozen to death.

The dog and the man share the same instincts but the man chooses to ignore his instincts. The dog also stays away from the man's reach when the man tries to set him on fire. Ultimately, when he dies of his own fool-hardiness and arrogance, the dog howls for a little while and then leaves the corpse to the elements and seeks a place of shelter and sustenance.

(c) The man's inadequate imagination frequently causes him to misjudge the supremacy of nature. He is also self-assured in his strength of endurance, which depends on artificial resources, and not natural instincts or abilities. The man is indifferent because he is devoid of imagination. The man is capable and practical, but he does not take into account that when nature strikes, no one can escape. He learns that no man can defeat nature. An arrogant man, unmindful of nature's might, will easily fall victim to its wrath.

He thought of the old-timer on Sulphur Creek, imagining him warm and comfortable and smoking a pipe, he said to himself, "You were right, old hoss; you were right".

Answer 14.

Miss Meadows sways through a gamut of emotions and that too quickly throughout the plot.

The author's amusing sarcastic lines in the beginning that Miss Meadows "stared at the Science Mistress. Everything about her was sweet, pale, like honey. You would not have been surprised to see a bee caught in the tangles of that yellow hair" sums up quite nicely Miss Meadow's character as one who leans toward envy. It is thought-provoking that Miss Meadows' disposition appears to have an undesirable effect on the singers in the music class. We can see it when Miss Meadows asks the girls to sing a 'lament'. 'Every note was a sigh, a sob, a groan of awful mournfulness.' This line is noteworthy in numerous ways, the girl's reaction to the song echoes what Miss Meadows is going through. The girls were obviously irritated with her but she said to herself that why would they care about anyway for a person who was standing "there bleeding to death, pierced to the heart, by such a letter".

Just as Miss Meadows is feeling melancholic, by selecting a lament and asking her class to belt it out, Miss Meadows seems to be transferring her own despondency to upset the girls in her singing class. Reflecting her feelings and raging emotions, the music moves along, the students picking up instinctively Miss Meadow's emotions themselves get irritable and afraid. Later she reprimands them for singing expressionlessly.

During her lesson, she's informed by another colleague that Basil, her fiancé, has sent a telegram for her. To this she says, "I hope it's not a bad news" This line states her reaction over the telegram as letters arriving in the school hours explained of a "very bad news, such as death or a serious accident or something to that effect."But to her surprise, when she reads in the letter that Basil asked her to ignore the first letter and that he had bought the hat-stand which they had been thinking of lately, indicating that their engagement is happening.

"On the wings of hope, of love, of joy" she returns to her class and says to her students "Don't look so doleful girls, It ought to be warm, joyful and eager." She asks them to sing a happy song for one of felicitations. She also urges her students to imagine 'flowers o'erladen; baskets of fruit and ribbons to boot.'

Miss Meadow's voice sung the loudest of all the voices, glowing with expression. This line shows that Miss Meadows, perhaps, is describing her own feelings, after Basil had reaffirmed their engagement. Just the way she had transferred her feelings to the girls and made them feel miserable by singing something that echoes her feelings, she now asks them to sing something lighter and happier reflecting her blissful thoughts.

Answer 15.

When Mrs. Mallard heard the news of Brently Mallard—her husband's death in a rail-road accident, she wept at once, overwhelmed with sadness. Then she headed towards her room to spend some time alone. Out of physical exhaustion, she sat in an armchair facing the window in her room.

From the window she observed the scenery outside of – the tops of the trees that were aquiver with new spring life; the delicious smell of rain in the air; the sight of a peddler crying his wares; the faint sound of a distant song; countless sparrows twittering in the eaves and patches of blue sky that showed through the clouds. All these evoked a different feeling, a new hope in her. A little whispered word came from her mouth. She said it over and over again, "Free, free, free!" A feeling of terror escaped her eyes and they became bright. Her pulses became fast and "the coursing blood warmed and relaxed every inch of her body."

Mrs. Mallard understands the ways of the world, to be bereft of her strongest tie; she knows that she will mourn when Brently's corpse will arrive, but beyond that the realisation of 'a long procession of years to come' which she will live independently awaits her. She may have loved her husband, but she also realises the prospects that wait for her, that life would be freer than it had been, "Free! Body and soul free!" she kept whispering.

Finally, she descended the stairs with her sister, with sadness manifested outwardly and her private thoughts of happiness within. Somebody opened the frontdoor with a latch key and Brently Mallard entered inside, 'travel-strained, composedly carrying his grip-sack and umbrella.' It was learnt that he had been far from the scene of accident and didn't even have the knowledge of it.Though Richard tried to screen him from his wife's view but it was too late. Being weak-hearted, Mrs. Mallard died of shock. The doctors said that she had died of heart disease–a "joy that kills." The ending is ironical as Mrs. Mallard does not die of happiness but of sadness and shock of being robbed of her new future.

Answer 16.

(a) 'The Darkling Thrush' begins with endings: the end of a year, the end of a day 'weakening eye of day' makes the poem to be set in dusk, even the end of the century. Leaning against a gate, he calls the "Frost was spectre-grey," sees the dregs of winter, "household fires", "frost and the shoots" which looked like a broken lyre's strings. This points at the season being winter. Everything is dark and depressing. All is cold, bleak and gray. He gives the description of the season. The century is now dying and its "corpse" visible to all and lying in the "crypt" of the "cloudy canopy" or the overcast grey skies.

(b) When the poet was contemplating on the turns of the century, a thrush arrived with evensong. It seemed as if an aged thrush which is described as "frail, gaunt and small" with blast-beruffled plume" had chosen to fling its soul into the enveloping darkness as it sang in full throated song with unlimited joy. The song was loud, musical and elevating. Positivity, hope and energy enters into the dull arena. Despite being old and weak, the bird's voice resounded through the pall. The song of the thrush may have dispelled the gloom of the place where the poet stood in observation, yet the poet feels no joy at all. "I was unaware"—shows he was too cynical of the world to feel hopeful. Yet the hope in the thrush's song does make him feel that the world could still be saved. The poem is rich in metaphor as the title is a metaphor itself. Darkling means "related to darkness" which shows the end of the day.

(c) The poem ends on a confusing note. The speaker says that he was unaware. The confusion lay in whether he was touched by the bird's song yet unaware of the reasons to be happy and motivated or he was completely unaware of the positivity in the bird's voice and was unimpressed by its singing and its implications. As a modernist, the poet may be ambiguous in his realisation. He wants to hope for things to get better but does not know the reasons to feed that hope. The thrush symbolised the spirit of resurrection of new life of joy and hope that lay in store in the future. Although the knowledge of hope is only available to the bird, Hardy accepts the bird's song as a sign that there is hope for the future. Some critics believe that The Darkling Thrush is an elegy as it laments symbolically on the ending of the era.

Answer 17.

'The Spider and the Fly' reveals the story of a cunning spider who lures a naive fly with his clever words and traps her in his web. The spider praises his web by using words like "tis the prettiest little parlor" with many "pretty things" and tries to tempt the fly with the prospect of beautiful things. However, the fly refuses him saying that whoever goes to his parlour, doesn't come back.

He fakes concern for the fly's exhaustion and offers his bed to her to rest. He says that his home has "pretty curtains drawn around, the sheets are fine and thin." He asks her sleep in the beautiful bed. To this the fly says that whoever sleeps in his bed never wakes up and does not fall into his trap.

The spider also uses the promises of the delicious food to attract the fly toward its home-cordially invites her to its pantry. "I have within my pantry good store of all that's nice"; he wants the fly to feast upon "all the nice" and tasty treats. Yet the fly swerves away from this offer and says I've heard what's in your pantry, and I do not wish to see."

However, the fly is cunning and smart enough to keep her head on her shoulders and refuses the spider once again. Yet, the spider does not for go his flattery and calls her witty and wise. Still, the fly thanks him profusely but does not relent saying that it would return some other day.

The interaction between the predator and the prey is written in a way to make this warning easier for children to understand. The poet has tried to show that flattery should never be allowed to get to the head because wicked people flatter innocent gullible ones to lure them into their traps and later use them for their own good.

Finally in a last attempt, the spider entices her by praising her wings and physical beauty. He calls her wings beautiful. He suggests to the fly that she can look at herself, her beauty in a small looking glass present at his home. The fly, losing her head, visits the spider's home later and finally gets caught in its trap. The spider having caught the fly in its web happily sings "pretty fly, with the pearl and silver wing". The poem ends with a small message to all the children. The poet cautions them against flattery by evil counselors and imposters. She asks them to take a lesson from the fly's foolishness resulting in her tragic end and warns them not to be deceived by false flattery.

Answer 18.

Tennyson's sickness and advanced age perhaps contributed to this personal and unforgettable account on death. Tennyson wrote this poem three years before his death. The poem speaks of his calm and compliant attitude toward dying.

The line "And one clear call for me!" clearly suggests that he has heard the clarion call from death. The poem clearly highlights his peace with his impending death.The "evening star" is pointing at his way, and he believes that the "clear call" of death has finally come. He is ready; the poem is layered with a sense of eagerness and acceptance.

With the use of the image of a sand bar, the poet tries to explain life and death. Another vital image is of "crossing," which suggests Christian implications: the suggestion of crossing can be both to go to the other world after death, or to the act of a Christian crossing himself as a gesture of faith and piety.

"And may there be no sadness of farewell, When I embark"; This is where the reaffirmation becomes clear where he says that he is about to embark upon the journey, not end his journey. As he is ready to die, he does not want his friends and family to cry at his parting. The placid acceptance of his death and his solace in the thought of meeting his 'Pilot' helps the poet conquer his fears. In the end, the time of death has arrived for the poet. He wants no sadness in his parting from the world. The reason to not grieve is that he hopes to greet his Pilot or God in the afterlife.

The religious importance of crossing was familiar to the poet. The cross was where Christ had died and now as the poet dies, he brings up the same image. This is a spiritual poem. The poet has faith in himself and his God and has made peace with his passing. The title means travelling calmly and serenely from this life to the next. The Pilot has been used as a metaphor for God and the poet hopes to meet Him face to face.

●●

ENGLISH-II 2005

Due to change in the present syllabus, the papers from 2018 to 2006 are not given.

QUESTIONS

Section – A

(Answer one question)

THE TEMPEST – *Shakespeare*

Question 1.

*Choose **three** of the passages **(a)** to **(d)** and answer as briefly as possible the questions that follow :*

(a) ***Miranda*** : *You have often*
Begun to tell me what I am, but stopp'd
And left me to a bootless inquisition,
Concluding, "Stay; not yet."

Prospero : *The hour's now come;*
The very minute bids thee ope thine ear;
Obey and be attentive...

(i) *Where are Prospero and Miranda at this moment ? What had Miranda seen, a little while ago, that had moved her emotionally ?*

(ii) *What had Prospero often tried to tell her earlier ?*

(iii) *Which 'hour' does Prospero refer to ? What does he want Miranda to do now ?*

(iv) *Explain the phrase—"bootless inquisition."*

(v) *What does Prospero ask Miranda immediately after he tells her to be attentive ?*

(b) ***Prospero*** : *I did say so,*
When first I rais'd the tempest
Say, my spirit,
How fares the king and's followers ?

Ariel : *Confin'd together*
In the same fashion as you gave in charge.
Just as you left them; all prisoners, sir,
In the line-grove which weather-fends your cell;
They cannot budge till your release.

(i) *Where are Prospero and Ariel ? What is Prospero dressed in ?*

(ii) *What does Prospero refer to when he says "I did say so" ? Who is the "king" referred to by Prospero ?*

(iii) *Comment on Prospero's use of the expression, "my spirit."*

(iv) *Explain : "In the line-grove... your cell".*

(v) *What reaction does Ariel expect from Prospero on seeing his prisoners' condition?*

(vi) *Mention one quality of Ariel as reflected in this extract.*

(c) ***Ferdinand*** : *Wherefore weep you ?*

Miranda : *At mine unworthiness, that dare not offer*
What I desire to give, and much less take
What I shall die to want. But this is trifling,
And all the more it seeks to hide itself,
The bigger bulk it shows. Hence, bashful cunning,
And prompt me, plain and holy innocence
I am your wife, if you will marry me;

(i) *Where are Ferdinand and Miranda at the present moment ? Who is the third person unseen by both of them ?*

(ii) *Why is Miranda crying ? What is it that she dared not to offer ?*

(iii) *Explain : "... and much less take*
What I shall die to want."

(iv) *What is "it" referred to in the line*
"And all the more it seeks to hide itself..."

(v) *What does "bashful cunning" mean ? What proposal does Miranda make?*

(vi) *What will she do if her proposal is rejected?*

(d) ***Sebastian*** : *Methinks, I do.*

Antonio : *And how does your content*
Tender your own good fortune?

Sebastian : *I remember*

Antonio : *You did supplant your brother*
Prospero.

Antonio : *True;*
And look how well my garments sit upon me,
Much feater than before; my brother's servants.
Were then my fellows, now they are my men.

(i) *Why does Sebastian say, "Methinks, I do" ?*

(ii) *What does Antonio mean by "And how does... own good fortune" ?*

(iii) *What is meant by the term "supplant" ? How did Antonio "supplant" his own brother?*

(iv) *Which "garments" does Antonio refer to ? How do they fit him ?*

(v) *Who are the "my men" referred to by Antonio? When were they his "fellows"? Comment on Antonio's tone in this speech.*

(vi) *How does Sebastian comment on this speech of Antonio ?*

Section – B

(Answer ***four*** *questions on at least* ***three*** *textbooks which may include The Tempest OR Saint Joan)*

THE TEMPEST—*Shakespeare*

Question 3.

With close reference to Act I, Scene II describe how Ferdinand is charmed and let to Prospero's cave and his first meeting with Miranda.

Comment on Prospero's apparently cruel attitude towards Ferdinand.

Question 4.

Though mischievous and playful, Ariel is most sincere in his devotion towards his master Prospero. Discuss.

ANSWERS

Section – A

Answer 1.

(a) (i) Prospero and Miranda are on the island at this moment.

Miranda has witnessed the scene of shipwreck which has moved her emotionally

(ii) Prospero had tried to tell her about the events of their lives that brought them to this remote island, *i.e.*, the story of the government by which he was overthrown and expelled from his dukedom.

(iii) Prospero refers to the maturity that Miranda has attained, *i.e.*, she is now matured enough to understand everything that Prospero is going to narrate at this moment. He wants Miranda to pay total attention to what he is going to state and explain.

(iv) "Bootless inquisition" means an unsuccessful enquiry. Here Miranda wants to say that she has never been satisfied by her father's answers to her queries about her past.

(v) Immediately after telling her to be attentive, Prospero asks Miranda to recall the time and incidents when she was just a child.

(b) (i) Prospero and Ariel are at a venture on the island where Prospero's alchemic experiments are in progress. Prospero is dressed in his magic robe.

(ii) Prospero refers to the moment for which he has been waiting since twelve years. The 'king' referred to by Prospero is his treacherous brother Antonio.

(iii) Prospero addresses Ariel as 'my spirit'. The expression suggests that Prospero is in a pleasant mood and is glad at the performance of Ariel.

(iv) In the line–"grove which weather-fends your cell" means that amongst the dense grove of lime, which protects Prospero's cell from bad weather. Here Ariel has made captive, the enemies of his master Prospero.

(v) Ariel expects that Prospero will have pity on seeing his prisoners' condition.

(vi) Ariel strictly carried out Prospero's orders as instructed by him. Therefore, we can say that Ariel was obedient to Prospero.

(c) (i) At the present moment Ferdinand is labouring as a slave of Prospero on the island and Miranda is there watching him. Prospero is the third person unseen by both of them.

(ii) Miranda is crying to see her love labouring as a log-man and she is quite helpless. She dares not to offer herself to Ferdinand as his wife.

(iii) The given expression shows Miranda's innocent and selfless love for Ferdinand.

(iv) In the given part of statement 'it' is referred to as 'true love' which cannot be hidden, because the more you try to hide it the more it is exposed.

(v) "Bashful cunning" means pretence of shyness. By her innocence Miranda proposes about marriage to Ferdinand.

(vi) Miranda will die a virgin if her proposal is rejected.

(d) (i) By saying "Methinks I do", Sebastian says that he fully understands what Antonio suggests when he asks, "For your advancement! Do you understand me"?

(ii) Antonio meant that Sebastian should be inclined to look favourably on his lucky chance for his advancement.

(iii) By the term 'supplant' it is meant that Antonio usurped the dukedom from his brother Prospero. Antonio treacherously took hold of the administration and usurped the dukedom of Milan from Prospero and left him and his three-year-old daughter in a broken boat at the mercy of the sea.

(iv) The garments referred to are the clothes that were provided for the wedding of the King's daughter Claribel at Tunis. The garments fit him more suitably.

(v) Men who were once the men of Prospero are referred to as "my men" by Antonio. They were his 'fellows' when Prospero was the duke of Milan. Antonio's tone, here, is of self-praise and contentment.

(vi) On Antonio's speech, Sebastian remarks, "But for your conscience", which means that whether Antonio doesn't feel guilty which suggests that his tone is sarcastic and satirical.

Section – B

Answer 3.

After the opening scene of the play, the second scene is of great importance and is an integral part of it. Ferdinand is brought on the scene by Ariel. He follows Ariel's charming song which refers to his drowned father, and thus comes within the view of Prospero and Miranda.

Prospero points him out to Miranda. On seeing the handsome youngman, Miranda expresses her unfeigned admiration for him. Ferdinand too is charmed by the beauty and grace of the young lady. Prospero is glad to see that they exchange glances.

In this very first meeting with Miranda, Ferdinand is so deeply charmed by the untouched beauty of the maiden that he offers to make Miranda, the Queen of Naples, since Ferdinand is the heir of his father, who, he believes has drowned in the tempest. Prospero, however, wants to be sure about Ferdinand's claim that he is the King of Naples, and calls him a traitor and spy and bids him to follow. On being insulted to this degree Ferdinand draws his sword, but finds himself unable to lift it.

Miranda, who has fallen in love with Ferdinand at first sight, begs her father not to be hard with him. Prospero discourages her admiration for Ferdinand and sternly bids her to keep her mouth shut. But Miranda declares that she has no ambition to see a 'goodlier man'. Ferdinand submits to Prospero's biddings by saying;

"Might I but through my prison once a day
Behold this maid. All corners else O'the' earth
Let liberty make use of – space enough
Have I in such a prison"

Though Prospero apparently looks very strict while treating Ferdinand, he actually tests the strength of Ferdinand's love for Miranda, because Prospero is also a father with a tender heart for his daughter, his "Cherubin."

Answer 4.

Many a supernatural character has been created by Shakespeare and other playwrights, but Ariel is unmatched. Unlike most supernatural characters, Ariel possesses many human qualities like love for freedom, gratitude, sympathy, desire for praise, dedication to his master and faculty of mischief. When examined deeply, two aspects of his character makes him intriguing. These are his latent faculty of mischief and his sincere devotion towards his master Prospero.

Ariel's playful nature with which he performs those duties enables him to practise his powers upon beings such as Stephano and Trinculo. The scene in which he successfully plants a discord among Stephano, Trinculo and Caliban is full of excellent befooling. The manner in which he gives an account of it to Prospero, shows that he enjoyed the fun of this foolery. There are many instances when Ariel successfully amuses the reader or the audience when the play is read or enacted.

Ariel is an ardent lover of freedom, yet he is bound to Prospero by a feeling of sincere devotion towards him. He performs those tasks and duties punctually and promptly which are imposed upon him by his master. Except for a single occasion, when he shows impatience with Prospero for assigning him so many tasks and postponing his liberation. Throughout the play he shows total devotion to his master Prospero which is very rarely found in supernaturals.

Thus, we can say that, Ariel is most sincere in his devotion towards his master though he is quite mischieveous and playful, yet he is true and not an exaggeration.

●●

ENGLISH-II

2004

QUESTIONS

Section – A

THE TEMPEST—*Shakespeare*

Question 1.

*Choose **three** of the passages **(a)** to **(d)** and answer as briefly as possible the questions that follow :*

(a) ***Prospero*** : *[To Ferdinand] Come on; obey:*
Thy nerves are in their infancy again,
And have no vigour in them.

Ferdinand : *So they are :*
My spirits, as in a dream, are all bound up.
My father's loss, the weakness which I feel,
The wreck of all my friends, nor this man's threats,
To whom I am subdued are but light to me,
Might I but through my prison once a day
Behold this maid. All corners else o'th' earth
Let liberty make use of—space enough
Have I in such a prison.

(i) *Where are Prospero and Ferdinand ? Mention two commands given by Prospero that Ferdinand has to obey.*

(ii) *Explain : "Thy nerves are in their infancy again."*

(iii) *What are "but light" to Ferdinand ?*

(iv) *Who is "this maid" referred to ? What had shocked the maid a little while ago ?*

(v) *Where will Ferdinand find "space enough" to live in? What should "liberty" make use of ?*

(b) ***Antonio*** : *O !*
If you but knew how you the purpose cherish
Whiles thus you mock it, how in stripping it
You more invest it, ebbing men, indeed,
Most often do so near the bottom run
By their own fear of sloth.

Sebastian : *Prithee, say on.*
The setting of thine eye and cheek proclaim
A matter from thee, and a birth, indeed,
Which throes thee much to yield.

(i) *Which "purpose" does Antonio refer to ? Where are Antonio and Sebastian at the present moment ?*

(ii) *How had Antonio mocked the "purpose" he cherished? What is implied by "stripping it" and "invest it" ?*

(iii) *What does Sebastian request Antonio to "say on" ?*

(iv) *How does Sebastian conclude that Antonio has much to convey ?*

(v) *About whom does Antonio speak just after Sebastian's speech ? What is his purpose in talking about him ?*

(c) ***Prospero*** : ***[aside]*** *Poor worm, thou art infected !*
This visitation shows it.

Miranda : *You look wearily.*

Ferdinand : *No, noble mistress, 'tis fresh morning with me*
When you are by at night. I do beseech you,—
Chiefly that I might set it in my prayers,—
What is your name ?

Miranda : *Miranda—O my father,*
I have broke your hest to say so!

Ferdinand : *Admir'd Miranda,*
Indeed the top of admiration; worth
What's dearest to the world! Full many a lady
I have eyed with best regard; and many a time
Th' harmony of their tongues hath into bondage
Brought my too diligent ear.

(i) *Who is the "poor worm" referred to ? What has the "worm' been "infected" with ?*

(ii) *What makes Ferdinand "look wearily" ? How does Ferdinand feel in Miranda's presence ?*

(iii) *Why does Ferdinand wish to know Miranda's name? What compliments does he bestow upon Miranda ?*

(iv) *Explain : "Th' harmony of their tongues hath into bondage. Brought my too diligent ear."*

(v) *What are Prospero's feelings on knowing about the couple's feelings for each other ?*

(d) ***Alonso*** : *Now all the blessings*
Of a glad father compass thee about !
Arise, and say how thou cam'st here.

Miranda : *O, Wonder !*
How many goodly creatures are there here !
How beauteous mankind is! O brave new world.
That has such people in't !

Prospero : *'Tis new to thee*

Alonso : *What is this maid with whom thou wast at play ?*
Your eld'st acquaintance cannot be three hours.
Is she the goddess that hath sever'd us,
And brought us thus together ?

(i) *Whom does Alonso ask to "arise" ? What is he glad "about" ?*

(ii) *What is Miranda wonderstruck about ?*

(iii) *Explain the terms, "eldest acquaintance" and "sever'd us".*

(iv) *What information does Alonso receive in response to his question, "Is she the goddess that hath sever'd us. And brought us together" ?*

(v) *Why is the world 'new' to Miranda ?*

Section – B

*(Answer **four** questions on at least **three** textbooks which may include The Tempest **OR** Saint Joan)*

THE TEMPEST—*Shakespeare*

Question 3.

Describe briefly but vividly the scene where Prospero reveals to Miranda, his brother's secret plot to kill them and usurp his dukedom.

What are your feelings for Prospero and Miranda's fate at the hands of a cruel brother?

Question 4.

Miranda's appeal lies in her sweet innocence, gentle nature and pure selfless love for Ferdinand. Referring closely to the text bring out the truth of the statement.

ANSWERS

Section – A

Answer 1.

(a) (i) Prospero and Ferdinand are on a lonely island where the latter is captivated by the former. The two commands given by Prospero that Ferdinand has to obey are :

(a) To carry and pile up the logs of wood.

(b) To be obedient always to his commands by following them.

(ii) Prospero treats Ferdinand cruelly, as he is under trial. Prospero thus speaks to him that all his muscles are limp and lifeless.

(iii) All the personal loss and threats of Prospero does not affect Ferdinand and he is ready to face it all.

(iv) Miranda is the "maid" referred to. She is shocked to find that her own father can be so cruel to someone.

(v) Ferdinand will enjoy enough space in a prison where he gets a single glance of Miranda each day. Though he will be in prison, liberty shall prevail in all corners of the earth.

(b) (i) Here, the desire of Sebastian is referred to by Antonio who wants to be the King of Naples. Antonio and Sebastian are presently at another part of the island.

(ii) By pretending as if he was talking in sleep, Antonio mocked the "purpose" he cherished. The two phrases mean that the more you put it off, the greater the importance you give to it.

(iii) Sebastian requests Antonio to "say on" the desire he had in his heart.

(iv) By reading the expressions on Antonio's face, Sebastian concludes that he has much to convey.

(v) After Sebastian's speech, Antonio speaks about Francisco's effort to convince Alonso that Ferdinand was dead fighting against the tempestuous waves.

The purpose in talking about the death of Ferdinand is that Sebastian should think seriously about becoming the King of Naples.

(c) (i) Miranda is referred to as the "poor worm", and is infected with love for Ferdinand, the crown price.

(ii) Tiredness from hard physical labour makes Ferdinand look weary. But the very presence of Miranda refreshes him.

(iii) Ferdinand wishes to know Miranda's name so that he could pray for her. "Admired Miranda—to of all admiration" is the compliment that he bestows upon her.

(iv) Ferdinand's attention was drawn towards the melodious manner of women talking among themselves.

(v) Prospero is filled with fatherly love when he comes to know about the feelings of love that the couple has for each other, and he wishes that the grace of Almighty may rain "on that which breeds between them."

(d) (i) Alonso asks his son Ferdinand to arise. Finding his son alive he is very glad.

(ii) So many "goodly creatures" are the cause of wonder for innocent Miranda.

(iii) Finding Ferdinand and Miranda close to each other, Alonso expresses his feeling that their acquaintance should be an old one and he thus calls it 'eld'st'. Alonso takes Miranda as the goddess of the island who has 'sever'd or separated them.

(iv) In response to his question, Alonso comes to know from Ferdinand that Miranda is not a goddess but a mortal human being.

(v) Miranda had only seen two men in her life, who were Caliban and her father Prospero. So, the appearance of many human beings suddenly is a 'new' world to her.

Section – B

Answer 3.

Miranda is sorry to see the shipwreck and thinks that so many "Noble Creatures" have been perished. She requests her father to stop the storm if he had raised it. Then Prospero, after removing his magical robe, addresses Miranda in the capacity of a father and tells her that twelve years since he was the duke of Milan, but was wrongly dethroned from his kingdom by his brother Antonio whom he loved and trusted blindly.

Prospero made Antonio the incharge of managing the kingdom's affairs as he was busy in his secret study of magic. This made Antonio ambitious and his evil mind came to work. Prospero tells that he believed that he was indeed the duke; out of the substitution, and executing the outward face of royalty, with all prerogative. He tells that Antonio became an expert in whom to promote and whom to thrash for gaining hierarchy. He compares Antonio to an 'ivy' that had hid his princely trunk and "suck'd his verdure out on't".

Antonio considered Prospero to be no longer capable of managing state affairs and thereby joined hands with Alonso, the king of Naples, who was his relentless enemy. Alonso agreed to help Antonio on two conditions *i.e.,* "to give him annual tribute, do him homage, subject his coronet to his crown, and bend the dukedom, yet unbow'd"

The opportunist King of Naples agreed to "extirpate" Prospero from Milan along with his three year old daughter, Miranda and "Confer fail Milan, with all honours" on Antonio. Thereby, in the darkness of the night Prospero and Miranda were carried on a ship with no sail or mast and were left in the sea. They were thus left in a miserable state to die.

Though a reader feels sympathy for both Prospero and Miranda who were deceived by a cruel brother and uncle, whom they trusted so much, but one may also feel that Prospero should have exercised his powers prudently and should not have left the affairs of the state to his treacherous brother in the first place.

Answer 4.

It is a true estimate of Miranda that her appeal lies in her sweet innocence, gentle nature and pure selfless love for Ferdinand. This is seen when Miranda believes that people have perished in the shipwreck and she says "O, I have suffered with those that I saw suffer." She further states, "O the cry did knock against my very heart!" and wishes if "I had been any god of power, I would have sunk the sea within the earth; or ere, it should the good ship so have swallow'd and the fraughting souls within her."

When Miranda meets Ferdinand for the first time, she says, "What is't a spirit ?" She says to her father "Believe me Sir, It carries a brave form; but tis a spirit", as she had so far only seen her father and Caliban. She takes Ferdinand to be a spirit but Prospero tells her that he is a human being and she falls in love, i.e., love at first sight. She finds Ferdinand to be noble, but is shocked to see her father behave in a strange manner and requests him to be kind and gentle towards him.

Miranda is convinced that Ferdinand is one in whom "Nothing" can dwell in such a temple. When Prospero orders Ferdinand to be obedient to his commands, Miranda pleads for him to her father, "Dear father, Make not too rash a trial of him, for He's gentle and not fearful." On being rebuked by her father, she implores mercy on behalf of Ferdinand and says, "Sir, have piti, : I'll be his surety". It is an example of her innocence.

She tries to convince Ferdinand by saying, "my father's of a better nature, sir, than he appears by speech." When her father says that there are better looking men in the world than Ferdinand, she says, "My affections are then most humble; I have no ambition to see a goodlier man!"

On seeing Ferdinand carrying logs of wood, against the orders of her father, she goes to meet him and is even willing to carry that for him. She finds the task so hard and laborious that she remorses by saying, "when this (logs) burns I will weep for having wearied you," and then she asks him to take some rest.

Miranda is an example of innocence and straight forwardness. Her conversation with Ferdinand lacks cunningness and shrewdness. She is overjoyed when told that Ferdinand also loves her, and she considers herself to be very lucky to have his love as a great reward.

●●

ENGLISH-II

2003

QUESTIONS

Section – A

THE TEMPEST – *Shakespeare*

Question 1.

*Choose **three** of the passages **(a)** to **(d)** and answer as briefly as possible the questions that follow :*

(a) ***Prospero*** : *Hear a little further,*
And then I'll bring thee to the present business
Which now's upon's; without the which this story
Were most impertinent.
Miranda : *Wherefore did they not*
That hour destroys us?
Prospero : *Well demanded, wench :*
My tale provokes that question. Dear, they trust not,

(i) *Which "hour" is Miranda referring to ?*

(ii) *According to Prospero why couldn't his enemies destroy them?*

(iii) *What did they do to Miranda and Prospero ?*

(iv) *What according to Prospero is the "present business which now's upon's" ?*

(v) *What was the main reason which gave Prospero courage to bear his troubles ?*

(b) ***Gonzalo*** : *Had I plantation of this isle, my lord—*
Antonio : *He'd sow't with nettle-seed.*
Sebastian : *Or docks, or mallows.*
Gonzalo : *—And were the king on't, what would I do ?*
Sebastian : *'Scape being drunk, for want of wine.*
Gonzalo : *I' th' commonwealth I would be contraries*
Execute all things,

(i) *Explain what Gonzalo means by "Had I plantation of this isle"? Whom does he address as "my lord" ?*

(ii) *Mention four things which Gonzalo says he will have in his "commonwealth".*

(iii) *Who are Antonio and Sebastian ? What kind of attitude do they have towards Gonzalo ?*

(iv) *Give one remark made earlier by Sebastian which shows his cold attitude.*

(v) *What is Gonzalo's motive in talking about a commonwealth ? Which characteristics of his stand out in this scene ?*

(c) ***Ferdinand*** : *This is strange. Your father's in some passion*
That works him strongly.
Miranda : *Never till this day*
Saw I him touch'd with anger, so distemper'd
Prospero : *[To Ferdinand] You do look, my son, in mov'd sort,*
As if you were dismay'd. Be cheerful, sir;
Our revels now are ended…

(i) *Why is Prospero "touch'd with anger so distemper'd"? Explain the phrase.*

(ii) *What "revels" is Prospero referring to ? Why were they held ?*

(iii) *What does Prospero go on to say about human beings and life in the same speech ?*

(iv) *What does Prospero now order Ariel to do ?*

(v) *"At this hour*
Lies at my mercy all mine enemies" says Prospero at the end of this scene. Show how far this is true.

(d) ***Prospero*** : *Mark but the badges of these men, my lords,*
Then say if they be true. This misshapen knave.
His mother was a witch, and one so strong
That could control the moon, make flows and ebbs,
And deal in her command without her power.
These three have robb'd me, and this demi-devil—
For he's a bastard one–had plotted with them.
To take my life. Two of these fellows you
Must know and own; this thing of darkness I
Acknowledge mine.

(i) *Who are the "my lords" Prospero is addressing ? Where are they?*

(ii) *Who are "these men" ? Why does Prospero want "my lords" to look at their "badges" ?*

(iii) *"This misshapen knave"—to whom is Prospero referring? What does Prospero say about this person's mother ?*

(iv) *Describe the condition and dress of "these men" ? How did they come to be in this state?*

(v) *How does Prospero deal with the three of them ? State what you think of Prospero's decision.*

Section – B

*(Answer **four** questions on at least **three** textbooks which may include The Tempest **OR** Saint Joan)*

THE TEMPEST—*Shakespeare*

Question 3.

Briefly but vividly describe the opening scene of the play "The Tempest", bringing out clearly the impression the scene makes on your mind and its importance in the play.

Question 4.

Antonio is an out and out villain. With close reference to what he says and does, show how far you agree with this statement.

ANSWERS

Section – A

Answer 1.

(a) (i) Miranda refers to the hour when with the help of Alonso and Sebastian, Antonio removed her and her father Prospero, from Milan.

(ii) According to Prospero, enemies could not destroy them because the people of Milan loved them very much and they would have revolted against Antonio.

(iii) Miranda and Prospero were placed on a deserted boat and were left in the midst of the ocean by the enemies to die.

(iv) The present business refers to Prospero's wish to tell Miranda the reason for raining the storm in the sea.

(v) The smile of Miranda gave courage to Prospero to bear the troubles.

(b) (i) Gonzalo means to say that if he would be the king of this island he would do many things. Alonso, the King of Naples, is addressed as 'my lord'.

(ii) In his commonwealth Gonzalo would :

(a) Have no magistrates.

(b) Have no occupation; men will be idle.

(c) Neither have riches nor poverty.

(d) Women will be innocent and pure.

(iii) Antonio is the brother of Prospero, who had usurped the dukedom of Milan from him. Sebastian is the brother of the King of Naples, Alonso. They are making fun of Gonzalo.

(iv) "T'was what a sweet marriage", was end of the remarks made earlier by Sebastian to Alonso. He spoke in an ironical manner and blamed Alonso for whatever happened to them.

(v) Gonzalo's motive in talking about a commonwealth is just to amuse his master Alonso, who is sorry and sad to have lost his son in the storm. In this scene, Gonzalo's characteristics like loyalty and sincerity stand out.

(vi) The cause of Prospero's anger is the treachery or wicked conspiracy hatched by Caliban against him with the help of Stephano and Trinculo. The phrase means that Miranda has never seen her father in such an angry state.

(c) (i) By 'revels' he refers to the masque in which various goddesses and spirits took part. It was held to bless the marriage of Miranda and Ferdinand.

(ii) By 'revels' he refers to the masque in which various goddesses and spirits took part. It was held to bless the marriage of Miranda and Ferdinand.

(iii) Prospero says that human beings are as unreal as the dreams and their brief existence is followed by death that puts an end to that existence.

(iv) Prospero orders Ariel to bring 'trumpery' *i.e.,* rich clothings, to deceive Caliban and the plotters against him.

(v) This is true to every extent as Alonso, Antonio, and Sebastian are in a fit of madness, whereas Trinculo, Stephano, and Caliban are chased away by spirits in disguise of dogs and hounds.

(d) (i) The "my lords" referred here by Prospero are Alonso, Sebastian, Gonzalo, Antonio, and others. They are in front of the cell of Prospero.

(ii) "These men" are Caliban, Stephano, and Trinculo. He wants to point them out to look at their 'badges' because the clothes belonged to Prospero which 'these men' had stolen.

(iii) Prospero is referring to Caliban. He says about Sycorax, the mother of Caliban, that she was a witch and was so powerful that she had power over the moon and could cause the sea to ebb and flow.

(iv) "These men" are heavily drunk and are wearing the clothes they had stolen. They are in this state because of Prospero's plan and are under the charm of his magical power.

(v) Prospero pardons them. It is a good decision as forgiveness is the keynote of the play.

Answer 3.

As the title of the play itself suggests, the opening scene of 'The Tempest' is a vivid description of a ship caught in a storm. It dramatically tells how a ship is wrecked and how people in it come to the island. It also introduces the audience to some of the important characters of the play.

As the scene unfolds, the captain of the ship is seen ordering his boatswain to call all men to duty as the ship is in danger. The boatswain instructs the mariners to adjust the topsail as he is apprehensive of the outcome of the storm. On board the ship are Alonso, the king of Naples, Sebastian his brother, Antonio, the duke of Milan, Alonso's son Ferdinand, Gonzalo, an honest councillor of Alonso, and other lords. Alonso shows deep concern for the ship but the boatswain tells them impatiently to go down to their cabins as they mar the crew's labour. When Gonzalo asks the boatswain to be patient, the boatswain retorts that he is hindering his work. He tells Gonzalo that, the waves do not care for the name of a king. He is rude towards Gonzalo. He commands him to go to his cabin and prepare himself for any misfortune that may fall on him. Gonzalo is annoyed but hopes that this man does not die by drowning but be hanged later on.

This scene also shows Shakespeare's profound knowledge of sea and seamanship.

The boatswain tries to keep the ship far from the seashore, but is irritated by Sebastian and Antonio who curses and speaks ill words to him. Gonzalo wanted everyone to join the king at prayers, but at the same time a confused noise comes :

"Mercy on us!–We split, we split!–Farewell,

my wife and children!–Farewell, brother!–We split, we split, we split!"

The scene ends with Gonzalo saying:

"The wills above be done, but I would fain die a dry death."

Though the scene is not very long yet it is very effective. The treatment of the scene compels us to think that no one on board the ship would survive. So, it can be said that the presentation of the opening scene is realistic and makes the readers fill with curiosity and interest about the happenings going to take place later in the play.

Answer 4.

Antonio is the most wicked and hateful character in the play. He does evil without remorse or sense of any guilt. He does not regret his deeds towards his brother Prospero. He is loyal to none. He even plots to kill Alonso and instigates Sebastian against him. Prospero calls him "perfidious", a "false uncle" while narrating his tale to Miranda.

Antonio betrayed Prospero's trust in him and usurped his dukedom. He cruelly put his brother and his three-year-old daughter in a deserted ship and left them at the mercy of the sea-god. He was not guilty or ashamed; instead boasts to Sebastian later :

"And look how well my garments sit upon me.
Much feater than before."

Antonio speaks ill about Gonzalo and calls him a talkative person. He instigates Sebastian to kill Alonso so that, he would be free from the tribute he has to pay to Alonso. He forgets all the help Alonso had extended to him in order to remove Prospero from Milan.

He laughs at Gonzalo for his loyalty and is ready to kill him and the King of Naples in their sleep. Prospero in the end remarks about Antonio—

"For you wicked Sir,
Whom to call a brother
Would even infect my mouth."

Thus, we see that Antonio is an out and out villain.

ENGLISH-II 2002

QUESTIONS

Section – A

THE TEMPEST—*Shakespeare*

Question 1.

Choose **three** *of the passages* **(a)** *to* **(d)** *and answer as briefly as possible the questions that follow :*

(a) ***Boatswain*** : *None that I love more than myself. You are a councilor;*
if you can command these elements to silence, and work
the peace of the present, we will not hand a rope more;
use your authority. If you cannot, give thanks you have
lived so long, and make yourself ready in your cabin for
the mischance of the hour, if it so hap. [To the mariners]–
Cheerly, good hearts! [To the courtiers]–Out of the way, I say!

(i) *To whom is the Boatswain speaking ? What does he mean in this way ?*

(ii) *What were the conditions that made the boatswain react in this way ?*

(iii) *What hope does the person speak taking from the attitude of the Boatswain?*

(iv) *How can they 'make yourself ready in your cabin' ? For what are they asked to make themselves ready ?*

(v) *What opinion do you form of the Boatswain from the above extract ?*

(b) ***Sebastian*** : *What a strange drowsiness posseses them!*
Antonio : *It is the quality o' th' climate*
Sebastian : *Why*
Doth it not then bur eyelids sink? I find not
Myself dispos'd to sleep.
Antonio : *Nor I : my spirits are nimble.*
They fell together all, as by consent;
They dropp'd, as by a thunder-stroke. What might.
Worthy Sebastian ? O what might– ? No more.–
Any yet methinks I see it in thy face
What thou should'st be. Th' occasion speaks thee, and
My strong imagination sees a crown
Dropping upon thy head.

(i) *Why did such a 'strange drowsiness' possess them ? Who are referred to as "them" ? Where are "they" ?*

(ii) *Who is Antonio ? How was he responsible for Prospero's present condition ?*

(iii) *Explain : "Th' occasion speaks thee".*

(iv) *What does Antonio ultimately convince Sebastian to do ? What role does Antonio himself intend to play ?*

(v) *How are Antonio and Sebastian prevented from doing what they had planned ?*

(c) ***Stephano*** : *How now shall this be compassed ? Canst*
thou bring me to the party
Caliban : *Yea, yea, my lord. I'll yield him thee asleep,*
Where thou may'st knock a nail into his head.
Ariel : *Thou liest, thou canst not.*
Caliban : *What a pied ninny's this! Thou scurvy patch!*
I do beseech thy greatness, give him blows,
And take his bottle from him : When that's gone
He shall drink naught but brine; for I'll not show him
Where the quick freshes are.

(i) *Who is Stephano ? What is the "this" that is to be "compassed" ?*

(ii) *Who is Caliban speaking about when he says "I'll yield him thee asleep" ?What is the reason for Caliban's hatred towards that person ?*

(iii) *Who is Caliban calling "a pied ninny" ? Explain the terms "a pied ninny" and "scurvy patch" ?*

(iv) *What is Ariel's intention in being a part of this conversation ?*

(v) *What is the plot Caliban outlines before Stephano a little later in the scene ?*

(d) ***Prospero*** : *Fairly spoke.*
Sit then and talk with her, she is thine own.

What, Ariel! my industrious servant Ariel!
Enter Ariel.

Ariel : *What would my potent master ? Here I am.*

Prospero : *Thou and thy meaner fellows your last service*
Did worthily perform, and I must use you
In such another trick. Go bring the rabble
O'er whom I give thee pow'r here to this place.

(i) *Who is Prospero addressing at the beginning of the extract ? How did that person come to be there ?*

(ii) *What was "fairly spoke" ?*

(iii) *What was the "last service" worthily performed by Ariel and his fellows ?*

(iv) *What is the "such another trick" Prospero wants to use Ariel for and for what purpose?*

(v) *After the performance of "such another trick" Prospero is disturbed for some reason. What is he reminded of ?*

Section – B

*(Answer **four** questions on at least **three** textbooks which may include The Tempest **OR** Saint Joan)*

THE TEMPEST—*Shakespeare*

Question 3.

Compare Caliban and Ariel as servants of Prospero.

Question 4.

'The fifth act sees the play moving smoothly to its expected closure'. Outline briefly but vividly the main events that take place in Act V.

ANSWERS

Section – A

Answer 1.

(a) (i) The boatswain addresses Gonzalo : "None that I love more than myself." By this he means that he loves himself more than he cared for the king.

(ii) As a severe storm was raging, the ship could be wrecked at any moment. The courtiers were interferring in the sailors' efforts who were trying their best to save the ship. These were the ghastly conditions that made the boatswain to react in this way.

(iii) The rude attitude of the boatswain convinces the person that the boatswain had no drowning mark upon him and that he will be hanged, if he is alive later.

(iv) They were asked to make themselves ready for death as the ship would certainly wreck and they would all drown. Thus they can make themselves ready to meet their death.

(v) The boatswain is a sincere worker though a bit coarse in manner. He is seriously concerned about the safety of all.

(b) (i) Ariel's magical music causes a strange drowsiness on them. 'Them' refers to Alonso, Gonzalo, Adrian and Francisco. They are on another part of the island.

(ii) Antonio is Prospero's younger brother and the usurping Duke of Milan. Antonio conspired with Alonso, the King of Naples, to overthrow Prospero. He had put Prospero and Miranda on a rotten carcass of a boat to perish in the open sea and declared himself the Duke of Milan.

(iii) The phrase means that this is the apt hour for Sebastian to act.

(iv) Antonio finally convinces Sebastian to kill Alonso and Gonzalo and become the King of Naples. Antonio himself intends to kill Alonso.

(v) Ariel at that moment sings in Gonzalo's ear and awakens him, thus preventing Antonio and Sebastian from doing what they had planned.

(c) (i) Stephano is Alonso's butler. "This" is the purpose of Stephano becoming the king of the island.

(ii) Caliban is speaking about Prospero. Caliban hates Prospero because he believes that he is the real owner of this island, being the son of Sycorax, the witch. Prospero had taken this island from him by his power of magic and Prospero also ill-treated Caliban and punished him in many ways.

(iii) Caliban calls Trinculo a 'pied ninny'. 'Pied ninny' means a fool who wears striped clothes and 'scurvy patch' means a diseased clown.

(iv) Ariel wants to deliberately cause a quarrel between Stephano and Trinculo. Then he wants to inform Prospero about the conspiracy that is being plotted against him.

(v) Caliban outlines the plot that Prospero goes to sleep in the afternoon; "There thou mayst brain him, having first seiz'd his books; for without them he's but a sot, as I am" or by seizing his books first he can be rendered helpless. Stephano can then kill Prospero easily, marry his daughter, and become the king of the island.

(d) (i) Prospero is addressing Ferdinand, who after being shipwrecked, had swum to the shore and was lured to Prospero's cell by Ariel's music.

(ii) The statement of Ferdinand was "fairly spoke" in which he says that he would not be tempted to do anything that destroys the sacredness of the marriage.

(iii) The last service worthily performed by Ariel and his fellows was to spread a banquet before Alonso and his companions. But before those hungry and tired men could eat, Ariel appeared like a harpy and made the banquet to vanish. He then reminded them of the evil they had done to Prospero and Miranda.

(iv) Prospero wants Ariel to present a masque before Ferdinand and Miranda. He wants to show them the potency of his art and also depict the Goddesses Iris, Juno and Ceres blessing them.

(v) Prospero is disturbed because he is aware of the conspiracy being hatched by Caliban, Stephano, and Trinculo against his life.

Answer 3.

Shakespeare is the master playwright who has crafted a multitude of characters unparalleled in English literature. His famous play 'The Tempest' is also a panorama of different types of characters who represent perfectly the identity imparted to them. Both Ariel and Caliban are servants of Prospero but they oppose each other. However there are some similarities too; both dislike hardwork, both are deeply desirous of freedom, and belong to different realms beyond human experience.

One is a super-human and the other is sub-human.

Ariel is a soulless being, dainty and delicate. He is associated with flowers and bees. He serves Prospero wholeheartedly.

"All hail, great master, grave sir, hail! I come to answer thy best pleasure he 't to fly.

To swim, to dive into the fire, to ride

On the curl'd clouds; to thy strong bidding tasks."

Before Prospero's arrival on the island, Ariel was in the service of Sycorax. When he refused to carry out her "earthy and abhorr'd commands" she confined him into a "cloven pine". He howled away twelve winters before Prospero arrived and freed him from the prison with the help of his magical powers. Ariel's gratitude to Prospero makes him a willing, responsible, and sincere servant. Prospero is able to raise the tempest; cause a magical shipwreck; lead his enemies to the island, and finally cause them to repent for their misdeeds with Ariel's help only.

Ariel fulfils Prospero's orders to the minutest detail causing no injury to any person on the ship, not even a minutest "hair pershed".

Though Ariel is an airy spirit, yet like a man he enjoys harsh work and treats his victims roughly. He is eager for praise and reports to Prospero about how he carried out his instructions.

"To every article

I boarded the king's ship. Now on the beak,

Now in the waist, the dech, in every cabin,

I flam'd amazement."

Ariel has the power over all the elements and that is why he controls every phenomenon in the play. He creates the storm, the thunder and lightning, mirages that deceive, the masque, the banquet, and the conflict between Stephano and Trinculo. He awakes Gonzalo and saves him and Alonso from being killed. He chastises Alonso, Antonio and Sebastian as 'three men of sin'. He causes Prospero to change his intention of taking revenge to that of forgiving his enemies.

Caliban, on the other hand is a cowardly sub-human creature, "a freckl'd whelp, hag-born" not honoured with a human shape. He is all earth, a "thing of darkness", a "dull thing", "tortoise", "poisonous slave", "filth", etc. Prospero teaches him language but he uses this language to curse Prospero.

"You taught me language and my profit on't

Is I know how to curse."

In the beginning, Prospero seems to be kind to Caliban and teaches him how to name the "bigger light and how the less" and then Caliban loves him, shows him all the "qualities of the isle." When Caliban is reduced to abject slavery by Prospero he begins to hate him. He bitterly realises that this island which was his is taken away by Prospero.

Caliban himself is confined in a hard rock and keeps away from the rest of the island after he tried to "violate the honour" of Miranda. However, he has no defence against the powerful magic of Prospero and has to make fire, fetch wood and serve in offices that profit his master. Prospero threatens him with dire punishment—"side stitches that shalt pen thy breath up", "urchins shall all exercise on thee" and "pinches as thick as honeycomb." Caliban is therefore obliged to obey Prospero because he is aware of his magical power that would control his mother's god Setebos.

Thus he too serves Prospero, but unwillingly. He longs to be free from him and this makes him conspire with Stephano against Prospero, his master. He curses Prospero and revels at having found a new master :

"A plague upon the tyrant that I serve!
I'll hear him no more sticks, but follow thee
Thou wondrous man"

He enjoys his drunken state and enjoys his new found freedom as he sings

"Ban,'ban Ca-Caliban
Has a new master–get a new man"

At the end he realises his mistake at the end when he says:

"What a thrice double–ass
Was I to take this drunkard for a god,
And worship this dull fool!"

Caliban too gets his freedom and his island back in the end. Caliban is though a sub-human yet wins our sympathy. He is obscene, but an attractive and interesting character.

Answer 4.

'The Tempest' is estimated as Shakespeare's last complete work and is interpreted by many as his will and testament. Prospero is the matured spirit of Shakespeare– a peaceful soul, seasoned to human weaknesses.

In the last Act, Prospero's project is nearly complete.

"Now does my project gather to ahead
My charms crack not, my spirit obey and time
Goes upright with his carriage."

Ariel has tortured and tormented all the courtiers with his mirages, illusions and magical powers that they are almost out of their minds. With grief and fear, they are held in a state of complete handicap. Ariel is moved to gaiety and tells Prospero :

"... if you now behold them your affections
would become tender"

Prospero is deeply moved by these words and he realises that

"Though with their high wrongs I am struck to th' quick.
Yet with my nobler reason against my fury
Do I take part. The rarer action is
In virtue than in vengeance..."

He prepares to release the prisoners but before that he surrenders his magical art and his soliloquy allows us a glimpse of the depth of his art.

"To the dread rattling thunder
Have I given fire...
"Graves at my command
Have wak'd their sleepers, op'd and let' emforth...

But this "rough magic" he decides to abjure;

"I'll break my staff,
Bury it certain fathoms in the earth,
And deeper than did ever plummet sound
I'll drown my book"

When Ariel brings his enemies into his presence, Prospero at first addresses Gonzalo as an honourable man and forgives Alonso and Sebastian and even Antonio.

"that entertain'd ambition
Expell'd remorse and nature"

Prospero gets back his dukedom and restores Ferdinand to Alonso. When he draws back the curtain of his cave we see Alonso, Ferdinand and Miranda playing chess. Alonso, overwhelmed with joy, exclaims.

"If this prove
A vision of the island, one dear son
Shall I twice lose."

Miranda is struck with amazement at the sight of so many goodly creatures (human beings)—

"... O wonder!
How many goodly creatures are there here !
How beauteous mankind, is ! O brave new world
that has such people in't !"

The boatswain is brought before this gathering by Ariel and he reports that the ship is "tight and yare" and "bravely rigg'd". His report makes confirmation that "these are not natural events, they now strengthen from strange to stranger."

Caliban and his companions are then dealt with amnesty, despite the fact that Caliban is a "demi devil" and plotted to kill Prospero. Caliban promises to be "wise hereafter" and "seek for grace." He realises his foolishness and there is hope for his realisation as well.

Prospero grants freedom to Ariel after he has brought the ships safely back to Naples. He promises to enlighten Alonso with his life's story and experiences on this island and prepares to return to the old world of Milan.

Prospero finally bids farewell to his audience, asking their forgiveness for his faults, and appreciation for his efforts to please and amuse them in the end.

●●

HINDI

2020

(Maximum Marks : 100)
(Time allowed : Three hours)
(Candidates are allowed additional 15 minutes for **only** reading the paper.
They must NOT start writing during this time).
Answer questions **1, 2** and **3** is Section **A** and **four** other questions from
Section **B** on **at least three** of the prescribed textbooks.
The intended marks for questions or parts of questions are given in brackets [ ].

QUESTIONS

SECTION – A

Language – 50 Marks

Question 1.

Write a composition in approximalely 400 words in Hindi on any ONE of the topics given below: [20]

किसी एक विषय पर निबन्ध लिखिए जो लगभग 400 शब्दों से कम न हो :

(i) *निःस्वार्थ भाव से की गई सहायता से असीम आनंद तथा संतोष प्राप्त होता है। किसी ऐसी ही एक घटना का वर्णन कीजिए जब आपने अपनी परेशानियों की परवाह किए बिना किसी जरूरतमंद व्यक्ति की मदद की थी। यह भी स्पष्ट कीजिए कि इस अनुभव से आपके जीवन पर क्या प्रभाव पड़ा?*

(ii) *"जल ही जीवन है। जल के बिना सुनहरे कल की कल्पना करना व्यर्थ है।" वर्तमान युग में जल संकट की समस्या किस प्रकार विकराल रूप लेती जा रही है? जल संरक्षण की आवश्यकता तथा इसके विभिन्न उपायों पर प्रकाश डालते हुए अपने विचार प्रस्तुत कीजिए।*

(iii) *आपके विद्यालयी जीवन का यह अन्तिम वर्ष है। आज आपका विदाई समारोह आयोजित किया गया है। इतने वर्षों का मित्रों एवं अध्यापकों का साथ छूटने वाला है। इन बीते वर्षों के न भूलने वाले खट्टे–मीठे अनुभव लिखिए।*

(iv) *"मनुष्य के नैतिक उत्थान का जिम्मेदार परिवार एवं समाज है"–विषय के पक्ष या विपक्ष में अपने विचार व्यक्त कीजिए।*

(v) *विश्व के मानचित्र पर भारत की एक नई पहचान उभर रही है, इसका कारण है "आज का जागरूक भारत" व्याख्या कीजिए।*

(vi) *निम्नलिखित में से किसी एक पर मौलिक कहानी लिखिए :*

(a) *"बीती ताहि बिसार दे आगे की सुध लेय।"*

(b) *एक मौलिक कहानी लिखिए जिसका अन्तिम वाक्य हो : और अपने घर सकुशल पहुँचने पर हमने चैन की साँस ली।*

Question 2.

Read the passage given below carefully and answer in Hindi the questions that follow, using your own words:

निम्नलिखित अवतरण को पढ़कर, अन्त में दिए गए प्रश्नों के उत्तर अपने शब्दों में लिखिए :

पुराने समय की बात है, एक गाँव में दो किसान रहते थे। दोनों ही बहुत गरीब थे, दोनों के पास थोड़ी–थोड़ी जमीन थी, दोनों उसमें ही मेहनत करके अपना और अपने परिवार का गुजारा करते थे।

अकस्मात् कुछ समय पश्चात दोनों की एक ही दिन, एक ही समय पर मृत्यु हो गयी। यमराज दोनों को एक साथ भगवान के पास ले गए। भगवान ने उन्हें देख के उनसे पूछा, "तुम्हारे इस जीवन में क्या कमी थी?" भगवान की बात सुनकर उनमें से एक किसान बड़े गुस्से से बोला, "हे भगवन्! आपने इस जन्म में मुझे बहुत घटिया जिन्दगी दी थी। आपने कुछ भी नहीं दिया था मुझे। पूरी ज़िन्दगी मैंने बैल की तरह खेतों में काम किया, जो कुछ भी कमाया वह सब पेट भरने में लगा दिया, न ही मैं कभी अच्छे कपड़े पहन पाया और न ही कभी अपने परिवार को अच्छा खाना खिला पाया। जो भी पैसे कमाता था, कोई आकर मुझसे लेकर चला जाता था और मेरे हाथ में कुछ भी नहीं आया। देखो, कैसी जानवरों जैसी ज़िन्दगी जी है मैंने।"

उसकी बात सुनकर भगवान कुछ समय मौन रहे और पुनः उस किसान से पूछा, "तो अब तुम क्या चाहते हो, इस जन्म में मैं तुम्हें क्या बनाऊँ ?"

भगवान का प्रश्न सुनकर वह किसान पुनः बोला, "भगवन्! आप कुछ ऐसा कर दीजिए कि मुझे कभी किसी को कुछ भी देना ना पड़े। मुझे तो केवल चारों तरफ से पैसा ही पैसा मिले।"

अपनी बात कहकर वह किसान चुप हो गया। भगवान ने उसकी बात सुनी और कहा, "तथास्तु! तुम अब जा सकते हो, मैं तुम्हें ऐसा ही जीवन दूँगा जैसा तुमने मुझसे माँगा है।"

उसके जाने के बाद भगवान ने दूसरे किसान से पूछा, "तुम बताओ, तुम्हारे जीवन में क्या कमी थी ?" उस किसान ने भगवान के सामने हाथ जोड़ते हुए कहा, "हे भगवन्। आपने मुझे सब कुछ दिया, मैं आपसे क्या माँगू। आपने मुझे एक अच्छा परिवार दिया, मुझे कुछ जमीन दी जिस पर मेहनत से काम करके मैंने अपने परिवार को एक अच्छा जीवन दिया। खाने के

लिए आपने मुझे और मेरे परिवार को भरपेट भोजन दिया। मैं और मेरा परिवार कभी भूखे पेट नहीं सोया। बस एक ही कमी थी मेरे जीवन में, जिसका मुझे पूरी ज़िन्दगी अफसोस रहा और आज भी है। मेरे दरवाज़े पर कभी कुछ भूखे–प्यासे लोग आते थे भोजन माँगने के लिए परन्तु कभी–कभी भोजन न होने के कारण मैं उन्हें खाना नहीं दे पाता था और वे मेरे द्वार से भूखे ही लौट जाते थे। ऐसा कहकर वह चुप हो गया।''

भगवान ने उसकी बात सुनकर उससे पूछा, ''तो अब क्या चाहते हो तुम, इस जन्म में मैं तुम्हें क्या बनाऊँ ?'' किसान ने हाथ जोड़ते हुए भगवान से विनती की, ''हे प्रभु ! आप कुछ ऐसा कर दें कि मेरे द्वार से कोई भूखा–प्यासा ना जाए।'' भगवान ने कहा, ''तथास्तु ! तुम जाओ तुम्हारे द्वार से कभी कोई भूखा–प्यासा नहीं जाएगा।''

अब दोनों का पुनः उसी गाँव में एक साथ जन्म हुआ। दोनों एक साथ बड़े हुए। पहला व्यक्ति जिसने भगवान से कहा था कि उसे चारों तरफ से केवल धन मिले उसे कभी किसी को कुछ देना ना पड़े, वह व्यक्ति उस गाँव का सबसे बड़ा भिखारी बना। अब उसे किसी को कुछ देना नहीं पड़ता था और जो कोई भी आता उसकी झोली में पैसे डालकर ही जाता था।

दूसरा व्यक्ति जिसने भगवान से कहा था कि उसे कुछ नहीं चाहिए, केवल इतना हो जाए कि उसके द्वार से कभी कोई भूखा–प्यासा न जाए, वह उस गाँव का सबसे अमीर आदमी बना।

ईश्वर ने जो दिया है उसी में संतुष्ट रहना बहुत जरूरी है। अक्सर देखा जाता है कि सभी लोगों को हमेशा दूसरों की चीज़ें ज्यादा पसंद आती हैं और इसके चक्कर में वे अपना जीवन भी अच्छे से नहीं जी पाते। हर बात के दो पहलू होते हैं–सकारात्मक और नकारात्मक, अब ये हमारी सोच पर निर्भर है कि हम चीजों को नकारात्मक रूप से देखते हैं या सकारात्मक रूप से। अच्छा जीवन जीना है, तो अपनी सोच को अच्छा बनाना होगा। चीज़ों में कमियाँ निकालने की बजाय भगवान ने जो दिया है उसका आनंद लेना और हमेशा दूसरों के प्रति सेवा भाव रखना होगा। जिस दिन हमारी सोच बदलेगी, जीवन के प्रति हमारा दृष्टिकोण भी बदल जाएगा।

(i) दोनों किसान कहाँ रहते थे? उन दोनों में क्या समानताएँ एवं क्या विषमताएँ थीं? [4]

(ii) पहले किसान को अपने जीवन से क्या शिकायत थी? वह दूसरे जन्म में क्या बनना चाहता था? [4]

(iii) दूसरे किसान ने भगवान से अपने लिए क्या माँगा और क्यों? [4]

(iv) दोनों किसानों का पुनर्जन्म किस रूप में हुआ? अब उनका जीवन कैसा था? [4]

(v) इस गद्यांश से हमें क्या शिक्षा मिलती है? [4]

Question 3.

(a) Correct the following sentences and rewrite: [5]

निम्नलिखित वाक्यों को शुद्ध करके लिखिए :

(i) श्याम तेजी से दौड़ता है।

(ii) वह मेरे शब्दों पर ध्यान नहीं देता।

(iii) उसने गीत की दो–चार लड़ियाँ गाईं।

(iv) हत्यारे को मृत्युदण्ड की सजा मिली।

(v) हम हमारे देश के लिए जान दे देंगे।

(b) Use the following idioms in sentences of your own to illustrate their meaning : [5]

निम्नलिखित मुहावरों को वाक्यों में प्रयुक्त कीजिए :

(i) हाथ तंग होना।

(ii) चुल्लू भर पानी में डूब मरना।

(iii) आसमान सिर पर उठाना।

(iv) कान भरना।

(v) इधर–उधर की हाँकना।

SECTION – B

Prescribed Textbooks – 50 Marks

*(Answer **four** questions from this Section on **at least three** of the prescribed textbooks)*

गद्य संकलन (Gadya Sanklan)

Question 4.

''जैसे भी हो, इस बार बेटू को अपने साथ लेकर ही जाना होगा। यही हाल रहा तो इसकी जिंदगी चौपट हो जाएगी। यह भी कोई ढंग है भला।''

(i) उक्त कथन कौन, किससे और किस संदर्भ में कह रहा है ? [1½]

(ii) श्रोता उक्त कथन सुनकर धर्म–संकट में क्यों था ? [3]

(iii) बेटू के आ जाने से अम्मा का जीवन किस तरह बीतता था ? [3]

(iv) 'मजबूरी' कहानी के माध्यम से कहानीकार पाठकों का ध्यान किस ओर आकृष्ट कर रही हैं? [5]

Question 5.

''म्लेच्छों ने मुझे मुलतान की लूट में पकड़ लिया। मैं उनकी कठोरता में जीवित रहकर बराबर उनका विरोध ही करती रही।'' कथन के आधार पर इरावती की व्यथा का वर्णन करते हुए उसका चरित्र–चित्रण कीजिए। [12½]

Question 6.

''गौरी एक चरित्र प्रधान कहानी है''। कहानी के आधार पर गौरी की देशभक्ति एवं त्याग का वर्णन करते हुए बताइए कि गौरी का योगदान सीताराम जी की तुलना में कहीं कम नहीं था। [12½]

काव्य मंजरी (Kavya Manjari)

Question 7.

क्या हवाएँ थीं कि उजड़ा प्यार का वह आशियाना,
कुछ ना आया काम तेरा, शोर करना गुल मचाना,
माना कि उन शक्तियों के साथ चलता जोर किसका
किन्तु ऐ निर्माण के प्रतिनिधि, तुझे होगा बताना
जो बसे हैं, वो उजड़ते हैं, प्रकृति के जड़ नियम से,
पर किसी उजड़े हुए को, फिर बसाना कब मना है ?
है अँधेरी रात पर दीवा जलाना कब मना है ?

(i) प्रस्तुत पद्यांश के कवि तथा कविता का नाम लिखिए। यह किस प्रकार की कविता है ? [1½]

(ii) 'प्यार का आशियाना' कैसे उजड़ गया? मनुष्य का शोरगुल मचाना काम क्यों नहीं आया ? [3]

(iii) *'निर्माण के प्रतिनिधि' किसे कहा गया है और क्यों ? 'प्रकृति का जड़ नियम' क्या है ? समझाइए।* **[3]**

(iv) *प्रस्तुत कविता से कवि क्या सन्देश देना चाहते हैं? समझाकर लिखिए।* **[5]**

Question 8.

'एक फूल की चाह' कविता के माध्यम से कवि सियारामशरण गुप्त जी ने छुआछूत जैसी सामाजिक कुरीति पर कुठाराघात किया है। – सिद्ध कीजिए। **[12½]**

Question 9.

"आः धरती कितना देती है" का मूल प्रतिपाद्य लिखिए। प्रस्तुत कविता द्वारा कवि ने क्या सन्देश दिया है ? **[12½]**

"सारा आकाश" (Saara Akash)

Question 10.

"तूने मुझे बचा लिया, वरना सच कहता हूँ कि पागल हो जाता। तू नहीं जानता, हमारे घर की हालत क्या है।" मेरी समझ में नहीं आ रहा था कि कैसे अपनी कृतज्ञता को व्यक्त करूँ। मेरी आँखें भर आईं।

(i) *उपन्यास तथा उपन्यासकार का नाम लिखिए। यह किस प्रकार का उपन्यास है ?* **[1½]**

(ii) *उपर्युक्त कथन का वक्ता कौन है? वक्ता किसके प्रति आभारी है और क्यों ?* **[3]**

(iii) *वक्ता ने श्रोता से कितने रुपये उधार लिए और उन रुपयों से किसके लिए क्या खरीदा? उसके बाद वक्ता जब घर पहुँचा तो घरवालों की क्या प्रतिक्रिया हुई ?* **[3]**

(iv) *श्रोता का चरित्र–चित्रण कीजिए।* **[5]**

Question 11.

"सारा आकाश" उपन्यास के आधार पर समर के बाबूजी का चरित्र–चित्रण कीजिए। **[12½]**

Question 12.

"सारा आकाश" राजेन्द्र यादव द्वारा लिखित एक उद्देश्यपूर्ण रचना है–उपन्यास के आधार पर इस कथन की व्याख्या कीजिए। **[12½]**

"आषाढ़ का एक दिन" (Aashad Ka Ek Din)

Question 13.

विलोम क्या है? एक असफल कालिदास। और कालिदास? एक असफल विलोम। हम कहीं एक–दूसरे के बहुत निकट पड़ते हैं।

(i) *वक्ता और श्रोता का परिचय दीजिए।* **[1½]**

(ii) *प्रस्तुत संवाद का प्रसंग स्पष्ट कीजिए।* **[3]**

(iii) *उपर्युक्त पंक्तियों के आधार पर वक्ता का दृष्टिकोण स्पष्ट कीजिए।* **[3]**

(iv) *उपर्युक्त संवाद के आधार पर बताइए कि विलोम और कालिदास के बीच कैसे संबंध थे ?* **[5]**

Question 14.

"अम्बिका भावनाओं में नहीं यथार्थ में जीती है।" आषाढ़ का एक दिन नाटक के आधार पर अम्बिका की चारित्रिक विशेषताओं का वर्णन कीजिए। **[12½]**

Question 15.

प्रियंगुमंजरी मल्लिका को अपने साथ चलने के लिए क्यों कहती है? मल्लिका की इस पर क्या प्रतिक्रिया थी ? **[12½]**

ANSWERS

Answer 1.

(i) **निःस्वार्थ सेवा**

सच है कि निःस्वार्थ भाव से की गई सहायता से असीम आनन्द तथा संतोष प्राप्त होता है। उत्कृष्ट धर्म है मानव सेवा अथवा मानवता धर्म।

गोस्वामी तुलसीदास ने रामचरितमानस में लिखा है–

"परहित सरिस धर्म नहिं भाई,
पर पीड़ा सम नहिं अधमाई।"

निःस्वार्थ भाव से सेवा करना। अपने हित की चिंता न करते हुए जब व्यक्ति दूसरों की सहायता करता है तो उसे आत्मिक शांति मिलती है और हृदय निर्मल हो जाता है।

आज भी जब मैं उस घटना को याद करता हूँ तो दिल को सुकून मिलता है। बात उन दिनों की है, जब मैं दसवीं कक्षा में था। हमारे परिवार में हम चार लोग थे। माताजी, पिताजी, मेरी बड़ी बहन आँचल और मैं। घर का काम–काज करने के लिए लीला बाई आती थीं। मुझे प्रति माह 500 रुपए पॉकेट मनी के रूप में मिलते थे। जीवन नियमित क्रम से चल रहा था। माता–पिता दोनों आई–टी कम्पनियों में कार्यरत थे।

स्कूल से आते ही माँ को आवाज़ न लगाकर लीला बाई को ही आवाज़ लगाता। वे मेरा अच्छा ख्याल रखती थीं। यह साल हमारे लिए बहुत मुख्य था। बोर्ड की परीक्षा की तैयारी करनी थी। मेरी आँखों पर मोटा चश्मा लगा था। एक बार यह चश्मा टूट गया। माँ ने पैसे भी दिए, नया चश्मा लाने के लिए। संयोग से लीला बाई का लड़का भी दसवीं कक्षा में था। उस दिन लीला बाई बहुत उदास थीं। मेरे पूछने पर उन्होंने बताया कि उनके लड़के को स्कूल से निकाल दिया गया था। पिछले तीन महीनों से वह फीस नहीं भर पाई थीं। सारा पैसा बीमार पति की देखभाल में लग जाता है।

उनकी बातें सुनकर मुझे बहुत बुरा लगा। वह माँ से रुपए माँगने में हिचकिचा रही थीं। पहले ही 20 हजार रुपए ले चुकी थीं। मैंने जल्दी से अपना पर्स टटोला। पाया कि उसमें नया चश्मा लाने के लिए जो पैसे मिले थे, वे वहीं पर थे। चश्मा लाना भी जरूरी था पर मैंने लीला बाई की मदद करने का निश्चय किया। मैंने उस टूटे चश्मे को जैसे–तैसे जोड़कर पहनना शुरू किया। माँ ने जो पैसे दिए और मेरी पॉकेट मनी से मैंने लीलाबाई के बेटे की फीस भरी। वह खुशी–खुशी स्कूल जाने लगा।

भले ही यह बात मैंने अपने पिताजी तथा माँ से छुपाई पर बोर्ड की परीक्षाओं के बाद सारी बात बताई, तो माँ खुश हो गईं और उज्ज्वल भविष्य के लिए आशीर्वाद भी दिया। उन्होंने मुझसे कहा कि दूसरों की सहायता करने से जैसा सुकून मिलता है, वैसी खुशी किसी और से नहीं। माँ ने ही मुझे नया चश्मा ला कर दिया। मैंने भी बात छिपाने की गलती के लिए माफी माँगी।

मित्रों, इस तरह लीला बाई की मदद करके मुझे असीम आनन्द की अनुभूति हुई, जीवन की सार्थकता का अनुभव हुआ। जीवन की सन्तुष्टि इसी में है। हम अपने छोटे–छोटे सुकृत्यों से दूसरों की मदद कर सकते हैं। आज मैं कक्षा बारह का छात्र हूँ। अपनी पढ़ाई के साथ–साथ चार गरीब बच्चों को पढ़ाता हूँ और उनकी फीस का इंतजाम भी करता हूँ। मेरे इस कार्य से प्रेरित होकर मेरा मित्र भी इस कार्य में मेरा सहयोग देता है।

सच है कि निस्वार्थ सेवा ही सच्ची सेवा है, उसके सामने सब सुख फीके हैं। परोपकार का कार्य करने के बाद जिस आनंद की अनुभूति होती है उस विषय में कवि रहीम ने ठीक ही लिखा है–

"यों रहीम सुख होत है उपकारी के संग।
बाटन वारे को लगे ज्यों मेंहदी को रंग।।"

(ii) जल ही जीवन है

"जल है तो कल है
जीवन का ये शुभ मंगल है।"

सम्पूर्ण सृष्टि का आधार जल है। हमारी धरती पर कुल क्षेत्रफल का 71% जल है। जल के कारण ही धरती पर जीवन सम्भव हो पाया है। समस्त प्राणियों व प्रकृति के लिए जल अमृत से कम नहीं है। मानव शरीर में लगभग 70% जल पाया जाता है। मनुष्य भोजन के बिना कई महीनों तक जीवित रह सकता है, लेकिन जल के बिना मात्र सात–आठ दिनों में उसकी मृत्यु हो जायेगी। जैसे साँस लेना जरूरी है, वैसे ही जीने के लिए जल जरूरी है।

वैसे तो धरती पर पीने लायक जल केवल दो प्रतिशत है। यह जल हमें नदियों, तालाबों और वर्षा के पानी से उपलब्ध होता है। भूमिगत जल भी पेयजल का एक प्रमुख स्रोत है। हमारी पृथ्वी पर सभी संसाधन सीमित मात्रा में हैं। वैसे ही जल भी सीमित मात्रा में ही उपलब्ध है।

दुर्भाग्य यह है कि आज जिस तरह से जल का दुरुपयोग हो रहा है, अगर ऐसा ही चलता रहा तो यह जल संकट की समस्या विकराल रूप धारण कर लेगी। आने वाली पीढ़ी को जल मिलेगा या नहीं, यह भी निश्चित रूप से नहीं कहा जा सकता है।

जल की सही कीमत वही लोग जानते हैं जो कि कई किलोमीटर दूर से पानी का एक घड़ा लेकर आते हैं। हमारे देश के कुछ हिस्से ऐसे हैं, जहाँ बरसों से सूखा पड़ा है और समस्त प्राणी प्रदूषित पानी पीने को मजबूर हैं।

हमारे देश में पेड़ों की भी अंधा–धुंध कटाई की जा रही है। जिसका प्रभाव वर्षा पर पड़ा है। वृक्ष विहीन क्षेत्रों में वर्षा की कमी के कारण जल का प्राकृतिक रूप से भंडारण नहीं हो पाता और जल आपूर्ति में कमी रहती है।

इसके साथ ही अनावश्यक रूप से बढ़ती जनसंख्या भी इसका एक कारण है। अधिक जनसंख्या के लिये अधिक संसाधन चाहिये और जल एक सीमित संसाधन है। जनसंख्या वृद्धि के अनुपात में बढ़ता शहरीकरण भी जल स्त्रोतों को खत्म कर रहा है। वृक्ष काटे जा रहे हैं, कुँओं–तालाबों को पाटा जा रहा है। शहरों में लगाई जा रही नई–नई फैक्ट्रियाँ जल स्त्रोतों को दूषित कर रही हैं। अत्यधिक दोहन के कारण पेयजल का अभाव होता जा रहा है। महानगरों में बढ़ती आबादी और जल के सदुपयोग पर ध्यान न देने के कारण मुँबई, चेन्नई जैसे महानगर आज जल संकट के मुहाने पर खड़े हैं। गुजरात और राजस्थान में भी स्थिति बहुत खराब है।

जल हमारी पृथ्वी की अमूल्य धरोहर है, अतः हर एक व्यक्ति को जागरुक होकर जल को बचाना होगा। हमें ऐसे कदम उठाने होंगे, जिनसे भूमिगत जल की मात्रा भी बढ़े। हमें अनावश्यक कार्यों में जल का उपयोग बंद करना होगा। वर्षा के जल को टंकी बना संग्रहित करना होगा। पेड़ों की कटाई पर रोक लगानी होगी, अधिक मात्रा में पेड़–पौधों लगाने होंगे। उद्योग–धंधों व कारखानों का दूषित जल नदियों में मिलने से रोकना होगा। हमें भी नदियों व तालाबों में कचरा नहीं फेंकना चाहिए। प्लास्टिक का उपयोग पूर्णतया बंद करना होगा। पूरे देश की नदियों को अगर जोड़ दिया जाए तो देश के किसी भी हिस्से में जल की कमी नहीं होगी। सरकार भी इस दिशा में प्रयास कर रही है।

जनता में जागरूकता फैलानी होगी कि जल संरक्षण कितना आवश्यक है और इसका क्या महत्व है। तीसरा विश्वयुद्ध संभवतः जल के लिए ही लड़ा जायेगा। यह मानव सभ्यता के लिए एक गम्भीर विषय है।

जल संरक्षण की जिम्मेदारी हर नागरिक की है। हमें चाहिये कि जल के महत्व को समझकर जल संरक्षण का संकल्प लें। कवि ने ठीक ही लिखा है–

"जल बचाने का करो जतन
जीवन का है ये बहुमूल्य रत्न"
"हर बच्चा, बूढ़ा और जवान
पानी को बचाकर बने महान"

(iii) विद्यालय का विदाई समारोह

छात्र जीवन किसी भी मनुष्य के लिये बहुत यादगार और महत्वपूर्ण समय होता है। छात्र विद्यालय में ज्ञान प्राप्ति के लिये जाता है और इस पूरे काल में वह जो कुछ भी सीखता है वही भविष्य में उसका मार्गदर्शन करता है।

मेरा विद्यालय में कक्षा नर्सरी से कक्षा बारह तक का सफर बहुत ही रोमांचक रहा। अध्यापकों के मार्गदर्शन में, विद्यालय प्रबंधन के सहयोग और मित्रों के साथ में बहुत कुछ सीखा–समझा और जाना। केवल पढ़ाई ही नहीं बल्कि अन्य क्षेत्रों में भी नई–नई चीजें सीखीं।

शुरूआत के दिनों में विद्यालय आने के नाम से ही रोना, माँ का मनाना सब आज याद आ रहा है। लेकिन कुछ समय बाद मित्रों और अध्यापकों के साथ इतना अपनापन हो गया कि विद्यालय घर जैसा लगने लगा। पढ़ाई और शरारतें करते–करते कब कक्षा बारह की विदाई समारोह का दिन आ गया पता ही नहीं चला। कक्षा ग्यारह के विद्यार्थियों ने हमें स्वयं के बनाये हुये निमंत्रण पत्रों के द्वारा विदाई समारोह में आने का निमंत्रण दिया। हम कक्षा बारह के विद्यार्थियों ने निश्चय किया कि हम भी उन सबके लिये कुछ उपहार लेकर जायेंगे ताकि हमारे जाने के बाद भी वे हमें याद करते रहें।

समारोह वाले दिन सभी लोग बहुत उत्साहित थे। हमारे प्रधानाचार्य जी ने कार्यक्रम की शुरूआत करते हुये हम सबको अग्रिम भविष्य के लिये शुभकामनायें दीं। उसके बाद अन्य कक्षाओं के छात्रों ने छोटे–छोटे कार्यक्रम प्रस्तुत किये जिसमें नृत्य–गीत और अन्य चीजें शामिल थीं।

हम सबसे विद्यालय से जुड़े कुछ यादगार किस्से सुनाने को कहा गया। मेरी बारी आने पर मैंने बताया कि कैसे एक बार मैं

और मेरे दोस्त विद्यालय की छुट्टी की घण्टी बजाकर भाग गये थे। अध्यापक बहुत पता लगाने पर भी जान नहीं पाये थे कि ये किसने किया। सभी अध्यापक यह जानकर बहुत हँसे।

अन्त तक सभी बहुत भावुक भी हो गये। अध्यापकों से अपनी शरारतों के लिये क्षमा याचना भी की। अध्यापकों ने भी आशीर्वाद देकर उज्जवल भविष्य की शुभकामनायें दी।

सभी सहपाठियों से भी भविष्य में मिलते रहने के वादे किये। साथ छूटने का गम सभी को था परंतु नये माहौल में जाने के लिये उत्साहित भी थे।

विदाई समारोह का यह कार्यक्रम मुझे आजीवन स्मरण रहेगा। सभी लोगों ने उन खूबसूरत पलों को तस्वीरों में कैद करके अपने पास रख लिया है। ताकि कभी–कभी उन्हें देखकर ये सुनहरे पल फिर से याद कर सकें।

मेरे इतने वर्षों के अनुभव को शब्दों में समेटना मुश्किल ही नहीं असम्भव है। मैं दिल से हर उस अध्यापक को कृतज्ञता ज्ञापित करता हूँ, जिन के अथक श्रम के कारण मैं अपने अनुभव साझा कर पाया हूँ।

(iv) नैतिक उत्थान में परिवार व समाज की भूमिका

सामान्य तौर पर नैतिकता का अर्थ हम सत्य, विनय, सेवा, आज्ञापालन, सहयोग, सहनशीलता, ईमानदारी, सम्मान, उत्तरदायित्व आदि गुणों से लगाते हैं। इन गुणों को अपनाकर व्यक्ति चाहे तो धरती पर स्वर्ग ला सकता है और इसके अभाव के कारण धरती पर ही नरक बना सकता है।

वर्तमान समय में इन मूल्यों का ह्रास होता जा रहा है व मनुष्य का नैतिक उत्थान की जगह नैतिक पतन होता जा रहा है। इस कारण मानव मनुष्यत्व से पशुत्व की ओर जाता जा रहा है। इसका कारण संयुक्त परिवार प्रणाली का समाप्त होना, आधुनिकता व अति स्वच्छंदता आदि है। अतः नैतिक उत्थान की बहुत आवश्यकता है।

परिवार वह पाठशाला है जहाँ बच्चा बड़ों के लालन–पालन के द्वारा बहुत कुछ सीखता है।

प – पवित्रता का विकास

रि – रिपु–शत्रुओं से सुरक्षा

वा – वात्सल्य भाव का पोषण

र – रचनात्मकता का सृजन

गीली मिट्टी को जिस आकार में ढालो वह ढल जाती है, परिवार वह संस्था है जो बच्चों में नैतिक गुणों को बचपन से ही ढालना शुरू कर देती है, क्योंकि यही सही उम्र होती है।

परिवार में कई सदस्य जैसे–दादा–दादी, माता–पिता, चाचा–चाची व चचेरे भाई–बहन आदि होते हैं। सभी सदस्य बालक का उत्थान करने में उस माली की तरह भूमिका अदा करते हैं जो अपनी बगिया के फूलों की देखभाल करता है। प्रत्येक परिवार के कुछ नियम होते हैं इसी कारण बच्चे दायरे में रहना सीखते हैं। परिवार में अलग–अलग रिश्ते समय पड़ने पर एक–दूसरे के काम आते हैं। इस कारण बच्चे सहयोग व जिम्मेदारी सीखते हैं। भाई–बहनों के साथ रहने से वह चीजों को बाँटने व मिल–जुलकर रहना सीखते हैं। बड़ों के साथ रहने से विनय व विवेक का गुण सीखते हैं। परिवार में रहने से बच्चा हर बात पर अपनी मनमानी नहीं कर सकता। अतः उसमें सहनशीलता का गुण आता है।

अतः यह कह सकते हैं कि परिवार एक मनुष्य के नैतिक उत्थान में बहुत बड़ा सहयोग देता है।

इसी तरह समाज भी मनुष्य के नैतिक उत्थान में अहम भूमिका अदा करता है। व्यक्ति से घर, घर से परिवार तथा परिवारों का समूह ही समाज कहलाता है।

व्यक्ति एक सामाजिक प्राणी है। उसका समाज के साथ सम्बन्ध अटूट होता है। समाज में भी देखा जाता है कि एक व्यक्ति, दूसरे व्यक्ति को व एक परिवार, दूसरे परिवार का सुख–दुख के समय साथ देते हैं। इसी तरह आपसी सहयोग व प्रेम की भावना का विकास होता है।

यदि समाज में एक व्यक्ति सक्षम है और दूसरा अक्षम है तो वह उसकी मदद करता है जिससे दान व सेवा की भावना मनुष्य में जागृत होती है।

प्रत्येक समाज के कुछ नियम होते हैं, समाज में रहने से व नियमों के अनुरूप ढालने से ही मनुष्य का नैतिक उत्थान हो सकता है। इसी प्रकार विद्यालय भी नैतिक मूल्यों को पल्लवित करने में महत्वपूर्ण भूमिका निभाता है। शिक्षक उन्हें नैतिकता की सीख देते हैं और स्वयं अपने नैतिकतापूर्ण क्रियाकलापों से उनके समक्ष आदर्श प्रस्तुत करते हैं।

अतः कहा जा सकता है कि परिवार, विद्यालय या समाज मनुष्य के नैतिक उत्थान के लिये जिम्मेदार होते हैं।

(v) आज का जागरुक भारत

भारत देश कृषि प्रधान देश है। अनेकता में एकता इसकी शान है। यहाँ की संस्कृति यहाँ के प्राण हैं। यहाँ की संस्कृति एवं परम्पराएँ पूरे विश्व में विख्यात हैं।

यहाँ विविध जाति, संप्रदाय व धर्म के लोग भाईचारे के साथ रहते हैं। प्राचीन काल में भारत को सोने की चिड़िया कहा जाता था। आधुनिक भारत विकासशील देशों में उन्नति के पथ पर अग्रसर सबसे पहला देश है। आज का भारत जागरुक है। आज प्रत्येक नागरिक देश के विकास के लिये प्रयत्नशील है।

''स्वच्छ भारत अभियान'' जो 2 अक्टूबर 2014 को देश के प्रधानमंत्री ने आरम्भ किया था। देश इस अभियान को सफल बनाने के लिए एकजुट है। जनता की जागरूकता प्रशंसनीय है। गरीबी व निरक्षरता को जड़ से मिटाने के लिए सरकार प्रयासरत है। कृषि और खाद्य प्रसंस्करण, शिक्षा व स्वास्थ्य सुरक्षा, सूचना व संचार तकनीक, भरोसेमंद इलेक्ट्रॉनिक पावर, औद्योगिकी में बढ़ते कदमों ने देश की आर्थिक स्थिति को मजबूत बनाया है।

भारत ने विकासशील देशों की आवाज़ बनने की कोशिश भी की है। आज जिस गति से भारत विकास पथ पर दौड़ रहा है, उससे भारत की बड़ी आर्थिक ताकत बनने की उम्मीद है। जलवायु परिवर्तन की समस्या से निपटने के लिए भी भारतवासियों को जागरूक किया जा रहा है।

''स्वच्छ भारत, स्वस्थ भारत'' के नारे के साथ नए भारत के निर्माण का आगाज़ हुआ है। वर्तमान में योग को शारीरिक, मानसिक व आत्मिक शांति के लिए पूरे विश्व में बड़े पैमाने पर अपनाया जा रहा है। 21 जून 2015 को प्रथम अन्तर्राष्ट्रीय योग दिवस मनाया गया। प्रथम बार विश्व योग दिवस के अवसर पर 1920 देशों में योग का आयोजन किया गया।

बढ़ते प्लास्टिक प्रदूषण ने हमारे पर्यावरण को काफी क्षति पहुँचाई है। इस ओर जनता को जागरूक करने के लिए स्वतन्त्रता दिवस पर प्रधानमंत्री नरेन्द्र मोदी ने प्लास्टिक मुक्त भारत अभियान का शुभारम्भ किया।

इस तरह से समय–समय पर देश व विश्व की उन्नति, अमन व शांति बनाए रखने के लिए भारत ने सशक्त व ठोस कदम उठाकर भारत के लोगों को ही नहीं बल्कि पूरे विश्व को जागरूक करने में अहम भूमिका निभाई है। पर्यावरण व जल संरक्षण व महिला सुरक्षा व साक्षरता के लिए उठाये गये कदम प्रशंसनीय है। अपने इन्हीं महत्वपूर्ण कार्यों व अभियानों द्वारा विश्व के मानचित्र पर भारत की नई पहचान उभर रही है।

इकबाल की पंक्तियाँ :

"सारे जहाँ से अच्छा, हिन्दोस्ताँ हमारा।
हम बुलबुले हैं इसकी ये गुलस्ताँ हमारा।

(vi) (a) "बीती ताहि बिसार दे, आगे की सुधि लेय" इस लोकोक्ति का अर्थ है मनुष्य को बीती बातों को भुला कर आगे बढ़ने की बात सोचनी चाहिए। रुका हुआ पानी गन्दगी से भरकर कीचड़ हो जाता है और बहता हुआ पानी निर्मल व स्वच्छ रहता है। ठीक उसी प्रकार मनुष्य को अपने भूतकाल की दुखद घटनाओं, हानियों व असफलता के बारे में सोचकर व्यथित नहीं होना चाहिए। निराशा से आशा की ओर कदम बढ़ाने होंगे। आगे क्या करना है यह सोचना होगा।

रोहन और मोहन अच्छे मित्र थे। एक साथ पढ़ते व खेलते थे। रोहन पढ़ाई में विशेष रुचि लेता था। वह नियमित समय पर गृहकार्य पूरा करता था। मोहन, रोहन से अलग था। पढ़ने से ज्यादा खेलने में समय व्यतीत करता था। मोहन के समझाने पर भी वह उसकी एक नहीं सुनता। धीरे–धीरे मोहन पढ़ाई में पीछे रहने लगा और परिणामतः वार्षिक परीक्षा में फेल हो गया।

रोहन आगे की पढ़ाई करने शहर चला गया और एक सफल डॉक्टर बना। मोहन के साथ विपरीत होने लगा। फेल होने के बाद मोहन गलत संगति में पड़ गया। पिता की सारी ज़ायदाद भी बिक गई, उसकी हालत किसी भिखारी से कम न थी।

कुछ सालों बाद रोहन गाँव वापस आया। गांव आकर वह मोहन की दयनीय स्थिति को देखकर काँप गया।

मोहन की ऐसी हालत देखकर वह बहुत दुखी होता है। मोहन को अपने साथ शहर ले जाता है।

रोहन उसे समझाता है कि वह बीती बातों को भूल जाए और एक नये जीवन की शुरूआत करे। रोहन उसे एक कम्पनी में सफाई कर्मचारी के रूप में लगा देता है। वह उसे समझाता है कि अब आगे उसे क्या करना है, अपना जीवन बेहतर कैसे बनाना है, इस बारे में गम्भीरता से सोचे। वह हमेशा उसके साथ है। जिंदगी आगे बढ़ने का नाम है, रुकने का नहीं। इसलिए ठीक ही कहा है–"बीती ताहि बिसार दे, आगे की सुध लेय"।

(b) पूरा परिवार उत्साहित था कि इस बार होली एक अलग ही ढंग से मनाई जायेगी। रंगों का त्यौहार होली का नाम लेते ही मन में अजीब सी स्फूर्ति आ जाती है। इस बार यह तय किया गया कि होली वाटर पार्क में मनाई जायेगी। बस एक ही तकलीफ थी कि ये वाटर पार्क घर से करीब 60 किमी. की दूरी पर था। हम कुल तेरह सदस्य थे। मैं, पिंकी, माताजी, पिताजी, चाचा, चाची, रोहन, बुआ और उनका परिवार।

इस बार होली सोमवार की थी। इसलिए मज़ा भी दुगुना था। रविवार और सोमवार दोनों ही दिन स्कूल की छुट्टी थी। पिताजी ने एक बड़ी गाड़ी बुक की थी। सब सुबह छः बजे निकल पड़े। माँ ने तो चार दिन पहले ही तैयारी शुरू कर दी। खाने का भी थोड़ा बहुत सामान साथ लिया।

यात्रा बड़ी सुखद व रोमांचक थी। बेसब्री से वॉटर पार्क पहुँचने का इंतजार हो रहा था। गाड़ी में हमने अंताक्षरी भी खेली। वॉटर पार्क पहुँचने से कुछ ही दूर पहले गाड़ी रुक गई। हमने सोचा ट्रैफिक जाम हुआ होगा। हमारे आगे दो और कारें खड़ी थीं।

अचानक हमने देखा कि कुछ लोग मुखौटा पहने हमारी गाड़ी की तरफ बढ़ रहे हैं। ड्राइवर के पास पिताजी बैठे थे। जैसे ही पिताजी ने उन्हें देखा, उन्होंने ड्राइवर से गाड़ी के शीशे बंद करने को कहा और कहा कि गाड़ी को पीछे की तरफ ले जायें। हमारी खुशनसीबी थी कि ड्राइवर भी सतर्क व सजग था। गाड़ी में बैठे सब लोग अत्यधिक भयभीत थे। सबकी साँसें तेज चल रही थीं। पता नहीं आगे क्या होगा? क्या आज हम बच पायेंगे या फिर ये खतरनाक लुटेरे हमें मार देंगे, कुछ नहीं कहा जा सकता था।

जैसे ही गाड़ी पीछे की तरफ चलने लगी, वे हमारी तरफ भागने लगे और गाड़ी को रोकने के लिए धमकाने लगे। ड्राइवर ने बड़े संयम से काम लिया। वह तेज़ गति से गाड़ी को पीछे ले गया और सही वक्त पर गाड़ी को मोड़कर रास्ता बदल दिया। वे भी कुछ देर तक हमारे पीछे भागे पर कुछ दूरी के बाद दिखाई नहीं दिये।

हम सबने भगवान को धन्यवाद दिया पर एक अनजाना सा डर मन में समा गया था। पिताजी ने नाश्ते के लिए गाड़ी को रोकने के लिए कहा तो हम सबने एक साथ मना कर दिया। हम सबकी हालत बहुत खराब थी। सब मौन थे। कोई कुछ नहीं बोल रहा था। सब जैसे–तैसे घर पहुँचना चाहते थे। कुछ ही देर में हम घर पहुँचने वाले थे। होली के त्यौहार का ये दिन हमें जिन्दगी भर याद रहेगा। गाड़ी घर के दरवाजे के बाहर रुकी और अपने घर सकुशल पहुँचने पर हमने चैन की साँस ली।

Answer 2.

(i) दोनों किसान गाँव में रहते थे। दोनों ही बहुत गरीब थे, दोनों के पास थोड़ी–थोड़ी ज़मीन थी, दोनों उसमें ही मेहनत करके अपना और अपने परिवार का गुज़ारा करते थे। दोनों की सोच में बहुत विषमताएँ थीं। एक किसान लालची और स्वार्थी था उसका दृष्टिकोण नकारात्मक था। जबकि दूसरे किसान का जीवन के प्रति रवैया सकारात्मक था। वह अपने जीवन से संतुष्ट एवं खुश था। उसमें परमार्थ की भावना थी।

(ii) पहले किसान का कहना था कि उसे बहुत घटिया ज़िन्दगी मिली थी। भगवान ने उसे कुछ नहीं दिया। पूरी ज़िन्दगी वह बैलों की तरह खेतों में काम करता रहा। जितना कमाता वह पेट भरने में लग जाता था। न अच्छे कपड़े पहने और न ही कभी अपने परिवार को अच्छा खाना खिला पाया। जो भी कमाता, कोई दूसरा उससे ले जाता था। उसकी ज़िन्दगी जानवरों के समान थी। वह भगवान से कुछ ऐसा करने की प्रार्थना करता है जिससे उसे कभी किसी को कुछ भी देना न पड़े। चारों तरफ से पैसा ही पैसा मिले।

(iii) दूसरे किसान ने भगवान से विनती की कि उसके द्वार से कोई भूखा–प्यासा न जाये।

उसे इस बात का पूरी ज़िन्दगी अफसोस रहा कि उसके दरवाज़े पर कभी कुछ भूखे और प्यासे लोग आते थे ,और भोजन माँगते तो, कभी–कभी भोजन न होने के कारण वह उन्हें खाना नहीं दे पाता था और वे उसके द्वार से भूखे ही लौट जाते थे। इसलिए उसने भगवान से विनती की कि उसके द्वार से कोई भूखा–प्यासा न जाये।

(iv) दोनों किसानों का पुनः उसी गाँव में एक साथ जन्म हुआ। दोनों एक साथ बड़े हुए। पहला किसान/व्यक्ति जिसने यह माँग की कि उसे चारों तरफ से पैसे मिलें, उसे कभी किसी को कुछ देना न पड़े, वह व्यक्ति उस गाँव का सबसे बड़ा भिखारी बना। जो कोई भी आता, वह उसकी झोली में पैसा डालकर ही जाता था। अब उसे किसी को कुछ देना नहीं पड़ता था। दूसरा व्यक्ति जिसने भगवान से यह प्रार्थना की कि उसके द्वार से कभी कोई भूखा–प्यासा न जाए, वह उस गाँव का सबसे अमीर आदमी बना। वह नहीं चाहता था कि उसके द्वार से कोई भूखा–प्यासा जाये। अमीर आदमी बनकर वह अपने परमार्थ वृत्ति को पूरा कर पाया।

(v) यह एक शिक्षाप्रद गद्यांश है। ईश्वर ने जो दिया है उसी में सन्तुष्ट रहना बहुत जरूरी है। अक्सर देखा जाता है कि सभी लोगों को हमेशा दूसरों की चीजें पसन्द आती हैं और इसके चक्कर में वे अपना जीवन भी अच्छी तरह से नहीं जी पाते। हर बात के दो पहलू होते हैं– सकारात्मक और नकारात्मक। अच्छा जीवन जीना है तो, अपनी सोच को अच्छा बनाना होगा, चीज़ों में कमियाँ निकालने की बजाय भगवान ने जो दिया है उसका आनन्द लेना और हमेशा दूसरों के प्रति सेवा–भाव रखना होगा। जिस दिन हमारी सोच बदलेगी, जीवन के प्रति हमारा दृष्टिकोण भी बदल जायेगा।

Answer 3.

(a) **(i)** श्याम तेज़ दौड़ता है।

(ii) वह मेरी बातों पर ध्यान नहीं देता।

(iii) उसने गीत की दो–चार कड़ियाँ गाईं।

(iv) हत्यारे को मृत्युदण्ड मिला।

(v) हम अपने देश के लिए जान दे देंगे।

(b) **(i)** नौकरी छूट जाने के कारण रमेश का हाथ तंग हो गया था।

(ii) राम चोरी करते हुए पकड़ा गया तो उसकी हालत ऐसी हो गई, जैसे वह चुल्लू भर पानी में डूब मरेगा।

(iii) अध्यापक के कक्षा में न आने पर बच्चों ने आसमान सिर पर उठा लिया।

(iv) वह अपने साथियों द्वारा कान भरने पर अपने ही भाई से लड़ बैठा।

(v) रमेश के इधर–उधर की हाँकने की आदत के कारण उस पर कोई विश्वास नहीं करता था।

SECTION – B

Answer 4.

(i) उक्त कथन रमा ने अपने पति रामेश्वर से कहा। दूसरी बार जब रमा गाँव आती है तो बेटू के चाल–ढाल और व्यवहार को देखकर चिन्तित हो जाती है उसे लगता है कि अम्मा का लाड़ प्यार बेटू में अच्छी आदतें नहीं बनने दे रहा इसलिये वह बेटू को साथ ले जाने की बात कहती है।

(ii) श्रोता रामेश्वर इस कथन को सुनकर धर्मसंकट में था क्योंकि अगर वे बेटू को अपने साथ ले गए तो अम्मा बहुत दुखी होगीं। बेटू को भी अम्मा से बहुत लगाव था। अम्मा को छोड़कर बेटू का रहना मुश्किल था। दूसरी तरफ पत्नी का कहना भी सही था कि बेटू को शहर ले जाकर उसकी ठीक से पढाई लिखाई कराई जाये।

अम्मा की भावनाओं का ख्याल और बेटू के भविष्य से समझौता ना कर पाने के कारण वह धर्मसंकट में था इसलिये सारी बात रमा पर छोड़कर वह अकेले बम्बई चला जाता है।

(iii) बेटू के आ जाने से अम्मा को एक नया जीवन मिलता है। वह पोते में अपने बेटे का बचपन देखती है। अम्मा का सारा दिन बेटू को खिलाने–पिलाने और देख–भाल में ही बीत जाता। वह सारा दिन बेटू को गोद में लिए घूमती, उसके साथ खेलती और अपने पास ही सुलाती। बेटू के आने से अम्मा अपने गठिया के दर्द को भी भूल जाती है। वह उसे लोरियाँ और कहानियाँ भी सुनाती। उसकी हर इच्छा पूरी करती। अपने पोते के लिए घड़ी में समय देखना तथा बोतल में दूध भरना सीखती है। बेटू के आने से ऐसा लगता है मानो पतझड़ भरे जीवन में बहार आ गई हो।

(iv) 'मजबूरी– कहानी के द्वारा कहानीकार मन्नू भंडारी ने नारी मन की कोमल भावनाओं, बुढ़ापे में आने वाली अकेलेपन की समस्या का मर्मस्पर्शी चित्रण किया है। जीवन की परिस्थितियों में एक स्त्री के सामने किस प्रकार की मजबूरियाँ आती हैं इसका संवेदनशील वर्णन किया गया है। वात्सल्यमयी माँ की ममता हृदयस्पर्शी है। कहानी में अम्मा अपने बेटे व पोते से बहुत प्रेम करती है। एक और संतान का जन्म होने वाला था, इसलिए रमा और रामेश्वर अपने पहले बेटे बेटू को अम्मा के पास छोड़ देते हैं। अम्मा बेटू में अपने पुत्र का बचपन देखती है। थोड़ा बड़ा होने पर बेटू भी मुम्बई चला जाता है और अपनी दादी को भूल जाता है। इस तरह कहानीकार ने 'मजबूरी'–कहानी के माध्यम से सभी पात्रों की मजबूरियों को दिखाने की कोशिश की है।

Answer 5.

हिन्दू कन्या इरावती को मुल्तान की लूट में पकड़ लिया गया और दासी बना दिया गया। मलेच्छों के कठोर और निर्दयी व्यवहार के बीच भी वह डरी नहीं न ही कमजोर पड़ी। काशी के एक महाजन के हाथों 500 दिरहम में उसे बेच दिया जाता है। इरावती स्त्री जाति की प्रतिष्ठा, सम्मान व गौरव पर आँच नहीं आने देती। बलराज के प्रेम–प्रदर्शन पर वह उसे बताती है कि वह अपवित्र हो चुकी है। इस तरह वह स्पष्टवादिता का परिचय देती है। महाजन के कहे अनुसार वह हर कुत्सित कार्य करती है। दासी के जीवन के कष्टों को सहनशीलता से सहती है। इरावती, बलराज और मालिक दोनों के प्रति ईमानदार है। इरावती अपने आत्मसम्मान के प्रति भी सजग है। वह बलराज के व्यवहार से दुखी होती है। वह बलराज की पत्नी बनने का प्रस्ताव अस्वीकार कर देती है। वह कहती है कि उसने सेठ को लिखकर दिया है कि वह उसके घर का कुत्सित से कुत्सित कर्म करेगी और कभी भी विद्रोह नहीं करेगी और न ही भागने की चेष्टा करेगी। उसने सिर पर तृण रखकर अपने को बेचने

की स्वीकृति दी है। इरावती स्पष्टवादी व वचनबद्ध और मालिक के लिये ईमानदार है। वह बलराज से कुछ नहीं छुपाती है। युद्ध में घायल होने पर वह बलराज को जल पिलाती है। तिलक भी अपनी बहन इरावती को पहचान लेता है। अन्त में बलराज जाटों का सरदार बनता है और इरावती उसकी रानी। संघर्ष और दुखों से भरे इरावती के जीवन में आखिरकार खुशियाँ दस्तक दे ही देती हैं।

Answer 6.

गौरी एक चरित्र प्रधान भावनात्मक कहानी है। गौरी एक सशक्त नारी पात्र है। गौरी, बाबू राधाकृष्णन की इकलौती उन्नीस वर्षीय सुन्दर कन्या थी, विवाह योग्य होने पर माता–पिता को उसके विवाह की चिंता हुई। दृढ़ निश्चयी व निर्भीक गौरी लज्जा और संकोच के कारण अपने माता–पिता से यह नहीं कह पाई कि वे उसका विवाह किसी से भी करवा दे। वह हर हाल में सुखी रहेगी।

विदुर सीताराम अपने दो छोटे बच्चों के साथ उसे देखने आते हैं। वे एक सच्चे, ईमानदार देशभक्त थे और देशभक्ति आन्दोलनों में कई बार जेल जा चुके थे परंतु राधाकृष्णन नहीं चाहते थे कि उनकी बेटी उनके दो बच्चों की धाय माँ बने। गौरी सीताराम की देशभक्ति व सादगी के विषय में जानकर मन ही मन उन्हें अपना पति मान लेती है। गौरी का देशभक्ति से प्रभावित होना उसके अन्तर्मन में छिपे देशप्रेम की भावनाओं को दर्शाता है। सीताराम जी के जेल जाने पर वह नौकर के साथ कानपुर जाकर बच्चों की देख–भाल करती है। हर महीने बच्चों के कुशल–क्षेम के समाचार भी जेल भिजवाती है। गौरी त्याग की मूर्ति है। धनी तहसीलदार साहब का रिश्ता ठुकराकर सीताराम और उसके दोनों बच्चों की माँ बनती है। जिस तरह गौरी ने एक देशभक्त, स्वतंत्रता सेनानी के बच्चों की माँ बनकर उन्हें गृहस्थी की जिम्मेदारियों से मुक्त कर, अपने कार्यक्षेत्र में स्वतंत्र रूप से कार्य करने का अवसर देकर जो त्याग किया है वह किसी मायने में युद्धभूमि में किये गये त्याग से कम नहीं है। अतः गौरी के सराहनीय त्याग, उसके साहस, उसकी देशभक्ति व कर्त्तव्यनिष्ठा का मूल्य सीताराम जी के योगदान से कम नहीं आँका जा सकता।

काव्य मंजरी (Kavya Manjari)

Answer 7.

(i) प्रस्तुत पद्यांश के कवि हरिवंश राय बच्चन हैं और कविता का नाम है 'अंधेरे का दीपक'। यह कविता सकारात्मकता का प्रतीक है और एक आशावादी कविता है।

(ii) समय की तेज़ आँधी से प्यार का आशियाना उजड़ गया है। कवि कहते हैं कि प्रकृति के नियम और शक्तियों के आगे किसी की नहीं चलती। मनुष्य के शोर मचाने से कुछ नहीं बदलने वाला। अतः मनुष्य को यथार्थवादी बनना होगा। घटनाओं व चुनौतियों से हार नहीं माननी चाहिए। शोर मचाने से दुख कम नहीं होगा। अतः हमें जाने वाले की याद में निराश, निष्क्रिय होकर नहीं बैठना चाहिये। प्रकृति में सृजन व संहार का क्रम चलता ही रहता है।

(iii) मनुष्य को निर्माण का प्रतिनिधि कहकर सम्बोधित किया गया है क्योंकि जब प्रकृति सब–कुछ तहस–नहस कर देती है तो मनुष्य ही सब कुछ बनाता और सँवारता है। मनुष्य अपने पुरुषार्थ और सामर्थ्य से उजड़े हुये को फिर से बसाता है। प्रकृति का जड़ नियम यही है कि जो बसा है वह अवश्य ही उजड़ेगा। इस संसार में प्रत्येक वस्तु नश्वर है। निर्माण के बाद विनाश और पुनः निर्माण का क्रम चलता रहता है। मनुष्य को यह शाश्वत सत्य समझकर परिवर्तन से विचलित होकर निराश नहीं होना चाहिये। उसे हमेशा अपने कर्म पथ पर अग्रसर रहना चाहिये।

(iv) कविता से कवि आशावाद का संदेश देना चाहते हैं। यह एक सकारात्मक कविता है। इस कविता का आधार आशा और विश्वास है। सुख–दुख तो मनुष्य के जीवन में आते–जाते रहते हैं। अंधेरी रात और दीपक का सन्दर्भ भी इसी कारण दिया गया है। निर्माण और ध्वंस तो प्रकृति का जड़ नियम है। इस कविता में कवि ने मनुष्य के जीवन में घटने वाली दुखद घटनाओं, व असफलताओं से निराश न होने और आत्मविश्वास बनाए रखने का संदेश दिया है। दुःखों के क्षण में भी मनुष्य को धैर्य बनाए रखना चाहिए। सृजन और विध्वंस तो निरन्तर क्रमानुसार चलते ही रहेंगे। विपरीत परिस्थितियों में निराश होकर नहीं बैठना चाहिए। हर परिस्थिति में सकारात्मक सोच रखनी चाहिए। आशा और विश्वास ही मनुष्य को जीवन का सच्चा सुख दे सकते हैं।

Answer 8.

'एक फूल की चाह' कविता के माध्यम से गुप्तजी ने छुआछूत जैसी कुरीति पर कुठाराघात किया है। छुआछूत भारतीय समाज की संकीर्ण विचारधारा पर प्रकाश डालता है। इसमें एक दलित पिता की पीड़ा को बताया गया है। उसकी बेटी महामारी की चपेट में आ जाती है। उसका शरीर ज्वर से तपने लगता है। ज्वर के कारण उसका शरीर और स्वर कमज़ोर हो जाता है। पिता बेटी को बचाने की भरसक कोशिश भी करता है। एक दिन बेटी पिता से देवी माँ का प्रसाद रूपी फूल लाने को कहती है। दलित होने के कारण पिता मंदिर में नहीं जा सकता था। लेकिन अपनी बेटी की अन्तिम इच्छा को पूरी करने के लिए देवी माँ के मन्दिर से फूल लेने जाता है। लाचार पिता को मंदिर में प्रवेश करने और पूजा का फूल लेने की धृष्टता के लिए खूब पीटा जाता है। दलितों को अछूत कहकर तिरस्कार करना और ऐसा मानना कि उसके मंदिर में प्रवेश करने से धर्म भ्रष्ट हो जायेगा। कुएँ से पानी निकालने पर जल अपवित्र हो जायेगा। ऐसी संकीर्ण सोच पर गुप्तजी ने तीखा प्रहार किया है। ऐसी कठोर निर्मम कुरीति के कारण पिता अपनी बेटी की अन्तिम इच्छा को भी पूरी नहीं कर पाता है।

जाति के आधार पर किसी से भेद–भाव करना सामाजिक अपराध है। छुआछूत के नाम पर दलितों को मन्दिर में प्रवेश करने से रोकना मानवता पर प्रहार करने के समान है। हमारे देश का संविधान प्रत्येक नागरिक को समानता का अधिकार देता है, अतः हमें उसका सम्मान करना चाहिए और भेद–भाव नहीं करना चाहिए।

Answer 9.

प्रस्तुत कविता ''आः धरती कितना देती है'' में कवि ने यह संदेश दिया है कि हम जैसा बोयेंगे वैसा ही पायेंगे। अच्छे कर्मों का फल अच्छा और बुरे कर्मों का फल बुरा ही होगा। बीज रूपी मणियाँ बोने से सेम रूपी फलियाँ ही उगेंगी तथा ममता, समता व भाईचारे के बीज बोयेंगे तो मानवता रूपी श्रेष्ठ फल प्राप्त होंगे।

भूगर्भ में खनिजों व रत्नों का व्यापक धन भरा हुआ है। धरती रत्नों की जन्मदात्री है। स्वार्थी मनुष्य उसे अपनी तृष्णा और कामना से सींचता है। पंत जी लालच में आकर बचपन में धरती में पैसे बो देते हैं। उस समय उन्होंने सोचा कि पैसे बोने से पैसों के पौधे उग आयेंगे। जब धरती से अंकुर नहीं फूटा, तब बाल कवि मन में सोचता है कि धरती बंजर है। कई साल बीतने पर भी पैसों की फसल नहीं उगती है। तब कवि मिट्टी खोद कर घर के आँगन में सेम के बीज दबा देते हैं। कवि बीज बोने के बाद, इस घटना को भूल जाते हैं। एक दिन संध्या के समय टहलते हुए कवि ने देखा कि सेम के अंकुर फूट गए हैं। धीरे–धीरे ये अंकुर पत्तों से लद गए। कुछ समय पश्चात आँगन में बेल फैल गई। उनके लताओं पर तारों जैसे फूल खिलने लगे। सेम की फलियों को खाने का आनन्द भी उठाया।

अपने इस अनुभव से कवि कहते हैं कि वसुन्धरा मनुष्य को रत्नों की खान देती है। मनुष्य लोभ और स्वार्थवश इसका हनन कर रहा है। धरती को समाज का प्रतीक मानकर कवि कहते हैं कि हमें ममता और समता के बीज बोने चाहिए।

''सारा आकाश'' (Saara Akash)

Answer 10.

(i) उपन्यास का नाम – सारा आकाश

उपन्यासकार का नाम – राजेंद्र यादव

यह एक सामाजिक एवं पारिवारिक उपन्यास है जिसमें मध्यमवर्गीय संयुक्त परिवार के युवक के संघर्ष, नारी शोषण, अंधविश्वास और रूढ़िवादी परम्पराओं का यथोचित वर्णन किया गया है।

(ii) उपर्युक्त कथन का वक्ता समर है। वह अपने मित्र दिवाकर के प्रति आभारी है। दिवाकर ने समर को अपने पिता के मित्र की प्रिंटिग प्रेस में नौकरी दिलाकर उसे बहुत सहयोग किया है। वहाँ उसे 75 रुपये तनख्वाह मिलेगी। प्रेस में उसे प्रूफ रीडिंग का काम करना था। समर को आगे की पढ़ाई के लिये पैसों की आवश्यकता थी। अतः यह नौकरी पाकर वह दिवाकर का अहसानमंद था कि अब वह बिना किसी बाधा के अपनी आगे की पढ़ाई पूरी कर पायेगा।

(iii) वक्ता समर ने श्रोता दिवाकर से 20 रुपए उधार लिए और उन रुपयों से अपनी पत्नी प्रभा के लिए धोती खरीदी। समर जब धोती लेकर घर पहुँचता है तो घर में कोहराम मच जाता है। समर की माँ बहुत दुखी होती है। पिताजी भी समर को खरी–खोटी सुनाते हैं। घर के किसी भी सदस्य को समर का प्रभा के लिए धोती लाना अच्छा नहीं लगा।

(iv) श्रोता – दिवाकर

दिवाकर समर का सच्चा व आदर्श मित्र है। वह इंटर का मेधावी छात्र है तथा विवाहित है। वह समर के घर के हालात से परिचित था। वह समर की मदद करता है। उसे प्रूफ रीडिंग की नौकरी लगवा देता है। दिवाकर आधुनिक विचारों वाला युवक था। वह अपनी पत्नी का सम्मान करता है और खुश भी रखता है। वह अपनी माँ और पत्नी किरण के बीच सन्तुलन बनाए रखता है। वह एक आदर्श पति है। वह रूढ़िवादी माँ का साथ न देकर अपनी पत्नी का साथ देता है। प्रभा के लिए धोती लाने के लिए वह अपने मित्र समर को उधार भी देता है। उसमें स्वार्थ भावना लेश मात्र भी नहीं है। वह अपने मित्र की हर सम्भव सहायता करता है और उसका उचित मार्गदर्शन भी करता है। अतः दिवाकर एक सज्जन पुरुष व सच्चा मित्र है।

Answer 11.

समर के बाबूजी सेवानिवृत्त हैं। उनका बहुत बड़ा परिवार है। उनके चार पुत्र क्रमशः धीरज, समर, अमर, कुंवर और एक पुत्री मुन्नी है। वे संयुक्त परिवार का समर्थन करते हैं। आर्थिक हालात कमजोर थे। वे क्रोधी स्वभाव के भी हैं। वे रूढ़िवादी परम्पराओं पर विश्वास करते थे। स्त्री शिक्षा को बढ़ावा नहीं देते। इसलिए मुन्नी की पढ़ाई बीच में ही रोक देते हैं और उसका विवाह कर देते हैं। पर्दा प्रथा का समर्थन करते हैं। इसी कारण प्रभा को भी घूँघट न निकालने पर डाँटते हैं। बाबूजी दहेज प्रथा के भी पक्षधर हैं। वे समर के मामाजी से शिकायत भी करते हैं कि समर के विवाह में उसके ससुराल वालों ने कुछ भी नहीं दिया। अमर के विवाह प्रस्ताव आने पर वे स्पष्ट कहते हैं कि लेन–देन की बात साफ होगी। संकीर्ण विचारधारा की मानसिकता रखने वाले बाबूजी की भाषा भी निम्न स्तर की थी। बाबूजी को पेंशन के रूप में पच्चीस रुपए मिलते। घर में हमेशा आर्थिक तंगी की समस्या रहती थी।

बाबूजी वात्सल्य भावना से परिपूर्ण थे। भले ही उनके व्यवहार में निष्ठुरता व संवादों में कठोरता थी। वे स्वभाव से चिड़चिड़े हैं। ठाकुर साहब चाहते हैं कि घर में सब कुछ उनके कहे अनुसार ही चले। संक्षेप में ठाकुर साहब रूढ़िवादी विचारों में विश्वास रखने वाले दहेज लोभी पिता हैं।

Answer 12.

"सारा आकाश" राजेन्द्र यादव द्वारा लिखित एक उद्देश्यपूर्ण रचना है। मध्यमवर्गीय परिवार के आर्थिक एवं सामाजिक जीवन के संघर्षों का यथार्थवादी चित्रण है। कहानी का प्रमुख पात्र समर किस प्रकार के ज्वलंत संघर्षों से गुजरता है, इसका मार्मिक चित्रण इस उपन्यास में देखा जा सकता है। मध्यम वर्ग की सामान्य अपरिपक्व मानसिकता का सजीव वर्णन करने में उपन्यासकार सफल हुए हैं। राजेन्द्र यादव जी ने इस उपन्यास को मर्मस्पर्शी बनाने के लिए अंधविश्वासों, परम्पराओं तथा संकीर्ण भावनाओं का हृदयस्पर्शी वर्णन किया है। राजेंद्र यादव अपने उपन्यास में बाल विवाह, महिला उत्पीड़न, दहेज प्रथा, नारी शिक्षा व निष्ठुर समाज जैसे महत्वपूर्ण बिन्दुओं पर पाठकों को सोचने व चिन्तन करने पर मजबूर करते हैं। मुन्नी के चरित्र के द्वारा बलपूर्वक विवाह करवाना, दुख देना व कम पढ़ी लिखी लड़की का चित्रण किया है। उपन्यास का मुख्य पात्र समर कमजोर मानसिकता का प्रतीक है। वह मानसिक अन्तर्द्वन्द्व से जूझता रहता है। वह भारत की प्राचीन परम्पराओं से बंधा है। वह महत्वाकांक्षी भी है। विवाह, परिवार के प्रति आक्रोश, पत्नी के प्रति अरुचि, परीक्षा की चिंता आदि ने समर के सपने को चूर–चूर कर दिया था और वह कुंठाओं से घिर जाता है। उपन्यास का हर पात्र लेखक के उद्देश्य को पाठकों तक पहुँचाने में सफल होता है। देश की, समाज की विशेष तौर पर मध्यमवर्गीय परिवार की आर्थिक स्थिति का यथार्थ व सजीव चित्रण करने में यह उपन्यास सफल हुआ है। लेखक राजेन्द्र यादव ने मानवीय जीवन के तनावों और संघर्षों को पूरी संवेदनशीलता के साथ यथार्थ स्थान दिया है।

''आषाढ़ का एक दिन'' (Aashad Ka Ek Din)

Answer 13.

(i) वक्ता विलोम है और श्रोता कालिदास है। विलोम और कालिदास कवि हैं। विलोम को मल्लिका का प्रेम नहीं मिलता। विलोम कालिदास की आलोचना करता है। विलोम और कालिदास एक ही चरित्र के दो पहलू हैं। कालिदास भावुक और तरल हैं, विलोम यथार्थवादी, आक्रामक और अन्तद्वन्द्वों को पचाये हुये हैं दोनों परस्पर विरोधी होते हुये भी एक दूसरे के पूरक हैं।

(ii) आचार्य वररुचि उज्जयिनी के राजा की तरफ से कालिदास को सम्मानित करने के लिए अपने साथ लेने आये हैं। कालिदास राजकवि का सम्मान स्वीकार नहीं करना चाहते थे। राजकीय सम्मान के प्रति उदासीन कालिदास को जगदम्बा के मन्दिर से मनाकर मल्लिका घर लाती है। घर पर विलोम भी रहता है। विलोम कालिदास का प्रतिद्वंद्वी कवि है। उसका घर पर रहना मल्लिका और कालिदास दोनों को अच्छा नहीं लगा। कालिदास विलोम पर व्यंग्य करते हुए कहता है कि विलोम का कुछ भी करना अप्रत्याशित नहीं है। कालिदास के व्यंग्य पर विलोम यथार्थवादी बनकर यह वाक्य कहता है कि विलोम क्या है? एक असफल कालिदास और कालिदास एक सफल विलोम।

(iii) वक्ता विलोम अपनी तुलना कालिदास से करता है। वह कहता है कि वह भी कालिदास की तरह गाँव का रहने वाला है और कवि है। वह भी मल्लिका से प्रेम करता है पर मल्लिका उससे नहीं। कालिदास को राजकीय सम्मान भी मिलता है, इसलिए वह अपने आपको असफल कालिदास कहता है। वह कालिदास को सफल विलोम कहता है, क्योंकि उसे मल्लिका का प्यार भी मिलता है और राजकीय सम्मान भी। वे दोनों एक जैसे होते हुए भी एक दूसरे से विपरीत हैं।

(iv) विलोम और कालिदास दोनों ही एक गाँव में रहने वाले कवि हैं। दोनों ही मल्लिका से प्यार करते थे। दोनों एक–दूसरे को नापसंद करते थे। दोनों के बीच प्यार और घृणा का अंतर्द्वंद्व देखा जा सकता है। विलोम कालिदास का आलोचक भी है। वह यथार्थवादी है इसलिए कालिदास के उज्जयिनी जाने से पूर्व वह उसे मल्लिका से विवाह करने की बात कहता है। कालिदास और विलोम के बीच का द्वंद ही भाव प्रेम और सौन्दर्य की पराजय का कारण बनता है। मल्लिका द्वारा विलोम के तिरस्कृत होने पर, विलोम इस तिरस्कार का मुख्य कारण कालिदास को ही मानता है। कालिदास की अपेक्षा विलोम स्पष्टवादी है।

Answer 14.

'अम्बिका' ''आषाढ़ का एक दिन'' नाटक का महत्वपूर्ण नारी पात्र है। वह एक ग्रामीण वृद्ध महिला है। वह मल्लिका की माँ है। अम्बिका भावनाओं में नहीं यथार्थ में जीती है। वह हमेशा अपनी बेटी मल्लिका के लिए चिंतित रहती है। अम्बिका कर्मठ व्यक्तित्व की परिचायक भी है। हर माँ की तरह वह भी अपनी बेटी के विवाह के लिए चिंतित रहती है।

अम्बिका दिल से कम दिमाग से ज्यादा सोचती है। वह भावनाओं को कम महत्व देती है। वह अपनी बेटी को समझाते हुए कहती भी है कि भावना एक छलावा है, आत्म प्रवंचना है। मात्र भावना में जीने से जीवन की आवश्यकताओं की पूर्ति नहीं हो सकती है। वह लोक नीति को महत्व देती है। वह कहती है कि कालिदास का उज्जयिनी जाने से सम्मान बढ़ेगा। कालिदास के विदाई के समय वह मल्लिका को वहाँ जाने से रोकती है। बहुत दुखी होकर मल्लिका जब टूट जाती है, तब वह अपनी बेटी को सांत्वना भी देती है।

हम निश्चित रूप से कह सकते हैं कि अम्बिका का हृदय वात्सल्य से ओत–प्रोत है, वह अपनी बेटी को सुखी देखना चाहती है। वह यथार्थवादी और आदर्श माँ हैं।

Answer 15.

''आषाढ़ का एक दिन'' नाटक में प्रियंगुमंजरी एक राजकन्या है। वह महाकवि कालिदास की पत्नी है। प्रियंगुमंजरी का पात्र उदारता व सहृदयता के गुण दर्शाता है। वह प्रकृति प्रेमी है। कालिदास के काव्य से प्रेम होने के कारण ही उसका विवाह कालिदास से होता है। उज्जयिनी में विवाहोपरान्त कालिदास का नाम मातृगुप्त रख दिया जाता है और वह कश्मीर का शासन भार सँभालता है।

प्रियंगुमंजरी मल्लिका से मिलने भी जाती है। वह जानती है कि मल्लिका कालिदास की प्रेयसी है और कालिदास भी उससे प्रेम करता है। कालिदास की रचनाओं की प्रेरणा स्रोत मल्लिका और ग्रामीण परिवेश है। वह मल्लिका के घर का परिसंस्कार करना चाहती है और वह चाहती थी कि मल्लिका किसी राजकर्मचारी से विवाह कर ले, उसके साथ उसकी संगिनी बन कर रहे। वहाँ की वनस्पतियों, पाषाणों, पशु–पक्षियों व कुछ गरीब बच्चों को अपने साथ ले जाना चाहती है, जिससे कालिदास को राजमहल में रहते हुए भी कभी अपने परिवेश का वियोग न खटके। मल्लिका इस प्रस्ताव को अस्वीकार कर देती है। मल्लिका स्वाभिमानी युवती है। वह प्रियंगुमंजरी से कहती है कि वह बहुत उदार है। परन्तु उसे भी ऐसे घर में रहने का अभ्यास है, इसलिए उसे कोई असुविधा नहीं होती। मल्लिका कालिदास से निस्वार्थ प्रेम करती है। कालिदास के लिए वह कोई भी त्याग करने के लिए तैयार थी। इस तरह नाटक के ये दो प्रमुख नारी पात्र प्रियंगुमंजरी और मल्लिका आदर्श गुणों की परिचायक हैं।

●●

HINDI

2019

QUESTIONS

SECTION – A
Language – 50 Marks

Question 1.

Write a composition in approximately 400 words in Hindi on any ONE of the topics given below: [20]

किसी एक विषय पर निबन्ध लिखिए जो लगभग 400 शब्दों से कम न हो :

(i) *'तकनीकी विकास ने मानव को सुविधाओं का दास बना दिया है'–इस विषय पर अपने विचार व्यक्त कीजिए।*

(ii) *'वर्तमान युग में आगे बढ़ने के लिए धन की आवश्यकता है न कि प्रतिभा की–इस विषय के पक्ष या विपक्ष में अपने विचार लिखिए।*

(iii) *पुस्तक एक सच्ची मित्र, गुरु और मार्गदर्शक का कार्य करके जीवन की धारा को बदल सकती है–'मेरी प्रिय पुस्तक' विषय पर अपने विचार प्रस्तुत कीजिए।*

(iv) *'निरन्तर अभ्यास करने से इच्छित कार्य में सफलता मिलती है।'–इस कथन को अपने जीवन के किसी निजी अनुभव के द्वारा विस्तारपूर्वक लिखिए।*

(v) *'सहशिक्षा के माध्यम से बालक–बालिका के मध्य मित्रता और समानता का भाव जागता है।'–इस विषय पर अपने विचार विस्तारपूर्वक लिखिए।*

(vi) *निम्नलिखित में से किसी एक विषय पर मौलिक कहानी लिखिए–*

(a) *"साँच को आँच नहीं।"*

(b) *एक ऐसी मौलिक कहानी लिखिए जिसका अन्तिम वाक्य हो।*

.......और इस तरह उन्होंने मुझे माफ कर दिया।

Question 2.

Read the passage given below carefully and answer in Hindi the questions that follow, using your own words :

निम्नलिखित अवतरण को पढ़कर, अन्त में दिए गए प्रश्नों के उत्तर अपने शब्दों में लिखिए :

एक राजा का दरबार लगा हुआ था। सर्दियों के दिन थे, इसलिए राजा का दरबार खुले स्थान पर लगा था। पूरी आम सभा सुबह की धूप में बैठी थी। महाराज ने सिंहासन के सामने एक मेज डलवा रखी थी। राजा के परिवार के सभी सदस्य, पंडित जन, दीवान आदि सभी दरबार में बैठे थे।

उसी समय एक व्यक्ति आया और राजा से दरबार में मिलने की आज्ञा माँगी। प्रवेश मिल गया, तो उसने कहा, "मेरे पास दो वस्तुयें हैं, बिल्कुल एक जैसी लेकिन एक नकली है और एक असली। मैं हर राज्य के राजा के पास जाता हूँ और उन्हें परखने का आग्रह करता हूँ लेकिन कोई परख नहीं पाता, सब हार जाते हैं और मैं विजेता बनकर घूम रहा हूँ। अब आपके नगर में आया हूँ।"

राजा ने उसे दोनों वस्तुओं को पेश करने का आदेश दिया, तो उसने दोनों वस्तुएँ मेज पर रख दीं। बिल्कुल समान रूप–रंग, समान प्रकाश, सब कुछ नख–शिख समान। राजा ने कहा, "ये दोनों वस्तुएँ एक हैं", तो उस व्यक्ति ने कहा, "हाँ दिखाई तो एक सी ही देती हैं लेकिन हैं भिन्न। इनमें से एक है बहुत कीमती हीरा और एक है काँच का टुकड़ा, लेकिन रूप रंग सब एक हैं। कोई आज तक परख नहीं पाया कि कौन सा हीरा है और कौन सा काँच। कोई परख कर बताए कि ये हीरा है या काँच। अगर परख खरी निकली, तो मैं हार जाऊँगा और यह कीमती हीरा मैं आपके राज्य की तिजोरी में जमा करवा दूँगा, यदि कोई न पहचान पाया तो इस हीरे की जो कीमत है उतनी धनराशि आपको मुझे देनी होगी। इसी प्रकार मैं कई राज्यों से जीतता आया हूँ।"

राजा ने कई बार उन दोनों वस्तुओं को गौर से देखकर परखने की कोशिश की और अंत में हार मानते हुए कहा–"मैं तो नहीं परख सकूँगा।"

दीवान बोले–"हम भी हिम्मत नहीं कर सकते, क्योंकि दोनों बिल्कुल समान हैं।" सब हारे, कोई हिम्मत नहीं जुटा पाया। हारने पर पैसे देने पड़ेंगे, इसका किसी को कोई मलाल नहीं था क्योंकि राजा के पास बहुत धन था लेकिन राजा की प्रतिष्ठा गिर जायेगी, इसका सबको भय था।

कोई व्यक्ति पहचान नहीं पाया। आखिरकार पीछे थोड़ी हलचल हुई। एक अंधा आदमी हाथ में लाठी लेकर उठा। उसने कहा, "मुझे महाराज के पास ले चलो, मैंने सब बातें सुनी हैं और यह भी सुना कि कोई परख नहीं पा रहा है। एक अवसर मुझे भी दो।" एक आदमी के सहारे वह राजा के पास पहुँचा, उसने राजा से प्रार्थना की "मैं तो जन्म से अंधा हूँ फिर भी मुझे एक अवसर दिया जाए जिससे मैं भी एक बार अपनी बुद्धि को परखूँ और हो सकता है कि सफल भी हो जाऊँ और यदि सफल न भी हुआ, तो वैसे भी आप तो हारे ही हैं।"

राजा को लगा कि इसे अवसर देने में कोई हर्ज नहीं है और राजा ने उसे अनुमति दे दी। उस अंधे आदमी के हाथ में दोनों वस्तुएँ दे दी गयीं और पूछा गया कि इनमें कौन सा हीरा है और कौन सा काँच?

उस आदमी ने एक मिनट में कह दिया कि यह हीरा है और यह काँच। जो व्यक्ति इतने राज्यों को जीतकर आया था, वह नतमस्तक हो गया और बोला, "सही है, आपने पहचान लिया, आप धन्य हैं! अपने वचन के मुताबिक यह हीरा मैं आपके राज्य की तिजोरी में दे रहा हूँ।" सब बहुत खुश हो गये। और जो व्यक्ति आया था वह भी बहुत प्रसन्न हुआ कि कम से कम कोई तो मिला असली और नकली को परखने वाला। राजा और अन्य

सभी लोगों ने उस अंधे व्यक्ति से एक ही जिज्ञासा जताई कि, "तुमने यह कैसे पहचाना कि यह हीरा है और वह काँच?"

उस अंधे ने कहा "सीधी सी बात है राजन्, धूप में हम सब बैठे हैं, मैंने दोनों को छुआ। जो ठण्डा रहा वह हीरा, जो गरम हो गया वह काँच।"

यही बात हमारे जीवन में भी लागू है, जो व्यक्ति बात–बात में अपना आपा खो देता है, गरम हो जाता है और छोटी से छोटी समस्याओं में उलझ जाता है वह काँच जैसा है और जो विपरीत परिस्थितियों में भी सुदृढ़ रहता है और बुद्धि से काम लेता है वही सच्चा हीरा है।

(i) राजा का दरबार कहाँ और क्यों लगा था? वहाँ कौन–कौन उपस्थित था? तभी वहाँ कौन, क्या लेकर आया? **[4]**

(ii) उस व्यक्ति ने उन वस्तुओं की क्या विशेषता बताई तथा राजा के सामने क्या शर्त रखी? **[4]**

(iii) कोई भी उन वस्तुओं को पहचान क्यों नहीं सका? उन्हें किस बात का डर था? **[4]**

(iv) अन्त में उन वस्तुओं की पहचान किसने एवं किस प्रकार की? सब लोगों की इस पर क्या प्रतिक्रिया थी? **[4]**

(v) इस कहानी से हमें क्या शिक्षा मिलती है? **[4]**

Question 3.

(a) Correct the following sentences and rewrite : [5]

निम्नलिखित वाक्यों को शुद्ध करके लिखिए :

(i) तेरे को अध्यापिका ने शीघ्र बुलाया है।

(ii) तुम अकारण बेकार में डर रहे हो।

(iii) प्रातः भ्रमण में कैसी आनन्द आती है।

(iv) इमारत के गिर जाने की आशा है।

(v) मुझे पानी का एक गर्म लोटा चाहिए।

(b) Use the following idioms in sentences of your own to illustrate their meaning : [5]

निम्नलिखित मुहावरों को वाक्यों में प्रयुक्त कीजिए :

(i) कलेजे का टुकड़ा। **(ii)** मुँह में पानी भरना।

(iii) रोड़ा अटकाना। **(iv)** बाग–बाग होना।

(v) हाथ मलना।

SECTION – B

Prescribed Textbooks – 50 Marks

*(Answer **four** questions from this Section on **at least three** of the prescribed textbooks)*

गद्य संकलन (Gadya Sanklan)

Question 4.

"हाँ, और लोग पीछे आते हैं। कई सौ आदमी साथ आये हैं। यहाँ तक आने में सैकड़ों उठ गये पर सोचता हूँ कि बूढ़े पिता की मुक्ति तो बन गई। धन और है ही किसलिए।"

(i) उपर्युक्त कथन किस पाठ से लिया गया है? इस कथन का वक्ता कौन है? यह कथन किस स्थान पर कहा जा रहा है? **[1½]**

(ii) वक्ता द्वारा यह कथन किस सन्दर्भ में कहा गया था? **[3]**

(iii) 'धन और है ही किसलिए।'–वक्ता ऐसा क्यों कहता है? **[3]**

(iv) वक्ता के सम्बन्ध में श्रोता को क्या–क्या पता चलता है, उसका प्रभाव श्रोता पर क्या पड़ता है? अन्ततः श्रोता क्या निर्णय लेता है? **[5]**

Question 5.

नीलम का परिचय देते हुए बताइए कि उसके परिवार में कौन–कौन था? वह किस धोखे में अपना जीवन अभी तक व्यतीत कर रही थी? उसे इसका आभास कैसे हुआ? स्पष्ट कीजिए। **[12½]**

Question 6.

'भक्तिन का जीवन संघर्ष एवं कर्मठता का जीवंत उदाहरण है'–उसके जीवन के विभिन्न अध्यायों का वर्णन करते हुए इस कथन की पुष्टि कीजिए। **[12½]**

काव्य मंजरी (Kavya Manjari)

Question 7.

किन्तु हम बहते नहीं हैं।
क्योंकि बहना रेत होना है।
हम बहेंगे, तो रहेंगे ही नहीं।
पैर उखड़ेंगे, प्लवन होगा, ढहेंगे, सहेंगे बढ़ जाएँगे,
और फिर हम पूर्ण होकर भी कभी क्या धार बन सकते?

(i) कौन बहते नहीं हैं? कविता के प्रसंग में बताइए। **[1½]**

(ii) 'बहना रेत होना' कैसे है? **[3]**

(iii) 'बहना' प्रक्रिया का क्या प्रभाव पड़ता है? कविता के संदर्भ में समझाइए। **[3]**

(iv) प्रस्तुत कविता का सामाजिक संदर्भ उजागर कीजिए। **[5]**

Question 8.

"कवि नागार्जुन ने हिमालय के वर्षाकालीन सौन्दर्य का मोहक चित्रण किया है।" पठित कविता 'बादल को घिरते देखा है' के आधार पर व्याख्या कीजिए। **[12½]**

Question 9.

महादेवी वर्मा पथिक को क्या प्रेरणा दे रही है और क्यों? 'जाग तुझको दूर जाना' कविता के आधार पर स्पष्ट कीजिए। **[12½]**

"सारा आकाश" (Saara Akash)

Question 10.

अपने काँपते हुए बेजान हाथों को निहायत डरते–डरते उसके कंधे पर रखकर भर्राए और खंडित स्वर में कहा, "प्रभा तुम मुझसे नाराज हो......?" साथ ही मुझे आश्चर्य हो रहा था कि यह कौन मेरे भीतर से बोल रहा है।

(i) प्रभा कौन थी और वह किससे नाराज थी? **[1½]**

(ii) उपन्यास के नायक का नाम लिखते हुए यह स्पष्ट कीजिए कि उसके आत्मविश्लेषण का क्या परिणाम निकला? **[3]**

(iii) पति ने अपने बदले स्वभाव का परिचय किस प्रकार दिया? **[3]**

(iv) पति को अपने बदले हुए व्यवहार पर कैसा अनुभव हुआ और इससे क्या स्पष्ट होता है? **[5]**

Question 11.

"सारा आकाश" उपन्यास में वर्णित संयुक्त परिवार की समस्याओं के बारे में शिरीष भाई साहब के विचारों को स्पष्ट कीजिए। **[12½]**

Question 12.

'सारा आकाश' उपन्यास में समर की भाभी मध्यमवर्गीय परिवार की भाभियों का प्रतिनिधित्व करती है।' इस कथन को ध्यान में रखते हुए भाभी की चारित्रिक विशेषताओं का वर्णन कीजिए। **[12½]**

"आषाढ़ का एक दिन" (Aashad Ka Ek Din)

Question 13.

"............हमारा शरीर कोमल है, तो क्या हुआ? हम पीड़ा सह सकते हैं। एक बाण प्राण ले सकता है, तो उँगलियों का कोमल स्पर्श प्राण दे भी सकता है।"

(i) *प्रस्तुत पंक्तियों में वक्ता किससे संवाद कर रहा है, सन्दर्भ सहित लिखिए।* **[1½]**

(ii) *'हमारा शरीर कोमल है, तो क्या हुआ? हम पीड़ा सह सकते हैं।' वक्ता द्वारा ऐसा कहने का प्रयोजन स्पष्ट कीजिए।* **[3]**

(iii) *'एक बाण प्राण ले सकता है, तो उँगलियों का कोमल स्पर्श प्राण दे भी सकता है।' पंक्ति का आशय स्पष्ट कीजिए।* **[3]**

(iv) *'आषाढ़ का एक दिन' नाटक की भाषा–शैली की विशेषताएँ बताइए।* **[5]**

Question 14.

'विलोम का चरित्र मोहन राकेश की एक अनुपम नाटकीय चरित्र–सृष्टि है' कथन के आधार पर विलोम की चारित्रिक विशेषताएँ लिखिए। **[12½]**

Question 15.

"यह पन्ने अपने हाथों से बनाकर सिये थे"— उक्त कथन किसका है और किससे कहा गया है? इन पन्नों के बारे में वक्ता–श्रोता के बीच क्या बातचीत हुई? उनकी बातचीत में छिपे भाव को भी स्पष्ट कीजिए। **[12½]**

ANSWERS

SECTION – A

Answer 1.

(i) मानव : सुविधाओं का दास

प्रकृति ने मनुष्य को बुद्धि देकर उसे पशुओं से भिन्न बनाया है। वह अपनी बुद्धि के बल पर नये–नये आविष्कार कर संसार में विकास के मार्ग को आलोकित करता है। सुविधापूर्ण जीवन जीने की ललक में मानव नित नई खोजें करता रहा, नित नई तकनीकों के द्वारा तरह–तरह के आविष्कार कर जीवन को आरामदायक बनाता रहा। सुबह जागने से लेकर रात में सोने तक मनुष्य तकनीकी में ही जीवन–यापन करता है।

वर्तमान युग विज्ञान और तकनीकी का युग है। आज तकनीकी उन्नति ने लोगों को बहुत लाभ पहुँचाया है। इसके कारण ही आज मनुष्य अच्छा जीवन व्यतीत कर रहा है।

दूसरे शब्दों में, यह कहा जा सकता है कि तकनीकी विकास ने मानव को सुविधाओं का दास बना दिया है। वह इस पर इतना निर्भर हो गया है कि आज वह उसके बिना जीने की कल्पना भी नहीं कर सकता। अतः इस विकास ने मानव के जीवन स्तर को बहुत प्रभावित किया है। प्राचीन समय में चलाई जाने वाली बैलगाड़ी तो आज संग्रहालय की शोभा बन कर रह गई है और उसका स्थान मोटर चलित वाहनों ने ले लिया है।

आज के युग में कई देशों में तकनीक इतनी विकसित हो चुकी है कि मनुष्य को केवल बटन मात्र को दबाने की आवश्यकता रह गई है। रेडियो, टेलीविजन का जमाना तो अब लगभग समाप्त–सा ही हो चुका है। उनका स्थान आज मोबाइल, कम्प्यूटर, लैपटॉप आदि ने ले लिया है। संचार साधनों ने विश्व के सभी अदृश्य कोनों को भी दृश्यवान बना दिया है। जरूरत की हर छोटी से बड़ी चीज तक में तकनीकी के प्रयोग ने मानव में आलस्यपूर्ण प्रवृति को जन्म दिया है। कम समय में काम निपटाने की प्रवृति के कारण मानव तकनीकी पर पूरी तरह निर्भर हो चुका है। वॉशिंग मशीन से लेकर मोबाइल फोन, खाना बनाने के उपकरण, साफ–सफाई के आधुनिक उपकरण आदि ने मानव शरीर को बिल्कुल निष्क्रिय बना दिया है। इससे स्वास्थ्य पर भी बहुत बुरा असर पड़ा है। मानसिक विकास भी अवरुद्ध हुआ है।

देश का आर्थिक विकास हो चाहें औद्योगिक विकास सभी में ही तकनीकी ने अपना महत्वपूर्ण योगदान दिया है। मानव जीवन को सुखी व निरोगी बनाने में विज्ञान और तकनीक का अपना अलग ही योगदान है।

वैसे यह कहना अतिशयोक्ति नहीं है कि तकनीकी ने मानव की आत्मनिर्भरता को समाप्त ही कर दिया है। वह इसका दास बन कर रह गया है। जहाँ पहले मनुष्य सभी कार्य स्वयं करता था वहीं आज वह अपने हर कार्य के लिए तकनीक पर निर्भर हो गया है। आज तो वह मिलने के लिए भी मोबाइल का ही सहारा लेता है इसलिए आज रिश्ते भी सिमट गये हैं। 'यदि मनुष्य ने आज स्वयं को आत्मनिर्भर नहीं बनाया तो वह दिन दूर नहीं जब मनुष्य केवल आलसी और अकर्मण्य बनकर रह जायेगा।' आधुनिक उपकरणों की इस भीड़ में मानव अस्तित्व कहीं खो जायेगा। किसी ने सच ही कहा है–

'सावधान मनुष्य, यदि तकनीक है तलवार की धार,
तो उसे दे फैंक और अपना ले आत्मनिर्भरता के द्वार।'

(ii) प्रतिभा का महत्व

आज का युग विज्ञान का युग है। इस युग में आधुनिक संसाधनों ने इतनी उन्नति की है कि जीवन अधिक सरल हो गया है और यह सब सम्भव हुआ है मनुष्य की प्रतिभा से। मनुष्य इतना अधिक प्रतिभाशाली है कि अपनी प्रतिभा के बल पर वह संसार में अपना लोहा मनवा सकता है। मनुष्य की इसी प्रतिभा ने उसे पशुओं से पृथक किया हुआ है। प्रतिभा किसी एक देश की परिरक्षित निधि नहीं है।

प्रतिभा के बल पर ही मनुष्य संसार में अपनी अलग पहचान बना सकता है। कुछ लोगों का मानना है कि आगे बढ़ने के लिए प्रतिभा की नहीं धन की आवश्यकता होती है। ऐसा मानना तर्कसंगत नहीं है। धन के बल पर हम अपने जीवन के सभी सुख खरीद सकते हैं पर जीवन की शान्ति इससे नहीं खरीदी जा सकती। संसार में जन्म लेने वाले प्रत्येक व्यक्ति के अन्दर कोई–न–कोई प्रतिभा छिपी हुई होती है। बस जरूरत होती है उसे सँवारने की, उसे निखारने की। अपनी उसी

प्रतिभा के बल पर व्यक्ति अपना जीवन आगे बढ़ा सकता है। यह सही है कि जीवन को खुशहाल बनाने के लिए धन की आवश्यकता होती है क्योंकि धन के माध्यम से ही हम अपनी दैनिक जरूरतों को पूरा कर सकते हैं पर इसका यह अर्थ कदापि नहीं कि आगे बढ़ने के लिए प्रतिभा का कोई मोल नहीं। मनुष्य अपनी प्रतिभा के बल पर ही स्वयं का जीवन जीने योग्य बना सकता है। संसार में अनेक ऐसे महापुरुषों ने जन्म लिया है जो धनी नहीं थे पर अपनी प्रतिभा के बल पर वे आगे बढ़ते गये। लाल बहादुर शास्त्री जी धनी परिवार से नहीं थे पर अपनी प्रतिभा के बल पर वह भारत के प्रधानमन्त्री बने। मध्यमवर्गीय परिवार के अब्दुल कलाम अपनी प्रतिभा के कारण ही देश के राष्ट्रपति के पद पर आसीन हुए। यदि मनुष्य अपनी प्रतिभा का प्रयोग नहीं करेगा तो वह नष्ट हो जायेगी।

आज ऐसा माना जाता है कि जिसके पास धन है वही सुखी और सुविधा–सम्पन्न है। प्रतिभा को धन से नहीं खरीदा जा सकता। किसी भी देश के मस्तिष्क को लगातार पुनः पूर्ति और उन्नति की आवश्यकता होती है। पैसे में यद्यपि बहुत आकर्षण होता है पर फिर भी वह प्रतिभा को बाँधकर नहीं रख सकता। एक प्रतिभाहीन व्यक्ति अपने लिए धन से सभी भौतिक संसाधन तो खरीद सकता है पर मन की शान्ति नहीं। वह जीवन के सफ़र में आगे तब तक नहीं बढ़ सकता जब तक वह अपनी प्रतिभा का प्रयोग नहीं करता हो। भौतिक सम्पदा में बहुत बड़ा आकर्षण होता है, परन्तु जिस आवश्यकता की पूर्ति लोग करना चाहते हैं उसका यह एक तुच्छ भाग है। अतः यह स्पष्ट है कि आगे बढ़ने के लिए प्रतिभा की आवश्यकता होती है न कि धन की।

(iii) मेरी प्रिय पुस्तक

पुस्तक मनुष्य की सच्ची मित्र और पथ–प्रदर्शक होती हैं। जीवन में सफलता प्राप्त करने में पुस्तकों की महत्वपूर्ण भूमिका है। मन को स्वस्थ रखने में पुस्तकें अहम् भूमिका निभाती हैं। इनके ज्ञान से मनुष्य की बुद्धि तो विकसित होती है बल्कि वह मनुष्य की कठिन से कठिन समस्याओं का निवारण कर उसके आत्मबल को बढ़ाती है। अतः पुस्तकें ज्ञानार्जन का प्रमुख स्रोत है।

लोकमान्य बाल गंगाधर तिलक ने सही कहा है–'मैं नरक में भी उत्तम पुस्तकों का आदर करूँगा क्योंकि इनमें नरक को भी स्वर्ग में बदलने की क्षमता है।'

रामचरितमानस मेरी प्रिय पुस्तक है। इसकी रचना गोस्वामी तुलसीदास जी ने 1631 में की थी। यह अवधी भाषा में लिखी हुई है। इसमें मर्यादा पुरुषोत्तम भगवान श्री राम के जीवन चरित्र को 7 काण्डों में प्रस्तुत किया गया है। रामचरितमानस की कथावस्तु का प्रमुख स्रोत महर्षि वाल्मीकि द्वारा रचित रामायण है। इसमें राम की रावण पर विजय दिखाते हुए प्रतीकात्मक रूप से सत्य, न्याय और धर्म की विजय को दिखाया गया है।

इस पुस्तक की प्रमुख विशेषताएँ हैं–

(1) राम का आदर्श चरित्र

(2) आदर्श लोकमंगल

(3) नीति, सदाचार और समन्वय की भावना

(4) रामराज्य के माध्यम से आदर्श राज्य की कल्पना।

रामचरित मानस हमें बताती है कि मर्यादित जीवन कैसा होता है। श्रीराम के चरित्र ने मुझे सबसे अधिक प्रभावित किया। ऐसे ही उन्हें मर्यादा पुरुषोत्तम नहीं कहा गया है। उन्होंने एक पुत्र, एक पति, एक उत्तम राजा होने की मर्यादा का पालन करने के लिये अपना पूरा जीवन लगा दिया। सभी सुख त्याग दिये। लक्ष्मण और शत्रुघ्न का राम के प्रति प्रेम भी आदर्श भाईचारे को दर्शाता है। माता सीता और उर्मिला पतिव्रता स्त्री थीं।

रामचरितमानस हिन्दी साहित्य का वह सुगन्धित फूल है जिसकी सुगन्ध से तन–मन में एक अनोखा उत्साह संचारित होता है। ये ग्रन्थ दोहा–चौपाई में लिखा हुआ है। इसके प्रत्येक काण्ड की भाषा शैली उत्कृष्ट है। संस्कृत के श्लोकों से आरम्भ हुई कथा फलागम की ओर बढ़ती है। अलंकारों का प्रयोग इसमें स्वाभाविक रूप से हुआ है। सभी रसों का समावेश इसमें है।

इस पुस्तक को पढ़ने से परिवार और समाज के दुःखों को दूर करने की प्रेरणा मिलती है और मन शान्त होता है। जो भी इसे एक बार पढ़ता है कि इसे बार–बार पढ़ने से अपने को रोक नहीं पाता है। इस ग्रन्थ को पवित्र गंगा की संज्ञा दी गई है क्योंकि एक बार इसमें डुबकी लगाने पर शरीर में जिस मधुर रस का संचार होता है वह अमर है।

अतः रामचरितमानस अपनी भव्यता की कहानी खुद ही कहती है इसलिए यह मेरे जीवन की अमूल निधि है।

(iv) अभ्यास : सफलता की कुंजी

"करत करत अभ्यास के, जड़मति होत सुजान,
रसरि आवत जात पै, सिल पर परत निसान।"

यह कथन पूर्णतः सत्य है। निरन्तर अभ्यास से असम्भव को सम्भव बनाया जा सकता है। नियमित अभ्यास एक व्यक्ति की बौद्धिक और सृजनात्मक क्षमताओं की ओर इंगित करता है। किसी भी कार्य को उचित योजना और अभ्यास के अनुसार करना एक व्यक्ति का पूर्ण प्रदर्शन की ओर नेतृत्व करता है। किसी भी कार्य को करने में गुणवत्ता लाने के साथ ही एक व्यक्ति को अन्य गुणों के लिए तैयार करना नियमित अभ्यास से ही सम्भव है।

अभ्यास कमियों को सुधार कर कार्य की पूर्णता में व्यक्ति की सहायता करता है। निश्चित ही ज्ञान बहुत बड़ा होता है पर अभ्यास के अभाव में उसका कोई अर्थ नहीं रह जाता। कपिल देव, सचिन तेन्दुलकर, मेरिकॉम आदि निरन्तर अभ्यास के बल पर ही एक सफल और प्रसिद्ध हस्ती बने।

प्रत्येक मनुष्य के अन्दर किसी न किसी प्रकार का भय व्याप्त रहता है परन्तु जब तक सोने को भी बार–बार तपाया नहीं जाता तब तक उसमें भी निखार नहीं आता है। अतः हमें मन का भय अभ्यास के माध्यम से दूर करना चाहिए।

इस सन्दर्भ में मुझे अपने पूर्व कक्षाओं में हुई परीक्षाओं के दिन याद आते हैं जिसमें अभ्यास के द्वारा ही मैंने उनका डटकर सामना किया।

मेरी अंग्रेजी की परीक्षा थी। चूँकि वह वार्षिक परीक्षा थी एवं तैयारी कुछ ठीक–ठाक नहीं थी इसलिए मुझे सोचकर घबराहट होती थी कि कहीं मैं फेल न हो जाऊँ परन्तु मेरे अध्यापक महोदय ने मुझे पूर्व परीक्षाओं के प्रश्न पत्र अधिक से अधिक करने की सलाह दी और मैंने अधिक प्रश्न पत्रों का अभ्यास किया जिसके परिणामस्वरूप मैं पास ही नहीं अपितु अच्छे अंकों

के साथ उत्तीर्ण हुआ। अभ्यास एक साधना है जो ध्यान लगाकर साधना करता है उसे सफलता अवश्य मिलती है। अभ्यास से जड़मति भी विद्वान हो जाते हैं। निरंतर चलते रहने से मंद गति वाले भी जीत जाते हैं।

(v) सहशिक्षा का महत्व

शिक्षा मानव के आन्तरिक गुणों को विकसित कर उसके व्यक्तित्व को निखारती है। शिक्षा के आलोक से जीवन में अंधकार नष्ट होता है। शिक्षा का प्रचलन वैदिक युग से है। संस्कृत में कहा है–**"विद्याविहीनः पशुः समानः"**

सहशिक्षा का अर्थ है बालक–बालिकाओं को एक साथ शिक्षा देना। एक साथ शिक्षा देने का अर्थ है कि विद्यालय से सम्बन्धित सभी गतिविधियों में बालक एवं बालिका का समान रूप से साथ–साथ भाग लेना। देश–विदेश में अनेक विद्वानों द्वारा इसका विरोध किया गया क्योंकि अनुसार स्त्री–पुरुष का एक साथ शिक्षा ग्रहण करना उचित नहीं है। इसके लिए उन्होंने यह मत दिया कि सहशिक्षा से विपरीत यौनाकर्षण विद्यार्थियों के मन को भ्रमित करता है और वे अध्ययन पर कम ध्यान देते हैं, परन्तु यह सर्वथा अनुचित है।

सहशिक्षण तो लड़के–लड़कियों की झिझक को समाप्त करने में अहम् भूमिका निभाता है। एक ही कक्षा में अध्ययनरत बालक–बालिकाओं के मध्य स्वस्थ मित्रवत सम्बन्ध स्थापित होते हैं और वे दोनों के लिये प्रत्येक परिस्थिति में सहायक होते हैं। आज समाज में जहाँ नजर जाती है वहाँ कोई भी स्थान ऐसा नहीं है जहाँ नर–नारी साथ काम न कर रहे हों। आज के समय की माँग नर–नारी की समानता है। इसके बिना तो मानव समाज की कल्पना भी नहीं की जा सकती।

शिक्षा जीवन के लिए मनुष्य को तैयार करती है और सहशिक्षा के माध्यम से इसे अनुशासित और नैतिक बनाकर अपने लक्ष्य को प्राप्त किया जा सकता है। यह तभी सम्भव हो सकता है जब सहशिक्षण को सामान्य रूप से देखा जाये क्योंकि यही एक माध्यम है जो स्त्री–पुरुष के मध्य स्वस्थ मित्रता और सामंजस्य स्थापित करने में सक्षम है।

एक ही कक्षा में एक साथ पढ़ने से उनके मध्य मित्रता के संस्कार विकसित और दृढ़ होते हैं। उनमें समानता का भाव आता है। वे स्वयं को एक समान ही मानते हैं और उनके मन में कोई भी भेदभाव नहीं रहता। सहशिक्षा बालक–बालिकाओं में आत्मविश्वास, आपसी सहयोग, स्पर्धा की भावना, एक दूसरे के प्रति सम्मान और मानसिक सामंजस्य बनाने में सहायता करती है। प्रत्येक बालक–बालिका के लिये यह आवश्यक है।

(vi) (a) साँच को आँच नहीं

"साँच को आँच नहीं" मुहावरे का अर्थ है कि सच्चाई को किसी सबूत की आवश्यकता नहीं होती, वह तो वक्त के साथ सामने आ जाती है। इस दुनिया में लोग अनेक कारणों के लिए झूठ के पथ को चुनने का फैसला करते हैं। वे इस बात से बेखबर होते हैं कि एक झूठ को छिपाने के लिए सैंकड़ों झूठ बोलने पड़ते हैं। जो लोग सच्चाई के पथ को चुनते हैं, वह जिन्दगी में खुश रहते हैं। इसका एक उदाहरण याद आ रहा है कि किसी कम्पनी में बोर्ड के साथ मनोज की मीटिंग थी। इस मीटिंग में सब लोग अपने सोच–विचार और सुझाव को पेश करने वाले थे और बाद में रिपोर्ट बनाकर बॉस को दी जानी थी। मनोज, मानव और मिहिर कॉन्फ्रेन्स रूम में पहुँचे और सब बॉस का इंतजार करने लगे। बॉस के न आने पर मीटिंग शुरू हो गई। मीटिंग की शुरूआत नये यन्त्र के प्रस्ताव से हुई। पहले मानव ने अपनी सोच और विचार बताये कि उन्हें किस तरह का यन्त्र बनाना चाहिए और उसके लिए कम्पनी कितना धन लगायेगी। फिर मिहिर ने एक यन्त्र का सुझाव दिया परन्तु अन्य सदस्यों को यह यन्त्र ज्यादा लाभकारी नहीं लगा। इसलिए उन्होंने मनोज की ओर देखा। मनोज के सुझाव व यन्त्र का डिजाइन देखकर बोर्ड के सभी सदस्यों के चेहरे पर मुस्कान आई और अन्त में उन्होंने रिपोर्ट तैयार कर बॉस के पास जाने का फैसला किया।

मनोज, मिहिर और मानव इस बात से बेखबर थे कि कॉन्फ्रेन्स रूम के बाहर कम्पनी के एक सदस्य राम ने उनकी सारी बातें सुन ली थीं। मीटिंग समाप्त होने के तुरन्त बाद राम कॉन्फ्रेन्स रूम में गया। उसने सारे सुझाव और यन्त्र के बारे में भी जानकारी एक बड़े से कागज पर लिखी। वह उस कागज को लेकर सीधा बॉस के पास गया। उसे पता था कि मनोज और उसके साथी तो बॉस को यन्त्र के बारे में कल बतायेंगे इसलिए उसने पहले जाकर बताने की सोची ताकि बॉस को लगे कि यह उसका प्रस्ताव है। राम के सुझाव जानकर बॉस खुश हो गये।

अगले दिन जब मनोज और उसके साथियों ने यन्त्र का सुझाव बॉस को बताया तो उन्होंने कहा कि यह तो राम ने कल ही उन्हें बताया था। यह सुनकर मनोज और उसके साथी हैरान हो गये। बॉस ने जब उनकी सारी बात सुनी तो वे काफी उलझन में पड़ गये कि यह सुझाव वास्तव में किसका है? आखिरकार उन्होंने राम को बुलवाया और उन्होंने राम से उस यन्त्र की उपयोगिता और उसमें लगने वाली लागत के बारे में कई सवाल–जवाब किये। राम इन सवाल–जवाबों के चक्कर में फँस गया और उससे कुछ कहते न बना। फिर बॉस ने वही सवाल–जवाब मनोज और उसके साथियों से किये तो उन्होंने पाया कि उन्होंने न केवल उनके सवालों के जवाब देकर उन्हें सन्तुष्ट किया बल्कि और भी अधिक जानकारी उन्हें दी। सच्चाई बॉस के सामने आ गई। सच ही है कि **"साँच को आँच नहीं"**

(b) ......इस तरह उन्होंने मुझे माफ किया

बात उस समय की है जब मैं कक्षा सात में पढ़ती थी। मुझे गणित विषय से बहुत डर लगता था या यूँ कहें कि मैं उससे बहुत घबराती थी। अक्टूबर का महीना था, दिवाली की छुट्टियाँ आने वाली थीं और हमें सभी विषयों में दिवाली का गृहकार्य मिला था जो विद्यालय खुलते ही हमें सम्बन्धित अध्यापक या अध्यापिका को दिखाना था। सब विद्यार्थियों के साथ–साथ मैं भी बहुत खुश थी।

दिवाली आने में अभी कुछ ही दिन शेष थे। अपने भाई–बहनों के साथ मिलकर मैं घर की साफ–सफाई में लगी हुई थी। हम सब इसके साथ ही विद्यालय का गृहकार्य भी पूरा करने में लगे हुए थे। सभी विषयों का काम मैं पूरा कर चुकी थी। बस गणित का कार्य यह सोचकर छोड़ दिया कि इसे दिवाली के बाद पूरा करूँगी। समय बीतता गया और विद्यालय खुलने का दिन भी आ पहुँचा। गणित का कार्य मैंने अभी भी नहीं किया था और अब मुझे यह चिन्ता थी कि मैं

अपने गणित के अध्यापक के सामने कार्य न करने का क्या बहाना बनाऊँगी।

अगले दिन विद्यालय जाना है यह सोचकर मुझे रात–भर नींद नहीं आई, पर विद्यालय तो जाना ही था। मैं नियत समय पर तैयार होकर विद्यालय पहुँची। जैसे ही गणित का कालांश आया और अध्यापक महोदय ने कक्षा में प्रवेश किया तो मेरे तो होश ही उड़ गये। उन्होंने सबसे कॉपी माँगी, सबने दे दी, बस मैंने ही नहीं दी। अध्यापक ने मुझसे कॉपी नहीं देने का कारण पूछा तो मैंने कह दिया कि मेरा थोड़ा–सा काम अभी बाकी है। यह सुनकर अध्यापक ने मेरी डायरी में कार्य न करने का नोट डाल दिया और उस पर पिताजी के हस्ताक्षर करवाने का आदेश दिया। मैंने हाँ में सर हिला दिया लेकिन घर पर आकर मैंने खुद ही उस पर हस्ताक्षर कर दिये और अगले दिन अध्यापक को दिखा दिये। अध्यापक देखते ही पहचान गये और घर से मेरे पिताजी को बुलवा लिया, मेरी स्थिति काटो तो खून नहीं जैसी हो गई। मुझे लगा कि अब तो मुझे कड़ी–से–कड़ी सजा मिलेगी, परन्तु मेरे पिताजी ने मामले की गम्भीरता को देखते हुए अध्यापक से कहा कि ये हस्ताक्षर उन्हीं के द्वारा किये गये हैं। यह सुनकर मैं हैरान रह गई।

घर आकर मैंने पिताजी से झूठ बोलने का कारण पूछा तो वे बोले कि आज यदि मैं तुम्हें इस गलती के लिए सजा दिलवाता तो मैं गलत होता। तुम्हें सजा दिलवाने की अपेक्षा समझाने की जरूरत है इसलिए मैंने ऐसा किया। यह सुनकर मेरी आँखों में आँसू आ गये और मैं पिताजी के चरणों में गिर पड़ी और इस तरह उन्होंने मुझे माफ कर दिया।

Answer 2.

(i) राजा का दरबार खुले स्थान पर लगा था क्योंकि सर्दियों के दिन थे और ठण्ड से बचने के लिए पूरी आम सभा वहाँ बैठी थी। वहाँ राजा के परिवार के सभी सदस्य, पण्डितजन, दीवान आदि सभी उपस्थित थे। तभी वहाँ एक व्यक्ति एक जैसी दिखने वाली दो वस्तुएँ लेकर आया। जिसमें एक असली थी और एक नकली थी।

(ii) उस व्यक्ति ने उन वस्तुओं की यह विशेषता बताई कि वे दोनों ही आकार, रंग–रूप सब में एक–समान हैं। दोनों ही देखने में बिल्कुल एक जैसी लगती हैं, परन्तु उनमें भिन्नता है। दोनों में से एक काँच है और एक हीरा है। अत्यधिक समानता के कारण कोई इन्हें परख नहीं पाता।

उसने राजा के सामने शर्त रखी कि यदि आपके राज्य में कोई मुझे सही परख कर बता दे कि इनमें से कौन–सा हीरा है और कौन–सा काँच तो मैं यह हीरा आपकी तिजोरी में जमा करवा दूँगा और आप नहीं परख पाये तो इस हीरे की कीमत के बराबर धनराशि आपको मुझे देनी होगी।

(iii) कोई भी उन वस्तुओं को इसलिए नहीं पहचान सका क्योंकि दोनों ही वस्तुएँ रंग–रूप और आकार–प्रकार में नख से शिख तक एक समान थीं। राजा, दीवान तथा सभी दरबारी उसे परखने में असमर्थ रहे और उन्होंने अपनी हार मान ली। हारने पर पैसे देने पड़ेंगे, इस बात का किसी कोई मलाल नहीं था पर राजा की प्रतिष्ठा गिर जायेगी इस बात का सबको डर था।

(iv) अंत में उन वस्तुओं की पहचान एक अन्धे व्यक्ति ने की। उसने उन दोनों को छूकर देखा। धूप में होने के कारण, जो गरम था वह काँच था और जो ठण्डा था वह हीरा था। सभी अंधे व्यक्ति के कार्य से खुश हुये, यहाँ तक कि जो व्यक्ति वहाँ आया था वह भी खुश हुआ कि उसे कोई तो मिला जो सही परख कर सका।

(v) इस कहानी से हमें यह शिक्षा मिलती है कि रूप–रंग आकार में समान होते हुए भी हर व्यक्ति एवं वस्तु की अपनी एक अलग पहचान होती है। जो व्यक्ति छोटी–छोटी बातों पर अपना आपा खो देता है और समस्याओं में उलझ जाता है वह काँच जैसा होता है। इसके विपरीत जो प्रत्येक परिस्थिति में भी अपने विवेक और बुद्धि से काम लेता है वह हीरे के समान ही निखर जाता है।

Answer 3.

(a) **(i)** तुम्हें अध्यापिका ने शीघ्र बुलाया है।

(ii) तुम अकारण डर रहे हो।

(iii) प्रातः भ्रमण में कैसा आनन्द आता है।

(iv) इमारत के गिर जाने की आशंका है।

(v) मुझे एक लोटा गर्म पानी चाहिए।

(b) **(i)** **कलेजे का टुकड़ा**–हर बच्चा अपने माता–पिता के कलेजे का टुकड़ा होता है।

(ii) **मुँह में पानी भरना**–रसगुल्ला देखते ही मेरे मुँह में पानी भर आता है।

(iii) **रोड़ा अटकाना**–सफलता के मार्ग में रोड़े अटकाने वालों की कमी नहीं है।

(iv) **बाग–बाग होना**–परीक्षा में प्रथम आने की खबर सुनकर मैं बाग–बाग हो गया।

(v) **हाथ मलना**–अपना शिकार जाते देखकर शेर हाथ मलता रह गया।

SECTION – B

गद्य संकलन (Gadya Sanklan)

Answer 4.

(i) उपर्युक्त कथन 'प्रेमचन्द' द्वारा रचित 'पुत्र–प्रेम' नामक पाठ से लिया गया है। इस कथन का वक्ता वह युवक है जिसके पिता की मृत्यु हो गई थी और उनकी शवयात्रा को गाजे–बाजे के साथ लाया जा रहा था। यह कथन मणिकर्णिका घाट पर कहा गया है।

(ii) वक्ता द्वारा यह कथन उस सन्दर्भ में है जब यह युवक अपने पिता की अर्थी लेकर मणिकर्णिका घाट पर पहुँचा और प्रभुदास की मृत्यु होने पर चैतन्यदास भी उसी घाट पर उपस्थित थे। तब उस युवक ने प्रसन्न होकर कहा "भैया रुपया पैसा हाथ का मैल है" क्योंकि उसका कहना था कि उसने अपने पिता की बीमारी में अपनी सारी धन–दौलत लुटा दी थी। साथ ही उनकी इच्छानुसार वह मृत्यु उपरांत उन्हें उसी घाट पर लाया था। वह खुश था कि उसने अपने पिता की सब इच्छाएँ पूरी की और प्रभुदास की मृत्यु होने पर चैतन्यदास का पुत्र प्रेम हावी हो रहा था और सोच रहे थे कि यदि तीन हजार रुपये खर्च करा दिये होते तो प्रभुदास जीवित होते। घाट पर उपस्थित युवक उन्हें बहुत बड़ी शिक्षा दे गया। मनुष्य के जीवन से अधिक धन नहीं है।

(iii) धन को समाप्त होने पर दुबारा कमाया जा सकता परन्तु यदि व्यक्ति चला जाये तो उसे वापिस नहीं लाया जा सकता। वक्ता

ने अपने पिता की देखभाल में कोई कसर नहीं छोड़ी, फिर भी वह उन्हें बचा नहीं पाया। वक्ता को इस बात का कोई मलाल नहीं था कि उसके पिता पर बहुत धन खर्च हुआ इसलिए उसने कहा कि 'धन और है ही किसलिए।'

(iv) वक्ता के सम्बन्ध में श्रोता को पता चलता कि अर्थी श्रोता के पिता की है, जिसके बीमार होने पर उनके पुत्र ने बहुत इलाज करवाया, जिसने जहाँ बताया वहाँ लेकर गये। वैद्यों ने जो कुछ भी कहा उसमें कोई कसर नहीं छोड़ी। घर की सारी पूँजी जमीन जायदाद भी पिता की सेवा में लगा दी।

इन बातों का श्रोता पर गहरा प्रभाव पड़ता है। वह अपनी हृदयहीनता, आत्महीनता और भौतिकता से दबे जा रहे थे। उनके अन्दर की ज्वाला उस चिता की ज्वाला से कम दग्धकारिणी नहीं थी अन्ततः श्रोता यह निर्णय लेता है कि वह अपने पुत्र की अन्त्येष्टि पर हजारों रुपये खर्च करेगा और संतप्त हृदय को शांत करेगा।

Answer 5.

नीलम, मालती जोशी द्वारा रचित कहानी 'आउटसाइडर' की प्रमुख पात्र है। वह अपने परिवार की सबसे बड़ी लड़की है। उसके पिता की आकस्मिक मृत्यु के बाद घर और परिवार की सारी जिम्मेदारी नीलम के ऊपर आ जाती है। वह नौकरी कर घर को चलाने की जिम्मेदारी निभाने का निर्णय लेती है और नौकरी करने लगती है। जिस उम्र में लड़की विवाह के सुन्दर स्वप्न संजोती है उस उम्र में वह अविवाहित रहने का निर्णय लेती है और अपने भाई–बहनों को काबिल बनाती है। वह एक कॉलेज में अध्यापन कार्य कर और अविवाहित रहकर अपने पूरे परिवार के भरण–पोषण का उत्तरदायित्व निभाती है।

उसके तीन भाई हैं–सबसे बड़ा सुजीत, मंझला सुदीप ओर छोटा सुमित। इन तीनों की पत्नियाँ क्रमशः सुषमा, अलका और नेहा हैं। तीन भाइयों के बाद उनकी एक छोटी बहिन भी है जिसका नाम पूनम है उसके पति का नाम नरेश है। अपने तीनों भाइयों को पढ़ा–लिखा कर उसने इस काबिल बना दिया कि वह अपनी जिम्मेदारी खुद उठा सकें। सुदीप कनाडा में काम करता है, सुजीत बैंक में काम करता है और सुमित मेडिकल की पढ़ाई कर रहा था। धीरे–धीरे समय के साथ–साथ सुजीत और सुदीप का भी विवाह हो जाता है। परिवार में अब छोटे भाई सुमित का विवाह हो रहा है। इस अवसर पर सभी घर पर एकजुट होते हैं और नीलम को विवाह करने के लिए मनाते हैं। अब उसकी उम्र पैंतालीस वर्ष हो जाती है। अब उसे लगने लगता है कि उसने अपनी सभी जिम्मेदारियों को पूरा कर लिया है इसलिए अब वह आराम से बाकी जीवन काटेगी।

अभी तक वह इस धोखे में जीवन व्यतीत कर रही थी कि उसने सबके लिए अपने जीवन की सभी खुशियाँ न्यौछावर कर दीं तो वे सब भी उसका पूरा ध्यान रखेंगे इसलिए अपने ट्रांसफर और प्रमोशन के प्रस्ताव को भी वह ठुकरा देती है। जब वह सबके कहने के बावजूद भी अपने विवाह प्रस्ताव को ठुकरा देती है तो सुदीप की पत्नी से सुजीत की पत्नी कहती है कि वह तो पहले से ही जानती थी कि दीदी शादी करने से मना कर देंगी क्योंकि इतने दिनों तक वह घर की बॉस बनी घूमती रहीं अब उनके लिए ससुराल के बन्धन में रहना मुमकिन नहीं होगा। उसे भ्रम था कि उसके भाई उसे अपने पास से दूर नहीं जाने देंगे पर उसका यह भ्रम तब चकनाचूर हो जाता है जब वह अपनी भाभी अलका को यह कहते हुए सुनती है कि जब तक नीलम दीदी इस घर में हैं, यह घर उसका नहीं हो सकता। उनके कारण उसे अपनी बहुत सी इच्छाओं का गला घोंटना पड़ता है। अब उसे अपनी बहन पूनम की कही हुई बात याद आती है कि यह घर आज भी तुम्हारे लिए पराया ही है। अभी तो तुम्हारे पास नौकरी है, जब तुम सेवानिवृत्त हो जाओगी तो तुम्हारी हैसियत इस घर में एक फालतू सामान जैसी हो जायेगी जिसे न फेंका जा सके और न रखा जा सके। यह सुनकर उसका सारा भ्रम टूट जाता है।

Answer 6.

संस्मरण लेखिका महादेवी वर्मा जी द्वारा रचित 'स्मृति की रेखाएँ' में संकलित 'भक्तिन' एक प्रसिद्ध संस्मरणात्मक रेखाचित्र है। इसमें उन्होंने अपनी सेविका भक्तिन के अतीत एवं वर्तमान का परिचय देते हुए उसके व्यक्तित्व, उसकी कर्मठता एवं उसके संघर्षपूर्ण जीवन का बड़ा ही जीवन्त शब्द चित्र प्रस्तुत किया है।

भक्तिन जिसका वास्तविक नाम लछमिन था, के जीवन के चार अध्यायों का महादेवी वर्मा ने बड़ा ही सुन्दर एवं सजीव चित्रण किया है। अपने जीवन के पहले अध्याय में वह एक अनाम धन्या गोपालिका की बेटी के रूप में कठिन परिश्रम के पश्चात् भी अभाव का दर्द सहने वाली, विमाता के सौतेलेपन का शिकार होकर कष्ट सहने वाली, छोटी उम्र में हँड़िया गाँव के एक गोपालक के पुत्र के साथ ब्याही और नौ वर्ष में ही गौने के बाद ससुराल आकर पूरा काम–काज सँभालने वाली एक कार्य कुशल बहू के रूप में दिखाई पड़ती है।" सम्पत्ति की रक्षा में सतर्क विमाता ने उसके पिता के मरणांतक रोग का समाचार उसे तब भेजा, जब वह मृत्यु की सूचना बन चुका था।" अपने पिता के अन्तिम दर्शन का सुख भी भक्तिन के भाग्य में नहीं बदा था।

अपने जीवन के दूसरे अध्याय में भी भाग्य की कुंचित रेखाओं से उसका पीछा नहीं छूटा। उसने एक के बाद एक तीन कन्याओं को जन्म दिया। सास और जिठानियों के लड़के होने के कारण उसे उनकी उपेक्षा ही नहीं सहनी पड़ी वरन् घर का सारा काम भी उसके जिम्मे आ पड़ा था, पर उसके स्वभाव और मेहनत का ही परिणाम था कि पति उससे बहुत प्यार करता था "अपने पति के प्यार के दम पर ही उसने परिवार से अलगौझा करके सबको अंगूठा दिखा दिया।

जीवन के तीसरे अध्याय में भी दुर्भाग्य ने उसका पीछा नहीं छोड़ा। उसकी बड़ी लड़की भी विधवा हो गई और सम्पत्ति के लालच में उसके जेठ के लड़के ने अपनी विधवा चचेरी बहन का विवाह अपने तीतर लड़ाने वाले साले से छल के साथ करवा दिया। दामाद की कामचोरी तथा पारिवारिक द्वेष के कारण घर की समृद्धि जाती रही और विवश होकर भक्तिन को कमाई के लिए शहर आना पड़ा।

जीवन के चौथे अध्याय में वह लेखिका के पास नौकरी के लिए पहुँची। उसके छोटे कद, दुबले शरीर, पतले होंठ और गले में कंठीमाला पहने देख लेखिका ने उसका नाम भक्तिन रख दिया क्योंकि वह अपना वास्तविक नाम (लछमिन) किसी को बताना नहीं चाहती थी। महादेवी वर्मा के यहाँ भी उनकी सेविका कम, अभिभाविका के रूप में भक्तिन ने अपने कठिन परिश्रम का परिचय दिया। पहले ही दिन कोयले की मोटी रेखा से चौके की सीमा निर्धारित कर खाना बनाना शुरु कर दिया और न चाहते

हुए भी लेखिका को पाक विद्या से समझौता करना पड़ा। इस देहाती वृद्धा से वे इतना प्रभावित हुई कि वे अपनी असुविधाएँ छिपाने लगी। सेवा धर्म में भक्तिन हनुमान से स्पर्द्धा करती थी।

भक्तिन में दुर्गुण भी थे। यहाँ–वहाँ पड़े पैसों को मटकी में छुपाकर रखना वह गलत नहीं मानती थी क्योंकि इसे वह अपना घर समझती थी। भक्तिन अपनी मालकिन के व्यक्तित्व को असाधारण मानती थी। स्वयं को महत्व देने के लिए वह कभी उनकी उत्तर पुस्तिकाओं को बाँधकर, कभी अधूरे चित्र को कोने में रखकर, कभी रंग की प्याली धोकर, कभी चटाई को आँचल से झाड़कर जो सहायता करती थी, उससे वह अपने अधिक बुद्धिमान होने का प्रमाण देना चाहती थी।

लेखिका के साथ वह उनकी छाया की तरह रहती थी। किसी भी परिस्थिति में उसे लेखिका का साथ छोड़ना स्वीकार नहीं था। युद्ध के समय जब सब लोग पलायन कर रहे थे, उस वक्त उसका तथा लेखिका का रिश्ता स्वामी और सेवक का न होकर आत्मीयता का दिखता है। लेखिका के जेल जाने पर वह भी उनके साथ वहाँ जाने को तैयार रहती है ताकि वह महादेवी वर्मा का काम कर सके। उसके इस प्यार को देखकर लेखिका इतनी भावुक हो जाती है और यह कल्पना भी नहीं कर पाती है कि उनका अन्तिम बुलावा आने पर उस देहातिन वृद्धा की क्या दशा होगी?

इस तरह अपने कठिन संघर्ष एवं परिश्रम के द्वारा भक्तिन अपने व्यक्तित्व की छाप न सिर्फ लेखिका के हृदय पर छोड़ती है वरन् पाठकों के हृदय में भी अपनी जगह बना लेती है।

काव्य मंजरी (Kavya Manjari)

Answer 7.

(i) नदी के द्वीप बहते नहीं हैं उनके मन में अपनी माँ नदी के प्रति पूर्ण समर्पण भाव है परन्तु वह बहते नहीं है। यदि वे नदी के साथ बह जाएँगे तो उनका कोई स्वयं का अस्तित्व नहीं रहेगा। कवि कहते हैं कि जिसका जो गुण है, उसको धारण करने से अस्तित्व बना रहता है।

(ii) बहने का अर्थ है अपने अस्तित्व को समाप्त करना। यदि द्वीप बहेंगे तो उनका अस्तित्व समाप्त हो जायेगा और वे रेत बनकर कहीं और चले जायेंगे। उनका रूप–स्वरूप नष्ट हो जाएगा, नदी की रेत से ही द्वीप का निर्माण हुआ है और उसका त्याग स्थिर रहकर सर्मपण करने में है, नदी की धारा के साथ बहने में नहीं।

(iii) द्वीप कभी भी नदी की प्रवाहित होने वाली धारा नहीं बन सकते। द्वीप का अस्तित्व स्थिर रहते हुए त्याग की भावना में है वे हमेशा ही नदी के द्वीप कहलाते हैं। वे नदी की गति के साथ नहीं चलते। यदि वे नदी के साथ प्रवाहित हो जायेंगे तो स्वयं रेत बन जायेंगे और उसका अस्तित्व समाप्त हो जायेगा। यदि ये अपने स्थान से डिग जायेंगे, तो सभी स्थान जल से पूर्ण हो जायेंगे। यदि ये पूर्ण हो जाये तो भी धारा नहीं बन सकते। ये द्वीप रेत बनकर बहकर जल को गन्दा कर देंगे और इस प्रकार का जल किसी भी काम का नहीं रहेगा।

(iv) प्रस्तुत कविता का सामाजिक सन्दर्भ यह है कि व्यक्ति का अस्तित्व समाज के अनुरूप चलने से है। समाज के विपरीत चलने में उसका कोई अस्तित्व नहीं है। यहाँ द्वीप, नदी और भूखण्ड को कमशः व्यक्ति, परम्परा और समाज के प्रतीक के रूप में चुना गया है। कवि का विचार है कि जिस प्रकार द्वीप 'भू' का ही एक खण्ड हैं, परन्तु नदी के कारण उसका अस्तित्व अलग है, उसी प्रकार व्यक्ति भी समाज का अंग है परन्तु सामाजिक परम्पराएँ उसे विशिष्ट व्यक्तित्व प्रदान करती है।

Answer 8.

कवि ने हिमालय के वर्षाकालीन सौन्दर्य का बड़ा ही मोहक चित्र प्रस्तुत किया है। हिमालय की ऊँची–ऊँची चोटियाँ बर्फ से ढकी हुई रहती हैं। उन हिमाच्छादित चोटियों पर जब धूप पड़ती है तो उन पर पड़ी हुई ओस की बूँदें मोती के समान चमकती हुई सी प्रतीत होती हैं। इस सौन्दर्य का कवि ने जी भरकर पान किया है। ऊँची–ऊँची चोटियों के मध्य स्वच्छ और निर्मल पानी से भरी हुई झीलें हैं जिनमें तैरते हुए हंस कमल के खट्टे–मीठे तन्तुओं का स्वाद लेने के लिए व्याकुल रहते हैं। प्रभात की सुन्दर बेला में धीमी–धीमी वायु मंद गति से चल रही है। उगते हुए सूर्य की कोमल किरणें चोटियों पर पड़ रही हैं। उस सुनहरी धूप में चकवा–चकवी का जोड़ा प्रेमालाप कर रहा है।

"मैंने तो भीषण जाड़ों में
नभ चुंबी कैलाश शीर्ष पर
महामेघ को झंझानिल से,
गरज–गरज भिड़ते देखा है
बादल को घिरते देखा है।"

बर्फ से ढकी हुई घाटियों में एक कस्तूरी मृग अपनी नाभि से आती हुई सुगन्ध को पाने के लिए घूम रहा है और जब वह उसे नहीं मिल पाता तो वह चिढ़ने लगता है। उसका चिढ़ना कवि को बहुत प्रिय लग रहा है। कवि को घिरते हुए बादलों का यह दृश्य बड़ा ही मनोरम लग रहा है।

हिमालय के बर्फीले और अतीव सुन्दर आँगन को देखकर कवि सोचता है कि कुबेर की अलकानगरी कहाँ चली गई और कालिदास की मेघदूत में वर्णित आकाशगंगा का भी कुछ पता नहीं है। कवि कहते हैं कि भीषण जाड़ों में आकाश को छूते हुए कैलाश पर्वत पर एक बड़े बादल को टकराते हुए देखा है।

हिमालय पर्वत पर स्थित देवदारु के जंगलों में कल–कल करते झरनों का स्वर कवि को स्वर्गिक आनन्द की अनुभूति करा रहा है। इस प्रकार इस सौन्दर्य को देखकर कवि इतना अभिभूत हो गया है कि वह प्रकृति के प्रति नत–मस्तक हो रहा है।

Answer 9.

महादेवी वर्मा पथिक को सांसारिक बाधाओं और विपदाओं के बीच में भी संकल्प पथ पर अपना विकास करते हुए निरन्तर आगे बढ़ने की प्रेरणा दे रही हैं। वह पथिक से कहती हैं कि साधना के कष्टों को स्वीकार करते हुए उसमें निरन्तर लीन रहकर अपने लक्ष्य की प्राप्ति में लगे रहो क्योंकि उसे अभी बहुत दूर जाना है। अतः उसके लिए आलस्य और प्रमाद उचित नहीं है क्योंकि ये विकास के मार्ग में बाधक हैं। महादेवी जी ने अजर–अमर जीवात्मा को पथिक की संज्ञा दी है। वह कहती हैं कि जीवात्मा अपनी छाया को वास्तविकता की संज्ञा देकर अपने लिए कारागार का निर्माण नहीं करे। जीवात्मा वस्तुतः अमर पुत्र है। अतः उसे माया के आकर्षण में नहीं बाँधना है और न ही मृत्यु को अपने हृदय में स्थान देना है। ये सब तो पथिक के मार्ग की कठिनाइयाँ हैं जो उसे लक्ष्य से भटकाने का प्रयास

करती हैं। पथिक को इन सब पर विजय प्राप्त करते हुए आगे बढ़ते जाना है।

कवयित्री का मानना है कि जीवन पथ बाधाओं से युक्त होने के साथ–साथ लंबा भी है। अतः पथिक को दृढ़ संकल्प के साथ आगे बढ़ना है। चाहे सदा अविचल रहने वाला हिमालय पर्वत कम्पित हो जाये, आकाश भले ही प्रलय वर्षा करने लगे, भले ही अंधकार की काली छाया प्रकाशवान सूर्य को निगल जाये और भयानक तूफान विद्युत की गति से चारों ओर फैल जाये फिर भी पथिक को निडर और अविचल रहकर अपने पथ पर अविराम चलते रहना है। इस मार्ग पर चलते हुए उसके मार्ग में अनेक आकर्षण और मधुर कठिनाइयाँ आयेंगी जो उसे मार्ग में डिगाने का प्रयास करेंगी। इन सबसे पथिक को निरन्तर सजग रहना है। सांसारिक सुख मानव के लिए सबसे बड़ी बाधा है। अतः इन मिथ्या सुखों का त्याग कर उसे अपने संकल्प–पथ पर बढ़ते जाना है। ये सभी सांसारिक सुख क्षणभंगुर होते हैं, वास्तव में इनका कोई मोल नहीं होता। हमें पतंगे की तरह दीपक के प्रेम में जलकर राख बनने एवं अपने प्रेम को अमर दीपक का हिस्सा बनाकर अमरतत्व प्राप्ति की ओर अग्रसर रहना चाहिए और साधना के पथ पर यानि अंगारों की शैय्या पर तपस्या रूपी कलियाँ बिछानी चाहिए एवं उस अज्ञात प्रियतम से मिलने के लिए प्रयत्नशील रहना चाहिए। अतः कवयित्री पथिक को यह प्रेरणा दे रही है कि उसे अपने कर्तव्य पर इतना आगे बढ़ना है जिसके आगे मार्ग ही न हो।

"सारा आकाश" (Saara Akash)

Answer 10.

(i) प्रभा समर की पत्नी और नाटक की प्रमुख नारी पात्र थी। वह समर से नाराज थी। प्रभा व समर के बीच लड़ाई का कारण था उन दोनों के बीच उत्पन्न हुई गलतफहमियाँ। समर अपने अहं के कारण प्रभा से बात भी नहीं करता था और उसने प्रभा को कसकर तमाचा भी मारा था।

(ii) उपन्यास के नायक का नाम समर है। समर आत्मविश्लेषण करता है कि उसने कभी भी प्रभा पर भरोसा नहीं किया। आज प्रभा को देखकर उसे यह विश्वास हो जाता है कि उसका अनुमान गलत था और उसे अपने किये पर पछतावा भी होता है। वह उठकर प्रभा के पास गया और बोला, "ओस पड़ रही है, बीमार पड़ सकती हैं, रोना बंद कीजिए और कमरे में जाकर सो जाइए।" इतनी–सी बात कहते–कहते उसका गला भर आया।

(iii) प्रभा का पति समर पिघल जाता है। वह अनुभव करने लगा कि उसकी सारी दृढ़ता, अकड़, संयम, आत्मनुशासन आदि समाप्त हो गये हैं और इन सबके स्थान पर अनुताप, ग्लानि, आत्म भर्त्सना के भाव लहरा रहे हैं और यह उसके भीतरी मन को मथ रहे हैं। शरीर से थका हुआ और मन से हारा हुआ वह उस क्षण खुद को टूटा हुआ अनुभव करने लगता है।

(iv) समर को लगा वह खड़ा नहीं रह सकता। वह प्रभा के सामने धरती पर बैठ गया। पश्चाताप व ग्लानि के दुर्वह भार को न सह सकने के कारण उसके मुख से यह शब्द निकल पड़े। "प्रभा क्या तुम मुझसे रूष्ट हो?" उसे स्वयं पर आश्चर्य हो रहा था अपने कठोरता की पाषाण मूर्ति से मोम की पुतली बन जाने पर वह स्वयं को ही नहीं पहचान पाया कि यह वही है जो इतनी मधुरता से अपनी पत्नी के दुःख को समझकर उसकी नाराजगी को दूर करने का प्रयास कर रहा था। इससे यह पता चलता है कि समर मूलतः भावुक, सहृदय युवक था, झूठे अहंभाव ने उसका मन दूषित कर दिया था।

Answer 11.

राजेन्द्र यादव जी द्वारा लिखित 'सारा आकाश' के पात्र शिरीष भाई साहब के विचारों द्वारा संयुक्त परिवार से होने वाली मुश्किलों, हताशा एवं अवसाद की स्थिति का बड़ा ही तर्कपूर्ण वर्णन करते हुए उसे 'अभिशाप' बताया गया है। परिवार के साथ मिलकर रहने की भावुकता कितने भयंकर परिणामों को जन्म देती है। 'रात–दिन हैरान–परेशान रहेंगे, लड़ेंगे–मरेंगे, सब होगा लेकिन रहेंगे साथ ही, इज्जत का सवाल, अलग चूल्हा न जलने की हठ" ऐसी ही न जाने कितनी बातें, जो भयंकर परिणाम हैं संयुक्त परिवार के।

शिरीष भाई दिवाकर के चचेरे भाई हैं। आधुनिक विचारों वाले शिरीष भाई का मानना है कि रूढ़ियों में बँधकर विकास करना सम्भव नहीं है। समय के साथ समाज एवं व्यक्ति विशेष की मानसिक विचारधारा में परिवर्तन होना आवश्यक है। शिरीष यहाँ अपनी बहन का इलाज कराने आए थे। उनसे जब समर एवं दिवाकर की बात होती है तब वे दोनों परिवारों की स्थिति जानकर इस निष्कर्ष पर पहुँचे कि लगातार चलने वाली तकरार एवं झगड़े संयुक्त परिवार के कारण होते हैं।

उनका कहना था या तो परिवार इतना समर्थ हो कि एक घर में रहते हुए भी सबको अलग–अलग सुविधाएँ मिलें और स्वाभिमान की रक्षा भी हो सके तो ही संयुक्त परिवार चल सकता है अन्यथा अर्थभाव और अशिक्षा में रहकर तो अशांति ही बनी रहेगी।

वास्तविक स्थिति जानने के पश्चात् शिरीष भाई समर की ओर संकेत करके कहते हैं–"इस संयुक्त परिवार परम्परा को तोड़ना ही होगा। अब आपके लिए ही कहूँ कि अगर आप खुद जिंदा रहना चाहते हैं या चाहते हैं कि आपकी पत्नी भी रहे तो इसके सिवाय कोई रास्ता नहीं है कि अलग रहिए। जैसा भी हो अलग रहिए........।"

शिरीष भाई ऐसा कहेंगे, यह समर ने सोचा ना था। यह सुनकर वह स्तब्ध रह गया क्योंकि यह सलाह उसके लिए अप्रत्याशित थी। शिरीष आगे समर से कहते हैं–............"सच्चाई यह है कि आपकी पत्नी की सारी आकांक्षाएँ और आपके सारे प्रयत्न इस वातावरण में बिल्कुल बेबस होकर घुट–घुटकर मर जाएँगे। एक के बाद एक मुसीबतें निकलती आएँगी और आप लोग शायद सुख की नींद के सपने में ही जिन्दगी बिता देंगे"।

शिरीष का मानना था कि संयुक्त परिवार में रहकर व्यक्ति को इच्छाओं का दमन करना पड़ता है। दूसरे महँगाई बढ़ने के कारण, परिवार के सदस्यों की संख्या अधिक होने के कारण आज की आर्थिक स्थिति में संयुक्त परिवार चलाना अत्यंत कठिन है।

"संयुक्त परिवार का शाब्दिक अर्थ चाहे कितना महान हो, उसका सबसे बड़ा नुकसान यह होता है कि परिवार का कोई भी सदस्य अपने व्यक्तित्व का विकास नहीं कर पाता। सारा समय तो समस्याएँ बनाने में या बनी–बनायी समस्याओं को सुलझाने में चला जाता है।" लड़ाई–झगड़ा, खींचतान, बदला, ग्लानि सब मिलाकर वातावरण ऐसा विषैला और दमघोंटू बना रहता

है जिसमें व्यक्ति दो हिस्से में बँट जाता है। वह बाहर आर्थिक लड़ाई से जूझता है और भीतर उसे अपनी पारिवारिक उलझनें भी सुलझानी पड़ती हैं। आदमी का सीधा–सरल दिमाग इन्हीं उलझनों में पिस जाता है।

संयुक्त परिवार का विरोध करते हुए आगे शिरीष भाई कहते हैं–संयुक्त परिवार में व्यक्ति को अपनी शक्तियों और प्रतिभा को सही दिशा में बढ़ाने का अवसर ही नहीं मिलता।......... बौद्धिक और मानसिक विकास के लिए आपके पास शक्ति ही नहीं बचती।''

इसका अर्थ यह कदापि नहीं कि हम परिवार से अलग होकर परिवार के प्रति अपने दायित्व को भूल जाएँ उसे विस्मृत कर दें। जो जिम्मेदारी या कर्तव्य परिवार के प्रति हमारा है उसे अलग रहकर भी पूरा किया जा सकता है। कर्तव्य के नाम पर संयुक्त परिवार में रहकर अपने जीवन का नाश करना बेवकूफी है।

स्पष्ट है शिरीष एकांकी परिवार के समर्थक एवं संयुक्त परिवार के विरोधी जान पड़ते हैं। समय के साथ समाज की इस पारिवारिक इकाई में यदि परिवर्तन की तनिक भी गुंजाइश है तो उसमें परिवर्तन स्वीकार्य होना ही चाहिए।

Answer 12.

उपन्यासकार राजेन्द्र यादव ने 'सारा आकाश' उपन्यास की रचना मध्यमवर्गीय पृष्ठभूमि को ध्यान में रखकर की है। ऐसा ही एक परिवार समर का है जिसमें कुल नौ सदस्य हैं। धीरज बड़ा बेटा है इसलिए उसकी पत्नी को सभी भाई–बहन भाभी कहते हैं। स्त्री पात्रों में प्रभा के बाद उपन्यास में भाभी का चरित्र ही सबसे महत्वपूर्ण है। 'सारा आकाश' की भाभी भी एक आम भारतीय परिवार की भाभी की तरह है जिसमें अच्छाइयाँ कम बुराइयाँ अधिक हैं। उसके चरित्र की विशेषताएँ निम्नलिखित हैं—

1. **ईर्ष्यालु प्रवृत्ति**—समर के विवाह के बाद से ही भाभी समर की पत्नी प्रभा से ईर्ष्या करने लगती है। प्रभा की पढ़ाई–लिखाई और सुन्दरता उसमें ईर्ष्या का भाव जगा देती है, इसीलिए प्रभा को नीचा दिखाने का कोई भी मौका वह नहीं चूकती है। जहाँ एक और वह प्रभा के प्रति अम्माजी के कान भरती है वहीं समर को भी उसके प्रति भड़काती रहती है।
2. **षड्यन्त्र रचने में निपुण**—समर की भाभी प्रभा के काम में हर समय कमी निकालती रहती थी। ताकि प्रभा की कोई तारीफ न करें। इसके लिए वह कोई न कोई षड्यन्त्र रचती थी। जैसे—प्रभा ने विवाह के बाद जब पहली बार खाना बनाया तो दाल में उसने नमक सही डाला था। क्योंकि प्रभा स्वयं चखकर देख चुकी थी, परन्तु मौका पाते ही भाभी ने दाल में इतना नमक डाल दिया कि दाल कड़वी हो गई और प्रभा को सबके ताने सुनने पड़े।
3. **निन्दा प्रेमी**—समर की भाभी को दूसरों की बुराई करने में बहुत आनन्द आता है। प्रभा के मायके चले जाने पर वह समर से प्रभा की तरह–तरह से बुराई करती है और यह जताना चाहती है कि प्रभा को अपनी पढ़ाई और सुन्दरता पर बहुत घमण्ड है। वह समर को बड़ी चतुराई से अपनी बातों के जाल में फँसाकर स्वयं को समर का हितैषी सिद्ध करते हुए जी भरकर प्रभा की निन्दा करती है।
4. **झगड़ालू प्रवृत्ति**—समर की भाभी स्वभाव से ही झगड़ा करने वाली महिला है। परिवार में छोटी से छोटी बात को बढ़ा–चढ़ाकर कहना भाभी के स्वभाव की विशेषता है। बात का बतंगड़ बना देना और बेवजह उल्टी–सीधी बातें सुनाना भाभी की एक अन्य विशेषता है।
5. **अन्धविश्वासी महिला**—समर की भाभी अशिक्षित महिला है वह अन्धविश्वासी भी है। वह भूत–प्रेत, तन्त्र–मन्त्र, गण्डा–ताबीज आदि में आँख मूँदकर भरोसा करती है। एक दिन अनजाने में मिट्टी समझकर जब प्रभा मिट्टी के गणेश जी से बर्तन माँज लेती है तो भाभी यह कहकर कि अब तो अनिष्ट होकर ही रहेगा, कोहराम मचा देती है।
6. **व्यंग्य बाण कसने वाली**—प्रभा पर व्यंग्यबाण छोड़ने और उस पर ताना कसने में भाभी कभी पीछे नहीं हटती है। उसे प्रभा से तीन शिकायतें थीं। पहली यह कि वह दहेज में कुछ नहीं लाई, दूसरी यह कि वह पढ़ाई का बहुत दिखावा करती है, तीसरी यह कि उसे अपनी खूबसूरती का बहुत घमण्ड है। वस्तुतः यह भाभी की हीनभावना थी जिसके कारण वह प्रभा को तंगकर, उसे छुपाने का प्रयास करती है। देवर के प्रति सहज भाव रखने वाली भाभी के रूप में वह समर की हितैषी बन उसे सलाह देती दिखाई देती है पर साड़ी एवं घर के कामकाज के सम्बन्ध में उन दोनों के बीच सहजता नहीं दिखाई देती है। समय–समय पर वह समर और प्रभा को साथ रहने की सलाह देकर परिवार के प्रति अपने प्रेम भाव को प्रकट करती है।

इस प्रकार लेखक ने 'सारा–आकाश' में जिस भाभी का चित्रण किया है वह आज भी हमारे परिवारों में देखी जा सकती है। नई बहू के प्रति असहयोगी व्यवहार, उसकी निन्दा और गलतियों को बढ़ा–चढ़ाकर प्रस्तुत करना, एक आम भारतीय भाभी का यही चरित्र है।

''आषाढ़ का एक दिन'' (Aashad Ka Ek Din)

Answer 13.

(i) प्रस्तुत पंक्ति मोहन राकेश के 'आषाढ़ का एक दिन' से ली गई है। यहाँ कालिदास एक हरिणशावक से संवाद कर रहे हैं। इसमें बाण से घायल हिरन को मल्लिका अपने हाथों से उठाकर कालिदास के साथ अपने घर लाती है।

(ii) वक्ता अर्थात कालिदास अत्यंत भावुक एवं संवेदनशील पुरुष है हरिणशावक की पीड़ा उसके अन्तर्मन को व्याकुल कर देती है वह न केवल उसके प्रति सहानुभूति प्रदर्शित करता है अपितु तन के साथ–साथ उसके मन की सुकोमलता, हरिणशावक के प्रति उसे सहानुभूति के भाव से भर देती है। इसी कारण वक्ता हरिणशावक को पीड़ा सह सकने का ढ़ाँढस बँधाने का प्रयत्न कर रहा है।

(iii) प्रस्तुत पंक्तियों का वक्ता कविहृदय एवं संवेदनशील होने के कारण संवेदना से ओत–प्रोत है। इसलिए वह प्रेम को सदैव हिंसा और घृणा से ऊपर पाता है। वह अपनी सुकोमल उँगलियों के स्पर्श से हरिणशावक को पीड़ा से राहत प्रदान करने का प्रयास करता है। वक्ता को पूर्ण विश्वास है कि यदि एक बाण प्राण हर लेने की क्षमता रखता है, तो संवेदनाओं के सम्मिश्रण से ओत–प्रोत प्रेम सहित किया गया उसकी उँगलियों का वह कोमल स्पर्श उसके प्राण बचाने की क्षमता से परिपूर्ण है।

(iv) ''आषाढ़ का एक दिन'' नाटक की भाषा विषयवस्तु को प्रभावशाली बनाने में पूरी तरह सफल रही है। भाषा पात्रानुकूल है। भाषा में प्रवाहमयता तथा बोधगम्यता है।

प्रस्तुत नाटक में भाषा की काव्यात्मकता विशेष प्रभाव उत्पन्न करने वाली है। कालिदास के रचनाकाल के अनुसार ही भाषा का प्रयोग हुआ है। शब्दों का चयन पात्रों के भावानुकूल है। जैसे—

"मातुल...मैं समझता हूँ कि जो कुछ मैं समझ पाता हूँ, सत्य सदा उसके विपरीत होता है और मैं जब उस विपरीत तक पहुँचने लगता हूँ तो सत्य उस विपरीत से विपरीत हो जाता है।"

उपर्युक्त कथन में जहाँ प्रवाहमयता है वहीं काव्यात्मकता एक विशिष्ट चमत्कार उत्पन्न करती प्रतीत हो रही है। प्रकोष्ठ, आतिथ्य, निष्ठुर, अभ्यर्थना आदि ऐसे शब्द हैं जो नाटक की भाषिक समृद्धता को प्रदर्शित करते हैं।

शैली की दृष्टि से नाटक में कुछ ऐसे बिम्ब हैं जो दर्शकों को बाँधे रखते हैं। आरम्भ में बिजली की चमक, अम्बिका का चिंतित होना, कालिदास के पैरों की आहट, अलग होते प्रेमी, मल्लिका का आँसुओं में डूबा हुआ चेहरा आदि दृश्य प्रभावकारी है।

अतः नाटककार द्वारा नाटक में भाषा–शैली का बड़ी सफलता एवं अनुकूलता के साथ प्रयोग किया गया है।

Answer 14.

'आषाढ़ का एक दिन' नाटक मोहन राकेश जी द्वारा लिखित एक प्रसिद्ध नाटक है। इसमें कालिदास व मल्लिका के अतिरिक्त विलोम भी एक पात्र है। पूरे नाटक में वह कालिदास के विपरीत ही रहता है। इसमें विलोम ग्राम्य परिवेश से है और वह खलनायक के रूप में है। मल्लिका कालिदास से प्रेम करती है और विलोम मल्लिका से विवाह करने के लिए प्रयास करता है। वह कालिदास और मल्लिका के विषय में मल्लिका की माँ के कान भरता है—

1. **कालिदास से ईर्ष्या करना**— "जानता हूँ अम्बिका! मल्लिका बहुत भोली है। वह लोक और जीवन के सम्बन्ध में कुछ नहीं जानती। वह यह नहीं चाहती कि मैं इस घर में आऊँ क्योंकि कालिदास नहीं चाहता, और कालिदास क्यों नहीं चाहता? क्योंकि मेरी आँखों में उसे अपने हृदय का सत्य झाँकता दिखाई देता है। उसे उलझन होती है...किन्तु तुम नहीं जानती अम्बिका, मेरा एकमात्र दोष यह है कि मैं जो भी अनुभव करता हूँ उसे कह देता हूँ।"
2. **अपनी उपस्थिति का भान कराना**—मल्लिका के न चाहने पर भी वह हर परिस्थिति में उपस्थित हो जाता है। वह मल्लिका से कहता है—"यह अनुभव करने की मैंने आवश्यकता नहीं समझी कि तुम मुझसे घृणा करती हो, मैं जानता हूँ परन्तु मैं तुमसे घृणा नहीं करता, मेरे यहाँ होने के लिए इतना ही पर्याप्त है।"
3. **अपनी विजय में मदहोश होना**—कालिदास के उज्जयिनी जाने पर उनका विवाह प्रियगुंमंजरी से हो जाता है। इधर मल्लिका की माँ का निधन हो जाता है। परिस्थितियाँ कुछ ऐसा मोड़ लेती हैं कि मल्लिका को विलोम के सामने आत्मसमर्पण करने के लिए विवश होना पड़ता है। अंत में जब कालिदास मल्लिका के पास आते हैं तो विलोम स्वयं को विजयी घोषित करता है।
4. **कालिदास के प्रति व्यंग्य बाण छोड़ना**—"तुम चाहते हो कि इस समय मैं यहाँ से चला जाऊँ। मैं चला जाता हूँ इसलिए नहीं कि तुम आदेश देते हो बल्कि इसलिए कि तुम यहाँ आज अतिथि के रूप में आये हो और अतिथि की इच्छा का मान करना चाहिए। देखना मल्लिका, आतिथ्य में कोई कमी न रहे। जो अतिथि वर्ष में एक बार आया है, वह आगे जाने आयेगा भी या नहीं।"
5. **नकारात्मक सोच वाला व्यक्ति**—विलोम नकारात्मक सोच वाला व्यक्ति है, वह कहता भी है—"विलोम क्या है? एक असफल कालिदास और कालिदास? एक सफल विलोम।" अपनी नकारात्मक सोच के कारण वह कहीं–कहीं ईर्ष्यालु प्रवृत्ति का भी दिखाई पड़ता है और समय–समय पर कालिदास को अपने कटाक्ष से आहत करता है—"राजधानी के वैभव में जाकर ग्राम प्रान्तर को भूल तो नहीं जाओगे?
6. **हठी व्यक्ति**—विलोम एक हठी व्यक्ति भी है। वह मल्लिका के न चाहते हुए भी उसके घर में बिन बुलाए अतिथि की तरह आता है, उसके जख्मों को कुरेदता है। मल्लिका और कालिदास के सम्बन्धों को लेकर सवाल करता है कि कालिदास अकेले ही उज्जयिनी जा रहा है या मल्लिका भी उसके साथ जा रही है? अम्बिका, मल्लिका और कालिदास, तीनों ही उसे वहाँ से जाने के लिए कहते हैं पर वह अपनी बात कहे बिना वहाँ से टलता नहीं है। यहाँ एक तरफ तो ऐसा लगता है मानो उसमें कोई आत्मसम्मान नहीं है पर दूसरी तरफ उसका अम्बिका के साथ सहानुभूति रखना, उसे मल्लिका के भोलेपन के बारे में समझाना उसके संवेदनशील व्यक्तित्व एवं शुभचिंतक रूप को प्रकट करता है।
7. **वैवाहिक जीवन सुखमय नहीं**—विलोम यथार्थवादी, आक्रामक और अर्न्तद्वन्द्वों के पचाए हुए हैं। कालिदास से प्रतिद्वन्द्विता उसे विजयी बना चुकी है, लेकिन भीतर ही भीतर वह एक हारा हुआ व्यक्ति है। यद्यपि वह मल्लिका का पति है पर मल्लिका के भावकोष्ठ में सदैव कालिदास विद्यमान रहता है। विलोम इसी विडम्बनापूर्ण त्रासदी को झेलता है। इसलिए उसका वैवाहिक जीवन भी सुखमय नहीं है।

उपर्युक्त विवेचन से स्पष्ट है कि खलनायक के रूप में विलोम की स्थिति सार्थक है। खलनायक के साथ–साथ वह कालिदास से भी अधिक विकसित पात्र है। उसकी विजय होने का अर्थ भावना पर यथार्थ की विजय है।

Answer 15.

मोहन राकेश द्वारा लिखित नाटक 'आषाढ़ का एक दिन' एक यथार्थवादी नाटक है जो पात्रों के आंतरिक व मानसिक अर्न्तद्वन्द्व तथा संवेदनशीलता को प्रकट करता है। प्रस्तुत पंक्ति नाटक की नायिका मल्लिका द्वारा कही गई है तथा श्रोता नायक कालिदास है।

कालिदास वर्षों बाद सत्ता और प्रभुता का मोह छोड़, मातृगुप्त के कलेवर से मुक्त होकर पुनः कालिदास के कलेवर में जीने की इच्छा लिए अपनी प्रेमिका (मल्लिका) के पास फिर से एक नया जीवन जीने की आस में आता है। अपनी इच्छा के विपरीत, न चाहते हुए भी वह अनमनेपन से उज्जयिनी चला गया था। वहाँ रहकर उसने जो कुछ लिखा, वह यहाँ के जीवन का ही संचय था, वह कहता है 'कुमार सम्भव' की पृष्ठभूमि यह हिमालय है और तपस्विनी उमा तुम हो। 'मेघदूत' के यक्ष की पीड़ा, मेरी पीड़ा है और विरह विमर्दिता यक्षिणी तुम हो। 'अभिज्ञान शाकुन्तलम्' में शकुन्तला के रूप में तुम्हीं मेरे सामने थीं। मैंने जब–जब लिखने का प्रयत्न किया तुम्हारे और अपने जीवन के इतिहास को फिर–फिर दोहराया। वह चाहता था कि मल्लिका सब कुछ पढ़े।

मल्लिका ने बताया कि उसने वह पढ़ा है और अपने हाथों से सीकर पन्ने बनाए थे, सोचा था कि जब कालिदास राजधानी से आएगा, तो वह उसे यह भेंट देगी और इन पृष्ठों

पर अपने सबसे बड़े महाकाव्य की रचना करने के लिए कहेगी परन्तु उस बार कालिदास वहाँ आकर भी मल्लिका से मिलने नहीं आया और यह भेंट यहीं पड़ी रही। अब तो ये पन्ने टूटने भी लगे थे और उसे यह कहते हुए संकोच हो रहा था कि ये उसकी रचना के लिए है।

कालिदास को आश्चर्य होता है कि जो पृष्ठ मल्लिका ने अपने हाथों से बनाए थे ताकि वह उस पर एक महाकाव्य रच सके, वे पृष्ठ तो अब कोरे नहीं है। उन पर तो बहुत कुछ लिखा जा चुका है। "स्थान–स्थान पर इन पर पानी की बूँदें पड़ी हैं जो निःसन्देह वर्षा की बूँदें नहीं है।" लगता है उसने अपनी आँखें से इन कोरे पृष्ठों पर बहुत कुछ लिखा है और आँखों से ही नहीं, स्थान–स्थान पर ये पृष्ठ स्वेद–कणों से मैले हुए हैं, कई स्थानों पर नखों से छीलने एवं दाँतों से काटने के निशान हैं, इसके अतिरिक्त ये ग्रीष्म की धूप के हल्के–गहरे रंग, हेमन्त की पत्रधूलि और इस घर की सीलन से ये पृष्ठ अब कोरे नहीं रहे। इन पर एक महाकाव्य की रचना हो चुकी है। अनन्त सर्गों का एक महाकाव्य इस पर लिखा जा चुका है– "इन पृष्ठों पर अब नया कुछ क्या लिखा जा सकता है?"

उसकी और मल्लिका की बातचीत का भाव यह था कि मल्लिका के दुःख भरे जीवन की दास्तान वह उन पन्नों में देख रहा था खुद उसने भी बहुत कुछ सहा था। अब वह चाहता था कि "इसके आगे भी तो जीवन शेष है", फिर अथ से इसे आरम्भ किया जा सकता है। तभी उसे बच्ची के रोने की आवाज सुनाई पड़ती है जो कि मल्लिका का वर्तमान था। कोरे पृष्ठों पर महाकाव्य लिखने का समय बीत चुका था। अथ से जीवन आरम्भ करने की इच्छा का वस्तुतः 'समय' के साथ द्वन्द्व था और वह पाता है कि समय अधिक शक्तिशाली है, क्योंकि वह प्रतीक्षा नहीं करता।

मल्लिका के साथ फिर से एक नई शुरुआत की आशा लिए कालिदास जिस क्षत–विक्षत हालत में भीगता हुआ वहाँ आता है उसी हालत में अपनी प्रेमिका, जो अब विलोम की पत्नी है, को टूटे हुए मन से छोड़कर चल देता है। मल्लिका भी उन कोरे पृष्ठों पर लिखे जा चुके अपने जीवन अध्ययन को अपनी नियति मानकर रूक जाती है, उसका वर्तमान उसे फिर से एक नए 'अथ' की शुरुआत रोक देता है।

●●

HINDI

2018

QUESTIONS

SECTION – A
Language – 50 Marks

Question 1.

Write a composition in approximately 400 words in Hindi on any ONE of the topics given below: **[20]**

किसी एक विषय पर निबन्ध लिखिए जो लगभग 400 शब्दों से कम न हो :

(i) *आज की भागदौड़ भरी जिन्दगी में आदमी मानसिक तनाव से ग्रस्त है, इसे दूर करने तथा जीवन को खुशहाल बनाने के तरीकों के बारे में अपने विचार प्रस्तुत कीजिए।*

(ii) *"कर्म ही प्रबल है, भाग्य नहीं"*—इस कथन के पक्ष या विपक्ष में अपने विचार प्रकट कीजिए।

(iii) *"आज के भौतिकवादी युग में त्यौहारों का रूप–स्वरूप बदल रहा है। त्यौहारों में व्यावसायिकता बढ़ती जा रही है" तथ्य की विवेचना कीजिए।*

(iv) *"जीवन में सफलता पाने के लिए कठिन संघर्ष की आवश्यकता होती है"*—इस कथन को अपने जीवन के किसी निजी अनुभव के द्वारा पुष्ट कीजिए।

(v) *"नारी घर और बाहर दोनों जगह अपनी भूमिका निभाते हुए नित नई चुनौतियों का सामना करती है।" विभिन्न क्षेत्रों में नारी के योगदान को ध्यान में रखते हुए इस विषय पर अपने विचार लिखिए।*

(vi) *निम्नलिखित में से किसी एक विषय पर मौलिक कहानी लिखिए—*

(a) "पर उपदेश कुशल बहुतेरे।"

(b) एक ऐसी कहानी लिखिए जिसका अन्तिम वाक्य हो—

..........काश! ऐसा पल मेरे जीवन में भी आया होता।

Question 2.

Read the passage given below carefully and answer in Hindi the questions that follow, using your own words :

निम्नलिखित अवतरण को पढ़कर, अन्त में दिए गए प्रश्नों के उत्तर अपने शब्दों में लिखिए:–

किसी नगर में एक नवयुवक रहता था जिसका नाम सुन्दर था। वह मेहनत करने से हमेशा बचता था। जब भी कोई काम उसके सामने आ जाता था जिसमें उसे मेहनत करनी हो, तो वह उस कार्य से दूर भागने लगता था। मेहनत को लेकर उसके मन में यह बात बैठ गयी थी कि वह कभी मेहनत नहीं कर सकता, लेकिन उसके अन्दर अच्छी बात यह थी कि वह अपने जीवन में सफल होना चाहता था। वह सोचता था क्या कोई ऐसा व्यक्ति है जो उसे सफलता का मन्त्र दे सके। इस प्रश्न को लेकर वह बहुत से लोगों और विद्वानों के पास गया। कोई कहता था कि माता–पिता की सेवा करना सफलता का मन्त्र है, तो कोई कहता था कि लोगों की मदद करना सफलता का मन्त्र है, लेकिन किसी का भी उत्तर उसे सन्तुष्ट नहीं कर पाता था। एक दिन जब वह अपने नगर की एक सड़क से गुजर रहा था, तो उसने एक साधु को देखा जिसे एक बहुत बड़ी भीड़ ने घेर रखा था। उस साधु को उसने पहले कभी अपने नगर में नहीं देखा था। साधु के बारे में पूछने पर पता चला कि वे साधु लोगों के प्रश्नों के बहुत सटीक उत्तर देते हैं, आज तक कोई भी व्यक्ति उनके उत्तर से असन्तुष्ट नहीं हुआ है। सुन्दर की आँखों में चमक आ गई। उसने सोचा कि क्यों न साधु से अपने प्रश्न का उत्तर जाना जाए। अगर उन्होंने मुझे सफलता का मन्त्र बता दिया तो मैं जरूर सफल हो जाऊँगा।

वह साधु के पास गया और उनसे पूछा, "साधु महाराज, मैं अपने जीवन में सफल होना चाहता हूँ, क्या आप मुझे सफलता का मन्त्र बता सकते हैं?" साधु के चेहरे पर मधुर मुस्कान आ गयी और उन्होंने कहा, "तुम्हारे इस प्रश्न के बारे में मैं तुम्हें अभी नहीं बताऊँगा। इस नगर में मुझे 10 दिन तक रुकना है। तुम कल आकर मुझसे मिलो।" अगले दिन साधु ने उसे एक बहुत बड़ी और मोटी किताब देते हुए कहा, "अगर तुम्हें सफलता का मन्त्र जानना है तो इसके लिए तुम्हें इस किताब को पढ़ना होगा। इस किताब के किसी एक पृष्ठ पर सफलता का मन्त्र दिया हुआ है। जैसे ही तुम उस पृष्ठ को पढ़ोगे, तो तुरन्त तुम्हें वह मन्त्र मिल जायेगा, लेकिन शर्त यह है कि इस किताब को तुम शुरू से पढ़ोगे, यदि तुमने इसे कहीं बीच में से पढ़ा, तो वह मन्त्र तुम्हें नहीं मिल पायेगा।"

सुन्दर किसी भी तरह सफलता का मन्त्र जानना चाहता था। अतः उसने साधु की शर्त मान ली और तुरन्त उस किताब को शुरू से पढ़ना प्रारम्भ कर दिया। वह जल्दी से जल्दी उस पृष्ठ पर पहुँचना चाहता था, जहाँ सफलता का मन्त्र लिखा हुआ था। अतः उसने किताब को लगातार पढ़ना जारी रखा। कब रात हुई और कब दिन, उसे बिल्कुल भी ध्यान नहीं था। वह खाना और पीना तक भूल गया था। हर समय किताब पढ़ता रहता था। नींद बहुत सताती, तो कुछ देर सो जाता, लेकिन उठते ही पढ़ने बैठ जाता। सात दिन बाद जब वह किताब के आखिरी पृष्ठ पर पहुँचा, तो उसे लगा कि यह तो किताब का आखिरी पृष्ठ है, यहाँ पर मुझे सफलता का मन्त्र मिलना तय है, लेकिन जब वह किताब की आखिरी लाइन पर पहुँचा तो उसमें लिखा था– "अगर तुम्हें सफलता का मन्त्र जानना है, तो इस किताब के आखिरी 'कवर' पृष्ठ की जिल्द हटा कर देखो।"

सुन्दर ने तुरन्त पिछले 'कवर' पृष्ठ की जिल्द को हटाया, तो कुछ लाइनें वहाँ लिखी हुई थीं। उन्हें पढ़ते ही वह खुशी से उछलने लगा और चिल्लाने लगा, "मुझे सफलता का मंत्र

मिल गया! मुझे सफलता का मन्त्र मिल गया।'' इतना कहकर वह फिर से उन लाइनों को पढ़ने लगा, जिनमें यह लिखा था–''जिस तरह तुमने इस किताब को पढ़ने के लिए अपने दिन और रात एक कर दिए, तुम्हें अपने खाने–पीने का भी ध्यान नहीं रहा, हर समय सफलता का मन्त्र खोजने के लिए लगातार किताब पढ़ते रहे, तुमने अपना हर पल इस किताब में सफलता का मन्त्र ढूँढ़ने में लगा दिया, किसी भी अन्य चीज के बारे में तुमने एक पल के लिए भी नहीं सोचा, लगातार उत्साह और लगन के साथ तुमने अपने प्रत्येक क्षण को मन्त्र पाने में डुबो दिया। यदि इसी ललक और दृढ़ इच्छा के साथ तुम दुनिया के किसी भी कार्य में सफलता प्राप्त करना चाहोगे, तो कोई भी तुम्हें सफल होने से नहीं रोक सकता।''

विवेकानन्द जी ने भी हमें सफलता का मन्त्र कुछ इस तरह बताया है–''अपना जीवन एक लक्ष्य पर निर्धारित करो। अपने पूरे शरीर को उस एक लक्ष्य से भर दो और एक दूसरे विचार को अपनी जिन्दगी से निकाल दो। यही सफलता की कुंजी है।''

(i) *सुन्दर किस चीज से घबराता था और क्यों? उसकी एक अच्छी बात क्या थी?* **[4]**

(ii) *सुन्दर साधु के पास क्यों गया? समझाकर लिखिए।* **[4]**

(iii) *साधु ने सुन्दर को सफलता का मन्त्र पाने के लिए क्या कहने को कहा? समझाकर लिखिए।* **[4]**

(iv) *सुन्दर को सफलता का मन्त्र कैसे मिला? समझाकर लिखिए।* **[4]**

(v) *इस गद्यांश से आपको क्या शिक्षा मिलती है?* **[4]**

Question 3.

(a) Correct the following sentences and rewrite : **[5]**

निम्नलिखित वाक्यों को शुद्ध करके लिखिए :

(i) *ममता गाने की कसरत कर रही है।*

(ii) *पिछले कुछ वर्षों के बीच भारत की आबादी बढ़ी है।*

(iii) *अपने बुरे दुष्कर्मों के कारण वह आज कंगाल है।*

(iv) *चोर सोमनाथ के घर पाँव दबाकर आया।*

(v) *स्वार्थी मित्र काम निकलते ही आँखें नीची कर लेते हैं।*

(b) Use the following idioms in sentences of your own to illustrate their meaning : **[5]**

निम्नलिखित मुहावरों को वाक्यों में प्रयुक्त कीजिए :

(i) *पापड़ बेलना।* **(ii)** *कंधे से कंधा मिलाना।*

(iii) *पीठ दिखाना।* **(iv)** *दाल में काला होना।*

(v) *फूला न समाना।*

SECTION – B

Prescribed Textbooks – 50 Marks

*(Answer **four** questions from this Section on **at least three** of the prescribed textbooks)*

गद्य संकलन (Gadya Sanklan)

Question 4.

''ये वे हाथ नहीं हो सकते, मैं मन में सोच रही थी, जो बच्चों को मीठी लोरी की थपकनें देकर सुलाते हैं, पति की कमीज में बटन टाँकते हैं या चिमटा–सनसी पकड़ते हैं।''

(i) *प्रस्तुत पंक्तियाँ किस पाठ से ली गई हैं? इस पाठ की लेखिका कौन हैं? उन्हें कहाँ से गाड़ी पकड़नी थी?* **[1½]**

(ii) *सफर में उस डिब्बे में कौन–कौनसी महिलाएँ थीं? उनका परिचय अत्यन्त संक्षेप में दीजिए।* **[3]**

(iii) *लेखिका ने उपर्युक्त कथन किस सन्दर्भ में कहा है?* **[3]**

(iv) *उपर्युक्त कथन जिस महिला के बारे में कहा गया है, वे कहाँ जा रही थीं और क्यों? उनका कौन–सा सामान उन्हें परेशान किए जा रहा था?* **[5]**

Question 5.

रज्जब कौन था? उसका पेशा क्या था? वह अपने पेशे से मिले धन को लेकर कहाँ जा रहा था? क्या वह अपने गन्तव्य स्थल पर पहुँच सका? यदि हाँ, तो कैसे? 'शरणागत' कहानी के आधार पर स्पष्ट कीजिए। **[12½]**

Question 6.

'क्या निराश हुआ जाए?' निबन्ध में निबन्धकार का क्या उद्देश्य है? किन दो घटनाओं के द्वारा निबन्धकार ने यह बताने का प्रयास किया कि मनुष्यता अब भी पूरी तरह समाप्त नहीं हुई? उन घटनाओं का संक्षेप में वर्णन कीजिए। **[12½]**

काव्य मंजरी (Kavya Manjari)

Question 7.

भीतर जो डर रहा छिपाए,
हाय! वही बाहर आया।
एक दिवस सुखिया के तन को
ताप–तप्त मैंने पाया।
ज्वर में विह्वल हो बोली वह,
क्या जानूँ किस डर–से–डर,
मुझको देवी के प्रसाद का,
एक फूल ही दो लाकर।

(i) *'मैंने' शब्द किसके लिए प्रयुक्त हुआ है? उसे किस बात का डर था?* **[1½]**

(ii) *सुखिया का स्वभाव कैसा था? उसके इस स्वभाव का क्या परिणाम निकला? उसने किससे, क्या इच्छा जाहिर की?* **[3]**

(iii) *क्या सुखिया की इच्छा पूरी हो सकी? कारण सहित लिखिए।* **[3]**

(iv) *इस कविता में कवि ने किस बुराई को किस प्रकार उजागर किया है?* **[5]**

Question 8.

''बाल लीला'' के आधार पर बताइए कि सूरदास जी ने श्रीकृष्ण के बाल स्वरूप का कैसा वर्णन किया है? श्रीकृष्ण अपने बड़े भाई बलराम की शिकायत किससे करते हैं और क्या शिकायत करते हैं? **[12½]**

Question 9.

'प्रकृति भाग्य–बल से नहीं, भुजबल से झुकती।'– उद्यमी नर' कविता के आधार पर सिद्ध कीजिए। **[12½]**

"सारा आकाश" (Saara Akash)

Question 10.

"मुझे पता होता तो मैं कभी भी नहीं करती। मैंने समझा कि कोई सादा मिट्टी का ढेला है।"

(i) *प्रस्तुत पंक्तियों के वक्ता और श्रोता कौन–कौन हैं? उनके बीच किस विषय पर चर्चा हो रही है?* **[1½]**

(ii) *वक्ता की बात सुनकर श्रोता ने गुस्से में क्या–क्या कहा?* **[3]**

(iii) उक्त घटना के सन्दर्भ में समर की क्या प्रतिक्रिया थी? **[3]**

(iv) *समर की आत्मग्लानि का अपने शब्दों में वर्णन कीजिए।* **[5]**

Question 11.

प्रभा के परदा न करने से परिवार में क्या प्रतिक्रिया हुई, उसका परदा न करना कहाँ तक उचित था, स्पष्ट कीजिए। **[12½]**

Question 12.

'सारा आकाश' उपन्यास के प्रमुख पात्र 'समर' का चरित्र–चित्रण कीजिए। **[12½]**

"आषाढ़ का एक दिन" (Aashad Ka Ek Din)

Question 13.

तुम जिसे भावना कहती हो वह केवल छलना और आत्म–प्रवंचना है।······भावना में भावना का वरण होता है!······मैं पूछती हूँ भावना में भावना का वरण क्या होता है?

(i) *उक्त कथन नाटक के किस अंक से लिया गया है तथा उक्त कथन किसने, किससे कहा है?* **[1½]**

(ii) *उक्त कथन किस सन्दर्भ में कहा गया है? स्पष्ट कीजिए।* **[3]**

(iii) *उक्त कथन के माध्यम से वक्ता श्रोता को क्या समझाना चाहती है?* **[3]**

(iv) *वक्ता की चारित्रिक विशेषताएँ लिखिए।* **[5]**

Question 14.

'आषाढ़ का एक दिन' नाटक के आधार पर कालिदास का चरित्र–चित्रण कीजिए। **[12½]**

Question 15.

'आषाढ़ का एक दिन' शीर्षक नाटक ऐतिहासिक होते हुए भी आधुनिक समस्याओं का आकलन प्रतीत होता है।— व्याख्या कीजिए। **[12½]**

ANSWERS

Answer 1.

(i) **मानसिक तनाव**

तनाव आज जीवन का अभिन्न अंग है। बच्चा, बूढ़ा, अमीर–गरीब, बुद्धिजीवी–श्रमशील, पुरुष–स्त्री सभी इससे ग्रसित हैं। आज के युग को यदि तनाव युग कहें तो कोई अतिशयोक्ति न होगी। तनाव सिद्धान्त के जनक मनोवैज्ञानिक हॉन सालाई के अनुसार, आज ऐसे व्यक्ति की कल्पना असम्भव है जो मानसिक तनाव को अनुभव न करता हो, तनाव एक महाव्याधि के रूप में मानव जाति को आतंकित किये है।

तनाव मनुष्य के लिए हानिकारक ही साबित हो ऐसी बात नहीं है। यह मनुष्य के विकास एवं प्रगति में कारक तत्व है। आपातकालीन परिस्थितियों का सामना करने में मन एवं शरीर को तनाव ही तैयार करता है। 'Overcoming Tension' के लेखक डॉ. केनेथ हेम्वली के अनुसार, प्रागैतिहासिक काल का आदिमानव इस तनाव द्वारा ही जंगली पशुओं के हमलों से बचने में समर्थ रहता था। तनाव की स्थिति में एड्रीनल ग्रन्थि से एड्रीनेलिन नामक हार्मोन स्रावित होता है, जो शरीर की प्रतिरोधक क्षमता बढ़ाता है व बाह्य दबाव सहन करने में सक्षम बनाता है। यदि तनाव न होता तो शायद आज का मनुष्य प्रागैतिहासिक आदिमानव की स्थिति में ही होता।

तनावपूर्ण स्थिति से निपटने की व्यवस्था प्रकृति ने मनुष्य के शरीर में दी है, लेकिन लम्बे समय तक तनावपूर्ण स्थिति के बने रहने के कारण यह व्यवस्था चरमरा जाती है क्योंकि इस अवस्था में हार्मोन का स्राव अनावश्यक एवं अत्यधिक मात्रा में होता है जो शरीर के विभिन्न अंग–अवयवों पर दुष्प्रभाव डालता है। माँसपेशियों में खिंचाव, पेट दर्द, कंधे व पैर में दर्द, थकान, सिरदर्द आदि शारीरिक लक्षण दिखाई पड़ते हैं। अधिक गम्भीर होने पर इसके दुष्परिणाम आँत के फोड़े (पेप्टिक अल्सर), हृदय रोग, उच्च रक्तचाप, अनिद्रा, चिड़चिड़ापन, एकाग्रता में कमी जैसे रोगों के रूप में प्रकट होते हैं। अत्यधिक तनाव शरीर व मन को तोड़ कर रख देता है व शीघ्र ही बुढ़ापे की ओर ले जाता है। यह असामयिक मृत्यु का कारण भी बन सकता है।

कारण—प्रगति की दौड़ में मनुष्य मानवीय मर्यादा, नैतिक एवं आध्यात्मिक मूल्यों को तिलांजलि दे चुका है। एक–दूसरे को नीचा दिखाने की प्रतिस्पर्धा, विलासी जीवन, अमर्यादित आचरण, उद्दण्ड व्यवहार एवं विकृत चिन्तन दिनचर्या के अंग बन गये हैं। परिणामतः बढ़ती असुरक्षा, हिंसा, ईर्ष्या, द्वेष, भय, चिन्ता, निराशा, खिन्नता जैसे मानसिक विकास मनुष्य के जीवन में कुहराम मचाये हुए हैं।

"The Stress Solution" के लेखक **लाहले एच. मिलर** एवं **अल्या डेल स्मिथ** के अनुसार, तनाव एक जटिल, बहुआयामी एवं वैयक्तिक समस्या है जो व्यक्ति के स्वास्थ्य, कार्यक्षमता एंव धन की तबाही कर देती है। तनाव के रूप में पड़ने वाले दबाव बाह्य परिस्थितियों पर आन्तरिक मनःस्थिति के कारण हो सकते हैं। तनाव के व्यक्तिगत, पारिवारिक, सामाजिक, आर्थिक, व्यावसायिक एवं वातावरणजन्य कई कारण हो सकते हैं।

तनाव से राहत के लिए आज तरह–तरह की औषधियाँ इज़ाद की जा रही हैं। तनाव से तत्काल मुक्ति के लिए लोग दर्दनिवारक औषधियों यहाँ तक कि शराब एवं अन्य नशीले पदार्थों का सेवन करते हैं। यह सब तनाव रूपी महाव्याधि का समाधान नहीं है।

लेखन, पेंटिंग, ड्राइंग, नाचना–कूदना एवं ड्रामा जैसी कलात्मक अभिव्यक्तियों द्वारा हम तनाव को शानदार ढंग से कम कर सकते हैं। भावनात्मक उद्वेग के क्षणों को कागज पर उतारने से तनाव कम हो जाता है। शोध तथा अध्ययनों में पाया गया

है कि ऐसा करने से व्यक्ति के रोग–प्रतिरोधी तत्व; जैसे टी सेल्स और क्रिल सेल्स में उल्लेखनीय वृद्धि हो जाती है। दिनभर की उथल–पुथल तक विस्तृत करने पर यह प्रक्रिया प्रकारान्तर में डायरी लेखन का रूप ले लेती है जिसकी महत्ता आत्मनिरीक्षण व व्यक्तित्व निर्माण के रूप में असंदिग्ध है।

नाचकूद में व्यायाम के साथ–साथ तनाव से राहत मिलती है। मनपसन्द संगीत भी तनाव से राहत का प्रभावशाली उपाय है। वैज्ञानिकों ने पाया कि सूर्यास्त के समय संगीत सुनने से एण्डोरफिन नामक रसायन पदार्थ निकलता है जो रोग प्रतिरोधक क्षमता को मजबूत करता है।

आज मनोविज्ञान में 'आर्ट थेरेपी' के अन्तर्गत क्रोध व अवसाद आदि से ग्रसित व्यक्ति को कागज, केनवास व कलम–ब्रुश थमाकर अपने गुस्से व अवसाद को कागज या केनवास पर उगलवा कर हल्का तथा शान्त किया जाता है।

हँसना तनाव की औषधि है। विशेषज्ञों का कहना है कि हँसते समय मस्तिष्क से एण्डोरफिन नामक रसायन निस्तृत होता है जो पीड़ा में कमी करता है व प्रतिरोधक क्षमता को बढ़ाता है।

जीवन की परेशानियों व कुण्ठाओं से उत्पन्न तनाव को हल्का करने का एक उपाय यह भी है कि अपने हृदय को किसी विश्वसनीय बन्धु के आगे खोलें। इससे मन हल्का हो जायेगा व समस्या का समाधान हो जायेगा। जब हम अपने लिए ही परेशान हों, तब दूसरों के लिए कुछ करने की सोचें। इससे हमें असीमित आनन्द मिलेगा।

(ii) कर्म की प्रधानता

वायु, जल, भोजन आदि मनुष्य की मूलभूत आवश्यकताएँ हैं। वह इनके बिना जीवित नहीं रह सकता है। इन सबके लिए श्रम की आवश्यकता होती है। श्रम से ही वह अपने कृत्यों का सुचारु रूप से संचालन कर सकता है तथा स्वयं को शारीरिक व मानसिक रूप से स्वस्थ रख सकता है।

वह श्रम से ही सम्पत्ति, यश सभी कुछ पा लेता है। मनुष्य का श्रम ही उसकी सफलता का मूलमंत्र है। समाज में उल्लेखनीय सफलता प्राप्त करने वाले व्यक्तियों को हम प्रायः 'भाग्यशाली' की संज्ञा देते हैं। यदि हम स्वयं उनसे यह प्रश्न करें अथवा उनकी जीवन शैली का अध्ययन करें तो उनकी सफलता के पीछे उनका अनवरत संघर्ष का पता चलेगा। परिश्रम एवं संघर्ष क्षमता से ही वे सफलता की मंजिल तक पहुँचने में सक्षम हो सके हैं। श्रम अथवा कर्म ही पूजा है। गीता में भगवान श्रीकृष्ण ने अर्जुन को महाभारत के युद्ध में कर्मयोग का ही उपदेश दिया था—

'कर्मण्येवाधिकारस्ते, मा फलेषु कदाचन्।

'कर्मयोग के उनके उपदेश से अर्जुन का मोह भंग हुआ था और वे महाभारत के युद्ध में विजयी हुए थे। जिस प्रकार पूजा, अर्चना व ध्यान से ईश्वर की प्राप्ति होती है उसी प्रकार मनुष्य को यश, वैभव व सफलता तभी मिलती है जब वह श्रम को 'पूजा' की तरह मानता है।

''मही नहीं जीवित है मिट्टी से डरने वालों से,
ये जीवित है इसे फूँक सोना करने वालों से।''

यदि कार्य मनोयोग से नहीं किया गया होता है तब वह जल्दी ही थकान एवं निराशा महसूस करना प्रारम्भ कर देता है और कार्य से उसकी अरुचि उत्पन्न हो जाती है। इतिहास में ऐसे समस्त लोग जिन्होंने सफलता के नए आयाम स्थापित किए वे सभी हमारी तरह पहले सामान्य लोग थे, परन्तु अपने श्रम व लगन के बल पर वे सफलता की सीढ़ियों पर चढ़ते चले गये और अन्ततः उन्होंने अपनी मंजिल प्राप्त कर ली। ऐसे लोग ही दूसरों के लिए आदर्श बने जिनका अनुसरण आज पूरा समाज, राष्ट्र व विश्व कर रहा है।

नेपोलियन राष्ट्रपति बनने से पूर्व तेंतालीस बार असफल हुए थे, उन्होंने कभी हार स्वीकार नहीं की। उनके अथक परिश्रम एवं लगन ने ही उन्हें सफलता के शिखर तक पहुँचा दिया। नेपोलियन की ही भाँति ऐसे असंख्य उदाहरण हैं जिन्होंने कर्म को ही अपनी पूजा व ध्येय माना और जीवन पथ पर सफलता के नए आयाम स्थापित किए।

वे सभी व्यक्ति जो अकर्मण्य होते हैं अर्थात् कर्म से दूर भागते हैं वे सभी भाग्यवादी हो जाते हैं।

कर्म को पूजा मानकर जीवन में सफलता के नए आयाम स्थापित किए जा सकते हैं। कर्म के माध्यम से ही मनुष्य स्वयं अपनी असीमित शक्तियों को पहचानकर उनका सदुपयोग कर सकता है। यह मनुष्य के श्रम का ही परिणाम है जब उसने ऊँचे पहाड़ों को काटकर वहाँ सड़कें बनाई हैं।

खुद को भाग्य के भरोसे वही छोड़ता है जो कर्म नहीं करना चाहता। कर्म करने से भाग्य बनता है। कर्म ही बड़ा होता है, इसीलिए ये नहीं कहते कि ''भाग्य ही पूजा'' है बल्कि ये कहते हैं कि ''कर्म ही पूजा है।''

आप देखेंगे कि अगर थॉमस अल्वा एडिसन और हरलेण्ड सांडर्स जैसे लोग भाग्य के भरोसे होते तो दो–तीन बार असफल होने पर ही हार मान लेते; लेकिन उन्होंने कर्म किया एडिसन 1000 और सांडर्स 1009 बार असफल होने के बाद सफल हुए।

जब **विल्मा रुडोल्फ** को पोलियो हुआ और केरोली टाक्कस जिस हाथ से शूटिंग करते थे उस हाथ में हथगोले का विस्फोट हो गया और वह हाथ बिल्कुल बेकार हो गया तो लोगों ने तो यही कहा होगा देखो इनका भाग्य कितना खराब है, पर दुनिया जानती है इन्होंने कैसे कर्म कर अपना भाग्य बनाया और दोनों ने आत्मशक्ति के बल पर अपने–अपने खेल में ओलंपिक गोल्ड मैडल जीता और कर्म की महानता को सिद्ध किया—

खुदी को कर बुलंद इतना कि हर तकदीर से पहले
खुदा बंदे से खुद पूछे बता तेरी रजा क्या है?

श्रम–शक्ति के द्वारा ही मनुष्य ने प्रकृति में अभूतपूर्व परिवर्तन ला दिया है। यह मनुष्य के श्रम व विचार–शक्ति का ही परिणाम है जिससे मनुष्य आज चन्द्रमा पर अपनी विजय पताका फहरा पाया है। अतः हम सभी को श्रम के महत्त्व को समझना चाहिए, क्योंकि श्रम ही सफलता की कुंजी है।

(iii) त्यौहार और उनका स्वरूप

भारत विभिन्न संस्कृतियों, संप्रदायों एवं भाषाओं का देश है। यहाँ अनेक प्रकार के उत्सव और त्यौहार मनाए जाते हैं। वसंतोत्सव, होली, वैसाखी, ईद, स्वतंत्रता दिवस, गणतंत्र दिवस, गाँधी जयंती, बाल दिवस, दशहरा, दीपावली, ओणम, पोंगल, गुरुपर्व, रक्षाबंधन, बड़ा दिन (क्रिसमस) आदि अनेक ऐसे त्योहार हैं जिनका संबंध संस्कृति, धर्म, देश या देश के महापुरुषों के साथ है।

आज के आधुनिक युग में राष्ट्रीय महत्त्व के त्यौहारों को छोड़ दिया जाए तो लगभग सभी त्यौहारों का रूप बदल गया है। नगरीकरण, व्यवसायीकरण, मशीनीकरण आदि संस्कृतियों ने अपना रंग जमाना शुरू कर दिया है। हमारे परंपरागत त्यौहार भी पाश्चात्य संस्कृति की चपेट में आते जा रहे हैं।

रक्षाबंधन की बात की जाए तो अभी कुछ वर्ष पूर्व यह एक सादा एवं पवित्र त्यौहार था। भाई–बहन के रिश्ते को रक्षाबंधन के दिन सूत (धागा) बाँधकर और सुदृढ़ करने की परंपरा निभाई जाती थी। परंतु आज के व्यवसायिक युग में इसका स्वरूप ही बदल गया है। उत्पादकों, वितरकों, विज्ञापनों तथा केबिल टी. वी. के चैनलों के कारण इस पवित्र त्यौहार में चकाचौंध, प्रदर्शन और कृत्रिमता का समावेश हो गया है। जहाँ घर में पड़े एक धागे और गुड़ से काम चल जाता था वहाँ अब एक–दूसरे से बढ़कर खर्च करने की भावना पनप रही है। एक रुपए से लेकर एक लाख तक की राखियाँ आज बाजार में उपलब्ध हैं। सौ रुपए से लेकर दो हजार तक के मूल्य के मिठाई के डिब्बे उपलब्ध हैं।

क्रिसमस, दीपावली, दशहरा, गुरुपर्व आदि उत्सवों को मनाने के ढंग भी बदल चुके हैं। नगरतोरण बनाए जाते हैं, अमीर लोग अपने घरों पर मात्र सजावटी प्रकाश करने में ही हजारों–लाखों खर्च कर देते हैं। पटाखों तथा दूसरे प्रकार की आतिशबाजी खर्चीली, भड़कीली और यांत्रिक होती जा रही है। ऐसे त्यौहारों पर आज दिखावा अधिक होने लगा है। पास धन हो या न हो, व्यर्थ प्रदर्शन करना आवश्यक है। लोग उधार लेकर उपहारों का आदान–प्रदान करने लगे हैं। प्रतिस्पर्धा बढ़ने लगी है।

पहले लोग घरों में बने व्यंजनों तथा उपकरणों तक सीमित रहते थे और आनंदपूर्वक इन उत्सवों का आनंद उठाते थे। आज वह बीती बात हो चुकी है। नववर्ष के संदर्भ में अनेक विज्ञापन पहले ही आने शुरू हो जाते हैं। क्लबों, होटलों, सभा–समितियों, आमोद–केन्द्रों की ओर से अपने–अपने ढंग से नववर्ष की पूर्व–संध्या के कार्यक्रमों की घोषणा होने लगती है। युवावर्ग विशेष रूप से पाश्चात्य संस्कृति की ओर आकृष्ट होता है। पहले लोग अपने घरों में ही ऐसे अवसरों का आनंद लेते और एक–दूसरे को शुभकामनाएँ देते थे। आजकल होटल रूपी व्यावसायिक केंद्रों में हजार रुपये से लेकर लाख रुपये तक ऐसे समारोहों के टिकट से प्रवेश पाने का प्रावधान होता है और सनक से भरे लोग वहाँ जाने से नहीं चूकते। नृत्य, मदिरा, नग्नता आदि का सहारा लेकर ऐसे व्यवसायी लोगों को खूब लूटने लगे हैं।

उपहारों, संदेशों, खाद्यानों और व्यंजनों पर खूब पैसा बहाया जाता है। लोग पदोन्नति के लाभ में अपने अधिकारियों को प्रसन्न करने के लिए ऐसे अवसरों पर अपनी सामर्थ्य से भी बढ़कर उपहार देने का प्रयत्न करते हैं। घरों में बच्चों और स्त्रियों की मानसिकता भी बदली है। वे भी ऐसे उत्सवों पर नए वस्त्रों, उपकरणों और वाहनों की माँग करने लगी हैं। जन्मदिन पर मध्यवर्ग का भरसक शोषण होता है।

उत्सवों में बढ़ती जा रही इस अंधाधुंध व्यावसायिकता को हम सब ही रोक सकते हैं। त्यौहारों को उनके मूल रूप में मनाना चाहिये। युवावर्ग को इसके लिए आगे आना चाहिए। वे ऐसे उपभोक्तावादी दृष्टिकोण पर नियंत्रण पाएँगे, तो ही उनका भविष्य उज्ज्वल हो पाएगा।

(iv) संघर्ष और सफलता

जीवन में कुछ पाने के लिए संघर्ष अनिवार्य है। बिना संघर्ष मात्र भाग्य के दम पर कुछ नहीं पाया जा सकता। संघर्ष सफलता की कुंजी है। इसको मैं अपने जीवन की एक घटना से बताने जा रहा हूँ।

मेरे पिता दिल्ली में सरकारी नौकरी में थे। कुछ परिस्थितियों के कारण मुझे अपनी प्रारम्भिक शिक्षा गाँव के जूनियर हाईस्कूल में करनी पड़ी। आठवीं करने के बाद नवीं कक्षा में, मेरे पिता ने मेरा प्रवेश दिल्ली के एक अच्छे विद्यालय में कराया।

मेरे लिए यह विद्यालय और वहाँ का पर्यावरण बिल्कुल नया था। पढ़ाई में मेरी गिनती ज्यादा अच्छे विद्यार्थियों में नहीं थी। कक्षा में मुझे सबसे पीछे बैठना अच्छा लगता था और मेरी मित्र–मण्डली में भी वही छात्र थे जिनका पढ़ाई के प्रति रुझान न के बराबर था।

प्रथम सत्र की परीक्षा का परिणाम घोषित हुआ। यह जानकर मुझे आश्चर्य नहीं था कि मुझे गणित में शून्य मिला और अंग्रेजी में एक अंक। यह बात मैंने घर में नहीं बताई।

उसके बाद छमाही परीक्षा में भी यही स्थिति रही। वार्षिक परीक्षा में मेरी उम्मीद के विपरीत मे तीन विषयों में पास था और दो में कम्पार्टमेंट। कम्पार्टमेंट का पेपर हुआ और मैं एक विषय अंग्रेजी में फेल हो गया।

मेरे फेल होने पर घर के सभी सदस्यों को आघात लगा और मुझे शर्मिन्दगी हुई। यही वह समय था जब मैंने कठोर प्रण किया कि मैं खुद को बदल दूँगा।

फेल होने की निराशा, सहपाठियों से बिछड़ जाने का दुख सब कुछ भुलाकर मैं सिर्फ पढ़ाई में जुट गया। पूरे साल पढ़ाई पर ध्यान न देने के कारण मैं पढ़ाई में काफी पिछड़ गया था। मैंने इस साल पूरी तरह मेहनत करने की ठानी। अंग्रेजी और गणित जो कि मेरे कमजोर विषय थे उन्हें अधिक समय दिया। सभी विषयों का पूरा पाठ्यक्रम पढ़ा और दोहराया।

मेरी यह मेहनत तब रंग ले आई जब मेरी दसवीं कक्षा का परीक्षाफल घोषित हुआ। मैंने परीक्षा 75% अंकों से उत्तीर्ण की और अंग्रेजी में 80 अंक लाया।

मेरी खुशी की सीमा ना थी। परिवार के सभी लोग खुश हुये और मुझे यकीन आ गया कि वाकई मेहनत से क्या हासिल नहीं हो सकता।

उसी दिन मैंने निश्चय कर लिया कि कभी मेहनत से जी नहीं चुराऊँगा और हमेशा मन लगाकर पढ़ाई करूँगा। असफलता मिलने पर हमें निराश नहीं होना चाहिये बल्कि दोबारा मेहनत करनी चाहिये सफलता अवश्यक मिलती है।

(v) भारतीय नारी के योगदान

प्राचीन काल से ही नारी का समाज में महत्त्वपूर्ण स्थान रहा है। तब नारी को बहुत आदर की दृष्टि से देखा जाता था। उसे पुरुष के समान ही अधिकार प्राप्त थे। मध्यकाल में नारी की स्थिति बिगड़ी परन्तु आधुनिक युग में नारी फिर से अपने पुराने रूप में वापस आई है। आज नारी पुरुष के साथ कंधे से कंधा मिलाकर चल रही है। वह एक माँ, बहन, पत्नी, बेटी सभी उत्तरदायित्वों को निभाते हुये भी बाहर की दुनिया से सामंजस्य बिठा रही है।

घर और बाहर दोनों तरफ सामंजस्य बिठाते हुये यह दोहरी भूमिका निभाना बहुत मुश्किल है। कामकाजी महिलाओं को अपने काम के साथ–साथ अपने घरेलू उत्तरदायित्व भी निभाने पड़ते हैं। पति–बच्चों और कार्यालय की जिम्मेदारी पूरी करते हुये वह अपने आराम और सुख को भी त्याग देती है। वह दिनभर अपनी जिम्मेदारी को पूरा करने में जुटी रहती है।

समय चाहे जितना भी बदल गया हो परन्तु आज भी कई जगह लोग महिलाओं के प्रति दूषित सोच रखते हैं। आज भी कई जगह रूढ़िवादी सोच का सामना कामकाजी और गैर–कामकाजी महिलाओं को करना पड़ता है। परन्तु भारतीय महिलायें इन सबसे लड़ते हुये भी अपने सपनों को पूरा करने में जुटी रहती हैं।

आज हर एक क्षेत्र में महिलाओं का योगदान सराहनीय है। राजनैतिक और सामाजिक सुधारों के क्षेत्र में महिलाओं ने महत्त्वपूर्ण योगदान दिया।

इन्दिरा गाँधी ने देश की महिला प्रधानमन्त्री के तौर पर शासन–व्यवस्था को बखूबी संभाला और दुर्गाबाई देशमुख ने महिलाओं के कल्याण के लिए केन्द्रीय समाज कल्याण बोर्ड की स्थापना की। पद्म भूषण डॉ. मुथुलक्ष्मी रेड्डी ने मद्रास में कैंसर अस्पताल की स्थापना की। वह मद्रास स्टेट सोशल वेलफेयर एडवाइजरी बोर्ड की 1957 तक अध्यक्षा रहीं। दिल्ली की समाज सुधारक रुस्तमजी फरोदोनजी ने महिलाओं के लिए नागरिक एवं राजनीतिक दोनों अधिकारों के लिए संघर्ष किया।

वर्तमान में भी देश के विभिन्न राज्यों के ग्रामीण क्षेत्रों में पंचायतों में भी महिलाएँ विभिन्न पदों पर आसीन हैं। भारतीय साहित्य के क्षेत्र में महादेवी वर्मा, सुभद्रा कुमारी चौहान से लेकर नई पीढ़ी की महिला रचनाकारों ने भी पहचान बनाई है। अरुंधती राय को उनके उपन्यास 'द गॉड ऑफ स्माल थिंग्स' के लिए बहुप्रतिष्ठित बुकर पुरस्कार मिला।

आज मीडिया, पत्रकारिता एवं जनसंचार के क्षेत्र में भी महिलाओं का वर्चस्व है। सेना, वायुयान उड़ान, पर्वतारोहण आदि विभिन्न क्षेत्रों में महिलाओं ने अपनी भागीदारी करके लैंगिक असमानता को दूर कर दिखाया है। शिक्षा, विज्ञान, खेल–कूद, व्यवसाय, सूचना–प्रौद्योगिकी, चिकित्सा आदि सभी क्षेत्रों में महिलाएँ पुरुषों से अधिक योग्य सिद्ध हो रही हैं।

आर्थिक स्तर पर महिलाओं ने स्वयं उद्यमी बनकर राष्ट्रीय विकास करते हुए अपनी जगतव्यापी पहचान बनाई है।

आज भारतीय महिला सजगता और जिम्मेदारी से कार्यालय और घर के कार्यों को संभाले हुए हैं। घर को चलाने, पति के भार को कम करने और सुख सुविधाओं से सम्पन्न जीवन जीने के प्रति भारतीय कामकाजी महिलाओं के प्रयत्न प्रशंसनीय हैं।

(vi) (a) पर उपदेश कुशल बहुतेरे

मानव सृष्टि का सर्वोत्कृष्ट प्राणी है। वह सुन्दर से सुन्दर, महान् से महान् जीवन के आदर्शों का अनुगमन करना चाहता है। उपदेश देना बहुत अच्छी बात है, पर यह भी आवश्यक है कि उपदेशक स्वयं अपने जीवन में उन उपदेशों को व्यावहारिक रूप में आत्मसात कर ले। आदेश देना या उपदेश देना बहुत आसान है, पर मन, वचन और कर्म से उनको निभाना बहुत कठिन है। गुरु द्रोणाचार्य ने जब अपने शिष्यों को 'सत्यं वद' का उपदेश दिया तो सभी पाण्डवों ने अगले दिन पाठ याद करके सुना दिया पर युधिष्ठिर ने जब तक इस सूत्र को व्यावहारिक जीवन में नहीं उतार लिया तब तक उन्होंने गुरु जी को पाठ नहीं सुनाया।

कथनी और करनी का अन्तर ही मनुष्य को महान् नहीं बनने देता।

हर व्यापारी, अधिकारी तथा नेता अपने नौकरों, कर्मचारियों तथा जनता को ईमानदारी तथा कर्मठता का उपदेश देता है, जबकि वह स्वयं भ्रष्टाचार के पथ पर बढ़ता रहता है। नेता मंच पर आकर कितनी सारगर्भित बातें कहते हैं, पर उनका आचरण हमेशा उनकी बातों के विपरीत होता है।

यही कारण है कि आजकल के नेताओं की बात पर कोई विश्वास नहीं करता और न ही उनके प्रति किसी के मन में श्रद्धा उत्पन्न होती है। शिक्षा के क्षेत्र में भी यह बात बड़ी महत्वपूर्ण है कि अध्यापक उन सिद्धान्तों पर आचरण करें जिनका प्रचार वह अपने भाषणों में करते हैं।

यही प्रवृत्ति हमें बहुत स्थानों पर दिखाई देती है कि हम दूसरों की कमियों का और अपने गुणों का बखान करते हैं। यह बहुत निरर्थक और अनैतिक है।

हमें दूसरों की कमजोरियाँ और अच्छाइयाँ सन्तुलित करके देखनी चाहिए। सभी को पूर्ण परिवेश में स्वीकार करना चाहिए और उपदेशक की बजाय दर्द बाँटने वाला सहृदय व संवेदनशील मित्र बनना चाहिए।

(b) हसरत

घर में आज सुबह से ही चहल–पहल मची हुई थी। साहब के बेटे का जन्मदिन था। सभी बहुत खुश थे। आज घर में कई तरह के पकवान बनाये जा रहे थे क्योंकि शाम को एक बड़ी पार्टी का आयोजन था। मुझे इस घर में आये अभी दो साल हुए थे। पिताजी काफी बीमार चल रहे थे इसलिए काम पर नहीं जा रहे थे। घर में कमाने वाला और कोई नहीं था। माँ मेरे पैदा होते ही चल बसी थी। घर में दो बहनें भी थीं। लड़की होने के कारण पिताजी उन्हें घर से बाहर जाने नहीं देते थे। घर की माली हालत देखकर मैंने भी अपनी पढ़ाई छोड़ दी और 14 वर्ष की आयु में ही काम पर लग गया था। जिसका आज जन्मदिन था उसका नाम रोहण था। वैसे तो वह मेरे हम उम्र ही था इसलिए हम दोनों में दोस्ती थी। घरवालों को हमारा इस तरह उठना–बोलना, खेलना पसन्द नहीं था। बात–बात पर वह रोहण को मुझसे मिलने पर डाँटते थे। आज जन्मदिन की रौनक में सब अपने–अपने काम में व्यस्त थे। घर में तरह–तरह के पकवान बनने की खुशबू आ रही थी। मैं भी घर के कामों हाथ बँटाने लगा। शाम को पाँच बजते ही रोहण के पिताजी जब घर आये तो उनके हाथ में एक बड़ा केक था। मेज पर उसे रखते ही रोहण का चेहरा अपना बर्थ–डे केक देखकर खिल गया। आज उसने बहुत सुन्दर कपड़े पहने थे। तभी मेरी नजर अपने कपड़ों पर गई वह बहुत ही गन्दे और फटे हुए थे। छह बजते ही पार्टी शुरू हो गई। सभी रिश्तेदार और रोहण के दोस्त सुन्दर–सुन्दर कपड़ों और उपहारों के साथ आने शुरू हो गये थे। सभी रोहण को उसके जन्मदिन पर बधाई दे रहे थे। रोहण अपने दोस्तों के साथ बहुत खुश था और उनके साथ मिलकर तस्वीरें खिंचवा रहा था। मैं दूर खड़ा यह नजारा देख रहा था

तभी रोहण की नजर मुझ पर पड़ी। वह मेरे पास आया और बोला—"चल ना, तू यहाँ अकेला क्या कर रहा है चल, तुझे मैं अपने दोस्तों से मिलवाऊँ।"

मैंने अपना हाथ झटकते हुए कहा, "नहीं, रोहण! तुम जाओ! ये बड़े लोगों की पार्टी है। मैं यहीं ठीक हूँ।"

तभी उसकी माँ की नजर हम पर पड़ गई और वह रोहण का हाथ पकड़कर उसे वहाँ से ले गई। मैं फिर वहीं कोने में खड़ा सबको देखने लगा। केक काटने का समय हो गया था। जैसे ही रोहण ने केक काटा सबने तालियाँ बजाई। मैंने भी खुश होकर जोर–जोर से तालियाँ बजाई। अब सब उसे केक खिला रहे थे और बधाई दे रहे थे। रोहण की माँ ने भी उसका माथा चूमते हुए उसे केक खिलाया और फिर गले लगा लिया। माँ–बेटे का ऐसा प्यार देखकर मेरी आँखों में अचानक आँसू आ गये। दिल पसीज गया। मन से आवाज आई—काश! ऐसा पल मेरे जीवन में भी कभी आया होता।

मैं वहाँ से अपने आँसुओं को पोंछता हुआ चला गया और घर के काम–काज में लग गया क्योंकि यही मेरे जीवन की सच्चाई थी। हम जैसे बच्चों के नसीब में केवल सपने ही हैं जो अपने हैं।

Answer 2.

(i) सुन्दर मेहनत करने से घबराता था क्योंकि मेहनत करना उसे अच्छा नहीं लगता था। उसकी अच्छी बात यह थी कि वह अपने जीवन में सफल होना चाहता था। वह हमेशा सफलता का मूलमंत्र पाने के लिए लोगों से मिलता और उनकी सलाह लेता, उस पर अमल भी करता था।

(ii) सुन्दर साधु के पास सफलता का मन्त्र जानने की इच्छा से गया था क्योंकि उसे पता चला कि यह साधु बहुत ज्ञानी हैं। वे किसी भी प्रश्न का बहुत सटीक उत्तर देते हैं और कोई भी व्यक्ति उनके उत्तर से कभी असन्तुष्ट नहीं हुआ है। इसी मंशा से अपने प्रश्न का उत्तर जानने के लिए वह उनके पास गया।

(iii) साधु ने सुन्दर को सफलता का मन्त्र प्राप्त करने के लिए एक मोटी किताब पढ़ने को दी एवं साथ ही उन्होंने यह शर्त रखी कि किताब को पूरा क्रमानुसार लगातार पढ़ना होगा तभी उसे अपने प्रश्न का उत्तर मिलेगा, बीच–बीच में से कोई पृष्ठ पढ़ने पर या छोड़ने पर तुम्हें सफलता का मन्त्र नहीं मिलेगा।

(iv) सुन्दर ने साधु की शर्त मानकर वह किताब पढ़नी शुरू कर दी। वह जल्द से जल्द उस पृष्ठ पर पहुँचना चाहता था जहाँ सफलता का मंत्र लिखा हुआ था। इस क्रम में वह खाना–पीना और सोना सब कुछ भूल गया था। दिन–रात वह सिर्फ किताब ही पढ़ता रहता था। जिस दिन वह अंतिम पृष्ठ पर पहुँचा, उस दिन उसे बहुत खुशी हो रही थी। वास्तव में, उसे अंतिम पृष्ठ पर ये लिखा हुआ मिला कि दृढ़ इच्छाशक्ति, लगन, निष्ठा और समय का सदुपयोग ही सफलता का मंत्र है।

(v) इस गद्यांश से हमें शिक्षा मिलती है कि यदि हम जीवन में सफल होना चाहते हैं, तो हमें किसी भी कार्य को पूरा मन लगाकर करना चाहिए, तब तक नहीं रूकना चाहिए जब तक हमें अपना लक्ष्य न मिल जाए। कार्य के प्रति ललक और दृढ़ इच्छा शक्ति होगी, तो सफलता प्राप्ति से कोई हमें नहीं रोक सकता।

Answer 3.

(a)
- **(i)** ममता गाने का अभ्यास कर रही है।
- **(ii)** पिछले कुछ वर्षों के दौरान भारत की आबादी बढ़ी है।
- **(iii)** अपने दुष्कर्मों के कारण वह आज कंगाल है।
- **(iv)** चोर सोमनाथ के घर दबे पाँव आया।
- **(v)** स्वार्थी मित्र काम निकलते ही आँखें फेर लेते हैं।

(b)
- **(i)** उसे नौकरी पाने के लिए बहुत पापड़ बेलने पड़े।
- **(ii)** आज स्त्री एवं पुरुष कंधे से कंधा मिलाकर राष्ट्र निर्माण में सहयोग कर रहे हैं।
- **(iii)** वीर पुरुष कभी भी रणभूमि में पीठ नहीं दिखाते।
- **(iv)** उसे अनजान व्यक्ति के साथ जाते समय दाल में कुछ काला लगा।
- **(v)** परीक्षा में मेरे प्रथम आने पर, मेरे पिताजी फूले नहीं समा रहे थे।

SECTION – B

गद्य संकलन (Gadya Sanklan)

Answer 4.

(i) प्रस्तुत पंक्तियाँ गद्य संकलन के पाठ 'सती' से ली गई हैं, जिसकी लेखिका 'शिवानी जी' हैं। उन्हें प्रयाग स्टेशन से गाड़ी पकड़नी थी।

(ii) सफर में उस डिब्बे में लेखिका शिवानी थीं जिनका पूरा नाम गौरापंत है। इनका जन्म 17 अक्टूबर, 1923 में गुजरात के राजकोट नामक स्थान पर हुआ था। इनकी कहानियाँ सामाजिक आडम्बरों और रूढ़िवादों पर करारा व्यंग्य करती हैं। दूसरी महिला एक पंजाबी थी जो एक समाज सेविका है जो पंजाब में विस्थापित स्त्रियों के लिए बनाये गये एक आश्रम की संचालिका थी। तीसरी महिला एक महाराष्ट्रीयन महिला थी जो कम बोलती थी। वह देखने में सुन्दर प्रौढ़ा, किताब पढ़ने में व्यस्त थी। चौथी महिला का नाम मदालसा था, जो प्रिटोरिया से अपने पति की मृत देह लेने आई थी और अपने पति के साथ सती होना चाहती थी।

(iii) जब लेखिका ने पंजाबी महिला को देखा जो खद्दर का सलवार, कमीज व दुपट्टा पहने हुई थी, उन्होंने महसूस किया कि पंजाबी महिला के गोरे रंग में भी रूखापन था। चेहरा रौबदार था, लेकिन चेहरे पर लावण्य नहीं था। ठुड्डी पर एक बड़ा सा तिल था जिस पर दो–तीन बाल लटक रहे थे जिन्हें वे अंगुली में लपेट कर छल्ले–सा घुमा रही थीं। उनके हाथ खुरदुरे थे। हाथों की बनावट मर्दानी थी जिसे देखकर लेखिका ने उपर्युक्त वाक्य कहे।

(iv) उपर्युक्त कथन पंजाबी महिला के बारे में कहा गया है जो शायद किसी मीटिंग में भाग लेने जा रही थीं क्योंकि उनके हाथ में एक मोटी–सी फाइल थी जिसे वह बीच–बीच में हिल–हिलकर कुछ आँकड़ों को पहाड़ों की भाँति रटने लगती थीं। अभी हाल में ही वे विदेश से लौटी थीं और लखनऊ किसी समाज कल्याण गोष्ठी में भाग लेने जा रही थीं। उनके पास एक सुराहीदान था जिसकी एक टाँग अधिकांश सुराहीदानों की भाँति कुछ छोटी थी, जिससे वह बार–बार लुढ़क रहा था। वे बेचारी चश्मा उतारकर रखती हाथ की जासूसी अंग्रेजी पुस्तक जिसे पढ़ने

में उन्हें पर्याप्त आनन्द आ रहा था औंधी कर बर्थ पर टिकातीं और झुंझलाकर सुराहीदान ठिकाने से लगा देतीं। परन्तु जैसे ही वे अपनी पुस्तक में रस लेने लगती, सुराहीदान फिर से लुढ़क जाता। यही बात उनको परेशान कर रही थी।

Answer 5.

रज्जब एक गरीब आदमी था। वह जाति से कसाई तथा ललितपुर का रहने वाला था। इस समय रोजगार के लिए वह बाहर गया था जहाँ से लगभग तीन सौ रुपये की कमाई करके ललितपुर जा रहा था। उसके साथ उसकी पत्नी थी जो बुखार से तप रही थी। लौटते में देर हो गयी थी इसलिए उसे रात बिताने के लिए दो हाथ जगह चाहिए थी।

रास्ता काफी सुनसान था जिसके कारण उसने यह तय किया कि मड़पुरा गाँव में रातभर के लिए रुक जाये, लेकिन उसकी समस्या यह थी कि गाँव के लोग उसे जानते थे कि वह कसाई है इसलिए किसी ने उसे आश्रय नहीं दिया। अन्त में उसने गाँव के एक ठाकुर के घर जाने का निर्णय किया। ठाकुर ने उसे अपने मकान के एक कोने में रहने की अनुमति दे दी। रज्जब अपनी पत्नी के साथ वहाँ सो जाता है। सुबह होते ही ठाकुर उसे बाहर निकल जाने का हुक्म दे देता है।

उस समय रज्जब की पत्नी एक कदम भी चलने के योग्य न थी। वह जाकर एक पेड़ के नीचे बैठ जाता है और यह सोचता है कि उसकी पत्नी शायद कुछ समय में ठीक हो जाये, परन्तु ऐसा न हुआ।

रज्ज़ब किराए पर एक बैलगाड़ी लेकर मड़पुरा से अपनी पत्नी के साथ देर से निकलता है। मार्ग में उसकी गाड़ी को चार–पाँच लठधारी रोकर उसे लूटने एवं मारने की कोशिश करते हैं। रज्जब कमर की गाँठ को एक हाथ से सँभालते हुए कहता है–

"मैं बहुत गरीब आदमी हूँ। मेरे पास कुछ नहीं है। मेरी औरत गाड़ी में बीमार पड़ी है। मुझे जाने दीजिए।"

लाठी मारने वालों में से एक गरीब ठाकुर भी था, जिसने रज्ज़ब को एक रात्रि के लिए शरण दी थी। जब उसने गाड़ीवान के मुख से ललितपुर के कसाई का नाम सुना तो उसकी लाठी ऊपर की ऊपर ही रह गई। वह शरणागत की रक्षा करना अपना धर्म समझता था अतः वह अपने साथियों से भी उसे छोड़ने के लिए कहता है। उसके साथी जब उसकी बात मानने को तैयार नहीं होते हैं तब गरीब ठाकुर कहता है–

"खबरदार, जो उसे छुआ। नीचे उतरो, नहीं तो तुम्हारा सिर चूर किए देता हूँ। वह मेरी शरण आया था।"

ठाकुर एक बुन्देला था और बुन्देला अपनी शरण में आए हुए के साथ विश्वासघात नहीं करता। वह गाड़ीवान से भी उसे सही सलामत गन्तव्य स्थल पर पहुँचा आने के लिए कह देता है।

Answer 6.

आचार्य हजारी प्रसाद द्विवेदी जी ने अपने निबन्ध 'क्या निराश हुआ जाये?' की रचना एक विशेष उद्देश्य से की है। इस निबन्ध की रचना के दौरान निबन्धकार का उद्देश्य स्पष्ट रूप से परिलक्षित होता है। निबन्ध के प्रारम्भ में ही लेखक ने देश की वर्तमान स्थिति की विवेचना की है। लेखक यह बताना चाहता है कि आज देश में एक ऐसे वातावरण का निर्माण हो गया है कि जिसे देख ऐसा प्रतीत होता है कि देश में कोई ईमानदार आदमी रह ही नहीं गया है। आज हर जगह दोष और भ्रष्टाचार ही दिखाई देता है। यदि इस पृष्ठभूमि पर विचार करें तो हर व्यक्ति दोषी दिखाई देता है। इसलिए यह चिन्तन का विषय है।

इस आधार पर यह कहना कि अब देश के लोगों में गुणों का अभाव है, यह अवश्य ही असंगत होगा। दोषों को बढ़ा–चढ़ाकर दिखाया जाता है तो गुण उसमें छिप जाते हैं और हमारा ध्यान उस ओर नहीं जा पाता है। ऐसी परिस्थिति में हमें चिन्तन करने की आवश्यकता है या निराश होने की?

लेखक ने अपने बस चालक और टिकट बाबू की ईमानदारी की घटना के उदाहरणों द्वारा यह स्पष्ट रूप से प्रमाणित किया है कि समाज में अभी सच्चाई और ईमानदारी बाकी है।

लेखक यह बताता है कि एक बार गलती से दस रुपये की जगह सौ रुपये का नोट देने के बाद वह आकर ट्रेन में बैठ गया, जबकि टिकट दस रुपये का था, थोड़ी ही देर में टिकट बाबू विनम्रता के साथ उसे नब्बे रुपये लौटा देता है। यह कहना कि अब ईमानदारी लुप्त हो गई है सही नहीं है। यह घटना अनेक बेईमानी की घटनाओं से अधिक शक्तिशाली है।

लेखक वहीं दूसरे प्रसंग में यह कहता है कि मैं अपनी पत्नी और तीन बच्चों के साथ जा रहा था, निर्धारित स्थान से पहले एक निर्जन स्थान पर बस ने जवाब दे दिया। रात के दस बजे का समय था। तब कण्डक्टर साइकिल लेकर चलता बना। लोगों ने सोचा कहीं वही डकैतों को न भेज दे। कुछ नौजवान ड्राइवर को मारने की योजना बना रहे थे। मैंने बड़ी मुश्किल से ड्राइवर को बचाया। उसी समय हम यह देखते हैं कि बस अड्डे से खाली बस लेकर कण्डक्टर आ रहा है। आते ही उसने कहा कि यह बस चलने लायक नहीं है इसलिए वह नई बस लाया है। साथ ही वह लेखक के बच्चों के लिए पानी और दूध भी लाया था। अन्त में लोगों ने अपने व्यवहार के लिए ड्राइवर से माफी माँगी।

इन उदाहरणों को देखकर यह नहीं कहा जा सकता कि मनुष्यता एकदम समाप्त हो गई है।

द्विवेदी जी ने यह समझाया कि सिर्फ कपटपूर्ण घटनाओं का हिसाब रखने से हमें कष्ट की अनुभूति होती है, वहीं दूसरी ओर जब हम ऐसी घटनाओं पर ध्यान देते हैं जिसमें लोगों ने अकारण ही सहायता की है, तो ऐसी घटनाएँ मन और आशा और हिम्मत देती है।

काव्य मंजरी (Kavya Manjari)

Answer 7.

(i) 'मैंने' शब्द लेखक सियारामशरण गुप्त ने सुखिया के पिता के लिए प्रयोग किया है। उन्हें डर था कि कहीं उनकी पुत्री को महामारी न लग जाये। इसलिए वह अपनी पुत्री सुखिया को बाहर खेलने नहीं जाने देते थे।

(ii) सुखिया बहुत चंचल स्वभाव की थी, उसे बाहर खेलना बहुत पसन्द था। उसके पिता उसे बाहर खेलने जाने से बहुत रोकते थे, परन्तु उसका खेलना नहीं रुकता था और वह पलभर भी घर में नहीं ठहरती थी। उसके स्वभाव का यह परिणाम निकला कि एक दिन खेलते–खेलते उसे बहुत तेज बुखार हो जाता है। वह अपने पिता से इच्छा जताती है कि वे देवी माँ के मन्दिर से प्रसाद का एक फूल उसके लिए ले आयें।

(iii) नहीं सुखिया की इच्छा पूरी नहीं हो सकी, क्योंकि जब उसके पिता फूल लाने के लिए मन्दिर गये तो पुजारी ने उसे हाथों में

दीप व फूल दे दिए परंतु जब वह फूल लेकर जाने ही वाला था कि अचानक लोगों का शोर सुनाई दिया कि यह अछूत मन्दिर में कैसे आ गया? सभी ने उसको मारा पीटा और जेल भेज दिया। जब वह छूटकर आया तो उसे पता चला कि उसकी पुत्री कई दिन पहले मर चुकी थी।

(iv) इस कविता में कवि ने अस्पृश्ता की बुराई को बड़े ही मार्मिक ढंग से प्रस्तुत किया गया है। उस समय समाज में अछूत एक अभिशाप था। उन दिनों छुआछूत जैसी सामाजिक बुराई का बड़ा बोल–बाला था। अछूत व्यक्तियों का मन्दिर में प्रवेश वर्जित था। इसी समस्या को एक पिता पुत्री के माध्यम से बताया गया है कि एक पिता अछूत होने के कारण अपनी पुत्री की अन्तिम इच्छा भी पूरी न कर सका। क्योंकि उसकी पुत्री ने एक फूल देवी माँ के मन्दिर से माँगा था, लेकिन अछूत होने के कारण लोगों ने उसे मारा–पीटा और जेल भेज दिया। जेल से छूटने के बाद उसे पता चला कि उसकी पुत्री का निधन हो चुका है।

Answer 8.

सूरदास ने अत्यन्त मनोयोग से कृष्ण की बाल–लीलाओं का वर्णन किया है। सूरदास जी कहते हैं कि किस प्रकार श्रीकृष्ण घुटनों के बल चलते हुए बहुत सुन्दर लग रहे हैं। उनका पूरा शरीर धूल में सना हुआ है। मुख पर दही लिपटा हुआ है। सूरदास ने कृष्ण की बाल–सुलभ चेष्टाओं का बड़ा सटीक, सुन्दर और रोचक वर्णन किया है। इनके पदों को पढ़कर पाठकों के सामने श्रीकृष्ण की लीलाओं का चित्र–सा उपस्थित हो जाता है जिससे उनके हृदय में श्रीकृष्ण के प्रति प्रेम और भक्ति का समावेश होता है। सूरदास बाल कृष्ण की लीलाओं का वर्णन करते हुए कहते हैं कि कृष्ण अपने छोटे हाथ में मक्खन लिए हुए अति शोभायमान लग रहे हैं। उनके गाल सुन्दर हैं और आँखें भी अति सुन्दर हैं। उनके माथे पर गोरोचन का तिलक भी शोभा को प्राप्त हो रहा है। उनकी लटकती हुई लटें उनके चेहरे पर ऐसे शोभायमान हैं जैसे काले भंवरों के झुंड मधुर रस का पान किसी सुन्दर फूल पर बैठकर कर रहे हों। वह गले में वज्र तथा शेर का नाखून लगा हुआ कठुला पहने हैं।

सूरदास का मन श्रीकृष्ण की बाल–लीलाओं के वर्णन में अधिक लगा है। उन्होंने बाल–लीलाओं का ऐसा सजीव चित्रण किया है जैसे सूर स्वयं ही उन लीलाओं को देख रहे हों। जैसे–

'शोभित कर नवनीत लिए
घुटरुन चलत रेनु–तन–मंडित,
मुख दधि–लेप किए

श्रीकृष्ण यशोदा के आंगन में घुटनों के बल ही चलते हैं। एक दिन उन्होंने ताजा निकला माखन एक हाथ में लिया और लीला करने लगे।

चारु कपोल लोल लोचन, गोरोचन तिलक किए।'
लट लटकनि मनु मत्त मधुप–गन, मादक मधुहिं पिए।''

श्रीकृष्ण के गाल सुंदर, नेत्र चंचल हैं। माथे पर गोरोचन का तिलक लगा हुआ है। मुख पर काले घुँघराले बालों की लट आ रही है, जो बहुत सुन्दर लग रही है मानो काले भौंरे मस्त करने वाले शहद को पी रहे हों।

'कहन लागे मोहन मैया–मैया।
नंद–महर सौं बाबा–बाबा, अरु हलधर सौं भैया।

श्रीकृष्ण जब बोलना सीखते हैं तो उनके मुख से निकले सुंदर शब्द 'मैया–मैया', 'बाबा–बाबा', 'भैया–भैया' सुनना बहुत ही आनन्ददायक लगता है।

''मैया मोहि दाऊ बहुत खिझायौ।
मौं सो कहत मोल को लीन्हों, तू जसुमति कब जायो।''

यशोदा मैया अपना वात्सल्य प्रकट करते हुए ऊँचे स्थान से कन्हैया का नाम पुकार कर उन्हें कहीं भी दूर जाने से मना करती हैं, ताकि किसी की गाय उन्हें नुकसान न पहुँचा दे। श्रीकृष्ण जब थोड़े और बड़े होते हैं, तो वे अपने बड़े भाई बलराम की शिकायत अपनी माँ यशोदा से करते हैं कि बलराम भैया उसे बहुत चिढ़ाते हैं–कि तुझे तो खरीदा गया है, तू यशोदा का पैदा किया हुआ बेटा नहीं है। तू देख! नंद बाबा और यशोदा माता दोनों ही गोरे हैं, पर तू तो काला है।

सूरदास जी आगे वर्णन करते हुए कहते हैं कि श्रीकृष्ण इसी बात पर गुस्सा हो जाते हैं और खेलने भी नहीं जाते हैं। वे अपनी माँ से कहते हैं कि तुम भी सिर्फ मुझे ही मारना सीखी हो, बलराम को तो कुछ नहीं कहती मोहन के मुख से गुस्से भरी बातें सुनकर यशोदा मैया मन ही मन में इतनी प्रसन्न होती हैं कि उसका वर्णन नहीं किया जा सकता।

अन्त में हम कह सकते हैं कि परम संत सूरदास जी के आराध्य देव श्रीकृष्ण थे। 'वात्सल्य रस' को उन्होंने अपने इष्टदेव की लीलाओं का वर्णन करते हुए प्रकट किया है। शैशव अवस्था के कृष्ण के रूप का वर्णन, उनका बोलना और थोड़े बड़े होने पर अपने बड़े भाई की शिकायत माँ यशोदा से करना अद्भुत दृश्य उपस्थित करते हैं, जो संसार में अद्वितीय है।

Answer 9.

उद्यमी नर ही संसार में इच्छित सुख–साधनों को भोगने में समर्थ हो पाता है। ईश्वर ने इस पृथ्वी के अन्दर सभी धन–धान्य छिपाकर रख दिये हैं। उद्यमी मनुष्य ही उन्हें प्राप्त करता है। उद्यमी नर कभी भी भाग्य के भरोसे रहकर पछताता नहीं है, वह तो कार्य को पूरा करके दिखा देता है। संत कबीर ने कहा है–

''कर बहियाँ बल आपणी, छाँड़ परायी आस।
जिनके आँगन है नदी, सो कत मरत पियास।''

इससे यह साबित होता है कि मनुष्य को अपनी भुजाओं के बल पर विश्वास रखना चाहिए। ईश्वर ने पूरा कोष प्रकृति के भीतर छिपाकर रख दिया है। इसे उद्यमी नर ही भोग सकता है। यदि अब भी कोई मनुष्य उन्हें परिश्रम करके प्राप्त नहीं करता है तो इसमें ईश्वर का दोष कहाँ है। तुलसीदास जी ने कहा है–

''सकल पदारथ है जग माहीं।
करमहीन नर पावत नाहीं।''

संसार के सभी पदार्थ कर्मशील मनुष्य के लिए हैं, कर्महीन के लिए नहीं। इतिहास इस बात का साक्षी है कि परिश्रमी और उद्यमी व्यक्ति को हमेशा सफलता मिलती है। गीता में श्रीकृष्ण ने अर्जुन को कर्मयोग का ही उपदेश दिया था। संस्कृत का यह श्लोक कि–

''उद्यमेन हि सिध्यन्ति कार्याणि न मनोरथैः।
न हि सुप्तस्य सिंहस्य प्रविशन्ति मुखे मृगाः।।

यह इस बात को प्रमाणित करता है कि बिना परिश्रम के मनुष्य को कुछ प्राप्त नहीं होता। ईश्वर ने जो–जो बहुमूल्य वस्तुएँ या

जीवन के लिए महत्वपूर्ण साधन पृथ्वी के अन्दर छिपाकर रख दिये हैं, वे सब मनुष्य के हितार्थ ही हैं, ईश्वर या प्रकृति के काम तो आने नहीं है। मनुष्य उन्हें अपने परिश्रम के बल से ही प्राप्त कर सकता है। आकाश पृथ्वी पर जल वर्षा कर अन्न उत्पन्न करने का साधन बना देता है, लेकिन मनुष्य को हल चलाकर बीज बोना पड़ेगा और उगती फसल की रक्षा करनी होगी तभी उसे खाने के लिए अन्न प्राप्त होगा। कवि के अनुसार परिश्रम के सम्मुख पृथ्वी, नभ–तल सभी झुके रहते हैं। वे परिश्रम करने वाले का विरोध नहीं करते, बल्कि सहयोग करते हैं।

भाग्य के भरोसे रहकर मनुष्य कुछ प्राप्त नहीं कर सकता। ब्रह्मा ने भी मनुष्य के भाग्य में ऐसा कुछ लिखा हो कि उसे बिना कुछ किये सब कुछ मिल जायेगा, ऐसा कोई प्रमाण नहीं मिलता। कर्मवीर न तो भाग्य के भरोसे रहता है और न ही भाग्य के भरोसे रहकर पश्चाताप करता है। प्रकृति कभी भाग्य के सामने झुकती नहीं है। भाग्य के भरोसे रहना मनुष्य की सबसे बड़ी भूल है। कोई भी मनुष्य ब्रह्मा से अपने भाग्य में सुख लिखाकर लाया हो, ऐसा नहीं है। भाग्य के भरोसे तो आलसी ही रहते हैं। यदि भाग्य ही प्रबल हो तो मनुष्य को पृथ्वी में संचित कोष अपने आप प्राप्त हो जायें, पर ऐसा नहीं होता।

इस सम्बन्ध में भाग्य उसकी कोई सहायता नहीं करता। कवि भाग्यवादी नहीं है, वह तो उद्यमी नर की ही प्रशंसा करने वाला व्यक्ति है। यहाँ एक प्रश्न अवश्य उठता है कि किसी का कमाया हुआ धन कोई भोग रहा है। कवि का कहना है कि पाप से जो संचित धन है उसे भाग्य के छल से कोई और भोगता है, यह एक धोखा है जो भाग्य का नाम लेकर धनी व्यक्ति के धन का उपभोग करता है। ऐसा भोग बहुत कम समय तक ही चलता है। कवि ने नर–समाज का भाग्य मनुष्य का परिश्रम और परिश्रम से बहाया हुआ पसीना बताया है।

इस कविता में कवि ने यही प्रेरणा दी है कि भाग्य के भरोसे न रहकर कर्मशील और परिश्रमी बनो। अपने कार्य से कभी विरक्त मत हो।

"सारा आकाश" (Saara Akash)

Answer 10.

(i) प्रस्तुत पंक्तियों की वक्ता प्रभा और श्रोता अम्माजी हैं। जब प्रभा ने मिट्टी के गणेश को मिट्टी का ढेला समझकर उससे बर्तन माँज दिये थे तो अम्मा जी उसे डाँट रही थीं, तब उपरोक्त वाक्य प्रभा ने अम्माजी से कहे।

(ii) वक्ता अर्थात् प्रभा की बात सुनकर श्रोता अर्थात् अम्माजी ने गुस्से में कहा कि हमारी पुरखा हमें जवाब दिये जा रही है। तूने समझा काहे को? अच्छी थी। दिखाई थोड़े दिया होगा कि उसमें कलावा बँधा था या नहीं। बात पर बात कहे चली जा रही है, जो एक भी बात बिना जवाब दिए छोड़ी हो। अच्छी सीख सिखाई है तेरे माँ–बाप ने बड़ों से जुबान लड़ाने की। कोई आये कोई जाये किसी की चिन्ता ही नहीं और ऊपर से कैंची जैसी जुबान चलावे। मेरी बेटी होती तो इतने जवाब पर खींचकर जिन्दा गाड़ देती।

(iii) प्रभा के बारे में अपनी भाभी से पूरी बात सुनकर समर का गुस्सा सातवें आसमान पर पहुँच गया और वह सीधे प्रभा के सामने पहुँच गया और पूरी ताकत से प्रभा के गाल पर एक चाँटा जड़ दिया। वह गाली देते हुए बोला यहाँ रहना हो तो ठीक से रहो, नहीं तो जहाँ दीखे अभी चली जाओ। बड़ी नास्तिक की बच्ची बनी है।

(iv) प्रभा को चाँटा मारने के बाद समर को बहुत ग्लानि हो रही थी। क्योंकि जिन्दगी में उसने पहली बार किसी स्त्री पर हाथ उठाया था। वह अपने आपको बहुत कोस रहा था कि उसे ऐसा नहीं करना चाहिए। पश्चाताप के कारण वह रातभर सो न सका और रोता रहा। एक बार भाभी को चाँटा मारने पर उसने अपने भाई को बहुत बुरा भला कहा था और आज स्वयं ही उसने यह घटिया काम कर डाला। उसे ऐसी ग्लानि हो रही थी कि वह अपने दोनों गालों पर जोर से तमाचे मारे जिससे गाल पर उँगलियाँ छप जायें।

Answer 11.

राजेन्द्र यादव द्वारा लिखित 'सारा आकाश' का मुख्य पात्र समर यूँ तो सैद्धान्तिक रूप से परदा प्रथा का विरोधी था परन्तु जब उसने प्रभा को अम्मा बाबूजी के सामने बिना सिर पर पल्ला लिए आते देखा तो उसे भी बुरा लगा था, वास्तव में वह प्रभा को अपने परिवार के संस्कारों के अनुसार चलाना चाहता था। उसकी पत्नी प्रभा पढ़ी लिखी थी। वह अपने ससुर व जेठ से परदा नहीं करती थी पर वह बड़ों के सामने सिर–झुकाए रहती थी। शुरु–शुरु में किसी ने इस बात पर ध्यान नहीं दिया कि वह परदा नहीं करती है। बाबूजी हमेशा खाँस कर भीतर आते। एक बार प्रभा मुँह खोले दिखाई दी तो उन्होंने समझा कि शायद बहू को उनके आने का पता ही नहीं चला। बाद में उन्हें पता चला कि प्रभा परदा न करके केवल अपना मुँह दूसरी ओर कर लेती है। बाबूजी के शब्दों में "फिर बेटी और बहू में फर्क ही क्या रह गया? बेटी भी मुँह खोले बाल बिखेरे घूमती है और बहू को भी चिंता नहीं है कि पल्ला कहाँ जा रहा है।"

परदा न करने पर बहुत हल्ला हुआ। भाभी ने खूब ताने कसे। अम्मा ने भी समझाया कि भले ही उसके घर में परदा न किया जाता हो पर इस घर की रीति उसे रखनी चाहिए। इस बात की पूरी बिरादरी में चर्चा है। नामकरण पर लोग आएँगे तो देखकर क्या कहेंगे कि बहू परदा भी नहीं करती। धीरज ने भी समर से एक–दो बार कहा कि वह बहू को क्यों नहीं रोकता है? अम्मा–बाबू को अच्छा नहीं लगता है, तो कम से कम उनकी इज्जत के लिए ही सही, जब तक वे दोनों हैं उनका कहना मान लेना चाहिए। घर में चलता आया है, इसमें क्या बुराई है। प्रभा ने कोई उत्तर नहीं दिया, पर परदा नहीं किया।

प्रभा, अम्मा–बाबूजी सबके सामने मुँह खोले रहती है यह व्यवहार हैरान करने वाला था। समर के लिए नया भी था। समर प्रभा से बोलता नहीं था। पर मुन्नी से जोर से एक–दो बार कहा–"मुन्नी तेरी यह भाभी बड़ी बदतमीज है। किसी ने आदर शिष्टाचार नहीं सिखाया?" पर प्रभा पर इस कटाक्ष का कोई प्रभाव नहीं पड़ा। घर में धीरे–धीरे विवाद समाप्त हो गया। जब कोई बाहर से आता तो ध्यान रखा जाता, कि प्रभा सामने न पड़े।

परदा प्रथा एक सम्मानजनक प्रथा नहीं है। बड़ों का सम्मान केवल परदा करने से नहीं होता। सिर्फ घूँघट कर लें और बड़ों का सम्मान न करें उनसे जबान लड़ाएँ यह उचित नहीं। प्रभा ने कभी अम्मा, बाबू जी, सेठ जी, तथा समर को जबाव नहीं दिया। प्रभा ने घूँघट नहीं किया, पर कभी मान–सम्मान पर आँच न आने दी। घूँघट केवल दिखावा मात्र है। दिखावे के लिए परदा करना उचित नहीं, आँखें नीचे करके एवं सिर झुकाकर बड़ों के

सामने विनम्रता से बातें करना ही सही मायने में सम्मान है न कि सिर पर पल्ला रखना।

Answer 12.

समर 'सारा आकाश' उपन्यास का नायक है और प्रभा का पति है। उसका अभी विद्यार्थी जीवन चल रहा है। उसकी इच्छा के खिलाफ शादी कर दी गई है, इसलिए वह अपनी नव–विवाहिता पत्नी से सामीप्य नहीं बना पाता। समर की चरित्रगत विशेषताएँ इस प्रकार हैं–

1. **मेधावी**—प्रारम्भ में वह एक पढ़ने वाले विद्यार्थी के रूप में सामने आता है। उसकी लगन पढ़ने में है इसीलिए वह नहीं चाहता था कि उसका विवाह कर दिया जाये। उसकी इच्छा के प्रतिकूल विवाह होने से उसकी पढ़ाई में बाधा पड़ती है। घर में रोज किसी न किसी बात पर चिक–चिक होती रहती है इसलिए वह अक्सर अपने मित्र दिवाकर के घर पर पढ़ने के लिए चला जाता है। पढ़ने में व्यवधान न पड़े इसलिए वह अपनी पत्नी से भी बोलचाल नहीं करता। वह इण्टर में पढ़ रहा है और कॉलेज जाता है। वार्षिक परीक्षा में वह द्वितीय श्रेणी में पास हुआ। इससे प्रकट होता है कि वह पढ़ने में होशियार है।
2. **शुरू में निराशावादी**—विवाह के बाद जो उत्साह और उमंग युवकों में होता है, उसका समर में अभाव है। वह अपने साथियों से बातचीत करने में कतराता, परिचित–अपरिचितों से आँख चुराता है कि वे उससे सुहागरात के सम्बन्ध में मजाक करेंगे। वह प्रारब्ध को बोझीला मानता है और कहता है कि उसमें भागने का साहस नहीं है। उसने अपनी डायरी में लिखा था, "मैं इतनी जल्दी अपनी हिम्मत क्यों हार जाता हूँ? अभी से क्यों सोचता हूँ कि मेरी सारी महत्वकांक्षाएँ मर गई?...... मुसीबतों से इतनी जल्दी घबरा जाना अच्छी आदत नहीं है।" भाभी के कथनानुसार कि लालाजी क्या सोच रहे हो? तुम्हारी तो शादी न हुई, दुनियाभर की चिन्ताओं का बोझ तुम्हारे ऊपर आ गया हो। इस कथन से उसकी निराशा मालूम पड़ती है और लड़कों की तरह वह हँस बोल नहीं पाता जैसे उसका मन ही मर गया हो। एक गहरी निराशा, सूक्ष्म किन्तु व्यापक थकान उसकी रग–रग में बैठ गई है।
3. **शान्त स्वभाव वाला**—वह घर में शान्ति चाहता है, लेकिन वह उसे घर में नहीं मिल पाती। घर में रोजाना किसी न किसी बात पर चिक–चिक होती रहती इसलिए वह पढ़ नहीं पाता। अक्सर पढ़ने के लिए वह दिवाकर के घर पर चला जाता है। अम्माजी और बाबूजी को तो रोज ही प्रभा के काम–काज के बारे में जली–कटी सुनानी होतीं और इस माहौल में वह अशान्त हो जाता। पढ़ाई के लिए शान्त वातावरण आवश्यक है जो समर को नहीं मिल पाता।
4. **1Ppk fe=**—वह दिवाकर का सच्चा मित्र है। उन दोनों में प्रगाढ़ता है। समर उसके घर में पढ़ने चला जाता है और कभी–कभी वहाँ खाना भी खा लेता है। वह उसका आदर करता है। जब कभी रुपयों की आवश्यकता होती है तो दिवाकर उसे उधार रुपये भी दे देता है। दिवाकर ने उसकी नौकरी की व्यवस्था भी की। उन दोनों में खूब गहरा प्रेम है। उसने कभी भी दिवाकर की आलोचना नहीं की। कभी–कभी मजाक में दिवाकर उसके पहनावे को जरूर टोक देता। कभी सिनेमा जाते तो साथ–साथ जाते। उसे दिवाकर के पास आकर सच्चा सुख मिलता है।
5. **पुराने विचारों और पहनावे वाला**—समर पुराने विचारों वाला व्यक्ति है और साधारण कपड़े पहनने वाला है। उसे चटक–मटक और फैशन वाले कपड़े पसन्द नहीं हैं और न ही वह इतना खर्च कर सकता है। दिवाकर ने एक दिन उससे कहा, "देशी और विदेशी की बात नहीं जानता, लेकिन देशीपने का नाम लेकर कार्टून जैसी सूरत बनाये रखना भी मुझे पसन्द नहीं है। मेरी बात का बुरा मत मानना, लेकिन अगर मैं पूछूँ कि ऊँची–ऊँची बिना क्रीज–इस्तरी की पतलून और गंदी सिकुड़ी कमीज या अंग्रेजी ढंग से कटे लटकती लट्ठ जैसी गाँठ वाली चोटी ही अगर देशीपना है तो हम इसे दूर से प्रणाम करते हैं।" समर ने उसे व्यक्तिगत पसन्द और विश्वास कहकर टाल दिया। उसे क्या पता कि उसके (समर) के पास बस यही कपड़े हैं।
6. **शुरू में पत्नी के खिलाफ, बाद में प्यार करने वाला**—प्रारम्भ में तो समर अपनी नवविवाहिता पत्नी के खिलाफ ही रहा क्योंकि उसकी मर्जी के खिलाफ घरवालों ने उसकी शादी कर दी थी। इसलिए वह बहुत दिनों तक उससे बोला भी नहीं। हमेशा उससे गुस्सा ही रहा। कभी उसने यह जानने की कोशिश भी नहीं की कि वह कैसे रह रही है, लेकिन उत्तरार्द्ध में वह उससे बोलने भी लगा और प्यार भी करने लगा। वह उसकी स्थिति और अधिक काम का भी ध्यान रखता और घरवालों से उसकी खातिर उलझ भी जाता।
7. **शिरीष भाई की बातों पर विश्वास करने वाला**—पहले तो शिरीष भाई की बातें समर की समझा में ही नहीं आई, पर धीरे–धीरे वह उन पर विश्वास करने लगा। उसने हिन्दू समाज के ढाँचे के बारे में बताया कि शादी–विवाह एक पवित्र बन्धन है। शिरीष भाई ने बताया कि ऋषि–मुनियों के बनाये सारे धर्म–नियम संग्रहालय में रखने लायक हैं, किसी काम में आने लायक नहीं हैं। धर्म के नाम पर पैसा कमाना, लोगों को धोखा देना ठीक नहीं है, उनके विचार से सच्चा हिन्दू सोचता या जाँचता कुछ नहीं, वह तो चाहे आध्यात्मिक ज्ञान हो या रिश्वत और ब्लैक का रुपया, सबको ग्रहण करना जानता है। शिरीष भाई के अनुसार तो मनुष्य का धर्म 'मानवता' है बाकी सब ढकोसला है। धीरे–धीरे समर भी उनकी बात मानने लगा था।

सब मिलाकर समर का चरित्र मेधावी, निराशावादी, स्वाभिमानी और अपने फैसले पर अडिग रहने वाला है। वह घर में परेशान रहता है और बाहर जाकर खुश रहता है। उसका सम्पूर्ण जीवन निराशा और कष्टों में बीतता है। अन्त में तो वह आत्महत्या तक करने को तैयार हो जाता है, लेकिन अपनी सूझ–बूझ से आत्महत्या करने से अपने को बचा लेता।

"आषाढ़ का एक दिन" (Aashad Ka Ek Din)

Answer 13.

(i) उक्त कथन नाटक के प्रथम अंक से लिया गया है। उक्त कथन एक वृद्धा अम्बिका ने अपनी पुत्री मल्लिका से कहा है।

(ii) जब अग्निमित्र एक परिवार में मल्लिका के विवाह की बात करने जाता है और उस परिवार वाले विवाह के लिए इंकार कर देते हैं क्योंकि उस परिवार तक भी कालिदास और मल्लिका की बातें पहुँच जाती हैं। अग्निमित्र के आने पर मल्लिका इसे खुशी–खुशी स्वीकार करती है। मल्लिका कहती हैं—माँ, मैं जानती हूँ। अपवाद होता है। फिर भी मुझे अपराध का अनुभव नहीं होता। मैंने भावना में एक भावना का वरण

किया है। मेरे लिए वह सम्बन्ध बड़ा है। मैं वास्तव में अपनी भावना से प्यार करती हूँ, जो पवित्र और कोमल है।

(iii) अम्बिका अपनी बेटी मल्लिका को समझाती है कि केवल भावनाओं के सहारे जीवन व्यतीत नहीं किया जा सकता। केवल भावनाएँ धोखा–मात्र हैं। ऐसा सोचना अपने प्रति अन्याय है। जीवन यथार्थ के सहारे जिया जाता है। भावना मानव की आवश्यकता को पूरा नहीं कर सकती। अम्बिका मल्लिका के विचारों से एक दम सहमत नहीं है। उसे अपने बेटी का कालिदास के साथ घूमना–फिरना बिल्कुल पसन्द नहीं है, वह जानती है कि कालिदास आत्मकेन्द्रित व्यक्ति है। वह उसकी भावनाओं को नहीं समझेगा और भावनाओं के सहारे जीवन नहीं जिया जा सकता है। इन पंक्तियों के माध्यम से नाटककार ने युवा पीढ़ी को जीवन की सच्चाई का आइना दिखाने का प्रयत्न किया है।

(iv) **1. कुशल माँ**—नाटक के प्रारम्भ में जब मल्लिका भीगकर आती है तब वह अपनी माँ से कहती है कि वह उसे डाँटे कि वह भीगकर क्यों आई है? तो इस कथन के प्रत्युत्तर में अम्बिका द्वारा यह कथन कहा गया है—"अम्बिका, मुझे सब अच्छा लगता है। तुम्हारे यहाँ रहते मैं अकेली नहीं होती।"

2. पुत्री के भविष्य को लेकर सशंकित—अम्बिका हमेशा मल्लिका के भविष्य को लेकर चिन्तित रहती है। उसे यह चिन्ता सताती रहती है कि कालिदास को लेकर तरह–तरह का अपवाद है जो मल्लिका के भविष्य को लेकर बाधक है। अम्बिका, मल्लिका का विवाह करना चाहती है, उसकी दृष्टि में कालिदास एक आत्मसीमित व्यक्ति है।

3. समझाने वाली—अम्बिका बार–बार अपनी पुत्री को समझाने का प्रयास करती है कि भावनाओं और कल्पनाओं से जीवन के यथार्थ का सामना नहीं किया जा सकता। वह यह समझाने का प्रयास करती है कि भावना और कल्पना से जीवनयापन नहीं किया जा सकता है।

4. कर्म में विश्वास रखने वाली—अम्बिका नाटक के प्रारम्भ से ही हमेशा अपने कर्म में प्रवृत्त रहती है। बीच में मल्लिका से बात करती रहती है। वह अपने कार्य पर विशेष ध्यान देती है। उसके चरित्र के इस गुण को देखकर यह लगता है कि वह कर्म में विश्वास रखने वाली है।

5. दूरदर्शी—अम्बिका, मल्लिका और कालिदास के बीच के सम्बन्ध को क्षणिक मानती है। वह कालिदास के व्यक्तित्व को सामान्य नहीं मानती है उसके अनुसार कालिदास के जीवन में मल्लिका का स्थान स्पष्ट नहीं है।

Answer 14.

'आषाढ़ का एक दिन' नाट्य लेखन के क्षेत्र में एक महत्वपूर्ण उपलब्धि है। वस्तुतः हिन्दी के जिन एक–दो लेखकों ने नाटकों में यथार्थपरकता और व्याख्यात्मकता का श्रीगणेश किया था, 'आषाढ़ का एक दिन' उसकी महत्वपूर्ण परिणति है। इस नाटक की प्रत्यक्ष विषय–वस्तु कालिदास के जीवन से सम्बन्धित है। सारी घटनाएँ आषाढ़ के एक दिन पर ही घटित होती हैं। प्रेयसी मल्लिका के आग्रह पर कालिदास उसे छोड़कर चला जाता है। उसे सारी सुख–सुविधाएँ मिल जाती हैं, परन्तु अपने मानसिक द्वन्द्व की वजह से उसे कहीं भी शान्ति नहीं मिलती है। इस प्रकार कालिदास का चरित्र ही नाटक के कथानक का आधार है, अतः कालिदास की चारित्रिक विशेषताएँ निम्नलिखित हैं—

1. दयालु—नाटक के प्रथम अंक में कालिदास का प्रवेश ही हरिणशावक के साथ होता है जो किसी शर (तीर) से आहत हुआ है। जिसे लेकर कालिदास मल्लिका के घर आता है और उससे कहता है कि अँगुलियों का कोमल स्पर्श तुम्हें प्राण दे सकता है। इस प्रकार कालिदास का मंच पर आगमन ही जीवों पर दया करने वाले एक कोमल हृदय वाले व्यक्तित्व से होता है, जो उसके चरित्र की एक महत्वपूर्ण विशेषता को उद्घाटित करता है।

2. निर्भीक—कालिदास में निर्भीकता भी है, जो उसके चरित्र के महत्वपूर्ण पक्ष को उजागर करती है। नाटक के प्रथम अंक में ही दंतुल जब कालिदास को प्राणों और शक्ति का भय दिखाते हुए यह कहता है कि उसे हरिणशावक को देना ही पड़ेगा अन्यथा परिणाम बुरा होगा। तब यह सुनने के बाद भी कालिदास किसी भी परिणाम की चिन्ता किये बिना हरिणशावक को लेकर चले जाते हैं। इस प्रकार यह प्रकट हो जाता है कि कालिदास के चरित्र में निर्भीकता और शरणागत–वत्सलता है जो उनके चरित्र के निर्माण में एक महत्वपूर्ण योगदान करता है।

3. सन्तोषी—कालिदास का वर्तमान बड़े संघर्ष में बीत रहा है, वह मातुल की गाय चराने के लिए जाता है। जीविकोपार्जन भी कालिदास के समक्ष एक बड़ी समस्या बन जाती है। ऐसे समय में आचार्य वररुचि उज्जयिनी के राजा की तरफ से कालिदास को सम्मानित करने के लिए अपने साथ ले जाने के लिए आये हैं। कालिदास को यह पता है कि उन्हें राजकवि का सम्मान दिया जायेगा। उनकी जीविकोपार्जन की समस्या का अन्त हो जायेगा। यह सब जानकर भी कालिदास उज्जयिनी जाने से इंकार कर देते हैं। मातुल के द्वारा कालिदास के सन्दर्भ में कहा गया यह कथन उनकी सन्तोषवृत्ति को दर्शाता है।

"………और वे वंशावतंस कहते हैं कि, मुझे यह सम्मान नहीं चाहिए। मैं राजकीय मुद्राओं से क्रीत होने के लिए नहीं हूँ।"

4. प्रतिभाशाली—कालिदास का व्यक्तित्व एक असाधारण प्रतिभा से सम्पन्न है। उनकी कृतियाँ और उन कृतियों की प्रसिद्धि कालिदास की असाधारण प्रतिभा को व्यक्त करती हैं। अनेक स्थलों पर भिन्न–भिन्न पात्रों द्वारा नाटककार ने यह बात बताई है। जब संगिनी के पूछने पर मल्लिका उसे यह बताती है कि हमारा जीवन हर दृष्टि से बहुत साधारण है तो इस कथन के जवाब में कहा गया संगिनी का यह कथन कालिदास की प्रतिभा का परिचायक है—

"यह मैं नहीं मान सकती। इस प्रदेश ने कालिदास जैसी असाधारण प्रतिभा को जन्म दिया है। यहाँ की तो प्रत्येक वस्तु असाधारण होनी चाहिए।"

5. मानसिक अंतर्द्वन्द्व का शिकार—कालिदास ही नाटक का वह पात्र है जिसे नाटककार ने बाह्य और आन्तरिक द्वन्द्व को प्रदर्शित करने के लिए चुना। कालिदास के माध्यम से राकेश जी ने एक ऐसे पात्र की सृष्टि की है जिसका जीवन यथार्थ और आदर्शवाद के बीच झूलता रहता है। अंक तीन में मल्लिका से बातचीत के दौरान कालिदास ने अपने कथन द्वारा अपने चरित्र के इस द्वन्द्व के सन्दर्भ में खुद ही कहा है—

"………मैं यह आशा लिए हुए चला गया कि कल एक ऐसा समय आयेगा जब मैं तुमसे यह सब कह सकूँगा और तुम्हें अपने मत के द्वन्द्व का विश्वास दिला सकूँगा।"

6. निर्णय लेने में अक्षम—कालिदास को जब भी किसी प्रकार का निर्णय लेना होता है तो पूरी तरह से एक निर्णय पर पहुँच नहीं पाता जिससे उसका सम्पूर्ण जीवन द्वन्द्व में बीतता है। पहली बार उज्जयिनी जाने के सन्दर्भ में निर्णय लेने में मल्लिका उसकी सहायता करती है, दूसरी बार वह आकर भी मल्लिका से नहीं मिलता, तीसरे अंक में कालिदास अथ से प्रारम्भ करने की बात तो करता है, परन्तु दूसरे ही क्षण वह मल्लिका को पुनः छोड़कर चला जाता है।

7. निष्कर्ष—अन्त में निष्कर्ष के रूप में यह कहा जा सकता है कि कालिदास का प्रभाव पूरे नाटक पर पड़ता है। जहाँ कालिदास अप्रतिम प्रतिभा का स्वामी है, वहीं दूसरी ओर उसके चरित्र में अनेक ऐसे गुण भी हैं जो उसके चरित्र की दुर्बलता को भी उद्‌घाटित करते हैं। कहीं–कहीं तो ऐसा प्रतीत होता है कि मल्लिका के इस कष्टपूर्ण जीवन का एकमेव उत्तरदायी कालिदास ही है। कालिदास के चाहे बिना मल्लिका का औदात्य आत्मदान अकेला छूट जाता है। कुछ भी हो कालिदास ने नाटक में अपने चरित्र का बड़ी ही सफलता से निर्वाह किया है।

Answer 15.

'आषाढ़ का एक दिन' ऐतिहासिक होते हुए भी आधुनिकता का परिचय देता है। इसको आधुनिक युग का नाटक भी कहा जाता है। यह नाटक महाकवि कालिदास के निजी जीवन पर आधारित है। आषाढ़ का एक दिन का परिवेश ऐतिहासिक है, परन्तु उसमें आज की आधुनिक समस्याओं को प्रस्तुत किया गया है। कालिदास कश्मीर जाने के मार्ग में मल्लिका के गाँव प्रदेश से गुजरता है पर उससे मिलता नहीं है। इससे पहले जब वह गाँव गया तो मल्लिका से भावना के स्तर पर जुड़ा हुआ था, लेकिन भौतिक स्तर पर वह मल्लिका से बहुत दूर चला जाता है। कहा जाता है कि आषाढ़ का एक दिन नाटक के कथासूत्र तथा पात्र भले ही ऐतिहासिक हों, किन्तु मोहन राकेश ने वर्तमान समस्या को पाठकों के सम्मुख प्रस्तुत किया है। आज भी गाँव में प्रेम विवाह को अच्छी नजर से नहीं देखा जाता, न ही समाज उसको स्वीकार करता है। ये नाटक ऐतिहासिक मुखौटे के बावजूद आधुनिक पुरुष की विवशता को बड़े ही स्पष्ट, यथार्थ और सार्थक धरातल पर प्रस्तुत करते हैं।

यह एक सुव्यवस्थित एवं यथार्थवादी नाटक है जिसमें बाह्य और अन्तर्द्वन्द्व को दर्शाया गया है। इस नाटक में पाठकों को जीवन की सच्चाइयों से रूबरू कराना है। मनुष्य कभी–कभी अपनी ही बनाई हुई परिस्थितियों में फँस जाता है और उसे आकांक्षाओं से अलग होकर रहना पड़ता है। चाहे उसने कभी ऐसा सोचा नहीं था। मल्लिका कालिदास को उज्जयिनी भेजती है। इसका परिणाम यह होता है कि उसे हमेशा के लिए खो देती है। वहाँ वह राजनीतिक कार्यों में उलझ जाता है। राजनीति में उसकी बिल्कुल भी रुचि नहीं थी इसलिए वह परेशान रहता है। मल्लिका और कालिदास दोनों अलग–अलग हो जाते हैं।

●●

HINDI 2017

QUESTIONS

SECTION – A
Language – 50 Marks

Question 1.

Write a composition in approximately 400 words in Hindi on any ONE of the topics given below: [20]

किसी एक विषय पर निबन्ध लिखिए जो लगभग 400 शब्दों से कम न हो :

(i) *स्कूल के प्रधानाध्यापक का पद बहुत ही आदरणीय तथा जिम्मेदारीपूर्ण है। यदि वह पद आपको प्राप्त हो जाए, तो आप उस पद की जिम्मेदारियों को किस प्रकार पूरा करेंगे?*

(ii) *समाज में फैली बुराइयों को दूर करने के लिए कड़े कानून की नहीं, बल्कि नैतिक मूल्यों को बढ़ावा देने की आवश्यकता है।*

(iii) *पर्यटन एक बड़ा उद्योग बन गया है जिसके कारण सांस्कृतिक धरोहरें नष्ट हो रही हैं। इनके बचाव के लिए सुझाव दीजिए ताकि वह उद्योग फलता–फूलता रहे। इस विषय के अन्तर्गत विस्तार से लिखें।*

(iv) *अनुशासित व्यक्ति सुखी और स्वस्थ जीवन जीता है।— विवेचना कीजिए।*

(v) *मनुष्य अनुभव से बहुत कुछ सीखता है। कभी–कभी अनुभव कड़वा भी हो जाता है। क्या आपको कभी कोई कड़वा अनुभव हुआ है? उस अनुभव का वर्णन करते हुए लिखें, आपने उससे क्या सीखा? उसका विस्तृत वर्णन करें।*

(vi) *निम्नलिखित में से किसी एक पर मौलिक कहानी लिखिए—*

(a) *मन के हारे हार है, मन के जीते जीत।*

(b) *एक ऐसी मौलिक कहानी लिखिए जिसका अन्तिम वाक्य हो:*

काश! मैंने माँ की बात मानी होती।

Question 2.

Read the passage given below carefully and answer in Hindi the questions that follow, using your own words :

निम्नलिखित अवतरण को पढ़कर, अन्त में दिए गए प्रश्नों के उत्तर अपने शब्दों में लिखिए:–

ज्ञान–प्राप्ति के अनन्तर महात्मा बुद्ध ने अपना उपदेश सारनाथ में दिया। उसमें उन्होंने कहा— "जीवन के दो मार्ग हैं। एक मार्ग पर चलकर लोग सुख–साधनों को जुटाते हैं और भोगों से लिपटे रहते हैं। यह जीवन को सुखी नहीं बनाता। दूसरा मार्ग त्याग का है, तपस्या का है। इस पर चलकर लोग शरीर को कठोर यातनाएँ देकर भी अपने–आपको सुखी नहीं बना पाते। अतः ज्ञान–प्राप्ति के लिए मानसिक सुख–शांति के लिए बीच का मार्ग ही श्रेयस्कर है। बुद्ध का यही सिद्धान्त मध्यम–मार्ग के नाम से प्रसिद्ध है। कहते हैं कि महात्मा बुद्ध जहाँ बोधिवृक्ष के नीच ध्यान–मग्न बैठे थे, उस मार्ग से एक भजन–मण्डली जा रही थी। गीत की स्वर–लहरी वातावरण को गुंजित कर रही थी। सहसा समाधिस्थ बुद्ध के कानों में गायिका के गीत के मधुर स्वर गूँज उठे— 'वीणा के तारों को ढीला न छोड़ो, नहीं तो उससे मधुर स्वर नहीं निकलेगा और न ही उन्हें इतना कसो कि वे टूट ही जायें।' इन शब्दों से बुद्ध को मध्यम मार्ग की प्रेरणा मिली। उन्होंने अनुभव किया कि न तो भोगों से पूर्णतया तृप्ति हो सकती है और न ही भोगों को पूर्ण रूप से त्यागा जा सकता है। अतः जीवन के लिए भोग और त्याग के बीच का मार्ग ही श्रेयस्कर है।

महात्मा बुद्ध का चिंतन सरल और स्वाभाविक था। उनके मत में ऊँच–नीच का भेदभाव न था। राजा से लेकर रंक तक उनके लिए समान थे। उन्होंने धर्म के बाह्य आडम्बरों की अपेक्षा आन्तरिक शुद्धि पर बल दिया। यज्ञ में दी जाने वाली पशु–बलि का उन्होंने विरोध किया। जन–साधारण की भाषा में उन्होंने अहिंसा और प्रेम का जो उपदेश दिया उसने जादू का काम किया और देखते ही देखते लाखों नर–नारी उनके अनुयायी बन गये। 45 वर्षों तक उन्होंने घूम–घूमकर अपने अमृतमयी उपदेशों से जन–कल्याण किया।

महात्मा बुद्ध जीवन के 80 वर्ष पार कर चुके थे। शरीर दिन–प्रतिदिन क्षीण होता जा रहा था। अन्त में उन्होंने कुशीनगर की ओर प्रस्थान किया। कुशीनगर पहुँचते–पहुँचते उनका शरीर निढाल हो गया। उनके प्रिय शिष्य भिक्षु आनन्द ने शैय्या तैयार की। वह उस पर लेट गए। शिथिलता बढ़ती जा रही थी। निर्वाण की घड़ियाँ आ पहुँची थीं। यह देख आनन्द की आँखों से आँसू फूट पड़े। बोला— "देव! अब हमारा मार्ग–दर्शन कौन करेगा?" बुद्ध ने आँखें खोलीं। बोले— "आनन्द! तुम्हारे सामने विशाल कर्म–क्षेत्र है। जरा, रोग और मृत्यु से त्रस्त मानवता को अहिंसा और प्रेम का मार्ग दिखाओ। क्यों भूलते हो संसार नश्वर है? जन्म के बाद मृत्यु और मृत्यु के बाद जन्म प्रकृति का अटल नियम है। रही मार्ग–दर्शन की बात। कौन किसका मार्ग–दर्शन करता है? कौन किसका मार्ग–दर्शन कर सकता है? "अप्प दीपो भव–अपना दीपक आप बनो।" यह कहते हुए तथागत सदा के लिए समाधिस्थ हो गए। वातावरण शोक से भर गया। तथागत के निष्पंद पार्थिव शरीर के चारों ओर खड़े भिक्षु आँसू बहा रहे थे। रह–रह कर उनके कानों में तथागत के शब्द गूँज रहे थे "भिक्षुओं! मुझ पर विश्वास मत करना। मैं जो कहता हूँ उस पर भी इसलिए विश्वास मत करना कि मैंने कहा है। सोचना, विचारना और अपने अनुभव की कसौटी पर जो खरा उतरे वही करना।"

(i) *मध्यम–मार्ग से आप क्या समझते हैं? महात्मा बुद्ध को मध्यम–मार्ग की प्रेरणा कैसे मिली?* **[4]**

(ii) महात्मा बुद्ध के मत की क्या–क्या विशेषताएँ थीं कि देखते ही देखते लाखों नर–नारी उनके अनुयायी बन गए? [4]

(iii) महात्मा बुद्ध ने शोकाकुल भिक्षु को क्या उपदेश दिया? इससे हमें क्या शिक्षा मिलती है? [4]

(iv) महात्मा बुद्ध के निष्पंद पार्थिव शरीर के चारों ओर खड़े शोकाकुल भिक्षुओं के कानों में क्या शब्द गूँज रहे थे? इन शब्दों से क्या प्रेरणा मिलती है? [4]

(v) 'अप्प दीपो भव' का क्या अर्थ है? ज्ञान–प्राप्ति के बाद महात्मा बुद्ध ने अपना उपदेश कहाँ दिया और उनका निर्वाण कहाँ हुआ? [4]

Question 3.

(a) Correct the following sentences and rewrite : [5]

निम्नलिखित वाक्यों को शुद्ध करके लिखिए :

(i) मैंने आज जाना है।

(ii) पुलिस को देखकर चोर आठ तीन ग्यारह हो गए।

(iii) सूर्यास्त के छिपते ही वातावरण में अँधेरा छा गया।

(iv) साहित्य और समाज का घोर संबंध है।

(v) यह कार्य तत्काल अभी पूर्ण करना है।

(b) Use the following idioms in sentences of your own to illustrate their meaning : [5]

निम्नलिखित मुहावरों को वाक्यों में प्रयुक्त कीजिए :

(i) गले का हार होना। (ii) रंग चढ़ना।

(iii) सिर उठाना। (iv) चाँदी होना।

(v) अपने मुँह मियाँ मिट्ठू बनना।

SECTION – B

Prescribed Textbooks – 50 Marks

*(Answer **four** questions from this Section on **at least three** of the prescribed textbooks)*

गद्य संकलन (Gadya Sanklan)

Question 4.

एक जगह गरम–गरम जलेबियाँ बन रही थीं। बच्चों के लिए थोड़ी–सी खरीद लीं। घर के दरवाजे पर पहुँचे। दरवाजा खुला था। घर के अन्दर पैर रखने में हृदय धड़कता था। न जाने बच्चे किस हालत में हों?

(i) किसने और कब जलेबियाँ खरीदीं? जलेबियाँ खरीदने वाले का बच्चों से क्या सम्बन्ध था? [1½]

(ii) जलेबियाँ खरीदने वाला व्यक्ति कहाँ और क्यों गया था? [3]

(iii) सीताराम जी की अनुपस्थिति में बच्चों की देखभाल किसे करनी पड़ती थी? इस बारे में सीताराम जी क्यों चिंतित थे? [3]

(iv) इस बार बच्चों की देखभाल किसने और क्यों की? [5]

Question 5.

"'मजबूरी' कहानी मातृत्व प्रेम से परिपूर्ण एक सरल वृद्धा की कहानी है।" इस कथन की पुष्टि कहानी की घटनाओं के आधार पर कीजिए। [12½]

Question 6.

संस्कृति की परिभाषा न देकर लेखक ने किन लक्षणों का उल्लेख कर संस्कृति को समझाने का प्रयत्न किया है? उदाहरण सहित स्पष्ट कीजिए। [12½]

काव्य मंजरी (Kavya Manjari)

Question 7.

जाउँ कहाँ तजि चरण तुम्हारे।
काको नाम पतित पावन जग केहि अति दीन पियारे।
कौने देव बराइ बिरद हित, हठि–हठि अधम उधारे।
खग, मृग, व्याध, पषान, विटप जड़, जवन–कवन सुत तारे।
देव, दनुज, मुनि, नाग, मनुज सब, माया–बिबस बिचारे।
तिनके हाथ दास तुलसी प्रभु, कहा अपुनपौ हारे।

(i) प्रस्तुत पंक्तियाँ कहाँ से ली गई हैं? इनके कवि कौन हैं? भक्त ने किसके प्रति अपनी आस्था प्रकट की है? [1½]

(ii) पतित पावन किसे कहा गया है और क्यों? [3]

(iii) 'माया–बिबस बिचारे' पंक्ति का भाव स्पष्ट कीजिए। [3]

(iv) "जाऊँ कहाँ तजि चरन तुम्हारे" शीर्षक पद के आधार पर कवि की भक्ति–भावना पर प्रकाश डालिए। [5]

Question 8.

साखी के आधार पर सिद्ध कीजिए कि कबीरदास जी एक सफल कवि एवं श्रेष्ठ उपदेशक थे, उन्होंने बाहरी आडम्बरों या पाखंडों का विरोध करके किस चीज पर अधिक ध्यान देने पर जोर दिया है? [12½]

Question 9.

'जैसा हम बोयेंगे वैसा ही पायेंगे।' पंक्ति को आधार बनाकर 'आः धरती कितना देती है' कविता की समीक्षा कीजिए। [12½]

"सारा आकाश" (Saara Akash)

Question 10.

"बाबूजी तुम मुझे अपने हाथ से जहर देकर मार डालो मेरा गला घोंट दो········ मुझे वहाँ मत भेजो········ ।।"

(i) इस कथन का वक्ता कौन है? उसका परिचय दीजिए। [1½]

(ii) उसे कहाँ भेजा गया और क्यों? समझाकर लिखिए। [3]

(iii) बेटी की दुर्दशा देखकर माता–पिता की क्या स्थिति थी? [3]

(iv) उपन्यास के आधार पर तत्कालीन नारी की दशा का वर्णन कीजिए। [5]

Question 11.

समर का मन आत्म–ग्लानि से कब भर गया और क्यों? [12½]

Question 12.

'सारा आकाश' उपन्यास के आधार पर नायिका प्रभा का चरित्र–चित्रण कीजिए। [12½]

"आषाढ़ का एक दिन" (Aashad Ka Ek Din)

Question 13.

"यह लोकनीति है मैं तो कहूँगा कि लोकनीति और मूर्खनीति दोनों का एक ही अर्थ है।"

(i) *उपर्युक्त कथन कौन, किससे और कब कह रहा है?* [1½]

(ii) *उपर्युक्त वाक्य किस सन्दर्भ में कहा गया है?* [3]

(iii) *वक्ता का संक्षिप्त परिचय दीजिए।* [3]

(iv) *'आषाढ़ का एक दिन' नाटक का उद्देश्य लिखिए।* [5]

Question 14.

मल्लिका 'आषाढ़ का एक दिन' नाटक की महत्वपूर्ण पात्र है, जिसके चरित्र ने सर्वाधिक प्रभावित किया है। उसकी चारित्रिक विशेषताओं का वर्णन कीजिए। [12½]

Question 15.

मोहन राकेश द्वारा लिखित 'आषाढ़ का एक दिन' नाट्य लेखन के क्षेत्र में एक महत्वपूर्ण उपलब्धि है। नाटक के शीर्षक की सार्थकता पर प्रकाश डालिए। [12½]

ANSWERS

Answer 1.

(i) **यदि मैं प्रधानाध्यापक बन जाऊँ तो**

विद्यालय एक परिवार है। जैसे परिवार में अनेक सदस्य होते हैं, परन्तु उसका एक प्रमुख होता है। उसी प्रकार से छात्र व अध्यापक के ऊपर भी एक पद होता है, वह प्रधानाध्यापक के नाम से जाना जाता है। विद्यालय का प्रमुख होने के कारण उसका स्थान आदरणीय व दायित्वपूर्ण होता है। इन्हीं विशेषताओं से प्रभावित होकर मेरे मन में भी इस पद के प्रति अभिलाषाएँ जाग्रत हो उठीं। एक दिन मैं अपने कमरे में बैठा था, तो मुझे अपने प्रधानाध्यापक के दण्ड, व्यवहार का ध्यान हो आया, किस छात्र को कैसे, कब दण्ड की आवश्यकता है वह अच्छी तरह जानते थे। किस समय सही शिक्षा देकर किसी छात्र को सुधारा जाये, इसे वह भलीभाँति जानते थे। मेरी भी शिक्षा पूर्ण होने को थी। मेरे मन में भी यह विचार आया कि यदि मैं प्रधानाध्यापक बन जाऊँ तो मेरा व्यवहार मेरे छात्रों के प्रति कैसा होगा?

मेरा प्रथम प्रयास होगा कि अनुशासन का पूर्ण ध्यान रखूँ क्योंकि आज विद्यालयों में सबसे बड़ी समस्या अनुशासन की है। छात्र व शिक्षक समय पर विद्यालय आयें इसके लिए मैं स्वयं पन्द्रह मिनट पहले विद्यालय पहुँचूँगा क्योंकि बड़े लोग जैसा आचरण करते हैं छोटे भी वैसा ही आचरण करते हैं। मुझे मालूम है कि हर व्यक्ति को मैं खुश नहीं कर सकता। इसलिए मैं 'नहीं' कहना सीख लूँगा। मैं हर कक्षा में चालीस से ज्यादा छात्र नहीं रखूँगा। कम छात्र होने से कक्षा में पढ़ाई का वातावरण अच्छा होगा जिससे छात्रों का सर्वांगीण विकास होगा।

छात्रों के स्वास्थ्य का ध्यान रखते हुए उनके खेलने का विशेष प्रबन्ध रखूँगा। विद्यालय के भवन को सुन्दर, आकर्षक व स्वच्छ रखूँगा। सुन्दर बगीचे में भाँति–भाँति के पेड़–पौधे व फूलों के पौधे होंगे। इसके अतिरिक्त एक तरणताल भी होगा। पीने के लिए स्वच्छ जल, बैठने की पूरी व्यवस्था, जनरेटर की व्यवस्था व साथ–साथ एक छोटे पुस्तकालय की भी व्यवस्था करूँगा।

विद्यालय में हर वर्ग के विद्यार्थी शिक्षा लेते हैं, इसलिए शुल्क कम रखना होगा। इसके अतिरिक्त विद्यालय में छात्रों का गणवेश भी एक सा रहेगा, क्योंकि इससे निर्धन व धनी छात्रों के बीच का फर्क समाप्त हो जायेगा। इसका एक फायदा यह होगा कि छात्रों में मैत्रीभाव विकसित होगा और उनमें भी हीनभावना व उच्च भावना जाग्रत नहीं होगी। इसके अतिरिक्त एक गरीब देश का नागरिक होने के नाते मैं निर्धन छात्रों के लिए शुल्क माफी का प्रावधान करूँगा। मेरा यह भी प्रयास रहेगा कि मैं छात्रों को उचित संस्कार दे सकूँ क्योंकि संस्कार सम्पन्नता ही कुल और देश का गौरव बढ़ाती है। इसके अतिरिक्त समय–समय पर ऐतिहासिक कथा–कहानियों के द्वारा छात्रों को चारित्रिक शिक्षा दूँगा। बाल युवा मंच की स्थापना कर हर हफ्ते में एक बार छात्रों को कहानियाँ, कविताएँ और विभिन्न विषयों पर भाषण देने का अभ्यास करवाऊँगा। इससे छात्रों में नैतिकता, सदाचार, विनम्रता तथा विवेक आदि सद्‌गुणों का विकास होगा। विद्यालय में व्यायाम और खेलकूद भी अनिवार्य होगा ताकि बौद्धिक विकास के साथ–साथ छात्रों का शारीरिक विकास भी हो सके।

विद्यालय के समस्त अध्यापक, छात्र व मैं स्वयं मिलकर सामाजिक उत्थान के लिए भी कार्य करेंगे। इसलिए हे प्रभु! मुझे यह सुअवसर जरूर दे, जिससे मैं प्रधानाचार्य बनकर अपने विद्यालय, समाज, देश और विद्यार्थियों की उन्नति में सहयोगी बन सकूँ।

(ii) **नैतिक मूल्यों को बढ़ावा देने की आवश्यकता**

किसी ने सच ही कहा है कि यदि मनुष्य का चरित्र चला गया, तो उसका सब कुछ चला गया। आज प्रतिदिन देखने में, सुनने में, समाचार पत्रों में पढ़ने से यह स्पष्ट है कि आम आदमी से लेकर आई. ए. एस. अधिकारी तक भ्रष्टाचार में डूबे हुए हैं। इसलिए समाज में फैली बुराइयों को दूर करने के लिए कड़े कानून की नहीं, नैतिक मूल्यों की आवश्यकता है। लालबहादुर शास्त्री, अटलजी ऐसे नेता थे जो नैतिक मूल्यों को महत्व देते थे। आज अन्तरात्मा की आवाज न सुनकर केवल भौतिक सुविधाओं की ओर ध्यान दिया जा रहा है।

हर खाद्य पदार्थ में मिलावट ही आ रही है। व्यापारी वर्ग अत्यधिक मुनाफे के प्रलोभन में नैतिक मूल्यों को तिलांजलि दे रहा है। आज नैतिक मूल्यों के अभाव में परिवार टूट रहे हैं, अपने स्वयं के बच्चे, पत्नी के अलावा अन्य सदस्यों पर ध्यान नहीं दिया जा रहा है। पहले नैतिक मूल्यों के कारण संयुक्त परिवार में परिवार के सभी सदस्य इकट्ठे रहते थे।

गाँधी जी ने जीवन भर नैतिक मूल्यों को अपने जीवन में स्थान दिया। वे हर कार्य अपनी अन्तरात्मा की आवाज (नैतिक मूल्यों) के अनुसार करते थे। यदि जीवन में उच्च आदर्शों, नैतिक मूल्यों को महत्व दिया जाये तो परिवार से लेकर समाज एवं समाज से लेकर राष्ट्र, हर क्षेत्र में चहुँमुखी विकास कर सकता है। आज न्यायालय भी नैतिक मूल्यों के अभाव में सही निर्णय देने में असमर्थ होते जा रहे हैं।

नैतिक मूल्यों के अभाव के कारण व्यक्ति के चरित्र में गिरावट आती जा रही है। आज अपराधों का ग्राफ बढ़ता जा रहा है। चोरी, डकैती, बलात्कार, हत्याएँ इसलिए हो रही हैं कि व्यक्ति

उच्च आदर्शों, नैतिक मूल्यों को जीवन में स्थान नहीं दे पा रहा है, इसलिए बच्चों को आज अच्छे संस्कारों की नितान्त आवश्यकता है। हमें अपने चरित्र से, व्यवहार से बच्चों के सामने नैतिक मूल्यों के उदाहरण प्रस्तुत करने चाहिए।

आज के बच्चे ही कल का भविष्य हैं। अतः शिक्षा के साथ-साथ नैतिक मूल्यों को जीवन में स्थान दें तथा हर कार्य अपनी अन्तरात्मा एवं उच्च आदर्शों को ध्यान में रखकर करें। परिवार बच्चों की प्रथम पाठशाला है। अतः हर माता-पिता को स्वयं का आचरण शुद्ध, सरल, पवित्र एवं मर्यादापूर्ण रखना चाहिए। आज जो संस्कार बच्चों में स्थापित होंगे वे ही आगे चलकर देश और समाज के, परिवार के विकास में सहायक होंगे।

ईमानदारी, सत्यता, विवेक, करुणा, प्रलोभनों से दूर रहना, पवित्रता, आदि नैतिक मूल्यों के आदर्श तत्व हैं। इन आदर्श सिद्धान्तों से जीवन में आध्यात्मिकता का जन्म होता है। एक आदर्श परिवार या देश के संचालन हेतु परिवार के सभी सदस्यों में शिष्टाचार, सदाचार, त्याग, मर्यादा, अनुशासन व परिश्रम की आवश्यकता होती है। यदि इन नैतिक मूल्यों का अभाव होगा तो राष्ट्र में स्वार्थ लोलुपता तथा पद लोलुपता बनी रहेगी, जिससे परिवार भी टूटेगा, राष्ट्र भी भ्रष्टाचार से प्रदूषित होता रहेगा। इसलिए अहंभाव तथा स्वार्थ त्यागकर प्रलोभनों से दूर रहकर अपने जीवन में विनम्रता, त्याग और परोपकार को जीवन में स्थान देना होगा, तभी हम एक आदर्श परिवार, आदर्श परिवार, आदर्श समाज और आदर्श राष्ट्र की कल्पना कर सकते हैं।

(iii) सांस्कृतिक धरोहरों को सहेजते हुए पर्यटन उद्योग के फलने-फूलने के लिए सुझाव

भारत एक विशाल देश है। इसका परिचय हमें यहाँ विद्यमान पर्यटन स्थलों से मिलता है। यद्यपि इस उद्योग में कुछ समय से मन्दी आ रही है। वस्तुतः प्रत्येक देश का सौन्दर्य वहाँ का प्राकृतिक वातावरण, ऐतिहासिक स्थल, संस्कृति एवं सभ्यता अधिक होता है और इसी आकर्षण के वशीभूत ही लोग विभिन्न देशों अथवा राज्यों में भ्रमण हेतु जाते हैं। पर्यटन को उद्योग का दर्जा देने के पश्चात् सरकार की मुख्य कोशिश पर्यटन को बढ़ावा देने की है। हमारे पास पर्यटकों को आकर्षित करने की तकनीक जरूर है, किन्तु इससे भविष्य में उपजने वाली अनेकानेक समस्याओं का सामना करने और उनसे निपटने की तकनीक उपलब्ध नहीं है।

अतिथि देवो भव की अवधारणा वाला भारत आज विश्वस्तर पर पर्यटन में लगातार पिछड़ रहा है। ख्याल यह भी है कि भारत में पर्यटन उद्योग को सिर्फ पर्यटकों के लिए आरक्षित एवं संरक्षित मान लिया गया है, जबकि देश में ऐसे लोगों की संख्या बढ़ती जा रही है जो अपनी संस्कृति से रूबरू होना चाहते हैं।

यही नहीं देश में पर्यटन का सर्वथा एकांगी अर्थ यह लगा लिया जाता है, जिसे हम भारतदर्शन का पर्याय मान लें तो गलत न होगा। यह अपने आप में कम आश्चर्य नहीं है कि सर्वाधिक मुद्रार्जन का स्रोत आज खिसककर दूसरे स्थान पर पहुँच गया है, जबकि तमाम सरकारी दावों के अनुसार आज देश में यह उद्योग दिन दूना-रात चौगुना बढ़ रहा है। प्राकृतिक सुषमा की पिटारी में जब कोई नयी चीज नहीं बची है। अच्छा माल उपलब्ध न होने के कारण यह पिछड़ता जा रहा है। भारत में हम बहुत सी बातों को रस्म अदायगी के तौर पर ही लेते हैं जिसके चलते पर्यटन का भी वही हश्र होता जा रहा है जो हमारी राष्ट्रभाषा का हो रहा है।

आज इस उद्योग को बढ़ावा देने के लिए बेहतर विपणन अथवा मार्केटिंग की जरूरत है। भारत में विदेशी पर्यटन में आ रहे ठहराव के लिए अक्सर आतंकवाद को जिम्मेदार ठहरा दिया जाता है, जबकि और देश भी इससे जूझ रहे हैं, फिर भी वे देश-पर्यटन को लेकर विशेष उत्साहित हैं।

पर्यटन मन्त्रालय के आय के पिटारे में इनका नाममात्र भी योगदान नहीं है। हमारे यहाँ पर्यटकों के लिए सुविधाएँ नाममात्र की है और ठगी से भरपूर हैं। इसी तरह परिवार सुविधाओं और विदेशी में समुचित प्रचार-प्रसार के अभाव जैसी कई कमियाँ भी हैं। बहुत योजनाएँ कागजों पर सिमट कर रह जाती हैं। पर्यटन धीरे-धीरे प्रदूषण में तब्दील होता जा रहा है क्योंकि अनुशासनहीन पर्यटन की बदौलत आज सांस्कृतिक धरोहरों पर जिस तेजी से कूड़ा करकट जमा हो रहा है वह एक दिन एवरेस्ट से भी ऊँचा हो जायेगा।

आज आवश्यकता इस बात की है कि पर्यटन तन्त्र को सांस्कृतिक धरोहरों की सुव्यवस्था और विकास पर जोर देना चाहिए। सभी विभागों में समन्वय व सामंजस्य स्थापित किया जाये। वर्तमान समय में पर्यटन के विकास में उड़ान सेवा और कम्प्यूटर की आवश्यकता बहुत बढ़ गयी है।

विदेशी पर्यटकों हेतु बाजार की खोज करनी होगी। पर्यटन संस्कृतियों को बढ़ावा देने के लिए स्वदेशी संस्कृति की रक्षा करना अति आवश्यक है। हम भारतीयों को 'अतिथि देवो भव' की परम्परा को याद रखना बहुत जरूरी है क्योंकि पर्यटन राष्ट्रीय और अन्तर्राष्ट्रीय सद्भाव को बढ़ाने का एक अनूठा माध्यम है।

इसलिए सांस्कृतिक धरोहरों को सहेजने के लिए पर्यटन के चेहरे को निखारने की पहल बहरहाल हम लोगों को ही करनी होगी।

(iv) अनुशासित व्यक्ति सुखी और स्वस्थ जीवन जीता है

प्रकृति के समस्त तत्व अनुशासन से प्रेरित होते हैं। संसार में सूर्य, चन्द्रमा, ऋतुएँ आदि का जो नियत क्रम चलता है, वह यह सिद्ध करता है कि प्रकृति के मूल में भी एक सुनियोजित व्यवस्था कार्य कर रही है।

अनुशासन का अर्थ (अनु + शासन) अर्थात् स्वयं पर शासन अथवा स्वयं के नियमों के अनुसार चलने की दृढ़ शक्ति। यह एक व्यापक तथ्य है जिससे जीवन में व्यवस्था आती है। अनुशासन के मार्ग में भटक जाने पर व्यक्ति चरित्रहीन, दुराचारी, पतित तथा निन्दनीय हो जाता है। समाज में उसका कोई सम्मान नहीं करता। अनुशासन के अभाव में उसका तन तथा मन दोनों दूषित हो जाता है।

विद्यार्थी अनुशासन में रहकर ही स्वास्थ्य, शिक्षा, व्यवहार तथा आचार ग्रहण कर सकता है। नियमित रूप से अध्ययन करना, विद्यालय जाना, व्यायाम करना, गुरुजनों से सद्व्यवहार करना ही विद्यार्थी जीवन का अनुशासन है।

बच्चे का प्रथम सम्पर्क अपने माता-पिता से होता है। बच्चे को अनुशासन की प्रथम शिक्षा घर से ही प्राप्त होती है। इसके

बाद बाह्य समाज तथा विद्यालय से अनुशासन की शिक्षा प्राप्त होती है।

अनुशासन के दो रूप होते हैं—एक बाह्य व एक आन्तरिक। बाह्य रूप हमें बचपन से सिखाया जाता है जैसे—पंक्ति में खड़े होना, बड़ों का आदर करना, सड़क के नियमों का पालन करना, कक्षा में शान्त रहना, कार्यालय में अपना कार्य ईमानदारी से करना आदि व आन्तरिक अनुशासन कर्तव्य या प्रेरणा के रूप में उभरता है।

जो व्यक्ति अनुशासित होता है, उसके स्वभाव में भी विनम्रता व नम्रता आ जाती है। इससे व्यक्ति अपनी शक्ति के दुरुपयोग को रोकेगा। वह सजग रहकर कर्तव्य का पालन करेगा। इसलिए यह सत्य ही है कि नियमित और नियन्त्रित जीवन व्यतीत करने वाला व्यक्ति सच्ची सुख–शान्ति का अनुभव करता है। अनुशासन में रहकर ही सच्ची शक्ति का संचार होता है। सेना अनुशासन में रहकर ही युद्ध कर पाती है, विद्यार्थी अनुशासन में रहकर ही ज्ञान प्राप्त कर सकता है। व्यापारी एवं उद्योगपति भी अनुशासन का पालन करके ही अपना उत्तरदायित्व निभा पाते हैं।

वास्तव में, बिना अनुशासन के कोई भी क्रियाकलाप नहीं चल सकता। चाहे वह विद्यार्थी जीवन हो या वह आम मनुष्य का जीवन हो। अनुशासन जीवन के हर मोड़ पर आवश्यक है। सफलता व असफलता का मापदण्ड ही अनुशासन है। सद्भावना का विस्तार करके विश्व जीवन को भव्यता प्रदान करने का एकमात्र साधन ही अनुशासन है। अनुशासनमय जीवन ही यथार्थ जीवन का परिचायक है व अनुशासित व्यक्ति ही जीवन में सुखी है, यह शाश्वत सत्य है।

(v) **मेरा कड़वा अनुभव**

मनुष्य अनुभव से ही सीखता है, परन्तु जिन्दगी में वही अनुभव जब कड़वा हो जाता है तब उससे दुःख तो बहुत होता है, परन्तु वह एक अच्छी सीख भी दे जाता है।

मैं कॉलोनी में नया–नया आया था। आस–पास जान–पहचान न होने के कारण बहुत परेशान था। धीरे–धीरे मेरी भी घनिष्ठता कई परिवारों से हो गई। एक परिवार का हमारे यहाँ आना–जाना भी काफी होने लगा। तीज–त्योहार, जन्मदिन सब साथ–साथ मनाते। वे समय–समय पर हमें उपहार देते, हम भी समय–समय पर उपहार लेते व देते थे। घर में आना–जाना बढ़ने के कारण उन पर मेरा विश्वास बहुत अधिक बढ़ गया। एक दिन उस परिवार से अपने घर के कुछ कामों के सिलसिले में रुपयों से सम्बन्धित बातें हुई। मैंने अपने घर में कुछ रुपयों के रखने के बारे में उन्हें जानकारी दी।

एक दिन हम सब उसी मैत्रीभाव को संजोये अपने घर में बैठे थे, सहसा मुझे कुछ रुपये कमरे में इधर–उधर गिरे हुए दिखाई दिये। मैं समझ गया कि आज तक जिसे मैं अपना मित्र समझ रहा था वही शायद आज धोखा दे रहा है। पूछने पर "यह रुपये किसके हैं?" कोई जवाब नहीं मिला। मैं तथा वह अलमारी व पैसे रखने को सभी जगह देख चुके थे जहाँ से वह रुपये गायब थे। वह क्षणभर का अनुभव मेरे लिए ऐसा था मानो मैं आसमान से धरती पर आ गिरा। मुझे विश्वास ही नहीं हो रहा था कि जो मेरा इतना अन्तरंग मित्र था वह मेरे साथ ऐसा भी कर सकता है। सच तो यह था कि पूछने या कहने की हिम्मत न उसमें थी न मुझमें थी, परन्तु सत्य तो आखिर सत्य ही होता है। उसे हमें स्वीकारना ही पड़ता है। वह दोस्ती का जो सुखद अनुभव मैं आज तक करता आया था वह शायद क्षण में ही समाप्त हो गया, परन्तु इस घटना ने मुझे जिन्दगी के उस कड़वे सत्य से परिचित करा दिया कि किसी पर विश्वास तो करो पर इतना विश्वास भी न करो कि कोई आपके साथ विश्वासघात कर जाये। वह दिन था और आज का दिन है, मुझे दोस्ती करना तो आ गया है, पर अब मैं दोस्त की पहचान अच्छे से करने लगा ताकि ऐसा कड़वा अनुभव मेरे जीवन में फिर न हो।

(vi) **(a) मन के हारे हार है, मन के जीते जीत**

"जो भी परिस्थितियाँ जीवन में मिलें,
काँटे चुभें या कलियाँ खिलें,
इंसान हारे नहीं,
यही सन्देश जीवन का सही।"

मनुष्य का मन बहुत चंचल और बलवान होता है। शारीरिक क्रियाएँ मन पर ही निर्भर करती हैं। यदि मन में शक्ति, उत्साह और उमंग का जोश हो तो शरीर भी तेजी से कार्य करता है। व्यक्ति की हार–जीत उसके मन की दुर्बलता–सबलता पर निर्भर करती है; अगर मनुष्य को जीवन में सफलता प्राप्त करनी है तो उसे दृढ़ संकल्पित होना आवश्यक है। वह कौन–सा बल था, जिसके आधार पर मुट्ठीभर हड्डियों वाले महात्मा गाँधी ने विश्व–विजयी अंग्रेजों को देश से बाहर का रास्ता दिखा दिया? निश्चय ही इसके पीछे मन की सबलता है। मन को सबल बनाने के लिए सत्य, न्याय और कल्याण के भाव का होना आवश्यक है। अगर मनुष्य के मन में सत्य की शक्ति नहीं है, तथा वह न्याय का पक्षधर नहीं है, तो उसका मन स्थिर नहीं हो सकता। मन की स्थिरता, दृढ़ता और धैर्य ऐसे गुण हैं जो व्यक्ति को विजय की ओर अग्रसर करते हैं। संकटों को सहर्ष झेलते हुए भी जो व्यक्ति कर्तव्य–मार्ग पर चलते रहते हैं, वे ही विजयी होते हैं।

यदि व्यक्ति आशावादी है और जीवन में कुछ पाने की इच्छा रखता है अर्थात् उज्ज्वल भविष्य की ओर देखता है तो उसे सकारात्मक विचारों के साथ प्रगति की ओर बढ़ते जाना चाहिए। इसके विपरीत निराशावादी व्यक्ति जीवन में दुर्बल मन के कारण अक्सर असफल रहते हैं। ऐसे व्यक्ति संघर्ष करने की भावना न होने के कारण पहले ही हार स्वीकार कर लेते हैं। सफल व्यक्ति कर्मवीर होते हैं, असफल व्यक्ति हीनभावना से ग्रसित होते हैं। हालांकि परिस्थितियाँ कभी–कभी इतनी भयावह होती हैं कि व्यक्ति चाहकर भी कुछ नहीं कर पाता है, परन्तु अगर वह दृढ़निश्चयी है, तो वह परिस्थितियों को अपने वश में कर अपने उत्थान की ओर अग्रसर रहता है।

इसलिए सच ही कहा है "मन के हारे हार है, मन के जीते जीत।"

(b) काश! मैंने माँ की बात मानी होती

यह उन दिनों की बात है जब हमने अपने जर्जर–से मकान को छोडकर एक नवनिर्मित कॉलोनी में किराए का घर लेना तय किया था। घर के सामान को बाँधा जाने लगा। रविवार को नए घर में जाने का निश्चय किया गया।

पिताजी कल शाम ही एस्टेट एजेन्ट से घर की चाबियाँ ले आए थे। अगले दिन हमें घर खाली करना था। मैंने मम्मी के

साथ सुबह ही उठकर दोपहर का भोजन पकाकर पूरी रसोई समेट दी थी। इतने में ट्रक वाले आए और समान लादने लगे। सामान लदने लगा तो पापा ने हमें चाबियाँ देकर कहा कि हम लोग ऑटोरिक्शा में जाकर घर को खोलें और झाड़ लगा दें। वे ट्रक के साथ आएंगे। भैया तुरन्त एक ऑटोरिक्शा ले लाए और हम उस पर दोपहर का टिफिन लिए चल दिये।

गाँधी पार्क, कोठी नं. 64 वाह! क्या भव्य कोठी है। हमने ऑटो वाले को किराया दिया और दरवाजा खोलने लगे परन्तु यह क्या, चाबी तो लग ही नहीं रही। बहुत प्रयास किया। मम्मी ने भी जोर लगाया। तभी पड़ोसन दिखाई दी। उसने पूछा हम लोग कौन हैं? जब उसे पूरा मामला समझाया गया तो वह बोली–वे लोग हमारे पास भी चाबियों का गुच्छा दे गये थे। आप उससे कोशिश कर लो। हो सकता है कि दलाल ने आपको किसी और घर की चाबियाँ भूल से दे दी हों।

हम पड़ोसिन द्वारा दी गई चाबियों को लगाते गए और हमारे सामने गेट, मुख्य द्वार, शयनकक्ष आदि सब खुलते गये। इतने कमरे! पापा ने तो उन लोगों को ठग ही लिया। तीन हजार में तो दो कमरों का मकान नहीं मिलता और यहाँ तीन शयनकक्ष, एक बैठक, एक बड़ी लॉबी, स्टोर, पोर्च और गैरेज थे। बाहर लॉन भी था।

हम बहुत उत्साहित थे। झाड़ लगाकर पूरे घर को पानी और फिनायल से धो दिया। पंखे खुले छोड़ दिये ताकि फर्श सूख जाए। तब खिडकियों दरवाजों के जाले पोंछने लगे। हम सब पापा का इंतजार कर रहे थे।

पापा नहीं आए। दोपहर हो आई। ट्रक नहीं आया। ट्रक की प्रतीक्षा में हम बाहर गेट पर आ गए। ट्रक का कोई अता–पता नहीं था। हमारे कपड़े मजदूरों जैसे मटियाहे और गंदे हो चुके थे। पापा की कोई खबर नहीं थी। मम्मी के चेहरे पर चिंता की रेखाएँ उभर रही थीं। साफ–सफाई में हम भूल ही गए थे कि ट्रक को यहाँ पहुँचने में दो घंटे से अधिक समय नहीं लगना चाहिए। हम तो पंद्रह मिनट में आ गए थे परन्तु बड़ी गाड़ियाँ बाई पास से होकर मुख्य मार्ग से इस कॉलोनी की ओर आ सकती हैं। फिर भी दो घंटे बहुत थे। यहाँ तो दोपहर भी बीत रही थी।

इसी बीच पड़ोसन चाय–नाश्ता ले आई। हमें भूख लग आई थी। हमने चाय तो नहीं पी, परन्तु पकौड़े भरपेट खाए। मम्मी ने केवल चाय ली। वे घबरा रही थी। पड़ोसिन ने सांत्वना के स्वर में कहा–आखिर मशीनरी है, खराब भी पड़ सकती है।

मम्मी ने इतना कहा–''मैं न कहती थी कि घर बदलने से पहले मोबाइल ले लो पर तुम मेरी सुनती ही नहीं हो। आज पास में मोबाइल होता तो झट से पता चल जाता।''

तभी मेरे दिमाग में एक युक्ति सूझी। मम्मी के पर्स में पापा का दिया कार्ड था। एस्टेट एजेंट का फोन व पता उसी कार्ड पर था। मैंने कार्ड देखा तो पता पास ही का था। मैं अपने भाई के साथ उधर निकल पड़ी। मम्मी में चलने की शक्ति नहीं थी। एस्टेट एजेंट ने सारी बात सुनकर हँसना शुरू कर दिया। मैं हैरान थी कि यह कैसा दलाल है जो अपने ग्राहक की मुसीबत में भी खिल्ली उड़ा रहा था। तभी उसने संयत होकर कहा–''गुड़िया रानी! आपको गाँधी नगर जाना था और आप आ गए गाँधी पार्क में। आपका ट्रक वहीं पहुँचा होगा। वह कॉलोनी शहर के पश्चिमी छोर पर है। जल्दी जाइए, आपका तो जून में भी अप्रैल फूल बन गया।'' हम वहाँ से ऑटो करके गाँधी नगर की ओर चल दिये। उस घटना को सुनाकर हम आज भी अपने परिचितों का मनोरंजन करते हैं। काश! हमने माँ की बात मानकर मोबाइल लिया होता।

Answer 2.

(i) महात्मा बुद्ध के समस्त उपदेशों का सार ही मध्यम मार्ग है। महात्मा बुद्ध के अनुसार जीवन–यापन के दो मार्ग हैं। जिनमें से पहला भोग–विलास, सुख–सम्पन्नता एवं भौतिकवादी जीवन की ओर जाता है और दूसरा अत्यन्त कष्ट व कठोर तपस्या से होते हुए त्याग की ओर अग्रसर है। किन्तु विडम्बना यह है कि इनमें से कोई भी मार्ग यथार्थ का बोध नहीं कराता और न ही वास्तविक सुख की अनुभूति प्रदान करता है। उनके अनुसार न तो भोगों से पूर्णतः तृप्ति हो सकती है और न ही भोगों को पूर्णरूप से त्यागा जा सकता है। अतः जीवन के लिए भोग और त्याग के बीच का मार्ग अर्थात् मध्यम मार्ग ही श्रेयस्कर है।

महात्मा बुद्ध को मध्यम मार्ग की प्रेरणा एक भजन–मण्डली के गीत से मिली। गीत के बोल थे–''वीणा के तारों को ढीला न छोड़ो नहीं तो उससे मधुर स्वर नहीं निकलेगा और न ही उन्हें इतना कसो कि वे टूट ही जायें।'' इस गीत के माध्यम से महात्मा बुद्ध के अन्तर्मन में मध्यम मार्ग की श्रेष्ठता का विचार अंकुरित हुआ।

(ii) महात्मा बुद्ध के जीवन काल में हिन्दू धर्म अनेक रूढ़िवादी मान्यताओं की जकड़ में था। जातिगत भेदभाव अपने चरम पर था तथा धार्मिक नीतियाँ इन कुरीतियों को और बढ़ावा देती थीं। ऐसे समय में महात्मा बुद्ध के द्वारा एक ऐसे मत की स्थापना की गई जिसकी विचारधारा सरल तथा स्वाभाविक थी। उनके मत में ऊँच–नीच का भेदभाव नहीं था। राजा से लेकर रंक तक उनके लिए समान थे। उन्होंने धर्म के बाह्य आडम्बरों की अपेक्षा आंतरिक शुद्धि पर बल दिया। यज्ञ में दी जाने वाली पशु बलि का विरोध किया। जनसाधारण की भाषा का प्रयोग करते हुए उन्होंने अहिंसा व प्रेम का ऐसा पाठ पढ़ाया कि लाखों नर–नारी श्रद्धास्वरूप उनके अनुयायी बन गये।

(iii) महात्मा बुद्ध ने उसे जीवन के कालचक्र से अवगत कराते हुए कहा कि जन्म और मृत्यु प्रकृति का अटल नियम है तथा संसार की नश्वरता ही परम सत्य है। जहाँ तक मार्गदर्शन का प्रश्न है तो इस संसार में कोई किसी का पथ–प्रदर्शक नहीं हो सकता। जीवन में वही व्यक्ति सच्चा ज्ञान प्राप्त कर सकता है जो अपने भीतर छिपे अंधकार को दूर कर स्वयं के मार्ग को प्रकाशित करे।

यह उपदेश केवल उनके शिष्यों के लिए नहीं वरन् हम सभी के लिए जीवन–यापन का मूलमन्त्र होना चाहिए।

अतः हमारा कर्तव्य बनता है कि हम स्वयं को इस योग्य बनायें कि बिना किसी बाह्य सहयोग के सुखपूर्वक अपना जीवन व्यतीत कर सकें।

(iv) पार्थिव शरीर के चारों ओर खड़े भिक्षुओं के कानों में तथागत के शब्द गूँज रहे थे कि भिक्षुओं को उन पर अंधे के समान विश्वास न करके अपने ज्ञान और अनुभव के आधार पर निर्णय लेने पर विचार करना चाहिये। इनसे हमें यह प्रेरणा मिलती है कि किसी का भी अंध–भक्त होने के स्थान पर आत्मचिंतन, आत्मज्ञान और अनुभव पर विश्वास करके ही कदम उठाना चाहिये।

(v) 'अप्प दीपो भव' का अर्थ है—अपना प्रकाश स्वयं बनो। यह छोटी सी पंक्ति मानव–जीवन की सार्थकता का सार समेटे हुए है। क्योंकि वास्तव में उसी व्यक्ति का जीवन सफल है जिसने अपने बल पर मार्ग में आने वाली कठिनाइयों पर विजय प्राप्त की है।

ज्ञानप्राप्ति के बाद महात्मा बुद्ध ने अपना उपदेश सारनाथ में दिया और उनका निर्वाण 80 वर्ष की उम्र में कुशीनगर में हुआ।

Answer 3.

(a) **(i)** मुझे आज जाना है।

(ii) पुलिस को देखकर चोर नौ दो ग्यारह हो गये।

(iii) सूर्यास्त होते ही वातावरण में अँधेरा छा गया।

(iv) साहित्य और समाज का घनिष्ठ सम्बन्ध है।

(v) यह कार्य तत्काल पूर्ण करना है।

(b) **(i)** **गले का हार होना**—प्यारा होना।

प्रयोग—मेधावी बच्चे गुरुओं व माता–पिता के गले का हार होते हैं।

(ii) **रंग चढ़ना**—अन्य व्यक्ति की तरह व्यवहार करना।

प्रयोग—अजय के साथ विजय पर भी उसका रंग चढ़ गया है।

(iii) **सिर उठाना**—प्रबल होना।

प्रयोग—मौका पाते ही अन्यायी व अत्याचारी सिर उठाने लगते हैं।

(iv) **चाँदी होना**—फायदा होना।

प्रयोग—रोहन की सरकारी नौकरी लगने के बाद उसके परिवार की तो चाँदी हो गई है।

(v) **अपने मुँह मियाँ मिट्ठू बनना**—अपनी तारीफ स्वयं करना।

प्रयोग—अपने मुँह मियाँ मिट्ठू बनने वाले समाज में हँसी के पात्र बनते हैं।

SECTION – B

गद्य संकलन (Gadya Sanklan)

Answer 4.

(i) जलेबियाँ सीताराम जी ने जेल से बाहर आने के बाद घर जाते हुए खरीदीं। जलेबियाँ खरीदने वाले बच्चों के पिता थे।

(ii) जलेबियाँ खरीदने वाला व्यक्ति अर्थात् सीताराम जी सत्याग्रह संग्राम में भागीदार थे। अतः उन्हें राजद्रोह के आरोप में एक वर्ष सश्रम कारावास की सजा हुई थी।

(iii) पहले तो सीताराम जी की बूढ़ी बुआ घर में रहा करती थी, जो उनकी अनुपस्थिति में बच्चों की देखभाल किया करती थी, किन्तु उनके जाने के बाद सीताराम जी के पास कहारिन के अलावा कोई और सहारा न था। सीताराम जी सत्याग्रह–संग्राम में बढ़–चढ़कर हिस्सा लेने वालों में से थे। वे जानते थे कि देश की परिस्थिति, गवर्नमेण्ट का रुख तथा महात्मा जी के वक्तव्य इस बात का संकेत थे कि सत्याग्रह संग्राम किसी भी पल शुरू हो सकता था। देश के प्रति उनका कर्तव्य उन्हें इस संग्राम का हिस्सा बनने के लिए प्रेरित करता था, परन्तु अपने भोले–भाले, मासूम, छोटे–छोटे बच्चों की चिन्ता भी उन्हें बराबर बेचैन करती रहती। जेल की सजा होने पर बच्चों को कहारिन के भरोसे छोड़ने को बाध्य थे। यद्यपि वे जानते थे कि कहारिन के भी तीन–चार बच्चे थे और ऐसे में वह उनके बच्चों का ठीक से ख्याल नहीं रख पाती होगी, फिर भी वे राष्ट्र–प्रेम के कारण बच्चों को अकेला छोड़ने के लिए विवश थे।

(iv) इस बार बच्चों की देखभाल गौरी ने की। गौरी राधाकृष्ण और कुन्ती की इकलौती संतान थी। उन्नीस साल की आयु हो जाने पर भी वे उसका विवाह न कर सके थे, यही चिन्ता उन्हें दिन–रात खाये जाती थी। वे गौरी के लिए योग्य वर चाहते थे, लेकिन दहेज देने की उनकी सामर्थ्य न थी। ऐसे में वे सीताराम जी से गौरी के विवाह के प्रस्ताव के लिए मिले। सीताराम जी विधुर थे और उनके दो बच्चे थे। उम्र में भी वे गौरी से काफी बड़े थे। इसलिए राधाकृष्ण जी इस रिश्ते को लेकर उत्साहित तो नहीं थे, लेकिन फिर भी वे उन्हें घर आकर लड़की देख लेने का प्रस्ताव दे आये थे। इसी सन्दर्भ में सीताराम जी राधाकृष्ण के घर पधारे तभी उनका गौरी से प्रथम परिचय हुआ।

सीतारामजी की सादगी एवं देश के प्रति उनके कर्तव्यभाव को देखकर गौरी उनसे अत्यधिक प्रभावित हुई। यद्यपि राधाकृष्ण जी ने इस रिश्ते को विवाह तक न पहुँचने दिया, परन्तु गौरी के मन में सीताराम जी की अमिट छाप पड़ चुकी थी।

इसी कारण जब दूल्हे के पिता की मृत्यु के कारण उसका विवाह टल गया था तब गौरी को सीताराम जी को गिरफ्तार करके जेल भेज देने की खबर मिली तो उसने तुरन्त कानपुर जाकर उनके बच्चों की देखभाल करने का निर्णय लिया। उसका यह दृढ़निश्चय देखकर उसकी माँ भी उसे रोकने का साहस न कर सकी और गौरी निश्चिन्त होकर बच्चों की देखभाल करने के लिए चली गई।

Answer 5.

प्रसिद्ध साहित्य लेखिका मन्नू भण्डारी द्वारा रचित कहानी 'मजबूरी' एक सरल वृद्धा के ममत्व को उकेरने वाली कथा है। बूढ़ी अम्मा के चरित्र का मर्मस्पर्शी चित्रण इस कहानी को और भी जीवंत बना देता है।

कहानी के केन्द्र में बूढ़ी अम्मा हैं। हाँ, वे सिर्फ अम्मा हैं—पहले रामेश्वर की अम्मा और फिर बेटू की अम्मा।

इस चरित्र का कोई नाम नहीं है और यही तथ्य इस चरित्र को सर्वव्यापी तथा कालजयी बनाता है। ये कहानी हर उस माँ की है जो अपने बच्चों के लालन–पालन में अपना सर्वस्व न्यौछावर कर देती है और फिर एक दिन अपने दिल पर पत्थर रखकर उन्हीं बच्चों को उनकी प्रगति की खातिर स्वयं से दूर कर देती है। अपने जीवन के संध्याकाल के झंझावतों में उन सुनहरे दिनों की यादों के चिराग को प्रज्ज्वलित रखने का भरसक प्रयास करती है।

बूढ़ी अम्मा का चरित्र ऐसी ही माताओं का प्रतिनिधित्व करता है। कहानी की शुरुआत में बूढ़ी अम्मा कुछ भूली–बिसरी लोरियाँ गुनगुनाती हुई नजर आती हैं। वे उत्साहित हैं क्योंकि पूरे तीन वर्ष बाद उनका बेटा रामेश्वर अपने परिवार के साथ उनसे मिलने आ रहा है। बूढ़ी अम्मा गठिया से पीड़ित हैं, लेकिन बेटे के आने की खुशी में भरी सर्दी में भी आँगन लीपने लग जाती हैं। अपने बेटे के मोह में वे इतनी पागल हैं कि उसके खिलाफ कुछ नहीं सुन सकतीं। इसीलिए जब नर्बदा

कहती है—''तीन बरस बाद आ रहा है तो मैं तो यही कहूँगी कि उनमें मोह–माया नहीं है। तुम यों ही मरी जाती हो उसके पीछे।'' इस बात पर बूढ़ी अम्मा उसे झिड़कर कहती हैं, ''देख नर्बदा मेरे रामेश्वर के लिए कुछ मत कहना।'' बेटे बहू के स्वागत के लिए वह सभी तैयारियाँ करके रखती है।

रामेश्वर के आते ही वह अपने पोते बहू को सीने से लगा लेती है।

जब उन्हें पता चलता है कि रमा बेटू को उनके पास छोड़कर जाना चाहती है तो वह खुशी से पागल हो जाती है और बहू को दुआएँ देती है—''······तुम्हारी सब साध पूरी हो, तुम बड़भागी होओ!'' वह पूरे गाँव भर में ढिंढोरा पीट देती है कि अब से बेटू उनके पास ही रहेगा। अम्मा का सारा दिन बेटू को खिलाने में बीत जाता है। कभी वह उसको नोन–राई करती तो कभी उसके साथ आँख–मिचौली खेलती तथा कभी वे बेटू का घोड़ा बनकर पूरे आँगन में दौड़ती फिरतीं। बेटू के पालन–पोषण के लिए उन्होंने नये–तरीके सीखने में भी कोताही नहीं बरती। उम्र के इस पड़ाव पर उन्होंने घड़ी में समय देखना तथा बोतल में दूध भरना सीखा। यहाँ तक कि जब शीशी में दूध पलटते समय जब थोड़ा दूध बाहर गिर जाता तो वे अपराधिनी की भाँति बहू के सामने गिड़गिड़ातीं—''बहुत जल्द सीख लूँगी, बहू! जरा–सा हाथ काँप गया था।'' इस ग्लानि में बेटू को खो देने का उनका डर साफ झलकता था।

जाने से पहले जब रामेश्वर ने अम्मा और पिताजी के लिए नये कपड़े बनवाये तो वे गर्व से उन्हें गाँव भर में दिखाती फिरीं। सबसे कहती—''अम्मा के पीछे तो बस रामेश्वर पागल है।······ उसका वश चले तो सारा घर ही मुझ पर लुटा दे।''

अगले साल जब रमा वापस आई तो बेटू के व्यवहार को देखकर अचम्भित रह गई। दादी के लाड़–प्यार में वह जिद्दी हो गया था। सारा दिन दादी का पल्लू पकड़े घूमता रहता था। रमा को क्रोध आया क्योंकि बेटू को देखकर लगता न था कि वह किसी पढ़ी–लिखी सभ्य लड़की का बच्चा है, किन्तु वह सब्र करके रह गई पर जब दो साल बाद भी बेटू के स्वभाव में कोई परिवर्तन नहीं आया तो उससे रहा न गया और उसने उसे अपने साथ ले जाने का निर्णय कर लिया।

रमा का यह निर्णय अम्मा के लिए वज्रपात के समान था, पर फिर भी बेटू की भलाई के लिए वह इस कष्ट को भी सहने के लिए तैयार हो गई—''ले जा बहू, ले जा। मेरा बेटू फूले–फले, पढ़–लिखकर लायक बने, इससे बढ़कर खुशी की बात मेरे लिए और क्या हो सकती है?''

पर जब अम्मा को खबर मिली कि बेटू ने उनकी याद में रो–रोकर बुरा हाल कर लिया है तथा उसे बुखार भी आ गया तो वे झटपट जाकर बहू के मायके से उसे ले आई। मातृत्व प्रेम एक बार फिर वास्तविकता पर हावी हो गया। वे जानती थीं कि रमा को यह अच्छा नहीं लगेगा पर बेटे के मोहपाश में ऐसी बँधी थीं कि डोर तोड़े न टूटती थी। इसी प्रकार एक साल और बीत गया। इस बार रमा बेटू को अपने साथ बम्बई ले गई। अम्मा ने जिद करके शिब्बू को उसके साथ भेजा क्योंकि उन्हें उम्मीद थी कि बेटू उनके बिना नहीं रह पायेगा और शिब्बू उसे वापस ले आयेगा।

वैसे तो अम्मा सबके सामने यह दर्शातीं कि वे बेटू को भेजकर खुश हैं तथा यही कामना करतीं कि बेटू उन्हें भूल जाये। जब शिब्बू ने लौटकर उन्हें बताया कि बेटू वहाँ हिलमिल गया है तो वे हतप्रभ–सी रह गई तथा बार–बार रटती रहीं—''बेटू मुझे भूल गया।'' इस एक वाक्य में उनके जीवन की सारी आकांक्षाएँ आकर विलीन हो गई, बिल्कुल वैसे जैसे अस्त होते सूर्य की लालिमा, रात्रि की सघन कालिमा में विलीन हो जाती हैं।

Answer 6.

''संस्कृति क्या है?'' नामक निबन्ध के माध्यम से रामधारी सिंह दिनकर जी ने अत्यन्त सरल, सहज एवं स्पष्ट भाषा द्वारा लोगों में 'संस्कृति' के प्रति फैली भ्रांतियों को दूर करने का सराहनीय प्रयास किया है।

निबन्ध के आरम्भ में ही दिनकर जी ने स्पष्ट किया है कि संस्कृति को परिभाषित करना दुःसाध्य है, कठिन है। य़द्यपि कुछ उदाहरणों की सहायता से उसे समझा जरूर जा सकता है।

साधारणतः हम सभ्यता और संस्कृति को एक–दूसरे के पर्याय के रूप में लेते हैं। सभ्यता और संस्कृति एक–दूसरे के पूरक हैं तथा एक–दूसरे पर निर्भर भी करते हैं, लेकिन कदाचित वे एक–दूसरे के पर्याय नहीं हैं। सभ्यता का सम्बन्ध भौतिकता से है, जबकि संस्कृति की पहचान विचारों से होती है। इसी आधार पर कह सकते हैं कि हर सभ्य व्यक्ति सुसंस्कृत ही होगा, यह आवश्यक नहीं है। बहुत–से ऐसे लोग हैं जो सड़े–गले झोंपड़ों में रहते हैं, जिनके पास पर्याप्त कपड़े भी नहीं होते और न ही कपड़े पहनने के अच्छे तरीके ही उन्हें मालूम होते हैं, लेकिन फिर भी उनमें विनय और सदाचार होता है। वे दूसरे के दुःख से दुःखी होते हैं तथा दूसरों का दुःख दूर करने के लिए वे खुद मुसीबत उठाने को भी तैयार रहते हैं।

दिनकर जी ने संस्कृति को विभिन्न लक्षणों के माध्यम से समझाने का प्रयास किया है। उनके अनुसार संस्कृति का पहला लक्षण विभिन्न मानवीय मूल्यों जैसे दया, सदाचार, सच्चाई, सद्भावना आदि की उपस्थिति होना है। इस तर्क के पक्ष में वे छोटा नागपुर की आदिवासी जनता का उदाहरण देते हैं जो भले ही सभ्य न हो, लेकिन सुसंस्कृत अवश्य है।

संस्कृति का दूसरा लक्षण है—सादगी। प्राचीन भारत में ऋषिगण जंगलों में निवास करते थे। फूस की झोंपड़ियों में वास करना, जंगलों के जीवों से दोस्ती और प्यार करना, किसी भी मोटे काम को अपने हाथ से करने में किसी भी तरह की हिचकिचाहट न दिखाना, आदि उनका जीवन था। भले ही वे यूरोपीय परिभाषा के अनुसार सभ्यता के मानकों पर खरे न उतरते हों, लेकिन वे स्वयं में केवल सुसंस्कृत ही नहीं बल्कि हमारी संस्कृति के निर्माता भी थे। सभ्यता एवं संस्कृति में यही मौलिक भेद है।

संस्कृति शाश्वत है। इसका निर्माण यकायक नहीं होता बल्कि यह एक सतत प्रक्रिया है, जिसकी रचना सौ–पचास वर्षों में ही की जा सकती है। इसके विपरीत सभ्यता नश्वर है क्योंकि इसका आधार भौतिक वस्तुएँ हैं जिनका विनाश निश्चित है।

संस्कृति जीवन जीने का तरीका है और यह तरीका सदियों से जमा होकर उस समाज में छाया रहता है, जिसमें हम जन्म लेते हैं। संस्कृति का साथ जन्म–जन्मांतर तक रहता है, जबकि सभ्यता की सामग्रियों से हमारा सम्बन्ध शरीर के साथ ही छूट जाता है क्योंकि संस्कृति आत्मा का गुण है तथा सभ्यता शरीर का।

सभ्यता और संस्कृति में सभ्यता दृष्टिगोचर होती है जबकि संस्कृति उसे संवारती और सजाती है। जिस प्रकार दूध में घी और फूल में गंध छिपी रहती है उसी प्रकार सभ्यता में संस्कृति छिपी होती है। संस्कृति के बिना सभ्यता बेमानी है। सभ्यता की वस्तुएँ टूट–फूट सकती हैं, पर संस्कृति का विनाश सरलता से नहीं होता।

सभ्यता के उपकरण बटोरे जा सकते हैं पर उनके उपयोग के लिए जिस संस्कृति की आवश्यकता होती है उसे अर्जित करने में समय लगता है। सभ्य व्यक्ति यदि विनयशील नहीं है तो वह हंसी का पात्र नहीं बन सकता। लोग उसे चढ़ाने लगते हैं क्योंकि सभ्यता के साधनों के उपयोग का भी अपना–अपना अंदाज होता है।

संस्कृति का स्वभाव है कि वह आदान–प्रदान से बढ़ती है। जब भी दो देश वाणिज्य, व्यापार अथवा शत्रुता या मित्रता के कारण आपस में मिलते हैं, तब उनकी संस्कृति एक–दूसरे को प्रभावित करने लगती है। संसार में शायद ही कोई देश यह दावा कर सकता है कि उस पर किसी अन्य देश की संस्कृति का प्रभाव नहीं पड़ा है। पहले एक देश में बुद्ध होते हैं तब दूसरे देश में ईसामसीह का जन्म होता है। जब भी दो जातियाँ मिलती हैं, उनके सम्पर्क या संघर्ष से जिन्दगी की एक नई धारा फूट निकलती है, जिसका प्रभाव दोनों पर पड़ता है। आदान–प्रदान की प्रक्रिया संस्कृति की जान है और इसी के सहारे वह स्वयं को जिन्दा रखती है।

काव्य मंजरी (Kavya Manjari)

Answer 7.

(i) प्रस्तुत पंक्तियाँ 'विनयपत्रिका' से ली गई हैं। इनके कवि तुलसीदास हैं। भक्त ने अपने आराध्य प्रभु श्रीराम के प्रति अपनी आस्था प्रकट की है।

(ii) यहाँ पतित पावन श्रीराम को कहा गया है। तुलसीदास जी का मानना है कि श्रीराम का जन्म दीन–दुःखियों के उद्धार व पापियों के संहार के लिए हुआ था। उनके अतिरिक्त किसी ने अपने भक्तों का उद्धार करने के लिए पापियों का संहार नहीं किया है।

(iii) इस पंक्ति का भाव यह है कि तुलसीदास श्रीराम के प्रति अपनी एकनिष्ठ भक्ति प्रकट करते हुए कहते हैं कि देवता, राक्षस, ऋषि–मुनि, सर्प तथा मनुष्य आदि सभी माया के वश में हैं। चारों ओर माया का ही साम्राज्य है। उनके समक्ष अपने आत्मसम्मान एवं स्वाभिमान को गिराकर स्वयं को चोट क्यों पहुँचाये। ऐसे में आपकी आराधना ही एकमात्र साधन है, जिससे सबका कल्याण हो सकता है।

(iv) तुलसीदास श्रीराम के अनन्य भक्त हैं। उनकी भक्ति में दास–भावना स्पष्ट दृष्टिगोचर होती है। उन्हें लगता है कि श्रीराम की उपासना एवं आराधना से समस्त सुखों को प्राप्त किया जा सकता है। राम भक्ति से परम सन्तोष प्राप्त होता है। राम ही शक्तिशाली तथा सौन्दर्य के अक्षय भण्डार हैं। वे श्रीराम के समक्ष ही अपनी दीनता स्वीकार करते हैं तथा उनके चरणों को त्यागकर अन्यत्र नहीं जाना चाहते। तुलसी के अनुसार श्रीराम ही दीनों पर दया करके उनका उद्धार करते हैं। उनका मानना है कि जो देवता पशु–पक्षी से लेकर मनुष्य तक का उद्धार करने की शक्ति रखता है, वही सर्वश्रेष्ठ है और वह है—मर्यादा पुरुषोत्तम श्रीराम। उनके अनुसार राम ही अत्यन्त उदार हृदय वाले हैं तथा समस्त कामनाओं को पूर्ण करने वाले हैं।

तुलसीदास की भक्ति भावना की प्रबलता उनके पद में परिलक्षित हो रही है। श्रीराम में ही उनकी अपार श्रद्धा एवं आस्था है।

Answer 8.

कबीरदास हिन्दी की संत काव्यधारा की ज्ञानाश्रयी शाखा के निर्गुण ब्रह्म के उपासक थे। धर्म के बाहरी आचार–व्यवहार तथा कर्मकाण्डों में उनकी लेशमात्र भी आस्था नहीं थी। समाज में व्याप्त संकीर्णताओं को देखकर उनका मन व्याकुल हो उठता था। उनकी व्यंग्यपूर्ण वाणी से विद्रोही स्वर मुखरित होने लगते थे। कबीर के तूफ़ानी तथा हुंकार भरे व्यक्तित्व के समक्ष रूढ़ि व परम्परा रूपी जर्जर दीवारें धराशायी हो जाती थीं। वे एकता एवं प्रेम की मजबूत नींव पर मानवता का विशाल दुर्ग खड़ा करना चाहते थे।

कबीर ने अपनी वाणी के माध्यम से तत्कालीन कर्मकाण्डों व बाह्य आडम्बरों का खुलकर विरोध किया। साथ ही उन्होंने साम्प्रदायिकता, जातिप्रथा तथा धार्मिक कुरीतियों की जमकर भर्त्सना की।

कबीरदास श्रेष्ठ कवि तो थे ही वे सफल उपदेशक भी थे। वे मनुष्य को किसी भी बात पर गर्व न करने का उपदेश देते थे। उनका मानना था कि ईश्वर किसी मन्दिर या मस्जिद में नहीं बल्कि सभी मनुष्यों के अन्तर्मन में विद्यमान है, लेकिन मनुष्य अहंकारवश उसके अस्तित्व का अनुभव नहीं कर पाता।

जब मैं था, तब हरि नहीं, अब हरि हैं मैं नाहीं

प्रस्तुत साखियों में कबीर ने निर्गुण ब्रह्म के महत्व पर प्रकाश डालते हुए उसकी उपासना करने का उपदेश दिया है। उनका कहना है कि मात्र वेदों के अध्ययन या संसार के बनाये नियमों का पालन करने से ईश्वर की प्राप्ति नहीं होती। उनकी प्राप्ति तब होती है जब सद्गुरु की शरण में जाकर सच्चे ज्ञानरूपी दीपक का प्रकाश हमारे अन्तर्मन पर पड़ता है।

कवि के शब्दों में—

कबीर बादल प्रेम का हम पर वर्षा आइ।
अंतरि भीगी आत्मा हरी भई बनराइ।।

मनुष्य के जीवन में प्रेम का महत्वपूर्ण स्थान है। यही प्रेम जब ईश्वर के प्रति हो तो जीवन सफल हो जाता है। उनका मानना है कि ईश्वर की प्रेमरूपी वर्षा से मनुष्य का तन ही नहीं उसकी आत्मा भी सराबोर हो जाती है।

ईश्वर से बिछड़कर मनुष्य अपना सुख–चैन खो देता है। उसके बाद सांसारिक वैभव भी हमें सुख की अनुभूति नहीं करा सकता। कबीर के अनुसार सच्चा सुख ईश्वर से 'ऐक्य' स्थापित करने में है।

वे कहते हैं कि यदि किसी व्यक्ति को पारस पत्थर की प्राप्ति तब हो जब वह लोहे के बदले लकड़ी ले चुका हो तो उसके लिए वह पारस पत्थर व्यर्थ है। ठीक उसी प्रकार यदि मृत्यु के बाद भी ईश्वर से मिलन न हो सके तो सारा जीवन व्यर्थ ही है; क्योंकि ईश्वर की प्राप्ति ही सुखदायक व कल्याणकारी है। कबीरदास हमें ब्राह्य आडम्बरों के प्रति सचेत करते हैं। उनका मानना है कि केवल बाहरी दिखावे के लिए ईश्वर का नाम जपकर हमें बहुमूल्य समय नहीं गँवाना चाहिए क्योंकि ऐसे आडम्बरों से ईश्वर की प्राप्ति नहीं होती।

ईश्वर की खोज में भटक रहे मनुष्यों को वे सन्देश देते हैं कि पर्वत–पर्वत भटकते हुए दर–दर की ठोकरें खाकर भी ईश्वर की प्राप्ति करना कठिन है क्योंकि ऐसी कोई बूटी या दवा नहीं है; जिसके सेवन से ईश्वर की प्राप्ति हो सके तथा जीवन सफल हो सके। उनका मानना है कि प्रभु की वास्तविक छवि की अनुभूति संतों के सत्संग द्वारा ही सम्भव है।

जिस ह्रदय में अभियान का वास होता है, उस ह्रदय में ईश्वर कभी नहीं बसता। यदि एक बार ईश्वर से सच्चा मिलन हो जाये तो अभिमान स्वतः ही ह्रदय से निकल जाता है। इन सभी बिन्दुओं के आधार पर हम कह सकते हैं कि कबीरदास सच्चे युगद्रष्टा व श्रेष्ठ समाज सुधारक थे, जिन्होंने आजीवन समाज में व्याप्त विकृत रूढ़ियों, परम्पराओं तथा आडम्बरों का जमकर विरोध किया। अतः वे निश्चित रूप से एक सफल कवि व श्रेष्ठ उपदेशक के रूप में वंदनीय हैं।

Answer 9.

प्रकृति के सुकुमार कवि सुमित्रानन्दन पंत द्वारा रचित कविता 'आः धरती कितना देती है' बेहद सरल एवं सहज भाषा में जीवन के गूढ़ सत्य को हमारे समक्ष प्रस्तुत करती है। 'जैसी करनी वैसी भरनी' यही हमारे जीवन का अटल सत्य है। हम जैसा कर्म करेंगे वैसा ही फल पायेंगे। यदि हम दूसरों के प्रति अपमानजनक व्यवहार करेंगे तो स्वयं भी सम्मान पाने का अधिकार खो देंगे।

जीवन के इस महत्वपूर्ण सन्देश को कवि ने प्रकृति के माध्यम से समझाने का प्रयास किया है। यह प्रयास सार्थक व सराहनीय है। कवि अपने बाल्यकाल में धरती में पैसे बो देते हैं तथा कल्पना करते हैं कि इस धरती से पैसों के विशाल वृक्ष उगेंगे। वृक्ष से रुपये–पैसों को प्राप्त करके वे अमीर हो जायेंगे, परन्तु उनकी कल्पना के विपरीत उस मिट्टी में पैसों का एक भी अंकुर नहीं फूटा। जो पैसे बोये थे वे भी मिटकर धूल हो गये। कवि हताशा में कई दिनों तक इंतजार करते रहे तब जाकर उन्हें अहसास हुआ कि यह उनकी नादानी थी।

कवि के शब्दों में—

मैं हताश हो, बाट जोहता रहा दिनों तक,
बाल कल्पना के अपलक पावड़े बिछाकर,
मैं अबोध था, मैंने गलत बीज बोए थे,
ममता को रोपा था, तृष्णा को सींचा था।

जीवन की अर्द्धशती गुजरने के दौरान उन्होंने जीवन के कई रूप और रंग देखे। फूलों को खिलते देखा तो पत्तों को झड़ते भी देखा।

फिर उन्होंने एक दिन बारिश के समय उत्सुकतावश आँगन के कोने में सेम के बीज बो दिये, लेकिन समय के साथ वे इस छोटी–सी बात को भूल गये। कुछ समय बाद उनके हर्ष की सीमा न रही जब उन्होंने देखा कि उन बीजों में से नये–नये, छोटे–छोटे अंकुर फूट रहे हैं और देखते ही देखते सेम की हरी–भरी बेल आँगन में लहरा उठी। उस समय कवि को यह अहसास हो गया कि कुछ समय पूर्व किया गया थोड़ा–सा श्रम आज बेल के रूप में उनके सामने उपस्थित था। अब कवि समझ चुके थे कि पृथ्वी रत्न–प्रसूता है, इसमें पैसा बोकर इसका अपमान नहीं करना चाहिए। इसमें तो ममता के बीज बोने चाहिए जिससे मानवता की सुनहरी फसलें लहलहाने लगें।

कवि के शब्दों में—

जिससे उगल सके फिर धूल सुनहली फसलें,
मानवता की—जीवन श्रम से हंसे दिशाएँ

कवि कहते हैं कि धरती माता में मानवता की सुनहरी फसलें उगाई जायेंगी। मनुष्य के परिश्रम का परिणाम ही चारों ओर प्रसन्नता भर देगा। प्रकृति का नियम है कि हम जैसा बोयेंगे वैसा ही फल प्राप्त करेंगे।

वस्तुतः इस कविता के माध्यम से कवि मानव–कल्याण का सन्देश देना चाहते हैं। उनका कहना है कि जब हम सभी परस्पर सहयोग, प्रेम, सद्भावना, सहानुभूति, त्याग व भाईचारे की भावना के साथ जीवन व्यतीत करेंगे तभी हमारे राष्ट्र का कल्याण सम्भव है।

इन सभी मानवीय गुणों को अपनाने से ही पृथ्वी पर सर्वत्र शान्ति एवं सुख की स्थापना सम्भव है।

"सारा आकाश" (Saara Akash)

Answer 10.

(i) इस कथन की वक्ता 'मुन्नी' है। मुन्नी उपन्यास के महिला–पात्रों में से एक है। वह समर की छोटी बहन है, जो पति द्वारा प्रताड़ित होने के बाद मायके में रह रही है। मुन्नी प्रभा की इकलौती शुभचिन्तक भी है।

(ii) मुन्नी की शादी अल्पायु में ही कर दी गई थी। उसका पति खराब चरित्र का हिंसक पुरुष था। वह मुन्नी को बहुत मारता था तथा उसके सामने ही एक दूसरी औरत को घर ले आया था। समाज के लोगों का भी उस पर कोई प्रभाव नहीं पड़ता था। उसके अत्याचारों से तंग आकर मजबूरन मुन्नी को मायके में आसरा लेना पड़ा, लेकिन दो साल बाद अचानक वह मुन्नी को लेने उसके मायके आ गया।

यूँ तो मुन्नी के घरवालों को उस पर तनिक भी भरोसा नहीं है; लेकिन आर्थिक परिस्थितियों तथा सामाजिक प्रतिष्ठा के दबाव में आकर परिवारवालों द्वारा मुन्नी को जबरन ससुराल भेज दिया गया।

(iii) जब मुन्नी अपने पति का घर छोड़कर आई थी, तब उसके पूरे शरीर पर चोटों तथा पति द्वारा किये गये अत्याचारों के गहरे निशान थे। घरवालों को उसके प्रति सहानुभूति थी; लेकिन वक्त बीतने के साथ ही जीवन की कड़वी सच्चाइयों ने उस सहानुभूति को जिम्मेदारियों में बदल दिया।

घर की आर्थिक स्थिति तो पहले से ही खराब थी और उस पर मुन्नी की जिम्मेदारी तथा समाज में बदनामी के डर ने माँ–बाप को परेशान कर रखा था, परन्तु जब मुन्नी का पति माफी माँगते हुए उसे घर ले जाने की कहने लगा तो उन्हें उस पर बिल्कुल

भी भरोसा नहीं हुआ। अम्मा ने तो मुन्नी को भेजने से साफ मना कर दिया पर बाबूजी बेटी को उसकी गृहस्थी में सुखी देखना चाहते थे।

जब मुन्नी बाबूजी के पैरों में लिपटकर, रो–रोकर उसे वहाँ न भेजने का आग्रह करने लगी तो बाबूजी और अम्मा दोनों ही फूट–फूटकर रोने लगे। परन्तु लोक–लाज के कारण उन्हें न चाहते हुए भी मुन्नी को उसकी ससुराल भेजना पड़ा।

(iv) राजेन्द्र यादव द्वारा रचित उपन्यास 'सारा आकाश' में कई महिला पात्र हैं तथा सभी की भूमिका अत्यन्त महत्वपूर्ण है, परन्तु जिन पात्रों द्वारा तत्कालीन नारी की दशा का मार्मिक एवं यथार्थ चित्रण हुआ है वे हैं—प्रभा और मुन्नी।

मुन्नी का पात्र पितृसत्तात्मक समाज की विसंगतियों का सजीव चित्रण करता है। वह अपने पति द्वारा प्रताड़ित व शोषित है; लेकिन फिर भी समाज में उसी की बदनामी होती है। जब वह परेशान होकर पिता के घर में आसरा पाती है तो वहाँ उसे बोझ समझा जाता है। अम्मा के लाख विरोध करने पर भी पिता द्वारा उसे ससुराल भेज दिया गया जिसकी परिणति थी उसकी मृत्यु।

दूसरा नारी चरित्र प्रभा का है। प्रभा इस उपन्यास की नायिका है। प्रभा शिक्षित नारी का प्रतिनिधित्व भी करती है। वह शिक्षित है, इस बात का उसके ससुराल वालों के लिए कोई महत्व नहीं है। बल्कि शिक्षा के नाम पर उसे बार–बार ताने दिये जाते हैं। समर उसकी शिक्षा के कारण ही उसे अहंकारी समझता है तथा विवाह के सालभर बाद भी वह उससे बोलता तक नहीं है। प्रभा यह सब बड़ी ही शान्ति व धैर्य के साथ सहती चली जाती है। इस पर भी वह घर की सारी जिम्मेदारियों का निर्वहन करती रहती है।

भाभी और अम्मा का चरित्र इस कथन को चरितार्थ करता है कि नारी ही नारी की पीड़ा का कारण होती है।

अतः यह उपन्यास नारी के निराश–हताश तथा आत्मनिर्भरता की कमी से जूझते जीवन को प्रदर्शित करता है।

Answer 11.

समर शुरू से ही विवाह के लिए सहमत नहीं था। उसके विचार दार्शनिक जैसे थे तथा उसके सपने बहुत ऊँचे थे। वह महान बनना चाहता था तथा बहुत पढ़ना चाहता था। लोकनीति का हवाला देते हुए उसका विवाह जबरदस्ती प्रभा के साथ करा दिया गया। पहले से ही खिन्न समर को प्रभा की खामोशी ने गुस्से में बदल दिया। उसे लगा कि प्रभा को अपनी पढ़ाई का घमण्ड है तथा उसी गुस्से में उसने प्रभा से जीवनभर बात न करने का निश्चय कर लिया।

प्रभा सभी घरवालों की प्रताड़ना सहती तथा चुपचाप घर का सारा काम करती रहती। किसी से कोई शिकायत न करती तब भी उसी को केन्द्र बनाकर हर दिन कोई न कोई महाभारत छिड़ी ही रहती।

समर जो कि पहले ही अपने घर के क्लेशों से तंग था, अब ज्यादातर समय घर से बाहर बिताने लगा था। जैसे ही वह घर लौटता कोई न कोई हंगामा चल रहा होता।

ऐसे ही जब एक दिन वह हारा–थका घर लौटा तो देखा कि अम्मा जी गुस्से से आग–बबूला हुई जा रही थीं। भाभी अन्दर जोर–जोर से रो रहीं थीं तथा प्रभा एक कोने में अपराधिनी की भाँति सिर झुकाये खड़ी थी।

जब समर ने हंगामे का कारण पूछा तो पता चला कि पण्डित जी ने जिन गणेश जी की पूजा की थी, उसे मिट्टी का ढेला समझकर प्रभा ने उससे बर्तन माँज दिये थे। अपने इस अपराध के लिए प्रभा बार–बार माफी माँग रही थी—'अम्मा जी मुझे पता होता तो मैं ऐसा कभी नहीं करती।' लेकिन फिर भी सारा घर प्रभा को दोषी ठहराने में तुला हुआ था।

समर यह सब सुनकर आग–बबूला हो गया तथा तमतमाते हुए एक करारा तमाचा प्रभा के गाल पर जड़ दिया। वह इतने गुस्से में था कि प्रभा को गाली देने से भी नहीं चूका तथा घर से निकल जाने तक के लिए कह डाला।

भयंकर आवेश में उसने यह सब कर तो दिया, लेकिन कमरे में आते ही उसका मन आत्म–ग्लानि से भर गया। वह जानता था कि उसने जो किया वह गलत था।

जब भी उसके भैया भाभी को मारते थे तो वह उनकी घोर निन्दा किया करता था, लेकिन आज उसने स्वयं अपनी पत्नी पर हाथ उठाया था। वह इस सच को बर्दाश्त न कर सका और बच्चों की तरह फूट–फूटकर रोने लगा।

Answer 12.

प्रभा इस उपन्यास की नायिका है। वह समकालीन स्त्री का प्रतिनिधित्व करती है। अपनी ससुराल की बाकी स्त्रियों के मुकाबले वह अधिक शिक्षित है, लेकिन उसे इस बात का अभिमान नहीं है। फिर भी उसकी इसी शिक्षा को लेकर उसके ससुराल वाले उसे तानों से छलनी कर देते हैं, परन्तु फिर भी सभी विषम परिस्थितियों से जूझते हुए प्रभा का चरित्र अविचल खड़ा नजर आता है। उसके चरित्र की प्रमुख विशेषताएँ इस प्रकार हैं—

1. **रूपवती**—प्रभा अत्यन्त रूपवती है। वह अत्यन्त सुन्दर है, यह बात बार–बार दोहराई जाती है। उसके इसी सौन्दर्य से जलकर शायद समर की भाभी उसके काम को लेकर नुक्ताचीनी करती रहती है।
2. **सौम्य स्वभाव की धनी**—प्रभा सौम्य स्वभाव की धनी है। ससुरालवालों द्वारा बार–बार ताने दिये जाने पर भी वह कुछ नहीं बोलती, न ही किसी का अपमान करती है। अकारण डाँट पड़ने पर भी वह अपमानसूचक शब्दों का प्रयोग नहीं करती। यहाँ तक कि अपने प्रति समर की बेरुखी को भी वह अपने तक सीमित रखती है। वह घरवालों से कोई शिकायत नहीं करती।
3. **कुशल गृहिणी**—भाभी के बच्चे के जन्म के बाद प्रभा पर जिम्मेदारियों का बोझ और भी बढ़ गया। शुरू में उसे मुन्नी से थोड़ा–बहुत सहारा जरूर मिलता है, पर उसके चले जाने के बाद सारा काम प्रभा अकेले ही करती है। दिन–रात काम में जुटे रहने के बाद भी उससे सहानुभूति रखने वाला कोई नहीं है। बल्कि उसके काम में मीन–मेख निकालकर हर रोज उसे नीचा दिखाकर डाँटा जाता; लेकिन वह कभी भी जिम्मेदारियों से पीछे नहीं हटती।
4. **स्वाभिमानी**—प्रभा शिक्षित है और शिक्षा एवं स्वाभिमान का चोली–दामन का साथ है। वह समर की बेरुखी को सहन कर लेती है, लेकिन उसके सामने गिड़गिड़ाकर उसका ध्यान अपनी ओर नहीं खींचना चाहती। इसीलिए जब भाभी उससे

समर के कमरे में सोने के लिए कहती हैं तो वह रुआंसी होकर कहती है—"मुझसे तो नहीं होता जेठानी जी, कि कोई दुत्कारता रहे और पूँछ हिलाते रहो, ठोकर मारता रहे और तलुए चाटते रहो।"

प्रभा अपने स्वाभिमान को दाँव पर लगाकर पति का प्रेम पाने की अभिलाषी नहीं है।

5. **चरित्रवान**—प्रभा उज्जवल एवं पवित्र चरित्र की धनी है। बाबूजी जब उसे छत पर जाने के लिए डाँटते हैं तो वह अपमान का घूँट पीकर रह जाती है तथा दोबारा ऐसा नहीं करती। पति की बेरुखी भी उसको चारित्रिक पतन की ओर नहीं जाने देती।

6. **साज–सज्जा के प्रति अनासक्ति**—नवविवाहिता होने पर भी प्रभा को सौन्दर्य–संसाधनों के प्रति आसक्ति नहीं है। उसका रहन–सहन सादा है। घर के कामों में वह इतनी उलझी रहती है कि फटी साड़ियों पर भी उसका ध्यान नहीं जाता। यूँ तो उसके पास कुछ जेवर हैं पर वह उन्हें पहनती नहीं है। समर की सहायता के लिए उन गहनों को बेचने के लिए सदैव तत्पर रहती है। कार्य की अधिकता के कारण वह अपने बालों को भी कई दिनों तक सुलझा नहीं पाती।

7. **पतिव्रता नारी**—प्रभा पतिव्रता है। वह अपने पति की इच्छाओं का सम्मान करती है। समर सालभर तक उससे बात नहीं करता; लेकिन फिर भी वह कोई शिकायत नहीं करती। यहाँ तक कि वह समर के लिए घण्टों भूखी बैठी रहती है। जब समर उससे बात करना शुरू कर देता है तब भी वह सारे काम उसकी इच्छानुसार ही करती है। उसके सपनों की सहभागिनी बनती है तथा जीवन की हर परिस्थिति में उसका साथ निभाने का वचन भी देती है।

8. **धैर्यवान एवं विवेकशील**—विषम से विषम परिस्थितियाँ भी कोई स्त्री तब झेल जाती है जब उसका पति उसके साथ हो; लेकिन प्रभा को यह सहारा भी न था। बल्कि उसका पति भी उसके कष्टों से विमुख ही रहता था। सारा घर उसे ताने देता, अभिमानी कहता था, लेकिन ऐसी कठिन परिस्थितियों में भी प्रभा ने संयम से काम लिया तथा उसने धैर्य नहीं छोड़ा। वह धैर्यवान होने के साथ विवेकशील नारी भी है। उसने कभी भी आवेश में आकर कोई निर्णय नहीं लिया तथा तकलीफों को सहने के बाद भी वह पूरे समर्पण के साथ अपनी ससुराल की जिम्मेदारियाँ निभाती गई।

प्रभा के चरित्र की इन विशेषताओं के आधार पर हम कह सकते हैं कि प्रभा एक आदर्श भारतीय नारी है जो नारीत्व के सभी गुणों से सम्पन्न है।

"आषाढ़ का एक दिन" (Aashad Ka Ek Din)

Answer 13.

(i) प्रस्तुत कथन के वक्ता मातुल हैं जो कि महाकवि कालिदास के मामा हैं। वे यह कथन अम्बिका से कह रहे हैं क्योंकि वह कालिदास द्वारा उज्जयिनी जाकर सम्मान न ग्रहण करने के निर्णय को लोकनीति का नाम दे रही है। मातुल इस निर्णय के खिलाफ हैं और इसीलिए वे इसे कालिदास की मूर्खता मानते हैं।

(ii) मातुल जब यह बताते हैं कि कालिदास को आचार्य वररूचि उज्जियनी ले जाने के लिये आये हैं लेकिन कालिदास जाने के लिये मना कर देते हैं। अम्बिका यह सब सुनने के बाद कहती है कि वह उज्जियनी अवश्य जायेगा क्योंकि सम्मान प्राप्त होने से पहले सम्मान के प्रति प्रकट की गई उदासीनता व्यक्ति के महत्व को बढ़ा देती है। वह मातुल से कहती है कि तुम्हें तो प्रसन्न होना चाहिये कि तुम्हारा भांजा लोकनीति में निपुण है।

मातुल अम्बिका का ये कथन सुनने के बाद कहता है कि सम्मान के प्रति कालिदास द्वारा व्यक्त की गई उदासीनता मूर्ख नीति कहलायेगी क्योंकि जो व्यक्ति सम्मान दे रहा है वह रूष्ट हो सकता है। सम्मान करने का निर्णय बदल सकता है।

(iii) वक्ता मातुल है। वे कालिदास के मामा हैं। वे स्वभाव से चापलूस हैं अतः किसी भी प्रकार से हाथ में आये सम्मान के मौके को गँवाना नहीं चाहते। वे कालिदास के सम्मान द्वारा अपने जीवन की कठिनाइयों को दूर करना चाहते हैं। उनके चरित्र के माध्यम से नाटककार ने यह दिखाने का प्रयास किया है कि सामाजिक क्रान्ति अपने ही मूल्यों के कारण अन्ततः पंगु हो जाती है।

(iv) 'आषाढ़ का एक दिन' एक सुव्यवस्थित यथार्थवादी नाटक है जिसमें बाह्य व आन्तरिक यथार्थ के अन्तर्द्वन्द्व को संवेदनशीलता के साथ दर्शाया गया है। इस नाटक का उद्देश्य पाठकों को जीवन की कठोर सच्चाइयों से रूबरू कराना है। मनुष्य कई बार अपने ही द्वारा पैदा की गई परिस्थितियों का शिकार बन जाता है और अन्ततः अपनी वास्तविक आकांक्षाओं से दूर होकर एक ऐसा जीवन जीने को बाध्य हो जाता है जिसकी उसने कभी कल्पना भी न की थी। मल्लिका अपनी इच्छा से कालिदास को उज्जयिनी भेजती है और अन्त में उसे हमेशा के लिए खो देती है। कालिदास बेमन चला तो जाता है, पर न चाहते हुए भी वह कवि से राजनीतिज्ञ बन जाता है तथा ऐसे कामों में उलझ जाता है जिनमें उसकी कोई दिलचस्पी नहीं है। फलस्वरूप वह अपने जीवन का सुख–चैन खो बैठता है।

इन्हीं परिस्थितिजन्य अन्तर्द्वन्द्व को यथार्थ के धरातल पर दिखाना इस नाटक का उद्देश्य है।

Answer 14.

नाटक 'आषाढ़ का एक दिन' पात्रों की दृष्टि से भी सफल नाटक है; क्योंकि पात्र–संयोजन कथानक के अनुसार है। मोहन राकेश के इस नाटक में दो प्रमुख पात्र हैं—कालिदास और मल्लिका। जिनमें मल्लिका का चरित्र नाटक के सन्देश को अपने में समेटे हुए है। नाटक का यही चरित्र सम्पूर्ण नाटक का 'प्राण' है जिसने कालिदास व अन्य पात्रों के सहयोग से नाटककार के उद्देश्य को पूरा करने में महत्वपूर्ण भूमिका का निर्वहन किया है। मल्लिका का चरित्र अनेक विशेषताओं का स्वामी है जो नाटक को प्रभावशाली बनाने में सहायक हैं।

मल्लिका के चरित्र की प्रमुख विशेषताएँ निम्नवत् हैं—

1. **प्रकृति–प्रेमी**—नाटक के प्रारम्भ में ही जब पाठकों का मल्लिका से पहली बार सामना होता है तब वह बारिश में भीगी हुई प्रकृति के सौन्दर्य का वर्णन अपनी माँ अम्बिका से कर रही होती है। वह मेघ–मालाओं से आच्छादित पर्वत–शिखरों के सौन्दर्य में खुद को ही भूल जाती है—"माँ आज के वे क्षण मैं कभी नहीं भूल सकती। सौन्दर्य का ऐसा साक्षात्कार मैंने कभी नहीं किया। जैसे वह सौन्दर्य अस्पृश्य होते हुए भी मांसल हो।"

2. **त्याग की अनुपम मूर्ति एवं निःस्वार्थी**—अभावग्रस्त जीवन में मल्लिका ने दुर्दशा के दिन भी देखे, लेकिन अपने किसी

भी निर्णय में उसने स्वार्थ को हावी नहीं होने दिया। यदि वह स्वार्थी होती तो कालिदास को अकेले उज्जयिनी नहीं भेजती बल्कि उस पर विवाह के लिए दबाव बनाती, परन्तु वह कालिदास की प्रतिभा को पूरी तरह से पल्लवित होने देना चाहती थी इसलिए समाज में हो रही अपनी बदनामी को दरकिनार करके उसने कभी कालिदास के सामने विवाह का प्रसंग तक नहीं छेड़ा।

कालिदास के प्रियंगुमंजरी से विवाह की बात सुनकर भी वह कालिदास के लिए अपने प्रेम व सम्मान को कम नहीं होने देती।

3. **भावुक हृदय की स्वामिनी**—मल्लिका भावुक हृदय की स्वामिनी है। वह सबके मन की भावनाओं को समझती है तथा उन्हें महत्व भी देती है।

   जब प्रियंगुमंजरी मल्लिका से घर–परिवार की कल्पना के विषय में प्रश्न करती है तो अम्बिका खुद कहती है—"इसके मन में यह कल्पना नहीं है क्योंकि यह भावना के स्तर पर जीती है।"

   मल्लिका स्वयं अपनी भावनाओं को अपनी माँ के समक्ष रखती है—"मैंने भावना में एक भावना का वरण किया है। मैं वास्तव में अपनी भावना से प्रेम करती हूँ जो पवित्र है, कोमल है, अनश्वर है……।"

4. **स्वाभिमानी**—मल्लिका जिसका व्यक्तित्व एक तरफ भावुक तथा कोमल भावनाओं से युक्त है वहीं दूसरी ओर उसने कभी भी अपने आत्मसम्मान को आहत नहीं होने दिया। विषम परिस्थितियों में भी वह प्रियंगुमंजरी के सहायता के प्रस्ताव को अस्वीकार कर देती है। मातुल का यह कथन उसके इस गुण को उद्घाटित करता है—'प्रियंगुमंजरी ने तुम्हारे लिए कुछ वस्त्र और स्वर्णमुद्राएँ भिजवाई थीं जो तुमने लौटा दीं?"

5. **सच्ची प्रेयसी**—मल्लिका सच्ची प्रेयसी के धर्म का निर्वहन करते हुए बिना किसी अपेक्षा के कालिदास की प्रतिभा को विकसित होने के लिए कहीं भी अपने प्रेम को बाधा नहीं बनने देती तथा समाज के सामने उसके हर निर्णय का समर्थन करती रहती हैं। उसके खिलाफ एक शब्द भी सुनना उसके लिए असह्य है।

6. **संघर्षशील**—मल्लिका का पूरा जीवन संघर्षशील रहा है। उसने पूरे नाटक में जीवन में संघर्ष किया तथा कभी–भी अपनी इस अवस्था का जिम्मेदार किसी और को नहीं ठहराया। अभावग्रस्त जीवन में भी वह अपने आत्मसम्मान तथा कालिदास के प्रति अपने प्रेम को बचाये रखने के लिए संघर्षरत रहती है।

   मल्लिका इस नाटक की धुरी है। वह ध्रुव तारे के समान नाटक के सभी चरित्रों के बीच अटल खड़ी नजर आती है।

Answer 15.

नाटकों में युगीन समस्याओं का चित्रण करते हुए उनका समाधान करने का भी निरन्तर प्रयास किया जाता है। मानव जीवन की सजीव, मूर्त एवं प्रत्यक्ष झाँकी, जैसी नाटक में दिखाई जाती है, वह दूसरी विधाओं से निश्चित रूप से कठिन है।

'आषाढ़ का एक दिन' शीर्षक की सार्थकता इस तथ्य से स्पष्ट होती है कि इस नाटक का आरम्भ एवं अन्त दोनों ही आषाढ़ के दिन होते हैं।

आषाढ़ का एक दिन मल्लिका तथा कालिदास के निःस्वार्थ व निष्फल प्रेम का प्रतीक है।

आषाढ़ के महीने में सघन मेघ–मालाओं से आच्छादित पर्वत श्रृंखलाओं के साये में ही मल्लिका व कालिदास का प्रेम पल्लवित होता है तथा उसी आषाढ़ के दिन मल्लिका कालिदास को उज्जयिनी जाने के लिए मना लेती है। आषाढ़ का एक दिन ही उनके बिछोह का प्रथम बार साक्षी बनता है।

दूसरे अंक में जब मल्लिका को कालिदास के वापस आने की सूचना मिलती है तो वह विचलित हो जाती है क्योंकि वह दिन आषाढ़ का दिन नहीं है। "आज वर्षों के बाद तुम लौटकर आये हो। सोचती थी तुम आओगे तो उसी तरह मेघ घिरे होंगे, वैसा ही अँधेरा–सा दिन होगा, वैसे ही एक बार फिर वर्षा में भीगूँगी और तुमसे कहूँगी कि देखो मैंने तुम्हारी सब रचनाएँ पढ़ी हैं……परन्तु आज तुम आये हो, तो सारा वातावरण ही और है।"

मल्लिका के इन विचारों से स्पष्ट होता है कि वह स्वयं 'आषाढ़ के एक दिन' को अपने प्रेम का प्रतीक मानती है।

तीसरे अंक में हम देखते हैं कि फिर से वही आषाढ़ का एक दिन है। वर्षा और मेघों का गर्जन है और कालिदास पुनः मल्लिका के जीवन में प्रवेश करते हैं। मल्लिका सब कुछ भूलकर उनके प्रेम में खो जाना चाहती हैं, लेकिन विलोम उसके जीवन की सच्चाई से कालिदास को रूबरू करा देता है। कालिदास की अथ से आरम्भ करने की इच्छा दम तोड़ देती है और एक बार फिर वह मल्लिका को छोड़कर चले जाते हैं।

अतः हम देखते हैं कि 'आषाढ़ का एक दिन' इस नाटक में इतना महत्वपूर्ण है कि कभी–कभी वह स्वयं में एक पात्र–सा प्रतीत होता है।

अतः शीर्षक 'आषाढ़ का एक दिन' पूर्णतः सार्थक है।

●●

HINDI

2016

QUESTIONS

SECTION – A
Language – 50 Marks

Question 1.

Write a composition in approximately 400 words in Hindi on any ONE of the topics given below: [20]

किसी एक विषय पर निबन्ध लिखिए जो लगभग 400 शब्दों से कम न हो :

(i) *"भूकम्प प्रकृति का वह विनाशकारी रूप है जिसकी कल्पना भी मनुष्य–मन को चोट पहुँचाती है।"–इस कथन को ध्यान में रखते हुए उस समय का वर्णन कीजिए जब आपके नगर में भूकम्प आया, उसका आम जनजीवन पर क्या प्रभाव पड़ा ? आपने तथा आपके साथियों ने राहत कार्यों में क्या योगदान दिया। विस्तार से लिखिए।*

(ii) *अपराधी को नहीं बल्कि अपराध को समाप्त करने से एक मजबूत राष्ट्र तैयार होता है। इस विषय के पक्ष या विपक्ष में अपने विचार लिखिए।*

(iii) *"बालश्रम समाज पर एक अभिशाप है।" इस विषय को ध्यान में रखते हुए उन बच्चों की जरूरतों और मजबूरियों पर एक प्रस्ताव लिखिए।*

(iv) *"अहंकार से वृद्धि रुक जाती है।" इस विचार को अपने जीवन की घटना के आधार पर स्पष्ट कीजिए।*

(v) *"योग स्वस्थ जीवन का आधार"–विवेचना कीजिए।*

(vi) *निम्नलिखित विषयों में से किसी **एक** पर मौलिक कहानी लिखिए–*

(a) *कहानी का अन्तिम वाक्य होगा........*
"इस घटना ने मेरा हृदय परिवर्तन कर दिया।"

(b) *"जहाँ चाह वहाँ राह।" कहावत को आधार बनाकर एक कहानी लिखिए।*

Question 2.

Read the passage given below carefully and answer in Hindi the questions that follow, using your own words:

निम्नलिखित अवतरण को पढ़कर, अन्त में दिए गए प्रश्नों के उत्तर अपने शब्दों में लिखिए :

विनम्रता तथा सरलता ऐसे गुण हैं जो हमें सच्चे अर्थों में मनुष्य बनाते हैं। विनम्रता मानव को उसके दिव्य स्वभाव से जोड़ती है। उसे दूसरों के प्रति सहृदय बनाती है। वह सरलता से प्रेरित होकर विपत्ति में पड़े मनुष्यों की मदद करता है। स्वयं अभाव में रहकर भी दूसरों की यथासंभव सहायता करता है। विनम्र व्यक्ति की वाणी बड़ी मधुर होती है। इनके मृदुवचन मन की कटुता को समाप्त करते हैं। वाणी की मिठास के कारण इनके अनेकानेक मित्र बन जाते हैं। लोग इस प्रकार के व्यक्तियों से चाहे जितने भी क्रोध में बात करें, विनम्रता का जादू क्षणभर में क्रोध को शांत कर देता है। अतः इस प्रकार विनम्रता बड़ी सहजता से क्रोध जैसे बड़े मनोविकार पर भी विजय प्राप्त कर लेती है। ये व्यक्ति समाज के लिए मार्गदर्शक बन जाते हैं। इनकी सरलता में जो स्वाभाविकता होती है वह लोगों को सहज ही अपनी ओर आकर्षित करती है। इनका अनुकरण करने का प्रयास सभी करते हैं। जिससे वे भी विनम्रता को अपने जीवन में अपनाकर अपना जीवन सफल बना सकें। ईर्ष्या, द्वेष, घृणा मनोविकार को पराजित कर पाएं।

अनेक महान विभूतियों का जीवन सरलता तथा विनम्रता का प्रत्यक्ष उदाहरण है। विश्वप्रसिद्ध दार्शनिक सुकरात शक्ल से अत्यंत ही कुरूप थे। एक दिन अकेले में वे दर्पण हाथ में लेकर अपना मुँह देख रहे थे, तभी उनका प्रिय शिष्य आया और सुकरात को दर्पण देखता पाकर बहुत आश्चर्य में पड़ गया। वह कुछ बोला नहीं मात्र मुस्कुराने लगा। सुकरात ने उससे कहा "शायद तुम सोच रहे हो कि मुझ जैसा असुंदर व्यक्ति आखिर शीशा क्यों देख रहा है ?" सुकरात ने उसे समझाया "वत्स शायद तुम नहीं जानते कि मैं यह शीशा क्यों देखता हूँ ? मैं कुरूप हूँ। इसलिये प्रतिदिन शीशा देखता हूँ। शीशा देखकर मुझे अपनी कुरूपता का ज्ञान होता है। मैं अपने रूप को जानता हूँ। इसलिए प्रतिदिन प्रयत्न करता हूँ कि ऐसे, अच्छे कार्य करूँ जिनसे मेरी यह कुरूपता ढक जाए।" यह सरलता शिष्य के हृदय को छू गई। उसे गहरा जीवन दर्शन बड़ी आसानी से उसके गुरु ने सिखा दिया।

हमें भी अपने जीवन में इस अच्छे गुण को अपनाना चाहिए। यह एक बहुत बड़ी शक्ति है। इससे हम अपने केन्द्र से जुड़ते हैं। हमारी जीवनदृष्टि विशाल तथा भव्य बनती है। हम मानवता को नई ऊँचाई प्रदान करते हैं। हमारी विनम्रता तथा सरलता एक स्वस्थ समाज को जन्म देती है।

(i) *विनम्रता किस प्रकार मनुष्य के स्वभाव को दिव्यता प्रदान करती है ?* **[4]**

(ii) *सरल तथा विनम्र मनुष्यों की वाणी की विशेषता तथा प्रभाव का वर्णन करें।* **[4]**

(iii) *सामान्य लोग सरल और विनम्र लोगों का अनुसरण करने का प्रयास क्यों करते हैं ?* **[4]**

(iv) *सुकरात की प्रतिदिन शीशा देखने वाली घटना का वर्णन अपने शब्दों में कीजिये।* **[4]**

(v) *इस गद्यांश से आपको क्या शिक्षा मिलती है ?* **[4]**

Question 3.

(a) Correct the following sentences and rewrite: [5]

निम्नलिखित वाक्यों को शुद्ध करके लिखिए :

(i) *मुझे आपसे अनेकों बात करनी हैं।*

(ii) *हमें अपनी मातृभूमि पर अहंकार है।*

(iii) *मेरी बात राकेश हँसी से टाल गया।*

(iv) जो काम करो वह पूरा जरूर करो।

(v) यह सारा कार्य मैंने करा है।

(b) *Use the following idioms in sentences of your own to illustrate their meaning :* [5]

निम्नलिखित मुहावरों को वाक्यों में प्रयुक्त कीजिए :

(i) आँच न आना।

(ii) कान खड़े होना।

(iii) घी के दीये जलाना।

(iv) चेहरे का रंग उड़ना।

(v) राई का पहाड़ बनाना।

Section B is not given due to change in Present Syllabus.

ANSWERS

Answer 1.

(i) भूकम्प : प्रकृति का विनाशकारी रूप

ऊपरी सतह से लेकर अन्तर्भाग तक पृथ्वी कई परतों से बनी हुई है। पृथ्वी की बाहरी सतह कई कठोर खण्डों या विवर्तनिक प्लेटों में विभाजित है जो क्रमशः कई लाख सालों की अवधि में पूरे सतह से विस्थापित होती है। पृथ्वी की आन्तरिक सतह एक अपेक्षाकृत ठोस भूपटल की मोटी परत से बनी हुई है और सबसे अन्दर होता है एक कोर, जो एक तरल बाहरी कोर और एक ठोस लोहे की आन्तरिक कोर बनी हुई है। बाहरी सतह की जो विवर्तनिक प्लेट हैं वे बहुत धीरे–धीरे गतिमान हैं।

ये प्लेट आपस में टकराते भी हैं और एक–दूसरे से अलग भी होते हैं। ऐसी स्थिति में घर्षण के कारण भूपटल पर अचानक दरारें फूट सकती हैं। इस अचानक तेज हलचल के कारण जो शक्ति उत्सर्जित होती है, वही भूकम्प के रूप में तबाही मचाती है। कभी–कभी तो यह स्थिति बहुत भयावह हो जाती है। इसके परिणामस्वरूप पृथ्वी के ऊपर स्थित जड़–चेतना, हर प्राणी और पदार्थ का या तो विनाश हो जाता है या फिर वह सर्वनाश की–सी स्थिति में पहुँच जाता है। जापान के विषय में तो प्रायः सुना जाता है कि वहाँ तो अक्सर भूकम्प आकर विनाशलीला प्रस्तुत करते ही रहते हैं। इस कारण लोग वहाँ लकड़ियों के बने घरों में रहते हैं। इसी प्रकार का एक भयानक भूकम्प बहुत वर्षों पहले अविभाजित भारत के क्वेटा नामक स्थान पर आया था। उसने शहर के साथ–साथ हजारों घर–परिवारों का नाम तक भी बाकी नहीं रहने दिया था।

अभी कुछ वर्षों पहले गढ़वाल और महाराष्ट्र के कुछ भागों को भूकम्पन के दिल दहला देने वाले हादसों का शिकार होना पड़ा था। प्रकृति की यह कैसी लीला है कि वह मानव–शिशुओं के घर–घरौंदों को तथा स्वयं उनको भी कच्ची मिट्टी के खिलौनों की तरह तोड़–मरोड़कर रख देती है। पहले यह भूकम्प गढ़वाल के पहाड़ी इलाकों में आया था, जहाँ इसने बहुत नुकसान पहुँचाया था। थोड़े दिनों पश्चात् महाराष्ट्र के लातूर में फिर एक भूकम्प आया जिसने वहाँ सब कुछ मटियामेट कर दिया था।

26 जनवरी, 2001 को गुजरात सहित पूरे भारत ने भूकम्प का कहर देखा। भुज सहित सम्पूर्ण गुजरात में भारी जान–माल का नुकसान हुआ। 8 अक्टूबर, 2005 को पाकिस्तान के कब्जे वाले कश्मीर और उससे सटे भारतीय कश्मीर में दिल दहला देने वाला भूकम्प आया। उसमें जहाँ एक लाख से अधिक लोग काल के गाल में समा गये, वहीं लाखों लोग घायल हुए तथा अरबों रुपये की सम्पत्ति की भी हानि हुई।

भूकम्प वैज्ञानिकों का कहना है कि अभी तक ऐसा कोई उपकरण या यन्त्र विकसित नहीं हुआ है जिससे यह बात पता चल सके कि अमुक–अमुक क्षेत्रों में भूकम्प आने वाला है। भूकम्प के आते समय 'रिक्टर स्केल' पर सिर्फ उसकी तीव्रता का ही माप लिया जा सकता है।

गुजरात के भूकम्प को देखकर वहाँ अफरा–तफरी मच गई। आँखों–देखी इस भयंकर दुर्घटना ने सभी लोगों के मन में एक प्रश्न खड़ा कर दिया कि आखिर गुजरात प्रान्त का क्या होगा ? वहाँ जन–हानि और धन हानि का अनुमान ही लगाना कठिन था। भूकम्प आने के बाद गुजरात प्रान्त की स्थिति एक श्मशान घाट या कब्रिस्तान की भाँति हो गई थी। हम पाँच छात्र किसी परीक्षा के लिए वहाँ गये थे। हम एक होटल में ठहरे थे। हमें जैसे ही पता लगा तो हम लोग बाहर निकल आये और उस स्थान की तरफ बढ़े जहाँ से चीख–पुकार या भयंकर शोर शराबा हो रहा था। लोग उधर दौड़े जा रहे थे। हम भी भागकर वहाँ पहुँचे और पीड़ित लोगों की मदद की। पुलिस, एम्बुलैंस और फायरब्रिगेड की गाड़ियाँ आ चुकी थीं। उस भयंकर दृश्य को देखकर प्राण सूख रहे थे। हम पाँचों छात्रों ने हिम्मत नहीं हारी और घायलों को एम्बुलैंस में बिठाया। दबे हुए लागों को मलबा हटाकर बाहर निकाला। मृतकों के ढेर देखकर करुणा भी रोने लगी। ऐसी त्रासदी भगवान किसी देश को न दे।

भूकम्प, भूस्खलन और हिम स्खलन पैदा कर सकता है, जो पहाड़ी और पर्वतीय इलाकों में क्षति का कारण हो सकता है। भूकम्प के कारण, किसी विद्युत लाइन के टूटने के कारण आग लग सकती है। भूकम्प के कारण मिट्टी द्रवीकृत हो सकती है जिससे इमारतों और पुलों को नुकसान पहुँच सकता है। समुद्र के अन्दर भूकम्प से सुनामी आ सकती है। भूकम्प से क्षतिग्रस्त बाँध के कारण बाढ़ आ सकती है। भूकम्प से जीवन की हानि, सम्पत्ति की हानि, मूलभूत आवश्यकताओं की कमी, रोग इत्यादि होते हैं।

यह एक प्राकृतिक आपदा है। हमें इससे अपनी रक्षा के उपाय करने चाहिए।

(ii) राष्ट्र विकास के लिए अपराधी नहीं, अपराध की समाप्ति

अपराधी को अपराध करने से पहले अपराध का बोध होता है और उसके दण्ड का भी आभास होता है। अपनी आत्मा, समाज और कानून व्यवस्था के विरुद्ध जो कार्य किया जाता है वह अपराध की श्रेणी में आता है। कोई इन्सान अच्छा बुरा नहीं होता, बल्कि हालात उसे अच्छा या बुरा बनाते हैं। कुछ अपराध ऐसे होते हैं जो धर्म के विरुद्ध किये जाते हैं, कुछ

अपराध समाज के विरुद्ध किये जाते हैं, कुछ व्यक्ति विशेष को हानि पहुँचाने के लिए किये जाते हैं। अपराध तो अपराध है वह चाहे किसी भी कारण से किया गया हो। कानून की दृष्टि में प्रत्येक अपराध के लिए दण्ड निहित है।

पक्ष : साधु–सन्तों और मनीषियों ने कहा है कि पापी से नहीं पाप से घृणा करो। कई ऐसी घटनाएँ हुई हैं जिनमें अपराधी ने अपराध छोड़कर समाज के हित में कार्य किये। अपराधी को समाज और कानून द्वारा सुधरने का अवसर प्रदान करना चाहिए। यदि वह जुर्म छोड़कर देश हित में कार्य करता है तो इससे राष्ट्र के निर्माण में सहायता मिलेगी। जो उच्चकोटि का अपराधी होता है उसमें मजबूत आत्मिक शक्ति होती है। यदि वह शक्ति देश के निर्माण में लग जाये तो कितना भला होगा राष्ट्र का। अतः अपराधी की स्थिति को समझकर उसकी कठिनाइयों को दूर करने का प्रयास करना चाहिए। उसे कानून से भी कुछ सुविधाएँ दी जायें और उसे समाज सेवा का बोध कराया जाये तो अवश्य ही उसका हृदय परिवर्तन होगा और वह जुर्म का मार्ग छोड़कर विकास के मार्ग पर चलेगा।

संत विनोबा भावे ने एक 'भूदान' आन्दोलन चलाया था जिसमें किसानों से थोड़ी–थोड़ी जमीन लेकर भूमिहीनों को दी जाती थी। यह आन्दोलन सफल रहा। इसी दौर में डाकुओं के आत्मसमर्पण का कार्य भी तेजी से चला। जितने भी नामी डाकू थे उन्होंने अपने साथियों के साथ आत्मसमर्पण किया। कानून ने उदारता से उनका साथ दिया और अपनी सजा काटकर जब वे आये तो उन्हें गाँवों में जमीन और घर दिये गये जिससे कि वे एक साधारण नागरिक की तरह अपना जीवन व्यतीत करने लगे। यह था विनाश से निर्माण की ओर आना। भगवान बुद्ध के समय एक अंगुलिमाल डाकू हुआ था जिसने एक हजार व्यक्तियों की हत्या करने का प्रण लिया था। संख्या गिनने के लिए वह जो आदमी मारता था उसकी एक अंगुली काटकर उसकी माला बनाकर अपने गले में पहन लेता था। जनता उसके डर के मारे त्राहि–त्राहि कर उठी थी। जिस जंगल में वह रहता था वहाँ राजा ने एक पहरेदार बैठा दिया था कि उस जंगल से होकर कोई यात्री न जाये। राजा ने भगवान बुद्ध से उसके आतंक से मुक्ति दिलाने की प्रार्थना की। भगवान बुद्ध उस जंगल में गये और अंगुलिमाल के सामने निडर होकर प्रसन्न मुद्रा में खड़े हो गये। अंगुलिमाल को यह देखकर आश्चर्य हुआ कि उसके सामने आते ही लोगों की घिग्घी बँध जाती है, लेकिन यह व्यक्ति निडर होकर खड़ा है। कहते हैं भगवान बुद्ध ने उसे उपदेश दिया जिसे सुनकर अपने अपनी कटार भगवान बुद्ध के चरणों में रख दी और उनका शिष्य बनकर अहिंसा के मार्ग पर चलने लगा। इससे हमें यह बोध होता है कि भगवान बुद्ध ने भी अपराधी को नष्ट नहीं किया, बल्कि उसके अपराध को नष्ट किया।

इन तथ्यों से पता चलता है कि अपराधी को नहीं बल्कि अपराध को समाप्त करने से मजबूत राष्ट्र तैयार होता है।

विपक्ष : यह कथन सत्यता से कोसों दूर है क्योंकि अपराधी अपने अपराध को कभी नहीं छोड़ता। यदि ऐसा होता, तो सजा काटकर आये अपराध फिर से अपराध नहीं करते, सजा काटकर आया हुआ व्यक्ति और अधिक जुर्म करता है। उसका हृदय इतना कठोर हो जाता है कि उसे बदलना मुश्किल नहीं असम्भव है।

साधु संन्यासियों की बात आज कौन सुनता और मानता है ? यदि अपराध समाप्त करने से अपराधी समाप्त हो जाता तो अब तक भारत में अपराध देखने को नहीं मिलता। भारत क्या सभी देशों में आज आतंकवादी माहौल के कारण साँस लेना भी दूभर हो रहा है। क्या आतंकवादियों का हृदय परिवर्तन सम्भव है ? कभी नहीं। प्राचीनकाल में अपराध के कठोर दण्ड नियत थे इसलिए अपराधों की संख्या न के बराबर थी। चोरी करने पर चोर के हाथ काट दिये जाते थे। अतः लोग अपने घरों में ताले नहीं लगाते थे। अपराधी को कठोर दण्ड देने से उसके साथ–साथ उसके साथियों द्वारा किया जाने वाला अपराध भी समाप्त हो जाता है।

प्रातःकाल जज साहब भ्रमण के लिए जा रहे थे तभी उन्होंने देखा कि एक व्यक्ति ने दूसरे व्यक्ति को चाकू मार दिया और भाग गया। कुछ हफ्तों बाद उसी हत्या का मुकदमा उन्हीं जज साहब के कोर्ट में आया। जज साहब को आश्चर्य हुआ कि खून के इल्जाम में पकड़ा गया आदमी कोई और व्यक्ति है, वह व्यक्ति नहीं जिसको जज साहब ने देखा था। सारे सबूत उस पकड़े गये व्यक्ति के खिलाफ थे। जज साहब ने फैसला देने से पहले उस पकड़े गये व्यक्ति को अकेले में बुलाया और उससे पूछा कि तुम इस केस में कैसे फँस गये? खून तुमने नहीं किसी और ने किया है, लेकिन सारे सबूत तुम्हारे खिलाफ हैं। उस व्यक्ति ने शान्तिपूर्ण ढंग से उत्तर दिया कि जज साहब आप सबूतों के आधार पर अपना निर्णय दें, वह व्यक्ति जिसने खून किया है वह इस बार तो बच गया है, लेकिन आगे पकड़ा जायेगा जैसे मैं कई बार बच गया हूँ, पर इस बार पकड़ा गया। यह सुनकर जज साहब अवाक् रह गये और ऊपर भगवान की ओर देखने लगे।

यदि हम इस आशा में कि अपराधी सुधर जायेंगे, उन पर दया कर उन्हें छोड़ दें तो यह कभी सम्भव नहीं है। हम अपना समय बर्बाद कर रहे हैं। अपराधियों के अपराध समाप्त करने के लिए उनके लिए कठोर दण्ड की व्यवस्था करनी होगी।

(iii) 'बालश्रम समाज पर एक अभिशाप है'

"बालश्रम समाज पर एक अभिशाप है।" यह कथन इस सत्य को प्रतिपादित करता है कि 'बालश्रम' समाज पर एक कलंक है। यह एक अपराध है जिसमें कि बच्चों को जबरदस्ती काम करने पर लगा दिया जाता है और उनको आर्थिक जिम्मेदारी बड़ों की तरह उठाने को मजबूर किया जाता है। 'इण्टरनेशनल लेबर ऑर्गनाइजेशन' (ILO) के अनुसार पन्द्रह वर्ष की उम्र से कम के बच्चों को धन कमाने के कार्यों में लगाना एक अपराध है। बालश्रम के चलते उनका भविष्य बर्बाद हो रहा है। इस पर दुनिया के सभी देश गम्भीरता से विचार कर रहे हैं।

बालश्रम की समस्या विकासशील देशों में अधिक पायी जाती है। विकासशील देशों में गरीबी के कारण बच्चों को कम उम्र में ही मजबूरी में काम आदि करना पड़ता है। जिसके कारण वे अपनी शिक्षा, स्वास्थ्य, सामाजिक स्वतन्त्रता तथा अपने बचपन से वंचित रह जाते हैं। दुनिया के सभी विकासशील देश इस समस्या को सुलझाने का भरसक प्रयत्न कर रहे हैं, लेकिन

चोरी–छिपे यह रोजगार धड़ल्ले से चल रहा है। समाज का बड़ा तबका बालश्रम का भरपूर फायदा ले रहा है। लाखों बच्चे बालश्रम की चपेट में आकर अपना जीवन चौपट कर रहे हैं। ये बच्चे ही भारत का भविष्य हैं जो बिगड़ता हुआ दिखाई दे रहा है। बालश्रम एक जुर्म है। ये जुर्म बच्चों के माता–पिता, कारखानों के मालिक, रेस्टोरेन्ट के मालिक और घर के मालिक, दुकानदार, दस्तकार, बुक बाइंडर आदि खुले रूप से कर रहे हैं। जिससे बच्चों का जीवन नष्ट हो रहा है और इसकी चिन्ता किसी को नहीं है।

बालश्रम बढ़ने के निम्न कारण हैं जो इस श्रम को कम नहीं होने देते–

1. गरीबी और बढ़ती हुई बेरोजगारी विकासशील देशों में बालश्रम को जन्म दे रही है।
2. विश्व संगठन की रिपोर्ट बताती है कि दुनिया के एक–चौथाई लोग अत्यन्त गरीब हैं।
3. शिक्षा की कमी है जिसे अधिकतर बच्चे प्राप्त नहीं कर पाते।
4. विकासशील देश के लोग बालश्रम के कानून को तोड़कर बालश्रम को बढ़ावा दे रहे हैं।
5. समाज का पूर्ण नियन्त्रण खेती पर काम करने वाले बच्चों पर और घर में काम करने वाले बच्चों पर नहीं है।
6. बालश्रम करने वाले बच्चे अपने परिवार की दो वक्त की रोटी जुटाने के लिए इसे महत्व देते हैं और इसके ऊपर वे कुछ और सोचते ही नहीं हैं।
7. व्यापारिक संगठन और गृह उद्योग छोटे बच्चों को नौकरी पर रख लेते हैं, ताकि कम कीमत में ज्यादा काम ले सकें। वे बच्चों को प्रोत्साहित करते हैं और छोटे बच्चे उसी को अपना भाग्य मानकर स्वीकार कर लेते हैं।

बालश्रम की समस्या को सुलझाने के लिए उपाय—

1. समाज में फैली इस कुप्रथा को बन्द करने के लिए हमें नये सिरे से प्रयास करने होंगे।
2. बालश्रम को समाप्त करने के लिए समाज में ऐसी इकाइयाँ बनें जो बालकों के श्रम पर ध्यान देकर बालकों को उस काम के प्रति निरुत्साहित करें और उनके पढ़ने की निःशुल्क व्यवस्था करें।
3. बच्चों के माता–पिता को उनकी पढ़ाई के लिए प्रोत्साहित करें।
4. समाज में जागृति उत्पन्न की जाये जिसका परिणाम बालश्रम से छुटकारा हो।
5. सरकार उन गरीब परिवारों की कुछ आर्थिक सहायता करे जिनके बच्चे गरीबी के कारण बालश्रम में लग जाते हैं।
6. सरकार बालश्रम समाप्ति के लिए कड़े कानून बनाये जिसमें माता–पिता और जहाँ बच्चे काम करते हैं उनको दण्ड देने का प्रावधान हो।

भारतीय सामाजिक कार्यकर्ता कैलाश सत्यार्थी जिन्होंने बालश्रम को रोकने के लिए अपना पूर्ण सहयोग दिया तथा इसके लिए उन्हें विश्व का सर्वश्रेष्ठ सम्मान 'नोबेल पुरस्कार' मिला। ये मध्य प्रदेश में पैदा हुए और अपनी शिक्षक की नौकरी त्यागकर 'बालश्रम' की समाप्ति के कार्य में लग गये। 1980 में इन्होंने 'बचपन बचाओ' आन्दोलन चलाया। 144 देशों में इन्होंने लगभग 83,000 बच्चों को बालश्रम से उबारा और उनकी शिक्षा आदि का भी सरकार के सहयोग से प्रबन्ध किया।

बालश्रम एक बड़ी सामाजिक समस्या है। इसको माता–पिता, अध्यापक, सामाजिक कार्यकर्ता और सरकार मिलकर सुलझा सकते हैं। बच्चों को शिक्षित बनाया जाये और उनकी कमाने वाली मानसिकता को सुधारा जाये।

(iv) 'अहंकार से वृद्धि रुक जाती है'

'अहंकार से वृद्धि रुक जाती है' यह उक्ति अक्षरशः सत्य है। अहंकार एक ऐसा अवगुण है जो मनुष्य के हृदय और मस्तिष्क पर हावी रहता है और उसे उसका भान भी नहीं होता। वह उसके मार्ग में प्रतिपग बाधा उत्पन्न करता है। अन्य मनुष्य ही समझ पाते हैं कि अमुक मनुष्य अहंकारी है। अतः वे उससे दूरी बनाकर रखते हैं। अहंकार न दिखाई देते हुए भी स्पष्ट परिलक्षित हो जाता है। अहंकार, अहंकारी व्यक्ति को भरोसा दिलाता है कि उसके सब काम अंहकार के कारण ही होंगे पर होता इसके विपरीत ही है। मनुष्य का जीवन अपने आस–पास के वातावरण से प्रभावित होता है। मूलरूप से मानव के विचारों और कार्यों को उसके संस्कार, वंश परम्पराएँ ही दिशा दे सकती हैं, अहंकार नहीं। यदि उसे अच्छा वातावरण मिलता है तो वह श्रेष्ठ कार्यों को सम्पादित करता है। यदि अपने अहंकार के कारण उसे उचित वातावरण सुलभ नहीं हो पाता है तो उसके कार्य भी उससे प्रभावित होते हैं। प्रत्येक व्यक्ति का व्यक्तित्व उसकी जीवन शैली और आचरण के रूप में प्रतिबिम्बित होता है। यदि मनुष्य का आचरण नैतिकता के अनुरूप है तो उस मनुष्य की सामाजिक प्रतिष्ठा अन्य लोगों के लिए प्रेरणा बन जाती है और वह व्यक्ति समाज में सम्मान पाता है। इसके विपरीत यदि मनुष्य का आचरण सामाजिक, सांस्कृतिक, वैधानिक, धार्मिक नियमों के प्रतिकूल और अहंकार से प्रभावित है तो वह समाज में अपमान का पात्र होता है। अतः मनुष्य का जीवन अहंकारहीन और सदाचारमय होना अधिक महत्व रखता है। सदाचार का जीवन में विशेष महत्व है। यदि मनुष्य का आचरण सत्य पर आधारित हो तो मनुष्य शारीरिक, मानसिक, बौद्धिक एवं भौतिक रूप से समृद्ध रहता है।

अहंकार और दम्भ मनुष्य को यह दर्शाते हैं कि वह सभी मनुष्यों से श्रेष्ठ है और अपने बलबूते पर ही सब काम कर सकता है। यह अभिमान उसके कार्यों में बाधा उत्पन्न करता है। ईश्वर भी अपने भक्तों में अभिमान के अंकुर फूटने नहीं देता। जहाँ अभिमान प्रतिष्ठापित है वहाँ ईश्वर का निवास नहीं होता; जैसे—

"जब मैं था तब हरि नहीं, अब हरि है मैं नाहि।"

बाली को अपने पर अभिमान था, पर श्रीराम ने उसे मारकर उसका अहंकार तोड़ दिया। एक बार गरुड़ को अपनी गति का अभिमान हुआ था। वह हनुमान जी को बुलाने गये और चाहते थे कि हनुमान जी उनकी पीठ पर बैठकर चलें तो शीघ्र रामजी के पास पहुँच जायेंगे। हनुमानजी ने उनके साथ जाने को मनाकर दिया और कहा कि 'मैं आता हूँ'। वापिस लौटने पर गरुड़ ने रामजी के पास जाकर देखा कि हनुमान जी रामजी से वार्ता करके लौट रहे हैं। तब वह अपनी गति के अभिमान पर लज्जित हुआ। इन प्रसंगों से पता चलता है कि अहंकार सबके कार्यों में बाधा डालता है।

मुझे भी कई बार यह अभिमान हुआ कि मैं सभी छात्रों से अधिक बुद्धिमान हूँ। इसका परिणाम यह हुआ कि मेरे मित्र भी मुझसे दूरी बनाने लगे। मुझे किसी का सहयोग प्राप्त नहीं हुआ। मैं अपनी कक्षा में भी पिछड़ने लगा। तब मेरे एक

हितैषी अध्यापक ने बड़े प्रेम से मुझे समझाया कि मेरे मन में बैठा अहंकार ही मेरे मार्ग में बाधा बन गया है। मैंने धीरे–धीरे अपने अन्दर बैठे अभिमान को दूर किया और पुनः अपने मित्रों का सहयोग मुझे मिलने लगा। कहाँ तो मैं समझता था कि मुझे सब कुछ आता है, लेकिन परीक्षा में बहुत कम अंक आने पर मुझे ज्ञात हुआ कि मेरा अभिमान ही मेरे विकास में बाधा डाल रहा था। अब मैं अभिमान से रहित होकर अपने पथ पर तीव्रगति से अग्रसर हो रहा हूँ।

तुलसीदास जी ने अभिमान को पाप का मूल बताया है—

''दया धर्म का मूल है, पाप मूल अभिमान।
तुलसी दया न छोड़िये, जब लगि घट में प्रान।''

(v) 'योग स्वस्थ जीवन का आधार है'

योग भारत की प्राचीन संस्कृति का गौरवमयी हिस्सा है जिसकी वजह से भारत सदियों तक विश्व गुरु रहा है। योग एक ऐसी सुलभ एवं प्राकृतिक पद्धति है जिससे स्वस्थ मन एवं शरीर के साथ अनेक आध्यात्मिक लाभ प्राप्त किये जा सकते हैं। योग को स्वामी रामदेव ने गुफाओं और कन्दराओं से निकालकर आम जन तक पहुँचाया है। भारतीय धर्म और दर्शन में योगा का अत्यधिक महत्व है। आध्यात्मिक उन्नति या शारीरिक और मानसिक स्वास्थ्य के लिए योग की आवश्यकता व महत्व को प्रायः सभी दर्शनों एवं धार्मिक सम्प्रदायों ने एकमत व मुक्तकंठ से स्वीकार किया है। आधुनिक युग में योग का महत्व बढ़ गया है। स्वास्थ्य रक्षा के लिए योग रामबाण दवा है। इसके सहारे मनुष्य स्वस्थ तो रहता ही है साथ ही साथ कई बीमारियों से छुटकारा भी पाता है।

योग भारत और नेपाल में एक आध्यात्मिक प्रक्रिया को कहते हैं जिसमें शरीर, मन और आत्मा को एक साथ लाने का प्रयास होता है। यह शब्द, प्रक्रिया और धारणा बौद्ध धर्म, जैन धर्म और हिन्दू धर्म में ध्यान प्रक्रिया से सम्बन्धित है। 'योग' शब्द भारत से बौद्ध धर्म के साथ चीन, जापान, तिब्बत, दक्षिण–पूर्व एशिया और श्रीलंका में भी फैल गया है और इस समय सारे सभ्य जगत में लोग इससे परिचित हैं। प्राचीन जीवन पद्धति लिए योग आज के परिवेश में हमारे जीवन को स्वस्थ और खुशहाल बना सकता है। आज के प्रदूषित वातावरण में योग एक ऐसी औषधि है जिसका कोई साइड इफैक्ट नहीं है, बल्कि योग के अनेक आसन जैसे शवासन हाई ब्लडप्रैशर को सामान्य करता है, जीवन के लिए संजीवनी है, अनुलोम–विलोम कपालभांति प्राणायाम, भ्रामरी प्राणायाम मन को शान्त करता है। वक्रासन हमें अनेक बीमारियों से बचाता है। आज कम्प्यूटर की दुनिया में दिनभर उसके सामने बैठे–बैठे काम करने से अनेक लोगों को कमर दर्द एवं गर्दन दर्द होना एक आम बात हो गई है, ऐसे में शलभासन तथा ताड़ासन हमें दर्द से छुटकारा दिलाते हैं। पवन मुक्तासन अपने नाम के अनुरूप पेट से गैस की समस्या को दूर करता है। गठिया की समस्या को मेरुदंडासन दूर करता है। योग में ऐसे अनेक आसन हैं जिनको जीवन में अपनाने से कई बीमारियाँ समाप्त हो जाती हैं और खतरनाक बीमारियों का असर भी कम हो जाता है। 24 घण्टे में से महज कुछ मिनट का ही समय यदि योग में उपयोग करते हैं तो अपनी सेहत को हम चुस्त–दुरुस्त रख सकते हैं। फिट रहने के साथ योग हमें सकारात्मक ऊर्जा भी देता है। योग से शरीर में रोग प्रतिरोधक क्षमता का विकास होता है। यह कहना अतिशयोक्ति न होगा कि योग हमारे लिए हर तरह से आवश्यक है। यह हमारे शारीरिक, मानसिक और आत्मिक स्वास्थ्य के लिए लाभदायक है। योग के माध्यम से आत्मिक सन्तुष्टि, शान्ति और ऊर्जावान चेतना की अनुभूति प्राप्त होती है जिससे हमारा जीवन तनावमुक्त तथा हर दिन सकारात्मक ऊर्जा के साथ आगे बढ़ता है। हमारे देश की ऋषि परम्परा से प्राप्त योग को आज विश्व भी अपना रहा है। गीता में योग के बारे में लिखा है, ''योग स्वयं की स्वयं के माध्यम से स्वयं तक पहुँचने की यात्रा है।''

योगासनों से कुछ निम्नलिखित लाभ हैं—

1. योग का प्रयोग शारीरिक, मानसिक और आध्यात्मिक लाभों के लिए हमेशा से होता रहा है। आज की चिकित्सा शोधों ने ये साबित कर दिया है कि योग शारीरिक और मानसिक रूप से मानव जाति के लिए वरदान है।

2. जहाँ 'जिम' आदि से शरीर के किसी खास अंग का ही व्यायाम होता है, वहीं योग से शरीर के समस्त अंग–प्रत्यंगों, ग्रन्थियों का व्यायाम होता है जिससे अंग–प्रत्यंग सुचारु रूप से काम करने लगते हैं।

3. योगाभ्यास से रोगों से लड़ने की शक्ति बढ़ती है। बुढ़ापे में भी स्वस्थ बने रह सकते हैं। त्वचा पर चमक आती है, शरीर स्वस्थ, निरोग और बलवान बनता है।

4. जहाँ एक तरफ योगासन माँसपेशियों को पुष्ट करता है जिससे दुबला–पतला व्यक्ति भी ताकतवर और बलवान बन जाता है वहीं दूसरी ओर योग के नित्य अभ्यास से शरीर से फैट भी कम हो जाता है। इस तरह योग कृश और स्थूल दोनों के लिए लाभदायक है।

5. योगासन के नित्य अभ्यास से माँसपेशियों का अच्छा व्यायाम होता है जिससे तनाव दूर होकर अच्छी नींद आती है, भूख अच्छी लगती है, पाचन सही रहता है।

योग आज एक थैरेपी के रूप में भी उपयोगी सिद्ध हो चुका है। हर एक रोग के लिए विशेष आसन है। रोग विशेष के उपचार के लिए विशेष आसन है। कई ऐसे रोग जिसके उपचार से डॉक्टर जवाब दे देता है उसके लिए भी योग लाभदायक हो सकता है। आसन करना किसी दक्ष व्यक्ति से ही सीखना चाहिए। रोगों से लड़ने के लिए हमारे शरीर में भी प्रतिरोधात्मक शक्ति होती है तथा इन शक्ति को योगासन द्वारा बढ़ाया जा सकता है।

(vi) (a) इस घटना ने मेरा हृदय परिवर्तन कर दिया

भारतीय इतिहास में अनेक ऐसे प्रसंग मिलते हैं जिनमें किसी घटना से प्रभावित होकर व्यक्ति का हृदय परिवर्तन हो गया। हृदय परिवर्तन किसी सद्मार्ग पर चलने के लिए हुआ हो या अपने को परोपकार में लगाने के लिए हुआ हो या फिर दान आदि के द्वारा धार्मिक प्रोत्साहन के लिए हुआ हो। मनुष्य का हृदय अत्यन्त कोमल, दयापूर्ण, प्रेम से भरा और मनोबल से परिपूर्ण होता है।

महर्षि वाल्मीकि अपनी युवावस्था में डाकू रत्नाकर के नाम से जाने जाते थे। एक बार ऋषियों ने अपने उपदेश से उनका हृदय परिवर्तन कर दिया और अपने कल्याण के लिए 'राम–राम' जपने का उपदेश दिया। वे 'राम–राम' का उल्टा 'मरा–मरा' जपने लगे और आगे चल कर महर्षि वाल्मीकि हुए

जिन्होंने संस्कृत में वाल्मीकी रामायण लिखी। इनके बारे में एक चौपाई इस प्रकार हैं—

^^mYVk uke tik tx tkuk
okYehfd Hk; cã lekuk**

दूसरा उदाहरण है सम्राट् अशोक का जिन्होंने 'कलिंग'देश के युद्ध में भीषण नरसंहार के बाद बौद्ध भिक्षु के उपदेश से अपना हृदय परिवर्तन कर लिया और आगे युद्ध न करने की प्रतिज्ञा की। उन्होंने अपनी तलवार रख दी और फिर कभी उसे नहीं उठाया। यह भी मानवता के कल्याण में किया गया 'हृदय परिवर्तन' था।

एक दिन छमाही परीक्षा के बाद मैंने बाजार में घूमने का निश्चय किया और अपनी साइकिल उठाकर बाजार की तरफ चल दिया। तभी मुझे याद आया कि पिताजी की कुछ दवा लेनी थी। इमरजेन्सी के पास एक मेडीकल की बड़ी दुकान है। मैं वहाँ जाकर खड़ा हुआ तो देखा कि गाँव का एक वृद्ध व्यक्ति डॉक्टर का पर्चा लेकर दुकान पर आया और बताया कि उसके बेटे का ऑपरेशन हो रहा है इसलिए उसे शीघ्र इन दवाइयों की आवश्यकता है।

दुकानदार ने दवा निकालकर दे दी और पैसे माँगे। उस वृद्ध के पास 50 रुपये कम थे। इस पर दुकानदार ने दवा वापस लेकर एक तरफ रख दी। वृद्ध गुहार करता रहा कि वह पैसे बाद में दे जायेगा इस समय ऑपरेशन चल रहा है। मुझे उस पर दया आई और उसके शब्दों में सत्यता का आभास हुआ। दवा लेकर जाना उसके लिए कितना आवश्यक था यह समझ में आ रहा था। उसने कातर दृष्टि से देखा पर किसी ने उसकी तरफ ध्यान नहीं दिया। वह रुआँसा हो गया। मैंने दुकानदार से कहा कि वह उसको दवा दे दे बाकी के पैसे मैं दे दूँगा। दुकानदार ने उसे दवा दे दी। उसने धन्यवाद सूचक दृष्टि से देखा और चला गया। मैंने अपनी जेब देखी तो उसमें उतने पैसे नहीं थे। अतः मैंने अपने हाथ की घड़ी उतारकर दुकानदार को दी। उसने आश्चर्य से मुझे देखा और घड़ी लेकर एक लिफाफे में रख दी। मैंने उससे कहा कि रुपये देकर घड़ी ले जाऊँगा।

मैं घर आ गया और यह घटना पिताजी को सुनाई। वे मेरे इस कृत्य पर बहुत खुश हुए। वे बड़े सहृदय हैं और इस प्रकार की मदद करते रहते हैं। मैंने पिताजी से 1050 रुपये माँगे। उन्होंने तुरन्त ही ग्यारह सौ रुपये मुझे दे दिये।

इस घटना ने मेरे हृदय को झकझोर कर रख दिया था कि किसी जरूरत के समय पैसा न होने से किस संकट से गुजरना पड़ सकता है। मैं उसी मेडीकल की दुकान पर गया और पचास रुपये देकर अपनी घड़ी वापस ले ली।

फिर उस दुकानदार को विश्वास में लेकर एक हजार रुपये उसके पास जमा कराये कि इस प्रकार का परेशान कोई भी व्यक्ति उसकी दुकान पर लेने आये उसके पास रुपये कम हों तो इन रुपयों में से उसकी पूर्ति कर दे। उसने हर्षित होकर मुझे धन्यवाद दिया। मैंने उसको अपना मोबाइल नम्बर दिया कि रुपये समाप्त हो जायें तो मुझे फोन कर दे। मैं तुरन्त ही पैसे जमा कर दिया करूँगा। इस प्रकार यह छोटी–सी मदद मेरी तरफ से चलती रहेगी।

इस घटना ने मेरा हृदय परिवर्तन कर दिया।

(b) 'जहाँ चाह वहाँ राह।'

मनुष्य के सभी कार्य मन से ही संचालित होते हैं। मन में मनन की शक्ति है। मननशीलता के कारण ही मनुष्य को चिन्तनशील प्राणी कहा जाता है। संस्कृत में एक कहावत है—'मन एवं मनुष्याणां कारणं बन्धमोक्षयो' अर्थात् मन ही मनुष्य के बन्धन और मोक्ष का कारण है। यदि मन न चाहे तो मनुष्य बड़े–से–बड़े बन्धनों की उपेक्षा कर सकता है। मन के चाहने से ही राह बनती है। शंकराचार्य ने कहा है 'जिसने मन को जीत लिया, उसने जगत को जीत लिया।' मन की संकल्पशक्ति से व्यक्ति को राह मिल जाती है अर्थात् वह अपने कार्य में सफल हो जाता है। अनेक ऐसे उदाहरण मिलते हैं जिसमें मन की संकल्पशक्ति के द्वारा व्यक्तियों ने अपनी हार को विजय में परिवर्तित कर दिया। महाभारत के युद्ध में पाण्डवों की जीत का कारण था कि श्रीकृष्ण ने उनके मनोबल को दृढ़ कर दिया था। नचिकेता ने चाहा कि वह मृत्यु पर विजय प्राप्त करे और आत्मा का रहस्य जान सके। यमराज के द्वारा उसे आत्मा का रहस्य समझाया गया। सावित्री ने अपने मनोबल की चाह से यमराज से अपने मरे हुए पति को आत्मा को वापस प्राप्त कर लिया। महाराणा प्रताप ने अपनी दृढ़ मनःशक्ति से अकबर की सेना से लोहा लिया। तेनसिंह चाहता था कि वह एवरेस्ट की चोटी पर चढ़कर विजय प्राप्त करे। उसने कोशिश की और एवरेस्ट की चोटी पर विजय प्राप्त कर ली।

मेवाड़ का इतिहास वीरता से भरा पड़ा है। वहाँ का क्रूर शासक बनवीर निष्कंटक राज करना चाहता था। उसने यह योजना बनाई कि शहर में एक रात दीपदान और नाच–गान के उत्सव का आयोजन करवाया जाये। उसका लक्ष्य था कि जब जनता इस उत्सव में व्यस्त होगी तब वह महाराणा और मेवाड़ के उत्तराधिकारी बालक उदयसिंह को मार देगा और निष्कंटक राज्य करेगा। शहर में यह उत्सव चल रहा था और लोग आनन्द में व्यक्त थे।

इधर राजमहल में उदयसिंह की धाय 'पन्ना' उसकी रक्षा और देखभाल के लिए थी। वह एक देशभक्त क्षत्राणी थी। वह बनवीर की क्रूरता से परिचित थी। पन्ना धाय का बेटा चन्दन उदयसिंह की आयु का ही था और वे दोनों उसी राजमहल में पन्ना की देखरेख में रहते थे। पन्ना को शक था कि दीपदान के उत्सव में बनवीर कुछ षड्यंत्र कर सकता है। तभी एक दासी ने उसे बताया कि बनवीर ने महाराणा का कत्ल कर दिया है और वह उदयसिंह के महल की तरफ आ रहा है। पन्ना उदयसिंह को बचाना चाहती थी। उसने अपने पुत्र चन्दन को उदयसिंह के राजसी वस्त्र पहनाकर कुँवर उदयसिंह के पलंग पर सुला दिया। तभी कीरत बारी एक बड़ी टोकरी लेकर जूठी पत्तल उठाने के लिए आया। वह भी मेवाड़ को बहुत प्रेम करता था। पन्ना ने उससे कहा कि वह मेवाड़ के कुँवर की रक्षा करे। तब पन्ना ने सोते हुए उदयसिंह को कीरत की टोकरी में रख दिया और ऊपर से पत्तलों से ढक दिया तथा उसे बताये हुए स्थान पर मिलने को कहा। महल के बाहर अनेक सैनिक तैनात थे, लेकिन कीरत को किसी ने टोका नहीं और वह वहाँ से आसानी से निकल गया। उसके बाद बनवीर हाथ में तलवार लेकर आया और पन्ना को लालच देकर उदयसिंह के बारे में पूछा। पन्ना ने

उससे बहुत गुहार लगाई, लेकिन वह नहीं माना। तब पन्ना ने उँगली के इशारे से पलंग की ओर संकेत किया जिस पर उसका पुत्र चन्दन सो रहा था। बनवीर ने एक ही बार में उस बालक का काम तमाम कर दिया। पन्ना चीत्कार करके वहीं गिर पड़ी। उसका ऐसा बलिदान इतिहास में स्वर्ण–अक्षरों में लिखा गया।

फिर पन्ना उस स्थान पर पहुँची जहाँ कीरत को पहुँचने को कहा था। फिर वह कुँवर उदयसिंह को लेकर पड़ोसी मित्र राज्य में चली गई जहाँ उदयसिंह का पालन हुआ। पन्ना ने सच्चे मन से जो चाहा वह पूरा हुआ। अतः हम कह सकते हैं— 'जहाँ चाह वहाँ राह।'

Answer 2.

(i) विनम्रता सच्चे अर्थ में मनुष्य को श्रेष्ठ बनाती है। वह मनुष्य को उसके दिव्य स्वभाव से जोड़ती है। विनम्र व्यक्ति में ही सहृदयता आती है जिससे वह दूसरे मनुष्यों की सहायता करता है। विनम्र व्यक्ति का कोई शत्रु नहीं होता। वह सबको अपना प्रिय बना लेता है। विनम्रता मनुष्य का सर्वश्रेष्ठ गुण है। विनम्र व्यक्ति ही सबका चहेता होता है।

(ii) सरल तथा विनम्र मनुष्यों की वाणी मधुर होती है। वे सामने वाले व्यक्ति के अनुरूप ही सद्व्यवहार करते हैं। सरल व्यक्ति किसी से धोखाधड़ी नहीं करता। विनम्र व्यक्ति सरल हृदय तो होता ही है साथ–साथ वह अपनी मधुर वाणी से सुनने वाले को सन्तोष और सुख प्रदान करता है। उसके मृदुवचन मन की कटुता को समाप्त करते हैं। वाणी की मधुरता के कारण इनके अनेकानेक मित्र बन जाते हैं। विनम्र व्यक्ति की मधुर वाणी क्रोधी को भी शान्त कर देती है।

(iii) सामान्य लोग सरल और विनम्र लोगों का अनुसरण करने का प्रयास इसलिए करते हैं क्योंकि वे सभी उन लोगों की तरह विनम्रता को अपनाकर अपने जीवन को महान और सफल बना सकें। विनम्रता से ईर्ष्या, द्वेष, घृणा जैसे मनोविकार धीरे–धीरे समाप्त हो जाते हैं और मनुष्य सद्गुणों की ओर प्रेरित होता है।

(iv) विश्वप्रसिद्ध दार्शनिक सुकरात शक्ल में अत्यन्त कुरूप थे। एक दिन वे दर्पण में अपना मुख देख रहे थे। तभी एक शिष्य वहाँ आया तो उसे आश्चर्य हुआ कि कुरूप होते हुए भी गुरुजी अपना मुख दर्पण में क्यों देख रहे हैं ? लेकिन वह कुछ बोला नहीं। गुरु ने उसके मन के विचारों को जानकर उससे कहा कि वत्स मैं कुरूप हूँ इसलिए शीशा देखकर मैं यह प्रयत्न करता हूँ कि ऐसे अच्छे कार्य करूँ जिससे मेरी कुरूपता ढक जाये। संसार में रूप का नहीं बल्कि विद्वता और सत्कर्मों का ही महत्व है। शिष्य की समझ में यह बात आ गई।

(v) इस गद्यांश से हमें यह शिक्षा मिलती है कि हमें अपने जीवन में अच्छे गुणों को अपनाना चाहिये। इससे हमारी जीवनदृष्टि विशाल तथा भव्य बनती है। हम मानवता को नई ऊँचाई प्रदान करते हैं। हमारी विनम्रता तथा सरलता एक स्वस्थ समाज को जन्म देती है। हमारे जीवन में सद्गुणों का विकास होता है और सत्पुरुषों का संग मिलता है।

Answer 3.

(a) **(i)** मुझे आपसे अनेक बातें करनी हैं।

(ii) हमें अपनी मातृभूमि पर गर्व है।

(iii) मेरी बात राकेश हँसी में टाल गया।

(iv) जो काम करो उसे पूरा जरूर करो।

(v) यह सारा कार्य मैंने किया है।

(b) **(i)** **आँच न आना**—कोई नुकसान न होना।

प्रयोग—शेयरों की उलटफेर में करीब–करीब सभी को घाटा हुआ, लेकिन रमेश भाग्यशाली रहा कि उसको कोई **आँच न आई**।

(ii) **कान खड़े होना**—सावधान हो जाना।

प्रयोग—जब से सुशील को चाँदी के काम में घाटा हुआ है तब से उसके **कान खड़े हो गये** और अब वह सावधानी से चलता है।

(iii) **घी के दिये जलाना**—बहुत खुशी मनाना।

प्रयोग—व्यापार में उम्मीद से ज्यादा मुनाफा होने पर सेठ लक्ष्मीचन्द के यहाँ **घी के दिये जल रहे हैं**।

(iv) **चेहरे का रंग उड़ना**—घबरा जाना।

प्रयोग—चोर ने चोरी के माल को ऐसी जगह छिपा दिया कि किसी को पता न चले; लेकिन जब पुलिस ने उसको पकड़ लिया तो उसके **चेहरे का रंग उड़ गया**।

(v) **राई का पहाड़ बनाना**—बड़ा–चढ़ा कर कहना।

प्रयोग—रमेश की तो आदत है कि वह हर घटना को **राई का पहाड़ बनाकर** बताता है।

●●

PHYSICAL EDUCATION 2020

(Paper-I)
(THEORY)
(Maximum Marks : 70)
(Time allowed : Three hours)
*(Candidates are allowed additional 15 minutes for **only** reading the paper. They must NOT start writing during this time.)*
*Answer **any five** questions from **Section A**.*
***Section B** comprises one question on each game, with five subparts (a), (b), (c) (d) and (e).*
*Select **any two** games from this Section.*
*Answer **any three** of the **five** subparts (a), (b), (c), (d) and (e) from **each** of the **two** games selected by you in this Section.*
The intended marks for questions or parts of questions are given in brackets [ ].

QUESTIONS

SECTION–A (40 Marks)

*Answer **any five** questions.*

Question 1.

Briefly explain the role of physical education in national integration [8]

Question 2.

What is meant by Sports Training ? Explain its importance. [8]

Question 3.

Write a short note on Netaji Subhas National Institute of Sports (NSNIS) and list its aims and objectives. [8]

Question 4.

Explain the advantages and disadvantages of knock-out tournaments. Name the other three types of tournaments. [8]

Question 5.

(a) *What is meant by drug abuse ?* [8]

(b) *Discuss the various objectives of health education.*

Question 6.

(a) *State the various causes of disability.* [8]

(b) *Discuss first-aid treatment for sprains and strains.*

Question 7.

(a) *Briefly discuss the importance of correct posture.* [8]

(b) *With regard to first-aid, what does RICE stand for ?*

SECTION–B (30 Marks)

*Select **any two** games from this Section.*

*Answer any **three** of the **five** subparts (a), (b), (c), (d) and (e) from **each** of the **two** games selected by you. Each subpart is for **5 marks**.*

FOOTBALL

Question 8. [3 × 5]

(a) *Explain any five offences committed by a football player for which a Red card is shown.* [5]

(b) *Explain the following terms associated with the game of football :* [5]

(i) *Corner kick* **(ii)** *Throw in*

(c) *What is time lost ? State the various situations under which time is lost during the game.* [5]

(d) *List any five occasions when the referee blows the whistle during the game.* [5]

(e) **(i)** *When is a penalty kick awarded in a game of football ?* [5]

(ii) *From which spot is a penalty kick taken ?*

CRICKET

Question 9. [3 × 5]

(a) *Mention the role of a match referee.* [5]

(b) *Draw any five fielding positions in a 30 yards circle.* [5]

(c) *Explain the five situations when a no ball is called by an umpire.* [5]

(d) *Explain the terms leg bye and leg before wicket.* [5]

(e) *Explain the Duckworth rule in the game of cricket.* [5]

HOCKEY

Question 10. [3 × 5]

(a) *Explain the following :* [5]

(i) *Flick* **(ii)** *Long corner*

(b) *List any five international tournaments of hockey.* [5]

(c) *Mention the basic compulsory equipment used by hockey players.* [5]

(d) *Write any four fundamental skills of hockey and briefly explain any one of them.* [5]

(e) *Write the full form of IHF and mention any three duties of IHF.* [5]

BASKETBALL

Question 11. [3 × 5]

(a) *What is the difference between charging and blocking?* [5]

(b) *List any five technical equipment used by the table officials.* [5]

(c) *Explain the Five Second Rule and the Eight Second Rule.* [5]

(d) *Explain the following terms with reference to the game of basketball :* [5]

(i) *Time out* (ii) *Pivot*

(e) *Differentiate between foul and team foul.* [5]

VOLLEYBALL

Question 12. [3 × 5]

(a) *Explain the following :* [5]

(i) *Round-arm service*

(ii) *Tennis service*

(b) *Differentiate between a smash and a serve.* [5]

(c) *Name five international tournaments of volleyball.* [5]

(d) *List five types of game delay.* [5]

(e) *Explain the following terms :* [5]

(i) *Rally* (ii) *Spike*

TENNIS

Question 13. [3 × 5]

(a) *List the various situations under which a player, while serving during a game of tennis may make foot fault.* [5]

(b) *What is the pattern of scoring in a tie-breaker ?* [5]

(c) *Mention any four fundamental skills required in the game of tennis and briefly explain any one of them.* [5]

(d) *Name the officials who are required to conduct a tennis match.* [5]

(e) (i) *When is an advantage given in a game of tennis?* [5]

(ii) *What is a cross court shot ?*

BADMINTON

Question 14. [3 × 5]

(a) *List the duties of the line judge and the service judge.* [5]

(b) *Mention any five faults a badminton player may commit during the game.* [5]

(c) *Describe the scoring system in a game of badminton.* [5]

(d) *State any five duties of a service judge.* [5]

(e) *Mention any five important tournaments of badminton.* [5]

SWIMMING

Question 15. [3 × 5]

(a) *Mention any three duties of the starter and any two duties of the chief inspector of turns.* [5]

(b) *Mention the rules related to the conduct of races.* [5]

(c) *List the order of strokes in the medley relay events.* [5]

(d) *Write a short note on SFI.* [5]

(e) *Name the events which start with a dive and the events which start from within the water.* [5]

ATHLETICS

Question 16. [3 × 5]

(a) *What is the diameter of a shot put circle ? What are the measurements of a stop board ?* [5]

(b) *Differentiate between sector and lane in athletics.* [5]

(c) *Draw a diagram of a javelin runway and sector with any three specifications.* [5]

(d) *Describe fundamental skills required for long jump.* [5]

(e) *Mention any five international competitions in athletics.* [5]

ANSWERS

SECTION–A

Answer 1.

In a country like India where there is so much diversity with regards to religion, caste, creed, language etc. physical education plays a very important role in bringing unity and in promoting national integration. Physical education, as well as sports, helps in integrating the body with the soul. The term integration means co-operation. It is an act by individuals towards achieving some common goal. Hence, national integration can be achieved through proper physical education. Physical education promotes peace, goodwill, friendship and universal brotherhood between the players during competitions.

Physical education provides a platform to act across the barriers of national boundaries. National events provide an opportunity of personal interaction between the sports persons of different states and bring them together and closer to share their experience.

Physical education helps much in national integration. It helps in the overall development of one's personality. It also helps to be physically fit, emotionally balanced and mentally sharp. It also help in adjusting socially well. Physical education is a type of physical activity that can help in the total personality development of a child and in the perfection of the body and mind.

Answer 2.

Sports Training : Sports training is a scientific and systematic way of preparation of sports persons for the highest level of sports performance. Sports training is the main component and the basic form of preparing the sportsperson. It includes all the basic types of preparation of the sportsperson. Sports training is a physical, technical, moral and intellectual preparation for the performance with the help of physical exercises.

According to Matwejew : Sports training is the basic form of preparation of sportsmen.

According to Hardial Singh (1993) : Sports training is a pedagogical process, based on scientific principles, aiming at preparing sportsmen for higher performances in sports competitions.

Importance of sports training :

1. **Personality development** : Due to sports training the personality of a sports person develops through participation in games and sports.

2. **Physical fitness development** : Due to sports training the physical fitness of a sports person develops through participation in different activities, exercises or sports and games.

3. **Skill or technique development** : Due to sports training the skill or technique of a sports person develops.

4. **Tactical development** : Tactical training is designed to improve various strategies and is based upon analysis of the tactics of opponents. Specific training in sports helps the sports persons to make the best use of their abilities and techniques so that the chances of success in competition are increased.

5. **Mental training** : Improvement of mental abilities is critical for high level performance in any sport. The training aims at development of positive attitude towards sports and competition, dedication and devotion towards particular sport or event, sincerity and honesty, self-confidence and optimum level of aspiration.

Answer 3.

Netaji Subhas National Institute of Sports commonly known as National Institute of Sports (NIS) is the academic wing of the Sports Authority of India (SAI) and Asia's largest sports institute located in the city of Patiala. Dr. K.L. Shrimali, the then Union Education Minister opened the National Institute of sports on 7th may, 1961. The NIS was renamed "Netaji Subhas National Institute of Sports" on the 76th birth anniversary of Netaji in January 1973.

The main function of NSNIS is to promote sports in the country and to produce high calibre coaches. NSNIS gives sports scholarships like the sports talent search scholarships scheme to school students (10+2), college or university students, national women championship, who are doing Ph.D, M.Phil, MS Diploma in sports coaching.

Aims and objectives of NSNIS

1. The main aim and objective of NSNIS is to give a scientific orientation to the coaching profession in the country.

2. Another main aim and objective of NSNIS is to provide systematic and scientific coaching to our national teams in various sports disciplines for participation in international competitions.

3. To provide scholarships to the players to encourage them towards sports.

4. To help 'Nehru Yuva Kendra' in order to encourage games and sports.

5. To train the teams for Olympic Games, Commonwealth games, and Asian games.

6. To organise the National Sports Championship for women.

7. To organise rural sports tournaments.

Answer 4.

The main advantages and disadvantages of the knock-out tournaments are :

Advantages of knock-out tournaments

1. There is economy of expenditure.

2. The tournament can be finished in a short period of time.

3. Minimum numbers of officials are required.

4. Helpful in enhancing standard of sports.

5. The competition is keen and intense because of the fear of elimination of a team from the tournament, the moment it is defeated.

Disadvantages of knock-out tournaments

1. There are many chances of elimination of good teams in preliminary rounds.

2. There are many chances of weak teams to enter in the final round.

3. Spectators may not have enough interest in the final round.

4. A winner of a particular round may have to wait to meet the winner of another match of the same round or the previous round.

Other three types of tournaments

1. **League or Round Robin tournament**
   - (i) Single league tournament.
   - (ii) Double league tournament.
2. **Combination tournaments**
   - (i) Knock out cum knock out tournament.
   - (ii) Knock out cum league tournament.
   - (iii) League cum league tournament.
   - (iv) League cum knock-out tournament.
3. **Challenge tournaments**
   - (i) Ladder tournament.
   - (ii) Pyramid tournament.

Answer 5.

(a) **Drug abuse :** Chemical substances that excite a person mentally or physically are called drugs. Substance abuse, also known as drug abuse, is use of a drug in amounts or by methods which are harmful to the individual or others. Drug abuse is the non-permissive consumption of certain substance that may lead to physical and psychological dependence. It is a form of substance-related disorder. A drug is something that affects our body. Drugs pass through the body and interfere with brain's neurotransmitters.

An ergogenic aid is defined as something that improves physical work performance. Drugs taken by an athlete helps him/her run faster, jump higher, and prolong the time to reach total fatigue. Physiologically drugs increase alertness, reduce sense of fatigue, increase blood pressure and heart rate, elevate blood sugar and increase muscle tension. Commonly abused drugs are marijuana, cocaine, heroin, alcohol, prescription drugs etc.

(b) Health education is a profession of educating people about health. Health education can be defined as the principle by which individuals and groups of people learn to behave in a manner conducive to the promotion, maintenance, or restoration of health. Health education is a matter of public relations stating the activities of health department, and to many others it provides knowledge about health and diseases.

Objectives of health education

1. To develop the health attitudes.
2. To educate people about the preservation and development of their physical, mental and emotional health.
3. To improve general conditions of living in the community.
4. To eradicate the diseases through health drive programmes.
5. To develop health consciousness in the school and in the community.
6. To influence the parents and other adults through the health educative programme for better habits and attitudes in children.
7. To educate people about superstitions and myths regarding safety measures of health.

Answer 6.

(a) A physical or mental condition that limits a person's movements, senses or activities is known as disability.

Various causes of disability are :

1. Illnesses like cancer, heart attack or diabetes cause the majority of long-term disabilities. Back pain, injuries, and arthritis are also significant causes.
2. Lifestyle choices and personal behaviour that lead to obesity are becoming major contributing factors.
3. Another main cause of disability is low quality of parental (early childhood) care.
4. Violence, conflict, traffic and occupational accidents cause injuries, which also lead to disability.
5. Another main cause of disability is accidents. Accidents cause a wide range of handicaps, including spinal damage and loss of limbs.

(b) **First aid treatment for sprains and strains**

The first aid treatment for sprain and strain is a five part programme. The programme of treatment is called "RICER".

1. **R** = REST (Rest is required after sprain and strain injury).
2. **I** = ICE (Ice-packs reduce swelling and pain).
3. **C** = COMPRESSION (It limits swelling, it is simply pressing the injured part with little force).
4. **E** = ELEVATION (Elevation of injured part above level of heart helps to drain excess fluid with force of gravity).
5. **R** = REFERRAL (it means refer patient to hospital in severe condition).

Answer 7.

(a) **Importance of correct posture**

1. Correct posture keeps bones and joints in the correct alignment so that muscles are being used properly.
2. Good posture of an individual conveys good impression of his well being.
3. Good posture reflects the alertness, activeness, agility and wholesomeness of an individual's personality.
4. Good posture allows the minimum or economical use of energy, effort and time.
5. Good posture enables an individual to follow good health habits so as to lead a healthy life.
6. Good posture helps in developing the strength, physical fitness and athletic ability in an individual.
7. A person having good posture is able to draw attention towards him and is able to gain social acceptability, recognition and social efficacy easily.
8. Correct posture helps to decrease the abnormal wearing of joint surfaces.

(b) In first aid treatment for any injury the RICE stands for :

1. **R** = REST (Rest is required after injury).
2. **I** = ICE (Ice-packs reduce swelling and pain).

3. C = COMPRESSION (It limits swelling; it is simply pressing the injured part with little force).
4. E = ELEVATION (elevation of injured part above level of heart helps to drain excess fluid with force of gravity).

SECTION–B

FOOTBALL

Answer 8.

(a) Five offences committed by a football player for which a Red card is shown are :

1. **Misconduct :** Misconduct will result in the player being dismissed (sent off) from the field indicated by a red card.
2. **Serious foul play :** Serious foul play is a foul committed using excessive force (*i.e.,* the player is deliberately or attempting to injure his/her opponent while challenging for the ball)
3. Using offensive, insulting or abusive language.
4. Deliberately doing handball during play.
5. Assaulting the referee during play.

(b) (i) **Corner kick :** A corner kick is awarded to the opposing team when a player kicks the ball over his own goal line. The corner is taken from within the quarter circle of the corner nearest to the spot where the ball went out of the field of play. For a corner kick, all opposing players must be at least 10 yards from the ball.

(ii) Throw in : A team is awarded a throw-in when the ball goes over the side line, last touched by an opponent. A player throws the ball in over his/her head, with both feet on the ground at the moment of release. At least part of each foot must be on or behind the sideline. If the throw-in is done incorrectly, the opposing team is awarded a throw-in.

(c) **Time lost :** It is the wastage of time during the game. Extra time in football is made up the time lost in activities other than playing football like substitutions, during fouls, during injuries and showing cards like yellow and red card etc. This is calculated by the match official outside the ground. The referee is the official timekeeper for the match. The referee during the match stops its watch every time the play stops, but the clock goes on and when the 90 minutes are reached in the referee watch, the match official outside play field shows extra time through LED board.

(d) Five occasions when the referee blows the whistle during the game are :

1. When the game starts.
2. During handball.
3. During throw-in.
4. During corner kick.
5. During penalty kick.

(e) (i) A team is awarded a penalty kick when an opposing player commits an intentional foul within their opponent's penalty area. Also penalty kick is awarded to the opponent's team whose any player had handball in opponent's penalty area. All players except the kicker and the goalkeeper must stand outside the penalty area, at least 10 yards from the ball.

(ii) In front of each goal is an area known as the penalty area. The penalty kick is taken from the penalty spot. The penalty kick is taken 11 metres from the centre of the goalpost which is known as penalty spot.

CRICKET

Answer 9.

(a) A match referee is an official appointed to oversee professional cricket matches. The match referee's responsibility is to ensure that the ICC cricket code of conduct is upheld during the game, to assess any breaches of the code, and to hand out penalties for any breaches. Match referees give ratings to the quality of the pitch after the match and submit it to the ICC.

A match referee remains off the field of play at all times during the actual play of the game, observing events from a spectator area. The referee makes no decisions of any relevance to the play or result of the game; such decisions are the sole responsibility of the appointed umpires. Following every game, the match referee composes and submits a match report to the ICC, nothing any events or actions by players or umpires that may be a concern in terms of the code of conduct or the laws of cricket. Match referees are frequently former cricket players who have had distinguished careers on the field.

(b)

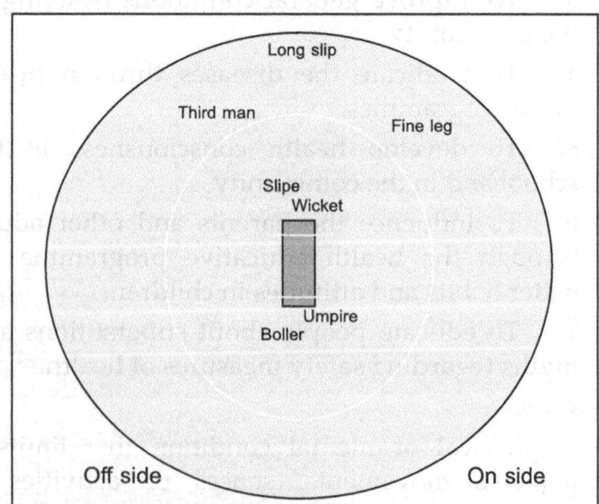

(c) The five situations when a no ball is called by an umpire are :

1. The ball is not bowled, but thrown.

2. The bowlers back foot does not land within the area bounded by bowling and return crease extended or it touches the return crease or its extension.

3. A part of the front foot, whether raised or grounded is not behind the popping crease at the time of delivering the ball.

4. The bowler does not indicate the change in his mode of delivery and whether he wants to bowl over or round the wicket.

5. Attempts to run out a striker or non-striker, before delivering a ball. If a non-striker attempts to steal a run during his run up before delivery, then he can do so.

(d) 1. **Leg bye :** A leg bye is a run scored by the batting team if the batsman has not hit the ball with their bat, but the ball has hit the batsman's body or protective gear. If the ball deflects off the batsman's body they can attempt to score runs in a similar manner as if they had hit the ball. The numbers of runs scored are scored as leg byes; they are added to the team's total, but not to the number of runs scored by the batsman or to the runs conceded by the bowler. Umpires signal a leg-bye by touching a raised knee or leg.

2. **Leg Before Wicket (LBW) :** When a delivered ball strikes the batsman's leg, pad or body, and the umpire judges that the ball would otherwise have struck the stumps.

Leg before wicket is one of the ways in which a batsman can be dismissed in the sport of cricket. Following an appeal by the fielding side, the umpire may rule a batsman out LBW if the ball would have struck the wicket, but was instead intercepted by any part of the batsman's body (except the hand holding the bat). The umpire's decision will depend on a number of criteria, including where the ball pitched, whether the ball hit in line with the wickets, and whether the batsman was attempting to hit the ball.

(e) The Duckworth-Lewis-Stern method (DLS) is a mathematical formulation designed to calculate the target score for the team batting second in a limited over's cricket match interrupted by weather or other circumstances. The method was devised by two English statisticians, Frank Duckworth and Tony Lewis and was formerly known as the Duckworth-Lewis method (D/L). it was introduced in 1997, and adopted officially by the ICC in 1999.

The DLS method is an attempt to set a statistically fair target for the second team's innings, which is the same difficulty as the original target. The basic principle is that each team in a limited-over's match has two resources available with which to score runs (over's to play and wickets remaining), and the target is adjusted proportionally to the change in the combination of these two resources.

HOCKEY

Answer 10.

(a) (i) **Flick :** This technique is used for penalty strokes. It is similar to push but the ball is lifted at a low height. The flick is an important movement in the field of hockey, in which player shoots or flicks the ball by elevating it with their stick. It is the move that a player uses to shoot penalties, but it is also very useful in normal game play. The way to learn how flick is, to first make sure your that player's body is positioned properly, and then performs the right movements with his body and stick to elevate the ball in certain direction.

(ii) **Long corner :** It is awarded to the attacking team after the ball goes over the end line (not between the goal post) from the stick of the defender. The ball is placed five yard away from the side line over the end line.

(b) **The five International tournaments of Hockey are :**

1. Olympic games
2. World Championship
3. Commonwealth games
4. Men's World Cup
5. Women's World Cup

(c) The basic compulsory equipment used by Hockey players are :

Equipments used by regular ground player are:

1. Proper kit
2. Hockey shoes
3. Long stocking
4. Shin guards
5. Hockey stick

Equipments used by goal keeper are :

1. Helmet
2. Face guard
3. Neck guard
4. Chest guard
5. Elbow guard
6. Hockey gloves
7. Abdominal guard
8. Thigh pads and Leg pads
9. Kicker and Hockey stick

(d) **The four fundamental skills of hockey are :**

1. Holding the stick
2. Dribble
3. Push
4. Hit

Push : This stroke is used to send the ball to shorter distance. Right hand is placed low on the stick and pushes the stick forward while the left hand holds the top of handle. This skill is mostly used to pass the ball to the own team player when she is close.

(e) IHF stands for the Indian Hockey Federation. Indian Hockey Federation was formed on 7th November 1925 in Gwalior. Hockey India is the new governing body of hockey in India. It was formed after Indian Hockey Federation was dismissed in 2008 by IOA. The main duties of Indian Hockey Federation are :

1. The main duty of Indian Hockey Federation is to develop and manage the national teams programs and to be in close coordination with the sports authority of India.

2. Another main duty of Indian Hockey Federation is to train players at sub junior, junior and senior level.

3. The governing body of IHF engages the national coaches with international standard certification programs, and educate them about equipments, technical officiating and train them for umpiring in international level. Indian Hockey Federation is also responsible for inbringing top international hockey events such as FIH, men's world cup in 2010, hockey India is dedicated to getting our National team both men and women into top three in the world.

BASKETBALL

Answer 11.

(a) 1. **Blocking :** It is an illegal personal contact which impedes the progress of opponent player with or without the ball. Blocking occurs when a defensive player legally deflects a field goal attempt from an offensive player to prevent a score. The defender is not allowed to make contact with the offensive player's hand (unless the defender is also in contact with the ball) or a foul is called.

2. **Charging :** It is a personal contact with or without ball, by pushing, illegal guarding the opponent. It is preventing defense from rear of players or back side of player. A charge, or player-control foul, occurs when a dribbler charges into a defender who has already established his position.

(b) Technical equipments used by the table officials are :

1. Game clock and stop watch.
2. Score sheet.
3. Whistle-operated stop clock systems.
4. 24 second device.
5. Team foul markers. (20 cm wide and 35 cm high) Team foul indicator.

(c) 1. **Five Second Rule :** A closely guarded player holding the ball should pass, roll, bounce or shoot the ball within five seconds. Failure to do so is a violation. In basketball, the five-second rule, or five second violation, is a rule that helps promote continuous play.

2. **Eight Second Rule :** When a player has the ball in their own half or backcourt, they have 8 seconds to move the ball over the halfway line into the front court. Otherwise they will lose possession. This refers to when the team with possession does not advance the ball out of the back court past the half-court in 8 seconds or less. Once the team inbounds the ball, they have no more than eight seconds to advance the ball past mid-court.

(d) (i) **Time-out :** One time out is allowed in the first, second and third periods and two time-outs in fourth period. Time out last for one full minute in basketball. In sports, a time-out or time-out is a halt in the play. This allows the coaches of either team to communicate with the team, example, to determine strategy or inspire morale, as well as to stop the game clock.

(ii) **Pivot :** A player holding the ball is allowed to step once or more than once in any direction with the same foot, while the other foot, which is called the pivot foot, is kept in contact with the ground during this action.

(e)

	Foul	Team Foul
1.	A player is assessed a personal foul every time he commits a defensive foul. Each player can commit up to 6 personal fouls per game, after which he will already be disqualified from the game.	Team fouls are accumulations of personal fouls (except offensive fouls) committed by players on a team.
2.	A foul is an infringement or breaking of a rule that involves a personal contact with an opponent or unsportsman, like behaviour. A personal foul occurs if a player blocks, holds, pushes, charges, trips, checks the progress of an opponent or uses rough tactics.	Each personal foul committed by a player is also counted against his team; when a team goes over the limit; its opponent is awarded a free-throw.

3. Blocking : A personal contact which checks the progress of an opponent.	In America's NBA, the limit is five fouls-personal or technical-in any one period, after which the opposing team get two free throws.
4. Charging : Pushing or contacting an opponent's torso.	When a team has committed seven players fouls, personal or technical in one half, then for all further player fouls one and one rule comes into operation.
5. Guarding from behind : That results in personal contact.	According to this one and one rule, for all subsequent player fouls, the player against whom the foul has been committed is given one free throw. If this free throw is successful he gets one more free throw. But if he fails to score in the first chance he is not given the second free throw and the game is continued as per rules.
6. Holding : It is holding or catching an opponent to check his movement.	
7. Illegal use of hands : In trying to play the ball, contacts an opponent with hands.	

VOLLEYBALL

Answer 12.

(a) (i) Round arm service : The server stands sideways with one foot ahead. The hitting arm is given rotation as blowing action of cricket. In this service ball is hit hard with open palm.

(ii) Tennis service : The server can stand any where behind the 9 meter service line. The player executes it, as a spike behind the service line. If it is hit well, the serve is a powerful offensive skill. Other variations of tennis serve are top spin, reverse spin or floating serve.

(b)

S. No.	Smash	Serve
1.	It is a powerful attack which is executed by the spiker who has an excellent sense of balance in the air and can perceive and anticipate the actions and positions of the opposing team members. Contact with the ball takes place above the net. A player specialized in smash is known as spiker.	Serve is the act of putting the ball in play by a player. It sounds simple, but actually it describes one of the most important plays in the game of volleyball.
2.	After lifting the ball when it comes down near the net and the ball is hit powerfully with open hand.	The ball is held in the left hand to the front and right side of the body so that it will be directly in line with the right hand.
3.	There are many types of smashing *i.e.,* one-man spike, two men spike or complex spike.	The service can be done by the following methods : **(i) Underhand service :** Mostly beginners use this technique. By taking one foot ahead, the ball is hit by the heel of the hand. **(ii) Round arm service :** The server stands behind the sideways with one foot ahead. The hitting arm is given rotation as blowing action of cricket. In this service, the ball is hit hard with open palm.

(c) Five international tournaments of volleyball are :

1. World Volleyball Championship.
2. Volleyball World Cup.
3. Olympic Volleyball Championship.
4. Asian Volleyball Championship.
5. Commonwealth Volleyball Championship.

(d) Five types of game delay are :

1. Delaying a substitution.
2. Prolonging other interruptions, after having been instructed to resume the game.
3. Requesting an illegal substitution.
4. Repeating an improper request.
5. Delaying the game by a team member.

(e) (i) Rally : A rally is a time between the serve and the end of the play. The team that wins a rally wins a point, moreover wins the right to serve at the same point. The rally ends when the ball touches the ground or when it is hit out of bounds or when the player does not return it legally.

(ii) Spike : It is powerful attacks, a player who execute the spike has an excellent sense of balance in the air and can perceive and anticipate the actions and positions of the opposing team members. Contact with the ball takes place above the net. A player specialized in smash is known as spiker.

TENNIS

Answer 13.

(a) The rarer multiple reasons under which a player, while serving during a game of tennis may make foot fault are :

1. A foot fault takes place when the server assumes an illegal position while serving.
2. Illegal release of the ball.
3. If the ball before bouncing, hits any object other than the net cord, the opponents racket or body, it is a fault.

(b) The first player to reach seven points wins the tie-break and the set. But if the score reaches six-points all, the winner is the first player to win two points in a row. The player whose turn it was to serve in the set serves the first point of the tie-break.

This is sometimes played instead of third set. A match tie-break is played like a regular tie-break but the winner must win ten points instead of seven. The tie-breaker consists of an odd number of points with two point margin required. But if the score reaches six points all, the winner is the first player to win two points in a row. The player whose turn it was to serve in the set serves the first point of the tie-break.

(c) Fundamental skills of tennis are given below :

1. Serve **2.** Ground stroke **3.** Volley
4. Lob **5.** Smash **6.** Grip

Serve : This is the stroke that starts the game. A serve or more formally a service in tennis is a shot to start a point. The serve is initiated by tossing the ball into the air and hitting it into the diagonally opposite service box without touching the net. The serve may be hit under or overhead. The server stands behind the baseline and has two tries to send the ball into the diagonally opposite service court. The first serve in the match is made from the right side of the court, and the players then change sides after each other.

(d) The officials who are required to conduct a tennis match are :

1. Line Umpires **2.** Chair Umpires
3. Community Official **4.** Referees
5. 1 to 4 ball boys

(e) (i) The final set of a tennis match is called advantage set. The set does not have a tie breaker instead; the player has to maintain difference of two games after winning 6 games. The set continues till one of them achieves the target.

(ii) A stroke in which ball goes diagonally in opponent's court. A cross court shot is a shot hit from the left or right side of one player's court to the left or right side of the other player's court from each players own point of view, so that it crosses the lengthwise centreline of the court.

BADMINTON

Answer 14.

(a) 1. Duties of line judge : Badminton line judge's role is to judge whether the shuttle has landed in or out of the court, if the shuttle lands near the lines they are assigned to. If the line judge is unsighted for instance of a player they should signal the umpire that, they did not see where the shuttlecock landed.

2. Duties of Service Judge : To check the service fault of the serving player, like waist fault, foot fault, line cut, service court error. He changes the shuttle on demand of player; he makes the cleaning of surface in case of sweat falling over the court. He observes the need of break for cleaning sweat from the racket. He assists match umpire in taking decision.

(b) Five faults a badminton player may commit during the game are :

1. The shuttle after being hit by that player or side, lands outside the boundaries of the court.
2. Shuttle is struck twice by a player or in a succession by two players of the same side.
3. Shuttle passes through or under the net or touches the person or dress of a player.

4. Touches the roof or ceiling.
5. If the shuttle is caught by hand or held on the racket.

(c) A match consists of the best of 3 games of 21 points. Every time there is a service, there is a point scored. The side winning a rally adds a point to its score. At 20 all, the side which gains 2 points lead first, wins that game. The side winning a game serves first in the next game.

A rally is lost if the shuttle is hit into the net, or over the net but outside of the opponent's court. A rally is also lost if the shuttle touches the player's clothing or body, or if it is hit before it crosses over the net.

(d) Duties of a service judge are :
1. Calling service faults.
2. Keeping hold of and handling out shuttles.
3. Checking shuttle speeds and the height of the net.
4. Making sure the position of the poles is correct.
5. Service judge watches over the server and checks whether the serves made are legal. He changes the shuttle on the demand of the player.

(e) Important tournaments of Badminton are :
1. Thomas Cup
2. Uber Cup
3. All England Championship
4. Asian Championship
5. World Championship

SWIMMING

Answer 15.

(a) **Duties of starter :**

1. Starter is responsible for sending swimmers off the block and making sure each swimmer leaves on time.
2. The starter shall have power to decide whether the start is fair, subject only to the decision of the referee.
3. The starter has the duty to penalize the swimmer for a false start.
4. The starter controls the swimmers from the time the referee turns them over to when the race begins.

Duties of chief inspector of turns :

1. The chief inspector of turns shall ensure that inspectors of turns fulfil their duties during the competition.
2. The main duty of chief inspector of turns is to observe the inspectors of turns at the relevant end of the pool.
3. Chief inspector of turns ensures swimmers comply with the rules for turning and also the starting and finishing rules. The rules differ for each stroke, so inspector of turns needs to know the rules for each swimming stroke.

(b) The rules related to the conduct of races are :
1. All individual races must be held as separate gender events.
2. A swimmer swimming over the course alone shall cover the whole distance to qualify.
3. The swimmer must remain and finish the race in the same lane in which he or she started.
4. Pulling on the lane rope is not allowed.
5. From the beginning of the first arm stroke after the start and each turn, the body shall be kept on the breast. Under water kicking on the side is allowed. It is not permitted to roll onto the back at any time.
6. At each turn and at the finish of the race, the touch shall be made with both hands simultaneously, at above or below the water surface.

(c) In Medley relay events, swimmers will cover the four swimming strokes in the following order :

1. **Backstroke :** When using a backstroke ledge at the start, the toes of the both feet must be in contact with the end wall or face of the touchpad. By that point, the head must have broken the surface.
2. **Breaststroke :** After the start and after each turn, the swimmer may take one arm stroke completely back to the legs during which the swimmer may be submerged.
3. **Butterfly :** Both arms shall be brought forward simultaneously over the water and brought backward simultaneously under the water through-out the race.
4. **Freestyle :** Freestyle means that in an event so designated the swimmer may swim and style, except that in individual medley or medley relay events, freestyle means any style other than backstroke, breaststroke or butterfly.

(d) The Swimming Federation of India (SFI) is the national governing body for aquatic sports in India. The SFI was formed by the manager of the National Swimming Association (NSA) and the Indian Swimming Federation (ISF) in 1948. Prior to the merger, the NSA and ISF had been engaged in disputes. While the Calcutta based NSA received affiliation from FINA in 1923-33, the ISF had the support of the Indian Olympic Association. The union government intervened to resolve the dispute by merging the two entities to form the SFI.

In may 2019, a new set of office-bearers took charge at the swimming federation of India (SFI) AGM held, with R.N. Jayaprakash of Tamil Nadu and Gujarat's D. Chokshi Monal being unanimously elected as President and

Secretary respectively. The SFI currently oversees competition in the sports of swimming, masters swimming, synchronized swimming, diving, high diving and water polo. It is affiliated to FINA (International Swimming Federation), and the Asian Swimming Federation (ASF).

SFI regularly conducts clinics and training programmes for its coaches and technical team. Supported by FINA global experts execute these clinics provided. SFI has provided FINA scholarships to swimmers across the years. The scholarships include international training and support for event participation.

(e) Events which start with dive :

1. **Freestyle :** 50 m, 100 m, 200 m, 400 m, 800 m and 1500 m.
2. **Backstroke :** 100 m and 200 m.
3. **Breaststroke :** 100 m and 200 m.
4. **Butterfly :** 100 m and 200 m.

Events which start within the water

1. 5 km events.
2. 10 km events.
3. 25 km events.

ATHLETICS

Answer 16.

(a) Competitors take their throw from inside a circle 7 feet (2.135 metres) in diameter, with a toe board approximately 10 cm (4 inches) high at the front of the circle. The distance thrown is measured from the inside of the circumference of the circle to where the shot lands at its nearest disturbance of the soil. The diameter of the shot for men should be 110 to 130 mm, and for women the diameter of shot should be 95 to 110 mm.

The shot put stop board or toe board should be 10 centimetres high or 98 to 102 mm high, and measure 1.21 to 1.23 meters in length and 0.112 meters in width or 112 mm to 300 mm wide.

(b) **1. Sector :** It is an area bounded by two radii and the included arc of a circle. The sector lines are laid from the centre point on the runway through the cross points of the throwing arc and the lines of the runway. The length of the sector is 100 m. At this distance the inner edges of the sector lines are 50 m apart. The shot must land in the legal sector (34.920) of the throwing area.

2. Lane : A standard (or regulation) athletic track has length of 400 m and 8 lanes. Lanes are bounded with lines; each lane should be 1.22 m wide and marked by lines 5 cm in width. All lanes should be of the same width. The runners should run between the lanes in which lane they start first, if runner will go outside the lane or disturb other runner he will be disqualified.

(c)

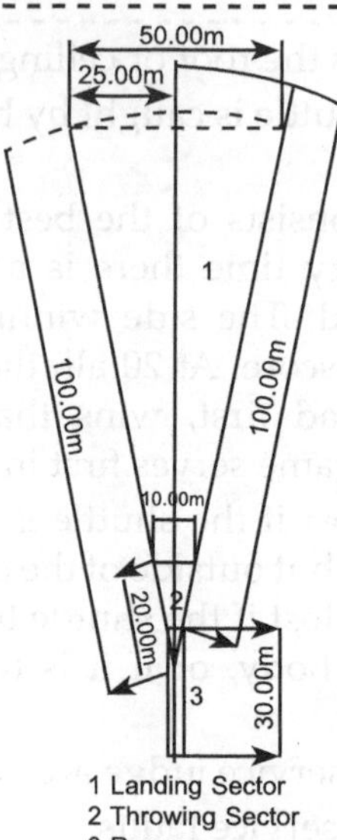

(d) In order to learn proper technique of long jump, the following skills or steps should be mastered :

1. **Check marks :** Marks or objects placed alongside the width of the runway for guidance in the approach run.
2. **Initial stance :** Position of the body at the marked starting point.
3. **Approach run :** The run from the starting point up to the take off board.
4. **Take off :** This means taking a jump in the air. Technically, it is the conversion of the momentum gained by a jumper into a thrust of the body upward and forward.
5. **Flight in the air :** It means the movements made by a jumper while he is air borne. These movements may be of paddling in the air, hang style or of a hitch kick style.
6. **Landing :** Falling in the pit.

(e) Five international competitions in athletics are :

1. **Olympic Games :** Although many games competitions are organised as a part of Olympics, track and field competitions are the star attraction of these games.
2. **Asian Games :** The first Asian games were held in New Delhi on 4th March, 1951. The then President of India Dr. Rajendra Prasad had inaugurated these games.
3. **Commonwealth Games :** The commonwealth games are an international multi-sport event involving athletes from the Commonwealth of Nations. The event was first started in 1930.
4. **World Championships :** The primary competitions where all nations may compete.
5. **IAAF Diamond League :** The diamond league is an annual series of elite track and field athletic competitions comprising 15 of the best invitational athletics meetings.

●●

PHYSICAL EDUCATION 2019

QUESTIONS

SECTION–A (40 Marks)

*Answer **any five** questions.*

Question 1.

What is meant by warming up ? Write the advantages of warming up. **[8]**

Question 2.

(a) *Explain the merits and demerits of league tournament.* **[8]**

(b) *Draw a league fixture for eight teams according to cyclic method.*

Question 3.

Discuss the importance of health education for adults and younger generation. **[8]**

Question 4.

Mention the causes and corrective measures for the following : **[8]**

(a) *Lordosis* **(b)** *Flat foot*

Question 5.

Write in detail the effects of insufficient sleep, on human body. **[8]**

Question 6.

List the various ill effects of alcohol on an individual. **[8]**

Question 7.

Explain what is contusion. Mention the steps to be followed for its prevention. **[8]**

SECTION–B (30 Marks)

*Select **any two** games from this Section.*

*Answer any **three** of the **five** subparts (a), (b), (c), (d) and (e) from **each** of the **two** games selected by you. Each subpart is for 5 **marks**.*

FOOTBALL

Question 8. **[3 × 5]**

(a) *List any five fundamental skills which a football player must posses.* **[5]**

(b) *What decision is to be taken by the referee if the ball bursts or becomes defective during the course of a match ?* **[5]**

(c) *Explain the difference between tackling and trapping.* **[5]**

(d) *Mention the basic compulsory equipment used by a referee while conducting a match.* **[5]**

(e) *Mention any five duties of Assistant Referees.* **[5]**

CRICKET

Question 9. **[3 × 5]**

(a) *Explain the following :* **[5]**

(i) *Dolly Catch* (ii) *Gully*

(iii) *Seam* (iv) *Hook shot*

(v) *Half-volley*

(b) *Differentiate between popping crease and bowling crease.* **[5]**

(c) *Explain the following terms :* **[5]**

(i) *Ball tampering* (ii) *Sweep*

(d) (i) *Write the full forms of ICC and BCCI.* **[5]**

(ii) *State the function of ICC and BCCI.*

(e) *Under which conditions can an umpire change the ball ?* **[5]**

HOCKEY

Question 10. **[3 × 5]**

(a) *Write a short note on Hockey Stick.* **[5]**

(b) *Mention any five duties of the captain of a Hockey team.* **[5]**

(c) *Write the equipment used by a Goal Keeper.* **[5]**

(d) *When is a penalty stroke awarded ?* **[5]**

(e) *State the difference between a scoop and a stroke.* **[5]**

BASKETBALL

Question 11. **[3 × 5]**

(a) *Explain Three Second Rule and Twenty-four Second Rule.* **[5]**

(b) *What is free throw ? State the steps to be followed for taking a free throw.* **[5]**

(c) *Mention the violations in the game of Basketball.* **[5]**

(d) *What is a double foul in the game of basketball ? What is the penalty for this foul ?* **[5]**

(e) *Explain technical foul and multiple foul.* **[5]**

VOLLEYBALL

Question 12. **[3 × 5]**

(a) *Write the duties of a Volleyball coach.* **[5]**

(b) *Write the service faults in Volleyball.* **[5]**

(c) (i) *What is the procedure adopted to start a game ?* **[5]**

(ii) *List any two National level tournaments in Volleyball.*

(d) *Differentiate between :* **[5]**

(i) *Service zone and substitution zone.*

(ii) *Positional fault and rotational fault.*

(e) *Mention the duties of a line judge.* **[5]**

TENNIS

Question 13. [3 × 5]

(a) *Explain the following :* [5]

(i) *Double fault* (ii) *Chip shot*

(b) *List any five types of strokes in the game of Tennis.* [5]

(c) *What are the duties of a coach in Tennis ?* [5]

(d) *Explain the following :* [5]

(i) *Set* (ii) *Ace*

(iii) *Slice* (iv) *Down the line*

(v) *Drive*

(e) *Explain how a forehand shot is taken.* [5]

BADMINTON

Question 14. [3 × 5]

(a) *Discuss the rule of a service in doubles.* [5]

(b) *What happens if the shuttlecock gets stuck on the net?* [5]

(c) (i) *What is meant by bird in the game of Badminton?* [5]

(ii) *Name the different types of strokes played by a badminton player.*

(d) *Explain the following terms :* [5]

(i) *Side out* (ii) *Waist fault*

(e) (i) *Explain the following shuttle grips :* [5]

(1) *Out of hand grip*

(2) *Mid grip*

(3) *Base grip*

(ii) *What is the weight of a shuttle cock ? How many feathers are there in a shuttle cock ?*

SWIMMING

Question 15. [3 × 5]

(a) *List the freestyle events for men.* [5]

(b) (i) *What is the position of backstroke turn indicators?* [5]

(ii) *Which lane is given to the best swimmer ?*

(c) *Explain the starting stance of a swimmer in the backstroke event.* [5]

(d) (i) *What is a Dolphin kick ?* [5]

(ii) *Describe any two duties of the judges of stroke.*

(e) *Explain the following terms :* [5]

(i) *Tapper*

(ii) *Peddler*

ATHLETICS

Question 16. [3 × 5]

(a) *Explain the event triple jump.* [5]

(b) *What is the difference between marking radius and running radius ?* [5]

(c) *Draw a neat diagram of a Shot-Put sector with all its specifications.* [5]

(d) *List the fundamental skills required by an athlete, for the event of Discus throw.* [5]

(e) *Explain the following terms :* [5]

(i) *Stagger*

(ii) *Athlete*

(iii) *Shoulder Shrug*

(iv) *Heat*

(v) *Lane*

ANSWERS

Answer 1.

Prevention is better than cure is an age old saying. Injury prevention is one of the best ways to ensure that performance doesn't dwindle or drop. There are many ways of preventing injuries. Warming up and cooling down helps in preventing sports injuries.

Warming up is preparing the athletes mentally and physically to adjust the sudden increase in the movement. Warming up is a process by which human body is brought to a condition at which it safely responds to the nerve impulses of the person for quick and efficient action. An insufficient warming up may contribute to muscle or tendon injuries. Components of warming up are jogging, running, stretching, calisthenics, flexibility exercise, games, sports or event related activities.

Muscle should be warmed up :

1. To increase blood circulation.
2. To enable oxygen to reach to the working muscle group.
3. To increase temperature of the muscle.
4. To increase metabolism of the muscle.

Warming up can be divided into two types, namely, Physiological warm up and Psychological warm up.

(i) Physiological Warm Up :

In this warm up a player is prepared physically for the activity. All the systems of the body are made ready to bear stress during the activity.

Psysiological warm up creates following impact on the body :

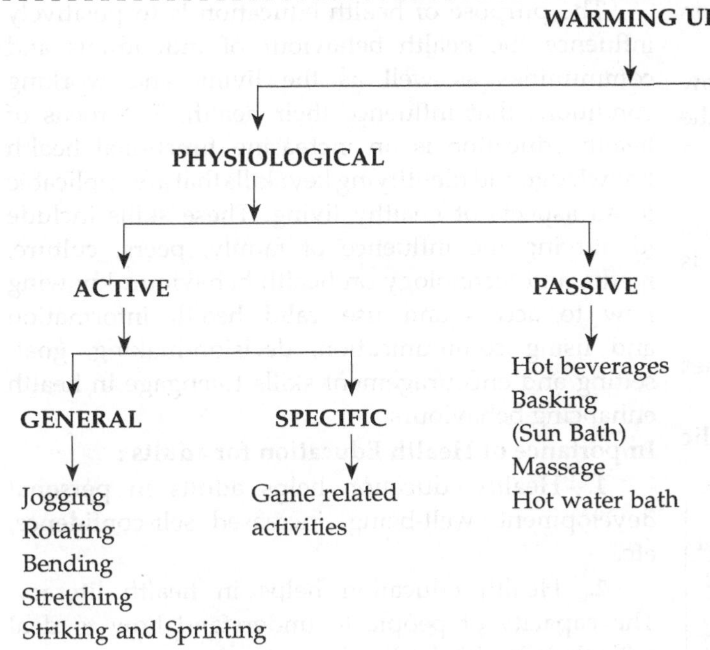

1. Body temperature is raised gradually which in turn increases the blood supply to the muscles. Warming up enhances the body core temperature. If the core temperature of the body is increased by 1°C then basal metabolic rate (BMR) is increased by 14% which enhances the physical work capacity.

2. Haemoglobin (RBC) carries more oxygen, which results in more oxygen exchange between blood or tissue and myoglobin.

3. Propagation of nerve impulses (signals) become faster which improves and sharpens the reaction time.

4. Rate of contraction of muscle becomes faster as the viscosity is lowered. Thus the chances of injuries or wear and tear are minimized.

5. Increases the stroke volume as per demand of muscles to be used in the activity.

6. Increases the lung ventilation, which supplies more oxygen. It results in the removal of lactic acid which helps in improving the endurance.

7. Improves agility, coordination and range of motion (ROM) in the joints.

8. It helps the athlete to enjoy the second wind at the earliest possible.

(ii) Psychological Warm Up :

In this warm up athletes are prepared mentally for the activity. It is very important for all games and sports specially combating games like judo, wrestling, boxing etc. Psychological warm up can be started many days before the competition and can be given during the actual competition. It reduces tension and nervousness. It improves the concentration required for the main task. Through psychological warm up a player is motivated by boosting his or her moral in following ways :

1. Telling the players about his or her capabilities.
2. Telling about the weakness of the opponent.
3. Clarifying his doubts.
4. Reminding about his or her past wins and defeats.
5. Through rewards or punishments.
6. Sometime provoking him.
7. Thumbs up signal.

Answer 2.

(a) In a league tournament each team competes with every other team participating in the tournament. Winners are decided on the basis of the points acquired after all the competitions. For each win a team may be awarded 2 points, for draw 1 point and for a loss no point is awarded. Formulae to find the number of matches in a league tournament = n (n – 1)/2, where n = number of teams participating in the tournament. For example : The number of matches in a league tournament of 8 teams = 8 (8 – 1)/2 = 28 matches. In a single league tournament, every team competes with every other team participating in the tournament once. In a double league tournament, every team competes with every other team participating in the tournament twice.

Advantages of league tournament :

1. Luck factor does not have a role to play as each team gets to play with all the teams.
2. If a good team has lost a match, it still can be winner of the tournament because it has other matches to play.

3. There is less pressure on players to perform well because each match is not crucial.

4. More number of matches for players to show their talent and also for the selectors to spot the talent.

Disadvantages of league tournament :

1. It takes more time to finish the tournament. Number of matches in a league tournament is more compared to knock-out tournament.

2. It is not economical.

3. Each match is not crucial. So, some matches may become boring for spectators.

(b) League fixture for 8 teams according to cyclic method:

1 Round	2 Round	3 Round	4 Round	5 Round	6 Round	7 Round
8 Vs 1	7 Vs 1	6 Vs 1	5 Vs 1	4 Vs 1	3 Vs 1	2 Vs 1
7 Vs 2	6 Vs 8	5 Vs 7	4 Vs 6	3 Vs 5	2 Vs 4	8 Vs 3
6 Vs 3	5 Vs 2	4 Vs 8	3 Vs 7	2 Vs 6	8 Vs 5	7 Vs 4
5 Vs 4	4 Vs 3	3 Vs 2	2 Vs 8	8 Vs 7	7 Vs 6	6 Vs 5

Answer 3.

According to National Conference of Preventive Medicine USA, Health education is a process that informs, motivates and helps people to adopt and maintain healthy practices and lifestyle, advocates environmental changes as needed to facilitate this goal and conducts professional training and research to the same end.

Lawrence Green defined Health Education as "a combination of learning experiences designed to facilitate voluntary actions conducive to health."

Health education is a social science that draws from the biological, environmental, psychological, physical and medical sciences to promote health and prevent disease, disability and premature death through education-driven voluntary behaviour change activities.

Health education is the process by which individuals and group of people learn to promote, maintain, restore health. It is a process aimed at stimulating the wish to be healthy, to know how to stay healthy, to do what they can individually and collectively do to maintain health, and to seek help when needed.

Health education as applied to health issues is a process with intellectual, psychological and social dimensions relating to activities which increase the abilities of people to make informed decisions affecting their personal, family and community well being. This process is based on scientific principles which facilitates learning and behavioural change in both health personnel and consumers, including children and youth.

The purpose of health education is to positively influence the health behaviour of individuals and communities as well as the living and working conditions that influence their health. The focus of health education is on increasing functional health knowledge and identifying key skills that are applicable to all aspects of healthy living. These skills include identifying the influence of family, peers, culture, media and technology on health behaviour; knowing how to access and use valid health information and using communication, decision-making, goal-setting and encouragement skills to engage in health enhancing behaviours.

Importance of Health Education for adults :

1. Health education helps adults in personal development, well-being, increased self-confidence, etc.

2. Health education helps in health literacy. The capacity of people to understand how to deal with their health (indications of a doctor, instruction leaflets on medicines, information leaflets on illnesses, etc.) helps in better health of adults.

3. Health education helps adult to have better understanding of the preventive measures and treatment for communicable and non-communicable diseases.

4. Health education helps in promotion of healthy lifestyles. Courses on nutrition, cooking, yoga and gym classes, etc., help the adults to have healthy concepts for having healthy lifestyle.

5. Helps in promoting the proper use of health services available for adults.

6. By focusing on prevention, health education reduces the costs of adults (both financial and human) that individuals would spend on medical treatment.

Importance of Health Education for young generation:

1. Helps in instilling behaviours into young generation to prevent or delay the onset of the leading causes of death.

2. Helps in promoting the proper use of health services available for young generation.

3. Health education builds students knowledge, skills, and positive attitudes about health. Health education teaches about physical, mental, emotional and social health. It motivates students to improve and maintain their health, prevent disease, and reduce risky behaviours.

4. Health education curriculum and instruction help students learn skills they will use to make healthy choices throughout their lifetime. It provides logical reasons for avoiding alcohol, tobacco, and other drugs.

5. It helps to inform the young generation regarding the importance of injury prevention, having good physical and emotional health, proper intake of nutrition, physical activity, prevention of diseases and sexuality and family life.

6. Health education promotes learning in other subjects. In general, healthy students learn better. Numerous studies have shown that healthier students tend to do better in school. They have higher attendance, have better grades, and perform better in tests.

Answer 4.

(a) Lordosis : Lordosis is a medical term used to describe an inward curvature of a portion of the vertebral column. It is an increased forward curvature in the lumber region and is the main cause of problems in walking and standing. It can affect the lower back and neck. This can lead to excess pressure on the spring causing pain and discomfort. Lordotic curves refer to the inward curve of the lumber spine (just above the buttocks).

Names of the yoga exercise which help in treatment of Lordosis : Paschimotanasana (Forward Bend Posture), Padahastasana (Standing Forward Bend Posture), Janu Sirsasana (Head to Knee Posture) Halasana (Plough Posture) Naukasana (Boat Posture) etc.

Following exercise can be performed as a remedy for Lordosis:

1. From standing position bend forward from hip level. Repeat this exercise ten times.

Causes	Symptoms	Treatment
1. Birth defect.	Stomach is outward.	1. Surgery.
2. Deficiency of the vitamin-D, calcium, etc., during childhood.	The lower back dips inward.	2. Following correct posture of standing, sitting and lying.
3. Misaligned healing of a spinal fracture or injury to the supporting ligaments of the spine.	Back pain or strain.	3. Treatment by exercise (a) Flexibility exercise for lower back muscle which is strong. (b) Strengthening exercise for muscle of stomach which is weak (forward bending exercise). (c) Abdominal crunches. (d) Place the pillow under the stomach while sleeping in a prone lying position.
4. Habits from everyday activities such as lifting bag, bucket filled with water etc. using one hand.		
5. Obesity can put uneven pressure on the spine, causing curvature.		
6. Pregnancy period can result in Lordosis.		
7. Abdominal muscle of the body is weak and lower back muscle is strong.		

2. Lie down on back, raise head and legs simultaneously for 10 times.

3. Perform sit-ups regularly.

4. Halasana should be performed regularly.

5. Lie down in supine position, *i.e.*, on back, then raise legs at 45 degree angle. Remain in this position for sometime.

6. Toe touching should be performed.

7. Sit down and extend your legs forward. Try to touch forehead to knees. Repeat this exercise 10 times.

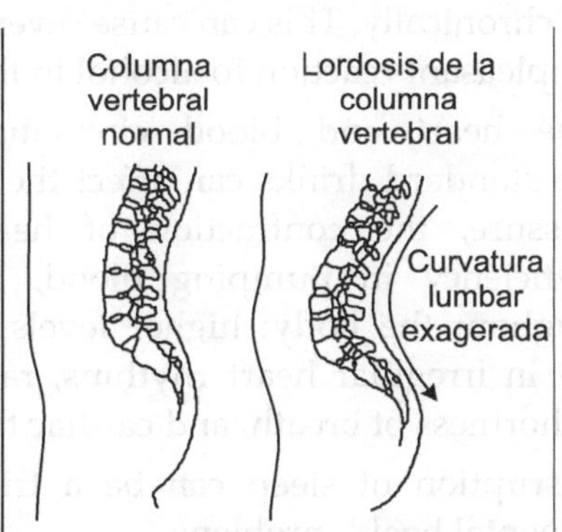

(b) Flat Foot : Flat foot is one physical deformity in which the medial longitudinal and transverse arches of foot are depressed and medial border of foot comes in contact with ground. Flat foot deformity can be diagnosed with a very simple test called as 'Wet Foot test'. In this deformity the arch of the foot collapses, with the entire sole of the foot coming into complete or near-complete contact with the ground.

Causes	Symptoms	Treatment
1. Birth defect.	Difficulty in running and walking.	1. Surgery.
2. Overweight or obesity Deficiency of the vitamin D, calcium, etc., during childhood.		2. Standing on toes.
		3. Walking on toes.
		4. Picking small stone with toes and placing it a little far.

Names of the yoga exercise which help in treatment of Flat foot : Tadasana (Tree posture), Utkatasana (Chair posture), etc.

Following exercises can be performed as a remedy for Flat Foot deformity :

1. Walking on heels.
2. Walking on inner and outer side of feet.
3. Walking on toes.
4. To perform up and down the heels.
5. Jumping on toes for sometime.
6. To skip on rope.
7. To perform Vajrasana, the yogic asana.

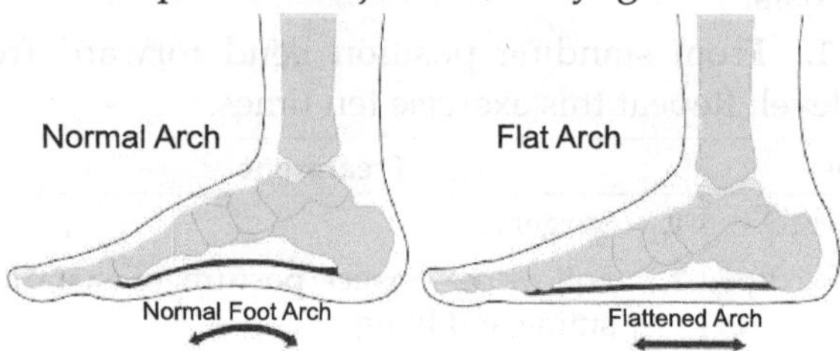

Flat foot

Answer 5.

Sleep plays an important role in the physical health of people. Sleep deprivation occurs when an individual fails to get enough sleep. The amount of sleep that a person needs varies from one person to another, but on an average most adults need about seven to eight hours of sleep every night in order to feel well rested and alert. Effects of insufficient sleep, on human body are as follows :

1. Increases the risk of heart disease, kidney disease, high blood pressure, diabetes, and stroke. Sleep deficiency also increases the menace of obesity. One study of teenagers showed that with each hour of sleep lost, the chances of becoming obese went up. Sleep deficiency increases the risk of obesity in other age groups as well.

2. Sleep helps to maintain a balance of the hormones that regulates appetite. Sleep also affects the hormone that controls the blood glucose level and its deficiency may increase the risk for diabetes.

3. Sleep also supports healthy growth and development. Deep sleep triggers the body to release the hormone that promotes normal growth in children and teens. This hormone also boosts muscle mass and helps repair cells and tissues in children, teens, and adults. Prolonged sleep deficiency can change the way the immune system responds.

4. Sleep disturbance can also be an indication of other underlying medical conditions or psychological disorder. It may, for instance, be a symptom of an Anxiety or Mood Disorder. Furthermore, sleep disturbance that is chronic or severe can bring on or exacerbate serious psychological disorder.

Answer 6.

Alcohol has some immediate or short-term physical effects on the body as a whole, including the brain, the gut and pancreas, the heart and circulation, sleep control and sexual functions. The most obvious and immediate effects of alcohol are on the brain. The effects of alcohol on the brain are felt within about five minutes of alcohol being swallowed. Alcohol slows down or blocks many of the brain's functions. The initial effect might be a reduction in tension or inhibitions, making the drinker feel more relaxed or excited. However, as the intake of alcohol increases, these effects are counter balanced by less pleasant effects, such as :

1. Drowsiness
2. Loss of balance
3. Poor coordination
4. Slower reaction times (critical when driving or operating machinery)
5. Slurred speech
6. Slowed thought processes
7. Nausea and vomiting

As more alcohol is consumed and the blood alcohol concentration (BAC) rises, performance and behaviour deteriorate progressively. The risk of accident, violence or injury increases rapidly with increasing blood alcohol levels. If the BAC reaches a high enough level, it can lead to life threatening events such as :

1. Unconsciousness.

2. Inhibition of normal breathing may be fatal, particularly as the person may vomit and can suffocate if the vomit is inhaled.

3. The pituitary gland at the base of the brain, which controls the body's fluid balance, can leave the person dehydrated and with a headache.

4. The gut, by irritating the lining of the gut as the level of alcohol is increased, results in diarrhoea as a possible outcome.

5. The pancreas, which may become inflamed, sometimes chronically. This can cause severe pain and a highly unpleasant reaction to alcohol in the future.

6. The heart and blood circulation, where one or two standard drinks can affect the heart rate, blood pressure, the contraction of heart muscle and its efficiency in pumping blood, and blood flow throughout the body; higher levels of alcohol may result in irregular heart rhythms, raised blood pressure, shortness of breath, and cardiac failure.

7. Disruption of sleep can be a trigger for a variety of mental health problems.

8. Sexual functioning, particularly in men, who may find it more difficult to get an erection after drinking alcohol beyond low-risk levels.

9. Alcohol can increase the level of risk in particular conditions or diseases, including: cancer; cirrhosis of the liver; heart disease and stroke; cognitive problems (including memory and reasoning); and dementia.

Answer 7.

Contusion is an injury caused by a forceful hit by a hard object. It is an internal injury. There is no external cut or bleeding over the skin.

Signs and Symptoms : Swelling, pain, bluish colour appears on the outer surface of the skin.

Management of Injuries

P – Protection : Protect the injured part from further injury. Do not attempt to put pressure or massage the injured part.

R – Rest : Give rest to the injured part and restrict the movement.

I – Ice : Apply ice to the injured part for 15-20 minutes and repeat it 3-4 times in a day if required for first 24-48 hours of injury, after that apply heat for the dispersion of clotted blood.

C – Compression : Compress the injured part with a crape bandage to restrict the movement.

E – Elevation : Elevate the injured part with slings if required.

S – Support : Support the injured part to avoid painful movement until medical assistance is provided.

Injuries are part of the sport and almost every athlete gets injured at one time or another. Most sports injuries can be predicted and prevented with understanding and immediate treatment.

Following ways helps to prevent contusion :

1. Avoiding workout routine mistakes may help to prevent injuries. Workout routine mistakes are like:

(a) Ignoring one's medical condition.

(b) Doing the same workout routine day after day.

(c) Never taking a rest day.

(d) Too much, too soon, too fast, too hard and too often use of muscle.

(e) Missing several workouts in a row and then adding additional mileage to catch up.

(f) Not using the proper technique.

(g) Not wearing the right protective gear.

(h) Not doing conditioning exercises before for strengthening the muscles used in play.

(i) Not performing stretching exercises before and after games or practice.

(j) Not drinking plenty of fluids before, during and after exercise or play. Drinking water prevents dehydration and keeps the muscle flexible.

(k) Not having a rest period during practice and games. Rest period can reduce injuries and prevent heat illness.

(l) Not stopping the activity if there is pain.

(m) Not wearing light clothing.

2. Having adequate amount of calorie and supplements like iron and calcium for the body helps to prevent injury.

3. The player should have a positive approach to game and should look into the game as a means of developing the quality of sportsmanship and hard work. Not having a positive approach to game can create emotional stress and this will affect the physiological process which results in lack of concentration leading to injuries.

4. Immediate first aid care should be provided to the athlete for preventing aggravation of injury. Treatment and rehabilitation can be done through various physiotherapy, treatments and modalities.

5. Proper Rehabilitation should be given to an injury for faster and maximum recovery.

6. Return to activity after full recovery helps to prevent the reoccurrence of injury as the muscles are tunes to take the load.

SECTION–B

FOOTBALL

Answer 8.

(a) Fundamental skills which a football player must possess are :

1. Controlling : To control the ball is to master it. Properly controlling the ball means that a move will be successful. The control movements to focus on are directed control and gathering the ball while moving these will introduce speed into the play.

2. Passing : This is the action of giving the ball to a team-mate. It is an essential part of team play. As the core of the game, passing allows a team to:

(i) Keep possession of the ball;

(ii) Set up attacks;

(iii) Change the direction of play;

(iv) Counter-attack;

(v) Provide a decisive or final pass.

Controlling ball	Passing	Shooting	Dribbling	Heading
Chest trap	High Pass	Instep kick	Zig-zag dribble	Attacking header
Thigh trap	Ground Pass	Out-step kick	Dribble and turn left side	Standing header
Feet trap		Bicycle kick	Dribble and turn right side	
Belly trap		Lofted Kick		

3. **Shooting :** This is an action with the objective of dispatching the ball into the opponent's goal. It is the logical conclusion, the culmination of an attack. It is what football is all about. Shooting requires technical qualities (striking the ball well, accuracy), physical qualities (power, coordination, balance) and mental qualities (determination, audacity, self-confidence).

4. **Dribbling :** This is how an individual moves with the ball when faced by opponents or obstacles. Dribbling allows the player in possession of the ball to eliminate one or more opponents by :

(i) Making a manoeuvre and taking individual risks;

(ii) Setting up a team move;

(iii) Gaining time to allow support from team-mates;

(iv) Deceiving an opponent (the concept of the feint).

(5) Headers : Headers can be associated with mastery of the ball (juggling and control) and passing or shooting the ball. Heading is an essential skill for defending and marking.

(b) Decision to be taken by the referee if the ball burst or becomes defective during the course of a match

1. Play is stopped and restarted by dropping the replacement ball where the original ball became defective.

2. If the ball becomes defective at a kick-off, goal kick, corner kick, free kick, penalty kick or throw-in, the restart is retaken.

3. If the ball becomes defective during a penalty kick or kicks from the penalty mark as it moves forward and before it touches a player, crossbar or goal posts, the penalty kick is retaken.

4. The ball may not be changed during the match without the referee's permission.

(c) 1. **Tackling:** Most forms of football have a move known as a tackle. The primary and important purposes of tackling are to dispossess an opponent of the ball, to stop the player from gaining ground towards goal or to stop them from carrying out what they intend.

2. **Trapping:** Trapping is to gain control over the ball. It is a skill applied to keep possession of the ball. It is a method of gaining control of the ball of a pass or a loose ball. Trapping the ball involves using the feet, thighs or chest to bring the ball to the ground while keeping it in the player's possession. It can include stopping the ball completely or redirecting the ball in a controlled manner. When trying to stop or slow the ball, players cushion the ball as it moves towards them.

(d) Referees must have the following equipment :

1. Whistle(s)
2. Watch(es)
3. Red and yellow cards
4. Notebook (or other means of keeping a record of the match).

(e) Five duties of Assistant Referee : A match official with a flag positioned on one half of each touchline to assist the referee particularly with offside situations and goal kick /corner kick / throw-in decisions:

They indicate when:

1. The whole of the ball leaves the field of play and which team is entitled to a corner kick, goal kick.

2. The whole of the ball leaves the field of play and which team is entitled to a throw-in.

3. Player in an offside position may be penalised.

4. A substitution is requested.

5. At penalty kicks, the goalkeeper moves off the goal line before the ball is kicked and if the ball crosses the line; if additional assistant referees have been appointed the assistant referee takes a position in line with the penalty mark. The assistant referee's assistance also includes monitoring the substitution procedure. The assistant referee may enter the field of play to help control the 9.15 m (10 yards) distance.

CRICKET

Answer 9.

(a) (i) Dolly Catch : A very easy catch taken by a fielder.

(ii) Gully : A close fielder near the slip fielders, at an angle to a line between the two sets of stumps of about 100 to 140 degrees.

(iii) Seam : The stitching on the ball.

(iv) Hook Shot : A shot, similar to a pull, but played so that the ball is struck when it is above the batsman's shoulder.

(v) Half Volley : A delivery that bounces just short of the block hole, usually easy to drive or glance away.

(b) Difference between Popping crease and Bowling crease : Crease is one of the several lines on the pitch near the stumps (the "popping crease", the "return crease" and the "bowling crease").

Note : Measuremets are in centimetres
100 cms = 1 metre
Lines in Black are imaginary only and do not exist as pitch markings during play

122 122
Return Crease
Bowling Crease
Wicket or Stumps
Popping or Batting Crease
366 minimum
51
152
264
152
51
1768
2012
2256

The popping crease has the same length, is parallel to the bowling crease and is 4 feet (1.22 m) in front of the wicket. Although it is considered to have unlimited length (in other words, runs across the entire field) the popping crease must be marked to at least 6 feet (1.83 metres) perpendicular to the pitch, on either side of the middle of the pitch.

The bowling crease lie 22 yards (66 feet or 20.12 m) away, and marks the other end of the pitch. A bowling crease is 8 feet 8 inches (2.64 m) long with the middle stump placed dead centre. For the fielding side, the crease defines whether there is a no-ball because a fielder has encroached on the pitch or the wicket-keeper has moved in front of the wicket before he is permitted to do so.

(c) **(i)** Ball tampering is the unlawful alteration of the surface or seam of a ball on the field, to affect its motion when bowled.

(ii) Sweep is a shot played to a good length slow delivery. The batsman gets down on one knee and "sweeps" the ball to the leg side.

(d) **(i)** ICC—International Cricket Council.

Headquarters : Dubai, United Arab Emirates.

BCCI—The Board of Control for Cricket in India.

Headquarters : Mumbai.

(ii) Functions of the International Cricket Council (ICC)

1. It is responsible for organizing major international tournaments of cricket, in which the "Cricket World Cup" is the main.

2. It employs all umpires and referees, who are responsible for the successful conduct of all Test Matches, One-Day International and Twenty-20 International tournaments.

3. It deals with the code of conduct for cricket, as well as the professional standards of discipline, action against corruption and match fixing.

4. It monitors the cricket playing conditions, the bowling action review, and the other rules of the ICC.

Functions of the Board of Control for Cricket in India (BCCI)

1. BCCI works as a money churning machine which controls and improves the quality and standard of cricket in India.

2. To select teams to represent India in Test Match, One-Day International and Twenty–20 and in any other format in India and abroad as BCCI may decide from time to time.

3. BCCI decides the sponsor, controls IPL, decides with whom and where India will play and also how much of what format will team play.

4. In consultation with various state cricket associations BCCI conducts Ranji, Duleep Trophy, Challenger Series and the other domestic tournaments.

5. To appoint representative in International Cricket Council.

(6) To appoint managers and other team officials for Indian team.

(e) Conditions in which Umpire can change the Ball are :

1. If ball is lost or has become unfit for playing. If during play, the ball cannot be found or recovered or the umpires agree that it has become unfit for playing, the umpires shall replace it with a ball which had wear comparable with which the previous ball had received before the need for its replacement. When the ball is replaced, the umpire shall inform the batsmen and the fielding captain.

2. If the umpires together agree that the condition of the ball has been unfairly changed by a member or members of the fielding side, the batsman at the wicket shall choose the replacement ball from a set of six other balls of various degrees of usage (including a new ball) and of the same brand as the ball in use prior to the breaking.

3. If the umpires together agree that the condition of the ball has been unfairly changed by a member or members of the bowling side, the umpires shall select and bring into use immediately, a ball which shall have wear comparable to that of the previous ball immediately prior to the breaking.

HOCKEY

Answer 10.

(a) Each player carries a stick and cannot take part in the game without it. The stick for an adult is usually in the range 89–95 cm (35–38") long. A maximum length of 105 cm (41") was stipulated from 2016. The maximum permitted weight is 737 grams. The majority of players use a stick in the range 19 oz to 22 oz (538 g - 623 g). Traditionally hockey sticks were made of hickory, ash or mulberry wood with the head of the sticks being hand carved. Sticks made of wood continue to be made but the higher grade sticks are now manufactured from composite materials which were first permitted in 1992. These sticks usually contain a combination of fibreglass, aramid fiber and carbon fibre in varying proportions according to the characteristics (flexibility; stiffness; resistance to impact and abrasion) required.

(b) Five Duties of the Captain in Hockey Team are :

1. Captain must wear a distinctive arm-band or similar distinguishing article on an upper arm or shoulder or over the upper part of a sock.
2. Captain is responsible for the behaviour of all players in their team.
3. To ensure that substitutions of players in their team are carried out correctly.
4. Understanding different game strategies and latest rules and regulation helps to make plays on the field.
5. Work hard to develop the relationships between the players and coach(es).

(c) Equipment used by Goalkeeper are :

Hand protectors :

1. Each have a maximum width of 228 mm and length of 355 mm when laid flat, palm upwards.
2. Must not have any additions to retain the stick when the stick is not held in the hand.

Leg guards : Each have a maximum width of 300 mm when on the leg of the goalkeeper.

(d) A penalty stroke is awarded :

1. For an offence by a defender in the circle which does not prevent the probable scoring of a goal.
2. For an intentional offence in the circle by a defender against an opponent who does not have possession of the ball or an opportunity to play the ball.
3. For an intentional offence by a defender outside the circle but within the 23 metres area they are defending.
4. For intentionally playing the ball over the back-line by a defender.

*(**Note:** goalkeepers are permitted to deflect the ball with their stick, protective equipment or any part of their body in any direction including over the back line.)*

5. For an offence by a defender in the circle which prevents the probable scoring of a goal.
6. If the ball hits a piece of equipment lying in the circle and a probable goal is prevented, a penalty stroke may be awarded.
7. For an intentional offence in the circle by a defender against an opponent who has possession of the ball or an opportunity to play the ball.
8. If there is another offence or misconduct before the awarded penalty has been taken.

(e) Scoop is a method of passing or shooting the ball by leveraging the stick underneath the ball to pick it up and fling it through the air while Penalty stroke is a free one-on-one shot from the penalty stroke line awarded when a foul prevents a likely goal. The goalkeeper must remain behind the goal line until the ball is hit.

BASKETBALL

Answer 11.

(a) 1. **Three Second Rule:** The three seconds rule is often termed as a lane violation requires that in basketball, a player shall not remain in the opponents' restricted area for more than three consecutive seconds while that player's team is in control of a live ball in the frontcourt and the game clock is running. The countdown starts when one foot enters the restricted area and resets when both feet leave the area.

2. **Twenty-four Second Rule :** A shot for a field goal is attempted near the end of the shot clock period and the shot clock signal sounds while the ball is in the air. If the ball does not touch the ring, a violation has occurred unless the opponents have gained an immediate and clear control of the ball. The ball shall be awarded to the opponents for the throw-in from the place nearest to where the game was stopped, except directly behind the backboard.

(b) A free throw is a shot that a player gets after they are fouled. It is a direct and unguarded shot from 15-feet away from the hoop. Player gets two free throws if shot does not go in the hoop after the foul, no basket is awarded. If the shot does go in after the foul, then the basket counts and one more free throw is awarded. If a player gets a technical foul (a foul for bad behaviour), then a player on the other team is given one free throw.

Steps to be followed for taking free throw are :

1. Get the body in position. If one is right handed then right foot should be lined up with the rim. Shoulders should be square and one should feel balanced - being off balanced will affect control.

2. Make sure to keep eyes on the front of the rim before the shot and while the shoot.

3. Keep shooting arm parallel to the rim when shoot the ball.

4. Bend knees. The strength of shot should come from legs rather than arms.

5. Shoot the ball off finger tips. If using palm, shot will be flat and not get good arc.

6. Make sure to follow through with shot! This will help to get arc, rotation, and consistency.

(c) **The violation in the game of Basketball are:**

1. **Walking or Travelling :** Taking more than 'a step and a half' without dribbling the ball is travelling. Moving pivot foot once one has stopped dribbling is travelling.

2. **Carrying or palming :** When a player dribbles the ball with hand too far to the side of or, sometimes, even under the ball.

3. **Double Dribble :** Dribbling the ball with both hands on the ball at the same time or picking up the dribble and then dribbling again is a double dribble.

4. **Held ball :** Occasionally, two or more opposing players will gain possession of the ball at the same time. In order to avoid a prolonged and/or violent tussle, the referee stops the action and awards the ball to one team or the other on a rotating basis.

5. **Goaltending :** If a defensive player interferes with a shot while it is on the way down toward the basket, while it is on the way up toward the basket after having touched the backboard, or while it's in the cylinder above the rim, it's goaltending and the shot counts. If committed by an offensive player, it's a violation and the ball is awarded to the opposing team for a throw-in.

6. **Backcourt violation :** Once the offense has brought the ball across the mid-court line, they cannot go back across the line during possession. If they do, the ball is awarded to the other team to pass inbounds.

7. **Time restrictions :** A player passing the ball inbounds has five seconds to pass the ball. If the player does not, then the ball is awarded to the other team. Other time restrictions include the rule that a player cannot have the ball for more than five seconds when being closely guarded and, in some states and levels, shot-clock restrictions requiring a team to attempt a shot within a given time frame.

(d) **Double foul :** A foul is considered to be a double foul, when both fouls have equal penalties.

Penalty for Double Foul : The penalties offset one another, so the offense retains the possession of the ball and inbounds the ball from the side-line. Both players, however, are charged with personal fouls.

(e) 1. **Technical Foul :** An official warning is given to a player for an action or behaviour which, if repeated, may lead to a technical foul. That warning shall also be communicated to the head coach of that team and shall apply to any member of that team for any similar actions for the remainder of the game. An official warning shall be given only when the ball becomes dead and the game clock is stopped.

2. **Multiple Foul :** A multiple foul occurs when 2 or more teammates commit personal fouls against the same opponent at approximately the same time. For all double, multiple and simultaneous fouls the clock must be stopped between fouls.

VOLLEYBALL

Answer 12.

(a) **The Duties of Volleyball Coach are :**

1. Duties of a volleyball coach is to **focus on the process of the match.**

2. **Help every player in the team :** Give every player equal attention, regardless of skill level. More skillful players often are given more opportunities to play, so pay closer attention to less-skilled players during practice.

3. **Separate the performance from the performer:** One of the most important things coaches can do is build every player's self-esteem, no matter what their skill level is. Make sure every comment is constructive, sandwiched between compliments and correction.

4. **Be demanding and disciplined, never demeaning :** The coaches style must be based on the laws of learning. If screaming, yelling, or putting down a player was the best way to teach, then all teachers in school classrooms would be using that method. Coaches must make sure they are teaching, not taking out their frustrations or setting unrealistic expectations. Coaches should never be demeaning their athletes.

5. **Coach smarter, not harder :** Experience is a great teacher, but attending coaching clinics and learning from other coaches is essential. Sharing and bouncing ideas off other coaches will lead to greater coaching wisdom.

6. **Be innovative, not imitative :** This is hard to do, but it's best to not just copy other successful coaches. For instance, don't just copy volleyball practice plans.

(b) **Service Faults in Volleyball are :** The service is the act of putting the ball into play, by the back-right player, placed in the service zone.

1. The ball shall be hit with one hand or any part of the arm after being tossed up or released from the hand(s).

2. Only one toss or release of the ball is allowed. Dribbling or moving the ball in the hands is permitted.

3. At the moment of the service hit or take-off for a jump service, the server must not touch the court (the end line included) or the floor outside the service zone.

4. After the hit, he/she may step or land outside the service zone, or inside the court.

5. The server must hit the ball within 8 seconds after the 1st referee whistles for service.

6. A service executed before the referee's whistle is cancelled and repeated.

(c) **(i) Procedure Adopted to Start the game are :**

1. Before the match, the 1st referee carries out a toss to decide upon the first service and the sides of the court in the first set.

2. If a deciding set is to be played, a new toss will be carried out.

3. The toss is taken in the presence of the two team captains.

4. The winner of the toss chooses either the right to serve or to receive the service, or the side of the court. The loser takes the remaining choice.

(ii) National Level Tournament in Volleyball are :

1. **Senior Nationals :** The Tournament is held for state teams all over the country. Few Government Institutions also participate in this annual tournament. The tournament follows a league-cum-knockout system, both for men and women teams. It is considered as the topmost volleyball league currently in India. Playing in this league leads to the Final step in selection towards the Senior National Team.

2. **Youth Nationals :** It is held for the U-21 Category state teams all over the country. Few Government Institutions also participate in this annual tournament. The tournament follows a league-cum-knockout system, both for boys and girls teams. Playing in this league leads to the Final step in selection towards the U-21 National Team.

(d) **(i) Difference between Service Zone and Substitution Zone :**

1. **Service Zone :** The service zone is a 9 m wide area behind each end line. It is laterally limited by two short lines, each 15 cm long, drawn 20 cm behind the end line as an extension of the side lines. Both short lines are included in the width of the service zone. In depth, the service zone extends to the end of the free zone.

2. **Substitution Zone:** This is the part of the free zone through which substitutions are carried out. The substitution zone is limited by the extension of both attack lines up to the scorer's table.

(ii) Difference between Positional Fault and Rotational Fault :

1. **Positional Fault :** The team commits a positional fault, if any player is not in his/her correct position at the moment the ball is hit by the server. When a player is in court through illegal substitution, and game restarts, this is counted as a positional fault with the consequences of an illegal substitution.

2. **Rotational Fault :** A rotational fault is committed when the service is not made according to the rotational order. It leads to the scorer stops play by the buzzer; the opponent gains a point and next service.

(e) **Duties of line judge are :**

1. To perform their functions by using flags (40 × 40 cm), to signal.

2. To check whether the ball is 'in' or 'out' whenever the ball lands near their line(s).

3. To check the touches of "out" balls by the team receiving the ball.

4. To check whether the ball is touching the antenna, the served ball and the third hit of the team crossing the net outside the crossing space, etc.

5. To check any player (except the server) stepping outside of his/her court at the moment of the service hit.

6. To check the foot faults of the server.

7. To check whether there is any contact with the top 80 cm of the antenna on their side of the court by any player during his/her action of playing the ball or interfering with the play.

8. To check whether the ball is crossing the net outside the crossing space into the opponent's court or touching the antenna on his/her side of the court.

9. To repeat the signal at the 1st referee's request.

TENNIS

Answer 13.

(a) Double Fault and Chip Shot

(i) Double Fault : Two missed serves in a row. The server will lose the point. The server has two attempts to get the ball in. If the ball lands outside the service box or does not clear the net or the net post, it is known as a 'fault'. After one fault the server may try again. If both tries result in faults, a 'double fault' is called and the opponent wins the point.

(ii) Chip Shot : A chip is an under-spin blocking motion used to hit a tennis ball. In other words, blocking a shot with backspin, creating a low trajectory. The aim of a chip, is to put the opponent under pressure.

(b) Five Types of Strokes in Tennis are:

1. Serve : The essentials of this stroke include fluid motion and rotation of the entire body into the shot. Turn to predominant side at the baseline with the front foot angled towards the net post. Bring the racquet back, while tossing the ball in the air slightly toward the other shoulder with free hand. Keeping eyes on the ball, quickly coil up at the waist and knees and release into an overhand motion that drives the ball at the opposite server box.

2. Forehand : This is usually the strongest shot, but requires footwork and focus. Remain in ready position with the knees and torso bent. Keep eyes on the ball as you reach back with the racquet. However, too much backswing may result in a late stroke that lacks power or direction. How much control one has over the return swing will depend on playing strategy and timing. Concentration, balance and practice will help to develop more control.

3. Backhand : It takes good footwork and timing to get into position for the backhand. Fluid body rotation is important. Beginning players tend to use a short backswing and lack follow-through, making their backhand shot weaker. Backhands can be delivered with one or both hands. The one-handed style will give more reach while the two-handed technique provides better power and control.

4. Volley : A volley occurs when one approach the net and hit the ball in the air. A half-volley is volleying after a bounce, not in the air. Making a strong volley involves a short backswing and hitting the ball when it's directly in front.

5. Overhead : An overhead stroke is any shot besides the hit that the ball hits the racket's long axis close to vertical. The point of contact between the ball and racket with this stroke is typically above the players head. An overhead stroke can be used to hit off an opposing player's lob stroke.

(c) Duties of Coach in Tennis are :

1. Learn how each student learns best : Some students are visual learners, since they emulate what they see. Other students are auditory learners who understand by listening. Some are kinesthetic learners, in that they need to perform a movement to understand it best. Make an assessment of the students' learning style and adjust the message accordingly.

2. Separate the performance from the performer: One of the most important things coaches can do is build every player's self-esteem, no matter what their skill level is, make sure every comment is constructive, sandwiched between compliments and correction.

3. Be demanding and disciplined, never demeaning : The coaches style must be based on the laws of learning. If screaming, yelling, or putting down a player was the best way to teach, then all teachers in school classrooms would be using that method. Coaches must make sure they are teaching, not taking out their frustrations or setting unrealistic expectations. Coaches should never be demeaning their athletes.

4. Coach smarter, not harder : Experience is a great teacher, but attending coaching clinics and learning from other coaches is essential. Sharing and bouncing ideas off other coaches will lead to greater coaching wisdom.

5. Be innovative, not imitative : This is hard to do, but it's best to not just copy other successful coaches. For instance, don't just copy practice plans.

6. Put the needs of the player first : The experience of coaching is first and foremost about providing a secure, productive, healthy and motivating environment for players to learn, develop and grow. It is not about satisfying the coaches or parents' needs. Do not exploit your players.

7. Be a role model : A coach should emphasise the importance of hard work, perseverance attentiveness, determination, health and fitness. The greatest impact however, is when coach personally embody and repeatedly display these attributes. "Do as I say" is not as powerful a message as "Do as I do!"

(d) (i) Set: A set consists of 6 games; a player that win 6 games with at least a two-game advantage wins the set.

(ii) Ace : When a serve is not returned by the receiving tennis player, resulting in a point for the player who served.

(iii) Slice : A type of shot with underspin.

(iv) Down the Line : A ball that is hit straight along the side-line of the tennis court.

(v) Drive: It is fairly a hard hit. Two major types of drives are forehand and backhand drive. Flat drive is hit hard and fairly low. Slice drive is a hit lower than flat and much lower than top spin. Top spin drive is hit hard and relatively low, but has more margin of clearance than flat.

(e) Steps for Taking Forehand Shot are: As with most strokes in tennis the forehand is hit on the move using a variety of different spins.

1. As the ball approaches stay in open stance position.
2. Rotates hips, shoulders and arms as part of the take back.
3. Now start unwinding body with dominant arm dropping down (to help generate top spin) as it trails the rest of the body forward towards the ball.
4. Continue rotating and swing the racket up, over and through the ball and make contact.
5. The front of the body and head should be facing the net.
6. Follow through by continuing the arms extension forward then across the body towards the non-dominant side.
7. Finally, complete the stroke with the weight shifted to the left leg (if right handed) and the racket head hovering around the left shoulder.

BADMINTON

Answer 14.

(a) Service Rule in Doubles

1. Service Boundaries for Badminton Doubles: Serve to the darkened area so that the shuttlecock falls inside.

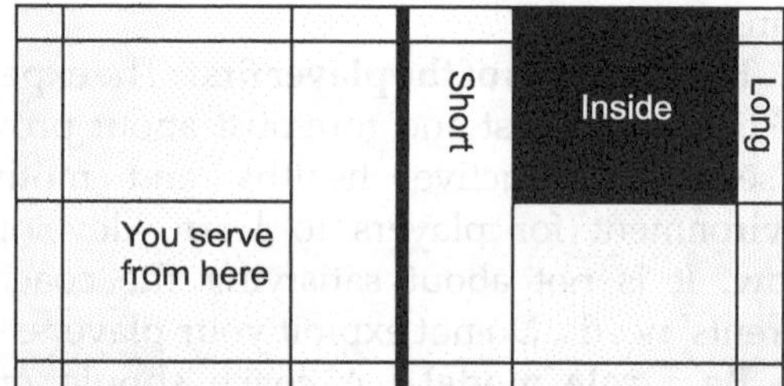

2. Serving and receiving courts

(i) A player of the serving side shall serve from the right service court when the serving side has not scored or has scored an even number of points in that game.

(ii) A player of the serving side shall serve from the left service court when the serving side has scored an odd number of points in that game.

(iii) The player of the receiving side who served last shall stay in the same service court from where he/she served last. The reverse pattern shall apply to the receiver's partner.

(iv) The player of the receiving side standing in the diagonally opposite service court to the server shall be the receiver.

(v) The players shall not change their respective service courts until they win a point when their side is serving.

(b) If shuttlecock gets struck in net : After the service is returned, the shuttle is caught on the net and remains suspended on its top, or after passing over the net, is caught in the net then the umpire shall call out let. When a "let" occurs, play since the last service shall not count and the player who served last shall serve again.

(c) (i) A shuttlecock (also called a bird or birdie) is a high-drag projectile used in the sport of badminton. It has an open conical shape formed by feathers (or a synthetic alternative) embedded into a rounded cork (or rubber) base. The shuttlecock's shape makes it extremely aerodynamically stable. Regardless of initial orientation, it will turn to fly cork first, and remain in the cork-first orientation.

(ii) Different Strokes Played by Badminton Player are :

1. Defensive Badminton Clear/Lob : A clear or lob is a good way to get yourself more organized while you're under pressure and buys some time to work back into the rally.

Overhead Defensive Clear

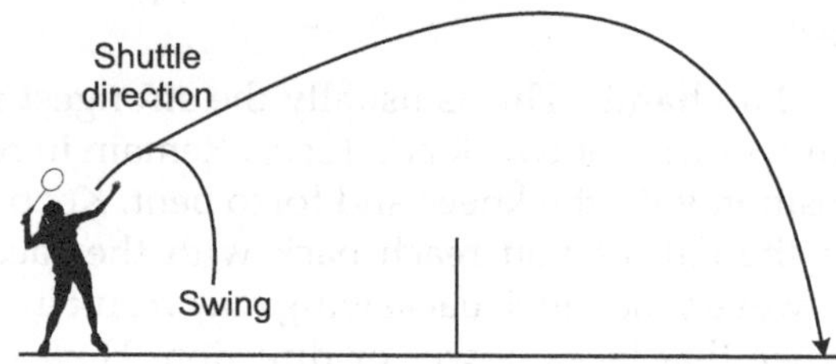

Underarm Defensive Clear

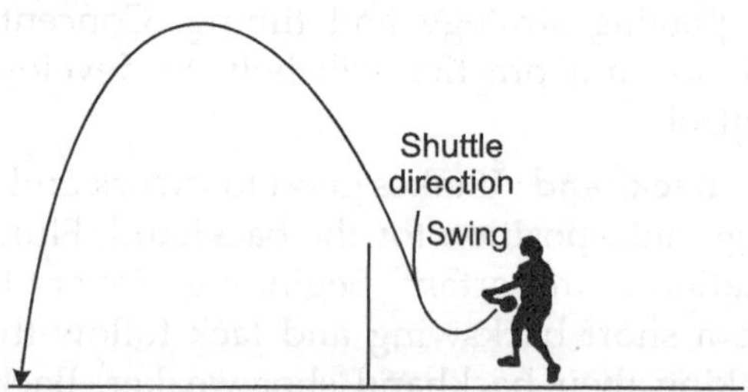

2. Badminton Drop Shot : The badminton drop shot is semi-offensive. It is performed from the back of the court and is usually used to maintain the pace of the game. A drop shot performed at a steep angle lands on the ground quicker, compared to the clear or lob. Opponent has to rush forward to retrieve it.

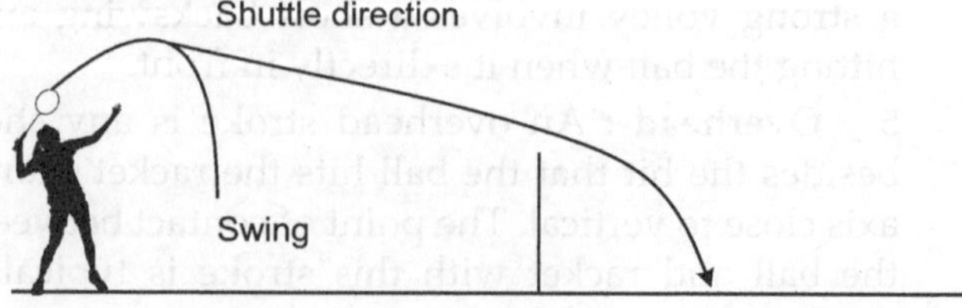

3. Badminton Smash : The smash is often known as the 'winning shot' in badminton. It's used with the intention to end a rally.

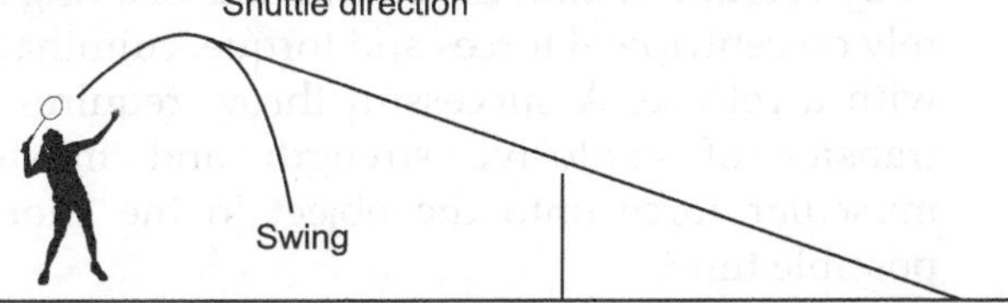

4. Badminton Jump Smash : This is a more powerful version of the badminton smash. It is also one of the hardest badminton shots to perform. To hit this shot, one has to jump and hit the smash in mid-air.

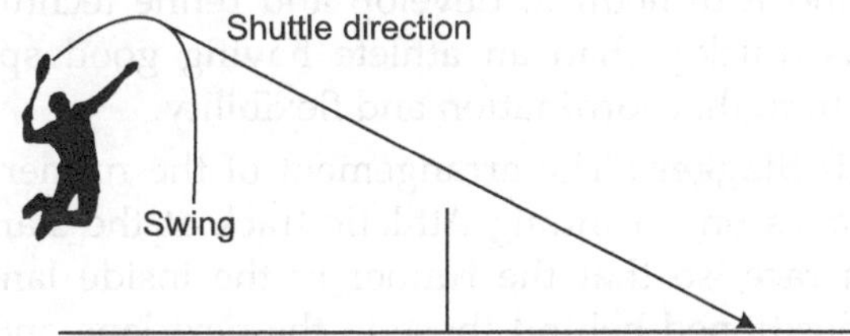

5. Badminton Net Shot (Tumbling / Spinning Net Shot) : Badminton shots along the net (known as 'net shots') are used to create opportunities and dominate the front of the court. A tumbling net shot enables the shuttle to tumble across just slightly above the net.

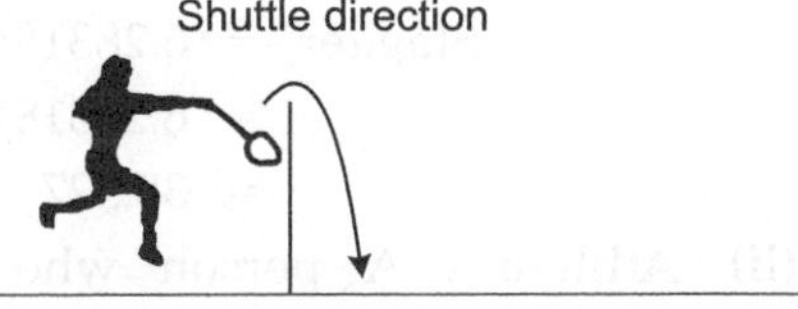

6. Badminton Drive : Drives are fast and furious badminton shots exchanged horizontally across the net. One of the main reasons to hit the drive is to increase the pace of the game.

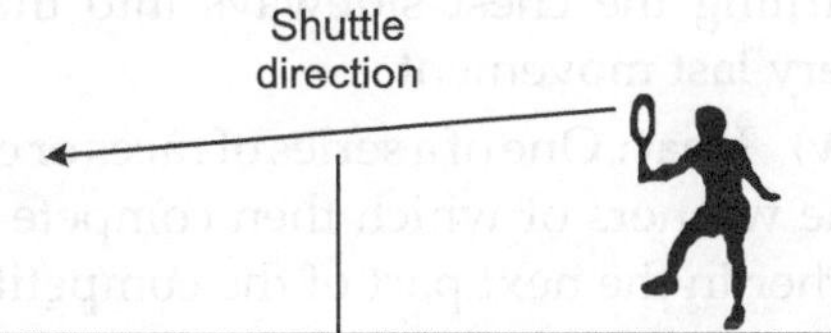

(d) (i) **Side Out :** This occurs when the side that is serving loses the serve and becomes the receiving team.

(ii) Waist Fault : The whole shuttle shall be below the server's waist at the instant of being hit by the server's racket. The waist shall be considered an imaginary line round the body, level with the lowest part of the server's bottom rib. This is an important rule: it's to prevent the player from smashing as he / she serves.

(e) (i) **Shuttle Grips : (1) Out of Hand Grip :** In this, the shuttle cock is held with tip of the thumb and index finger.

(2) Mid Grip : In this, the shuttle cock is held with tip of the thumb and two finger at base of the shuttle.

(3) Base Grip : In this, the shuttle cock is held with tip of the thumb and fingers around the shuttle at the shafts of its feathers and the cork.

(ii) The feathered shuttle shall weigh from 4.74 to 5.50 grams and shall have 16 feathers fixed at the base.

SWIMMING

Answer 15.

(a) Freestyle events for men are 50, 100, 200, 400, 800 and 1500 metres.

Freestyle Relays 4 × 100 and 4 × 200 metres.

(b) **(i) Position of Turn Backstroke Indicator:** Backstroke turn indicator posts should be installed 5 metres from each end of the pool and backstroke flags should be 1.8 metres to 2.5 metres above the water.

(ii) Lane for Best Swimmer : Except for 50 metre events in 50 metre pools, assignment of lanes shall be(number 1 lane being on the right side of the pool (0 when using pools with 10 lanes) when facing the course from the starting end) by placing the fastest swimmer or team in the centre lane in pool with an odd number of lanes, or in lane 3 or 4 respectively in pool shaving 6 or 8 lanes. In pools using 10 lanes, the fastest swimmer shall be placed in lane 4. The swimmer having the next fastest time is to be placed on his left, then alternating the others to right and left in accordance with the submitted times. Swimmers with identical times shall be assigned their lane positions by draw within the aforesaid pattern.

(c) **Swimmers Starting Stance in Backstroke Event:** Prior to the starting signal, the swimmers shall line up in the water facing the starting end, with both hands holding the starting grips. Standing in or on the gutter or bending the toes over the lip of the gutter is prohibited. When using a backstroke ledge at the start, the toes of both feet must be in contact with the end wall or face of the touchpad. Bending the toes over the top of the touchpad is prohibited.

(d) (i) **Dolphin Kick :** In the dolphin kick, both legs do a simultaneous whipping movement, with the feet pointed. The dolphin kick is a bi-directional (functioning in two directions) kick that requires abdominal, hip, lower back, and leg strength. Also, it requires an amazing amount of ankle flexibility and anaerobic conditioning.

(ii) Two Duties of the Judges of the Stroke are :

1. Judge of stroke shall ensure that the rules related to the style of swimming designated for the event are being observed, and shall observe the turns and the finishes to assist the inspectors of turns.

2. Judges of Stroke shall report to the referee any violation on signed cards detailing the event, lane number, and the infraction.

(e) (i) **Tapper :** A tapper in blind swimming is a person that taps a swimmer on the head when they are coming to a wall. Tappers may assist swimmers by running from one end of the pool to the other.

(ii) **Peddler** : It is an equipment which assist in learning swimming skill. It assists in swimming.

ATHLETICS

Answer 16.

(a) **Triple Jump Event :** The triple jump, sometimes referred to as the hop, step and jump or the hop, skip and jump, is a track and field event, similar to the long jump. It is a horizontal jump. The competitor runs down the track and performs a hop, a bound and then a jump into the sand pit. In hop, the athlete takes off and lands on the same foot. In step, athlete lands on the other foot. In jump, athlete lands in any manner, usually with both feet together. The triple jump was inspired by the ancient Olympic Games and has been a modern Olympics event since the Games inception in 1896.

(b) Running distance radius is an imaginary line through which the athlete is supposed to run, what is actually used for marking the track is called curve distance radius or marking distance radius.

(c) **Diagram of Shot-Put Sector with all its specifications :**

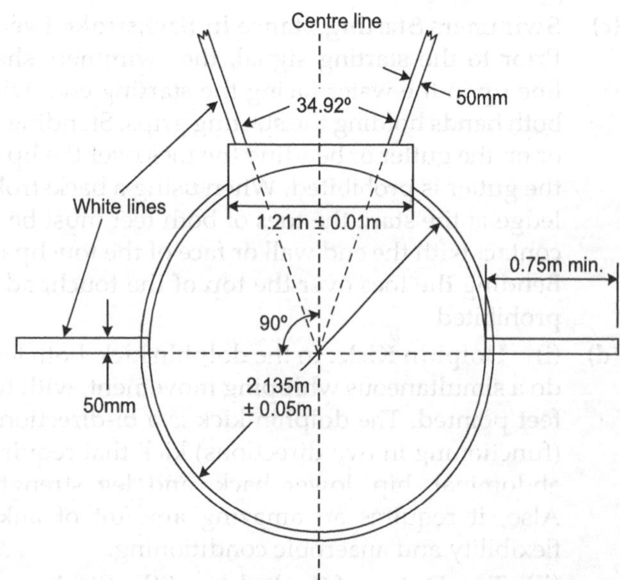

(d) **Fundamental Skills Required by an Athlete for Discus Throw :** The discus throw consists of three movements: the glide or rotation, the standing power position, and the release. Movements involved in the discus throw include body rotation within the boundaries of a ring and rely on centripetal forces and torque, culminating with a release. A successful throw requires the transfer of explosive strength and maximal muscular force onto the object in the shortest possible time.

The importance of muscular strength drives the requirement for throwers to develop leg strength in the early stages of their careers. Strength is very important for a thrower. Athlete with weaker speed, strength, coordination and flexibility find it difficult to develop and refine technique as quickly than an athlete having good speed strength, coordination and flexibility.

(e) (i) **Stagger :** The arrangement of the runners in lanes on a running Athletic track at the start of a race, so that the runner in the inside lane is positioned behind those in the next lane and so on until the outside lane. This is done to place the runners to run the equal distance for a specified distance race.

Stagger for 400 m track

$$= 2 - (1.22 \times (\text{Lane} - 1))$$

For example, applying the formula to lane six:

$$\text{Stagger} = 6.283184\ (1.22 \times 5)$$
$$= 6.283184 \times 6.1$$
$$= 38.327$$

(ii) **Athlete :** A person who takes part in competitive track and field events (athletics).

(iii) **Shoulder Shrug :** Shoulder shrugging technique is a type of finishing in which athlete shrugs his opposite shoulder to the tape by turning the chest sideways into the tape at the very last movement.

(iv) **Heat:** One of a series of races or competitions, the winners of which then compete against each other in the next part of the competition.

(v) **Lane :** A narrow marked section of a track or that is used by one person taking part in a race. The track should have a minimum of 6 and a maximum of 8 lanes separated and bounded on both sides by white lines 50 mm wide. The lanes shall all be 1.22 m ± 0.01 m wide including the lane line on the right.

Note : For all tracks constructed before 1 January 2004, the lanes may have a width of maximum 1.25m. However, when such a track is fully resurfaced or replaced, the lane width shall comply with this rule.

●●

PHYSICAL EDUCATION 2018

QUESTIONS

SECTION–A (40 Marks)

*Answer **any five** questions.*

Question 1.

Explain the role of physical education in the development of personality of an individual. **[8]**

Question 2.

Explain briefly what is meant by interval training ? Write the method of conducting interval training. **[8]**

Question 3.

Write a note on the Sports Authority of India (SAI) giving its objectives. **[8]**

Question 4.

What is meant by a tournament ? Explain the various types of tournaments. Draw a fixture of 11 teams on the knock-out basis. **[8]**

Question 5.

Explain the term Disability. State the various services available for the rehabilitation of the disabled, in India. **[8]**

Question 6.

What is meant by soft-tissue injury ? Explain the role of an individual in the prevention of sports related accidents. **[8]**

Question 7.

Write short notes on the following : **[8]**

(a) *Splint* (b) *Personal hygiene*

SECTION–B (30 Marks)

*Select **any two** games from this Section.*

*Answer any **three** of the **five** subparts (a), (b), (c), (d) and (e) from **each** of the **two** games selected by you. Each subpart is for **5 marks**.*

FOOTBALL

Question 8. **[3 × 5]**

(a) *What is meant by kick-off, in the game of football ? State four situations of a kick-off in the game.* **[5]**

(b) *Explain the following terms :* **[5]**

(i) *Offside* (ii) *Dropped-ball*

(c) *Write any five duties of a referee during the game.* **[5]**

(d) *In the game of football, what is meant by added time ? State the difference between added time and extra time.* **[5]**

(e) *What is meant by throw-in ? Explain the process of throw-in.* **[5]**

CRICKET

Question 9. **[3 × 5]**

(a) *Explain follow-on with reference to the game of cricket.* **[5]**

(b) *Mention any five occasions when the umpire calls or signals for a no ball.* **[5]**

(c) *Explain the following :* **[5]**

(i) *Sledgin* (ii) *IPL*

(d) *Explain the techniques of wicket keeping.* **[5]**

(e) *Mention five front foot drives played in cricket.* **[5]**

HOCKEY

Question 10. **[3 × 5]**

(a) *Draw a neat diagram of a hockey field, with all its dimensions.* **[5]**

(b) *When is a penalty corner awarded ? Write the procedure followed for taking a penalty corner.* **[5]**

(c) *Write the procedure followed for the start and the restart of a match of hockey.* **[5]**

(d) *Explain the following terms :* **[5]**

(i) *Penalty corner* (ii) *Dribbling*

(e) *Mention any five duties of a referee during the game.* **[5]**

BASKETBALL

Question 11. **[3 × 5]**

(a) *List any five duties of a referee.* **[5]**

(b) *Explain the following terms :* **[5]**

(i) *Travelling* (ii) *Bounce pass*

(c) *Mention any five occasions when the game clock is stopped during the game.* **[5]**

(d) *List the equipment used in the game of basketball.* **[5]**

(e) *Write any five skills a basketball player should possess.* **[5]**

VOLLEYBALL

Question 12. **[3 × 5]**

(a) *Draw a neat diagram of a volleyball court. Mark the zones and the positions of the players.* **[5]**

(b) *Write the basic equipment used by volleyball players.* **[5]**

(c) *Explain the following terms :* **[5]**

(i) *Libero* (ii) *Screening*

(d) *Write five duties of the captain of a volleyball team.* [5]

(e) *Explain the procedure to be followed when a match is interrupted by bad weather or dim light.* [5]

TENNIS

Question 13. [3 × 5]

(a) *Draw a neat diagram of a tennis racket with all its specifications.* [5]

(b) *Write any five duties of a referee.* [5]

(c) *Explain the following :* [5]

(i) *Deuce* (ii) *Half volley*

(d) *State any five instances when a player loses a point during the game.* [5]

(e) (i) *Name four important tournaments of tennis.*

(ii) *Expand the following :* [5]

(1) *ATF* (2) *NLTA*

BADMINTON

Question 14. [3 × 5]

(a) *Draw a neat diagram of a badminton court with all its specifications.* [5]

(b) *Explain the scoring system in the game of badminton.* [5]

(c) *Explain the following terms :* [5]

(i) *Short service line* (ii) *Back gallery*
(iii) *Foot fault* (iv) *Love all*
(v) *Rotation*

(d) *Write a short note on each of the following:* [5]

(i) *Thomas Cup* (ii) *Uber Cup*

(e) *Expand the following :* [5]

(i) *BWF* (ii) *SGFI* (iii) *BAI*
(iv) *IBF* (v) *IBL*

SWIMMING

Question 15. [3 × 5]

(a) *List the equipment used in swimming competitions.* [5]

(b) *What is false start rope ?* [5]

(c) *List any five swimming events.* [5]

(d) *Explain the following :* [5]

(i) *Control room supervisor*

(ii) *Individual medley*

(e) *Write a short note on FINA.* [5]

ATHLETICS

Question 16. [3 × 5]

(a) *List the events of Decathlon and Heptathlon, in athletics.* [5]

(b) *Draw a neat diagram of a discuss throw sector, with all its specifications.* [5]

(c) *Explain the event high jump. Write the different techniques used in high jump.* [5]

(d) *Mention five important duties of a referee, in Athletics.* [5]

(e) *Explain the following terms :* [5]

(i) *Baton* (ii) *Marathon*
(iii) *Hurdling* (iv) *Take off area*
(v) *Approach run*

ANSWERS

Answer 1.

Personality represents those characteristics of the person that account for consistent patterns of behaviour. Broadly, four factors influence how an individual respond in any given situation : the genetic make-up, past experience, the nature of the situation in which he/she find themselves and the free will. Personality development means the process in which an individual's attitude, discipline, helpfulness, team spirit, patience, unity, friendship etc., grows.

(i) **Attitude : 1.** Physical Education helps an individual to develop the attitude of competition against themselves for their betterment.

2. Physical Education helps the individual to develop an attitude of learning from their failure and success rather than getting too much elated about their success and depressed by their failure.

3. Physical Education helps in developing an attitude of equanimity at the time of success and failure and also develops an attitude to pursue excellence.

4. Physical Education develops an attitude to respect sports, other participants, coaches, official and themselves.

(ii) **Discipline : 1.** Physical Education helps in developing effective skills for dealing with conflict, difficult opponents, and other people when they are negative or oppositional during sports competition and thereby an individual gets disciplined.

2. Individual realizes through sports and games the value of being systematic, consistent and punctual as these qualities help the individual to develop his sporting ability. Accepting these qualities help the individual to become disciplined.

(iii) **Helpfulness : 1.** While playing sports and games an individual learns the importance of helping others. For developing bonding one has to assist, motivate and cooperate.

2. The senior players get respect from junior players, when they help the junior players to develop their skill by sharing their experience.

3. The quality of helpfulness is required during training session and sharing of equipment.

4. It is required during competition when an individual's performance is getting affected.

5. Through sports and games an individual understands one's success is not only the result of one's effort but also due to the help received from coach, team members, family etc.

(iv) Team Spirit : 1. Nature of sports is such that there is one common goal to be achieved by each member of a team and that goal can be called as team goal and that is winning. This common goal develops the quality of team spirit.

2. To win as a team a sense of respect for each member of team is required. Apart from sense of respect for team member, there is a requirement of giving priority to team's achievement than individuals. This thought and act of one while playing sports and game develops the quality of team spirit.

(v) Patience : 1. Individual while playing sports and games understands the importance of patience. One needs to regularly do hard work and has to be patient for the development of performance.

2. One needs to adjust with the team members. In order to adjust with others one needs to have patience. Applying strategies requires patience. For developing the quality and consistency one needs to be patient.

(vi) Unity : 1. Common goal develops unity. The team which lacks unity has more chances to get defeated than the team which has unity.

2. Each player in the team may have different responsibility and each member getting assistance and motivation to fulfill their responsibility from other team members develops a sense of unity among team members.

(vii) Friendship : 1. Quality of helpfulness, sacrifice, assistance, motivation, support and respect among the players develops the quality of friendship.

2. Matches between countries help in developing the quality of friendship among players of different country.

Answer 2.

Interval Training : This training is useful for the development of endurance. There are repeated sessions or interval of relatively intense exercise. Each interval is followed by a period of incomplete recovery. If one is running 400m intervals on a track and it takes approximately 90 seconds to complete each run then the recovery period between the intervals should be at least 90 seconds. This training develops cardio respiratory endurance. In high intensity interval training there is a risk of an athlete getting injured. During the interval period, the heart rate may reach to 180 beats/minutes followed by a recovery period. When the heart rate in recovery period comes down to 120 beats/minute then the next interval period should start.

Variables which decides the effect of interval method are :

1. Speed of work
2. Duration of work
3. Duration of recovery
4. Number of repetition
5. Nature of recovery

Interval method is of two type :

(a) Intensive Interval Method : When the intensity of the activity is 80-90% of the maximum speed at which an individual can run, then it is called intensive interval method.

(b) Extensive Interval Method : When the intensity of the activity is 70-80% of the maximum speed at which an individual can run, then it is called extensive interval method.

On the basis of duration interval training may be classified into three types : Short interval method, medium interval method and long interval method.

Short	Medium	Long
15 seconds – 2 minutes	2 minutes – 8 minutes	8 minutes – 15 minutes

Method of Conducting Interval Method :

(a) Before starting interval method one should have undergone a period of continuous running.

(b) Maximum heart rate of the individual doing interval training should be noted. Generally maximum heart rate is found by the formula 220 minus age. Maximum heart rate will help to find the target heart rate that the individual has to run during the interval.

(c) Number of repetition of intervals should be decided after taking into account the athletes age and level of practice.

(d) Improvement in performance can be done by changing the variable. It would be better to change one variable at a time.

(e) See that the surface of running is flat.

(f) Warm up and cooling down should be done properly before starting interval training.

(g) If the heart rate does not come back to 120 beats/minutes within a period of time it is expected *i.e.,* around 90 seconds then it should be understood that workload is difficult for the individual and the workload should be adjusted.

(h) Have a heart rate monitor as it will help in increasing and decreasing intensity according to heart rate, depending on which interval one is.

(i) Do not over exert while training.

Answer 3.

The Sports Authority of India (SAI), was set up on 25th January, 1984 as a Society registered under Societies Act, 1860 with the objective of promotion of Sports and Games. It comes under the Department of Sports, Govt. of India. SAI head office is in Jawaharlal Nehru Sports Complex, New Delhi.

SAI is responsible for the management of the following :

1. One high altitude training centre called Netaji Subhash High Altitude Training Centre (Shilaroo, Himachal Pradesh).

2. Two SAI Sports Academic Institutions (NSNIS Patiala and LNCPE Thiruvananthapuram).

3. Five stadiums at New Delhi (Jawaharlal Nehru Sports Complex, Indira Gandhi Sports Complex, Major Dhyan Chand National Stadium, Dr. Shyama Prasad Mukherjee Swimming Pool Complex, Dr. Karni Singh Shooting Ranges).

4. Ten SAI Regional Centres (SRC) are located at (Chandigarh, Sonipat, Lucknow, Guwahati, Imphal, Kolkata, Bhopal, Bengaluru, Mumbai and Gandhinagar).

5. Fourteen "Centre of Excellences" at different parts of India.

6. Twenty "Special Area Games" (SAG) in different parts of India.

7. Fifty six "Sports Training Centres" (STC) in different parts of India.

Objectives of SAI :

1. Talent Scouting at micro level and nurturing talent towards excellence.

2. Training and International Exposure.

3. Support Training with Scientific and Sports equipment and scientific personnel.

4. Monitoring and enhancing performance with a scientific evaluation system.

5. Training and preparation of national teams.

6. Sports Infrastructure development and maintenance.

7. Maintenance and upgradation of 4 stadia complexes and a shooting range in Delhi.

8. To produce coaches and physical educationists of high caliber in different disciplines of sports to broad base sports.

Answer 4.

Tournament : It is a series of competition between different teams. It helps to decide the positions like first, second, third etc. For a tournament to take place there should be minimum two teams.

Various Types of Tournaments : Tournaments are of different types like Knock-out, League, and League cum knock-out, challenge etc.

1. Knock-out Tournament : In a knock-out tourna-ment if a team loses a single match it will be out of the tournament. Only the winners move to the next round of matches. Formula to find the number of matches in a knock-out tournament = n – 1 where n = number of teams participating in the tournament. The number of matches in a knock-out tournament of 17 teams = 17 – 1 = 16 matches.

2. League Tournament : In a league tournament every team participating in the tournament has to compete with every other team. Winners are decided on the basis of the points acquired after all the competition. For each win a team may be awarded 2 point, for draw 1 point and for a loss no point is awarded. Formula to find the number of matches in a league tournament = n (n – 1)/2.Where n = number of teams participating in the tournament. The number of matches in a league tournament of 9 teams = 9 (9 – 1) /2 = 36 matches.

3. League cum Knock-out Tournament : In a league cum knock-out tournament competition is conducted in two phase. In the first phase, competition is conducted in the form of a league and through this competition a pre-decided number of teams are selected on the basis of points secured from the top. In the second phase, pre-decided number of teams selected from the league matches will participate in a knock-out tournament and from this, first and second teams are decided.

Fixture of 11 teams on the knock-out basis.

Number of Byes in Lower Half

$$= \frac{nb+1}{2} = \frac{5+1}{2} = 3$$

Number of Byes in Upper Half

$$= \frac{nb-1}{2} = \frac{5-1}{2} = 2$$

IR IIR IIIR IVR

Upper Half: 1 Bye; 2, 3 → 1; 4, 5 → 2; 6 Bye; 4; 5; 8

Lower Half: 7 Bye; 8, 9 → 3; 10 Bye; 11 Bye; 6; 7; 9

10

Number of Bye = 16 – 11 = 5 Bye

Answer 5.

Disability means an inability to do work. It is a state of an individual when he/she is not able to work efficiently. It may be present from birth, or occur during a person's lifetime. It may be physical, cognitive, sensory, emotional, developmental, or combination of these. Disabilities cover impairments, activity limitations, and participation restrictions. Impairment is a problem in body function or structure; an activity limitation is a difficulty encountered by an individual in executing a task or action; while a participation restriction is a problem experienced by an individual in involvement with the life situations. Thus, disability is a complex phenomenon, reflecting an interaction between features of a person's body and the environment of the society where he or she lives.

Various services available for the rehabilitation of the disabled India :

1. Rehabilitation Council of India, New Delhi: Regulation of training policies and programs in Person with disability (PWD) rehabilitation.

2. National Institute of Rehabilitation Training and Research, Cuttack, Odisha: Coordinates training for rehabilitation personnel, service delivery programs for rehabilitation, vocational training, placement and rehabilitation of persons with mobility disability.

3. Society for Rehabilitation and Research of the Handicapped, New Delhi : A organisation of Person with disability (PWD) rehabilitation, Co-education school for age group 16-30 with hearing impairment.

4. Association for Advancement and Rehabilitation of Handicapped, New Delhi : A organisation of Person with disability (PWD) rehabilitation, social awareness and occupational centre for persons with intellectual disability.

5. Amar Joyti Research and Rehabilitation Center, New Delhi formed in 1981: A organisation for Person with disability.

6. Indian Red Cross Society established in 1920. It has around 400 branches all over India.

7. All India Blind Relief Society formed in 1946 coordinates with various institutions of blind. It works for the welfare of blind.

8. Artificial limb Manufacturing Corporation in Kanpur provides artificial limbs to disabled.

9. Kamayani School for Mentally Handicapped, Pune, formed in 1964 provides work and training for handicapped persons.

10. National Institute for the Mentally Handicapped, Secunderabad, provides training and rehabilitation for persons with learning disability.

Answer 6.

A soft tissue injury is the damage of muscles, ligaments and tendons throughout the body. Common soft tissue injuries usually occur from a sprain, strain, a one off blow resulting in a contusion or overuse of a particular part of the body. Soft tissue injuries can result in pain, swelling, bruising and loss of function.

Role of various individuals in prevention of sports related accidents

To prevent sports related accidents in a school, the players and physical education teacher has a major role.

Things to be performed by a player to prevent sports related accidents :

1. Warm up properly before entering the playground.

2. Use safety equipments prescribed for a game.

3. Play the game with concentration and not carelessly.

4. Coordinate and cooperate with your team members so that you have a good rapport with your team members. Unsportsman like behaviour should not be shown against the opponents.

5. Have right frame of mind to play the game. If you are sad and you do not wish to play then do not enter the court.

6. If there is an injury, give proper rest and later rehabilitation exercise till the injury has healed completely and then start playing.

7. Learn the correct skill and execute it in a right manner.

8. Do not try to exert more than your limitation.

Things to be performed by Physical Education teacher or coach to prevent sports related accidents :

1. Have medical checkup for students before including them for training.

2. Test their performance and decide their schedule.

3. Start with simple skills and then move to advanced skills.

4. Ensure that ground is neatly marked and levelled, surroundings are neat and clean before starting training.

5. Divide the class into small groups and engage them in variety of sports to prevent overcrowding in a particular ground.

6. Ensure that first aid kit is carried to the ground to attend to an injury as quickly as possible.

7. Ensure that students are playing in proper playing kit and using the desired safety equipment.

Basis	Sprain	Strain	Contusion
Meaning	It is an injury to a ligament. According to the severity it may be classified as mild, moderate or severe.	It is an injury to a muscle or tendon. According to the severity it may be classified as mild, moderate or severe.	It is an injury caused due to forceful hit by a hard object. It is an internal injury. There is no external cut or bleeding over the skin.
Signs and Symptoms	Pain, swelling, loss of movement, lack of strength at the place of injury.	Pain, swelling, loss of movement, lack of strength at the place of injury.	Swelling, pain, bluish colour appears on the outer surface of the skin.
Treatment	**P – Protection:** Protect the injured part from further injury. Do not attempt to pressure or massage the injured part. **R – Rest:** Give rest to the injured part and restrict the movement. **I – Ice:** Apply ice to the injured part for 15-20 minutes and repeat it 3-4 times in a day if required for first 24-48 hours of injury, after that apply heat for the dispersion of clotted blood. **C – Compression:** Compress the injured part with a crape bandage to restrict the movement. **E – Elevation:** Elevate the injured part with the slings if required. **S – Support:** Support the injured part to avoid painful movement until medical assistance is provided. For severe strain/sprain treatment is surgery.		

Answer 7.

(a) **Splint :** A splint is a device used for support or immobilisation of a limb or the spine. It can be used in multiple situations, including temporary immobilisation of potentially broken bones or damaged joints and support for joints during activity. Splinting immobilises injured extremities and prevents further injury, decreases pain and bleeding and allows healing to begin. Splint is most commonly used during fractures, dislocations and sprains.

Advantages :

1. Stabilisation of an extremity fracture with a splint will preserve proper bone alignment and alleviate discomfort.
2. After reduction of a dislocated joint, anatomical positioning is maintained by application of a splint.
3. Splints may be used to give desired position at rest and functional positioning to the involved active joint.

Temporary splints can be used to alleviate pain and to prevent additional injury during transport to an emergency department or while awaiting evaluation by a specialist.

Contraindication :

1. For injuries that are prone to severe swelling, even a splint can become restrictive as the swelling increases. Extra padding may be placed under the splint if extensive swelling is anticipated, allowing room for expansion.
2. When neurovascular compromise is present as a result of a fracture or dislocation, splints should not be placed without first attempting immediate reduction.
3. Use appropriate universal precautions for potential exposure to body fluids when open wounds are present.
4. Whenever possible, use sheets or pads to protect the patient's clothing during splint application.

(b) **Personal Hygiene :** It is the basic concept of cleaning, grooming and it is the first step to good health. It helps to protect ourselves and maintain good health. Personal hygiene is learnt from the family and the society one lives in.

Advantages of having Personal Hygiene :

Personal hygiene includes different habits like washing hands and brushing teeth which keep bacteria, viruses and fungus far away from our bodies. These habits helps to protect mental health and to have a good feeling about one self.

Disadvantages of not looking after Personal Hygiene :

People who do not take care of their personal hygiene like wearing dirty clothes or having bad breath, may find it difficult to get acceptance by the society and may get prone to many diseases (physical and mental).

Some of the Personal Hygiene practices are :

1. Washing and combing hair.
2. Washing carefully around the eyes and avoiding rubbing of eyes, particularly with dirty hands or clothes.

3. Washing skin frequently to remove sweat, dirt and dead skin cells. Use soap to remove dirty matter from skin.

4. Washing hands well, particularly before preparing food, before eating, after excreting, after handling dirty objects, before and after cleaning and treating sores and wounds, after handling animals and after handling chemicals at work or in the home.

5. Keeping nails clean and trim.

6. Wash the feet and removing the dirt and sweat from between the toes.

SECTION–B

FOOTBALL

Answer 8.

(a) Kick-off is the method of starting and, in some cases, restarting play in the game of football.

Four situations of kick-off in a game of football are :

1. At the start of the match.
2. After a goal has been scored.
3. At the start of the second half of the match.
4. At the start of each period of extra time.

(b) **(i) Offside :** Offside is one of the laws of football association which states that, a player is in an offside position, if any of their body part except the hands and arms is in the opponents' half of the pitch and closer to the opponents' goal line than both the ball and the second-last opponent (not necessarily the goalkeeper). Being in an offside position is not an offence in itself; a player who was in an offside position at the moment the ball last touched, or was played by a team mate, must then become involved in active play in the opinion of the referee, in order for an offence to occur. When the offside offence occurs, the referee stops play and awards an indirect free kick to the defending team from the place where the offending player became involved in active play.

(ii) Dropped-Ball : It is a method of restarting the play. It is designed to offer no advantage to either side, generally being awarded when play has been stopped due to reasons other than normal game-play or misconduct.

The referee drops the ball at the position where it was when play was stopped, unless play was stopped inside the goal area, in which the ball is dropped on the goal area line, which is parallel to the goal line at the point nearest to where the ball was when play was stopped. The ball is in play when it touches the ground. Any number of players may contest a dropped ball (including the goalkeepers); the referee cannot decide who may contest a dropped ball or its outcome.

(c) **Five duties of referee :**

1. Enforces the laws of the game.
2. Control the match in co-operation with the assistant referees.
3. Acts as time keeper and keep a record of the match.
4. Restart the match after it has been stopped.
5. Takes disciplinary action against player, guilty of cautionable and sending off offenses.

(d) **Added Time :** Time allowed at the end of each half for time 'lost' because of substitutions, injuries, disciplinary action, goal celebration etc.

Extra Time	Added Time
1. A method of trying to decide the outcome of a match involving two equal additional periods of play not exceeding 15 minutes each.	Time allowed at the end of each half for time 'lost' because of substitutions, injuries, disciplinary action, goal celebration etc.
2. Extra time starts with a kick off.	Additional time does not start with kick off.
3. During extra time, a short drinks break is permitted at half-time of extra time.	There is no drinks break.
4. Extra time has two equal half with teams changing ends in between.	Duration of additional time is not fixed and there is no division or change of court during additional time.
5. Extra time is given for deciding the winner by the referee after the end of normal duration of play and additional time.	Fourth official will indicate the minimum duration of additional time the referee intends to play at the end of each half (including extra time).

(e) A throw-in is a method of restarting the play. A throw-in is awarded to the opponents of the player who last touched the ball when the whole of the ball passes over the touchline, either on the ground or in the air.

A goal cannot be scored directly from a throw-in:

1. If the ball enters the opponents' goal – a goal kick is awarded.
2. If the ball enters the thrower's goal – a corner kick is awarded.

Procedure of Throw-in :

At the moment of delivering the ball, the thrower must :

1. Face the field of play.

2. Have part of each foot on the touchline or on the ground outside the touchline.

3. Throw the ball with both hands from behind and over the head from the point where it left the field of play.

CRICKET

Answer 9.

(a) Follow-on is a term used in test matches. The side which bats first, if leads by at least 200 runs then they shall have the option of asking the other side to follow their innings. The same option shall be available in matches of shorter duration with the minimum leads as follows : 150 runs in a match of 3 or 4 days; 100 runs in a 2-days match; 75 runs in a 1-day match.

When all the innings are completed, the team with the most runs wins. If there is a tie, then the result stands as TIE means no win no lose.

If by the end of the final day's play all the innings are not completed, the game is a draw, no matter who appeared to be "winning". Thus, the strategic importance of sometimes declaring an innings closed, in order to have enough time to dismiss the other team and so win the game.

(b) Refer to ISC Paper 2020, Answer 9 (c)

(c) **(i) Sledging :** It is a term used to describe the practice by which some players seek to gain an advantage by insulting or verbally intimidating the opposing player. The purpose is to weaken the opponent's concentration and thereby making the opponent to make mistakes or underperform. It can be effective because the batsman stands within hearing range of the bowler and certain close fielders and vice versa. The insults may be direct or feature in conversations among fielders designed to be overheard.

(ii) IPL : Indian Premier League is a professional Twenty-20 cricket league in India contested generally during April and May of every year by teams representing Indian cities. The league was founded by the Board of Control for Cricket in India in 2007, and is regarded as the brainchild of Lalit Modi, the founder and former commissioner of the league. The IPL is the most-attended cricket league in the world and in 2014 ranked sixth by average attendance among all sports leagues. In 2010, the IPL became the first sporting event in the world to be broadcast live on YouTube.

(d) **1. Footwork :** Footwork is the most significant thing for a wicket keeper. If our feet are right, all things moves with it. Many say that poor wicket keepers have to dive a lot. This mean that a keeper having good footwork probably don't have to dive very often. Taking the ball efficiently means, right footwork so that the head is behind the ball on impact. Footwork of keeper consists of frequent, small, precise adjustments just like a dancer's or a tennis player's which are made in response to the trajectory and movement of the cricket ball so that the head remains in the optimum position.

2. Catching the Ball : For a wicket keeping, catching the ball is the key point. Judge the position where the elevation of the ball begins to decrease, that would be the ideal place for a keeper to stand. After positioning according to the speed of the bowler, one should be ready to catch the ball. When the ball reaches, catch the ball with relaxed but firm hands. Keep a steady head with eyes on the ball all the way into the gloves.

3. Diving : Diving often compensates for bad footwork so diving a lot should remind us to move our feet. However, diving becomes essential for the keeper when the ball is extremely out of his standing range. A thick edge on the ball would make ball go away from the keeper so he has to dive to take the ball. When keeper is standing up from the crouch position, weight equally distributed, it permit him to execute a dive easily.

(e) **1. Off Drive :** Played to a ball pitched just around the off stump. The bat should come down in a straight line in the direction of mid-off, to make sure that the ball is propelled towards that direction.

2. Cover Drive : Played to a ball pitched slightly wider outside the off stump. The bat should come down in the direction of extra cover, so as to hit the ball in that area.

3. Front Foot Cut : Played to a ball pitched wider outside the off stump, in the direction square of the wickets.

4. On Drive : One of the most difficult drives to play, which is being eradicated from the modern game gradually, as not many exponents of it are left. Played between mid-on and mid-wicket.

5. Straight Drive : The most eye-catching of all drives, played straight down the ground.

The batsman may choose which drive to play on a particular good length ball, but emphasis definitely lies on getting to the pitch of the ball.

HOCKEY

Answer 10.

(a)

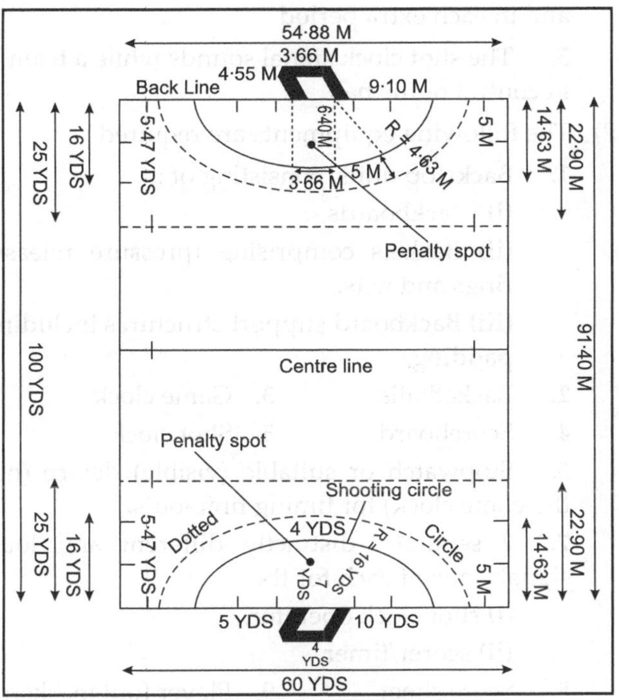

Hockey Field

(b) A Penalty Corner is Awarded :

1. For an offence by a defender in the striking circle which does not prevent the probable scoring of a goal.
2. For an intentional offence in the circle by a defender against an opponent who does not have possession of the ball or an opportunity to play the ball.
3. For an intentional offence by a defender outside the circle but within the 23 metres area they are defending.
4. For intentionally playing the ball over the back-line by a defender.
5. When the ball becomes lodged in a player's clothing or equipment while in the circle they are defending.

Procedure for taking Penalty Corner : An attacker pushes a ball from a mark which is 10 yards away from the goal post and other player stops the ball with hockey stick at the top of the shooting circle, then another player hits the ball in the goal post. It can be taken from both sides right and left. Meanwhile, the opposing team tries to stop the goal.

(c) 1. The team which wins the toss has to start match with a centre pass.

2. After a goal scored, match will start by centre pass by the team which received the goal.

3. After half time, the opposing team who lost the toss will restart the game with centre pass.

4. A bully is taken to restart the match when match stops due to an injury or any other reason and no penalty is awarded.

5. A free hit is taken by a defender 15 metres in front of the centre of the goal-line to restart a match when a penalty stroke has been completed and no goal has been scored.

(d) (i) Penalty Corner : A penalty corner is awarded to an offensive team when the defensive team commits a foul inside the shooting circle. It is also awarded when a defender commits an intentional foul outside the circle within the 25 yard (23 metres) defensive area. A penalty corner is taken on the backline inside the circle from a mark at least 10 metres from the goal post, from whichever side of the goal the attacking team prefers. The ball can be hit or pushed, but the attacker must have at least one foot outside the field of play.

(ii) Dribbling : Dribbling is to move with the ball when the ball is in possession using a hockey stick. While dribbling, protect the ball and dribble into the space. Different types of dribbling are Straight Dribble, Loose Dribble, Indian Dribble, Dribbling Push Back, One Handed – Right Hand Dribble, One Handed- Reverse Side Dribble.

(e) Duties of Referee during the game are as follows :

1. The Referee shall impose and report to the scorekeeper such penalties as are prescribed by the playing rules. He shall allow the goals scored.
2. The Referee may consult with the Linesmen, Goal Judge and Video Goal Judge, if available, in matters of disputed goals before making his decision, which shall be final.
3. He shall report to the scorekeeper the number of the goal scorer and any player(s) entitled to assists.
4. At the first stoppage of play, he shall cause to be announced over the public address system the reason for not allowing a goal.
5. He shall measure any equipment at his own discretion or if requested by the Captain of either team.

BASKETBALL

Answer 11.

(a) 1. The referee shall inspect and approve all equipments, including all the signals used by the table officials and their assistant during the game.

2. He shall designate the official game clock, twenty-four second device, stopwatch and recognise the table officials.

3. He shall not permit any player to wear objects which may cause injury to other players.

4. Make the final decision whenever necessary, or when the officials disagree. To make a final decision he may consult the umpire, the commissioner and/or the table officials.

5. At the end of each half and of each extra period or at anytime he feels necessary, he shall carefully examine the score sheet and approve the score and confirm the time that remains to be played.

(b) (i) Travelling : Travelling is a violation in basketball and is penalized by awarding the ball to the opposing team out of bounds closest to where the travel occurred. The travel rule is intended to prevent players from gaining an advantage by moving with the ball without dribbling.

The basic concept of travelling is based on the "pivot foot." Once a player receives the ball or picks up his dribble, he is allowed to move one foot, while the other foot has to remain on the floor as a pivot foot. The foot which is in first contact with the ground while the ball is caught becomes the Pivot foot. In case the player has caught the ball in air and landed on both the foot then the foot which is moved first without dribbling will not be called Pivot foot and the other foot will be considered as Pivot foot. Pivot foot is allowed to rotate, as long as the ball of the foot remains on the floor at all times.

A travel occurs when the player lifts or drags the pivot foot from the floor before releasing the ball for dribbling.

A travel occurs when the player lifts the pivot foot and the other foot and returns to the floor before releasing the ball on a pass or a shot.

(ii) Bounce Pass : In bounce pass, the holding of the ball, body positions and passing action is approximately the same as in two hand chest pass. However, in bounce pass, the ball is thrown in such a way that the ball should bounce at specific place, so that it may bounce up to the waist level of the receiver easily. This type of pass can be performed with one hand only. There are two types of bounce pass :

1. One handed bounce pass.

2. Two handed bounce pass.

(c) Game clock is stopped when :

1. Time expires at the end of a period, if not stopped automatically by the game clock itself.

2. An official blows his whistle while the ball is live.

3. A Basket is scored against a team which has requested a time-out.

4. A Basket is scored when the game clock shows 2:00 minutes or less in the fourth period and in each extra period.

5. The shot clock signal sounds while a team is in control of the ball.

(d) The following equipments are required :

1. Backstop units, consisting of :

(i) Backboards.

(ii) Baskets comprising (pressure release) rings and nets.

(iii) Backboard support structures including padding.

2. Basketballs **3.** Game clock

4. Scoreboard **5.** Shot clock

6. Stopwatch or suitable (visible) device (not the game clock) for timing time-outs.

7. 2 separate, distinctly different and loud signals, one of each for the

(i) shot clock operator,

(ii) scorer/timer.

8. Score sheet **9.** Player foul markers

10. Team foul markers

11. Alternating possession indicator

12. Playing floor **13.** Playing court

14. Adequate lighting.

Following are the five skills which a basketball player must possess:

(e) 1. Dribbling : In this skill the player advances with the ball by bouncing it on the ground. This skill allows to move up and down the court and dodge past the defenders and execute plays. Proper dribbling requires ball-handling skills and knowledge of how to spread your fingers for ball control. It is best to dribble equally well with both hands. Examples of different skills of dribbling are—in between legs dribble, behind the back dribble etc.

2. Shooting : In order to score points in basketball, you need to shoot the ball into the hoop. This requires the ability to properly hold and throw the ball into the air towards the basket while avoiding defenders. A proper shot requires precise aiming, arm extension and lift from the legs. Examples of different shooting skills are—jump shots, layups, hook shot etc.

3. Running : Running is a big part of basketball. In a full-court game, one needs to run back and forth during the transitions from offense to defense. On offense running will help to avoid defenders and get to the basket quicker. On defense, running is needed to run after the opponent, especially during fast breaks.

4. **Passing** : Passing means the movement of the ball from one player to another. Passing is the best way to keep possession of the basketball and is a faster way of moving the ball up the court than dribbling. Examples of different passing skills are—Bounce Pass, Chest Pass, Overhead Pass etc.

5. **Jumping** : Jumping is another skill that can define how good a basketball player is. Jumping skill is required during the jump ball in the beginning, while taking shots, trying to catch a pass and rebound. On defensive jumping ability is required to block a shot or a pass by opponent. Jumping skill is required for collecting rebounds too.

VOLLEYBALL

Answer 12.

(a)

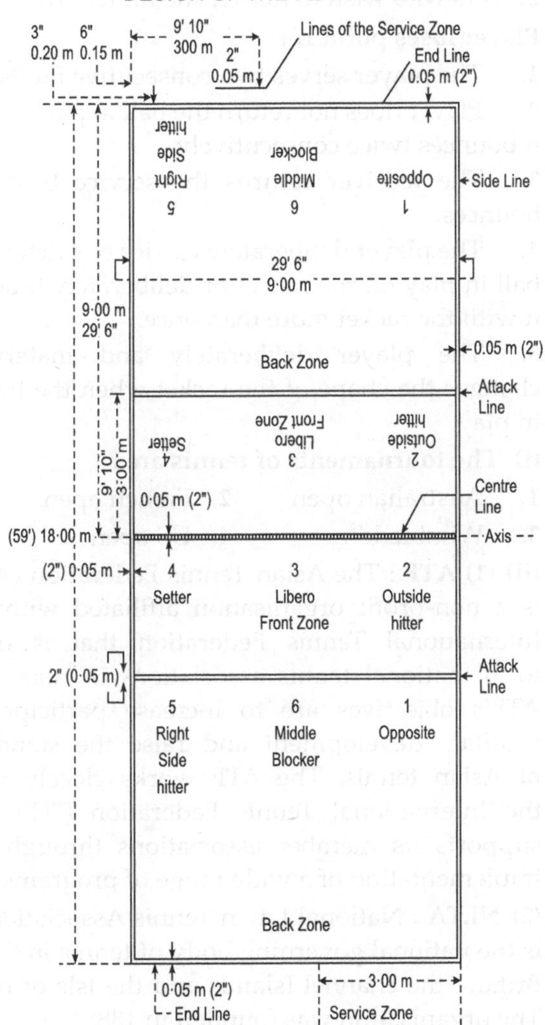

(b) A player's equipment consists of a jersey, shorts, socks (the uniform), sport shoes, protective gear like knee pads, elbow pads, shin guards, mouth guards and protective cups for men.

The colour and the design for the jerseys, shorts and socks must be uniform for the team (except for the Libero). The uniforms must be clean.

1. The shoes must be light and pliable with rubber or composite soles without heels.

2. The number must be placed on the jersey at the centre of the front and of the back. The colour and brightness of the numbers must contrast with the colour and brightness of the jerseys.

3. The team captain must have on his/her jersey a stripe of 8 × 2 cm underlining the number on the chest.

4. It is forbidden to wear uniforms of a colour different from that of the other players (except for the Liberos), and/or without official numbers.

(c) **(i) Libero** : The Libero player is a defensive player and must wear a uniform which has a different dominant colour from any colour of the rest of the team. Libero replacements are not counted as substitutions. Libero should not serve, block or attempt to block. The Libero is allowed to replace any player in a back row position. He/she is restricted to perform as a back row player and is not allowed to complete an attack hit from anywhere (including playing court and free zone) if, at the moment of the contact the ball is entirely higher than the top of the net.

(ii) Screening : Screening is said to have happened when :

1. The players of the serving team prevents their opponent, through individual or collective screening, from seeing the server and the flight path of the ball.

2. A player or a group of players of the serving team make(s) a screen by waving arms, jumping or moving sideways during the execution of the service, or by standing in a group, and in so doing hides both the server and the flight path of the ball until the ball reaches the vertical plane of the net.

If referee calls screening against a player or a team then the point will be awarded to opposite team.

(d) Five duties of a captain of a volleyball team are as follows :

1. Prior to the match the team captain signs the score sheet and represents his/her team in the toss.

2. During the match and while on the court, the team captain is the game captain. When the

team captain is not on the court, the coach or the team captain must assign another player on the court, but not the Libero, to assume the role of game captain.

3. In the absence of the coach, to request time-outs and substitutions to referee.

4. At the end of the game, thanks the referees and signs the score sheet to ratify the result.

5. Captain may request for a change of ball to the referee.

(e) The procedure to be followed when the match is interrupted by bad weather or dim light is :

1. The first referee, the organiser and the control committee, if there is one, shall decide the measures to be taken to re-establish normal conditions.

2. If the match is resumed on the same playing court, the interrupted set shall continue normally with the same score, players (except expelled or disqualified ones) and positions. The sets already played will keep their scores.

3. If the match is resumed on another playing court, the interrupted set is cancelled and replayed with the same team members and the same starting line-ups (except expelled or disqualified ones) and the record of all sanctions will be maintained. The sets already played will keep their scores.

4. Should one or several interruptions occur, exceeding 4 hours in total, the whole match shall be replayed.

TENNIS

Answer 13.

(a)

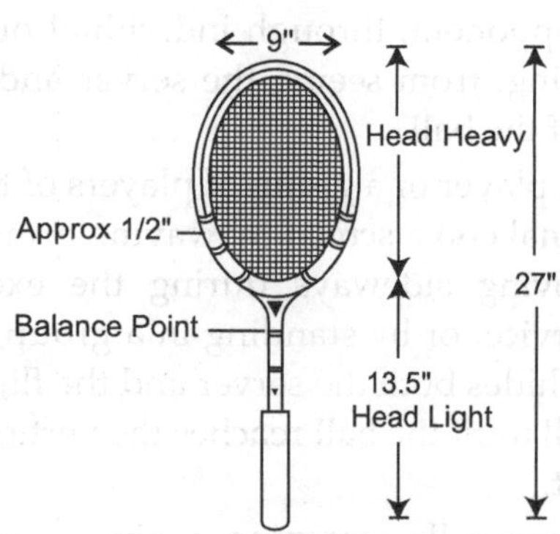

Diagram of a Tennis Racket

(b) The five duties of a referee :

1. A referee settles disputes when two players disagree.

2. A referee gives the final verdict on whether the court is good for playing tennis. This includes making sure the playing surface is safe, the net is of the correct height, the lighting is adequate, the court is dry or that the temperature is not too cold or hot.

3. A referee enforces match start times, issuing penalties and disqualifications for late starts.

4. Referees should wear clothing that identifies them as officials. They should introduce themselves to players.

5. Referee should not yell instructions or warnings to players.

(c) **(i) Deuce :** Deuce is a term used in scoring. A deuce occurs in tennis when the score is 40-40 (both players have 40 points). To win the game, a player need to get the advantage and then get an extra point. The game can come back to deuce if the player that has got a point after deuce (advantage) loses a point in the next.

(ii) Half-Volley : The half volley is not actually a volley, but rather a forehand or backhand groundstroke hit immediately after the ball bounces onto your side of the court. A half volley in tennis is a shot that is hit immediately after the ball bounces but before it reaches the apex of its bounce. It is a fundamental shot for all doubles players who wish to master the midcourt.

(d) Player loses point if :

1. The player serves two consecutive faults.

2. Player does not return the ball in play before it bounces twice consecutively.

3. The receiver returns the service before it bounces.

4. The player deliberately carries or catches the ball in play on the racket or deliberately touches it with the racket more than once.

5. The player deliberately and materially changes the shape of the racket when the ball is in play.

(e) **(i) The tournaments of tennis are :**

1. Australian open 2. French open

3. Wimbledon 4. US open

(ii) (1) ATF : The Asian Tennis Federation (ATF) is a non-profit organisation affiliated with the International Tennis Federation that is open to all national tennis associations in Asia. The ATF's objectives are to increase participation facilitate development and raise the standard of Asian tennis. The ATF works closely with the International Tennis Federation (ITF) and supports its member associations through the implementation of a wide range of programs.

(2) NLTA : National Lawn Tennis Association. It is the national governing body of tennis in Great Britain, the channel Islands and the Isle of man. The organisation was founded in 1888.

BADMINTON

Answer 14.

(a)

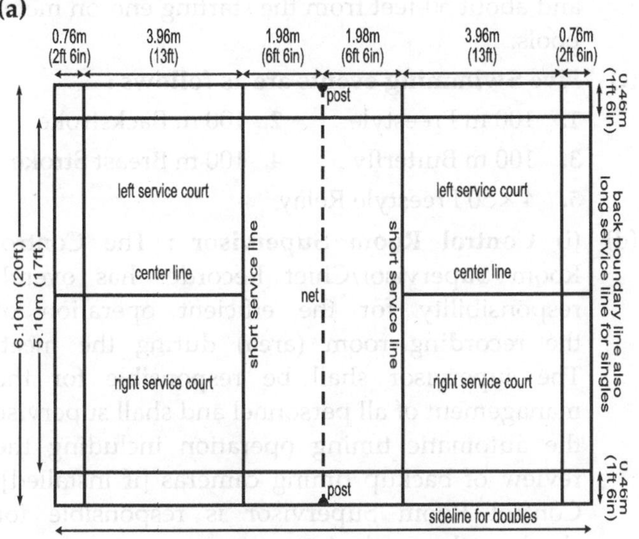

Diagram of Badminton Court

COURSE OF ACTION/EXPLANATION	SCORE	SERVICE FROM SERVICE COURT	SERVER AMD WINNER OF RECEIVER THE RALLY		
	Love All	Right service court. Being the score of the serving side is even.	A serves to C, A and C are the initial server and receiver.	A & B	C D B A
A and B win a point. A and B will change service courts. A serves again from left service court. C and D will stay in the same service courts.	1 – 0	Left service court. Being the score of the serving side is odd.	A serves to D	C & D	C D A B
C and D win a point and also right to serve. Nobody will change their respective service courts.	1 – 1	Let service court. Being the score of the serving side is odd.	D serves to A	A & B	C D A B
A and B win a point and also right to serve. Nobody will change their respective service courts.	2 – 1	Right service court. Being the score of the serving side is even.	B serves to C	C & D	C D A B
C and D win a point and also right to serve. Nobody will change their respective service courts.	2 – 2	Right service court. Being the score of the serving side is even.	C serves to B	C & D	C D A B
C and D win a point. C and D will change service courts. C serves from left service court. A and B will stay in the same service courts.	3 – 2	Left service court. Being the score of the serving side is odd.	C serves to A	A & B	D C A B
A and B win a point and also right to serve. Nobody will change their respective service courts.	3 – 3	Left service court. Being the score of the serving side is odd.	A serves to C	A & B	D C A B
A and B win a point. A and B will change service courts. A serves again from right service court. C and D will stay in the same service courts.	4 – 3	Right service court. Being the score of the serving side is even.	A serves to D	C & D	D C B A

(b) Every time there is a serve – there is a point scored. A rally is won, when a shuttle is hit over the net and onto the floor of the opponent's court. A rally is lost, if the shuttle is hit into the net, or over the net but outside of the opponent's court. A rally is also lost, if the shuttle touches the player's clothing or body, or if it is hit before it crosses over the net. The side winning a rally adds a point to its score.

In Badminton World Federation statutes book 2018-19 it is written that, it is permissible to play by prior arrangement, either.

1. One game of 21 points.
2. The best of 3 games of 15 points.
3. The best of five games of 11 points.

(c) **(i) Short Service Line :** It is a line 1.98 m from the net, which the serve must reach to be legal. This line marks the non-volley zone or fore-court of the playing area.

(ii) Back Gallery : The area in between the two last lines in the back area of the court *i.e.,* the second last line which is parallel to the centre line from the back end of the court *i.e.,* the line which is considered the end line for doubles serve and the last line which is parallel to the centre line from the back end of the court *i.e.,* the line which is considered end line in singles serve.

(iii) Foot Fault : According to the service rule of badminton, some part of both feet of the server and the receiver shall remain in contact with the surface of the court in a stationary position from the start of the service to until the service is delivered if it does not happen a foot fault will be called by the referee and the point will be awarded to the opponent of the foot faultier.

(iv) Love all : In badminton the game starts with love all. It means the score is 0-0.

(v) Rotation : Rotation in badminton may signify following :

1. Rotation of shoulder or elbow while executing various skills of badminton.
2. It may signify the order of receiving and order of serving in a doubles badminton match.

(d) **(i) Thomas Cup :** The Thomas Cup or World Men's Team Championships was first staged in 1948-49. Thomas Cup is named after Sir George Alan Thomas from England. The Thomas Cup was played every three years in the beginning. Since 1982, the Tournament has been held every two years with a format of 3 singles and 2 doubles. Since 1984, Uber Cup and Thomas cup are held together.

(ii) Uber Cup : Uber Cup or Women's World Team Championships was added for the first time in 1956-57. Uber Cup is named after Mrs. H.S. Uber from England. The Uber Cup was played every three years in the beginning. Since 1984, the tournament has been held every two years with the format of three singles and two doubles. Since 1984, Uber Cup and Thomas cup are held together.

(e) **(i) BWF :** Badminton World Federation.

(ii) SGFI : School Games Federation of India.

(iii) BAI : Badminton Federation of India.

(iv) IBF : International Badminton Federation.

(v) IBL : Indian Badminton League.

SWIMMING

Answer 15.

(a) 1. Swimsuit 2. Earplugs/Nose Clips 3. Goggles 4. Training tools 5. Swimming cap 6. Fins, Paddles and Boards 7. Stop watch

(b) A false start/recall rope is a rope that stretches across the width of competitive racing pools. It stops swimmers who were unaware of a false start. They're specifically designed to get the attention of swimmers, to prevent them from exerting unnecessary energy in the case of a false start. Typically, the rope is made of some sort of polymer plastic to prevent deterioration from the water, and includes two floats and two quick-snap connectors with weights. The rope is typically located about halfway on yard pools and about 50-feet from the starting end on meter pools.

(c) **Five Swimming events are as follows :**

1. 100 m Freestyle.
2. 100 m Backstroke
3. 100 m Butterfly
4. 100 m Breast Stroke
5. 4 × 50 Freestyle Relay

(d) **(i) Control Room Supervisor :** The Control Room Supervisor/Chief Recorder has overall responsibility for the efficient operation of the recording room (area) during the meet. The supervisor shall be responsible for the management of all personnel and shall supervise the automatic timing operation including the review of backup timing cameras [if installed]. Control Room Supervisor is responsible for checking the results from computer printouts.

(ii) Individual Medley : In individual medley events, the swimmer covers the four swimming strokes in the following order : butterfly, backstroke, breaststroke and freestyle. Each of the strokes must cover one quarter (1/4) of the distance.

(e) **The Federation International de Natation (FINA)** was founded in London July 19, 1908 during the Olympic Games in London (GBR). Eight national federations were responsible for the formation of FINA : Belgium, Denmark, Finland, France, Germany, Great Britain, Hungary and Sweden. Some of the objectives of FINA are to promote and encourage the development of aquatics in all possible regards throughout the world; to provide fair and drug free sport; to promote and encourage the development of international relations, to encourage participation in aquatics disciplines at all levels throughout the world regardless of age, gender or race etc.

FINA controls the development of the following aquatic events :

1. **Swimming :** freestyle, backstroke, breast-stroke, butterfly, medley, relays, mixed relays.
2. **Diving :** Springboard (1 m, 3 m), platform (10 m), synchronized diving (3 m and 10 m), mixed and team events.
3. **High Diving :** Men (27m) and women (20m).
4. **Water Polo :** Men and women.
5. **Synchronized Swimming :** Solo, duet, mixed duet, team and free combination.
6. **Open Water Swimming :** 5 km, 10 km, 25 km, team event.

ATHLETICS

Answer 16.

(a) The heptathlon consists of seven events and is a standard part of men and women's indoor track and field competitions, but only women compete in the heptathlon outdoors and in the heptathlon in the Olympics. Men's indoor track events often include the heptathlon. For women, the heptathlon is composed of the 100 m hurdles, long jump, javelin throw, 200 m high jump, 800m and shot put. For men's indoor competitions, the heptathlon consists of the 60 m, 1,000 m, shot put, 60 m hurdles, high jump, long jump and pole vault.

The decathlon is a 10-discipline combined track and field event that is primarily a men's event. A decathlon is sometimes held in women's track and field events, but not at major international competitions such as the Olympics. The events in the men's decathlon include the 100 m long jump, shot put, high jump, 400 m, 110 m hurdles, discuss throw, pole vault, javelin throw and 1,500m. The women's decathlon includes the same events.

(b)

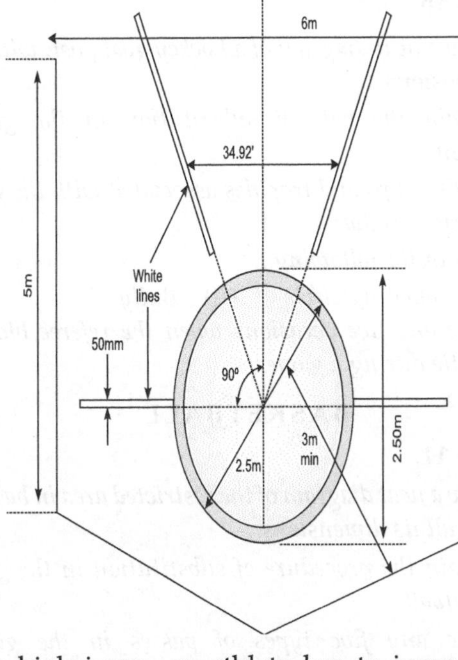

(c) In high jump, an athlete has to jump unaided and take off from one foot over a four-metre long horizontal bar. High jumper seek to clear the greatest height without knocking the bar to the ground. An athlete has three attempts per height, although athlete can elect to 'pass', *i.e.*, advance to a greater height despite not having cleared the current one. Three consecutive failures at the same height, or combination of heights, cause an athlete's elimination.

High jump technique can be mainly divided into three approach (Running towards the high jump bar) take-off (Movement of lifting the body up in air and flight (what is done in air to landing on mat). Major high jump techniques are :

1. **Scissor Jump :** Legs are lifted over the bar in alteration one after the other.
2. **Eastern Cut-off :** Athlete rotates the trunk into a horizontal position at the peak of the jump.
3. **Western Roll :** In this the athlete clears the bar on his/her side, with the take-off leg tucked under the rest of the body.
4. **Straddle Jump :** The athlete cleared the bar face-down, with the body stretched along the bar.
5. **Fosbury-flop :** The athlete runs up in a curve, jumps by taking off his outer foot and twists his body to clear the bar with his back. He finishes the movement by lifting his legs over the bar and landing on a mattress.

(d)

1. Referee shall have authority to warn or exclude from competition, any athlete guilty of acting in an unsporting or improper manner.
2. The referee shall not act as a judge or umpire but may take any action or decision according to the rules based on his own observations and may overrule a decision of a judge.
3. The appropriate referee shall check all final results.
4. If, in the opinion of the appropriate referee, circumstances arise at any competition such that the justice demands that any event or any part of an event should be contested again, he shall have the authority to declare the event or any part of an event void and that it shall be held again, either on the same day or on some future occasion, as he shall decide.
5. The referee may reconsider a decision (whether made in the first instance or in considering a protest) on the basis of any available evidence, provided the new decision is still applicable.

(e) **(i) Baton :** A short light stick that one member of a team in a relay race passes to the next person to run.

(ii) Marathon : A long running race of 42.195 km.

(iii) Hurdling : In hurdling a runner races over a series of obstacles called hurdles, which are set at a fixed distance apart.

(iv) Take off Area : The area marked on the ground from where the jumper takes off for a jump.

(v) Approach Run : The approach run is a buildup run by the athlete before moving the stage of taking off for a jump.

●●

PHYSICAL EDUCATION 2017

QUESTIONS

SECTION–A (40 Marks)

Answer any five questions.

Question 1.

Discuss the role of Physical Education in the development of an individual through the medium of games and sports. **[8]**

Question 2.

Explain the procedure of conducting circuit training, with the help of a diagram. Explain any four benefits of circuit training. **[8]**

Question 3.

Explain the meaning of weight training. Write any eight weight training exercises. **[8]**

Question 4.

What is International Olympic Committee (IOC) ? Write any five functions of the IOC. **[8]**

Question 5.

What is meant by correct posture ? Explain any two common postural deformities. **[8]**

Question 6.

Explain the meaning of Health Education. Mention any five principles of health education. **[8]**

Question 7.

Write short notes on the following : **[8]**

(a) Cardiopulmonary resuscitation.

(b) Stimulants.

SECTION-B (30 Marks)

*Select **any two** games from this Section*

*Answer any **three** of the **five** subparts (a), (b), (c), (d) and (e) from **each** of the **two** games selected by you. Each subpart is for **5 marks**.*

FOOTBALL

Question 8. **[3 × 5]**

(a) *Draw a neat diagram of a football field, showing all its specifications.* **[5]**

(b) *State any five ways of trapping the ball by various body parts.* **[5]**

(c) *Mention the basic compulsory equipment used by a football player.* **[5]**

(d) *Explain goal-kick. When is a goal-kick awarded ?* **[5]**

(e) *Explain the following :* **[5]**

(i) IFAB *(ii)* AIFF

CRICKET

Question 9. **[3 × 5]**

(a) *State any five pre-match duties of the umpire.* **[5]**

(b) *Name any five different variations of spin bowling, in cricket.* **[5]**

(c) *Specify the measurement of the following :* **[5]**

(i) *Length and width of a bat*

(ii) *Circumference and weight of a ball*

(iii) *Length of wickets above the ground.*

(d) *Write any five instances when a ball becomes dead.* **[5]**

(e) *Explain the following terms :* **[5]**

(i) *An appeal* *(ii)* *Bump ball*

HOCKEY

Question 10. **[3 × 5]**

(a) *Draw a neat diagram of a hockey goal post, with all its dimensions.* **[5]**

(b) *Explain the rule of substitution in the game of hockey.* **[5]**

(c) *List five cups and trophies associated with the game of hockey in India.* **[5]**

(d) *Explain the following :* **[5]**

(i) *Shooting circle* *(ii)* *Bully*

(e) *Write any five occasions when the referee blows the whistle during a game.* **[5]**

BASKETBALL

Question 11. **[3 × 5]**

(a) *Draw a neat diagram of the restricted area in basketball with all its dimensions.* **[5]**

(b) *Explain the procedure of substitution in the game of basketball.* **[5]**

(c) *Name any five types of passes in the game of basketball.* **[5]**

(d) *Mention any five duties of a scorer and an assistant scorer.* **[5]**

(e) *Explain the following terms :* **[5]**

(i) *Screening*

(ii) *Three Second Rule*

VOLLEYBALL

Question 12. **[3 × 5]**

(a) *Write any five skills a volleyball player should possess.* **[5]**

(b) *Explain the following :* [5]

(i) Blocking faults

(ii) Antennae

(c) *List any five important duties of a volleyball referee.* [5]

(d) *State any five violations of rules that are punished by the referee.* [5]

(e) *State the dimensions of the following :* [5]

(i) Circumference of the ball

(ii) Height of the net for men and women

(iii) Distance of attack line from the centre

(iv) Length of the net

(v) Length and breadth of the volleyball court.

TENNIS

Question 13. [3 × 5]

(a) *Draw a neat diagram of the tennis court, with its dimensions.* [5]

(b) *Explain the pattern of scoring in the game of tennis.* [5]

(c) *Explain the following :* [5]

(1) Let *(2) Foot fault*

(d) *State any five conditions when the service is considered to be a fault service.* [5]

(e) *Write the instances when a player hinders the opponent. What are the penalties for it ?* [5]

BADMINTON

Question 14. [3 × 5]

(a) *List any five duties of an umpire.* [5]

(b) *Explain the method of testing the speed of a shuttle.* [5]

(c) *List any five basic skills required by a player, in the game of badminton.* [5]

(d) *Define the following :* [5]

(i) Smash shot *(ii) Drop shot*

(iii) Drive shot *(iv) Clear shot*

(v) Toss or lob shot

(e) *State the following dimensions :* [5]

(i) Height of the net at the centre

(ii) Length and width of the racket

(iii) Weight of the shuttle

(iv) Number of feathers in a shuttle

(v) Width of the court lines.

SWIMMING

Question 15. [3 × 5]

(a) *Explain synchronised swimming and marathon swimming.* [5]

(b) *State the duties of the following :* [5]

(i) Judges of stroke

(ii) Recorders

(c) *Classify the different types of races in swimming.* [5]

(d) *Name any five important swimming tournaments.* [5]

(e) *Explain the following :* [5]

(i) Lane ropes

(ii) Automatic officiating equipment.

ATHLETICS

Question 16. [3 × 5]

(a) *Draw a neat diagram of standard 400 m track, with all its dimensions.* [5]

(b) *List the fundamental skills required by an athlete, for hammer throw.* [5]

(c) *What is false start ? Explain any two types of finishing techniques.* [5]

(d) *Write any five duties of the call-room manager.* [5]

(e) *Explain the following :* [5]

(i) Exchange zone *(ii) Starting blocks.*

ANSWERS

SECTION–A (40 Marks)

Answer 1.

1. Games and sports help not only in the physical fitness but also the psychological fitness. A combination of physical and psychological fitness leads to an overall development of an individual.
2. Games and sports help an individual to maintain a good shape and size of the body.
3. They also help in the improvement of the physiological capacities of the various systems of the body. Games and sports affect not only the physical and physiological aspects of the body but also the psychological fitness.
4. They help to improve perception, emotional stability, motivation and intelligence of an individual.
5. They help the individual to bear stress and tension better.

Therefore, we can see that the games and sports helps in the overall development of an individual.

Answer 2.

Circuit training is a form of body conditioning or resistance training using high-intensity aerobics. It targets strength building or muscular endurance. An exercise "circuit" is one completion of all prescribed exercises in the program.

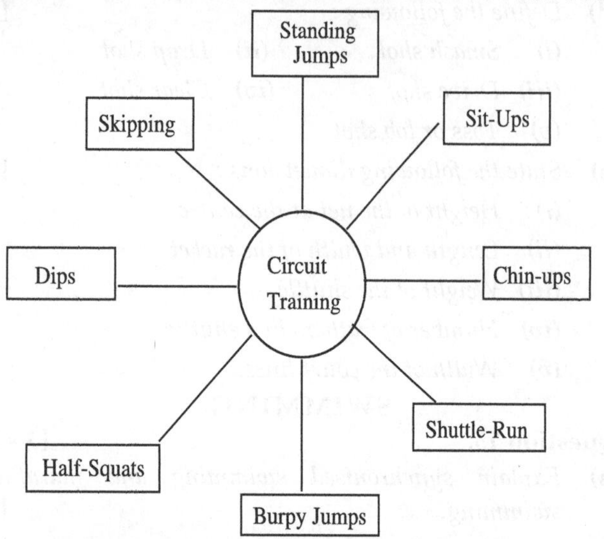

Four benefits of Circuit Training :

1. **Get Maximum Results in Minimum Time :** Circuit training is the perfect workout for the time-crunched fitness. You can set up as many or as few stations as you like and then continue through the circuit until your time runs out. You can fit it in effective, total-body workout whether you have 15 min or an hour.
2. **Send your Metabolism Soaring :** Circuit training kicks your fat-burning furnace into high gear. You'll torch more calories in 20 minutes than you would in an entire hour on the elliptical reading a magazines. Plus, the afterburn effect will have you torching calories for up to 48 hours after your workout ends.
3. **Never Waste Another Minute :** It is easy to swap out one station for a different exercise and still get the same great workout.
4. **Beat Gym Boredom :** Circuit training is such a flexible format, the options in how you set up your workout are limitless. Whether you like to stick to one routine for a while to measure your progress or prefer mixing it up every time you go to the gym, having a plan will make you more confident and efficient.

Answer 3.

Weight training is a common type of strength training for developing the strength and size of skeletal muscles. It utilizes the force of gravity in the form of weighted bars, dumbbells or weight stacks in order to oppose the force generated by muscle through concentric or eccentric contraction. Weight training uses a variety of specialized equipment to target specific muscle groups and types of movement.

1. **Bodybuilding :** Bodybuilders train in a specific manner to achieve a specific goal making their muscles bigger. They tend to lift in the eight- to 12-rep range and only train one muscle group per day, per week. This type of training is the best for making your muscles larger but not necessarily stronger.
2. **Power Lifting :** Power lifting is the best type of training for maximal strength - what many people would consider "brute" strength. A power lifter's goal isn't appearance, it's strength. Power lifters focus on lifting an extremely heavy amount of weight for only a few repetitions.
3. **Circuit Weight Training :** Circuit weight training is when you do a number of exercises in quick succession, usually lifting a light amount of weight for a high number of repetitions. For example, doing 20 bench presses, 20 squats and shoulder presses with little or no rest in between. This type of training is ideal for burning fat, increasing endurance and making some gains in strength.
4. **Isometric Weight Training :** Isometric weight training is when you hold a weight up in one position for a given amount of time without going through a range of motion. For example, holding a dumbbell straight in front of you for 30 seconds. This type of training increases your stamina and makes you stronger in the positions you hold.
5. **High-Volume Training :** High volume training is similar to body building in the sense that you only work your muscle groups once a week. This type of training places a strong emphasis on increasing muscular endurance one exercise at a time. Not only does it makes your muscles larger, it also makes them better at keeping up a certain amount of resistance for a long period of time. It is done by performing one exercise at a time, lifting the same amount of weight 10 times for 10 sets. Each week, you try to increase the amount of weight slightly.
6. **Resistance Bands :** With resistance bands, you can strengthen and tone all the major muscle groups. The advantage of resistance bands is that, you can adjust the resistance, using less resistance when needed or increasing the resistance to increase the challenge.
7. **Pull-Ups :** A pull-up is when your hands are facing away from you. This will work your back and biceps.
8. **Back Squat:** This is the squat movement where the biggest amount of weight is moved. This is the one and original squat movement which has been around for as long as people have been lifting weights.

Answer 4.

The body that governs the Olympic Games is the International Olympic Committee (IOC). This is a non-governmental organization that was created on June 23rd, 1894, Paris, France based in Lausanne Switzerland, by Baron Pierre De Coubertin. IOC is the governing body of Modern Olympic Games.

Main Role—Functions of International Olympic Committee (IOC) : As IOC being the governing body, it has various roles and functions. The roles and functions are as follows :

1. To boost support and promote the ethics in games and sports.
2. To encourage and support the organisation, its progress, coordination of sports and sporting competition.
3. To act in contradiction of any form of discrimination that disturbs the Olympic Movement.
4. To execute the principle of equality between men and women by encouraging and supporting in elevating the women in sports at all level.
5. To fight against illegal practices by atheletes (doping) in sport.

Answer 5.

Correct posture means the balancing of body in accurate and proper manner while sitting, standing, reading and writing, walking, running or during any other action.

Postural deformities imply not having proper alignment of body parts. An individual who has postural deformities cannot perform his work efficiently. Some postural deformities are kyphosis, flat foot, knock knee etc.

1. **Lordosis :** Lordosis denotes or indicates increased anterior curvature of spine. It is an exaggeration or increase in the amount of the normal concavity of the lumber region of the spine, *i.e.*, an increase in the extension curve of the lumber region beyond normal region. It is an inward curvature of the spine with protruding abdomen resulting in hollow back.
2. **Scoliosis :** The term 'scoliosis' is derived from a Greek word, which means twisting or bending. Scoliosis means a lateral bending of spine to one or more sides from midline. It is an exaggerated lateral curvature *i.e.* sideward curvature or deviation of the spine, with the shoulder lower and hip higher on one side. It may be present in a very small part of spine or it can involve the whole or major part of spine. Usually, there is a single curve in one direction only and called 'C' curve scoliosis. When two or more than two lateral curves are present in the spine, then it is called 'S' curve scoliosis or compound and double scoliosis.

Answer 6.

Health education is one strategy of health promotion and is focused on helping individuals learn and use health-enhancing skills. Health education helps to know about the causes and occurrence of various diseases that mainly is the result of sedentary lifestyles and lack of activities. So, knowing of these may help people to participate in physical activities. It will also helpful to know about the nutritional facts, healthy and harmful food and suitable physical activity or exercise for an individual.

Principles of Health Education :

1. Supported by the latest knowledge from research.
2. A systematic, comprehensive and consistent activity.
3. Respects environment or culture of an individual.
4. Encourages personal investment of an individual.
5. Adapted to age, gender, education and particular health, mental or social problems of an individual or community.

Answer 7.

(a) **Cardiopulmonary resuscitation :** Cardiopulmonary resuscitation (CPR) is a procedure to support and maintain breathing and circulation for an infant, child, or adolescent who has stopped breathing (respiratory arrest) and/or whose heart has stopped (cardiac arrest).

CPR is performed to restore and maintain breathing and circulation and to provide oxygen and blood flow to the heart, brain and other vital organs. CPR can be performed by trained people or healthcare professionals on infants, children, adolescents and adults.

(b) **Stimulants :** A stimulant is one of the classes of doping that comprises types of drugs that increases alertness, reduce fatigue and also enhances competitiveness, and hostility. Using of this, also produces loss of judgements that may lead to accidents to others in some sports. Some of the examples are like amphetamine, cocaine, pentetetrazol, phentermine, ephedrine etc. It improves the performances by providing a psychological and physical stimulus. It increases the heart and breathing rates, the functions of the brain and aggressiveness. They are taken orally, by sniffing, by smoking or by injecting in the body.

SECTION-B (30 Marks)

FOOTBALL

Answer 8.

(a)

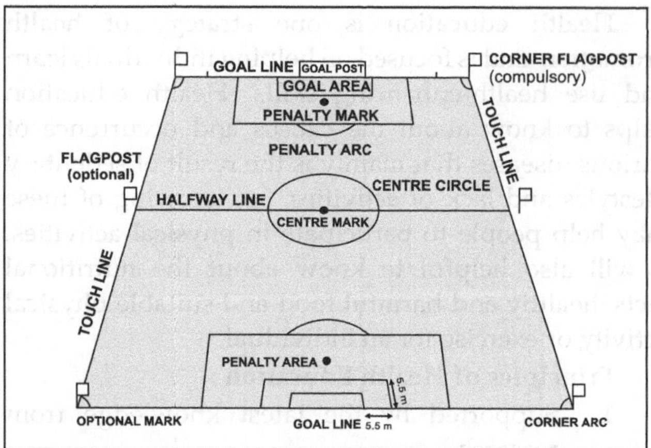

Football field

(b) **1.** You can trap both high and low balls with your foot in many different ways.

2. You can trap balls on the ground, in the air and during or just after the bounce.

3. You can trap the ball with the inside, outside, underside and top of the foot.

4. There are two basic ways of trapping a high ball.

5. One is to allow the ball to bounce up a little when trapping it.

6. You can then play the ball on before it reaches the ground.

(c) **The basic compulsory equipments required by a football player are :**

1. A jersey or shirt with sleeves.
2. **Shorts :** If thermal undershorts are worn, they are of the same main colour as the shorts.
3. Stockings.
4. **Shinguards :** These are covered entirely by the stockings. These are made of a suitable material (rubber, plastics or similar substances). They provide a reasonable degree of protection.
5. Footwear

(d) **Goal-kick :** When the whole of the ball passes over the goal-line excluding that portion between the goal posts, either in the air or on the ground, having last been played by one of the attacking team, it shall be kicked directly into play beyond the penalty area from a point within that half of the goal-area nearest to the point where it crossed the line by a player of the defending team. A goal-keeper shall not receive the ball into his hands from a goal-kick in order that he may thereafter kick it into play. If the ball is not kicked beyond the penalty area *i.e.* directly into play, the kick shall be retaken. The kicker shall not play the ball a second time until it has touched or been played by another player. A goal shall not be scored directly from such a kick. Players of the opposite team shall remain outside the penalty-area until the ball has been kicked out of the penalty area.

A goal kick is awarded to the defending team when the ball goes out of the field of play by crossing either on the ground or in the air, the goal line, without a goal being scored, when the last person to touch the ball was from the attacking team.

(e) **(i) IFAB :** The International Football Association Board.

It is the body that determines the laws of the game of association football. IFAB was founded in 1886 to agree standardised laws for international competition, and has since acted as the "guardian" of the internationally used Laws; since its establishment FIFA in 1904, the sport's top governing body, has recognised IFAB's jurisdiction over the Laws.

IFAB is known to take a highly conservative attitude regarding changes to the laws of the game.

(ii) AIFF : The AIFF is a National Association registered with the Registrar of Societies, Mumbai, under the Societies Registration Act, 1860. The headquarters of the AIFF is in Dwarka, New Delhi. AIFF is a member of FIFA, AFC and Indian Olympic Association and is self obliged to respect the Statues, Regulations, Directives and Decisions of FIFA and AFC and to ensure that these are likewise respected by its members. Its main objective is to improve and popularise the game of football constantly and promote it throughout India in the light of its unifying, educational, cultural and humanitarian values, particularly through youth and development programmes to urban, rural and remote areas including schools, colleges and universities. Its responsibility is to draw up regulations and provisions and ensure their enforcement.

CRICKET

Answer 9.

(a)

Umpire Duties Before the Match

Day to Previous Day | Day Before the Match | The Match Day

(i) **Day to Previous Day : 1.** Acknowledge the appointment and seek particulars relevant to the match. (Venue, Date, Teams, Partners, Scorers etc.)

2. Connect with partner(s) and agree on the time to meet the venue on previous day.
3. Revise laws and connect to public information relevant to the match or tournament.

(ii) Day Before the match : 1. Meet partner as planned and jointly meet match and ground authority at the venue–aim to establish report.

2. Enquire about weather, ground and light conditions affecting the match.
3. Check facilities like drainage, cover at disposal etc.

(iii) The Match Day : 1. Umpire to be present on the ground and report to ground authority at least 45 mins before scheduled start to the match.

2. Umpire have to inspect the playing field and satisfy that the things are fix or fit for play to start.
3. Meet the scorers etc.

(b) 1. Off spin or off break.

2. Leg spin or leg break.
3. Top spin.
4. Googly.
5. Arm ball (Right arm off-spin)

(c) (i) Length 38 inches width 4·25 inches.

(ii) Circumference–22·4 to 22·9 cms and Weight : 155·9 gms to 163 gms.

(iii) 28 inches.

(d) A ball is dead in cricket when :

1. It is finally settled in the hands of the wicket keeper.
2. A boundary is scored.
3. A batsman is dismissed.
4. Lost ball.
5. Whether played or not, it lodges in the clothing or equipment of a batsman or the clothing of an umpire.

(e) **(i) An appeal :** The umpires shall not give a batsman out unless appealed to by the other side which shall be done prior to the bowler beginning his next delivery.

An appeal "How's that" ? shall cover all ways of being out. Answering appeals, the umpire at the bowler's wicket shall answer appeals before the other umpire in all cases, except those arising out of 35 (Hit wicket) or 39 (Stumped) or 38 (Run out) when this occurs at the striker's wicket.

(ii) Bump ball : A delivery that bounces very close to the batsman's foot, after he has played a shot, such that it appears to have come directly from the bat without ground contact. The result is often the appearances of a clean catch.

HOCKEY

Answer 10.

(a) Goal Dimensions

Code	*Metre*	*Code*	*Metre*
A	3·66	E	Minimum 0·90
B	2·14	F	0·050
C	0·46	G	0·050 to 0·075
D	Minimum 1·20		

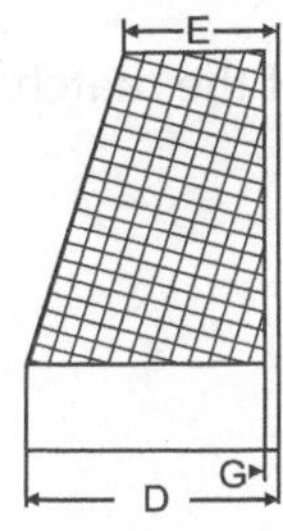

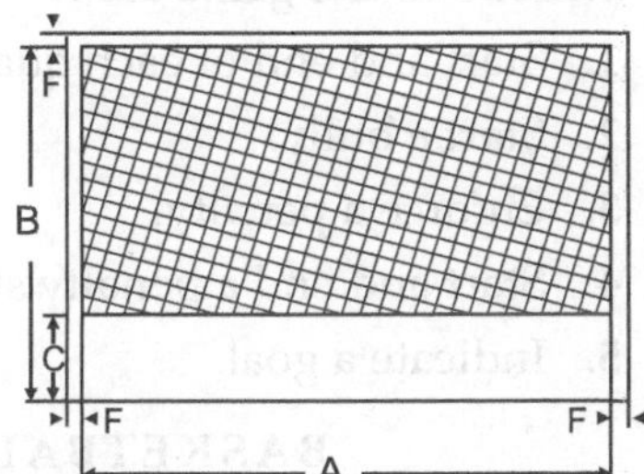

(b) The Rules of Substitution : 1. There is no limit to the number of players who may be substituted at the same time nor to the number of times any player may substitute or be substituted.

2. Substitution of a player may be made only after a player from the same team has left the field.
3. Time shall not be stopped for substitutions except for goalkeepers.
4. Substitutes shall not be permitted for suspended players during their suspension.
5. After completing a suspension, a player may be substituted without first returning to the field.
6. Players leaving or entering the field shall do so at the centre-line or such other place on the side of the field decided by the umpires before the game.

(c)

Cups	Trophies
1. Agha Khan Cup	1. Dhyan Chand Trophy
2. Scindia Gold Cup	2. Nehru Trophy
3. Bombay Gold Cup	3. MCC Trophy
4. Rangaswami Cup	4. Gurmeet Trophy
5. Indira Gandhi Gold Cup	5. Guru Nanak Trophy

(d) **(i) Shooting Circle :** A semi-circle is drawn on the ground 16 yards in front of the goal, which is referred to as the shooting circle in Hockey.

1. A goal is only awarded, when the ball is touched inside the circle by an offender or defender and the goal is converted.
2. The severity of the foul will be in higher degree when it is committed inside the circle.
3. An intentional foul inside the circle will be awarded with a penalty stroke.

4. An unintentional foul inside the circle will be awarded with a penalty corner.

(ii) Bully : It is a call used to start or restart play. The referee puts the ball between two opposing players. Two players of both the teams stand in front of each other. They touch this stick to the ground three times and alternately their opponent's stick after that ball is put in play.

(e) Five occasions when the umpire shall blow the whistle in the game are :

1. Start and end of each quarter of the match.
2. Start a bully.
3. Enforce a penalty.
4. Start and end a penalty stroke.
5. Indicate a goal.

BASKETBALL

Answer 11.

(a)

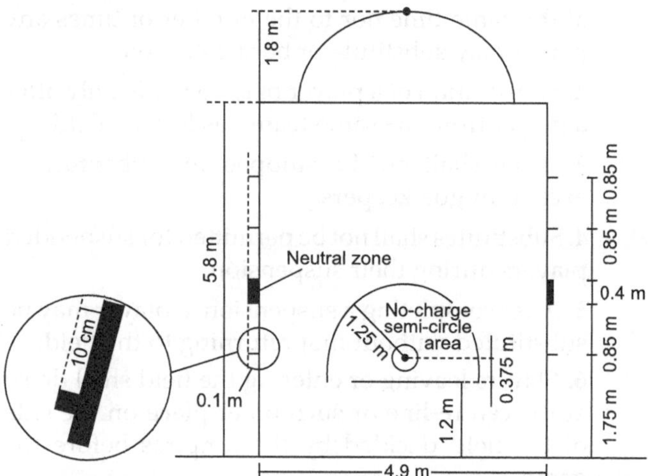

Restricted Area

(b) The coach should send the substitute who must be ready to play to the Scorer. After reporting to the Scorer, the substitute must sit on the seat provided until the Scorer sounds his signal. He should then stand up and indicate to the nearer Official that he wishes to enter the court. He should not enter the court until beckoned to do so by an Official. If substitution is required during a charged time-out, the substitute must report to the Scorer and the nearer Official before entering the court.

(c) The five types of passes, in the game of basket ball are as follows :

1. Chest Pass
2. Bounce Pass
3. Baseball Pass
4. Outlet Pass
5. No-look pass

(d) Duties of the Scorer :

1. The scorer shall keep a record of the names and number of players who are to start the game and of all substitute who enter the game.
2. The scorer shall keep a chronological running summary of the points scored, he shall record the Baskets made and free throws made or missed.
3. He shall record the time-outs charged to each team and shall notify the coach through an official when he has taken a second time-out in each half or has no more time-outs left in a half or extra period.
4. He shall also indicate the number of fouls committed by each player by raising, in a manner visible to both coaches and by using the numbered markers as provided.
5. Effect substitutions.

Duties of Assistant Scorer : The assistant scorer shall note when each half is to start and shall notify to the referee more than three minutes before this time so that he may notify the teams, or cause them to be notified, at least three minutes before the quarter is to start. He shall signal the scorer two minutes before starting time.

(e) (i) Screening : Screening occurs when a player attempt to prevent an opponent who does not control the ball from reaching a desired position on the court. Screening may be legal or illegal.

Legal screening occurs when the player who is attempting to screen an opponent is not in motion but standing still and has both feet on the floor. Illegal screening occurs, when the player who is attempting to screen an opponent is moving when contact occurs with the player who is being screened. If that occurs, the screener is guilty of blocking and blocking is a foul.

(ii) Three Second Rule : A player shall not remain in the opponents restricted area for more than three consecutive seconds while his team is in control of a live ball in the front court and the game clock is running.

VOLLEYBALL

Answer 12.

(a) Five fundamental skills required by a player :

1. The service.
2. The Net Recovery.
3. The Block
4. The Set-up
5. The Attack (smash)

(b) (i) Blocking Faults : A blocker commits a faults when :

1. He blocks outside the antennae and touches the ball in the opponent's space.

2. He interferes with the play of an opponent in the opponent, space, either before or during the action.

3. As a back line player, he is effecting or participating in a block.

4. He blocks the opponent's service. (he may neither block nor attack in the front zone from higher than the top of the net).

(ii) **Antennae :** The antennae are two flexible rods made or fiber glass or similar material, 1.80 m long and 10 mm in diameter.
The antennae shall be marked with 10 cm bands of contrasting colours, preferably red and white. Each is fastened on the outer edge of each vertical side band extending net and on opposite sides of the net. They are considered as parts of the net and mark its side limits.

(c) 1. The referee initiates service, and controls the tempo of the game.

2. The referee is primarily responsible for judgements regarding ball handling.

3. The referee is only responsible for rotation of serving team.

4. The referee has the duty of communication with umpire, captains and statistician.

5. The referee performs the toss up for service and court.

(d) 1. **Consecutive Contacts :** A player may not hit the ball 2 times consecutively except when blocking, provided the contacts take place simultaneously. However, a player may make consecutive contacts on the teams first team contact, provided the contacts occur during one action.

2. **Four Hits :** It is a violation for a team to hit the ball 4 times before returning it.

3. **Assisted Hit :** It is illegal for a player or any object to assist a teammate in playing the ball.

4. **Catch :** It is illegal to catch or throw the ball. The ball must rebound from the hit.

5. **Ball Crossing the Net Outside the Crossing Space :** The ball must cross the net within the crossing space. The ball must cross over the net, between the antennae and their imaginary extension, and without contacting the ceiling.

(e) (i) 65 cm to 67 cm.

(ii) 2·43 m for Men.
2·24 m for women.

(iii) 3 m

(iv) 9·50 m

(v) Length : 18 meters.
Width : 9 meters.

TENNIS

Answer 13.

(a)

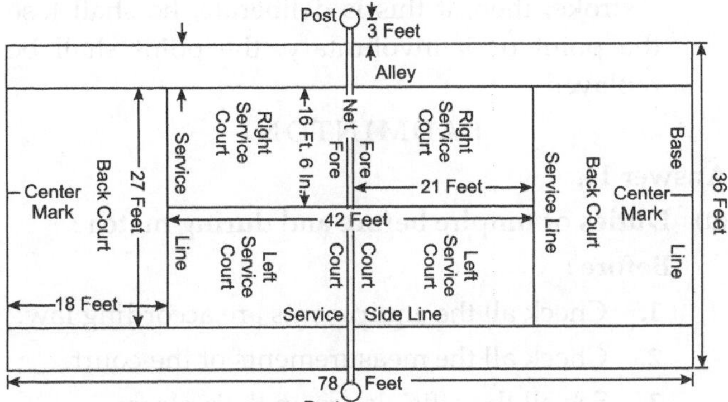

(b) The tennis scoring system is the standard system by which tennis matches and tournaments are generally operated and scored. A tennis tournament is organised into matches between players (for singles tournaments) or teams of two players (for doubles tournaments). The matches of the tournaments are grouped into rounds. In round 1, all players (or teams) are paired and play against each other in matches. The losers are said to leave, or be out. They no longer compete in the tournament (this is single elimination). The winner of the final round will go further for the other matches of the tournament. A tennis match is composed of points, games, and sets. Traditionally, matches are either a best of three sets or best of five sets format. The best of five set format is typically only played in the men's singles or doubles matches at Majors and Davis Cup matches.

(c) 1. **Let :** A serve is called a let when the ball hits the net cord but still lands in the service court. Such a serve is not considered a fault and the server may repeat the service attempt. A ball that hits the net cord but lands outside the service box is still a fault.

2. **Foot Fault :** Foot faults occurs when the player is serving and he steps on or over the line before making contact with the ball.

(d) **Five conditions when the service is considered to be a fault are :**

1. When the server touches inside the court while serving.

2. Two consecutive service faults leading to lose on point.

3. If the players miss the ball in attempting to strike it.

4. If the ball goes in the side gallery.

5. If the ball served touches the permanent fixture other than the net, strap or band.

(e) **Player Hinders Opponent :** If a player commits any act which hinders his opponent in making a stroke, then, if this is deliberate, he shall lose the point or if involuntary, the point shall be replayed.

BADMINTON

Answer 14.

(a) **Duties of umpire before and during match :**

Before :

1. Check all the equipments are according law.
2. Check all the measurements of the court.
3. See all the officials are on their places.

During :

1. Uphold and enforce the laws of game.
2. Give a decision on any appeal.
3. Appoint or remove the line judges.
4. Record and report to the referee all the matters.

(b) To test a shuttle, a player shall use a full underhand stroke which makes contact with the shuttle over the back boundary line. The shuttle shall be hit at an upward angle and in a direction parallel to the side lines.

A shuttle of correct speed will land not less than 530 mm and not more than 990 mm short of the other back boundary line as in Diagram B.

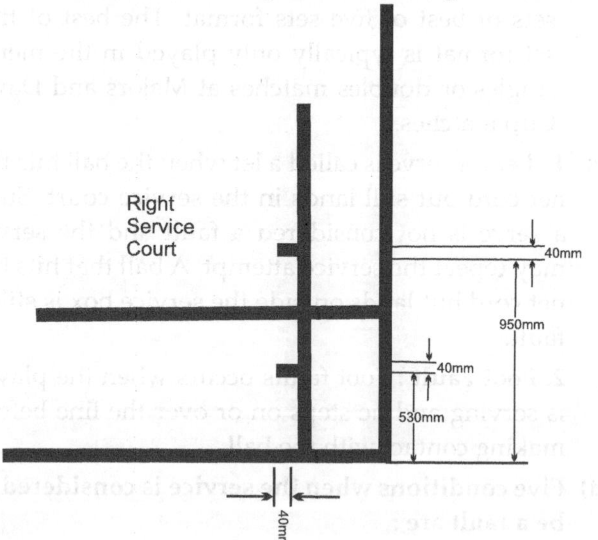

(c) **Fundamental skills required by a player :**

1. Correct footwork, 2. Service
3. Lob 4. Drop
5. Smash 6. Back hand
7. Net Play

(d) **(i) Smash shot :** It is a powerful stroke which pushes the shuttle cock to move downwards sharply. This is the most important and point winning stroke.

(ii) Drop Shot : It is a very soft shot that can help in winning points if played well. This stroke can be executed from forehand and backhand sides in the frontcourt of the opponent. Drop-shots are of two types—fast and slow drop-shots.

(iii) Drive Shot : It is a stroke in which the shuttle crosses over a net in a horizontal or flat trajectory. It can be played from both forehand and backhand sides. It is prominently used in the game of doubles a player intends to keep the shuttle as low as possible.

(iv) Clear Shot : This is the most common and important of all badminton strokes that can be played overhead or underarm and can be executed from both forehand and backhand sides of the court. The clear can be of attacking (flat) and defensive (long and high) in nature.

(v) Toss or Lob Shot : Prior to the start of a game, there will be a toss. The winner of the toss makes a decision to choose an end of the court or to serve first. If the player chooses the court, the opponent player will serve first. On the other hand, lobbing or lob shot is a stroke executed at the baseline, hitting the ball far to the back of the opponent court. Any good lob should be over the opponent's reach at mid court, hence pushing them to the back of the baseline.

(e) (i) 5 Ft high in the centre.

(ii) L = 680 mm.

W = 230 mm.

(iii) 4·75 —5·50 gms.

(iv) 16

(v) 6·1 m full width of the court.

1·98 m × from the centre line dividing the width of the court.

SWIMMING

Answer 15.

(a) 1. **Synchronised Swimming :** Synchronised swimming is a hybrid of swimming performing a synchronised routine of elaborate move in the water accompanied by music. It demands advanced water skills and requires great strength, endurance etc. Synchronised swimming is both an individual and team sport.

2. **Marathon Swimming :** Marathon swimming is a continuous open water swimming of at least 10km. This distance was chosen for its equivalence in time-to-completion to a 26·2 mile marathon run.

(b) (i) Judges of Strokes : At the start, the stroke Judge is responsible for observing the swimmers at the 15 meter mark for all strokes except the breaststroke. After the start, the Stroke judge walks the sides of the pool, preferably abreast of swimmers during all strokes except freestyle. The breaststroke and butterfly are best judged by walking slightly behind the swimmers. Judges should walk at a pace that is sufficient to stay with the swimmers. If the field spreads out, the stroke judge should maintain a position slightly ahead of the trailing swimmers while maintaining contact with the lead swimmer(s). If there are two (2) stroke judges working one side of the pool (usually in long course meets), a "lead-lag" observation pattern could be utilized as the field spreads out.

(ii) Recorders : Record withdrawls from the competition, enter results on official forms and maintain record for team awards as appointment.

(c) Types of races in swimming are as follows :

1. Freestyle
2. Back stroke
3. Breast stroke
4. Butterfly
5. Individual medley
6. Relays.

(d) Five important swimming tournaments are :

1. FINA Swimming World Cup
2. European Masters Swimming
3. Swedish Swimming
4. Australian Swimming.
5. World Open Water Championships.

(e) (i) Lane Ropes : They shall extend the full length of the course, secured at each end wall to anchor brackets recessed into the end walls. Each lane rope will consists of floats placed end to end having a minimum diameter of 5 cm to a maximum of 11 cm. The colour of the floats extending for a distance of 5·0 m from each end of the pool shall be distinct from the rest of the floats. There shall not be more than one lane rope between each lane.

(ii) Automatic Officiating Equipments : Automatic Officiating Device is used in any competition. The placings and timings so determined shall have precedence over the decisions of human judges and timers.

ATHLETICS

Answer 16.

(a)

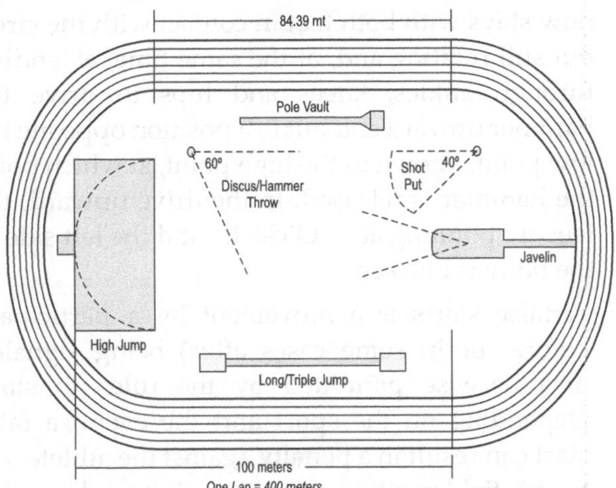

400-meter Run : One full lap around an outdoor track. Runners remain in their assigned lanes and receive staggered starts. Although competitors begin in starting blocks and the 400 is technically considered a sprint the runners must pace themselves a bit. Come from behind victories.

(b) Hammer throw is a sport in athletics (track and field) in which a hammer is hurled for distance, using two hands with a throwing circle.

1. **Grip :** The athlete's left hand, wearing the glove, grips the handle between the 2nd and 3rd joints of the fingers and the right hand is then placed inside the handle on top of the left hand.

2. **Swings :** The thrower stands at the rear of the circle, feet parallel, with back towards the throwing sector. The preliminary swings are used to commence momentum of the throw.

3. **Entry :** At the end of the swings, the athelete is about to enter the turning phase. With the hammer directly in front of the thrower, the hammer is pushed to the left and the turns are started.

4. **Turns :** The number of turns performed is a personal choice, but in general 3 or 4 are used. These are a series of movements performed to increase the speed of the hammer using balance through the feet, and power from the legs and hips. The feet works together throughout the throw. The left foot is rotated into the first turn. The left foot is rotated 180∞ on the heel towards the direction of the throw, and the right foot rotates 180∞ on the toe. The thrower then picks up the right foot, brings it fully around and places it on the ground parallel to the left. This completes one turn. The elite throwers usually perform 3 or 4 turns, whereas the beginners perform one or two turns.

5. Releasing of Hammer : After the turns performed by the thrower have accelerated, the hammer head to its maximal velocity, the thrower now stays with both feet in contact with the circle but still rotating and, at the same time, extending through ankles, knees and hips to drive the hammer upward and out to a position opposite the low point, known as the high point, at which point the hammer is released. In the drive upward, the feet are pointing at "9 O'clock" and the left side of the body is blocked.

(c) A false starts is a movement by a participant before (or in some cases after) being signaled or otherwise permitted by the rules to start. Depending on the sport and the event, a false start can result in a penalty against the athlete's or team's field position, a warning that a subsequent false start will result in disqualification, or immediate disqualification of the athlete from further competition.

Generally, the athletes use three types of finishing techniques which are mention below :

1. Run through : This technique is generally adopted by novice athletes. In this technique, the finishing line is crossed with the full speed by the athletes.

2. Lunging : In this technique, the athlete bends his 'torso' towards the finishing line in the last stride. This technique is adopted by well experienced athletes.

3. Shoulder shrug : An athlete tends to bend his one shoulder and the complete action is performed in the last stride near the finishing line by the extraordinary athletes.

(d) Duties of call-room manager are :

1. Organise the call-room area.
2. Brief the team.
3. Attend the technical meeting.
4. Organise the team.
5. Receive and relay any messages.

(e) **(i) Exchange Zone :** It is an area of the track where the baton is passed from one runner to the other during a relay race.

(ii) Starting Block : It is a device used in the sports of track and field by sprint athletes to hold their feet at the start of a race so they do not slip as they push out at the sound of the gun.

●●

PHYSICAL EDUCATION 2016

QUESTIONS

SECTION–A (40 Marks)

*Answer **any five** questions.*

Question 1.

Explain how games and sports are helpful in promoting national integration among the players. **[8]**

Question 2.

What is meant by sports training ? Explain the importance of sports training in the field of games. **[8]**

Question 3.

List the various career options available in the field of physical education. Discuss any two of them in detail. **[8]**

Question 4.

Mention the importance of intramural competitions and extramural competitions. Write any two objectives of each. **[8]**

Question 5.

What is meant by drugs and drug abuse ? Name any two drugs that are banned by the World Anti-Drugs Agency (WADA). Explain why these drugs are banned. **[8]**

Question 6.

What is scoliosis ? State the causes and corrective measures for scoliosis. **[8]**

Question 7.

What is meant by bruises ? Mention five measures that can be taken by an individual to prevent sports related accidents. **[8]**

SECTION B (30 Marks)

*Select **any two** games from this Section*

*Answer any **three** of the **five** subparts (a), (b), (c), (d) and (e) from **each** of the **two** games selected by you. Each subpart is for **5 marks**.*

FOOTBALL

Question 8. **[3 × 5]**

(a) *What is meant by wall formation in the game of football ? With the help of a diagram show any one defensive formation and any one offensive formation in the game.* **[5]**

(b) *State the importance of technical area in the game of football. Also, give its specifications.* **[5]**

(c) *Mention any five duties of the fourth official in the game.* **[5]**

(d) *What is a direct free kick in the game of football ? State any three situations when a referee awards a direct free kick.* **[5]**

(e) *Explain the following :* **[5]**

(i) Heading

(ii) Goal area

CRICKET

Question 9. **[3 × 5]**

(a) *Draw a neat labelled diagram of a cricket field and show any ten fielding positions of the players.* **[5]**

(b) *Explain the term power play in the game of cricket. What are the restrictions imposed on the players during the power play ?* **[5]**

(c) *What is meant by tampering of ball ? Mention any two ways by which a ball is tampered.* **[5]**

(d) *What are the duties of the third umpire ? How does the third umpire convey his decision during the game ?* **[5]**

(e) *Explain the following terms :* **[5]**

(i) Bouncer *(ii) Lost ball*

HOCKEY

Question 10. **[3 × 5]**

(a) *Mention the protective equipment used by the goalkeeper during the game.* **[5]**

(b) *What is Astroturf ? State its importance in the game of hockey.* **[5]**

(c) *When is a free hit awarded in the game of hockey ? Write any two famous tournaments associated with the game.* **[5]**

(d) *Explain the following :* **[5]**

(i) Push pass

(ii) Team formation

(e) *Write any five basic skills a hockey player should possess.* **[5]**

BASKETBALL

Question 11. **[3 × 5]**

(a) *Write a short note on FIBA. Define the term violation in basketball.* **[5]**

(b) *Differentiate between a personal foul and a foul.* **[5]**

(c) *Mention any five duties of a time-keeper during the game.* **[5]**

(d) *Explain the term jump shot in the game of basketball. Write any four types of passes used in the game.* **[5]**

(e) *Explain the following terms :* **[5]**

(i) Technical foul

(ii) Pivoting

VOLLEYBALL

Question 12. [3 × 5]

(a) *Explain the system of rotation in the game of volleyball. Mention any four rotation faults a player may commit during the game.* [5]

(b) *Explain the following :* [5]
(i) Assisted hit *(ii) Penetration fault*

(c) *(i) Who is a Spiker in the game of volleyball ?* [2]
(ii) List any three blocking faults the players may commit during the game. [3]

(d) *What is the method of scoring in the game of volleyball? State the various types of delays in the game.* [5]

(e) *Expand the following :* [5]
(i) FIVB *(ii) VFI*
(iii) AVF *(iv) UVA*
(v) JVA

TENNIS

Question 13. [3 × 5]

(a) *(i) Explain the rule of advantage in the game of tennis.* [5]
(ii) What is meant by the term ace ?

(b) *List the duties of a chair umpire in the game of tennis.* [5]

(c) *(i) Give detailed specifications of a tennis ball with reference to its :* [4]
(1) Material *(2) Weight*
(3) Circumference *(4) Colour*
(ii) Mention the width of the strap of the net used in Tennis. [1]

(d) *Write a short note on Grand Slam Tournament. Also, name any two other important tournaments associated with the game.* [5]

(e) *Explain any five basic shots used by the players in the game.* [5]

BADMINTON

Question 14. [3 × 5]

(a) *Name any five important tournaments associated with the game of badminton.* [5]

(b) *Mention any five faults a player may commit during the game.* [5]

(c) *Draw a neat diagram of a shuttle with all specifications.* [5]

(d) *Define the following :* [5]
(i) Bird *(ii) Let*
(iii) Drive *(iv) Double hit*
(v) Service fault

(e) *(i) Explain the procedure to be followed for performing a correct service in the game of badminton.* [3]
(ii) When do the players change ends in the game ? [2]

SWIMMING

Question 15. [3 × 5]

(a) *Draw a neat diagram of a standard swimming pool with its dimensions.* [5]

(b) *List the officials required for conducting a swimming competition.* [5]

(c) *With regard to swimming, explain the techniques of the following :* [5]
(i) Breast-stroke *(ii) Free style*

(d) *State the warming-up exercises used in swimming.* [5]

(e) *Explain the following:* [5]
(i) Spring board
(ii) Marathon swimming

ATHLETICS

Question 16. [3 × 5]

(a) *What is meant by stagger distance ? For a 400 m track race, state the stagger distance for each lane of a standard track.* [5]

(b) *Draw a neat diagram of a javelin runway and its sectors with all dimensions.* [5]

(c) *List the track and field events in athletics.* [5]

(d) *State any five duties of a physiotherapist in athletics.* [5]

(e) *Explain the following :* [5]
(i) Wind-gauge operator
(ii) Steeple-chase race

ANSWERS

SECTION–A (40 Marks)

Answer 1.

National integration means unity of nation. Physical Education plays an important role in building it. National integration includes feeling of brotherhood, friendship and oneness. When the people of different countries participate in games and sports and visit other district/ state / country they come across the different ways of living, customs, religion, traditions, languages, eating habits, beliefs, etc. In sports camp, players come from different states, live together, practice to participate in one game by living and working together and exchange ideas, habits etc.

Games and sports attract people from all over the world or nation. Social qualities such as patience, cooperation, sympathy etc. are developed. It satisfies the basic needs such as recognition, and self respect. Sports is a universal language. At its best, it can bring people together, no matter what their origin, background, religious beliefs or economic status.

Physical Education provides a common platform where they forget all their differences. It plays very important role in bringing about national unity.

Answer 2.

Refer to ISC Paper 2020, Answer 2

Answer 3.

Career options available in the field of physical education :

1. Professional Sportsman.
2. Sports Manager.
3. Sports Teacher or Lecturer.
4. Sports Coach.
5. Gym Instructor.
6. Sports Officials.
7. Sports Event Co-ordinator.
8. Sports Journalist.
9. Sports Engineer.
10. Marketing and Manufacturing Sports.
11. Sports Photographer.
12. Physical Therapist.

Sports Teacher or Lecturer : Physical education has a major role to play in the school and college system. Physical education occupies a dignified position in the higher education sector. Apart from being taught in training colleges it has been included as an elective subject in degree course. Physical education professionals whosoever qualifies NET, M.Phil, B.P.Ed., M.P.Ed., Ph.D. degree may get job as Teacher / Lecturer in the school/colleges.

Sports Journalist : Sports Journalist write about and report on amateur and professional sports. As a sports journalist, you can expect a variety of job duties such as reporting game statistics, interviewing coaches and players and offering game commentary. One can work in a variety of media, including radio, television and print.

Answer 4.

Importance of Intramural Competitions :

1. Intramurals are very significant for physical, mental, emotional and social development of students.

2. They instill and inculcate discipline among the students.

3. They provide maximum recreation to the students.

4. They develop healthy competition among the participants.

5. They help in improving the personality of the students.

Importance of Extramural Competitions :

1. Provide opportunity to schools to show their sports capabilities.

2. Enhance technical knowledge.

3. Boost personal efficacy and self esteem.

4. Increased level of self confidence.

Objectives of Intramurals :

1. Provide opportunities for development of positive attitudes.

2. Allow the development of sports and knowledge and strategies through participation in a safe and competitive situation.

Objective of Extramurals :

1. To improve the standard of sports.

2. To aid in the development of sportsmanship and fraternity.

Answer 5.

Drugs : Chemical substances that excite a person mentally or physically are called drugs. Drugs are powerful stimulant of the brain they can produce an ecstatic sensation of both physical and mental power with temporary disappearance of feeling of fatigues and hunger.

Drug abuse : It is the excessive self-damaging use of drugs, leading to addiction or dependence, serious physiological injury, psychological harm or even death.

Two drugs banned by the World Anti-Drugs Agency are :

1. Stimulants are banned because they may artificially stimulate the mind and body, thereby improving an athlete's performance and giving them an unfair advantage.

2. Anabolic agents are banned because they are prescribed for medical use only. Use of anabolic agents may enhance an athlete's performance. Another possibility is serious medical side effects on the user.

Answer 6.

Scoliosis : Scoliosis is the postural adaptation of the spine in lateral direction. It is a medical condition in which a persons spinal axis has a three-dimensional deviation. When viewed from the side, the spine should show a mild roundness in the upper back and shows a degree of swayback (inward curvature) in the lower back.

Causes of Scoliosis :

1. Hereditary defects.
2. One side paralysis of spinal muscles.
3. Short leg of one side.
4. One side flat foot.
5. One side vision and hearing defects.

Corrective Measures : This can be cured through voluntary efforts :

1. By hanging oneself from the hands.
2. Using body weight.
3. By developing strength in spinal extensors through corrective exercises.
4. Swim by using backstroke technique.

Answer 7.

Bruises : Bruises are injuries caused by the hitting of a foreign agent on the skin without producing external wound. Bruise can be defined as an injury or mark where the skin has not been broken but it is darker in colour, often as a result of being hit by something.

Five measures taken by an individual to prevent sports related accidents.

1. Every individual should have pre-participation physical examination.
2. Sports person should focus on developing muscular strength and endurance as well as cardiovascular fitness and flexibility.
3. Limit workout and practices to maximum two hours.
4. All sports persons should use appropriate that fits properly in practices as well as competitions.
5. Ice should be available on the sidelines of every game and practice to apply to appropriate injuries.

SECTION B (30 Marks)

FOOTBALL

Answer 8.

(a) Wall formation : A line of players organized by a team to stand shoulder by shoulder and try to prevent a free kick being scored. A wall is designed to obstruct the kicker's view of goal and restrict the target he has to shoot at. Therefore, we can say that the wall is primarily used to cut off the shooter's angle to the post nearest to the spot of the free kick.

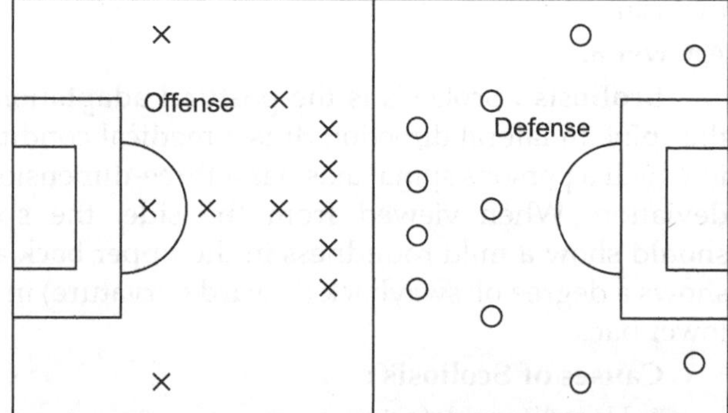

(b) Technical area :

1. In this area team coach, manager and players (extra) sit together and make the strategies of game.
2. Only one person at a time is authorized to convey tactical instruction and he must return to his position after the instructions.
3. The coach and other occupants of the technical area must behave in a responsible manner.

The technical area extends 1 m on either side of the designated seated area and extends forward upto a distance of 1 m from the touch lines.

(c) Duties of fourth official :

1. Assists the administrative functions.
2. Assess the players.
3. Ensues that the substitutions are conducted in an orderly manner.
4. Serves as a replacement official in the event.
5. Acts as the contact point between the match official and any non- participants.

(d) Direct free kick : A **direct free kick** is a method of restarting play in a game of football following a foul. A goal may be scored directly against the opposing side without the ball having first touched another player. A direct free kick is awarded to the opposing team if a player commits any of the following offenses :

1. Kicks or attempts to kick an opponent.
2. Charges an opponent.
3. Handles the ball deliberately (except for the goalkeeper within his own penalty area).

(e) (i) Heading : Heading is an important skill which is used by players. To head a ball accurately, a player should judge the flight of the ball and wait for the proper jump and hit the ball with the head.

(ii) Goal area : A goal area is defined at each end of the field. Two lines are drawn at right angles to the goal line 5·5 m from the inside of each goal post. These lines extend into the field of play for a distance of 5·5 m and are joined by a line drawn parallel with the goal line.

Answer 9.

(a)

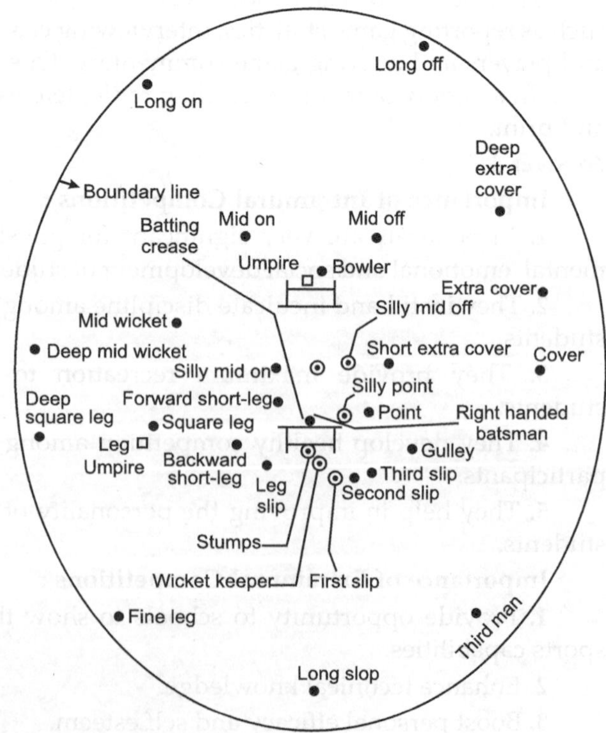

(b) Power play is the name for the fielding restrictions in limited-over and Twenty- 20 matches in cricket. It is of 30 yard circle in the centre of field.

Restrictions imposed on the players during the power play :

1. Mandatory powers play (1-10 overs) : Only two fielders are allowed outside. the 30-yard circle.

2. Between over 11 and 40, a maximum of four fielders are allowed outside the 30-yard circle.

3. In the final 10 overs (41-50), a maximum of five fielders will be allowed to field outside the restricted area.

(c) Ball tampering : Unlawful alteration of the surface or seam of a ball on the field to effect its motion is known as Ball Tampering. The two ways by which a ball is tampered are :

1. Rubbing the ball on the ground.

2. Picking the threads of the main seam.

(d) Third umpire duties : The third umpire is also known as off-field umpire who usually gives his decision when the on-field umpires are unsure. The third umpire looks at various TV replays from different angles and comes to a conclusion by pressing the appropriate signal. The field umpire can use his discretion to refer a close decision to the third umpire regarding the dismissal of a batsman, catches or boundaries via a wireless set or a signal system. Third umpire conveys his decision through radio link and television replay. A red light indicates that the batsman is out and a green one otherwise.

(e) (i) Bouncer : In a game of cricket, a bouncer is a type of delivery in which a ball pitched short by a fast bowler so that, it rises up to the batsman's chest or head (or even higher).

(ii) Lost ball : The ball which is hit by the batsman is gone out of the field and fielder is not able to find it is called lost ball.

Answer 10.

(a) 1. Hand protector
2. Helmet
3. Gloves
4. Leg guards.
5. Chest guards
6. Elbow pads
7. Hockey stick

(b) Astroturf is a surface of synthetic fiber made to look like natural grass.

Importance : Astroturf is water base, due to this the water layer in the hockey pitch (field) has LSR (Low Side Resistance) and shock absorbency properties for excellent players, safety unlike some sand based hockey pitches, the surface will remain soft even at low temperatures thereby, reducing the risk of friction burn.

(c) Free hit : Free hits are awarded when offences are committed outside the scoring circles. The ball may be hit, pushed or lifted in any direction by the team offended against.

Tournaments : 1. Champions Trophy
2. Azlan Shah Cup
3. Agha Khan Cup
4. Indira Gold Cup

(d) (i) Push pass is used for short and accurate pass when there is no time. This pass can be executed while dribbling. While implementing the pass the left shoulder should be pointed in the direction of the pass with the left foot forward. The right hand remains 6 to 8 inches lower than in the fundamental position.

(ii) Team formation is the position assigned to the player play on the field to protect the goal from opponent team and score goal for the team. There are :

1. Three forwards
2. Three mid fielders
3. Three defenders
4. One sweeper
5. One goalie.

(e) Five skills of a hockey player are :

1. Dribbling
2. Dodge
3. Holding of Hockey stick
4. Hitting a stationary ball
5. Penalty stroke

Answer 11.

(a) FIBA : The International Basketball Federation is more commonly known as FIBA. It defines the international rules of basketball, specifies the facilities required, regulates the transfer of players across countries and controls the appointment of international referees. This world governing body of basketball was founded in 1932. It brings together 214 National Basketball Federations from all over the world.

Violation : A violation is an infraction of the rules and the most minor class of illegal action. Most variations are committed by the team with the possession of the ball, when a player mishandles the ball or makes an illegal move.

(b) Personal foul : A personal foul is committed by the contact foul with an opponent, whether the ball is live or dead.

Foul : A Foul is an infraction of the rules more serious than a violation.

(c) Duties of Timekeeper :

1. To set the clock at the beginning of each quarter.

2. To keep score on the clock for each team during the game.

3. To start and stop the clock as indicated by the referee.

4. To indicate the appropriate quarters for the game.

5. To keep team fouls and indicate if possible.

(d) **Jump shot :** Jump shot is an attempt to score a basket by jumping usually straight up and in mid jump, propelling the ball in an arc. It is accomplished by the player bringing his or her elbow up until it is aligned with the hoops then sent towards the hoop in a high arc into the basket.

Four types of passes :

(1) Chest pass, (2) Bounce pass,

(3) Under hand pass (4) Overhead pass.

(e) (i) **Technical foul :** An infraction by a player or a coach that is deemed unsportsman like conduct is called Technical Foul. The penalty for this infraction is one free throw for the opposing team and possession.

(ii) **Pivoting :** The action a player will use while in possession of the ball to make in any direction while still keeping one foot in contact with the floor is called pivoting. It is very useful for getting rid of an opponent.

Answer 12.

(a) **Rotation :** It is a process of changing the positions of the players within a court in a clockwise direction after getting a chance to serve.

Rotation fault :

1. When the player is not being positioned correctly on the court at service.

2. When the middle front player is being overlapped with the right front player.

3. When middle back player is being overlapped with the middle front player.

4. When determining whether the player in position 1 and player in position 6 are overlapped.

(b) (i) **Assisted hit :** A player takes support from a team mate or any object in order to reach the ball within the playing area.

(ii) **Penetration fault :** It is permitted for a player to step under the net into the opponents court provided, that he does not interfere with the opponent's play and if it interferes the play of opponents it is a fault.

(c) (i) **Spiker :** The title given to the player who finishes the attack by hitting the ball across the net fast and hard.

(ii) **Blocking fault :**

1. When the blocker blocks an opponent's serve.

2. Blocker is not given a chance to attack the opponent.

3. The ball is deflected out of play off a block.

4. A back row player or libro completes a block.

(d) **There are two scoring methods in Volleyball :**

1. **Rally scoring :** In rally scoring team scores a point when the other team fails to return the ball over the net and hit the ball out of bounds commits an infraction or makes a service error.

2. **Side out scoring :** In side out scoring the service team scores a point when the opponents fail to return the ball over the net hit the ball out of bounds.

Ways of delays in game :

1. Delaying a substitution.

2. Request for an illegal substitution.

3. Repeating an improper request.

4. Delaying the game by a team member.

(e) (i) FIVB—Federation International de Volleyball.

(ii) VFI—Volleyball Federation of India.

(iii) AVF—Asian Volleyball Federation.

(iv) UVA—United Volleyball Association.

(v) JVA—Japan Volleyball Association.

Answer 13.

(a) (i) **Advantage :** A player, who scores a point after a deuce is said to have gained an advantage.

(ii) **Ace :** When a receiver fails to control the ball while receiving a service the server is said to have served an ace.

(b) **Duties of a chair umpire :**

1. The chair umpire has the right to over rule a line or a net umpire.

2. The chair umpire is responsible for calling any line or net umpire where no line umpire or net umpire is assigned.

3. Play may be stopped or suspended at anytime the chair umpire decides.

4. The chair umpire calls the score of the match to the players and spectators.

5. The chair umpire is responsible for completing a score card.

(c) (i)

(1) Material	—	Rubber
(2) Weight	—	56·7—60·24 gram
(3) Circumference	—	6·35 — 6·67 gram
(4) Colour	—	Yellow or white.

(ii) Width of the strap of the net used in Tennis—5 cm (2 inch).

(d) The Grand Slam Tournaments are considered to be the most prestigious Tennis in the world. They are held annually. There are four tournaments included in the category of Grand Slam—The Australian Open, French Open, Wimbledon and US Open are the four Grand Slam Tournament.

These are held at four different venues in Australia, France, The United Kingdom and United States of America.

Two other important tournaments :

1. Federation Open
2. Asian Tennis Championship.

(e) Drive : An offensive shot hit with an extra force.

Forehand : The stroke hit on the right side of a right-handed player.

Chip shot : An under-cutting stroke with a forehand or backhand action.

Top spin : The clockwise rotation of the ball at right angle.

Volley : When a player strikes the ball without allowing it to touch the ground.

Answer 14.

(a) Five important tournaments :

1. Thomas Cup 2. Uber Cup
3. Wills World Cup 4. Asia Cup
5. All England Badminton Championship

(b) Five faults :

1. When a player touches the net or its supports with racket or in person.
2. When a player invades an opponents court under the net with racket.
3. A player deliberately distracts an opponent by action such as shouting.
4. If the shuttle is hit by a player and the player's partner successively.

(5) When the player hits the shuttle twice.

(c)

$2\frac{1}{8} - 2\frac{1}{2}$ inches

$2\frac{1}{2} - 2\frac{2}{3}$ inches

Feather : 0.054 mm – 0.064 mm

Weight : 4.74 grams – 5.50 grams

$1 - 1\frac{1}{8}$ inches

(d) (i) Bird : It is another name for the shuttle cock. It is made of natural and synthetic materials when a 'let' occurs, play since the last service shall not count and the player what should last shall serve again.

(ii) Let : The stopping of a play owing to any outside interference is called let.

(iii) Drive : It is a flat stroke in which the shuttle is sent over the net, just clearing it and parallel to the floor.

(iv) Double hit : When a player hits the shuttle cock twice in row.

(v) Service fault : In attempting to serve the server misses the shuttle.

(e) (i) Procedure of correct service :

1. Once the players are ready for the service, the first forward movement of the server's racket head shall be the start of the service.
2. Once started, the service is delivered when the shuttle is hit by the server's racket or in attempting to serve, the server misses the shuttle.
3. The server shall not serve before the receiver is ready. However, the receiver shall be considered to have been ready if a return of the service is attempted.
4. In doubles, during the delivery of service, the partners may take up any positions within then respective courts, which do not in sight the opposing server or receiver.

(ii) Change of ends : Player will change ends at the end of the first game, at the end of the second game. If there is the third game or if the leading score reaches 11 points then also the ends will be changed.

Answer 15.

(a)

20 mts 10 mts 20 mts

CASCAE

Toddlers and warmuppool

Divingwell

AREA

20 mts

25 mts

50 mts

Standard Pool

(b) List of officials required is as follows :

1.	Referee	1
2.	Control room supervisor	1
3.	Judges of stroke	4
4.	Starters	2
5.	Chief inspectors of turns	2
6.	Inspectors of turns— (1 at each end of each lane)	

7.	Chief recorder	1
8.	Clerk of course	2
9.	False start rope personnel	1
10.	Announcer	1

(c) (i) **Breast stroke :** The breast stroke starts with the swimmer lying in the water, face down, arms extended straight forward and legs extended straight to the back, arm with move out sweep, in sweep and recovery, the movement start with the out sweep and the leg will move like frog, kick in this breathing is usually done during the beginning of the in sweep phase of arm through the mouth.

(ii) **Free style :** In this, leg kick should be performed with extended ankles kicks in the legs up and down from the hip with toes turned in and loose knees and arm forward, in front of face pull down under the water steadily until the hand leaves the water at high level, little figure upper most bend and left the elbow to clear the water and swing the arm forward. Then repeat and turn the head side ways breath only as needed.

(d) **Warm up exercises :**

1. Shoulder stretch 2. Tricep stretching
3. Muscles stretching
4. Lower pectorals muscle stretch
5. Forward lunge 6. Quadricep stretch
7. Hamstring stretch 8. Calf stretches
9. Neck stretches 10. Long Hip adductor

(e) (i) **Spring board :** A spring board is used for diving and is a board that is itself a spring. Spring Boards are commonly fixed by a hinge at one end and the other end usually hangs over a swimming pool with a point midway between the hinge and the end resting on adjustable fulcrum.

(ii) **Marathon swimming :** Marathon swimming is a class of open water swimming defined by long distance at least 10 km. Unlike marathon foot races which have a specifically defined distance, Marathon swims vary in distance.

Answer 16.

(a) **Stagger distance :** When the runner runs in lanes around curves then they will run, unequal distance. In order to make the running distance same for all runners, we give them staggered start (plus distance reduce from the curve) so that they cover the same distance from start to finish line.

Stagger for 400 mts Trace :

Lane No.	Distance
1st lane	00 : 00 mts
2nd lane	07 : 04 mts
3rd lane	14 : 71 mts
4th lane	22 : 38 mts
5th lane	30 : 05 mts
6th lane	37 : 72 mts
7th lane	45 : 39 mts
8th lane	53 : 06 mts

(b)

(c) **Track events :**

Sprint races	—	100 mts, 200 mts, 400 mts, 110 mts hurdles 4 × 100 mts 4 × 400 mts relays
Middle races	—	800 mts, 1500 mts
Long distance	—	3000 mts, 5000 mts 10000 mts, 3000 steeplechase
Walking events	—	20 km, 50 km
Cross country	—	12 km
Marathon race	—	26 miles and 385 yards or 42 km and 195 mts

Field events :

Shot put	Discus Throw
Hammer Throw	Javelin Throw
Long Jump	Triple Jump
High Jump	Pole Vault

(d) **Five duties of a physiotherapist in athletics :**

1. Protecting the injury to prevent further damage.

2. Provide proper exercise to the injured part.

3. Return functional movement and use by restoring flexibility and strength in and around the injured area.

4. Pain management is also a part of sports physiotherapist's responsibility.

5. Deciding when you are ready to return to the sports.

(e) (i) **Wind gauge operator :** The wind gauge operator ensures that the gauge is placed as per rule. He ascertains the velocity of the wind in the running direction in appropriate event and then records and signs the result obtained and communicates it to the Competition Secretary.

(ii) **Steeple-chase race :** The standard distance is 3000 mts, there are 28 hurdles jumps and 7 water jump included in the 3000 mts event. In this event, there shall be five jumps in each lap, after the finish line has been passed for the first time with the water jump as the fourth.

MATHEMATICS 2020

(Maximum Marks : 100)
(Time allowed : Three hours)

*(Candidates are allowed additional 15 minutes for **only** reading the paper. They must NOT start writing during this time.)*

This Question Paper consists of three sections A, B and C.

*Candidates are required to attempt all questions from **Section A** and all questions **EITHER** from **Section B OR Section C.***

***Section A :** Internal choice has been provided in three questions of four marks each and two questions of six marks each.*

***Section B :** Internal choice has been provided in two questions of four marks each.*

***Section C :** Internal choice has been provided in two questions of four marks each.*

All working, including rough work, should be done on the same sheet as, and adjacent to the rest of the answer.

The intended marks for questions or parts of questions are given in brackets [ ].

Mathematical tables and graph papers are provided.

QUESTIONS

SECTION – A (80 Marks)

Question 1. **[10 × 2]**

(i) *Determine whether the binary operation $*$ on R defined by $a * b = |a - b|$ is commutative. Also, find the value of $(-3) * 2$.*

(ii) *Prove that :*

$\tan^2(\sec^{-1} 2) + \cot^2(\operatorname{cosec}^{-1} 3) = 11.$

(iii) *Without expanding at any stage, find the value of the determinant :*

$$\Delta = \begin{vmatrix} 20 & a & b+c \\ 20 & b & a+c \\ 20 & c & a+b \end{vmatrix}$$

(iv) *If* $\begin{pmatrix} 2 & 3 \\ 5 & 7 \end{pmatrix}\begin{pmatrix} 1 & -3 \\ -2 & 4 \end{pmatrix} = \begin{pmatrix} -4 & 6 \\ -9 & x \end{pmatrix}$*, find x.*

(v) *Find* $\frac{dy}{dx}$ *if* $x^3 + y^3 = 3axy$

(vi) *The edge of a variable cube is increasing at the rate of 10 cm/sec. How fast is the volume of the cube increasing when the edge is 5 cm long ?*

(vii) *Evaluate :* $\int_4^5 |x - 5|\, dx$

(viii) *Form a differential equation of the family of the curves* $y^2 = 4ax$.

(ix) *A bag contains 5 white, 7 red and 4 black balls. If four balls are drawn one by one with replacement, what is the probability that none is white ?*

(x) *Let A and B be two events such that*

$$P(A) = \frac{1}{2},\ P(B) = p \text{ and } P(A \cup B) = \frac{3}{5}$$

find 'p' if A and B are independent events.

Question 2. **[4]**

If the function $f : R \to R$ *be defined as*

$f(x) = \frac{3x+4}{5x-7}, \left(x \neq \frac{7}{5}\right)$ *and* $g : R \to R$ *be defined as*

$g(x) = \frac{7x+4}{5x-3}, \left(x \neq \frac{3}{5}\right)$ *show that* $(g \circ f)(x) = (f \circ g)(x)$.

Question 3. **[4]**

(a) *If* $\cos^{-1}\frac{x}{2} + \cos^{-1}\frac{y}{3} = \theta$*, then prove that*

$$9x^2 - 12xy\cos\theta + 4y^2 = 36\sin^2\theta$$

OR

(b) *Evaluate :* $\cos(2\cos^{-1}x + \sin^{-1}x)$ *at* $x = \frac{1}{5}$.

Question 4. **[4]**

Using properties of determinants, show that

$$\begin{vmatrix} x & p & q \\ p & x & q \\ q & q & x \end{vmatrix} = (x-p)(x^2 + px - 2q^2)$$

Question 5. **[4]**

Verify Rolle's theorem for the function, $f(x) = -1 + \cos x$ *in the interval* $[0, 2\pi]$.

Question 6. [4]

If $y = e^{m \sin^{-1} x}$, *prove that*

$$(1-x^2)\frac{d^2y}{dx^2} - x\frac{dy}{dx} = m^2 y$$

Question 7. [4]

(a) *The equation of tangent at (2, 3) on the curve* $y^2 = px^3 + q$ *is* $y = 4x - 7$. *Find the values of 'p' and 'q'.*

OR

(b) *Using L' Hospital's rule, evaluate :*

$$\lim_{x \to 0} \frac{xe^x - \log(1+x)}{x^2}$$

Question 8. [4]

(a) *Evaluate :* $\int \frac{dx}{\sqrt{5x - 4x^2}}$

OR

(b) *Evaluate :* $\int \sin^3 x \cos^4 x \, dx$

Question 9. [4]

Solve the differential equation

$$(1 + x^2)\frac{dy}{dx} = 4x^2 - 2xy$$

Question 10. [4]

Three persons A, B and C shoot to hit a target. Their probabilities of hitting the target are $\frac{5}{6}, \frac{4}{5}$ *and* $\frac{3}{4}$ *respectively. Find the probability that :*

(i) *Exactly two persons hit the target.*

(ii) *At least one person hits the target.*

Question 11. [6]

Solve the following system of linear equations using matrices:

$$x - 2y = 10,\ 2x - y - z = 8,\ -2y + z = 7$$

Question 12. [6]

(a) *Show that the radius of a closed right circular cylinder of given surface area and maximum volume is equal to half of its height.*

OR

(b) *Prove that the area of right-angled triangle of given hypotenuse is maximum when the triangle is isosceles.*

Question 13. [6]

(a) *Evaluate :*

$$\int \tan^{-1}\sqrt{\frac{1-x}{1+x}}\,dx$$

OR

(b) *Evaluate :* $\int \frac{2x+7}{x^2 - x - 2}dx$

Question 14. [6]

The probability that a bulb produced in a factory will fuse after 150 days of use is 0·05.

Find the probability that out of 5 such bulbs:

(i) *None will fuse after 150 days of use.*

(ii) *Not more than one will fuse after 150 days of use.*

(iii) *More than one will fuse after 150 days of use.*

(iv) *At least one will fuse after 150 days of use.*

SECTION – B (20 Marks)

Question 15. [3×2]

(a) *Write a vector of magnitude of 18 units in the direction of the vector* $\hat{i} - 2\hat{j} - 2\hat{k}$.

(b) *Find the angle between the two lines :*

$$\frac{x+1}{2} = \frac{y-2}{5} = \frac{z+3}{4} \text{ and } \frac{x-1}{5} = \frac{y+2}{2} = \frac{z-1}{-5}$$

(c) *Find the equation of the plane passing through the point (2, –3, 1) and perpendicular to the line joining the points (4, 5, 0) and (1, –2, 4).*

Question 16. [4]

(a) *Prove that*

$$\vec{a}\,.[(\vec{b} + \vec{c}) \times (\vec{a} + 3\vec{b} + 4\vec{c})] = [\vec{a}\ \vec{b}\ \vec{c}]$$

OR

(b) *Using vectors, find the area of the triangle whose vertices are:*

A (3, –1, 2), B (1, –1, –3) and C (4, –3, 1)

Question 17. [4]

(a) *Find the image of the point (3, –2, 1) in the plane* $3x - y + 4z = 2$

OR

(b) *Determine the equation of the line passing through the point (–1, 3, –2) and perpendicular to the lines :*

$$\frac{x}{1} = \frac{y}{2} = \frac{z}{3} \text{ and } \frac{x+2}{-3} = \frac{y-1}{2} = \frac{z+1}{5}$$

Question 18. [6]

Draw a rough sketch of the curves $y^2 = x$ *and* $y^2 = 4 - 3x$ *and find the area enclosed between them.*

SECTION – C (20 Marks)

Question 19. [3 × 2]

(a) *The selling price of a commodity is fixed at ₹ 60 and its cost function is* $C(x) = 35x + 250$

(i) *Determine the profit function.*

(ii) *Find the break even points.*

(b) *The revenue function is given by* $R(x) = 100x - x^2 - x^3$. *Find*

(i) *The demand function.*

(ii) *Marginal revenue function.*

(c) *For the lines of regression $4x - 2y = 4$ and $2x - 3y + 6 = 0$, find the mean of 'x' and the mean of 'y'.*

Question 20. **[4]**

(a) *The correlation coefficient between x and y is 0.6. If the variance of x is 225, the variance of y is 400, mean of x is 10 and mean of y is 20, find*

(i) *the equations of two regression lines.*

(ii) *the expected value of y when $x = 2$.*

OR

(b) *Find the regression coefficients b_{yx}, b_{xy} and correlation coefficient 'r' for the following data :*

(2, 8), (6, 8), (4, 5), (7, 6), (5, 2).

Question 21. **[4]**

(a) *The marginal cost of the production of the commodity is $30 + 2x$, it is known that fixed costs are ₹ 200, find*

(i) *The total cost.*

(ii) *The cost of increasing output from 100 to 200 units.*

OR

(b) *The total cost function of a firm is given by $C(x) = \frac{1}{3}x^3 - 5x^2 + 30x - 15$ where the selling price per unit is given as ₹ 6. Find for what value of x will the profit be maximum.*

Question 22. **[6]**

A company uses three machines to manufactue two types of shirts, half sleeves and full sleeves. The number of hours required per week on machine M_1, M_2 and M_3 for one shirt of each type is given in the following table :

	M_1	M_2	M_3
Half sleeves	1	2	8/5
Full sleeves	2	1	8/5

None of the machines can be in operation for more than 40 hours per week. The profit on each half sleeve shirt is ₹ 1 and the profit on each full sleeve shirt is ₹ 1·50. How many of each type of shirts should be made per week to maximize the company's profit ?

SOLUTIONS

SECTION – A

Solution 1.

(i) We have,

$$a * b = |a - b| = |-(b-a)| = b - a = |b - a| = b * a$$

$\Rightarrow \quad a * b = b * a$

So, $*$ is commutative on R.

Now, $\quad (-3) * 2 = |-3 - 2| = |-5| = 5$ **Ans.**

(ii) To prove : $\tan^2(\sec^{-1} 2) + \cot^2(\text{cosec}^{-1} 3) = 11$

Let $\sec^{-1} 2 = \theta$

$\Rightarrow \sec\theta = 2$

$\therefore \tan^2\theta = \sec^2\theta - 1 = 2^2 - 1 = 3$

Also, let $\text{cosec}^{-1} 3 = \alpha$

$\Rightarrow \text{cosec}\,\alpha = 3$

$\therefore \cot^2\alpha = \text{cosec}^2\alpha - 1 = 3^2 - 1 = 8$

Now, L.H.S. $= \tan^2(\sec^{-1} 2) + \cot^2(\text{cosec}^{-1} 3)$

$= \tan^2\theta + \cot^2\alpha$

$= 3 + 8 = 11 =$ R.H.S. **Hence Proved.**

(iii) Given :

$$\Delta = \begin{vmatrix} 20 & a & b+c \\ 20 & b & a+c \\ 20 & c & a+b \end{vmatrix}$$

Taking 20 common from C_1,

$$\Rightarrow \quad \Delta = 20\begin{vmatrix} 1 & a & b+c \\ 1 & b & a+c \\ 1 & c & a+b \end{vmatrix}$$

Applying $C_2 \to C_2 + C_3$,

$$\Delta = 20\begin{vmatrix} 1 & a+b+c & b+c \\ 1 & a+b+c & a+c \\ 1 & a+b+c & a+b \end{vmatrix}$$

Taking $(a + b + c)$ common from C_2,

$$\Delta = 20(a+b+c)\begin{vmatrix} 1 & 1 & b+c \\ 1 & 1 & a+c \\ 1 & 1 & a+b \end{vmatrix}$$

$\because \quad C_1 = C_2$

$\therefore \quad \Delta = 0$ **Ans.**

(iv) Given :

$$\begin{pmatrix} 2 & 3 \\ 5 & 7 \end{pmatrix}\begin{pmatrix} 1 & -3 \\ -2 & 4 \end{pmatrix} = \begin{pmatrix} -4 & 6 \\ -9 & x \end{pmatrix}$$

$$\Rightarrow \begin{pmatrix} 2\times1+3\times(-2) & 2\times(-3)+3\times4 \\ 5\times1+7\times(-2) & 5\times(-3)+7\times4 \end{pmatrix} = \begin{pmatrix} -4 & 6 \\ -9 & x \end{pmatrix}$$

$$\Rightarrow \begin{pmatrix} -4 & 6 \\ -9 & 13 \end{pmatrix} = \begin{pmatrix} -4 & 6 \\ -9 & x \end{pmatrix}$$

$\therefore \quad x = 13$ **Ans.**

(v) Given : $\quad x^3 + y^3 = 3axy$

Differentiating w.r.t. x, we get

$$3x^2 + 3y^2\frac{dy}{dx} = 3a\left(1.y + x\frac{dy}{dx}\right)$$

$$\Rightarrow \quad x^2 + y^2 \frac{dy}{dx} = ay + ax\frac{dy}{dx}$$

$$\Rightarrow \quad \frac{dy}{dx}(y^2 - ax) = ay - x^2$$

$$\Rightarrow \quad \frac{dy}{dx} = \frac{ay - x^2}{y^2 - ax}$$ **Ans.**

(vi) Let the edge of a cube be x units.

$\therefore \quad \frac{dx}{dt} = 10$ cm/sec [Given]

Now, Volume of cube $(V) = x^3$

Differentiating w.r.t. t, we get

$$\frac{dV}{dt} = 3x^2 \frac{dx}{dt}$$

$$\Rightarrow \quad \frac{dV}{dt} = 3 \times 5^2 \times 10 \quad [\because x = 5 \text{ cm}]$$

$$\Rightarrow \quad \frac{dV}{dt} = 750$$

Hence, the volume of the cube is increasing at the rate 750 cm³/sec. **Ans.**

(vii) $$\int_4^5 |x-5|\,dx = \int_4^5 -(x-5)\,dx$$

$$= \int_4^5 (5-x)\,dx$$

$$= \left[5x - \frac{x^2}{2}\right]_4^5$$

$$= \left[5\times5 - \frac{5\times5}{2}\right] - \left[5\times4 - \frac{4\times4}{2}\right]$$

$$= \left[25 - \frac{25}{2}\right] - \left[20 - \frac{16}{2}\right]$$

$$= \frac{50-25}{2} - 12$$

$$= \frac{25}{2} - 12$$

$$= \frac{25-24}{2} = \frac{1}{2}$$ **Ans.**

(viii) Given : $y^2 = 4ax$

Differentiating both sides w.r.t. x, we get

$$2y\frac{dy}{dx} = 4a$$

$$\Rightarrow \quad 2y\frac{dy}{dx} = \frac{y^2}{x} \quad \left[\because 4a = \frac{y^2}{x}\right]$$

$$\Rightarrow \quad 2\frac{dy}{dx} = \frac{y}{x}$$

$$\Rightarrow \quad 2x\frac{dy}{dx} - y = 0,$$

which is the required differential equation. **Ans.**

(ix) Total number of balls = 5 + 7 + 4 = 16

Total number of non-white balls = 7 + 4 = 11

$\therefore$ Probability of drawing a non-white ball $= \frac{11}{16}$

Since, balls are drawn one by one with replacement, So, probability of drawing non-white balls 4 times

$$= \left(\frac{11}{16}\right)^4 = \frac{14641}{65536}$$ **Ans.**

(x) Given : $P(A) = \frac{1}{2}$, $P(B) = p$ and $P(A \cup B) = \frac{3}{5}$.

We know, $P(A \cup B) = P(A) + P(B) - P(A \cap B)$

$$\Rightarrow \quad \frac{3}{5} = \frac{1}{2} + p - P(A \cap B)$$

$$\Rightarrow \quad P(A \cap B) = \frac{1}{2} - \frac{3}{5} + p = p + \frac{1}{2} - \frac{3}{5}$$

$$= p + \frac{5-6}{10}$$

$$= p - \frac{1}{10}$$

Now, A and B are independent events.

$$\therefore \quad P(A \cap B) = P(A) \cdot P(B)$$

$$\Rightarrow \quad p - \frac{1}{10} = \frac{1}{2} \times p$$

$$\Rightarrow \quad p - \frac{p}{2} = \frac{1}{10}$$

$$\Rightarrow \quad \frac{p}{2} = \frac{1}{10}$$

$$\therefore \quad p = \frac{2}{10} = \frac{1}{5}$$ **Ans.**

Solution 2.

Given : $f(x) = \frac{3x+4}{5x-7}$, $x \neq \frac{7}{5}$ and $g(x) = \frac{7x+4}{5x-3}$, $3x \neq \frac{3}{5}$

Now, $(g\,o\,f)(x) = g\{f(x)\}$

$$= g\left(\frac{3x+4}{5x-7}\right)$$

$$= \frac{7\left(\frac{3x+4}{5x-7}\right) + 4}{5\left(\frac{3x+4}{5x-7}\right) - 3}$$

$$= \frac{\frac{21x+28+20x-28}{5x-7}}{\frac{15x+20-15x+21}{5x-7}}$$

$$= \frac{41x}{41} = x$$

Also, $(f \circ g)(x) = f\{g(x)\}$

$$= f\left(\frac{7x+4}{5x-3}\right)$$

$$= \frac{3\left(\frac{7x+4}{5x-3}\right)+4}{5\left(\frac{7x+4}{5x-3}\right)-7}$$

$$= \frac{\frac{21x+12+20x-12}{5x-3}}{\frac{35x+20-35x+21}{5x-3}}$$

$$= \frac{41x}{41} = x$$

$\because \quad (g \circ f)(x) = (f \circ g)(x) = x$

Thus, $(g \circ f)(x) = (f \circ g)(x)$ **Hence Proved.**

Solution 3.

(a) Given : $\cos^{-1}\frac{x}{2} + \cos^{-1}\frac{y}{3} = \theta$

$$\Rightarrow \cos^{-1}\left[\frac{x}{2}\times\frac{y}{3} - \sqrt{1-\frac{x^2}{4}}.\sqrt{1-\frac{y^2}{9}}\right] = \theta$$

$$\left[\because \cos^{-1}a + \cos^{-1}b = \cos^{-1}\left(ab - \sqrt{1-a^2}.\sqrt{1-b^2}\right)\right]$$

$$\Rightarrow \frac{xy}{6} - \sqrt{1-\frac{x^2}{4}}.\sqrt{1-\frac{y^2}{9}} = \cos\theta$$

$$\Rightarrow \frac{xy}{6} - \cos\theta = \sqrt{1-\frac{x^2}{4}}.\sqrt{1-\frac{y^2}{9}}$$

Squaring both sides, we get

$$\Rightarrow \left(\frac{xy}{6} - \cos\theta\right)^2 = \left(1-\frac{x^2}{4}\right)\left(1-\frac{y^2}{9}\right)$$

$$\Rightarrow \frac{x^2y^2}{36} + \cos^2\theta - \frac{2xy}{6}\cos\theta = 1 - \frac{x^2}{4} - \frac{y^2}{9} + \frac{x^2y^2}{36}$$

$$\Rightarrow \frac{x^2}{4} + \frac{y^2}{9} - \frac{xy\cos\theta}{3} = 1 - \cos^2\theta$$

$$\Rightarrow \frac{9x^2 + 4y^2 - 12xy\cos\theta}{36} = \sin^2\theta$$

$$\Rightarrow 9x^2 + 4y^2 - 12xy\cos\theta = 36\sin^2\theta$$

$$\Rightarrow 9x^2 - 12xy\cos\theta + 4y^2 = 36\sin^2\theta$$

Hence Proved.

OR

(b) $\cos(2\cos^{-1}x + \sin^{-1}x) = \cos[\cos^{-1}x + \cos^{-1}x + \sin^{-1}x]$

$= \cos[\cos^{-1}x + (\cos^{-1}x + \sin^{-1}x)]$

$$= \cos\left[\cos^{-1}x + \frac{\pi}{2}\right]$$

$$= \cos\left(\frac{\pi}{2} + \cos^{-1}x\right)$$

$$= -\sin(\cos^{-1}x) \qquad \left[\because \cos\left(\frac{\pi}{2}+\theta\right) = -\sin\theta\right]$$

$$= -\sin\left(\sin^{-1}\sqrt{1-x^2}\right) \qquad \left[\because \cos^{-1}x = \sin^{-1}\sqrt{1-x^2}\right]$$

$$= -\sqrt{1-x^2}$$

$$= -\sqrt{1-\left(\frac{1}{5}\right)^2} \qquad \left[\because x = \frac{1}{5}\text{(Given)}\right]$$

$$= -\sqrt{1-\frac{1}{25}} = -\sqrt{\frac{25-1}{25}}$$

$$= -\sqrt{\frac{24}{25}} = -\frac{2\sqrt{6}}{5}$$ **Ans.**

Solution 4.

To prove :

$$\begin{vmatrix} x & p & q \\ p & x & q \\ q & q & x \end{vmatrix} = (x-p)(x^2+px-2q^2)$$

Let $\Delta = \text{LHS} = \begin{vmatrix} x & p & q \\ p & x & q \\ q & q & x \end{vmatrix}$

Applying $C_1 \to C_1 - C_2$, we get

$$\Delta = \begin{vmatrix} x-p & p & q \\ p-x & x & q \\ q-q & q & x \end{vmatrix}$$

Taking $(x-p)$ as common from C_1, we get

$$\Delta = (x-p)\begin{vmatrix} 1 & p & q \\ -1 & x & q \\ 0 & q & x \end{vmatrix}$$

Now, expanding along R_1, we get

$$\Delta = (x-p)[1(x^2-q^2) - p(-x-0) + q(-q-0)]$$

$$= (x-p)[x^2 - q^2 + px - q^2)$$

$$= (x-p)(x^2+px-2q^2) = \text{RHS}.$$

Hence Proved.

Solution 5.

Given : $f(x) = -1 + \cos x$ in $[0, 2\pi]$

1. Since, $\cos x$ is continuous in $[0, 2\pi]$, so, $f(x)$ is a continuous function in $[0, 2\pi]$.
2. $f(x)$ is differentiable on $(0, 2\pi)$

3. $\because \quad f(0) = -1 + \cos 0 = -1 + 1 = 0$

and $\quad f(2\pi) = -1 + \cos(2\pi)$

$= -1 + \cos(2\pi - 0)$

$= -1 + \cos 0$

$= -1 + 1 = 0$

$\therefore \quad f(0) = f(2\pi)$

Thus, $f(x)$ satisfies all conditions of Rolle's theorem.

So, there exists a value of x, say $c \in (0, 2\pi)$ such that $f'(c) = 0$

Now, $\quad f(x) = -1 + \cos x$

$\therefore \quad f'(x) = 0 - \sin x$

$\Rightarrow \quad f'(c) = 0 - \sin c$

$\because \quad f'(c) = 0$

$\therefore \quad -\sin c = 0$

or $\quad \sin c = 0$

$\therefore \quad c = 0, \pi, 2\pi, ...$

Thus, $c = \pi \in (0, 2\pi)$, such that $f'(c) = 0$

Hence, Rolle's theorem is verified. **Hence Proved.**

Solution 6.

Given : $y = e^{m \sin^{-1} x}$

Differentiating w.r.t. x, we get

$$\frac{dy}{dx} = e^{m \sin^{-1} x} \frac{d}{dx}(m \sin^{-1} x)$$

$$\Rightarrow \quad \frac{dy}{dx} = y \times \frac{m}{\sqrt{1-x^2}} \quad ...(i)$$

or $$\sqrt{1-x^2}\frac{dy}{dx} = my$$

Squaring both LHS and RHS,

$$(1-x^2)\left(\frac{dy}{dx}\right)^2 = m^2y^2$$

Differentiating w.r.t. x,

$$\Rightarrow \quad (1-x^2) \times 2\left(\frac{dy}{dx}\right)\left(\frac{d^2y}{dx^2}\right) + \left(\frac{dy}{dx}\right)^2(-2x) = m^2\sqrt{2y}\frac{dy}{dx}$$

$$\Rightarrow \quad (1-x^2)\frac{d^2y}{dx^2} + \frac{dy}{dx}(-x) = m^2y$$

$$\Rightarrow \quad (1-x^2)\frac{d^2y}{dx^2} - x\frac{dy}{dx} = m^2y$$ **Hence Proved.**

Solution 7.

(a) Given : Equation of curve is,

$$y^2 = px^3 + q$$

Differentiating both sides w.r.t. x, we get

$$2y\frac{dy}{dx} = 3px^2$$

$$\Rightarrow \quad \frac{dy}{dx} = \frac{3px^2}{2y}$$

$$\therefore \quad \left(\frac{dy}{dx}\right)_{(2,3)} = \frac{3p \times 2^2}{2 \times 3}$$

or $$m = \left(\frac{dy}{dx}\right)_{(2,3)} = \frac{12p}{6} = 2p$$

Since, $y = 4x - 7$ is the tangent to the curve at point (2, 3).

So, on comparing with $y = mx + c$, we get

$m = 4$

Now, $\quad 2p = 4$

$\Rightarrow \quad p = \frac{4}{2} = 2$

Since, point (2, 3) lies on the curve,

$\therefore \quad 3^2 = p \times 2^3 + q$

$\Rightarrow \quad 9 = 2 \times 8 + q \quad [\because p = 2]$

$\Rightarrow \quad 9 - 16 = q$

$\Rightarrow \quad q = -7$

Hence, $\quad p = 2$ and $q = -7$ **Ans.**

OR

(b) $$\lim_{x\to 0}\frac{xe^x - \log(1+x)}{x^2} \quad \left[\frac{0}{0}\text{ form}\right]$$

Differentiating w.r.t. x, on numerator and denominator

$$= \lim_{x\to 0}\frac{1.e^x + xe^x - \frac{1}{1+x}}{2x} \quad \left[\frac{0}{0}\text{ form}\right]$$

Again, differentiating w.r.t. x, on numerator and denominator,

$$= \lim_{x\to 0}\frac{e^x + 1e^x + xe^x + \frac{1}{(1+x)^2}}{2}$$

Using limits, $x = 0$

$$\frac{e^0 + e^0 + 0.e^0 + \frac{1}{(1+0)^2}}{2} = \frac{1+1+1}{2} = \frac{3}{2}$$ **Ans.**

Solution 8.

(a) $$\int \frac{dx}{\sqrt{5x - 4x^2}} = \frac{1}{2}\int \frac{dx}{\sqrt{\frac{5}{4}x - x^2}}$$

$$= \frac{1}{2}\int \frac{dx}{\sqrt{\frac{5}{4}x - x^2 + \left(\frac{5}{8}\right)^2 - \left(\frac{5}{8}\right)^2}}$$

$$= \frac{1}{2}\int \frac{dx}{\sqrt{\left(\frac{5}{8}\right)^2 - \left(x^2 - \frac{5}{4}x + \left(\frac{5}{8}\right)^2\right)}}$$

$$= \frac{1}{2}\int \frac{dx}{\sqrt{\left(\frac{5}{8}\right)^2 - \left(x - \frac{5}{8}\right)^2}}$$

$$= \frac{1}{2}\sin^{-1}\left(\frac{x - \frac{5}{8}}{\frac{5}{8}}\right) + c$$

$$\left[\because \int \frac{dx}{\sqrt{a^2 - x^2}} = \sin^{-1}\frac{x}{a} + c\right]$$

$$= \frac{1}{2}\sin^{-1}\left(\frac{8x-5}{5}\right) + c \quad \textbf{Ans.}$$

OR

(b) Let $I = \int \sin^3 x \cos^4 x\, dx$

$= \int \sin^2 x . \cos^4 x (\sin x\, dx)$

$= \int (1 - \cos^2 x) \cos^4 x (\sin x\, dx)$

Let $\cos x = t$

$\Rightarrow \quad -\sin x\, dx = dt$

or $\quad \sin x\, dx = -dt$

$\therefore \quad I = \int (1 - t^2)\, t^4 (-dt)$

$= -\int (t^4 - t^6)\, dt$

$= \int (t^6 - t^4)\, dt$

$= \frac{t^7}{7} - \frac{t^5}{5} + c$

$= \frac{\cos^7 x}{7} - \frac{\cos^5 x}{5} + c$ **Ans.**

Solution 9.

Given differential equation is,

$$(1 + x^2)\frac{dy}{dx} = 4x^2 - 2xy$$

$$\Rightarrow \quad \frac{dy}{dx} = \frac{4x^2}{1+x^2} - \frac{2xy}{1+x^2}$$

$$\text{or} \quad \frac{dy}{dx} + \frac{2xy}{1+x^2} = \frac{4x^2}{1+x^2}$$

This is of the form $\frac{dy}{dx} + Py = Q$

where, $P = \frac{2x}{1+x^2}$ and $Q = \frac{4x^2}{1+x^2}$

Thus, the solution of given differential equation is,

$$y \times \text{I.F.} = \int (Q \times \text{I.F.})\, dx + c$$

where, $\text{I.F.} = e^{\int P dx}$

$= e^{\int \frac{2x}{1+x^2}dx}$

$= e^{\log(1+x^2)}$

$$\left[\because \int \frac{f'(x)}{f(x)}dx = \log[f(x)] + c\right]$$

$= 1 + x^2 \quad [\because e^{\log(a)} = a]$

$\therefore \quad y \times (1 + x^2) = \int \left[\frac{4x^2}{1+x^2} \times (1+x^2)\right] dx + c$

$\Rightarrow \quad (1 + x^2)\, y = \int 4x^2\, dx + c$

$\Rightarrow \quad (1 + x^2)\, y = \frac{4x^3}{3} + c$ **Ans.**

Solution 10.

Given : $P(A) = \frac{5}{6}$, $P(B) = \frac{4}{5}$ and $P(C) = \frac{3}{4}$

So, $P(\overline{A}) = 1 - P(A) = 1 - \frac{5}{6} = \frac{1}{6}$

$P(\overline{B}) = 1 - P(B) = 1 - \frac{4}{5} = \frac{1}{5}$

$P(\overline{C}) = 1 - P(C) = 1 - \frac{3}{4} = \frac{1}{4}$

(i) Probability (exactly two persons hit the target)

$= P(A) \cdot P(B) \cdot P(\overline{C}) + P(A) \cdot P(\overline{B}) \cdot P(C) + P(\overline{A}) \cdot P(B) \cdot P(C)$

$= \frac{5}{6} \times \frac{4}{5} \times \frac{1}{4} + \frac{5}{6} \times \frac{1}{5} \times \frac{3}{4} + \frac{1}{6} \times \frac{4}{5} \times \frac{3}{4}$

$= \frac{1}{6} + \frac{1}{8} + \frac{1}{10}$

$= \frac{20+15+12}{120} = \frac{47}{120}$ **Ans.**

(ii) Probability (at least one person hits the target)

$= 1 -$ No one will hit the target

$= 1 - P(\overline{A}\overline{B}\overline{C})$

$= 1 - \left(\frac{1}{6} \times \frac{1}{5} \times \frac{1}{4}\right)$

$= 1 - \frac{1}{120}$

$= \frac{120-1}{120}$

$= \frac{119}{120}$ **Ans.**

Solution 11.

Given equations are,

$$x-2y = 10$$
$$2x-y-z = 8$$
and
$$-2y+z = 7$$

The given equations can be arranged in matrix form as,

$$\begin{bmatrix}1 & -2 & 0\\2 & -1 & -1\\0 & -2 & 1\end{bmatrix}\begin{bmatrix}x\\y\\z\end{bmatrix} = \begin{bmatrix}10\\8\\7\end{bmatrix}$$

Let $A = \begin{bmatrix}1 & -2 & 0\\2 & -1 & -1\\0 & -2 & 1\end{bmatrix}, X = \begin{bmatrix}x\\y\\z\end{bmatrix}$ and $B = \begin{bmatrix}10\\8\\7\end{bmatrix}$

$\therefore \quad AX = B \quad$ ...(i)

Now,
$$|A| = \begin{vmatrix}1 & -2 & 0\\2 & -1 & -1\\0 & -2 & 1\end{vmatrix}$$
$$= 1(-1-2)+2(2+0) + 0(-4+0)$$
$$= -3+4 = 1 \neq 0$$

So, A is invertible.

Now, the minors of the elements of | A | are

$$M_{11} = -3, M_{12} = 2, M_{13} = -4$$
$$M_{21} = -2, M_{22} = 1, M_{23} = -2$$
$$M_{31} = 2, M_{32} = -1, M_{33} = 3$$

Co-factors of the elements of | A | are

$$A_{11} = -3, A_{12} = -2, A_{13} = -4$$
$$A_{21} = 2, A_{22} = 1, A_{23} = 2$$
$$A_{31} = 2, A_{32} = 1, A_{33} = 3$$

We know, $\quad \text{adj. A} = \begin{bmatrix}A_{11} & A_{12} & A_{13}\\A_{21} & A_{22} & A_{23}\\A_{31} & A_{32} & A_{33}\end{bmatrix}^T$

$$\therefore \quad \text{adj. A} = \begin{bmatrix}-3 & -2 & -4\\2 & 1 & 2\\2 & 1 & 3\end{bmatrix}^T$$
$$= \begin{bmatrix}-3 & 2 & 2\\-2 & 1 & 1\\-4 & 2 & 3\end{bmatrix}$$

$$\therefore \quad A^{-1} = \frac{1}{|A|}(\text{adj. A})$$
$$= 1.\begin{bmatrix}-3 & 2 & 2\\-2 & 1 & 1\\-4 & 2 & 3\end{bmatrix}$$

From (i), we have

$$AX = B$$

$\Rightarrow \quad A^{-1}AX = A^{-1}B$ [Multiplying A^{-1} on both sides]

$\Rightarrow \quad IX = A^{-1}B \quad [AA^{-1} = I]$

$\Rightarrow \quad X = A^{-1}B$

$$\Rightarrow \quad \begin{bmatrix}x\\y\\z\end{bmatrix} = \begin{bmatrix}-3 & 2 & 2\\-2 & 1 & 1\\-4 & 2 & 3\end{bmatrix}\begin{bmatrix}10\\8\\7\end{bmatrix}$$

$$\Rightarrow \quad \begin{bmatrix}x\\y\\x\end{bmatrix} = \begin{bmatrix}-30+16+14\\-20+8+7\\-40+16+21\end{bmatrix}$$

$$\Rightarrow \quad \begin{bmatrix}x\\y\\z\end{bmatrix} = \begin{bmatrix}0\\-5\\-3\end{bmatrix}$$

$\therefore \quad x = 0, y = -5$ and $z = -3$. **Ans.**

Answer 12.

(a) Let S be the given surface area of a closed right circular cylinder whose radius is r and height is h.

$$\therefore \quad S = 2\pi rh + 2\pi r^2$$

$$\Rightarrow \quad h = \frac{S-2\pi r^2}{2\pi r} \quad \text{...(i)}$$

Now, volume of cylinder (V) = $\pi r^2 h$

$$\therefore \quad V = \pi r^2\left(\frac{S-2\pi r^2}{2\pi r}\right) \quad \text{[From equation (i)]}$$

$$\Rightarrow \quad V = \frac{Sr-2\pi r^3}{2}$$

$$\therefore \quad \frac{dV}{dr} = \frac{S-6\pi r^2}{2} \quad \text{[differentiating w.r.t. } r.] \text{...(ii)}$$

and
$$\frac{d^2V}{dr^2} = \frac{-12\pi r}{2} = -6\pi r \quad \text{[Again diff. w.r.t. } r.] \quad \text{...(iii)}$$

For maximum and minimum value,

Put $\quad \frac{dV}{dr} = 0 \quad$ [from eq. (ii)]

$$\Rightarrow \quad \frac{S-6\pi r^2}{2} = 0$$

$$\Rightarrow \quad S = 6\pi r^2$$

Putting the value of S in equation (i), we get

$$h = \frac{6\pi r^2 - 2\pi r^2}{2\pi r}$$
$$= \frac{4\pi r^2}{2\pi r} = 2r$$

or $\quad r = \frac{h}{2}$

and $\quad \left[\frac{d^2V}{dr^2}\right]_{r=h/2} = -6\pi \times \frac{h}{2} \quad$ [from eq. (iii)]

$$= -3\pi h < 0$$

[$\because h$ cannot be negative]

So, volume is maximum when $r = \frac{h}{2}$.

Hence Proved.

OR

(b) Let h be the hypotenuse of the right-angled triangle and x be its altitude.

So, base $= \sqrt{h^2 - x^2}$ [using Pythagoras theorem]

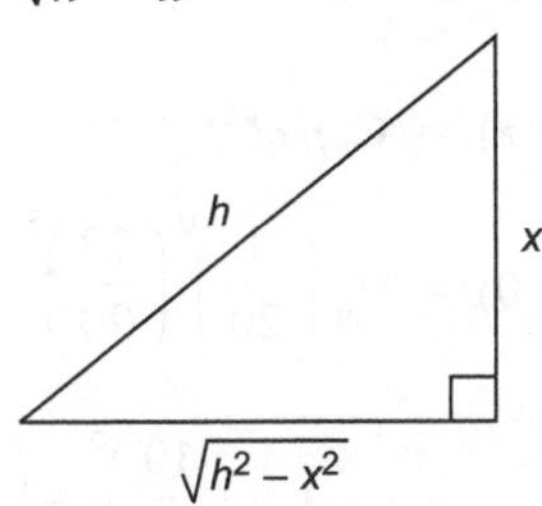

Now, area of triangle (A) $= \frac{1}{2} \times \sqrt{h^2 - x^2} \times x$

$$\therefore \frac{dA}{dx} = \frac{1}{2}\left[x\left\{\frac{1}{2}(h^2 - x^2)^{-1/2} \times (-2x)\right\} + \sqrt{h^2 - x^2}\right]$$

$$= \frac{1}{2}\left[\frac{-x^2}{\sqrt{h^2 - x^2}} + \sqrt{h^2 - x^2}\right]$$

$$= \frac{1}{2}\left[\frac{-x^2 + h^2 - x^2}{\sqrt{h^2 - x^2}}\right]$$

$$= \frac{1}{2}\left[\frac{h^2 - 2x^2}{\sqrt{h^2 - x^2}}\right]$$

and
$$\frac{d^2A}{dx^2} = \frac{1}{2}\left[\frac{\sqrt{h^2 - x^2}.(-4x) - (h^2 - 2x^2)\left\{\frac{1}{2}(h^2 - x^2)^{-1/2}(-2x)\right\}}{\left(\sqrt{h^2 - x^2}\right)^2}\right]$$

$$= \frac{1}{2}\left[\frac{-4x\left(\sqrt{h^2 - x^2}\right) + \frac{(h^2 - 2x^2)x}{\sqrt{h^2 - x^2}}}{(h^2 - x^2)}\right]$$

$$= \frac{1}{2}\left[\frac{-4x(h^2 - x^2) + (h^2 - 2x^2)x}{(h^2 - x^2).\sqrt{h^2 - x^2}}\right]$$

$$= \frac{1}{2}\left[\frac{-4xh^2 + 4x^3 + h^2x - 2x^3}{(h^2 - x^2)^{3/2}}\right]$$

$$= \frac{1}{2}\left[\frac{2x^3 - 3xh^2}{(h^2 - x^2)^{3/2}}\right]$$

For maximum and minimum value,

Put $\frac{dA}{dx} = 0$

$$\Rightarrow \frac{1}{2}\left[\frac{h^2 - 2x^2}{\sqrt{h^2 - x^2}}\right] = 0$$

$$\Rightarrow h^2 - 2x^2 = 0$$

$$\Rightarrow x = \frac{h}{\sqrt{2}}$$

$$\therefore \left(\frac{d^2A}{dx^2}\right)_{x=\frac{h}{\sqrt{2}}} = \frac{1}{2}\left[\frac{2 \times \frac{h^3}{2\sqrt{2}} - 3 \times \frac{h}{\sqrt{2}} \times h^2}{\left(h^2 - \frac{h^2}{2}\right)^{3/2}}\right]$$

$$= \frac{1}{2}\left[\frac{h^3 - 3h^3}{\left(\frac{h^2}{2}\right)^{3/2}\sqrt{2}}\right]$$

$$= \frac{-2h^3}{2 \times \frac{h^3}{2\sqrt{2}} \times \sqrt{2}}$$

$$= -2 < 0$$

Thus, A is maximum at $x = \frac{h}{\sqrt{2}}$

Now, Base $= \sqrt{h^2 - x^2}$

$$= \sqrt{h^2 - \frac{h^2}{2}} = \sqrt{\frac{h^2}{2}} = \frac{h}{\sqrt{2}}$$

and Altitude $= x = \frac{h}{\sqrt{2}}$

Base $= \sqrt{h^2 - x^2}$

$$= \sqrt{h^2 - \left(\frac{h}{\sqrt{2}}\right)^2} = \frac{h}{\sqrt{2}}$$

Since, Base = Altitude $= \frac{h}{\sqrt{2}}$

Hence, the triangle is isosceles. **Hence Proved.**

Answer 13.

(a) Let $I = \int \tan^{-1}\sqrt{\frac{1-x}{1+x}}dx$

Let $x = \cos t$

$\Rightarrow \quad dx = -\sin t\, dt$

$\therefore \quad I = \int \tan^{-1}\sqrt{\dfrac{1-\cos t}{1+\cos t}}\,(-\sin t\, dt)$

$= \int \tan^{-1}\sqrt{\dfrac{2\sin^2\frac{t}{2}}{2\cos^2\frac{t}{2}}}\,(-\sin t)\, dt$

$= \int \tan^{-1}\left(\tan\dfrac{t}{2}\right)(-\sin t)\, dt$

$= -\int \dfrac{t}{2}\sin t\, dt$

$= -\dfrac{1}{2}\int t\sin t\, dt$

$= -\dfrac{1}{2}\left[t(-\cos t) - \int 1.(-\cos t)\, dt\right]$

[Integrating by parts]

$= \dfrac{1}{2}t\cos t - \dfrac{1}{2}\sin t + c$

$= \dfrac{1}{2}t\cos t - \dfrac{1}{2}\sqrt{1-\cos^2 t} + c$

$= \dfrac{1}{2}x\cos^{-1}x - \dfrac{1}{2}\sqrt{1-x^2} + c$ **Ans.**

OR

(b) $\int \dfrac{2x+7}{x^2-x-2}dx = \int \dfrac{2x+7}{x^2-2x+x-2}dx$

$= \int \dfrac{2x+7}{(x-2)(x+1)}dx$

Let $\dfrac{2x+7}{(x-2)(x+1)} = \dfrac{A}{x-2} + \dfrac{B}{x+1}$

$\Rightarrow \quad \dfrac{2x+7}{(x-2)(x+1)} = \dfrac{A(x+1)+B(x-2)}{(x-2)(x+1)}$

$\Rightarrow \quad 2x+7 = (A+B)x + A - 2B$

Comparing both sides, we get

$A + B = 2$ and $A - 2B = 7$

Solving these equations for A and B, we get

$A = \dfrac{11}{3}$ and $B = -\dfrac{5}{3}$

Now, $\int \dfrac{2x+7}{(x-2)(x+1)}dx = \dfrac{11}{3}\int \dfrac{1}{(x-2)}dx - \dfrac{5}{3}\int \dfrac{1}{x+1}dx$

$= \dfrac{11}{3}\log|x-2| - \dfrac{5}{3}\log|x+1| + c$ **Ans.**

Solution 14.

Let p be the probability that the bulb will fuse after 150 days of use.

[Since it is a case of binomial distribution]

So, $p = 0.05 = \dfrac{5}{100} = \dfrac{1}{20}$

$\therefore \quad q = 1 - p = 1 - \dfrac{1}{20} = \dfrac{19}{20}$

and $n = 5$

We know, $P(x = r) = {}^nC_r p^r q^{n-r}$

(i) $P(x = 0) = {}^5C_0\left(\dfrac{1}{20}\right)^0\left(\dfrac{19}{20}\right)^5$

$= 1 \times 1 \times \left(\dfrac{19}{20}\right)^5 = \left(\dfrac{19}{20}\right)^5$ **Ans.**

(ii) $P(x \le 1) = P(x = 0) + P(x = 1)$

$= \left(\dfrac{19}{20}\right)^5 + {}^5C_1\left(\dfrac{1}{20}\right)^1\left(\dfrac{19}{20}\right)^4$

[Putting value of P from (i)]

$= \left(\dfrac{19}{20}\right)^5 + 5 \times \dfrac{1}{20} \times \left(\dfrac{19}{20}\right)^4$

$= \left(\dfrac{19}{20}\right)^4\left(\dfrac{19}{20} + \dfrac{1}{4}\right)$

$= \left(\dfrac{19}{20}\right)^4\left(\dfrac{19+5}{20}\right)$

$= \left(\dfrac{19}{20}\right)^4 \dfrac{24}{20}$

$= \dfrac{6}{5}\left(\dfrac{19}{20}\right)^4$ **Ans.**

(iii) $P(x > 1) = 1 - P(x \le 1)$

$= 1 - \dfrac{6}{5}\left(\dfrac{19}{20}\right)^4$ **Ans.**

(iv) $P(x \ge 1) = 1 - P(x = 0)$

$= 1 - \left(\dfrac{19}{20}\right)^5$

[Putting value of P from (i)]

$= 1 - \left(\dfrac{19}{20}\right)^5$ **Ans.**

SECTION – B

Solution 15.

(a) Let $\vec{a} = \hat{i}-2\hat{j}-2\hat{k}$

Then, $|\vec{a}| = \sqrt{1^2+(-2)^2+(-2)^2}$

$= \sqrt{1+4+4} = \sqrt{9} = 3$

$\therefore$ Unit vector, $\hat{a} = \dfrac{\vec{a}}{|\vec{a}|}$

$= \dfrac{1}{3}\hat{i} - \dfrac{2}{3}\hat{j} - \dfrac{2}{3}\hat{k}$

Hence, the required vector of magnitude 18 units is $18\hat{a}$.

$\therefore$ Required vector $= \dfrac{18}{3}\hat{i} - \dfrac{36}{3}\hat{j} - \dfrac{36}{3}\hat{k}$

$= 6\hat{i}-12\hat{j}-12\hat{k}$ **Ans.**

(b) Given lines are,

$$\frac{x+1}{2} = \frac{y-2}{5} = \frac{z+3}{4} \quad \text{...(i)}$$

and

$$\frac{x-1}{5} = \frac{y+2}{2} = \frac{z-1}{-5} \quad \text{...(ii)}$$

$\therefore$ Direction ratios of the (i), $\langle a_1, b_1, c_1\rangle = \langle 2, 5, 4\rangle$ and, direction ratios of line (ii), $\langle a_2, b_2, c_2\rangle = \langle 5, 2, -5\rangle$.

The angle between two lines is given as,

$$\cos\theta = \frac{a_1a_2+b_1b_2+c_1c_2}{\sqrt{a_1^2+b_1^2+c_1^2}.\sqrt{a_2^2+b_2^2+c_2^2}}$$

$$= \frac{2\times5+5\times2+4\times(-5)}{\sqrt{2^2+5^2+4^2}.\sqrt{5^2+2^2+(-5)^2}}$$

$$= \frac{10+10-20}{\sqrt{4+25+16}.\sqrt{25+4+25}}$$

$$= \frac{0}{\sqrt{45}.\sqrt{54}}$$

$\Rightarrow \quad \cos\theta = 0$

$\therefore \quad \theta = \dfrac{\pi}{2} = 90^\circ$

Hence, the angle between the two lines is $\dfrac{\pi}{2}$. **Ans.**

(c) The general equation of a plane is,

$$a(x-x_1)+b(y-y_1)+c(z-z_1) = 0$$

So, the equation of the plane passing through $(2, -3, 1)$ is,

$$a(x-2)+b(y+3)+c(z-1) = 0 \quad \text{...(i)}$$

The direction ratios of the line joining the points (4, 5, 0) and (1, –2, 4) are (1 – 4, –2 – 5, 4 – 0) *i.e.*, (–3, – 7, 4).

Since, the plane is perpendicular to the line whose direction ratios are – 3, – 7 and 4.

So, direction ratios of the normal to the plane are – 3, –7 and 4.

Putting these values in equation (i), we get

$$-3(x-2)+(-7)(y+3)+4(z-1) = 0$$

$$-3x+6-7y-21+4z-4 = 0$$

$\Rightarrow \quad -3x-7y+4z-19 = 0$

or $\quad 3x+7y-4z+19 = 0$

which is the required equation of plane. **Ans.**

Solution 16.

(a) Consider,

$$\text{L.H.S.} = \vec{a}.[(\vec{b}+\vec{c})\times(\vec{a}+3\vec{b}+4\vec{c})]$$

$$= \vec{a}.[\vec{b}\times\vec{a}+3(\vec{b}\times\vec{b})+4(\vec{b}\times\vec{c})+\vec{c}\times\vec{a} + 3(\vec{c}\times\vec{b})+4(\vec{c}\times\vec{c})]$$

$$= \vec{a}.[\vec{b}\times\vec{a}+0+4(\vec{b}\times\vec{c})+\vec{c}\times\vec{a}+3(\vec{c}\times\vec{b})+0]$$

$$= \vec{a}.(\vec{b}\times\vec{a})+4[\vec{a}.(\vec{b}\times\vec{c})]+\vec{a}.(\vec{c}\times\vec{a}) + 3[\vec{a}.(\vec{c}\times\vec{b})]$$

$$= 0+4[\vec{a}\ \vec{b}\ \vec{c}]+0+3[\vec{a}\ \vec{c}\ \vec{b}]$$

$$= 4[\vec{a}\ \vec{b}\ \vec{c}]-3[\vec{a}\ \vec{b}\ \vec{c}]$$

$$= [\vec{a}\ \vec{b}\ \vec{c}] = \text{R.H.S.}$$

Hence Proved.

OR

(b) Given : Vertices of ΔABC are A (3, –1, 2), B (1, –1, –3) and C (4, –3, 1).

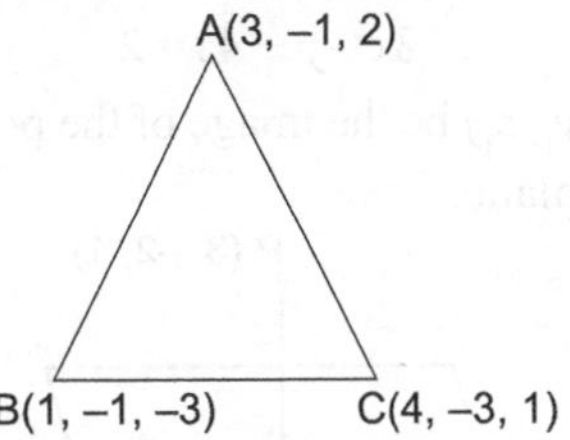

So, Position vector of A $= (3\hat{i}-\hat{j}+2\hat{k})$

Position vector of B $= (\hat{i}-\hat{j}-3\hat{k})$

Position vector of C $= (4\hat{i}-3\hat{j}+\hat{k})$

$\therefore \quad \vec{AB}$ = (p.v. of B) – (p.v. of A)

$= (\hat{i}-\hat{j}-3\hat{k})-(3\hat{i}-\hat{j}+2\hat{k})$

$= -2\hat{i}+0-5\hat{k}$

and $\vec{AC}$ = (p.v. of C) – (p.v. of A)

$= (4\hat{i}-3\hat{j}+\hat{k})-(3\hat{i}-\hat{j}+2\hat{k})$

$= \hat{i}-2\hat{j}-\hat{k}$

We know, area of $\Delta ABC = \frac{1}{2}|\vec{AB} \times \vec{AC}|$

Now, $\vec{AB} \times \vec{AC} = \begin{vmatrix} \hat{i} & \hat{j} & \hat{k} \\ -2 & 0 & -5 \\ 1 & -2 & -1 \end{vmatrix}$

$= (0-10)\hat{i} - (2+5)\hat{j} + (4-0)\hat{k}$

$= -10\hat{i} - 7\hat{j} + 4\hat{k}$

$\therefore \quad |\vec{AB} \times \vec{AC}| = \sqrt{(-10)^2+(-7)^2+(4)^2}$

$= \sqrt{100+49+16}$

$= \sqrt{165}$

So, Area of $\Delta ABC = \frac{1}{2}|\vec{AB} \times \vec{AC}|$

$= \frac{1}{2} \times \sqrt{165}$

$= \frac{\sqrt{165}}{2}$ sq. units **Ans.**

Solution 17.

(a) Given: Equation of plane is,

$3x - y + 4z = 2$ ...(i)

Let Q (x_1, y_1, z_1) be the image of the point P (3, –2, 1) in the given plane.

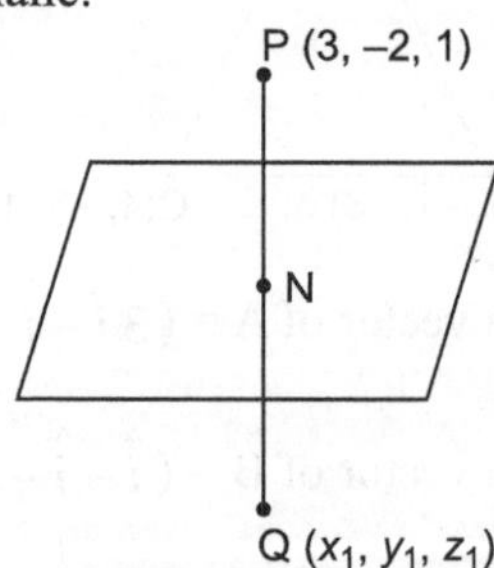

The equation of a line passing through (3, –2, 1) and perpendicular to the given plane is,

$\frac{x-3}{3} = \frac{y+2}{-1} = \frac{z-1}{4} = k$ (say)

So, the coordinates of a general point on this line are $(3k + 3, -k - 2, 4k + 1)$.

If N is the foot of the perpendicular from P to the given plane, then N lies on the plane.

Let the coordinates of N be $(3k + 3, -k - 2, 4k + 1)$.

$\therefore 3(3k + 3) - (-k - 2) + 4(4k + 1) - 2 = 0$

[Putting N in eq. (i)]

$\Rightarrow \quad 9k + 9 + k + 2 + 16k + 4 - 2 = 0$

$\Rightarrow \quad 26k + 13 = 0$

$\Rightarrow \quad k = -\frac{13}{26}$

$= -\frac{1}{2}$

So, coordinates of N $= \left[3\times\left(-\frac{1}{2}\right)+3, -\left(-\frac{1}{2}\right)-2, 4\times\left(-\frac{1}{2}\right)+1\right]$

$= \left(\frac{3}{2}, -\frac{3}{2}, -1\right)$

Since, N is the mid-point of PQ,

$\therefore$ By mid-point formula,

$\frac{3+x_1}{2} = \frac{3}{2}$

$\Rightarrow \quad x_1 = 3 - 3 = 0$

Also, $\frac{-2+y_1}{2} = -\frac{3}{2}$

$\Rightarrow \quad y_1 = -3 + 2 = -1$

and $\frac{1+z_1}{2} = -1$

$\Rightarrow \quad 1 + z_1 = -2$

$\Rightarrow \quad z_1 = -2 - 1 = -3$

Hence, the required image of the point P is Q(0, –1, –3). **Ans.**

OR

(b) Let the equation of required line be,

$\frac{x-x_1}{a} = \frac{y-y_1}{b} = \frac{z-z_1}{c}$

$\because$ The line passes through (–1, 3, –2).

$\therefore \quad \frac{x+1}{a} = \frac{y-3}{b} = \frac{z+2}{c}$ ...(i)

Now, given lines are

$\frac{x}{1} = \frac{y}{2} = \frac{z}{3}$

and $\frac{x+2}{-3} = \frac{y-1}{2} = \frac{z+1}{5}$

Since, the required line is perpendicular to these two lines

$\therefore \quad a \times 1 + b \times 2 + 3 \times c = 0$

$\Rightarrow \quad a + 2b + 3c = 0 \quad ...(ii)$

and $\quad a \times (-3) + b \times 2 + c \times 5 = 0$

$\Rightarrow \quad -3a + 2b + 5c = 0 \quad ...(iii)$

Solving equations (ii) and (iii) using cross-multiplication method,

$$\frac{a}{10-6} = \frac{b}{-9-5} = \frac{c}{2+6}$$

$$\Rightarrow \quad \frac{a}{4} = \frac{b}{-14} = \frac{c}{8} = k \text{ (say)}$$

$\Rightarrow \quad a = 4k, b = -14k$ and $c = 8k.$

Putting these values in equation (i), we get

$$\frac{x+1}{4k} = \frac{y-3}{-14k} = \frac{z+2}{8k}$$

$$\Rightarrow \quad \frac{x+1}{2} = \frac{y-3}{-7} = \frac{z+2}{4},$$

which is the required equation of line. **Ans.**

Solution 18.

Given equation of curves are,

$$y^2 = x \quad ...(i)$$

and $$y^2 = 4 - 3x \quad ...(ii)$$

These are the equations of parabola.

Solving equations (i) and (ii), we get

$$x = 4 - 3x$$

$\Rightarrow \quad 4x = 4$

$\Rightarrow \quad x = 1$

$\therefore \quad y^2 = 1$

$\Rightarrow \quad y = \pm 1$

So, (1, 1) and (1, –1) are the points of intersection of the two parabolas.

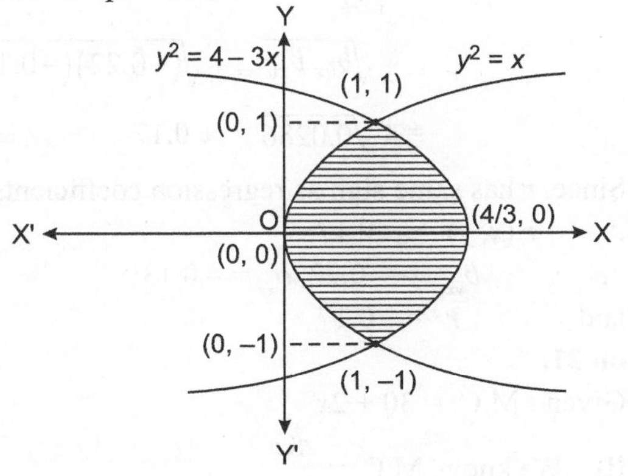

$$\therefore \text{ Required area} = \int_{-1}^{1} x_{II}\, dy - \int_{-1}^{1} x_I dy$$

$$= \int_{-1}^{1} \frac{4-y^2}{3} dy - \int_{-1}^{1} y^2 dy \qquad \left[x_{II} = \frac{4-y^2}{3},\ x_I = y^2 \right]$$

$$= \frac{1}{3}\left[4y - \frac{y^3}{3}\right]_{-1}^{1} - \left[\frac{y^3}{3}\right]_{-1}^{1}$$

$$= \frac{1}{3}\left[\left(4\times1 - \frac{1^3}{3}\right) - \left(4\times(-1) - \frac{(-1)^3}{3}\right)\right] - \left[\frac{1^3}{3} - \frac{(-1)^3}{3}\right]$$

$$= \frac{1}{3}\left[4 - \frac{1}{3} + 4 - \frac{1}{3}\right] - \left[\frac{1}{3} + \frac{1}{3}\right]$$

$$= \frac{1}{3}\left[8 - \frac{2}{3}\right] - \frac{2}{3}$$

$$= \frac{1}{3} \times \frac{22}{3} - \frac{2}{3}$$

$$= \frac{22}{9} - \frac{2}{3}$$

$$= \frac{22-6}{9}$$

$$= \frac{16}{9} \text{ sq. units.}$$ **Ans.**

SECTION – C

Solution 19.

(a) Given : $\quad C(x) = 35x + 250$

and $\quad S(x) = 60x$

(i) $\quad P(x) = S(x) - C(x)$

$= 60x - 35x - 250$

$= 25x - 250$ **Ans.**

(ii) For break even points,

$P(x) = 0$

$\Rightarrow \quad 25x - 250 = 0 \quad$ [Putting P(x) from (i)]

$\Rightarrow \quad x = \frac{250}{25} = 10$

$\therefore \quad x = 10$ **Ans.**

(b) Given : $\quad R(x) = 100x - x^2 - x^3$

(i) We know, $R(x) = P(x) \times x$

$$\therefore \quad P(x) = \frac{R(x)}{x}$$

$$= \frac{100x - x^2 - x^3}{x}$$

$= 100 - x - x^2$ **Ans.**

(ii) We know $\text{M.R.} = \frac{dR}{dx}$

$$= \frac{d}{dx}(100x - x^2 - x^3)$$

$= 100 - 2x - 3x^2$ **Ans.**

(c) Given : Lines of regression are,

$4x - 2y = 4 \quad ...(i)$

and $\quad 2x - 3y + 6 = 0$

$\Rightarrow \quad 2x - 3y = -6 \quad ...(ii)$

Multiplying equation (ii) by 2 and then subtracting equation (i) from it,

$$4x - 6y = -12$$
$$4x - 2y = 4$$
$$- \quad + \quad -$$
$$-4y = -16$$

$\Rightarrow$ $y = \frac{16}{4} = 4$

Putting $y = 4$ in equation (i), we get $4x - 2 \times 4 = 4$

$\Rightarrow$ $4x = 4 + 8 = 12$

$\Rightarrow$ $x = \frac{12}{4} = 3$

$\therefore$ $\bar{x} = 3$ and $\bar{y} = 4$ **Ans.**

Solution 20.

(a) Given : $r = 0{\cdot}6$, $\bar{x} = 10$, $\bar{y} = 20$, var. $(x) = (\sigma_x^2)$ $= 225$ and var. $(y) = \sigma_y^2 = 400$.

(i) $\sigma_x = \sqrt{225} = 15$ and $\sigma_y = \sqrt{400} = 20$

$$b_{yx} = r\frac{\sigma_y}{\sigma_x} = 0{\cdot}6 \times \frac{20}{15} = \frac{4}{5}$$

and $$b_{xy} = r\frac{\sigma_x}{\sigma_y} = 0{\cdot}6 \times \frac{15}{20} = \frac{9}{20}$$

So, line of regression of y on x is

$$y - \bar{y} = b_{yx}(x - \bar{x})$$

$\Rightarrow$ $y - 20 = \frac{4}{5}(x - 10)$

$\Rightarrow$ $5y - 100 = 4x - 40$

$\Rightarrow$ $4x - 5y + 60 = 0$ ...(i)

Line of regression of x on y is

$$x - \bar{x} = b_{xy}(y - \bar{y})$$

$\Rightarrow$ $x - 10 = \frac{9}{20}(y - 20)$

$\Rightarrow$ $20x - 200 = 9y - 180$

$\Rightarrow$ $20x - 9y - 20 = 0$ **Ans.**

(ii) At $x = 2$

$4 \times 2 - 5y + 60 = 0$ [Putting $x = 2$ in eq. (i)]

$\Rightarrow$ $5y = 68$

$\therefore$ $y = \frac{68}{5}$ **Ans.**

OR

(b)

x	x^2	y	y^2	xy
2	4	8	64	16
6	36	8	64	48
4	16	5	25	20
7	49	6	36	42
5	25	2	4	10
$\Sigma x = 24$	$\Sigma x^2 = 130$	$\Sigma y = 29$	$\Sigma y^2 = 193$	$\Sigma xy = 136$

$$b_{yx} = \frac{\Sigma xy - \frac{1}{n}\Sigma x \Sigma y}{\Sigma x^2 - \frac{1}{n}(\Sigma x)^2}$$

$$= \frac{136 - \frac{1}{5}(24)(29)}{130 - \frac{1}{5}(24)^2}$$

$$= \frac{\frac{136 \times 5 - 696}{5}}{\frac{130 \times 5 - 576}{5}}$$

$$= \frac{680 - 696}{650 - 576}$$

$$= \frac{-16}{74} = -0.22$$

$$b_{xy} = \frac{\Sigma xy - \frac{1}{n}\Sigma x \Sigma y}{\Sigma y^2 - \frac{1}{n}(\Sigma y)^2}$$

$$= \frac{136 - \frac{1}{5}(24)(29)}{193 - \frac{1}{5}(29)^2}$$

$$= \frac{136 \times 5 - 696}{193 \times 5 - 841}$$

$$= \frac{680 - 696}{965 - 841}$$

$$= \frac{-16}{124} = -0.13$$

$$r = \pm\sqrt{b_{yx}\, b_{xy}} = \pm\sqrt{(-0.22)(-0.13)}$$

$$= \pm\sqrt{0.0286} = \pm 0.17$$

Since, r has same sign as regression coefficients

$\therefore$ $r(x, y) = -0.17$

$\therefore$ $b_{yx} = -0.22$, $b_{xy} = -0.13$

and $r = -0.17$ **Ans.**

Solution 21.

(a) Given : M.C. $= 30 + 2x$

(i) We know, M.C. $= \frac{dC}{dx}$

$\Rightarrow$ $\frac{dC}{dx} = 30 + 2x$

Intergrating both sides w.r.t. x

$$\int \frac{dC}{dx}\,dx = \int (30 + 2x)dx$$

$\Rightarrow$ $C = 30x + \frac{2x^2}{2} + k$

$\Rightarrow$ $C = 30x + x^2 + k$

It is given that,

When $x = 0$, $C = 200$

$\therefore$ $200 = 0 + 0 + k$

$\Rightarrow$ $k = 200$

$\therefore$ Total cost $C(x) = 30x + x^2 + 200$. **Ans.**

(ii) Cost of increasing output from 100 to 200

$$= \int_{100}^{200} (30+2x)\,dx$$

$$= \left[30x + x^2\right]_{100}^{200}$$

$$= [30 \times 200 + (200)^2] - [30 \times 100 + (100)^2]$$

$$= [6{,}000 + 40{,}000] - [3{,}000 + 10{,}000]$$

$$= 46{,}000 - 13{,}000$$

$= ₹\,33{,}000$ **Ans.**

OR

(b) Given : $C(x) = \frac{1}{3}x^3 - 5x^2 + 30x - 15$

$[\because P = 6 \therefore S(x) = 6x]$

$\therefore$ $P(x) = S(x) - C(x)$

$$= 6x - \frac{1}{3}x^3 + 5x^2 - 30x + 15$$

$$= 5x^2 - \frac{1}{3}x^3 - 24x + 15$$

Now, $\frac{dP}{dx} = 10x - \frac{3x^2}{3} - 24$

$$= 10x - x^2 - 24$$

For maximum and minimum profit,

Put $\frac{dP}{dx} = 0$

$\Rightarrow$ $10x - x^2 - 24 = 0$

$\Rightarrow$ $x^2 - 10x + 24 = 0$

$\Rightarrow$ $(x-6)(x-4) = 0$

$\Rightarrow$ $x = 6$ or $x = 4$

Now, $\frac{d^2P}{dx^2} = 10 - 2x$

$\therefore$ $\left(\frac{d^2P}{dx^2}\right)_{x=6} = 10 - 2 \times 6 = -2 < 0$

So, at $x = 6$, $P(x)$ is maximum.

and $\left(\frac{d^2P}{dx^2}\right)_{x=4} = 10 - 2 \times 4 = 2 > 0$

So, at $x = 4$, $P(x)$ is minimum.

Hence, at $x = 6$, profit is maximum. **Ans.**

Solution 22.

Let the number of half sleeve shirts made per week be x and the number of full sleeve shirts made per week be y.

So, the objective function is,

Maximize $Z = x + 1.50y$

Subject to constraints,

$x + 2y \leq 40$ ...(i)

$2x + y \leq 40$ ...(ii)

$\frac{8}{5}x + \frac{8}{5}y \leq 40$

or $x + y \leq 25$ ...(iii)

and $x \geq 0$ and $y \geq 0$

From inequation (i), we have

$x + 2y = 40$

x	0	40
y	20	0

From inequation (ii), we have

$2x + y = 40$

x	0	20
y	40	0

From inequation (iii), we have

$x + y = 25$

x	0	25
y	25	0

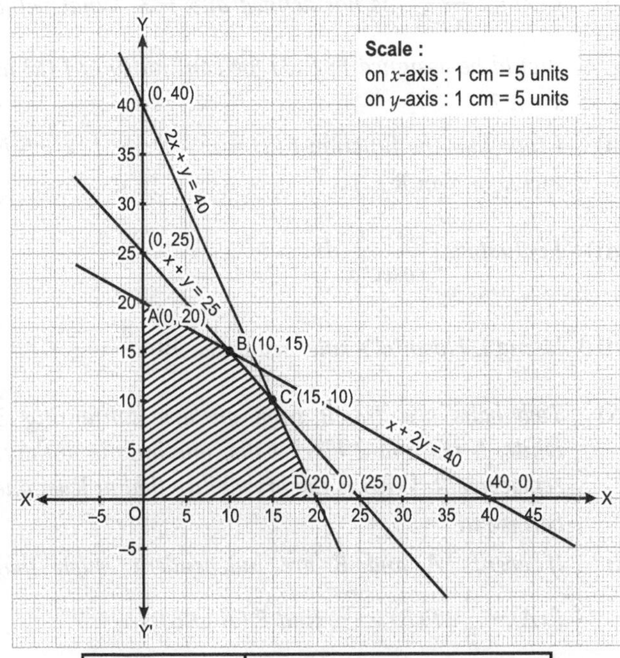

Points	$Z = x + 1.50y$
A (0, 20)	$0 + 1.50 \times 20 = ₹\,30$
B (10,15)	$10 + 1.50 \times 15 = ₹\,32.50$
C (15, 10)	$15 + 1.50 \times 10 = ₹\,30$
D (20, 0)	$20 + 1.50 \times 0 = ₹\,20$
O (0, 0)	0

So, Z is maximum at B (10, 15).

Hence, 10 half sleeves shirts and 15 full sleeve shirts should be made per week for maximum profit. **Ans.**

MATHEMATICS 2019

QUESTIONS

SECTION—A (80 Marks)

Question 1. [10 × 2]

(i) *If $f : R \to R$, $f(x) = x^3$ and $g : R \to R$, $g(x) = 2x^2 + 1$, and R is the set of real numbers, then find fog (x) and gof (x).*

(ii) *Solve : $\sin (2 \tan^{-1} x) = 1$.*

(iii) *Using determinants, find the values of k, if the area of triangle with vertices (– 2,0), (0, 4) and (0, k) is 4 square units.*

(iv) *Show that (A + A′) is symmetric matrix, if $A = \begin{bmatrix} 2 & 4 \\ 3 & 5 \end{bmatrix}$.*

(v) *$f(x) = \dfrac{x^2 - 9}{x - 3}$ is not defined at x = 3. What value should be assigned to f (3) for continuity of f (x) at x = 3 ?*

(vi) *Prove that the function $f(x) = x^3 - 6x^2 + 12x + 5$ is increasing on R.*

(vii) *Evaluate : $\int \dfrac{\sec^2 x}{\operatorname{cosec}^2 x} dx$.*

(viii) *Using L'Hospital's rule, evaluate : $\lim\limits_{x \to 0} \dfrac{8^x - 4^x}{4x}$.*

(ix) *Two balls are drawn from an urn containing 3 white, 5 red and 2 black balls, one by one without replacement. What is the probability that at least one ball is red ?*

(x) *If events A and B are independent, such that $P(A) = \dfrac{3}{5}$, $P(B) = \dfrac{2}{3}$, find $P(A \cup B)$.*

Question 2. [4]

If $f: A \to A$ and $A = R - \left\{\dfrac{8}{5}\right\}$, show that the function $f(x) = \dfrac{8x+3}{5x-8}$ is one-one onto. Hence, find f^{-1}.

Question 3. [4]

(a) *Solve for x :*

$$\tan^{-1}\left(\frac{x-1}{x-2}\right) + \tan^{-1}\left(\frac{x+1}{x+2}\right) = \frac{\pi}{4}$$

OR

(b) *If $\sec^{-1} x = \operatorname{cosec}^{-1} y$, show that $\dfrac{1}{x^2} + \dfrac{1}{y^2} = 1$.*

Question 4. [4]

Using properties of determinants prove that :

$$\begin{vmatrix} x & x(x^2+1) & x+1 \\ y & y(y^2+1) & y+1 \\ z & z(z^2+1) & z+1 \end{vmatrix} = (x-y)(y-z)(z-x)(x+y+z)$$

Question 5. [4]

(a) *Show that the function $f(x) = |x - 4|$, $x \in R$ is continuous, but not differentiable at x = 4.*

OR

(b) *Verify the Lagrange's mean value theorem for the function :*

$$f(x) = x + \frac{1}{x} \text{ in the interval } [1, 3].$$

Question 6. [4]

If $y = e^{\sin^{-1} x}$ and $z = e^{-\cos^{-1} x}$, prove that $\dfrac{dy}{dz} = e^{\pi/2}$.

Question 7. [4]

A 13 m long ladder is leaning against a wall, touching the wall at a certain height from the ground level. The bottom of the ladder is pulled away from the wall, along the ground, at the rate of 2 m/s. How fast is the height on the wall decreasing when the foot of the ladder is 5 m away from the wall ?

Question 8. [4]

(a) *Evaluate : $\int \dfrac{x(1+x^2)}{1+x^4} dx$.*

OR

(b) *Evaluate : $\int_{-6}^{3} |x+3|\, dx$*

Question 9. [4]

Solve the differential equation : $\dfrac{dy}{dx} = \dfrac{x+y+2}{2(x+y)-1}$

Question 10. [4]

Bag A contains 4 white balls and 3 balck balls, while Bag B contains 3 white balls and 5 black balls. Two balls are drawn from bag A and placed in bag B. Then, what is the probability of drawing a white ball from bag B ?

Question 11. **[6]**

Solve the following system of linear equations using matrix method :

$$\frac{1}{x}+\frac{1}{y}+\frac{1}{z} = 9$$

$$\frac{2}{x}+\frac{5}{y}+\frac{7}{z} = 52$$

$$\frac{2}{x}+\frac{1}{y}-\frac{1}{z} = 0$$

Question 12. **[6]**

(a) *The volume of a closed rectangular metal box with a square base is 4096 cm³. The cost of polishing the outer surface of the box is ₹ 4 per cm². Find the dimensions of the box for the minimum cost of polishing it.*

OR

(b) *Find the point on the straight line $2x + 3y = 6$, which is closest to the origin.*

Question 13. **[6]**

Evaluate : $\int_0^{\pi} \frac{x \tan x}{\sec x + \tan x} dx$

Question 14. **[6]**

(a) *Given three identical Boxes A, B and C, Box A contains 2 gold and 1 silver coin, Box B contains 1 gold and 2 silver coins and Box C contains 3 silver coins. A person chooses a Box at random and takes out a coin. If the coin drawn is of silver, find the probability that it has been drawn from the Box which has the remaining two coins also of silver.*

OR

(b) *Determine the binomial distribution where mean is 9 and standard deviation is $\frac{3}{2}$. Also, find the probability of obtaining at most one success.*

SECTION—B (20 Marks)

Question 15. **[3 × 2]**

(a) *If $\vec{a}$ and $\vec{b}$ are perpendicular vectors, $|\vec{a}+\vec{b}| = 13$ and $|\vec{a}| = 5$, find the value of $|\vec{b}|$.*

(b) *Find the length of the perpendicular from origin to the plane $\vec{r}\cdot(3\hat{i} - 4\hat{j} - 12\hat{k}) + 39 = 0$.*

(c) *Find the angle between the two lines $2x = 3y = -z$ and $6x = -y = -4z$.*

Question 16. **[4]**

(a) *If $\vec{a} = \hat{i} - 2\hat{j} + 3\hat{k}$, $\vec{b} = 2\hat{i} + 3\hat{j} - 5\hat{k}$, prove that $\vec{a}$ and $\vec{a} \times \vec{b}$ are perpendicular.*

OR

(b) *If $\vec{a}$ and $\vec{b}$ are non-collinear vectors, find the value of x such that the vectors $\vec{\alpha} = (x-2)\vec{a} + \vec{b}$ and $\vec{\beta} = (3 + 2x)\vec{a} - 2\vec{b}$ are collinear.*

Question 17. **[4]**

(a) *Find the equation of the plane passing through the intersection of the planes $2x + 2y - 3z - 7 = 0$ and $2x + 5y + 3z - 9 = 0$, such that the intercepts made by the resulting plane on the X-axis and the Z-axis are equal.*

OR

(b) *Find the equation of the lines passing through the point (2, 1, 3) and perpendicular to the lines $\frac{x-1}{1} = \frac{y-2}{2} = \frac{z-3}{3}$ and $\frac{x}{-3} = \frac{y}{2} = \frac{z}{5}$.*

Question 18. **[6]**

Draw a rough sketch and find the area bounded by the curve $x^2 = y$ and $x + y = 2$.

SECTION—C (20 Marks)

Question 19. **[3 × 2]**

(a) *A company produces a commodity with ₹24,000 as fixed cost. The variable cost estimated to be 25% of the total revenue received on selling the product, is at the rate of ₹8 per unit. Find the break-even point.*

(b) *The total cost function for a production is given by*

$$C(x) = \frac{3}{4}x^2 - 7x + 27.$$

Find the number of units produced for which M.C. = A.C. (M.C. = Marginal Cost and A.C. = Average Cost)

(c) *If $\bar{x} = 18$, $\bar{y} = 100$, $\sigma_x = 14$, $\sigma_y = 20$ and correlation coefficient $r_{xy} = 0{\cdot}8$, find the regression equation of y on x.*

Question 20. **[4]**

(a) *The following results were obtained with respect to two variables x and y.*

$\Sigma x = 15, \Sigma y = 25, \Sigma xy = 83, \Sigma x^2 = 55, \Sigma y^2 = 135$ *and* $n = 5$

(i) *Find the regression coefficient b_{xy}.*

(ii) *Find the regression equaiton of x on y.*

OR

(b) *Find the equation of the regression line of y on x, if the observations (x, y) are as follows :*

(1, 4), (2, 8), (3, 2), (4, 12), (5, 10), (6, 14), (7, 16), (8, 6), (9, 18)

Also, find the estimated value of y when x = 14.

Question 21. **[4]**

(a) *The cost function of a product is given by* $C(x) = \frac{x^3}{3} - 45x^2 - 900x + 36$ *where x is the number of units produced. How many units should be produced to minimise the marginal cost ?*

OR

(b) *The marginal cost function of x units of a product is given by MC = $3x^2 - 10x + 3$. The cost of producing one unit is ₹7. Find the total cost function and average cost function*

Question 22. **[6]**

A carpenter has 90, 80 and 50 running feet respectively of teak wood, plywood and rosewood which is used to produce product A and product B. Each unit of product A requires 2, 1 and 1 running feet and each unit of product B requires 1, 2 and 1 running feet of teak wood, plywood and rosewood respectively. If product A is sold for ₹ 48 per unit and product B is sold for ₹ 40 per unit, how many units of product A and product B should be produced and sold by the carpenter, in order to obtain the maximum gross income ?

Formulate the above as a Linear Programming Problem and solve it, indicating clearly the feasible region in the graph.

SOLUTIONS

SECTION—A

Solution 1.

(i) Given, $f(x) = x^3$ and $g(x) = 2x^2 + 1$

$\therefore$ $f(g(x))$ and $g(f(x))$ both are defined over common domain.

$$fog(x) = f(g(x)) = f(2x^2+1) = (2x^2+1)^3 = (2x^2)^3 + (1)^3 + 3\times(2x^2)^2 + 3\times 2x^2 = 8x^6 + 12x^4 + 6x^2 + 1$$ **Ans.**

$$gof(x) = g(f(x)) = g(x^3) = 2(x^3)^2 + 1 = 2x^6 + 1$$ **Ans.**

(ii) Given, $\sin(2\tan^{-1}x) = 1$

$\Rightarrow\ 2\tan^{-1}x = \sin^{-1}1$

$\Rightarrow\ 2\tan^{-1}x = \frac{\pi}{2}$

$\Rightarrow\ \tan^{-1}x = \frac{\pi}{4}$

$\Rightarrow\ x = \tan\frac{\pi}{4}$

$\Rightarrow\ x = 1$ **Ans.**

(iii) Given, Area of $\Delta ABC = 4$

$$\Rightarrow \frac{1}{2}\begin{vmatrix} x_1 & y_1 & 1 \\ x_2 & y_2 & 1 \\ x_3 & y_3 & 1 \end{vmatrix} = 4$$

$$\Rightarrow \frac{1}{2}\begin{vmatrix} -2 & 0 & 1 \\ 0 & 4 & 1 \\ 0 & k & 1 \end{vmatrix} = 4$$

$$\Rightarrow \frac{1}{2}\,|[-2(4-k)]| = 4$$

[$\because$ Area cannot be negative]

$\Rightarrow\ |-8+2k| = 8$

$\Rightarrow\ -8+2k = \pm 8$

$-8+2k = 8$ or $-8+2k = -8$

$\Rightarrow\ 2k = 16$ or $2k = 0$

$\Rightarrow\ k = 8$ or $k = 0$ **Ans.**

(iv) Given, $A = \begin{bmatrix} 2 & 4 \\ 3 & 5 \end{bmatrix}$

Then, $A' = \begin{bmatrix} 2 & 3 \\ 4 & 5 \end{bmatrix}$

$$\therefore\ A + A' = \begin{bmatrix} 2 & 4 \\ 3 & 5 \end{bmatrix} + \begin{bmatrix} 2 & 3 \\ 4 & 5 \end{bmatrix} = \begin{bmatrix} 4 & 7 \\ 7 & 10 \end{bmatrix}$$

Now, $(A+A')' = \begin{bmatrix} 4 & 7 \\ 7 & 10 \end{bmatrix}$

$\because\ (A+A') = (A+A')'$

Hence, $(A + A')$ is a symmetric matrix.

Hence Proved.

(v) Given, $f(x) = \frac{x^2-9}{x-3}$

$\Rightarrow\ f(x) = \frac{(x-3)(x+3)}{x-3}$

$\Rightarrow\ \lim_{x\to 3} f(x) = \lim_{x\to 3}(x+3)$

$$= \lim_{x \to 3} (x+3)$$

$$f(3) = 6$$

Hence, $f(x)$ to be continuous at $x = 3$, $f(3) = 6$.

Ans.

(vi) Given, $f(x) = x^3 - 6x^2 + 12x + 5$

$$\therefore \quad f'(x) = 3x^2 - 12x + 12$$

$$= 3(x^2 - 4x + 4)$$

$$= 3(x-2)^2$$

$$\Rightarrow \quad 3(x-2)^2 \geq 0, \text{ for all } x \in R$$

$$\Rightarrow \quad f'(x) \geq 0, \text{ for all } x \in R.$$

$\therefore f(x)$ is increasing for $x \in R$. **Hence Proved.**

(vii)

$$\int \frac{\sec^2 x}{\operatorname{cosec}^2 x} dx = \int \frac{1}{\cos^2 x} \times \frac{\sin^2 x}{1} dx$$

$$= \int \tan^2 x \, dx$$

$$= \int (\sec^2 x - 1) \, dx$$

$$= \int \sec^2 x \, dx - \int 1 \, dx$$

$$= \tan x - x + C$$

Ans.

(viii) Applying L, Hospital's rule,

$$= \lim_{x \to 0} \frac{8^x - 4^x}{4x} \quad \left[\frac{0}{0} \text{ form}\right]$$

$$= \lim_{x \to 0} \frac{\frac{d}{dx}(8^x - 4^x)}{\frac{d}{dx}(4x)}$$

$$= \lim_{x \to 0} \frac{8^x \log 8 - 4^x \log 4}{4}$$

$$\left\{\because \frac{d(a^x)}{dx} = a^x \log a\right\}$$

$$= \frac{8^0 \log 8 - 4^0 \log 4}{4}$$

$$= \frac{\log 8 - \log 4}{4}$$

$$= \frac{1}{4} \log \frac{8}{4} = \frac{1}{4} \log 2.$$

Ans.

(ix) Given, Number of red balls = 5

Number of black balls = 2

Number of white balls = 3

P (at least one ball is red)

= 1 – P (no ball is red)

$$= 1 - \frac{^5C_2}{^{10}C_2}$$

$$= 1 - \frac{5 \times 4}{10 \times 9}$$

$$= 1 - \frac{2}{9} = \frac{7}{9}$$

$$= 0{\cdot}777 = 0{\cdot}78$$

Ans.

(x) Given, $P(A) = \frac{3}{5}$ and $P(B) = \frac{2}{3}$

$\because$ A and B are independent

$$\therefore P(A \cap B) = P(A).\,P(B) = \frac{3}{5} \times \frac{2}{3} = \frac{2}{5}$$

$$\because \quad P(A \cup B) = P(A) + P(B) - P(A \cap B)$$

$$= \frac{3}{5} + \frac{2}{3} - \frac{2}{5}$$

$$= \frac{9 + 10 - 6}{15}$$

$$= \frac{13}{15} = 0{\cdot}8666$$

Ans.

Solution 2.

Given, $f(x) = \dfrac{8x+3}{5x-8}$

For one-one, Let $x_1, x_2 \in A$ such that,

$$f(x_1) = f(x_2)$$

$$\therefore \quad \frac{8x_1 + 3}{5x_1 - 8} = \frac{8x_2 + 3}{5x_2 - 8}$$

$$\Rightarrow \quad (8x_1 + 3)(5x_2 - 8) = (8x_2 + 3)(5x_1 - 8)$$

$$\Rightarrow 40x_1x_2 - 64x_1 + 15x_2 - 24 = 40x_1x_2 - 64x_2 + 15x_1 - 24$$

$$\Rightarrow \quad -64x_1 - 15x_1 = -64x_2 - 15x_2$$

$$\Rightarrow \quad -79x_1 = -79x_2$$

$$\Rightarrow \quad x_1 = x_2$$

$\therefore f(x)$ is one-one **Hence Proved.**

For onto,

$$\text{Let } f(x) = y$$

$$\Rightarrow \quad y = \frac{8x+3}{5x-8}$$

$$\Rightarrow \quad y(5x - 8) = 8x + 3$$

$$\Rightarrow \quad 5xy - 8y = 8x + 3$$

$$\Rightarrow \quad 5xy - 8x = 8y + 3$$

$$\Rightarrow \quad x(5y - 8) = 8y + 3$$

$$\Rightarrow \quad x = \frac{8y+3}{5y-8}, \text{ when } y \in A$$

$$\therefore \quad 5y - 8 \neq 0$$

$$y \neq \frac{8}{5}$$

Range of $f = R - \left\{\frac{8}{5}\right\} = A$

$\therefore$ Codomain = Range

$\therefore f(x)$ is onto function. **Hence Proved.**

$\because$ f is one one onto $\Rightarrow f$ is invertible

$$\Rightarrow f(x) = y$$

$$\Rightarrow f^{-1}(y) = x$$

$$\Rightarrow f^{-1}(y) = \frac{8y+3}{5y-8}$$

i.e., $f^{-1}(x) = \frac{8x+3}{5x-8}$ **Ans.**

Solution 3.

(a) Given, $\tan^{-1}\left(\frac{x-1}{x-2}\right) + \tan^{-1}\left(\frac{x+1}{x+2}\right) = \frac{\pi}{4}$

$$\Rightarrow \tan^{-1}\left[\frac{\left(\frac{x-1}{x-2}\right)+\left(\frac{x+1}{x+2}\right)}{1-\left(\frac{x-1}{x-2}\right)\left(\frac{x+1}{x+2}\right)}\right] = \frac{\pi}{4}$$

$$\Rightarrow \tan^{-1}\left[\frac{(x-1)(x+2)+(x+1)(x-2)}{(x-2)(x+2)-(x-1)(x+1)}\right] = \frac{\pi}{4}$$

$$\Rightarrow \tan^{-1}\left[\frac{(x^2+x-1)+(x^2-x-2)}{(x^2-4)-(x^2-1)}\right] = \frac{\pi}{4}$$

$$\Rightarrow \left[\frac{(x^2+x-2)+(x^2-x-2)}{(x^2-4)-(x^2-1)}\right] = \tan\frac{\pi}{4}$$

$$\Rightarrow \frac{x^2+x-2+x^2-x-2}{x^2-4-x^2+1} = 1$$

$$\Rightarrow \frac{2x^2-4}{-3} = 1$$

$$\Rightarrow 2x^2 - 4 = -3$$

$$\Rightarrow 2x^2 = -3+4$$

$$\Rightarrow 2x^2 = 1$$

$$\Rightarrow x^2 = \frac{1}{2}$$

$$\Rightarrow x = \pm\frac{1}{\sqrt{2}}$$ **Ans.**

OR

(b) Given, $\sec^{-1} x = \operatorname{cosec}^{-1} y$

$$\Rightarrow \cos^{-1}\frac{1}{x} = \sin^{-1}\frac{1}{y}$$

$$\Rightarrow \cos^{-1}\frac{1}{x} = \frac{\pi}{2} - \cos^{-1}\frac{1}{y}$$

$$\Rightarrow \cos^{-1}\frac{1}{x} + \cos^{-1}\frac{1}{y} = \frac{\pi}{2}$$

$$\Rightarrow \cos^{-1}\left(\frac{1}{x}\times\frac{1}{y} - \sqrt{1-\frac{1}{x^2}}\sqrt{1-\frac{1}{y^2}}\right) = \frac{\pi}{2}$$

$$\Rightarrow \frac{1}{xy} - \sqrt{\frac{x^2-1}{x^2}}\sqrt{\frac{y^2-1}{y^2}} = \cos\frac{\pi}{2}$$

$$\Rightarrow \frac{1-\sqrt{x^2-1}\sqrt{y^2-1}}{xy} = 0$$

$$\Rightarrow 1-\sqrt{x^2-1}\sqrt{y^2-1} = 0$$

$$\Rightarrow 1 = \sqrt{x^2-1}\sqrt{y^2-1}$$

On squaring both sides, we get

$$(x^2-1)(y^2-1) = 1$$

$$\Rightarrow x^2y^2 - x^2 - y^2 + 1 = 1$$

$$\Rightarrow x^2 + y^2 = x^2y^2$$

$$\Rightarrow \frac{x^2}{x^2y^2} + \frac{y^2}{x^2y^2} = 1$$

[Dividing by x^2y^2 on both sides]

$$\Rightarrow \frac{1}{x^2} + \frac{1}{y^2} = 1$$

Hence Proved.

Alternatively,

$$\sec^{-1} x = \operatorname{cosec}^{-1} y$$

$$\Rightarrow \cos^{-1}\frac{1}{x} = \sin^{-1}\frac{1}{y}$$

$$\Rightarrow \cos^{-1}\frac{1}{x} = \cos^{-1}\sqrt{1-\frac{1}{y^2}}$$

$$\left[\because \sin^{-1}\alpha = \cos^{-1}\sqrt{1-\alpha^2}\right]$$

$$\Rightarrow \frac{1}{x} = \sqrt{1-\frac{1}{y^2}}$$

On squaring both sides, we get

$$\frac{1}{x^2} = 1 - \frac{1}{y^2}$$

$$\Rightarrow \frac{1}{x^2} + \frac{1}{y^2} = 1$$ **Hence Proved.**

Solution 4.

$$\text{L.H.S.} = \begin{vmatrix} x & x(x^2+1) & x+1 \\ y & y(y^2+1) & y+1 \\ z & z(z^2+1) & z+1 \end{vmatrix}$$

Applying $R_2 \to R_2 - R_1$ and $R_3 \to R_3 - R_1$

$$\Rightarrow \begin{vmatrix} x & x(x^2+1) & x+1 \\ y-x & y^3+y-x^3-x & y-x \\ z-x & z^3+z-x^3-x & z-x \end{vmatrix}$$

$$\Rightarrow \begin{vmatrix} x & x(x^2+1) & x+1 \\ (y-x) & (y-x)(y^2+x^2+xy+1) & y-x \\ (z-x) & (z-x)(z^2+x^2+zx+1) & z-x \end{vmatrix}$$

Taking $(y-x)$ and $(z-x)$ common from R_2 and R_3, respectively.

$$\Rightarrow (y-x)(z-x)\begin{vmatrix} x & x(x^2+1) & x+1 \\ 1 & y^2+x^2+xy+1 & 1 \\ 1 & z^2+x^2+zx+1 & 1 \end{vmatrix}$$

Now, applying $R_2 \to R_2 - R_3$,

$$\Rightarrow (y-x)(z-x)\begin{vmatrix} x & x(x^2+1) & x+1 \\ 0 & (y-z)(x+y+z) & 0 \\ 1 & z^2+x^2+zx+1 & 1 \end{vmatrix}$$

Taking $(y-z)$ common from R_2,

$$\Rightarrow (y-z)(y-x)(z-x)\begin{vmatrix} x & x(x^2+1) & x+1 \\ 0 & x+y+z & 0 \\ 1 & z^2+x^2+zx+1 & 1 \end{vmatrix}$$

Now, Expanding along R_2,

$$\Rightarrow (y-x)(y-z)(z-x)(x+y+z)\begin{vmatrix} x & x+1 \\ 1 & 1 \end{vmatrix}$$

$$\Rightarrow (y-x)(y-z)(z-x)(x+y+z)(x-x-1)$$

$$\Rightarrow (x-y)(y-z)(z-x)(x+y+z)$$ **Hence Proved.**

Solution 5.

(a) $$f(x) = |x-4| = \begin{cases} x-4, x \geq 4 \\ -(x-4), x < 4 \end{cases}$$

$$\text{R.H.L.} = \lim_{x\to 4^+} f(4) = \lim_{h\to 0} f(4+h)$$

$$= \lim_{h\to 0}(4+h-4)$$

$$= \lim_{h\to 0} h = 0$$

$$\text{L.H.L.} = \lim_{x\to 4^-} f(x) = \lim_{h\to 0} f(4-h)$$

$$= \lim_{h\to 0} -(4-h-4)$$

$$= \lim_{h\to 0} h = 0$$

$$f(4) = 4-4 = 0$$

$\because$ R.H.L. = L.H.L. $= f(4) = 0$

$\therefore f(x)$ is continuous at $x = 4$

Now, $$\text{R.H.D.} = \lim_{x\to 4^+} f'x = \lim_{h\to 0}\frac{f(4+h)-f(4)}{h}$$

$$= \lim_{h\to 0}\frac{(4+h-4)-(4-4)}{h}$$

$$= \lim_{h\to 0}\frac{h}{h} = \lim_{h\to 0} 1$$

$$= 1$$

and $$\text{L.H.D.} = \lim_{x\to 4^-} f'x = \lim_{h\to 0} f\frac{(4-h)-f(4)}{-h}$$

$$= \lim_{h\to 0}\frac{-(4-h-4)-(4-4)}{-h}$$

$$= \lim_{h\to 0}\frac{h}{-h} = -1$$

$\because$ R.H.D. $\neq$ L.H.D.

$\therefore$ $f(x)$ is not differentiable at $x = 4$.

Hence, $f(x)$ is continuous at $x = 4$ but not differentiable at $x = 4$. **Hence Proved.**

OR

(b) Given, $f(x) = x + \frac{1}{x'}$ $x \in [1, 3]$

(i) $f(x)$ is continuous in [1, 3]

(ii) $f(x)$ is differentiable for $x \in (1, 3)$.

$\therefore$ Lagrange's mean value theorem is applicable.

$\therefore$ There must atleast be one value of x, say c, in the open interval (1, 3), such that

$$f'(c) = \frac{f(b)-f(a)}{b-a} \quad \text{...(i)}$$

$$\because \quad f(x) = x + \frac{1}{x}$$

$$\therefore \quad f'(x) = 1 - \frac{1}{x^2} \quad \text{...(ii)}$$

$$\Rightarrow \quad 1 - \frac{1}{c^2} = \frac{\left(3+\frac{1}{3}\right)-\left(1+\frac{1}{1}\right)}{3-1}$$

[from eq. (i) and (ii)]

$$\Rightarrow \quad 1 - \frac{1}{c^2} = \frac{\frac{10}{3}-2}{2}$$

$$\Rightarrow \quad 1 - \frac{1}{c^2} = \frac{10-6}{6}$$

$$\Rightarrow \quad 1-\frac{1}{c^2} = \frac{4}{6}=\frac{2}{3}$$

$$\Rightarrow \quad 1-\frac{2}{3} = \frac{1}{c^2}$$

$$\Rightarrow \quad \frac{1}{3} = \frac{1}{c^2}$$

$$\Rightarrow \quad c = \pm\sqrt{3}$$

$= 1.73$ lies in the open interval (1, 3)

$\because \quad 1 < \sqrt{3} < 3.$

Hence, Lagrange's mean value theorem is verified.

Hence Proved.

Solution 6.

Given, $y = e^{\sin^{-1}x}$ and $z = e^{-\cos^{-1}x}$

$$\frac{dy}{dx} = e^{\sin^{-1}x} \times \frac{1}{\sqrt{1-x^2}} \text{ and } \frac{dz}{dx} = e^{-\cos^{-1}x} \times \frac{1}{\sqrt{1-x^2}}$$

$$\Rightarrow \quad \frac{dy}{dz} = \frac{dy}{dx}\times\frac{dx}{dz}$$

$$\Rightarrow \quad \frac{dy}{dz} = \frac{e^{\sin^{-1}x}}{\sqrt{1-x^2}}\times\frac{\sqrt{1-x^2}}{e^{-\cos^{-1}x}}$$

$$\Rightarrow \quad \frac{dy}{dz} = \frac{e^{\sin^{-1}x}}{e^{-\cos^{-1}x}} = e^{\sin^{-1}x}.e^{\cos^{-1}x}$$

$$\Rightarrow \quad \frac{dy}{dz} = e^{\sin^{-1}x+\cos^{-1}x}$$

$$\Rightarrow \quad \frac{dy}{dz} = e^{\pi/2}$$ **Hence Proved.**

Solution 7.

In ΔABC,

$x^2 + y^2 = (13)^2$ (By Pythagoras theorem)

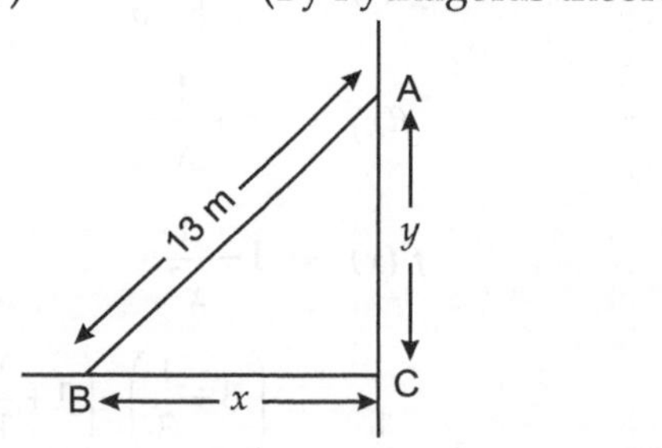

$$\Rightarrow \quad x^2 + y^2 = 169 \quad \text{...(i)}$$

Differentiating w.r.t. time (t),

$$2x\frac{dx}{dt} + 2y\frac{dy}{dt} = 0$$

$$\Rightarrow \quad \frac{dy}{dt} = -\frac{x}{y}\frac{dx}{dt} \quad \text{...(ii)}$$

when $x = 5m$, $\frac{dx}{dt} = 2$m/s

$y^2 = 169 - x^2$ [using (i)]

$\Rightarrow \quad y^2 = 169 - 25$

$\Rightarrow \quad y^2 = 144$

$\Rightarrow \quad y = \sqrt{144} = \pm 12$

$\Rightarrow \quad y = 12$m

[$\because$ Length cannot be negative]

From equation (ii), we get

$$\frac{dy}{dt} = \frac{-5}{12}\times 2$$

$$\Rightarrow \quad \frac{dy}{dt} = \frac{-5}{6} \text{ m/s.}$$

Negative sign indicates the height is decreasing.

Ans.

Solution 8.

(a) $$\int\frac{x(1+x^2)}{1+x^4}dx = \int\frac{x}{1+x^4}dx + \int\frac{x^3}{1+x^4}dx$$

Put $x^2 = t$ in I[st] Integral	Put $1 + x^4 = v$ in IInd integral
$2x\,dx = dt$	$4x^3\,dx = dv$
$x\,dx = \frac{dt}{2}$	$x^3\,dx = \frac{dv}{4}$

$$= \frac{1}{2}\int\frac{dt}{1+t^2} + \frac{1}{4}\int\frac{dv}{v}$$

$$= \frac{1}{2}\tan^{-1}t + \frac{1}{4}\log|v| + C$$

$$= \frac{1}{2}\tan^{-1}x^2 + \frac{1}{4}\log|1+x^4| + C$$ **Ans.**

OR

(b) $$\int_{-6}^{3}|x+3|\,dx = \int_{-6}^{-3}|x+3|\,dx + \int_{-3}^{3}|x+3|\,dx$$

$$= -\int_{-6}^{-3}(x+3)dx + \int_{-3}^{3}(x+3)dx$$

$$= -\left[\frac{x^2}{2}+3x\right]_{-6}^{-3} + \left[\frac{x^2}{2}+3x\right]_{-3}^{3}$$

$$= -\left[\left(\frac{9}{2}-9\right)-\left(\frac{36}{2}-18\right)\right] + \left[\left(\frac{9}{2}+9\right)-\left(\frac{9}{2}-9\right)\right]$$

$$= \frac{9}{2} + 0 + 18 = \frac{45}{2}$$ **Ans.**

Solution 9.

Given, $\frac{dy}{dx} = \frac{x+y+2}{2(x+y)-1}$ ...(i)

Let $x+y=v$

$$1+\frac{dy}{dx} = \frac{dv}{dx}$$

$$\frac{dy}{dx} = \frac{dv}{dx}-1$$

From equation (i),

$$\frac{dv}{dx}-1 = \frac{v+2}{2v-1}$$

$$\Rightarrow \frac{dv}{dx} = \frac{v+2}{2v-1}+1$$

$$\Rightarrow \frac{dv}{dx} = \frac{v+2+2v-1}{2v-1}$$

$$\Rightarrow \frac{dv}{dx} = \frac{3v+1}{2v-1}$$

$$\Rightarrow \frac{2v-1}{3v+1}dv = dx$$

On integrating both sides,

$$\int\frac{2v-1}{3v+1}dv = \int dx$$

2v – 1 can be written as :

$$2v-1 = A(3v+1)+B$$

$$2v-1 = 3Av+A+B$$

$$3A = 2 \quad \text{and} \quad A+B = -1$$

$$A = \frac{2}{3}, \quad \frac{2}{3}+B = -1$$

$$B = -1-\frac{2}{3}$$

$$B = \frac{-5}{3}$$

$$\therefore 2v-1 = \frac{2}{3}(3v+1)-\frac{5}{3}$$

$$\therefore \int\frac{\frac{2}{3}(3v+1)-\frac{5}{3}}{3v+1}dv = \int dx$$

$$\Rightarrow \frac{2}{3}\int\frac{3v+1}{3v+1}dv-\frac{5}{3}\int\frac{dv}{3v+1} = x+C$$

$$\Rightarrow \frac{2}{3}v-\frac{5}{3}\times\frac{1}{3}\log|3v+1| = x+C$$

$$\Rightarrow \frac{2}{3}(x+y)-\frac{5}{9}\log|3(x+y)+1| = x+C$$

$$\Rightarrow \frac{2}{3}y-\frac{5}{9}\log|3(x+y)+1| = x-\frac{2x}{3}+C$$

$$\Rightarrow \frac{2}{3}y-\frac{5}{9}\log|3(x+y)+1| = \frac{x}{3}+C \quad \textbf{Ans.}$$

Solution 10.

Bag A	Bag B
4 White	3 White
3 Black	5 Black

Let E_1 be the event when 2 white balls transferred from bag A to bag B.

E_2 be the event when 2 black balls transferred from bag A to bag B.

E_3 be the event 1 white ball transferred from bag A to bag B.

A be the event when ball is drawn from bag B.

$$P(E_1) = \frac{^4C_2}{^7C_2} = \frac{2}{7}$$

$$P(E_2) = \frac{^3C_2}{^7C_2} = \frac{1}{7}$$

$$P(E_3) = \frac{^4C_1\times{}^3C_1}{^7C_2} = \frac{4}{7}$$

$$P\left(\frac{A}{E_1}\right) = \frac{5}{10} = \frac{1}{2}$$

$$P\left(\frac{A}{E_2}\right) = \frac{3}{10}$$

$$P\left(\frac{A}{E_3}\right) = \frac{4}{10}$$

$$P(A) = P(E_1)\cdot P\left(\frac{A}{E_1}\right) + P(E_2)\cdot P\left(\frac{A}{E_2}\right) + P(E_3)\cdot P\left(\frac{A}{E_3}\right)$$

$$= \frac{2}{7}\times\frac{1}{2}+\frac{1}{7}\times\frac{3}{10}+\frac{4}{7}\times\frac{4}{10}$$

$$= \frac{1}{7}+\frac{3}{70}+\frac{16}{70}$$

$$= \frac{10+3+16}{70} = \frac{29}{70} \quad \textbf{Ans.}$$

Solution 11.

Let $\frac{1}{x} = a$, $\frac{1}{y} = b$ and $\frac{1}{z} = c$.

∴ The given equations become,

$$a+b+c = 9$$
$$2a+5b+7c = 52$$
$$2a+b-c = 0$$
$$AX = B$$

Let $A=\begin{bmatrix}1&1&1\\2&5&7\\2&1&-1\end{bmatrix}$ $B=\begin{bmatrix}9\\52\\0\end{bmatrix}$ and $X=\begin{bmatrix}a\\b\\c\end{bmatrix}$

$$|A| = 1(-5-7)-1(-2-14)+1(2-10)$$
$$= -12+16-8$$
$$= -4$$

Cofactors of A are

$A_{11} = -12$ $A_{12} = 16$ $A_{13} = -8$

$A_{21} = 2$ $A_{22} = -3$ $A_{23} = 1$

$A_{31} = 2$ $A_{32} = -5$ $A_{33} = 3$

$$\text{adj } A = \begin{bmatrix}-12&16&-8\\2&-3&1\\2&-5&3\end{bmatrix}^T$$

$$= \begin{bmatrix}-12&2&2\\16&-3&-5\\-8&1&3\end{bmatrix}$$

$$A^{-1} = \frac{1}{|A|}\text{adj } A$$

$$= \frac{1}{-4}\begin{bmatrix}-12&2&2\\16&-3&-5\\-8&1&3\end{bmatrix}$$

$$X = A^{-1}B$$

$$\begin{bmatrix}a\\b\\c\end{bmatrix} = \frac{1}{-4}\begin{bmatrix}-12&2&2\\16&-3&-5\\-8&1&3\end{bmatrix}\begin{bmatrix}9\\52\\0\end{bmatrix}$$

$$\begin{bmatrix}a\\b\\c\end{bmatrix} = \frac{1}{-4}\begin{bmatrix}-108+104+0\\144-156-0\\-72+52+0\end{bmatrix}$$

$$\begin{bmatrix}a\\b\\c\end{bmatrix} = \frac{1}{-4}\begin{bmatrix}-4\\-12\\-20\end{bmatrix}$$

$$\therefore a = \frac{-1}{4}\times(-4) = 1, b = \frac{-12}{-4} = 3 \text{ and } c = \frac{-20}{-4} = 5$$

$$\therefore x = \frac{1}{a} = 1, y = \frac{1}{b} = \frac{1}{3} \quad z = \frac{1}{c} = \frac{1}{5}$$

$$\Rightarrow x = 1, y = \frac{1}{3} \text{ and } z = \frac{1}{5}$$ **Ans.**

Solution 12.

(a) Let Length of the box $= x$ cm

Height of the box $= h$ cm

$\therefore$ Breadth of the box $= x$ cm [$\because$ square base]

$\because$ Volume of the box $= 4096$ cm³

$$x^2h = 4096$$

$$h = \frac{4096}{x^2} \quad \text{...(i)}$$

Surface area of outer surface $= 2[lb+bh+lh]$

$$= 2[x^2+xh+hx]$$
$$= 2x^2+4hx$$

Cost of polishing the surface = ₹ 4 per cm²

So, Total cost, $C = 4(2x^2+4hx)$

$$\Rightarrow C = 8x^2+16xh$$

$$\Rightarrow C = 8x^2+16x\left[\frac{4096}{x^2}\right]$$

[From (i)]

$$\Rightarrow C = 8x^2+\frac{65536}{x}$$

Differentiating w.r.t. x, we get

$$\frac{dC}{dx} = 16x - \frac{65536}{x^2}$$

$$\frac{d^2C}{dx^2} = 16+\frac{65536\times 2}{x^3}$$

$$\frac{d^2C}{dx^2} > 0$$

$\because$ Cost of polishing the surface is minimum,

$$\frac{dC}{dx} = 0$$

$$\Rightarrow 16x - \frac{65536}{x^2} = 0$$

$$\Rightarrow 16x^3 - 65536 = 0$$

$$\Rightarrow x^3 = \frac{65536}{16}$$

$$\Rightarrow x^3 = \frac{16\times 4096}{16}$$

$$\Rightarrow x^3 = 4096$$

$$\Rightarrow x = 16 \text{ cm}$$

$$\therefore h = \frac{4096}{16\times 16} = 16 \text{ cm}$$

$\therefore$ Length = 16 cm, breadth = 16 cm and height = 16 cm. **Ans.**

OR

(b) The given equation of the line is $2x + 3y = 6$. Let the point on the straight line be P (x, y) and O(0, 0) be the origin.

$$\therefore \quad OP = \sqrt{(x-0)^2 + (y-0)^2}$$

$$OP = \sqrt{x^2 + y^2}$$

$$\therefore \quad OP^2 = x^2 + y^2$$

Now, OP is minimum when OP^2 is minimum

Let $\quad OP^2 = f(x)$

$$\therefore \quad f(x) = x^2 + y^2$$

$$\because \quad 2x + 3y = 6$$

$$\therefore \quad y = \frac{6-2x}{3}$$

$$\Rightarrow f(x) = x^2 + \left(\frac{6-2x}{3}\right)^2$$

Differentiating w.r.t. x,

$$\Rightarrow \quad f'(x) = 2x + \frac{2}{9}(6-2x)(-2)$$

For $f(x)$ to be minimum,

$$f'(x) = 0$$

$$\Rightarrow 2x - \frac{4(6-2x)}{9} = 0$$

$$\Rightarrow \quad 2x = \frac{4(6-2x)}{9}$$

$$\Rightarrow \quad 9x = 12 - 4x$$

$$\Rightarrow \quad 13x = 12$$

$$\Rightarrow \quad x = \frac{12}{13}$$

$$\therefore \quad y = \frac{6-2x}{3} = \frac{6-2\times\frac{12}{13}}{3}$$

$$y = \frac{54}{39} = \frac{18}{13}$$

$\therefore$ The point on the straight line P(x, y) is $\left(\frac{12}{13}, \frac{18}{13}\right)$

Now, $\quad f''(x) = 2 - \frac{4}{9} \times (-2)$

$$= 2 + \frac{8}{9} = \frac{26}{9}$$

$$\because \quad f''(x) > 0$$

$\therefore$ P $\left(\frac{12}{13}, \frac{18}{13}\right)$ is closest to the origin. **Ans.**

Solution 13.

Let $I = \int_0^{\pi} \frac{x \tan x}{\sec x + \tan x} dx \quad \ldots(i)$

Using $\int_0^a f(x)dx = \int_0^a f(a-x)dx$

$$I = \int_0^{\pi} \frac{(\pi - x)\tan(\pi - x)}{\sec(\pi - x) + \tan(\pi - x)} dx$$

$$I = \int_0^{\pi} \frac{(\pi - x)(-\tan x)}{(-\sec x) + (-\tan x)} dx$$

$$\Rightarrow \quad I = \int_0^{\pi} \frac{(\pi - x)\tan x}{\sec x + \tan x} dx \quad \ldots(ii)$$

Adding equations (i) and (ii), we get

$$2I = \int_0^{\pi} \frac{\pi \tan x}{\sec x + \tan x} dx$$

$$\Rightarrow \quad 2I = \pi \int_0^{\pi} \frac{\tan x}{\sec x + \tan x} dx$$

$$\Rightarrow \quad 2I = \pi \int_0^{\pi} \frac{\tan x}{\sec x + \tan x} \times \frac{(\sec x - \tan x)}{(\sec x - \tan x)} dx$$

$$\Rightarrow \quad 2I = \pi \int_0^{\pi} \frac{\tan x (\sec x - \tan x)}{\sec^2 x - \tan^2 x} dx$$

$$\Rightarrow \quad 2I = \pi \int_0^{\pi} \frac{\sec x \tan x - \tan^2 x}{1} dx$$

$$[\because \sec^2 x - \tan^2 x = 1]$$

$$\Rightarrow \quad 2I = \pi \int_0^{\pi} \sec x \tan x - \sec^2 x + 1\, dx$$

$$[\because \tan^2 x = \sec^2 x - 1]$$

$$\Rightarrow \quad 2I = \pi \left[\sec x - \tan x + x\right]_0^{\pi}$$

$$= \pi[-1 - 0 + \pi - (1 - 0 + 0)]$$

$$\Rightarrow \quad = \pi[\pi - 2] \Rightarrow I = \frac{\pi}{2}[\pi - 2]$$ **Ans.**

Solution 14.

(a) Let E_1, E_2 and E_3 be the events to choose the box A, B and C, respectively.

$$\therefore P(E_1) = P(E_2) = P(E_3) = \frac{1}{3}$$

Let A be the event that coin drawn is silver

$$\therefore P\left(\frac{A}{E_1}\right) = \frac{1}{3}, P\left(\frac{A}{E_2}\right) = \frac{2}{3}, P\left(\frac{A}{E_3}\right) = \frac{3}{3} = 1$$

Using Bayes' theorem,

$$P\left(\frac{E_3}{A}\right) = \frac{P(E_3).P\left(\frac{A}{E_3}\right)}{P(E_1).P\left(\frac{A}{E_1}\right)+P(E_2).P\left(\frac{A}{E_2}\right)+P(E_3).P\left(\frac{A}{E_3}\right)}$$

$$= \frac{\frac{1}{3}\times 1}{\frac{1}{3}\times\frac{1}{3}+\frac{1}{3}\times\frac{2}{3}+\frac{1}{3}\times 1}$$

$$= \frac{\frac{1}{3}}{\frac{1}{9}+\frac{2}{9}+\frac{1}{3}}$$

$$= \frac{\frac{1}{3}}{\frac{1+2+3}{9}} = \frac{1}{3}\times\frac{9}{6}$$

$$= \frac{1}{2}$$ **Ans.**

OR

(b) Given, mean $= np = 9$ ...(i)

and standard deviation $= \sqrt{npq} = \frac{3}{2}$

$\Rightarrow \quad npq = \frac{9}{4}$

$\Rightarrow \quad 9q = \frac{9}{4}$

[Putting $np = 9$ from eq. (i)]

$\Rightarrow \quad q = \frac{1}{4}$

$\because \quad p+q = 1$

$\Rightarrow \quad p = 1-\frac{1}{4} = \frac{3}{4}$

$\because \quad npq = \frac{9}{4}$

$\Rightarrow \quad n\times\frac{3}{4}\times\frac{1}{4} = \frac{9}{4}$

$\Rightarrow \quad n = 12$

Binomial distribution $\rightarrow \left(\frac{1}{4}+\frac{3}{4}\right)^{12}$

P (at most one success) $= p\,(X = 0) + p\,(X = 1)$

$= {}^{12}C_0 p^0 q^{12} + {}^{12}C_1 p^1 q^{12-1}$

$= {}^{12}C_0\left(\frac{3}{4}\right)^0\times\left(\frac{1}{4}\right)^{12} + {}^{12}C_1\left(\frac{3}{4}\right)^1\times\left(\frac{1}{4}\right)^{12-1}$

$= \left(\frac{1}{4}\right)^{12} + 12\times\frac{3}{4}\times\left(\frac{1}{4}\right)^{11}$

$= \left(\frac{1}{4}\right)^{11}\left[\frac{1}{4}+12\times\frac{3}{4}\right]$

$= \left(\frac{1}{4}\right)^{11}\left[\frac{1}{4}+9\right]$

$= \frac{37}{(4)^{12}}$ **Ans.**

SECTION—B

Solution 15.

(a) Given, vectors $\vec{a}$ and $\vec{b}$ are perpendicular.

$\therefore \quad \vec{a}\cdot\vec{b} = 0$

Given, $|\vec{a}+\vec{b}| = 13$

On squaring both sides, we get

$|\vec{a}+\vec{b}|^2 = (13)^2$

$\Rightarrow \quad (\vec{a}+\vec{b}).(\vec{a}+\vec{b}) = 169$

$\Rightarrow \quad |\vec{a}|^2 + |\vec{b}|^2 + 2\,\vec{a}\cdot\vec{b} = 169$

$\Rightarrow \quad 25 + |\vec{b}|^2 + 0 = 169$

[$\because$ Given $|\vec{a}| = 5$]

$\Rightarrow \quad |\vec{b}|^2 = 169 - 25$

$\Rightarrow \quad |\vec{b}|^2 = 144$

$\Rightarrow \quad |\vec{b}| = \sqrt{144}$

$\Rightarrow \quad |\vec{b}| = 12$ **Ans.**

(b) Given, equation of the plane is

$\vec{r}\cdot(3\hat{i}-4\hat{j}-12\hat{k}) + 39 = 0$

Let $\vec{r} = (x\hat{i}+y\hat{j}+z\hat{k})$

$\therefore \quad (x\hat{i}+y\hat{j}+z\hat{k})\cdot(3\hat{i}-4\hat{j}-12\hat{k}) + 39 = 0$

$\Rightarrow \quad 3x - 4y - 12z + 39 = 0$

Perpendicular distance from the origin to plane

$$= \left|\frac{39}{\sqrt{9+16+144}}\right|$$

$$= \left|\frac{39}{\sqrt{169}}\right|$$

$$= \left|\frac{39}{13}\right| = 3 \text{ units.}$$ **Ans.**

(c) Given equation of lines are

$$2x = 3y = -z$$

$$\therefore \quad \frac{x}{\frac{1}{2}} = \frac{y}{\frac{1}{3}} = \frac{z}{-1}$$

Direction Ratio of this line $= \langle a_1, b_1, c_1\rangle$

$$= \left\langle \frac{1}{2}, \frac{1}{3}, -1\right\rangle$$

$$= \langle 3, 2, -6\rangle$$

and $6x = -y = -4z$

$$\frac{x}{\frac{1}{6}} = \frac{y}{-1} = \frac{z}{\frac{1}{-4}}$$

Direction Ratio of this line $= \langle a_2, b_2, c_2\rangle$

$$= \left\langle \frac{1}{6}, -1, -\frac{1}{4}\right\rangle$$

$$= \langle 2, -12, -3\rangle$$

$$= \langle -2, 12, 3\rangle$$

Angle between the given lines :

$$\cos\theta = \frac{a_1a_2 + b_1b_2 + c_1c_2}{\sqrt{a_1^2+b_1^2+c_1^2}\sqrt{a_2^2+b_2^2+c_2^2}}$$

$$= \frac{3\times(-2) + 2\times 12 + (-6)\times 3}{\sqrt{9+4+36}\sqrt{4+144+9}}$$

$$= \frac{-6+24-18}{\sqrt{49}\sqrt{157}}$$

$$\cos\theta = 0$$

$$\cos\theta = \cos\frac{\pi}{2}$$

$$\theta = \frac{\pi}{2} \text{ or } 90°.$$ **Ans.**

Solution 16.

(a) Given vectors are $\vec{a} = \hat{i} - 2\hat{j} + 3\hat{k}$

and $\vec{b} = 2\hat{i} + 3\hat{j} - 5\hat{k}$

$$\vec{a}\times\vec{b} = \begin{vmatrix} \hat{i} & \hat{j} & \hat{k} \\ 1 & -2 & 3 \\ 2 & 3 & -5 \end{vmatrix}$$

$$= \hat{i}(10-9) - \hat{j}(-5-6) + \hat{k}(3+4)$$

$$= \hat{i} + 11\hat{j} + 7\hat{k}$$

Now,

$$\vec{a}\cdot\left(\vec{a}\times\vec{b}\right) = (\hat{i}-2\hat{j}+3\hat{k})\cdot(\hat{i}+11\hat{j}+7\hat{k})$$

$$= 1 - 22 + 21 = 0$$

$$\because \quad \vec{a}\cdot\left(\vec{a}\times\vec{b}\right) = 0$$

$\therefore$ $\vec{a}$ and $\left(\vec{a}\times\vec{b}\right)$ are perpendicular.

Hence Proved

OR

(b) Since $\vec{\alpha}$ and $\vec{\beta}$ are collinear

$$\therefore \quad \vec{\alpha} = \lambda\vec{\beta} \text{ \{where '}\lambda\text{' is any scalar\}}$$

$$(x-2)\vec{a} + \vec{b} = \lambda[(3+2x)\vec{a} - 2\vec{b}]$$

$$(x-2)\vec{a} + \vec{b} = (3+2x)\lambda\vec{a} - 2\lambda\vec{b}$$

On comparing the coefficients of $\vec{a}$ and $\vec{b}$

$$(3+2x)\lambda = x - 2 \quad \ldots\text{(i)}$$

and $1 = -2\lambda$

$$\lambda = -\frac{1}{2}$$

Put $\lambda = -\frac{1}{2}$ in equation (i), we get

$$(3+2x)\times\left(-\frac{1}{2}\right) = x - 2$$

$$\Rightarrow \quad -3 - 2x = 2x - 4$$

$$\Rightarrow \quad -3 - 2x = 2x - 4$$

$$\Rightarrow \quad -3 + 4 = 2x + 2x$$

$$\Rightarrow \quad 4x = 1$$

$$\Rightarrow \quad x = \frac{1}{4}$$ **Ans.**

Solution 17.

(a) Equation of the required plane passing through the intersection of the given plane is :

$$(2x + 2y - 3z - 7) + \lambda(2x + 5y + 3z - 9) = 0$$

$$x(2 + 2\lambda) + y(2 + 5\lambda) + z(3\lambda - 3) - 7 - 9\lambda = 0 \quad \ldots\text{(i)}$$

For x-intercept, put $y = 0$ and $z = 0$.

we get

$$(2 + 2\lambda)\,x - 7 - 9\lambda = 0$$

$$x = \frac{7+9\lambda}{2+2\lambda}$$

For z- intercept, put $x = 0$ and $y = 0$

we get

$$z\,(3\lambda - 3) - 7 - 9\lambda = 0$$

$$z = \frac{7+9\lambda}{3\lambda-3}$$

$\because$ x - intercept = z-intercept

$$\frac{7+9\lambda}{2+2\lambda} = \frac{7+9\lambda}{3\lambda-3}$$

$\Rightarrow$ $3\lambda - 3 = 2 + 2\lambda$

$\Rightarrow$ $3\lambda - 2\lambda = 2 + 3$

$\Rightarrow$ $\lambda = 5$

put $\lambda = 5$ in equation (i), we get

$(2 + 10)\,x + (2 + 25)\,y + (15 - 3)\,z - 7 - 45 = 0$

$\Rightarrow$ $12x + 27\,y + 12z - 52 = 0$

This is the required equation of the plane. **Ans.**

OR

(b) The given line is passing through the point (2, 1, 3) and perpendicular to the lines

$$\frac{x-1}{1} = \frac{y-2}{2} = \frac{z-3}{3} \text{ and } \frac{x}{-3} = \frac{y}{2} = \frac{z}{5}$$

Let $\vec{a} = \hat{i}+2\hat{j}+3\hat{k}$

$\vec{b} = -3\hat{i}+2\hat{j}+5\hat{k}$

$$\vec{a}\times\vec{b} = \begin{vmatrix} \hat{i} & \hat{j} & \hat{k} \\ 1 & 2 & 3 \\ -3 & 2 & 5 \end{vmatrix}$$

$$= \hat{i}\,(10 - 6) - \hat{j}\,(5 + 9) + \hat{k}\,(2 + 6)$$

$$= 4\hat{i} - 14\hat{j} + 8\hat{k}$$

$\therefore$ D.R's of required line = $\langle 4, -14, 8\rangle$ or $\langle 2, -7, 4\rangle$

$\therefore$ Required equation of line is

$$\frac{x-2}{2} = \frac{y-1}{-7} = \frac{z-3}{4}$$ **Ans.**

Solution 18.

Given equation of curve is $x^2 = y$ and equation of line is $x + y = 2$

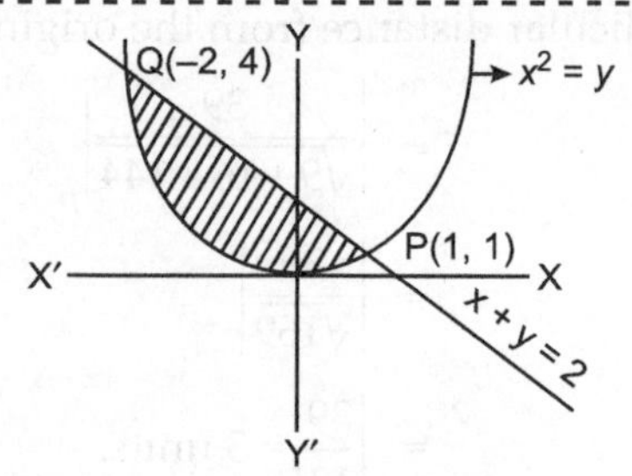

$\because$ $x^2 = y$ and $y = 2 - x$

$\therefore$ $x^2 = 2 - x$

$\Rightarrow$ $x^2 + x - 2 = 0$

$\Rightarrow$ $x^2 + 2x - x - 2 = 0$

$\Rightarrow$ $x\,(x + 2) - (x + 2) = 0$

$\Rightarrow$ $(x + 2)\,(x - 1) = 0$

$x = -2$ and $x = 1$

when $x = -2$,

$y = 4$ (Putting $x = -2$ in curve $y = x^2$)

and when $x = 1$,

$y = 1$ (Putting $x = 1$ in curve $y = x^2$)

$\therefore$ The points are (– 2, 4) and (1, 1).

$$\text{Required area} = \int_{-2}^{1}[(2-x)-x^2]\,dx$$

$$= \left[2x - \frac{x^2}{2} - \frac{x^3}{3}\right]_{-2}^{1}$$

$$= \left[2-\frac{1}{2}-\frac{1}{3}\right] - \left[-4-2+\frac{8}{3}\right]$$

$$= \frac{7}{6} + \frac{10}{3}$$

$$= \frac{9}{2} \text{ sq. units.}$$ **Ans.**

SECTION—C

Solution 19.

(a) Given, fixed cost = ₹24,000

Let number of units be 'x'

$\therefore$ $R(x) = 8x$

[$\because$ Given selling price per unit, P = ₹8]

Variable cost = 25% of R (x)

$$= \frac{25}{100} \times 8x$$

$$= 2x$$

$C(x) = 2x + 24{,}000$

$\therefore$ Break-even point,

$R(x) = C(x)$

$2x + 24{,}000 = 8x$

$6x = 24{,}000$

$x = 4{,}000$ units. **Ans.**

(b) Given,
$$C(x) = \frac{3}{4}x^2 - 7x + 27$$
$$\text{M.C.} = \frac{d}{dx}C(x) = \frac{3}{4} \times 2x - 7$$
$$= \frac{3}{2}x - 7$$
$$\text{A.C.} = \frac{C(x)}{x} = \frac{3}{4}x - 7 + \frac{27}{x}$$
$\because$ M.C. = A.C. (given)
$$\frac{3}{2}x - 7 = \frac{3}{4}x - 7 + \frac{27}{x}$$
$$\Rightarrow \frac{3}{2}x - \frac{3}{4}x = \frac{27}{x}$$
$$\Rightarrow \frac{6x - 3x}{4} = \frac{27}{x}$$
$$\Rightarrow \frac{3x}{4} = \frac{27}{x}$$
$$\Rightarrow x^2 = 36$$
$\Rightarrow x = \pm 6$, but x cannot be negative

$\therefore$ Number of units = 6. **Ans.**

(c) Given, $\bar{x} = 18$, $\bar{y} = 100$, $\sigma_x = 14$, $\sigma_y = 20$, $r_{xy} = 0{\cdot}8$
$$byx = r_{xy}\left(\frac{\sigma_y}{\sigma_x}\right)$$
$$= 0{\cdot}8\left(\frac{20}{14}\right) = \frac{8}{7}$$
Now, Regression equation of y on x is
$$y - \bar{y} = b_{yx}(x - \bar{x})$$
$$\Rightarrow y - 100 = \frac{8}{7}(x - 18)$$
$$\Rightarrow 7y - 700 = 8x - 144$$
$$\Rightarrow 8x - 7y + 556 = 0$$
Ans.

Solution 20.

(a) Given, $\Sigma x = 15$, $\Sigma y = 25$, $\Sigma xy = 83$, $\Sigma x^2 = 55$, $\Sigma y^2 = 135$ and $n = 5$

(i)
$$b_{xy} = \frac{\Sigma xy - \frac{1}{n}\Sigma x\Sigma y}{\Sigma y^2 - \frac{1}{n}(\Sigma y)^2}$$
$$= \frac{83 - \frac{1}{5} \times 15 \times 25}{135 - \frac{1}{5} \times (25)^2}$$
$$= \frac{83 - 75}{135 - 125} = \frac{8}{10} = \frac{4}{5}$$
$$= 0{\cdot}8$$
Ans.

(ii)
$$\bar{x} = \frac{\Sigma x}{n} = \frac{15}{5} = 3$$
$$\bar{y} = \frac{\Sigma y}{n} = \frac{25}{5} = 5$$
$\therefore$ Equation of regression line x on y,
$$x - \bar{x} = b_{xy}(y - \bar{y})$$
$$x - 3 = \frac{4}{5}(y - 5)$$
$$5x - 15 = 4y - 20$$
$$5x - 4y + 5 = 0$$
Ans.

OR

(b)

x	y	xy	x^2
1	4	4	1
2	8	16	4
3	2	6	9
4	12	48	16
5	10	50	25
6	14	84	36
7	16	112	49
8	6	48	64
9	18	162	81
$\Sigma x = 45$	$\Sigma y = 90$	$\Sigma xy = 530$	$\Sigma x^2 = 285$

Here, $n = 9$
$$b_{yx} = \frac{\Sigma xy - \frac{1}{n}\Sigma x\Sigma y}{\Sigma x^2 - \frac{1}{n}(\Sigma x)^2}$$
$$b_{yx} = \frac{530 - \frac{1}{9} \times 45 \times 90}{285 - \frac{1}{9} \times 45 \times 45}$$
$$b_{yx} = \frac{530 - 450}{285 - 225}$$
$$b_{yx} = \frac{80}{60} = \frac{4}{3}$$
$$\bar{x} = \frac{\Sigma x}{n}$$
$$= \frac{45}{9} = 5$$
$$\bar{y} = \frac{\Sigma y}{n} = \frac{90}{9} = 10$$

$\therefore$ Regression equation of y on x is

$$y - \bar{y} = b_{yx}(x - \bar{x})$$

$$\Rightarrow \quad y - 10 = \frac{4}{3}(x - 5)$$

$$\Rightarrow \quad 3y - 30 = 4x - 20$$

$$\Rightarrow \quad 4x - 3y + 10 = 0$$ **Ans.**

Now, when $x = 14$

$$4 \times 14 - 3y + 10 = 0$$

$$\Rightarrow \quad 3y = 56 + 10$$

$$\Rightarrow \quad 3y = 66$$

$$\Rightarrow \quad y = 22$$ **Ans.**

Solution 21.

(a) $$C(x) = \frac{x^3}{3} - 45x^2 - 900x + 36$$

$$\text{M.C.} = \frac{d}{dx}C(x) = \frac{3x^2}{3} - 90x - 900$$

$$= x^2 - 90x - 900$$

Now, the marginal cost is to be minimised *i.e.*

$$\frac{d}{dx}(\text{M.C}) = 2x - 90$$

$$\frac{d}{dx}(\text{M.C.}) = 0$$

$$2x - 90 = 0$$

$$x = 45$$

$$\frac{d^2}{dx^2}(\text{M.C.}) = 2$$

$$\because \quad \frac{d^2}{dx^2}(\text{M.C.}) > 0$$

So, marginal cost is minimum.

$\therefore$ Number of units produced = 45 **Ans.**

OR

(b) Given $$\text{M.C.} = 3x^2 - 10x + 3$$

$$C(x) = \int(\text{M.C.})\,dx$$

$$= \int(3x^2 - 10x + 3)\,dx$$

$$= 3\frac{x^3}{3} - 10\frac{x^2}{2} + 3x + k$$

where k is constant

$$= x^3 - 5x^2 + 3x + k$$

$\because$ Cost of one unit is ₹ 37 *i.e.*,

$$x = 1$$

$$\therefore \quad C(1) = 7$$

$$\Rightarrow \quad 1 - 5 + 3 + k = 7$$

$$\Rightarrow \quad -4 + 3 + k = 7$$

$$\Rightarrow \quad -1 + k = 7$$

$$\Rightarrow \quad k = 7 + 1$$

$$\Rightarrow \quad k = 8$$

$$\therefore \quad C(x) = x^3 - 5x^2 + 3x + 8$$

and

$$\text{Average cost, A.C.} = \frac{C(x)}{x}$$

$$= \frac{x^3 - 5x^2 + 3x + 8}{x}$$

$$= x^2 - 5x + 3 + \frac{8}{x}$$ **Ans.**

Solution 22.

Let x units of product A and y unit of product B should be produced.

Total running feet teak wood available = 90

Total running feet plywood available = 80

Total running feet rosewood available = 50

For teak wood

$$2x + y \le 90 \quad \ldots\text{(i)}$$

For plywood

$$x + 2y \le 80 \quad \ldots\text{(ii)}$$

For rosewood

$$x + y \le 50 \quad \ldots\text{(iii)}$$

$$x \ge 0, y \ge 0$$

To maximise the gross income

$$\text{Maximise } Z = 48x + 40y$$

From equation (i)

$$2x + y = 90$$

x	0	45
y	90	0

From equation (ii)

$$x + 2y = 80$$

x	0	80
y	40	0

From equation (iii)

$$x + y = 50$$

x	0	50
y	50	0

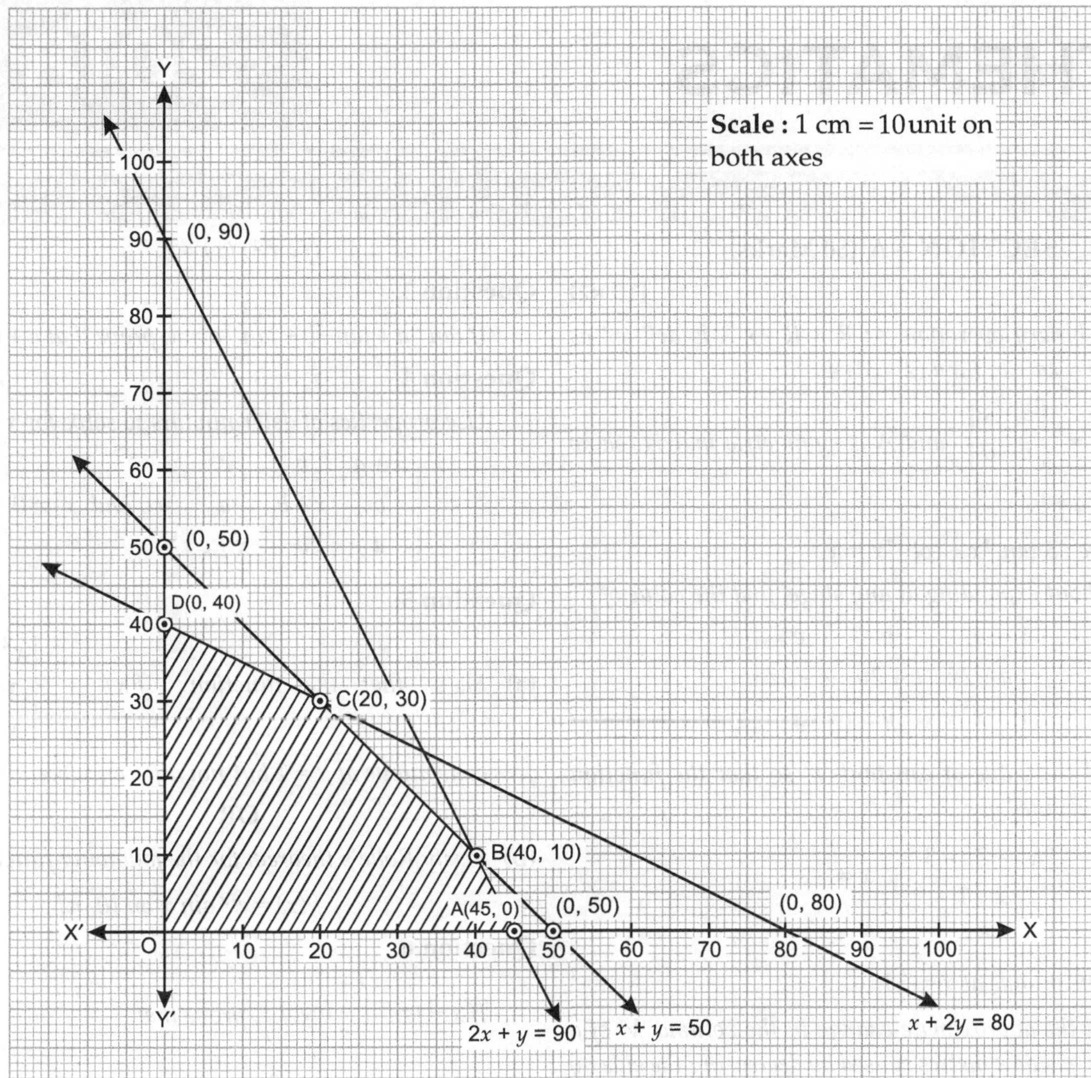

Corner points are O (0, 0), A (45, 0), B (40, 10), C (20, 30) and D (0, 40)

For gross profit

$$Z = 48x + 40y$$

For point O, $Z = 0$

For point A,

$$Z = 48 \times 45 + 40 \times 0$$
$$= 2160 + 0$$
$$= ₹32160$$

For point B,

$$Z = 48 \times 40 + 40 \times 10$$
$$= 1920 + 400$$
$$= ₹32320$$

For point C,

$$Z = 48 \times 20 + 40 \times 30$$
$$= 960 + 1200$$
$$= ₹32160$$

For point D,

$$Z = 48 \times 0 + 40 \times 40$$
$$= ₹31600$$

$\therefore$ Z is maximum at B (40, 10).

Hence, 40 units of product A and 10 units of product B must be sold to get gross income of ₹32320. **Ans.**

LL

MATHEMATICS 2018

QUESTIONS

SECTION—A (80 Marks)

Question 1. [10 ×2]

(i) *The binary operation* $* : R \times R \to R$ *is defined as* $a * b = 2a + b$. *Find* $(2 * 3) * 4$.

(ii) *If* $A = \begin{bmatrix} 5 & a \\ b & 0 \end{bmatrix}$ *and A is symmetric matrix, show that* $a = b$.

(iii) *Solve :* $3 \tan^{-1}x + \cot^{-1}x = \pi$.

(iv) *Without expanding at any stage, find the value of :*

$$\begin{vmatrix} a & b & c \\ a+2x & b+2y & c+2z \\ x & y & z \end{vmatrix}$$

(v) *Find the value of constant 'k' so that the function f (x) defined as :*

$$f(x) = \begin{cases} \dfrac{x^2 - 2x - 3}{x+1} & , x \neq -1 \\ k & , x = -1 \end{cases}$$

is continuous at $x = -1$.

(vi) *Find the approximate change in the volume 'V' of a cube of side x metres caused by decreasing the side by 1%.*

(vii) *Evaluate :*

$$\int \frac{x^3 + 5x^2 + 4x + 1}{x^2} dx$$

(viii) *Find the differential equation of the family of concentric circles* $x^2 + y^2 = a^2$.

(ix) *If A and B are events such that* $P(A) = \frac{1}{2}$, $P(B) = \frac{1}{3}$ *and* $P(A \cap B) = \frac{1}{4}$, *then find :*

(a) $P(A/B)$ **(b)** $P(B/A)$

(x) *In a race, the probabilities of A and B winning the race are* $\frac{1}{3}$ *and* $\frac{1}{6}$ *respectively. Find the probability of neither of them winning the race.*

Question 2. [4]

If the function $f(x) = \sqrt{2x-3}$ *is invertible then find its inverse. Hence prove that*

$$(f o f^{-1})(x) = x.$$

Question 3. [4]

If $\tan^{-1}a + \tan^{-1}b + \tan^{-1}c = \pi$, *prove that* $a + b + c = abc$

Question 4. [4]

Use properties of determinants to solve for x :

$$\begin{vmatrix} x+a & b & c \\ c & x+b & a \\ a & b & x+c \end{vmatrix} = 0 \text{ and } x \neq 0.$$

Question 5. [4]

(a) *Show that the function* $f(x) = \begin{cases} x^2 & , x \leq 1 \\ \dfrac{1}{x} & , x > 1 \end{cases}$ *continuous at* $x = 1$ *but not differentiable.*

OR

(b) *Verify Rolle's theorem for the following function :*

$$f(x) = e^{-x} \sin x \text{ on } [0, \pi]$$

Question 6. [4]

If $x = \tan\left(\frac{1}{a}\log y\right)$, *prove that*

$$(1 + x^2)\frac{d^2y}{dx^2} + (2x - a)\frac{dy}{dx} = 0.$$

Question 7. [4]

Evaluate : $\int \tan^{-1}\sqrt{x}\, dx$.

Question 8. [4]

(a) *Find the points on the curve* $y = 4x^3 - 3x + 5$ *at which the equation of the tangent is parallel to the X-axis.*

OR

(b) *Water is dripping out from a conical funnel of semi-verticle angle* $\frac{\pi}{4}$ *at the uniform rate of* $2\ cm^2/sec$ *in the surface, through a tiny hole at the vertex of the bottom. When the slant height of the water level is 4 cm, find the rate of decrease of the slant height of the water.*

Question 9. [4]

(a) *Solve :* $\sin x \frac{dy}{dx} - y = \sin x . \tan \frac{x}{2}$

OR

(b) *The population of a town grows at the rate of 10% per year. Using differential equation, find how long will it take for the population to grow 4 times.*

Question 10. **[6]**

(a) *Using matrices, solve the following system of equations :*

$$2x - 3y + 5z = 11$$
$$3x + 2y - 4z = -5$$
$$x + y - 2z = -3$$

OR

(b) *Using elementary transformation, find the inverse of the matrix :*

$$\begin{bmatrix} 0 & 1 & 2 \\ 1 & 2 & 3 \\ 3 & 1 & 1 \end{bmatrix}$$

Question 11. **[4]**

A speaks truth in 60% of the cases, while B in 40% of the cases. In what percent of cases are they likely to contradict each other in stating the same fact?

Question 12. **[6]**

A cone is inscribed in a sphere of radius 12 cm. If the volume of the cone is maximum, find its height.

Question 13. **[6]**

(a) *Evaluate :* $\int \frac{x-1}{\sqrt{x^2 - x}} dx$

OR

(b) *Evaluate :* $\int_0^{\pi/2} \frac{\cos^2 x}{1 + \sin x \cos x} dx$

Question 14. **[6]**

From a lot of 6 items containing 2 defective items, a sample of 4 items are drawn at random. Let the random variable X denote the number of defective items in the sample. If the sample is drawn without replacement, find :

(a) *The probability distribution of X.*

(b) *Mean of X.*

(c) *Variance of X.*

SECTION—B (20 Marks)

Question 15. **[3 × 2]**

(a) *Find* λ *if the scalar projection of* $\vec{a} = \lambda\hat{i} + \hat{j} + 4\hat{k}$ on $\vec{b} = 2\hat{i} + 6\hat{j} + 3\hat{k}$ is 4 units.

(b) *The Cartesian equation of a line is :* $2x - 3 = 3y + 1 = 5 - 6z$. *Find the vector equation of a line passing through* $(7, -5, 0)$ *and parallel to the given line.*

(c) *Find the equation of the plane through the intersection of the planes.*

$\vec{r} \cdot (\hat{i} + 3\hat{j} - \hat{k}) = 9$ *and* $\vec{r} \cdot (2\hat{i} - \hat{j} + \hat{k}) = 3$ *and passing through the origin.*

Question 16. **[4]**

(a) *If A, B, C are three non-collinear points with position vectors* $\vec{a}, \vec{b}, \vec{c}$ *respectively, then show that the length of the perpendicular from C on AB is :*

$$\frac{\left|(\vec{a} \times \vec{b}) + (\vec{b} \times \vec{c}) + (\vec{b} \times \vec{a})\right|}{\left|\vec{b} - \vec{a}\right|}$$

OR

(b) *Show that the four points A, B, C and D with position vectors* $4\hat{i} + 5\hat{j} + \hat{k}, -\hat{j} - \hat{k}, 3\hat{i} + 9\hat{j} + 4\hat{k}$ *and* $4(-\hat{i} + \hat{j} + \hat{k})$ *respectively, are coplanar.*

Question 17. **[4]**

(a) *Draw a rough sketch of the curve and find the area of the region bounded by curve* $y^2 = 8x$ *and the line* $x = 2$.

OR

(b) *Sketch the graph of* $y = |x + 4|$. *Using integration, find the area of the region bounded by the curve* $y = |x + 4|$ *and* $x = -6$ *and* $x = 0$.

Question 18. **[6]**

Find the image of a point having position vector : $3\hat{i} - 2\hat{j} + \hat{k}$ *in the plane*

$$\vec{r} \cdot (3\hat{i} - \hat{j} + 4\hat{k}) = 2.$$

SECTION—C (20 Marks)

Question 19. **[3 × 2]**

(a) *Given the total cost function for x units of a commodity as :*

$$C(x) = \frac{1}{3}x^3 + 3x^2 - 16x + 2.$$

Find :

(i) *Marginal cost function*

(ii) *Average cost function*

(b) *Find the coefficient of correlation from the regression lines :*

$$x - 2y + 3 = 0 \text{ and } 4x - 5y + 1 = 0.$$

(c) *The average cost function associated with producing and marketing x units of an item is given by*

$AC = 2x - 11 + \frac{50}{x}$. Find the range of values of the output x, for which AC is increasing.

Question 20. **[4]**

(a) *Find the line of regression of y on x from the following table :*

x	1	2	3	4	5
y	7	6	5	4	3

Hence, estimate the value of y when x = 6.

OR

(b) *From the given data :*

Variable	x	y
Mean	6	8
Standard Deviation	4	6

and correlation coefficient is $\frac{2}{3}$. Find :

(i) *Regression coefficients b_{yx} and b_{xy}*

(ii) *Regression line x on y*

(iii) *Most likely value of x when y = 14.*

Question 21. **[4]**

(a) *A product can be manufactured at a total cost $C(x) = \frac{x^2}{100} + 100x + 40$, where x is the number of units produced. The price at which each unit can be sold is given by $P = \left(200 - \frac{x}{400}\right)$ Determine the production level x at which the profit is maximum. What is the price per unit and total profit at the level of production ?*

OR

(b) *A manufacturer's marginal cost function is $\frac{500}{\sqrt{2x+25}}$ Find the cost involved to increase production from 100 units to 300 units.*

Question 22. **[6]**

A manufacturing company makes two types of teaching aids A and B of Mathematics for Class X. Each type of A requires 9 labour hours for fabricating and 1 labour hour for finishing. Each type of B requires 12 labour hours for fabricating and 3 labour hours for finishing. For fabricating and finishing, the maximum labour hours available per week are 180 and 30 respectively. The company makes a profit of ₹ 80 on each piece of type A and ₹ 120 on each piece of type B. How many pieces of type A and type B should be manufactured per week to get a maximum profit ? Formulate this as Linear Programming Problem and solve it. Identify the feasible region from the rough sketch.

SOLUTIONS

SECTION—A

Solution 1.

(i) Given binary operation * R × R → R defined as

$$a * b = 2a + b$$
$$(2 * 3) = 2 \times 2 + 3$$

So, $(2 * 3) * 4 = (2 \times 2 + 3) * 4$ (Multiplying 4 on both sides)

$$= 7 * 4$$
$$= 2 \times 7 + 4$$
$$= 14 + 4 = 18.$$ **Ans.**

(ii) We have, $A = \begin{bmatrix} 5 & a \\ b & 0 \end{bmatrix}$

∵ A is symmetric matrix

So, $A^T = A$

$$\begin{bmatrix} 5 & b \\ a & 0 \end{bmatrix} = \begin{bmatrix} 5 & a \\ b & 0 \end{bmatrix}$$

On comparing the elements of both matrix, we get

$a = b$ **Hence Proved.**

(iii) We have,

$$3\tan^{-1}x + \cot^{-1}x = \pi$$
$$2\tan^{-1}x + \tan^{-1}x + \cot^{-1}x = \pi$$
$$2\tan^{-1}x + \frac{\pi}{2} = \pi$$
$$\left[\because \tan^{-1}x + \cot^{-1}x = \frac{\pi}{2}\right]$$
$$2\tan^{-1}x = \pi - \frac{\pi}{2}$$
$$2\tan^{-1}x = \frac{\pi}{2}$$
$$\tan^{-1}x = \frac{\pi}{4}$$
$$x = \tan\frac{\pi}{4}$$
$$x = 1$$ **Ans.**

(iv) Let $D = \begin{vmatrix} a & b & c \\ a+2x & b+2y & c+2z \\ x & y & z \end{vmatrix}$

On applying $R_1 \to R_1 + 2R_3$, we get

$$D = \begin{vmatrix} a+2x & b+2y & c+2z \\ a+2x & b+2y & c+2z \\ x & y & z \end{vmatrix}$$

$\because$ R_1 and R_2 are identical.

So, $D = 0$ **Ans.**

(v) Given,

$$f(x) = \begin{cases} \dfrac{x^2-2x-3}{x+1} & , x \neq -1 \\ k & , x = -1 \end{cases}$$

Since $f(x)$ is continuous at $x = -1$

$$\lim_{x \to -1} f(x) = f(-1)$$

$$\Rightarrow \lim_{x \to -1} \frac{x^2-2x-3}{x+1} = k$$

$$\Rightarrow \lim_{x \to -1} \frac{x^2-3x+x-3}{x+1} = k$$

$$\Rightarrow \lim_{x \to -1} \frac{(x-3)(x+1)}{x+1} = k$$

$$\Rightarrow \lim_{x \to -1} (x-3) = k$$

$$\Rightarrow -1-3 = k$$

$$\Rightarrow -4 = k$$

$$\therefore k = -4$$ **Ans.**

(vi) Given, Side of cube $= x$ metre

and side is decreasing by 1%

i.e.,
$$\frac{dx}{dt} = -1\% \text{ of } x$$

$$= \frac{-x}{100}$$

Volume of cube $= x^3\,m^3$

$V = x^3$ cubic metre

Differentiate with respect to 't',

$$\frac{dV}{dt} = 3x^2 \frac{dx}{dt}$$

$$= 3x^2 \left(\frac{-x}{100}\right)$$

$$= \frac{3}{100}(-x^3)$$

$= -3\%$ volume of cube

So, volume of cube decreases 3% of its volume.

Ans.

(vii) We have, $\displaystyle\int \frac{x^3+5x^2+4x+1}{x^2}\,dx$

$$= \int \frac{x^3}{x^2}dx + \int \frac{5x^2}{x^2}dx + 4\int \frac{x}{x^2}dx + \int \frac{1}{x^2}dx$$

$$= \int x\,dx + 5\int dx + 4\int \frac{1}{x}dx + \int x^{-2}\,dx$$

$$= \frac{x^2}{2} + 5x + 4\log x - \frac{1}{x} + c,$$

where c is constant of integration. **Ans.**

(viii) Given, equation of family of circle is

$$x^2 + y^2 = a^2$$

Differentiate with respect to x.

$$2x + 2y\frac{dy}{dx} = 0$$

Required differential equation is,

$$x + y\frac{dy}{dx} = 0$$ **Ans.**

(ix) Given, $P(A) = \dfrac{1}{2}$, $P(B) = \dfrac{1}{3}$ and $P(A \cap B) = \dfrac{1}{4}$

(a)
$$P(A/B) = \frac{P(A \cap B)}{P(B)}$$

$$= \frac{\frac{1}{4}}{\frac{1}{3}}$$

$$= \frac{3}{4}$$ **Ans.**

(b)
$$P(B/A) = \frac{P(A \cap B)}{P(A)}$$

$$= \frac{\frac{1}{4}}{\frac{1}{2}}$$

$$= \frac{2}{4} = \frac{1}{2}$$ **Ans.**

(x) Let E and F be the event of winning the game by A and B respectively.

Here, $P(E) = \dfrac{1}{3}$

So, $P(\overline{E}) = 1 - \dfrac{1}{3}$

$$= \frac{2}{3}$$

and $\quad P(F) = \frac{1}{6}$

So, $\quad P(\bar{F}) = 1 - \frac{1}{6} = \frac{5}{6}$

P(neither of them winning the game)

$$= P(\bar{E}) \cdot P(\bar{F})$$

$$= \frac{2}{3} \times \frac{5}{6} = \frac{5}{9}$$ **Ans.**

Solution 2.

Given, $\quad f(x) = \sqrt{2x-3}$

Let, $\quad f(x) = y$

So, $\quad y = \sqrt{2x-3}$

Squaring both sides, we get

$$y^2 = 2x - 3$$

$$y^2 + 3 = 2x$$

So, $\quad x = \frac{y^2+3}{2} = f^{-1}(y)$

So, inverse of f

i.e., $\quad f^{-1}(x) = \frac{x^2+3}{2}$

Now to prove,

$$(f o f^{-1})(x) = x$$

We take, $\quad$ L.H.S. $= (f o f^{-1}) x$

$$= f(f^{-1}(x))$$

$$= \sqrt{2\left(\frac{x^2+3}{2}\right) - 3}$$

$$\left[\because f(x) = \sqrt{2x-3}\right]$$

$$= \sqrt{x^2+3-3}$$

$(f o f^{-1}) x = x$ **Hence Proved.**

Solution 3.

Given, $\tan^{-1} a + \tan^{-1} b + \tan^{-1} c = \pi$

$\Rightarrow \quad \tan^{-1} a + \tan^{-1} b = \pi - \tan^{-1} c$

$\Rightarrow \quad \tan^{-1}\left(\frac{a+b}{1-ab}\right) = \pi - \tan^{-1} c$

$\Rightarrow \quad \frac{a+b}{1-ab} = \tan(\pi - \tan^{-1} c)$

$\Rightarrow \quad \frac{a+b}{1-ab} = -\tan(\tan^{-1} c)$

$\Rightarrow \quad \frac{a+b}{1-ab} = -c$

$\Rightarrow \quad a + b = -c(1 - ab)$

$\Rightarrow \quad a + b = -c + abc$

$\Rightarrow \quad a + b + c = abc$ **Hence Proved.**

Solution 4.

We have,

$$\begin{vmatrix} x+a & b & c \\ c & x+b & a \\ a & b & x+c \end{vmatrix} = 0$$

On applying $C_1 \to C_1 + C_2 + C_3$, we get

$$\begin{vmatrix} x+a+b+c & b & c \\ x+a+b+c & x+b & a \\ x+a+b+c & b & x+c \end{vmatrix} = 0$$

Taking $(x + a + b + c)$ common from C_1

$$(x+a+b+c)\begin{vmatrix} 1 & b & c \\ 1 & x+b & a \\ 1 & b & x+c \end{vmatrix} = 0$$

On applying $R_2 \to R_2 - R_1$, $R_3 \to R_3 - R_1$, we get

$$(x+a+b+c)\begin{vmatrix} 1 & b & c \\ 0 & x & a-c \\ 0 & 0 & x \end{vmatrix} = 0$$

On expanding along C_1, we get

$$(x + a + b + c)(x^2 - 0) = 0$$

$\therefore \quad x + a + b + c = 0$

or $\quad x^2 = 0 \quad [\because x \neq 0]$

$\Rightarrow \quad x = -(a + b + c)$ **Ans.**

Solution 5.

(a) Given, $f(x) = \begin{cases} x^2, & x \leq 1 \\ \frac{1}{x}, & x > 1 \end{cases}$

For continuity,

L.H.L. at $(x = 1) = \lim_{x \to 1^-} f(x) = \lim_{x \to 1^-} x^2$

$\lim_{h \to 0} (1 - h)^2$ (Put $x = 1 - h$)

On putting $h = 0$,

$$(1)^2 = 1$$

R.H.L. at $(x = 1) = \lim_{x \to 1^+} f(x) = \lim_{x \to 1} \frac{1}{x}$

$\lim_{h \to 0} \frac{1}{1+h}$ (Put $x = 1 + h$)

On putting $h = 0$,

$$\frac{1}{1} = 1$$

$$f(1) = \lim_{x \to 1} x^2$$
$$= (1)^2$$
$$= 1$$

$\because$ L.H.L. at $(x = 1)$ = R.H.L. at $(x = 1) = f(1)$

So, $f(x)$ is continuous at $x = 1$.

For differentiability,

$$\text{L.H.D. at } (x = 1) = \lim_{h \to 0^-} \frac{f(1-h) - f(1)}{-h}$$
$$= \lim_{h \to 0} \frac{(1-h)^2 - 1}{-h}$$
$$= \lim_{h \to 0} \frac{1 + h^2 - 2h - 1}{-h}$$
$$= \lim_{h \to 0} \frac{h^2 - 2h}{-h}$$
$$= \lim_{h \to 0} \frac{h(h-2)}{-h} = 2$$

$$\text{R.H.D. at } x = 1 = \lim_{h \to 0^+} \frac{f(1+h) - f(1)}{h}$$
$$= \lim_{h \to 0} \frac{\frac{1}{1+h} - 1}{h}$$
$$= \lim_{h \to 0} \frac{1 - 1 - h}{h(1+h)}$$
$$= \lim_{h \to 0} \frac{-h}{h(1+h)}$$
$$= \lim_{h \to 0} \frac{-1}{1+h}$$
$$= -1$$

$\because$ L.H.D. at $(x = 1) \neq$ R.H.D. at $(x = 1)$.

$\Rightarrow$ $f(x)$ is not differentiable at $x = 1$.

Hence Proved.

OR

(b) To verify Rolle's theorem,

$$f(x) = e^{-x} \sin x \text{ on } [0, \pi]$$

Since $\sin x$ and e^{-x} are both continuous and differentiable everywhere so

(i) $f(x)$ is continuous in $[0, \pi]$

(ii) $f(x)$ is differentiable in $(0, \pi)$

(iii)
$$f(0) = e^{-0} \sin 0$$
$$= 0 \qquad (\because \sin 0 = 0)$$
$$f(\pi) = e^{-\pi} \sin \pi$$
$$= 0 \qquad (\because \sin \pi = 0)$$

$$\therefore \quad f(0) = f(\pi)$$

$\because$ All the three conditions of Rolle's theorem are satisfied, there exists at least one value of x say c, in $(0, \pi)$ such that $f'(c) = 0$.

$$f'(x) = \frac{e^x \cos x - \sin x e^x}{(e^x)^2}$$
$$= \frac{e^x(\cos x - \sin x)}{(e^x)^2}$$
$$f'(x) = \frac{\cos x - \sin x}{e^x}$$

Now, $$f'(c) = \frac{\cos c - \sin c}{e^c}$$

$\because$ $$f'(c) = 0$$

$\Rightarrow$ $e^{-c}(\cos c - \sin c) = 0$

$\Rightarrow$ $\cos c - \sin c = 0$ $\qquad [\because e^{-c} \neq 0]$

$\Rightarrow$ $\sin c = \cos c$

$\Rightarrow$ $\tan c = 1$

$\Rightarrow$ $\tan c = \tan \frac{\pi}{4}$

$\Rightarrow$ $c = \frac{\pi}{4}$

Clearly, this belongs to the open interval $(0, \pi)$.

Hence, Rolle's theorem verified. **Hence Verified.**

Solution 6.

Given, $x = \tan\left(\frac{1}{a}\log y\right)$

$$\tan^{-1} x = \frac{1}{a}\log y$$

Differentiating with respect to x,

$$\frac{1}{1+x^2} = \frac{1}{a} \times \frac{1}{y} \cdot \frac{dy}{dx}$$
$$ay = \frac{dy}{dx}(1+x^2)$$

$\Rightarrow$ $$(1+x^2)\frac{dy}{dx} = ay$$

Again, differentiating with respect to x,

$$(1+x^2)\frac{d^2y}{dx^2} + \frac{dy}{dx} \times 2x = a\frac{dy}{dx}$$

$$\Rightarrow (1+x^2)\frac{d^2y}{dx^2}+2x\frac{dy}{dx}-a\frac{dy}{dx}=0$$

$$\Rightarrow (1+x^2)\frac{d^2y}{dx^2}+(2x-a)\frac{dy}{dx}=0$$ **Hence Proved.**

Solution 7.

We have, $\int \tan^{-1}\sqrt{x}\ dx$

Let, $x = t^2$

$\Rightarrow \sqrt{x} = t$ and $dx = 2t\,dt$

$$\Rightarrow \int \tan^{-1}\sqrt{x}\ dx = \int \tan^{-1} t \times 2t\,dt$$

$$= 2\int \underset{(II)}{t}\ \underset{(I)}{\tan^{-1} t}\ dt$$

$$= 2\left[\tan^{-1} t \times \frac{t^2}{2} - \int \frac{1}{1+t^2}\times\frac{t^2}{2}dt\right]$$

(Integration by parts)

$$= 2\left[\frac{t^2}{2}\tan^{-1} t - \frac{1}{2}\int\frac{t^2}{1+t^2}dt\right]$$

$$= 2\left[\frac{t^2}{2}\tan^{-1} t - \frac{1}{2}\int\frac{t^2+1-1}{1+t^2}dt\right]$$

$$= t^2\tan^{-1} t - \int\frac{t^2+1}{1+t^2}dt + \int\frac{1}{1+t^2}dt$$

$$= t^2\tan^{-1} t - \int dt + \int\frac{1}{1+t^2}dt$$

$$= t^2\tan^{-1} t - t + \tan^{-1} t + c$$

(c = constant of integration)

$$\Rightarrow \int \tan^{-1}\sqrt{x}\ dx = x\tan^{-1}\sqrt{x} - \sqrt{x} + \tan^{-1}\sqrt{x} + c$$

(Substituting the value of t)

$$\Rightarrow \int \tan^{-1}\sqrt{x}\ dx = (x+1)\tan^{-1}\sqrt{x} - \sqrt{x} + c$$ **Ans.**

Solution 8.

(a) Given, Equation of curve is

$$y = 4x^3 - 3x + 5 \quad \ldots(i)$$

Differentiate with respect to x,

$$\frac{dy}{dx} = 12x^2 - 3$$

$\because$ Tangent is parallel to X-axis

So, $$\frac{dy}{dx} = 0$$

$$\Rightarrow 12x^2 - 3 = 0$$

$$\Rightarrow x^2 = \frac{1}{4}$$

$$\Rightarrow x = \pm\frac{1}{2}$$

So, when $x = \frac{1}{2}$

$$\Rightarrow y = 4\left(\frac{1}{2}\right)^3 - 3\times\frac{1}{2}+5 \quad \text{[using (i)]}$$

$$= 4\times\frac{1}{8}-\frac{3}{2}+5$$

$$= \frac{1}{2}-\frac{3}{2}+5$$

$$= -\frac{2}{2}+5$$

$$= -1+5 = 4$$

So, the point on the curve is $\left(\frac{1}{2}, 4\right)$

and when $x = -\frac{1}{2}$

$$\Rightarrow y = 4\left(-\frac{1}{2}\right)^3 - 3\left(-\frac{1}{2}\right)+5$$

$$y = 4\left(-\frac{1}{8}\right)+\frac{3}{2}+5$$

$$\Rightarrow y = -\frac{1}{2}+\frac{3}{2}+5$$

$$\Rightarrow y = 1+5 = 6$$

So, the point on the curve is $\left(-\frac{1}{2}, 6\right)$

So, the required points on the curve are $\left(\frac{1}{2}, 4\right)$ and $\left(-\frac{1}{2}, 6\right)$. **Ans.**

OR

(b) At any time 't' water forms a right circular cone of semi-vertical angle $\theta = \frac{\pi}{4}$ Let 'l' be the slant height and 'r' is the radius of cone at that instant.

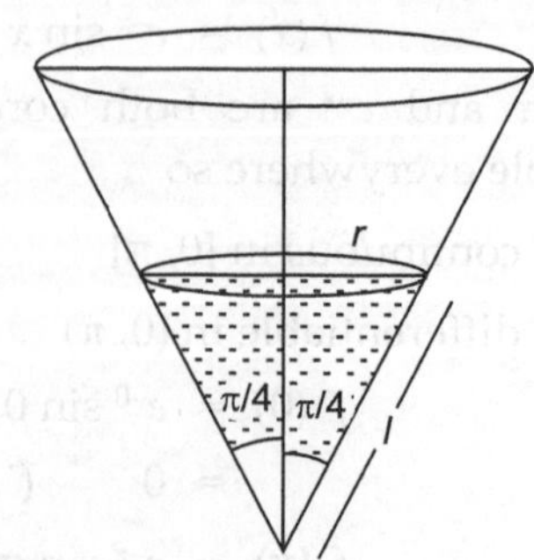

So, $\frac{r}{l} = \sin\theta \left[\sin\theta = \frac{\text{perpendicular}}{\text{hypotenuse}}\right]$

$\Rightarrow \quad \frac{r}{l} = \sin\frac{\pi}{4}$

$\Rightarrow \quad \frac{r}{l} = \frac{1}{\sqrt{2}}$

$\Rightarrow \quad r = \frac{l}{\sqrt{2}}$

$\because$ Curved surface area of cone,

$$A = \pi r l$$

$$A = \pi \times \frac{l}{\sqrt{2}} \times l$$

$$A = \pi \times \frac{1}{\sqrt{2}} \times l^2$$

Differentiating with respect to 't',

$$\frac{dA}{dt} = \pi \times \frac{1}{\sqrt{2}} \times 2l\frac{dl}{dt}$$

$\because$ Water is leaking (dripping out) at the surface.

i.e., $\frac{dA}{dt} = -2 \text{ cm}^2/\text{sec.}$

So, $-2 = \pi \times \frac{1}{\sqrt{2}} \times 2l\frac{dl}{dt}$

$$\frac{dl}{dt} = \frac{-\sqrt{2}}{\pi l}$$

$\because \quad l = 4 \text{ cm}$

So, $\frac{dl}{dt} = \frac{-\sqrt{2}}{4\pi} \text{ cm/s}$

Hence, the slant height is decreasing at the rate of $\frac{\sqrt{2}}{4\pi}$ cm/s. **Ans.**

Solution 9.

(a) Given, differential equation is,

$$\sin x\frac{dy}{dx} - y = \sin x \cdot \tan\frac{x}{2}$$

Divide by $\sin x$ on both sides

$$\frac{dy}{dx} - \frac{y}{\sin x} = \tan\frac{x}{2}$$

It is a linear differential equation, where

$$P = \frac{-1}{\sin x} = -\operatorname{cosec} x$$

$$Q = \tan\frac{x}{2}$$

Integrating factor

$$e^{\int P dx} = e^{-\int \operatorname{cosec} x\, dx}$$

$$= e^{-\log\tan x/2} = e^{\log(\tan x/2)^{-1}}$$

$$= y \cdot \frac{1}{\tan\frac{x}{2}} \qquad (\because e^{\log x} = x)$$

Complete solution of differential equation is,

$$y \cdot \frac{1}{\tan\frac{x}{2}} = \int \tan\frac{x}{2} \times \frac{1}{\tan\frac{x}{2}} dx + c$$

$$\frac{y}{\tan\frac{x}{2}} = \int dx + c$$

$$\frac{y}{\tan\frac{x}{2}} = x + c$$

$y \cot\frac{x}{2} = x + c$ **Ans.**

OR

(b) Let P_0 be the initial population and let the population after 't' years be P. Then,

$$\frac{dP}{dt} = \frac{10P}{100}$$

$$\frac{dP}{dt} = \frac{P}{10}$$

$$\frac{dP}{P} = \frac{dt}{10}$$

Integrating both sides,

$$\int\frac{dP}{P} = \int\frac{dt}{10}$$

$$\log P = \frac{1}{10}t + c \qquad \ldots\text{(i)}$$

At $t = 0$, $P = P_0$

$$\log P_0 = \frac{1}{10} \times 0 + c$$

So, $c = \log P_0$

So, equation (i) will become :

$$\log P = \frac{1}{10}t + \log P_0$$

$$\log P - \log P_0 = \frac{1}{10}t$$

$$\log\frac{P}{P_0} = \frac{1}{10}t$$

$$t = 10\log\left(\frac{P}{P_0}\right)$$

Since population grows 4 times,

$$P = 4P_0$$

So,
$$t = 10\log\frac{4P_0}{P_0}$$

$$t = 10\log 4 = 20\log 2$$

Thus, the population will grow 4 times in 20 log 2 years. **Ans.**

Solution 10.

(a) Given system of equations is,

$$2x - 3y + 5z = 11$$
$$3x + 2y - 4z = -5$$
$$x + y - 2z = -3$$

which can be written as AX = B where,

$$A = \begin{bmatrix} 2 & -3 & 5 \\ 3 & 2 & -4 \\ 1 & 1 & -2 \end{bmatrix} B = \begin{bmatrix} 11 \\ -5 \\ -3 \end{bmatrix}, X = \begin{bmatrix} x \\ y \\ z \end{bmatrix}$$

So,
$$X = A^{-1}B.$$

$$|A| = 2\begin{vmatrix} 2 & -4 \\ 1 & -2 \end{vmatrix} + 3\begin{vmatrix} 3 & -4 \\ 1 & -2 \end{vmatrix} + 5\begin{vmatrix} 3 & 2 \\ 1 & 1 \end{vmatrix}$$
$$= 2(-4+4) + 3(-6+4) + 5(3-2)$$
$$= 0 + 3(-2) + 5 \times 1$$
$$= -6 + 5$$
$$= -1$$

$\because$ $|A| \neq 0$ so A^{-1} exist.

Cofactors of A

$$A_{11} = 0, \quad A_{12} = 2, \quad A_{13} = 1$$
$$A_{21} = -1, \quad A_{22} = -9, \quad A_{23} = -5$$
$$A_{31} = 2, \quad A_{32} = 23, \quad A_{33} = 13$$

$$\text{adj. } A = \begin{bmatrix} 0 & 2 & 1 \\ -1 & -9 & -5 \\ 2 & 23 & 13 \end{bmatrix}^T$$

$$= \begin{bmatrix} 0 & -1 & 2 \\ 2 & -9 & 23 \\ 1 & -5 & 13 \end{bmatrix}$$

Now,
$$A^{-1} = \frac{1}{|A|}\text{adj. } A$$

$$= -\begin{bmatrix} 0 & -1 & 2 \\ 2 & -9 & 23 \\ 1 & -5 & 13 \end{bmatrix}$$

$$= \begin{bmatrix} 0 & 1 & -2 \\ -2 & 9 & -23 \\ -1 & 5 & -13 \end{bmatrix}$$

$$X = A^{-1}B$$

$$= \begin{bmatrix} 0 & 1 & -2 \\ -2 & 9 & -23 \\ -1 & 5 & -13 \end{bmatrix}\begin{bmatrix} 11 \\ -5 \\ -3 \end{bmatrix}$$

$$\begin{bmatrix} x \\ y \\ z \end{bmatrix} = \begin{bmatrix} 1 \\ 2 \\ 3 \end{bmatrix}$$

So, $x = 1, y = 2, z = 3$ **Ans.**

OR

(b) Let
$$A = \begin{bmatrix} 0 & 1 & 2 \\ 1 & 2 & 3 \\ 3 & 1 & 1 \end{bmatrix}$$

Then,
$$A = I_3A$$

$$\begin{bmatrix} 0 & 1 & 2 \\ 1 & 2 & 3 \\ 3 & 1 & 1 \end{bmatrix} = \begin{bmatrix} 1 & 0 & 0 \\ 0 & 1 & 0 \\ 0 & 0 & 1 \end{bmatrix}A$$

On applying $R_1 \leftrightarrow R_2$, we get

$$\begin{bmatrix} 1 & 2 & 3 \\ 0 & 1 & 2 \\ 3 & 1 & 1 \end{bmatrix} = \begin{bmatrix} 0 & 1 & 0 \\ 1 & 0 & 0 \\ 0 & 0 & 1 \end{bmatrix}A$$

On applying $R_3 \to R_3 - 3R_1$, we get

$$\begin{bmatrix} 1 & 2 & 3 \\ 0 & 1 & 2 \\ 0 & -5 & -8 \end{bmatrix} = \begin{bmatrix} 0 & 1 & 0 \\ 1 & 0 & 0 \\ 0 & -3 & 1 \end{bmatrix}A$$

On applying $R_3 \to R_3 + 5R_2$, we get

$$\begin{bmatrix} 1 & 2 & 3 \\ 0 & 1 & 2 \\ 0 & 0 & 2 \end{bmatrix} = \begin{bmatrix} 0 & 1 & 0 \\ 1 & 0 & 0 \\ 5 & -3 & 1 \end{bmatrix}A$$

On applying $R_1 \to R_1 - 2R_2$, we get

$$\begin{bmatrix} 1 & 0 & -1 \\ 0 & 1 & 2 \\ 0 & 0 & 2 \end{bmatrix} = \begin{bmatrix} -2 & 1 & 0 \\ 1 & 0 & 0 \\ 5 & -3 & 1 \end{bmatrix}A$$

On applying $R_2 \to R_2 - R_3$, we get

$$\begin{bmatrix} 1 & 0 & -1 \\ 0 & 1 & 0 \\ 0 & 0 & 2 \end{bmatrix} = \begin{bmatrix} -2 & 1 & 0 \\ -4 & 3 & -1 \\ 5 & -3 & 1 \end{bmatrix} A$$

On applying $R_3 \to \frac{R_3}{2}$, we get

$$\begin{bmatrix} 1 & 0 & -1 \\ 0 & 1 & 0 \\ 0 & 0 & 1 \end{bmatrix} = \begin{bmatrix} -2 & 1 & 0 \\ -4 & 3 & -1 \\ 5/2 & -3/2 & 1/2 \end{bmatrix} A$$

On applying $R_1 \to R_1 + R_3$, we get

$$\begin{bmatrix} 1 & 0 & 0 \\ 0 & 1 & 0 \\ 0 & 0 & 1 \end{bmatrix} = \begin{bmatrix} 1/2 & -1/2 & 1/2 \\ -4 & 3 & -1 \\ 5/2 & -3/2 & 1/2 \end{bmatrix} A$$

So, $$A^{-1} = \begin{bmatrix} 1/2 & -1/2 & 1/2 \\ -4 & 3 & -1 \\ 5/2 & -3/2 & 1/2 \end{bmatrix}$$

Ans.

Solution 11.

Let E be the event that A speaks the truth and F be the event that B speaks the truth. Then $\bar{E}$ will be the event that A speaks lie and $\bar{F}$ will be the event that B speaks a lie. They will contradict each other if A speaks the truth and B does not speak the truth or if A does not speak the truth and B speaks the turth.

Now, $$P(E) = 60\% = \frac{60}{100}$$

$$P(\bar{E}) = 1 - P(E) = 1 - \frac{60}{100} = \frac{40}{100}$$

$$P(F) = 40\% = \frac{40}{100}$$

$$P(\bar{F}) = 1 - P(F) = 1 - \frac{40}{100} = \frac{60}{100}$$

P(A and B contradict each other)

$$= P(E)\,P(\bar{F}) + P(F)\,P(\bar{E})$$

$$= \frac{60}{100} \times \frac{60}{100} + \frac{40}{100} \times \frac{40}{100}$$

$$= \frac{36}{100} + \frac{16}{100} = \frac{52}{100} = 52\%$$

Hence, A and B will contradict each other in 52% cases. **Ans.**

Solution 12.

Let height of cone be 'h' cm, and its radius is 'r' cm.

Also, given radius of sphere = 12 cm.

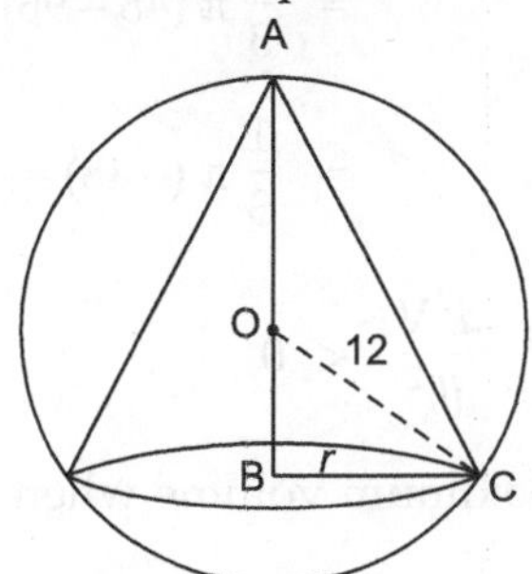

So,
$$OA = OC = 12 \text{ cm}$$
$$BC = r \text{ cm}$$
$$AB = h \text{ cm}$$

In Δ ABC,
$$12 + OB = h$$
$$OB = h - 12$$

In Δ OBC,
$$OC^2 = OB^2 + BC^2$$
$$(12)^2 = (h-12)^2 + (r)^2$$
$$144 = h^2 + 144 - 24h + r^2$$
$$24h - h^2 = r^2$$

Now, volume of cone,

$$V = \frac{1}{3}\pi r^2 h$$
$$= \frac{1}{3}\pi (24h - h^2)\,h$$
$$\Rightarrow \quad V = \frac{1}{3}\pi (24h^2 - h^3)$$

Differentiate with respect to h,

$$\frac{dV}{dh} = \frac{1}{3}\pi (48h - 3h^2)$$

Again, differentiating with respect to h,

$$\frac{d^2V}{dh^2} = \frac{1}{3}\pi (48 - 6h)$$

For maximum and minimum value,

$$\frac{dV}{dh} = 0$$

$$\Rightarrow \frac{1}{3}\pi (48h - 3h^2) = 0$$

$$\Rightarrow \quad 48h = 3h^2$$

$$\Rightarrow \quad h = 16 \text{ cm} \qquad [\because h \neq 0]$$

$$\left.\frac{d^2V}{dh^2}\right|_{h=16} = \frac{1}{3}\pi(48 - 6 \times 16)$$

$$= \frac{1}{3}\pi(48-96)$$

$$= \frac{1}{3}\pi(-48) = -16\pi$$

$$\frac{d^2V}{dh^2} < 0$$

So, it has maximum volume when height of cone = 16 cm. **Ans.**

Solution 13.

(a) We have, $\int \frac{x-1}{\sqrt{x^2-x}}dx$

Let, $x-1 = A\frac{d}{dx}(x^2-x)+B$

$$x-1 = A(2x-1)+B$$

$$x-1 = 2Ax - A + B$$

Comparing the coefficient of both sides, we get

$$2A = 1$$

$$A = \frac{1}{2}$$

and $-A+B = -1$

$$-\frac{1}{2}+B = -1$$

$$B = -\frac{1}{2}$$

$$\Rightarrow \quad x-1 = \frac{1}{2}(2x-1)-\frac{1}{2}$$

Let, $I = \int \frac{x-1}{\sqrt{x^2-x}}dx$

$$= \int \frac{\frac{1}{2}(2x-1)-\frac{1}{2}}{\sqrt{x^2-x}}dx$$

$$= \frac{1}{2}\int \frac{2x-1}{\sqrt{x^2-x}}dx - \frac{1}{2}\int \frac{dx}{\sqrt{x^2-x}}$$

Let, $x^2 - x = t$

On differentiating both sides

$$(2x-1)dx = dt$$

$$I = \frac{1}{2}\int \frac{dt}{\sqrt{t}} - \frac{1}{2}\int \frac{dx}{\sqrt{x^2-x+\frac{1}{4}-\frac{1}{4}}}$$

$$= \frac{1}{2}\times 2\sqrt{t} - \frac{1}{2}\int \frac{dx}{\sqrt{\left(x-\frac{1}{2}\right)^2-\left(\frac{1}{2}\right)^2}}$$

$$= \sqrt{t} - \frac{1}{2}\log\left|\left(x-\frac{1}{2}\right)+\sqrt{\left(x-\frac{1}{2}\right)^2-\left(\frac{1}{2}\right)^2}\right| + c$$

(where c is constant of integration) **Ans.**

$$= \sqrt{x^2-x} - \frac{1}{2}\log\left|x-\frac{1}{2}+\sqrt{x^2-x}\right| + c$$

OR

(b) Let, $I = \int_0^{\pi/2} \frac{\cos^2 x}{1+\sin x\cos x}dx$ ...(i)

Using property

$$\int_0^a f(x)dx = \int_0^a f(a-x)dx$$

$$I = \int_0^{\pi/2} \frac{\cos^2\left(\frac{\pi}{2}-x\right)}{1+\sin\left(\frac{\pi}{2}-x\right)\cdot\cos\left(\frac{\pi}{2}-x\right)}dx$$

$$I = \int_0^{\pi/2} \frac{\sin^2 x}{1+\cos x\cdot\sin x}dx \quad \text{...(ii)}$$

Adding equations (i) and (ii),

$$2I = \int_0^{\pi/2} \frac{dx}{1+\cos x\cdot\sin x}$$

($\because \sin^2 x + \cos^2 x = 1$)

Divide by $\cos^2 x$ in numerator and denominator,

$$2I = \int_0^{\pi/2} \frac{\sec^2 x\,dx}{\sec^2 x + \tan x}$$

$$= \int_0^{\pi/2} \frac{\sec^2 x}{\tan^2 x + \tan x + 1}dx$$

($\tan^2 x + 1 = \sec^2 x$)

Let $\tan x = t$, $\sec^2 x\,dx = dt$ and $x = 0$

$$\Rightarrow \quad t = 0 : x = \frac{\pi}{2} \Rightarrow t = \infty$$

$$2I = \int_0^{\infty} \frac{dt}{t^2+t+1}$$

$$= \int_0^{\infty} \frac{dt}{t^2+t+\frac{1}{4}+\frac{3}{4}}$$

$$= \int_0^{\infty} \frac{dt}{\left(t+\frac{1}{2}\right)^2+\left(\frac{\sqrt{3}}{2}\right)^2}$$

$$\Rightarrow \quad 2I = \frac{2}{\sqrt{3}}\left[\tan^{-1}\frac{t+\frac{1}{2}}{\frac{\sqrt{3}}{2}}\right]_0^{\infty}$$

$$\Rightarrow \quad 2I = \frac{2}{\sqrt{3}}\left[\tan^{-1}\left(\frac{2t+1}{\sqrt{3}}\right)\right]_0^{\infty}$$

$$\Rightarrow \quad 2I = \frac{2}{\sqrt{3}}\left[\tan^{-1}(\infty)-\tan^{-1}\left(\frac{1}{\sqrt{3}}\right)\right]$$

$$\Rightarrow \quad 2I = \frac{2}{\sqrt{3}}\left(\frac{\pi}{2}-\frac{\pi}{6}\right)$$

$$\Rightarrow \quad 2I = \frac{2}{\sqrt{3}}\times\frac{2\pi}{6}$$

$$\Rightarrow \quad I = \frac{\pi}{3\sqrt{3}}$$ **Ans.**

Solution 14.

Here, X denotes the number of defective items in the sample.

So, X can take value 0, 1, 2.

$$P(X=0) = \frac{{}^4C_4}{{}^6C_4} = \frac{1}{15}$$

$$P(X=1) = \frac{{}^4C_3\times{}^2C_1}{{}^6C_4} = \frac{8}{15}$$

$$P(X=2) = \frac{{}^4C_2\times{}^2C_2}{{}^6C_4} = \frac{2}{5}$$

So,

(a) Probability distribution of X :

X	0	1	2
P(X)	$\frac{1}{15}$	$\frac{8}{15}$	$\frac{2}{5}$

(b) For mean :

X_i	P_i	P_iX_i	X_i^2	$P_iX_i^2$
0	$\frac{1}{15}$	0	0	0
1	$\frac{8}{15}$	$\frac{8}{15}$	1	$\frac{8}{15}$
2	$\frac{2}{5}$	$\frac{4}{5}$	4	$\frac{8}{5}$

$$\text{Mean} = \Sigma P_iX_i$$
$$= \frac{8}{15}+\frac{4}{5}$$
$$= \frac{20}{15} = \frac{4}{3}$$ **Ans.**

(c) Variance $= \Sigma P_iX_i^2 - (\Sigma P_iX_i)^2$

$$= \left(\frac{8}{15}+\frac{8}{5}\right)-\left(\frac{4}{3}\right)^2$$

$$= \frac{32}{15}-\frac{16}{9} = \frac{16}{45}$$ **Ans.**

SECTION—B

Solution 15.

(a) Projection of $\vec{a}$ on $\vec{b}$ $= \frac{\vec{a}\cdot\vec{b}}{|\vec{b}|}$

$$= \frac{(\lambda\hat{i}+\hat{j}+4\hat{k})\cdot(2\hat{i}+6\hat{j}+3\hat{k})}{\sqrt{4+36+9}}$$

$$4 = \frac{2\lambda+6+12}{\sqrt{49}}$$

$$\Rightarrow \quad 4\times 7 = 2\lambda+18$$
$$\Rightarrow \quad 28-18 = 2\lambda$$
$$\Rightarrow \quad 2\lambda = 10$$
$$\Rightarrow \quad \lambda = 5$$ **Ans.**

(b) Given equation of line is,

$$2x-3 = 3y+1 = 5-6z$$

$$\frac{x-\frac{3}{2}}{\frac{1}{2}} = \frac{y+\frac{1}{3}}{\frac{1}{3}} = \frac{z-\frac{5}{6}}{-\frac{1}{6}}$$

Direction ratio : $\left\langle\frac{1}{2},\frac{1}{3},-\frac{1}{6}\right\rangle$

$$= \langle 3,2,-1\rangle$$

Normal vector

$$\bar{n} = 3\hat{i}+2\hat{j}-\hat{k}$$

Vector equation of line through (7, – 5, 0)

i.e., $\bar{a} = 7\hat{i}-5\hat{j}$

So, $\vec{r} = \vec{a}+\lambda\vec{n}$

$$\vec{r} = 7\hat{i}-5\hat{j}+\lambda(3\hat{i}+2\hat{j}-\hat{k})$$ **Ans.**

(c) Equation of plane through the intersection of given plane is,

$$\vec{r}\cdot(\hat{i}+3\hat{j}-\hat{k})-9+\lambda[\vec{r}\cdot(2\hat{i}-\hat{j}+\hat{k})-3] = 0$$

$$\vec{r}\cdot[(1+2\lambda)\hat{i}+\hat{j}(3-\lambda)+\hat{k}(-1+\lambda)]-9-3\lambda = 0$$

$\because$ Plane passes through origin (0, 0, 0), then vector equation

$$= 0\cdot\hat{i}+0\cdot\hat{j}+0\cdot\hat{k}$$

It satisfy the equation of plane

$$-9-3\lambda = 0$$
$$3\lambda = -9$$
$$\lambda = -3$$

So, equation of required plane

$$\vec{r}(-5\hat{i}+6\hat{j}-4\hat{k}) = 0$$

Ans.

Solution 16.

(a) Given, A, B, C are three non-collinear points with position vectors $\vec{a}, \vec{b}$ and $\vec{c}$ respectively.

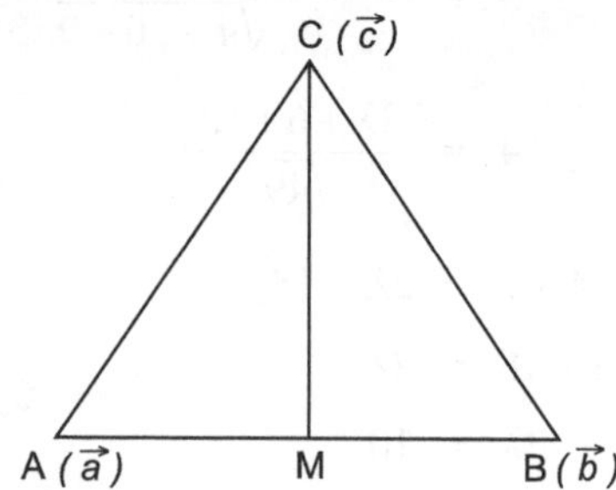

i.e. $\vec{OA} = \vec{a}, \vec{OB} = \vec{b}, \vec{OC} = \vec{c}$

$$\vec{AB} = \vec{OB}-\vec{OA}$$
$$= \vec{b}-\vec{a}$$
$$\vec{AC} = \vec{OC}-\vec{OA}$$
$$= \vec{c}-\vec{a}$$

CM is length of perpendicular from C on AB.

$$\text{Area of } \Delta \text{ ABC} = \frac{1}{2}\left|\vec{AB}\times\vec{AC}\right|$$

and $\text{Area of } \Delta \text{ ABC} = \frac{1}{2}\vec{AB}\times\vec{CM}$

$$\text{So, } \frac{1}{2}\vec{AB}\times\vec{CM} = \frac{1}{2}\left|\vec{AB}\times\vec{AC}\right|$$

$$\vec{AB}\times\vec{CM} = \left|(\vec{b}-\vec{a})\times(\vec{c}-\vec{a})\right|$$

$$= \left|\vec{b}\times\vec{c}-\vec{b}\times\vec{a}-\vec{a}\times\vec{c}+\vec{a}\times\vec{a}\right|$$

$$(\because \vec{a}\times\vec{a}=0)$$

$$\vec{AB}\times\vec{CM} = \left|\vec{b}\times\vec{c}+\vec{a}\times\vec{b}+\vec{c}\times\vec{a}\right|$$

$$\left(\begin{array}{l}\because -\vec{b}\times\vec{a}=\vec{a}\times\vec{b} \\ -\vec{a}\times\vec{c}=\vec{c}\times\vec{a}\end{array}\right)$$

$$\vec{CM} = \frac{\left|\vec{a}\times\vec{b}+\vec{b}\times\vec{c}+\vec{c}\times\vec{a}\right|}{\left|\vec{AB}\right|}$$

$$\vec{CM} = \frac{\left|\vec{a}\times\vec{b}+\vec{b}\times\vec{c}+\vec{c}\times\vec{a}\right|}{\left|\vec{b}-\vec{a}\right|}$$

Hence Proved.

OR

(b) Given the position vector of four points.

$$\overline{OA} = 4\hat{i}+5\hat{j}+\hat{k}$$
$$\overline{OB} = -\hat{j}-\hat{k}$$
$$\overline{OC} = 3\hat{i}+9\hat{j}+4\hat{k}$$
$$\overline{OD} = -4\hat{i}+4\hat{j}+4\hat{k}$$

To show that the four points are coplanar, we have to show,

$$\left[\overline{AB}\ \overline{AC}\ \overline{AD}\right] = 0$$

$$\overline{AB} = \overline{OB} - \overline{OA}$$
$$= -\hat{j}-\hat{k}-4\hat{i}-5\hat{j}-\hat{k}$$
$$= -4\hat{i}-6\hat{j}-2\hat{k}$$
$$\overline{AC} = \overline{OC}-\overline{OA}$$
$$= 3\hat{i}+9\hat{j}+4\hat{k}-4\hat{i}-5\hat{j}-\hat{k}$$
$$= -\hat{i}+4\hat{j}+3\hat{k}$$
$$\overline{AD} = \overline{OD}-\overline{OA}$$
$$= -4\hat{i}+4\hat{j}+4\hat{k}-4\hat{i}-5\hat{j}-\hat{k}$$
$$= -8\hat{i}-\hat{j}+3\hat{k}$$

$$\left[\overline{AB}\ \overline{AC}\ \overline{AD}\right] = \begin{vmatrix} -4 & -6 & -2 \\ -1 & 4 & 3 \\ -8 & -1 & 3 \end{vmatrix}$$

$$= -4(12+3)+6(-3+24)-2(1+32)$$
$$= -60+126-66$$
$$= -126+126 = 0$$

Hence, given points are coplanar. **Hence Proved.**

Solution 17.

(a) Given equation of parabola $y^2 = 8x$ with vertex (0, 0) and line $x = 2$.

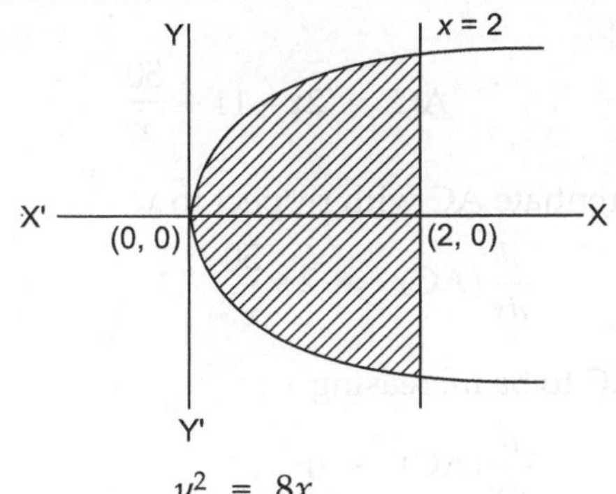

$$y^2 = 8x$$

$$y = \sqrt{8x}$$

$$= 2\sqrt{2}\sqrt{x}$$

$$\text{Required area} = 2\int_0^2 2\sqrt{2}\sqrt{x}dx$$

$$= 4\sqrt{2}\left[\frac{x^{3/2}}{3/2}\right]_0^2$$

$$= \frac{4\sqrt{2}\times 2}{3}(2)^{3/2}$$

$$= \frac{8}{3}(2)^2 = \frac{32}{3} \text{ sq. units} \qquad \textbf{Ans.}$$

OR

(b) Given, $y = |x+4| = \begin{cases} (x+4) \text{ if } x+4 \geq 0 \\ -(x+4) \text{if } x+4 < 0 \end{cases}$

$$= \begin{cases} (x+4) \text{ if } x \geq -4 \\ -x-4 \text{ if } x < -4 \end{cases}$$

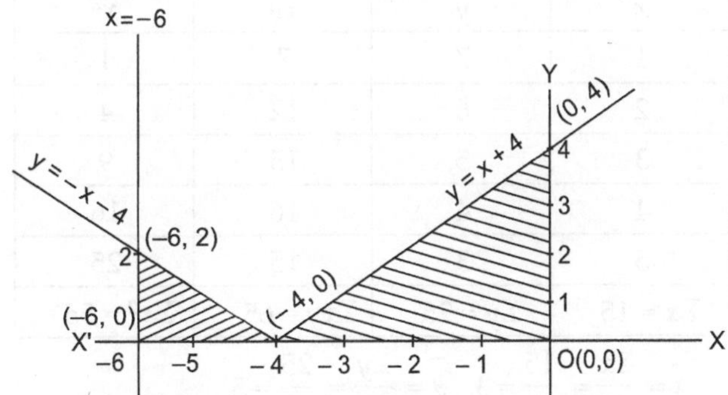

$$\text{Required area} = \int_{-6}^{0}|x+4|\,dx$$

$$= \int_{-6}^{-4}|x+4|\,dx + \int_{-4}^{0}|x+4|\,dx$$

$$= \int_{-6}^{-4}-(x+4)dx + \int_{-4}^{0}(x+4)dx$$

$$= -\left[\frac{x^2}{2}+4x\right]_{-6}^{-4} + \left[\frac{x^2}{2}+4x\right]_{-4}^{0}$$

$$= -\left[\left(\frac{16}{2}-16\right)-\left(\frac{36}{2}-24\right)\right] + \left[-\left(\frac{16}{2}-4\times 4\right)\right]$$

$$= -[(8-16)-(18-24)] + [-8+16]$$

$$= -(-8+6)+(8)$$

$$= 2+8 = 10 \text{ sq. unit.} \qquad \textbf{Ans.}$$

Solution 18.

Let Q be the image of the point having position vector P $(3\hat{i}-2\hat{j}+\hat{k})$ in the plane

$$\vec{r}\cdot(3\hat{i}-\hat{j}+4\hat{k}) = 2$$

PQ is normal to the plane.

So, equation of line PQ

$$\vec{r} = (3\hat{i}-2\hat{j}+\hat{k})+\lambda(3\hat{i}-\hat{j}+4\hat{k})$$

∵ Q lie on the line PQ so let the position vector of point Q

$$= (3+3\lambda)\hat{i}+(-2-\lambda)\hat{j}+(1+4\lambda)\hat{k}$$

∵ 'R' is the mid-point of PQ.

So, position vector of R

$$= \frac{(3+3\lambda)\hat{i}+(-2-\lambda)\hat{j}+(1+4\lambda)\hat{k}+(3\hat{i}-2\hat{j}+\hat{k})}{2}$$

$$= \frac{(6+3\lambda)\hat{i}+(-4-\lambda)\hat{j}+(2+4\lambda)\hat{k}}{2}$$

∵ 'R' lie in the plane $\vec{r}(3\hat{i}-\hat{j}+4\hat{k}) = 2$

So, $\left[\left(\frac{6+3\lambda}{2}\right)\hat{i}+\left(\frac{-4-\lambda}{2}\right)\hat{j}+\left(\frac{2+4\lambda}{2}\right)\hat{k}\right]$

$$(3\hat{i}-\hat{j}+4\hat{k}) = 0$$

$$\Rightarrow 3\left(\frac{6+3\lambda}{2}\right)-1\left(\frac{-4-\lambda}{2}\right)+4\left(\frac{2+4\lambda}{2}\right)-2 = 0$$

$$\Rightarrow \frac{18+9\lambda}{2}+\frac{4+\lambda}{2}+\frac{8+16\lambda}{2} = 2$$

$$\Rightarrow 30+26\lambda = 4$$

$$\Rightarrow 26\lambda = -26$$

$$\Rightarrow \lambda = -1$$

So, position vector of Q

$$= (3+3(-1))\hat{i}+(-2+1)\hat{j}+(1+4(-1))\hat{k}$$

$$= 0-\hat{j}-3\hat{k}$$

i.e., $-\hat{j}-3\hat{k}$

$\Rightarrow x = 0, y = -1, z = -3.$

Thus, the image is (0, – 1, – 3). **Ans.**

SECTION—C

Solution 19.

(a) Given cost function is,

$$C(x) = \frac{1}{3}x^3 + 3x^2 - 16x + 2$$

(i)
$$\text{M.C.} = \frac{d}{dx}[C(x)]$$
$$= \frac{1}{3} \times 3x^2 + 6x - 16$$
$$= x^2 + 6x - 16 \quad \textbf{Ans.}$$

(ii) Average cost function $= \dfrac{C(x)}{x}$

$$= \frac{1}{3}x^2 + 3x - 16 + \frac{2}{x} \quad \textbf{Ans.}$$

(b) Given lines of regression are

$x - 2y + 3 = 0$ and $4x - 5y + 1 = 0$

Let, $x - 2y + 3 = 0$ is equation of regression line y on x.

So,
$$2y = x + 3$$
$$y = \frac{1}{2}x + \frac{3}{2}$$
$$b_{yx} = \frac{1}{2}$$

and $4x - 5y + 1 = 0$ is regression equation of line x on y.

$$4x = 5y - 1$$
$$\Rightarrow \quad x = \frac{5}{4}y - \frac{1}{4}$$
$$b_{xy} = \frac{5}{4}$$
$$b_{yx} \cdot b_{xy} = \frac{1}{2} \times \frac{5}{4} = \frac{5}{8} < 1$$

So, our assumption is true

$$r^2 = b_{xy} \cdot b_{yx}$$
$$r^2 = \frac{1}{2} \times \frac{5}{4}$$
$$r = \pm\sqrt{\frac{5}{8}}$$
$$r = \frac{\pm\sqrt{5}}{2\sqrt{2}} = 0{\cdot}79$$

But signs of b_{yx} and b_{xy} are +ve so

$$r = \frac{1}{2}\sqrt{\frac{5}{2}} = 0{\cdot}79 \quad \textbf{Ans.}$$

(c) Given,

$$AC = 2x - 11 + \frac{50}{x}$$

Differentiate AC with respect to x.

$$\frac{d}{dx}(AC) = 2 - \frac{50}{x^2}$$

For AC to be increasing

$$\frac{d}{dx}(AC) > 0$$
$$\Rightarrow \quad 2 - \frac{50}{x^2} > 0$$
$$2 > \frac{50}{x^2}$$
$$2x^2 > 50$$
$$x^2 > 25$$
$$x^2 - 25 > 0$$
$$(x+5)(x-5) > 0$$

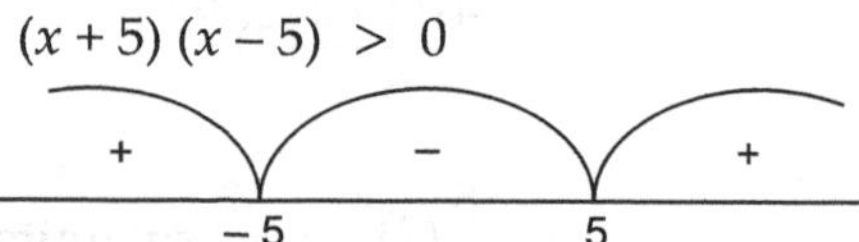

For AC to be increasing range of x is

$$(-\infty, -5) \cup (5, \infty)$$

But, $x \neq$ negative,

$\therefore$ AC is increasing in range of x is $(5, \infty)$. **Ans.**

Solution 20.

(a)

x	y	xy	x^2
1	7	7	1
2	6	12	4
3	5	15	9
4	4	16	16
5	3	15	25
$\Sigma x = 15$	$\Sigma y = 25$	$\Sigma xy = 65$	$\Sigma x^2 = 55$

$$\bar{x} = \frac{\Sigma x}{N} = \frac{15}{5} = 3, \quad \bar{y} = \frac{\Sigma y}{N} = \frac{25}{5} = 5$$

$$b_{yx} = \frac{\Sigma xy - \frac{1}{N}\Sigma x \Sigma y}{\Sigma x^2 - \frac{1}{N}(\Sigma x)^2}$$
$$= \frac{65 - \frac{1}{5}15 \times 25}{55 - \frac{1}{5}15 \times 15}$$

$$= \frac{65-75}{55-45} = \frac{-10}{10} = -1$$

So, regression equation of y on x.

$$y - \bar{y} = b_{yx}(x - \bar{x})$$
$$y - 5 = -1\,(x-3)$$
$$y = -x + 3 + 5$$
$$y = -x + 8$$

When $x = 6$,

$$y = -6 + 8$$
$$y = 2 \qquad \textbf{Ans.}$$

OR

(b) Given, $\bar{x} = 6$, $\bar{y} = 8$

$$\sigma_x = 4, \sigma_y = 6, r = \frac{2}{3}$$

(i)
$$b_{yx} = r\frac{\sigma_y}{\sigma_x}$$
$$= \frac{2}{3} \times \frac{6}{4} = 1$$
$$b_{xy} = r\frac{\sigma_x}{\sigma_y}$$
$$= \frac{2}{3} \times \frac{4}{6} = \frac{8}{18} = \frac{4}{9} \qquad \textbf{Ans.}$$

(ii) Regression line of x on y

$$x - \bar{x} = b_{xy}(y - \bar{y})$$
$$x - 6 = \frac{4}{9}(y-8)$$
$$9x - 54 = 4y - 32$$
$$9x - 4y - 22 = 0 \qquad \textbf{Ans.}$$

(iii) When $y = 14$,

So,
$$9x = 4y + 22$$
$$9x = 4 \times 14 + 22$$
$$9x = 56 + 22$$
$$9x = 78$$
$$x = \frac{78}{9} = 8{\cdot}67 \qquad \textbf{Ans.}$$

Solution 21.

(a) Given, cost function

$$C(x) = \frac{x^2}{100} + 100x + 40$$
$$R(x) = P \cdot x$$
$$R(x) = 200x - \frac{x^2}{400}$$
$$P(x) = R(x) - C(x)$$
$$= 200x - \frac{x^2}{400} - \frac{x^2}{100} - 100x - 40$$
$$= 100x - \frac{5x^2}{400} - 40$$
$$\frac{d}{dx}[P(x)] = 100 - \frac{10x}{400}$$
$$\frac{d^2}{dx^2}[P(x)] = \frac{-10}{400} = \frac{-1}{40}$$
$$\frac{d^2}{dx^2}P(x) < 0$$

So, it has maximum profit.

For maximum and minimum profit.

$$\frac{d}{dx}P(x) = 0$$
$$100 - \frac{10x}{400} = 0$$
$$100 = \frac{10x}{400}$$
$$x = 4000$$

Total profit at the level of production.

$$Px = 4000 = 100 \times 4000 - \frac{5 \times (4000)^2}{400} - 40$$
$$= 400000 - \frac{4000 \times 4000}{80} - 40$$
$$= 400000 - 200000 - 40$$
$$Px = 4000 = ₹\,1{,}99{,}960$$
$$\text{Price per unit} = 200 - \frac{4000}{400} \quad [\text{Taking } x = 4000]$$
$$= 200 - 10 = ₹\,190 \qquad \textbf{Ans.}$$

OR

(b) Given marginal cost function $= \dfrac{500}{\sqrt{2x+25}}$

Total increased cost when x increases from 100 to 300 units.

$$C(x) = \int_{100}^{300} \frac{500}{\sqrt{2x+25}}\,dx$$
$$= 500\int_{100}^{300} \frac{dx}{\sqrt{2x+25}}$$
$$= 500\int_{100}^{300} (2x+25)^{-1/2}\,dx$$
$$= 500\left[\frac{\sqrt{2x+25}}{2 \times \frac{1}{2}}\right]_{100}^{300}$$

$= 500\left[\sqrt{2\times300+25} - \sqrt{2\times100+25}\right]$

$= 500 \times 10 = ₹\ 5000$ **Ans.**

Solution 22.

Let x and y be the number of teaching aids of type A and B respectively. Then the LPP is,

Maximize $Z = 80x + 120y$

Subject to constraints :

$$9x + 12y \leq 180$$
$$x + 3y \leq 30$$

and $x \geq 0, y \geq 0,$

First we draw the lines AB and CD whose equations are

$$9x + 12y = 180$$

$\Rightarrow$ $3x + 4y = 60$ ...(i)

x	20	0
y	0	15

and $x + 3y = 30$ ...(ii)

x	30	0
y	0	10

Let P be the point of intersection of the lines

$$3x + 4y = 60$$

and $x + 3y = 30$

Solving these equations, we get point P(12, 6).

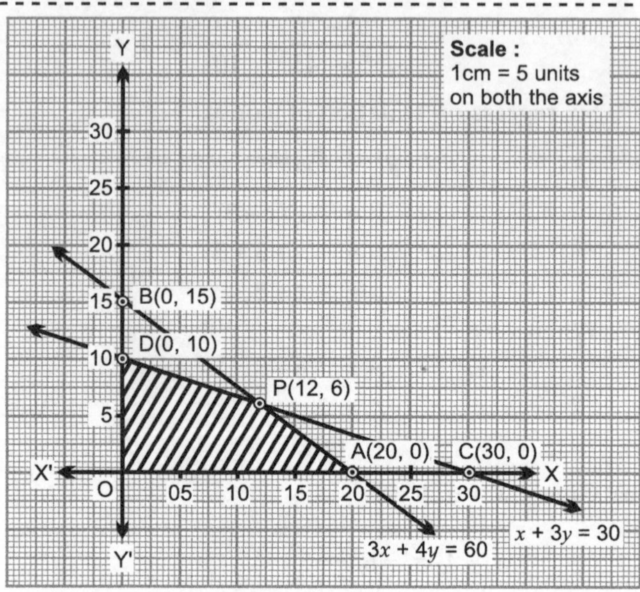

The feasible region is OAPDO which is shaded in the figure.

The vertices of the feasible region are O (0, 0), A (20, 0), P (12, 6) and D (0, 10).

The value of objective function :

$Z = 80x + 120y$ as follows :

Corner Points	Maximize $Z = 80x + 120y$
At O (0, 0)	0
At A (20, 0)	1600
At P (12, 6)	960 + 720 = 1680 maximum
At D (0, 10)	1200

$\therefore$ The profit is maximum at P(12, 6) *i.e.,* when the teaching aids of types A and B are 12 and 6 respectively.

Also maximum profit = ₹ 1680 per week. **Ans.**

MATHEMATICS 2017

QUESTIONS

SECTION—A (80 Marks)

Question 1. **[10 × 3]**

(i) *If the matrix* $\begin{pmatrix} 6 & -x^2 \\ 2x-15 & 10 \end{pmatrix}$ is symmetrix; find the value of x.

(ii) *If* $y - 2x - k = 0$ *touches the conic* $3x^2 - 5y^2 = 15$, *find the value of k.***

(iii) *Prove that* $\frac{1}{2}\cos^{-1}\left(\frac{1-x}{1+x}\right) = \tan^{-1}\sqrt{x}$

(iv) *Using L' Hospital's rule, evaluate :*

$$\lim_{x\to\pi/2}\left(x\tan x - \frac{\pi}{4}\cdot\sec x\right)$$

(v) *Evaluate :* $\int \frac{1}{x^2}\sin^2\left(\frac{1}{x}\right)dx$

(vi) *Evaluate :*

$$\int_0^{\pi/4}\log(1+\tan\theta)d\theta$$

(vii) *By using the data* $\overline{x} = 25$, $\overline{y} = 30$, $b_{yx} = 1{\cdot}6$ *and* $b_{xy} = 0{\cdot}4$, *find :*

(a) *The regression equation y on x.*

(b) *What is the most likely value of y when* $x = 60$?

(c) *What is the coefficient of correlation between x and y ?*

(viii) *A problem is given to three students whose chances of solving it are* $\frac{1}{4}, \frac{1}{5}$ *and* $\frac{1}{3}$ *respectively. Find the probability that the problem is solved.*

(ix) *If* $a + ib = \frac{x+iy}{x-iy}$ *prove that* $a^2 + b^2 = 1$ *and* $\frac{b}{a} = \frac{2xy}{x^2-y^2}$.**

(x) *Solve :*

$$\frac{dy}{dx} = 1 - xy + y - x$$

Question 2.

(a) *Using properties of determinants, prove that :* **[5]**

$$\begin{vmatrix} a & b & b+c \\ c & a & c+a \\ b & c & a+b \end{vmatrix} = (a+b+c)(a-c)^2$$

(b) *Given that :*

$$A = \begin{bmatrix} 1 & -1 & 0 \\ 2 & 3 & 4 \\ 0 & 1 & 2 \end{bmatrix} \text{ and } B = \begin{bmatrix} 2 & 2 & -4 \\ -4 & 2 & -4 \\ 2 & -1 & 5 \end{bmatrix}, \text{ find } AB.$$ **[5]**

Using this result, solve the following system of equation :

$$x - y = 3,\ 2x + 3y + 4z = 17 \text{ and } y + 2z = 7.$$

Question 3.

(a) *Solve the equation for x :* **[5]**

$$\sin^{-1}x + \sin^{-1}(1-x) = \cos^{-1}x,\ x \neq 0$$

(b) *If A, B and C are the elements of Boolean algebra, simplify the expression* $(A' + B')(A + C') + B'(B + C)$. *Draw the simplified circuit.*** **[5]**

Question 4.

(a) *Verify Langrange's mean value theorem for the function :* **[5]**

$f(x) = x(1 - \log x)$ *and find the value of 'c' in the interval [1, 2].*

(b) *Find the coordinates of the centre, foci and equation of directrix of the hyperbola.*

$$x^2 - 3y^2 - 4x = 8.^{**}$$ **[5]**

Question 5.

(a) *If* $y = \cos(\sin x)$, *show that :* **[5]**

$$\frac{d^2y}{dx^2} + \tan x\frac{dy}{dx} + y\cos^2 x = 0$$

(b) *Show that the surface area of a closed cuboid with square base and given volume is minimum when it is a cube.* **[5]**

Question 6.

(a) *Evaluate :* **[5]**

$$\int \frac{\sin 2x}{(1+\sin x)(2+\sin x)}dx$$

(b) *Draw a rough sketch of the curve* $y^2 = 4x$ *and find the area of the region enclosed by the curve and the line* $y = x$. **[5]**

** Answer is not given due to change in the present syllabus.

Question 7.

(a) *Calculate the Spearman's rank correlation coefficient for the following data and interpret the result :*** **[5]**

x	35	54	80	95	73	73	35	91	83	81
y	40	60	75	90	70	75	38	95	75	70

(b) *Find the line of best fit for the following data, treating x as dependent variable (Regression equation x on y):* **[5]**

x	14	12	13	14	16	10	13	12
y	14	23	17	24	18	25	23	24

Hence, estimate the value of x when y = 16.

Question 8.

(a) *In a class of 60 students, 30 opted for Mathematics, 32 opted for Biology and 24 opted for both Mathematics and Biology. If one of these students is selected at random, find the probability that :* **[5]**

(i) *The student opted for Mathematics or Biology.*

(ii) *The student has opted neither Mathematics nor Biology.*

(iii) *The student has opted Mathematics but not Biology.*

(b) *Bag A contains 1 white, 2 blue and 3 red balls. Bag B contains 3 white, 3 blue and 2 red balls. Bag C contains 2 white, 3 blue and 4 red balls. One bag is selected at random and then two balls are drawn from the selected bag. Find the probability that the balls drawn are white and red.* **[5]**

Question 9.

(a) *Prove that locus of z is circle and find its centre and radius if jj is purely imaginary.*** **[5]**

(b) *Solve :* **[5]**

$$(x^2 - yx^2)\, dy + (y^2 + xy^2)\, dx = 0$$

SECTION—B (20 Marks)

Question 10.

(a) *If $\vec{a}, \vec{b}, \vec{c}$ are three mutually perpendicular vectors of equal magnitude, prove that $\left(\vec{a}+\vec{b}+\vec{c}\right)$ is equally inclined with vectors $\vec{a}$, $\vec{b}$ and $\vec{c}$.* **[5]**

(b) *Find the value of λ for which the four points with position vectors $6\hat{i}-7\hat{j}, 16\hat{i}-19\hat{j}-4\hat{k}, \lambda\hat{j}-6\hat{k}$ and $2\hat{i}-5\hat{j}+10\hat{k}$ are coplanar.* **[5]**

** Answer is not given due to change in the present syllabus.

Question 11.

(a) *Show that the lines $\frac{x-4}{1} = \frac{y+3}{-4} = \frac{z+1}{7}$ and $\frac{x-1}{2} = \frac{y+1}{-3} = \frac{z+10}{8}$ intersect. Find the coordinates of their point of intersection.* **[5]**

(b) *Find the equation of the plane passing through the point (1, – 2, 1) and perpendicular to the line joining the points A(3, 2, 1) and B(1, 4, 2).* **[5]**

Question 12.

(a) *A fair die is rolled. If face 1 turns up, a ball is drawn from Bag A. If face 2 or 3 turns up a ball is drawn from Bag B. If face 4 or 5 or 6 turns up, a ball is drawn from Bag C. Bag A contains 3 red and 2 white balls, Bag B contains 3 red and 4 white balls and Bag C contains 4 red and 5 white balls. The die is rolled, a Bag is picked up and a ball is drawn. If the drawn ball is red, what is the probability that it is drawn from Bag B ?* **[5]**

(b) *An urn contains 25 balls of which 10 balls are red and the remaining green. A ball is drawn at random from the urn, the colour is noted and the ball is replaced. If 6 balls are drawn in this way, find the probability that :* **[5]**

(i) *All the balls are red.*

(ii) *Not more than 2 balls are green.*

(iii) *Number of red balls and green balls are equal.*

SECTION—C (20 Marks)

Question 13.

(a) *A machine costs ₹ 60,000 and its effective life is estimated to be 25 years. A sinking fund is to be created for replacing the machine at the end of its life time when its scrap value is estimated as ₹ 5,000. The price of the new machine is estimated to be 100% more than the price of the present one. Find the amount that should be set aside at the end of each year, out of the profits, for the sinking fund if it accumulates at an interest of 6% per annum compounded annually.*** **[5]**

(b) *A farmer has a supply of chemical fertilizer of type A which contains 10% nitrogen and 6% phosphoric acid and of type B which contains 5% nitrogen and 10% phosphoric acid. After soil test, it is found that at least 7kg of nitrogen and same quantity of phosphoric acid is required for a good crop. The fertilizer of type A costs ₹ 5.00 per kg and the type B costs ₹ 8.00 per kg. Using Linear programming, find how many kilograms of each type of the fertilizer should be bought to meet the requirement and for the cost to be minimum. Find the feasible region in the graph.* **[5]**

Question 14.

(a) *The demand for a certain product is represented by the equation $p = 500 + 25x - \frac{x^2}{3}$ in rupees where x is the number of units and p is the price per unit. Find :* **[5]**

(i) Marginal revenue function.

(ii) The marginal revenue when 10 units are sold.

(b) *A bill of ₹ 60,000 payable 10 months after date was discounted for ₹ 57,300 on 30th June, 2007. If the rate of interest was $11\frac{1}{4}$ % per annum, on what date was the bill drawn ? *** **[5]**

Question 15.**

(a) *The price relatives and weights of a set of commodities are given below :* **[5]**

Commodity	*A*	*B*	*C*	*D*
Price relatives	*125*	*120*	*127*	*119*
Weights	*x*	*2x*	*y*	*y + 3*

If the sum of the weights is 40 and weighted average of price relatives index number is 122, find the numerical values of x and y.

(b) *Construct 3 yearly moving averages from the following data and show on a graph against the original data :* **[5]**

Year	Annual sales (in lakhs)
2000	18
2001	22
2002	20
2003	26
2004	30
2005	22
2006	24
2007	28
2008	32
2009	35

** Answer is not given due to change in the present syllabus.

SOLUTIONS

SECTION—A

Solution 1.

(i) Given, the matrix $\begin{pmatrix} 6 & -x^2 \\ 2x-15 & 10 \end{pmatrix}$ is symmetric.

So, $$\begin{pmatrix} 6 & -x^2 \\ 2x-15 & 10 \end{pmatrix}^T = \begin{pmatrix} 6 & -x^2 \\ 2x-15 & 10 \end{pmatrix}$$

$$\begin{pmatrix} 6 & 2x-15 \\ -x^2 & 10 \end{pmatrix} = \begin{pmatrix} 6 & -x^2 \\ 2x-15 & 10 \end{pmatrix}$$

On comparing both sides, we get

$$\Rightarrow \quad 2x - 15 = -x^2$$
$$\Rightarrow \quad x^2 + 2x - 15 = 0$$
$$\Rightarrow \quad x^2 + 5x - 3x - 15 = 0$$
$$\Rightarrow \quad x(x+5) - 3(x+5) = 0$$
$$(x+5)(x-3) = 0$$
$$x = -5 \text{ or } 3$$ **Ans.**

(iii) $$\text{L.H.S.} = \frac{1}{2}\cos^{-1}\left(\frac{1-x}{1+x}\right)$$

Putting $x = \tan^2\theta$

$$\text{L.H.S.} = \frac{1}{2}\cos^{-1}\left(\frac{1-\tan^2\theta}{1+\tan^2\theta}\right)$$

$$= \frac{1}{2}\cos^{-1}\left(\frac{1-\tan^2\theta}{\sec^2\theta}\right)$$

$$= \frac{1}{2}\cos^{-1}(\cos^2\theta - \sin^2\theta)$$

$$= \frac{1}{2}\cos^{-1}(\cos 2\theta)$$

$$= \frac{2\theta}{2} = \theta$$

$$\because \quad \tan\theta = \sqrt{x}$$

$$\Rightarrow \quad \theta = \tan^{-1}\sqrt{x} = \text{R.H.S.}$$

Hence Proved.

(iv) $$\underset{x\to\pi/2}{\text{Lt}}\left(x\tan x - \frac{\pi}{2}\sec x\right) \qquad [\infty = \infty]$$

$$\Rightarrow \underset{x\to\pi/2}{\text{Lt}}\left[\frac{x\sin x}{\cos x} - \frac{\pi}{2}\cdot\frac{1}{\cos x}\right]$$

$$\Rightarrow \underset{x\to\pi/2}{\text{Lt}}\left[\frac{x\sin x - \frac{\pi}{2}}{\cos x}\right] \qquad \left[\frac{0}{0}\text{ form}\right]$$

Differentiating both numerator and denominator w.r.t. x,

$$\Rightarrow \lim_{x\to\pi/2}\left[\frac{x\cos x + \sin x}{-\sin x}\right] = \frac{1}{-1} = -1$$ **Ans.**

(v) Let $I = \int \frac{1}{x^2} \sin^2\left(\frac{1}{x}\right) dx$

Put $\frac{1}{x} = t$

$-\frac{1}{x^2} dx = dt$

$\Rightarrow \quad I = -\int \sin^2 t \, dt$

$= -\int \frac{(1-\cos 2t)dt}{2}$

$= \frac{1}{2}\int \cos 2t dt - \frac{1}{2}\int dt$

$= \frac{1}{2}\left(\frac{\sin 2t}{2}\right) - \frac{t}{2} + c$

(where c is constant of integration)

$= \frac{1}{4}\sin\left(\frac{2}{x}\right) - \frac{1}{2x} + c$ **Ans.**

(vi) We have, $\int_0^{\pi/4} \log(1+\tan\theta)\, d\theta$

Let $I = \int_0^{\pi/4} \log(1+\tan\theta)\, d\theta$ ...(i)

Also, $I = \int_0^{\pi/4} \log\left(1+\tan\left(\frac{\pi}{4}-\theta\right)\right) d\theta$

$\left[\because \int_b^a f(x)\,dx = \int_b^a f(a-x)\,dx\right]$

$\Rightarrow \quad I = \int_0^{\pi/4} \log\left(1+\frac{1-\tan\theta}{1+\tan\theta}\right) d\theta$

$\Rightarrow \quad I = \int_0^{\pi/4} \log\left(\frac{1+\tan\theta+1-\tan\theta}{1+\tan\theta}\right) d\theta$

$\Rightarrow \quad I = \int_0^{\pi/4} \log\left(\frac{2}{1+\tan\theta}\right) d\theta$ ...(ii)

Adding (i) and (ii), we get

$2I = \int_0^{\pi/4} \log(1+\tan\theta)\, d\theta + \int_0^{\pi/4} \log\frac{2}{1+\tan\theta}\, d\theta$

$\Rightarrow \quad 2I = \int_0^{\pi/4} \log\left[\frac{2(1+\tan\theta)}{1+\tan\theta}\right] d\theta$

$\Rightarrow \quad 2I = \int_0^{\pi/4} \log 2\, d\theta$

$\Rightarrow \quad I = \frac{1}{2}\log 2[\theta]_0^{\pi/4}$

$\Rightarrow \quad I = \frac{1}{2}\log 2\left[\frac{\pi}{4}-0\right]$

$\therefore \quad I = \frac{\pi}{8}\log 2$ **Ans.**

(vii) Given, $\bar{x} = 25$, $\bar{y} = 30$, $b_{yx} = 1{\cdot}6$ and $b_{xy} = 0{\cdot}4$.

(a) The regression equation of y on x is

$y - \bar{y} = b_{yx}(x - \bar{x})$

$y - 30 = 1{\cdot}6\,(x - 25)$

$\Rightarrow \quad y = 1{\cdot}6x - 40 + 30$

$\Rightarrow \quad y = 1{\cdot}6x - 10$ **Ans.**

(b) When $x = 60$, $y = 1{\cdot}6\,(60) - 10$

$y = 96 - 10$

$\therefore \quad y = 86$ **Ans.**

(c) Since, b_{yx} and b_{xy} are both positive, so r should be + ve.

$r = \sqrt{b_{yx} \cdot b_{xy}}$

$\Rightarrow \quad r = \sqrt{1{\cdot}6 \times 0{\cdot}4}$

$\therefore \quad r = 0{\cdot}8$

Hence, the coefficient of correlation between x and y is 0·8. **Ans.**

(viii) Let A, B, C be the events that the problem is solved by three students. Then,

$P(A) = \frac{1}{4}$, $P(B) = \frac{1}{5}$, $P(C) = \frac{1}{3}$

$P(\bar{A}) = 1 - (A)$,

$P(\bar{B}) = 1 - P(B)$,

$P(\bar{C}) = 1 - P(C)$

$P(\bar{A}) = \frac{3}{4}$, $P(\bar{B}) = \frac{4}{5}$, $P(\bar{C}) = \frac{2}{3}$

Now, A, B and C are independent events.

$\therefore$ Required probability that the problem is solved

$= 1 - P(\bar{A}) \cdot P(\bar{B}) \cdot P(\bar{C})$

$= 1 - \frac{3}{4} \cdot \frac{4}{5} \cdot \frac{2}{3}$

$= 1 - \frac{2}{5} = \frac{3}{5}$ **Ans.**

(x) We have, $\frac{dy}{dx} = 1 - xy + y - x$

$\frac{dy}{dx} = 1 + y - x - xy$

$$\frac{dy}{dx} = (1+y) - x(1+y)$$

$$\frac{dy}{dx} = (1+y)(1-x)$$

$$\frac{dy}{1+y} = (1-x)\,dx$$

Integrating both sides, we get

$$\int \frac{1}{(1+y)}\,dy = \int (1-x)\,dx$$

$$\log(1+y) = x - \frac{x^2}{2} + c$$

(where c is constant of integration) **Ans.**

Solution 2.

(a) Let $\Delta = \begin{vmatrix} a & b & b+c \\ c & a & c+a \\ b & c & a+b \end{vmatrix}$ = L.H.S.

Applying $R_1 \to R_1 + R_2 + R_3$

$$\Delta = \begin{bmatrix} a+b+c & a+b+c & 2(a+b+c) \\ c & a & c+a \\ b & c & a+b \end{bmatrix}$$

$$\Delta = (a+b+c)\begin{bmatrix} 1 & 1 & 2 \\ c & a & c+a \\ b & c & a+b \end{bmatrix}$$

(Taking out $(a+b+c)$ as common)

Applying $C_1 \to C_1 - \frac{1}{2}C_3$ and $C_2 \to C_2 - \frac{1}{2}C_3$

$$\Delta = (a+b+c)\begin{bmatrix} 0 & 0 & 2 \\ (c-a)/2 & (a-c)/2 & c+a \\ (b-a)/2 & c-(a+b)/2 & a+b \end{bmatrix}$$

Now, expanding along R_1

$$\Delta = (a+b+c)\cdot 2\left[\frac{(c-a)}{2}\cdot\frac{(2c-a-b)}{2} - \frac{(a-c)}{2}\frac{(b-a)}{2}\right]$$

$$= (a+b+c)\cdot\frac{1}{2}[2c^2 - ac - bc - 2ac$$

$$+ a^2 + ab - ab + a^2 + bc - ac]$$

$$= (a+b+c)\,\frac{1}{2}(2a^2 + 2c^2 - 4ac)$$

$$= (a+b+c)(a-c)^2 = \text{R.H.S.}$$ **Hence Proved.**

(b) We have,

$$A = \begin{bmatrix} 1 & -1 & 0 \\ 2 & 3 & 4 \\ 0 & 1 & 2 \end{bmatrix} \text{ and } B = \begin{bmatrix} 2 & 2 & -4 \\ -4 & 2 & -4 \\ 2 & -1 & 5 \end{bmatrix}$$

Then,

$$A\cdot B = \begin{bmatrix} 1\times2+(-1)(-4)+0 & 1\times2+(-1)\times2+0 & 1\times(-4)+(-1)(-4)+0 \\ 2\times2+3\times(-4)+4\times2 & 2\times2+3\times2+4\times(-1) & 2\times(-4)+3\times(-4)+4\times5 \\ 0+1\times(-4)+2\times2 & 0+1\times2+2\times(-1) & 0+1\times(-4)+2\times5 \end{bmatrix}$$

$$\therefore \quad AB = \begin{bmatrix} 6 & 0 & 0 \\ 0 & 6 & 0 \\ 0 & 0 & 6 \end{bmatrix} = 6I_3$$

$$\Rightarrow \quad \frac{1}{6}AB = I$$

$$\Rightarrow \quad A^{-1} = \frac{1}{6}\cdot B \qquad \text{...(i)}$$

Now, writing the given system of equations in the matrix form AX = C as

$$\begin{bmatrix} 1 & -1 & 0 \\ 2 & 3 & 4 \\ 0 & 1 & 2 \end{bmatrix}\begin{bmatrix} x \\ y \\ z \end{bmatrix} = \begin{bmatrix} 3 \\ 17 \\ 7 \end{bmatrix}$$

where, $A = \begin{bmatrix} 1 & -1 & 0 \\ 2 & 3 & 4 \\ 0 & 1 & 2 \end{bmatrix}, X = \begin{bmatrix} x \\ y \\ z \end{bmatrix}$

and $C = \begin{bmatrix} 3 \\ 17 \\ 7 \end{bmatrix}$

$$\Rightarrow \quad AX = C$$

$$\Rightarrow \quad X = A^{-1}C$$

$$\Rightarrow \quad X = \frac{1}{6}BC \qquad \text{[from eq. (i)]}$$

$$= \frac{1}{6}\begin{bmatrix} 2 & 2 & -4 \\ -4 & 2 & -4 \\ 2 & -1 & 5 \end{bmatrix}\begin{bmatrix} 3 \\ 17 \\ 7 \end{bmatrix}$$

$$= \frac{1}{6}\begin{bmatrix} 2\times3+2\times17+(-4)\times7 \\ (-4)\times3+2\times17+(-4)\times7 \\ 2\times3+(-1)\times17+5\times7 \end{bmatrix}$$

$$\begin{bmatrix} x \\ y \\ z \end{bmatrix} = \frac{1}{6}\begin{bmatrix} 12 \\ -6 \\ 24 \end{bmatrix} = \begin{bmatrix} 2 \\ -1 \\ 4 \end{bmatrix}$$

Hence, $x = 2$, $y = -1$ and $z = 4$. **Ans.**

Solution 3.

(a) We have, $\sin^{-1}x + \sin^{-1}(1-x) = \cos^{-1}x,\ x \neq 0$

$$\sin^{-1}\left[x\sqrt{1-(1-x)^2} + (1-x)\sqrt{1-x^2}\right] = \cos^{-1}x$$

$$\Rightarrow \sin^{-1}\left[x\sqrt{1-1-x^2+2x}+(1-x)\sqrt{1-x^2}\right]$$
$$= \sin^{-1}\left[\sqrt{1-x^2}\right]$$

$$\Rightarrow x\sqrt{2x-x^2}+(1-x)\sqrt{1-x^2} = \sqrt{1-x^2}$$

$$x\sqrt{2x-x^2} = \sqrt{1-x^2}-(1-x)\sqrt{1-x^2}$$

$$x\sqrt{2x-x^2} = \sqrt{1-x^2}(1-1+x)$$

$$x\sqrt{2x-x^2} = \sqrt{1-x^2}\,(x)$$

$$\sqrt{2x-x^2} = \sqrt{1-x^2} \qquad [x \neq 0]$$

On squaring both sides, we get

$$2x - x^2 = 1 - x^2$$
$$2x = 1$$
$$x = \frac{1}{2}$$ **Ans.**

Solution 4.

(a) Given, $f(x) = x\ (1 - \log x)$ in [1, 2] ...(i)

The function is continuous in the closed interval [1, 2] as x and $\log x$ are continuous in [1, 2].

$$\text{Now, } f'(x) = 1 - \log x + x\left(-\frac{1}{x}\right)$$
$$= 1 - \log x - 1$$
$$f'(x) = -\log x \qquad \text{...(ii)}$$

which exists for all values in (1, 2).

So, $f(x)$ is differentiable in (1, 2).

By Lagrange's mean value theorem, we have

$$f'(c) = \frac{f(2)-f(1)}{2-1}$$

$$-\log c = \frac{2(1-\log 2)-1(1-\log 1)}{2-1}$$

[from eq. (i) and (ii)]

$$-\log c = \frac{2-2\log 2-1}{1} \qquad (\because \log 1 = 0)$$

$$-\log c = 1 - \log 2^2$$
$$\log c = \log 4 - \log e \qquad (\because \log e = 1)$$
$$\log c = \log\left(\frac{4}{e}\right)$$
$$c = \frac{4}{e} \text{ which lies in the interval } (1, 2)$$

Hence, the theorem is verified. **Hence Verified.**

Solution 5.

(a) Given, $y = \cos(\sin x)$ ...(i)

On differentiating both sides, we get

$$\frac{dy}{dx} = -\sin(\sin x)\cos x \qquad \text{...(ii)}$$

Again, differentiating both sides

$$\frac{d^2y}{dx^2} = -\cos(\sin x)\cos^2 x + \sin(\sin x)\sin x$$

$$\frac{d^2y}{dx^2} = -y\cos^2 x + \frac{\sin x}{(-\cos x)}\frac{dy}{dx}$$

[Using (i) and (ii)]

$$\frac{d^2y}{dx^2} + \tan x\frac{dy}{dx} + y\cos^2 x = 0$$ **Hence Proved.**

(b) Let h be the height and x be the side of the square base of the closed cuboid. Then,

$$\text{Surface area} = 2(x^2 + 2hx)$$
$$A = 2x^2 + 4hx \qquad \text{...(i)}$$

and
$$\text{Volume (V)} = x^2 h$$
$$\Rightarrow h = \frac{V}{x^2} \qquad \text{...(ii)}$$

Now,
$$A = 2x^2 + 4x\left(\frac{V}{x^2}\right)$$

[Putting h from eq. (ii) in eq. (i)]

$$A = 2x^2 + \frac{4V}{x}$$

On differentiating w.r.t. x,

$$\frac{dA}{dx} = 4x - \frac{4V}{x^2} \qquad \text{...(iii)}$$

For surface area (A) to be minimum, $\frac{dA}{dx} = 0$.

$$4x - \frac{4V}{x^2} = 0$$
$$4x = \frac{4V}{x^2}$$
$$V = x^3$$
$$x^2 h = x^3 \quad \text{[Putting V from eq. (ii)]}$$
$$h = x$$

Differentiating (iii) w.r.t. x, we get

$$\frac{d^2A}{dx^2} = 4 + \frac{8V}{x^3}$$
$$= 4 + \frac{8x^2h}{x^3}$$

[Putting V from eq. (ii)]

$$= 4 + \frac{8h}{x}$$

$$\left(\frac{d^2A}{dx^2}\right)_{h=x} = 4 + 8 = 12 > 0$$

Hence, the surface area is minimum when length = breadth = height = x.
i.e. when it is a cube. **Hence Proved.**

Solution 6.

(a) Let, $$I = \int \frac{\sin 2x}{(1+\sin x)(2+\sin x)}dx$$

$$I = \int \frac{2\cdot\sin x\cos x}{(1+\sin x)(2+\sin x)}dx$$

[Put $\sin x = t$, $\cos x\, dx = dt$]

$$I = \int \frac{2t\, dt}{(1+t)(2+t)}$$

Let, $$\int \frac{2t}{(1+t)(2+t)} = \int \frac{A}{(1+t)} + \frac{B}{(2+t)}$$

$$2t = A(2+t) + B(1+t)$$

putting $t = -1$ and $t = -2$, we get

$$-2 = A(2-1)$$
$$-2 = A$$
and $$-4 = B(1-2)$$
$$-4 = -B$$
$$\Rightarrow \quad A = -2 \text{ and } B = 4$$

Now, $$I = -2\int\frac{dt}{1+t} + 4\int\frac{dt}{2+t}$$

$$\Rightarrow \quad I = -2\log|1+t| + 4\log|2+t| + c$$
$$\Rightarrow \quad I = -2\log|1+\sin x| + 4\log|2+\sin x| + c$$
$$\Rightarrow \quad I = \log|1+\sin x|^{-2} + \log|2+\sin x|^4 + c$$
$$\Rightarrow \quad I = \log\left|\frac{(2+\sin x)^4}{(1+\sin x)^2}\right| + c$$

Ans.

(b) Given equation of the curves are

$$y^2 = 4x \quad \ldots(i)$$
and $$y = x \quad \ldots(ii)$$

Solving (i) and (ii), we get $x = 0, 4$

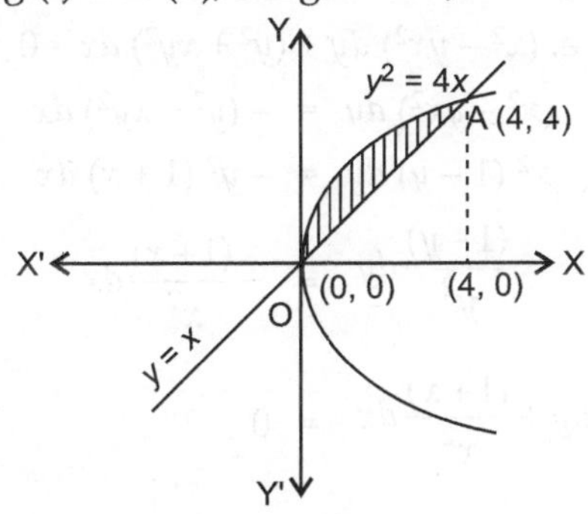

Equation (i) gives $y = 2\sqrt{x}$

Point of intersection of (i) and (ii) are (0, 0) and (4, 4).

$\therefore$ Required area $= \int_0^4 2\sqrt{x}dx - \int_0^4 x\, dx$

$$= 2\frac{[x^{3/2}]_0^4}{3/2} - \frac{[x^2]_0^4}{2}$$
$$= \frac{4}{3}(4^{3/2} - 0^{3/2}) - \frac{1}{2}(4^2 - 0^2)$$
$$= \frac{4}{3}\times 8 - \frac{1}{2}\times 16 = \frac{32}{3} - 8$$
$$= \frac{32-24}{3} = \frac{8}{3} \text{ sq. units.}$$ **Ans.**

Solution 7.

(b) Let us construct the table :

x	y	xy	y^2
14	14	196	196
12	23	276	529
13	17	221	289
14	24	336	576
16	18	288	324
10	25	250	625
13	23	299	529
12	24	288	576
$\Sigma x = 104$	$\Sigma y = 168$	$\Sigma xy = 2154$	$\Sigma y^2 = 3644$

$$\bar{x} = \frac{\Sigma x}{n} = \frac{104}{8} = 13$$

$$\bar{y} = \frac{\Sigma y}{n} = \frac{168}{8} = 21$$

Now, $$b_{xy} = \frac{\Sigma xy - \frac{\Sigma x \Sigma y}{n}}{\Sigma y^2 - \frac{1}{n}(\Sigma y)^2}$$

$$= \frac{2154 - \frac{104\times 168}{8}}{3644 - \frac{(168)^2}{8}}$$

$$b_{xy} = \frac{2154-2184}{3644-3528}$$

$$b_{xy} = \frac{-30}{116}$$

The regression equation of x on y is

$$x - \bar{x} = b_{xy}(y - \bar{y})$$
$$x - 13 = \frac{-30}{116}(y-21)$$
$$116x - 1508 = -30y + 630$$
$$116x + 30y = 1508 + 630$$

$$116x + 30y = 2138$$
$$58x + 15y = 1069$$

Value of x when $y = 16$

$$\Rightarrow \quad 58x + 15 \times 16 = 1069$$
$$58x = 1069 - 240$$
$$58x = 829$$
$$x = \frac{829}{58}$$
$$x = 14{\cdot}2931 \qquad \textbf{Ans.}$$

Solution 8.

(a) Let M : Students who opted Mathematics and B : Students who opted Biology

Given, $P(M) = \frac{30}{60} = \frac{1}{2}$

$$P(B) = \frac{32}{60} = \frac{8}{15}$$
$$P(M \cap B) = \frac{24}{60} = \frac{2}{5}$$

(i) Probability of student opted for Mathematics or Biology

$$= P(M \cup B)$$
$$= P(M) + P(B) - P(M \cap B)$$
$$= \frac{1}{2} + \frac{8}{15} - \frac{2}{5}$$
$$= \frac{15+16-12}{30} = \frac{19}{30} \qquad \textbf{Ans.}$$

(ii) Probability of student opted for neither Mathematics nor Biology $= P(M' \cap B')$

$$= 1 - P(M \cup B)$$
$$= 1 - \frac{19}{30}$$
$$= \frac{11}{30} \qquad \textbf{Ans.}$$

(iii) Probability of student opted for Mathematics but not Biology $= P(M \cap B')$

$$= P(M) - P(M \cap B)$$
$$= \frac{1}{2} - \frac{2}{5}$$
$$= \frac{5-4}{10} = \frac{1}{10} \qquad \textbf{Ans.}$$

(b) Let E_1, E_2, E_3 and A be the events defined as follows :

E_1 = bag A is chosen

E_2 = bag B is chosen

E_3 = bag C is chosen

and A = ball drawn are white and red.

Since, there are three bags and one of the three bags is chosen at random.

Therefore, $P(E_1) = P(E_2) = P(E_3) = \frac{1}{3}$

Now, $P(A/E_1)$ = Probability of drawing a white and red ball when first bag A has been chosen

$$= \frac{{}^1C_1 \times {}^3C_1}{{}^6C_2} = \frac{1 \times 3 \times 2}{6 \times 5} = \frac{1}{5}$$

Similarly, $P(A/E_2) = \frac{{}^3C_1 \times {}^2C_1}{{}^8C_2}$

$$= \frac{3 \times 2 \times 2}{8 \times 7} = \frac{3}{14}$$

and $P(A/E_3) = \frac{{}^2C_1 \times {}^4C_1}{{}^9C_2}$

$$= \frac{2 \times 4 \times 2}{9 \times 8} = \frac{2}{9}$$

By the law of total probability, we get

$$P(A) = \frac{1}{3} \times \frac{1}{5} + \frac{1}{3} \times \frac{3}{14} + \frac{1}{3} \times \frac{2}{9}$$
$$= \frac{1}{15} + \frac{1}{14} + \frac{2}{27}$$
$$= \frac{126+135+140}{1890}$$
$$= \frac{401}{1890} = 0{\cdot}212 \qquad \textbf{Ans.}$$

Solution 9.

(b) We have, $(x^2 - yx^2)\, dy + (y^2 + xy^2)\, dx = 0$

$$(x^2 - yx^2)\, dy = -(y^2 + xy^2)\, dx$$
$$x^2 (1 - y)\, dy = -y^2 (1 + x)\, dx$$
$$\frac{(1-y)}{y^2} dy = -\frac{(1+x)}{x^2} dx$$
$$\frac{(1-y)}{y^2} dy + \frac{(1+x)}{x^2} dx = 0$$

On integrating both sides, we get

$$\int \frac{(1-y)}{y^2} dy + \int \frac{(1+x)}{x^2} dx = c$$
$$\int \frac{1}{y^2} dy - \int \frac{1}{y} dy + \int \frac{1}{x^2} dx + \int \frac{1}{x} dx = c$$

$$-\frac{1}{y}-\log|y|-\frac{1}{x}+\log|x| = c$$

$$\log\left|\frac{x}{y}\right|-\frac{1}{x}-\frac{1}{y} = c$$

which is the required solution. **Ans.**

SECTION—B

Solution 10.

(a) Since, $\vec{a}, \vec{b}, \vec{c}$ are three mutually perpendicular vectors of equal magnitude, therefore.

$$\vec{a}\cdot\vec{b} = \vec{b}\cdot\vec{c} = \vec{c}\cdot\vec{a} = 0 \text{ and } |\vec{a}| = |\vec{b}| = |\vec{c}| = x$$

Let $\quad \vec{d} = \vec{a}+\vec{b}+\vec{c}$

$$\text{Then, } |\vec{d}|^2 = |\vec{a}+\vec{b}+\vec{c}|^2$$

$$= (\vec{a}+\vec{b}+\vec{c})(\vec{a}+\vec{b}+\vec{c})$$

$$|\vec{d}|^2 = \vec{a}.\vec{a}+\vec{a}.\vec{b}+\vec{a}.\vec{c}+\vec{b}.\vec{a}$$

$$+\vec{b}.\vec{b}+\vec{b}.\vec{c}+\vec{c}.\vec{a}+\vec{c}.\vec{b}+\vec{c}.\vec{c}$$

$$= x^2+0+0+0+x^2+0+0+0+x^2 = 3x^2$$

$$|\vec{d}| = \sqrt{3}x$$

Let, the angles between $\vec{d}$ and $\vec{a}, \vec{b}, \vec{c}$ be α, β, γ respectively. Then,

$$\cos\alpha = \frac{\vec{d}.\vec{a}}{|\vec{d}|.|\vec{a}|}$$

$$= \frac{(\vec{a}+\vec{b}+\vec{c})+\vec{a}}{\sqrt{3}x\cdot|\vec{a}|}$$

$$= \frac{\vec{a}.\vec{a}+\vec{b}.\vec{a}+\vec{c}.\vec{a}}{\sqrt{3}x|\vec{a}|}$$

$$= \frac{x^2+0+0}{\sqrt{3}x\cdot|\vec{a}|} = \frac{x^2}{\sqrt{3}x\cdot|\vec{a}|}$$

$$= \frac{x^2}{\sqrt{3}x\cdot x} = \frac{x^2}{\sqrt{3}x^2} = \frac{1}{\sqrt{3}}$$

Similarly, $\cos\beta = \frac{1}{\sqrt{3}}$ and $\cos\gamma = \frac{1}{\sqrt{3}}$

Thus, $\cos\alpha = \cos\beta = \cos\gamma$

i.e., $\quad \alpha = \beta = \gamma = \cos^{-1}\left(\frac{1}{\sqrt{3}}\right)$ **Hence Proved.**

(b) Let A, B, C, D be the given points. Then,

$\vec{AB}$ = Position vector of B – Position vector of A

$$= (16\hat{i}-19\hat{j}-4\hat{k})-(6\hat{i}-7\hat{j})$$

$$= 10\hat{i}-12\hat{j}-4\hat{k}$$

$\vec{AC}$ = Position vector of C – Position vector of A

$$= (\lambda\hat{j}-6\hat{k})-(6\hat{i}-7\hat{j})$$

$$= -6\hat{i}+(\lambda+7)\hat{j}-6\hat{k}$$

$\vec{AD}$ = Position vector of D – Position vector of A

$$= (2\hat{i}-5\hat{j}+10\hat{k})-(6\hat{i}-7\hat{j})$$

$$= -4\hat{i}+2\hat{j}+10\hat{k}$$

Given points are coplanar, if vectors $\vec{AB}, \vec{AC}, \vec{AD}$ are coplanar.

i.e. $\quad \left[\vec{AB}, \vec{AC}, \vec{AD}\right] = 0$

$$\text{Now, } \left[\vec{AB}, \vec{AC}, \vec{AD}\right] = \begin{vmatrix} 10 & -12 & -4 \\ -6 & \lambda+7 & -6 \\ -4 & 2 & 10 \end{vmatrix}$$

$$\Rightarrow \quad 10(10\lambda+70+12)+12(-60-24)$$

$$-4(-12+4\lambda+28) = 0$$

$$10(10\lambda+82)+12(-84)-4(16+4\lambda) = 0$$

$$100\lambda+820-1008-64-16\lambda = 0$$

$$\Rightarrow \quad 84\lambda = 252$$

$$\Rightarrow \quad \lambda = \frac{252}{84}$$

$$= 3$$ **Ans.**

Solution 11.

(a) Let $\quad \frac{x-4}{1} = \frac{y+3}{-4} = \frac{z+1}{7} = r$

and $\quad \frac{x-1}{2} = \frac{y+1}{-3} = \frac{z+10}{8} = r_1$

Direction ratios of these lines are 1, – 4, 7 and 2, – 3, 8.

Since, $1 : -4 : 7 \neq 2 : -3 : 8$

∴ Lines are not parallel.

Now, $(r+4, -4r-3, 7r-1)$ and $(2r_1+1, -3r_1-1, 8r_1-10)$ are the general coordinates of points on the two lines respectively.

The lines will intersect if the points coincide.

i.e., if $r + 4 = 2r_1 + 1$,

$$-4r - 3 = -3r_1 - 1,\ 7r - 1 = 8r_1 - 10$$

i.e., $r - 2r_1 + 3 = 0,\ -4r + 3r_1 - 2 = 0,\ 7r - 8r_1 + 9 = 0$

Solving the equations, we get

$$r_1 = 2 \text{ and } r = 1$$

which also satisfies the third equation.

Hence, the given lines intersect and their point of intersection is

$(1 + 4, -4(1) - 3, 7(1) - 1) = (5, -7, 6)$ **Ans.**

(b) Equation of line AB is

$$\frac{x-3}{1-3} = \frac{y-2}{4-2} = \frac{z-1}{2-1}$$

$$\frac{x-3}{-2} = \frac{y-2}{2} = \frac{z-1}{1} = r \text{ (say)}$$

Direction ratios of the line are $-2, 2, 1$. Since the line is perpendicular to the required plane. Therefore, direction ratios of a normal to the plane are proportional to $-2, 2, 1$. Hence the equation of the plane passing through the point $(1, -2, 1)$ and having the line as normal is,

$$-2(x-1) + 2(y+2) + 1(z-1) = 0$$

$$\Rightarrow \quad -2x + 2 + 2y + 4 + z - 1 = 0$$

$$\Rightarrow \quad -2x + 2y + z + 5 = 0$$

$$\text{or} \quad 2x - 2y - z - 5 = 0$$

is the required equation of plane. **Ans.**

Solution 12.

(a) Let E_1, E_2, E_3 and A be the events defined as follows :

E_1 = Bag A is chosen

E_2 = Bag B is chosen

E_3 = Bag C is chosen

and A = Red ball is drawn

Then, $P(E_1) = \frac{1}{6}$

$P(E_2) = \frac{2}{6}$

$P(E_3) = \frac{3}{6}$

and $P(A/E_1) = \frac{3}{5}$

$P(A/E_2) = \frac{3}{7}$

and $P(A/E_3) = \frac{4}{9}$

By Bayes' theorem. $P(E_2/A)$

$$= \frac{P(E_2)P(A/E_2)}{P(E_1)P(A/E_1)+P(A/E_2)+P(E_3)P(A/E_3)}$$

$$= \frac{\frac{2}{6}\times\frac{3}{7}}{\frac{1}{6}\times\frac{3}{5}+\frac{2}{6}\times\frac{3}{7}+\frac{3}{6}\times\frac{4}{9}}$$

$$= \frac{1/7}{\frac{1}{10}+\frac{1}{7}+\frac{2}{9}} = \frac{1/7}{\frac{63+90+140}{630}}$$

$$= \frac{1/7}{\frac{293}{630}} = \frac{630}{7\times 293} = \frac{90}{293}$$ **Ans.**

(b) Given, probability of drawing a red ball $= \frac{10}{25} = \frac{2}{5}$ and the probability of drawing a green ball $= \frac{15}{25} = \frac{3}{5}\ \left(\text{or } 1-\frac{2}{5}=\frac{3}{5}\right)$

$P(X = r)$ = Probability of drawing r balls of red colour

$$= {}^6C_r\left(\frac{2}{5}\right)^r\left(\frac{3}{5}\right)^{6-r},\ r = 0, 1, 2, \ldots\ldots 6$$

(i) Probability of drawing all red balls

$$= {}^6C_6\left(\frac{2}{5}\right)^6\left(\frac{3}{5}\right)^0 = \left(\frac{2}{5}\right)^6 = \frac{64}{15625}$$ **Ans.**

(ii) Probability of drawing not more than 2 green balls.

i.e. Probability of drawing atleast 4 red balls.

$$= P(X \geq 4)$$

$$= P(X = 4) + P(X = 5) + P(X = 6)$$

$$= {}^6C_4\left(\frac{2}{5}\right)^4\left(\frac{3}{5}\right)^2 + {}^6C_5\left(\frac{2}{5}\right)^5\left(\frac{3}{5}\right)^1 + {}^6C_6\left(\frac{2}{5}\right)^6\left(\frac{3}{5}\right)^0$$

$$= \left(\frac{2}{5}\right)^4\left(\frac{6\times5}{2\times1}\times\frac{3\times3}{5\times5}+\frac{6\times3\times2}{5\times5}+\frac{2\times2}{5\times5}\right)$$

$$= \left(\frac{2}{5}\right)^4\left(\frac{175}{25}\right)$$

$$= \frac{7\times2\times2\times2\times2}{5\times5\times5\times5} = \frac{112}{625}$$ **Ans.**

(iii) Probability of drawing equal no. of red and green balls = $P(X = 3)$

$$= {}^6C_3\left(\frac{2}{5}\right)^3\left(\frac{3}{5}\right)^{6-3} = {}^6C_3\left(\frac{2}{5}\right)^3\left(\frac{3}{5}\right)^3$$

$$= \frac{6\times5\times4}{3\times2\times1}\times\frac{2\times2\times2}{5\times5\times5}\times\frac{3\times3\times3}{5\times5\times5}$$

$$= \frac{4320}{15625} = \frac{864}{3125}$$ **Ans.**

SECTION—C

Solution 13.

(b) Let x kg be the quantity of fertilizer of type A and y kg be the quantity of fertilizer of type B.

Now, x kg of fertilizer of type A contains 10% nitrogen and y kg of fertilizer of type B contains 5% nitrogen.

$$\frac{10}{100}x+\frac{5}{100}y \geq 7$$

Also, x kg of fertilizer of type A and y kg of fertilizer of type B contains 6% and 10% of phosphorus respectively.

$$\frac{6}{100}x+\frac{10}{100}y \geq 7$$

The cost of fertilizer of x kg of type A and y kg of type B will be ₹ $(5x + 8y)$

Hence, the given L.P.P. is

Minimize, $Z = 5x + 8y$

$$\frac{10}{100}x+\frac{5}{100}y \geq 7$$

or $$2x + y \geq 140$$

$$\frac{6}{100}x+\frac{10}{100}y \geq 7$$

or $$3x + 5y \geq 350$$

$$x \geq 0,\ y \geq 0$$

To solve this graphically draw the graphs of the equations.

$2x + y = 140$

x	0	70
y	140	0

and $3x + 5y = 350$

x	0	115	350/3
y	70	1	0

Both the lines

i.e., $2x + y = 140$ and $3x + 5y = 350$ intersect at point (50, 40)

Clearly the coordinates of the vertices of the feasible region are A (0, 140), B (50, 40), C (350/3, 0)

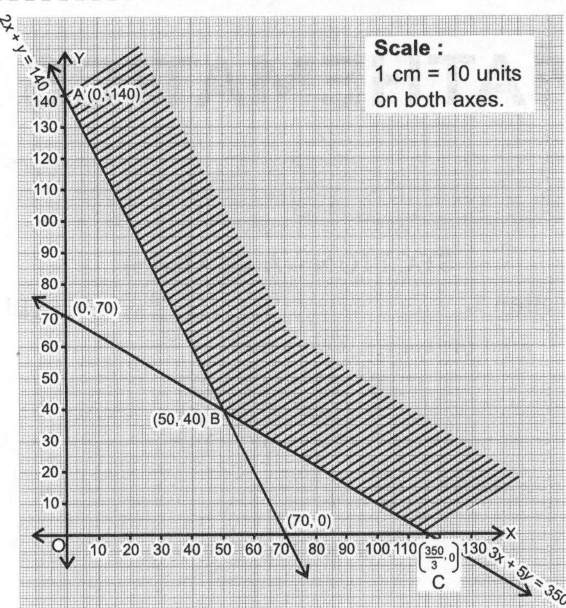

The values of the objective function at these corner points are given in the following table :

Corner Point	Objective Function $Z = 5x + 8y$
A (0, 140)	$Z = 0 + 8 \times 140 = 1120$
B (50, 40)	$Z = 5 \times 50 + 8 \times 40 = 570$ minimum
C (350/3, 0)	$Z = \frac{5\times350}{3}+0 = \frac{1750}{3} = 583{\cdot}33$

From the above, we find that Z is minimum at (50, 40) *i.e.*, when $x = 50$ and $y = 40$.

Hence, the farmer should use 50 kg of fertilizer of type A and 40 kg of fertilizer of type B to meet the requirement and the minimum cost is ₹ 570. **Ans.**

Solution 14.

(a) We have, $p = 500 + 25\,x - \frac{x^2}{3}$

(i) Now, total Revenue function (R) = px

$$R = 500\,x + 25x^2 - \frac{x^3}{3}$$

$\therefore$ Marginal Revenue function (MR) = $\frac{dR}{3}$

$$= 500 + 50\,x - x^2$$ **Ans.**

(ii) Marginal Revenue when 10 units are sold *i.e.*, when $x = 10$,

$$MR = 500 + 50\,(10) - (10)^2$$
$$= 500 + 500 - 100$$
$$= 1000 - 100$$
$$(MR)_{x=10} = 900$$ **Ans.**

●●

MATHEMATICS 2016

QUESTIONS

SECTION—A (80 Marks)

Question 1. [10 ×3]

(i) *Find the matrix X for which :*

$$\begin{bmatrix} 5 & 4 \\ 1 & 2 \end{bmatrix} X = \begin{bmatrix} 1 & -2 \\ 1 & 3 \end{bmatrix}$$

(ii) *Solve for x, if :*

$$\tan(\cos^{-1} x) = \frac{2}{\sqrt{5}}$$

(iii) *Prove that the line $2x - 3y = 9$ touches the conics $y^2 = -8x$. Also, find the point of contact.***

(iv) *Using L' Hospital's rule, evaluate :*

$$\lim_{x \to 0} \left(\frac{1}{x^2} - \frac{\cot x}{x} \right)$$

(v) *Evaluate :* $\int \tan^3 x \, dx$

(vi) *Using properties of definite integrals, evaluate :*

$$\int_0^{\pi/2} \frac{\sin x - \cos x}{1 + \sin x \cos x} dx$$

(vii) *The two lines of regressions are $x + 2y - 5 = 0$ and $2x + 3y - 8 = 0$ and the variance of x is 12. Find the variance of y and the coefficient of correlation.*

(viii) *Express $\frac{2+i}{(1+i)(1-2i)}$ in the form of $a + ib$. Find its modulus and argument.***

(ix) *A pair of dice is thrown. What is the probability of getting an even number on the first die or a total of 8 ?*

(x) *Solve the differential equation :*

$$x \frac{dy}{dx} + y = 3x^2 - 2$$

Question 2.

(a) *Using properties of determinants, prove that :* [5]

$$\begin{vmatrix} b+c & a & a \\ b & a+c & b \\ c & c & a+b \end{vmatrix} = 4abc$$

(b) *Solve the following system of linear equations using matrix method :* [5]

$$3x + y + z = 1,\ 2x + 2z = 0,\ 5x + y + 2z = 2$$

Question 3.

(a) *If $\sin^{-1} x + \tan^{-1} x = \frac{\pi}{2}$, prove that :* [5]

$$2x^2 + 1 = \sqrt{5}$$

(b) *Write the Boolean function corresponding to the switching circuit given below :* [5]

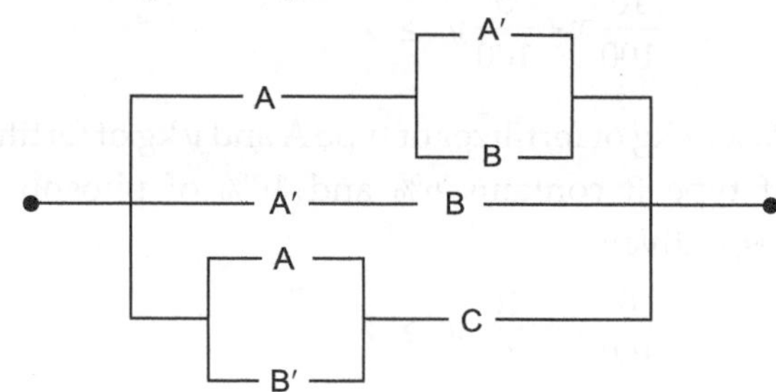

*A, B and C represent switches in 'on' position and A', B' and C' represent them in 'off' position. Using Boolean algebra, simplify the function and construct an equivalent switching circuit.***

Question 4.

(a) *Verify the conditions of Rolle's theorem for the following function :* [5]

$$f(x) = \log(x^2 + 2) - \log 3 \text{ on } [-1, 1]$$

Find a point in the interval, where the tangent to the curve is parallel to X-axis.

(b) *Find the equation of the standard ellipse, taking its axes as the coordinate axes, whose minor axis is equal to the distance between the foci and whose length of latus rectum is 10. Also, find its eccentricity.*** [5]

Question 5.

(a) *If $\log y = \tan^{-1} x$, prove that :* [5]

$$(1 + x^2) \frac{d^2y}{dx^2} + (2x - 1) \frac{dy}{dx} = 0$$

(b) *A rectangle is inscribed in a semicircle of radius r with one of its sides on the diameter of the semicircle. Find the dimensions of the rectangle to get maximum area. Also, find the maximum area.* [5]

Question 6.

(a) *Evaluate :* [5]

$$\int \frac{\sin x + \cos x}{\sqrt{9 + 16 \sin 2x}} dx$$

(b) *Find the area of the region bound by the curves $y = 6x - x^2$ and $y = x^2 - 2x$.* [5]

** Answer is not given due to change in the present syllabus.

Question 7.

(a) *Calculate Karl Pearson's coefficient of correlation between x and y for the following data and interpret the result :*** **[5]**

(1, 6), (2, 5), (3, 7), (4, 9), (5, 8), (6, 10), (7, 11), (8, 13), (9, 12).

(b) *The marks obtained by 10 candidates in English and Mathematics are given below :* **[5]**

Marks in English	Marks in Mathematics
20	17
13	12
18	23
21	25
11	14
12	8
17	19
14	21
19	22
15	19

Estimate the probable score for Mathematics if the marks obtained in English are 24.

Question 8.

(a) *A committee of 4 persons has to be chosen from 8 boys and 6 girls, consisting of at least one girl. Find the probability that the committee consists of more girls than boys.* **[5]**

(b) *An urn contains 10 white and 3 black balls while another urn contains 3 white and 5 black balls. Two balls are drawn from the first urn and put into the second urn and then a ball is drawn from the second urn. Find the probability that the ball drawn from the second urn is a white ball.* **[5]**

Question 9.

(a) *Find the locus of a complex number, $z = x + iy$, satisfying the relation $\left|\frac{z-3i}{z+3i}\right| \le \sqrt{2}$. Illustrate the locus of z in the Argand plane.*** **[5]**

(b) *Solve the following differential equation :* **[5]**

$x^2\,dy + (xy + y^2)\,dx = 0$, *when x = 1 and y = 1*

SECTION—B

Question 10.

(a) *For any three vectors $\vec{a}, \vec{b}, \vec{c}$, show that $\vec{a}-\vec{b}, \vec{b}-\vec{c}, \vec{c}-\vec{a}$ are coplanar.* **[5]**

(b) *Find a unit vector perpendicular to each of the vectors $\vec{a}+\vec{b}$ and $\vec{a}-\vec{b}$ where $\vec{a} = 3\hat{i}+2\hat{j}+2\hat{k}$ and $\vec{b} = \hat{i}+2\hat{j}-2\hat{k}$* **[5]**

** Answer is not given due to change in the present syllabus.

Question 11.

(a) *Find the image of the point (2, – 1, 5) in the line $\frac{x-11}{10} = \frac{y+2}{-4} = \frac{z+8}{-11}$.* **[5]**

Also, find the length of the perpendicular from the point (2, – 1, 5) to the line.

(b) *Find the Cartesian equation of the plane, passing through the line of intersection of the planes : $\vec{r}.\left(2\hat{i}+3\hat{j}-4\hat{k}\right)+5 = 0$ and $\vec{r}.\left(\hat{i}-5\hat{j}+7\hat{k}\right)+2 = 0$ and intersecting Y-axis at (0, 3).* **[5]**

Question 12.

(a) *In an automobile factory, certain parts are to be fixed into the chassis in a section before it moves into another section. On a given day, one of the three persons A, B and C carries out this task. A has 45% chance, B has 35% chance and C has 20% chance of doing the task. The probability that A, B and C will take more than the allotted time is $\frac{1}{6}, \frac{1}{10}$ and $\frac{1}{20}$ respectively. If it is found that the time taken is more than the allotted time, what is the probability that A has done the task ?* **[5]**

(b) *The difference between mean and variance of a binomial distribution is 1 and the difference of their squares is 11. Find the distribution.*** **[5]**

SECTION—C

Question 13.

(a) *A man borrows ₹ 20,000 at 12% per annum, compounded semi-annually and agrees to pay it in 10 equal semi-annual installments. Find the value of each installment, if the first payment is due at the end of two years.*** **[5]**

(b) *A company manufactures two types of products A and B. Each unit of A requires 3 grams of nickel and 1 gram of chromium, while each unit of B requires 1 gram of nickel and 2 grams of chromium. The firm can produce 9 grams of nickel and 8 grams of chromium. The profit is ₹ 40 on each unit of product of type A and ₹ 50 on each unit of type B. How many units of each type should the company manufacture so as to earn maximum profit ? Use linear programming to find the solution.* **[5]**

Question 14.

(a) *The demand function is $x = \frac{24-2p}{3}$ where x is the number of units demanded and p is the price per unit. Find :* **[5]**

(i) The revenue function R in terms of p.

(ii) The price and the number of units demanded for which the revenue is maximum.

(b) *A bill of ₹ 1,800 drawn on 10th September, 2010 at 6 months was discounted for ₹ 1,782 at a bank. If the rate of interest was 5% per annum, on what date was the bill discounted ?*** **[5]**

Question 15.**

(a) *The index number by the method of aggregates for the year 2010, taking 2000 as the base year, was found to be 116. If sum of the prices in the year 2000 is ₹ 300, find the values of x and y in the data given below :* **[5]**

Commodity	A	B	C	D	E	F
Price in the year 2000 (₹)	50	x	30	70	116	20
Price in the year 2010 (₹)	60	24	y	80	120	28

** Answer is not given due to change in the present syllabus.

(b) *From the details given below, calculate the five yearly moving averages of the number of students who have studied in a school. Also, plot these and original data on the same graph paper.* **[5]**

Year	Number of Students
1993	332
1994	317
1995	357
1996	392
1997	402
1998	405
1999	410
2000	427
2001	405
2002	438

SOLUTIONS

SECTION—A

Solution 1.

(i) Let $X = \begin{bmatrix} a & b \\ c & d \end{bmatrix}$

Then, $\begin{bmatrix} 5 & 4 \\ 1 & 1 \end{bmatrix}\begin{bmatrix} a & b \\ c & d \end{bmatrix} = \begin{bmatrix} 1 & -2 \\ 1 & 3 \end{bmatrix}$

$\Rightarrow \begin{bmatrix} 5a+4c & 5b+4d \\ a+c & b+d \end{bmatrix} = \begin{bmatrix} 1 & -2 \\ 1 & 3 \end{bmatrix}$

$\Rightarrow 5a + 4c = 1, 5b + 4d = -2, a + c = 1, b + d = 3.$

Solving these equations, we get

$a = -3, c = 4,$

$b = -14, d = 17$

Hence, $X = \begin{bmatrix} -3 & -14 \\ 4 & 17 \end{bmatrix}$ **Ans.**

(ii) Given, $\tan(\cos^{-1} x) = \dfrac{2}{\sqrt{5}}$

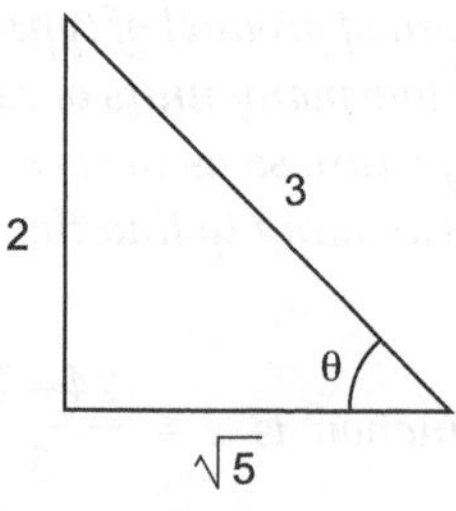

Let $\cos^{-1} x = \theta$ ...(i)

$\Rightarrow x = \cos\theta$

$\Rightarrow \tan\theta = \dfrac{2}{\sqrt{5}}$ [from (i)]

which gives $\cos\theta = \dfrac{\sqrt{5}}{3}$

Hence, $x = \dfrac{\sqrt{5}}{3}$ **Ans.**

(iv) We have, $\lim\limits_{x\to 0}\left(\dfrac{1}{x^2} - \dfrac{\cot x}{x}\right)$

$= \lim\limits_{x\to 0}\left(\dfrac{1}{x^2} - \dfrac{\cos x}{x\sin x}\right)$ [$\infty - \infty$ Form]

$= \lim\limits_{x\to 0}\dfrac{\sin x - x\cos x}{x^2 \sin x}$

$= \lim\limits_{x\to 0}\dfrac{\sin x - x\cos x}{x^3}\cdot\lim\limits_{x\to 0}\dfrac{x}{\sin x}$

$= \lim\limits_{x\to 0}\dfrac{\sin x - x\cos x}{x^3}\cdot 1$ $\left[\because \lim\limits_{x\to 0}\dfrac{\sin x}{x} = 1\right]$

$= \lim\limits_{x\to 0}\dfrac{\cos x - (\cos x - x\sin x)}{3x^3}$

$\left[\dfrac{0}{0}\text{ form, applying L' Hospital's rule}\right]$

$= \lim\limits_{x\to 0}\dfrac{x\sin x}{3x^2} = \lim\limits_{x\to 0}\dfrac{\sin x}{3x}$

$= \dfrac{1}{3}\lim\limits_{x\to 0}\dfrac{\sin x}{x}$ $\left[\because \lim\limits_{x\to 0}\dfrac{\sin x}{x} = 1\right]$

$= \dfrac{1}{3}$ **Ans.**

(v) Here, $\int \tan^3 x\,dx = \int \tan x \cdot \tan^2 x\,dx$

$= \int \tan x(\sec^2 x - 1)\,dx$

$$\int \tan^3 x\, dx = \int \tan x \cdot \sec^2 x\, dx - \int \tan x\, dx \quad ...(i)$$

In the first integral, put $u = \tan x$ so that,

$$\frac{du}{dx} = \sec^2 x$$

$$\Rightarrow \quad du = \sec^2 x\, dx \quad ...(ii)$$

$$\Rightarrow \int \tan^3 x\, dx = \int u \cdot du - \log|\sec x| + c$$

[From (i) and (ii)]

$$= \frac{u^2}{2} - \log|\sec x| + c$$

$$\therefore \int \tan^3 x\, dx = \frac{\tan^2 x}{2} - \log|\sec x| + c$$

Ans.

(vi) Let $I = \int_0^{\pi/2} \frac{\sin x - \cos x}{1 + \sin x \cos x} dx$

$$I = \int_0^{\pi/2} \frac{\sin x}{1 + \sin x \cos x} dx - \int_0^{\pi/2} \frac{\cos x}{1 + \sin x \cos x} dx$$

Here, $a + b - x = \frac{\pi}{2} - x$, so using the property

$$\int_b^a f(x)dx = \int_b^a f(a+b-x)dx$$

in first integral, we have

$$I = \int_0^{\pi/2} \frac{\sin\left(\frac{\pi}{2} - x\right)}{1 + \sin\left(\frac{\pi}{2} - x\right)\cos\left(\frac{\pi}{2} - x\right)} dx - \int_0^{\pi/2} \frac{\cos x}{1 + \sin x \cos x} dx$$

$$I = \int_0^{\pi/2} \frac{\cos x}{1 + \cos x \sin x} dx - \int_0^{\pi/2} \frac{\cos x}{1 + \sin x \cos x} dx$$

$$I = 0$$

Ans.

(vii) The given equation of the lines of regression are

$$x + 2y - 5 = 0 \quad ...(i)$$

and

$$2x + 3y - 8 = 0 \quad ...(ii)$$

Rewriting the equations (i) and (ii), we have

From equation (i),

$$y = \frac{-x}{2} + \frac{5}{2}$$

$$y = -0{\cdot}5\,x + 2{\cdot}5$$

(Regression line of y on x)

$$b_{yx} = r\frac{\sigma_y}{\sigma_x} = -0{\cdot}5 \quad ...(iii)$$

From equation (ii),

$$x = \frac{-3}{2}y + \frac{8}{2}$$

$$x = -1{\cdot}5\,y + 4$$

(Regression line of x on y)

$$b_{xy} = r\frac{\sigma_x}{\sigma_y} = -1{\cdot}5$$

$$\therefore \quad r^2 = b_{xy} \times b_{yx} = (-0{\cdot}5) \times (-1{\cdot}5) = 0{\cdot}75$$

$$\therefore \quad r = \sqrt{0{\cdot}75} = \pm 0{\cdot}866$$

But b_{xy} and b_{yx} being both – ve therefore, r is also – ve.

Correlation coefficient $(r) = -0{\cdot}866$.

Variance of x *i.e.,* $\sigma^2_x = 12$

$$\therefore \quad \sigma_x = \sqrt{12}$$

From equation (iii),

$$r\frac{\sigma_y}{\sigma_x} = -0{\cdot}5$$

$$-0{\cdot}866 \cdot \left(\frac{\sigma_y}{\sqrt{12}}\right) = -0{\cdot}5$$

$$\sigma_y = \frac{0{\cdot}5 \times \sqrt{12}}{0{\cdot}866} = 2$$

$\therefore$ Variance of y *i.e.* $\sigma_y^2 = 4$. **Ans.**

(ix) The total number of possible outcomes is 36.

i.e., $n(S) = 36$

Let A : getting a total of 8.

The favourable outcomes are : (2, 6), (4, 4), (6, 2), (3, 5), (5, 3)

$$\therefore \quad n(A) = 5$$

$$\therefore \quad P(A) = \frac{n(A)}{n(S)} = \frac{5}{36}$$

Ans.

B : getting an even number on the first die

The favourable outcomes are

(2, 1), (2, 2), (2, 3), (2, 4), (2, 5), (2, 6), (4, 1), (4, 2), (4, 3), (4, 4), (4, 5), (4, 6), (6, 1), (6, 2), (6, 3), (6, 4), (6, 5), (6, 6).

$$P(B) = \frac{n(B)}{n(S)} = \frac{18}{36} = \frac{1}{2}$$

Probability that both events will occur simultanously,

$$P(A \cap B) = \frac{n(A \cap B)}{n(S)} = \frac{3}{36} = \frac{1}{12}$$

$$P(A \cap B) = P(A) + P(B) - P(A \cup B)$$

$$= \frac{5}{36} + \frac{1}{2} - \frac{1}{12}$$

$$= \frac{5 + 18 - 3}{36} = \frac{20}{36} = \frac{5}{9}$$

Ans.

(x) Given, $x\frac{dy}{dx} + y = 3x^2 - 2$

Dividing by x on both sides, we get

$$\frac{dy}{dx} + \frac{1}{x} \cdot y = 3x - \frac{2}{x} \quad ...(i)$$

Here, $P = \frac{1}{x}$ and $Q = 3x - \frac{2}{x}$

$$\text{I. F.} = e^{\int P dx} = e^{\int \frac{1}{x} dx} = e^{\log x} = x$$

Multiplying equation (i) by I.F.

$$x\frac{dy}{dx} + \frac{y}{x}\cdot x = \left(3x - \frac{2}{x}\right)\cdot x$$

$$x\frac{dy}{dx} + y = 3x^2 - 2$$

On integrating, we get

$$xy = \int(3x^2 - 2)dx + c$$

$$xy = x^3 - 2x + c$$

which is the required solution. **Ans.**

Solution 2.

(a) Let $\Delta = \begin{vmatrix} b+c & a & a \\ b & a+c & b \\ c & c & a+b \end{vmatrix}$

Operating $R_1 \to R_1 + R_2 + R_3$, we get

$$\Delta = \begin{vmatrix} 2(b+c) & 2(a+c) & 2(a+b) \\ b & a+c & b \\ c & c & a+b \end{vmatrix}$$

Taking 2 as common from R_1, we get

$$\Delta = 2\begin{vmatrix} b+c & a+c & a+b \\ b & a+c & b \\ c & c & a+b \end{vmatrix}$$

Applying $R_2 \to R_2 - R_1$ and $R_3 \to R_3 - R_1$

$$\Delta = \begin{vmatrix} b+c & a+c & a+b \\ -c & 0 & -a \\ -b & -a & 0 \end{vmatrix}$$

Now, expanding along R_1, we get

$$\Delta = 2[(b+c)(-a^2) - (a+c)(0-ab) + (a+b)(ac)]$$
$$= 2[-a^2b - a^2c + a^2b + abc + a^2c + abc]$$
$$= 2\cdot 2abc = 4abc$$

Hence Proved.

(b) Writing the given equation in the matrix form AX = B as

$$\begin{bmatrix} 3 & 1 & 1 \\ 2 & 0 & 2 \\ 5 & 1 & 2 \end{bmatrix}\begin{bmatrix} x \\ y \\ z \end{bmatrix} = \begin{bmatrix} 1 \\ 0 \\ 2 \end{bmatrix}$$

where, $A = \begin{bmatrix} 3 & 1 & 1 \\ 2 & 0 & 2 \\ 5 & 1 & 2 \end{bmatrix}$, $X = \begin{bmatrix} x \\ y \\ z \end{bmatrix}$ and $B = \begin{bmatrix} 1 \\ 0 \\ 2 \end{bmatrix}$

$$\therefore \quad |A| = \begin{vmatrix} 3 & 1 & 1 \\ 2 & 0 & 2 \\ 5 & 1 & 2 \end{vmatrix}$$

$$= 3(0-2) - 1(4-10) + 1(2-0)$$
$$= 2 \neq 0$$

∴ A is non-singular.

∴ The system has the unique solution $X = A^{-1}B$.

$A_{11} = -2, A_{12} = 6, A_{13} = 2,$

$A_{21} = -1, A_{22} = 1, A_{23} = 2,$

$A_{31} = 2, A_{32} = -4, A_{33} = -2.$

$$\therefore \quad \text{adj } A = \begin{bmatrix} A_{11} & A_{21} & A_{31} \\ A_{12} & A_{22} & A_{32} \\ A_{13} & A_{23} & A_{33} \end{bmatrix} = \begin{bmatrix} -2 & -1 & 2 \\ 6 & 1 & -4 \\ 2 & 2 & -2 \end{bmatrix}$$

$$A^{-1} = \frac{\text{adj}A}{|A|} = \frac{1}{2}\begin{bmatrix} -2 & -1 & 2 \\ 6 & 1 & -4 \\ 2 & 2 & -2 \end{bmatrix}$$

$$= \begin{bmatrix} -1 & -1/2 & 1 \\ 3 & 1/2 & -2 \\ 1 & 1 & -1 \end{bmatrix}$$

Now, $AX = B \Rightarrow X = A^{-1}B$.

$$\begin{bmatrix} x \\ y \\ z \end{bmatrix} = \begin{bmatrix} -1 & -1/2 & 1 \\ 3 & 1/2 & -2 \\ 1 & 1 & -1 \end{bmatrix}\begin{bmatrix} 1 \\ 0 \\ 2 \end{bmatrix} = \begin{bmatrix} 1 \\ -1 \\ -1 \end{bmatrix}$$

$$\therefore \quad x = 1, y = -1, z = -1$$

Ans.

Solution 3.

(a) Given, $\sin^{-1} x + \tan^{-1} x = \frac{\pi}{2}$

$$\tan^{-1} x = \frac{\pi}{2} - \sin^{-1} x$$

$$\Rightarrow \quad x = \tan\left(\frac{\pi}{2} - \sin^{-1} x\right)$$

$$\Rightarrow \quad x = \cot(\sin^{-1} x)$$

$$\Rightarrow \quad x = \cot\left(\cot^{-1}\frac{\sqrt{1-x^2}}{x}\right)$$

$$\left[\because \sin^{-1} x = \cot^{-1}\frac{\sqrt{1-x^2}}{x}\right]$$

$$x = \frac{\sqrt{1-x^2}}{x}$$

$$x^2 = \sqrt{1-x^2}$$

$$x^4 = 1 - x^2$$

[On squaring both LHS & RHS]

$$\Rightarrow \quad x^4 + x^2 - 1 = 0$$

On solving, we get

$$x^2 = \frac{-1 \pm \sqrt{1-4(-1)}}{2}$$

$$\Rightarrow \quad x^2 = \frac{-1 \pm \sqrt{5}}{2}$$

$$x^2 \neq \frac{-1-\sqrt{5}}{2}$$

because x^2 cannot be negative

$$x^2 = \frac{-1+\sqrt{5}}{2}$$

$$\Rightarrow \quad 2x^2 + 1 = \sqrt{5}.$$ **Hence Proved.**

Solution 4.

(a) 1. Since $x^2 + 2 > 0$, $\forall\, x \in [-1, 1]$, therefore, the function $f(x) = \log(x^2+2) - \log(3)$ is continuous in $[-1, 1]$

2. $f'(x) = \frac{1}{x^2+2} \cdot 2x = \frac{2x}{x^2+2}$ which exists for all x. Thus, the function $f(x)$ is differentiable in the open interval $(-1, 1)$.

3. $f(-1) = \log(1+2) - \log 3 = 0$,
$f(1) = \log(1+2) - \log 3 = 0$
$\therefore \quad f(-1) = f(1) = 0$

Thus, $f(x)$ satisfies all the conditions of Rolle's theorem.

$\therefore$ There must exists atleast one value of x, say c, in the open interval $(-1, 1)$ such that $f'(c) = 0$.

Now, $$f'(c) = \frac{2c}{c^2+2} = 0$$

$$\Rightarrow \quad c = 0 \in (-1, 1)$$

Hence, Rolle's theorem is verified.

When $c = 0$,

$$f(0) = \log 2 - \log 3 = \log\left(\frac{2}{3}\right)$$

So, there exist a point $\left[0, \log\left(\frac{2}{3}\right)\right]$ on the given curve $f(x) = \log(x^2+2) - \log 3$ where the tangent is parallel to the X-axis. **Ans.**

Solution 5.

(a) Given, $$\log y = \tan^{-1} x$$

Differentiating both sides, w.r.t. x, we get

$$\frac{1}{y}\frac{dy}{dx} = \frac{1}{1+x^2}$$

$$(1+x^2)\frac{dy}{dx} - y = 0$$

Again, differentiating w.r.t. x, we get

$$2x\frac{dy}{dx} + (1+x^2)\frac{d^2y}{dx^2} - \frac{dy}{dx} = 0$$

$$(1+x^2)\frac{d^2y}{dx^2} + (2x-1)\frac{dy}{dx} = 0$$ **Hence Proved.**

(b) Let ABCD be the rectangle of sides x and $2y$ inscribed in the semicircle with centre O and radius r. Let OA = OB = y then AB = $2y$.

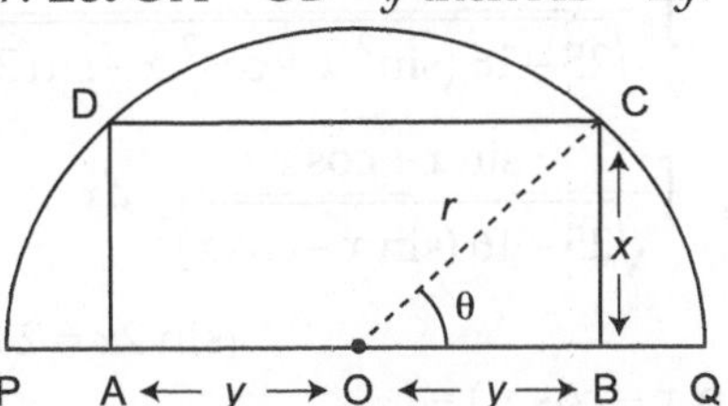

Let θ be the angle between radius of semicircle and side OB of rectangle, then $x = r\sin\theta$, $y = r\cos\theta$.

If A is the area of rectangle ABCD, then

$$A = 2y\,x = 2r\cos\theta . r\sin\theta$$

$$A = r^2 \sin 2\theta,$$

Differentiating w.r.t. θ, we get

$$\therefore \quad \frac{dA}{d\theta} = 2r^2\cos 2\theta \quad \ldots(i)$$

For max. or min. $\frac{dA}{d\theta} = 0 \Rightarrow 2r^2\cos 2\theta = 0$

$$\cos 2\theta = 0 = \cos\frac{\pi}{2}$$

$$\Rightarrow \quad \theta = \frac{\pi}{4}$$

Again, differentiating equation (i) w.r.t. θ, we get

$$\frac{d^2A}{d\theta^2} = -4r^2\sin 2\theta$$

$$\left(\frac{d^2A}{d\theta^2}\right)_{\theta=\frac{\pi}{4}} = -4r^2\sin\frac{\pi}{2} = -4r^2 < 0$$

Hence, area A is maximum when $\theta = \frac{\pi}{4}$

$\therefore$ For maximum area, $\theta = \frac{\pi}{4}$ and $x = r\sin\theta = r\sin\frac{\pi}{4} = \frac{r}{\sqrt{2}}$

$$2y = 2r\cos\theta = 2r\cos\frac{\pi}{4} = 2r \times \frac{1}{\sqrt{2}} = r\sqrt{2}$$

Hence, the dimensions of the rectangle are $\frac{r}{\sqrt{2}}$ and $r\sqrt{2}$.

and the area $= \frac{r}{\sqrt{2}} \times r\sqrt{2} = r^2$ units. **Ans.**

Solution 6.

(a) Let $I = \int \frac{\sin x + \cos x}{\sqrt{9 + 16\sin 2x}}\,dx$

$$= \int \frac{\sin x + \cos x}{\sqrt{9 + 16\sin 2x + 16 - 16}}\,dx$$

$$= \int \frac{\sin x + \cos x}{\sqrt{25 - 16\,(1 - \sin 2x)}}\,dx$$

$$= \int \frac{\sin x + \cos x}{\sqrt{25 - 16\,(\sin^2 x + \cos^2 x - \sin 2x)}}\,dx$$

$$\therefore\ I = \int \frac{\sin x + \cos x}{\sqrt{25 - 16\,(\sin x - \cos x)^2}}\,dx$$

$[\sin 2x = 2 \sin x \cos x]$

Let, $(\sin x - \cos x) = t$

$(\cos x + \sin x)\,dx = dt$

$$\Rightarrow I = \int \frac{dt}{\sqrt{(5)^2 - (4t)^2}} = \frac{1}{4}\sin^{-1}\left(\frac{4t}{5}\right) + c$$

$$\left[\because \int \frac{1}{\sqrt{a^2 - x^2}}\,dx = \sin^{-1}\left(\frac{x}{a}\right) + c\right]$$

Putting the value of t, we get

$$I = \frac{1}{4}\sin^{-1}\left[\frac{4(\sin x - \cos x)}{5}\right] + c$$ **Ans.**

(b) Given equations of the curves are

$y = 6x - x^2$ ...(i)

$y = x^2 - 2x$ ...(ii)

Solving equations (i) and (ii), we get $x = 0, 4$.

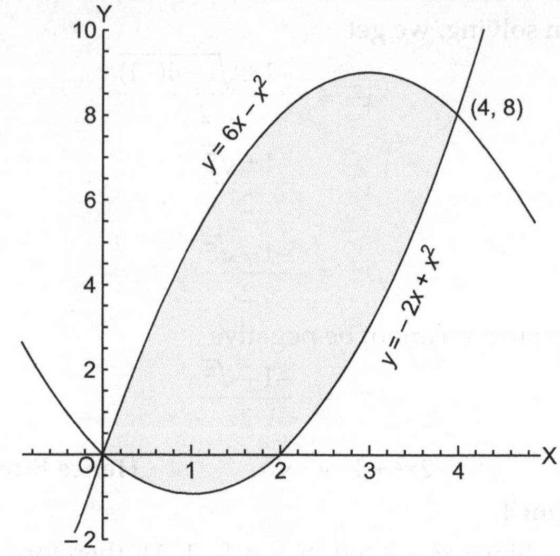

$$\text{Required area} = \int_0^4 (6x - x^2)dx - \int_0^4 (x^2 - 2x)dx$$

$$= \int_0^4 (6x - x^2 - x^2 + 2x)dx$$

$$= \int_0^4 (8x - 2x^2)dx$$

$$= \left[4x^2 - \frac{2x^3}{3}\right]_0^4 = \left(64 - \frac{128}{3}\right)$$

$$= \frac{64}{3} = 21\frac{1}{3} \text{ sq. units.}$$ **Ans.**

Solution 7.

(b) Let x denote the marks in English and y in Mathematics.

x	y	$dx = x - 16$	$dy = y - 18$	$dx \cdot dy$	$(dx)^2$	$(dy)^2$
20	17	4	−1	−4	16	1
13	12	−3	−6	18	9	36
18	23	2	5	10	4	25
21	25	5	7	35	25	49
11	14	−5	−4	20	25	16
12	8	−4	−10	40	16	100
17	19	1	1	1	1	1
14	21	−2	3	−6	4	9
19	22	3	4	12	9	16
15	19	−1	1	−1	1	1
$\Sigma x = 160$	$\Sigma y = 180$			$\Sigma dx\,dy = 125$	$\Sigma(dx)^2 = 110$	$\Sigma(dy)^2 = 254$

$\bar{x} = \frac{160}{10} = 16$ and $\bar{y} = \frac{180}{10} = 18.$

$$\therefore \text{A line of best fit is } y - \bar{y} = \frac{\Sigma dxdy}{\Sigma(dx)^2}(x - \bar{x})$$

$$y - 18 = \frac{125}{110}(x - 16)$$

when $x = 24$, then $y - 18 = \frac{125}{110}(24 - 16)$

$$y - 18 = \frac{25}{22} \times 8$$
$$22y - 396 = 200$$
$$22y = 396 + 200$$
$$y = \frac{596}{22}$$
$$= 27{\cdot}09 \text{ or } 27 \text{ marks (nearly)}$$ **Ans.**

Solution 8.

(a) Let A : Number of girls is more than number of boys.

B : The 4 member committee consists of atleast one girl.

Then, required probability = P(A/B)

$$= \frac{P(A \cap B)}{P(B)} = \frac{n(A \cap B)/n(S)}{n(B)/n(S)} = \frac{n(A \cap B)}{n(B)}$$

Now, $n(A \cap B) = {}^6C_4 . {}^8C_0 + {}^6C_3 . {}^8C_1 = 15 + 160 = 175$

$n(B) = {}^6C_1 . {}^8C_3 + {}^6C_2 . {}^8C_2 + {}^6C_3 . {}^8C_1 + {}^6C_4 . {}^8C_0 = 931$

$$\therefore \quad P(A/B) = \frac{n(A \cap B)}{n(B)}$$
$$= \frac{175}{931} = 0{\cdot}19$$ **Ans.**

(b) There are three mutually exclusive and exhaustive ways in which 2 balls can be transferred from first bag to second bag.

First way : Two white balls are transferred from first bag to second bag, probability of which is $\frac{{}^{10}C_2}{{}^{13}C_2}$.

In the second bag we now have 5 white and 5 black balls and probability of getting a white ball is $\frac{5}{10}$.

$$\therefore \quad \text{Required probability} = \frac{{}^{10}C_2}{{}^{13}C_2} \times \frac{5}{10}$$
$$= \frac{45}{78} \times \frac{5}{10} = \frac{225}{780}$$

Second way : Two black balls are transferred from first bag to the second bag probability of which is $\frac{{}^{3}C_2}{{}^{13}C_2}$.

Now, we have 3 white and 7 black balls in second bag and probability of getting a white ball is $\frac{3}{10}$.

$$\therefore \quad \text{Required probability} = \frac{{}^{3}C_2}{{}^{13}C_2} \times \frac{3}{10}$$
$$= \frac{3}{78} \times \frac{3}{10} = \frac{9}{780}.$$

Third way : One black and one white ball are transferred from first bag to second bag, the probability of which is $\frac{{}^{10}C_1 \times {}^3C_1}{{}^{13}C_2}$

In the second bag, there are now 4 white and 6 black balls and the probability of drawing a white ball is $\frac{4}{10}$.

$$\therefore \quad \text{Required probability} = \frac{{}^{10}C_1 \times {}^3C_1}{{}^{13}C_2} \times \frac{4}{10}$$
$$= \frac{30}{78} \times \frac{4}{10} = \frac{120}{780}$$

Since, these three cases are mutually exclusive, the required probability of drawing a white ball is

$$\frac{225}{780} + \frac{9}{780} + \frac{120}{780} = \frac{354}{780}$$
$$= \frac{177}{390} = \frac{59}{130}$$ **Ans.**

Solution 9.

(b) The given differential equation is,

$$x^2\,dy + (xy + y^2)\,dx = 0$$
$$\Rightarrow \quad \frac{dy}{dx} = -\frac{y(x+y)}{x^2}$$

which is homogeneous as each of the functions in the numerator and the denominator is a homogenous function of degree 2.

Putting $y = vx$ and $\frac{dy}{dx} = v + x\frac{dv}{dx}$, the given equation becomes

$$v + x\frac{dv}{dx} = \frac{-vx(x+vx)}{x^2}$$
$$\Rightarrow \quad x\frac{dv}{dx} = -v(1+v) - v$$
$$= -v(v+2)$$
$$\Rightarrow \quad \frac{dv}{v(v+2)} = -\frac{1}{x}dx$$
$$\therefore \quad \int \frac{dv}{v(v+2)} = -\int \frac{dx}{x}$$
$$\int \frac{1}{2}\left(\frac{1}{v} - \frac{1}{v+2}\right) dv = -\int \frac{dx}{x}$$

$$\Rightarrow \frac{1}{2}\log|v| - \frac{1}{2}\log|v+2| = -\log|x| + \log c$$

$$\Rightarrow \log\left|\frac{x\sqrt{v}}{\sqrt{v+2}}\right| = \log c$$

$$\Rightarrow \frac{x\sqrt{v}}{\sqrt{v+2}} = c$$

$$\Rightarrow \frac{x\sqrt{y}}{\sqrt{y+2x}} = c \qquad \left(\text{Putting } v = \frac{y}{x}\right)$$

$$\Rightarrow x^2y = A(y+2x) \qquad \ldots\text{(i)}$$

where $A = c^2$.

when $x = 1$ and $y = 1$, equation (i) becomes.

$$1 = A(1+2)$$

$$\Rightarrow A = \frac{1}{3}$$

Putting the value of A in equation (i), we get

$$3x^2y = (y+2x) \qquad \textbf{Ans.}$$

SECTION—B

Solution 10.

(a) $\left[\vec{a}-\vec{b}, \vec{b}-\vec{c}, \vec{c}-\vec{a}\right]$

$$= \left\{(\vec{a}-\vec{b})\times(\vec{b}-\vec{c})\right\}\cdot(\vec{c}-\vec{a})$$

$$= (\vec{a}\times\vec{b}-\vec{a}\times\vec{c}-\vec{b}\times\vec{b}+\vec{b}\times\vec{c})\cdot(\vec{c}-\vec{a})$$

$$= (\vec{a}\times\vec{b}+\vec{c}\times\vec{a}+\vec{b}\times\vec{c})\cdot(\vec{c}-\vec{a})$$

$$\left[\because \vec{b}\times\vec{b} = \vec{0}\right]$$

$$= (\vec{a}\times\vec{b})\cdot\vec{c}-(\vec{a}\times\vec{b})\cdot\vec{a}+(\vec{c}\times\vec{a})\cdot\vec{c}-(\vec{c}\times\vec{a})\cdot\vec{a}+(\vec{b}\times\vec{c})\cdot\vec{c}-(\vec{b}\times\vec{c})\cdot\vec{a}$$

[by distributive law]

$$= \left[\vec{a}\;\vec{b}\;\vec{c}\right]-\left[\vec{a}\;\vec{b}\;\vec{c}\right]+\left[\vec{c}\;\vec{a}\;\vec{c}\right]-\left[\vec{c}\;\vec{a}\;\vec{a}\right]+\left[\vec{b}\;\vec{c}\;\vec{c}\right]-\left[\vec{b}\;\vec{c}\;\vec{a}\right]$$

$$= \left[\vec{a}\;\vec{b}\;\vec{c}\right]-\left[\vec{b}\;\vec{c}\;\vec{a}\right]$$

[other scalar products having two equal vectors being 0.]

$$\left\{\because \left[\vec{b}\;\vec{c}\;\vec{a}\right] = \left[\vec{a}\;\vec{b}\;\vec{c}\right]\right\}$$

$$= \left[\vec{a}\;\vec{b}\;\vec{c}\right]-\left[\vec{a}\;\vec{b}\;\vec{c}\right] = 0$$

$\Rightarrow \vec{a}-\vec{b}, \vec{b}-\vec{c}, \vec{c}-\vec{a}$ are coplanar. **Hence Proved.**

(b) Given, $\vec{a} = 3\hat{i}+2\hat{j}+2\hat{k}$

and $\vec{b} = \hat{i}+2\hat{j}-2\hat{k}$

Now, $\vec{a}+\vec{b} = (3\hat{i}+2\hat{j}+2\hat{k})+(\hat{i}+2\hat{j}-2\hat{k})$

$$= 4\hat{i}+4\hat{j}+0\hat{k}$$

$\vec{a}-\vec{b} = (3\hat{i}+2\hat{j}+2\hat{k})-(\hat{i}+2\hat{j}-2\hat{k})$

$$= 2\hat{i}+0\hat{j}+4\hat{k}$$

We know that the unit vector perpendicular to both $\vec{a}+\vec{b}$ and $\vec{a}-\vec{b}$ is given by

$$\hat{n} = \frac{(\vec{a}+\vec{b})\times(\vec{a}-\vec{b})}{\left|(\vec{a}+\vec{b})\times(\vec{a}-\vec{b})\right|}$$

Here, $(\vec{a}+\vec{b})\times(\vec{a}-\vec{b}) = \begin{vmatrix}\hat{i} & \hat{j} & \hat{k}\\ 4 & 4 & 0\\ 2 & 0 & 4\end{vmatrix}$

$$= \hat{i}(16-0)-\hat{j}(16-0)+\hat{k}(0-8)$$

$$= 16\hat{i}-16\hat{j}-8\hat{k}$$

Now, $\left|(\vec{a}+\vec{b})\times(\vec{a}-\vec{b})\right|$

$$= \sqrt{(16)^2+(16)^2+(8)^2}$$

$$= \sqrt{256+256+64}$$

$$= \sqrt{576} = 24$$

Hence, the required unit vector

$$\hat{n} = \frac{(\vec{a}+\vec{b})\times(\vec{a}-\vec{b})}{\left|(\vec{a}+\vec{b})\times(\vec{a}-\vec{b})\right|}$$

$$= \frac{16\hat{i}-16\hat{j}-8\hat{k}}{24}$$

$$= \frac{2}{3}\hat{i}-\frac{2}{3}\hat{j}-\frac{1}{3}\hat{k} \qquad \textbf{Ans.}$$

Solution 11.

(a) Let P (2, – 1, 5) be the given point and AB the given line

$$\frac{x-11}{10} = \frac{y+2}{-4} = \frac{z+8}{-11} = r \text{ (say)}$$

P (2, – 1, 5)
A
B
M (1, 2, 3)
P′ (α, β, γ)

Draw PM ⊥ AB. Produce PM to P′ such that PM = MP′. Then P′ is image of P in AB.

Any point on AB is M $(10r + 11, -4r - 2, -11r - 8)$ ...(i)

Then, D.R′.s of MP are $10r + 11 - 2, -4r - 2 - (-1), -11r - 8 - 5$.

i.e., $10r + 9, -4r - 1, -11r - 13$.

and the d.c.′s of AB are proportional to 10, – 4, –11.

$\therefore$ MP ⊥ AB

$\therefore\ 10(10r + 9) - 4(-4r - 1) - 11(-11r - 13) = 0$

$\Rightarrow\ 100r + 90 + 16r + 4 + 121r + 143 = 0$

$\Rightarrow\ r = -1$

From equation (i), M ≡ (1, 2, 3)

Let P′ be the point (α, β, γ).

Since M (1, 2, 3) is the mid-point of PP′

$\therefore\ 1 = \dfrac{\alpha+2}{2},\ 2 = \dfrac{\beta-1}{2},\ 3 = \dfrac{\gamma+5}{2}$

$\Rightarrow\ \alpha = 0, \beta = 5, \gamma = 1$

$\therefore$ P′ ≡ (0, 5, 1)

Thus, P′ ≡ (0, 5, 1) is the image of P (2, – 1, 5) in the line AB.

Also, $MP = \sqrt{(1-2)^2 + (2+1)^2 + (3-5)^2}$

$= \sqrt{(1+9+4}$

$= \sqrt{14}$ units. **Ans.**

(b) Any plane passing through the intersection of the given planes is

$$\left[\vec{r}.\left(2\hat{i}+3\hat{j}-4\hat{k}\right)+5\right]+\lambda\left[\vec{r}.\left(\hat{i}-5\hat{j}+7\hat{k}\right)+2\right]=0$$

$$\vec{r}\left[(2+\lambda)\hat{i}+(3-5\lambda)\hat{j}+(-4+7\lambda)\hat{k}\right] = -5-2\lambda \quad \text{...(i)}$$

Given equation (i) intersects the Y-axis at (0, 3) so at Y-axis

Coordinate of x = 0

Coordinate of y = 3

Coordinate of z = 0

As at Y-axis coordinate of x and z will be 0.

$\therefore\ \vec{r} = 0\hat{i}+3\hat{j}+0\hat{k}$

$$\left(0\hat{i}+3\hat{j}+0\hat{k}\right)\cdot\left[(2+\lambda)\hat{i}+(3-5\lambda)\hat{j}+(-4+7\lambda)\hat{k}\right] = -5-2\lambda$$

$$0(2+\lambda)+3(3-5\lambda)+0(-4+7\lambda)+5+2\lambda = 0$$

$$\lambda = \frac{14}{13}$$

$\therefore$ The required plane is

$$\left(2+\frac{14}{13}\right)x+\left(3-5\times\frac{14}{13}\right)y+\left(-4+7\times\frac{14}{13}\right)z = -5-2\times\left(\frac{14}{13}\right)$$

$$\frac{(26+14)}{13}x+\left(\frac{39-70}{13}\right)y+\frac{(-52+98)}{13}z+\frac{65+28}{13}=0$$

or $40x - 31y + 46z + 93 = 0$ which is the required equation. **Ans.**

Solution 12.

(a) Let E_1, E_2, E_3 denote the events of carrying out the task by A, B and C respectively.

Let H denote the event of taking more time.

Then, $P(E_1) = 0{\cdot}45$, $P(E_2) = 0{\cdot}35$, $P(E_3) = 0{\cdot}20$

$P(H/E_1) = \dfrac{1}{6}$,

$P(H/E_2) = \dfrac{1}{10}$,

$P(H/E_3) = \dfrac{1}{20}$

Then, using Bayes' theorem

$$P(E_1/H) = \frac{P(E_1)\cdot P(H/E_1)}{P(E_1)P(H/E_1)+P(E_2)P(H/E_2)+P(E_3)\cdot P(H/E_3)}$$

$$= \frac{\frac{45}{600}}{\frac{45}{600}+\frac{35}{1000}+\frac{20}{2000}}$$

$$= \frac{\frac{3}{40}}{\frac{3}{40}+\frac{7}{200}+\frac{1}{100}}$$

$$= \frac{3}{40}\times\frac{200}{24} = \frac{5}{8}$$ **Ans.**

SECTION—C

Solution 13.

(b)

Type of Product	Total items: 9 gms Nickel (in gm)	8 gms Chromium (on gm)	Profit per unit (in ₹)
A	3	1	₹ 40
B	1	2	₹ 50

Let x = Number of units of type A

y = Number of units of type B.

Then, our linear programming problem will be,

$$\text{Maximize } Z = 40x + 50y$$

Subject to the constraints,

$$3x + y \le 9$$

$$x + 2y \le 8$$

and $x, y \ge 0$

Consider the equation,

$$3x + y = 9$$

x	0	3
y	9	0

and $x + 2y = 8$

x	0	8
y	4	0

Both the equations $3x + y = 9$ and $x + 2y = 8$ intersect at point (2, 3)

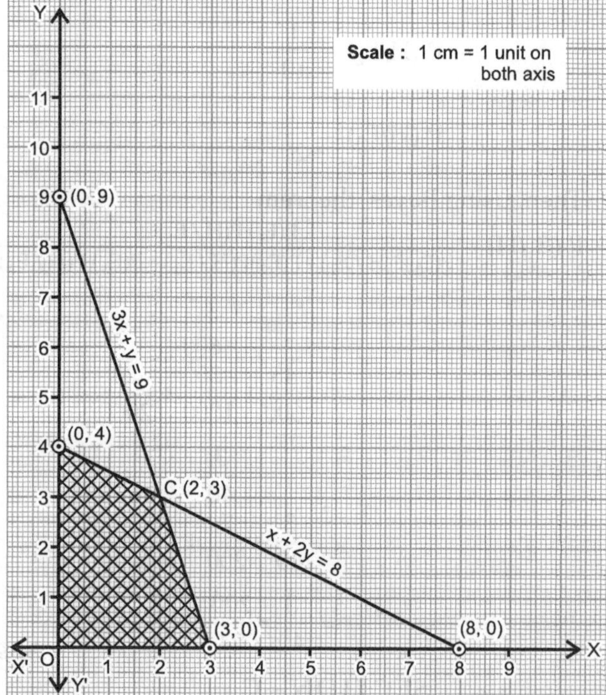

The solution set of this system is the shaded region in the diagram.

Now, we can determine the maximum value of Z by evaluating the value of Z at the four points (vertices) as shown below.

Vertices	$Z = 40x + 50y$
(0, 0)	$Z = 40 \times 0 + 50 \times 0 =$ ₹ 0
(3, 0)	$Z = 40 \times 3 + 50 \times 0 =$ ₹ 120
(0, 4)	$Z = 40 \times 0 + 50 \times 4 =$ ₹ 200
(2, 3)	$Z = 40 \times 2 + 50 \times 3 =$ ₹ 230 maximum

From graph,

Maximum profit, Z = ₹ 230

∴ Number of units of type A is 2 and number of units of type B is 3. **Ans.**

Solution 14.

(a) Given, $x = \dfrac{24 - 2p}{3}$, $p = 12 - \dfrac{3}{2}x$

(i) Revenue function $R(p) = px = p\left(\dfrac{24 - 2p}{3}\right)$

$$= 8p - \frac{2}{3}p^2 \quad \textbf{Ans.}$$

(ii) For maximum revenue,

$$\frac{dR(p)}{dp} = 8 - \frac{4}{3}p = 0 \Rightarrow p = 6$$

and $$\frac{d^2R(p)}{dp^2} = -\frac{4}{3} \ (< 0)$$

$\Rightarrow R(p)$ is maximum at $p = 6$.

For maximum revenue, price per unit is 6

$$\therefore \quad x = \frac{24 - 2 \times 6}{3}$$

$$= \frac{24 - 12}{3} = \frac{12}{3} = 4$$

∴ Price is ₹ 6 per unit and no. of units is 4 for which the revenue is maximum. **Ans.**

MATHEMATICS 2015

QUESTIONS

SECTION—A

Question 1. **[10 × 3]**

(i) *Find the value of k if* $M = \begin{bmatrix} 1 & 2 \\ 2 & 3 \end{bmatrix}$ *and* $M^2 - kM - I_2 = 0$.

(ii) *Find the equation of an ellipse whose latus rectum is 8 and eccentricity is* $\frac{1}{3}$. **

(iii) *Solve :* $\cos^{-1}(\sin \cos^{-1} x) = \frac{\pi}{6}$.

(iv) *Using L'Hospital's rule, evaluate :* $\lim_{x \to 0} \frac{x - \sin x}{x^2 \sin x}$.

(v) *Evaluate :* $\int \frac{2y^2}{y^2 + 4} dy$.

(vi) *Evaluate :* $\int_0^3 f(x)dx$.

where $f(x) = \begin{cases} \cos 2x, 0 \le x \le \frac{\pi}{2} \\ 3, \quad \frac{\pi}{2} \le x \le 3. \end{cases}$

(vii) *The two lines of regressions are* $4x + 2y - 3 = 0$ *and* $3x + 6y + 5 = 0$. *Find the correlation coefficient between x and y.*

(viii) *A card is drawn from a well shuffled pack of playing cards. What is the probability that it is either a spade or an ace or both ?*

(ix) *If 1, ω and ω² are the cube roots of unity, prove that* $\frac{a + b\omega + c\omega^2}{c + a\omega + b\omega^2} = \omega^2$**

(x) *Solve the differential equation :*

$$\sin^{-1}\left(\frac{dy}{dx}\right) = x + y.$$

Question 2.

(a) *Using properties of determinants, prove that :* **[5]**

$$\begin{vmatrix} 1 + a^2 - b^2 & 2ab & -2b \\ 2ab & 1 - a^2 + b^2 & 2a \\ 2b & -2a & 1 - a^2 - b^2 \end{vmatrix} = (1 + a^2 + b^2)^3.$$

(b) *Given two matrices A and B* **[5]**

$$A = \begin{vmatrix} 1 & -2 & 3 \\ 1 & 4 & 1 \\ 1 & -3 & 2 \end{vmatrix} \text{ and } B = \begin{vmatrix} 11 & -5 & -14 \\ -1 & -1 & 2 \\ -7 & 1 & 6 \end{vmatrix}$$

find AB and use this result to solve the following system of equations :

$x - 2y + 3z = 6, \quad x + 4y + z = 12, \quad x - 3y + 2z = 1.$

Question 3.

(a) *Solve the equation for x :*

$$\sin^{-1}\frac{5}{x} + \sin^{-1}\frac{12}{x} = \frac{\pi}{2}, x \neq 0.$$ **[5]**

(b) *A, B and C represent switches in 'on' position and A', B' and C' represent them in 'off' position. Construct a switching circuit representing the polynomial ABC + AB'C + A'B'C. Using Boolean Algebra, prove that the given polynomial can be simplified to C (A + B'). Construct an equivalent switching circuit.*** **[5]**

Question 4.

(a) *Verify Lagrange's mean value theorem for the following function :*

$$f(x) = 2 \sin x + \sin 2x \text{ on } [0, \pi].$$ **[5]**

(b) *Find the equation of the hyperbola whose foci are* $(0, \pm\sqrt{10})$ *and passing through the point (2, 3).*** **[5]**

Question 5.

(a) *If* $y = e^{m \cos^{-1} x}$, *prove that :*

$$(1 - x^2)\frac{d^2y}{dx^2} - x\frac{dy}{dx} = m^2y$$ **[5]**

(b) *Show that the rectangle of maximum perimeter which can be inscribed in a circle of radius 10 cm is a square of side* $10\sqrt{2}$ *cm.* **[5]**

Question 6.

(a) *Evaluate :* $\int \frac{\sec x}{1 + \mathrm{cosec}\, x} dx$ **[5]**

(b) *Find the smaller area enclosed by the circle* $x^2 + y^2$ *and the line* $x + y = 2$. **[5]**

Question 7.

(a) *Given that the observations are :* **[5]**

(9, – 4), (10, – 3), (11, – 1), (12, 0), (13, 1), (14, 3), (15, 5), (16, 8).

** Answer is not given due to change in the present syllabus.

Find the two lines of regression and estimate the value of y when $x = 13{\cdot}5$.

(b) *In a contest the competitors are awarded marks out of 20 by two judges. The scores of the 10 competitors are given below. Calculate Spearman's rank correlation.*** **[5]**

Competitors	Judge A	Judge B
A	2	6
B	11	11
C	11	16
D	18	9
E	6	14
F	5	20
G	8	4
H	16	3
I	13	13
J	15	17

Question 8.

(a) *An urn contains 2 white and 2 black balls. A ball is drawn at random. If it is white, it is not replaced into the urn. Otherwise, it is replaced with another ball of the same colour. The process is repeated. Find the probability that the third ball drawn is black.* **[5]**

(b) *Three persons A, B and C shoot to hit a target. If A hits the target four times in five trials, B hits it three times in four trials and C hits it two times in three trials, find the probability that :* **[5]**

(i) *Exactly two persons hit the target.*

(ii) *At least two persons hit the target.*

(iii) *None hit the target.*

Question 9.

(a) *If* $z = x + iy$, $\omega = \dfrac{2-iz}{2z-i}$ *and* $|\omega| = 1$, *find the locus of z and illustrate it in the Argand Plane.*** **[5]**

(b) *Solve the differential equation :* **[5]**

$$e^{x/y}\left(1-\frac{x}{y}\right) + (1 + e^{x/y})\frac{dx}{dy} = 0 \text{ when } x = 0, y = 1.$$

SECTION—B

Question 10.

(a) *Using vectors, prove that angle in a semicircle is a right angle.* **[5]**

(b) *Find the volume of a parallelopiped whose edges are represented by the vectors :* **[5]**

$$\vec{a} = 2\hat{i} - 3\hat{j} - 4\hat{k},\ \vec{b} = \hat{i} + 2\hat{j} - \hat{k},$$

and $\vec{c} = 3\hat{i} + \hat{j} + 2\hat{k}$.

** Answer is not given due to change in the present syllabus.

Question 11.

(a) *Find the equation of the plane passing through the intersection of the planes :* **[5]**

$x + y + z + 1 = 0$ *and* $2x - 3y + 5z - 2 = 0$ *and the point* $(-1, 2, 1)$.

(b) *Find the shortest distance between the lines*

$$\vec{r} = \hat{i} + 2\hat{j} + 3\hat{k} + \lambda(2\hat{i} + 3\hat{j} + 4\hat{k})$$

and $\vec{r} = 2\hat{i} + 4\hat{j} + 5\hat{k} + \mu(4\hat{i} + 6\hat{j} + 8\hat{k})$. **[5]**

Question 12.

(a) *Box I contains two white and three black balls. Box II contains four white and one black balls and box III contains three white and four black balls. A dice having three red, two yellow and one green face, is thrown to select the box. If red face turns up, we pick up box I, if a yellow face turns up we pick up box II, otherwise, we pick up box III. Then, we draw a ball from the selected box. If the ball drawn is white, what is the probability that the dice had turned up with a red face ?* **[5]**

(b) *Five dice are thrown simultaneously. If the occurrence of an odd number in a single dice is considered a success, find the probability of maximum three successes.* **[5]**

SECTION—C

Question 13.

(a) *Mr. Nirav borrowed ₹50,000 from the bank for 5 years. The rate of interest is 9% per annum compounded monthly. Find the payment he makes monthly if he pays back at the beginning of each month.*** **[5]**

(b) *A dietician wishes to mix two kinds of food X and Y in such a way that the mixture contains at least 10 units of vitamin A, 12 units of vitamin B and 8 units of vitamin C. The vitamin contents of one kg food is given below :* **[5]**

Food	Vitamin A	Vitamin B	Vitamin C
X	1 unit	2 units	3 units
Y	2 units	2 units	1 unit

One kg of food X costs ₹24 and one kg of food Y costs ₹36. Using Linear Programming, find the least cost of the total mixture which will contain the required vitamins.

Question 14.

(a) *A bill for ₹7,650 was drawn on 8th March, 2013, at 7 months. It was discounted on 18th May, 2013 and the holder of the bill received ₹7,497. What is the rate of interest charged by the bank ?*** **[5]**

(b) *The average cost function, AC for a commodity is given by* $AC = x + 5 + \dfrac{36}{x}$, *in terms of output x.*

Find :

(i) *The total cost, C and marginal cost, MC as a function of x.*

(ii) *The outputs for which AC increases.* **[5]**

Question 15.**

(a) *Calculate the index number for the year 2014, with 2010 as the base year by the weighted aggregate method from the following data :* **[5]**

Commodity	Price in ₹		Weight
	2010	2014	
A	2	4	8
B	5	6	10
C	4	5	14
D	2	2	19

(b) *The quarterly profits of a small scale industry (in thousands of rupees) is as follows :* **[5]**

Year	Quarter 1	Quarter 2	Quarter 3	Quarter 4
2012	39	47	20	56
2013	68	59	66	72
2014	88	60	60	67

Calculate four quarterly moving averages. Display these and the original figures graphically on the same graph sheet.

** Answer is not given due to change in the present syllabus.

SOLUTIONS

SECTION—A

Solution 1.

(i) Given, $M = \begin{bmatrix} 1 & 2 \\ 2 & 3 \end{bmatrix}$

$$M^2 = M \cdot M = \begin{bmatrix} 1 & 2 \\ 2 & 3 \end{bmatrix}\begin{bmatrix} 1 & 2 \\ 2 & 3 \end{bmatrix}$$

$$= \begin{bmatrix} 1\times1+2\times2 & 1\times2+2\times3 \\ 2\times1+3\times2 & 2\times2+3\times3 \end{bmatrix}$$

$$= \begin{bmatrix} 1+4 & 2+6 \\ 2+6 & 4+9 \end{bmatrix} = \begin{bmatrix} 5 & 8 \\ 8 & 13 \end{bmatrix}$$

Now, $M^2 - kM - I_2 = 0$

$$\Rightarrow \quad M^2 - I_2 = kM$$

$$\Rightarrow \quad \begin{bmatrix} 5 & 8 \\ 8 & 13 \end{bmatrix} - \begin{bmatrix} 1 & 0 \\ 0 & 1 \end{bmatrix} = k\begin{bmatrix} 1 & 2 \\ 2 & 3 \end{bmatrix}$$

$$\Rightarrow \quad \begin{bmatrix} 4 & 8 \\ 8 & 12 \end{bmatrix} = k\begin{bmatrix} 1 & 2 \\ 2 & 3 \end{bmatrix}$$

$$\Rightarrow \quad 4\begin{bmatrix} 1 & 2 \\ 2 & 3 \end{bmatrix} = k\begin{bmatrix} 1 & 2 \\ 2 & 3 \end{bmatrix}$$

On comparing both sides, we get

$$\Rightarrow \quad k = 4.$$ **Ans.**

(iii) Given, $\cos^{-1}(\sin \cos^{-1} x) = \dfrac{\pi}{6}$

$$\Rightarrow \quad \sin(\cos^{-1} x) = \cos\frac{\pi}{6}$$

$$\Rightarrow \quad \sin(\cos^{-1} x) = \sin\frac{\pi}{3}$$

$$\left[\because \cos\frac{\pi}{6} = \cos\left(\frac{\pi}{2} - \frac{\pi}{3}\right) = \sin\frac{\pi}{3}\right]$$

$$\Rightarrow \quad \cos^{-1} x = \frac{\pi}{3}$$

$$\Rightarrow \quad x = \cos\frac{\pi}{3}$$

Hence, $x = \dfrac{1}{2}$. **Ans.**

(iv) Given,

$$\lim_{x\to0}\frac{x - \sin x}{x^2 \sin x} = \lim_{x\to0}\frac{x-\sin x}{x^3}\cdot\frac{x}{\sin x}$$

$$= \lim_{x\to0}\frac{x-\sin x}{x^3}\cdot 1 \quad \left[\because \lim_{x\to0}\frac{\sin x}{x} = 1\right]$$

$$= \lim_{x\to0}\frac{x-\sin x}{x^3} \quad \left[\text{form } \frac{0}{0}\right]$$

Applying L'Hospital's rule, we have

$$= \lim_{x\to0}\frac{1-\cos x}{3x^2} \quad \left[\text{form } \frac{0}{0}\right]$$

Applying L'Hospital's rule, we have

$$= \lim_{x\to0}\frac{\sin x}{6x}$$

$$= \frac{1}{6}\lim_{x\to0}\frac{\sin x}{x} = \frac{1}{6}.$$

$$\left[\because \lim_{x\to0}\frac{\sin x}{x} = 1\right]$$ **Ans.**

(v) Given, $\displaystyle\int \frac{2y^2}{y^2+4}\,dy$

$$= 2\int \frac{y^2+4-4}{y^2+4}\,dy$$

$$= 2\int dy - 8\int \frac{dy}{y^2+4}$$

$$= 2y - 8\cdot\frac{1}{2}\tan^{-1}\left(\frac{y}{2}\right) + c$$

$$\left[\because \int \frac{dx}{x^2+a^2} = \frac{1}{a}\tan^{-1}\left(\frac{x}{a}\right)\right]$$

$$= 2y - 4\tan^{-1}\left(\frac{y}{2}\right) + c.$$ **Ans.**

(vi) $$\int_0^3 f(x)dx = \int_0^{\pi/2} f(x)dx + \int_{\pi/2}^3 f(x)dx$$

$$= \int_0^{\pi/2} \cos 2x\, dx + \int_{\pi/2}^3 3\, dx$$

$$= \left[\frac{\sin 2x}{2}\right]_0^{\pi/2} + [3x]_{\pi/2}^3$$

$$= \left[\frac{\sin 2\times\pi/2}{2} - \frac{\sin 2\times 0}{2}\right] + [3\times 3 - 3\pi/2]$$

$$= \frac{\sin\pi}{2} - 0 + 9 - \frac{3\pi}{2}$$

$$= 9 - \frac{3\pi}{2}$$ $(\because \sin\pi = 0)$ **Ans.**

(vii) Let $4x + 2y - 3 = 0$ be the regression equation of y on x and $3x + 6y + 5 = 0$ the regression equation of x on y then

$$2y = -4x + 3$$

$$\Rightarrow \quad y = -2x + \frac{3}{2}$$

$$\therefore \quad b_{yx} = -2$$

and $$3x = -6y - 5$$

$$\Rightarrow \quad x = -\frac{6}{3}y - \frac{5}{3}$$

$$\Rightarrow \quad x = -2y - \frac{5}{3}$$

$$\therefore \quad b_{xy} = -2$$

$$r^2 = b_{yx} \times b_{xy}$$

$$= -2 \times (-2) = 4 > 1$$

which is impossible.

So regression equation of y on x is

$$3x + 6y + 5 = 0 \quad \ldots(i)$$

and regression equation of x on y is

$$4x + 2y - 3 = 0 \quad \ldots(ii)$$

From (i), we get

$$6y = -3x - 5$$

$$y = -\frac{1}{2}x - \frac{5}{6}$$

$$\therefore \quad b_{yx} = -\frac{1}{2}$$

From (ii), we get

$$4x = -2y + 3$$

$$x = -\frac{1}{2}y + \frac{3}{4}$$

$$\therefore \quad b_{yx} = -\frac{1}{2}$$

$\therefore$ Correlation coefficient is given by

$$r = \sqrt{b_{yx}\cdot b_{xy}}$$

$$= \sqrt{\left[-\frac{1}{2}\right]\times\left(-\frac{1}{2}\right)} = \frac{1}{2}$$

$$\Rightarrow \quad r = \pm\frac{1}{2}$$

But r has the same sign as regression coefficient

$$\therefore \quad r = -\frac{1}{2}.$$ **Ans.**

(viii) Let A = getting a spade

B = getting an ace

$$P(A) = \frac{13}{52} = \frac{1}{4},\ P(B) = \frac{4}{52} = \frac{1}{13}$$

$$P(\overline{A}) = 1 - \frac{1}{4} = \frac{3}{4}$$

$$P(\overline{B}) = 1 - \frac{1}{13} = \frac{12}{13}$$

Required probability $= 1 - P(\overline{A})\,P(\overline{B})$

$$= 1 - \frac{3}{4}\times\frac{12}{13}$$

$$= 1 - \frac{9}{13} = \frac{4}{13}$$ **Ans.**

(x) Given, $$\sin^{-1}\left(\frac{dy}{dx}\right) = x + y$$

$$\frac{dy}{dx} = \sin(x+y) \quad \ldots(i)$$

Put $v = x + y$, we get

$$\frac{dv}{dx} = 1 + \frac{dy}{dx} \Rightarrow \frac{dy}{dx} = \frac{dv}{dx} - 1$$

From (i), we have

$$\frac{dy}{dx} = \sin(x+y)$$

Putting the value of $x+y$ and $\frac{dy}{dx}$, we get

$$\frac{dv}{dx} - 1 = \sin v$$

$$\Rightarrow \quad \frac{dv}{dx} = \sin v + 1$$

$$\Rightarrow \quad dx = \frac{1}{1+\sin v}\, dv$$

On integrating both sides, $\int dx = \int \frac{1}{1+\sin v}\, dv$

$$x = \int \frac{1}{1+\sin v} \times \frac{1-\sin v}{1-\sin v}\, dv$$

$$= \int \frac{1-\sin v}{1-\sin^2 v}\, dv$$

$$= \int \frac{1-\sin v}{\cos^2 v}\, dv$$

$$= \int (\sec^2 v - \sec v \tan v)\, dv$$

$$\therefore \quad x = \tan v - \sec v + c.$$

$$x = \tan(x+y) - \sec(x+y) + c$$ **Ans.**

Solution 2.

(a)

$$\Delta = \begin{vmatrix} 1+a^2-b^2 & 2ab & -2b \\ 2ab & 1-a^2+b^2 & 2a \\ 2b & -2a & 1-a^2-b^2 \end{vmatrix}$$

Operating $R_1 \to R_1 + bR_3$,

$$= \begin{vmatrix} 1+a^2+b^2 & 0 & -b(1+a^2+b^2) \\ 2ab & 1-a^2+b^2 & 2a \\ 2b & -2a & 1-a^2-b^2 \end{vmatrix}$$

Taking $(1+a^2+b^2)$ common from R_1,

$$= (1+a^2+b^2)\begin{vmatrix} 1 & 0 & -b \\ 2ab & 1-a^2+b^2 & 2a \\ 2b & -2a & 1-a^2-b^2 \end{vmatrix}$$

Operating $R_2 \to R_2 - aR_3$,

$$= (1+a^2+b^2)\begin{vmatrix} 1 & 0 & -b \\ 0 & 1+a^2+b^2 & a(1+a^2+b^2) \\ 2b & -2a & 1-a^2-b^2 \end{vmatrix}$$

Taking $(1+a^2+b^2)$ common from R_2,

$$= (1+a^2+b^2)^2\begin{vmatrix} 1 & 0 & -b \\ 0 & 1 & a \\ 2b & -2a & 1-a^2-b^2 \end{vmatrix}$$

Now, on expanding along R_1, we get

$$= (1+a^2+b^2)^2\,[(1-a^2-b^2+2a^2)+(2b^2)]$$

$$= (1+a^2+b^2)^2 \cdot (1+a^2+b^2)$$

$$= (1+a^2+b^2)^3.$$ **Hence Proved.**

(b) Given, $A = \begin{bmatrix} 1 & -2 & 3 \\ 1 & 4 & 1 \\ 1 & -3 & 2 \end{bmatrix}$ and $B = \begin{bmatrix} 11 & -5 & -14 \\ -1 & -1 & 2 \\ -7 & 1 & 6 \end{bmatrix}$

$$AB = \begin{bmatrix} 1 & -2 & 3 \\ 1 & 4 & 1 \\ 1 & -3 & 2 \end{bmatrix}\begin{bmatrix} 11 & -5 & -14 \\ -1 & -1 & 2 \\ -7 & 1 & 6 \end{bmatrix}$$

$$= \begin{bmatrix} 11+2-21 & -5+2+3 & -14-4+18 \\ 11-4-7 & -5-4+1 & -14+8+6 \\ 11+3-14 & -5+3+2 & -14-6+12 \end{bmatrix}$$

$$= \begin{bmatrix} -8 & 0 & 0 \\ 0 & -8 & 0 \\ 0 & 0 & -8 \end{bmatrix} = -8\begin{bmatrix} 1 & 0 & 0 \\ 0 & 1 & 0 \\ 0 & 0 & 1 \end{bmatrix}$$

$$\therefore \quad AB = -8\, I_3$$

$$-\frac{1}{8}(B) = A^{-1}$$

The given system of equations can be written as,

$$\begin{bmatrix} 1 & -2 & 3 \\ 1 & 4 & 1 \\ 1 & -3 & 2 \end{bmatrix}\begin{bmatrix} x \\ y \\ z \end{bmatrix} = \begin{bmatrix} 6 \\ 12 \\ 1 \end{bmatrix}$$

or $\quad AX = C$

where, $\quad X = \begin{bmatrix} x \\ y \\ z \end{bmatrix}$ and $C = \begin{bmatrix} 6 \\ 12 \\ 1 \end{bmatrix}$

As A exists, the given system of equations has a unique solution

$$\therefore \quad X = A^{-1}C$$

$$\Rightarrow \quad X = \left(-\frac{1}{8}B\right)C$$

$$= -\frac{1}{8}\begin{bmatrix} 11 & -5 & -14 \\ -1 & -1 & 2 \\ -7 & 1 & 6 \end{bmatrix}\begin{bmatrix} 6 \\ 12 \\ 1 \end{bmatrix}$$

$$= -\frac{1}{8}\begin{bmatrix} 66-60-14 \\ -6-12+2 \\ -42+12+6 \end{bmatrix}$$

$$\begin{bmatrix} x \\ y \\ z \end{bmatrix} = -\frac{1}{8}\begin{bmatrix} -8 \\ -16 \\ -24 \end{bmatrix}$$

$$\begin{bmatrix} x \\ y \\ z \end{bmatrix} = \begin{bmatrix} 1 \\ 2 \\ 3 \end{bmatrix}$$

Hence, the solution of equations is,

$x = 1, y = 2, z = 3.$ **Ans.**

Solution 3.

(a) The given equation is

$$\sin^{-1}\frac{5}{x}+\sin^{-1}\frac{12}{x} = \frac{\pi}{2}$$

$$\Rightarrow \quad \sin^{-1}\frac{12}{x} = \frac{\pi}{2}-\sin^{-1}\left(\frac{5}{x}\right)$$

$$\Rightarrow \quad \frac{12}{x} = \sin\left(\frac{\pi}{2}-\sin^{-1}\frac{5}{x}\right)$$

$$\Rightarrow \quad \frac{12}{x} = \cos\left(\sin^{-1}\frac{5}{x}\right)$$

$$\Rightarrow \quad \frac{12}{x} = \sqrt{1-\left(\frac{5}{x}\right)^2}$$

$$\left[\because \cos(\sin^{-1}x)=\sqrt{1-x^2}\right]$$

Squaring both LHS and RHS

$$\Rightarrow \quad \left(\frac{12}{x}\right)^2 = 1-\frac{25}{x^2}$$

$$\Rightarrow \quad \frac{144}{x^2}+\frac{25}{x^2} = 1$$

$$\Rightarrow \quad \frac{169}{x^2} = 1$$

$$\Rightarrow \quad x^2 = 169$$

$$\Rightarrow \quad x = \pm\sqrt{169}$$

$$\Rightarrow \quad x = 13, -13$$

But $x = -13$ doesn't satisfy the given equation.

Hence the solution of the given equation is $x = 13$. **Ans.**

Solution 4.

(a) Given, $f(x) = 2\sin x + \sin 2x,\ x \in [0, \pi]$

$f(x)$ is continuous in $[0, \pi]$

$f(x)$ is differentiable in $(0, \pi)$

Thus, both the conditions of Lagrange's mean value theorem are satisfied by the function $f(x)$ in $[0, \pi]$, therefore, there exists at least one real number c in $(0, \pi)$ such that

$$f'(c) = \frac{f(\pi)-f(0)}{\pi-0}$$

Now, $\quad f(\pi) = 2\sin\pi + \sin 2\pi = 0$

$\quad f(0) = 2\sin 0 + \sin 0 = 0$

Differentiating $f(x)$ w.r.t. x, we get

$$f'(x) = 2\cos x + 2\cos 2x$$

Now, $\quad f'(c) = \frac{0}{\pi}$

$$\Rightarrow \quad 2\cos c + 2\cos 2c = 0$$

$$\Rightarrow \quad \cos c + \cos 2c = 0$$

$$\Rightarrow \quad 2\cos^2 c + \cos c - 1 = 0$$

$$(\because \cos 2x = 2\cos^2 x - 1)$$

$$\Rightarrow \quad 2\cos^2 c + 2\cos c - \cos c - 1 = 0$$

$$\Rightarrow \quad 2\cos c(\cos c + 1) - 1(\cos c + 1) = 0$$

$$\Rightarrow \quad (2\cos c - 1)(\cos c + 1) = 0$$

$$\Rightarrow \quad 2\cos c - 1 = 0 \text{ or } \cos c + 1 = 0$$

$$\Rightarrow \quad 2\cos c = 1 \text{ or } \cos c = -1$$

$$\Rightarrow \quad \cos c = \frac{1}{2} \text{ or } \cos c = -1$$

$$\Rightarrow \quad c = \frac{\pi}{3}, \pi$$

$$\therefore \quad c = \frac{\pi}{3} \qquad \left[\because \frac{\pi}{3} \in (0, \pi)\right]$$

Thus Lagrange's mean value theorem is verified.

Solution 5.

(a) Given, $\quad y = e^{m\cos^{-1}x}$ ...(i)

Differentiating (i) w.r.t. x, we get

$$\frac{dy}{dx} = e^{m\cos^{-1}x}\cdot m\left(\frac{-1}{\sqrt{1-x^2}}\right)$$

$$\Rightarrow \quad \sqrt{1-x^2}\frac{dy}{dx} = -my \qquad \text{[Using (i)]}$$

on squaring both LHS and RHS,

$$\Rightarrow \quad (1-x^2)\left(\frac{dy}{dx}\right)^2 = m^2y^2$$

Differentiating again w.r.t. x, we get

$$(1-x^2)\,2\frac{dy}{dx}\cdot\frac{d^2y}{dx^2}+\left(\frac{dy}{dx}\right)^2\cdot(-2x) = m^2\,2y\,\frac{dy}{dx}$$

$$\Rightarrow (1-x^2)\frac{d^2y}{dx^2} - x\frac{dy}{dx} = m^2y.$$ **Hence Proved.**

(b) Let ABCD be a rectangle inscribed in a circle of radius 10 cm with centre at O, then DB = 20 cm.

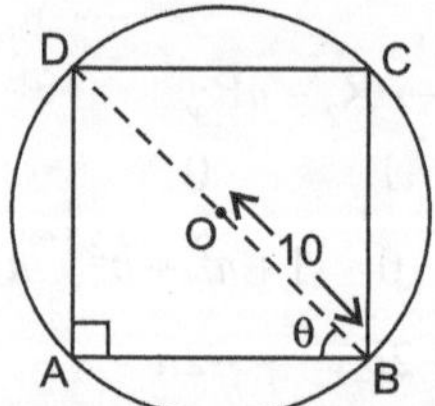

Let $\angle OBA = \theta$, $(0 < \theta < \pi/2)$ and θ is in radian.

Then, $\quad AB = 20\cos\theta$

$\quad AD = 20\sin\theta$

Let p be the perimeter of the rectangle ABCD, then

$$p = 2AB + 2AD$$

$$= 2(20\cos\theta + 20\sin\theta)$$

$$p = 40(\cos\theta + \sin\theta) \quad \text{...(i)}$$

Differentiating (i) w.r.t. θ, we get

$$\frac{dp}{d\theta} = 40(-\sin\theta + \cos\theta) \quad \ldots\text{(ii)}$$

Again, differentiating, we get

$$\frac{d^2p}{d\theta^2} = -40(\cos\theta + \sin\theta) \quad \ldots\text{(iii)}$$

Now, for maximum value $\frac{dp}{d\theta} = 0$

$$\Rightarrow \quad 40(-\sin\theta + \cos\theta) = 0$$

$$\Rightarrow \quad \cos\theta = \sin\theta$$

$$\Rightarrow \quad \tan\theta = 1$$

$$\theta = \tan^{-1}(1)$$

$$\Rightarrow \quad \theta = \frac{\pi}{4}$$

Also, $$\left(\frac{d^2p}{d\theta^2}\right)_{\theta=\pi/4} = -40\left(\frac{1}{\sqrt{2}} + \frac{1}{\sqrt{2}}\right)$$

$$= -40 \times \frac{2}{\sqrt{2}}$$

$$= -40\sqrt{2} < 0 \ (-\text{ve})$$

$\Rightarrow p$ is maximum when $\theta = \frac{\pi}{4}$.

Therefore, p is maximum when

$$AB = 20\cos\frac{\pi}{4} = 20 \times \frac{1}{\sqrt{2}} = 10\sqrt{2}$$

$$AD = 20\sin\frac{\pi}{4} = 20 \times \frac{1}{\sqrt{2}} = 10\sqrt{2}$$

i.e., when adjacent sides are equal and each of them is $10\sqrt{2}$ cm.

That is, a rectangle which can be inscribed in a circle of radius 10 cm is a square of side $10\sqrt{2}$ cm.

Hence Proved.

Solution 6.

(a) We have,

$$\int \frac{\sec x}{1+\operatorname{cosec} x}dx = \int \frac{1}{\cos x\left(1+\frac{1}{\sin x}\right)}dx$$

$$= \int \frac{\sin x}{\cos x(\sin x + 1)}dx$$

$$= \int \frac{\sin x \cos x}{\cos^2 x(1+\sin x)}dx$$

[Multiplying both numerator and denominator with $\cos x$]

$$= \int \frac{\sin x \cos x}{(1-\sin^2 x)(1+\sin x)}dx$$

$$= \int \frac{\sin x \cos x}{(1-\sin x)(1+\sin x)(1+\sin x)}dx$$

$$= \int \frac{\sin x \cos x}{(1-\sin x)(1+\sin x)^2}dx$$

Now, let $\sin x = t$, on differentiating, we get

$\cos x\, dx = dt$

$$\therefore \quad \int \frac{\sec x}{1+\operatorname{cosec} x}dx = \int \frac{t}{(1-t)(1+t)^2}dt$$

Let $$\frac{1}{(1-t)(1+t)^2} = \frac{A}{1-t} + \frac{A}{1+t} + \frac{C}{(1+t)^2}$$

$$\Rightarrow \quad t = A(1+t)^2 + B(1-t)(1+t) + C(1-t) \quad \ldots\text{(i)}$$

On putting $t = 1$ and $t = -1$, we get

$$1 = A \cdot 4 \quad \text{and} \quad -1 = C \cdot 2$$

$$\Rightarrow \quad A = \frac{1}{4} \quad \text{and} \quad C = -\frac{1}{2}$$

On equating coefficient of t^2 on both sides of (i), we get

$$0 = A - B$$

$$\Rightarrow \quad B = A = \frac{1}{4}$$

$$\therefore \int \frac{\sec x}{1+\operatorname{cosec} x}dx = \frac{1}{4}\int \frac{1}{1-t}dt + \frac{1}{4}\int \frac{dt}{1+t} - \frac{1}{2}\int \frac{dt}{(1+t)^2}$$

$$= \frac{1}{4}\frac{\log|1-t|}{-1} + \frac{1}{4}\log|1+t| - \frac{1}{2}\frac{(1+t)^{-1}}{-1} + c$$

$$= \frac{1}{4}\log\left|\frac{1+t}{1-t}\right| + \frac{1}{2(1+t)} + c$$

$$= \frac{1}{4}\log\left|\frac{1+\sin x}{1-\sin x}\right| + \frac{1}{2(1+\sin x)} + c. \quad \textbf{Ans.}$$

(b) Note : As per point of intersection equation of circle is taken as

$$x^2 + y^2 = 4$$

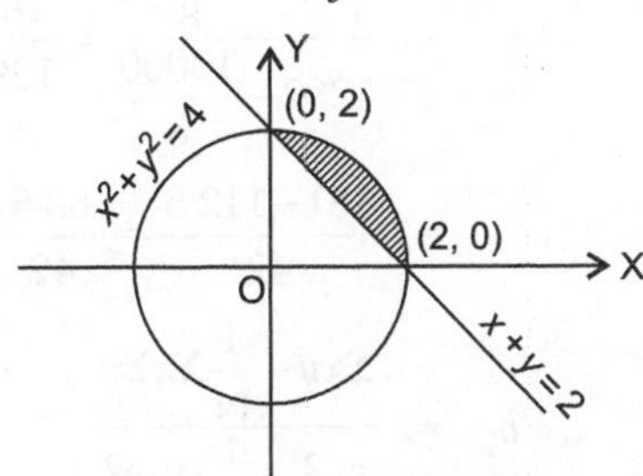

Equations of the circle and the line are

$$x^2 + y^2 = 4 \quad \text{and} \quad x + y = 2.$$

The curves intersect at (2, 0) and (0, 2).

Shaded area is the smaller area enclosed by the two curves and is

$$\int_0^2 \sqrt{4-x^2}\,dx - \int_0^2 (2-x)dx \qquad \ldots\text{(i)}$$

$$= \left[\frac{x\cdot\sqrt{4-x^2}}{2} + \frac{4}{2}\sin^{-1}\frac{x}{2}\right]_0^2 - \left[2x - \frac{x^2}{2}\right]_0^2$$

$$= \left[\frac{x\cdot\sqrt{4-x^2}}{2} + 2\sin^{-1}\frac{x}{2} - 2x + \frac{x^2}{2}\right]_0^2$$

$$= \frac{2\cdot\sqrt{4-4}}{2} + 2\sin^{-1}(1) - 4 + 2$$

$$= 2\cdot\frac{\pi}{2} - 2$$

$$= (\pi - 2) \text{ sq. unit.}$$ **Ans.**

Solution 7.

(a)

x	x^2	y	y^2	xy
9	81	−4	16	−36
10	100	−3	9	−30
11	121	−1	1	−11
12	144	0	0	0
13	169	1	1	13
14	196	3	9	42
15	225	5	25	75
16	256	8	64	128
$\Sigma x = 100$	$\Sigma x^2 = 1292$	$\Sigma y = 9$	$\Sigma y^2 = 125$	$\Sigma xy = 181$

$$\bar{x} = \frac{\Sigma x}{N} = \frac{100}{8} = 12{\cdot}5 \quad (\because N = 8)$$

$$\bar{y} = \frac{\Sigma y}{N} = \frac{9}{8} = 1{\cdot}125$$

$$b_{yx} = \frac{\Sigma xy - \frac{1}{N}\Sigma x\Sigma y}{\Sigma x^2 - \frac{1}{N}(\Sigma x)^2}$$

$$= \frac{181 - \frac{9\times 100}{8}}{1292 - \frac{10000}{8}} = \frac{181 - 112{\cdot}5}{1292 - 1250}$$

$$= \frac{181 - 112{\cdot}5}{42} = \frac{68{\cdot}5}{42} = 1{\cdot}63$$

$$b_{xy} = \frac{\Sigma xy - \frac{1}{N}\Sigma x\Sigma y}{\Sigma y^2 - \frac{1}{N}(\Sigma y)^2}$$

$$= \frac{181 - \frac{900}{8}}{125 - \frac{81}{8}}$$

$$= \frac{181 - 112{\cdot}5}{125 - 10{\cdot}125}$$

$$= \frac{68{\cdot}5}{114{\cdot}875} = 0{\cdot}596$$

Regression line of y on x,

$$y - \bar{y} = b_{yx}(x - \bar{x})$$

$$y - 1{\cdot}125 = 1{\cdot}63\,(x - 12{\cdot}5)$$

$$\Rightarrow \quad y - 1{\cdot}125 = 1{\cdot}63x - 20{\cdot}375$$

$$\Rightarrow \quad y = 1{\cdot}63x - 19{\cdot}25$$

Regression line of x on y,

$$x - \bar{x} = b_{xy}(y - \bar{y})$$

$$x - 12{\cdot}5 = 0{\cdot}596\,(y - 1{\cdot}125)$$

$$x - 12{\cdot}5 = 0{\cdot}596y - 0{\cdot}67$$

$$x = 0{\cdot}596y + 11{\cdot}83$$

Now, $\quad y = 1{\cdot}63x - 19{\cdot}25$

For $x = 13{\cdot}5$

The estimated value of $y = 1{\cdot}63 \times 13{\cdot}5 - 19{\cdot}25$

$$= 22{\cdot}005 - 19{\cdot}25$$

$$= 2{\cdot}755 \simeq 2{\cdot}8$$ **Ans.**

Solution 8.

(a) Here, there are four possibilities,

First Ball	Second Ball	Event
White	White	E_1
White	Black	E_2
Black	White	E_3
Black	Black	E_4

Let E denote the event of drawing a black ball in the third attempt

P (E/E_1) = 1, because there is no white ball to be selected.

P (E/E_2) = 3/4, because there are 3 black balls and 1 white ball

P (E/E_3) = 3/4, because again there are 3 black and 1 white ball

P (E/E_4) = 4/6 = 2/3, because there are 4 black and 2 white balls

$$\therefore \quad P(E) = P(E_1)\,P(E/E_1) + P(E_2)\,P(E/E_2) + P(E_3)\,P(E/E_3) + P(E_4)\,P(E/E_4)$$

$$= \frac{1}{6}\times 1 + \frac{1}{3}\times\frac{3}{4} + \frac{1}{5}\times\frac{3}{4} + \frac{3}{10}\times\frac{2}{3}$$

$$= \frac{1}{6} + \frac{1}{4} + \frac{3}{20} + \frac{1}{5}$$

$$= \frac{10+15+9+12}{60} = \frac{46}{60} = \frac{23}{30}. \quad \textbf{Ans.}$$

(b) Given, $P(A) = \frac{4}{5}, P(B) = \frac{3}{4}, P(C) = \frac{2}{3}$

$$\therefore P(\bar{A}) = 1 - \frac{4}{5} = \frac{1}{5}, P(\bar{B}) = 1 - \frac{3}{4} = \frac{1}{4}, P(\bar{C}) = 1 - \frac{2}{3} = \frac{1}{3}$$

(i) P (exactly two persons hit the target)

$$= P(A \cap B \cap \bar{C}) + P(A \cap \bar{B}) \cap C) + P(\bar{A} \cap B \cap C)$$

$$= P(A) \cdot P(B) \cdot P(\bar{C}) + P(A) \cdot P(\bar{B}) \cdot P(C) + P(\bar{A}) \cdot P(B) \cdot P(C)$$

$$= \frac{4}{5} \cdot \frac{3}{4} \cdot \frac{1}{3} + \frac{4}{5} \cdot \frac{1}{4} \cdot \frac{2}{3} + \frac{1}{5} \cdot \frac{3}{4} \cdot \frac{2}{3}$$

$$= \frac{12+8+6}{60} = \frac{26}{60} = \frac{13}{30}. \quad \textbf{Ans.}$$

(ii) P (at least two persons hit the target)

= P (all three persons hit the target) + P (exactly two persons hit the target)

$$= P(A \cap B \cap C) + \frac{13}{30} \quad \text{[Using (i)]}$$

$$= P(A)\, P(B)\, P(C) + \frac{13}{30}$$

$$= \frac{4}{5} \cdot \frac{3}{4} \cdot \frac{2}{3} + \frac{13}{30}$$

$$= \frac{2}{5} + \frac{13}{30} = \frac{12+13}{30} = \frac{25}{30} = \frac{5}{6}. \quad \textbf{Ans.}$$

(iii) P (None hit the target)

$$= P(\bar{A} \cap \bar{B} \cap \bar{C})$$

$$= P(\bar{A}) \cdot P(\bar{B}) \cdot P(\bar{C})$$

$$= \frac{1}{5} \cdot \frac{1}{4} \cdot \frac{1}{3} = \frac{1}{60}. \quad \textbf{Ans.}$$

Solution 9.

(b) Given diffferential equation is,

$$e^{x/y}\left(1 - \frac{x}{y}\right) + (1 + e^{x/y})\frac{dx}{dy} = 0$$

$$\Rightarrow \frac{dx}{dy} = \frac{-e^{x/y}(1-x/y)}{1+e^{x/y}} \quad \text{...(i)}$$

Since (i) is of the form $\frac{dx}{dy} = g(x/y)$. It is a homogeneous differential equation.

Put $x = vy$

$$\Rightarrow \frac{dx}{dy} = v + y\frac{dv}{dy}$$

Substituting the values of x and $\frac{dx}{dy}$ in (i), we get

$$v + y\frac{dv}{dy} = \frac{-e^v(1-v)}{1+e^v}$$

$$\Rightarrow y\frac{dv}{dy} = \frac{-e^v(1-v)}{1+e^v} - v$$

$$= \frac{-v-e^v}{1+e^v}$$

$$\Rightarrow \frac{1+e^v}{v+e^v}dv = -\frac{dy}{y}$$

On integrating both sides, we get

$$\log|v + e^v| = -\log|y| + \log c$$

$$\Rightarrow \log|y(v + e^v)| = \log c$$

$$\Rightarrow y\left(\frac{x}{y} + e^{x/y}\right) = c$$

$$\Rightarrow x + y\,e^{x/y} = c \quad \text{... (i)}$$

Where, A is arbitrary constant.

when $x = 0, y = 1 \Rightarrow c = 1$

$$[\because 0 + 1\,e^{0/1} = 0 + 1 \times 1 = 1]$$

Substituting this value of A in equation (i), we get

$$x + y\,e^{x/y} = 1,$$

which is the required particular solution. **Ans.**

SECTION—B

Solution 10.

(a) Let O be the centre of the semicircle and BA be the diameter. Let P be any point on the circumference of the semicircle.

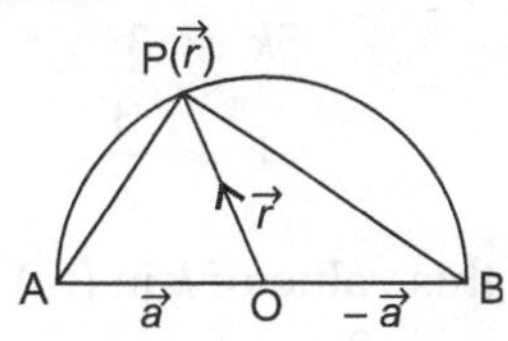

Let $\vec{OA} = \vec{a}$ then $= \vec{OB} = -\vec{a}$

Let $\vec{OP} = \vec{r}$

$$\therefore \vec{AP} = \vec{OP} - \vec{OA} = \vec{r} - \vec{a}$$

$$\vec{BP} = \vec{OP} - \vec{OB}$$

$$= \vec{r} - (-\vec{a}) = \vec{r} + \vec{a}$$

$$\vec{AP} \cdot \vec{BP} = (\vec{r} - \vec{a}) \cdot (\vec{r} + \vec{a})$$

$= \vec{r}^2 - \vec{a}^2$

$= \vec{a}^2 - \vec{a}^2$ [$\because r = a$ as, OP = OA]

$= 0$

$\because \overrightarrow{AP}$ is perpendicular to $\overrightarrow{BP}$.

$\Rightarrow \quad \angle APB = 90°.$ **Hence Proved.**

(b) Given, $\vec{a} = 2\hat{i} - 3\hat{j} - 4\hat{k}$

$\vec{b} = \hat{i} + 2\hat{j} - \hat{k}$

$\vec{c} = 3\hat{i} + \hat{j} + 2\hat{k}$

$$[\vec{a}\,\vec{b}\,\vec{c}] = \begin{vmatrix} 2 & -3 & -4 \\ 1 & 2 & -1 \\ 3 & 1 & 2 \end{vmatrix}$$

Expanding along first row, we get

$$\Rightarrow 2\begin{vmatrix} 2 & -1 \\ 1 & 2 \end{vmatrix} - (-3)\begin{vmatrix} 1 & -1 \\ 3 & 2 \end{vmatrix} + (-4)\begin{vmatrix} 1 & 2 \\ 3 & 1 \end{vmatrix}$$

$= 2(4 + 1) + 3(2 + 3) - 4(1 - 6)$

$= 2 \times 5 + 3 \times 5 - 4(-5)$

$= 10 + 15 + 20 = 45$ cu units **Ans.**

Solution 11.

(a) The equations of the given planes are

$x + y + z + 1 = 0$ ...(i)

$2x - 3y + 5z - 2 = 0$ ...(ii)

The equation of any plane passing through the line of intersection of the planes (i) and (ii) is

$x + y + z + 1 + k(2x - 3y + 5z - 2) = 0$ ...(iii)

where k is a parameter.

It passes through the point $(-1, 2, 1)$ if

$-1 + 2 + 1 + 1 + k(2 \times (-1) - 3(2) + 5 \times 1 - 2) = 0$

$\Rightarrow \quad 3 + k(-10 + 5) = 0$

$\Rightarrow \quad 3 - 5k = 0$

$\Rightarrow \quad 5k = 3$

$\Rightarrow \quad k = \frac{3}{5}$

Substituting this value of k in (iii),

$\Rightarrow x + y + z + 1 + \frac{3}{5}(2x - 3y + 5z - 2) = 0$

$\Rightarrow 5x + 5y + 5z + 5 + 6x - 9y + 15z - 6 = 0$

$\Rightarrow \quad 11x - 4y + 20z - 1 = 0$

which is the equation of the required plane.

Ans.

(b) Let l_1 and l_2 be the given lines whose equations are

$\vec{r} = \vec{a_1} + \lambda \vec{b}$

and $\vec{r} = \vec{a_2} + 2\mu \vec{b}$

Through the points $\vec{a_1} = \hat{i} + 2\hat{j} + 3\hat{k}$ and $\vec{a_2} = 2\hat{i} + 4\hat{j} + 5\hat{k}$ are parallel to $\vec{b} = 2\hat{i} + 3\hat{j} + 4\hat{k}$.

Hence the distance between the lines using the formula :

$$\frac{|\vec{b} \times (a_2 - a_1)|}{|\vec{b}|} = \frac{|(2\hat{i}+3\hat{j}+4\hat{k}) \times (\hat{i}+2\hat{j}+2\hat{k})|}{|\vec{b}|}$$

$$= \frac{|(2\hat{i}+3\hat{j}+4\hat{k}) \times (\hat{i}+2\hat{j}+2\hat{k})|}{|2\hat{i}+3\hat{j}+4\hat{k}|}$$

$$= \begin{vmatrix} \hat{i} & \hat{j} & \hat{k} \\ 2 & 3 & 4 \\ 1 & 2 & 2 \end{vmatrix}$$

$$= \sqrt{4+9+16}$$

$$= \frac{|-2\hat{i}+\hat{k}|}{\sqrt{29}} = \frac{\sqrt{5}}{\sqrt{29}} = 0{\cdot}415$$ **Ans.**

[**Note :** b_1 and b_2 are parallel as $b_2 = 2b_1$ and hence in the aforesaid formula b can be substituted by either of the two and answer remains unaltered.]

Solution 12.

(a) Let E_1, E_2, E_3 and A be the events defined as follows :

E_1 : red face turns up

E_2 : yellow face turns up

E_3 : green face turns up

A : white ball is drawn

We wish to calculate the probability that dice had turn up with red face when event A has occured (white ball is selected)

i.e., P (E_1/A)

By Bayes' theorem,

$$P(E_1/A) = \frac{P(E_1) \times P(A/E_1)}{P(E_1) \times P(A/E_1) + P(E_2) \times P(A/E_2) + P(E_3) \times P(A/E_3)}$$

$$= \frac{\frac{1}{2} \times \frac{2}{5}}{\frac{1}{2} \times \frac{2}{5} + \frac{1}{3} \times \frac{4}{5} + \frac{1}{6} \times \frac{3}{7}}$$

$$= \frac{\frac{1}{5}}{\frac{1}{5} + \frac{4}{15} + \frac{1}{14}}$$

$$= \frac{\frac{1}{5}}{\frac{113}{210}} = \frac{1}{5} \times \frac{210}{113}$$

$$= \frac{42}{113} = 0{\cdot}37$$ **Ans.**

(b) Let A be the event occurring an odd number and it is success.

$$p = P(A) = \frac{1}{2}, q = 1 - p = 1 - \frac{1}{2} = \frac{1}{2}$$

$\therefore$ P(at most 3 success) or P(maximum of 3 success)

$$= \sum_{r=0}^{3} {}^nC_r p^r q^{n-r} \text{ where } n \text{ is 5.}$$

$$= {}^5C_0\left(\frac{1}{2}\right)^5 + {}^5C_1\left(\frac{1}{2}\right)\left(\frac{1}{2}\right)^4 + {}^5C_2\left(\frac{1}{2}\right)^2\left(\frac{1}{2}\right)^3 + {}^5C_3\left(\frac{1}{2}\right)^3\left(\frac{1}{2}\right)^2$$

$$= \frac{1}{32} + \frac{5}{32} + \frac{10}{32} + \frac{10}{32}$$

$$= \frac{26}{32} = \frac{13}{16} = 0{\cdot}81$$ **Ans.**

SECTION—C

Solution 13.

(b) Let x be the number of units of food X and Y be the number of units of food Y be mixed to obtain the desired diet. Then LPP of the given problem is :

Minimize $Z = 24x + 36y$

Subject to the constraints,

$$x + 2y \geq 10,$$
$$2x + 2y \geq 12 \text{ i.e., } x + y \geq 6$$
$$3x + y \geq 8$$
$$x \geq 0, y \geq 0$$

We draw the lines $x + 2y = 10$, $x + y = 6$, $3x + y = 8$ and obtain the feasible region (unbounded and convex) is shown in the figure.

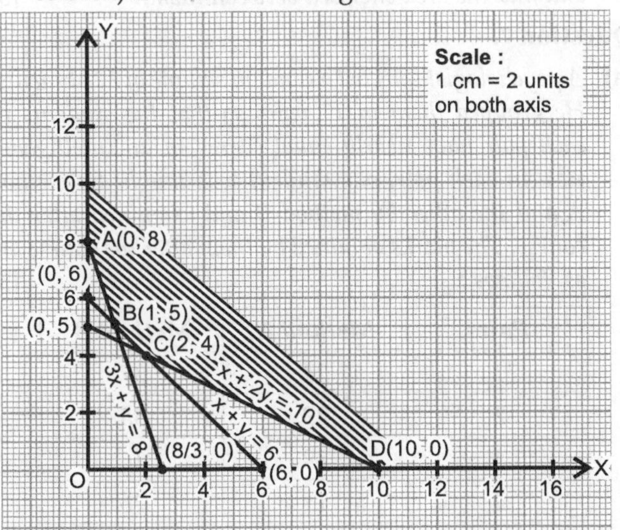

Thus, corner points are A (0, 8), B (1, 5), C (2, 4) and D (10, 0).

The values of Z (in R) at these points are given in the following table :

Corner Point	Objective Function $Z = 24x + 36y$
A (0, 8)	$Z = 24 \times 0 + 36 \times 8 = 288$
B (1, 5)	$Z = 24 \times 1 + 36 \times 5 = 204$
C (2, 4)	$Z = 24 \times 2 + 36 \times 4 = 192$ minimum
D (10, 0)	$Z = 24 \times 10 + 36 \times 0 = 240$

Therefore, Z has the minimum value and the minimum value is ₹192.

It occurs at C (2, 4).

i.e., when 2 kg of food X and 4 kg of food Y are mixed to get the desired diet. **Ans.**

Solution 14.

(b) Given, $\quad AC = x + 5 + \frac{36}{x}$

(i) $\quad$ Total cost (C) $= (AC)\,x$

$$= \left(x + 5 + \frac{36}{x}\right) \cdot x$$

$$= x^2 + 5x + 36$$

$$MC = \frac{d}{dx}(C)$$

$$= \frac{d}{dx}(x^2 + 5x + 36)$$

$$= 2x + 5.$$ **Ans.**

(ii) AC increases when $\frac{d}{dx}(AC) > 0$

$$\frac{d}{dx}\left(x + 5 + \frac{36}{x}\right) > 0$$

$$1 - \frac{36}{x^2} > 0$$

$$\Rightarrow \quad x^2 - 36 > 0$$

$$\Rightarrow \quad (x + 6)(x - 6) > 0$$

$$\Rightarrow \quad x < -6, x > 6$$

But x is positive therefore AC increases when $x > 6$. **Ans.**

LL

MATHEMATICS 2014

QUESTIONS

SECTION—A

Question 1. **[10 × 3]**

(i) *If* $A = \begin{bmatrix} 3 & 1 \\ 7 & 5 \end{bmatrix}$, *find the values of x and y such that* $A^2 + xI = yA$.

(ii) *Find the eccentricity and the coordinates of foci of the hyperbola* $25x^2 - 9y^2 = 225$.**

(iii) *Evaluate :* $\tan\left[2\tan^{-1}\frac{1}{2} - \cot^{-1}3\right]$

(iv) *Using L'Hospital's rule, evaluate :*

$$\lim_{x\to 0}(1 + \sin x)^{\cot x}$$

(v) *Evaluate :* $\int e^x \frac{(2 + \sin 2x)}{\cos^2 x}dx$

(vi) *Using properties of definite integrals, evaluate*

$$\int_0^{\pi/2} \frac{\sqrt{\sin x}}{\sqrt{\sin x} + \sqrt{\cot x}}dx$$

(vii) *For the given lines of regression,* $3x - 2y = 5$ *and* $x - 4y = 7$, *find :*

(a) *regression coefficients* b_{yx} *and* b_{xy}

(b) *coefficient of correlation* $r(x, y)$

(viii) *Express the complex number* $\frac{(1+\sqrt{3}i)^2}{\sqrt{3}-1}$ *in the form of a + ib. Hence, find the modulus and argument of the complex number.***

(ix) *A bag contains 20 balls numbered from 1 to 20. One ball is drawn at random from the bag. What is the probability that the ball drawn is marked with a number which is multiple of 3 or 4 ?*

(x) *Solve the differential equation :*

$$(x + 1)\,dy - 2xy\,dx = 0$$

Question 2.

(a) *Using properties of determinants, prove that :* **[5]**

$$\begin{vmatrix} a^2+1 & ab & ac \\ ba & b^2+1 & bc \\ ca & cb & c^2+1 \end{vmatrix} = a^2 + b^2 + c^2 + 1$$

(b) *Using matrix method, solve the following system of equation :* **[5]**

$$x - 2y = 10,\ 2x + y + 3z = 8 \text{ and } -2y + z = 7$$

Question 3.

(a) *If* $\cos^{-1}x + \cos^{-1}y + \cos^{-1}z = \pi$, *prove that* $x^2 + y^2 + z^2 + 2xyz = 1$ **[5]**

(b) *P, Q and R represent switches in on position and P′, Q′ and R′ represent switches in off position. Construct a switching circuit representing the polynomial PR + Q (Q′ + R) (P + QR). Using Boolean Algebra, simplify the polynomial expression and construct the simplified circuit.*** **[5]**

Question 4.

(a) *Verify Rolle's theorem for the function* $f(x) = e^x(\sin x - \cos x)$ *on* $\left[\frac{\pi}{4}, \frac{5\pi}{4}\right]$. **[5]**

(b) *Find the equation of the parabola with latusrectum joining points (4, 6) and (4, – 2)*** **[5]**

Question 5.

(a) *If* $y = \frac{x\sin^{-1}x}{\sqrt{1-x^2}}$, *prove that :* $(1 - x^2)\frac{dy}{dx} = x + \frac{y}{x}$ **[5]**

(b) *A wire of length 50 m is cut into two pieces. One piece of the wire is bent in the shape of a square and the other in the shape of a circle. What should be the length of each piece so that the combined area of the two is minimum ?* **[5]**

Question 6.

(a) *Evaluate :* $\int \frac{x + \sin x}{1 + \cos x}dx$. **[5]**

(b) *Sketch the graphs of the curves* $y^2 = x$ *and* $y^2 = 4 - 3x$ *and find the area enclosed between them.* **[5]**

Question 7.

(a) *A psychologist selected a random sample of 22 students. He grouped them in 11 pairs so that the students in each pair have nearly equal scores in an intelligence test. In each pair, one student was taught by method A and the other by method B and examined after the course. The marks obtained by them after the course are as follows :* **[5]**

Pairs	Method A	Method B
1	24	37
2	29	35
3	19	16
4	14	26

** Answer is not given due to change in the present syllabus.

5	30	23
6	19	27
7	27	19
8	30	20
9	20	16
10	28	11
11	11	21

*Calculate Spearman's Rank correlation.***

(b) *The coefficient of correlation between the values denoted by X and Y is 0·5. The mean of X is 3 and that of Y is 5. Their standard deviations are 5 and 4 respectively. Find :*

(i) *the two lines of regression.*

(ii) *the expected value of Y, when X is given 14.*

(iii) *the expected value of X, when Y is given 9.* **[5]**

Question 8.

(a) *In a college, 70% students pass in Physics, 75% pass in Mathematics and 10% students fail in both. One student is chosen at random. What is the probability that :* **[5]**

(i) *He passes in Physics and Mathematics.*

(ii) *He passes in Mathematics given that he passes in Physics.*

(iii) *He passes in Physics given that he passes in Mathematics.*

(b) *A bag contains 5 white and 4 black balls and another bag contains 7 white and 9 black balls. A ball is drawn from the first bag and two balls drawn from the second bag. What is the probability of drawing one white and two black balls ?* **[5]**

Question 9.

(a) *Using De. Moivre's theorem, find the least positive integer n such that* $\left(\frac{2i}{1+i}\right)^n$ *is a positive integer.*** **[5]**

(b) *Solve the following differential equation :*

$$(3xy + y^2)\,dx + (x^2 + xy)\,dy = 0$$ **[5]**

SECTION—B

Question 10.

(a) *In a triangle ABC, using vectors, prove that* $c^2 = a^2 + b^2 - 2ab\cos C$. **[5]**

(b) *Prove that :*

$$\vec{a}\cdot(\vec{b}+\vec{c})\times(\vec{a}+2\vec{b}+3\vec{c}) = [\vec{a}\ \vec{b}\ \vec{c}\,].$$ **[5]**

** Answer is not given due to change in the present syllabus.

Question 11.

(a) *Find the equation of a line passing through the points P (– 1, 3, 2) and Q (– 4, 2, – 2). Also, if the point R (5, 5, λ) is collinear with the points P and Q, then find the value of λ.* **[5]**

(b) *Find the equation of the plane passing through the points (2, – 3, 1) and (– 1, 1, – 7) and perpendicular to the plane* $x - 2y + 5z + 1 = 0$. **[5]**

Question 12.

(a) *In a bolt factory, three machines A, B and C manufacture 25%, 35% and 40% of the total production respectively. Of their respective outputs, 5%, 4% and 2% are defective. A bolt is drawn at random from the total production and it is found to be defective. Find the probability that it was manufactured by machine C.* **[5]**

(b) *On dialling certain telephone numbers, assume that on an average, one telephone number out of five is busy, ten telephone numbers are randomly selected and dialled. Find the probability that at least three of them will be busy.* **[5]**

SECTION—C

Question 13.

(a) *A person borrows ₹ 68,962 on the condition that he will repay the money with compound interest at 5% per annum in 4 equal annual instalments, the first one being payable at the end of the first year. Find the value of each instalment.*** **[5]**

(b) *A company manufactures two types of toys A and B. A toy of type A requires 5 minutes for cutting and 10 minutes for assembling. A toy of type B requires 8 minutes for cutting and 8 minutes for assembling. There are 3 hours available for cutting and 4 hours available for assembling the toys in a day. The profit is ₹ 50 each on a toy of type A and ₹ 60 each on a toy of type B. How many toys of each type should the company manufacture in a day to maximize the profit ? Use linear programming to find the solution.* **[5]**

Question 14.

(a) *A firm has the cost function* $C = \frac{x^3}{3} - 7x^2 + 111x + 50$ *and demand function* $x = 100 - p$.

(i) *Write the total revenue function in terms of x.*

(ii) *Formulate the total profit function P in terms of x.*

(iii) *Find the profit maximising level of output x.* **[5]**

(b) *A bill of ₹ 5050 is drawn on 13th April 2013. It was discounted on 4th July 2013 at 5% per annum. If the banker's gain on the transaction is ₹ 0·50, find the nominal date of the maturity of the bill.*** **[5]**

SOLUTIONS

Solution 1.

(i) Given, $A = \begin{bmatrix} 3 & 1 \\ 7 & 5 \end{bmatrix}$

$$A^2 = A \cdot A = \begin{bmatrix} 3 & 1 \\ 7 & 5 \end{bmatrix}\begin{bmatrix} 3 & 1 \\ 7 & 5 \end{bmatrix}$$

$$= \begin{bmatrix} 9+7 & 3+5 \\ 21+35 & 7+25 \end{bmatrix}$$

$$= \begin{bmatrix} 16 & 8 \\ 56 & 32 \end{bmatrix}$$

$$A^2 + xI^2 = yA$$

$$\begin{bmatrix} 16 & 8 \\ 56 & 32 \end{bmatrix} + x\begin{bmatrix} 1 & 0 \\ 0 & 1 \end{bmatrix} = y\begin{bmatrix} 3 & 1 \\ 7 & 5 \end{bmatrix}$$

$$\begin{bmatrix} 16+x & 8 \\ 56 & 32+x \end{bmatrix} = \begin{bmatrix} 3y & y \\ 7y & 5y \end{bmatrix}$$

$\Rightarrow \quad 7y = 56$

$\Rightarrow \quad y = 8$

and $\quad 16 + x = 3y$

$\therefore \quad x = 3 \times 8 - 16$

$= 24 - 16 = 8$

Thus, required value of x is 8 and y is 8. **Ans.**

(iii) $\tan\left[2\tan^{-1}\frac{1}{2} - \cot^{-1}3\right]$

$$= \tan\left[\tan^{-1}\left(\frac{2\times\frac{1}{2}}{1-\left(\frac{1}{2}\right)^2}\right) - \cot^{-1}3\right]$$

$$= \tan\left[\tan^{-1}\left\{\frac{1}{1-\frac{1}{4}}\right\} - \tan^{-1}\frac{1}{3}\right]$$

$$= \tan\left[\tan^{-1}\frac{4}{3} - \tan^{-1}\frac{1}{3}\right]$$

$$= \tan\left[\tan^{-1}\left(\frac{\frac{4}{3}-\frac{1}{3}}{1+\frac{4}{9}}\right)\right]$$

$$= \tan\left[\tan^{-1}\left(\frac{\frac{4-1}{3}}{\frac{9+4}{9}}\right)\right]$$

$$= \tan\left[\tan^{-1}\left(\frac{1}{13/9}\right)\right]$$

$$= \tan\left[\tan^{-1}\left(\frac{9}{13}\right)\right]$$

$$= \frac{9}{13}$$ **Ans.**

(iv) $\lim_{x\to 0}(1+\sin x)^{\cot x}$

Let $\quad L = \lim_{x\to 0}(1+\sin x)^{\cot x}$

$$\log L = \lim_{x\to 0}\log(1+\sin)^{\cot x}$$

$$\log L = \lim_{x\to 0}\cot x \log(1+\sin x)$$

$$= \lim_{x\to 0}\frac{\log(1+\sin x)}{\tan x}$$

$$\left[\frac{0}{0}\text{ form}\right]$$

Using L' Hospital's rule,

$$= \lim_{x\to 0}\frac{\frac{1}{1+\sin x}\times(\cos x)}{\sec^2 x}$$

$$= \lim_{x\to 0}\frac{\cos^3 x}{1+\sin x}$$

On using limits $x = 0$,

$$\log L = \frac{1}{1+0} = 1$$

$$\log L = 1$$

$$L = e^1 = e'$$ **Ans.**

(v) Let $\quad I = \int e^x\left(\frac{2}{\cos^2 x} + \frac{2\sin x\cos x}{\cos^2 x}\right)$

$$= \int e^x(2\sec^2 x + 2\tan x)\,dx$$

$$= 2\int e^x(\sec^2 x + \tan x)\,dx$$

$$= 2\left[\int e^x\sec^2 x\,dx + \int e^x\tan x\,dx\right]$$

$$= 2\left[e^x\int\sec^2 x\,dx - \int\left\{\frac{d}{dx}e^x\int\sec^2 x\,dx\right\}dx + \int e^x\tan x\,dx\right]$$

$$= 2\left[e^x\tan x - \int e^x\tan x\,dx + \int e^x\tan x\,dx\right]$$

$$= 2\,e^x\tan x + c$$ **Ans.**

(vi) Let $$I = \int_0^{\pi/2} \frac{\sqrt{\sin x}}{\sqrt{\sin x} + \sqrt{\cos x}}\, dx \quad ...(i)$$

$$= \int_0^{\pi/2} \frac{\sqrt{\sin\left(\frac{\pi}{2} - x\right)}}{\sqrt{\sin\left(\frac{\pi}{2} - x\right)} + \sqrt{\cos\frac{\pi}{2} - x}}\, dx$$

(Using property $\int_0^a f(x)\,dx = \int_0^a f(a-x)\,dx$)

$$I = \int_0^{\pi/2} \frac{\sqrt{\cos x}}{\sqrt{\cos x} + \sqrt{\sin x}}\, dx \quad ...(ii)$$

Adding equations (i) and (ii),

$$2I = \int_0^{\pi/2} \frac{\sqrt{\sin x} + \sqrt{\cos x}}{\sqrt{\sin x} + \sqrt{\cos x}}\, dx$$

$$2I = \int_0^{\pi/2} 1\, dx = [x]_0^{\pi/2}$$

$$I = \frac{1}{2}\left[\frac{\pi}{2} - 0\right]$$

$$I = \frac{\pi}{4}$$ **Ans.**

(vii) (a) The two regression lines are

$$3x - 2y = 5 \quad ...(i)$$

$$x - 4y = 7 \quad ...(ii)$$

From equation (i),

$$3x - 2y = 5$$

$$3x = 2y + 5$$

$$x = \frac{2}{3}y + \frac{5}{3}$$

$\therefore \quad b_{xy} = \frac{2}{3}$ (Regression coefficient of x on y)

From equation (ii)

$$x - 4y = 7$$

$$4y = x - 7$$

$$y = \frac{x}{4} - \frac{7}{4}$$

$\therefore \quad b_{yx} = \frac{1}{4}$ (Regression coefficient of y on x)

Now $r^2 = b_{xy} \cdot b_{yx} = \frac{2}{3} \times \frac{1}{4} = \frac{1}{6} < 1$, therefore, our assumption is correct.

$\therefore$ (i) is the regression line of x on y and (ii) is the regression line of y on x. **Ans.**

(b) Coefficient of correlation $r(x, y) = |r|$

$$= \sqrt{b_{xy} \cdot b_{yx}}$$

$$= \sqrt{\frac{2}{3} \times \frac{1}{4}} = \sqrt{\frac{1}{6}}$$

$$= \frac{1}{\sqrt{6}}$$ **Ans.**

(ix) Number of balls = 20

$\therefore \quad n(S) = 20$

Number of balls marked with multiple of 3 = 3, 6, 9, 12, 15, 18

$\therefore \quad n(A) = 6$

Number of balls marked with multiple of 4 = 4, 8, 12, 16, 20

$n(B) = 5$

Number of ball multiple of both 3 and 4 = 12

$n(A \cap B) = 1$

The required probability $P(A \cup B)$

$$= P(A) + P(B) - P(A \cap B)$$

$$= \frac{6}{20} + \frac{5}{20} - \frac{1}{20}$$

$$= \frac{10}{20} = \frac{1}{2}$$ **Ans.**

(x) Given differential equation is

$$(x + 1)\, dy - 2xy\, dx = 0$$

$$\frac{dy}{dx} = \frac{2xy}{x+1}$$

$$\frac{1}{y}\, dy = \left(\frac{2x}{x+1}\right) dx$$

$$\frac{1}{y}\, dy = \left(\frac{2x+2-2}{x+1}\right) dx$$

$$\frac{1}{y}\, dy = \left[\frac{2(x+1)}{(x+1)} - \frac{2}{(x+1)}\right] dx$$

$$\therefore \quad \int \frac{1}{y}\, dy = \int 2dx - \int \frac{2}{(x+1)}\, dx$$

$$\log |y| = 2x - 2\log |x + 1| + \log c$$

$$\log |y| + 2\log |x + 1| - \log c = 2x$$

$$\log\left(\frac{y(x+1)^2}{c}\right) = 2x$$

$$\frac{y(x+1)^2}{c} = e^{2x}$$

$$y = \frac{ce^{2x}}{(x+1)^2}$$ **Ans.**

Solution 2.

(a) Let $\Delta = \begin{vmatrix} a^2+1 & ab & ac \\ ba & b^2+1 & bc \\ ca & cb & c^2+1 \end{vmatrix}$

Applying $R_1 \to \frac{1}{a} R_1, R_2 \to \frac{1}{b} R_2, R_3 \to \frac{1}{c} R_3$

$$\Delta = \begin{vmatrix} a+\frac{1}{a} & b & c \\ a & b+\frac{1}{b} & c \\ a & b & c+\frac{1}{c} \end{vmatrix}$$

Applying $C_1 \to aC_1, C_2 \to bC_2, C_3 \to cC_3$

$$\Delta = \begin{vmatrix} a^2+1 & b^2 & c^2 \\ a^2 & b^2+1 & c^2 \\ a^2 & b^2 & c^2+1 \end{vmatrix}$$

Applying $C1 \to C_1 + C_2 + C_3$

$$\Delta = \begin{vmatrix} 1+a^2+b^2+c^2 & b^2 & c^2 \\ 1+a^2+b^2+c^2 & b^2+1 & c^2 \\ 1+a^2+b^2+c^2 & b^2 & c^2+1 \end{vmatrix}$$

Taking $(1 + a^2 + b^2 + c^2)$ as common

$$\Delta = (1+a^2+b^2+c^2)\begin{vmatrix} 1 & b^2 & c^2 \\ 1 & b^2+1 & c^2 \\ 1 & b^2 & c^2+1 \end{vmatrix}$$

Applying $R_2 \to R_2 - R_1$ and $R_3 \to R_3 - R_1$

$$\Delta = 1+a^2+b^2+c^2 \begin{vmatrix} 1 & b^2 & c^2 \\ 0 & 1 & 0 \\ 0 & 0 & 1 \end{vmatrix}$$

Expanding along C_1, we get

$$\Delta = 1+a^2+b^2+c^2 \begin{vmatrix} 1 & 0 \\ 0 & 1 \end{vmatrix}$$

$$= 1+a^2+b^2+c^2.$$ **Hence Proved.**

(b) Given system of equation is,

$$x - 2y = 10 \quad \ldots\text{(i)}$$
$$2x + y + 3z = 8 \quad \ldots\text{(ii)}$$
$$-2y + z = 7 \quad \ldots\text{(iii)}$$

$$A = \begin{bmatrix} 1 & -2 & 0 \\ 2 & 1 & 3 \\ 0 & -2 & 1 \end{bmatrix}, X = \begin{bmatrix} x \\ y \\ z \end{bmatrix},$$

and $B = \begin{bmatrix} 10 \\ 8 \\ 7 \end{bmatrix}$

$$|A| = 1(1+6) + 2(2-0) + 0$$
$$= 7 + 4 = 11 \neq 0$$

$\therefore A^{-1}$ exists.

$\therefore$ The system has the unique solution $X = A^{-1}B$.

$$A_{11} = \begin{vmatrix} 1 & 3 \\ -2 & 1 \end{vmatrix} = (1+6) = 7,$$

$$A_{12} = -\begin{vmatrix} 2 & 3 \\ 0 & 1 \end{vmatrix} = -2,$$

$$A_{13} = \begin{vmatrix} 2 & 1 \\ 0 & -2 \end{vmatrix} = -4$$

$$A_{21} = -\begin{vmatrix} -2 & 0 \\ -2 & 1 \end{vmatrix} = -(-2) = 2,$$

$$A_{22} = \begin{vmatrix} 1 & 0 \\ 0 & 1 \end{vmatrix} = 1,$$

$$A_{23} = -\begin{vmatrix} 1 & -2 \\ 0 & -2 \end{vmatrix} = -(-2) = 2$$

$$A_{31} = \begin{vmatrix} -2 & 0 \\ 1 & 3 \end{vmatrix} = -6,$$

$$A_{32} = -\begin{vmatrix} 1 & 0 \\ 2 & 3 \end{vmatrix} = -(3) = -3,$$

$$A_{33} = \begin{vmatrix} 1 & -2 \\ 2 & 1 \end{vmatrix} = 1 + 4 = 5$$

$$\therefore \quad \text{adj } A = \begin{bmatrix} 7 & -2 & -4 \\ 2 & 1 & 2 \\ -6 & -3 & 5 \end{bmatrix}^T$$

$$= \begin{bmatrix} 7 & 2 & -6 \\ -2 & 1 & -3 \\ -4 & 2 & 5 \end{bmatrix}$$

$$A^{-1} = \frac{\text{adj } A}{|A|}$$

$$= \frac{1}{11}\begin{bmatrix} 7 & 2 & -6 \\ -2 & 1 & -3 \\ -4 & 2 & 5 \end{bmatrix}$$

Now, $AX = B \Rightarrow X = A^{-1}B$

$$\begin{bmatrix} x \\ y \\ z \end{bmatrix} = \frac{1}{11}\begin{bmatrix} 7 & 2 & -6 \\ -2 & 1 & -3 \\ -4 & 2 & 5 \end{bmatrix}\begin{bmatrix} 10 \\ 8 \\ 7 \end{bmatrix}$$

$$\begin{bmatrix} x \\ y \\ z \end{bmatrix} = \frac{1}{11}\begin{bmatrix} 70+16-42 \\ -20+8-21 \\ -40+16+35 \end{bmatrix}$$

$$= \frac{1}{11}\begin{bmatrix} 44 \\ -33 \\ 11 \end{bmatrix}$$

$\Rightarrow \quad x = 4, y = -3, z = 1.$ **Ans.**

Solution 3.

(a) Given, $\cos^{-1}x + \cos^{-1}y + \cos^{-1}z = \pi$

$$\cos^{-1}x + \cos^{-1}y = \pi - \cos^{-1}z$$

$$\cos^{-1}\left(xy - \sqrt{1-x^2}\sqrt{1-y^2}\right) = \pi - \cos^{-1}z$$

$$xy - \sqrt{1-x^2}\sqrt{1-y^2} = \cos(\pi - \cos^{-1}z)$$

$$xy - \sqrt{1-x^2}\sqrt{1-y^2} = -\cos(\cos^{-1}z)$$

$$xy - \sqrt{1-x^2}\sqrt{1-y^2} = -z$$

$$xy + z = \sqrt{1-x^2}\sqrt{1-y^2}$$

Squaring both L.H.S. and R.H.S., we get

$$(xy+z)^2 = (1-x^2)(1-y^2)$$

$$x^2y^2 + z^2 + 2xyz = 1 - x^2 - y^2 + x^2y^2$$

$$x^2 + y^2 + z^2 + 2xyz = 1$$ **Hence Proved.**

Solution 4.

(a) Given function is,

$$f(x) = e^x(\sin x - \cos x), x \in \left[\frac{\pi}{4}, \frac{5\pi}{4}\right]$$

Sine, cosine and exponential functions are always continuous.

$\therefore$ Given function is continuous in $\left[\frac{\pi}{4}, \frac{5\pi}{4}\right]$

Differentiating w.r.t. x, we get

$$f'(x) = e^x(\cos x + \sin x) + (\sin x - \cos x)e^x$$

$$= e^x[\cos x + \sin x + \sin x - \cos x]$$

$$= 2e^x \sin x$$

which exists for all x in $\left(\frac{\pi}{4}, \frac{5\pi}{4}\right)$

$$f(\pi/4) = e^{\pi/4}\left(\frac{1}{\sqrt{2}} - \frac{1}{\sqrt{2}}\right) = 0$$

and $f(5\pi/4) = e^{5\pi/4}\left(-\frac{1}{\sqrt{2}} + \frac{1}{\sqrt{2}}\right) = 0$

$\therefore \quad f(\pi/4) = f(5\pi/4) = 0$

$\therefore$ The given function satisfies all three condition of Rolle's theorem.

For maxima or minima,

$$f'(c) = 0$$

$$2e^c \sin c = 0$$

$$\sin c = 0$$

$$c = n\pi + (-1)^n(0)$$

$$c = n\pi$$

$$c = \pi \quad \text{(for } n = 1\text{)}$$

π lies between $\left[\frac{\pi}{4}, \frac{5\pi}{4}\right]$ so Rolle's theorem is verified.

Solution 5.

(a) Given, $y = \frac{x\sin^{-1}x}{\sqrt{1-x^2}}$

$$y \cdot \sqrt{1-x^2} = x\sin^{-1}x \quad \ldots\text{(i)}$$

Differentiating both sides w.r.t. x, we get

$$y \cdot \frac{(-2x)}{2\sqrt{1-x^2}} + \sqrt{1-x^2}\frac{dy}{dx} = x \cdot \frac{1}{\sqrt{1-x^2}} + \sin^{-1}x$$

$$= \frac{x + \sqrt{1-x^2}\sin^{-1}x}{\sqrt{1-x^2}}$$

$$-xy + (1-x^2)\frac{dy}{dx} = x + \sqrt{1-x^2} \cdot \frac{y}{x}\sqrt{1-x^2} \quad \text{[from (i)]}$$

$$-xy + (1-x^2)\frac{dy}{dx} = x + \frac{y}{x}(1-x^2)$$

$$-xy + (1-x^2)\frac{dy}{dx} = x + \frac{y}{x} - yx$$

$$(1-x^2)\frac{dy}{dx} = x + \frac{y}{x}$$ **Hence Proved.**

(b) Length of wire = 50 m (Given)

Let, length of one piece for shape of square = x m

$\therefore$ Length of other piece for shape of circle

$$= (50 - x) \text{ m}$$

Now, perimeter of square = $4a = x$

$$\Rightarrow \quad a = \frac{x}{4}$$

and circumference of circle = $2\pi r = 50 - x$

$$\Rightarrow \quad r = \frac{50-x}{2\pi}$$

Combined area = $a^2 + \pi r^2$

$$= \frac{x^2}{16} + \pi\cdot\left(\frac{50-x}{2\pi}\right)^2$$

$$= \frac{x^2}{16} + \pi\cdot\frac{(50-x)^2}{4\pi^2}$$

$$A = \frac{x^2}{16} + \frac{(50-x)^2}{4\pi}$$

Differentiating w.r. t. x, we get

$$\frac{dA}{dx} = \frac{2x}{16} + \frac{2(50-x)(-1)}{4\pi}$$

$$\frac{dA}{dx} = \frac{x}{8} + \frac{(x-50)}{2\pi}$$

$$= \frac{\pi x + 4x - 200}{8\pi}$$

$$= \frac{x(4+\pi) - 200}{8\pi}$$

For maxima or minima,

$$\frac{dA}{dx} = 0$$

$$\therefore \quad x(4+\pi) - 200 = 0$$

$$x = \frac{200}{4+\pi}$$

since $\quad \frac{d^2A}{dx^2} > 0$

$\Rightarrow$ A is minimum at $x = \frac{200}{4+\pi}$

$\therefore$ Length of square wire, $x = \frac{200}{4+\pi}$ m

and length of circle wire = $50 - x$

$$= 50 - \frac{200}{4+\pi}$$

$$= \frac{50\pi}{4+\pi} \text{ m}$$ **Ans.**

Solution 6.

(a) Let $I = \int \frac{x + \sin x}{1 + \cos x}\, dx$

$$= \int \frac{x + \sin x}{2\cos^2 x/2}\, dx$$

$$= \int \frac{x}{2\cos^2 x/2}\, dx + \int \frac{2\sin(x/2)\cos(x/2)}{2\cos^2 x/2}\, dx$$

$$= \frac{1}{2}\int x\sec^2 x/2\, dx + \int \tan x/2\, dx$$

$$= \frac{1}{2}\left[x.\int \sec^2 x/2\, dx - \int\left(\frac{d}{dx}(x).\int \sec^2 x/2\, dx\right)dx\right] + \int \tan x/2\, dx$$

$$= \frac{1}{2}\left[x.\frac{\tan x/2}{1/2} - \int \frac{\tan x/2\, dx}{1/2}\right] + \int \tan x/2\, dx + c$$

$$= x\tan x/2 - \int \tan x/2\, dx + \int \tan x/2\, dx + c$$

$$= x\tan x/2 + c$$ **Ans.**

(b) Given curves are $y^2 = -3x + 4$

$$y^2 = -3\left(x - \frac{4}{3}\right)$$

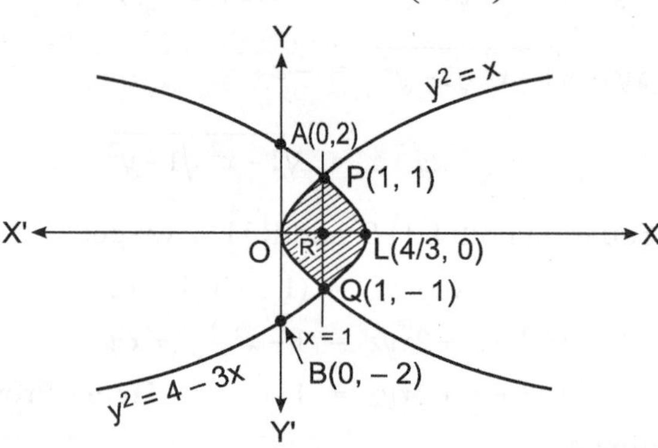

The vertex L of this parabola is $\left(\frac{4}{3}, 0\right)$. It cuts the Y-axis at A (0, 2) and B (0, – 2).

The points of intersection of these two parabolas are given by the equation

$$x = -3x + 4 \Rightarrow x = 1$$

Then $\quad y^2 = 1 \Rightarrow y = \pm 1$

Thus, the points of intersection are P (1, 1) and Q (1, – 1). Let PQ cut the X-axis at R.

$\therefore$ Total area of POQLP = 2 area of OPLRO

$$= 2\left[\int_0^1 \sqrt{x}\,dx + \int_1^{4/3} \sqrt{4-3x}\,dx\right]$$

$$= 2\left[\left(\frac{x^{3/2}}{3/2}\right)_0^1 + \left(\frac{2(4-3x)^{3/2}}{(-3)\times 3}\right)_1^{4/3}\right]$$

$$= 2\left[\left(\frac{2}{3}-0\right)-\frac{2}{9}(0-1)\right]$$

$$= 2\left[\frac{2}{3}+\frac{2}{9}\right] = 2\left[\frac{6+2}{9}\right]$$

$$= 2\left(\frac{8}{9}\right) = \frac{16}{9} \text{ sq. units}$$ **Ans.**

Solution 7.

(b) Given, $\bar{x} = 3, \bar{y} = 5, \sigma_x = 5, \sigma_y = 4, r = 0{\cdot}5$

$$b_{yx} = r \cdot \frac{\sigma_y}{\sigma_x} = 0{\cdot}5 \times \frac{4}{5} = 0{\cdot}4,$$

$$b_{xy} = r \cdot \frac{\sigma_x}{\sigma_y} = 0{\cdot}5 \times \frac{5}{4} = 0{\cdot}625$$

(i) Regression equation of y on x is

$$y - \bar{y} = b_{yx}(x - \bar{x})$$
$$y - 5 = (0{\cdot}4)(x - 3)$$
$$y = 0{\cdot}4x + 3{\cdot}8$$

Regression equation of x on y is

$$x - \bar{x} = b_{xy}(y - \bar{y})$$
$$x - 3 = (0{\cdot}625)(y - 5)$$
$$x = 0{\cdot}625\,y - 0{\cdot}125$$ **Ans.**

(ii) When $x = 14$

$$y = 0{\cdot}4 \times 14 + 3{\cdot}8$$
$$= 5{\cdot}6 + 3{\cdot}8 = 9{\cdot}4$$ **Ans.**

(iii) When $y = 9$

$$x = 0{\cdot}625 \times 9 - 0{\cdot}125$$
$$= 5{\cdot}625 - 0{\cdot}125 = 5{\cdot}5$$ **Ans.**

Solution 8.

(a) Let x% students pass in both Maths and Physics

Students who pass in Physics = 70%

$$\Rightarrow \quad P(A) = \frac{70}{100}$$

Students who pass in Maths = 75%

$$\Rightarrow \quad P(B) = \frac{75}{100}$$

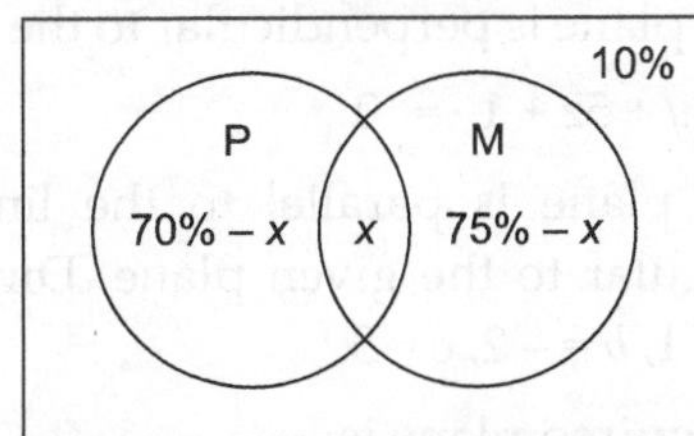

Students who fail in both $= P(\bar{A} \cap \bar{B})$

$$= 10\%$$

$$70\% - x + x + 75\% - x = 90\%$$

$$x = 55\% \Rightarrow P(B \cap A)$$
$$= \frac{55}{100}$$

(i) P (Passes in Physics and Mathematics)

$$= \frac{55}{100} = \frac{11}{20}$$ **Ans.**

(ii)

$$P(B/A) = \frac{P(A \cap B)}{P(A)}$$
$$= \frac{55/100}{70/100}$$
$$= \frac{55}{70} = \frac{11}{14}$$ **Ans.**

(iii)

$$P(A/B) = \frac{P(A \cap B)}{P(B)}$$
$$= \frac{55/100}{75/100}$$
$$= \frac{55}{75} = \frac{11}{15}$$ **Ans.**

(b)

Bag I	Bag II
4B	7W
5W	9B

Let A : the event of getting 1 black ball from Bag I

B : the event of getting 1 black and 1 white ball from Bag II

C : the event of getting 1 white ball from Bag I

D : the event of getting 2 black balls from Bag II

Then, $P(A) = \frac{{}^4C_1}{{}^9C_1} = \frac{4}{9}$, $P(B) = \frac{{}^9C_1 \times {}^7C_1}{{}^{16}C_2} = \frac{21}{40}$

$P(C) = \frac{{}^5C_1}{{}^9C_1} = \frac{5}{9}$, $P(D) = \frac{{}^9C_2}{{}^{16}C_2} = \frac{3}{10}$

∴ Required probability

$$= P(A)\,P(B) + P(C)\,P(D)$$
$$= \frac{4}{9} \times \frac{21}{40} + \frac{5}{9} \times \frac{3}{10}$$
$$= \frac{21+15}{90} = \frac{36}{90}$$
$$= \frac{2}{5}$$ **Ans.**

Solution 9.

(b) Given differential equation is,

$$(3xy + y^2)dx + (x^2 + xy)dy = 0$$

$$\frac{dy}{dx} = -\left(\frac{3xy + y^2}{x^2 + xy}\right) \quad \text{(Homogeneous)}$$

Put $y = vx \Rightarrow \frac{dy}{dx} = v + x \cdot \frac{dv}{dx}$

$$\frac{dv}{dx} \cdot x + v = -\left(\frac{3x.vx + v^2x^2}{x^2 + x.vx}\right)$$

$$\frac{dv}{dx} \cdot x = \frac{-3v - v^2}{1+v} - v$$

$$x \cdot \frac{dv}{dx} = \frac{-3v - v^2 - v - v^2}{1+v}$$

$$x \cdot \frac{dv}{dx} = \frac{-2v^2 - 4v}{1+v}$$

On integrating both sides

$$\int \frac{1+v}{2v^2+4v} dv = \int -\frac{1}{x} dx$$

or $$\frac{1}{4}\int \frac{2+2v}{2v+v^2} = -\int \frac{1}{x}$$

$$\frac{1}{4} \log |v^2 + 2v| = -\log |x| + \log c$$

$$\frac{1}{4} \log \left(\frac{y^2}{x^2} + 2\frac{y}{x}\right) = \log \frac{c}{x}$$

$$\log \left(\frac{y^2}{x^2} + \frac{2y}{x}\right) = 4 \log \frac{c}{x}$$

$$\log \left(\frac{y^2}{x^2} + \frac{2y}{x}\right) = \log \frac{c^4}{x^4}$$

$$\frac{y^2}{x^2} + \frac{2y}{x} = \frac{c^4}{x^4}$$

$$x^2y^2 + 2x^3y = c^4$$ **Ans.**

SECTION—B

Solution 10.

(a) $$\vec{a} + \vec{b} + \vec{c} = \vec{0}$$

$$\vec{a} + \vec{b} = -\vec{c}$$

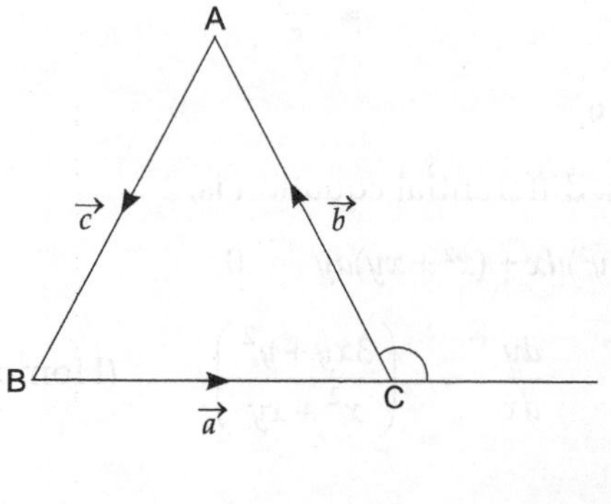

Squaring on both sides,

$$(\vec{a} + \vec{b})(\vec{a} + \vec{b}) = (-\vec{c})(-\vec{c})$$

$$b^2 + a^2 + 2ab \cos (\pi - C) = c^2$$

$$b^2 + a^2 + 2ab (-\cos C) = c^2$$

$$c^2 = a^2 + b^2 - 2ab \cos C$$

Hence Proved.

(b) L.H.S. $= \vec{a} \cdot (\vec{b} + \vec{c}) \times (\vec{a} + 2\vec{b} + 3\vec{c})$

$$= \vec{a} .(\vec{b} \times \vec{a} + \vec{b} \times 2\vec{b} + \vec{b} \times 3\vec{c} + \vec{c} \times \vec{a} + \vec{c} \times 2\vec{b} + \vec{c} \times 3\vec{c})$$

$$= \vec{a} .(\vec{b} \times \vec{a} + 3\vec{b} \times \vec{c} + \vec{c} \times \vec{a} + 2\vec{c} \times \vec{b})$$

$$= \vec{a} .(\vec{b} \times \vec{a}) + \vec{a} .(3\vec{b} \times \vec{c}) + \vec{a} .(\vec{c} \times \vec{a}) + \vec{a} .(2\vec{c} \times \vec{b})$$

$$= 0 + 3[\vec{a} .(\vec{b} \times \vec{c}) + 0 + 2(\vec{a} .(\vec{c} \times \vec{b}))]$$

$$= 3[\vec{a}\ \vec{b}\ \vec{c}] + 2[\vec{a}\ \vec{c}\ \vec{b}]$$

$$= 3[\vec{a}\ \vec{b}\ \vec{c}] - 2[\vec{a}\ \vec{b}\ \vec{c}]$$

$$= [\vec{a}\ \vec{b}\ \vec{c}] = \text{R.H.S.}$$ **Hence Proved.**

Solution 11.

(a) Equation of the line passing through the points P (– 1, 3, 2) and Q (– 4, 2, – 2) is

$$\frac{x+1}{-1+4} = \frac{y-3}{3-2} = \frac{z-2}{2+2}$$

$$\frac{x+1}{3} = \frac{y-3}{1} = \frac{z-2}{4}$$

$\because$ Point R (5, 5, λ) lies on it,

$$\frac{5+1}{3} = \frac{5-3}{1} = \frac{\lambda-2}{4}$$

$\Rightarrow \lambda - 2 = 8$ *i.e.*, $\lambda = 10$ **Ans.**

(b) Required plane is perpendicular to the given plane

$$x - 2y + 5z + 1 = 0$$

Required plane is parallel to the line which is perpendicular to the given plane. Direction ratio of line $a = 1, b = -2, c = 5$.

Hence, required plane is

$$\begin{vmatrix} x - x_1 & y - y_1 & z - z_1 \\ x_2 - x_1 & y_2 - y_1 & z_2 - z_1 \\ a & b & c \end{vmatrix} = 0$$

$$\Rightarrow \begin{vmatrix} x-2 & y+3 & z-1 \\ -1-2 & 1+3 & -7-1 \\ 1 & -2 & 5 \end{vmatrix} = 0$$

$$\Rightarrow \begin{vmatrix} x-2 & y+3 & z-1 \\ -3 & 4 & -8 \\ 1 & -2 & 5 \end{vmatrix} = 0$$

$$\Rightarrow (x-2)(20-16)-(y+3)(-15+8)+(z-1)(6-4) = 0$$

$$\Rightarrow (x-2)4-(y+3)(-7)+(z-1)2 = 0$$

$$\Rightarrow \quad 4x+7y+2z+11 = 0$$ **Ans.**

Solution 12.

(a) Let E_1 : the event that bolt is produced by machine A.

E_2 : the event that bolt is produced by machine B.

E_3 : the event that bolt is produced by machine C.

Here, E_1, E_2 and E_3 are mutually exclusive and exhaustive events.

We have,

$$P(E_1) = 25\% = \frac{25}{100}$$

$$P(E_2) = 35\% = \frac{35}{100}$$

$$P(E_3) = 40\% = \frac{40}{100}$$

Let E : the event that bolt chosen is found to be defective.

$\therefore \quad P(E/E_1) = 5\% = \frac{5}{100}$, $P(E/E_2) = 4\% = \frac{4}{100}$ and $P(E/E_3) = 2\% = \frac{2}{100}$

P (defective bolt is produced by machine C).

$$P(E_3/E) = \frac{P(E_3)\cdot P(E/E_3)}{P(E_1)\cdot P(E/E_1)+P(E_2)\cdot P(E/E_2)+P(E_3)\cdot P(E/E_3)}$$

$$= \frac{\frac{40}{100}\times\frac{2}{100}}{\frac{25}{100}\times\frac{5}{100}+\frac{35}{100}\times\frac{4}{100}+\frac{40}{100}\times\frac{2}{100}}$$

$$= \frac{80}{125+140+80}$$

$$= \frac{80}{345} = 0{\cdot}23$$ **Ans.**

(b) Probability of one phone number out of five is busy $p = \frac{1}{5}$, then $q = 1-\frac{1}{5} = \frac{4}{5}$ and $n = 10$

$\therefore$ Required probability (at least three phones are busy)

= 1 – (Probability of maximum two phones are busy)

= 1 – [P(0) + P(1) + P (2)]

$$= 1-\left[{}^{10}C_0\left(\frac{1}{5}\right)^0\left(\frac{4}{5}\right)^{10}+{}^{10}C_1\left(\frac{1}{5}\right)^1\left(\frac{4}{5}\right)^9 + {}^{10}C_2\left(\frac{1}{5}\right)^2\left(\frac{4}{5}\right)^8\right]$$

$$= 1-\left[1\times1\times\left(\frac{4}{5}\right)^{10}+10\times\frac{1}{5}\times\left(\frac{4}{5}\right)^9+45\times\left(\frac{1}{5}\right)^2\left(\frac{4}{5}\right)^8\right]$$

$$= 1-\left(\frac{4}{5}\right)^8\left[\frac{16}{25}+\frac{8}{5}+\frac{9}{5}\right]$$

$$= 1-\left(\frac{4}{5}\right)^8\left(\frac{16+40+45}{25}\right)$$

$$= 1-0{\cdot}1678\left(\frac{101}{25}\right)$$

$$= 1-0{\cdot}1678\times4{\cdot}04$$

$$= 1-0{\cdot}678 = 0{\cdot}322$$ **Ans.**

SECTION—C

Solution 13.

(b)

	Toy A	Toy B	Time in a day
Cutting time	5 min	8 min	180 min
Assembling time	10 min	8 min	240 min
Profit	50	60	
Assumed quantity	x	y	

$$\text{Profit function } Z = 50x+60y$$

$$x \geq 0, y \geq 0$$

$$5x+8y \leq 180$$

and $10x+8y \leq 240$ or $5x+4y \leq 120$

$5x+8y = 180$

	A	B
x	0	36
y	22·5	0

$5x+4y = 120$

	C	D
x	0	24
y	30	0

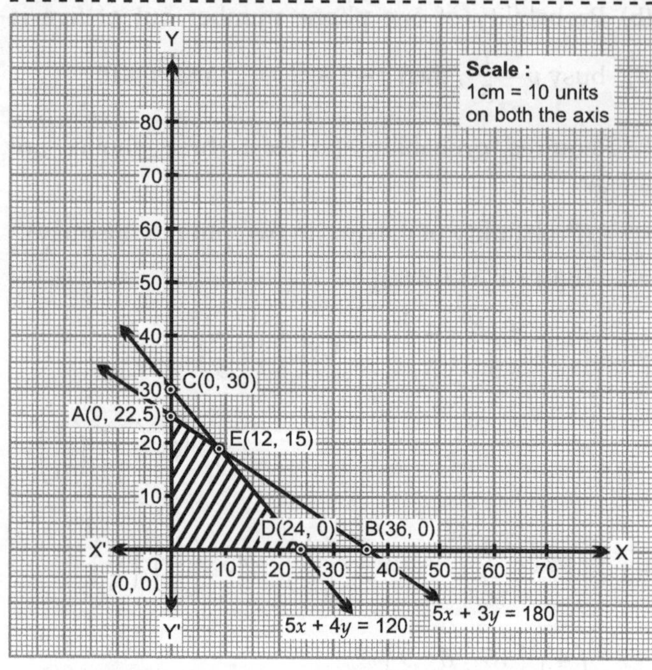

Corner point	Objective Function $Z = 50x + 60y$
At O (0, 0)	$Z = 50 \times 0 + 60 \times 0 = 0$
At D (24, 0)	$Z = 50 \times 24 + 60 \times 0 = 1200$
At E (12, 15)	$Z = 50 \times 12 + 60 \times 15 = 1500$ maximum
At A (0, 22.5)	$Z = 50 \times 0 + 60 \times 22{\cdot}5 = 1350$

Hence, the maximum profit is ₹ 1500 when 12 units of type A and 15 units of type B are produced.

Ans.

Solution 14.

(a) Cost function is

$$C(x) = \frac{x^3}{3} - 7x^2 + 111x + 50$$

Demand function is $x = 100 - p$

$\Rightarrow \quad p = 100 - x$

(i) Revenue function, $R(x) = p.x = x(100 - x)$

$$= 100x - x^2 \quad \ldots(i)$$

(ii) Profit function, P (x) = Revenue function − Cost function

$$= R(x) - C(x)$$

$$= 100x - x^2 - \frac{x^3}{3} + 7x^2 - 111x - 50$$

$$= -\frac{x^3}{3} + 6x^2 - 11x - 50 \quad \ldots(ii)$$

(iii) For profit function to be maximum.

Differentiating equation (ii) w.r. t. x, we get

$$\frac{dP}{dx} = -x^2 + 12x - 11 \quad \ldots(iii)$$

Now, $\frac{dP}{dx} = 0$

$\Rightarrow \quad -x^2 + 12x - 11 = 0$

$\Rightarrow \quad x^2 - 12x + 11 = 0$

$\Rightarrow \quad x^2 - 11x - x + 11 = 0$

$\Rightarrow \quad x(x - 11) - 1(x - 11) = 0$

$\Rightarrow \quad (x - 1)(x - 11) = 0$

$\Rightarrow \quad x = 1, 11$

Again, differentiating equation (iii) w.r.t. x, we get

$$\frac{d^2P}{dx^2} = 12 - 2x \quad \ldots(iv)$$

at $x = 1$ $\quad \frac{d^2P}{dx^2} = 10$

$\Rightarrow \quad \frac{d^2P}{dx^2} > 0$ (Minimum value)

at $x = 11$ $\quad \frac{d^2P}{dx^2} = -10$

$\Rightarrow \quad \frac{d^2P}{dx^2} < 0$ (Maximum value)

Hence, the profit maximising level of output at $x = 11$.

Ans.

MATHEMATICS 2013

QUESTIONS

SECTION—A

Question 1. [10 × 3]

(i) *If (A − 2I) (A − 3I) = 0, where* $A = \begin{bmatrix} 4 & 2 \\ -1 & x \end{bmatrix}$ *and* $I = \begin{bmatrix} 1 & 0 \\ 0 & 1 \end{bmatrix}$, *find the value of x.*

(ii) *Find the value(s) of k so that the line* $2x + y + k = 0$ *may touch the hyperbola :*

$$3x^2 - y^2 = 3.^{**}$$

(iii) *Prove that :* $\tan^{-1}\frac{1}{4} + \tan^{-1}\frac{2}{9} = \frac{1}{2}\sin^{-1}\frac{4}{5}$.

(iv) *Using L' Hospital's rule, evaluate :*

$$\lim_{x \to 0}\left(\frac{e^x - e^{-x} - 2x}{x - \sin x}\right)$$

(v) *Evaluate :* $\int \frac{1}{x + \sqrt{x}}\,dx$.

(vi) *Evaluate :* $\int_0^1 \log\left(\frac{1}{x} - 1\right)dx$

(vii) *Two regression lines are represented by* $4x + 10y = 9$ *and* $6x + 3y = 4$. *Find the line of regression of y on x.*

(viii) *If 1, ω and* ω^2 *are the cube roots of unity, evaluate* $(1 - \omega^4 + \omega^8)(1 - \omega^8 + \omega^{16})$.**

(ix) *Solve the differential equation :*

$$\log\left(\frac{dy}{dx}\right) = 2x - 3y$$

(x) *If two balls are drawn from a bag containing three red balls and four blue balls, find the probability that :*

(a) *They are of the same colour.*

(b) *They are of different colours.*

Question 2.

(a) *Using properties of determinants, prove that :* [5]

$$\begin{vmatrix} x & y & z \\ x^2 & y^2 & z^2 \\ y+z & z+x & x+y \end{vmatrix} = (x-y)(y-z)(z-x)(x+y+z)$$

(b) *Find* A^{-1}, *where* $A = \begin{vmatrix} 4 & 2 & 3 \\ 1 & 1 & 1 \\ 3 & 1 & -2 \end{vmatrix}$

Hence, solve the following system of linear equations :

$$4x + 2y + 3z = 2$$
$$x + y + z = 1$$
$$3x + y - 2z = 5$$ [5]

Question 3.

(a) *Solve for x :* $\sin^{-1} x + \sin^{-1}(1 - x) = \cos^{-1} x$. [5]

(b) *Construct a circuit diagram for the following Boolean Function :* [5]

$$(BC + A)(A'B' + C') + A'B'C'$$

*Using laws of Boolean Algebra, simplify the function and draw the simplified circuit.***

Question 4.

(a) *Verify Lagrange's mean value theorem for the function* $f(x) = \sqrt{x^2 - x}$ *in the interval [1, 4].* [5]

(b) *From the following information, find the equation of the hyperbola and the equation of its transverse axis :***
Focus : (− 2, 1), Directrix : $2x - 3y + 1 = 0$, $e = \frac{2}{\sqrt{3}}$. [5]

Question 5.

(a) *If* $y = (\cot^{-1} x)^2$, *show that*

$$(1 + x^2)^2 \frac{d^2y}{dx^2} + 2x(1 + x^2)\frac{dy}{dx} = 2.$$ [5]

(b) *Find the maximum volume of the cylinder which can be inscribed in a sphere of radius* $3\sqrt{3}$ *cm. (Leave the answer in terms of π)* [5]

Question 6.

(a) *Evaluate :* $\int \frac{\cos^{-1} x}{x^2}\,dx$ [5]

(b) *Find the area bounded by the curve* $y = 2x - x^2$ *and the line* $y = x$. [5]

Question 7.

(a) *Find the Karl Pearson's coefficient of correlation between x and y for the following data :*** [5]

** Answer is not given due to change in the present syllabus.

x	16	18	21	20	22	26	27	15
y	22	25	24	26	25	30	33	14

(b) *The following table shows the mean and standard deviation of the marks of Mathematics and Physics scored by the students in a school :* **[5]**

	Mathematics	Physics
Mean	84	81
Standard Deviation	7	4

The correlation coefficient between the given marks is 0·86. Estimate the likely marks in Physics if the marks in Mathematics are 92.

Question 8.

(a) *Bag A contains three red and four white balls; bag B contains two red and three white balls. If one ball is drawn from bag A and two balls from bag B, find the probability that :* **[5]**

(i) *One ball is red and two balls are white;*

(ii) *All the three balls are of the same colour.*

(b) *Three persons, Aman, Bipin and Mohan attempt a Mathematics problem independently. The odds in favour of Aman and Mohan solving the problem are 3 : 2 and 4 : 1 respectively and the odds against Bipin solving the problem are 2 : 1. Find :*

(i) *The probability that all the three will solve the problem.*

(ii) *The probability that problem will be solved.* **[5]**

Question 9.

(a) *Find the locus of the complex number $z = x + iy$, satisfying relations $\arg(z - 1) = \frac{\pi}{4}$ and $|z - 2 - 3i| = 2$. Illustrate the locus on the Argand plane.*** **[5]**

(b) *Solve the following differential equation :*

$ye^y\, dx = (y^3 + 2xe^y)\, dy$, *given that* $x = 0, y = 1$. **[5]**

SECTION—B

Question 10.

(a) *If $\vec{a}$ and $\vec{b}$ are unit vectors and θ is the angle between them, then show that*

$$|\vec{a} - \vec{b}| = 2 \sin\frac{\theta}{2}.$$ **[5]**

(b) *Find the value of λ for which the four points A, B, C, D with position vectors $-\hat{j} - \hat{k}$; $4\hat{i} + 5\hat{j} + \lambda\hat{k}$; $3\hat{i} + 9\hat{j} + 4\hat{k}$ and $-4\hat{i} + 4\hat{j} + 4\hat{k}$ are coplanar.* **[5]**

** Answer is not given due to change in the present syllabus.

Question 11.

(a) *Find the equation of a line passing through the point (– 1, 3, – 2) and perpendicular to the lines : $\frac{x}{1} = \frac{y}{2} = \frac{z}{3}$ and $\frac{x+2}{-3} = \frac{y-1}{2} = \frac{z+1}{5}$.* **[5]**

(b) *Find the equation of planes parallel to the plane $2x - 4y + 4z = 7$ and which are at a distance of five units from the point (3, – 1, 2).* **[5]**

Question 12.

(a) *If the sum and the product of the mean and variance of a binomial distribution are 1·8 and 0·8 respectively, find the probability distribution and the probability of at least one success.* **[5]**

(b) *For A, B and C, the chances of being selected as the manager of a firm are 4 : 1 : 2 respectively. The probabilities for them to introduce a radical change in the marketing strategy are 0·3, 0·8 and 0·5 respectively. If a change takes place; find the probability that it is due to the appointment of B.* **[5]**

SECTION—C

Question 13.

(a) *If Mr. Nirav deposits ₹ 250 at the beginning of each month in an account that pays an interest of 6% per annum compounded monthly, how many months will be required for the deposit to amount to at least ₹ 6,390 ?*** **[5]**

(b) *A mill owner buys two types of machines A and B for his mill. Machine A occupies 1,000 sq m of area and requires 12 men to operate it; while machine B occupies 1,200 sq m of area and requires 8 men to operate it. The owner has 7,600 sq m of area available and 72 men to operate the machines. If machine A produces 50 units and machine B produces 40 units daily, how many machines of each type should he buy to maximise the daily output ? Use Linear Programming to find the solution.* **[5]**

Question 14.

(a) *A bill of ₹ 60,000 was drawn on 1st April, 2011 at 4 months and discounted for ₹ 58,560 at a bank. If the rate of interest was 12% per annum, on what date was the bill discounted ?*** **[5]**

(b) *A company produces a commodity with ₹ 24,000 fixed cost. The variable cost is estimated to be 25% of the total revenue recovered on selling the product at a rate of ₹ 8 per unit. Find the following :* **[5]**

(i) *Cost function*

(ii) *Revenue function*

(iii) *Break-even point.*

Question 15.**

(a) *The price index for the following data for the year 2011 taking 2001 as the base year was 127. The simple average of price relatives method was used. Find the value of x :* **[5]**

Items	A	B	C	D	E	F
Price (₹ per unit) in year 2001	80	70	50	20	18	25
Price (₹ per unit) in year 2011	100	87·50	61	22	x	32·50

(b) *The profits of a paper bag manufacturing company (in lakhs of rupees) during each month of a year are :* **[5]**

Month	Profit
Jan.	1·2
Feb.	0·8
Mar.	1·4
Apr.	1·6
May	2·0
June	2·4
July	3·6
Aug.	4·8
Sept.	3·4
Oct.	1·8
Nov.	0·8
Dec.	1·2

Plot the given data on a graph sheet. Calculate the four monthly moving averages and plot these on the same graph sheet.

** Answer is not given due to change in the present syllabus.

SOLUTIONS

Solution 1.

(i) Given,

$$(A-2I)(A-3I)=0$$

$$\left\{\begin{bmatrix}4 & 2\\-1 & x\end{bmatrix}-\begin{bmatrix}2 & 0\\0 & 2\end{bmatrix}\right\}\left\{\begin{bmatrix}4 & 2\\-1 & x\end{bmatrix}-\begin{bmatrix}3 & 0\\0 & 3\end{bmatrix}\right\}=\begin{bmatrix}0 & 0\\0 & 0\end{bmatrix}$$

$$\Rightarrow \begin{bmatrix}2 & 2\\-1 & x-2\end{bmatrix}\begin{bmatrix}1 & 2\\-1 & x-3\end{bmatrix}=\begin{bmatrix}0 & 0\\0 & 0\end{bmatrix}$$

$$\Rightarrow \begin{bmatrix}0 & 2x-2\\-x+1 & x^2-5x+4\end{bmatrix}=\begin{bmatrix}0 & 0\\0 & 0\end{bmatrix}$$

Equating corresponding elements of equal matrices.

$$-x+1=0,\quad x=1 \qquad 2x-2=0,\quad x=1 \qquad x^2-5x+4=0,\quad x=1,4$$

But $x = 4$ does not satisfy all above equations.

$\therefore$ $x = 1$. **Ans.**

(iii) L.H.S. $= \tan^{-1}\frac{1}{4}+\tan^{-1}\frac{2}{9}$

$$= \tan^{-1}\left(\frac{\frac{1}{4}+\frac{2}{9}}{1-\frac{1}{4}\cdot\frac{2}{9}}\right)$$

$$= \tan^{-1}\left(\frac{\frac{17}{36}}{\frac{17}{18}}\right)=\tan^{-1}\frac{1}{2}$$

$$= \frac{1}{2}\left(2\tan^{-1}\frac{1}{2}\right)$$

$$= \frac{1}{2}\sin^{-1}\frac{2\left(\frac{1}{2}\right)}{1+\left(\frac{1}{2}\right)^2}$$

$$= \frac{1}{2}\sin^{-1}\frac{4}{5} = \text{R. H. S.}$$

Hence Proved.

(iv) Given,

$$\lim_{x\to 0}\left(\frac{e^x-e^{-x}-2x}{x-\sin x}\right), \qquad \left[\frac{0}{0}\right] \text{ form}$$

Using L' Hospital's rule,

$$= \lim_{x\to 0}\frac{e^x+e^{-x}-2}{1-\cos x}, \qquad \left[\frac{0}{0}\right] \text{ form}$$

Again using L' Hospital's rule,

$$= \lim_{x\to 0}\frac{e^x-e^{-x}}{\sin x}, \qquad \left[\frac{0}{0}\right] \text{ form}$$

Once again using L' Hospital's rule,

$$\Rightarrow \lim_{x\to 0}\frac{e^x+e^{-x}}{\cos x}=\frac{1+1}{1}=2$$ **Ans.**

(v) Given, $\int\frac{1}{x+\sqrt{x}}\,dx$

Put $\sqrt{x}=t \Rightarrow x=t^2 \Rightarrow dx=2t\,dt$

$$\therefore \int\frac{dx}{x+\sqrt{x}} = \int\frac{2t\,dt}{t^2+t}=2\int\frac{1}{t+1}\,dt$$

$$= 2\log|t+1|+c$$

$$= 2\log|\sqrt{x}+1|+c.$$ **Ans.**

(vi) Let,

$$I = \int_0^1 \log\left(\frac{1}{x}-1\right)dx$$

$$= \int_0^1 \log\left(\frac{1-x}{x}\right)dx \quad ...(i)$$

Using property,

$$\int_0^a f(x)\,dx = \int_0^a f(a-x)\,dx$$

$$I = \int_0^1 \log\left(\frac{1-(1-x)}{1-x}\right)dx$$

$$= \int_0^1 \log\left(\frac{x}{1-x}\right)dx$$

$$= \int_0^1 \log\left(\frac{1-x}{x}\right)^{-1}dx$$

$$= \int_0^1 (-1)\log\left(\frac{1-x}{x}\right)dx$$

$$I = -\int_0^1 \log\left(\frac{1-x}{x}\right)dx$$

$$I = -I \quad \text{[Using (i)]}$$

$$2I = 0$$

$$\therefore \quad I = 0$$

$$\therefore \quad \int_0^1 \log\left(\frac{1}{x}-1\right)dx = 0.$$ **Ans.**

(vii) Assuming $4x + 10y = 9$ to be regression line of y on x.

$$10y = -4x + 9$$

$$y = -\frac{4}{10}x + \frac{9}{10}$$

$$\Rightarrow \quad b_{yx} = \frac{-4}{10} = \frac{-2}{5}$$

Then, $6x + 3y = 4$

$$6x = -3y + 4$$

$$x = -\frac{3}{6}y + \frac{4}{6}$$

$$\Rightarrow \quad b_{xy} = \frac{-3}{6} = -\frac{1}{2}$$

Now, $b_{yx} \times b_{xy} = -\frac{2}{5} \times \left(-\frac{1}{2}\right) = \frac{1}{5} < 1$

which is true. Hence, our assumption is true, *i.e.*, regression line of y on x is

$$4x + 10y = 9.$$ **Ans.**

(ix) Given differential equation is,

$$\log\left(\frac{dy}{dx}\right) = 2x - 3y$$

$$\Rightarrow \quad \frac{dy}{dx} = e^{2x-3y}$$

$$\Rightarrow \quad \frac{dy}{dx} = \frac{e^{2x}}{e^{3y}}$$

$$\Rightarrow \quad e^{3y}\,dy = e^{2x}\,dx$$

Integrating both sides,

$$\int e^{3y}\,dy = \int e^{2x}\,dx$$

$$\frac{e^{3y}}{3} = \frac{e^{2x}}{2} + c$$

$$2e^{3y} - 3e^{2x} = K$$ **Ans.**

(x) Total number of ways of drawing 2 balls $= {}^7C_2$.

(a) P (Balls are of same colour)

$$= \frac{{}^3C_2}{{}^7C_2} + \frac{{}^4C_2}{{}^7C_2} = \frac{{}^3C_2 + {}^4C_2}{{}^7C_2}$$

$$= \frac{\frac{3!}{2!\cdot 1!} + \frac{4!}{2!\cdot 2!}}{\frac{7!}{2!\cdot 5!}} = \frac{3+6}{7\times 3} = \frac{9}{21} = \frac{3}{7}$$ **Ans.**

(b) P (Balls are of different colour)

$$= \frac{{}^3C_1 \times {}^4C_1}{{}^7C_2}$$

$$= \frac{\frac{3!}{1!\cdot 2!} + \frac{4!}{1!\cdot 3!}}{\frac{7!}{2!\cdot 5!}}$$

$$= \frac{3\times 4}{7\times 3} = \frac{4}{7}$$ **Ans.**

Solution 2.

(a) Let,

$$\Delta = \begin{vmatrix} x & y & z \\ x^2 & y^2 & z^2 \\ y+z & x+z & x+y \end{vmatrix}$$

Applying $R_3 \to R_3 + R_1$, we get

$$\Delta = \begin{vmatrix} x & y & z \\ x^2 & y^2 & z^2 \\ x+y+z & x+y+z & x+y+z \end{vmatrix}$$

$$= (x+y+z)\begin{vmatrix} x & y & z \\ x^2 & y^2 & z^2 \\ 1 & 1 & 1 \end{vmatrix}$$

(taking out $(x + y + z)$ as common)

Applying $C_1 \to C_1 - C_2$ and $C_2 \to C_2 - C_3$, we get

$$\Delta = (x+y+z)\begin{vmatrix} x-y & y-z & z \\ x^2-y^2 & y^2-z^2 & z^2 \\ 0 & 0 & 1 \end{vmatrix}$$

$$= (x+y+z)(x-y)(y-z)\begin{vmatrix} 1 & 1 & z \\ x+y & y+z & z^2 \\ 0 & 0 & 1 \end{vmatrix}$$

Expanding along R_3, we get

$$= (x+y+z)(x-y)(y-z)(y+z-x-y)$$

$$= (x-y)(y-z)(z-x)(x+y+z)$$

Hence Proved.

(b) Here, $|A| = \begin{vmatrix} 4 & 2 & 3 \\ 1 & 1 & 1 \\ 3 & 1 & -2 \end{vmatrix}$

$$= 4(-2-1) - 2(-2-3) + 3(1-3)$$

$$= -12 + 10 - 6 = -8 \neq 0$$

$\Rightarrow A^{-1}$ exists and $A^{-1} = \frac{1}{|A|}$ adj A.

$$A_{11} = \begin{vmatrix} 1 & 1 \\ 1 & -2 \end{vmatrix} = -3,$$

$$A_{12} = -\begin{vmatrix} 1 & 1 \\ 3 & -2 \end{vmatrix} = 5,$$

$$A_{13} = \begin{vmatrix} 1 & 1 \\ 3 & 1 \end{vmatrix} = -2$$

$$A_{21} = -\begin{vmatrix} 2 & 3 \\ 1 & -2 \end{vmatrix} = 7,$$

$$A_{22} = \begin{vmatrix} 4 & 3 \\ 3 & -2 \end{vmatrix} = -17,$$

$$A_{23} = -\begin{vmatrix} 4 & 2 \\ 3 & 1 \end{vmatrix} = 2$$

$$A_{31} = \begin{vmatrix} 2 & 3 \\ 1 & 1 \end{vmatrix} = -1,$$

$$A_{32} = -\begin{vmatrix} 4 & 3 \\ 1 & 1 \end{vmatrix} = -1,$$

$$A_{33} = \begin{vmatrix} 4 & 2 \\ 1 & 1 \end{vmatrix} = 2$$

$$\text{adj. A} = \begin{bmatrix} -3 & 5 & -2 \\ 7 & -17 & 2 \\ -1 & -1 & 2 \end{bmatrix}^T$$

$$= \begin{bmatrix} -3 & 7 & -1 \\ 5 & -17 & -1 \\ -2 & 2 & 2 \end{bmatrix}$$

$$\therefore \quad A^{-1} = \frac{1}{|A|} \text{adj. A}$$

$$= -\frac{1}{8}\begin{bmatrix} -3 & 7 & -1 \\ 5 & -17 & -1 \\ -2 & 2 & 2 \end{bmatrix}$$

$$= \frac{1}{8}\begin{bmatrix} 3 & -7 & 1 \\ -5 & 17 & 1 \\ 2 & -2 & -2 \end{bmatrix}$$ **Ans.**

The given system of equation is

$$4x + 2y + 3z = 2$$

$$x + y + z = 1$$

$$3x + y - 2z = 5$$

This system can be written as AX = B

Where, $A = \begin{bmatrix} 4 & 2 & 3 \\ 1 & 1 & 1 \\ 3 & 1 & -2 \end{bmatrix}$, $X = \begin{bmatrix} x \\ y \\ z \end{bmatrix}$ and $B = \begin{bmatrix} 2 \\ 1 \\ 5 \end{bmatrix}$

As $|A| \neq 0$, the given system has a unique solution $X = A^{-1}B$

$$\Rightarrow \quad X = \frac{1}{8}\begin{bmatrix} 3 & -7 & 1 \\ -5 & 17 & 1 \\ 2 & -2 & -2 \end{bmatrix}\begin{bmatrix} 2 \\ 1 \\ 5 \end{bmatrix}$$

$$= \frac{1}{8}\begin{bmatrix} 6 & -7 & +5 \\ -10 & +17 & +5 \\ 4 & -2 & -10 \end{bmatrix} = \frac{1}{8}\begin{bmatrix} 4 \\ 12 \\ -8 \end{bmatrix}$$

$$\Rightarrow \quad \begin{bmatrix} x \\ y \\ z \end{bmatrix} = \frac{1}{2}\begin{bmatrix} 1 \\ 3 \\ -2 \end{bmatrix} = \begin{bmatrix} 1/2 \\ 3/2 \\ -1 \end{bmatrix}$$

Hence, the solution of the given system of equation is

$$x = \frac{1}{2}, y = \frac{3}{2}, z = -1.$$ **Ans.**

Solution 3.

(a) The given equation is

$$\sin^{-1} x + \sin^{-1}(1-x) = \cos^{-1} x$$

$$\Rightarrow \sin^{-1} x + \sin^{-1}(1-x) = \frac{\pi}{2} - \sin^{-1} x$$

$$\Rightarrow \quad \sin^{-1}(1-x) = \frac{\pi}{2} - 2\sin^{-1} x \quad \text{...(i)}$$

Let, $\sin^{-1} x = y$

$\Rightarrow \quad x = \sin y \quad$ ...(ii)

Therefore, from (i), we get

$$\sin^{-1}(1-x) = \frac{\pi}{2} - 2y$$

$$\Rightarrow \quad 1-x = \sin\left(\frac{\pi}{2}-2y\right)$$

$$\Rightarrow \quad 1-x = \cos 2y$$

$$\Rightarrow \quad 1-x = 1-2\sin^2 y$$

$$\Rightarrow \quad 1-x = 1-2x^2 \quad \text{[from (ii)]}$$

$$\Rightarrow \quad 2x^2 - x = 0$$

$$\Rightarrow \quad x(2x-1) = 0$$

$$\Rightarrow \quad x = 0, \frac{1}{2}$$

Since, both these values satisfy the given equation. Hence, the solutions of the given equation are $x = 0, \frac{1}{2}$. **Ans.**

Solution 4.

(a) Given, $f(x) = \sqrt{x^2 - x},\ x \in [1, 4]$

Since $x^2 - x$ is continuous on R

and $\sqrt{x^2-x}$ exists in [1, 4].

$\Rightarrow f(x)$ is continuous in [1, 4].

Differentiating the given function w.r.t. x

$$f'(x) = \frac{1}{2}(x^2-x)^{-1/2}.(2x-1)$$

$$= \frac{2x-1}{2\sqrt{x^2-x}}$$

which exists $\forall\ x \in$ R.

$\therefore f(x)$ is differentiable in (1, 4).

Thus, both the conditions of Lagrange's mean value theorem is satisfied therefore, $\exists\ c$ in (1, 4).

Such that $f'(c) = \dfrac{f(4)-f(1)}{4-1}$

$$\frac{2c-1}{2\sqrt{c^2-c}} = \frac{\sqrt{12}-0}{3}$$

$$3(2c-1) = 2\sqrt{c^2-c}.\sqrt{12}$$

Squaring both sides, we get

$$9(4c^2-4c+1) = 48(c^2-c)$$

$$\Rightarrow \quad 3(4c^2-4c+1) = 16(c^2-c)$$

$$\Rightarrow \quad 12c^2-12c+3 = 16c^2-16c$$

$$\Rightarrow \quad 4c^2-4c-3 = 0$$

$$\Rightarrow \quad (2c-3)(2c+1) = 0$$

$$c = \frac{3}{2}, \frac{-1}{2}$$

Clearly $c = \frac{3}{2}$ lies in the interval (1, 4).

Hence, Lagrange's mean value theorem is verified and $c = \frac{3}{2}$. **Hence Proved.**

Solution 5.

(a) Given, $y = (\cot^{-1} x)^2$

Differentiating w.r.t. x,

Now, $$\frac{dy}{dx} = 2\cot^{-1}x\,\frac{d}{dx}(\cot^{-1}x)$$

$$\Rightarrow \quad \frac{dy}{dx} = 2\cot^{-1}x\left(\frac{-1}{1+x^2}\right)$$

$$\Rightarrow \quad (1+x^2)\frac{dy}{dx} = -2\cot^{-1}x$$

Differentiating once again w.r.t. x,

$$(1+x^2)\frac{d^2y}{dx^2} + 2x\frac{dy}{dx} = -2\cdot\frac{(-1)}{1+x^2}$$

$$\Rightarrow \quad (1+x^2)\frac{d^2y}{dx^2} + 2x\frac{dy}{dx} = \frac{2}{1+x^2}$$

$$\Rightarrow \quad (1+x^2)^2\frac{d^2y}{dx^2} + 2x(1+x^2)\frac{dy}{dx} = 2.$$ **Hence Proved.**

(b) Let h units be the height of the cylinder and R unit be the radius of the cylinder. Given $3\sqrt{3}$ cm is the radius of the sphere.

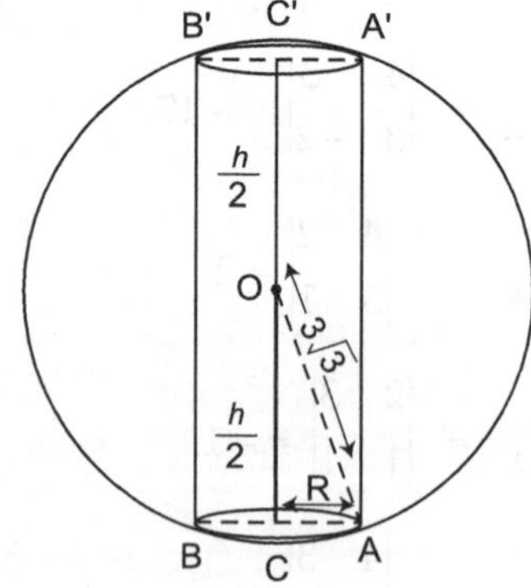

If V is the volume of the cylinder, then

$$V = \pi R^2 h \quad \text{...(i)}$$

Let O be the centre of the sphere and C′OC ⊥ B′A′ as well as BA.

From right angled ΔOCA,

$$(3\sqrt{3})^2 = \left(\frac{h}{2}\right)^2 + R^2$$

$$\Rightarrow \quad R^2 = 27 - \frac{h^2}{4} \quad \text{...(ii)}$$

$$\therefore \quad V = \pi\left(27 - \frac{h^2}{4}\right) \cdot h$$

$$V = 27\pi h - \pi\frac{h^3}{4}$$

$$\therefore \quad \frac{dV}{dh} = 27\pi - \frac{3\pi h^2}{4}$$

and $$\frac{d^2V}{dh^2} = -\frac{6\pi h}{4}$$

For maxima or minima,

$$\frac{dV}{dh} = 0$$

$$\therefore \quad 27\pi = \frac{3\pi h^2}{4}$$

$$h^2 = \frac{4\times27}{3}$$

$$h^2 = 36$$

$$h = 6, -6$$

and $$\left(\frac{d^2y}{dh^2}\right)_{h=6} = 6 = \frac{-6\pi\times6}{4} < 0$$

$\therefore$ V is maximum when $h = 6$, putting $h = 6$ in equation (ii).

$$R^2 = 27 - \frac{36}{4}$$

$$R^2 = 18$$

$\therefore$ From (i),

$$V = \pi R^2 h$$

$$= \pi \times 18 \times 6$$

$$= 108\pi \text{ cm}^3$$ **Ans.**

Solution 6.

(a) Given, $\int \frac{\cos^{-1}x}{x^2}\,dx$

Put $\cos^{-1}x = t$

$$\Rightarrow \quad x = \cos t$$

$$\Rightarrow \quad dx = -\sin t\,dt$$

$$\therefore \int \frac{\cos^{-1}x}{x^2}\,dx = \int \frac{t}{\cos^2 t}(-\sin t)\,dt$$

$$= -\int \underset{\text{I}}{t}\,\underset{\text{II}}{(\sec t\tan t)}\,dt$$

$$= \left(t.\sec t - \int 1.\sec t\,dt\right)$$

$$= -t\sec t + \log|\sec t + \tan t| + c$$

$$= \frac{-t}{\cos t} + \log\left|\frac{1+\sin t}{\cos t}\right| + c$$

Now, putting the value of 't', we get

$$\int \frac{\cos^{-1}x}{x^2}\,dx = \frac{-\cos^{-1}x}{x} + \log\left|\frac{1+\sin(\cos^{-1}x)}{x}\right| + c$$

$$= \frac{-\cos^{-1}x}{x} + \log\left|\frac{1+\sin\left(\sin^{-1}\left(\sqrt{1-x^2}\right)\right)}{x}\right| + c$$

$$= \frac{-\cos^{-1}x}{x} + \log\left|\frac{1+\sqrt{1-x^2}}{x}\right| + c$$ **Ans.**

(b)

Given curve is $y = 2x - x^2$

$$-y = x^2 - 2x$$

$$-y + 1 = x^2 - 2x + 1$$

$$-(y-1) = (x-1)^2 \quad \text{...(i)}$$

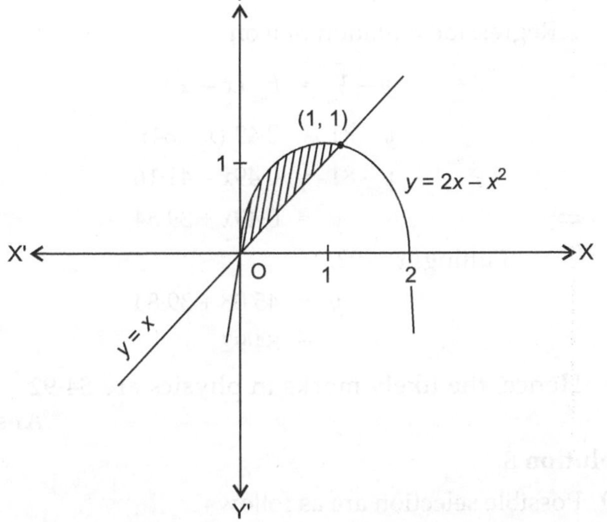

which represents a downward parabola with vertex at (1, 1).

To find point of intersection of the parabola and the line $y = x$

Put $y = x$ in equation (i)

$$-(x-1) = (x-1)^2$$

$$\Rightarrow \quad -x + 1 = x^2 - 2x + 1$$

$$\Rightarrow \quad x^2 - x = 0$$

$$x(x-1) = 0$$

$$\Rightarrow \quad x = 0, 1$$

$\therefore$ Points of intersection are (0, 0) and (1, 1).

$\therefore$ The area enclosed between the curve $y = 2x - x^2$ and the line $y = x$ is

$$\int_0^1 (2x - x^2 - x)\,dx = \int_0^1 (x - x^2)\,dx$$

$$= \left[\frac{x^2}{2} - \frac{x^3}{3}\right]_0^1$$

$$= \left(\frac{1}{2} - \frac{1}{3}\right) - (0 - 0)$$

$$= \frac{1}{6} \text{ sq. unit.}$$ **Ans.**

Solution 7.

(b) Given, mean marks in mathemtics, $\bar{x} = 84$,

Mean marks in physics, $\bar{y} = 81$

$$\sigma_x = 7, \sigma_y = 4$$

$$r = 0{\cdot}86$$

$$\therefore \quad b_{yx} = r \cdot \frac{\sigma_y}{\sigma_x}$$

$$= 0{\cdot}86 \times \frac{4}{7} = 0{\cdot}49$$

$\therefore$ Regression equation of y on x

$$y - \bar{y} = b_{yx}(x - \bar{x})$$

$$y - 81 = 0{\cdot}49(x - 84)$$

$$y - 81 = 0{\cdot}49x - 41{\cdot}16$$

$$\Rightarrow \quad y = 0{\cdot}49x + 39{\cdot}84$$

Putting $x = 92$,

$$y = 45{\cdot}08 + 39{\cdot}84$$

$$= 84{\cdot}92$$

Hence, the likely marks in physics are 84·92.

Ans.

Solution 8.

(a) Possible selection are as follows :

(i) 1 red ball from bag A, 2 white ball from bag B

1 white ball from bag A, 1 white ball from bag B, 1 red ball from B.

$\therefore$ P (one ball is red and two balls are white)

$$= \frac{^3C_1}{^7C_1} \times \frac{^3C_2}{^5C_2} + \frac{^4C_1}{^7C_1} \times \frac{^3C_1 \times {^2C_1}}{^5C_2}$$

$$= \frac{3}{7} \times \frac{3}{10} + \frac{4}{7} \times \frac{3\times 2}{10}$$

$$= \frac{9}{70} + \frac{24}{70} = \frac{33}{70}.$$ **Ans.**

(ii) Possible selections are as follows :

1 red ball from bag A, 2 red ball from bag B.

1 white ball from bag A, 2 white ball from bag B.

$\therefore$ P (All the three balls are of the same colour)

$$= \frac{^3C_1}{^7C_1} \times \frac{^2C_2}{^5C_2} + \frac{^4C_1}{^7C_1} \times \frac{^3C_2}{^5C_2}$$

$$= \frac{3}{7} \times \frac{1}{10} + \frac{4}{7} \times \frac{3}{10}$$

$$= \frac{3}{70} + \frac{12}{70} = \frac{15}{70} = \frac{3}{14}.$$ **Ans.**

(b) Odd against $= \dfrac{1-p}{p}$

Odd in favour $= \dfrac{p}{1-p}$

Let A : Event Aman solves problem

B : Event Bipin solves problem

C : Event Mohan solves problem

Aman	Mohan	Bipin
$\frac{3}{2} = \frac{p}{1-p}$	$\frac{4}{1} = \frac{p}{1-p}$	$\frac{2}{1} = \frac{1-p}{p}$
$\Rightarrow \quad p = \frac{3}{5}$	$\Rightarrow \quad p = \frac{4}{5}$	$\Rightarrow \quad p = \frac{1}{3}$
$\Rightarrow$ P (A) $= \frac{3}{5}$	$\Rightarrow$ P (C) $= \frac{4}{5}$	$\Rightarrow$ P (B) $= \frac{1}{3}$

(i) $P(A \cap B \cap C) = P(A).\,P(B).\,P(C)$

Probability that all three will solve the problem

$$= \frac{3}{5} \times \frac{1}{3} \times \frac{4}{5} = \frac{4}{25}$$ **Ans.**

(ii) Probability that the problem is not solved = the probability that all the three fail to solve the problem

$$\therefore \quad P(\bar{A} \cap \bar{B} \cap \bar{C}) = P(\bar{A}).\,P(\bar{B}).\,P(\bar{C})$$

$$= \left(1 - \frac{3}{5}\right) \cdot \left(1 - \frac{1}{3}\right) \cdot \left(1 - \frac{4}{5}\right)$$

$$= \frac{2}{5} \cdot \frac{2}{3} \cdot \frac{1}{5} = \frac{4}{75}$$

$\therefore$ The probability that the problem will be solved

$$= 1 - \frac{4}{75} = \frac{71}{75}.$$ **Ans.**

Solution 9.

(b) Given, $\quad y\,e^y\,dx = (y^3 + 2x\,e^y)\,dy$

$$\Rightarrow \quad y\,e^y \frac{dx}{dy} = y^3 + 2x\,e^y$$

$$\Rightarrow \quad \frac{dx}{dy} = \frac{2}{y} \cdot x + y^2 e^{-y}$$

$$\Rightarrow \quad \frac{dx}{dy} - \frac{2}{y} \cdot x = y^2 e^{-y} \quad \ldots(i)$$

$$\because \quad \frac{dx}{dy} - P_x = Q$$

$$\therefore \quad P = -\frac{2}{y}, Q = y^2 e^{-y}$$

It is a linear differential equation in x.

$$\text{I.F.} = e^{\int P dy} = e^{\int -\frac{2}{y}.dy}$$
$$= e^{-2 \log |y|}$$
$$= e^{\log |y|^{-2}}$$
$$= |y|^{-2} = \frac{1}{y^2}$$

Therefore, general solution of (i) is

$$x \times \text{I.F.} = \int Q \times \text{I.F.}\, dy + c$$

$$x \cdot \frac{1}{y^2} = \int y^2 e^{-y} . \frac{1}{y^2} \, dy + c$$

$$x \cdot \frac{1}{y^2} = \int e^{-y} \, dy + c$$

$$\Rightarrow \quad x \cdot \frac{1}{y^2} = \frac{e^{-y}}{-1} + c \quad \ldots(ii)$$

Given that $x = 0$ when $y = 1$

$$\Rightarrow \quad 0 = -e^{-1} + c$$
$$\Rightarrow \quad c = e^{-1}$$

Substituting the value of c in (ii), the required particular solution of the given equation is

$$\frac{x}{y^2} = -e^{-y} + e^{-1}$$

i.e., $\quad x = y^2(-e^{-y} + e^{-1}).$ **Ans.**

SECTION—B

Solution 10.

(a) $\vec{a} . \vec{b} = |\vec{a}| |\vec{b}| \cos\theta$

$$= 1.1 \cos\theta$$
$$= \cos\theta$$

Now, $|\vec{a} - \vec{b}|^2 = (\vec{a} - \vec{b})^2$

$$= |\vec{a}|^2 + |\vec{b}|^2 - 2\vec{a} . \vec{b}$$
$$= 1 + 1 - 2\cos\theta$$
$$= 2(1 - \cos\theta)$$

$$\Rightarrow \quad |\vec{a} - \vec{b}|^2 = 2 \cdot 2 \sin^2 \frac{\theta}{2}$$

$$\Rightarrow \quad |\vec{a} - \vec{b}| = 2 \sin \frac{\theta}{2}$$ **Hence Proved.**

(b) Let A, B, C and D be the given points whose position vectors are $-\hat{j} - \hat{k}$, $4\hat{i} + 5\hat{j} + \lambda\hat{k}$, $3\hat{i} + 9\hat{j} + 4\hat{k}$ and $-4\hat{i} + 4\hat{j} + 4\hat{k}$ respectively.

$$\vec{AB} = 4\hat{i} + 5\hat{j} + \lambda\hat{k} - (-\hat{j} - \hat{k})$$
$$= 4\hat{i} + 6\hat{j} + (\lambda + 1)\hat{k}$$

$$\vec{AC} = 3\hat{i} + 9\hat{j} + 4\hat{k} - (-\hat{j} - \hat{k})$$
$$= 3\hat{i} + 10\hat{j} + 5\hat{k}$$

$$\vec{AD} = -4\hat{i} + 4\hat{j} + 4\hat{k} - (-\hat{j} - \hat{k})$$
$$= -4\hat{i} + 5\hat{j} + 5\hat{k}$$

Since the points A, B, C and D are coplanar, vectors $\vec{AB}$, $\vec{AC}$ and $\vec{AD}$ are coplanar.

$$\Rightarrow \quad [\vec{AB}\ \vec{AC}\ \vec{AD}] = 0$$

$$\Rightarrow \quad \begin{bmatrix} 4 & 6 & \lambda+1 \\ 3 & 10 & 5 \\ -4 & 5 & 5 \end{bmatrix} = 0$$

$$\Rightarrow 4(50 - 25) - 3(30 - 5 - 5\lambda) - 4(30 - 10 - 10\lambda) = 0$$
$$\Rightarrow \quad 100 - 75 + 15\lambda - 80 + 40\lambda = 0$$
$$\Rightarrow \quad -55 + 55\lambda = 0$$
$$\Rightarrow \quad -1 + \lambda = 0$$
$$\Rightarrow \quad \lambda = 1$$ **Ans.**

Solution 11.

(a) Any line through the point $(-1, 3, -2)$ is

$$\frac{x-(-1)}{a} = \frac{y-3}{b} = \frac{z-(-2)}{c} \quad \ldots(i)$$

where a, b, c are its d.r.'s.

It is perpendicular to the lines

$$\frac{x}{1} = \frac{y}{2} = \frac{z}{3} \quad \ldots(ii)$$

and $$\frac{x+2}{-3} = \frac{y-1}{2} = \frac{z+1}{5} \quad \ldots(iii)$$

then $$1.a + 2.b + 3.c = 0$$

$$\Rightarrow \quad a + 2b + 3c = 0 \quad \ldots(iv)$$

and $$-3.a + 2.b + 5.c = 0$$

$$\Rightarrow \quad -3a + 2b + 5c = 0 \quad \ldots(v)$$

Subtracting (v) from (iv), we get

$$4a - 2c = 0$$

$$\Rightarrow \quad c = 2a \quad \ldots(vi)$$

and hence from (iv),

$$a + 2b + 3.2a = 0$$

$$\Rightarrow \quad 2b = -7a$$

$$\Rightarrow \quad b = \frac{-7a}{2} \quad \ldots\text{(vii)}$$

From (vi) and (vii),

$$a : b : c = a : -\frac{7a}{2} : 2a$$

$$\Rightarrow \quad a : b : c = 2 : -7 : 4$$

Putting these values in (i), the equation of the required lines are

$$\frac{x+1}{2} = \frac{y-3}{-7} = \frac{z+2}{4} \quad \textbf{Ans.}$$

(b) The given plane is $2x - 4y + 4z = 7$ ...(i)

Equation of any plane parallel to (i) is

$$2x - 4y + 4z + k = 0 \quad \ldots\text{(ii)}$$

Now, (ii) is at a distance of 5 units from the point (3, – 1, 2) if

$$\frac{|2\times3 - 4.(-1) + 4\times2 + k|}{\sqrt{(2)^2 + (-4)^2 + (4)^2}} = 5$$

$$\frac{|18+k|}{6} = 5$$

$$\Rightarrow \quad |18 + k| = 30$$

$$\Rightarrow \quad 18 + k = 30$$

$$\text{or} \quad 18 + k = -30$$

$$\Rightarrow \quad k = 12$$

$$\text{or} \quad k = -48$$

Substituting these values of k in (ii), the equations of the required planes are

$$2x - 4y + 4z + 12 = 0$$

$$\text{and} \quad 2x - 4y + 4z - 48 = 0$$

$$\text{or} \quad x - 2y + 2z + 6 = 0$$

$$\text{and} \quad x - 2y + 2z - 24 = 0 \quad \textbf{Ans.}$$

Solution 12.

(a) According to question, we have

$$np + npq = 1{\cdot}8$$

$$\Rightarrow \quad np\,(1 + q) = 1{\cdot}8 \quad \ldots\text{(i)}$$

$$\text{and} \quad np.npq = 0{\cdot}8$$

$$\Rightarrow \quad n^2p^2q = 0{\cdot}8 \quad \ldots\text{(ii)}$$

Dividing the square of (i) by (ii), we get

$$\frac{n^2p^2(1+q)^2}{n^2p^2q} = \frac{1{\cdot}8\times1{\cdot}8}{0{\cdot}8}$$

$$\Rightarrow \quad \frac{(1+q)^2}{q} = \frac{3{\cdot}24}{0{\cdot}8}$$

$$\Rightarrow \quad \frac{(1+q)^2}{q} = \frac{324}{80}$$

$$\Rightarrow \quad \frac{(1+q)^2}{q} = \frac{81}{20}$$

$$\Rightarrow \quad 20 + 2\times 20q + 20q^2 = 81q$$

$$\Rightarrow \quad 20q^2 - 41q + 20 = 0$$

$$\Rightarrow \quad 20q^2 - 25q - 16q + 20 = 0$$

$$\Rightarrow \quad 5q\,(4q - 5) - 4\,(4q - 5) = 0$$

$$\Rightarrow \quad (4q - 5)\,(5q - 4) = 0$$

$$q = \frac{5}{4} \text{ or } q = \frac{4}{5}$$

but $q \neq \frac{5}{4}$ as $\frac{5}{4} > 1$

$$\therefore \quad q = \frac{4}{5}$$

$$\Rightarrow \quad p = \frac{1}{5}$$

From (i), $n \times \frac{1}{5}\left(1 + \frac{4}{5}\right) = 1{\cdot}8$

$$\Rightarrow \quad n \times \frac{1}{5} \times \frac{9}{5} = 1{\cdot}8$$

$$\Rightarrow \quad 9n = 25 \times 1{\cdot}8$$

$$\Rightarrow \quad n = \frac{25\times1{\cdot}8}{9}$$

$$\Rightarrow \quad n = \frac{45}{9}$$

$$\therefore \quad n = 5$$

Hence, the binomial distribution is $(p + q)^n = \left(\frac{1}{5} + \frac{4}{5}\right)^5$

Probability of getting atleast 1 success = P (X ≥ 1)

$$= 1 - P(X = 0)$$

$$= 1 - {}^5C_0\left(\frac{1}{5}\right)^0\left(\frac{4}{5}\right)^5$$

$$= 1 - \frac{1024}{3125}$$

$$= \frac{2101}{3125} = 0{\cdot}67 \quad \textbf{Ans.}$$

(b) Let E_1, E_2, E_3 and E be the events as defined below :

E_1 = A is selected as Manager

E_2 = B is selected as Manager

E_3 = C is selected as Manager

and E = radical change occurs in marketing strategy

$$P(E_1) = \frac{4}{4+1+2} = \frac{4}{7}$$

$$P(E_2) = \frac{1}{4+1+2} = \frac{1}{7}$$

$$P(E_3) = \frac{2}{4+1+2} = \frac{2}{7}$$

Given, $P(E/E_1) = 0{\cdot}3$

$P(E/E_2) = 0{\cdot}8$

$P(E/E_3) = 0{\cdot}5$

We want to find the probability that the radical change in marketing strategy occured due to the appointment of B *i.e.*, we want to find $P(E_2/E)$.

By Bayes' theorem,

$$P(E_2/E) = \frac{P(E_2).\,P(E/E_2)}{P(E_1)P(E/E_1)+P(E_2).\,P(E/E_2)+P(E_3)P(E/E_3)}$$

$$= \frac{\frac{1}{7}\times 0{\cdot}8}{\frac{4}{7}\times 0{\cdot}3+\frac{1}{7}\times 0{\cdot}8+\frac{2}{7}\times 0{\cdot}5}$$

$$= \frac{0{\cdot}8}{1{\cdot}2+0{\cdot}8+1}$$

$$= \frac{0{\cdot}8}{3} = \frac{8}{30}$$

$$= \frac{4}{15}$$ **Ans.**

SECTION—C

Solution 13.

(b) The data given in the problem are as under :

	Machine A	Machine B	Maximum available
Area needed	1,000 sq. m	1,200 sq.m	7,600 sq.m
Labour force	12	8	72
Daily output	50 units	40 units	—

Let x and y be the number of machines A and B respectively.

Subject to constraints

$$1{,}000x + 1{,}200y \leq 7{,}600$$

or $$5x + 6y \leq 38$$

$$12x + 8y \leq 72$$

or $$3x + 2y \leq 18$$

$$x \geq 0, y \geq 0$$

Total output $Z = 50x + 40y$

or the problem is to maximise $Z = 50x + 40y$

Subject to constraints :

$$5x + 6y \leq 38$$

$$3x + 2y \leq 18$$

$$x \geq 0, y \geq 0$$

Now, we have the lines,

$3x + 2y = 18$

x	6	0	4
y	0	9	3

$5x + 6y = 38$

x	38/5	0	4
y	0	19/3	3

The graph of these in equation is as shown below :

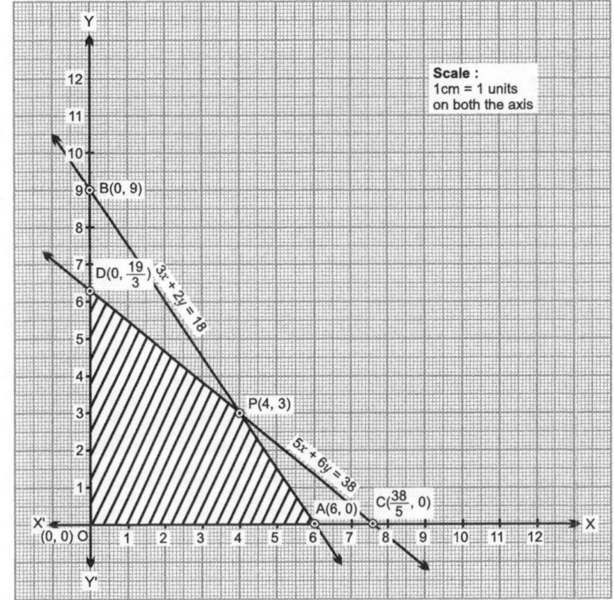

The vertices of the feasible region OAPD are O (0, 0), A (6, 0), P (4, 3) and D $\left(0, \frac{19}{3}\right)$.

Corner Points	$Z = 50x + 40y$
At O (0, 0),	$Z = 50 \times 0 + 40 \times 0 = 0$
At A (6, 0),	$Z = 50 \times 6 + 40 \times 0 = 300$
At P (4, 3),	$Z = 50 \times 4 + 40 \times 3 = 320$ maximum
At D $\left(0, \frac{19}{3}\right)$,	$Z = 50 \times 0 + 40 \times \frac{19}{3}$ $= \frac{760}{3} = 253{\cdot}33$

Thus, we see that Z is maximum at (4, 3).

$\therefore$ Number of machine A = 4

Number of machine B = 3. **Ans.**

Solution 14.

(b) Suppose that x number of units are produced and sold.

Given, fixed cost = ₹ 24,000

Revenue R (x) = ₹ $8x$

(i) As each unit's variable cost is 25% of revenue.

∴ The variable cost of x units = 25% of ₹ $8x$

= ₹ $2x$

∴ Total cost of producing x units

C (x) = Total fixed cost (TFC) + Total variable cost (TVC)

= ₹ $(24{,}000 + 2x)$ **Ans.**

(ii) Price of one unit = ₹ 8

∴ Total revenue on selling x units = R (x)

= ₹ $8x$ **Ans.**

(iii) At break-even values

$$\begin{aligned} C(x) &= R(x) \\ 24{,}000 + 2x &= 8x \\ 24{,}000 &= 6x \\ x &= 4{,}000 \end{aligned}$$

Hence, the break-even point is 4,000. **Ans.**

●●

MATHEMATICS 2012

QUESTIONS

SECTION—A

Question 1. **[10 × 3]**

(i) *Solve for x and y if* $\begin{bmatrix} x^2 \\ y^2 \end{bmatrix} + 2\begin{bmatrix} 2x \\ 3y \end{bmatrix} = 3\begin{bmatrix} 7 \\ -3 \end{bmatrix}$

(ii) *Prove that* $\sec^2(\tan^{-1} 2) + \text{cosec}^2(\cot^{-1} 3) = 15$.

(iii) *Find the equation of the hyperbola whose transverse and conjugate axes are the x and y axes respectively, given that the length of conjugate axis is 5 and distance between the foci is 13.***

(iv) *From the equations of the two regression lines,* $4x + 3y + 7 = 0$ *and* $3x + 4y + 8 - 0$*, find :*

(a) *Mean of x and y.*

(b) *Regression coefficients.*

(c) *Coefficient of correlation.*

(v) *Evaluate :* $\int e^x (\tan x + \log \sec x)\, dx$.

(vi) *Evaluate:* $\lim\limits_{x \to \pi/2} (\cos x \cdot \log \tan x)$.

(vii) *Find the locus of the complex number,* $Z = x + iy$**

given $\left|\dfrac{x + iy - 2i}{x + iy + 2i}\right| = \sqrt{2}$

(viii) *Evaluate :* $\int_1^2 \dfrac{\sqrt{x}}{\sqrt{3-x} + \sqrt{x}}\, dx$.

(ix) *Three persons A, B and C shoot to hit a target. If in trials, A hits the target 4 times in 5 shots, B hits 3 times in 4 shots and C hits 2 times in 3 trials. Find the probability that :*

(a) *Exactly two persons hit the target.*

(b) *At least two persons hit the target.*

(x) *Solve the differential equation :*
$(xy^2 + x)\, dx + (x^2 y + y)\, dy = 0$

Question 2.

(a) *Using properties of determinants, prove that :* **[5]**

$$\begin{vmatrix} a & a+b & a+b+c \\ 2a & 3a+2b & 4a+3b+2c \\ 3a & 6a+3b & 10a+6b+3c \end{vmatrix} = a^3$$

** Answer is not given due to change in the present syllabus.

(b) *Find the product of the matrices A and B where :* **[5]**

$$A = \begin{bmatrix} -5 & 1 & 3 \\ 7 & 1 & -5 \\ 1 & -1 & 1 \end{bmatrix}, B = \begin{bmatrix} 1 & 1 & 2 \\ 3 & 2 & 1 \\ 2 & 1 & 3 \end{bmatrix}$$

Hence, solve the following equations by matrix method :

$$\begin{aligned} x + y + 2z &= 1 \\ 3x + 2y + z &= 7 \\ 2x + y + 3z &= 2 \end{aligned}$$

Question 3.

(a) *Prove that :* $\cos^{-1}\dfrac{63}{65} + 2\tan^{-1}\dfrac{1}{5} = \sin^{-1}\dfrac{3}{5}$ **[5]**

(b) (i) *Write the Boolean expression corresponding to the circuit given below :*** **[5]**

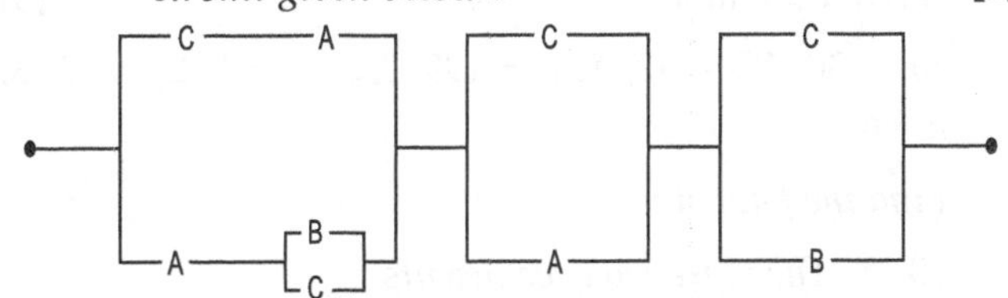

(ii) *Simplify the expression using laws of Boolean Algebra and construct the simplified circuit.*** **[5]**

Question 4.

(a) *Verify Rolle's theorem for the function :* **[5]**

$f(x) = \log\left\{\dfrac{x^2 + ab}{(a+b)x}\right\}$ *in the interval [a, b] where,* $0 \notin [a, b]$.

(b) *Find the equation of the ellipse with its centre at (4, –1) focus at (1, –1) and given that it passes through (8, 0).*** **[5]**

Question 5.

(a) *If,* $e^y(x + 1) = 1$*, then show that :* **[5]**

$$\frac{d^2y}{dx^2} = \left(\frac{dy}{dx}\right)^2$$

(b) *A printed page is to have a total area of 80 sq. cm with a margin of 1 cm at the top and on each side and a margin of 1·5 cm at the bottom. What should be the dimensions of the page so that the printed area will be maximum ?* **[5]**

Question 6.

(a) *Evaluate :* $\int \dfrac{dx}{x\{6(\log x)^2 + 7\log x + 2\}}$ **[5]**

(b) *Find the area of the region bounded by the curves* $x = 4y - y^2$ *and the Y-axis.* **[5]**

Question 7.

(a) *Ten candidates received percentage marks in two subjects as follows :*** **[5]**

Candidate	Mathematics Marks	Statistics Marks
A	80	72
B	88	84
C	76	90
D	74	66
E	68	54
F	65	50
G	40	54
H	43	38
I	40	30
J	80	43

Calculate Spearman's rank correlation coefficient and interpret your result.

(b) *The following results were obtained with respect to two variable x and y :* **[5]**

$\Sigma x = 30$, $\Sigma y = 42$, $\Sigma xy = 199$, $\Sigma x^2 = 184$, $\Sigma y^2 = 318$, $n = 6$.

Find the following :

(i) The regression coefficients.

(ii) Correlation coefficient between x and y.

(iii) Regression equation of y on x.

(iv) The likely value of y when $x = 10$.

Question 8.

(a) *A bag contains 8 red and 5 white balls. Two successive draws of 3 balls are made at random from the bag without replacements. Find the probability that the first draw yields 3 white balls and the second draw 3 red balls.* **[5]**

(b) *A box contains 30 bolts and 40 nuts. Half of the bolts and half of the nuts are rusted. If two items are drawn at random from the box, what is the probability that either both are rusted or both are bolts ?* **[5]**

Question 9.

(a) *Using De. Moivre's theorem prove that :*** **[5]**

$$\left(\frac{1+\cos\theta+i\sin\theta}{1+\cos\theta-i\sin\theta}\right)^n = \cos n\theta + i\sin n\theta, \text{ where } i=\sqrt{-1}$$

(b) *Solve the differential equation :* **[5]**

$$\frac{dy}{dx} - 3y\cot x = \sin 2x, \text{ given } y = 2, \text{ when } x = \frac{\pi}{2}$$

** Answer is not given due to change in the present syllabus.

SECTION—B

Question 10.

(a) *For any three vectors $\vec{a}, \vec{b}, \vec{c}$ prove :* **[5]**

$$\left[\vec{a}-\vec{b} \quad \vec{b}-\vec{c} \quad \vec{c}-\vec{a}\right] = 0.$$

(b) *In any triangle ABC, prove by vector method :* **[5]**

$$\frac{a}{\sin A} = \frac{b}{\sin B} = \frac{c}{\sin C}$$

Question 11.

(a) *Find the shortest distance between the lines :* **[5]**

$$\frac{x-8}{3} = \frac{y+9}{-16} = \frac{z-10}{7}$$

and $$\frac{x-15}{3} = \frac{y-29}{8} = \frac{5-z}{5}$$

(b) *Find the equation of the plane passing through the line of intersection of the planes $x + 2y + 3z - 5 = 0$ and $3x - 2y - z + 1 = 0$ and cutting off equal intercepts on the X and Z-axes.* **[5]**

Question 12.

(a) *In a class of 75 students, 15 are above average, 45 are average and the rest below average achievers. The probability that an above average achieving student fails is 0·005, that an average achieving student fails is 0·05 and the probability of a below average achieving student failing is 0·15. If a student is known to have passed, what is the probability that he is a below average achiever?* **[5]**

(b) *The probability that a bulb produced by a factory will fuse in 100 days of use is 0·05. Find the probability that out of 5 such bulbs, after 100 days of use :* **[5]**

(i) None fuse.

(ii) Not more than one fuses.

(iii) More than one fuses.

(iv) At least one fuses.

SECTION—C

Question 13.

(a) *Two tailors P and Q earn ₹ 150 and ₹ 200 per day respectively. P can stitch 6 shirts and 4 trousers a day, while Q can stitch 10 shirts and 4 trousers per day. How many days should each work to produce at least 60 shirts and 32 trousers at minimum labour cost ?* **[5]**

(b) *A machine costs ₹ 97,000 and its effective life is estimated to be 12 years. If scrap realises ₹ 2,000 only, what amount should be retained out of profits at the end of each year to accumulate at compound interest of 5% per annum in order to buy a new machine after 12 years? (use $1{\cdot}05^{12} = 1{\cdot}769$)*** **[5]**

Question 14.

(a) *A bill of ₹ 1,000 drawn on 7th May, 2011 for six months was discounted on 29th August, 2011 for cash payment of ₹ 988. Find the rate of interest charged by the bank.*** **[5]**

(b) *If total cost function is given by $C = a + bx + cx^2$, where x is the quantity of output. Show that :* **[5]**

$$\frac{d}{dx}(AC) = \frac{1}{x}(MC - AC),$$ *where MC is the marginal cost and AC is the average cost.*

Question 15.**

(a) *Find the consumer price index number for the year 2010 using year 2000 as the base year by using method of weighted aggregates :* **[5]**

Commodity	A	B	C	D	E
Year 2000 price (in ₹ per unit)	16	40	0·50	5·12	2·00
Year 2010 price (in ₹ per unit)	20	60	0·50	6·25	1·50
Weights	40	25	5·00	20·00	10·00

(b) *Calculate the 5 yearly moving averages of the number of students in a College from the following data and plot them on a graph paper :* **[5]**

Year	Number
1981	332
1982	317
1983	357
1984	392
1985	402
1986	405
1987	510
1988	427
1989	405
1990	438

** Answer is not given due to change in the present syllabus.

SOLUTIONS

Solution 1.

(i) Given, $\begin{bmatrix} x^2 \\ y^2 \end{bmatrix} + 2\begin{bmatrix} 2x \\ 3y \end{bmatrix} = 3\begin{bmatrix} 7 \\ -3 \end{bmatrix}$

$\Rightarrow \begin{bmatrix} x^2 \\ y^2 \end{bmatrix} + \begin{bmatrix} 4x \\ 6y \end{bmatrix} = \begin{bmatrix} 21 \\ -9 \end{bmatrix}$

$\Rightarrow \begin{bmatrix} x^2+4x \\ y^2+6y \end{bmatrix} = \begin{bmatrix} 21 \\ -9 \end{bmatrix}$

Equating corresponding elements of equal matrices, we get

$x^2 + 4x = 21 \qquad y^2 + 6y = -9$

$\Rightarrow x^2 + 4x - 21 = 0 \qquad y^2 + 6y + 9 = 0$

$\Rightarrow x^2 + 7x - 3x - 21 = 0 \qquad (y+3)^2 = 0$

$\Rightarrow x(x+7) - 3(x+7) = 0 \qquad y + 3 = 0$

$\Rightarrow (x+7)(x-3) = 0 \qquad y = -3$

$\Rightarrow x = -7, 3$ **Ans.**

(ii) L. H. S. $= \sec^2(\tan^{-1} 2) + \text{cosec}^2(\cot^{-1} 3)$

$= \sec^2\left(\sec^{-1}\sqrt{5}\right) + \text{cosec}^2\left(\text{cosec}^{-1}\sqrt{10}\right)$

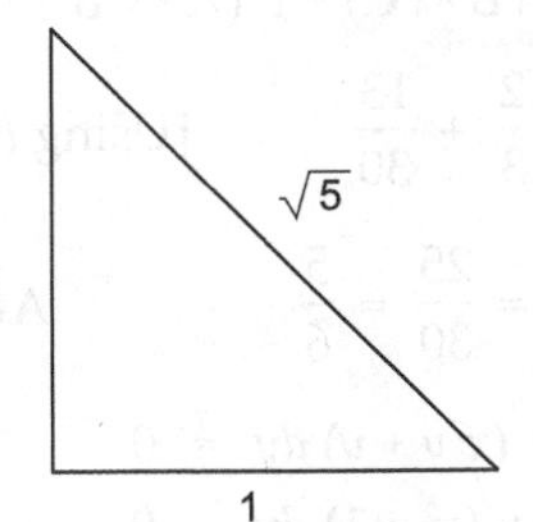

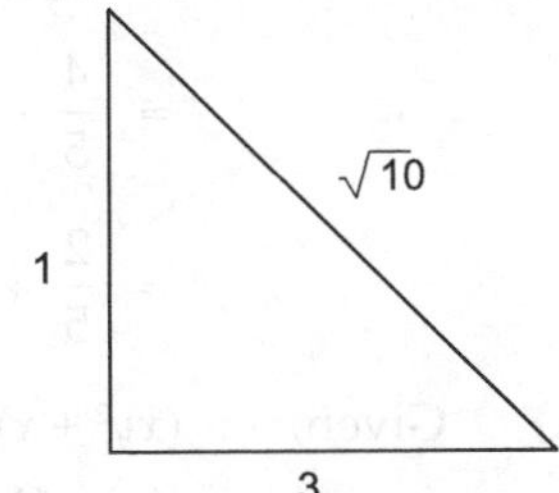

$= \left\{\sec\left(\sec^{-1}\sqrt{5}\right)\right\}^2 + \left\{\text{cosec}\left(\text{cosec}^{-1}\sqrt{10}\right)\right\}^2$

$= \left(\sqrt{5}\right)^2 + \left(\sqrt{10}\right)^2 = 5 + 10$

$= 15 =$ R. H. S. **Hence Proved.**

(iv) (a) The equations of the two lines of regression are

$4x + 3y + 7 = 0$ ...(i)

$3x + 4y + 8 = 0$ ...(ii)

Solving (i) and (ii) simultaneously, we get

$x = -4/7$ and $y = -11/7$

$\Rightarrow \bar{x} = -4/7$ and $\bar{y} = -11/7$ **Ans.**

(b) Let equation (i) be the regression equation of y on x and (ii) the regression equation of x on y then from (i)

$$y = -\frac{4}{3}x - \frac{7}{4}$$

$$\Rightarrow b_{yx} = -\frac{4}{3}$$

and from (ii),

$$x = -\frac{4}{3}y - \frac{8}{3}$$

$$\Rightarrow b_{xy} = -\frac{4}{3}$$

but $r^2 = b_{yx} \cdot b_{xy} = \left(-\frac{4}{3}\right) \times \left(-\frac{4}{3}\right) > 1$

which is impossible. Hence, equation (i) is the regression equation of x on y and equation (ii) is the regression equation of y on x. From (i)

$$x = -\frac{3}{4}y - \frac{7}{4}$$

$$\therefore \quad b_{xy} = -\frac{3}{4}$$

From (ii),

$$y = -\frac{3}{4}x - 2$$

$$\therefore \quad b_{yx} = -\frac{3}{4}$$ **Ans.**

(c) Coefficient of correlation

$$|r| = \sqrt{b_{yx}b_{xy}}$$

$$= \sqrt{\frac{-3}{4} \times \frac{-3}{4}}$$

$$\Rightarrow \quad r = \pm\frac{3}{4}$$

But r has the same sign as regression coefficients.

$$\therefore \quad r = \frac{-3}{4}$$ **Ans.**

(v) $\int e^x (\tan x + \log \sec x)\, dx$

$= \int e^x \tan x\, dx + \int e^x \log \sec x\, dx$
II I

$= \int e^x \tan x\, dx + \log (\sec x) \int e^x\, dx$

$-\int \left\{ \frac{d}{dx}(\log \sec x) . \int e^x dx \right\} dx$

$= \int e^x \tan x\, dx + \log (\sec x)\, e^x - \int e^x \tan x\, dx$

$= e^x \log \sec x + c$ **Ans.**

(vi) Given, $\lim_{x\to\pi/2} (\cos x \cdot \log \tan x)$ $[0 \cdot \infty \text{ form}]$

$$= \lim_{x\to\pi/2} \frac{\log \tan x}{\sec x} \quad \left[\frac{\infty}{\infty} \text{form}\right]$$

Using L' Hospital's rule,

$$= \lim_{x\to\pi/2} \frac{\frac{1}{\tan x} \times \sec^2 x}{\sec x . \tan x}$$

$$= \lim_{x\to\pi/2} \frac{\sec^2 x}{\tan^2 x . \sec x}$$

$$= \lim_{x\to\pi/2} \frac{\sec x}{\tan^2 x}$$

$$= \lim_{x\to\pi/2} \frac{1}{\cos x} \times \frac{\cos^2 x}{\sin^2 x}$$

$$= \lim_{x\to\pi/2} \frac{\cos x}{\sin^2 x}$$

$$= \frac{\cos \pi/2}{\sin^2 \pi/2} = \frac{0}{1} = 0$$ **Ans.**

(viii) Let, $I = \int_1^2 \frac{\sqrt{x}}{\sqrt{3-x}+\sqrt{x}}\, dx$ ...(i)

Using properties,

$$\int_a^b f(x)\, dx = \int_a^b f(a+b-x)\, dx$$

$$I = \int_1^2 \frac{\sqrt{3-x}}{\sqrt{3-(3-x)}+\sqrt{3-x}}\, dx$$

$$I = \int_1^2 \frac{\sqrt{3-x}}{\sqrt{x}+\sqrt{3-x}}\, dx \quad \text{...(ii)}$$

Adding equations (i) and (ii),

$$2I = \int_1^2 \frac{\sqrt{x}+\sqrt{3-x}}{\sqrt{x}+\sqrt{3-x}}\, dx$$

$$= \int_1^2 dx = [x]_1^2 = 2 - 1 = 1$$

$$\therefore \quad I = \frac{1}{2}$$ **Ans.**

(ix) Here, $P(A) = \frac{4}{5}$, $P(B) = \frac{3}{4}$, $P(C) = \frac{2}{3}$

$P(\bar{A}) = \frac{1}{5}$, $P(\bar{B}) = \frac{1}{4}$, $P(\bar{C}) = \frac{1}{3}$

(a) P (Exactly two hit the target)

$= P(A \cap B \cap \bar{C}) + (P\ \bar{A} \cap B \cap C)$

$+ P(A \cap \bar{B} \cap C)$

$[\because$ A,B,C are independent events$]$

$$= \frac{4}{5}.\frac{3}{4}.\frac{1}{3} + \frac{1}{5}.\frac{3}{4}.\frac{2}{3} + \frac{4}{5}.\frac{1}{4}.\frac{2}{3}$$

$$= \frac{1}{5} + \frac{1}{10} + \frac{2}{15} = \frac{13}{30}$$ **Ans.**

(b) P (at least two hit the target)

$= P(A \cap B \cap C) + P(A \cap B \cap \bar{C})$

$+ P(\bar{A} \cap B \cap C) + P(A \cap \bar{B} \cap C)$

$$= \frac{4}{5} \cdot \frac{3}{4} \cdot \frac{2}{3} + \frac{13}{30} \quad \text{[Using (a)]}$$

$$= \frac{2}{5} + \frac{13}{30} = \frac{25}{30} = \frac{5}{6}$$ **Ans.**

(x) Given, $(xy^2 + x)\, dx + (x^2y + y)\, dy = 0$

$$\Rightarrow \quad x(y^2+1)\, dx + y(x^2+1)\, dy = 0$$

$\Rightarrow \quad x(y^2+1)\,dx = -y(x^2+1)\,dy$

$\Rightarrow \quad \frac{x}{x^2+1}\,dx = \frac{-y}{y^2+1}\,dy$

Multiplying 2 on both LHS and RHS, we get

$\Rightarrow \quad \frac{2x}{x^2+1}\,dx = \frac{-2y}{y^2+1}\,dy$

On integrating both sides, we get

$\Rightarrow \quad \log|x^2+1| = -\log|y^2+1| + c$

$\Rightarrow \log|x^2+1| + \log|y^2+1| = c$

$\Rightarrow \quad |x^2+1|\,|y^2+1| = e^c$

$\Rightarrow \quad (x^2+1)(y^2+1) = \pm e^c$ **Ans.**

Solution 2.

(a) Let $D = \text{LHS} = \begin{vmatrix} a & a+b & a+b+c \\ 2a & 3a+2b & 4a+3b+2c \\ 3a & 6a+3b & 10a+6b+3c \end{vmatrix}$

Operating $R_2 \to R_2 - 2R_1$ and $R_3 \to R_3 - 3R_1$, we get

$$D = \begin{vmatrix} a & a+b & a+b+c \\ 0 & a & 2a+b \\ 0 & 3a & 7a+3b \end{vmatrix} = a\begin{vmatrix} a & 2a+b \\ 3a & 7a+3b \end{vmatrix}$$

(expanding along 1st column)

$= a\,[a\,(7a+3b) - 3a\,(2a+b)]$

$= a\,[7a^2 + 3ab - 6a^2 - 3ab]$

$= a \cdot a^2 = a^3 = \text{RHS}.$ **Hence Proved.**

(b) Given, $A = \begin{bmatrix} -5 & 1 & 3 \\ 7 & 1 & -5 \\ 1 & -1 & 1 \end{bmatrix}$, $B = \begin{bmatrix} 1 & 1 & 2 \\ 3 & 2 & 1 \\ 2 & 1 & 3 \end{bmatrix}$

$$\therefore AB = \begin{bmatrix} -5+3+6 & -5+2+3 & -10+1+9 \\ 7+3-10 & 7+2-5 & 14+1-15 \\ 1-3+2 & 1-2+1 & 2-1+3 \end{bmatrix}$$

$$AB = \begin{bmatrix} 4 & 0 & 0 \\ 0 & 4 & 0 \\ 0 & 0 & 4 \end{bmatrix}$$

$$AB = 4\begin{bmatrix} 1 & 0 & 0 \\ 0 & 1 & 0 \\ 0 & 0 & 1 \end{bmatrix}$$

$$AB = 4I_3$$

$\Rightarrow \quad B\left(\frac{1}{4}A\right) = I_3$

$\Rightarrow \quad B^{-1} = \frac{1}{4}A$

$\Rightarrow \quad B^{-1} = \frac{1}{4}\begin{bmatrix} -5 & 1 & 3 \\ 7 & 1 & -5 \\ 1 & -1 & 1 \end{bmatrix}$ ...(i)

The given system of equation is

$x + y + 2z = 1$

$3x + 2y + z = 7$

$2x + y + 3z = 2$

This system can be written as

$$\begin{bmatrix} 1 & 1 & 2 \\ 3 & 2 & 1 \\ 2 & 1 & 3 \end{bmatrix}\begin{bmatrix} x \\ y \\ z \end{bmatrix} = \begin{bmatrix} 1 \\ 7 \\ 2 \end{bmatrix} \text{ or } BX = C$$

where $X = \begin{bmatrix} x \\ y \\ z \end{bmatrix}$, $C = \begin{bmatrix} 1 \\ 7 \\ 2 \end{bmatrix}$

As B^{-1} exists, the given system has a unique solution

$X = B^{-1}C$

$\Rightarrow \quad X = \frac{1}{4}\begin{bmatrix} -5 & 1 & 3 \\ 7 & 1 & -5 \\ 1 & -1 & 1 \end{bmatrix}\begin{bmatrix} 1 \\ 7 \\ 2 \end{bmatrix}$ [Using (i)]

$\Rightarrow \quad \begin{bmatrix} x \\ y \\ z \end{bmatrix} = \frac{1}{4}\begin{bmatrix} -5+7+6 \\ 7+7-10 \\ 1-7+2 \end{bmatrix}$

$\Rightarrow \quad \begin{bmatrix} x \\ y \\ z \end{bmatrix} = \begin{bmatrix} 2 \\ 1 \\ -1 \end{bmatrix}$

$\Rightarrow \quad x = 2, y = 1, z = -1$

Hence, the solution of given system of equation is $x = 2, y = 1, z = -1.$ **Ans.**

Solution 3.

(a) To prove, $\cos^{-1}\frac{63}{65} + 2\tan^{-1}\frac{1}{5} = \sin^{-1}\frac{3}{5}$

Taking L.H.S. $= \sin^{-1}\sqrt{1-\left(\frac{63}{65}\right)^2} + \sin^{-1}\frac{2\cdot\frac{1}{5}}{1+\left(\frac{1}{5}\right)^2}$

$$\left[\text{Using, } \cos^{-1} = \sin^{-1}\sqrt{1-x^2},\ 2\tan^{-1}x = \sin^{-1}\frac{2x}{1+x^2}\right]$$

$= \sin^{-1}\sqrt{\frac{65^2-63^2}{65^2}} + \sin^{-1}\frac{2/5}{26/25}$

$= \sin^{-1}\sqrt{\frac{(65-63)(65+63)}{(65)^2}} + \sin^{-1}\frac{5}{13}$

$[\because a^2 - b^2 = (a-b)(a+b)]$

$$= \sin^{-1}\sqrt{\frac{2\times 128}{(65)^2}} + \sin^{-1}\frac{5}{13}$$

$$= \sin^{-1}\frac{16}{65} + \sin^{-1}\frac{5}{13}$$

[Using $\sin^{-1}x + \sin^{-1}y = \sin^{-1}\left(x\left\{\sqrt{1-y^2}\right\}+y\left\{\sqrt{1-x^2}\right\}\right)$]

$$= \sin^{-1}\left\{\frac{16}{65}\sqrt{1-\left(\frac{5}{13}\right)^2}+\frac{5}{13}\sqrt{1-\left(\frac{16}{65}\right)^2}\right\}$$

$$= \sin^{-1}\left(\frac{16}{65}\times\frac{12}{13}+\frac{5}{13}\times\frac{63}{65}\right)$$

$$= \sin^{-1}\left(\frac{16\times 12+315}{13\times 65}\right)$$

$$= \sin^{-1}\left(\frac{507}{65\times 13}\right) = \sin^{-1}\frac{3}{5}$$

Hence Proved.

Solution 4.

(a) Given function is,

$$f(x) = \log\left(\frac{x^2+ab}{(a+b)x}\right)$$

$$= \log(x^2+ab) - \log(a+b) - \log x \quad \ldots\text{(i)}$$

Since $a > 0$ and $\log x$ is continuous for all $x > 0$

$\because x \,\varepsilon\, [a, b]$, therefore, $f(x)$ is continuous in $[a, b]$.

$f(x)$ is differentiable in (a, b) and

$$f(a) = \log\left(\frac{a^2+ab}{(a+b)a}\right) = \log 1 = 0$$

$$f(b) = \log\left(\frac{b^2+ab}{(a+b)b}\right) = \log 1 = 0$$

$\Rightarrow \quad f(a) = f(b)$.

Thus, all the three conditions of Rolle's theorem are satisfied. Therefore, there exists at least one real number c in (a, b) such that $f'(c) = 0$.

Differentiating (i) w.r.t. x, we get

$$f'(x) = \frac{1}{x^2+ab}\cdot 2x - 0 - \frac{1}{x}$$

$$= \frac{2x}{x^2+ab} - \frac{1}{x}$$

$$= \frac{x^2-ab}{x(x^2+ab)}$$

Now, $f'(c) = 0$

$$\Rightarrow \quad \frac{c^2-ab}{c(c^2+ab)} = 0$$

$$\Rightarrow \quad c^2 - ab = 0$$

$$\Rightarrow \quad c^2 = ab$$

$$\Rightarrow \quad c = \pm\sqrt{ab}$$

$0 \notin [a, b] \Rightarrow -\sqrt{ab} \notin [a, b]$

$$\therefore \quad c = \sqrt{ab}$$

and $a < \sqrt{ab} < b$

So, there exists a real number

$c = \sqrt{ab}$ in (a, b) such that

$$f'(c) = 0$$

Hence, Rolle's theorem is verified and $c = \sqrt{ab}$.

Ans.

Solution 5.

(a) Given, $e^y(x+1) = 1$.

Taking logarithm of both sides, we get

$$\log(e^y(x+1)) = \log 1$$

$$y\log e + \log(x+1) = \log 1$$

$$\Rightarrow \quad y + \log(x+1) = 0$$

$$\Rightarrow \quad y = -\log(x+1)$$

Differentiating w.r.t. x, we get

$$\frac{dy}{dx} = -\frac{1}{x+1} \quad \ldots\text{(i)}$$

Differentiating (i) again w.r.t. x, we get

$$\frac{d^2y}{dx^2} = \frac{1}{(x+1)^2}$$

$$= \left(\frac{-1}{x+1}\right)^2$$

$$\Rightarrow \quad \frac{d^2y}{dx^2} = \left(\frac{dy}{dx}\right)^2 \quad \text{[From (i)]}$$

Hence Proved.

(b) Let x cm and y cm be the dimensions of the printed page.

$$\Rightarrow \quad xy = 80 \text{ sq cm.} \quad \ldots\text{(i)}$$

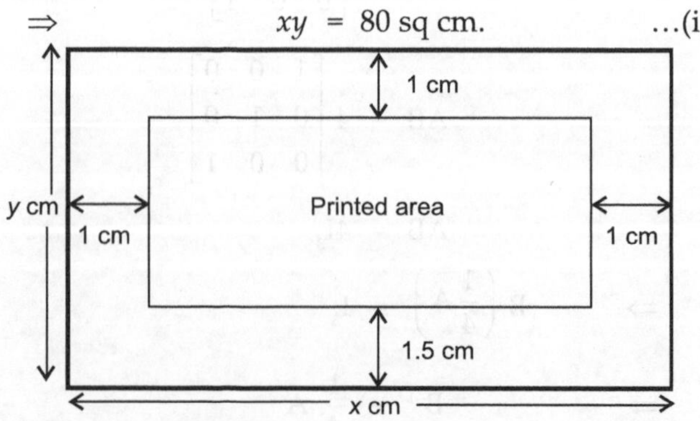

Length of printed material = $(x - 1 - 1)$ cm

$= (x - 2)$ cm.

Breadth of printed material = $(y - 1 - 1{\cdot}5)$ cm.

$= (y - 2{\cdot}5)$ cm.

Let A (x sq. cm) be the printed area then,

$$A = (x-2)(y-5/2)$$

$$A = xy - \frac{5x}{2} - 2y + 5$$

$$A = 80 - \frac{5x}{2} - \frac{2\times 80}{x} + 5 \text{ [using (i)]}$$

$$A = 85 - \frac{5x}{2} - \frac{160}{x}$$

Differentiating above equation w.r.t. x, we get

$$\frac{dA}{dx} = \frac{-5}{2} + \frac{160}{x^2}$$

Differcntiating above equation once again w.r.t. x

$$\frac{d^2A}{dx^2} = \frac{-320}{x^3}$$

For maximum and minimum area,

$$\therefore \quad \frac{dA}{dx} = 0$$

$$\frac{-5}{2} + \frac{160}{x^2} = 0$$

$$\Rightarrow \quad \frac{160}{x^2} = \frac{5}{2}$$

$$\Rightarrow \quad x^2 = 64$$

$$\Rightarrow \quad x = \pm 8$$

But, $x \neq 8$, as length cannot be – ve

$$\text{Also,} \left(\frac{d^2A}{dx^2}\right)_{x=8} = -\frac{320}{512} < 0$$

$\Rightarrow$ A is maximum at $x = 8$

From (i), $\quad xy = 80$

$$\Rightarrow \quad 8 \times y = 80$$

$$\Rightarrow \quad y = 10$$

Therefore, the dimensions of the printed page are 8 cm and 10 cm. **Ans.**

Solution 6.

(a) Let, $\log x = t \Rightarrow \frac{1}{x}\,dx = dt$

$$\therefore \int \frac{dx}{x\{6(\log x)^2 + 7\log x + 2\}} = \int \frac{dt}{6t^2 + 7t + 2}$$

$$6t^2 + 7t + 2 = 6t^2 + 3t + 4t + 2 = (3t+2)(2t+1)$$

$$= \int \frac{dt}{(3t+2)(2t+1)}$$

Let, $$\frac{1}{(3t+2)(2t+1)} = \frac{A}{3t+2} + \frac{B}{2t+1}$$

$$1 = A(2t+1) + B(3t+2)$$

On putting $t = -\frac{2}{3}$ and $t = -\frac{1}{2}$ respectively, we get

$$\therefore \quad 1 = A\left(2\times\left(-\frac{2}{3}\right)+1\right)$$

$$A = -3$$

$$1 = B\left(3\times\left(-\frac{1}{2}\right)+2\right)$$

$$\Rightarrow \quad B = 2$$

$$\int \frac{dt}{(3t+2)(2t+1)} = \int \frac{-3}{3t+2}\,dt + \int \frac{2}{2t+1}\,dt$$

$$= -3\cdot\frac{1}{3}\log|3t+2| + 2\cdot\frac{1}{2}\log|2t+1| + c$$

$$= -\log|3t+2| + \log|2t+1| + c$$

$$= \log\frac{|2t+1|}{|3t+2|} + c$$

$$= \log\left|\frac{2\log x + 1}{3\log x + 2}\right| + c.$$ **Ans.**

(b) Given curve is,

$$x = 4y - y^2$$

i.e., $$y^2 - 4y = -x$$

$$y^2 - 4y + 4 = -x + 4$$

$$(y-2)^2 = -(x-4)$$

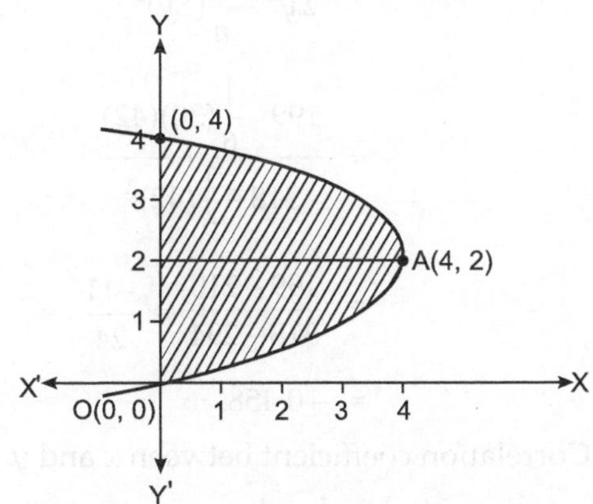

which represents a parabola with vertex at A (4, 2).

The parabola meets Y-axis *i.e.,* $x = 0$

$$4y - y^2 = 0$$

i.e., at $y = 0, 4.$

The area enclosed between the curve and the Y-axis is

$$\int_0^4 x\,dy = \int_0^4 (4y - y^2)\,dy$$

$$= \left[4\cdot\frac{y^2}{2} - \frac{y^3}{3}\right]_0^4$$

$$= \left[32 - \frac{64}{3}\right] - [0 - 0]$$

$$= \left|\frac{32}{3}\right| \text{ sq. unit}$$

$$= \frac{32}{3} \text{ sq. unit}$$

Area of the region bounded by the curves $x = 4y - y^2$ and Y-axis is $\frac{32}{3}$ sq. unit. **Ans.**

Solution 7.

(b) (i) Given, $\Sigma x = 30$, $\Sigma y = 42$, $\Sigma xy = 199$, $\Sigma x^2 = 184$, $\Sigma y^2 = 318$, $n = 6$

$$b_{yx} = \frac{\Sigma xy - \frac{1}{n}\Sigma x\cdot\Sigma y}{\Sigma x^2 - \frac{1}{n}(\Sigma x)^2}$$

$$= \frac{199 - \frac{1}{6}(30)(42)}{184 - \frac{1}{6}(30)^2}$$

$$= \frac{199 - 210}{184 - 150} = \frac{-11}{34} = -0\cdot323$$

$$b_{xy} = \frac{\Sigma xy - \frac{1}{n}\Sigma x\cdot\Sigma y}{\Sigma y^2 - \frac{1}{n}(\Sigma y)^2}$$

$$= \frac{199 - \frac{1}{6}(30)(42)}{318 - \frac{1}{6}(42)^2}$$

$$= \frac{199 - 210}{318 - 294} = \frac{-11}{24}$$

$$= -0\cdot458.$$ **Ans.**

(ii) Correlation coefficient between x and y.

$$r^2 = b_{yx}\cdot b_{xy}$$

$$= (-0\cdot323)\times(-0\cdot458)$$

$$= 0\cdot1479$$

$$\therefore \quad r = \pm\sqrt{0\cdot1479} = -0\cdot384$$

Since b_{xy} and b_{yx} have the same sign, that is why we have used negative sign with r. **Ans.**

(iii) $\bar{x} = \frac{\Sigma x}{6} = \frac{30}{6} = 5$

$$\bar{y} = \frac{\Sigma y}{6} = \frac{42}{6} = 7$$

Regression line of y on x.

$$y - \bar{y} = b_{yx}(x - \bar{x})$$

$$y - 7 = \frac{-11}{34}(x - 5)$$

$$34y - 238 = -11x + 55$$

$$11x + 34y - 293 = 0$$ **Ans.**

(iv) $x = 10$

$$11\times 10 + 34y - 293 = 0$$

$$\Rightarrow \quad 34y = 183$$

$$\Rightarrow \quad y = \frac{183}{34}$$

$$\Rightarrow \quad y = 5\cdot382$$ **Ans.**

Solution 8.

(a) Let A : Event that 3 balls in the first draw are all white

B : Event that 3 balls in the second draw are all red.

3 balls can be drawn out of 13 in ${}^{13}C_3$ and 3 white ball can be drawn out of 5 in ${}^{5}C_3$ ways.

$$P(A) = \frac{{}^{5}C_3}{{}^{13}C_3} = \frac{5}{143}$$

Since 3 balls drawn are not replaced before the second draw, we are left with 8 red and 2 white balls.

Now, 3 balls can be drawn in ${}^{10}C_3$ ways and 3 red balls can be drawn in ${}^{8}C_3$ ways.

$$P(B/A) = \frac{{}^{8}C_3}{{}^{10}C_3} = \frac{7}{15}$$

$$\therefore \quad P(A \cap (B/A)) = P(A)\cdot P(B/A)$$

$$= \frac{5}{143}\cdot\frac{7}{15}$$

$$= \frac{7}{429}$$ **Ans.**

(b) The number of bolts and nuts are 30 and 40 respectively. The number of rusted bolts and rusted nuts are 15 $\left(=\frac{30}{2}\right)$ and 20 $\left(=\frac{40}{2}\right)$ respectively.

Total number of items = 30 + 40 = 70.

Total number of rusted items = 15 + 20 = 35.

Total number of ways of drawing 2 items = ${}^{70}C_2$.

Let R and B be the events that the both items drawn are rusted items and both are bolts respectively.

R and B are not mutually exclusive events because there are 15 rusted bolts.

∴ P (items are both rusted or both bolts)

$$= P(R \cup B)$$

$$= P(R) + P(B) - P(R \cap B)$$

$$= \frac{{}^{35}C_2}{{}^{70}C_2} + \frac{{}^{30}C_2}{{}^{70}C_2} - \frac{{}^{15}C_2}{{}^{70}C_2}$$

$$= \frac{1\times 2}{70\times 69}\left(\frac{35\times 34}{1\times 2}+\frac{30\times 29}{1\times 2}-\frac{15\times 14}{1\times 2}\right)$$

$$= \frac{1850}{70\times 69} = \frac{185}{483}$$ **Ans.**

Solution 9.

(b) Given, differential equation is,

$$\frac{dy}{dx} - 3y\cot x = \sin 2x$$

This is a linear equation of the form $\frac{dy}{dx} + Py = Q$

where $P = -3\cot x,\ Q = \sin 2x$

$$\therefore \quad \text{I.F.} = e^{\int p\,dx}$$

$$\text{I.F.} = e^{\int -3\cot x\,dx} = e^{-3\int \cot x\,dx}$$

$$= e^{-3\log \sin x} = e^{\log(\sin x)^{-3}}$$

$$= (\sin x)^{-3} = \frac{1}{\sin^3 x}$$

∴ Its solution is $y\cdot$(I.F.)

$$= \int Q\,(\text{I.F.})\,dx + c$$

$$y\cdot\frac{1}{\sin^3 x} = \int \sin 2x\cdot\frac{1}{\sin^3 x}\,dx + c$$

$$\frac{y}{\sin^3 x} = \int \frac{2\sin x.\cos x}{\sin^3 x}\,dx + c$$

$$\frac{y}{\sin^3 x} = 2\int \operatorname{cosec} x.\cot x\,.\,dx + c$$

$$\frac{y}{\sin^3 x} = -2\operatorname{cosec} x + c$$

Putting $y = 2,\ x = \pi/2$,

$$\frac{2}{\sin^3 \pi/2} = -2\operatorname{cosec}(\pi/2) + c$$

$$2 = -2 + c \Rightarrow c = 4$$

$$\therefore \quad \frac{y}{\sin^3 x} = -\frac{2}{\sin x} + 4$$

$$y = -2\sin^2 x + 4\sin^3 x$$ **Ans.**

SECTION—B

Solution 10.

(a) LHS $= [\vec{a}-\vec{b}\ \ \vec{b}-\vec{c}\ \ \vec{c}-\vec{a}]$

$$= (\vec{a}-\vec{b})\times(\vec{b}-\vec{c})\cdot(\vec{c}-\vec{a})$$

$$= (\vec{a}\times\vec{b} - \vec{a}\times\vec{c} - \vec{b}\times\vec{b} + \vec{b}\times\vec{c})\cdot(\vec{c}-\vec{a})$$

$$= (\vec{a}\times\vec{b} - \vec{a}\times\vec{c} + \vec{b}\times\vec{c})\cdot(\vec{c}-\vec{a})$$

$(\because \vec{b}\times\vec{b} = 0)$

$$= (\vec{a}\times\vec{b})\cdot\vec{c} - (\vec{a}\times\vec{c})\cdot\vec{c} + (\vec{b}\times\vec{c})\cdot\vec{c}$$

$$-(\vec{a}\times\vec{b})\cdot\vec{a} + (\vec{a}\times\vec{c})\cdot\vec{a} - (\vec{b}\times\vec{c})\cdot\vec{a}$$

$$= (\vec{a}\times\vec{b})\cdot\vec{c} - 0 + 0 - 0 + 0 - (\vec{b}\times\vec{c})\cdot\vec{a}$$

$$= [\vec{a}\ \vec{b}\ \vec{c}] - [\vec{b}\ \vec{c}\ \vec{a}] \quad (\because [\vec{b}\ \vec{c}\ \vec{a}] = [\vec{a}\ \vec{b}\ \vec{c}])$$

$$= [\vec{a}\ \vec{b}\ \vec{c}] - [\vec{a}\ \vec{b}\ \vec{c}]$$

= 0 = RHS. **Hence Proved**

(b) In Δ ABC, let $\vec{BC} = \vec{a}$, $\vec{CA} = \vec{b}$ and $\vec{AB} = \vec{c}$, then

$$a = |\vec{BC}|,\ b = |\vec{CA}| \text{ and } c = |\vec{AB}|$$

From Δ ABC, we get

$$\Rightarrow \quad \vec{BC} + \vec{CA} = \vec{BA}$$

$$\Rightarrow \quad \vec{BC} + \vec{CA} + \vec{AB} = 0$$

$$\Rightarrow \quad \vec{a} + \vec{b} + \vec{c} = 0$$

$$\Rightarrow \quad \vec{a}\times(\vec{a}+\vec{b}+\vec{c}) = \vec{a}\times\vec{0}$$

$\Rightarrow \vec{a} \times \vec{a} + \vec{a} \times \vec{b} + \vec{a} \times \vec{c} = 0$

$\Rightarrow \quad \vec{a} \times \vec{b} = -\vec{a} \times \vec{c}$

$\Rightarrow \quad \vec{a} \times \vec{b} = \vec{c} \times \vec{a}$ ...(i)

Similarly, $\quad \vec{a} \times \vec{b} = \vec{b} \times \vec{c}$ ...(ii)

From (i) and (ii), we get

$$\vec{a} \times \vec{b} = \vec{b} \times \vec{c} = \vec{c} \times \vec{a}$$

$$|\vec{a} \times \vec{b}| = |\vec{b} \times \vec{c}| = |\vec{c} \times \vec{a}|$$

$$ab \sin(\pi - C) = bc \sin(\pi - A) = ca \sin(\pi - B)$$

$$ab \sin C = bc \sin A = ca \sin B$$

Dividing by abc throughout, we get

$$\frac{\sin C}{c} = \frac{\sin A}{a} = \frac{\sin B}{b}$$

$$\Rightarrow \quad \frac{a}{\sin A} = \frac{b}{\sin B} = \frac{c}{\sin C}$$

Hence Proved.

Solution 11.

(a) Given lines are : $\frac{x-8}{3} = \frac{y+9}{-16} = \frac{z-10}{7}$

$= t$ (say) ...(i)

and $\quad \frac{x-15}{3} = \frac{y-29}{8} = \frac{z-5}{-5} = s$ (say) ...(ii)

Any point on (i) is M $(8 + 3t, -9 - 16t, 10 + 7t)$
and any point on (ii) is N $(15 + 3s, 29 + 8s, 5 - 5s)$

$\therefore$ Direction numbers of the line MN are

$(15 + 3s - 8 - 3t, 29 + 8s + 9 + 16t, 5 - 5s - 10 - 7t)$

$\Rightarrow (7 + 3s - 3t, 38 + 8s + 16t, -5 - 5s - 7t)$

Now, MN will be the shortest distance between (i) and (ii).

If MN is perpendicular to both.

i.e., if $3(7 + 3s - 3t) + (-16)(38 + 8s + 16t) + 7(-5 - 5s - 7t) = 0$

and $3(7 + 3s - 3t) + 8(38 + 8s + 16t) + (-5)(-5 - 5s - 7t) = 0$

and $-154s - 314t - 622 = 0$

i.e., $\quad 77s + 157t + 311 = 0$ ...(iii)

and $\quad 98s + 154t + 350 = 0$

i.e., $\quad 49s + 77t + 175 = 0$ ...(iv)

On solving (iii) and (iv) simultaneously, we get

$t = -1$ and $s = -2$

$t = -1$ gives M (5, 7, 3)

$s = -2$ gives N (9, 13, 15)

$\therefore$ The shortest distance between the given lines

$$\text{MN} = \sqrt{(9-5)^2 + (13-7)^2 + (15-3)^2}$$

$$= \sqrt{196} = 14 \text{ units.}$$ **Ans.**

(b) The equation of any plane passing through the lines of intersection of the given planes

$x + 2y + 3z - 5 = 0$ and $3x - 2y - z + 1 = 0$ is

$(x + 2y + 3z - 5) + k(3x - 2y - z + 1) = 0$ ...(i)

For the intercept on X-axis, on putting $y = 0$ and $z = 0$, we get

$$x + 3kx - 5 + k = 0$$

$$\Rightarrow \quad x = \frac{5-k}{3k+1}$$

For the intercept on Z-axis, on putting $x = 0, y = 0$, we get

$$3z - kz - 5 + k = 0$$

$$\Rightarrow \quad z = \frac{5-k}{3-k}$$

$\therefore$ Intercepts on X-axis and Z-axis made by the plane (i) are $\frac{5-k}{3k+1}$ and $\frac{5-k}{3-k}$ respectively.

Since the intercepts on X-axis and Z-axis are equal.

$$\frac{5-k}{3k+1} = \frac{5-k}{3-k}$$

$\Rightarrow \quad -3k^2 + 14k + 5 = -8k + k^2 + 15$

$\Rightarrow \quad -4k^2 + 22k - 10 = 0$

$\Rightarrow \quad 4k^2 - 22k + 10 = 0$

$\Rightarrow \quad (4k - 2)(k - 5) = 0$

$\Rightarrow \quad k = \frac{1}{2}, 5$

On putting $k = 5$ in (i), we notice that the plane passes through origin and hence, it cannot make intercepts on axes. Therefore, $k = 1/2$ is the only admissible value.

Substituting $k = 1/2$ in equation (i), the equation of the required plane is

$x + 2y + 3z - 5 + 1/2(3x - 2y - z + 1) = 0$

$\Rightarrow \quad 2x + 4y + 6z - 10 + 3x - 2y - z + 1 = 0$

$\Rightarrow \quad 5x + 2y + 5z - 9 = 0$ **Ans.**

Solution 12.

(a) Let, E_1 : Event that student is above average

E_2 : Event that student is average

E_3 : Event that student is below average.

A : Event that student is known to have passed

$$P(E_1) = \frac{1}{5}, P(E_2) = \frac{3}{5}, P(E_3) = \frac{1}{5}$$

$$P(A/E_1) = 1 - 0{\cdot}005 = 0{\cdot}995$$

$$P(A/E_2) = 1 - 0{\cdot}05 = 0{\cdot}95$$

$$P(A/E_3) = 1 - 0{\cdot}15 = 0{\cdot}85$$

Using Bayes' theorem,

$$P(E_3/A) = \frac{P(E_3)\cdot P(A/E_3)}{P(E_1)\cdot P(A/E_1) + P(E_2)\cdot P(A/E_2) + P(E_3)\cdot P(A/E_3)}$$

$$= \frac{\frac{1}{5}\times\frac{85}{100}}{\frac{1}{5}\times\frac{995}{1000}+\frac{3}{5}\times\frac{95}{100}+\frac{1}{5}\times\frac{85}{100}}$$

$$= \frac{\frac{17}{100}}{\frac{199}{1000}+\frac{57}{100}+\frac{17}{100}}$$

$$= \frac{\frac{17}{100}}{\frac{199+570+170}{1000}}$$

$$= \frac{\frac{17}{100}}{\frac{939}{1000}} = \frac{17\times1000}{100\times939} = \frac{170}{939}$$

$= 0{\cdot}1810.$ **Ans.**

(b) Here, $p = 0{\cdot}05 = \frac{1}{20}, \; q = \frac{19}{20}$

(i) P (None fuse) = $P(0) = {}^5C_0\, p^0 q^{5-0}$

$$= \left(\frac{1}{20}\right)^0 \left(\frac{19}{20}\right)^5 = \left(\frac{19}{20}\right)^5$$ **Ans.**

(ii) P (Not more than one fuses) = P (0) + P (1)

$$= \sum_{r=0}^{1} {}^nC_r\, p_r\, q^{n-r}$$

$$= {}^5C_0 \left(\frac{1}{20}\right)^0 \left(\frac{19}{20}\right)^5 + {}^5C_1 \left(\frac{1}{20}\right)\left(\frac{19}{20}\right)^4$$

$$= \left(\frac{19}{20}\right)^4 \left(\frac{19}{20}+\frac{5}{20}\right)$$

$$= \left(\frac{19}{20}\right)^4 \left(\frac{24}{20}\right)$$

$$= \frac{6}{5}\left(\frac{19}{20}\right)^4$$ **Ans.**

(iii) P (More than one fuses)

= 1 – P (Not more than one fuse)

= 1 – [P(0) + P (1)]

$$= 1 - \frac{6}{5}\left(\frac{19}{20}\right)^4$$ **Ans.**

(iv) P (At least one fuses)

= 1 – P (None fuses)

= 1 – P (0)

$$= 1 - \left(\frac{19}{20}\right)^5$$ **Ans.**

SECTION—C

Solution 13.

(a) Let the tailor P work for x days and the tailor Q work for y days respectively.

$$Z = 150x + 200y$$

Here, the problem can be formulated as an L.P.P. as follows :

$$\text{Minimize } Z = 150x + 200y$$

Subject to the constraints :

$$6x + 10y \geq 60$$

or $\quad 3x + 5y \geq 30$

$$4x + 4y \geq 32$$

or $\quad x + y \geq 8$

and $\quad x \geq 0, y \geq 0$

Converting them into equations we obtain the following equations :

$3x + 5y = 30$, $\Rightarrow y = \frac{30-3x}{5}$,	$x + y = 8$, $y = 8 - x$.

x	0	10	5
y	6	0	3

x	0	8	5
y	8	0	3

The lines are shown on the graph paper and the feasible region (Unbounded convex) is shown shaded.

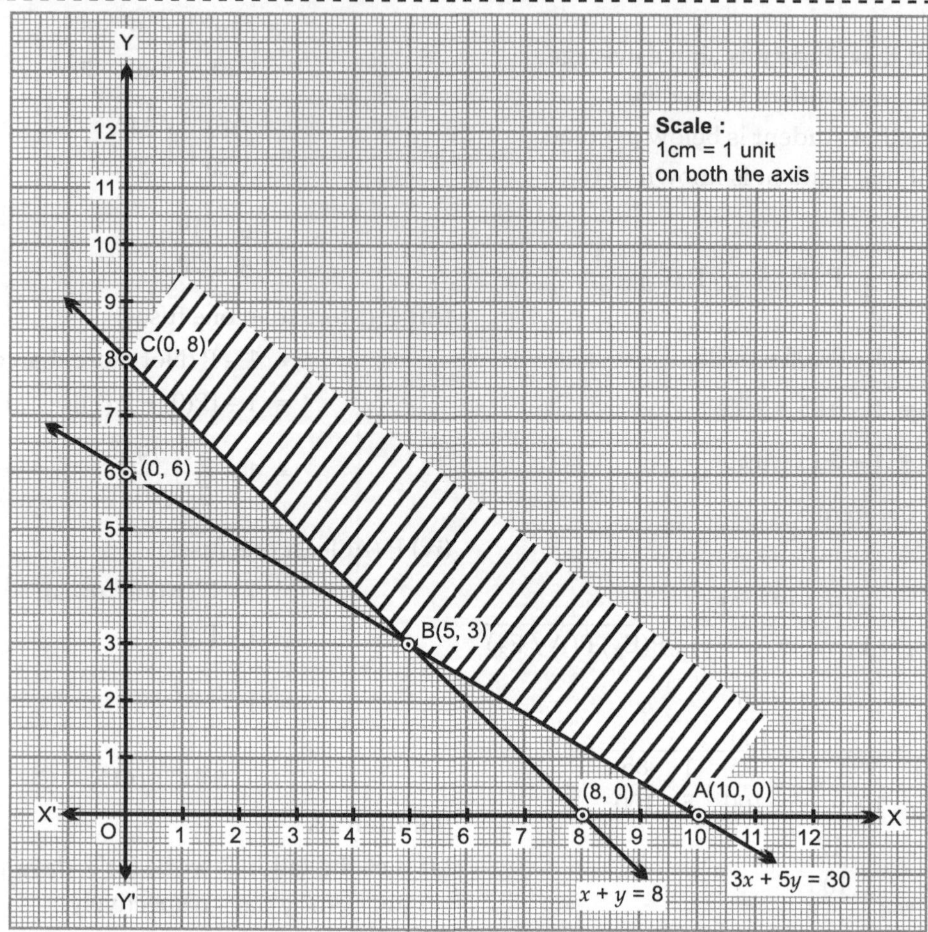

The corner points are
A (10, 0), B (5, 3) and C (0, 8).

Corner Points	Objective Function $Z = 150x + 200y$
At A (10, 0)	$Z = 150 \times 10 + 200 \times 0 = 1500$
At B (5, 3)	$Z = 150 \times 5 + 200 \times 3 = 1350$ minimum
At C (0, 8)	$Z = 150 \times 0 + 200 \times 8 = 1600$

Minimum value of Z is ₹ 1350 and it occurs at the point B (5, 3).

Hence, the labour cost is ₹ 1350 when P works for 5 days and Q works for 3 days. **Ans.**

Solution 14.

(b) Here, cost function $(C) = a + bx + cx^2$.

Quantity of the output $= x$.

$$\therefore \quad \text{Average cost (AC)} = \frac{C}{x} = \frac{a+bx+cx^2}{x} = \frac{a}{x} + b + cx$$

and Marginal cost (MC) $= \frac{d}{dx}(C)$

$$= \frac{d}{dx}(a + bx + cx^2) = b + 2cx.$$

$$\therefore \quad \frac{1}{x}(\text{MC} - \text{AC}) = \frac{1}{x}\left(b + 2cx - \frac{a}{x} - b - cx\right) = \frac{1}{x}\left(cx - \frac{a}{x}\right)$$

$$\therefore \quad \frac{1}{x}(\text{MC} - \text{AC}) = c - \frac{a}{x^2} \quad \ldots\text{(i)}$$

$$\frac{d(\text{AC})}{dx} = \frac{d}{dx}\left(\frac{a}{x} + b + cx\right)$$

$$\frac{d}{dx}(\text{AC}) = -\frac{a}{x^2} + c = c - \frac{a}{x^2} \quad \ldots\text{(ii)}$$

From (i) and (ii), we get

$$\frac{d}{dx}(\text{AC}) = \frac{1}{x}(\text{MC} - \text{AC}).$$

Hence Proved.

MATHEMATICS 2011

QUESTIONS

SECTION—A

Question 1. [10 × 3]

(i) *If* $A = \begin{bmatrix} 3 & -2 \\ 4 & -2 \end{bmatrix}$, *find x such that* $A^2 = xA - 2I$. *Hence find* A^{-1}.

(ii) *Find the values of k, if the equation* $8x^2 - 16xy + ky^2 - 22x + 34y = 12$ *represents an ellipse.***

(iii) *Solve for x :* $\sin(2\tan^{-1} x) = 1$.

(iv) *Two regression lines are represented by* $2x + 3y - 10 = 0$ *and* $4x + y - 5 = 0$. *Find the line of regression of y on x.*

(v) *Evaluate:* $\int \frac{\operatorname{cosec} x}{\log \tan\left(\frac{x}{2}\right)} dx$.

(vi) *Evaluate:* $\lim_{y \to 0} \frac{y - \tan^{-1} y}{y - \sin y}$.

(vii) *Evaluate:* $\int_0^1 \frac{x e^x}{(1+x)^2} dx$.

(viii) *Find the modulus and argument of the complex number* $\frac{2+i}{4i + (1+i)^2}$.**

(ix) *A word consists of 9 different alphabets, in which there are 4 consonants and 5 vowels. Three alphabets are chosen at random. What is the probability that more than one vowel will be selected?*

(x) *Solve the differential equation:* $\frac{dy}{dx} = e^{x+y} + x^2 e^y$.

Question 2. [10 × 3]

(a) *Using properties of determinants, show that* $p\alpha^2 + 2q\alpha + r = 0$, *given that:*

p, q and r are not in G.P. and

$$\begin{vmatrix} 1 & \frac{q}{p} & \alpha + \frac{q}{p} \\ 1 & \frac{r}{q} & \alpha + \frac{r}{q} \\ p\alpha + q & q\alpha + r & 0 \end{vmatrix} = 0.$$ [5]

(b) *Solve the following system of equations using matrix method:* [5]

$$\frac{2}{x} + \frac{3}{y} + \frac{10}{z} = 4$$

$$\frac{4}{x} - \frac{6}{y} + \frac{5}{z} = 1$$

$$\frac{6}{x} + \frac{9}{y} - \frac{20}{z} = 2$$

Question 3. [10 × 3]

(a) *Prove that:* $2\tan^{-1}\frac{1}{5} + \cos^{-1}\frac{7}{5\sqrt{2}} + 2\tan^{-1}\frac{1}{8} = \frac{\pi}{4}$. [5]

(b) *P, Q and R represent switches in 'ON position' and P', Q' and R' represent switches in 'OFF position'. Construct a switching circuit representing the polynomial:***

$$P(P + Q)\, Q(Q + R')$$

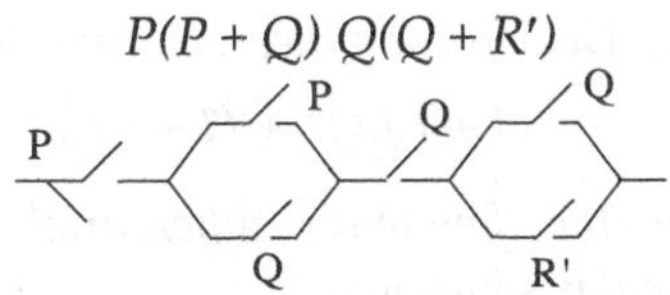

Using Boolean Algebra to show that the above circuit is equivalent to a switching circuit in which when P and Q are in 'ON position', the light is on. [5]

Question 4. [10 × 3]

(a) *Verify Lagrange's mean value theorem for the function* $f(x) = \sin x - \sin 2x$ *in the interval* $[0, \pi]$. [5]

(b) *Find the equation of the hyperbola whose foci are* $(0, \pm 13)$ *and the length of the conjugate axis is 20.***

Question 5. [10 × 3]

(a) *Evaluate:* $\int \frac{x^2 - 5x - 1}{x^4 + x^2 + 1} dx$. [5]

(b) *Draw a rough sketch of the curves* $y = (x - 1)^2$ *and* $y = |x - 1|$. *Hence, find the area of the region bounded by these curves.* [5]

Question 6. [10 × 3]

(a) *If the sum of the lengths of the hypotenuse and a side of a right angled triangle is given, show that the area of the triangle is maximum when the angle between them is* $\frac{\pi}{3}$. [5]

(b) *If* $y = x^x$, *prove that:* $\frac{d^2y}{dx^2} - \frac{1}{y}\left(\frac{dy}{dx}\right)^2 - \frac{y}{x} = 0$. [5]

Question 7. [10 × 3]

(a) *The following observations are given:*

(1, 4), (2, 8), (3, 2), (4, 12), (5, 10), (6, 14), (7, 16), (8, 6), (9, 18).

** Answer is not given due to change in the present syllabus.

Estimate the value of y when the value of x is 10 and also estimate the value of x when the value of y = 5.

(b) *Compute Karl Pearson's Coefficient of Correlation between sales and expenditures of a firm for six months.*** **[5]**

Sales (in lakhs of ₹)	18	20	27	20	21	29
Expenditures (in lakhs of ₹)	23	27	28	28	29	30

Question 8. **[10 × 3]**

(a) *A purse contains 4 silver and 5 copper coins. A second purse contains 3 silver and 7 copper coins. If a coin is taken out at random from one of the purses, what is the probability that it is a copper coin?* **[5]**

(b) *Aman and Bhuvan throw a pair of dice alternately. In order to win, they have to get a sum of 8. Find their respective probabilities of winning if Aman starts the game.* **[5]**

Question 9. **[10 × 3]**

(a) *Using De Moivre's theorem, find the value of:*

$$(1+i\sqrt{3})^6 + (1-i\sqrt{3})^6.^{**}$$ **[5]**

(b) *Solve the following differential equation for a particular solution:*

$$y - x\frac{dy}{dx} = x + y\frac{dy}{dx}, \text{ when } y = 0 \text{ and } x = 1.$$ **[5]**

SECTION—B

Question 10.

(a) *Prove that:*

$$[\vec{a}+\vec{b}\ \ \vec{b}+\vec{c}\ \ \vec{c}+\vec{a}] = 2[\vec{a}\ \vec{b}\ \vec{c}].$$ **[5]**

(b) *If D, E, F are mid-points of the sides of a triangle ABC, prove by vector method that:*

$$\text{Area of } \Delta DEF = \frac{1}{4}(\text{Area of } \Delta ABC).$$ **[5]**

Question 11.

(a) *Find the vector equation of the line passing through the point (– 1, 2, 1) and parallel to the line* $\vec{r} = 2\hat{i} + 3\hat{i} - \hat{k} + \lambda(\hat{i} - 2\hat{j} + \hat{k})$. *Also, find the distance between these lines.* **[5]**

(b) *Find the equation of the plane passing through the points A(2, 1, – 3), B(– 3, – 2, 1) and C(2, 4, – 1).*

Question 12.

(a) *A box contains 4 red and 5 black marbles. Find the probability distribution of the red marbles in a random draw of three marbles. Also find the mean and standard deviation of the distribution.* **[5]**

(b) *Bag A contains 2 white, 1 black and 3 red balls, Bag B contains 3 white, 2 black and 4 red balls and Bag C contains 4 white, 3 black and 2 red balls. One bag is chosen at random and 2 balls are drawn at random from that Bag. If the randomly drawn balls happen to be red and black, what is the probability that both balls come from Bag B?* **[5]**

SECTION—C

Question 13.

(a) *The price of a tape recorder is ₹ 1,661. A person purchases it by making a cash payment of ₹ 400 and agrees to pay the balance with due interest in 3 half-yearly equal instalments. If the dealer charged interest at the rate of 10% per annum compounded half-yearly equal instalments. If the order charged interest at the rate of 10% per annum compounded half-yearly, find the value of the instalments.*** **[5]**

(b) *A manufactures two types of tea-cups, A and B. Three machines are needed for manufacturing the tea cups. The time in minutes required for manufacturing each cup on the machines is given below:* **[5]**

Type of Cup	*Time in minutes*		
	Machine I	*Machine II*	*Machine III*
A	12	18	6
B	6	0	9

Each machine is available for a maximum of six hours per day. If the profit on each cup of type A is ₹ 1·50 and that on each cup of type B is ₹ 1·00, find the number of cups of each type that should be manufactured in a day to get maximum profit.

Question 14.

(a) *If the difference between Banker's discount and True discount of a bill for 73 days at 5% per annum is ₹ 10, find (i) the amount of the bill (ii) the Banker's discount.***

(b) *Given that the total cost function of x units of a commodity is:*

$$C(x) = \frac{x^3}{3} + 3x^2 - 7x + 16$$

(i) Find the Marginal Cost (MC).

(ii) Find the Average Cost (AC).

(iii) Prove that:

$$\text{Marginal Average Cost (MAC)} = \frac{x(MC) - C(x)}{x^2}.$$

Question 15.**

(a) *The price quotations of four different commodities for 2001 and 2009 are as given below. Calculate the index number for 2009 with 2001 as the base year by using weighted average of price relative method.* **[5]**

** Answer is not given due to change in the present syllabus.

Commodity	*Weight*	*Price (in ₹)*	
		2009	*2001*
A	*10*	*9·00*	*4·00*
B	*49*	*4·40*	*5·00*
C	*36*	*9·00*	*6·00*
D	*4*	*3·60*	*2·00*

(b) *The profit of a soft drink firm (in thousands of rupees) during each month of the year is as given below:*

Months	*Profit (in thousands of Rupees)*
January	*3·6*
February	*4·3*
March	*4·3*
April	*3·4*
May	*4·4*
June	*5·4*
July	*3·4*
August	*2·4*
September	*3·4*
October	*1·8*
November	*0·8*
December	*1·2*

Calculate the four monthly moving averages and plot these and the original data on a graph sheet. **[5]**

SOLUTIONS

SECTION—A

Solution 1.

(i) Given, $A = \begin{bmatrix} 3 & -2 \\ 4 & -2 \end{bmatrix}$, such that $A^2 = xA - 2I$

$$A^2 = \begin{bmatrix} 3 & -2 \\ 4 & -2 \end{bmatrix}\begin{bmatrix} 3 & -2 \\ 4 & -2 \end{bmatrix}$$

$$= \begin{bmatrix} 3\times 3 + (-2)\times 4 & 3\times(-2) + (-2)\times(-2) \\ 4\times 3 + (-2)\times 4 & 4\times(-2) + (-2)\times(-2) \end{bmatrix}$$

$$= \begin{bmatrix} 9-8 & -6+4 \\ 12-8 & -8+4 \end{bmatrix} = \begin{bmatrix} 1 & -2 \\ 4 & -4 \end{bmatrix}$$

Now, $A^2 = xA - 2I$

$$\Rightarrow \begin{bmatrix} 1 & -2 \\ 4 & -4 \end{bmatrix} = x\begin{bmatrix} 3 & -2 \\ 4 & -2 \end{bmatrix} - 2\begin{bmatrix} 1 & 0 \\ 0 & 1 \end{bmatrix}$$

$$\Rightarrow \begin{bmatrix} 1 & -2 \\ 4 & -4 \end{bmatrix} = \begin{bmatrix} 3x & -2x \\ 4x & -2x \end{bmatrix} - \begin{bmatrix} 2 & 0 \\ 0 & 2 \end{bmatrix}$$

$$\Rightarrow \begin{bmatrix} 1 & -2 \\ 4 & -4 \end{bmatrix} = \begin{bmatrix} 3x-2 & -2x \\ 4x & -2x-2 \end{bmatrix}$$

$\Rightarrow$ $3x - 2 = 1$

$\Rightarrow$ $3x = 3$ or $x = 1$

Now, $A^2 = xA - 2I$

$\Rightarrow$ $A^2 = 1\cdot A - 2I$

$\Rightarrow$ $A^2\cdot A^{-1} = AA^{-1} - 2I.A^{-1}$

[Multiplying both sides by A^{-1}]

$\Rightarrow$ $A(AA^{-1}) = I - 2A^{-1}$

$\Rightarrow$ $AI = I - 2A^{-1}$ $\quad [\because AA^{-1} = I]$

$\Rightarrow$ $2A^{-1} = I - AI$

$$= \begin{bmatrix} 1 & 0 \\ 0 & 1 \end{bmatrix} - \begin{bmatrix} 3 & -2 \\ 4 & -2 \end{bmatrix}\begin{bmatrix} 1 & 0 \\ 0 & 1 \end{bmatrix}$$

$$2A^{-1} = \begin{bmatrix} 1 & 0 \\ 0 & 1 \end{bmatrix} - \begin{bmatrix} 3 & -2 \\ 4 & -2 \end{bmatrix}$$

$$= \begin{bmatrix} 1-3 & 0+2 \\ 0-4 & 1+2 \end{bmatrix} = \begin{bmatrix} -2 & 2 \\ -4 & 3 \end{bmatrix}$$

$$\therefore A^{-1} = \frac{1}{2}\begin{bmatrix} -2 & 2 \\ -4 & 3 \end{bmatrix} \text{ or } \begin{bmatrix} -1 & 1 \\ -2 & 3/2 \end{bmatrix}.$$ **Ans.**

(iii) Given, $\sin(2\tan^{-1}x) = 1$

$$\Rightarrow \sin\left[\sin^{-1}\left(\frac{2x}{1+x^2}\right)\right] = 1$$

$$\left[\text{As, } 2\tan^{-1}x = \sin^{-1}\left(\frac{2x}{1+x^2}\right)\right]$$

$$\Rightarrow \frac{2x}{1+x^2} = 1$$

$\Rightarrow$ $2x = 1 + x^2$

or $x^2 - 2x + 1 = 0$

$\Rightarrow$ $(x-1)^2 = 0$

$\Rightarrow$ $x - 1 = 0$

$\Rightarrow$ $x = 1.$ **Ans.**

(iv) Let the line of regression of y on x is

$2x + 3y - 10 = 0$

$\Rightarrow$ $3y = -2x + 10$

or $y = \frac{-2}{3}x + \frac{10}{3}$

$\Rightarrow$ $b_{xy} = -\frac{2}{3}$

Then, the line of regression of x on y is

$4x + y - 5 = 0$

$\Rightarrow$ $4x = -y + 5$ or $x = -\frac{1}{4}y + \frac{5}{4}$

$$\Rightarrow \quad b_{xy} = -\frac{1}{4}$$

Now, for such cases $b_{yx} \cdot b_{xy} < 1$

$$\Rightarrow \quad b_{yx} \cdot b_{xy} = \left(\frac{-2}{3}\right)\left(-\frac{1}{4}\right)$$

$$= \frac{1}{6} < 1$$

which is true.

$\therefore$ Line of regression of y on x is

$$2x + 3y - 10 = 0. \qquad \textbf{Ans.}$$

(v) Given, $\displaystyle\int \frac{\operatorname{cosec} x}{\log \tan\left(\frac{x}{2}\right)} dx$

Put $\quad \log \tan\left(\frac{x}{2}\right) = t$

$$\frac{1}{\tan\frac{x}{2}} \cdot \frac{1}{2} \sec^2 \frac{x}{2}\, dx = dt$$

$$\frac{1}{2} \cdot \frac{\cos\frac{x}{2}}{\sin\frac{x}{2} \cos^2 \frac{x}{2}} dx = dt$$

or $$\frac{1}{2 \sin\frac{x}{2} \cos\frac{x}{2}} dx = dt$$

$$\Rightarrow \quad \frac{1}{\sin x} dx = dt,$$

or $$\operatorname{cosec} x\, dx = dt$$

$$\therefore \quad \int \frac{\operatorname{cosec} x\, dx}{\log \tan\left(\frac{x}{2}\right)} = \int \frac{dt}{t}$$

$$= \log t + c$$

$$= \log\left(\log \tan \frac{x}{2}\right) + c. \qquad \textbf{Ans.}$$

(vi) Given, $\displaystyle\lim_{y \to 0} \frac{y - \tan^{-1} y}{y - \sin y}$

On applying L' Hospital's rule, we get

$$= \lim_{y \to 0} \frac{1 - \frac{1}{1+y^2}}{1 - \cos y} \qquad \left[\frac{0}{0} \text{ form}\right]$$

Again, applying L' Hospital's rule, we get

$$= \lim_{y \to 0} \frac{0 - (-1)(1+y^2)^{-2}.2y}{\sin y}$$

$$= \lim_{y \to 0} \frac{2y}{(1+y^2)^2 \cdot \sin y}$$

$$= \lim_{y \to 0} \frac{2}{(1+y^2)^2} \cdot \left(\frac{y}{\sin y}\right)$$

$$= \frac{2}{(1+0)^2} \cdot 1 = 2. \qquad \textbf{Ans.}$$

(vii) Given,

$$\int_0^1 \frac{xe^x}{(1+x)^2} dx = \int_0^1 \frac{(x+1)-1}{(1+x)^2} \cdot e^x\, dx$$

$$= \int_0^1 \frac{(x+1)e^x}{(x+1)^2} dx - \int_0^1 \frac{e^x}{(x+1)^2} dx$$

$$= \int_0^1 \frac{e^x}{x+1} dx - \int_0^1 \frac{e^x}{(x+1)^2} dx$$

$$= \left[e^x \times \frac{1}{x+1}\right]_0^1 - \int_0^1 e^x \left[-\frac{1}{(x+1)^2}\right] dx - \int_0^1 \frac{e^x}{(x+1)^2} dx$$

[on integrating]

$$= \left[\frac{1}{2} e^1 - 1 \cdot e^0\right] + \int_0^1 \frac{1}{(x+1)^2} \cdot e^x\, dx$$

$$- \int_0^1 \frac{1}{(x+1)^2} \cdot e^x\, dx$$

$$= \frac{1}{2} e - 1 = \frac{e}{2} - 1. \qquad \textbf{Ans.}$$

(ix) Number of alphabets in the word = 9.

Number of consonants (C) = 4

Number of vowels (V) = 5

Three alphabets are chosen at random.

$\therefore$ Probability that more than one vowel will be selected

$$= P(2V \;\&\; 1C) + P(3V)$$

$$= \frac{{}^5C_2 \times {}^4C_1}{{}^9C_3} + \frac{{}^5C_3}{{}^9C_3}$$

$$= \frac{40}{84} + \frac{10}{84} = \frac{50}{84} = \frac{25}{42}. \qquad \textbf{Ans.}$$

(x) Given differential equation is

$$\frac{dy}{dx} = e^{x+y} + x^2 e^y$$

$$= e^x \cdot e^y + x^2 \cdot e^y$$

$$\Rightarrow \quad \frac{dy}{dx} = e^y (e^x + x^2)$$

$$\Rightarrow \quad \int \frac{dy}{e^y} = \int (e^x + x^2)\, dx$$

$$\Rightarrow \quad \int e^{-y}\,dy = \int e^{x}\,dx + \int x^2\,dx$$

$$\Rightarrow \quad -e^{-y} = e^{x} + \frac{x^3}{3} + c$$

$$\Rightarrow \quad -\frac{1}{e^{y}} = e^{x} + \frac{x^3}{3} + c$$

or $e^{x+y} + \dfrac{e^{y}\cdot x^3}{3} + ce^{y} + 1 = 0.$ **Ans.**

Solution 2.

(a) Given,
$$\begin{vmatrix} 1 & \frac{q}{p} & \alpha+\frac{q}{p} \\ 1 & \frac{r}{q} & \alpha+\frac{r}{q} \\ p\alpha+q & q\alpha+r & 0 \end{vmatrix} = 0$$

Applying $R_1 \to R_1 - R_2$, we get

$$\Rightarrow \quad \begin{vmatrix} 0 & \frac{q^2-pr}{pq} & \frac{q^2-pr}{pq} \\ 1 & \frac{r}{q} & \alpha+\frac{r}{q} \\ p\alpha+q & q\alpha+r & 0 \end{vmatrix} = 0$$

Applying $C_2 \to C_2 - C_3$, we get

$$\Rightarrow \quad \begin{vmatrix} 0 & 0 & \frac{q^2-pr}{pq} \\ 1 & -\alpha & \alpha+\frac{r}{q} \\ p\alpha+q & q\alpha+r & 0 \end{vmatrix} = 0$$

$$\Rightarrow \quad \frac{q^2-pr}{pq}[(q\alpha+r)+(p\alpha+q)\alpha] = 0$$

[expanding w.r.t. R_1]

$$\Rightarrow \quad \frac{(q^2-pr)}{pq}[q\alpha+r+p\alpha^2+q\alpha] = 0$$

or
$$\frac{(q^2-pr)}{pq}[p\alpha^2+2q\alpha+r] = 0$$

Given that p, q and r not in G.P.

$$\Rightarrow \quad \frac{p}{q} \neq \frac{q}{r} \text{ i.e., } pr \neq q^2 \text{ or } q^2 - pr \neq 0$$

$$\Rightarrow \quad p\alpha^2 + 2q\alpha + r = 0.$$ **Ans.**

(b) Let $\dfrac{1}{x} = p$, $\dfrac{1}{y} = q$ and $\dfrac{1}{z} = r$

$\therefore$
$$2p + 3q + 10r = 4 \quad \ldots\text{(i)}$$
$$4p - 6q + 5r = 1 \quad \ldots\text{(ii)}$$
$$6p + 9q - 20r = 2 \quad \ldots\text{(iii)}$$

Writing the given system of equations in matrix form.

$$\begin{bmatrix} 2 & 3 & 10 \\ 4 & -6 & 5 \\ 6 & 9 & -20 \end{bmatrix}\begin{bmatrix} p \\ q \\ r \end{bmatrix} = \begin{bmatrix} 4 \\ 1 \\ 2 \end{bmatrix}$$

or $\quad AX = B$

Here,
$$A = \begin{bmatrix} 2 & 3 & 10 \\ 4 & -6 & 5 \\ 6 & 9 & -20 \end{bmatrix}, B = \begin{bmatrix} 4 \\ 1 \\ 2 \end{bmatrix}, X = \begin{bmatrix} p \\ q \\ r \end{bmatrix}$$

$$\therefore \quad |A| = 2(120-45) - 3(-80-30) + 10(36+36)$$
$$= 150 + 330 + 720 = 1200$$

$\therefore \quad |A| \neq 0$

$\therefore$ A^{-1} exists.

Now co-factors of A are,

$$A_{11} = 75, A_{12} = 110, A_{13} = 72$$
$$A_{21} = 150, A_{22} = -100, A_{23} = 0$$
$$A_{31} = 75, A_{32} = 30, A_{33} = -24$$

$$\text{adj } A = \begin{bmatrix} 75 & 110 & 72 \\ 150 & -100 & 0 \\ 75 & 30 & -24 \end{bmatrix}^T$$

$$= \begin{bmatrix} 75 & 150 & 75 \\ 110 & -100 & 30 \\ 72 & 0 & -24 \end{bmatrix}$$

$$\therefore \quad A^{-1} = \frac{\text{adj } A}{|A|}$$

$$= \frac{1}{1200}\begin{bmatrix} 75 & 150 & 75 \\ 110 & -100 & 30 \\ 72 & 0 & -24 \end{bmatrix}$$

Now, $\quad X = A^{-1}B$

$$= \frac{1}{1200}\begin{bmatrix} 75 & 150 & 75 \\ 110 & -100 & 30 \\ 72 & 0 & -24 \end{bmatrix}\begin{bmatrix} 4 \\ 1 \\ 2 \end{bmatrix}$$

$$= \frac{1}{1200}\begin{bmatrix} 600 \\ 400 \\ 240 \end{bmatrix} = \begin{bmatrix} 1/2 \\ 1/3 \\ 1/5 \end{bmatrix}$$

$$\therefore \quad p = \frac{1}{2}, q = \frac{1}{3}, r = \frac{1}{5}$$

$$\Rightarrow \quad \frac{1}{x} = \frac{1}{2}, \frac{1}{y} = \frac{1}{3}, \frac{1}{z} = \frac{1}{5}$$

$$\Rightarrow \quad x = 2, y = 3, z = 5.$$ **Ans.**

Solution 3.

(a) L.H.S. $= 2\tan^{-1}\frac{1}{5} + \cos^{-1}\frac{7}{5\sqrt{2}} + 2\tan^{-1}\frac{1}{8}$

$$= 2\left(\tan^{-1}\frac{1}{5} + \tan^{-1}\frac{1}{8}\right) + \cos^{-1}\frac{7}{5\sqrt{2}}$$

$$= 2\tan^{-1}\left[\frac{\frac{1}{5}+\frac{1}{8}}{1-\frac{1}{5}\times\frac{1}{8}}\right] + \cos^{-1}\frac{7}{5\sqrt{2}}$$

$$\left[\because \tan^{-1}x + \tan^{-1}y = \tan^{-1}\left(\frac{x+y}{1-xy}\right)\right]$$

$$= 2\tan^{-1}\left[\frac{13/40}{39/40}\right] + \tan^{-1}\left[\frac{\sqrt{1-\frac{49}{50}}}{7/5\sqrt{2}}\right]$$

$$\left[\because \cos^{-1}x = \tan^{-1}\left(\frac{\sqrt{1-x^2}}{x}\right)\right]$$

$$= 2\tan^{-1}\frac{1}{3} + \tan^{-1}\frac{1}{7}$$

$$= \tan^{-1}\frac{1}{3} + \left(\tan^{-1}\frac{1}{3} + \tan^{-1}\frac{1}{7}\right)$$

$$= \tan^{-1}\frac{1}{3} + \tan^{-1}\frac{\left(\frac{1}{3}+\frac{1}{7}\right)}{\left(1-\frac{1}{3}\cdot\frac{1}{7}\right)}$$

$$= \tan^{-1}\frac{1}{3} + \tan^{-1}\frac{1}{2}$$

$$= \tan^{-1}\left(\frac{\frac{1}{3}+\frac{1}{2}}{1-\frac{1}{3}\cdot\frac{1}{2}}\right)$$

$$= \tan^{-1}\left(\frac{\frac{5}{6}}{\frac{5}{6}}\right) = \tan^{-1}1 = \frac{\pi}{4}.$$

Hence Proved.

Solution 4.

(a) The function $f(x) = \sin x - \sin 2x$ is differentiable for all values of x and hence differentiable in $[0, \pi]$. Also

$$f'(x) = \cos x - 2\cos 2x$$

By Lagrange's mean value theorem

We have, $$f'(c) = \frac{f(\pi) - f(0)}{\pi - 0}$$

$$\Rightarrow \quad \cos c - 2\cos 2c = \frac{0-0}{\pi} = 0$$

$$\Rightarrow \cos c - 2(2\cos^2 c - 1) = 0$$

$$[\because \cos 2x = 2\cos^2 x - 1]$$

$$\Rightarrow \quad 4\cos^2 c - \cos c - 2 = 0$$

$$\Rightarrow \quad \cos c = \frac{1 \pm \sqrt{1+32}}{8} = \frac{1 \pm \sqrt{33}}{8}$$

or $$\cos c = \frac{1 \pm 5{\cdot}7}{8}$$

$$\Rightarrow \quad \cos c = \frac{1 + 5{\cdot}7}{8}$$

or $$\cos c = \frac{1 - 5{\cdot}7}{8}$$

$$\Rightarrow \quad \cos c = 0{\cdot}84$$

or $$\cos c = -0{\cdot}59$$

$$\Rightarrow \quad c = 33°$$

or $c = 126{\cdot}2°$ (approx.)

Both values of c lie in the interval $[0, \pi]$.

Hence, the Mean value theorem is verified.

Hence Proved.

Solution 5.

(a) Given, $\int \frac{x^2 - 5x - 1}{x^4 + x^2 + 1}\,dx$

$$= \int \frac{x^2 - 1}{x^4 + x^2 + 1}\,dx - 5\int \frac{x}{x^4 + x^2 + 1}\,dx$$

$$= I_1 - 5I_2 \qquad \ldots\text{(i)}$$

$$I_1 = \int \frac{x^2 - 1}{x^4 + x^2 + 1}\,dx$$

$$= \int \frac{\left(1 - \frac{1}{x^2}\right)}{\left(x^2 + \frac{1}{x^2} + 1\right)}\,dx$$

$$= \int \frac{\left(1 - \frac{1}{x^2}\right)}{\left(x + \frac{1}{x}\right)^2 - 2 + 1}\,dx$$

$$= \int \frac{\left(1 - \frac{1}{x^2}\right)}{\left(x + \frac{1}{x}\right)^2 - 1}\,dx$$

put $$x + \frac{1}{x} = t$$

$$\left(1 - \frac{1}{x^2}\right)dx = dt$$

$$= \int \frac{dt}{t^2 - 1} = \frac{1}{2\times 1}\log\left|\frac{t-1}{t+1}\right| + c_1$$

$$= \frac{1}{2}\log\left|\frac{x + \frac{1}{x} - 1}{x + \frac{1}{x} + 1}\right| + c_1$$

$$= \frac{1}{2}\log\left|\frac{x^2 - x + 1}{x^2 + x + 1}\right| + c_1$$

Let $I_2 = \int \frac{x}{x^4 + x^2 + 1} dx$ put $x^2 = t$

$$= \int \frac{dt}{2(t^2 + t + 1)} \qquad 2x\,dx = dt$$

$$= \frac{1}{2}\int \frac{dt}{\left(t^2 + 2\times t\times\frac{1}{2} + \left(\frac{1}{2}\right)^2\right) + \left(\frac{\sqrt{3}}{2}\right)^2}$$

$$= \frac{1}{2}\int \frac{dt}{\left(t + \frac{1}{2}\right)^2 + \left(\frac{\sqrt{3}}{2}\right)^2}$$

$$= \frac{1}{2}\times\frac{1}{\sqrt{3/2}}\tan^{-1}\left(\frac{t + \frac{1}{2}}{\sqrt{3/2}}\right) + c_2$$

$$= \frac{1}{\sqrt{3}}\tan^{-1}\left(\frac{2t + 1}{\sqrt{3}}\right) + c_2$$

$$= \frac{1}{\sqrt{3}}\tan^{-1}\left(\frac{2x^2 + 1}{\sqrt{3}}\right) + c_2$$

Putting the values of I_1 and I_2 in equation (i)

$$I = \frac{1}{2}\log\left|\frac{x^2 - x + 1}{x^2 + x + 1}\right| + c_1 - \frac{5}{\sqrt{3}}\tan^{-1}\left(\frac{2x^2 + 1}{\sqrt{3}}\right) - 5c_2$$

$$= \frac{1}{2}\log\left|\frac{x^2 - x + 1}{x^2 + x + 1}\right| - \frac{5}{\sqrt{3}}\tan^{-1}\left(\frac{2x^2 + 1}{\sqrt{3}}\right) + c. \textbf{ Ans.}$$

Where, $c = c_1 - 5c_2$

(b) $y = (x - 1)^2$ is a parabola whose axis is a line $x = 1$ and vertex is at (1, 0).

$$y = (x - 1)^2$$

x	0	1	2
y	1	0	1

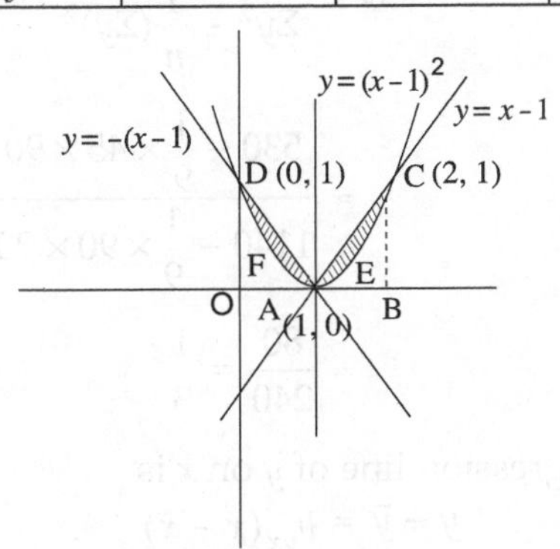

Also, $y = |x - 1| = \begin{cases} x - 1 \text{ if } x - 1 \geq 0 \\ -(x - 1) \text{ if } x - 1 < 0 \end{cases}$

$$= \begin{cases} x - 1 \text{ if } x \geq 1 \\ -(x - 1) \text{ if } x < 0 \end{cases}$$

$$y = x - 1$$

x	0	1	2
y	– 1	0	1

Also, $y = -(x - 1)$

x	1	0
y	0	1

Here, we have to find the area enclosed by

$$y = (x - 1)^2$$

and the lines, $y = (x - 1)$ and $y = -(x - 1)$

$\therefore$ Required Area = Area ABC – Area ABCEA + Area AOD – Area AODFA

$$= \int_1^2 (\text{Line } y = x - 1)\, dx - \int_1^2 [\text{Curve } y = (x - 1)^2]\, dx + \int_0^1 (\text{Line } y = -x - 1)\, dx - \int_0^1 [\text{Curve } y = (x - 1)^2]\, dx$$

$$= \int_1^2 (x - 1)\, dx - \int_1^2 (x - 1)^2\, dx + \int_0^1 -(x - 1)\, dx - \int_0^1 (x - 1)^2\, dx$$

$$= \left[\frac{x^2}{2} - x\right]_1^2 - \left[\frac{x^3}{3} - x^2 + x\right]_1^2 - \left[\frac{x^2}{2} - x\right]_0^1 - \left[\frac{x^3}{3} - x^2 + x\right]_0^1$$

$$= 0 - \left(-\frac{1}{2}\right) - \left[\frac{8}{3} - 4 + 2 - \frac{1}{3} + 1 - 1\right] - \left[\frac{1}{2} - 1\right] - \left[\frac{1}{3} - 1 + 1\right]$$

$$= \frac{1}{2} - \frac{1}{3} + \frac{1}{2} - \frac{1}{3} = 1 - \frac{2}{3} = \frac{1}{3} \text{ sq. units.}$$ **Ans.**

Solution 6.

(a) Let 'l' be the length of the hypotenuse of the given right angled ΔABC at B and $\angle CAB = \theta$ (in radian measure) and $0 < \theta < \pi/2$.

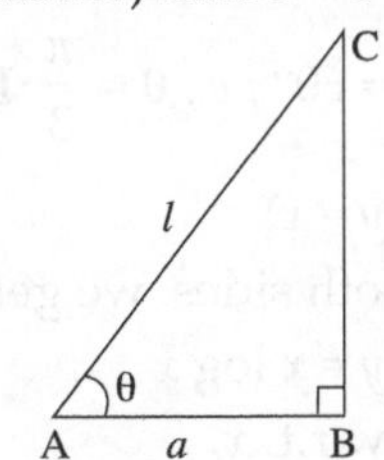

Then, $a = AB = l\cos\theta$ and $BC = l\sin\theta$.

Given the sum of hypotenuse and a side.

Let $S = l + a$

$= l + l\cos\theta = l(1+\cos\theta)$

or $l = \dfrac{S}{1+\cos\theta}$

Now, Area of the $\Delta ABC = \dfrac{1}{2}\cdot AB\cdot BC$

$= \dfrac{1}{2}(l\cos\theta)\cdot l\sin\theta$

$= \dfrac{1}{2}l^2\sin\theta\cos\theta$

$= \dfrac{1}{4}l^2\, 2\sin\theta\cos\theta$

or $A = \dfrac{\sin 2\theta}{4}\cdot\left[\dfrac{S^2}{(1+\cos\theta)^2}\right]$

$\left[\because l = \dfrac{S}{1+\cos\theta}\right]$

$$\frac{dA}{d\theta} = \frac{S^2}{4}\,\frac{\left[(1+\cos\theta)^2\frac{d}{d\theta}\sin 2\theta - \sin 2\theta\frac{d}{d\theta}(1+\cos\theta)^2\right]}{(1+\cos\theta)^4}$$

Now, $\dfrac{dA}{d\theta} = 0$

$\Rightarrow (1+\cos\theta)^2\cdot 2\cos 2\theta + 2\sin 2\theta\,(1+\cos\theta)\cdot\sin\theta = 0$

$\Rightarrow 2(1+\cos\theta)\,[(1+\cos\theta)\cdot\cos 2\theta + 2\sin\theta\cos\theta\cdot\sin\theta] = 0$

or $(1+\cos\theta)\,[(1+\cos\theta)(1-2\sin^2\theta) + 2\sin^2\theta\cos\theta] = 0$

$\Rightarrow (1+\cos\theta)\,[1-2\sin^2\theta+\cos\theta-2\sin^2\theta\cos\theta + 2\sin^2\theta\cos\theta] = 0$

$\Rightarrow (1+\cos\theta)\,[1-2(1-\cos^2\theta)+\cos\theta] = 0$

$(1+\cos\theta)\,[2\cos^2\theta+\cos\theta-1] = 0$

$\Rightarrow (1+\cos\theta)\,[2\cos^2\theta+2\cos\theta-\cos\theta-1] = 0$

$(1+\cos\theta)\,[2\cos\theta(\cos\theta+1)-1(\cos\theta+1)] = 0$

$\Rightarrow (1+\cos\theta)(1+\cos\theta)(2\cos\theta-1) = 0$

$\Rightarrow \cos\theta = -1$ (not possible) or $\cos\theta = \cos\theta = \cos 60°$

or $\theta = 60°$ *i.e.*, $\theta = \dfrac{\pi}{3}$ · **Hence Proved.**

(b) Given, $y = x^x$

Taking log on both sides, we get

$\log y = x\log x$

Differentiating w.r.t. x,

$$\frac{1}{y}\frac{dy}{dx} = x\cdot\frac{1}{x} + \log x$$

$$\frac{dy}{dx} = y(1+\log x) \quad \ldots(i)$$

Differentiating again w.r.t. x,

$$\frac{d^2y}{dx^2} = y\left(0+\frac{1}{x}\right) + (1+\log x)\frac{dy}{dx}$$

$$\frac{d^2y}{dx^2} = \frac{y}{x} + \frac{1}{y}\frac{dy}{dx}\cdot\frac{dy}{dx} \quad \text{[from (i)]}$$

$$\Rightarrow \frac{d^2y}{dx^2} - \frac{1}{y}\left(\frac{dy}{dx}\right)^2 - \frac{y}{x} = 0.$$ **Hence Proved.**

Solution 7.

(a) Here, we have

x	x^2	y	y^2	xy
1	1	4	16	4
2	4	8	64	16
3	9	2	4	6
4	16	12	144	48
5	25	10	100	50
6	36	14	196	84
7	49	16	256	112
8	64	6	36	48
9	81	18	324	162
$\Sigma x = 45$	$\Sigma x^2 = 285$	$\Sigma y = 90$	$\Sigma y^2 = 1140$	$\Sigma xy = 530$

$$\bar{x} = \frac{45}{9} = 5,\ \bar{y} = \frac{90}{9} = 10$$

$$b_{yx} = \frac{\Sigma xy - \frac{1}{n}(\Sigma x\Sigma y)}{\Sigma x^2 - \frac{1}{n}(\Sigma x)^2}$$

$$= \frac{530 - \frac{1}{9}\times 45\times 90}{285 - \frac{1}{9}\times 45\times 45} = \frac{80}{60} = \frac{4}{3}$$

$$b_{xy} = \frac{\Sigma xy - \frac{1}{n}(\Sigma x\Sigma x)}{\Sigma y^2 - \frac{1}{n}(\Sigma y)^2}$$

$$= \frac{530 - \frac{1}{9}\times 45\times 90}{1140 - \frac{1}{9}\times 90\times 90}$$

$$= \frac{80}{240} = \frac{1}{3}$$

$\therefore$ Regression line of y on x is

$$y - \bar{y} = b_{yx}(x - \bar{x})$$

$$y - 10 = \frac{4}{3}(x - 5)$$

or $$3y - 30 = 4x - 20$$

$\Rightarrow$ $$3y = 4x + 10$$

Given, $x = 10$, we get

$$3y = 4 \times 10 + 10 = 50$$

$$y = \frac{50}{3} = 16{\cdot}67$$

Again regression line of x on y is

$$x - \bar{x} = b_{xy}(y - \bar{y})$$

or $$x - 5 = \frac{1}{3}(y - 10)$$

or $$3x - 15 = y - 10$$

$$3x = y + 5$$

Given, $y = 5$, we get $3x = 5 + 5$

$$x = \frac{10}{3} = 3{\cdot}33.$$ **Ans.**

Solution 8.

(a) Probability of selecting 1st purse = Probability of selecting 2nd purse $= \frac{1}{2}$

Now, probability of taking out copper coin from 1st purse

$$= \frac{1}{2} \times \frac{5}{9} \times \frac{5}{18}$$

and probability of taking out copper coin from 2nd purse

$$= \frac{1}{2} \times \frac{7}{10} = \frac{7}{20}$$

Hence, the probability of taking out copper coin from either of the purse

$$= \frac{5}{18} + \frac{7}{20}$$

$$= \frac{50 + 63}{180} = \frac{113}{180}.$$ **Ans.**

(b) Success of Aman and Bhuvan is getting a total of 8 in a pair of dice.

$\therefore$ Favourable cases = {(2, 6), (3, 5), (4, 4), (5, 3)}, (6, 2)}

Let p = Probability of getting a total of 8

and q = Probability of not getting a total of 8

$\therefore$ $$p = \frac{5}{36} \text{ and } q = \frac{31}{36}$$

If Aman starts the game, he can win it in the first throw, 3rd throw, 5th throw, 7th throw and so on.

Now, probability of Aman's winning in the first throw $= p = \frac{5}{36}$.

Probability of Aman's winning in the third throw,

$$= qqp$$

$$= \frac{31}{36} \times \frac{31}{36} \times \frac{5}{36}$$

$$= \left(\frac{31}{36}\right)^2 \times \frac{5}{36}$$

Probability of Aman's winning in 5th throw

$$= qqqqp$$

$$= \left(\frac{31}{36}\right)^2 \times \frac{5}{36} \text{ and so on}$$

Since, all cases are mutually exclusive, therefore the chances of Aman's winning the game first

$$= p + qqp + qqqqp \ldots \infty$$

$$= p[1 + q^2 + q^4 + \ldots \infty]$$

$$= p\left[\frac{1}{1 - q^2}\right] \left[\text{In a G.P. } S_\infty = \frac{a}{1 - r}\right]$$

$$= \frac{5}{36}\left[\frac{1}{1 - \left(\frac{31}{36}\right)^2}\right]$$

$$= \frac{5}{36}\left[\frac{36 \times 36}{36 \times 36 - 31 \times 31}\right]$$

$$= \frac{5}{36} \times \frac{36 \times 36}{(5 \times 67)} = \frac{5 \times 36}{5 \times 67} = \frac{36}{67}$$

Since, Aman and Bhuvan are throwing the pair of dice alternately, hence the events are mutually exclusive and therefore, the sum of their probabilities of winning is 1.

$\therefore$ Probability of Bhuvan's winning

$$= 1 - \frac{36}{67} = \frac{31}{67}.$$ **Ans.**

Solution 9.

(b) Given differential equation is,

$$y - x\frac{dy}{dx} = x + y\frac{dy}{dx},$$

when $y = 0$ and $x = 1$

$$y - x = (x + y)\frac{dy}{dx}$$

or $$\frac{dy}{dx} = \frac{y - x}{y + x} \quad \ldots\text{(i)}$$

Put $$y = vx$$

$$\Rightarrow \quad \frac{dy}{dx} = v + x\frac{dv}{dx}$$

(i) $$v + x\frac{dv}{dx} = \frac{vx - x}{vx + x} = \frac{x(v-1)}{x(v+1)}$$

or $$v + x\frac{dv}{dx} = \frac{v-1}{v+1}$$

$$x\frac{dv}{dx} = \frac{v-1}{v+1} - \frac{v}{1}$$

$$x\frac{dv}{dx} = \frac{v - 1 - v^2 - v}{v+1}$$

$$= \frac{-(v^2+1)}{v+1}$$

or $$\frac{v+1}{v^2+1}\,dv = -\frac{dx}{x}$$

On integrating both sides,

$$\int \frac{v+1}{v^2+1}\,dv + \int \frac{dx}{x} = c$$

or $$\frac{1}{2}\int \frac{2v}{v^2+1}\,dv + \int \frac{1}{v^2+1}\,dv + \int \frac{dx}{x} = c$$

$$\Rightarrow \quad \log(v^2+1) + \tan^{-1} v + \log|x| = c$$

$$\Rightarrow \quad \frac{1}{2}\log\left(\frac{y^2}{x^2}+1\right) + \tan^{-1}\left(\frac{y}{x}\right) + \log|x| = c$$

$$\left[\text{replacing } v = \frac{y}{x}\right]$$

Now, when $y = 0$ and $x = 1$

$$c = \frac{1}{2}\log 1 + \tan^{-1} 0 + \log 1$$

$$c = 0 \qquad [\because \log 1 = 0]$$

$\therefore$ Required solution is

$$\frac{1}{2}\log\left(\frac{y^2}{x^2}+1\right) + \tan^{-1}\left(\frac{y}{x}\right) + \log|x| = 0$$

or $$\log\left(\frac{x^2+y^2}{x^2}\right) + 2\tan^{-1}\left(\frac{y}{x}\right) + 2\log x = 0$$

or $$\log\left(\frac{x^2+y^2}{x^2}\right) + \log x^2 + 2\tan^{-1}\left(\frac{y}{x}\right) = 0$$

or $$\log\left(\frac{x^2+y^2}{x^2}\cdot x^2\right) + 2\tan^{-1}\left(\frac{y}{x}\right) = 0$$

or $$\log(x^2+y^2) + 2\tan^{-1}\frac{y}{x} = 0.$$

Ans.

SECTION—B

Solution 10.

(a) We have

$$\text{L.H.S.} = [\vec{a}+\vec{b} \quad \vec{b}+\vec{c} \quad \vec{c}+\vec{a}]$$

$$= (\vec{a}+\vec{b})\cdot[(\vec{b}+\vec{c})\times(\vec{c}+\vec{a})]$$

$$= (\vec{a}+\vec{b})\cdot[(\vec{b}+\vec{c})+(\vec{b}+\vec{a})+(\vec{c}\times\vec{c})] + (\vec{c}\times\vec{a})]$$

$$= (\vec{a}+\vec{b})\cdot[(\vec{b}\times\vec{c})-(\vec{a}\times\vec{b})+(\vec{c}\times\vec{a})]$$

$$\left[\begin{array}{l}\because \quad \vec{c}\times\vec{c} = 0 \\ \text{and } (\vec{b}\times\vec{a}) = -(\vec{a}\times\vec{b})\end{array}\right]$$

$$= \vec{a}\cdot(\vec{b}\times\vec{c}) - \vec{a}\cdot(\vec{a}\times\vec{b}) + \vec{a}\cdot(\vec{c}\times\vec{a}) + \vec{b}\cdot(\vec{b}\times\vec{c}) - \vec{b}\cdot(\vec{a}\times\vec{b}) + \vec{b}\cdot(\vec{c}\times\vec{a})$$

$$= [\vec{a}\,\vec{b}\,\vec{c}] - [\vec{a}\,\vec{a}\,\vec{b}] + [\vec{a}\,\vec{c}\,\vec{a}] + [\vec{b}\,\vec{b}\,\vec{c}] - [\vec{b}\,\vec{a}\,\vec{b}] + [\vec{b}\,\vec{c}\,\vec{a}]$$

$$= [\vec{a}\,\vec{b}\,\vec{c}] + [\vec{b}\,\vec{c}\,\vec{a}]$$

[Other scalar triple products having two equal vectors being 0]

$$= [\vec{a}\,\vec{b}\,\vec{c}] + [\vec{a}\,\vec{b}\,\vec{c}]$$

$$= 2[\vec{a}\,\vec{b}\,\vec{c}] = \text{R.H.S.}$$

Hence Proved.

(b) We consider A as origin then the position vectors of B and C may be taken as $\vec{b}$ and $\vec{c}$.

$\therefore$ Position vector of mid point D of BC

$$= \frac{\vec{b}+\vec{c}}{2}$$

Position vector of E of AC $= \dfrac{\vec{c}}{2}$

Position vector of F of AB $= \dfrac{\vec{b}}{2}$

Hence $\vec{DE}$ = Position vector of E – Position vector of D

$$= \frac{\vec{c}}{2} - \frac{\vec{b}+\vec{c}}{2} = -\frac{\vec{b}}{2}$$

and $\vec{DF}$ = Position vector of F − Position vector of D

$$= \frac{\vec{b}}{2} - \frac{\vec{b}+\vec{c}}{2} = -\frac{\vec{c}}{2}$$

Now, area of ΔDEF

$$= \frac{1}{2}(\vec{DE} \times \vec{DF})$$

$$= \frac{1}{2}\left(\frac{-\vec{b}}{2}\right) \times \left(\frac{-\vec{c}}{2}\right) = \frac{1}{8}\vec{b} \times \vec{c}$$

$$= \frac{1}{4}\left(\frac{1}{2}\vec{b} \times \vec{c}\right)$$

$$= \frac{1}{4}[\text{Area of } \Delta ABC]$$

∴ Area of ΔDEF = $\frac{1}{4}$ (Area of ΔABC).

Hence Proved.

Solution 11.

(a) The given line is

$$\vec{r} = 2\hat{i} + 3\hat{j} - \hat{k} + \lambda(\hat{i} - 2\hat{j} + \hat{k}) \quad \ldots(i)$$

It passes through the point A_1 with position vector

$\vec{a_1} = 2\hat{i} + 3\hat{j} - \hat{k}$ and is parallel to the vector

$\vec{b} = \hat{i} - 2\hat{j} + \hat{k}$

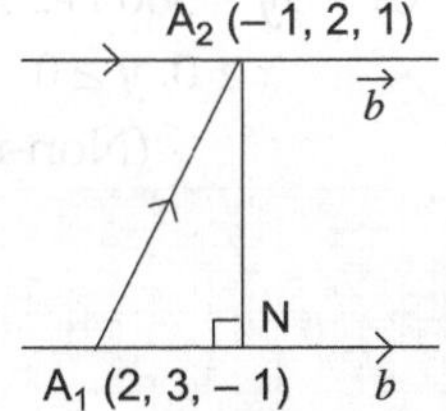

The equation of the line passing through the point A_2 (– 1, 2, 1) with position vector

$\vec{a_2} = -\hat{i} + 2\hat{j} + \hat{k}$

and parallel to the line (i) is

$$\vec{r} = -\hat{i} + 2\hat{j} + \hat{k} + \mu(\hat{i} - 2\hat{j} + \hat{k}) \quad \ldots(ii)$$

Also, $\bar{a}_2 - \bar{a}_1 = -3\hat{i} - \hat{j} + 2\hat{k}$

$$|\bar{b}| = \sqrt{(1)^2 + (2)^2 + (1)^2} = \sqrt{6}$$

and $\vec{b} \times (\vec{a_2} - \vec{a_1}) = \begin{vmatrix} \hat{i} & \hat{j} & \hat{k} \\ 1 & -2 & 1 \\ -3 & -1 & 2 \end{vmatrix}$

$$= \hat{i}(-4+1) - \hat{j}(2+3) + \hat{k}(-1-6)$$

$$= -3\hat{i} - 5\hat{j} - 7\hat{k}$$

$$\therefore \; |\vec{b} \times (\vec{a_2} - \vec{a_1})| = \sqrt{(-3)^2 + (-5)^2 + (-7)^2}$$

$$= \sqrt{9 + 25 + 49} = \sqrt{83}$$

∴ The distance between the parallel lines (i) and (ii)

$$= \frac{|\vec{b} \times (\vec{a_2} - \vec{a_1})|}{|\vec{b}|}$$

$$= \frac{\sqrt{83}}{\sqrt{6}} = \frac{\sqrt{83}}{6} \text{ units.}$$ **Ans.**

(b) Equation of any plane passing through A(2, 1, – 3) is

$$a(x-2) + b(y-1) + c(z+3) = 0 \quad \ldots(i)$$

since it passes through B(– 3, – 2, 1) and C(2, 4, – 1) we get

$$a(-3-2) + b(-2-1) + c(1, 3) = 0$$

i.e., $\quad -5a - 3b + 4c = 0$

$\Rightarrow \quad 5a + 3b - 4c = 0 \quad \ldots(ii)$

and $\quad a(2-2) + b(4-1) + c(-1+3) = 0$

i.e., $\quad 3b + 2c = 0 \quad \ldots(iii)$

Solving (ii) and (iii) $a : b : c = 18 : -10 : 15$ replacing in (i) required equation is

$$18(x-2) - 10(y-1) + 15(z+3) = 0$$

$$\Rightarrow \quad 18x - 36 - 10y + 10 + 15z + 45 = 0$$

$$\Rightarrow \quad 18x - 10y + 15z + 19 = 0.$$ **Ans.**

Solution 12.

(a) Let X denote the number of red marbles drawn from the box. Since there are 4 red marbles therefore X can take the values 0, 1, 2 and 3.

Now, P(X = 0) = Probability of getting no red marbles

= Probability of 3 marbles drawn are black

$$= \frac{^5C_3}{^9C_3} = \frac{10}{84} = \frac{5}{42}$$

P(X = 1) = Probability of getting 1 red marble

$$= \frac{^4C_1 \times {^5C_2}}{^9C_3}$$

$$= \frac{4 \times 10}{84} = \frac{10}{21}$$

$P(X = 2)$ = Probability of getting 2 red marbles

$$= \frac{{}^4C_2 \times {}^5C_1}{{}^9C_3} = \frac{6 \times 5}{84} = \frac{5}{14}$$

$P(X = 3)$ = Probability of getting 3 red marbles

$$= \frac{{}^4C_3}{{}^9C_3} = \frac{4}{84} = \frac{1}{21}$$

∴ The probability distribution of X is given as:

X	0	1	2	3
P(X)	5/42	10/21	5/14	1/21

Again

X_i	$P_i(x)$	$P_i \times X_i$	X_i^2	$P_i \times X_i^2$
0	5/42	0	0	0
1	10/21	10/21	1	10/21
2	5/14	5/7	4	10/7
3	1/21	1/7	9	3/7
		$\Sigma P_iX_i = 4/3$		$\Sigma P_i \times X_i^2 = 7/3$

Now, Mean $= \mu = \Sigma P_iX_i = \frac{4}{3}$

Standard deviation

$$= \sigma = \sqrt{\Sigma P_iX_i^2 - \mu^2}$$

$$= \sqrt{\frac{7}{3} - \frac{16}{9}}$$

$$= \sqrt{\frac{21 - 16}{9}}$$

$$= \frac{\sqrt{5}}{3}$$

$= 0{\cdot}745$. **Ans.**

(b) Given,

Bag A	Bag B	Bag C
2 W	3 W	4 W
1 B	2 B	3 B
3 R	4 R	2 R

Let E_1, E_2 and E_3 be the events of choosing bag A, bag B and bag C respectively.

∴ $$P(E_1) = P(E_2) = P(E_3) = \frac{1}{3}$$

Let E be the event of drawing 1 red and 1 black ball from the bags, then

$P(E/E_1)$ = Probability that 1 red and 1 black balls are drawn from bag A

$$= \frac{{}^3C_1 \times {}^1C_1}{{}^6C_2} = \frac{3}{15} = \frac{1}{5}$$

Similarly, $$P(E/E_2) = \frac{{}^4C_1 \times {}^2C_1}{{}^9C_2} = \frac{4 \times 2}{36} = \frac{2}{9}$$

and $$P(E/E_3) = \frac{{}^2C_1 \times {}^3C_1}{{}^9C_2} = \frac{2 \times 3}{36} = \frac{1}{6}$$

Required probability = Probability that 2 bals (1 red, 1 black) are drawn from bag B is given by $P(E_2/E)$

$$= \frac{P(E_2) \cdot P(E/E_2)}{P(E_1) \cdot P(E/E_1) + P(E_2) \cdot P(E/E_2) + P(E_3) \cdot P(E/E_3)}$$

$$= \frac{\frac{1}{3} \times \frac{2}{9}}{\frac{1}{3} \times \frac{1}{5} + \frac{1}{3} \times \frac{2}{9} + \frac{1}{3} \times \frac{1}{6}}$$

$$= \frac{\frac{1}{3} \times \frac{2}{9}}{\frac{1}{3}\left[\frac{1}{5} + \frac{2}{9} + \frac{1}{6}\right]}$$

$$= \frac{\frac{2}{9}}{\frac{18 + 20 + 15}{90}}$$

$$= \frac{2 \times 90}{9 \times 53} = \frac{20}{53}.$$ **Ans.**

SECTION—C

Solution 13.

(b) Let x be the number of A type tea cups and y be the number of B type tea cups.

The problem can be formulated as

Maximize the profit $P = 1{\cdot}50x + 1{\cdot}00y$

i.e., $P = 1{\cdot}50x + y$

Subject to the contraints:

$12x + 6y \le 360$ *i.e.*, $2x + y \le 60$

$18x + 0{\cdot}y \le 360$ *i.e.*, $x \le 20$

$6x + 9y \le 360$ *i.e.*, $2x + 3y \le 120$

and $x \ge 0, y \ge 0$

(Non-negative constraints)

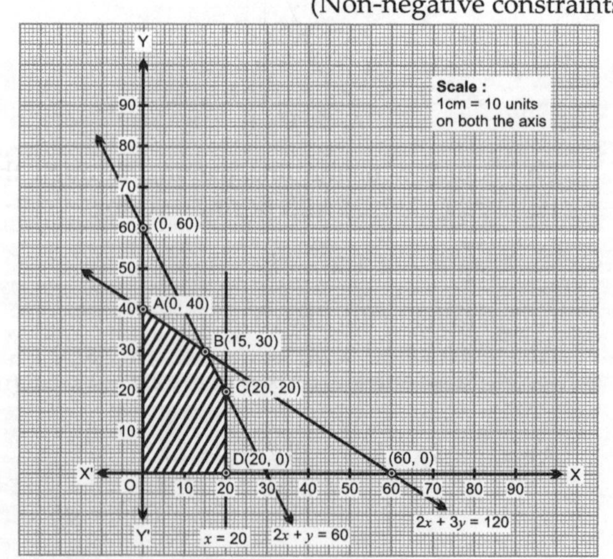

We draw the straight lines for given inequalities by converting it into equations.

$$2x + y = 60$$

$x = 20$, $2x + 3y = 120$, $x = 0$ and $y = 0$.

The shaded portion shows the feasible region.

The points B and C are points of intersection of lines $2x + y = 60$, with $2x + 3y = 120$ and $x = 20$ with $2x + y = 60$ respectively.

Thus, the 5 corner points of the feasible region are O(0, 0), A(0, 40), B(15, 30), C(20, 20) and D(20, 0).

Corner Points	Objective function $Z = 1{\cdot}50x + y$
At O(0, 0)	$Z = 1{\cdot}50 \times 0 + 1 \times 0 = 0$
At A(0, 40)	$Z = 1{\cdot}50 \times 0 + 1 \times 40 = 40$
At B(15, 30)	$Z = 1{\cdot}50 \times 15 + 1 \times 30 = 52{\cdot}5$ maximum
At C(20, 20)	$Z = 1{\cdot}50 \times 20 + 1 \times 20 = 50$
At D(20, 0)	$Z = 1{\cdot}50 \times 20 + 1 \times 0 = 30$

Clearly the maximum profit is at $x = 15$ and $y = 30$

Thus, the manufacturer should manufacture 15 cups of type A and 30 cups of type B to get maximum profit in a day. **Ans.**

Solution 14.

(b) Given $C(x) = \dfrac{x^3}{3} + 3x^2 - 7x + 16$

(i) Marginal Cost (M.C.)

$$= \frac{d}{dx}\left[\frac{x^3}{3} + 3x^2 - 7x + 16\right]$$

$$= x^2 + 6x - 7.$$ **Ans.**

(ii) Average Cost (A.C.)

$$= \frac{\text{Total cost}}{\text{Output}} = \frac{C(x)}{x}$$

$$= \frac{\frac{x^3}{3} + 3x^2 - 7x + 16}{x}$$

$$= \frac{x^2}{3} + 3x - 7 + \frac{16}{x}$$ **Ans.**

(iii) Marginal Average Cost (M.A.C.)

$$= \frac{d}{dx}\left[\frac{x^2}{3} + 3x - 7 + \frac{16}{x}\right]$$

$$= \frac{2x}{3} + 3 - \frac{16}{x^2} \quad \text{...(i)}$$

Again, $\dfrac{x(MC) - C(x)}{x^2}$

$$= \frac{x(x^2 + 6x - 7) - \frac{x^3}{3} - 3x^2 + 7x - 16}{x^2}$$

$$= \frac{1}{x^2}\left[x^3 + 6x^2 - 7x - \frac{x^3}{3} - 3x^2 + 7x - 16\right]$$

$$= \frac{1}{x^2}\left[x^3 - \frac{x^3}{3} + 3x^2 - 16\right]$$

$$= \frac{1}{x^2}\left[\frac{2}{3}x^3 + 3x^2 - 16\right]$$

$$= \frac{2}{3}x + 3 - \frac{16}{x^2} \quad \text{...(ii)}$$

From equation (i) and (ii), we get

⇒ Marginal average cost (M.A.C.)

$= x\,(\text{M.C.}) - C(x)\,x^2.$ **Hence Proved.**

COMPUTER SCIENCE 2020

(Paper-I)
(THEORY)
(Maximum Marks : 70)
(Time allowed : Three hours)

*(Candidates are allowed additional 15 minutes for **only** reading the paper. They must NOT start writing during this time.)*

*Answer **all** questions in Part-I (compulsory) and **six** questions from Part-II, choosing **two** questions from Section-A, **two** from Section-B and **two** from Section-C.*

All working, including rough work, should be done on the same sheet as the rest of the answer.

The intended marks for questions or parts of questions are given in brackets [ ].

QUESTIONS

PART–I (20 Marks)

*Answer **all** questions.*

While answering questions in this Part, indicate briefly your working and reasoning, wherever required.

Question 1.

(a) *State the properties of zero in Boolean algebra.* **[1]**

(b) *Find the complement of the following Boolean expression using De Morgan's law :* **[1]**

$F(P,Q,R) = P + (Q'.R)$

(c) *Find the dual of : $(A' + 0) .(B' + 1) = A'$* **[1]**

(d) *State whether the following proposition is a tautology, contradiction or a contingency :* **[1]**

$F = (P => Q) \vee (Q =>\sim P)$

(e) *Study the diagram given below and answer the questions that follow :* **[1]**

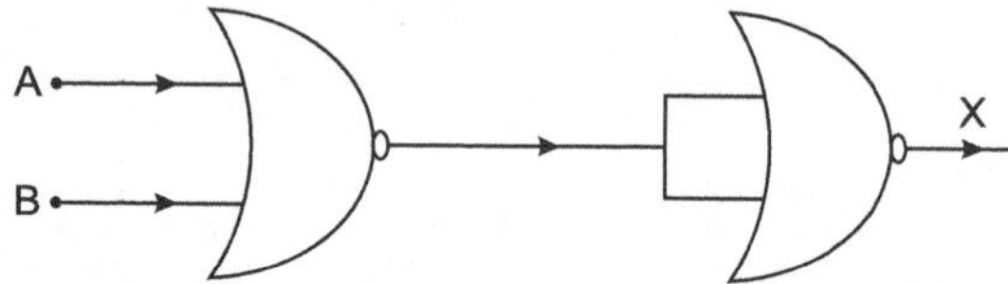

(i) *Name the basic gate which is represented by the diagram.*

(ii) *What will be the value of X when A = 1 and B = 0 ?*

Question 2.

(a) *State the difference between a Binary Tree structure and a Single Linked List.* **[2]**

(b) *A matrix B[10][20] is stored in the memory with each element requiring 2 bytes of storage. If the base address at B[2][1] is 2140, find the address of B[5][4] when the matrix is stored in* **Column Major Wise.** **[2]**

(c) *Convert the following infix notation to prefix form :* **[2]**

(X + Y) / (Z * W / V)

(d) *State the best case and the worst case complexity for bubble sort algorithm.* **[2]**

(e) *What is the significance of the keyword* **'new'** *in Java? Mention the areas where it is used.* **[2]**

Question 3.

The following function check() is a part of some class. What will the function check() return when the value of **(i)** *n = 25 and* **(ii)** *n = 10. Show the dry run/working.* **[5]**

```
int check(int n)
{   if(n<=1)
return 1;
if(n%2 == 0)
return 1 + check(n/2);
else
return 1 + check(n/2 + 1);
}
```

PART–II (50 Marks)

*Answer **six** questions in this part, choosing **two** questions from Section-A, **two** from Section-B and **two** from Section-C.*

SECTION–A

*Answer **any two** questions.*

Question 4.

(a) *Given the Boolean function :*

F(A, B, C, D) = ∑ (0, 1, 2, 3, 4, 5, 8, 9, 10, 11, 12, 13, 14).

(i) *Reduced the above expression by using 4-variable Karnaugh map, showing the various groups (i.e., octal, quads and pairs).* **[4]**

(ii) *Draw the logic gate diagram for the reduced expression. Assume that the variable and their complements are available as inputs.* **[1]**

(b) *Given the Boolean function :*

F(A, B, C, D,) = π (3, 4, 6, 9, 11, 12, 13, 14, 15).

(i) *Reduce the above expression by using 4-variable Karnaugh map, showing the various groups (i.e., octal, quads and pairs).* **[4]**

(ii) *Draw the logic gate diagram for the reduced expression. Assume that the variables and their complements are available as inputs.* **[1]**

Question 5.

(a) *Draw the logic circuit diagram for an octal to binary encoder and explain its working when a particular digit is pressed. Also, state the difference between encoders and decoders.* **[5]**

(b) *Draw the circuit of a two input XOR gate with the help of NOR gates.* **[3]**

(c) *Convert the following expression to its cardinal SOP form :* **[2]**

F(P, Q, R) = P'.Q'.R + P'.Q.R + P.Q'.R' + P.Q.R'

Question 6.

(a) *A company intends to develop a device to show the high status power load for a household invertor depending on the criteria given below:* **[5]**

- *If Air conditioner and Geyser are on.*

OR

- *If Air conditioner is off, but Geyser and Refrigerator are on.*

OR

- *If Geyser is off, but Air conditioner and Water purifier are on.*

OR

- *When all are on.*

The inputs are :

INPUTS	
A	*Air conditioner is on*
G	*Geyser is on*
R	*Refrigerator is on*
W	*Water purifier is on*

(In all the above cases 1 indicates yes and 0 indicates no.)

Output : *X [1 indicates high power, 0 indicates low power for all cases.]*

Draw the truth table for the inputs and outputs given above and write the ***SOP*** *expression for* ***X(A, G, R, W).***

(b) *Draw the truth table and derive an SOP expression for sum and carry for a full adder. Also, draw the logic circuit for the carry of a full adder.* **[3]**

(c) *Simplify the following expression using Boolean laws:* **[2]**

F = [(X' + Y) . (Y' + Z)]' + (X' + Z)

SECTION–B

Answer ***any two*** *questions.*

Each program should be written in such a way that it clearly depicts the logic of the problem. This can be achieved by using mnemonic names and comments in the program.

(Flowcharts and Algorithms are ***not*** *required.)*

The programs must be written in Java.

Question 7.

Design a class **Convert** *to find the date and the month from a given day number for a particular year.* **[10]**

Example : If day number is 64 and the year is 2020, then the corresponding date would be :

March 4, 2020 *i.e., (31 + 29 + 4 = 64)*

Some of the members of the class are given below :

Class name: ***Convert***

Data members/
instance variables:

n	:	*integer to store the day number*
d	:	*integer to store the day of the month (date)*
m	:	*integer to store the month*
y	:	*integer to store the year*

Methods/Member functions:

Convert()	:	*constructor to intitialise the data members with legal initial values*
void accept()	:	*to accept the day number and the year*
void day_to_date()	:	*converts the day number to its corresponding date for a particular year and stores the date in 'd' and the month in 'm'*
void display()	:	*displays the month name, date and year*

Specify the class **Convert** *giving details of the* **constructor(),** ***void accept(), void day_to_date()*** *and* ***void display().*** *Define a* ***main()*** *function to create an object and call the functions accordingly to enable the task.*

Question 8.

Design a class **BinSearch** *to search for a particular value in an array.* **[10]**

Some of the members of the class are given below :

Class name : ***BinSearch***

Data members/
instance variables:

arr[ ]	:	*to store integer elements*
n	:	*integer to store the size of the array*

Member functions/
Methods:

BinSearch(int nn) : *parameterised constructor to initialise n=nn*

void fillarray() : *to enter elements in the array*

void sort() : *sorts the array elements in ascending order using any standard sorting technique*

int bin_search (int l, int u, int v) : *searches for the value 'v' using* ***binary search*** *and* ***recursive technique*** *and returns its location if found otherwise return -1*

Define the class ***BinSearch*** *giving details of the* ***constructor(), void fillarray(), void sort()*** *and* ***int bin_search(int,int,int)****. Define the* ***main()*** *function to create an object and call the functions accordingly to enable the task.*

Question 9.

A class ***Mix*** *has been defined to mix two words, character by character, in the following manner :* **[10]**

The first character of the first word is followed by the first character of the second word and so on. If the words are of different length, the remaining characters of the longer word are put at the end.

Example : If the first word is "JUMP" and the second word is "STROLL", then the required word will be "JSUTMRPOLL".

Some of the members of the class are given below :

Class name *Mix*

Data member/instance variable :

wrd : *to store a word*

len : *to store the length of the word*

Member functions/methods :

Mix() : *default constructor to initialise the data members with legal initial values*

void feedword() : *to accept the word in UPPER case*

void mix_word (Mix P, Mix Q) : *mixes the words of objects P and Q as stated above and stores the resultant word in the current object*

void display() : *displays the word*

Specify the class ***Mix*** *giving the details of the* ***constructor(), void feedword(), void mix_word(Mix, Mix)*** *and* ***void display()****. Define the* ***main()*** *function to create objects and call the functions accordingly to enable the task.*

SECTION–C

Answer ***any two*** *questions.*

Each program should be written in such a way that it clearly depicts the logic of the problem stepwise.

This can be achieved by using comments in the program and mnemonic names or pseudo codes for algorithms. The programs must be written in Java and the algorithms must be written in general/standard form, wherever required/specified.

(Flowcharts are ***not*** *required.)*

Question 10.

A Circular queue is a linear data structure which works on the principle of FIFO, enables the user to enter data from the rear end and remove data from the front end with the rear end connected to the front end to form a circular pattern. Define a class ***CirQueue*** *with the following details :*

Class name ***CirQueue***

Data members/instance variables :

cq[ ] : *array to store the integers*

cap : *stores the maximum capacity of the array*

front : *to point the index of the front end*

rear : *to point the index of the rear end*

Member functions:

CirQueue(int max) : *constructor to initialise the data member cap=max, front=0 and rear=0*

void push(int n) : *to add integer in the queue from the rear end if possible, otherwise display the message "QUEUE IS FULL"*

int pop() : *removes and returns the integer from the front end of the queue if any, else returns -9999*

void show() : *displays the queue elements*

(a) *Specify the class* ***CirQueue*** *giving details of the functions* ***void push(int)*** *and* ***int pop()****. Assume that the other functions have been defined.* **[4]**

The main function and algorithm need NOT be written.

(b) *How is a linear queue structure different from a circular queue structure?* **[1]**

Question 11.

An interface ***Data*** *is defined with a data member and a method volume() which returns the volume of the implementing shape. A super class* ***Base*** *has been defined to contain the radius of a geometrical shape. Define a sub-class* ***CalVol*** *which uses the properties of the interface* ***Data*** *and the class* ***Base*** *and calculates the volume of a cylinder.* **[5]**

The details of the members of the interface and both the classes are given below

Interface name : Data

Data members:

double pi : initialise pi = 3.142

Member functions/methods :

double volume()

Class name : Base

Data member/instance variable:

rad : to store the radius in decimal

Member functions/methods:

Base(...) : parameterised constructor to initialise the data member

void show() : displays the radius with an appropriate message

Class name : CalVol

Data member/instance variable:

ht : to store the height in decimal

Member functions/methods:

CalVol() : parameterised constructor to initialise the data members of both the classes

double volume() : calculates the volume of a sphere by using the formula (pi x radius^2 x height)

void show() : displays the data members of both the classes and the volume of the sphere with appropriate message

*Assume that the interface **Data** and the super class **Base** has been defined. Using the **concept of inheritance,** specify the class **CalVol** giving the details of the **constructor(...), double volume()** and **void show()**.*

***The interface, super class, main function and algorithm** need NOT be written.*

Question 12.

(a) *A linked list is formed from the objects of the class **Node**. The class structure of the Node is given below :* **[2]**

```
class Node
{
int n;
Node next;
}
```

Write an algorithm OR a method to find the product of the integer numbers from an existing linked list.

The method declaration is as follows :

void Product_Node(Node str)

(b) *Answer the following questions from the diagram of a Binary Tree given below :*

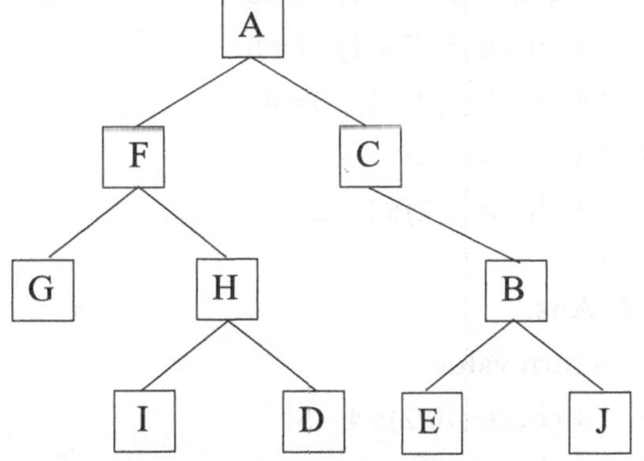

(i) *Write the post-order traversal of the left subtree of the above structure.* **[1]**

(ii) *State the degree of the Nodes E and H.* **[1]**

(iii) *Mention the external nodes of the right subtree.* **[1]**

ANSWERS

PART–I

Answer 1.

(a) Properties of zero is :

(i) $P + 0 = P$

(ii) $P . 0 = 0$

(b) $(P + (\overline{Q} . R))' = \overline{P} \cdot (\overline{Q} \cdot R)' = \overline{P} \cdot (Q + \overline{R})$

(c) $(\overline{A} + 0) . (\overline{B} + 1) = \overline{A}$

Dual is : $(\overline{A} . 1) + (\overline{B} . 0) = \overline{A}$

(d) $F = (P \rightarrow Q) + (Q \rightarrow \overline{P})$ [Conversion or as +]

$= \overline{P} + Q + \overline{Q} + \overline{P}$ [Apply implies rule]

$= \overline{P} + 1 = 1$ (Hence, it is a tautology)

(e) (i) OR Gate

(ii) $X = 1$

Answer 2.

(a)

	Binary Tree	Single Linked List
1.	It is a recursive data structure.	It is a linear data structure.
2.	It allocates memory randomly non-linearly.	It allocates memory sequentially.

(b) R = 10, C = 20

ES = 2, BA = 2140

LR = 2, LC = 1

I = 5, J = 4

In column major wise formulae :

B [I, J] = BA + ES ((I – LR) + R (J – LC))

B [5, 4] = 2140 + 2((5 – 2) + 10 (4 – 1))

= 2140 + 2 (3 + 10 * 3)

$= 2140 + 2 * 33$

$= 2140 + 66$

$= 2206$

(c) $(X + Y) / (Z * W / V)$

$= (+XY) / ((*ZW)/(V))$

$= (+XY) / (/*ZWV)$

$= /+XY/*ZWV$

(d) Best case complexity = O (1)

Worst case complexity = O (N^2)

(e) **new** keyword in java is used to dynamically allocate memory. It is used to create objects or arrays during run time.

Answer 3.

(i) n return value

25 1 + check (25/2 + 1) 5 = 6

13 1 + check (13/2 + 1) 4 = 5

7 1 + check (7/2 + 1) 3 = 4

4 1 + check (4/2) 2 = 3

2 1 + check (2/2) 1 = 2

1 1

= 6 **Ans.**

(ii) n return value

10 1 + check (10/2) 4 = 5

5 1 + check (5/2 + 1) 3 = 4

3 1 + check (3/2 + 1) 2 = 3

2 1 + check (2/2) 1 = 2

1 1

= 5

PART–II

SECTION–A

Answer 4.

(a) (i)

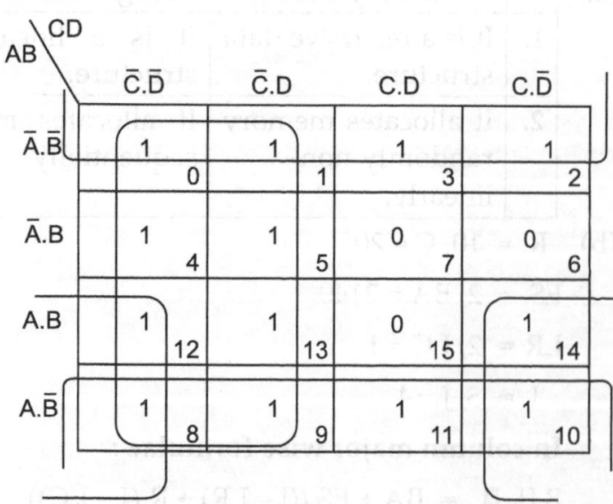

Octet 1 = $m_0 + m_1 + m_2 + m_3 + m_8 + m_9 + m_{10} + m_{11} = \bar{B}$

Octet 2 = $m_0 + m_1 + m_4 + m_5 + m_8 + m_9 + m_{12} + m_{13} = \bar{C}$

Quad 1 = $m_8 + m_{10} + m_{12} + m_{14} = A\bar{D}$

F = $\bar{B} + \bar{C} + A.\bar{D}$

(ii) The logic gate Diagram is :

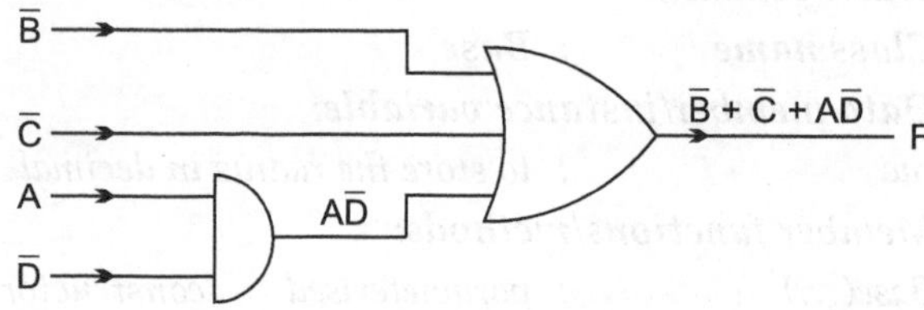

(b) (i)

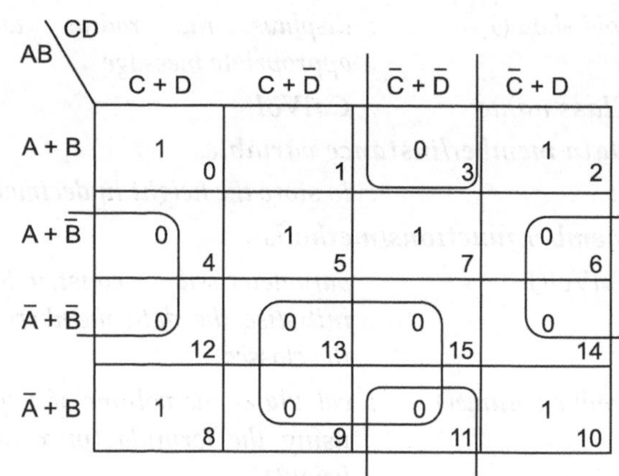

Quad 1 = $M_9 . M_{11} . M_{13} . M_{15} = \bar{A} + \bar{D}$

Quad 2 = $M_4 . M_6 . M_{12} . M_{14} = \bar{B} + D$

Pair 1 = $M_3 . M_{11} = B + \bar{C} + \bar{D}$

$= (\bar{A} + \bar{D}) . (\bar{B} + D) . (B + \bar{C} + \bar{D})$

(ii) The Logic Gate Diagram is :

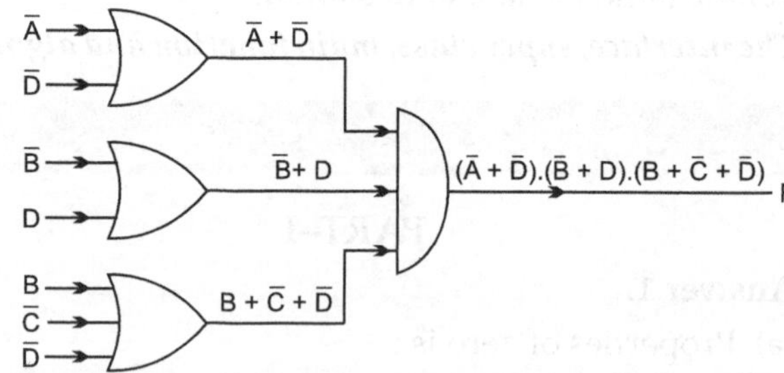

Answer 5.

(a) Encoder is a circuit which converts any signal to binary form while decoder is a circuit which converts binary data into signal form.

Octal to Binary (Octanary) Encoder

Truth Table :

F_2	F_1	F_0	Octal Numbers
0	0	0	0
0	0	1	1
0	1	0	2
0	1	1	3

1	0	0	4
1	0	1	5
1	1	0	6
1	1	1	7

Logic Circuit Diagram :

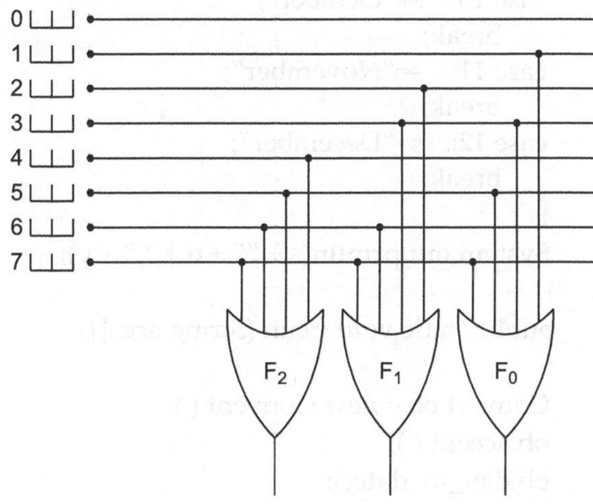

Octanary Encoder

(b) $X \oplus Y = X'Y + XY'$ [SOP Expression]

$F_{(X, Y)} = \Sigma(1, 2)$

$F_{(X, Y)} = \pi(0, 3)$

$X \oplus Y = (X + Y) . (X' + Y')$ [POS Expression]

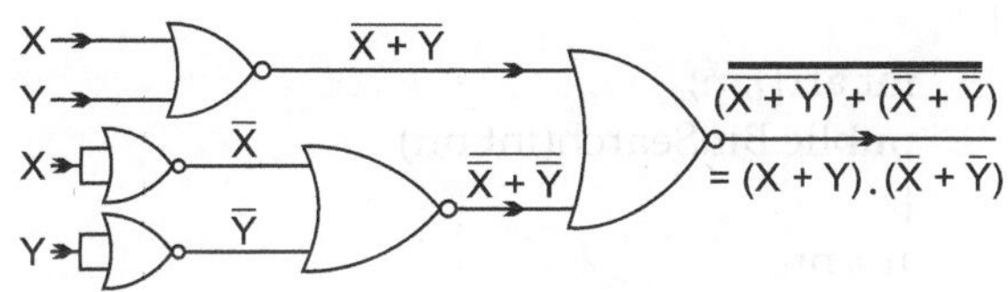

(c) $F(P.Q.R) = P'.Q'.R + P'.Q.R + P.Q'.R' + P.Q.R'$

$= 001 + 011 + 100 + 110$

$= 1 + 3 + 4 + 6$

$F(P.Q.R) = \Sigma (1, 3, 4, 6)$

Answer 6.

(a)

A	G	R	W	X	
0	0	0	0	0	0
0	0	0	1	0	1
0	0	1	0	0	2
0	0	1	1	0	3
0	1	0	0	0	4
0	1	0	1	0	5
0	1	1	0	1	6
0	1	1	1	1	7
1	0	0	0	0	8
1	0	0	1	1	9
1	0	1	0	0	10
1	0	1	1	1	11
1	1	0	0	1	12
1	1	0	1	1	13
1	1	1	0	1	14
1	1	1	1	1	15

$X(A, G, R, W) = \Sigma (6, 7, 9, 11, 12, 13, 14, 15)$

$\bar{A}GR\bar{W} + \bar{A}GRW + A\bar{G}\bar{R}W + A\bar{G}RW +$

$AG\bar{R}\bar{W} + AG\bar{R}W + AGR\bar{W} + AGRW$

(b)

X	Y	Z	Carry		Sum	
0	0	0	0		0	
0	0	1	0		1	$\bar{X}\bar{Y}Z$
0	1	0	0		1	$\bar{X}Y\bar{Z}$
0	1	1	1	$\bar{X}YZ$	0	
1	0	0	0		1	$X\bar{Y}\bar{Z}$
1	0	1	1	$X\bar{Y}Z$	0	
1	1	0	1	$XY\bar{Z}$	0	
1	1	1	1	XYZ	1	XYZ

$Sum = \bar{X}\bar{Y}Z + \bar{X}Y\bar{Z} + X\bar{Y}\bar{Z} + XYZ$

$= X \oplus Y \oplus Z$

$Carry = \bar{X}YZ + X\bar{Y}Z + XY\bar{Z} + XYZ$

$= XY + YZ + XZ$

Logic Circuit Diagram is :

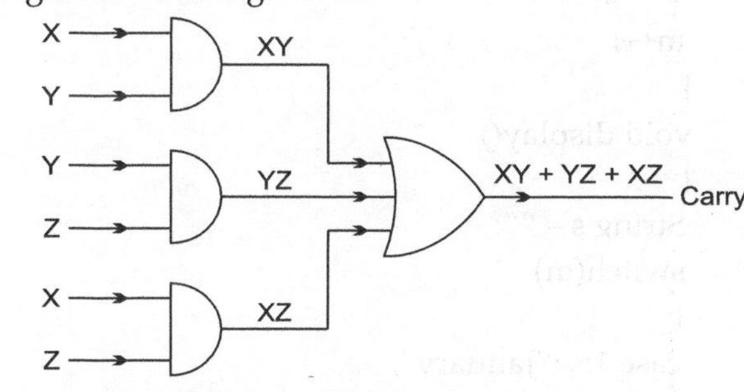

(c) $[(\bar{X} + Y).(\bar{Y} + Z)]' + \bar{X} + Z$

$= (\bar{X} + Y)' + (\bar{Y} + Z)' + \bar{X} + Z$

$= X.\bar{Y} + Y.\bar{Z} + \bar{X} + Z$ (Demorgan's Law)

$= \bar{X} + X\bar{Y} + Z + Y.\bar{Z}$ (Commutative Law)

$= (X + \bar{X}).(\bar{X} + \bar{Y}) + (Y + Z).(Z + \bar{Z})$

(Distributive Rule)

$= \bar{X} + \bar{Y} + Y + Z$ (Complementary Law)

$= 1 + \bar{X} + Z = 1$ (Properties of 1)

SECTION–B

Answer 7.

```
import java.util.*;
class Convert
{
    int n, d, m, y;
    public Convert()
    {
     n = d = m = y = 0;
    }
    void accept()
    {
     Scanner sc=new Scanner(System.in);
     System.out.println("Enter day no. and year");
     n = sc.nextInt();
     y = sc.nextInt();
    }
    void day_to_date()
    {
     int A[] = {31,28,31,30,31,30,31,31,30,31, 30,31};
     if(y%4 == 0)
     {
      A[1] + = 1;
     }
     while(n > A[m])
     {
     n = n – A[m];
     m++;
     }
     d = n;
     m++;
    }
    void display()
    {
     String s = "";
     switch(m)
     {
     case 1:s="January";
         break;
     case 2:s="February";
         break;
     case 3:s="March";
         break;
     case 4:s="April";
         break;
     case 5:s="May";
         break;
     case 6:s="June";
         break;
     case 7:s="July";
         break;
     case 8:s="August";
         break;
     case 9:s="September";
         break;
     case 10:    s="October";
         break;
     case 11:    s="November";
         break;
     case 12:    s="December";
         break;
     }
     System.out.println(s + "" + d + "," + y);
    }
    public static void main (String args[])
    {
     Convert ob = new Convent ( );
     ob.accept ( );
     ob.day_to_date();
     ob.display ( );
    }
}
```

Answer 8.

```
import java.util.*;
class BinSearch
{
    int arr[], n;
    public BinSearch(int nn)
    {
    n = nn;
    arr = new int[n];
    }
    void fillarray()
    {
    Scanner sc=new Scanner(System.in);
    int i;
    for(i=0; i<n; i++)
    {
    System.out.println("Enter a no.");
    arr[i]=sc.nextInt();
    }
        }
    }
    void sort()
    int i, j, t;
    for(i=0; i<n; i++)
    {
        for(j=i+1; j<n; j++)
        {
        if(arr[i]>arr[j])
        {
        t = arr[i];
```

```
        arr[i] = arr[j];
        arr[j] = t;
        }
        }
    }
    }
    int bin_search(int l, int u, int v)
    {
        int m;
        if(l > v)
        {
         return(- 1);
     }
     else
     {
        m = (l + u)/2;
        if(v > arr[m])
        {
         return(bin_search(m + 1, u, v));
     }
     else if(v < arr[m])
     {
        return(bin_search(l, m - 1, v));
     }
     else
     {
        return(m);
        }
     }
    }
    public static void main(String args[])
    {
        BinSearch ob = new BinSearch(10);
        ob.fillarray();
        ob.sort();
        int r = ob.bin_search(0,9,35);
        if(r = = - 1)
        {
         System.out.println("No. not Found");
     }
     else
     {
     System.out.println("No. Found at index" + r);
        }
     }
}
```

Answer 9.

```
import java.util.*;
class Mix
{
    String wrd;
int len;
public Mix()
{
 wrd = "RITESH";
 len = wrd.length();
}
void feedword()
{
 Scanner sc = new Scanner(System.in);
 System.out.println("Enter a string in upper case");
 wrd = sc.next();
 len = wrd.length();
}
void mix_word(Mix P, Mix Q)
{
 int l1, l2, l, i;
 String big;
 wrd = "";
 l1 = P.wrd.length();
 l2 = Q.wrd.length();
 if(l1 < l2)
 {
 l = l1;
 big = Q.wrd;
 }
 else
 {
 l = l2;
 big = P.wrd;
 }
 for(i=0; i<l; i++)
 {
 wrd = wrd + P.wrd.charAt(i) + Q.wrd. charAt(i);
 }
 wrd = wrd + big.substring(i);
 len = wrd.length();
}
void display()
{
System.out.println(wrd);
}
public static void main(String args[])
{
     Mix M1 = new Mix();
     Mix M2 = new Mix();
     Mix M = new Mix();
     M1.feedword();
     M2.feedword();
     M.mix_word(M1, M2)
```

```
        M.display();
    }
}
```

SECTION–C

Answer 10.

(a)
```
class CirQueue
{
    int cq[], cap, front, rear;
    public CirQueue(int max)
    {
    cap = max;
    cq = new int[cap];
    front = rear = 0;
    }
    void push(int n)
    {
    if((front == 1 && rear == cap – 1) | | (front
                                        == rear + 1))
    {
        System.out.println("QUEUE IS FULL");
    }
    else
    {
        if(rear == 0)
        {
        front = 1;
        rear = 1;
        }
        else if(rear == cap – 1)
        {
        rear = 1;
        }
        else
        {
        rear = rear + 1;
        }
        cq[rear] = n;
    }
    }
    int pop()
    {
    if(front==0)
    {
        return(– 9999);
    }
    else
    {
        int v = cq[front];
        if(front == rear)
        {
        front = rear = 0;
        }
        else if(front == cap – 1)
        {
        front = 1;
        }
        else
        {
        front = front + 1;
        }
        return(v);
        }
    }
}
```

(b) In linear queue we can waste some memory while circular queue is used for optimal utilisation of memory.

Answer 11.

```
class CalVol extends Base implements Data
{
    double ht;
    public CalVol(double r, double h)
    {
    super(r);
    ht = h;
    }
    double volume()
    {
    return(pi * rad * rad * ht);
    }
    void show()
    {
    super.show();
    System.out.println("Height is"+ht);
    System.out.println("Volume is" + volume());
    }
}
```

Answer 12.

(a)
1. Let ptr = str, P = 1
2. Repeat steps 3, 4 until ptr = null
3. P = P * ptr.n
4. ptr = ptr.next
5. Print P
6. End

(b) (i) G, I, D, H, F

(ii) Degree of E = 0 and H = 2

(iii) E and J

COMPUTER SCIENCE 2019

QUESTIONS

PART–I (20 Marks)

Answer ***all*** *questions.*

While answering questions in this Part, indicate briefly your working and reasoning, wherever required.

Question 1.

(a) *Name and draw the logic gate represented by the following truth table, where A and B are inputs and X is the output.* **[1]**

A	B	X
0	0	0
0	1	1
1	0	1
1	1	0

(b) *Write the canonical POS expression of :* **[1]**

$F(P, Q) = \Pi\ (0, 2)$

(c) *Find the dual of :* **[1]**

$X.Y + X.Y' = X + 0$

(d) *If* $F(A, B, C) = A'.B'.C' + A'.B.C'$ *then find F' using De Morgan's Law.* **[1]**

(e) *If A = "It is cloudy" and B = "It is raining", then write the proposition for :* **[1]**

(i) *Contrapositive*

(ii) *Converse*

Question 2.

(a) *What is an Interface ? How is it different from a class?* **[2]**

(b) *A matrix ARR[–4...6, 3...8] is stored in the memory with each element requiring 4 bytes of storage. If the base address is 1430, find the address of ARR[3][6] when the matrix is stored in* **Row Major Wise.** **[2]**

(c) *Convert the following infix notation to postfix form :* **[2]**

$(A + B * C) - (E * F/H) + J$

(d) *Compare the two complexities* $O(n^2)$ *and* $O(2^n)$*, also state which is better and why ?* **[2]**

(e) *State the difference between internal nodes and external nodes of a binary tree structure.* **[2]**

Question 3.

The following function **Mystery( )** *is a part of some class. What will the function Mystery() return when the value of* **num = 43629, x = 3 and y = 4** *respectively ? Show the dry run / working.* **[5]**

```
int Mystery(int num, int x, int y)
{
    if(num<10)
        return num;
    else
    {
        int z = num % 10;
        if(z % 2 == 0)
            return z*x + Mystery(num/10, x, y);
        else
            return z*y + Mystery(num/10, x, y);
    }
}
```

PART–II (50 Marks)

Answer ***six*** *questions in this part, choosing* ***two*** *questions from Section A,* ***two*** *from Section B and* ***two*** *from Section C.*

SECTION–A

Answer ***any two*** *questions.*

Question 4.

(a) *Given the Boolean function :*

$F(A, B, C, D) = \Sigma\ (0, 2, 3, 4, 5, 8, 10, 11, 12, 13).$

(i) *Reduce the above expression by using a 4-variable Karnaugh map, showing the various groups (i.e., octal, quads and pairs).* **[4]**

(ii) *Draw the logic gate diagram for the reduced expression using only NAND gates. Assume that the variables and their complements are available as inputs.* **[1]**

(b) *Given the Boolean function :*

$F(P, Q, R, S) = \pi\ (0, 1, 2, 8, 9, 11, 13, 15).$

(i) *Reduce the above expression by using a 4-variable Karnaugh map, showing the various groups (i.e. octal, quads and pairs).* **[4]**

(ii) *Draw the logic gate diagram for the reduced expression using only NOR gates. Assume that the variables and their complements are available as inputs.* **[1]**

Question 5.

(a) *How is a decoder different from a multiplexer ? Write the truth table and draw the logic circuit diagram for a 3 to 8 decoder and explain its working.* **[5]**

(b) *From the logic circuit diagram given below, derive the Boolean expression and simplify it to show that it represents a logic gate. Name and draw the logic gate.* **[3]**

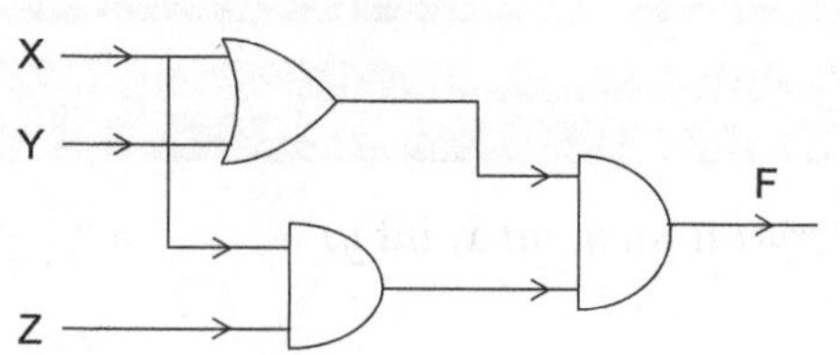

(c) *Using a truth table, state whether the following proposition is a Tautology, Contradiction or Contingency :* **[2]**

$\sim(P \Rightarrow Q) \Leftrightarrow (\sim P / Q)$

Question 6.

(a) *The owner of a company pays bonus to his salesman as per the criteria given below :* **[5]**

- *If the salesman works overtime for more than 4 hours but does not work on off days/holidays.*

OR

- *If the salesman works when festival sales are on and updates showroom arrangements.*

OR

- *If the salesman works on an off day/ holiday when the festival sales are on.*

The inputs are :

Inputs	
O	*Works overtime for more than 4 hours*
F	*Festival sales are on*
H	*Working on an off day/holiday*
U	*Updates showroom arrangements*

(In all the above cases 1 indicates yes and 0 indicates no).

Output : **X** *[1 indicates yes, 0 indicates no for all cases].*

Draw the truth table for the inputs and outputs given above and write the **POS** *expression for* **X(O, F, H, U).**

(b) *What is a half adder ? Write the truth table and derive an SOP expression for sum and carry for a half adder.* **[3]**

(c) *Simplify the following expression, using Boolean laws:* **[2]**

$(X + Z) . (X.Y + Y.Z') + X.Z + Y$

SECTION–B

Answer ***any two*** *questions.*

Each program should be written in such a way that it clearly depicts the logic of the problem.

This can be achieved by using mnemonic names and comments in the program.

(Flowcharts and Algorithms are ***not*** *required.)*

The programs must be written in Java.

Question 7.

Design a class **ArmNum** *to check if a given number is an Armstrong number or not.* **[10]**

[A number is said to be Armstrong if sum of its digits raised to the power of length of the number is equal to the number]

Example : $371 = 3^3 + 7^3 + 1^3$

$1634 = 1^4 + 6^4 + 3^4 + 4^4$

$54748 = 5^5 + 4^5 + 7^5 + 4^5 + 8^5$

Thus, 371, 1634 and 54748 are all examples of Armstrong numbers.

Some of the members of the class are given below :

Class Name : ***ArmNum***

Data members/ /instance variables:

n	:	*to store the number*
l	:	*to store the length of the number*

Methods/ Member functions:

ArmNum (int nn)	:	*parameterised constructor to initialise the data member n=nn*
int sum_ pow(int i)	:	*returns the sum of each digit raised to the power of the length of the number using recursive technique e.g., 34 will return* $3^2 + 4^2$ *(as the length of the number is 2)*
void is Armstrong( )	:	*checks whether the given number is an Armstrong number by invoking the function sum_pow ( ) and displays the result with an appropriate message*

Specify the class **ArmNum** *giving details of the* ***constructor(), int sum_pow(int) and void is Armstrong ()****. Define a* **main( )** *function to create an object and call the functions accordingly to enable the task.*

Question 8.

Design a class **MatRev** *to reverse each element of a matrix.* **[10]**

Example :

Matrix A

72	371	5
12	6	426
5	123	94

becomes

Matrix B

27	173	5
21	6	624
5	321	49

Some of the members of the class are given below :

Class Name	:	***MatRev***
Data members/ instance variables:		
arr[] []	:	*to store integer elements*
m	:	*to store the number of rows*
n	:	*to store the number of columns*
Member functions/ methods :		
MatRev *(int mm, int nn)*	:	*parameterised constructor to initialise the data members m = mm and n = nn*
void fillarray()	:	*to enter elements in the array*
int reverse (int x)	:	*returns the reverse of the number x*
void revMat (MatRev P)	:	*reverses each element of the array of the parameterised object and stores it in the array of the current object*
void show()	:	*displays the array elements in matrix form*

Define the class ***MatRev*** *giving details of the* ***constructor(), void fillarray(), int reverse(int), void revMat(MatRev)*** *and* ***void show( ).*** *Define the* ***main( )*** *function to create objects and call the functions accordingly to enable the task.*

Question 9.

A class ***Rearrange*** *has been defined to modify a word by bringing all the vowels in the word at the beginning followed by the consonants.* **[10]**

Example : ORIGINAL becomes OIIARGNL

Some of the members of the class are given below :

Class Name	:	***Rearrange***
Data members/ instances variable:		
wrd	:	to store a word
newwrd	:	to store the rearranged word
Member functions/ methods :		
Rearrange( )	:	default constructor
void readword ( )	:	to accept the word in UPPER case
void freq_vow_ con ( )	:	finds the frequency of vowels and consonants in the word and displays them with an appropriate message
void arrange( )	:	rearranges the word by bringing the vowels at the beginning followed by consonants
void display( )	:	displays the original word along with the rearranged word

Specify the class ***Rearrange,*** *giving the details of the* ***constructor( ), void readword( ), void freq_vow _ con( ), void arrange( )*** *and* ***void display( ).*** *Define the* ***main( )*** *function to create an object and call the functions accordingly to enable the task.*

SECTION–C

Answer any ***two*** *questions.*

Each program should be written in such a way that it clearly depicts the logic of the problem stepwise. This can be achieved by using comments in the program and mnemonic names or pseudo codes for algorithms. The programs must be written in Java and the algorithms must be written in general/ standard form, wherever required / specified. (Flowcharts are ***not*** *required)*

Question 10.

A super class ***Record*** *contains names and marks of the students in two different single dimensional arrays. Define a sub class* ***Highest*** *to display the names of the students obtaining the highest mark.* **[5]**

The details of the members of both the classes are given below :

Class Name	:	***Record***
Data member/ instance variable :		
n[]	:	array to store names
m[]	:	array to store marks
size	:	to store the number of students
Member functions/ methods :		
Record(int cap)	:	parameterised constructor to initialise the data member size = cap
void readarray()	:	to enter elements in both the arrays
void display()	:	displays the array elements
Class name	:	Highest
Data members/ instance variables :		
ind	:	to store the index

Member functions/ methods :

Highest(...)	:	parameterised constructor to initialise the data members of both the classes
void find( )	:	*finds the index of the student obtaining the highest mark and assign it to 'ind'*
void display( )	:	displays the array elements along with the names and marks of the students who have obtained the highest mark

*Assume that the super class Record has been defined. Using the **concept of inheritance**, specify the class **Highest** giving the details of the **constructor(...), void find()** and **void display()**.*

The super class, main function and algorithm need NOT be written.

Question 11.

*A linear data structure enables the user to add address from rear end and remove address from front. Define a class **Diary** with the following details :*

Class Name	:	***Diary***
Data member/ instance variable :		
Q[]	:	*array to store the addresses*
size	:	*stores the maximum capacity of the array*
start	:	*to point the index of the front end*
end	:	*to point the index of the rear end*
Member functions :		
Diary (int max)	:	*constructor to initialise the data member size = max, start = 0 and end = 0*
void pushadd (String n)	:	*to add address in the diary from the rear end if possible, otherwise display the message "NO SPACE"*
String popadd ( )	:	*removes and returns the address from the front end of the diary if any, else returns "?????"*
void show()	:	*displays all the addresses in the diary*

(a) *Specify the class **Diary** giving details of the functions **void pushadd(String)** and **String popadd( )**. Assume that the other functions have been defined.* **[4]**

The main function and algorithm need NOT be written.

(b) *Name the entity used in the above data structure arrangement.* **[1]**

Question 12.

(a) *A linked list is formed from the objects of the class **Node**. The class structure of the Node is given below:* **[2]**

```
class Node
{
    int num;
    Node link;
}
```

*Write an Algorithm **OR** a Method to find and display the sum of even integers from an existing linked list.*

The method declaration is as follows :

void SumEvenNode(Node str)

(b) *Answer the following questions from the diagram of a Binary Tree given below :*

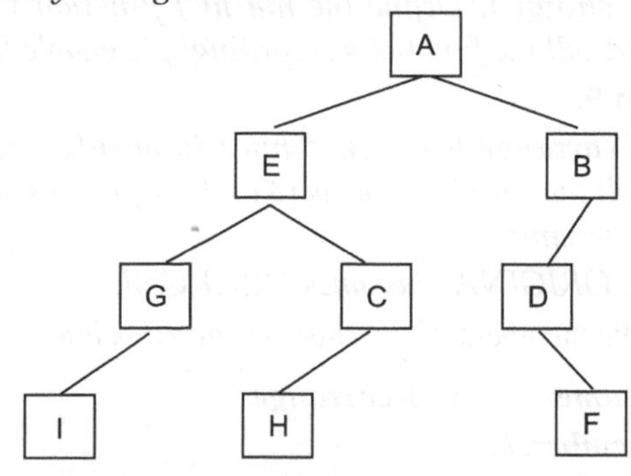

(i) *Write the pre-order traversal of the above tree structure.* **[1]**

(ii) *State the size of the tree.* **[1]**

(iii) *Name the siblings of the nodes E and G.* **[1]**

ANSWERS

PART–I

Answer 1.

(a) XOR gate :

Symbol of the gate :

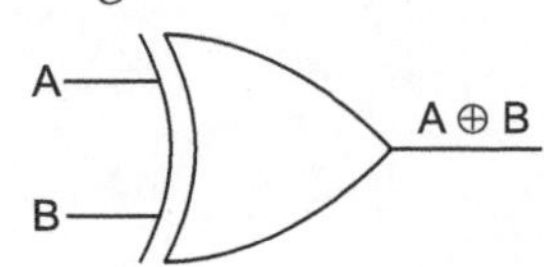

(b) The canonical POS expression is :

$$F(P, Q) = (P + Q).(P' + Q)$$

(c) Dual = (X + Y).(X + Y') = X.1

(d)

$$F' = (A'B'C' + A'BC')'$$
$$= (A'B'C')' \cdot (A'BC')'$$
$$= (A + B + C).(A + B' + C)$$

(e) **(i)** Contrapositive : (~B′ ⇒ ~A′) *i.e.,* if it is not raining then it is not cloudy.

(ii) Converse : (B ⇒ A) *i.e.,* if it is raining then it is cloudy.

Answer 2.

(a) An interface is a non-primitive data type which has static and final data members and prototype of functions (*i.e.,* functions are not defined).

Difference : Interface supports multiple inheritance whereas a class does not support multiple inheritance.

(b) **Row major order formulae :**

ARR[i] [j] = BA + W[(i – l_r)*cal + (j – l_c)]

B = 1430
W = 4
C = (8 – 3) + 1 = 6
I, J = 3, 6
I_0, J_0 = – 4, 3
= 1430 + 4 * ((3 – (– 4)) * 6 + (6 – 3))
= 1430 + 4 * (7 * 6 + 3)
= 1430 + 4 * 45
= 1430 + 180
= 1610

(c) Infix to Post fix :

= (A + B * C) – (E * F/H) + J
= ((A + (B * C)) – ((E * F)/H)) + J
= ((A + (BC*)) – (EF*)/H)) + J
= ((ABC* +) – (EF * H/)) + J
= (ABC* + EF * H/–) + J
= ABC* + EF * H/ – J +

(d)

S. No.	$O(n^2)$	$O(2^n)$
1.	It is the complexity of Quadratic time.	It is the complexity of Exponential time.
2.	It is used mostly for nested loops.	It is used mostly for tree recursion.

Since, $O(n^2)$ is less complex in comparison to $O(2^n)$, it is much better.

(e) Internal nodes are all the parent nodes except the root node in a tree whereas external nodes are the leaf nodes of the tree.

Answer 3.

Mystery(43629, 3, 4) // num = 43629, x = 3, y = 4
Z = 9
9 * 4 + Mystery(4362, 3, 4)
Mystery(4362, 3, 4) // num = 4362, x = 3, y = 4
Z = 2
2 * 3 + Mystery(436, 3, 4)
Mystery(436, 3, 4) // num = 436, x = 3, y = 4
Z = 6
6 * 3 + Mystery(43, 3, 4)
Mystery(43, 3, 4) // num = 43, x = 3, y = 4
Z = 3
3 * 4 + Mystery(4, 3, 4)
Mystery(4, 3, 4) // num = 4, x = 3, y = 4
Z = 4
36 + 6 + 18 + 12 + 4 = 76

PART–II

SECTION–A

Answer 4.

(a) **(i)** F(A, B, C, D) = Σ(0, 2, 3, 4, 5, 8, 10, 11, 12, 13)

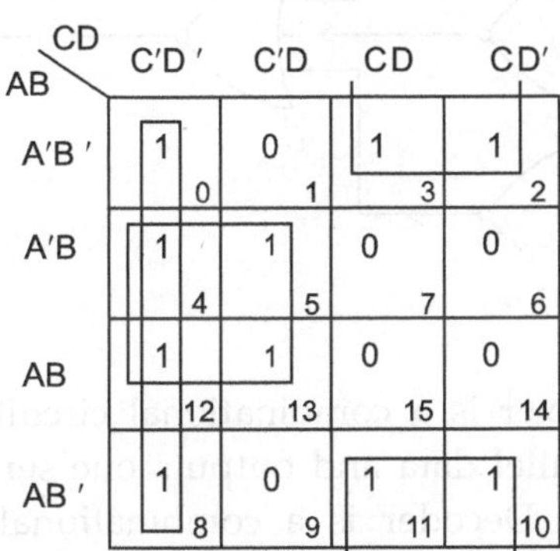

There are three quads :
Quad 1 (m0 + m4 + m8 + m12) = C′.D′
Quad 2 (m4 + m5 + m12 + m13) = B.C′
Quad 3 (m2 + m3 + m10 + m11) = B′.C
Hence, Reduced expression is :
F = C′D′ + BC′ + B′C

(ii) Logic Gate Diagram :

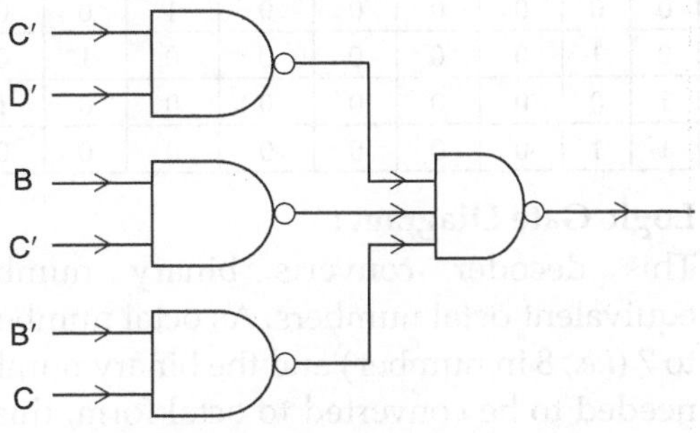

(b) (i) F(P, Q, R, S) = π(0, 1, 2, 8, 9, 11, 13, 15)

PQ \ RS	R + S	R + S′	R′ + S′	R′ + S
P + Q	0 (0)	0 (1)	1 (3)	0 (2)
P + Q′	1 (4)	1 (5)	1 (7)	1 (6)
P′ + Q′	1 (12)	0 (13)	0 (15)	1 (14)
P′ + Q	0 (8)	0 (9)	0 (11)	1 (10)

There are two quads and one pair :

Quad (M0.M1.M.8.M9) = Q + R

Quad (M9.M11.M13.M15) = P′ + S′

Pair 1 (M0, M2) = P + Q + S

Hence, Reduced expression is :

F = (Q + R).(P + Q + S).(P′ + S′)

(ii) Logic Gate Diagram :

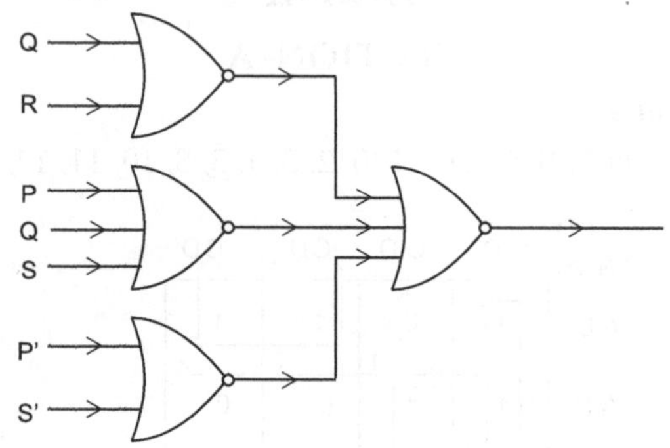

Answer 5.

(a) A multiplexer is a combinational circuit which inputs parallel data and outputs one serial data whereas, a Decoder is a combinational circuit which inputs n lines and outputs 2^n or fewer lines.

Truth table for 3 to 8 decoder :

BINARY (INPUT)			OCTAL (OUTPUT)							
B2	B1	B0	Oc0	Oc1	Oc2	Oc3	Oc4	Oc5	Oc6	Oc7
0	0	0	1	0	0	0	0	0	0	0
0	0	1	0	1	0	0	0	0	0	0
0	1	0	0	0	1	0	0	0	0	0
0	1	1	0	0	0	1	0	0	0	0
1	0	0	0	0	0	0	1	0	0	0
1	0	1	0	0	0	0	0	1	0	0
1	1	0	0	0	0	0	0	0	1	0
1	1	1	0	0	0	0	0	0	0	1

Logic Gate Diagram :

This decoder converts binary numbers to equivalent octal numbers. As octal numbers are 0 to 7 (*i.e.*, 8 in number) and the binary numbers are needed to be converted to octal form, this circuit requires 3 input lines and 8 output lines. Thus, this circuit is also called 3 to 8 decoder circuit.

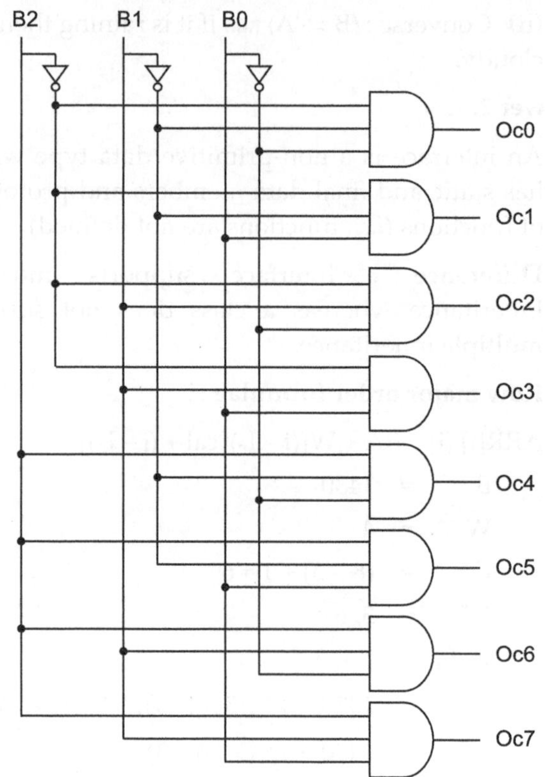

(b) Expression :

$$\begin{aligned} F(X, Y, Z) &= (X + Y).(X.Z) \\ &= X.X.Z + X.Y.Z \\ &= X.Z + X.Y.Z \\ &= X.Z.(1 + Y) \\ &= X.Z \end{aligned}$$

X → Z → [AND] → X.Z

It represents an AND gate.

(c) Truth table :

P	Q	P′	P ⇒ Q	~ (P ⇒ Q)	~ P ∨ Q	~ (P ⇒ Q) ⇔ (~ P/Q)
0	0	1	1	0	1	0
0	1	1	1	0	1	0
1	0	0	0	1	0	0
1	1	0	1	0	1	0

Thus, the proposition is a contradiction.

Answer 6.

(a)

O	F	H	U	X	Max Terms
0	0	0	0	0	O + F + H + U
0	0	0	1	0	O + F + H + U′
0	0	1	0	0	O + F + H′ + U

0	0	1	1	0	O + F + H′ + U′
0	1	0	0	0	O + F′ + H + U
0	1	0	1	1	
0	1	1	0	1	
0	1	1	1	1	
1	0	0	0	1	
1	0	0	1	1	
1	0	1	0	0	O′ + F + H′ + U
1	0	1	1	0	O′ + F + H′ + U′
1	1	0	0	1	
1	1	0	1	1	
1	1	1	0	1	
1	1	1	1	1	

POS Expression :

X (O, F, H, U) = (O + F + H + U).
(O + F + H + U′).(O + F + H′ + U).
(O + F + H′ + U′).(O + F′ + H + U).
(O′ + F + H′ + U).(O′ + F + H′ + U′)

(b) It is a logic circuit that adds two binary bits. It produces the output as SUM and CARRY.

Truth table for half adder :

A	B	SUM	CARRY	Minterm
0	0	0	0	A′.B′
0	1	1	0	A′.B
1	0	1	0	A.B′
1	1	0	1	A.B

From the truth table, now we can derive the Boolean expression as :

SUM = A′.B + A.B′ = A ⊕ B

CARRY = A.B

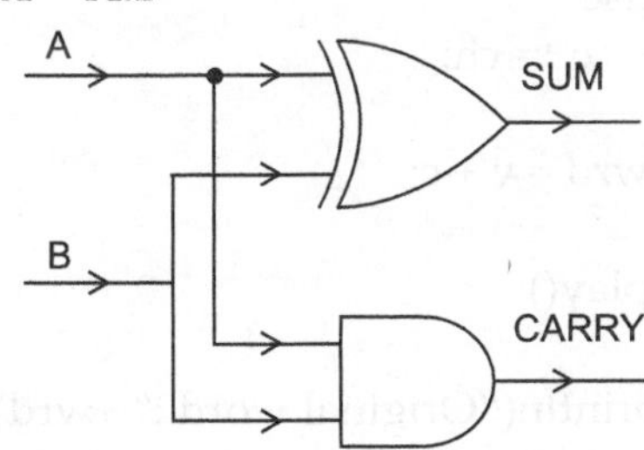

(c) Simplify :

(X + Z).(X.Y + Y.Z′) + X.Z + Y

= X.X.Y + X.Y.Z′ + X.Y.Z + Y.Z.Z′ + X.Z + Y

= X.Y + X.Y.Z′ + X.Y.Z + X.Z + Y

(X.X = X and Z.Z′ = 0, Idempotent law and Complement law)

= X.Y (1 + Z′ + Z) + X.Z + Y

(1 + Z + Z′ = 1, Property of 1)

= (X.Y + Y) + X.Z

(Absorption law, A + A.B = A)

= Y + X.Z

SECTION–B

Answer 7.

```
import java.util.Scanner;
class ArmNum
{
    int n;
    int 1;
    ArmNum(int nn)
    {
        n = nn;
        1 = 0
    }
    int sum_pow(int i)
    {
    if(i > 0)
    return(int)Math.pow(i % 10, 1) + sum_pow(i/10);
        else
            return 0;
    }
    void isArmstrong()
    {
        for(int i = n; i > 0; i = 1/10)
            ++l;
        int sum = sum_pow(n);
        if(sum == n)
        System.out.println("Armstrong no.");
        else
        System.out.println("Not an Armstrong no.");
    }
    public static void main(String ar[])
    {
        ArmNum obj = new ArmNum(371);
        obj.isArmstrong();
    }
}
```

Answer 8.

```
import java.util.Scanner;
class MatRev
{
    int arr[][];
    int m, n;
    MatRev(int mm, int nn)
    {
        m = mm;
        n = nn;
            arr = new int[m][n];
        }
        void fillarray()
        {
            Scanner sc = new Scanner(System.in);
            for(int i = 0; i < m; i++)
```

```
            for(int j = 0; j < n; j++)
            {
                System.out.println("Enter a no :");
                arr[i][j] = sc.nextInt();
            }
        }
    int reverse(int x)
    {
        int r = 0;
        while(x > 0)
        {
            r = r * 10 + x % 10;
            x = x / 10;
        }
        return r;
    }
    void revMat(MatRev P)
    {
        for(int i = 0; i < m; i++)
            for(int j = 0; j < n; j++)
                this.arr[i][j] = reverse(P.arr[i][j]);
    }
    void show()
    {
        for(int i = 0; i < m; i++)
        {
            for(int j = 0; j < n; j++)
            {
                System.out.println(arr[i][j] + "");
            }
            System.out.println():
        }
    }
    public static void main(String ar[])
    {
        MatRev ob = new MatRev(4, 4);
        ob.fillarray();
        MatRev xy = new MatRev(4, 4);
        xy.revMat(ob);
        xy.show();
    }
}
```

Answer 9.

```
import java.util.Scanner;
class Rearrange
{
    String wrd, newwrd;
    Rearrange()
    {
        wrd = newwrd = "";
    }
    void readword()
    {
    Scanner sc = new Scanner(System.in);
    System.out.ptintln("Accept a word in UPPER case");
        wrd = sc.next();
    }
    void freq_vow_con()
    {
        int vc = 0, cc = 0;
        int l = wrd.length();
        for(int i = 0; i < l; i++)
        {
    char ch = wrd.charAt(i);
    if(ch == 'A' || ch == 'E' || ch == 'I' ||
                                ch == 'O' || ch == 'U')
            vc++;
        else
            cc++;
        }
    System.out.println("Frequency of vowel :" + vc);
    System.out.println("Frequency of consonant :" +cc);
    }
    void arrange()
    {
        String v = "", c = "";
        int l = wrd.length();
        for(int i = 0; i < l; i++)
        {
char ch = wrd.charAt(i);
if(ch == 'A' || ch == 'E' || ch == 'I' || ch == 'O' || ch ==
                                                        'U')
            v += ch;
        else
            c += ch;
        }
        newwrd = v + c;
    }
    void display()
    {
System.out.println("Original word :" +wrd);
System.out.prtintln("Rearranged word :" +newwrd);
    }
    public static void main(String ar[])
    {
        Rearrange obj = new Rearrange();
        obj.readword();
        obj.freq_vow_con();
        obj.arrange();
        obj.display();
    }
}
```

SECTION–C

Answer 10.

```
class Highest extends Record
{
    int ind;
    Highest(int a)
    {
        super(a);
        ind = – 1;
    }
    void find()
    {
        int high = 0;
        for(int i = 0; i < size; i++)
        {
            if(m[i] > high)
            {
                high = m[i];
                ind = 1;
            }
        }
    }
    void display()
    {
        super.display();
        System.out.println(n[ind] + "got the
                        highest marks" + m[ind]);
    }
}
```

Answer 11.

(a) class Diary

```
{
    String Q[];
    int size, start, end;
    Diary(int max)
    {
        size = max;
        start = end = 0;
        Q = new int[size];
    }
    void pushadd(String n)
    {
        if(end < size – 1)
            Q[end++] = n;
        else
            System.out.println("NO SPACE);
    }
    String popadd()
    {
        if(start == end)
        {
            start = end = 0;
            return "?????";
        }
        else
            return Q[front++];
    }
}
```

(b) The entity being used in the above data structure arrangement is a Queue.

Answer 12.

(a) Algorithm :

Step 1 : Start

Step 2 : Declare a temporary node, p←str and declare c←0

Step 3 : Repeat steps 4 and 5 until P reaches null

Step 4 : If p.num % 2 = 0 then c←c + p.num

Step 5 : move p to next node

Step 6 : Print "Sum of even integers" is c

Step 7 : End

Method :

```
void SumEvenNode(Node str)
{
Node p = str;
int c = 0;
while(p != null)
{
if(p.num%2 == 0)
c += p.num;
p = p.next;
}
System.out.println("Sum of even integers :" +c);
}
```

(b) **(i)** A E G I C H B D F

(ii) 9

(iii) Sibling of node E is B and sibling of node G is C.

COMPUTER SCIENCE 2018

QUESTIONS

PART–I (20 Marks)

*Answer **all** questions.*

While answering questions in this Part, indicate briefly your working and reasoning, wherever required.

Question 1.

(a) *State the Commutative law and prove it with the help of a truth table.* **[1]**

(b) *Convert the following expression into its canonical POS form :* **[1]**

$$F(X, Y, Z) = (X + Y') \cdot (Y' + Z)$$

(c) *Find the dual of :* **[1]**

$$(A' + B) \cdot (1 + B') = A' + B$$

(d) *Verify the following proposition with the help of a truth tablé :* **[1]**

$$(P \wedge Q) \vee (P \wedge \sim Q) = P$$

(e) *If $F(A, B, C) = A'(BC' + B'C)$, then find F'.* **[1]**

Question 2.

(a) *What are Wrapper classes ? Give any two examples.* **[2]**

(b) *A matrix A[m][m] is stored in the memory with each element requiring 4 bytes of storage. If the base address at A[1][1] is 1500 and the address of A[4][5] is 1608, determine the order of the matrix when it is stored in* ***Column Major Wise.*** **[2]**

(c) *Convert the following infix notation to postfix form :* **[2]**

$$A + (B - C * (D/E) * F)$$

(d) *Define Big 'O' notation. State the two factors which determine the complexity of an algorithm.* **[2]**

(e) *What is exceptional handling ? Also, state the purpose of finally block in a try catch statement.* **[2]**

Question 3.

The following is a function of some class which checks if a positive integer is a Palindrome number by returning true or false. (A number is said to be palindrome if the reverse of the number is equal to the original number). The function does not use modulus (%) operator to extract digit. There are some places in the code marked by ?1?, ?2?, ?3?, ?4?, ?5?, which may be replaced by a statement expression so that the function works properly. **[5]**

```
boolean PalindromeNum(int N)
{
    int rev=?1?;
    int num=N;
    while(num>0)
    {
        int f=num/10;
        int s=?2?;
        int digit = num–?3?;
        rev = ?4? + digit;
        num/=?5?;
    }
    if(rev==N)
        return true;
    else
        return false;
}
```

(i) *What is the statement or expression at ?1?* **[1]**

(ii) *What is the statement or expression at ?2?* **[1]**

(iii) *What is the statement or expression at ?3?* **[1]**

(iv) *What is the statement or expression at ?4?* **[1]**

(v) *What is the statement or expression at ?5?* **[1]**

PART–II (50 Marks)

*Answer **six** questions in this part, choosing **two** questions from Section A, **two** from Section B and **two** from Section C.*

SECTION–A

*Answer **any two** questions.*

Question 4.

(a) *Given the Boolean function:*

$$F(A, B, C, D) = \Sigma\ (0, 2, 4, 8, 9, 10, 12, 13).$$

(i) *Reduce the above expression by using a 4-variable Karnaugh map, showing the various groups (i.e., octal, quads and pairs).* **[4]**

(ii) *Draw the logic gate diagram for the reduced expression. Assume that the variables and their complements are available as inputs.* **[1]**

(b) *Given the Boolean function:*

$$F(A, B, C, D) = \pi\ (3, 4, 5, 6, 7, 10, 11, 14, 15)$$

(i) *Reduce the above expression by using a 4-variable Karnaugh map, showing the various groups (i.e. octal, quads and pairs).* **[4]**

(ii) *Draw the logic gate diagram for the reduced expression. Assume that the variables and their complements are available as inputs.* **[1]**

Question 5.

(a) *A training institute intends to give scholarship to its students as per the criteria given below:* **[5]**

- *The student has excellent academic record but is financially weak.*

OR

- *The student does not have an excellent academic record and belongs to a backward class.*

OR

- *The student does not have an excellent academic record and is physically impaired.*

The inputs are :

Inputs	
A	Has excellent academic record
F	Financially sound
C	Belongs to a backward class
I	Is physically impaired

(In all the above cases 1 indicates yes and 0 indicates no).

Output : X [1 indicates yes, 0 indicates no for all cases].

Draw the truth table for the inputs and outputs given above and write the SOP expression for X (A, F, C, I).

(b) *Using the truth table, state whether the following proposition is a tautology, contingency or a contradiction :* **[3]**

$$\sim (A \wedge B)\ V\ (\sim A \Rightarrow B)$$

(c) *Simplify the following expression, using Boolean laws:* **[2]**

$$A\ .\ (A' + B)\ .\ C\ .\ (A + B)$$

Question 6.

(a) *What is an Encoder ? Draw the Encoder circuit to convert A-F hexadecimal numbers to binary. State an application of a Multiplexer.* **[5]**

(b) *Differentiate between Half Adder and Full Adder. Draw the logic circuit diagram for a Full Adder.* **[3]**

(c) *Using only NAND gates, draw the logic circuit diagram for A' + B.* **[2]**

SECTION–B

*Answer **any two** questions.*

Each program should be written in such a way that it clearly depicts the logic of the problem.

This can be achieved by using mnemonic names and comments in the program.

*(Flowcharts and Algorithms are **not** required.)*

The programs must be written in Java.

Question 7.

*Design a class **Perfect** to check if a given number is a perfect number or not. [A number is said to be perfect if sum of the factors of the number excluding itself is equal to the original number.]* **[10]**

Example : 6 = 1 + 2 + 3 (where 1, 2 and 3 are factors of 6, excluding itself)

Some of the members of the class are given below :

Class Name	:	***Perfect***
Data members/ instance variables:		
Num	:	*to store the number*
Perfect (int nn)	:	*parameterised constructor to initialise the data member num = nn*
int sum_of_factors(int i)	:	*return the sum of the factors of the number (num), excluding itself, using recursive technique*
void check( )	:	*checks whether the given number is perfect by invoking the function sum_of_factors( ) and displays the result with an appropriate message.*

*Specify the class **Perfect** giving details of the **constructor()**, **int sum_of_factors (int)** and **void check()**. Define a **main()** function to create an object and call the functions accordingly to enable the task.*

Question 8.

Two matrices are said to be equal if they have the same dimension and their corresponding elements are equal. **[10]**

For example : the two matrices A and B given below are equal :

Matrix A

1	2	3
2	4	5
3	5	6

Matrix B

1	2	3
2	4	5
3	5	6

*Design a class **EqMat** to check if two matrices are equal or not. Assume that the two matrices have the same dimension.*

Some of the members of the class are given below :

Class Name	:	***EqMat***
Data members/ instance variables:		
a[ ] [ ]	:	*to store integer elements*
m	:	*to store the number of rows*
n	:	*to store the number of columns*
Member functions/ methods :		
EqMat(int mm, int nn)	:	*parameterised constructor to initialise the data members m = mm and n = nn*
void readarray()	:	*to enter elements in the array*

int check(EqMat P, EqMat Q)	:	*checks if the parameterised objects P and Q are equal and returns 1 if true, otherwise returns 0*
void print()	:	*displays the array elements*

*Define the class **EqMat** giving details of the **constructor(), void readarray(), int check(EqMat, EqMat) and void print()**. Define the **main()** function to create objects and call the functions accordingly to enable the task.*

Question 9.

*A class **Capital** has been defined to check whether a sentence has words beginning with a capital letter or not.* **[10]**

Some of the members of the class are given below :

Class Name	:	***Capital***
Data members/ instances variable:		
sent	:	*to store a sentence*
freq	:	*stores the frequency of words beginning with a capital letter*
Member functions/ methods :		
Capital( )	:	*default constructor*
void input( )	:	*to accept the sentence*
boolean isCap (String w)	:	*checks and returns true if word begins with a capital letter, otherwise returns false*
void display( )	:	*displays the sentence along with the frequency of the words beginning with a capital letter*

*Specify the class **Capital,** giving the details of the **constructor(), void input(), boolean isCap(String)** and **void display()**. Define the **main()** function to create an object and call the functions accordingly to enable the task.*

SECTION–C

*Answer any **two** questions.*

Each program should be written in such a way that it clearly depicts the logic of the problem stepwise.

This can be achieved by using comments in the program and mnemonic names or pseudo codes for algorithms. The programs must be written in Java and the algorithms must be written in general/ standard form, wherever required / specified.

*(Flowcharts are **not** required)*

Question 10.

*A super class **Number** is defined to calculate the factorial of a number. Define a sub-class **Series** to find the sum of the series S = 1! +2! + 3! + 4! + …… + n!* **[5]**

The details of the members of both the classes are given below :

Class Name	:	***Number***
Data member/ instance variable:		
n	:	*to store an integer number*
Member functions/ methods :		
Number(int nn)	:	*parameterised constructor to initialise the data member n=nn*
int factorial(int a)	:	*returns the factorial of a number (factorial of n = 1 × 2 × 3 × …….. × n)*
void display()	:	*displays the data members*
Class name	:	***Series***
Data member/ instance variable :		
Sun	:	*to store the sum of the series*
Member functions / methods :		
Series(…)	:	*parameterised constructor to initialise the data members of both the classes*
void calsum( )	:	*calculates the sum of the given series*
void display( )	:	*displays the data members of both the classes*

*Assume that the super class **Number** has been defined. Using the **concept of inheritance,** specify the class **Series** giving the details of the **constructor(…), void calsum()** and **void display()**.*

The super class, main function and algorithm need NOT be written.

Question 11.

***Register** is an entity which can hold a maximum of 100 names. The register enables the user to add and remove names from the top most end only.*

*Define a class **Register** with the following details :*

Class Name	:	***Register***
Data members/ instance variables:		
stud[ ]	:	*array to store the names of the students*
cap	:	*stores the maximum capacity of the array*
top	:	*to point the index of the top end*

Member functions :

Register (int max)	:	*constructor to initialise the data member cap = max, top = –1 and create the string array*
void push (String n)	:	*to add names in the register at the top location if possible, otherwise display the message "OVERFLOW"*
String pop( )	:	*removes and returns the names from the top most location of the register if any, else returns "$$"*
void display( )	:	*displays all the names in the register*

(a) *Specify the class* ***Register*** *giving details of the functions* ***void push (String) and String pop()****. Assume that the other functions have been defined.* **[4]**

The main function and algorithm need NOT be written.

(b) *Name the entity used in the above data structure arrangement.* **[1]**

Question 12.

(a) *A linked list is formed from the objects of the class* ***Node****. The class structure of the Node is given below:* **[2]**

```
class Node
{
    int n;
    Node link;
}
```

Write an Algorithm ***OR*** *a Method to search for a number from an existing linked list.*

The method declaration is as follows :

void FindNode(Node str, int b)

(b) *Answer the following questions from the diagram of a Binary Tree given below :*

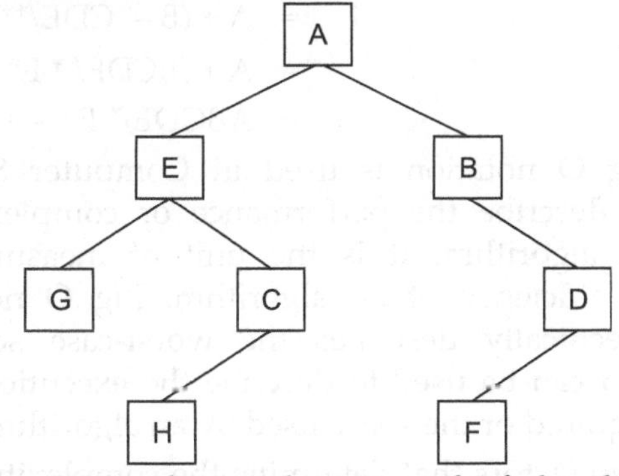

(i) *Write the inorder traversal of the above tree structure.* **[1]**

(ii) *State the height of the tree, if the root is at level 0 (zero).* **[1]**

(iii) *List the leaf nodes of the tree.* **[1]**

ANSWERS

PART–I

Answer 1.

(a) Commutative law states that :

1. X + Y = Y + X

2. X . Y = Y . X

Prooving with logic table :

X	Y	X+Y	Y+X	X.Y	Y.X
0	0	0	0	0	0
0	1	1	1	0	0
1	0	1	1	0	0
1	1	1	1	1	1

(b) $F(X, Y, Z) = (X + Y') . (Y' + Z)$

$= (X + Y' + 0) . (0 + Y' + Z)$

$= (X + Y' + Z.Z'). (X.X' + Y' + Z)$

$= (X + Y' + Z) . (X + Y' + Z') . (X + Y' + Z) . (X' + Y' + Z)$

$= (X + Y' + Z) . (X + Y' + Z') . (X' + Y' + Z)$

(c) Equation : $(A' + B) \cdot (1 + B') = A'B$

Dual : $A'. B + 0 . B' = A'. B$

(d) Verify : $(P \wedge Q) \vee (P \wedge \sim Q) = P$

P	Q	P∧Q	~Q	P∧~Q	(P∧Q) ∨ (P∧~Q)
0	0	0	1	0	0
0	1	0	0	0	0
1	0	0	1	1	1
1	1	1	0	0	1

(e) $F(A,B,C) = A'. (BC' + B'C)$

$F' = (A'. (BC' + B'C))'$

$= (A')' + (BC' + B'C)'$

$= A + (B.C')' . (B'.C)'$

$= A + (B' + C) . (B + C')$

Answer 2.

(a) A Wrapper class is a class whose object wraps or contains a primitive data type. When we create an object to a wrapper class, it contains a field and in this field, we can store a primitive data type. In other words, we can wrap a primitive value into a wrapper class object.

Two examples of wrapper classes : Integer for int data type, Character for char data type.

(b) Formula for the address in Column Major Wise :

$A = B + W \times ((I - I_0) + R \times (J - J_0))$

$A = 1608; B = 1500; I_0, J_0 = 1, 1; I, J = 4, 5; W = 4$

$1608 = 1500 + 4 \times ((4 - 1) + m \times (5 - 1))$

$1608 = 1500 + 4 \times (3 + 4m)$

$1608 = 1500 + 12 + 16m$

$16m = 1608 - (1500 + 12)$

$16m = 96$

$m = 6$

(c) Infix to Postfix :

A + (B – C * (D/E) * F) = A + (B – C * (DE/) * F)

= A + (B – CDE/**F)

= A + (BCDE/ * F* –)

= ABCDE/* F * – +)

(d) Big O notation is used in Computer Science to describe the performance or complexity of an algorithm. It is the unit of measurement of efficiency of an algorithm. Big O notation specifically describes the worst-case scenario and can be used to describe the execution time required or the space used by an algorithm.

Two factors that determine the complexity of an algorithm are Space and Time.

(e) Exception handling is the process of handling run time error during the execution of a program. Exception Handling in java is one of the most powerful mechanisms to handle the runtime errors so that normal flow of the application can be maintained.

Purpose of *finally* block in a try catch statement is to execute the statements placed in the finally block irrespective of the errors occurring. It's a good practice to use finally to handle program crashes.

Answer 3.

(i) 0 **(ii)** f * 10 **(iii)** s

(iv) rev * 10 **(v)** ´10

PART–II

SECTION–A

Answer 4.

(a) (i) F (A,B,C,D) = Σ (0, 2, 4, 8, 9, 10, 12, 13)

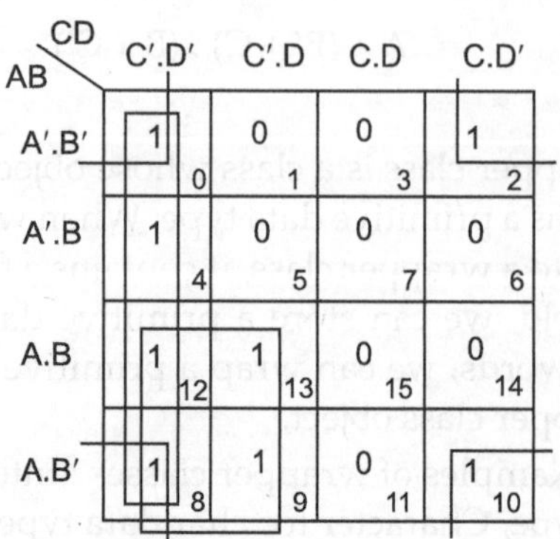

There are three quads :

Quad 1 (m0 + m2 + m8 + m10) = B′.D′

Quad 2 (m0 + m4 + m8 + m12) = C′.D′

Quad 3 (m8 + m9 + m12 + m13) = A.C′

Reduced expression is : F(A,B,C,D) = B′D′ + C′D′ + AC′

(ii) Logic Gate Diagram :

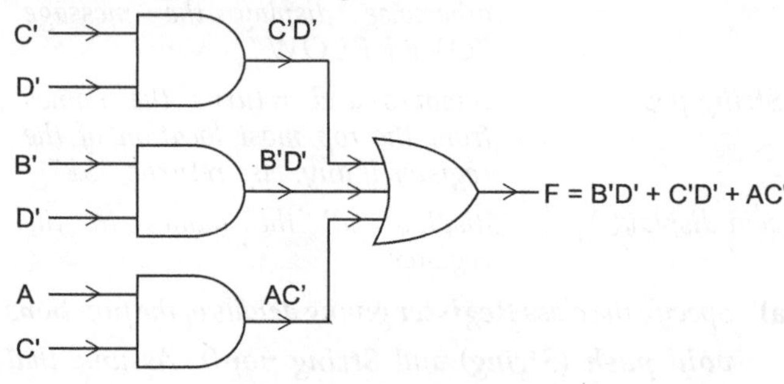

(b) (i) F(A,B,C,D) = π (3, 4, 5, 6, 7, 10, 11, 14, 15)

AB \ CD	C + D	C + D′	C′ + D′	C′ + D
A + B	1 (0)	1 (1)	0 (3)	1 (2)
A + B′	0 (4)	0 (5)	0 (7)	0 (6)
A′ + B′	1 (12)	1 (13)	0 (15)	0 (14)
A′ + B	1 (8)	1 (9)	0 (11)	0 (10)

There are three quads :

Quad 1 (M4 . M5 . M6 . M7) = A + B′

Quad 2 (M3 . M7 . M11 . M15) = C′+ D′

Quad 3 (M10 . M11 . M14 . M15) = A′+ C′

Reduced expression is :

F(A, B, C, D) = (A + B′) . (C′+ D′) . (A′ + C′)

(ii) Logic Gate Diagram :

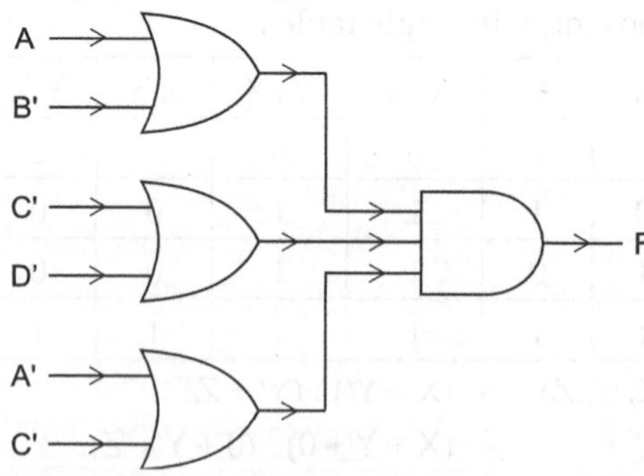

Answer 5.

(a) Truth table :

A	F	C	I	X	Minterm	Designation
0	0	0	0	0		0
0	0	0	1	1	A′.F′.C′.I	1
0	0	1	0	1	A′.F′.C.I′	2

0	0	1	1	1	A′.F′.C.I	3
0	1	0	0	0		4
0	1	0	1	1	A′.F.C′.I	5
0	1	1	0	1	A′.F.C.I′	6
0	1	1	1	1	A′.F.C.I	7
1	0	0	0	1	A.F′.C′.I′	8
1	0	0	1	1	A.F′.C′.I	9
1	0	1	0	1	A.F′.C.I′	10
1	0	1	1	1	A.F′.C.I	11
1	1	0	0	0		12
1	1	0	1	0		13
1	1	1	0	0		14
1	1	1	1	0		15

SOP Expression :

$$X(A,F,C,I) = A'F'C'I + A'F'CI' + A'F'CI + A'FC'I + A'FCI' + A'FCI + AF'C'I' + AF'C'I + AF'CI' + AF'CI$$

$$= \Sigma(1, 2, 3, 5, 6, 7, 8, 9, 10, 11)$$

(b) $\sim(A \wedge B) \vee (\sim A \Rightarrow B)$ is a tautology or contradiction.

A	B	A∧B	~(A∧B)	~ A	~ A⇒B	~ (A∧B) ∨ (~ A ⇒ B)
0	0	0	1	1	0	1
0	1	0	1	1	1	1
1	0	0	1	0	1	1
1	1	1	0	0	1	1

Hence, it is a tautology.

(c) Simplify :

$= (A.A' + A.B) . C.(A + B)$

$= (0 + A.B). (A.C + B.C)$ [by complement's law]

$= A.B.A.C + A.B.B.C$

$= A.B.C + A.B.C$ [by idempotent law]

$= ABC$ [by idempotent law]

Answer 6.

(a) An encoder is a combinational circuit that converts input in one form to another form. It is a combinational circuit which inputs 2^n or fewer lines and outputs n lines. Generally, encoders are used to convert decimal or octal or hexadecimal inputs into their equivalent binary form.

Encoder circuit to convert A-F hexadecimal inputs into their binary equivalent :

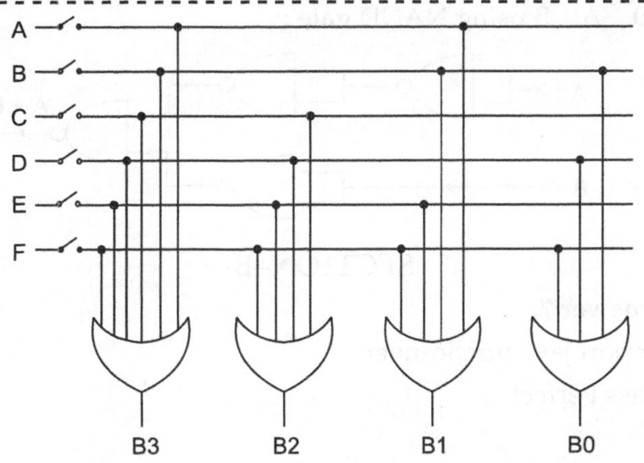

Application of Multiplexer: Multiplexers are used in various fields where multiple data needs to be transmitted using a single line such as routing of signals, data transmission, telephone exchanges, TV, etc.

(b) Half adder adds two input binary digits and outputs two binary bits, whereas Full adder adds three input binary bits and outputs two binary bits.

Truth table of a Full adder :

A	B	C	S_{out}	C_{out}
0	0	0	0	0
0	0	1	1	0
0	1	0	1	0
0	1	1	0	1
1	0	0	1	0
1	0	1	0	1
1	1	0	0	1
1	1	1	1	1

SUM $= A'B'C + A'BC' + AB'C' + ABC$

$= A \oplus B \oplus C$

CARRY $= A'BC + AB'C + ABC' + ABC$

$= AB + BC + AC$

Logic gate diagram of a Full adder :

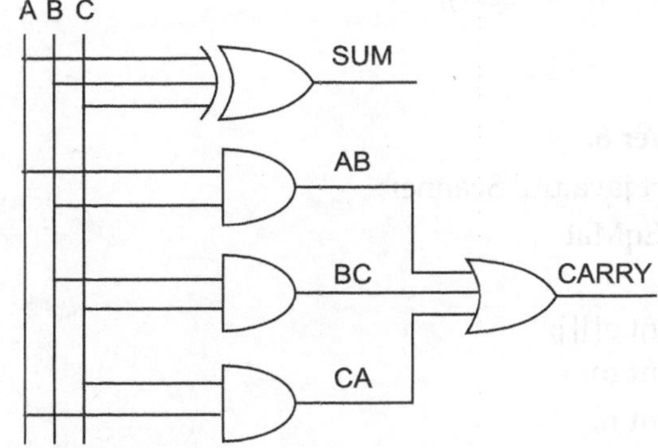

(c) A'+ B using NAND gate :

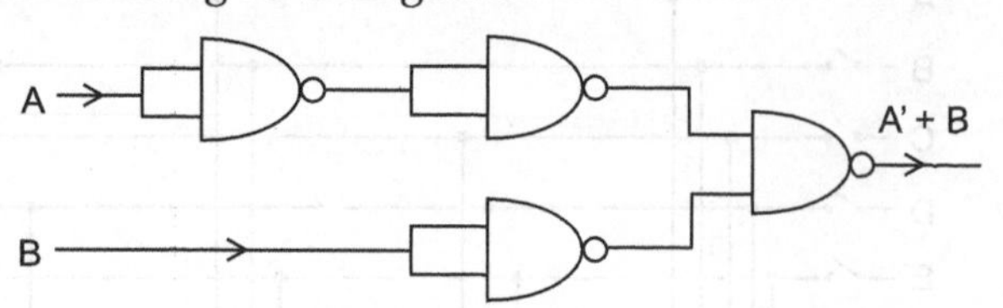

SECTION–B

Answer 7.

```
import java.util.Scanner;
class Perfect
{
    int num;
    public Perfect(int nn)
    {
        num = nn;
    }
    public int sum_of_factors(int i)
    {
        if(i == num)
            return 0;
        else if(n % i == 0)
            return i + sum_of_factors(i + 1);
        else
            return sum_of_factors(i + 1);
    }
    void check()
    {
        int s = sum_of_factors(1);
        if(s == num)
System.out.println(num+" is a perfect number");
        else
System.out.println(num+"is not a perfect number");
    }
    public static void main(String args[])
    {
        Scanner sc = new Scanner(System.in);
        System.out.println("Enter a number");
        int n = sc.nextInt();
        Perfect obj = new Perfect(n);
        obj.check();
    }
}
```

Answer 8.

```
import java.util.Scanner;
class EqMat
{
    int a[][];
    int m;
    int n;
    static Scanner SC = new Scanner(System.in);
    public EqMat(int mm, int nn)
    {
        m = mm;
        n = nn;
        a = new int[m][n];
    }
    public void readarray()
    {
        Scanner sc = new Scanner(System.in);
        System.out.println("Enter the elements
                                    for the array");
        for(int i = 0; i < m; i++)
        {
            for(int j = 0; j < n; j++)
            {
                a[i][j] = sc.nextInt();
            }
        }
    }
    public int check(EqMat P, EqMat Q)
    {
        for(int i = 0; i < P.m; i++)
        {
            for(int j = 0; j < P.n; j++)
            {
                if(P.a[i][j] != Q.a[i][j])
                    return 0;
            }
        }
        return 1;
    }
    public void print()
    {
        for(int i = 0; i < m; i++)
        {
            System.out.println();
            for(int j = 0; j < n; j++)
                System.out.print(a[i][j]+" " );
        }
    }
    public static void main(String args[])
    {
        System.out.println("Enter the row
                    and column size for the array");
        int a = sc.nextInt();
        int b = sc.nextInt();
        EqMat A = new EqMat(a,b);
```

```
    EqMat B = new EqMat(a,b);
    A.readarray();
    B.readarray();
    A.print();
    B.print();
    if(A.check(A, B) == 1)
    System.out.println("Both are equal matrix");
    else
    System.out.println("Both are unequal matrix");
    }
}
```

Answer 9.

```
import java.util.Scanner;
class Capital
{
    String sent;
    int freq;
    public Capital()
    {
        sent = "";
        freq = 0;
    }
    public void input()
    {
        Scanner sc = new Scanner(System.in);
        System.out.println("Enter a string");
        sent = sc.nextLine();
    }
    public boolean isCap(String w)
    {
        if(Character.isUpperCase(w.charAt(0)))
            return true;
        else
            return false;
    }
    public void display()
    {
        System.out.println("Given string :"+sent);
        sent += "";
        int len = sent.length();
        String word = "";
        for(int i = 0; i < len; i++)
        {
            char c = sent.charAt(i);
            if(c != ' ')
                word + = c;
            else
            {
                if(isCap(word))
                    freq ++;
                word = "";
            }
        }
        System.out.println("Number of words
            beginning with capital letters :"+freq);
    }
    public static void main(String args[])
    {
        Capital ob = new Capital();
        ob.input();
        ob.display();
    }
}
```

SECTION–C

Answer 10.

```
class Series extends Number
{
    int sum;
    public Series(int a)
    {
        super(a);
        sum = 0;
    }
    public void calsum()
    {
        for(int i = 1; i <= n; i++)
            sum += super.factorial(i);
    }
    public void display()
    {
        super.display();
        System.out.println(sum);
    }
}
```

Answer 11.

(a)

```
class Register
{
    String stud[];
    int cap;
    int top;
    Register(int max)
    {
        cap = max;
        top = -1;
        stud = new String[cap];
    }
    public void push(String n)
    {
```

```
        top++;
        if(top == cap)
        {
            System.out.println ("OVERFLOW");
            top--;
        }
        else
            stud[top] = n;
    }
    public String pop()
    {
        if(top == -1)
          return "$$";
        else
          return stud[top--];
    }
}
```

(b) Stack is used in the above data structure arrangement.

Answer 12.

(a) Algorithm :

Step 1: Start

Step 2: Declare a temporary node, t ← str

Step 3: If str = null then Print "Empty list" and return

Step 4: Repeat steps 5 to 6 while t ! = null

Step 5: If t.n = b then Print "Number matched" and quit the loop

Step 6: move t to next node

Step 7: if t = null then Print "Number not matched"

Step 8: End

Method definition :

```
void FindNode(Node str, int b)
{
    Node t = str;
    if(str == null)
    {
    System.out.println("Empty list");
    }
    while(t! = null)
    {
    if(t.n == b)
    {
     System.out.println("Number matched");
    }
        break;
    }
    t = t.link;
    if(t == null)
    System.out.println("Number not matched");
}
```

(b) **(i)** Inorder traversal : G, E, H, C, A, B, F, D

(ii) 3

(iii) G, H and F

●●

COMPUTER SCIENCE 2017

QUESTIONS

PART–I (20 Marks)

*Answer **all** questions.*

While answering questions in this Part, indicate briefly your working and reasoning, wherever required.

Question 1.

(a) *State the law represented by the following proposition and prove it with the help of a truth table :* **[1]**

$$P \vee P = P$$

(b) *State the Principle of Duality.* **[1]**

(c) *Find the complement of the following Boolean expression using De Morgan's law :* **[1]**

$$F(a, b, c) = (b' + c) + a$$

(d) *Draw the logic diagram and truth table for a 2 input XNOR gate.* **[1]**

(e) *If (~ P => Q) then write its :* **[1]**

(i) *Inverse* **(ii)** *Converse*

Question 2.

(a) *What is an interface ? How is it different from a class?* **[2]**

(b) *Convert the following infix expression to postfix form:* **[2]**

$$P * Q/R + (S + T)$$

(c) *A matrix P [15] [10] is stored with each element requiring 8 bytes of storage. If the base address at P[0] [0] is 1400, determine the address at P [10][7] when the matrix is stored in **Row Major Wise**.* **[2]**

(d) **(i)** *What is the worst case complexity of the following code segment :* **[2]**

```
for (int x = 1; x = a; x++)
{
    statements;
}
for (int y =1; y <= b; y++)
{
    for (int z = 1; z <= c; z++)
    {
        statement;
    }
}
```

(ii) *How would the complexity change if all the three loops went to N instead of a, b and c ?*

(e) *Differentiate between a constructor and a method of a class.* **[2]**

Question 3.

*The following function **magicfun()** is a part of some class. What will the function magicfun() return, when the value of n = 7 and n = 10, respectively ? Show the dry run / working :* **[5]**

```
int magicfun(int n)
{
    if (n == 0)
        return 0;
    else
        return magicfun(n / 2) * 10 + (n % 2);
}
```

PART–II (50 Marks)

*Answer **six** questions in this part, choosing **two** questions from Section A, **two** from Section B and **two** from Section C.*

SECTION–A

*Answer **any two** questions.*

Question 4.

(a) *Given the Boolean function :*

$$F(A, B, C, D) = \Sigma\,(2, 3, 4, 5, 6, 7, 8, 10, 11)$$

(i) *Reduce the above expression by using 4-variable Karnaugh map, showing the various groups (i.e., octal, quads and pairs).* **[4]**

(ii) *Draw the logic gate diagram for the reduced expression. Assume that the variables and their complements are available as inputs.* **[1]**

(b) *Given the Boolean function :*

$$F(P, Q, R, S) = \pi\,(0, 1, 2, 4, 5, 6, 8, 10)$$

(i) *Reduce the above expression by using 4-variable Karnaugh map, showing the various groups (i.e., octal, quads and pairs).* **[4]**

(ii) *Draw the logic gate diagram for the reduced expression. Assume that the variables and their complements are available as inputs.* **[1]**

Question 5.

(a) *A school intends to select candidates for an Inter-School Essay Competition as per the criteria given below :* **[3]**

- *The student has participated in an earlier competition and is very creative.*

OR

- *The student is very creative and has excellent general awareness, but has not participated in any competition earlier.*

OR

- *The student has excellent general awareness and has won prize in an inter-house competition.*

The inputs are :

Inputs	
A	*participated in a competition earlier*
B	*is very creative*
C	*won prize in an inter-house competition*
D	*has excellent general awareness*

(In all the above cases 1 indicates yes and 0 indicates no).

Output : X [1 indicates yes, 0 indicates no for all cases]

Draw the truth table for the inputs and outputs given above and write the POS expression for X (A, B, C, D).

(b) *State the application of a Half Adder. Draw the truth table and circuit diagram for a Half Adder.* **[4]**

(c) *Convert the following Boolean expression into its canonical POS form :* **[3]**

$$F(A,B,C) = (B + C')\cdot(A' + B)$$

Question 6.

(a) *What is a Multiplexer ? How is it different from a decoder ? Draw the circuit diagram for a 8 : 1 Multiplexer.* **[5]**

(b) *Prove the Boolean expression using Boolean laws. Also, mention the law used at each step.* **[3]**

$$F = (x' + z) + [(y' + z)\cdot(x' + y)]' = 1$$

(c) *Define maxterms and minterms. Find the maxterm and minterm when :* **[2]**

$$P = 0, Q = 1, R = 1 \text{ and } S = 0$$

SECTION–B

Answer ***any two*** *questions.*

Each program should be written in such a way that it clearly depicts the logic of the problem.

This can be achieved by using mnemonic names and comments in the program.

(Flowcharts and Algorithms are ***not*** *required.)*

The programs must be written in Java.

Question 7.

A class **Palin** *has been defined to check whether a positive number is a Palindrome number or not.* **[10]**

The number 'N' is palindrome if the original number and its reverse are same. Some of the members of the class are given below :

Class Name	:	***Palin***
Data members/ instance variables:		
num	:	*integer to store the number*
revnum	:	*integer to store the reverse of the number*
Methods/ Member functions :		
Palin( )	:	*constructor to initialise data members with legal initial values*
void accept( )	:	*to accept the number*
int reverse (int y)	:	*reverses the parameterised argument 'y' and stores it in 'revnum' using recursive technique*
void check( )	:	*Checks whether the number is a Palindrome by invoking the function reverse( ) and display the result with an appropriate message.*

Specify the class ***Palin*** *giving the details of the* ***constructor****(), void accept(), int reverse(int) and* ***void check****() . Define the* ***main****() function to create an object and call the functions accordingly to enable the task.*

Question 8.

A class **Adder** *has been defined to add any two accepted time.* **[10]**

Example :

Time A – 6 hours 35 minutes

Time B – 7 hours 45 minutes

Their sum is – 14 hours 20 minutes (where 60 minutes = 1 hour)

The details of the members of the class are given below :

Class Name	:	***Adder***
Data member/ instance variable:		
a[ ]	:	*integer array to hold two elements (hours and minutes)*
Member functions / methods:		
Adder( )	:	*constructor to assign 0 to the array elements*
void readtime( )	:	*to enter the elements of the array*
void addtime (Adder X, Adder Y)	:	*adds the time of the two parameterised objects X and Y and stores the sum in the current calling object*
void disptime( )	:	*displays the array elements with an appropriate message (i.e. hours = and minutes =)*

*Specify the class **Adder** giving details of the **constructor()**, **void readtime()**, **void addtime (Adder, Adder)** and **void disptime()**. Define the **main()** function to create objects and call the functions accordingly to enable the task.*

Question 9.

*A class **SwapSort** has been defined to perform string related operations on a word input. Some of the members of the class are as follows :* **[10]**

Class Name	:	***SwapSort***
Data member/ instance variable:		
wrd	:	*to store a word*
len	:	*integer to store length of the word*
swapwrd	:	*to store the swapped word*
sortwrd	:	*to store the sorted word*
Member functions/ methods :		
SwapSort( )	:	*default constructor to initialise data members with legal initial values*
void readword( )	:	*to accept a word in UPPER CASE*
Void swapchar( )	:	*to interchange/swap the first and last characters of the word in 'wrd' and stores the new word in **swapwrd**'*
Void sortword()	:	*Sorts the characters of the original word in alphabetical order and stores it in 'sortwrd'.*
void display( )	:	*displays the original word, swapped word and the sorted word*

*Specify the class **SwapSort** giving the details of the **constructor()**, **void readword()**, **void swapchar()**, **void sortword()** and **void display()**. Define the **main()** function to create an object and call the functions accordingly to enable the task.*

SECTION–C

*Answer any **two** questions.*

Each program should be written in such a way that it clearly depicts the logic of the problem stepwise.

This can be achieved by using comments in the program and mnemonic names or pseudo codes for algorithms. The programs must be written in Java and the algorithms must be written in general/standard form, wherever required / specified.

*(Flowcharts are **not** required)*

Question 10.

*A super class **Product** has been defined to store the details of a product sold by a wholesaler to a retailer. Define a sub-class **Sales** to compute the total amount paid by the retailer with or without fine along with service tax.* **[5]**

Some of the members of both the classes are given below :

Class Name	:	***Product***
Data member/ instance variable :		
name	:	*stores the name of the product code*
code	:	*integer to store the product code*
amount	:	*stores the total sale amount of the product (in decimals)*
Member functions/ methods :		
product (String n, int c, double p)	:	*parameterised constructor to assign data members name = n, code = c and amount = p*
void show( )	:	*displays the details of the data members*
Class name :		***Sales***
Data member/ instance variable :		
day	:	*stores number of days taken to pay the sale amount*
tax	:	*to store the service tax (in decimals)*
totamt	:	*to store the total amount (in deci mals)*
Member functions / methods :		
Sales (...)	:	*parameterised constructor to assign values to data members of both the classes*
void compute( )	:	*calculates the services tax @ 12.4% of the actual sale amount* *calculates the fine @ 2.5% of the actual sale amount only if the amount paid by the retailer to the wholesaler exceeds 30 days.* *calculates the total amount paid by the retailer as (actual sale amount + service tax + fine)*
void show( )	:	*displays the data members of super class and the total amount*

*Assume that the super class **Product** has been defined. Using the **concept of inheritance,** specify the class **Sales** giving the details of the **constructor(...)**, **void compute()** and **void show( ).***

The super class, main function and algorithm need NOT be written.

Question 11.

Queue is an entity which can hold a maximum of 100 integers. The queue enables the user to add integers from the rear and remove integers from the front.

Define a class **Queue** *with the following details :*

Class Name	:	***Queue***
Data members/ instance variables:		
Que[]	:	*array to hold the integer elements*
size	:	*stores the size of the array*
front	:	*to point the index of the front*
rear	:	*to point the index of the rear*
Member functions :		
Queue (int mm)	:	*constructor to initialise the data size = mm, front = 0, rear = 0*
void addele(int v)	:	*to add integer from the rear if possible else display the message "Overflow"*
int delele()	:	*returns elements from front if present, otherwise displays the message "Underflow" and returns – 9999*
void display()	:	*displays the array elements*

Specify the class **Queue** *giving details of* **ONLY** *the functions* **void addele(int)** *and* **int delele()***. Assume that the other functions have been defined.*

The main function and algorithm need NOT be written. **[5]**

Question 12.

(a) *A linked list is formed from the objects of the class* **Node***. The class structure of the Node is given below:* **[2]**

```
class Node
{
    int num;
    Node next;
}
```

Write an Algorithm **OR** *a Method to count the nodes that contain only odd integers from an existing linked list and returns the count.*

The method declaration is as follows :

int CountOdd(Node startPtr)

(b) *Answer the following questions from the diagram of a Binary Tree given below :*

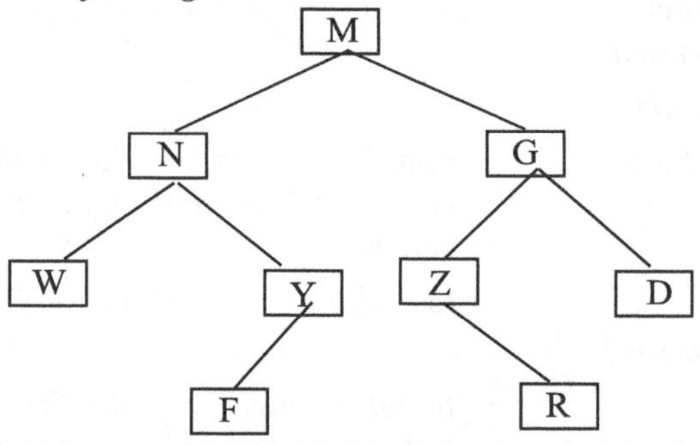

(i) *Write the postorder traversal of the above tree structure.* **[1]**

(ii) *State the level numbers of the nodes* **N** *and* **R** *if the root is at 0 (zero) level.* **[1]**

(iii) *List the internal nodes of the right subtree.* **[1]**

ANSWERS

PART–I

Answer 1.

(a) It is Idempotet law. It states that $P \vee P = P$ and $P \wedge P = P$

Truth Table :

P	P	P ∨ P
0	0	0
1	1	1
	R.H.S.	L.H.S.

L.H.S. = R.H.S.

(b) According to the principle of Duality, from one Boolean expression, another Boolean expression can be derived by changing :

1. All '.' into '+' and '+' into '.'
2. All 0 into 1 and 1 into 0
3. Complements remain unchanged.

(c) Complement of :

$$F(a, b, c) = (b' + c) + a)$$
$$= (b' + c) \cdot a'$$
$$= b'' \cdot c' \cdot a'$$
$$= b \cdot c' \cdot a'$$

(d)

X	Y	X ⊙ Y
0	0	1
0	1	0
1	0	0
1	1	1

$X \odot Y = \overline{X}.\overline{Y} + X.Y$

XNOR gate :

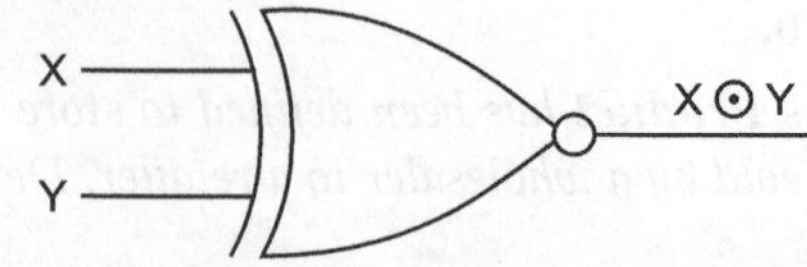

(e) (~ P ⇒ Q)

(i) Inverse : $P \Rightarrow \bar{Q}$

(ii) Converse : $Q \Rightarrow \sim P$

Answer 2.

(a) Interface is a non-primitive data type which has static and final data members and prototype of functions (*i.e.,* functions are not defined).

Interface	Class
Interface supports multiple inheritance.	Class does not support multiple inheritance.

(b) Infix to Postfix :

= P * Q/R + (S + T)

= P * Q/R + ST +

= PQ* /R + ST +

= PQ * R + ST+

= PQ*R/ST++

(c) R = 15, C = 10, ES = 8; LR = 0; LC = 0; BA = 1400, i = 10; j = 7

Row Major wise :

P [i] [j] = BA + ES (C (i – LR) + (j – LC))

P [10] [7] = 1400 + 8 (10 (10 – 0) + (7 – 0))

= 1400 + 856

= 2256

(d) (i) Total time taken :

$C_0 * a + C_1 (b * c)$

$= O(a + bc)$

(ii) Total time taken :

$O(N) + O(N)^2$

$O(N^2)$ taking the dominant term.

(e)

	Constructor	Method
1.	It does not have a return type, not even void.	It always has a return type.
2.	It has the same name as that of class name.	It does not have the same name as that of class name.

Answer 3.

1. Output — Dry Run

111 — $n = 7$

magicfun (3) * 10 + (1) } III

magicfun (1) * 10 + (1) } II

magicfun (0) * 10 + (1) } I

0

2. Output — Dry Run

1010 — $n = 10$

magicfun (5) * 10 + (0) } IV

magicfun (2) * 10 + (1) } III

magicfun (1) * 10 + (0) } II

magicfun (0) * 10 + (1) } I

0

PART–II

SECTION–A

Answer 4.

(a) (i) F(A,B,C,D) = Σ (2, 3, 4, 5, 6, 7, 8, 10, 11)

AB \ CD	$\bar{C}\bar{D}$	$\bar{C}D$	CD	$C\bar{D}$
$\bar{A}\bar{B}$	0 (0)	0 (1)	1 (3)	1 (2)
$\bar{A}B$	1 (4)	1 (5)	1 (7)	1 (6)
AB	0 (12)	0 (13)	0 (15)	0 (14)
$A\bar{B}$	1 (8)	0 (9)	1 (11)	1 (10)

There are two quads and one pair :

Quad 1 = $m_4 + m_5 + m_6 + m_7$ = A′B

Quad 2 = $m_2 + m_3 + m_{10} + m_{11}$ = B′C

Pair = $m_8 + m_{10}$ = AB′D′

Hence, Reduced Expression

= A′B′ + B′C+ AB′D′

(ii)

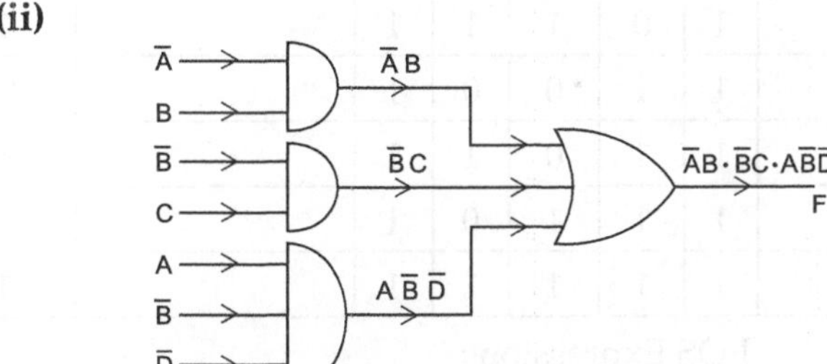

(b) (i) F (P, Q, R, S) = π (0, 1, 2, 4, 5, 6, 8, 10)

PQ \ RS	R + S	R + S̄	R̄ + S̄	R̄ + S
P + Q	0 (0)	0 (1)	1 (3)	0 (2)
P + Q̄	0 (4)	0 (5)	1 (7)	0 (6)
P̄ + Q̄	1 (12)	1 (13)	1 (15)	1 (14)
P̄ + Q	0 (8)	1 (9)	1 (11)	0 (10)

There are three quads :

Quad 1 = $M_0M_2M_4M_6$ = P + S

Quad 2 = $M_0M_1M_4M_5$ = P + R

Quad 3 = $M_0M_2M_8M_{10}$ = (Q + S)

Reduced Expression = (P + S).(P + R).(Q + S)

(ii)

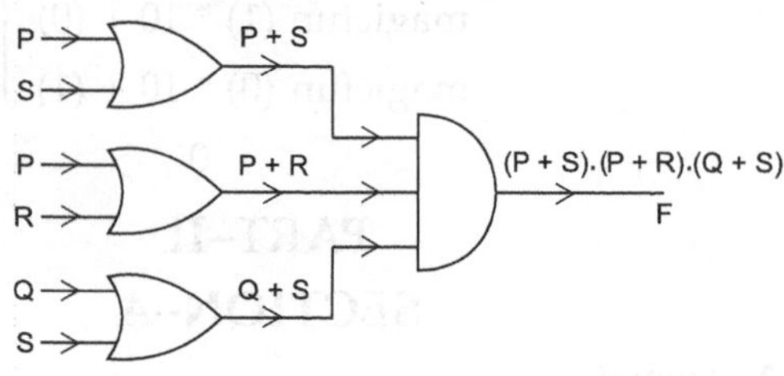

Answer 5.

(a)

A	B	C	D	X	Maxterm	Designation
0	0	0	0	0	A + B + C + D	0
0	0	0	1	0	$A + B + C + \overline{D}$	1
0	0	1	0	0	$A + B + \overline{C} + D$	2
0	0	1	1	1		3
0	1	0	0	0	$A + \overline{B} + C + D$	4
0	1	0	1	1		5
0	1	1	0	0	$A + \overline{B} + \overline{C} + D$	6
0	1	1	1	1		7
1	0	0	0	0	$\overline{A} + B + C + D$	8
1	0	0	1	0	$\overline{A} + B + C + \overline{D}$	9
1	0	1	0	0	$\overline{A} + B + \overline{C} + D$	10
1	0	1	1	1		11
1	1	0	0	1		12
1	1	0	1	1		13
1	1	1	0	1		14
1	1	1	1	1		15

POS Expression :

X (A, B, C, D) = π (0, 1, 2, 4, 6, 8, 9, 10)

X (A, B, C, D,) = (A + B + C + D) . $(A + B + C + \overline{D})$.

$(A + B + \overline{C} + D)$. $(A + \overline{B} + C + D)$.

$(A + \overline{B} + \overline{C} + D)$. $(\overline{A} + B + C + D)$.

$(\overline{A} + B + C + \overline{D})$. $(\overline{A} + B + \overline{C} + D)$

(b) Half Adder is a Digital electronic circuit which is used to add or manipulate 2 bits.

A	B	Sum	Carry
0	0	0	0
0	1	1	0
1	0	1	0
1	1	0	1

Sum : $\overline{A}.B + A.\overline{B} = A \oplus B$

Carry : A.B

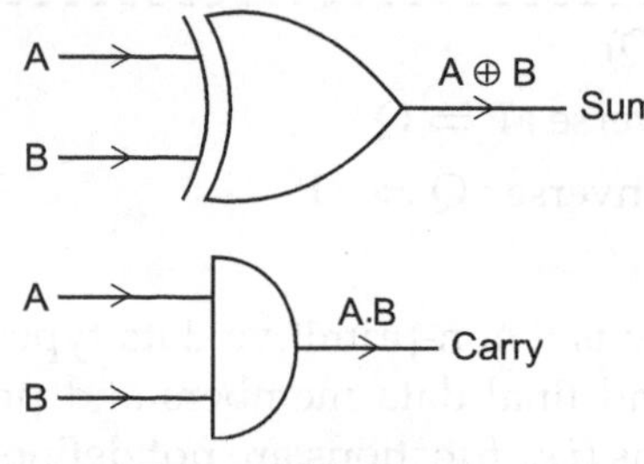

(c) Convert to canonical form :

F (A, B, C) = (B + C′) (A′ + B)

= (B + C′ + 0) (A′ + B + 0)

= ((B + C′) + AA′)) ((A′ + B) + CC′)

= (A + B + C′) (A′ + B + C′) (A′ + B + C) (A′ + B + C′)

= (A + B + C′) (A′ + B + C′) (A′ + B + C)

Answer 6.

(a) A multiplexer is a digital electronic/ combinational circuit which takes input from one of the many input lines and directs it towards one output line.

Multiplexer	Decoder
It takes input from one of the many input lines and directs it towards one output line by using select lines.	It takes input from n lines and output 2^n or fewer lines.

Circuit diagram of a 8 : 1 Multiplexer :

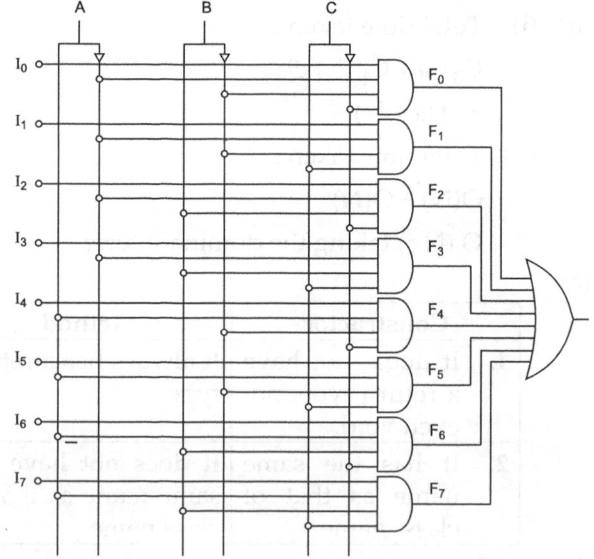

(b) **Prove :**

F = $(x' + z) + [(y' + z) . (x' + y)]' = 1$

= $(x' + z) + (y' + z)' + (x' + y)'$ [De-Morgan's law]

= $x' + z + y.z' + xy'$ [Distributive law]

= $(x' + x) . (x' + y') + (z + z') + (z + y)$

= $x' + y' + z + y$

= 1 (as $y + y' = 1$) [Identity law]

Hence Proved

(c) Maxterms refers to the sum of all the literals with or without bar present within a logic system.

Minterms refer to the product of all the literals with or without bar present within a logic system.

P = 0, Q = 1, R = 1, S = 0

Minterm = PQRS

Maxterm = P + Q + R + S

SECTION–B

Answer 7.

```
import java.util.*;
public class Palin
{
    int num, revnum;
    public Palin()
    {
        num = 0;
        revnum = 0;
    }
    void accept()
    {
        Scanner sc = new Scanner(System.in);
        System.out.println("Enter a no.");
        num = sc.nextInt();
    }
    int reverse(int y)
    {
        if(y > 0)
        {
            revnum = revnum * 10 + y %10;
            return(reverse(y/10));
        }
        else
            return revnum;
    }
    void check()
    {
        int p = num;
        if(reverse(P) == num)
        {
            System.out.println("Palindrome");
        }
        else
        {
            System.out.println("Not a Palindrome");
        }
    }
    public static void main(String args[])
    {
    Palin ob = new Palin();
        ob.accept();
        ob.check();
    }
}
```

Answer 8.

```
import java.util.*;
public class Adder
{
    int a[];
    public Adder()
    {
        a = new int[2];
        a[0] = 0;
        a[1] = 0;
    }
    void readtime()
    {
        Scanner sc = new Scanner(System.in);
        System.out.println("Enter hours and mins");
        a[0] = sc.nextInt();
        a[1] = sc.nextInt();
    }
    void addtime(Adder X, Adder Y)
    {
        a[1] = X.a[1] + Y.a[1];
        a[0] = a[1]/60;
        a[1] = a[1] % 60;
        a[0] + = X.a[0] + Y.a[0];
    }
    void disptime()
    {
        System.out.println("hours = " + a[0] + " and
                                minutes = " + a[1]);
    }
    public static void main(String args[])
    {
        Adder T1 = new Adder();
        Adder T2 = new Adder();
        Adder T3 = new Adder();
        T1.readtime();
        T2.readtime();
        T3.addtime(T1, T2);
        T3.disptime();
    }
}
```

Answer 9.

```
import java.util.*;
class SwapSort
{
    String wrd, swapwrd, sortwrd;
    int len;
    public class SwapSort()
    {
        wrd = "";
        len = 0;
        swapwrd = "";
        sortwrd = "";
    }
    void readword()
    {
        Scanner sc = new Scanner(System.in);
        System.out.println("Enter a word in
                                    Upper Case");
        wrd = sc.next();
        len = wrd.length();
    }
    void swapchar()
    {
        char ch1, ch2;
        ch1 = wrd.charAt(0);
        ch2 = wrd.charAt(len – 1);
        swapwrd = ch2 + wrd.substring
                                    (1, len – 1) + ch1;
    }
    void sortword()
    {
        int i, j;
        for(i = 65; i <= 90; i++)
        {
            for(j = 0; j < len; j++)
            {
                if(wrd.charAt(j) == i)
                {
                    sortwrd = sortwrd + wrd.charAt(j);
                }
            }
        }
    }
    void display()
    {
        System.out.println("Original word " + wrd);
        System.out.println("Swapped word
                                    " + swapwrd);
        System.out.println("Sorted word " + sortwrd);
    }
    public static void main(String args[])
    {
        SwapSort ob = new SwapSort();
        ob.readword();
        ob.swapchar();
        ob.sortword();
        ob.display();
    }
}
```

SECTION–C

Answer 10.

```
class Sales extends Product
{
    int day;
    double tax, totamt;
    public Sales(String nl, int c1, double p1, int d)
    {
        super(n1, c1, p1);
        day = d;
    }
    void compute()
    {
        double fine = 0.0;
        tax = 12.4 / 100 * amount;
        if(day > 30)
        {
            fine = 2.5 / 100 * amount;
        }
        else
        {
            fine = 0;
        }
        totamt = amount + tax + fine;
    }
    void show()
    {
        super.show();
        System.out.println(day);
        System.out.println(tax);
        System.out.println(totamt);
    }
}
```

Answer 11.

```
public class Queue
{
    int Que[], size, front, rear;
    public Queue()//optional
```

```
{
    size = 100;
    Que = new int[size];
    front = rear = 0;
}
public Queue(int mm) //optional
{
    size = mm;
    Que = new int[size];
    front = rear = 0;
}
void addele(int v)
{
    if(rear == size – 1)
    {
        System.out.println("Overflow");
    }
    else if(rear == 0)
    {
        front = rear = 1;
        Que[rear] = v;
    }
    else
    {
        Que[++rear] = v;
    }
}
int delele()
{
    int n;
    if(rear == 0)
    {
        n = –9999;
    }
    else if(front == rear)
    {
        n = Que[front];
        front = rear = 0;
    }
    else
    {
        n = Que[front++];
    }
    return(n);
}
}
```

Answer 12.

(a) Algorithm :

Step 1 : Start

Step 2 : let ptr = start, c = 0

Step 3 : Repeat steps 3 to 5 until ptr = NULL

Step 4 : let rem = ptr.num mod 2

Step 5 : if rem = 1 then

```
{
    c = c + 1
}
```

Step 6 : ptr = ptr.next

Step 7 : return

Step 8 : End

OR

Method :

```
int countodd(Node startPtr)
{
    int count = 0;
    node temp = new Node (startPtr);
    while(tem != null)
    {
        if(tem.num % 2 != 0)
        {
            count ++;
        }
        temp = temp.next;
    }
    return count;
}
```

(b) (i) W F Y N R Z D G M

(ii) Level of N = 1 and level of R = 3

(iii) G, Z

●●

COMPUTER SCIENCE 2016

QUESTIONS

PART–I (20 Marks)

*Answer **all** questions.*

While answering questions in this Part, indicate briefly your working and reasoning, wherever required.

Question 1.

(a) *State Involution law and prove it with the help of a truth table.* **[1]**

(b) *Show that* $X \vee \sim (Y \wedge X)$ *is a tautology.* **[1]**

(c) *Find the dual of :* **[1]**

$$Y.X + X' + 1 = 1$$

(d) *Write the maxterm and minterm, when the inputs are A = 0, B = 1, C = 1 and D = 0.* **[1]**

(e) *Draw the logic circuit of a NAND gate using NOR gates only.* **[1]**

Question 2.

(a) *Define the term fall through condition with reference to switch ( ) case.* **[2]**

(b) *Convert the following infix expression to postfix form:* **[2]**

$$A + B / C * (D / E * F)$$

(c) *A matrix A [m] [n] is stored with each element requiring 4 bytes of storage. If the base address at A [1] [1] is 1500 and the address at A [4] [5] is 1608. Determine the number of rows of the matrix when the matrix is stored in* ***Column Major Wise.*** **[2]**

(d) *From the class declaration given below, state the nature of the identifiers A, B, C and D :* **[2]**

class A extends B implements C, D

(e) *State one advantage and one disadvantage of using recursion over iteration.* **[2]**

Question 3.

The following function Check () is a part of some class. What will the function Check ( ) return when the values of both 'm' and 'n' are equal to 5 ? Show the dry run/working. **[5]**

```
int Check(int m, int n)
  {
      if(n == 1)
          return – m – –;
      else
          return ++m + Check(m, – –n);
  }
```

PART–II (50 Marks)

*Answer **six** questions in this part, choosing **two** questions from Section A, **two** from Section B and **two** from Section C.*

SECTION–A

*Answer any **two** questions.*

Question 4.

(a) *Given the Boolean function* **F (A, B, C, D) = Σ (1, 3, 5, 7, 8, 9, 10, 11, 14, 15).**

(i) *Reduce the above expression by using 4-variable Karnaugh map, showing the various groups (i.e. octal, quads and pairs).* **[4]**

(ii) *Draw the logic gate diagram for the reduced expression. Assume that the variables and their complements are available as inputs.* **[1]**

(b) *Given the Boolean function :*

F (A, B, C, D) = π (4, 6, 7, 10, 11, 12, 14, 15)

(i) *Reduce the above expression by using 4-variable Karnaugh map, showing the various groups (i.e. octal, quads and pairs).* **[4]**

(ii) *Draw the logic gate diagram for the reduced expression. Assume that the variables and their complements are available as inputs.* **[1]**

Question 5.

(a) *What is a decoder ? Draw the logic diagram for a binary to octal (3 to 8) decoder.* **[3]**

(b) *How is a half adder different from a full adder? Draw the truth table and derive the SUM and CARRY expression for a full adder. Also, draw the logic diagram for a full adder.* **[4]**

(c) *State whether the following expression is a Tautology, Contradiction or a Contingency, with the help of a truth table :* **[3]**

$$(X => Z) \vee \sim [(X => Y) \wedge (Y => Z)]$$

Question 6.

(a) *A passenger is allotted a window seat in an aircraft, if he/she satisfies the criteria given below :*

- *The passenger is below 15 years and is accompanied by an adult.*

OR

- *The passenger is a lady and is not accompanied by an adult.*

OR

- *The passenger is not below 15 years, but is traveling for the first time.*

The inputs are :

INPUTS	
A	The passenger is below 15 years age.
C	The passenger is accompanied by an adult.
L	The passenger is a lady.
F	The passenger is traveling for the first time.

(In all the above cases 1 indicates yes and 0 indicates no).

Output : W - Denotes the passenger is allotted a window seat (1 indicates yes and 0 indicates no).

Draw the truth table for the inputs and outputs given above and write the **SOP** expression for **W (A, C, L, F)**. **[5]**

(b) State the complement properties. Find the complement of the following Boolean expression using De Morgan's law : **[3]**

$$AB' + A' + BC$$

(c) Differentiate between Canonical form and Cardinal form of expression. **[2]**

SECTION–B

*Answer any **two** questions.*

Each program should be written in such a way that it clearly depicts the logic of the problem.

This can be achieved by using mnemonic names and comments in the program.

*(Flowcharts and Algorithms are **not** required.)*

The programs must be written in Java.

Question 7.

A disarium number is a number in which the sum of the digits to the power of their respective position is equal to the number itself. **[10]**

Example : $135 = 1^1 + 3^2 + 5^3$

Hence, 135 is a disarium number.

Design a class **Disarium** to check if a given number is a disarium number or not. Some of the members of the class are given below :

Class Name : **Disarium**

Data members / instance variables:

int num	:	stores the number
int size	:	stores the size of the number

Methods/ Member functions :

Disarium (int nn)	:	parameterised constructor to initialise the data members num = nn and size = 0
void count Digit ( )	:	counts the total number of digits and assigns it to size
int sumOf-Digits (int n, int p)	:	returns the sum of the digits of the number (n) to the power of their respective positions (p) using recursive technique
void check ()	:	checks whether the number is a disarium number and displays the result with an appropriate message

Specify the class **Disarium** giving the details of the **constructor (), void count Digit (), int sumOfDigits(int, int) and void check ()**. Define the main() function to create an object and call the functions accordingly to enable the task.

Question 8.

A class **Shift** contains a two dimensional integer array of order (m × n) where the maximum values of both m and n is 5. Design the class **Shift** to shuffle the matrix (i.e., the first row becomes the last, the second row becomes the first and so on). The details of the members of the class are given below : **[10]**

Class Name : **Shift**

Data member/ instance variable:

mat[] []	:	stores the array element
m	:	integer to store the number of rows
n	:	integer to store the number of columns

Member functions / methods :

Shift (int mm, int nn)	:	parameterised constructor to initialise the data members m = mm and n = nn
void input( )	:	enters the elements of the array
void cyclic (Shift P)	:	enables the matrix of the object (P) to shift each row upwards in a cyclic manner and store the resultant matrix in the current object
void display( )	:	displays the matrix elements

Specify the class **Shift** giving details of the **constructor()**, **void input** () void cyclic (Shift) and **void display** () Define the main () function to create an object and call the methods accordingly to enable the task of shifting the array elements.

Question 9.

*A class **Cons Change** has been defined with the following details :* **[10]**

Class Name	:	***ConsChange***
Data members/ instance variables:		
word	:	*stores the word*
len	:	*stores the length of the word*
Member functions/ methods :		
ConsChange()	:	*default constructor*
void readword ( )	:	*accepts the word in lowercase*
void shiftcons ( )	:	*shift all the consonants of the word at the beginning followed by the vowels (e.g. spoon becomes spnoo)*
void change-word ( )	:	*change the case of all occurring consonants of the shifted word to uppercase, for e.g., (spnoo becomes SPNoo)*
void show( )	:	*displays the original word, shifted word and the changed word*

*Specify the class **ConsChange** giving the details of the **constructor(), void readword(), void shift cons(), void changeword()** and **void show()**. Define the main() function to create an object and call the functions accordingly to enable the task.*

SECTION–C

*Answer any **two** questions.*

Each program should be written in such a way that it clearly depicts the logic of the problem stepwise.

This can be achieved by using comments in the program and mnemonic names or pseudo codes for algorithms. The programs must be written in Java and the algorithms must be written in general/standard form, wherever required / specified.

*(Flowcharts are **not** required)*

Question 10.

*A super class **Bank** has been defined to store the details of a customer. Define a sub-class **Account** that enables transactions for the customer with the bank. The details of both the classes are given below :* **[5]**

Class Name	:	***Bank***
Data member/ instance variable :		
name	:	*stores the name of the customer*
accno	:	*stores the account number*
p	:	*stores the principal amount in decimals*
Member functions/ methods :		
Bank(...)	:	*parameterised constructor to assign values to the instance variables*
void displays ( )	:	*displays the details of the customer*
Class name	:	*Account*
Data member/ instance variable :		
amt	:	*stores the transaction amount in decimals*
Member functions/ methods :		
Account(...)	:	*parameterised constructor to assign values to the instance variables of both the classes*
void deposit( )	:	*accepts the amount and updates the principal as p=p+amt*
void withdraw ( )	:	*accepts the amount and updates the principal as p=p–amt* *If the withdrawal amount is more than the principal amount, then display the message "INSUFFICIENT BALANCE".* *If the principal amount after withdrawal is less than 500, then a penalty is imposed by using the formula p = p – (500 – p)/10*
void display( )	:	*displays the details of the customer*

*Assume that the super class **Bank** has been defined. Using the **concept of Inheritance**, specify the class **Account** giving details of the **constructor(...), void deposit(), void withdraw()** and **void display()***

The super class and the main function need not to be written.

Question 11.

A bookshelf is designed to store the books in a stock with LIFO (Last in First Out) operation. Define a class Book with the following specifications : **[5]**

Class Name	:	***Book***
Data members/ instance variables:		
name [ ]	:	*stores the names of the books*

point	:	*stores the index of the topmost book*
	:	*stores the maximum capacity of the bookshelf*
Methods/ Member functions :		
Book(int cap)	:	*constructor to initialise the data members = cap and point = – 1*
void tell( )	:	*displays the name of the book which was last entered in the shelf. If there is no book left in the shelf, displays the message SHELF EMPTY'*
void add (string v)		*adds the name of the book to the shelf it possible, otherwise displays the message "SHELF FULL"*
void display( )	:	*displays all the names of the books available in the shelf*

*Specify the class **Book** giving the details of **ONLY** the functions **void tell ( )** and **void add (String)**. Assume that the other functions have been defined.*

The main function need not be written.

Question 12.

(a) *A linked list is formed from the objects of the class Node. The class structure of the Node is given below:* **[2]**

```
class Node
{
  String name;
  Node next :
}
```

*Write an Algorithm **OR** a Method to search for a given name in the linked list. The method of declaration is given below :*

boolean searchName (Node start, String v)

(b) *Answer the following questions from the diagram of a Binary Tree given below :*

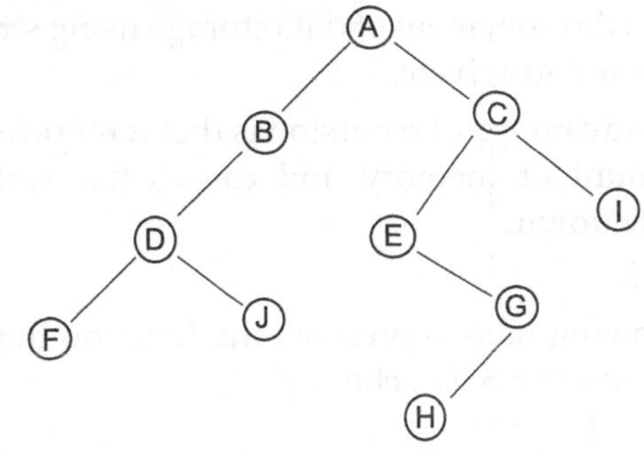

(i) *Write the inorder traversal of the above tree structure.* **[1]**

(ii) *Name the parent of the nodes B and G.* **[1]**

(iii) *Name the leaves of the right sub-tree.* **[1]**

ANSWERS

PART–I

Answer 1.

(a) Involution law states : $\overline{\overline{X}} = X$

X	$\overline{X}$	$\overline{\overline{X}}$
0	1	0
1	0	1

(b) $X \vee \sim (Y \wedge X)$

X	Y	$Y \wedge X$	$\sim (Y \wedge X)$	$X \vee \sim (Y \wedge X)$
0	0	0	1	1
0	1	0	1	1
1	0	0	1	1
1	1	1	0	1

Hence, it is a Tautology.

(c) Dual of : $Y.X + X' + 1 = 1$

$= (Y + X) . \overline{X} . 0 = 0$

(d) Maxterm : $A + \overline{B} + \overline{C} + A$

Minterm : $\overline{A} .B.C. \overline{D}$

(e) Logic circuit of a NAND gate using NOR gates :

NAND Expression = A′ + B′ or (A.B)′

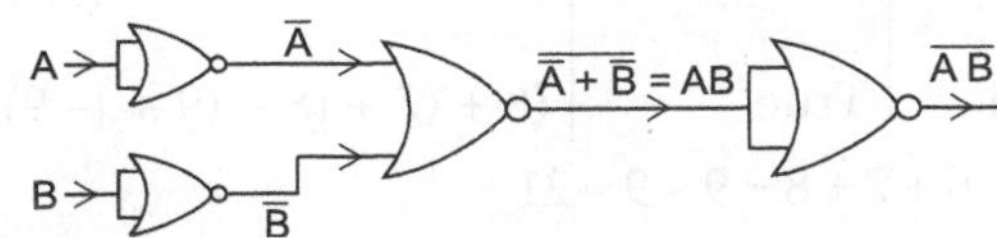

Answer 2.

(a) When break statement is missing after a case then the statements present in the following cases will get executed, such a situation is called fall through.

(b) A + B/C * (D/E * F)

= A + BC/ * (DE / * F)

= A + BC/ * DE/F *

= A + BC/DE/F * *

= ABC/DE /F * * +

(c) ES = 4, BA = 1500, LR = 1, LC = 1, I = 4, J = 5, R = *m*, *c* = *n*

In column major wise :

A [I, J] = BA + ES ((I – LR) + R (J – LC))

$$1608 = 1500 + 4\,((4-1) + m\,(5-1))$$
$$1608 - 1500 = 4\,(3 + 4m)$$
$$108 = 4\,(3 + 4m)$$
$$4m = 27 - 3$$
$$m = 24/4 = 6$$

(d) Here,

A is a child class/sub-class

B is a parent class/super class

C,D are interface

(e) Advantage of recursion is that it makes complex programs simpler and easy to understand by making the code reduced than iteration.

e.g., : Arrangement of data storage using structure tree and graph etc.

Disadvantage of recursion is that it requires large amount of memory and causes the system to slow down.

Answer 3.

The following table represents the function iterations for $n = 5$ and $m = 5$. *i.e.,* check (5, 5).

m	n	Condition ($n = 1$)	Return Value
5, 6	5, 4	False	(6 + check (6, 4))
6, 7	4, 3	False	(6 + (7 + check (7, 3)))
7, 8	3, 2	False	(6 + (7+ (8 + check (8, 2))))
8, 9	2, 1	False	(6+ (7 + (8 + (9 + check (9, 1)))))
9, 8	1	True	(6 + (7 + (8 + (9 + (– 9)))))

So, 6 + 7 + 8 + 9 – 9 = 21

PART–II

SECTION–A

Answer 4.

(a) F (A, B, C, D) = Σ (1, 3, 5, 7, 8, 9, 10, 11, 14, 15)

(i)

AB \ CD	$\bar{C}.\bar{D}$	$\bar{C}.D$	$C.D$	$C.\bar{D}$
$\bar{A}.\bar{B}$	0 (0)	1 (1)	1 (3)	0 (2)
$\bar{A}.B$	0 (4)	1 (5)	1 (7)	0 (6)
$A.B$	0 (12)	0 (13)	1 (15)	1 (14)
$A.\bar{B}$	1 (8)	1 (9)	1 (11)	1 (10)

There are three quads :

Quad 1 = $m_1 + m_3 + m_5 + m_7 = \bar{A}.D$

Quad 2 = $m_8 + m_9 + m_{10} + m_{11} = A.\bar{B}$

Quad 3 = $m_{10} + m_{11} + m_{14} + m_{15} = A.C$

So, the reduced expression is,

$= \bar{A}.D + A.\bar{B} + A.C$

(ii)

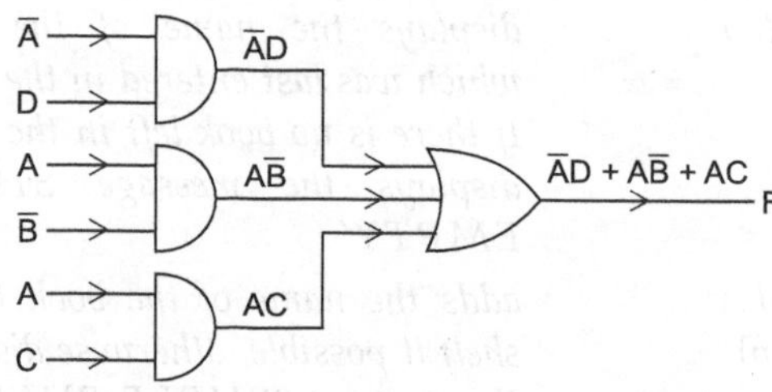

(b) **(i)** F (A, B, C, D) = π (4, 6, 7, 10, 11, 12, 14, 15)

AB \ CD	C + D	$C + \bar{D}$	$\bar{C} + \bar{D}$	$\bar{C} + D$
A + B	1 (0)	1 (1)	1 (3)	1 (2)
$A + \bar{B}$	0 (4)	1 (5)	0 (7)	0 (6)
$\bar{A} + \bar{B}$	0 (12)	1 (13)	0 (15)	0 (14)
$\bar{A} + B$	1 (8)	1 (9)	0 (11)	0 (10)

There are three quads :

Quad 1 = $M_4.M_6.M_{12}.M_{14} = \bar{B} + D$

Quad 2 = $M_6.M_7.M_{14}.M_{15} = \bar{B} + \bar{C}$

Quad 3 = $M_{10}.M_{11}.M_{14}.M_{15} = \bar{A} + \bar{C}$

So the reduced expression is,

$= (\bar{B} + D).(\bar{B} + \bar{C}).(\bar{A} + \bar{C})$

(ii)

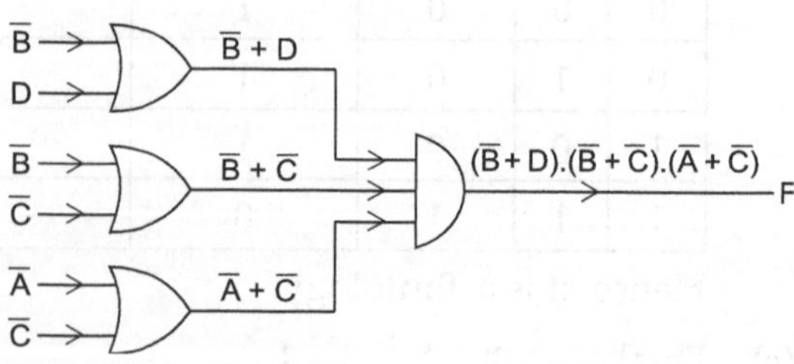

Answer 5.

(a) A decoder is a combinational circuit that converts binary information from n coded inputs to a maximum of 2^n unique outputs.

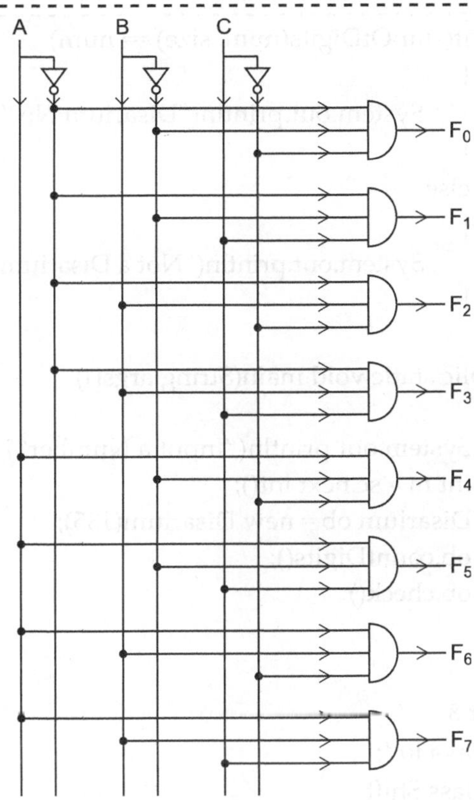

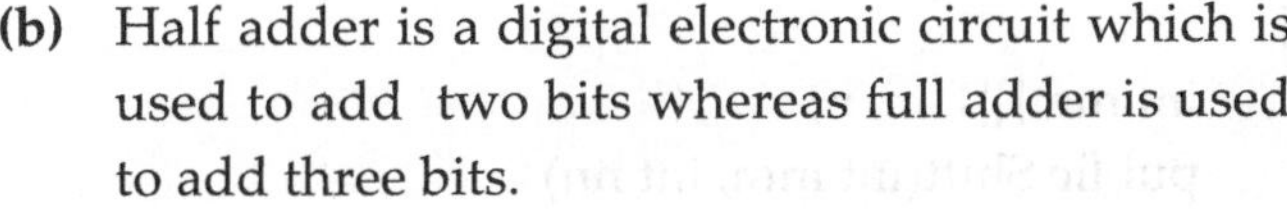

(b) Half adder is a digital electronic circuit which is used to add two bits whereas full adder is used to add three bits.

Half adder performs partial addition whereas a full adder performs total addition.

Truth table for a full adder :

X	Y	Z	Carry		Sum	
0	0	0	0		0	
0	0	1	0		1	$\bar{X}\,\bar{Y}\,Z$
0	1	0	0		1	$\bar{X}\,Y\,\bar{Z}$
0	1	1	1	$\bar{X}\,Y\,Z$	0	
1	0	0	0		1	$X\,\bar{Y}\,\bar{Z}$
1	0	1	1	$X\,\bar{Y}\,Z$	0	
1	1	0	1	$X\,Y\,\bar{Z}$	0	
1	1	1	1	$X\,Y\,Z$	1	$X\,Y\,Z$

$$\text{Sum} = \bar{X}\bar{Y}Z + \bar{X}Y\bar{Z} + X\bar{Y}\bar{Z} + XYZ$$
$$= X \oplus Y \oplus Z$$
$$\text{Carry} = \bar{X}YZ + X\bar{Y}Z + XY\bar{Z} + XYZ$$
$$= XY + YZ + XZ$$

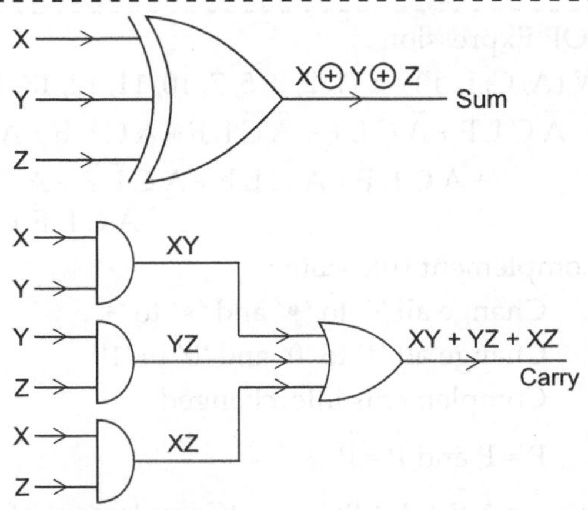

(c) The given expression E is :

$$(X \Rightarrow Z) \vee \sim [(X \Rightarrow Y) \wedge (Y \Rightarrow Z)]$$

X	Y	Z	X⇒Z A	X⇒Y B	Y⇒Z C	A.C D	D′	A + D′
0	0	0	1	1	1	1	0	1
0	0	1	1	1	1	1	0	1
0	1	0	1	1	0	0	1	1
0	1	1	1	1	1	1	0	1
1	0	0	0	0	1	0	1	1
1	0	1	1	0	1	0	1	1
1	1	0	0	1	0	0	1	1
1	1	1	1	1	1	1	0	1

Hence, it is a **tautology.**

Answer 6.

(a) Truth table for output expression W :

A	C	L	F	W	Maxterm	Designation
0	0	0	0	0		0
0	0	0	1	1	$\bar{A}\,\bar{C}\,\bar{L}\,F$	1
0	0	1	0	1	$\bar{A}\,\bar{C}\,L\,\bar{F}$	2
0	0	1	1	1	$\bar{A}\,\bar{C}\,L\,F$	3
0	1	0	0	0		4
0	1	0	1	1	$\bar{A}\,C\,\bar{L}\,F$	5
0	1	1	0	0		6
0	1	1	1	1	$\bar{A}\,C\,L\,F$	7
1	0	0	0	0		8
1	0	0	1	0		9
1	0	1	0	1	$A\,\bar{C}\,L\,\bar{F}$	10
1	0	1	1	1	$A\,\bar{C}\,L\,F$	11
1	1	0	0	1	$A\,C\,\bar{L}\,\bar{F}$	12
1	1	0	1	1	$A\,C\,\bar{L}\,F$	13
1	1	1	0	1	$A\,C\,L\,\bar{F}$	14
1	1	1	1	1	$A\,C\,L\,F$	15

SOP Expression :

$$W(A, C, L, F) = \Sigma(1, 2, 3, 5, 7, 10, 11, 12, 13, 14, 15)$$

$$= \bar{A}\,\bar{C}\,L\,F + \bar{A}\,\bar{C}\,L\,\bar{F} + \bar{A}\,\bar{C}\,L\,F + \bar{A}\,C\,\bar{L}\,F + \bar{A}\,C\,L\,F + A\,\bar{C}\,L\,\bar{F} + A\,\bar{C}\,L\,F + A\,C\,\bar{L}\,\bar{F} + A\,C\,\bar{L}\,F + A\,C\,L\,\bar{F} + ACLF$$

(b) Complement rule state :

1. Change all '+' to '•' and '•' to '+'
2. Change all '1' to '0' and '0' to '1'
3. Complements interchanged

$P = \bar{P}$ and $\bar{P} = P$

$= A\,\bar{B} + \bar{A} + BC$ (Complement of given expression)

$= (\bar{A} + B).A.(\bar{B} + \bar{C})$

(c) An expression which is entirely made up of either min. term or max. term is called canonical form of an expression whereas cardinal form of an expression is a complete system which clearly depicts the inputs and outputs and returns the cardinal numbers where output has occurred *e.g.*:

Canonical form : $\bar{A}.\bar{B}.\bar{C} + \bar{A}.B.\bar{C} + A.\bar{B}.\bar{C} + A.B.C$

Cardinal Form : $F(A, B, C) = \Sigma(0, 2, 4, 7)$

SECTION–B

Answer 7.

```
Public class Disarium
{
    int num, size, s = 0;
    static Scanner sc = new Scanner(System.in);
    public Disarium(int nn)
    {
        num = nn;
        size = 0;
    }
    void countDigit()
    {
        int t = num;
        while(t > 0)
        {
            size++;
            t = t / 10;
        }
    }
    int sumOfDigits(int n, int p)
    {
        return(n == 0) ? 0 : sumOfDigits(n/10, p – 1) +
                              (int)Math.pow(n%10, P);
    }
    void check()
    {
        countDigit();
        if(sumOfDigits(num, size) == num)
        {
            System.out.println("Disarium No.");
        }
        else
        {
            System.out.println("Not a Disarium No.");
        }
    }
    public static void main(String args[])
    {
        System.out.println("Input a Number");
        int m = sc.next int();
        Disarium ob = new Disarium(135);
        ob.countDigits();
        ob.check();
    }
}
```

Answer 8.

```
import java.io.*;
Public class Shift
{
    int mat[][], m, n;
    public Shift(int mm, int nn)
    {
        m = mm;
        n = nn;
        mat = new int[m][n];
    }
    void input()throws IOException
    {
        int i, j;
        Scanner sc = new Scanner(System.in));
        for(i = 0; i < m; i++)
        {
            for(j = 0; j < n; j++)
            {
                System.out.println("Enter a No.");
                mat[i][j] = sc.nextInt();
            }
        }
    }
    void cyclic(Shift P)
    {
        int i, j;
        for(i = 0; i < m; i++)
        {
            for(j = 0; j < n; j++)
            {
                if(i != 0)
```

```
                mat[i – 1][j] = P.mat[i][j];
            else
                mat[m – 1][j] = P.mat[0][j];
        }
    }
}
void display()
{
    int i, j;
    for(i = 0; i < m; i++)
    {
        System.out.println();
        for(j = 0; j < n; j++)
        {
            System.out.print(mat[i][j] + " ");
        }
    }
}
public static void main(String args[])
{
    Shift A = new Shift(5, 5);
    Shift B = new Shift(5, 5);
    A.input();
    B.cyclic(A);
    A.display();
    B.display();
}
}
```

Answer 9.

```
import java.io.*;
Public class ConsChange
{
    String word;
    int len;
    public ConsChange()
    {
        word = "";
        len = 0;
    }
    void readword()
    {
     Scanner sc = new Scanner(System.in);
     System.out.println("Enter a word in lower case");
        word = sc.next();
        len = word.length();
    }
    void shiftcons()
    {
        String v = "", c = "";
        int i;
        char ch;
        for(i = 0; i < len; i++)
        {
        ch = word.charAt(i);
            if(ch == 'a' || ch == 'e' || ch == 'i' || ch ==
                                                'o' || ch == 'u')
            {
                v = v + ch;
            }
            else
            {
                c = c + ch;
            }
        }
        word = c + v;
        System.out.println("shifted word is" + word);
    }
    void changeword()
    {
        int i;
        char ch;
        String s = "";
        for(i = 0; i < len; i++)
        {
            ch = word.charAt(i);
            if(ch ! = 'a' && ch ! = 'e' && ch ! = 'i'
            && ch ! = 'o' && ch ! = 'u')
            {
                s + = (char)(ch – 32);
            }
            else
                s = s + ch;
        }
        System.out.println("changed word is "+ s);
    }
    void show()
    {
        System.out.println("Original word =" + word);
        shiftcons();
        changeword();
```

```
    }
    public static void main(String args[])
                                    throws IOException
    {
        ConsChange ob = new ConsChange();
        ob.readword();
        ob.show();
    }
}
```

SECTION–C

Answer 10.

```
import java.io.*;
Public class Account extends Bank
{
    double amt;
    Scanner sc = new Scanner(System.in);
    Public Account(String n, int ano double p1,
                                          double a)
    {
        super(n, ano, p1);
        amt = 0.0;
    }
    void deposit()
    {
        System.out.println("Enter amount");
        amt = sc.nextDouble();
        p = p – amt;
    }
    void withdraw()
    {
        System.out.println("Enter amount");
        amt = Double.parseDouble(br.readLine());
        if(amt > p)
        {
    System.out.println("INSUFFICIENT BALANCE");
        }
        else
        {
            p = p – amt;
            if(p < 500)
            {
                p = p – (500 – p) / 10.0;
            }
        }
    }
    void display()
    {
        super.display();
        System.out.println(amt);
    }
}
```

Answer 11.

```
Public class Book
{
    String name[];
    int point, max;
    Book(int cap)
    {
        max = cap;
        point = –1;
        name = new String[max];
    }
    void tell()
    {
        if(point < 0)
        {
            System.out.println("SHELF EMPTY");
        }
        else
        {
            System.out.println(name[point]);
        }
    }
    void add(String v)
    {
        if(point < max – 1)
        {
            name[++point] = s;
        }
        else
        {
            System.out.println("SHELF FULL");
        }
    }
    void display()
    {
        for(int i = point; i >= 0; i––)
            System.out.println("\n" + name[i]);
    }
}
```

Answer 12.

(a) ALGORITHM

Step 1 : Let Ptr = start

Step 2 : Repeat steps 3, 4 until ptr = null

Step 3 : If Ptr. name = V then

```
{
        print "FOUND"
        break
}
```

Step 4 : Ptr = Ptr. next

Step 5 : If Ptr = null then

```
{
 Print "NOT FOUND"
}
```

Step 6 : End

OR

METHOD

```
boolean searchName(Node start, String v)
{
    Node temp = start;
    while(temp != null)
    {
        if(v.equals(temp.name) == true)
            return true;
    }
    temp = temp. next;
    return false;
}
```

(b) **(i)** F, D, J, B, A, E, H, G, C, I

(ii) Parent of B is A and parent of G is E

(iii) H, I

COMPUTER SCIENCE 2015

QUESTIONS

PART–I

*Answer **all** questions.*

While answering questions in this Part, indicate briefly your working and reasoning, wherever required.

Question 1.

(a) *Simplify:*
$(A+C).(A+A.D)+A.C+C$ **[2]**

(b) *Draw a logic circuit for :*
$(A+B).(C+D).C$ **[2]**

(c) *Verify the following proposition with the help of a truth table:*
$P \vee (\sim P \wedge Q) = P \vee Q$ **[2]**

(d) *State De Morgan's law and verify it, using a truth table.* **[2]**

(e) *Answer the questions related to the circuit given below:* **[2]**

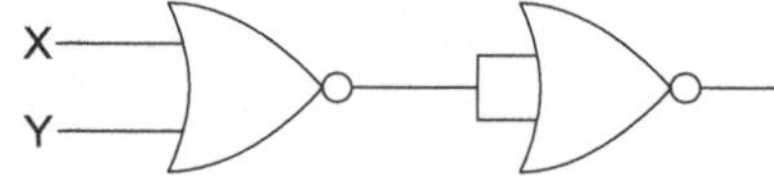

(i) *Give the output if, X = 1 and Y = 0*

(ii) *Name the basic gate represented by the above diagram.*

Question 2.

(a) *Define computational complexity. Calculate the complexity using Big 'O' notation for the following code segment:* **[2]**

```
for(int k = 0; k < n; k++)
s += k;
```

(b) *Convert the following infix notation into postfix form:*
$X + (Y - Z) + ((W + E) * F) / J$ **[2]**

(c) *Differentiate between **this keyword** and **super keyword.*** **[2]**

(d) *The array D[–2...10][3...8] contains double type elements. If the base address is 4110, find the address of D[4][5], when the array is stored in **Column Major Wise.*** **[2]**

(e) *State any two characteristics of a Binary tree.* **[2]**

Question 3.

(a) *The following function is a part of some class. Assume 'x' and 'y' are positive integers, greater than 0. Answer the given questions along with dry run /working.*

```
void someFun(int x, int y)
{
    if (x > 1)
    {
        if (x % y == 0)
        {
            System.out.print(y +"");
        }
        someFun(x/y, y);
    }
    else
    {
        someFun(x, y + 1);
    }
}
```

(i) *What will be returned by **someFun (24, 2)** ?* **[2]**

(ii) *What will be returned by **some Fun (84, 2)** ?* **[2]**

(iii) *State in one line what does the function **someFun()** do, apart from recursion?* **[1]**

(b) *The following is a function of some class which checks if a positive integer is an Armstrong number by returning true or false. (A number is said to be Armstrong if the sum of the cubes of all its digits is equal to the original number). The function does not use modulus (%) operator to extract digit. There are some places in the code marked by ?1?, ?2?, ?3?, ?4?, ?5? which may be replaced by a statement/expression so that the function works properly.*

```
boolean ArmstrongNum(int N)
{
    int sum = ?1?;
    int num = N;
    while(num > 0)
    {
        int f = num / 10;
        int s = ?2?;
        int digit = num - s;
        sum + = ?3?;
        num = ?4?;
    }
    if( ?5?)
        return true;
    else
        return false;
}
```

(i) *What is the statement or expression at ?1?* [1]

(ii) *What is the statement or expression at ?2?* [1]

(iii) *What is the statement or expression at ?3?* [1]

(iv) *What is the statement or expression at ?4?* [1]

(v) *What is the statement or expression at ?5?* [1]

PART–II

*Answer **seven** questions in this part, choosing **three** questions from Section A, **two** from Section B and **two** from Section C.*

SECTION–A

*Answer any **three** questions.*

Question 4.

(a) *Given the Boolean function F (A, B, C, D) = π (0, 1, 2, 3, 5, 7, 8, 9, 10, 11).*

(i) *Reduce the above expression by using 4-variable Karnaugh map, showing the various groups (i.e., octal, quads and pairs).* [4]

(ii) *Draw the logic gate diagram for the reduced expression. Assume that the variables and their complements are available as inputs.* [1]

(b) *Given the Boolean function:*

P (A, B, C, D) = ABC'D' + A'BC'D' + A'BC'D + ABC'D + A'BCD + ABCD

(i) *Reduce the above expression by using 4-variable Karnaugh map, showing the various groups (i.e., octal, quads and pairs).* [4]

(ii) *Draw the logic gate diagram for the reduced expression. Assume that the variables and their complements are available as inputs.* [1]

Question 5.

A person is allowed to travel in a reserved coach of the train, if he/she satisfies the criteria given below : [10]

- *The person has a valid reservation ticket and a valid ID proof.*

OR

- *The person does not have a valid reservation ticket, but holds a valid pass issued by the Railway department with a valid ID proof.*

OR

- *The person is disabled person and holds a valid pass issued by the Railway department along with a valid ID proof.*

The inputs are :

INPUTS

R	*The person has a valid reservation ticket.*
P	*The person holds a valid pass issued by the Railway department.*
D	*The person has a valid ID proof.*
H	*The person is disabled person.*

(In all of the above cases 1 indicates yes and 0 indicates no)

*Output : **T** - Denotes allowed to travel [1 indicates yes and 0 indicates no in all cases.]*

(a) *Draw the truth table for the inputs and outputs given above and write the **POS** expression for **T(R, P, D, H)**.* [5]

(b) *Reduce **T(R, P, D, H)** using Karnaugh map.* [5]

*Draw the logic gate diagram for the reduced **POS** expression for **T(R, P, D, H)** using only **NOR** gates. You may use gates with two or more inputs. Assume that the variable and their complements are available as inputs.*

Question 6.

(a) *Draw the truth table and logic gate diagram for an Octal to Binary encoder.* [4]

(b) *What is a Multiplexer? State an application of a Multiplexer. Also draw the logic diagram of a 4:1 Multiplexer.* [4]

(c) *Verify the following expression using Boolean laws. Also mention the law used at each step of simplification.* [2]

$$X.Y.Z + X.Y'.Z + X.Y.Z' = X.(Y + Z)$$

Question 7.

(a) *Derive a Boolean expression for the logic circuit given below and reduce the derived expression, using Boolean laws :* [3]

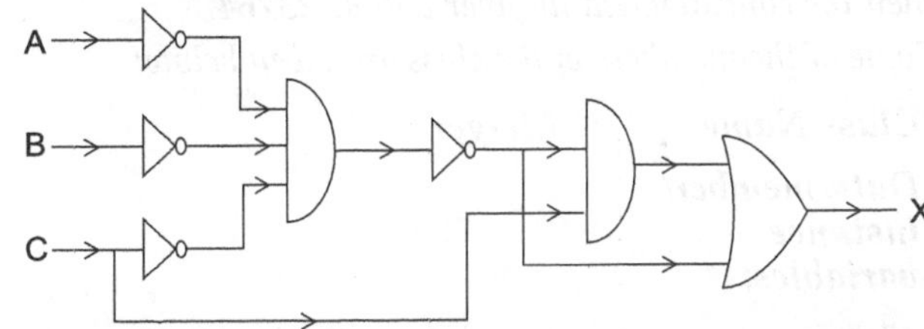

(b) *What are universal gates ? Construct a logic circuit using NAND gates only for the expression :*

A· (B + C) [3]

(c) *Define Half Adders. Draw the circuit diagram and the truth table for a Half Adder.* [4]

SECTION–B

*Answer any **two** questions.*

Each program should be written in such a way that it clearly depicts the logic of the problem.

This can be achieved by using mnemonic names and comments in the program.

*(Flowcharts and Algorithms are **not** required.)*

The programs must be written in Java.

Question 8.

*A class **Admission** contains the admission numbers of 100 students. Some of the data members / member functions are given below:* [10]

Class Name	:	***Admission***
Data member/ instance variable:		
Adno[ ]	:	*integer array to store admission numbers*
Member functions/ methods:		
Admission( )	:	*constructor to initialise the array elements*
void fillArray( )	:	*to accept the elements of array in ascending order*
int binSearch (int l, int u, int v)	:	*to search for a particular admission number (v) using* ***binary search*** *and* ***recursive technique*** *and returns 1 if found otherwise returns – 1*

Specify the class ***Admission*** *giving details of the* ***constructor, void fill Array()*** *and* ***int binSearch (int, int, int).*** *Define the* ***main()*** *function to create an object and call the functions accordingly to enable the task.*

Question 9.

A class ***Merger*** *concatenates two positive integers that are greater than 0 and produces a new merged integer.* **[10]**

Example *: If the first number is 23 and the second is* ***764****, then the concatenated number will be* ***23764****.*

Some of the members of the class are given below:

Class Name	:	***Merger***
Data member/ instance variables:		
n1	:	*long integer to store first number*
n2	:	*long integer to store second number*
mergNum	:	*long integer to store the merged number*
Member functions:		
Merger( )	:	*constructor to initialise the data members*
void readNum ( )	:	*to accept the values of data members n1 and n2*
void JoinNum ( )	:	*to concatenate the numbers n1 and n2 and store it in merg Num*
void show( )	:	*to display the original numbers and the merged number with appropriate messages*

Specify the class ***Merger****, giving the details of the* ***constructor, void readNum (), void JoinNum()*** *and* ***void show()****. Define the* ***main()*** *function to create an object and call the functions accordingly to enable the task.*

Question 10.

A class ***TheString*** *accepts a string of a maximum of 100 characters with only one blank space between the words.* **[10]**

Some of the members of the class are as follows:

Class Name	:	***TheString***
Data member/ instance variable:		
str	:	*to store a string*
len	:	*integer to store the length of the string*
wordcount	:	*integer to store the number of words*
cons	:	*integer to store the number of consonants*
Member functions/ methods:		
TheString( )	:	*default constructor to initialise the data members*
TheString (String ds)	:	*parameterised constructor to assign str=ds*
void count Freq ( )	:	*to count the number of words and the number of consonants and store them in word count and cons respectively*
void Display( )	:	*to display the original string, alongwith the numbers of consonants*

Specify the class ***TheString*** *giving the details of the* ***constructors, void count Freq()*** *and* ***void Display ().*** *Define the* ***main()*** *function to create an object and call the functions accordingly to enable the task.*

SECTION–C

Answer any ***two*** *questions.*

Each program should be written in such a way that it clearly depicts the logic of the problem stepwise. This can be achieved by using comments in the program and mnemonic names or pseudo codes for algorithms. The programs must be written in Java and the algorithms must be written in general/standard form, wherever required / specified.

(Flowcharts are ***not*** *required)*

Question 11.

WordPile *is an entity which can hold maximum of 20 characters. The restriction is that a character can be added or removed from one end only.*

Some of the members of classes are given below:

Class Name	:	***WordPile***
Data members/ instance variables:		
ch[ ]	:	*character array to hold the character elements*

capacity	:	*integer variable to store the maximum capacity*
top	:	*to point to the index of the topmost element*

Method/ Member functions:

WordPile (int cap)	:	*constructor to initialise the data member capacity = cap, top = -1 and create the WordPile*
void pushchar (char v)	:	*adds the character to the top of WordPile if possible, otherwise output a message "WordPile is full"*
char popChar()	:	*returns the deleted character from the top of the WordPile if possible, otherwise it returns '\\'*

(a) *Specify the class* ***WordPile*** *giving details of the* ***constructor, void pushchar(char)*** *and* ***Char popChar(). The main function and algorithm need not be written.*** **[8]**

(b) *What is the name of the entity described above and state one of its applications.* **[2]**

Question 12.

*A line on a plane can be represented by coordinates of the two-end points p1 and p2 as p1(x1, y1) and p2(x2, y2).***[10]**

A super class ***Plane*** *is defined to represent a line and a sub class* ***Circle*** *to find the length of the radius and the area of circle by using the required data members of super class. Some of the members of both the classes are given below:*

Class Name	:	***Plane***

Data members/ instance variables:

x1	:	*to store the x-coordinate of the first end point*
y1	:	*to store the y-coordinate of the first end point*

Member functions/ methods:

Plane(int nx, int ny)	:	*parameterised constructor to assign the data members x1 = nx and y1 = ny*
void Show( )	:	*to display the coordinates*
Class Name	:	*Circle*

Data members/ instance variables:

x2	:	*to store the x-coordinate of the second end point*
y2	:	*to store the y-coordinate of the second end point*
radius	:	*double variable to store the radius of the circle*
area	:	*double variable to store the area of the circle*

Member functions/ methods:

Circle(...)	:	*parameterised constructor to assign values to data members of both the classes*
void find Radius( )	:	*to calculate the length of radius using the formula:* $\left(\sqrt{(x_2-x_1)^2+(y_2-y_1)^2}\,/\,2\right)$ *assuming that* x_1, x_2, y_1, y_2 *are the coordinates of the two end of the diameters of a circle.*
void findArea( )	:	*to find the area of circle using formula:* $\pi * r * r$. *The value of pie* (π) *is 22/7 or 3.14*
void Show( )	:	*to display both the coordinates along with the length of radius and area of the circle*

Specify the class ***Plane,*** *giving details of the* ***constructor*** *and* ***void Show().*** *Using the* ***concept of inheritance,*** *specify the class* ***Circle*** *giving details of the* ***constructor, void findRadius(), void find Area()*** *and* ***void Show(). The main function and algorithm need not be written.***

Question 13.

(a) *A linked list is formed from the objects of the class:* **[4]**

```
class nodes
{
    int num;
    Nodes next;
}
```

Write an algorithm ***OR*** *a Method to print the sum of nodes that contains only odd integers of an existing linked list.*

The method declaration is as follows:

void NodesCount(Nodes starPtr)

(b) (i) *Give the meaning of the following common expression in Big O notation:* **[1]**

O(N)

$O(N^2)$

(ii) *List any two cases to analyse algorithm complexities.* **[1]**

(c) *Answer the following questions from the diagram of the Binary Tree given below :*

(i) *Name the leaf nodes of the right sub-tree.* **[1]**

(ii) *Write postorder traversal of the left sub-tree of node B including itself.* **[1]**

(iii) *State the level number of nodes R and M when the root is at level 0.* **[1]**

(iv) *Name the internal nodes of the tree.* **[1]**

ANSWERS

PART–I

Answer 1.

(a) Simplify :

$(A + C) . (A + A . D) + A . C + C$

$= (A + C) . (A (1 + D)) + C (A + 1)$

$= (A + C) . (A . 1) + C . 1$

$= A (A + C) + C$

$= (A . A) + A . C + C$ (by idempotent law)

$= A + A . C + C$

$= A . (1 + C) + C$

$= A . 1 + C$

$= A + C$

(b) $(A + B) . (C + D) . C$

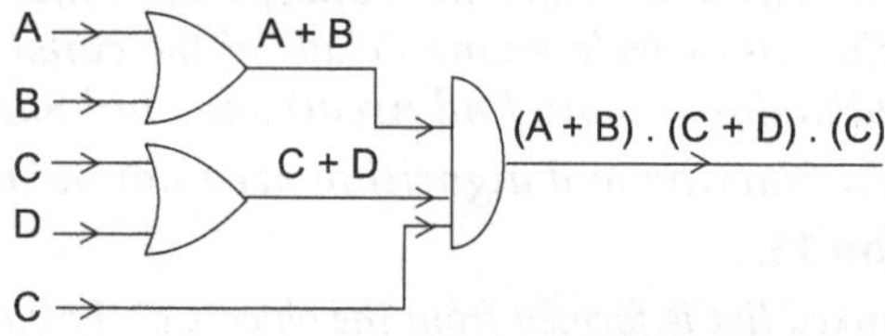

(c)

P	Q	~P	~P∧Q	P∨(~P∧Q)	P∨Q
0	0	1	0	0	0
0	1	1	1	1	1
1	0	0	0	1	1
1	1	0	0	1	1
				L.H.S.	R.H.S.

L.H.S. = R.H.S. **Hence, verified.**

(d) De Morgan's rule states that the complement of the sum of the variables is same as the product of the individual complements of the variables and vice-versa.

1. $\overline{P + Q} = \overline{P} \cdot \overline{Q}$

2. $\overline{P \cdot Q} = \overline{P} + \overline{Q}$

The following truth table verifies the above De Morgan's rule :

1. Verification of $\overline{A + B} = \overline{A} \cdot \overline{B}$

A	B	A + B	$\overline{A + B}$	$\overline{A}$	$\overline{B}$	$\overline{A} \cdot \overline{B}$
0	0	0	1	1	1	1
0	1	1	0	1	0	0
1	0	1	0	0	1	0
1	1	1	0	0	0	0
			L.H.S.			R.H.S.

L.H.S. = R.H.S. **Hence Proved.**

2. Verification of $\overline{A \cdot B} = \overline{A} + \overline{B}$

A	B	A · B	$\overline{A \cdot B}$	$\overline{A}$	$\overline{B}$	$\overline{A} + \overline{B}$
0	0	0	1	1	1	1
0	1	0	1	1	0	1
1	0	0	1	0	1	1
1	1	1	0	0	0	0
			L.H.S.			R.H.S.

L.H.S. = R.H.S. **Hence Proved.**

(e) **(i)** 1

(ii) It represents an OR gate.

Answer 2.

(a) Computational complexity is the study of how much of a given resource, a program uses. The resource is usually either space (how much memory) or time (how much basic operations).

for(int k = 0; k < n; k++)

s + = k;

Considering the loop execution in all the cases is n time.

∴ Let time taken be CO

∴ Total time = CO ∗ n

Complexity is linear so it is O(n).

(b) X + (Y – Z) + ((W + E) * F) / J

= X + (YZ –) + (WE + * F) / J

= X + (YZ –) + (WE + F *)/ J

= X + (YZ –) + (WE + F * J /)

= (XYZ – +) + (WE + F * J /)

= XYZ – + WE + F * J / +

(c)

This Keyword	Super Keyword
1. It is a reference variable, which referes to the current object.	It is a Keyword, used to access member of base/ super class, from the sub-class.
2. This Keyword is used to access global/ instance variable, when local variable and instance variables have the same name.	Super Keyword used to access members of super class from the sub-class.
3. It can access instance variables of current object.	It can access constructors, methods and instance variable of base class.

4. class TA
{
int a;
void fnTA1()
{
TA ob = new TA();
ob.a = this.a;
}
}
	class TB extends TA
	{
	int b;
	void fnTA1()
	{
	super.fnTA1();
	b = 10;
	System.out.print ("work done");
	}
	}

(d)

$$R = 10-(-2)+1 = 13$$
$$C = 8-3+1 = 6$$
$$LR = -2,\ LC = 3$$
$$BA = 4110,\ ES = 8$$
$$I = 4,\ J = 5$$
$$\because\ D[I, J] = BA + ES((I-LR)+R(J-LC))$$
$$D[4,5] = 4110 + 8((4-(-2))+13(5-3))$$
$$= 4110 + 8(6+26)$$
$$= 4110 + 8 \times 32$$
$$= 4110 + 256$$
$$= 4366$$

(e) 1. Each node can have at the most two children.
2. A unique path exists from the root to every other node.

Answer 3.

(a) (i)

x	y	Outer IF (X > 1)	Inner IF (X%Y = = 0)	Output
24	2	True	True	2
12	2	True	True	2
6	2	True	True	2
3	2	True	False	
3	3	True	True	3
1	3	False		

Output will be : 2 2 2 3

(ii)

X	Y	Outer IF (X >1)	Inner IF (X% Y= = 0)	Output
84	2	True	True	2
42	2	True	True	2
21	2	True	False	
21	3	True	True	3
7	3	True	False	
7	4	True	False	
7	5	True	False	
7	6	True	False	
7	7	True	True	7
1	7	False		

Output will be : 2 2 3 7

(iii) It displays the prime factors of X.

(b) (i) int sum = 0;

(ii) int s = *f* * 10;

(iii) sum + = digit * digit * digit;

(iv) num = num /10;

(v) if(sum = = N)

PART–II

SECTION–A

Answer 4.

(a) (i) F (A, B, C, D) = π (0, 1, 2, 3, 5, 7, 8, 9, 10, 11)

AB \ CD	C + D 00	C + $\bar{D}$ 01	$\bar{C}$ + $\bar{D}$ 11	$\bar{C}$ + D 10
A + B 00	0 (0)	0 (1)	0 (3)	0 (2)
A + $\bar{B}$ 01	1 (4)	0 (5)	0 (7)	1 (6)
$\bar{A}$ + $\bar{B}$ 11	1 (12)	1 (13)	1 (15)	1 (14)
$\bar{A}$ + B 10	0 (8)	0 (9)	0 (11)	0 (10)

Octal 1 = $M_0\ M_1\ M_2\ M_3$
$M_8\ M_9\ M_{10}\ M_{11}$
= B

Quad 1 = $M_1\ M_3\ M_5\ M_7$
= $A + \bar{D}$

So, the reduced expression is
= $B.(A + \bar{D})$

(ii) Logic Gate Diagram :

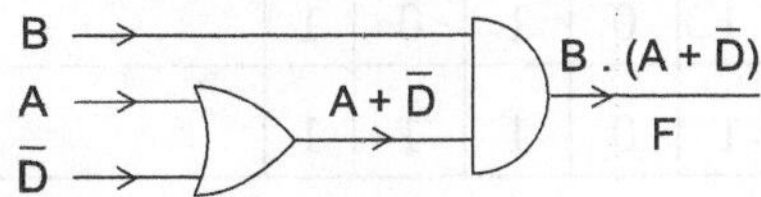

(b) (i) P (A, B, C, D)

$$= A B \overline{C} \overline{D} + \overline{A} B \overline{C} \overline{D} + \overline{A} B \overline{C} D + A B \overline{C} D + \overline{A} B C D + A B C D$$

AB \ CD	$\overline{C}.\overline{D}$	$\overline{C}.D$	$C.D$	$C.\overline{D}$
$\overline{A}.\overline{B}$	0 (0)	0 (1)	0 (3)	0 (2)
$\overline{A}.B$	1 (4)	1 (5)	1 (7)	0 (6)
$A.B$	1 (12)	1 (13)	1 (15)	0 (14)
$A.\overline{B}$	0 (8)	0 (9)	0 (11)	0 (10)

Quad 1 $= m_4 + m_5 + m_{12} + m_{13}$

$= B\overline{C}$

Quad 2 $= m_5 + m_7 + m_{13} + m_{15}$

$= BD$

So, the reduced expression is

$= B\overline{C} + BD$

(ii) Logic Gate Diagram :

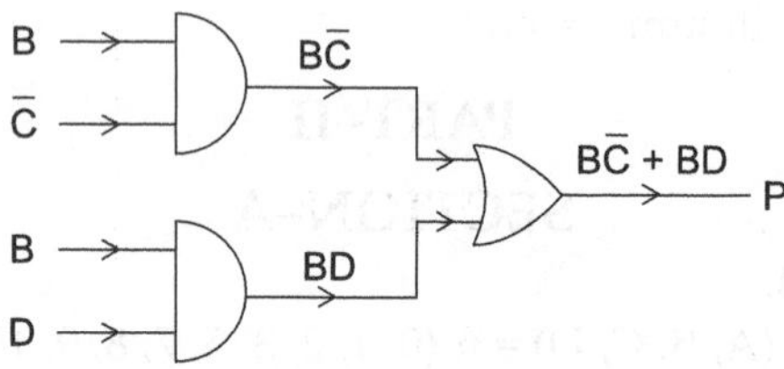

Answer 5. (a)

Designation	R	P	D	H	T	Maxterm
0	0	0	0	0	0	$R + P + D + H$
1	0	0	0	1	0	$R + P + D + \overline{H}$
2	0	0	1	0	0	$R + P + \overline{D} + H$
3	0	0	1	1	0	$R + P + \overline{D} + \overline{H}$
4	0	1	0	0	0	$R + \overline{P} + D + H$
5	0	1	0	1	0	$R + \overline{P} + D + \overline{H}$
6	0	1	1	0	1	
7	0	1	1	1	1	
8	1	0	0	0	0	$\overline{R} + P + D + H$
9	1	0	0	1	0	$\overline{R} + P + D + \overline{H}$
10	1	0	1	0	1	
11	1	0	1	1	1	
12	1	1	0	0	0	$\overline{R} + \overline{P} + D + H$
13	1	1	0	1	0	$\overline{R} + \overline{P} + D + \overline{H}$
14	1	1	1	0	1	
15	1	1	1	1	1	

T (R, P, D, H) = π (0, 1, 2, 3, 4, 5, 8, 9, 12, 13)

(b)

RP \ DH	D + H (00)	D + $\overline{H}$ (01)	$\overline{D}$ + $\overline{H}$ (11)	$\overline{D}$ + H (10)
R + P (00)	0 (0)	0 (1)	0 (3)	0 (2)
R + $\overline{P}$ (01)	0 (4)	0 (5)	1 (7)	1 (6)
$\overline{R}$ + $\overline{P}$ (11)	0 (12)	0 (13)	1 (15)	1 (14)
$\overline{R}$ + P (10)	0 (8)	0 (9)	1 (11)	1 (10)

Octat 1 $= M_0 . M_1 . M_4 . M_5 . M_8 . M_9 . M_{12} . M_{13}$

$= D$

Quad 1 $= M_0 . M_1 . M_2 . M_3$

$= R + P$

So, the reduced expression is

$= D . (R + P)$

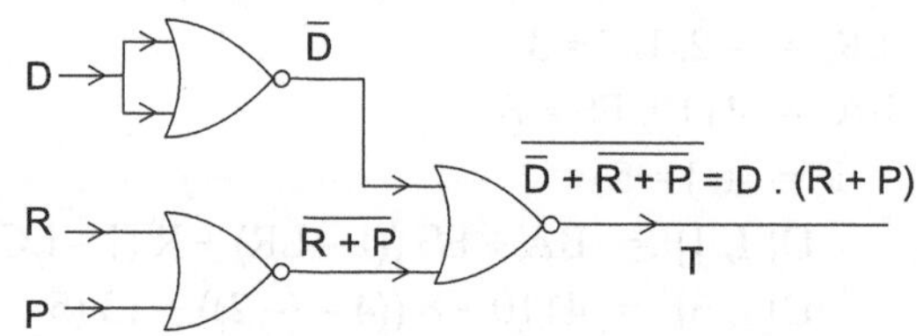

Answer 6.

(a)

Inputs	Outputs		
	F_2	F_1	F_0
0	0	0	0
1	0	0	1
2	0	1	0
3	0	1	1
4	1	0	0
5	1	0	1
6	1	1	0
7	1	1	1

Logic diagram for Octal to Binary Encoder :

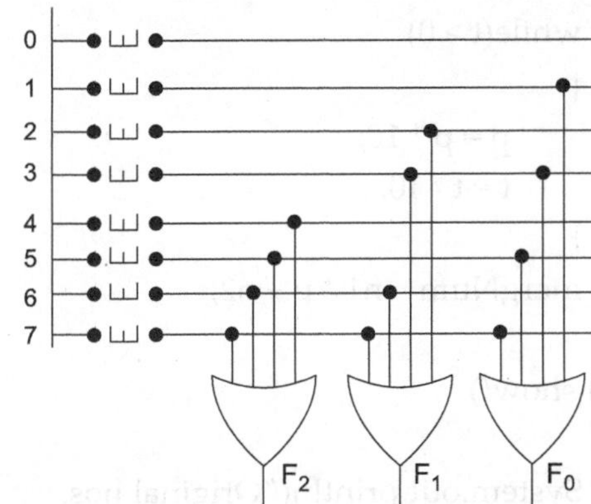

(b) Multiplexer is a combinational circuit that selects binary information from one of many input lines and directs it to a single output line. It is used to transmit a large number of information units over a single channel or line.

Application :

(i) It is used as a data selector.

(ii) It is used for routing of signals.

(iii) It is used for data transmission.

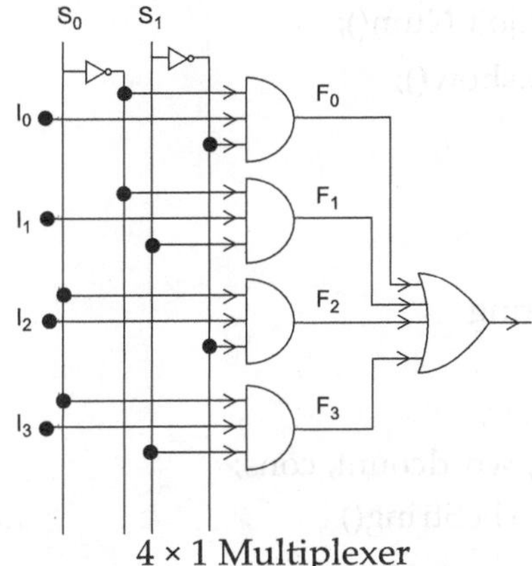

4 × 1 Multiplexer

(c) To prove :

Proof LHS

$X.Y.Z + X.\overline{Y}.Z + X.Y.\overline{Z} = X.(Y + Z)$

$= X.Z.(Y + \overline{Y}) + X.Y.\overline{Z}$ (Complementary Rule)

$= X.Z.1 + X.Y.\overline{Z}$ (Properties of 1)

$= X.Z + X.Y.\overline{Z}$

$= X.(Z + Y.\overline{Z})$ (Distributive Rule)

$= X.((Z + Y).(Z + Z'))$ (Complementary Rule)

$= X.((Y + Z).(1))$ (Properties of 1)

$= X.(Z + Y)$

Answer 7.

(a) $= \overline{(\overline{A}.\overline{B}.\overline{C})}.C + \overline{(\overline{A}.\overline{B}.\overline{C})}$

$= \overline{(\overline{A}.\overline{B}.\overline{C})}.(C + 1)$

$= \overline{\overline{A}.\overline{B}.\overline{C}}$

$= A + B + C$

(b) Universal gates can represent all the three basic gates— AND, OR and NOT. Thus all types of logic circuits are possible using the universal gates—NAND and NOR.

e.g., $= A.(B + C)$

$= A.B + A.C$

A, B → $\overline{AB}$; A, C → $\overline{AC}$; $\overline{\overline{AB}.\overline{AC}} = AB + AC$

(c) Refer to ISC Paper 2017, Answer 5 (b).

SECTION–B

Answer 8.

```
import java.io.*;
class Admission
{
    int Adno[];
    public Admission()
    {
        Adno[] = new int[100];
    }
    void fillArray()
    {
        int i;
        Scanner x = new Scanner(System.in);
        for(i = 0; i < 100; i++)
        {
            System.out.println("Enter a no.");
            Adno[i] = x.nextInt();
        }
    }
    int binSearch(int l, int u, int v)
    {
        if(l > u)
        {
            return(–1);
        }
        else
        {
            int mid = (l + u) / 2;
            if(v > Adno[mid])
            {
                return(binSearch(mid + 1, u, v));
            }
            else if(v < Adno[mid])
            {
```

```
            return(binSearch(l, mid –1, v));
        }
        else
        {
            return(1);
        }
    }
}
public static void main(String args[])
{
    Admission ob = new Admission();
    ob.fillArray();
    int v = x.nextInt();
    int x = ob.binSearch(0, ob.Adno.length – 1, v);
    if(x == 1)
    {
        System.out.println("Found");
    }
    else
    {
        System.out.println("Not Found");
    }
}
}
```

Answer 9.

```
import java.io.*;
class Merger
{
    long n1, n2, mergNum;
    public Merger()
    {
        n1 = 0;
        n2 = 0;
        mergNum = 0;
    }
    void readNum()
    {
        Scanner x = new Scanner(System.in);
        System.out.println("Enter 2 nos.");
        n1 = x.nextLong();
        n2 = x.nextLong();
    }
    void joinNum()
    {
        int p = 1;
        long t = n2;
        while(t > 0)
        {
            p = p * 10;
            t = t / 10;
        }
        mergNum = n1 * p + n2;
    }
    void show()
    {
        System.out.println("Original nos.
                    are " + n1 + " and " + n2);
        System.out.println("Merged no. is "
                    + mergNum);
    }
    public static void main(String args[])
    {
        Merger ob = new Merger();
        ob.readNum();
        ob.joinNum();
        ob.show();
    }
}
```

Answer 10.

```
class TheString
{
    String str;
    int len, wordcount, cons;
    public TheString()
    {
        str = "";
        wordcount = 0;
        cons = 0;
    }
    public TheString(String ds)
    {
        str = ds;
        len = str.length();
    }
    void countFreq()
    {
        int i;
        char ch;
        str = str.toLowerCase();
        for(i = 0; i < len; i++)
        {
            ch = str.charAt(i);
```

```
            if(ch == ' ')
            {
                wordcount++;
            }
            if(ch != 'a' && ch != 'e' && ch !=
            'i' && ch != 'o' &&ch != 'u')
            {
                cons++;
            }
        }
    }
    void Display()
    {
        System.out.println("Original String"+ str);
        System.out.println("Total words" +
                                wordcount + 1);
        System.out.println("Total Consonants" +
                                cons – wordcount);
    }
    public static void main(String args[])
    {
        TheString ob = new TheString
                        ("India is my country");
        ob.countFreq();
        ob.Display();
    }
}
```

SECTION–C

Answer 11.

(a) class WordPile

```
{
    char ch[] = new char[20];
    int capacity, top;
    public WordPile(int cap)
    {
        capacity = cap;
        top = –1;
        ch = new char[capacity];
    }
    void pushChar(char v)
    {
        if(top == capacity – 1)
        {
        System.out.println("WordPile is Full");
        }
        else
        {
            ch[++top] = v;
        }
    }
    char popChar()
    {
        if(top == –1)
        {
            return('\ \');
        }
        else
        {
            char a = ch[top];
            top– –;
            return(a);
        }
    }
}
```

(b) It is also known as STACK. It is used for LIFO, Recursion, infix to postfix.

Answer 12.

```
class Plane
{
        protected double x1, y1;
        public Plane(double nx,double ny)
        {
            x1 = nx;
            y1 = ny;
        }
        public void Show()
        {
            System.out.println(x1 + " ," + y1);
        }
}
class Circle extends Plane
{
double x2, y2, radius, area;
public Circle(double a, double b, double c, double d)
        {
            super(a, b);
            x2 = c;
            y2 = d;
        }
        void findRadius()
        {
            radius = Math.sqrt(Math.pow
```

```
        (x2 – x1, 2) + Math.pow(y2 – y1, 2)) / 2;
    }
    void findArea()
    {
        area = 3.14 * radius * radius;
    }
    public void Show()
    {
        super.show();
        System.out.println("Second x
        coordinate =" + x2 +"," + "Second y
                    coordinate =" + y2);
        System.out.println("Radius is" +
                    radius + "Area is" + area);
    }
}
```

Answer 13.

(a) Algorithm :

Step 1 : Start

Step 2 : Set temporary pointer (sayStr) to starPtr.

Step 3 : Set sum to 0

Step 4 : Repeat step 5 to step until pointer reaches null

Step 5 : if Str.num is Odd then add it to sum

Step 6 : move the pointer to next node

Step 7 : Print sum

Step 8 : End

OR

Method :

```
void NodesCount(Nodes strPtr)
{
    Nodes Str = new Nodes(starPtr);
    int sum = 0;
    while(Str != null)
    {
        if(Str.num % 2 != 0)
        sum += Str.num;
        Str = Str.next;
    }
    System.out.println(sum);
}
```

(b) (i) O(N) complexity means linear time complexity. It means that algorithm must have a single loop running 'N' times.

$O(N^2)$ complexity means quadratic time complexity. It means that algorithm must have nested loops running upto N times.

(ii) Best case, worst case are the two cases used to analyse algorithm complexities.

(c) (i) E, P

(ii) Post order traversal of left sub tree is :

R, N, M, C, B

(iii) Level 2

(iv) C, F, M, H

●●

COMPUTER SCIENCE 2014

QUESTIONS

PART–I

Answer all questions.

While answering questions in this Part, indicate briefly your working and reasoning, wherever required.

Question 1.

(a) *From the logic circuit diagram given below, find the output 'F' and simplify it. Also, state the law represented by the logic diagram.* **[2]**

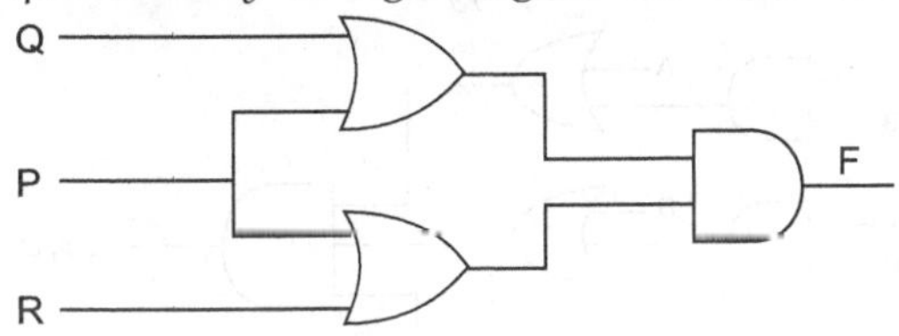

(b) *Write the truth table for a 2-input conjunction and disjunction in a proposition.* **[2]**

(c) *Find the complement of* $XY'Z + XY + YZ'$. **[2]**

(d) *Convert the following expression into its canonical POS form :* **[2]**

$$F(A, B) = (A + B) . A'$$

(e) *Minimise the following Boolean expression using the Karnaugh map :* **[2]**

$$F(A, B, C) = \overline{A}\overline{B}\overline{C} + \overline{A}BC + AB\overline{C} + ABC$$

Question 2.

(a) *State two advantages of using the concept of inheritance in Java.* **[2]**

(b) *An array AR [– 4...6, – 2...12], stores elements in* ***Row Major Wise,*** *with the address AR [2][3] as 4142. If each element requires 2 bytes of storage, find the Base address.* **[2]**

(c) *State the sequence of traversing a binary tree in :*

(i) *preorder*

(ii) *postorder* **[2]**

(d) *Convert the following infix expression into its postfix form :* **[2]**

$$(A/B + C) * (D/ (E – F))$$

(e) *State the difference between the functions* ***int nextInt*** *() and* ***boolean has NextInt***(). **[2]**

Question 3.

(a) *The following functions are part of some class:*

```
void fun1 (char s[], int x)
{ System.out.println(s);
  char temp;
  if (x < s.length/2)
  { temp=s[x];
    s[x]=s[s.length-x-1];
    s[s.length-x-1]=temp;
    fun1 (s, x + 1);
  }
}
void fun2(String n)
{ char c[] = new char[n.length()];
  for(int i = 0; i < c.length; i++)
    c[i] = n.charAt(i);
  fun1(c, 0);
}
```

(i) *What will be the output of* ***fun1()*** *when the value of s[] = {'j', 'U', 'N', 'E'} and x=1 ?* **[2]**

(ii) *What will be the output of* ***fun2()*** *when the value of n ="SCROLL" ?* **[2]**

(iii) *State in one line what does the function* ***fun1()*** *do apart from recursion.* **[1]**

(b) *The following is a function of some class which sorts an integer array a[] in ascending order using* ***selection sort*** *technique. There are some places in the code marked by* ***?1?, ?2?, ?3?, ?4?, ?5?*** *which may be replaced by a statement/ expression so that the function works properly :*

```
void selectsort(int []a)
{
  int i, j, t, min, minpos;
  for(i=0;i<?1?;i++)
  {
    min=a[i];
    minpos=i;
    for(j=?2?;j<a.length;j++)
    {
      if(min>a[j])
      {
        ?3? = j;
        min = ?4?;
      }
    }
    t=a[minpos];
    a[minpos]=a[i];
    a[i]=?5?;
  }
```

```
    for(int k=0;k<a.length;k++)
    System.out.println(a[k]);
}
```

(i) *What is the expression or statement at ?1?* [1]

(ii) *What is the expression or statement at ?2?* [1]

(iii) *What is the expression or statement at ?3?* [1]

(iv) *What is the expression or statement at ?4?* [1]

(v) *What is the expression or statement at ?5?* [1]

PART–II

*Answer **seven** questions in this part, choosing **three** questions from Section A, **two** from Section B and **two** from Section C.*

SECTION–A

*Answer any **three** questions.*

Question 4.

(a) *Given the Boolean function :*

F (A, B, C, D) = Σ (0, 1, 2, 3, 5, 6, 7, 10, 13, 14, 15)

(i) *Reduce the above expression by using 4-variable Karnaugh map, showing the various groups (i.e., octal, quads and pairs).* [4]

(ii) *Draw the logic gate diagram for the reduced expression. Assume that the variables and their complements are available as inputs.* [1]

(b) *Given the Boolean function :*

P (A, B, C, D) = π (0, 1, 2, 3, 5, 6, 7, 10, 13, 14, 15)

(i) *Reduce the above expression by using 4-variable Karnaugh map, showing the various groups (i.e., octal, quads and pairs).* [4]

(ii) *Draw the logic gate diagram for the reduced expression. Assume that the variables and their complements are available as inputs.* [1]

Question 5.

A school intends to select candidates for the Inter-School Athletic Meet, as per the criteria given below :

- *The candidate is from the Senior School and has participated in an Inter-School Athletic Meet earlier.*

OR

- *The candidate is not from the Senior School, but the height is between 5 ft. and 6 ft. and weight is between 50 kg and 60 kg.*

OR

- *The candidate is from the senior school and has height between 5 ft. and 6 ft., but the weight is not between 50 kg and 60 kg.*

The inputs are :

Inputs	
S	*Student is from the Senior School.*
W	*Weight is between 50 kg. and 60 kg.*
H	*Height is between 5 ft. and 6 ft.*
A	*Taken part in Inter-School Athletic Meet earlier.*

(In all of the above cases 1 indicates yes and 0 indicates no)

*Output : **X**-Denotes the selection criteria [1 indicates selected and 0 indicates rejected in all cases.]*

(a) *Draw the truth table for the inputs and outputs given above and write the **SOP** expression for X (S, W, H, A).* [5]

(b) *Reduce **X (S, W, H, A)** using Karnaugh map.* [5]

*Draw the logic gate diagram for the reduced **SOP** expression for X (S, W, H, A) using AND and OR gate. You may use gates with two or more inputs. Assume that the variable and their complements are available as inputs.*

Question 6.

(a) *With the help of a logic diagram and a truth table explain a **Decimal to Binary** encoder.* [4]

(b) *Derive a Boolean expression for the logic diagram given below and simplify it.* [3]

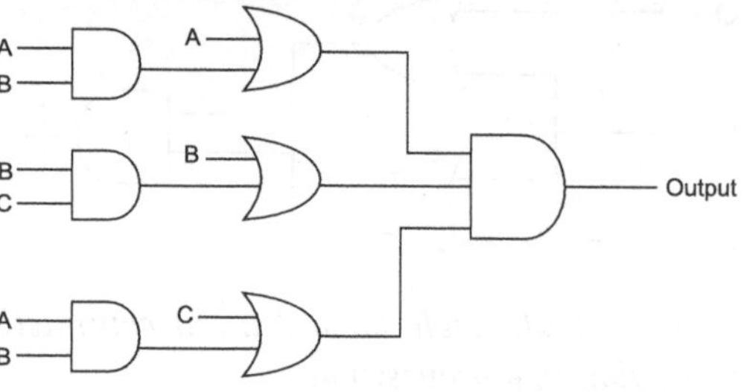

(c) *Reduce the following expression using Boolean laws :* [3]

F (A, B, C, D) = (A′ + C) (A′ + C′) (A′ + B + C′D)

Question 7.

(a) *Differentiate between XNOR and XOR gates. Draw the truth table and logic diagrams of 3 inputs XNOR gate.* [4]

(b) *Differentiate between a **proposition** and **wff**.* [2]

(c) *Define **Half Adder**. Construct the truth table and a logic diagram of a Half-Adder.* [4]

SECTION–B

*Answer any **two** questions.*

Each program should be written in such a way that it clearly depicts the logic of the problem.

This can be achieved by using mnemonic names and comments in the program.

*(Flowcharts and Algorithms are **not** required.)*

The programs must be written in Java.

Question 8.

*A class **Mixer** has been defined to merge two sorted integer arrays in ascending order. Some of the members of the class are given below :* [10]

Class name: **Mixer**

Data members/ instance variables :

int arr[] : *to store the elements of an array*

int n : *to store the size of the array*

Member functions :

Mixer(int nn) :	*constructor to assign n=nn*
void accept() :	*to accept the elements of the array in ascending order without any duplicates*
Mixer : mix(Mixer A)	*to merge the current object array elements with the parameterised array elements and return the resultant object*
void display() :	*to display the elements of the array*

Specify the class ***Mixer****, giving details of the* ***constructor (int), void accept(), Mixer mix(Mixer) and void display ()****. Define the* ***main()*** *function to create an object and call the function accordingly to enable the task.*

Question 9.

A class ***SeriesSum*** *is designed to calculate the sum of the following series :*

$$Sum = \frac{x^2}{1!} + \frac{x^4}{3!} + \frac{x^6}{5!} + \ldots \frac{x^n}{(n-1)!}$$

Some of the members of the class are given below :

Class name: ***SeriesSum***

Data Members/ instance variables :

x :	*to store an integer number*
n :	*to store number of terms*
sum :	*double variable to store the sum of the series*

Member functions :

seriesSum(int: xx,intnn)	*constructor to assign x=xx and n=nn*
double find fact(int m)	*to return the factorial of m using recursive technique.*
double findpower (int: x, int y)	*to return x raised to the power of y using recursive technique.*
void : calculate()	*to calculate the sum of the series by invoking the recursive functions respectively*
void display() :	*to display the sum of the series*

(a) *Specify the class* ***SeriesSum****, giving details of the* ***constructor(int, int), double findfact(int), double findpower (int, int), void calculate()*** *and* ***void display()****. Define the* ***main()*** *function to create an object and call the functions accordingly to enable the task.* **[8]**

(b) *State the* ***two*** *differences between* ***iteration*** *and* ***recursion****.* **[2]**

Question 10.

A sequence of fibonacci strings is generated as follows: **[10]**

S_0 = *"a"*, S_1 = *"b"*, $S_n = S_{(n-1)} + S_{(n-2)}$ where '+' denotes concatenation. Thus the sequence is : a, b, ba, bab, babba, babbabab, ...*n* terms.

Design a class ***FiboString*** *to generate fibonacci strings. Some of the members of the class are given below :*

Class name : ***FiboString***

Data Members/ instance variables :

x :	*to store the first string*
y :	*to store the second string*
z :	*to store the concatenation of the previous two strings*
n :	*to store the number of terms*

Member functions/ methods:

FiboString() :	*constructor to assign x = "a", y = "b" and z = "ba"*
void accept() :	*to accept the number of terms 'n'*
void generate() :	*to generate and print the fibonacci strings. The sum of ('+' i.e., concatenation) first two strings is the third string. E.g. "a" is first string, "b" is second string then the third will be "ba", and fourth will be "bab" and so on.*

Specify the class ***FiboString****, giving details of the* ***constructor(), void accept()*** *and* ***void generate()****. Define the* ***main()*** *function to create an object and call the functions accordingly to enable the task.*

SECTION–C

Answer any ***two*** *questions.*
Each Program should be written in such a way that it clearly depicts the logic of the problem stepwise.
This can be achieved by using comments in the program and mnemonic names or pseudo codes for algorithms.
The programs must be written in Java and the algorithms must be written in general/standard form, wherever required / specified.
(Flowcharts are ***not*** *required)*

Question 11.

A super class ***Stock*** *has been defined to store the details of the stock of a retail store. Define a subclass* ***Purchase*** *to store the details of the items purchased with the new rate and updates the stock. Some of the members of the classes are given below:* **[10]**

Class name : ***Stock***

Data Members/ instance variables :

item :	*to store the name of the item*
qty :	*to store the quantity of an item in stock*

rate :	*to store the unit price of an item*
amt :	*to store the net value of the item in stock*

Member functions :

Stock(...)	*parameterised constructor to assign values to the data members*
void display() :	*to display the stock details*

Class name : *Purchase*

Data members/ instance variables :

pqty :	*to store the purchased quantity*
prate :	*to store the unit price of the purchased item*

Member functions/ methods :

Purchase(...) :	*parameterised constructor to assign values to the data members of both classes*
void update( ) :	*to update stock by adding the previous quantity by the purchased quantity and replace the rate of the item if there is a difference in the purchase rate. Also update the current stock value as : (quantity * unit price)*
void display( ) :	*to display the stock details before and after updation.*

*Specify the class **Stock**, giving details of the **constructor()** and **void display()**. Using **concept of inheritance**, specify the class **Purchase**, giving details of the **constructor()**, **void update()** and **void display()**.*

The main function and algorithm need not be written.

Question 12.

A stack is a linear data structure which enables the user to add remove integers from one end only, using the concept of LIFO (Last in First Out). An array containing the marks of 50 students in ascending order is to be pushed into the stack.

*Define a class **Array_to_Stack** with the following details:* **[10]**

Class name: ***Array_to_stack***

Data members/ instance variables:

m[ ] :	*to store the marks*
st[ ] :	*to store the stack elements*
cap :	*maximum capacity of the array and stack*
top:	*to point the index of the topmost element of the stack*

Methods/ Member functions :

Array_to_Stack (int n) :	*parameterised constructor to initialise cap = n and top = – 1*
void input_ marks() :	*to input the marks from user and store it in the array m [ ] in ascending order and simultaneously push the marks into the stack st[] by invoking the function pushmarks().*
void pushmarks (int v) :	*to push the marks into the stack at top location if possible, otherwise, display "not possible"*
int popmarks() :	*to return marks from the stack if possible, otherwise, return-999*
void display() :	*To display the stack elements.*

*Specify the class **Array_to_Stack,** giving the details of the **constructor(int). void input_marks(), void pushmarks(int), int popmarks()** and **void display().***

The main function and the algorithm need not be written.

Question 13.

(a) *A linked list is formed from the objects of the class:* **[4]**

```
class Node
    {
        int    number;
        Node  nextNode;
    }
```

*Write an **algorithm OR a Method** to add a node at the end of an existing linked list. The method declaration is as follows :*

void addnode (Node start, int num)

(b) *Define the terms **complexity** and **big 'O' notation.*** **[2]**

(c) *Answer the following from the diagram of the Binary Tree given below :*

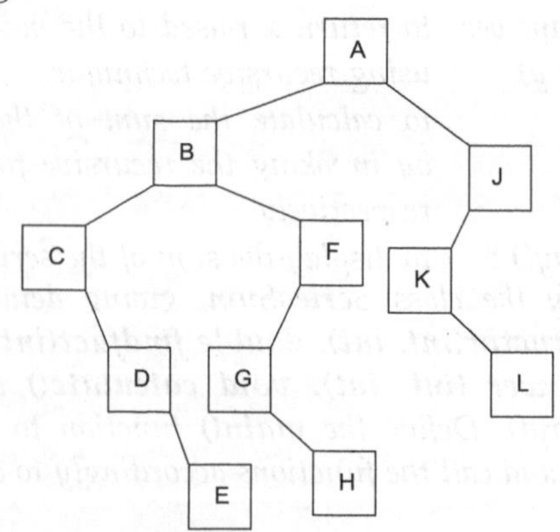

(i) *Root of the tree* **[1]**

(ii) *Left subtree* **[1]**

(iii) *Inorder traversal of the tree* **[1]**

(iv) *Size of the tree* **[1]**

ANSWERS

PART–I

Answer 1.

(a) $F = (P + Q) . (P + R) = P + QR$

It is distributive law.

(b)

X	Y	X . Y	X + Y
0	0	0	0
0	1	0	1
1	0	0	1
1	1	1	1

(c) $(X\bar{Y}Z + XY + Y\bar{Z})'$

$= (X\bar{Y}Z)' . (XY)' . (Y\bar{Z})'$

$= (\bar{X} + Y + \bar{Z}) . (\bar{X} + \bar{Y}) . (\bar{Y} + Z)$

(d) $(A + B) . \bar{A}$

$(A + B) . (\bar{A} + 0)$ [Properties of 0]

$(A + B) . (\bar{A} + B \cdot \bar{B})$ [$\because B . \bar{B} = 0$, complementrity]

$(A + B) . (\bar{A} + B) . (\bar{A} + \bar{B})$ [Distributive law]

(e) $F (A, B, C) = \bar{A}B\bar{C} + \bar{A}BC + AB\bar{C} + ABC$

A \ BC	$\bar{B}.\bar{C}$	$\bar{B}.C$	$B.C$	$B.\bar{C}$
$\bar{A}$	0 (0)	0 (1)	1 (3)	1 (2)
A	0 (4)	0 (5)	1 (7)	1 (6)

Quad 1 = $m_2 + m_3 + m_6 + m_7 = B$

Answer 2.

(a) Advantages of inheritance in Java :

1. It facilitates reusability of classes with additional functionality.
2. Its transitive nature helps to extend a program.
3. It exhibits real world concept.

(b) AR[2][3] = 4142

LR = – 4, LC = – 2

I = 2, J = 3, E.S = 2

R = 6 – (– 4) + 1 = 11, C = 12 – (– 2) + 1 = 15

∵ AR [I, J] = B.A + E.S [C(I – LR) + (J – LC)]

AR [2, 3] = B.A + 2 [(15 (2 – (– 4))) + 3 – (– 2)]

4142 = B.A + 2 [15 × 6 + 5]

4142 = B.A + 190

B.A = 4142 – 190

= 3952

(c) **(i)** Pre order is Root, Left, Right

(ii) Post order is Left, Right, Root.

(d) (A/B + C) * [D/(E – F)]

= (AB/ + C) * (D/EF–)

= (AB/C) + * (DEF –/)

= AB/C + DEF–/*

(e)

nextInt()	has nextInt()
Returns next token as an integer.	Returns true if the input has another token and that token can be interpreted as an integer.

Answer 3.

(a) **(i)** J U N E

J N U E

(ii) S C R O L L

L C R O L S

L L R O C S

L L O R C S

(iii) It reverses the part of a string from the specified location.

(b) **(i)** a.length –1

(ii) i + 1

(iii) minpos

(iv) a[j]

(v) t

PART–II

SECTION–A

Answer 4.

(a) **(i)** F (A, B, C, D)

$= \Sigma (0, 1, 2, 3, 5, 6, 7, 10, 13, 14, 15)$

AB \ CD	$\bar{C}.\bar{D}$	$\bar{C}.D$	$C.D$	$C.\bar{D}$
$\bar{A}.\bar{B}$	1 (0)	1 (1)	1 (3)	1 (2)
$\bar{A}.B$	0 (4)	1 (5)	1 (7)	1 (6)
$A.B$	0 (12)	1 (13)	1 (15)	1 (14)
$A.\bar{B}$	0 (8)	0 (9)	0 (11)	1 (10)

Quad1 = $m_0 + m_1 + m_2 + m_3 = \bar{A}\,\bar{B}$

Quad2 = $m_5 + m_7 + m_{13} + m_{15} = BD$

Quad3 = $m_2 + m_6 + m_{10} + m_{14} = C\bar{D}$

$= \bar{A} . \bar{B} + B . D + C . D$

(ii) Logic Gate Diagram :

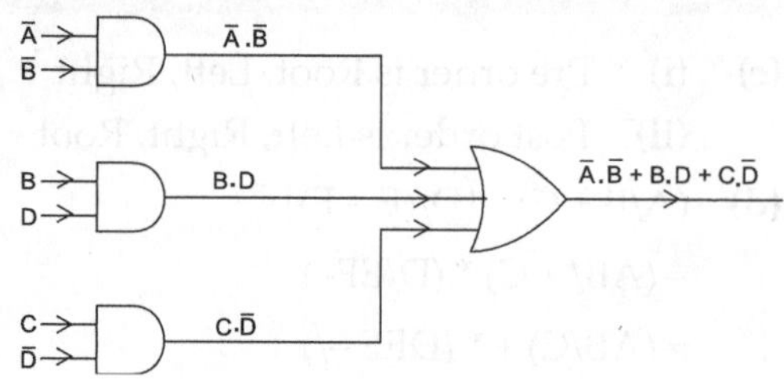

(b) (i) P (A, B, C, D) = π (0, 1, 2, 3, 5, 6, 7, 10, 13, 14,15)

AB \ CD	C + D	C + $\overline{D}$	$\overline{C}$ + $\overline{D}$	$\overline{C}$ + D
A + B	0 (0)	0 (1)	0 (3)	0 (2)
A + $\overline{B}$	1 (4)	0 (5)	0 (7)	0 (6)
$\overline{A}$ + $\overline{B}$	1 (12)	0 (13)	0 (15)	0 (14)
$\overline{A}$ + B	1 (8)	1 (9)	1 (11)	0 (10)

Quad1 = $M_0 . M_1 . M_2 . M_3 = A + B$

Quad2 = $M_5 . M_7 . M_{13} . M_{15} = \overline{B} + \overline{D}$

Quad3 = $M_2 . M_6 . M_{10} . M_{14} = \overline{C} + D$

$= (A + B) . (\overline{B} + \overline{D}) . (\overline{C} + D)$

(ii) Logic Gate Diagram :

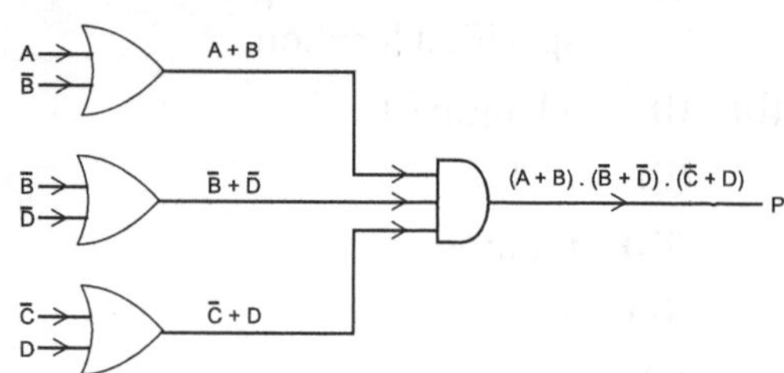

Answer 5.

(a)

S	W	H	A	X	
0	0	0	0	0	
0	0	0	1	0	
0	0	1	0	0	
0	0	1	1	0	
0	1	0	0	0	
0	1	0	1	0	
0	1	1	0	1	$\overline{S}.W.H.\overline{A}$
0	1	1	1	1	$\overline{S}.W.H.A$
1	0	0	0	0	
1	0	0	1	1	$S.\overline{W}.\overline{H}.A$
1	0	1	0	1	$S.\overline{W}.H.\overline{A}$
1	0	1	1	1	$S.\overline{W}.H.A$
1	1	0	0	0	
1	1	0	1	1	$S.W.\overline{H}.A$
1	1	1	0	0	
1	1	1	1	1	$S.W.H.A$

SOP expression :

$= \overline{S}.W.H.\overline{A} + \overline{S}.W.H.A + S.\overline{W}.\overline{H}.A + S.\overline{W}.H.\overline{A} + S.\overline{W}.H.A + S.W.\overline{H}.A + S.W.H.A$

$= X(S, W, H, A) = \Sigma (6, 7, 9, 10, 11, 13, 15)$

(b)

SW \ HA	$\overline{H}.\overline{A}$	$\overline{H}.A$	H.A	H.$\overline{A}$
$\overline{S}.\overline{W}$	0 (0)	0 (1)	0 (3)	0 (2)
$\overline{S}$.W	0 (4)	0 (5)	1 (7)	1 (6)
S.W	0 (12)	1 (13)	1 (15)	0 (14)
S.$\overline{W}$	0 (8)	1 (9)	1 (11)	1 (10)

Quad 1 $= m_9 + m_{11} + m_{13} + m_{15} = S.A$

Pair 1 $= m_6 + m_7 = \overline{S}.W.H$

Pair 2 $= m_{10} + m_{11} = S.\overline{W}.H$

$= S.A + \overline{S}.W.H + S.\overline{W}.H$

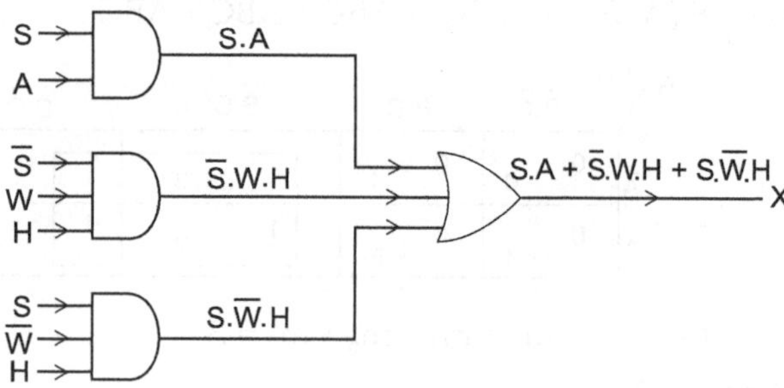

Answer 6.

(a) Decimal to binary encoder is used to convert decimal numbers into its equivalent binary form.

	F_3	F_2	F_1	F_0
0	0	0	0	0
1	0	0	0	1
2	0	0	1	0
3	0	0	1	1
4	0	1	0	0
5	0	1	0	1
6	0	1	1	0
7	0	1	1	1
8	1	0	0	0
9	1	0	0	1

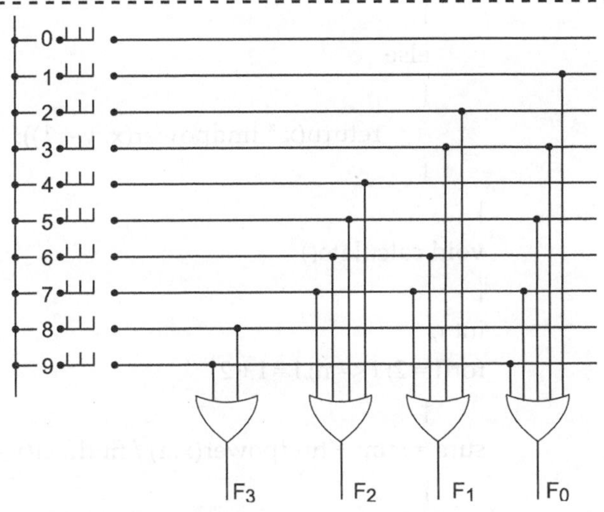

(b) $(AB + A)(BC + B)(AB + C)$

$= (A(B+1))(B(C+1))(AB+C)$ (By Distributive law)

$= (A)(B)(AB+C)$ (By Properties of 1)

$= AB(AB+C)$

$= AB.AB + ABC$ (By Distributive law)

$= AB + ABC$ (By Idempotance law)

$= AB(1+C)$ (By Distributive law)

$= AB$ (By Properties of 1)

(c) $F(A, B, C, D)$

$= (A'+C)(A'+C')(A'+B+C'\bullet D)$

$= (A'A' + A'C' + A'C + CC')(A'+B'+C'D)$ (By Distributive law)

$= (A' + A'C' + A'C)(A'+B'+C'D)$ (By Complementary law)

$= A'(1+C+C')(A'+B'+C'D)$ (By Distributive law)

$= A'(A'+B'+C'D)$ (Properties of 1) $(1+C+C'=1)$

$= (A' + A'B' + A'C'D)$ (By Distributive law)

$= A'(1+B'+C'D)$ (By Distributive law)

$= A'$ (Properties of 1) $(C'D+B'+1=1)$

Answer 7.

(a)

	XNOR	XOR
1.	Logically equivalent to an inverted XOR, *i.e.*, XOR gate followed by a NOT gate.	Logically equivalent to an exclusive OR. *i.e.*, if both inputs are 1 then output is 0 (reverse of inclusive OR).
2.	Produces 1 (high) as output when the input combination has even number of 0s.	Produces 1 (high) as output when the input combination has odd number of 1s.

Truth table of 3 input XNOR gate :

X	Y	Z	X ⊙ Y ⊙ Z
0	0	0	1
0	0	1	0
0	1	0	0
0	1	1	1
1	0	0	0
1	0	1	1
1	1	0	1
1	1	1	0

Logic gate diagram of 3 input XNOR gate :

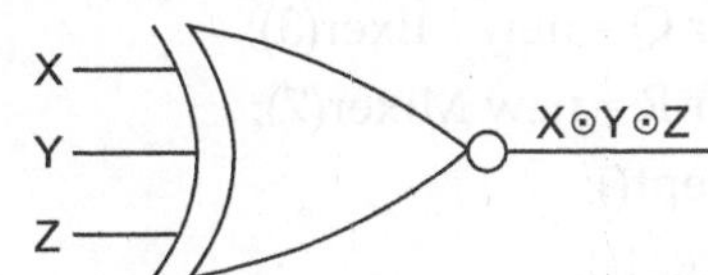

(b)

Proposition	wff
A proposition is an elementary atomic sentence that returns either true or false.	*wff's* are well formed formula which is a proposition that is satisfiable or valid.

(c) Refer to ISC Paper 2017, Answer 5 (b).

SECTION–B

Answer 8.

```
import java.util.*;
class Mixer
{
    int arr[];
    int n;
    static Scanner sc = new Scanner(System.in);
    Mixer(int nn)
    {
        n = nn;
        arr = new int[n];
    }
    void accept()
    {
        System.out.println("Enter" + n+"elements in
                                        ascending order");
        for(int i = 0; i < n; i++)
            arr[i] = sc.nextInt();
    }
    Mixer mix(Mixer A);
    {
        Mixer B = new Mixer(n + A.n);
        int x = 0;
        for(int i = 0; i < n; i++)
```

```
        B.arr[x++] = arr[i];
            for(int j = 0; j < A.n; j++)
                B.arr[x++] = A.arr[j];
        return B;
    }
    void display
    {
        for(int i = 0; i < n; i++)
            System.out.println(arr[i]);
    }
    static void main()
    {   Mixer P = new Mixer(4);
        Mixer Q = new Mixer(3);
        Mixer R = new Mixer(7);
        P.accept();
        Q.accept();
        R = P.mix(Q);
        R.display();
    }
}
```

Answer 9.

(a)

```
class SeriesSum
{
    int x, n;
    double sum;
    public SeriesSum(int xx, int nn)
    {
        x = xx;
        n = nn;
        sum = 0.0;
    }
    double findfact(int m)
    {
        if(m == 1)
        {
            return(1);
        }
        else
        {
            return(m * findfact(m – 1));
        }
    }
    double findpower(int x, int y)
    {
        if(y == 0)
        {
            return(x);
        }
        else
        {
            return(x * findpower(x, y – 1));
        }
    }
    void calculate()
    {
    int i;
    for(i = 2; i <= n; i = i + 2)
        {
    sum = sum + findpower(x, i) / findfact(i – 1);
        }
    }
    void display()
    {
        System.out.println("Sum is : " + sum);
    }
    public static void main(String args[])
    {
        SeriesSum ob = newSeriesSum(2, 6);
        ob.calculate();
        ob.display();
    }
}
```

(b) **1.** Iteration takes less amount of memory whereas recursion takes large amount of memory.

2. Iteration can work on slower CPU whereas recursion can only work on system having high speed and RAM.

Answer 10.

```
import java.io.*;
class FiboString
{
    String x, y, z;
    int n;
    public FiboString()
    {
        x = "a";
        y = "b";
        z = "ba";
    }
    void accept()
    {
    Scanner sc = new Scanner(System.in));
    System.out.println("Enter no. of terms");
```

```
    n = sc.nextInt();
    }
    void generate()
    {
        int i;
        System.out.print(x + " " + y + " ");
        for(i = 3; i <= n; i++)
        {
            z = y + x;
            x = y;
            y = z;
            System.out.print(z + " ");
        }
    }
    public static void main(String args[])
    {
        FiboString ob = new FiboString();
        ob.accept();
        ob.generate();
    }
}
```

SECTION–C

Answer 11.

```
class Stock
{
    protected String item;
    protected int qty;
    protected double rate, amt;
    public Stock(String it, int q, double r)
    {
        item = it;
        qty = q;
        rate = r;
        amt = qty * rate;
    }
    public void display()
    {
        System.out.println(item + " " + qty +
                               " " + rate + " " + amt);
    }
}
class Purchase extends Stock
{
    int pqty;
    double prate;
    public Purchase(String it1, int q1, double r1,
        double pr, int pq)
    {
        super(it1, q1, r1);
        pqty = pq;
        prate = pr;
    }
    void update()
    {
        qty += pqty;
        if(rate != prate)
        {
            rate = prate;
        }
        amt = qty * rate;
    }
    public void display()
    {
        System.out.println("Details Before Updation ");
        super.display();
        System.out.println("Details After Updation ");
        update();
        super.display();
    }
}
```

Answer 12.

```
import java.io.*;
class Array_to_Stack
{
    int m[], st[], cap, top;
    public Array_to_Stack(int n)
    {
        cap = n;
        top = -1;
        m = new int[cap];
        st = new int[cap];
    }
    void input_marks()
    {
        int i;
        Scanner sc = new Scanner(System.in);
        for(i = 0; i < cap; i++)
        {
            System.out.println("Enter a no.");
            m[i] = sc.nextInt();
            pushmarks(m[i]);
        }
    }
```

```
    void pushmarks(int v)
    {
        if(top == cap – 1)
        {
            System.out.println("Stack is full");
        }
        else
        {
            st[++top] = v;
        }
    }
    int popmarks()
    {
        if(top == –1)
        {
            return(–999);
        }
        else
        {
            return(st[top– –]);
        }
    }
    void display()
    {
        int i;
        for(i = top; i >= 0; i– –)
        {
            System.out.println(st[i]);
        }
    }
}
```

Answer 13.

(a) Algorithm :

Step 1 : Start

Step 2 : Set temporary pointer to start node

Step 3 : Repeat steps 4 until it reaches null

Step 4 : Move pointer to the next node

Step 5 : Creat new node, assign num to number and link to temporary node

Step 6 : End

Method :

```
void addnode(Node start, int num)
{
    Node A = new Node(start);
    while(A != null)
    A = A nextNode;
    Node C = new Node();
    C.number = num;
    C.nextNode = null;
    A.next = C;
}
```

(b) Complexity refers to the measure of the performance of an algorithm with respect to time and space whereas Big-O notation is a unit of measurement of an algorithm or represents complexity.

(c) (i) A

(ii)

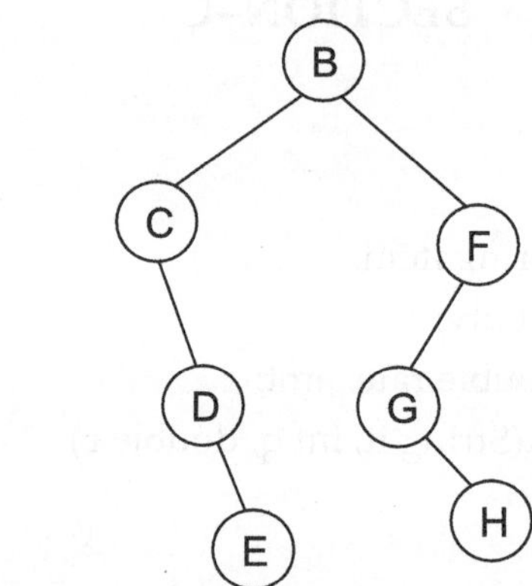

(iii) C, D, E, B, G, H, F, A, K, L, J

(iv) Size of the tree is 11.

COMPUTER SCIENCE 2013

QUESTIONS

PART–I

*Answer **all** questions.*
While answering questions in this Part, indicate briefly your working and reasoning, wherever required.

Question 1.

(a) *State the Principle of duality. Write the dual of:* **[2]**

$$(P + Q').R.1 = P.R + Q'.R$$

(b) *Minimize the expression using Boolean laws :* **[2]**

$$F = (A + B').(B + CD)'$$

(c) *Convert the following cardinal form of expression into its canonical form :* **[2]**

$$F(P, Q, R) = \pi(1, 3)$$

(d) *Using a truth table verify :* **[2]**

$$(\sim p \Rightarrow q) > p = (p \wedge \sim q) / (p \wedge q)$$

(e) *If A = 1 and B = 0, then find :* **[2]**

(i) $(A' + 1).B$

(ii) $(A + B')'$

Question 2.

(a) *Differentiate between **throw** and **throws** with respect to exception handling.* **[2]**

(b) *Convert the following infix notation to its postfix form:*

$E * (F / (G - H) * I) + J$ **[2]**

(c) *Write the algorithm for push operation (to add elements) in an array based stack.* **[2]**

(d) *Name the File Stream classes to :* **[2]**

(i) *Write data to a file in binary form.*

(ii) *Read data from a file in text form.*

(e) *A square matrix M[ ][ ] of size 10 is stored in the memory with each element requiring 4 bytes of storage. If the base address at M [0][0] is 1840, determine the address at M[4][8] when the matrix is stored in **Row Major Wise.*** **[2]**

Question 3.

(a) *The following function **Recur ()** is a part of some class. What will be the output of the function **Recur ()** when the value of n is equal to 10. Show the dry run / working.* **[5]**

```
void Recur (int n)
{
    if (n > 1)
    {
        System.out.print (n + "");
        if (n%2 != 0)
        {
            n = 3 * n + 1;
            System.out.print(n+ "");
        }
        Recur (n/2);
    }
}
```

(b) *The following function is a part of some class. Assume 'n' is a positive integer. Answer the given questions along with dry run / working.*

```
int unknown (int n)
{
    int i, k;
    if (n%2==0)
    {
        i = n/2;
        k=1;
    }
    else
    {
        k=n;
        n--;
        i=n/2;
    }
    while (i > 0)
    {
        k=k*i*n;
        i--;
        n--;
    }
    return k;
}
```

(i) *What will be returned by unknown (5)?* **[2]**

(ii) *What will be returned by unknown (6)?* **[2]**

(iii) *What is being computed by unknown (int n)?* **[1]**

PART–II

*Answer **seven** questions in this part, choosing **three** questions from Section A, **two** from Section B and **two** from Section C.*

SECTION–A

*Answer any **three** questions.*

Question 4.

(a) *Given the Boolean function :*

$$F(A, B, C, D) = \Sigma(0, 2, 4, 5, 8, 9, 10, 12, 13).$$

(i) *Reduce the above expression by using 4-variable K-Map, showing the various groups (i.e., octal, quads and pairs).* **[4]**

(ii) *Draw the logic gate diagram of the reduced expression. Assume that the variables and their complements are available as inputs.* **[1]**

(b) *Given the Boolean function :*

$F(P, Q, R, S) = \pi (0, 1, 3, 5, 7, 8, 9, 10, 11, 14, 15)$

(i) *Reduce the above expression by using 4–variable K-Map, showing the various groups (i.e., octal, quads and pairs).* **[4]**

(ii) *Draw the logic gate diagram of the reduced expression. Assume that the variables and their complements are available as inputs.* **[1]**

Question 5.

A Football Association coach analyses the criteria for a win/ draw of his team depending on the following conditions.

- *If the Centre and Forward players perform well but Defenders do not perform well.*

OR

- *If Goal keeper and Defenders perform well but the Centre players do not perform well.*

OR

- *If all the players perform well.*

The inputs are :

Inputs	
C	*Centre players perform well.*
D	*Defenders perform well.*
F	*Forward players perform well.*
G	*Goalkeeper performs well.*

(In all of the above cases 1 indicates yes and 0 indicates no)

Output : **X** *- Denotes the win/ draw criteria [1 indicates win/ draw and 0 indicates defeat in all cases.]*

(a) *Draw the truth table for the inputs and outputs given above and write the* **POS** *expression for X (C, D, F, G).* **[5]**

(b) *Reduce* **X** *(C, D, F, G) using Karnaugh's Map.*

Draw the logic gate diagram for the reduced **POS** *expression for X (C, D, F, G) using AND and OR gate. You may use gates with two or more inputs. Assume that the variable and their complements are available as inputs.* **[5]**

Question 6.

(a) *In the following truth table x and y are inputs and B and D are outputs :* **[3]**

INPUT		*OUTPUT*	
x	*y*	B	D
0	0	0	0
0	1	1	1
1	0	0	1
1	1	0	0

Answer the following questions:

(i) *Write the SOP expression for D.*

(ii) *Write the POS expression for B.*

(iii) *Draw a logic diagram for the SOP expression derived for D, using only NAND gates.*

(b) *Using a truth table, verify if the following proposition is valid or invalid :*

$(a \Rightarrow b) \bullet (b \Rightarrow c) = (a \Rightarrow c)$ **[3]**

(c) *From the logic circuit diagram given below, name the outputs (1), (2) and (3). Finally derive the Boolean expression and simplify it to show that it represents a logic gate. Name and draw the logic gate.* **[4]**

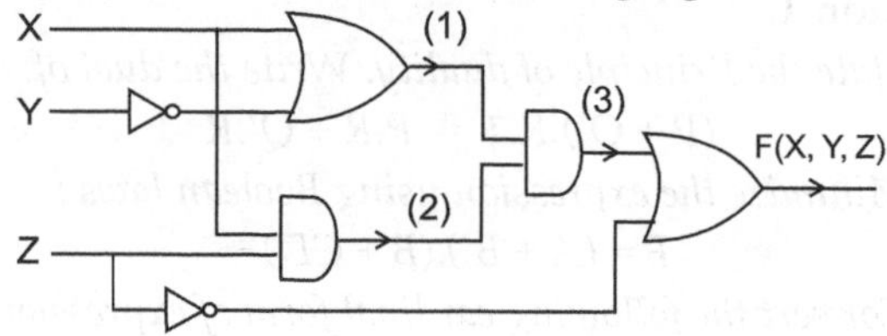

Question 7.

(a) *What are Decoders ? How are they different from Encoders ?* **[2]**

(b) *Draw the truth table and a logic gate diagram for a 2 to 4 Decoder and briefly explain its working.* **[4]**

(c) *A combinational logic circuit with three inputs P, Q, R produces output 1 if and only if an odd number of 0's are inputs.* **[4]**

(i) *Draw its truth table.*

(ii) *Derive a canonical SOP expression for the above truth table.*

(iii) *Find the complement of the above derived expression using De Morgan's theorem and verify if it is equivalent to its POS expression.*

SECTION–B

Answer any **two** *questions.*

Each program should be written in such a way that it clearly depicts the logic of the problem.

This can be achieved by using mnemonic names and comments in the program.

(Flowcharts and Algorithms are **not** *required)*

The Programs must be written in Java.

Question 8.

An emirp number is a number which is prime backwards and forwards. Example : 13 and 31 are both prime numbers. Thus, 13 is an emirp number.

Design a class Emirp to check if a given number is Emirp number or not. Some of the members of the class are given below : **[10]**

Class name: *Emirp*

Data members/ instance variables :

n : *stores the number*

rev :	*stores the reverse of the number*
f :	*stores the divisor*

Member functions:

Emirp (int nn) :	*to assign n = nn, rev = 0 and f = 2*
int isprime (int x) :	*check if the number is prime using the recursive technique and return 1 if prime otherwise return 0*
void isEmirp () :	*reverse the given number and check if both the original number and the reverse number are prime, by invoking the function isprime(int) and display the result with an appropriate message*

*Specify the class **Emirp** giving details of the **constructor (int), int isprime(int)** and void isEmirp(). Define the **main() function** to create an object and call the methods to check for Emirp number.* **[10]**

Question 9.

*Design a class **Exchange** to accept a sentence and interchange the first alphabet with the last alphabet for each word in the sentence, with single letter word remaining unchanged. The words in the input sentence are separated by a single blank space and terminated by a full stop.* **[10]**

Example :

Input : It is a warm day.

Output : tI si a marw yad.

Class name : ***Exchange***

Data Members/ instance variables :

sent :	*stores the sentence*
rev :	*to store the new sentence*
size :	*stores the length of the sentence*

Member functions :

Exchange() :	*default constructor*
void read-sentence() :	*to accept the sentence*
void exfirstlast() :	*extract each word and interchange the first and last alphabet of the word and form a new sentence rev using the changed words*
void display() :	*display the original sentence along with the new changed sentence.*

*Specify the class **Exchange** giving details of the **constructor(), void read-sentence(), void exfirst last()** and **void display()**. Define the **main()** function to create an object and call the functions accordingly to enable the task.*

Question 10.

*A class **Matrix** contains a two dimensional integer array of order [m × n]. The maximum value possible for both 'm' and 'n' is 25. Design a class **Matrix** to find the difference of the two matrices. The details of the members of the class are given below :* **[10]**

Class name : ***Matrix***

Data Members/ instance variables :

arr[ ][ ] :	*stores the matrix element*
m :	*integer to store the number of rows*
n :	*integer to store the number of columns*

Member functions :

Matrix(int mm, int nn) :	*to initialize the size of the matrix m = mm and n = nn*
void fillarray() :	*to enter the elements of the matrix*
Matrix SubMat (Matrix A) :	*subtract the current object from the matrix of parameterised object and return the resulting object*
void display() :	*display the matrix elements*

*Specify the class **Matrix** giving details of the **constructor(int,int), void fillarray(), Matrix Sub Mat(Matrix)** and **void display**. Define the **main()** function to create objects and call the methods accordingly to enable the task.*

SECTION–C

*Answer any **two** questions.*

Each Program should be written in such a way that it clearly depicts the logic of the problem stepwise. This can be achieved by using comments in the program and mnemonic names or pseudo codes for algorithms.

The programs must be written in Java.

The Algorithms must be written in general/standard form, wherever required / specified.

*(Flowcharts are **not** required)*

Question 11.

*A super class **Perimeter** has been defined to calculate the perimeter of a parallelogram. Define a subclass **Area** to compute the area of the parallelogram by using the required data members of the super class. The details are given below:* **[10]**

Class name: ***Perimeter***

Data Members/ instance variables :

a :	*to store the length in decimal*
b :	*to store the breadth in decimal*

Member functions :

Perimeter(...) :	*parameterised constructor to assign values to data members*
double Calculate() :	*calculate and return the perimeter of a parallelogram as 2* (length + breadth)*
void show( ) :	*to display the data members alongwith the perimeter of the parallelogram*

Class name :	Area
Data members/ instance variables :	
h :	*to store the height in decimal*
area :	*to store the area of the parallelogram*
Member functions :	
Area(...) :	*parameterised constructor to assign values to data members of both the classes*
void doarea( ) :	*compute the area as (breadth * height)*
void show( ) :	*display the data members of both classes along with the area and perimeter of the parallelogram.*

Specify the class **Perimeter** *giving details of the* **constructor(...), double Calculate() and void show().** *Using the* **concept of inheritance,** *specify the class* **Area** *giving details of the* **constructor (...), void doarea() and void show().**

Question 12.

A doubly queue is a linear data structure which enables the user to add and remove integers from either ends, i.e., from front or rear. Define a class **Dequeue** *with the following details :* **[10]**

Class name :	**Dequeue**
Data members/ instance variables :	
arr[ ] :	*array to hold up to 100 integer elements*
lim :	*stores the limit of the dequeue*
front :	*to point to the index of front end*
rear :	*to point to the index of the rear end*
Member functions :	
Dequeue (int 1) :	*constructor to initialise the data members lim=1; front=rear=0*
void addfront (int val) :	*to add integer from the front if possible else display the message ("Overflow from front")*
void addrear (int val) :	*to add integer from the rear if possible else display the message ("Overflow from rear")*
int popfront( ) :	*returns element from front, if possible otherwise returns – 9999*
int poprear( ) :	*returns element from rear, if possible otherwise returns – 9999*

Specify the class **Dequeue** *giving details of the* **constructor(int), void addfront (int), void addrear (int), int popfront( ) and int poprear( ).**

The main function and algorithm need not be written.

Question 13.

(a) *A linked list is formed from the objects of the class,* **[4]**

```
class node
{
    int item;
    Node next;
}
```

Write an algorithm **OR** *a Method to count the number of nodes in the linked list. The method declaration is given below :*

int count(Node ptr_start)

(b) *What is the Worst Case complexity of the following code segment :* **[2]**

(i)
```
for (int p = 0; p < N; p++)
{
    for (int q = 0; q < M; q++)
    {
        Sequence of statements;
    }
}
for (int r = 0; r < X; r++)
{
    Sequence of statements;
}
```

(ii) *How would the complexity change if all the loops went upto the same limit N ?*

(c) *Answer the following from the diagram of a Binary Tree given below :*

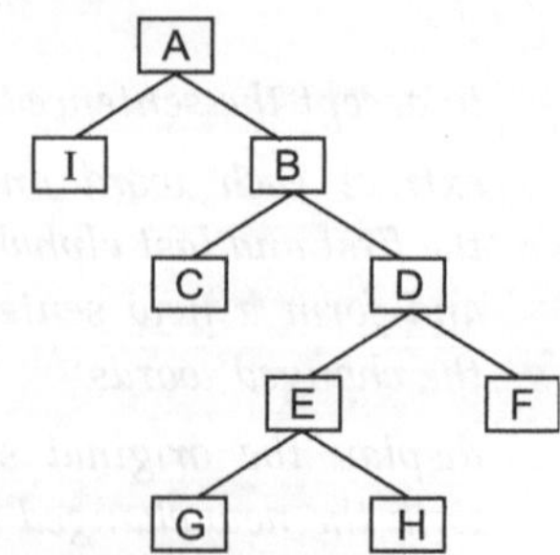

(i) *Preorder Transversal of tree.* **[1]**

(ii) *Children of node E.* **[1]**

(iii) *Left subtree of node D.* **[1]**

(iv) *Height of the tree when the root of the tree is at level 0.* **[1]**

ANSWERS

PART–I

Answer 1.

(a) Principal of Duality : For each Boolean equation there exists another equation which is dual of the previous equation. This is done by changing AND's to OR's and vice-versa, 0's to 1's and vice-versa, but compliments remain unchanged.

Dual of $P.\overline{Q}+R+0$ is $(P+R).(\overline{Q}+R)$

(b) $F = (A+B').(B+CD)'$

$= (A+\overline{B}).(\overline{B+CD})$

$= (A+\overline{B})[\overline{B}.\overline{CD}]$ (using De Morgan's law)

$= (A+\overline{B})[\overline{B}.(\overline{C}+\overline{D})]$

$= (A+\overline{B})(\overline{B}.\overline{C}+\overline{B}.\overline{D})$

(using distributive law)

$= A.\overline{B}.\overline{C}+A.\overline{B}.\overline{D}+\overline{B}.\overline{B}.\overline{C}+\overline{B}.\overline{B}.\overline{D}$

$= A.\overline{B}.\overline{C}+A.\overline{B}.\overline{D}+\overline{B}.\overline{C}+\overline{B}.\overline{D}$

$= \overline{B}.\overline{C}(A+1)+\overline{B}.\overline{D}(A+1)$

$= \overline{B}.\overline{C}+\overline{B}.\overline{D}$

$= \overline{B}(\overline{C}+\overline{D})$

(c) $F(P, Q, R) = \pi(1, 3)$

Canonical expression will be :

$(P+Q+\overline{R}).(P+\overline{Q}+\overline{R})$

(d) $(\sim p \Rightarrow q) \wedge p = (p \wedge \sim q) \vee (p \wedge q)$

p	q	~p	~q	~p ⇒ q	(~p ⇒q) ∧ p	p∧~q	p∧q	(p ∧ ~q) ∨(p ∧ q)
0	0	1	1	0	**0**	0	0	**0**
0	1	1	0	1	**0**	0	0	**0**
1	0	0	1	1	**1**	1	0	**1**
1	1	1	0	1	**1**	0	1	**1**
					L.H.S.			**R.H.S.**

Hence Proved.

(e) (i) 0 (ii) 0

Answer 2.

(a)

Throws	Throw
Throws is used to forward an exception to the calling function which can be handled later.	Throw is used to generate a user defined exception.

(b) E * (F/(G – H) * I) + J

= E * (F/GH – * I) + J

= E * (FGH – / * I) + J

= E * FGH – / I * + J

= EFGH – / ** + J

= EFGH – / I ** J +

(c) Step 1 – check if the stack is full.

Step 2 – If the stack is full, print "stack is full" and exit.

Step 3 – If the stack is not full, increments top to point next empty space.

Step 4 – Adds data element to the stack location, where top is pointing.

Step 5 – Returns Success.

(d) **(i)** File outputstream

(ii) File Reader

(e) BA = 1840, R = 10, C = 10, ES = 4, LR = 0, LC = 0, I = 4, J = 8

In row major :

$M[I, J] = BA + ES(C(I-LR)+(J-LC))$

$M[4, 8] = 1840 + 4(10(4-0)-(8-0))$

$= 1840 + 4 \times 48$

$= 1840 + 192$

$= 2032$

Answer 3.

(a) Output

Step 1	10					
Step 2	10	5	16			
Step 3	10	5	16	8		
Step 4	10	5	16	8	4	
Step 5	10	5	16	8	4	2

Finally output will be :

10 5 16 8 4 2

(b) **(i)** When *n* if-condition *k* *n* *i*

5 false 5 4 2

In loop

n	i	condition	k = 5
4	2	true	5 × 4 × 2 = 40
3	1	true	40 × 31 × 1 = 120
2	0	false	

= 120

(ii) When *n* if-condition *k* *n* *i*

6 true 1 6 3

In loop

n	*i*	condition	*k* = 1
6	3	true	1 × 6 × 3 = 18
5	2	true	18 × 5 × 2 = 180
4	1	true	180 × 4 × 1 = 720
3	0	false	

= 720

(iii) It is computing, factorial of *n*.

PART–II

SECTION–A

Answer 4.

(a) (i) $F(A, B, C, D) = \Sigma(0, 2, 4, 5, 8, 9, 10, 12, 13)$

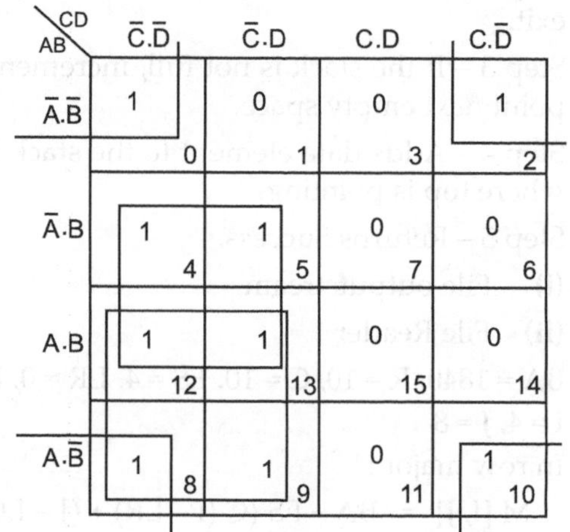

Quad 1 = m0 + m2 + m8 + m10 $= \overline{B}.\overline{D}$

Quad 2 = m4 + m5 + m12 + m13 $= B.\overline{C}$

Quad 3 = m8 + m9 + m12 + m13 $= A.\overline{C}$

$= A.\overline{C} + B.\overline{C} + \overline{B}.\overline{D}$

(ii) Logic Gate Diagram :

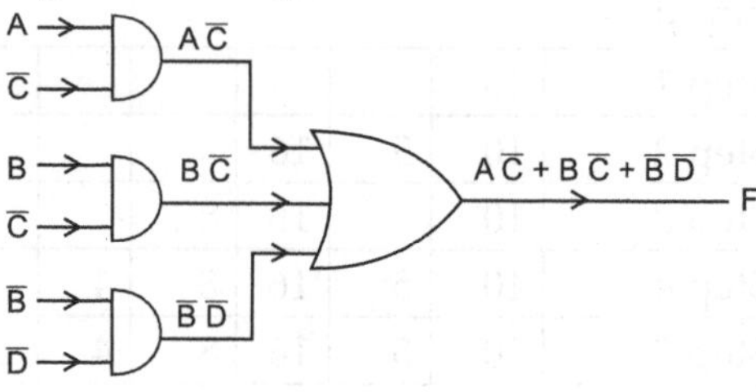

(b) (i) $F(P, Q, R, S) = \pi(0, 1, 3, 5, 7, 8, 9, 10, 11, 14, 15)$

PQ \ RS	$R+S$	$R+\overline{S}$	$\overline{R}+\overline{S}$	$\overline{R}+S$
$P+Q$	0 (0)	0 (1)	0 (3)	1 (2)
$P+\overline{Q}$	1 (4)	0 (5)	0 (7)	1 (6)
$\overline{P}+\overline{Q}$	1 (12)	1 (13)	0 (15)	0 (14)
$\overline{P}+Q$	0 (8)	0 (9)	0 (11)	0 (10)

Quad 1 = M0.M1.M8.M9

$= Q + R$

Quad 2 = M1.M3.M5.M7

$= P + \overline{S}$

Quad 3 = M10.M11.M14.M15

$= \overline{P} + \overline{R}$

$= (Q + R) . (P + \overline{S}) . (\overline{P} + \overline{R})$

(ii) Logic Gate Diagram :

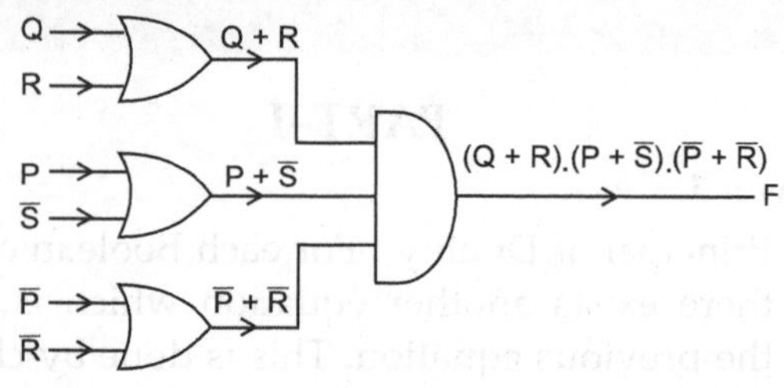

Answer 5.

(a)

C	D	F	G	X	Maxterm	Designation
0	0	0	0	0	$C+D+F+G$	0
0	0	0	1	0	$C+D+F+\overline{G}$	1
0	0	1	0	0	$C+D+\overline{F}+G$	2
0	0	1	1	0	$C+D+\overline{F}+\overline{G}$	3
0	1	0	0	0	$C+\overline{D}+F+G$	4
0	1	0	1	1		5
0	1	1	0	0	$C+\overline{D}+\overline{F}+G$	6
0	1	1	1	1		7
1	0	0	0	0	$\overline{C}+D+F+G$	8
1	0	0	1	0	$\overline{C}+D+F+\overline{G}$	9
1	0	1	0	1		10
1	0	1	1	1		11
1	1	0	0	0	$\overline{C}+\overline{D}+F+G$	12
1	1	0	1	0	$\overline{C}+\overline{D}+F+\overline{G}$	13
1	1	1	0	0	$\overline{C}+\overline{D}+\overline{F}+G$	14
1	1	1	1	1		15

POS expression :

$X(C, D, F, G) = \pi(0, 1, 2, 3, 4, 6, 8, 9, 12, 13, 14)$

$(C+D+F+G) . (C+D+F+\overline{G}) . (C+D+\overline{F}+G).$
$(C+D+\overline{F}+\overline{G}) . (C+\overline{D}+F+G) . (C+\overline{D}+\overline{F}+G).$
$(\overline{C}+D+F+G) . (\overline{C}+D+F+\overline{G}) . (\overline{C}+\overline{D}+F+G).$
$(\overline{C}+\overline{D}+F+\overline{G}) . (\overline{C}+\overline{D}+\overline{F}+G)$

(b)

CD \ FG	$F+G$	$F+\overline{G}$	$\overline{F}+\overline{G}$	$\overline{F}+G$
$C+D$	0 (0)	0 (1)	0 (3)	0 (2)
$C+\overline{D}$	0 (4)	1 (5)	1 (7)	0 (6)
$\overline{C}+\overline{D}$	0 (12)	0 (13)	1 (15)	0 (14)
$\overline{C}+D$	0 (8)	0 (9)	1 (11)	1 (10)

Quad 1 = M0.M1.M2.M3

= C + D

Quad 2 = M4.M6.M12.M14

= $\overline{D}$ + G

Quad 3 = M8.M9.M12.M13

= $\overline{C}$ + F

= (C + D) . ($\overline{D}$ + G) . ($\overline{C}$ + F)

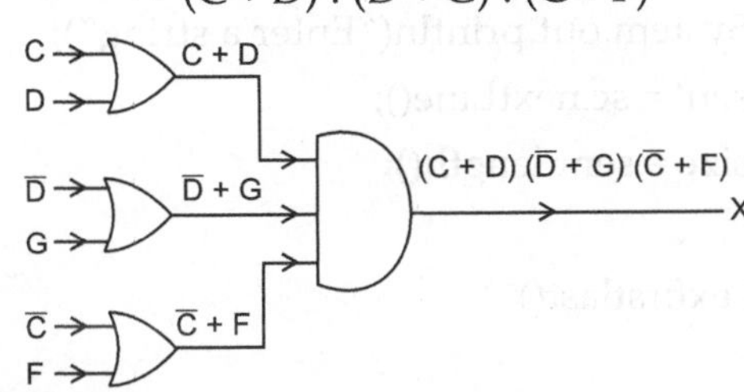

Answer 6.

(a) (i) $\overline{X}.Y + X.\overline{Y}$

(ii) $(X + Y) . (\overline{X} + Y) . (\overline{X} + \overline{Y})$

(iii)

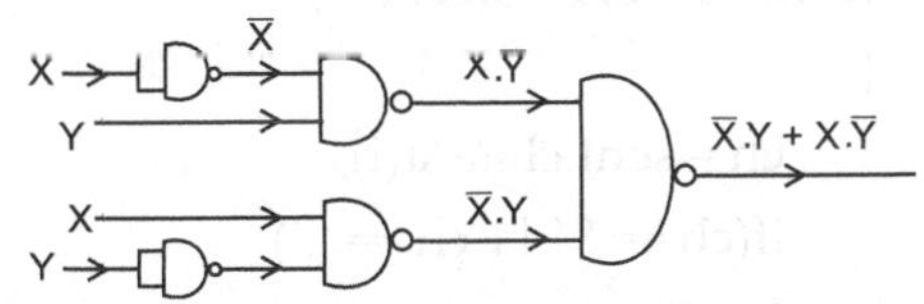

(b)

a	b	c	a ⇒ b	b ⇒ c	(a ⇒ b) . (b ⇒ c)	a ⇒ c
0	0	0	1	1	1	1
0	0	1	1	1	1	1
0	1	0	1	0	0	1
0	1	1	1	1	1	1
1	0	0	0	1	0	0
1	0	1	0	1	0	1
1	1	0	1	0	0	0
1	1	1	1	1	1	1

Hence, the following proposition is invalid.

(c) (1) $X + \overline{Y}$

(2) X.Z

(3) $X.Z\,(X + \overline{Y})$

Final expression = $X.Z.\,(X + \overline{Y}) + \overline{Z}$

= $X.Z + X.\overline{Y}.Z + \overline{Z}$

= $X.Z\,(1 + \overline{Y}) + \overline{Z}$

= $X.Z + \overline{Z}$ (Distributive Law)

= $(X + \overline{Z}).\,(Z + \overline{Z})$

= $X + \overline{Z}$

= $\overline{X}.Z$ (De Morgan's Law)

So, it represents the NAND Gate X.Z

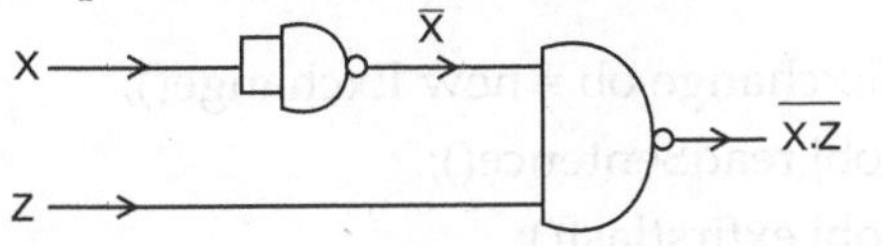

Answer 7.

(a) A decoder is a digital electronic circuit which converts a binary signal into another form whereas encoder is a digital electronic circuit which converts any signal to binary.

(b) A 2 to 4 decoder :

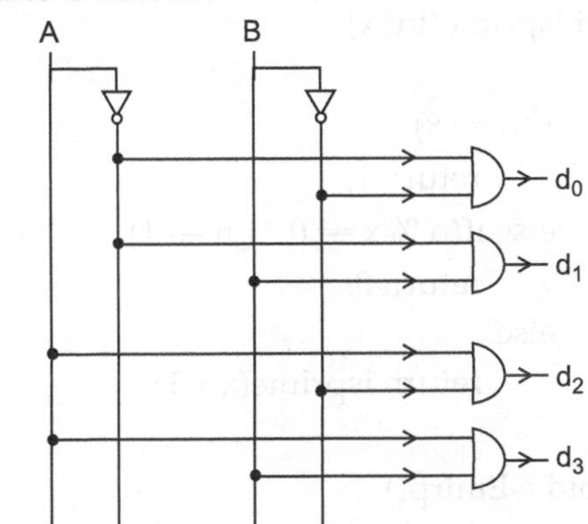

A	B	d_0	d_1	d_2	d_3
0	0	1	0	0	0
0	1	0	1	0	0
1	0	0	0	1	0
1	1	0	0	0	1

In a decoder if there are 2 input lines, it well be capable of producing 4 outputs, one for each of its state of given input.

(c) (i)

P	Q	R	Output
0	0	0	1
0	0	1	0
0	1	0	0
0	1	1	1
1	0	0	0
1	0	1	1
1	1	0	1
1	1	1	0

(ii) SOP expression is :

$\overline{P}\,\overline{Q}\,\overline{R} + \overline{P}\,Q\,R + P\,\overline{Q}\,R + P\,Q\,\overline{R}$

(iii) $(\overline{P}\,\overline{Q}\,\overline{R} + \overline{P}\,Q\,R + P\,\overline{Q}\,R + P\,Q\,\overline{R})'$

= $(P + Q + R) . (P + \overline{Q} + \overline{R})$.

$(\overline{P} + Q + \overline{R}) . (\overline{P} + \overline{Q} + R)$

Hence, it is equivalent to POS expression.

SECTION–B

Answer 8.

```
import java.util.*;
public class Emirp
{
        int n, rev, f;
        Emirp(int nn)
```

```
        {
            n = nn;
            rev = 0;
            f = 2;
        }
        int isprime(int x)
        {
            if(n == x)
                return 1;
            else if(n % x == 0 || n == 1)
                return 0;
            else
                return isprime(x + 1);
        }
        void isEmirp()
        {
                int x = n;
                while(x != 0)
                {
                    rev = rev * 10 + x % 10;
                    x = x / 10;
                }
                int ans1 = isprime(f);
                n = rev;
                f = 2;
                int ans2 = isprime(f);
                if(ans1 == 1 && ans2 == 1)
        System.out.println(n+" is an Emirp no.");
                else
        System.out.println(n+"is not an Emirp no.");
        }
        public static void main()
        {
                Scanner sc = new Scanner System.in);
                System.out.println("Enter a no.");
                int x = sc.nextInt();
                Emirp obj = new Emirp(x);
                Obj.isEmirp();
        }
}
```

Answer 9.

```
import java.util.*;
public class Exchange
{
    String sent, rev;
    int size;
    Exchange()
    {
        sent = "";
        rev = "";
    }
    void readSentence()
    {
        Scanner sc = new Scanner(System.in);
        System.out.println("Enter a string");
        sent = sc.nextLine();
        size = sent.length();
    }
    void exfirstlast()
    {
        int p = 0;
        char ch;
        String b;
        for(int i = 0; i < size; i++)
        {
            ch = sent.charAt(i);
            if(ch == ' ' || ch == '.')
            {
                b = sent.substring(p, i);
                if(b.length() != 1)
                {
            rev += b.charAt(b.length() - 1);
            rev += b.substring (1, b.length() - 1);
            rev += b.charAt(0);
                }
                else
                {
                    rev += b;
                    rev = rev + " ";
                }
            }
        }
    }
    void display()
    {
        System.out.println("\nInput :" +sent);
        System.out.println("\nOutput :" +rev);
    }
    public static void main()
    {
        Exchange ob = new Exchange();
        obj.readSentence();
        obj.exfirstlast();
```

```
        obj.display();
    }
}
```

Answer 10.

```
import java.io.*;
class Matrix
{
    int arr[][], m, n, i, j;
    public Matrix()
    {
        m = n = 25;
        arr = new int[m][n];
    }
    public Matrix(int mm, int nn)
    {
        m = mm;
        n = nn;
        arr = new int[m][n];
    }
    void fillarray()throws IOException
    {
    BufferedReader br = new BufferedReader
    (new InputStreamReader (System.in));
        for(i = 0; i < m;  i++)
        {
            for(j = 0;  j < n;  j++)
            {
            System.out.println("Enter a number");
            arr[i][j] = Integer.parseInt (br.readLine());
            }
        }
    }
    Matrix SubMat(Matrix A)
    {
        Matrix S = new Matrix(m,n);
        for(i = 0; i < m; i++)
        {
            for(j = 0; j < n; j++)
            {
                S.arr[i][j] = A.arr[i][j] – arr[i][j];
            }
        }
        return(S);
    }
    void display()
    {
        for(i = 0; i < m; i++)
        {
            for(j = 0; j < n; j++)
            {
                System.out.print(arr[i][j]);
            }
            System.out.println();
        }
    }
    public static void main(String args[])
                                        throws IOException
    {
    int r, c;
    BufferedReader br = new BufferedReader
                (new InputStreamReader(System.in));
    System.out.println("Enter rows and columns");
    r = Integer.parseInt(br.readLine());
    c = Integer.parseInt(br.readLine());
    Matrix M1 = new Matrix(r, c);
    Matrix M2 = new Matrix(r, c);
    Matrix M3 = new Matrix(r, c);
    M1.fillarray();
    M2.fillarray();
    M3 = M1.SubMat(M2);
    M3.display();
    }
}
```

SECTION–C

Answer 11.

```
class Perimeter
{
    protected double a, b;
    public Perimeter(double a1,double b1)
    {
        a = a1;
        b = b1;
    }
    public double Calculate()
    {
        double p = 2 * (a + b);
        return(p);
    }
    public void show()
    {
        System.out.println("Length : "+a);
        System.out.println("Breadth : "+b);
        System.out.println("Perimeter:"+Calculate());
    }
}
```

```
class Area extends Perimeter
{
    double h, area;
    public Area(int x, int y, int z)
    {
        super(x, y);
        h = z;
    }
    void doarea()
    {
        area = b * h;
    }
    void show()
    {
        super.show();
        System.out.println("Height : "+h);
        System.out.println("Area : "+area);
    }
}
```

Answer 12.

```
class Dequeue
{
    int arr[], lim, front, rear;
    public Dequeue()
    {
        arr = new int[100];
        front = rear = 0;
    }
    public Dequeue(int l)
    {
        lim = l;
        front = rear = 0;
        arr = new int[l];
    }
    void addrear(int val)
    {
        if(rear < lim – 1)
        {
            rear++;
            arr[rear] = val;
        }
        else
        {
            System.out.println("Overflow from rear");
        }
    }
    void addfront(int val)
    {
        if(front == 0 || front == 1)
        {
            System.out.println("Overflow from front");
        }
        else
        {
            front––;
            arr[front] = val;
        }
    }
    int popfront()
    {
        if(front <= rear)
        {
            if(front == 0)
            {
                front++;
            }
            int t = arr[front];
            front++;
            return(t);
        }
        else
        {
            return(– 9999);
        }
    }
    int poprear()
    {
        if(front > rear || front == 0 || rear == 0)
        {
            return(– 9999);
        }
        else
        {
            int t = arr[rear];
            rear––;
            return(t);
        }
    }
}
```

Answer 13.

(a) Algorithm :

Step 1 : Start.

Step 2 : Set temporary pointer to 1st node and count to 0.

Step 3 : Repeat step 4 and 5 until the pointer reaches null.

Step 4 : Increase the counter by 1.

Step 5 : Move temporary pointer to the next node.

Step 6 : Return the counter value.

Step 7 : End.

OR

Method :

```
int count_node(Node ptr_start)
{
    int ctr = 0;
    while(ptr_start != null)
    {
        ctr++;
        ptr_start = ptr_start.next;
    }
    return ctr;
}
```

(b) (i) O (M × N)

(ii) O (N^2)

(c) (i) A, I, B, C, D, E, G, H, F

(ii) G, H

(iii)

```
  E
 / \
G   H
```

(iv) 4

COMPUTER SCIENCE 2012

QUESTIONS

PART–I

*Answer **all** questions.*
While answering questions in this Part, indicate briefly your working and reasoning, wherever required.

Question 1.

(a) *Using a truth table, verify the following expression:* **[2]**

$X + (Y + Z) = (X + Y) + Z$

Also state the law.

(b) *Given,* $F(X, Y, Z) = (X' + Y') . (Y + Z')$ **[2]**

write the function in canonical product-of-sum form.

(c) *Draw the truth table and logic circuit for a 2-input XNOR gate.* **[2]**

(d) *Find the complement of the following expression :* **[2]**

$X' + XY'$

(e) *If* $(X \Rightarrow Y)$ *then write its :* **[2]**

(i) *Converse* **(ii)** *Contra positive*

Question 2.

(a) *Differentiate between the keywords* ***extends*** *and* ***implements.*** **[2]**

(b) *State how a binary tree is a recursive data structure.* **[2]**

(c) *A matrix B [10][7] is stored in the memory with each element requiring 2 bytes of storage. If the base address at B [x][1] is 1012 and the address at B [7][3] is 1060, determine the value 'x' where the matrix is stored in* ***Column Major wise.*** **[2]**

(d) *Convert the following infix notation to its postfix form:* **[2]**

$A + ((B + C) + (D + E) * F)/G$

(e) *What is a constructor ? State* ***one*** *difference between a constructor and any other member function of a class.* **[2]**

Question 3.

(a) *The following function is a part of some class which computes and sorts an array arr [ ] in ascending order using the* ***bubble sort technique.*** *There are some places in the code marked by ?1?, ?2?, ?3?, ?4?, ?5? which must be replaced by a statement/expression so that the function works properly :* **[5]**

```
void bubblesort(int arr[ ])
{
    int i, j, k, tmp;
    for (i = 0; ?1?; i++)
    {
        for(j = 0; ?2?;j++)
        {
            if(arr[j]>?3?)
            {
                tmp = arr[j];
                ?4? = arr[j+1];
                arr[j+1] = ?5?;
            }
        }
    }
}
```

(i) *What is the expression or statement at ?1?*

(ii) *What is the expression or statement at ?2?*

(iii) *What is the expression or statement at ?3?*

(iv) *What is the expression or statement at ?4?*

(v) *What is the expression or statement at ?5?*

(b) *The following function* ***witty()*** *is a part of some class. What will be the output of the function witty () when the value of n is 'SCIENCE' and the value of p is 5. Show the dry run/working :*

```
void witty(String n, int p)
{
    if (p < 0)
        System.out.println("");
    else
    {   System.out.println(n.char At(p) + ". ");
        witty (n, p-1);
        System.out.print(n.charAt(p));
    }
}
```

PART–II

Answer ***seven*** *questions in this part, choosing* ***three*** *questions from Section A,* ***two*** *from Section B and* ***two*** *from Section C.*

SECTION–A

Answer any ***three*** *questions*

Question 4.

(a) *Given the Boolean function :*

F (A, B, C, D) = Σ (4, 6, 7, 10, 11, 12, 14, 15).

(i) *Reduce the above expression by using 4 – variable K-Map, showing the various groups (i.e., octal, quads and pairs).* **[4]**

(ii) *Draw the logic gate diagram of the reduced expression. Assume that the variables and their complements are available as inputs.* **[1]**

(b) *Given the Boolean function :*

F (P, Q, R, S) = π (0, 5, 7, 8, 10, 12, 13, 14, 15)

(i) *Reduce the above expression by using 4 – variable K-Map, showing the various groups (i.e., octal, quads and pairs).* **[4]**

(ii) *Draw the logic gate diagram of the reduced expression. Assume that the variables and their complements are available as inputs.* **[1]**

Question 5.

(a) *The Principal of a school intends to select students for admission to class XI on the following criteria :* **[5]**

- *Student is of the same school and has passed the class X Board Examination with more than 60% marks.*

OR

- *Student is of the same school, has passed the class X Board Examination with less than 60% marks but has taken active part in co-curricular activities.*

OR

- *Student is not from the same school but has either passed the class X Board Examination with more than 60% marks or has participated in sports at the National level.*

The inputs are :

Inputs	
S	*Student of the same school.*
P	*Has passed the class X Board Examination with more than 60% marks.*
C	*Has taken active part in co-curricular activities.*
T	*Has participated in sports at the National Level.*

Output : **X** *- Denotes admission status [1 indicates granted and 0 indicates refused in all the cases.]*

(a) *Draw the truth table for the inputs and outputs given above and write the* ***SOP*** *expression.*

(b) *Reduce* **X** *(S, P, C, T) using Karnaugh's Map.*

Draw the logic gate diagram for the reduced ***SOP*** *expression for X (S, P, C, T) using AND and OR gate. You may use gates with two or more inputs. Assume that the variable and their complements are available as inputs.*

Question 6.

(a) *Verify algebraically if,* **[3]**

X'Y'Z'+X'Y'Z+X'YZ+X'YZ'+XY'Z'+XY'Z = X'+ Y'

(b) *Represent the Boolean expression X + YZ' with the help of NOR gates only.* **[3]**

(c) *Define the terms* ***Contingency, Contradiction*** *and* ***Tautology.*** **[4]**

(d) *Consider the following truth table where A and B are two inputs and X is the output :*

A	*B*	*X*
0	0	0
0	1	1
1	0	1
1	1	0

(i) *Name and draw the logic gate for the given truth table.*

(ii) *Write the POS of X (A, B).*

Question 7.

(a) *Define Multiplexer and state* ***one*** *of its uses. Draw the logic diagram for a 4 : 1 Multiplexer.* **[4]**

(b) *State how a* ***Half Adder*** *is different from a* ***Full Adder.*** *Also give their respective uses.* **[3]**

(c) *Minimize the following expression using Boolean laws:* **[3]**

$Q \cdot (Q' + P) \cdot R \cdot (Q + R)$

Also draw the logic gate for the reduced expression.

SECTION–B

Answer any ***two*** *questions*

Each program should be written in such a way that it clearly depicts the logic of the problem.

This can be achieved by using mnemonic names and comments in the program.

(Flowcharts and Algorithms are ***not*** *required)*

The Programs must be written in Java

Question 8.

A class Combine contains an array of integers which combines two arrays into a single array including the duplicate elements, if any, and sorts the combined array. Some of the members of the class are given below : **[10]**

Class name : *Combine*

Data Members/ instance variables :

com [ ] : *integer array*

size : *size of the array*

Member functions/ methods :

Combine(int nn): *parameterised constructor to assign size = nn*

void inputarray () : *to accept the array elements.*

void sort () : *sorts the elements of combined array in ascending order using the selection sort technique.*

void mix(Combine A, Combine B) :	*combines the parameterised object arrays and stores the result in the current object array along with duplicate elements, if any.*
void display () :	*displays the array elements*

Specify the class Combine giving details of the ***constructor (int), void inputarray(), void sort(), void mix (Combine, Combine)*** *and* ***void display*** *(). Also define the* ***main ()*** *function to create an object and call the methods accordingly to enable the task.*

Question 9.

Design a class VowelWord to accept a sentence and calculate the frequency of words that begin with a vowel. The words in the input string are separated by a single blank space and terminated by a full stop. The description of the class is given below :

Class name :	***VowelWord***
Data Members/ instance variables :	
str :	*to store a sentence*
freq :	*store the frequency of the words beginning with a vowel.*
Member functions :	
VowelWord () :	*constructor to initialise data members to legal initial value*
void read-str() :	*to accept a sentence*
void freq_vowel () :	*counts the frequency of the words that begin with a vowel*
void display () :	*to display the original string and the frequency of the words that begin with a vowel.*

Specify the class VowelWord giving details of the ***constructor (), void readstr (), void freq_vowel () and void display ().*** *Also define the* ***main ()*** *function to create an object and call the methods accordingly to enable the task.*

Question 10.

A happy number is a number in which the eventual sum of the square of the digits of the number is equal to 1. **[10]**

Example :

$$28 = (2)^2 + (8)^2 = 4 + 64 = 68$$
$$68 = (6)^2 + (8)^2 = 36 + 64 = 100$$
$$100 = (1)^2 + (0)^2 + (0)^2 = 1 + 0 + 0 = 1$$

Hence, 28 is a happy number.

Example :

$$12 = (1)^2 + (2)^2 = 1 + 4 = 5$$

Hence, 12 is not a happy number.

Design a class Happy to check if a given number is a happy number. Some of the members of the class are given below:

Class name :	***Happy***
Data Members/ instance variables :	
n :	*stores the number*
Member functions :	
Happy( ) :	*constructor to assign 0 to n.*
void getnum (int nn) :	*to assign the parameter value to the number n = nn.*
int sum_sq_digits(int x) :	*returns the sum of the square of the digits of the number x, using the recursive technique*
void ishappy( ) :	*checks if the given number is a happy number by calling the function sum_sq_digits (int) and displays an appropriate message.*

Specify the class Happy giving details of the ***constructor ( ), void getnum (int), int sum_sq_digits(int)*** *and* ***void ishappy ().*** *Also define a* ***main ( )*** *function to create an object and call the methods to check for happy number.*

SECTION–C

Answer any ***two*** *questions*

Each Program / Algorithm should be written in such a way that it clearly depicts the logic of the problem step wise.

This can also be achieved by using pseudo codes.

(Flowcharts are ***not*** *required)*

The programs must be written in Java.

The Algorithm must be written in general/standard form, wherever required.

Question 11.

Link is an entity which can hold a maximum of 100 integers. Link enables the user to add elements from the rear end and remove integers from the front end of the entity. Define a class Link with the following details : **[10]**

Class name :	***Link***
Data Members/ instant variables :	
lnk [ ] :	*entity to hold the integer elements.*
max :	*stores the maximum capacity of the entity*
begin :	*to point to the index of the front end.*

end	:	to point to the index of the rear end.

Member functions :

Link (int mm) :	constructor to initialise max = mm, begin = 0, end = 0.
void addlink (int v) :	to add an element from the rear index if possible otherwise display the message "OUT OF SIZE..."
int dellink ( ) :	to remove and return an element from the front index, if possible otherwise display the message "EMPTY..." and return –99
void display( ) :	displays the elements of the entity.

(a) *Specify the class Link giving details of the* **constructor (int), void add-link (int), int dellink ( )** *and* **void display ( ).** **[9]**

The main function and algorithm need not be written.

(b) *What type of data structure is the above entity ?* **[1]**

Question 12.

A super class Detail has been defined to store the details of a customer. Define a sub-class Bill to compute the monthly telephone charge of the customer as per the chart given below :

Number of Calls	Rate
1–100	Only rental charge
101–200	60 paisa per call + rental charge
201–300	80 paisa per call + rental charge
Above 300	1 rupee per call + rental charge

The details of both the classes are given below :

Class name: **Detail**

Data members/ instance variables :

name	:	to store the name of the customer
address	:	to store the address of the customer
telno	:	to store the phone number of the customer
rent	:	to store the monthly rental charge

Member functions :

Detail(...) :	parameterised constructor to assign values to data members
void show () :	to display the details of the customer

Class name Bill

Data members/ instance variables :

n	:	to store the number of calls
amt	:	to store the amount to be paid by the customer.

Member functions :

Bill (...) :	parameterised constructor to assign values to data members of both classes and to initialise amt = 0.0
void cal() :	calculate the monthly telephone charge as per the chart given above
void show () :	displays the details of the customer and amount to be paid.

Specify the class Details giving details of the **constructor ()** *and* **void show ().** *Using the* **concept of inheritance,** *specify the class Bill giving details of the* **constructor ( ), void cal () and void show ().**

The main () function and algorithm need not be written.

Question 13.

(a) *A linked list is formed from the objects of the class,* **[4]**

```
class node
{
    int p;
    String n;
    node next;
}
```

Write an **Algorithm OR** *a* **Method** *to search for a name and display the contents of that node. The method declaration is given below :*

void search (node start, string b)

(b) *What is the role of constants in complexity ? Explain briefly with an example.* **[2]**

(c) *Answer the following from the diagram of a Binary. Tree given below :*

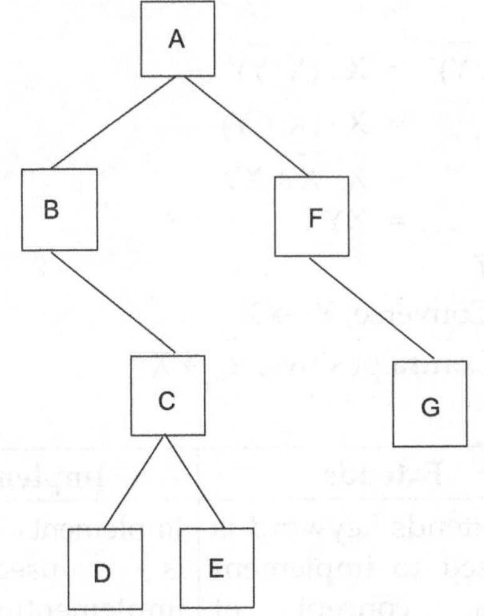

(i) *External nodes of the tree.* **[1]**

(ii) *Parent of node D.* **[1]**

(iii) *Inorder traversal of the tree.* **[1]**

(iv) *Right subtree of Node B.* **[1]**

ANSWERS

PART-I

Answer 1.

(a) $X+(Y+Z) = (X+Y)+Z$

x	y	z	x + y	(x + y) + z	y + z	X + (Y + Z)
0	0	0	0	0	0	0
0	0	1	0	1	1	1
0	1	0	1	1	1	1
0	1	1	1	1	1	1
1	0	0	1	1	0	1
1	0	1	1	1	1	1
1	1	0	1	1	1	1
1	1	1	1	1	1	1

It is associative law.

(b) $(\overline{X}+\overline{Y})\cdot(Y+\overline{Z})$

$= (\overline{X}+\overline{Y}+0)\cdot(Y+\overline{Z}+0)$ [Properties of 0]

$= (\overline{X}+\overline{Y}+Z\cdot\overline{Z})\cdot(Y+\overline{Z}+X\cdot\overline{X})$ [Complementrity]

$= (\overline{X}+\overline{Y}+Z)(\overline{X}+\overline{Y}+\overline{Z})\cdot(X+Y+\overline{Z})\cdot(\overline{X}+Y+\overline{Z})$

[Distributive law]

(c)

X	Y	X ⊙ Y
0	0	1
0	1	0
1	0	0
1	1	1

$X \odot Y = \overline{x\overline{y}+\overline{x}y}$

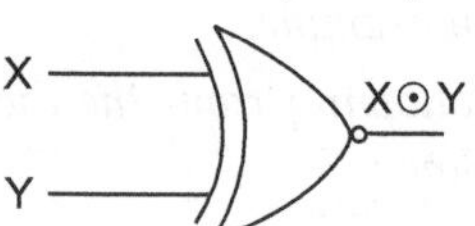

(d) $(\overline{X}+X\overline{Y})' = X.(X.\overline{Y})'$

$= X\cdot(\overline{X}+Y)$

$= X.\overline{X}+XY$

$= XY$

(e) $X \to Y$

(i) Converse, $Y \to X$

(ii) Contra positive, $\overline{Y} \to \overline{X}$

Answer 2.

(a)

Extends	Implements
1. Extends keyword is used to implement the concept of inheritance.	Implements keyword is used for implementing an interface by a class.
2. Extends is used when creating a sub-class.	Implements is used when implementing an interface.

(b) A tree inherits a recursive data structure because each node has reference to its other sub-trees. *e.g.*, the following binary tree :

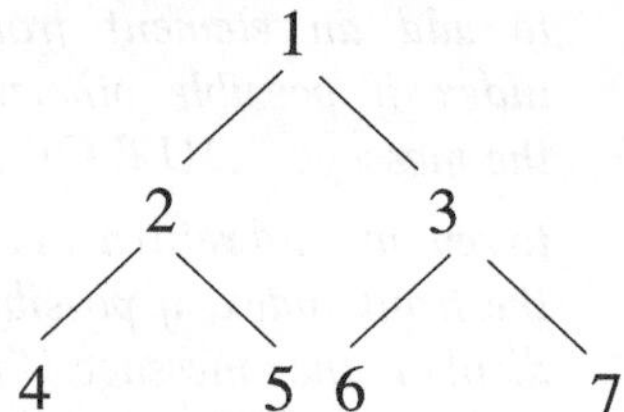

has two binary trees as its children, which are :

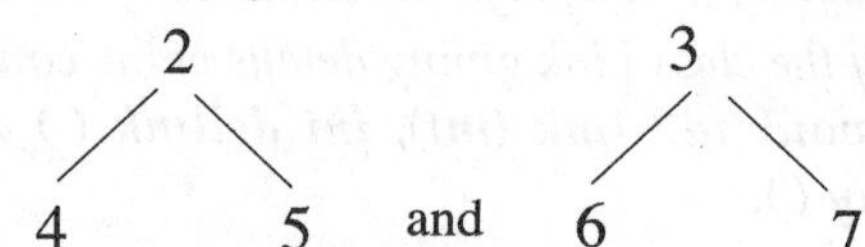

This property of binary tree is applicable at all levels of binary trees.

(c) BA = 1012, ES = 2, R = 10, I = 7, J = 3, LR = X, LC =1

Address of B [I, J] = BA + ES [(I – LR) + R (J – LC)]

B [7, 3] = 1012 + 2 [(7 – X) + 10 (3 – 1)]

1060 = 1012 + 2 [(7 – X) + 10 (2)]

1060 = 1012 + 2 [(7 – X) + 20)]

1060 – 1012 = 2 [(7 – X) + 20)]

48 = 2 [(7 – X) + 20)]

$(7 - X) = \frac{48}{2} - 20$

7 – X = 4

X = 7 – 4 = 3.

(d) A + ((BC +) + (DE +) * F)/G

⇒ A + ((BC +) + (DE +) F*))/G

⇒ A + (BC + DE + F* +)/G

⇒ A + (BC + DE + F* +G/)

⇒ ABC + DE + F* + G/ +

(e) Constructor is a member function used to create and initialise the object with legal set of values. Constructors have same name as that of class while functions have different names.

Answer 3.

(a) **(i)** i < arr. length

(ii) j < arr.length – 1 – i

(iii) arr [j + 1]

(iv) arr [j]

(v) tmp

(b) C.

N.

E.

I.

C.

S.

SCIENCE.

PART–II

SECTION–A

Answer 4.

(a) **(i)** F (A, B, C, D) = Σ (4, 6, 7, 10, 11, 12, 14, 15)

AB \ CD	$\overline{C}\,\overline{D}$	$\overline{C}$ D	C D	C $\overline{D}$
$\overline{A}\,\overline{B}$	0	1	3	2
$\overline{A}$ B	1 (4)	5	1 (7)	1 (6)
A B	1 (12)	13	1 (15)	1 (14)
A $\overline{B}$	8	9	1 (11)	1 (10)

Quad 1 = m6 + m7 + m14 + m15

= BC

Quad 2 = m10 + m11 + m14 + m15

= AC

Quad 3 = m4 + m6 + m12 + m14

= $B\overline{D}$

(ii) So the expession is

= $AC + BC + B\overline{D}$

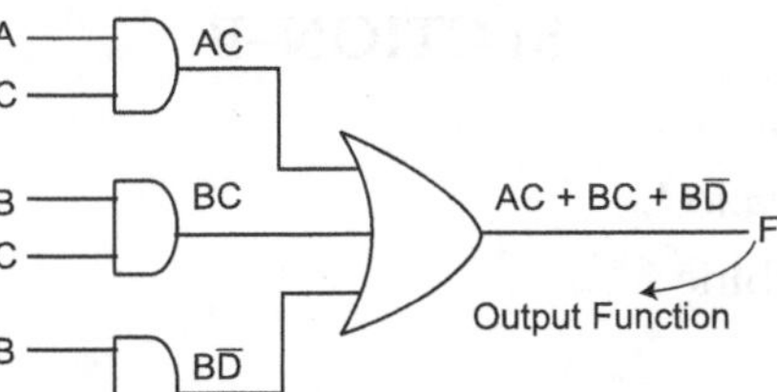

(b) **(i)** F (P, Q, R, S) = π (0, 5, 7, 8, 10, 12, 13, 14, 15)

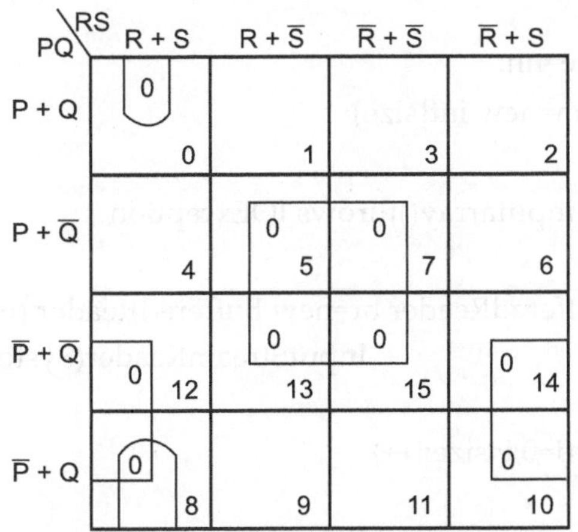

Quad 1 = m5 · m7 · m13 · m15

= $\overline{Q}+\overline{S}$

Quad 2 = m8· m10 · m12 · m14

= $\overline{P}+S$

Pair1 = m0 · m8

= Q + R + S

= $(\overline{P}+S)\cdot(\overline{Q}+\overline{S})\cdot(Q+R+S)$

(ii)

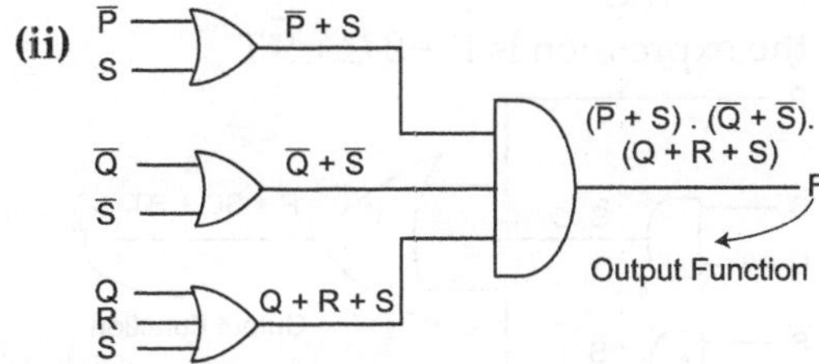

Answer 5.

(a)

S	P	C	T	X	
0	0	0	0	0	0
0	0	0	1	1	1
0	0	1	0	0	2
0	0	1	1	1	3
0	1	0	0	1	4
0	1	0	1	1	5
0	1	1	0	1	6
0	1	1	1	1	7
1	0	0	0	0	8
1	0	0	1	0	9
1	0	1	0	1	10
1	0	1	1	1	11
1	1	0	0	1	12
1	1	0	1	1	13
1	1	1	0	1	14
1	1	1	1	1	15

X (S, P, C, T) = Σ (1, 3, 4, 5, 6, 7, 10, 11, 12, 13, 14, 15)

$\overline{S}\,\overline{P}\,\overline{C}\,T+\overline{S}\,\overline{P}\,C\,T+\overline{S}\,P\,\overline{C}\,\overline{T}+\overline{S}\,P\,\overline{C}\,T+\overline{S}\,P\,C\,\overline{T}+$

$\overline{S}\,P\,C\,T+S\,\overline{P}\,C\,\overline{T}+S\,\overline{P}\,C\,T+S\,P\,\overline{C}\,\overline{T}+S\,P\,\overline{C}\,T+$

$S\,P\,C\,\overline{T}+S\,P\,C\,T.$

(b)

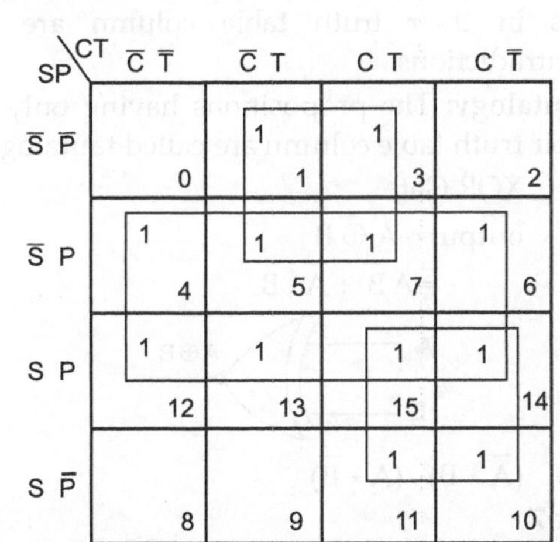

Octal 1 = m4 + m5 + m6 + m7 + m12 + m13 + m14 + m15

= P

Quad1 $= m1 + m3 + m5 + m7$

$= \overline{S}\,T$

Quad2 $= m10 + m11 + m14 + m15$

$= S\,C$

So the expression is $P + S\,C + \overline{S}\,T$.

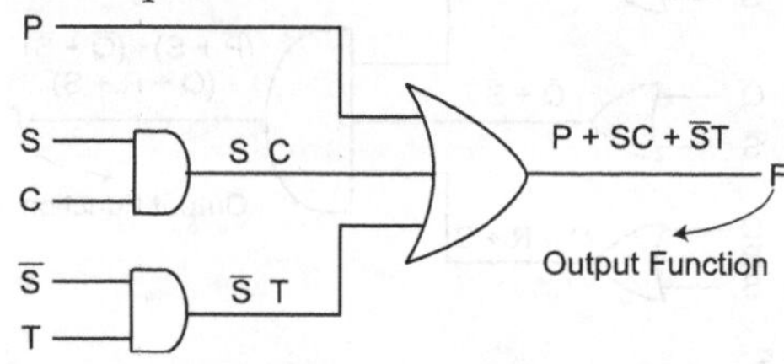

Answer 6.

(a) $\overline{X}\,\overline{Y}\,\overline{Z} + \overline{X}\,\overline{Y}Z + \overline{X}Y\overline{Z} + \overline{X}YZ + X\overline{Y}\,\overline{Z} + X\overline{Y}Z$

$= \overline{X}\,\overline{Y}\,(\overline{Z} + Z) + \overline{X}Y\,(Z + \overline{Z}) + X\overline{Y}\,(\overline{Z} + Z)$ [Distributive law]

$= \overline{X}\,\overline{Y} + \overline{X}Y + X\overline{Y}$ [Properties of 1]

$= \overline{X}\,(\overline{Y} + Y) + X\overline{Y}$ [Distributive law]

$= \overline{X} + X\overline{Y}$

$= (\overline{X} + X)\,.\,(\overline{X} + \overline{Y})$ [Distributive law]

$= 1\,.\,(\overline{X} + \overline{Y})$ [Property of 1]

$= \overline{X} + \overline{Y}$

(b) $X + Y\overline{Z}$

$= (X + Y) \cdot (X + \overline{Z})$

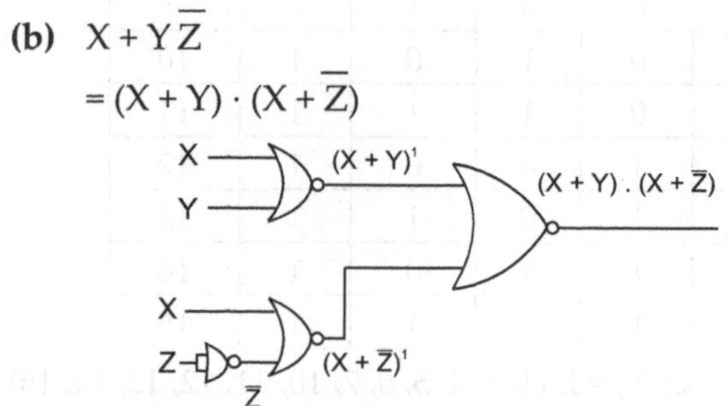

(c) **Contingency:** The propositions that have some combination of O's and 1's in their truth table column are called contingency.

Contradiction: The propositions having only O's in their truth table column are called contradictions.

Tautology: The propositions having only 1's in their truth table column are called tautologies.

(d) **(i)** XOR Gate

output $= A \oplus B$

$= A\overline{B} + \overline{A}\,.\,B$

(ii) $(\overline{A} + B)\,.\,(A + \overline{B})$

Answer 7.

(a) A multiplexer is a combinational circuit that selects binary information from more than one input lines and directs it to a single output line. It is used in parallel to serial conversion.

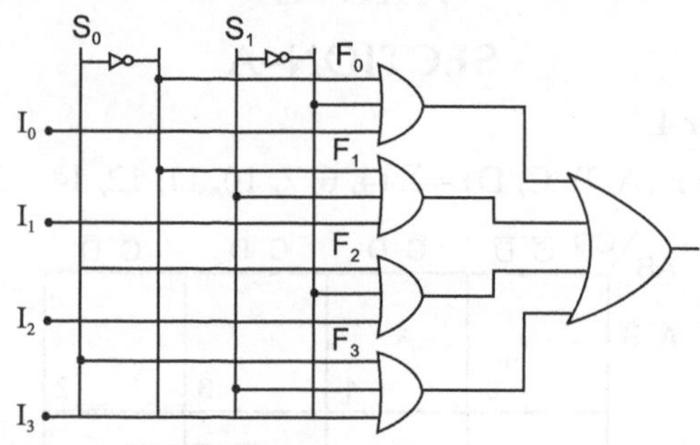

(b) Half adder is used to add two bits while full adder is used to add three bits.

Adder are used in ALU to sum the given binary number.

(c) $Q \cdot (\overline{Q} + P) \cdot R \cdot (Q + R)$

$= (Q \cdot \overline{Q} + PQ) \cdot R \cdot (Q + R)$

$= (O + PQ) \cdot R\,(Q + R)$

$= PQR \cdot (Q + R)$

$= (PQR \cdot Q + PQR \cdot R)$

$= PQR + PQR$

$= PQR$

SECTION–B

Answer 8.

```
import.java.io.*;
class Combine
{
 int com[],size;
  public Combine(int nn)
  {
    size=nn;
    com=new int[size];
  }
  void inputarray()throws IOException
  {
    BufferedReader br=new BufferedReader (new
                    InputStreamReader(System.in));
    int i;
    for(i=0;i<size;i++)
    {
      System.out.println("Enter a no.");
       com[i]=Integer.parseInt(br.readLine());
    }
  }
```

```
void sort()
{
  int i,j,t;
  for(i=0;i<size;i++)
  {
    for(j=i+1;j<size;j++)
    {
      if(com[i]>com[j])
      {
        t=com[i];
        com[i]=com[j];
        com[j]=t;
      }
    }
  }
}
void mix(Combine A, Combine B)
{
  int i,j;
  for(i=0,j=0;i<A.size;i++,j++)
  {
    com[j]=A.com[i];
  }
  for(i=0;i<B.size;i++,j++)
  {
    com[j]=B.com[i];
  }
}
void display()
{
  int i;
  for(i=0;i<size;i++)
  {
    System.out.println(com[i]);
  }
}
public static void main(String args[])throws
                                  IOException
{
  Combine c1=new Combine(5);
  Combine c2=new Combine(3);
  Combine c3=new Combine(8);
  c1.inputarray();
  c2.inputarray();
  c3.mix(c1,c2);
  c3.sort();
  c3.display();
}
}
```

Answer 9.

```
import java.io.*;
class VowelWord
{
 String str;
 int freq;
 public VowelWord()
 {
   str = " ";
   freq=0;
 }
 void readStr()throws IOException
 {
BufferedReader br=new
BufferedReader(new InputStreamReader(System.in));
System.out.println('Enter a Sentence.');
str=br.readLine();
 }
 void freq_vowel()
 {
  String w;
  int i,l;
  char ch;
  l=str.length();
  for(i=0;i<l;i++)
  {
    w="";
    while(str.charAt(i)!=' ' && str.charAt(i)!='.')
    {
      w=w+str.charAt(i);
      i++;
    }
    ch = w. char At (0);
    if (ch = = 'a' || ch = = 'A' || ch = = 'e' || = = 'E' ||
    ch = = 'i' || ch = = 'I' || ch = = 'o' || c = ='O' || ch
    = = 'u' || ch = ='U')
  {
    freq + +;
  }
 }
 void display()
 {
  System.out.println("Original String : " + str);
  System.out.println("Frequency of Words Beginning
                            with vowel are : "+ freq);
 }
 public static void main(String args[])throws
                                         IOException
 {
```

```
  VowelWord ob=new VowelWord();
  ob.readStr();
  ob.freq_vowel();
  ob.display();
 }
}
```

Answer 10.

```
Class Happy
{
  int n,s,d;
  public Happy()
  {
    n=0;
  }
  void getnum(int nn)
  {
    n=nn;
  }
  int sum_sq_digits(int x)
  {
    if(x>0)
    {
     d=x%10;
     s=s+d*d;
     x=x/10;
     return(sum_sq_digits(x));
    }
    else
    {
      return s;
    }
  }
  void ishappy()
  {
   int a;
    while(n>9)
    {
      s=0;
     n=sum_sq_digits(n);
    }
    if(n==1)
     System.out.println("Happy No.");
     else
     System.out.println("Not a Happy  No.");
  }
}
```

SECTION–C

Answer 11.

(a)

```
class Link
{
  int lnk[]=new int[100];
  int begin,end,max;
  public Link(int mm)
  {
    max=mm;
    begin=end=0;
  }
  void addlink(int v)
  {
   if(end==0)
   {
    end=begin=1;
    lnk[end]=v;
   }
   else if(end==max)
   {
    System.out.println('List is Full');
   }
   else
   {
    lnk[++end]=v;
   }
  }
  int dellink()
  {
    int a;
    if(begin==0)
    {
     System.out.println("Empty...");
      return(-99);
    }
    else if(begin==end)
    {
      a=lnk[begin];
      begin=end=0;
      return a;
    }
    else
    {
      a=lnk[begin];
      begin++;
      return(a);
    }
  }
  void display()
  {
    int i;
    for(i=begin;i<=end;i++)
    {
```

```
        System.out.println(lnk[i]);
      }
    }
  }
```

(b) It is a queue.

Answer 12.

```
Class Detail
{
  protected String name, add;
  protected long telno;
  protected double rent;
  public Detail(String n, String a, long t, double r)
  {
    name=n;
    add=a;
    telno=t;
    rent=r;
  }
  void show()
  {
    System.out.println(name + " " + add + " " + telno
                                      + " " + rent);
  }
}
class Bill extends Detail
{
  int n;
  double amt;
  public Bill(String nm,String ad,long tn,double
rn,int n1)
  {
    super(nm,ad,tn,rn);
    n=n1;
    amt=0.0;
  }
  void cal()
  {
    if(n>=1 && n<=100)
    {
      amt=rent;
    }
    else if(n>100 && n<=200)
    {
      amt=rent + (n-100)*0.60;
    }
    else if(n>200 && n<=300)
    {
      amt=rent + (100 * 0.60) + (n- 200)*0.80;
    }
    else if(n>300)
    {
      amt=rent + 100*0.60 + 100*0.80 + (n-300)*1;
    }
  }
  void show()
  {
    super.show();
    System.out.println(amt);
  }
}
```

Answer 13.

(a) Method

```
void search (node start, string b)
{
    boolean found = false;
    node tmp = start;
    while (tmp ! = nul)
    if ((tmp. n). equals (b) = = true)
    System out.println ("Item found" + tmp. p);
    System.out,println (tmp.n);
    found = true;
break;
}
    tmp = tmp. next;
}
if (found = = false)
    System.out.print ("Item not found");
}
```

Algorithm

Step 1 : Input b, the string to search

Step 2 : Let tmp = start, a temporary node created.

Step 3 : If (tmp.n = b) then print "found", tmp.p, tmp.n, Goto End
Else Goto steple

Step 4 : tmp = tmp-next, move tmp to the next node.

Step 5 : if tmp ! = null , then Goto step 3 else Goto step 6

Step 6 : Print "Not Found"

Step 7 : End

(b) Constants are used in calculating time complexity which can differ from one processor to another *e.g.*,

```
for (i = 1; i < N; i++)
}
System.out.print ln (i);suppose it takes
                              constant time C.
}
```

Total time taken by loop = C*n = Cn

(c) **(i)** D, E, G **(ii)** C

(iii) B, D, C, E, A, F, G.

(iv)

```
    C
   / \
  D   E
```

●●

COMPUTER SCIENCE 2011

PART—I

*Answer **all** questions*

While answering questions in this Part, indicate briefly your working and reasoning, wherever required.

Question 1.

(a) *State the two absorption laws. Verify any one of them using truth table.* **[2]**

(b) *Reduce the following expression :* **[2]**

F(A, B, C) = Σ (0,1,2,3,4,5,6,7)

Also find the complement of the reduced expression.

(c) *Name the logic gate for the following circuit diagram and write its truth table.* **[2]**

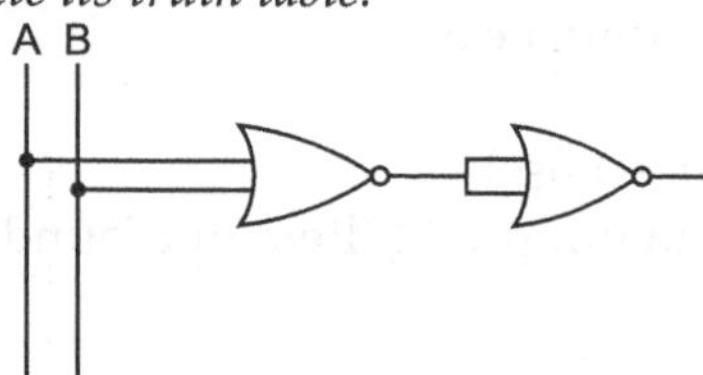

(d) *Using truth table, verify whether the following is true or false :* **[2]**

$(p \Rightarrow q) = (q \Rightarrow p)$

(e) *If A = 1, B = 0, C = 1 and D = 1 find its :* **[2]**

(i) Maxterm *(ii) Minterm*

Question 2.

(a) *How can we override a method in inheritance ?* **[2]**

(b) *A square matrix A [m × m] is stored in the memory with each element requiring 2 bytes of storage. If the base address A [1] [1] is 1098 and the address at A [4] [5] is 1144, determine the order of the matrix A [m × m] when the matrix is stored* ***Column Major*** *wise.* **[2]**

(c) *What is Big O notation ?* **[2]**

(d) *What is an exception ?* **[2]**

(e) *Convert the following infix expression to its postfix form :* **[2]**

*a + b * c – d/e*

Question 3.

(a) *The following is a part of some class. What will be the output of the function* ***mymethod( )*** *when the value of counter is equal to 3 ? Show the dry run / working.* **[5]**

```
void mymethod (int counter)
{
    if (counter == 0)
        System.out.println(" ");
    else
    {
        System.out.println
            ("Hello"+counter);
        mymethod(-- counter);
        System.out.println
            (" "+counter);
    }
}
```

(b) *The following function is a part of some class which computes and returns the greatest common divisor of any two numbers. There are some places in the code marked by ?1?, ?2?, ?3?, ?4? and ?5? which must be replaced by statement/expression so that the function works correctly.*

```
int gcd(int a, int b)
{
    int r;
    while(?1?)
    {
        r=?2?;
        b=?3?;
        a=?4?;
    }
    if(a= =0)
        reurn ?5?;
    else
        return – 1;
}
```

(i) What is the expression or statement at ?1? **[1]**

(ii) What is the expression or statement at ?2? **[1]**

(iii) What is the expression or statement at ?3? **[1]**

(iv) What is the expression or statement at ?4? **[1]**

(v) What is the expression or statement at ?5? **[1]**

PART—II

*Answer **seven** questions in this part, choosing **three** questions from Section A, **two** from Section B and **two** from Section C.*

SECTION—A

*Answer any **three** questions*

Question 4.

(a) *State the principle of Duality. Give the dual of the following :* **[3]**

(A′.B) + (C.1) = (A′ + C).(B + C)

(b) *Reduce the Boolean expressions to their simplest forms:* **[4]**

(i) $\{(C.D)' + A\} + A + C.D + A.B.$

(ii) $A.\{B + C\,(A.B + A.C)'\}$

(c) *Verify using a truth table if :* **[3]**

$(A \odot B \odot C)' = A \oplus B \oplus C$

Question 5.

(a) *Given* $F\,(P, Q, R, S) = \pi\,(2, 3, 6, 7, 9, 11, 12, 13, 14, 15)$ **[5]**

Reduce the above expression by using ***four*** *variable Karnaugh's Map. Draw the logic gate diagram of the reduced expression using NOR gate only.*

(b) *Given* $F\,(A, B, C, D) = \overline{A}\overline{B}\overline{C}\overline{D} + \overline{A}\overline{B}\overline{C}D + \overline{A}B\overline{C}D + A\overline{B}\overline{C}D + \overline{A}B\overline{C}\overline{D} + \overline{A}BC\overline{D}.$ **[5]**

Reduce the above expression by using ***four*** *variable Karnaugh's Map. Draw the logic gate dia-gram of the reduced expression using NAND gate only.*

Question 6.

(a) *Show with the help of a logic diagram how a NAND gate is equivalent to an OR gate.* **[3]**

(b) *Verify if the following is valid :* **[3]**

$(a \Rightarrow b) \wedge (a \Rightarrow c) = a \Rightarrow b \wedge c)$

(c) *What is a Decoder? Draw the truth table and logic circuit diagram for a 2 to 4 Decoder.* **[4]**

Question 7.

(a) *What is a Full Adder ? Draw the truth table for a Full Adder. Also derive SOP expression for the Full Adder and draw its logic circuit.* **[4]**

(b) *State how a Decoder is different from a Multiplexer. Also state one use of each.* **[3]**

(c) *Convert the following cardinal expression into its canonical form and reduce it using Boolean laws :* **[3]**

$F\,(L, M, O, P) = \pi\,(0, 2, 8, 10)$

SECTION—B

Answer any ***two*** *questions*

Each program should be written in such a way that it clearly depicts the logic of the problem.

This can be achieved by using mnemonic names and comments in the program.

(Flowcharts and Algorithms are ***not*** *required)*

The Programs must be written in Java

Question 8.

Input a sentence from the user and count the number of times, the words "an" and "and" are present in the sentence. Design a class **Frequency** *using the description given below :* **[10]**

Class name : ***Frequency***

Data Members/ variables :

text : *stores the sentence*

countand : *to store the frequency of the word "and"*

countan : *to store the frequency of the word "an"*

len : *stores the length of the string*

Member functions/ methods :

Frequency() : *constructor to initialise the instance variables*

void accept (String n): *to assign n to text, where the value of the parameter n should be in lower case.*

void checkandfreq( ) : *to count the frequency of "and"*

void checkanfreq( ) : *to count the frequency of "an"*

void display () : *to display the number of "and" and "an" with appropriate messages.*

Specify the class **Frequency** *giving details of the* **constructor (), void accept (String), void checkandfreq(), void checkanfreq( )** *and* **void display ().** *Also define the* **main ( )** *function to create an object and call methods accordingly to enable the task.*

Question 9.

A class **DeciOct** *has been define to convert a decimal number into its equivalent octal number. Some of the members of the class are given below :*

Class name : ***DeciOct***

Data Members/ instance variables :

n : *stores the decimal number*

oct : *stores the octal equivalent number*

Member functions :

DeciOct () : *constructor to initialise the data members n =0, oct 0.*

void getnum (int nn) : *assign nn to n*

void deci_oct () : *calculates the octal equivalent of 'n' and stores it in oct using the* ***recursive technique***

void show () : *displays the decimal number 'n', calls the function deci_oct() and displays its octal equivalent.*

(a) *Specify the class* **DeciOct,** *giving details of the* **constructor (), void getnum (int), void deci_oct () and void show ().** *Also define a* **main ( )** *function to create an object and call the functions accordingly to enable the task.* **[8]**

(b) *State any two disadvantages of using recursion.* **[2]**

Question 10.

You are given a sequence of N integers, which are called as pseudo arithmetic sequences (sequences that are in arithmetic progression). **[10]**

Sequence of N integers : 2, 5, 6, 8, 9, 12.

We observe that 2 + 12 = 5 + 9 = 6 + 8 = 14.

The sum of the above sequence can be calculated as 14 × 3 = 42.

For sequence containing an odd number of elements the rule is to double the middle element, for example 2, 5, 7, 9, 12 = 2 +12 = 5 + 9 = 7 + 7 = 14.

14 × 3 = 42 [middle element = 7]

*A class **Pseudoarithmetic** determines whether a given sequence is a pseudo-arithmetic sequence.*

The details of the class are given below :

Class name	:	***Pseudoarithmetic***
Data Members/instance variables	:	
n	:	*to store the size of the sequence*
a [ ]	:	*integer array to store the sequence of numbers*
ans, flag	:	*store the status*
sum	:	*store the sum of sequence of numbers*
r	:	*store the sum of the two numbers*
Member functions	:	
Pseudoarithmetic ( )	:	*default constructor*
void accept (int nn)	:	*to assign nn to n and to create an integer array. Fill in the elements of the array.*
boolean check ( )	:	*return true if the sequence is a pseudoarithmetic sequence otherwise returns false*

*Specify the class **Pseudoarithmetic**, giving the details of the **constructor ( ), void accept (int)** and **boolean check ( )**. Also define a **main ( )** function to create an object and call the member functions accordingly to enable the task.*

SECTION – C

*Answer any **two** questions*

Each Program / Algorithm should be written in such a way that it clearly depicts the logic of the problem step wise. This can also be achieved by using pseudo codes.

*(Flowcharts are **not** required)*

The programs must be written in Java.

The Algorithm must be written in general/standard form.

Question 11.

*A super class **Record** has been defined to store the names and ranks of 50 students. Define a sub-class **Rank** to find the highest rank along with the name. The details of both classes are given below :* **[10]**

Class name	:	***Record***
Data Members/instance variables	:	
name [ ]	:	*to store the name of students*
rnk [ ]	:	*to store the ranks of students*
Member functions	:	
Record ( )	:	*constructor to initialise data members*
void readvalues( )	:	*to store the names and ranks*
void display()	:	*displays the names and the corresponding ranks*

Class name	:	***Rank***
Data members/instance variables	:	
index	:	*integer to store the index of the topmost rank*
Member functions	:	
Rank( ) :		*constructor to invoke the base class constructor and to initialise index = 0*
void highest( ) :		*finds the index/location of the topmost rank and stores it in index without sorting the array.*
void display( ) :		*displays the names and ranks along with the name having the top-most rank.*

*Specify the class **Record** giving details of the **constructor( ) void readvalues( )** and **void display ( )**. Using the **concept of in-heritance,** specify the class **Rank** giving details of **constructor ( ), void highest ( )** and **void display ( ).***

THE MAIN( ) FUNCTION AND ALGO-RITHM NEED NOT BE WRITTEN.

Question 12.

Stack is a kind of data structure which can store elements with the restriction that an element can be added or removed from the top only.

*The details of the class **Stack** is given below :*

Class name	:	***Stack***
Data members/instance variables	:	
st[ ]	:	*the array to hold the names*
size	:	*the maximum capacity of the string array*
top	:	*the index of the top-most element of the stack*
ctr	:	*to count the number of elements of the stack*
Member functions	:	
Stack ( )	:	*default constructor*

Stack (int cap) :		*constructor to initialize size = cap and top = – 1.*
void pushname (String n) :		*to push a name into the stack. If the stack is full, display the message "OVERFLOW".*
String popname ( )	:	*removes a name from the top of the stack and returns it. If the stack is empty, display the message "UNDER-FLOW"*
void display ( )	:	*Display the elements of the stack.*

(a) *Specify class* ***Stack*** *giving details of the* ***constructors ( ), void pushname (String n), String popname ()*** *and* ***void display ( ).*** **[8]**

THE MAIN() FUNCTION AND ALGORITHM NEED NOT BE WRITTEN.

(b) *Under what Principle does the above entity work ?* **[2]**

Question 13.

(a) *A linked list is formed from the objects of the class,* **[4]**

```
class Node
    {
        int info;
        Node link;
    }
```

Write an algorithm ***OR*** *a Method for deleting a node from a linked list.*

The method declaration is given below :

void deletenode (Node start)

(b) *Distinguish between worst-case and best case complexity of an algorithm.* **[2]**

(c) *Answer the following from the diagram of a Binary Tree given below :*

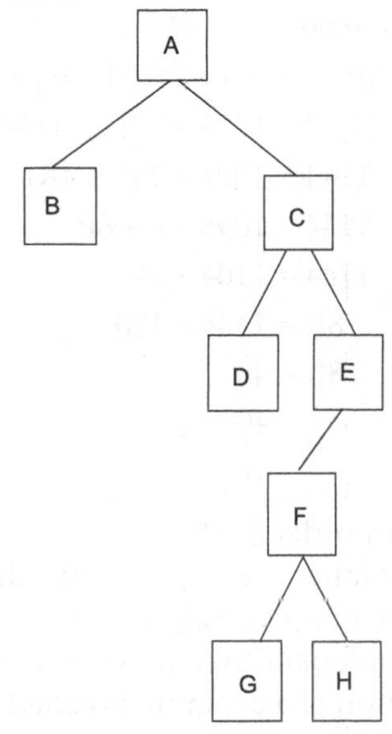

(i) *Write the Postorder tree traversal.* **[1]**

(ii) *Name the Leaves of the tree.* **[1]**

(iii) *Height of the tree.* **[1]**

(iv) *Root of the tree.* **[1]**

ANSWERS

PART–I

Answer 1.

(a) 1. $x + xy = x$

2. $x.(x + y) = x$

x	y	xy	$x + y$	$x + xy$	$x.(x+y)$
0	0	0	0	0	0
0	1	0	1	0	0
1	0	0	1	1	1
1	1	1	1	1	1

(b) F (A, B, C) = Σ (0, 1, 2, 3, 4, 5, 6, 7)

A \ BC	$\bar{B}\bar{C}$	$\bar{B}C$	BC	$B\bar{C}$
$\bar{A}$	1 (0)	1 (1)	1 (3)	1 (2)
A	1 (4)	1 (5)	1 (7)	1 (6)

F (A,B,C) = 1, since all its terms are present, so it is forever true expression.

F′ = 0

(c) OR will be formed using NOR gates

A	B	A + B	$\overline{A+B}$	$\overline{\overline{A+B}}$
0	0	0	1	0
0	1	1	0	1
1	0	1	0	1
1	1	1	0	1

(d) $P \Rightarrow q = \bar{q} \Rightarrow \bar{p}$

P	q	$p \Rightarrow q$	$\bar{q}$	$\bar{p}$	$\bar{q} \Rightarrow \bar{p}$
0	0	1	1	1	1
0	1	1	0	1	1
1	0	0	1	0	0
1	1	1	0	0	1

Hence both are equal.

(e) (i) $\bar{A}+\bar{B}+\bar{C}+\bar{D}$ (ii) $A\bar{B}CD$

Answer 2.

(a) When a function having same signature is defined in parent class as well as in child class, then the child class functions over ride the parent class functions and the parent class method can be used in child class with super keyword.

(b) BA = 1098
A[4][5] = 1144
ES = 2
lr = 1, lC = 1

In column major

A[I, J] = BA + ES [(I – lr) + R (J – lC)]
A [4, 5] = 1098 + 2 [(4 – 1) + R (5 – 1)]
1144 = 1098 + 2 (3 + 4R)
1144 = 1098 + 6 + 8R
1144 = 1104 + 8R
8R = 1144 – 1104
8R = 40

$R = \frac{40}{8} = 5.$

Matrix is of order 5 × 5.

(c) Big O notation is a way to determine the complexity of a given algorithm.

(d) The anamolous situation which occurs during the execution of a program is called an exception.

(e) *a + b* c – d/e*
*a + bc * – de/*
*abc * + – de/*
*abc * + de /-*

Answer 3.

(a) Hello 3
Hello 2
Hello 1

Working
mymethod (3)
mymethod (2)
mymethod (1)
mymethod (0)

(b) int gcd (int a, int b)
(i) a * b! = 0
(ii) b
(iii) a
(iv) a % b
(v) r

PART–II
SECTION–A

Answer 4.

(a) Principle of duality states :
1. change all 0 to 1 and 1 to 0
2. change all · to + and + to·

$\overline{A}B + C{\cdot}1 = (\overline{A}+C){\cdot}(B+C)$

Dual $(\overline{A}+B){\cdot}(C+0) = \overline{A}C + BC.$

(b) **(i)** {(CD)' + A} + A + CD' + AB.

[Complementarity Rule]

= 1 + A + A + AB = 1

[Properties of 1's]

(ii) A.{B + C.(AB + AC)'} = A. {(B + C.(A. (B + C))'}

……distributive law

$= A.\{(B + C.(\overline{A} + \overline{B+C}))\}$

$= A\{(B + C.\ (\overline{A} + \overline{B}.\overline{C})\}$

……De Morgan's law

$= A.\{(B + \overline{A}.C + C.B.\overline{C})\}$

$= A.\{(B + \overline{A}.C + 0)\}$

$= A.B + A.\overline{A}.C = A.B$

(c) 1. $A \oplus B = A\overline{B} + \overline{A}.B \quad A \odot B = \overline{A \oplus B}$

A	B	C	A ⊕ B ⊕ C	A⊙ B⊙ C	A⊙ B⊙ C
0	0	0	0	1	0
0	0	1	1	0	1
0	1	0	1	0	1
0	1	1	0	1	0
1	0	0	1	0	1
1	0	1	0	1	0
1	1	0	0	1	0
1	1	1	1	0	1

Answer 5.

(a) F(P, Q, R, S) = π (2, 3, 6, 7, 9, 11, 12, 13, 14, 15)

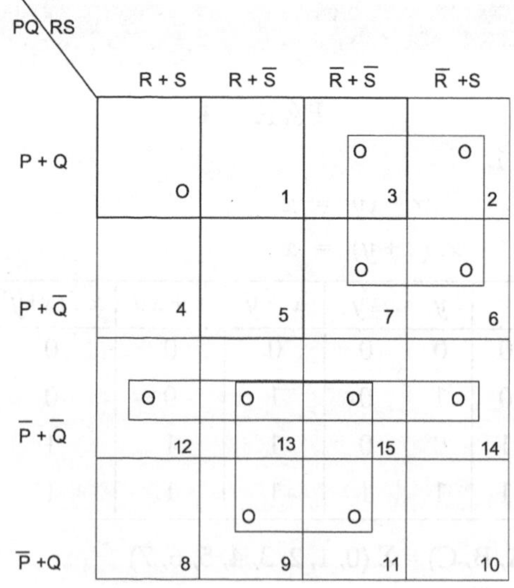

Quad 1 = $M_2 \cdot M_3 \cdot M_6 \cdot M_7$

$= P + \overline{R}$

Quad 2 = $M_{12} \cdot M_{13} \cdot M_{14} \cdot M_{15}$

$= \overline{P} + \overline{Q}$

Quad 3 = $M_9 \cdot M_{11} \cdot M_{13} \cdot M_{15}$

$= \overline{P} + \overline{S}$

$\Rightarrow \quad (P + \overline{R}) \cdot (\overline{P} + Q) \cdot (\overline{P} + S)$

(b) $\overline{A}\,\overline{B}\,\overline{C}\,\overline{D} + \overline{A}\,\overline{B}\,\overline{C}D + A\overline{B}\,\overline{C}\,\overline{D} + A\overline{B}\,\overline{C}D$

$+ \overline{A}B\overline{C}\,\overline{D} + \overline{A}BC\overline{D}$

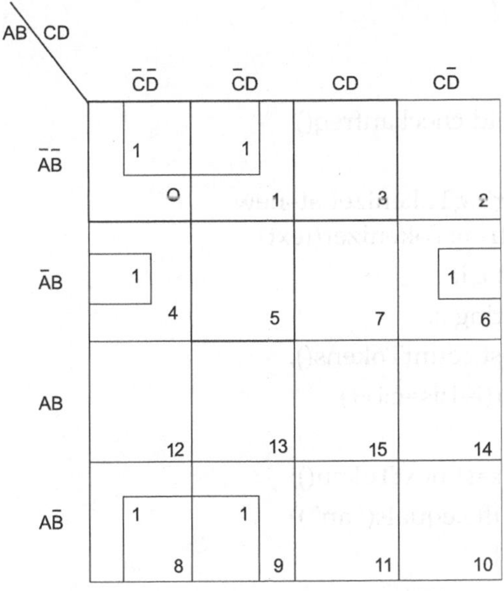

Quad 1 $= M_0 + M_1 + M_8 + M_9 = \overline{B}\,\overline{C}$

Pair 1 $= M_4 + M_6 = \overline{A}B\overline{D}$

So expression $\overline{B}\,\overline{C} + \overline{A}B\overline{D}$

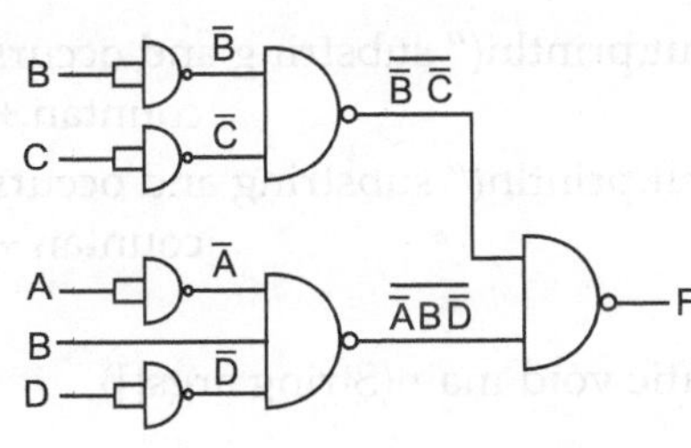

Answer 6.

(a)

(b) We can write the expression,

$(A \to B) \cdot (A \to C) = A \to (B \cdot C)$

A	B	C	BC	A→BC (RHS)	A→B	A→C	A → B· A → C (LHS)
0	0	0	0	1	1	1	1
0	0	1	0	1	1	1	1
0	1	0	0	1	1	1	1
0	1	1	1	1	1	1	1
1	0	0	0	0	0	0	0
1	0	1	0	0	0	1	0
1	1	0	0	0	1	0	0
1	1	1	1	1	1	1	1

(c) Decoder is a digital electronic circuit which converts binary signal to any other form.

A	B		
0	0	— F_0	(Zero)
0	1	— F_1	(One)
1	0	— F_2	(Two)
1	1	— F_3	(Three)

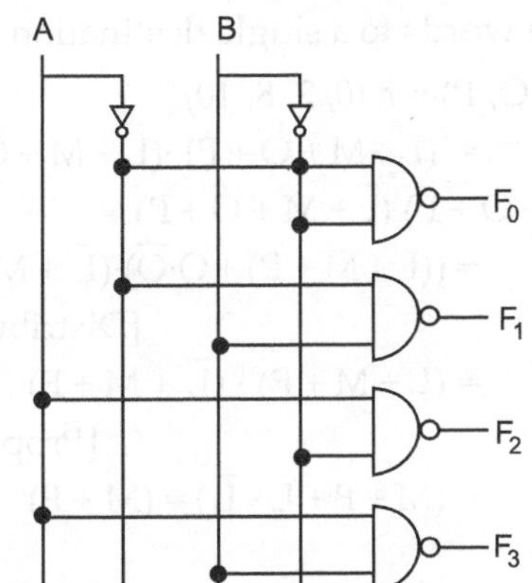

Answer 7.

(a) Full adder is a digital electronic circuit which is used to add 3 bits.

x	y	z	Carry		Sum	
0	0	0	0		0	
0	0	1	0		1	$\overline{x}\,\overline{y}z$
0	1	0	0		1	$\overline{x}y\overline{z}$
0	1	1	1	$\overline{x}yz$	0	
1	0	0	0		1	$x\overline{y}\,\overline{z}$
1	0	1	1	$x\overline{y}z$	0	
1	1	0	1	$xy\overline{z}$	0	
1	1	1	1	xyz	1	xyz

$$\text{Sum} = \overline{x}\,\overline{y}z + \overline{x}y\overline{z} + x\overline{y}\,\overline{z} + xyz$$

$$= x \oplus y \oplus z$$

$$\text{Carry} = \bar{x}yz+x\bar{y}z+xy\bar{z}+xyz$$
$$= xy + yz + xz.$$

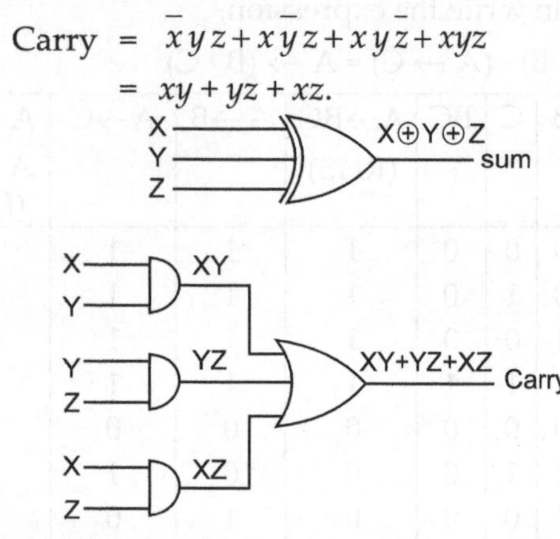

(b) Decoder is a combinational circuit which is used to convert binary signals to any other form. While multiplexer is a special form of decoder, which selects input from more than one line, and direct, it to the signals output channel.

Decoders are used in converting binary signals to any form, while multiplexers are used in parallel to serial conversion or when he wants to direct multiple words to a single destination.

(c) F (L, M, O, P) = π (0, 2, 8, 10)

$$= (L+M+O+P)\cdot(L+M+\bar{O}+P)$$
$$(\bar{L}+M+O+P)\,(\bar{L}+M+\bar{O}+P).$$
$$= ((L+M+P)+O\cdot\bar{O})\cdot(\bar{L}+M+P+O\cdot\bar{O}$$

[Distributive Law]

$$= (L+M+P)\cdot(\bar{L}+M+P)$$

[Properties of 0]

$$= (M+P+L\cdot\bar{L}) = (M+P)$$

Answer 8.

```
import java. io.*;
class Frequency
{
  String text;
  int countand,countan,len;
  public Frequency()
  {
    text="";
    countand=0;
    countan=0;
    len=0;
  }
  void accept(String n)
  {
    n=n.toLowerCase();
    text=n;
  }
    void checkandfreq()
  {
    StringTokenizer st=new
     StringTokenizer(text);
    int c,i;
    String s;
    c=st.countTokens();
    for(i=1;i<=c;i++)
    {
      s=st.nextToken();
      if(s.equals("and"))
      {
        countand++;
      }
    }
  }
    void checkanfreq()
  {
    StringTokenizer st=new
     StringTokenizer(text);
    int c,i;
    String s;
    c=st.countTokens();
    for(i=1;i<=c;i++)
    {
      s=st.nextToken();
      if(s.equals("an"))
      {
        countan++;
      }
    }
  }
  void display()
  {
  System.out.println(" substring and occurs " +
                              countan + " times");
  System.out.println(" substring and occurs " +
                              countan + " times");
  }
  public static void main(String args[])
  {
    Frequency ob= new Frequency();
    ob.accept("ritesh sahu, destination point jhansi");
    ob.checkandfreq();
    ob.checkanfreq();
    ob.display();
  }
}
```

Answer 9.

(a)
```
class DeciOct
{
  int n,oct;
   public DeciOct()
  {
    n=0; oct=0;
  }
  Void getnum(int nn)
  {
    n=nn;
  }
  void deci_oct()
  {
   if (n < = 0)
   return;
  }
   n = n/8;
   deci_oct ( );
   oct = oct * 10 + n% 8;
   }
   void show ( )
   {
    System.out. println ("Decimal number : "+
n);
   deci_oct ( );
   System out. print ("Octal Equivalent : "+oct);
   }
   void main ( )
   {
    DeciOct ob = new DeciOct ( );
    ob.get num (102);
    ob.show ( );
   }
}
```

(b) Program using recursion needs :

1. Large amount of memory.
2. Higher processor.

Answer 10.

```
public Pseudoarithmetic
{
int n, a [ ], sum, r;
boolean ans, flag;
Pseudoarithmetic ( )
{
  n = 0; sum = 0; r = 0;
  ans = true; // assume aus to state
                    psendo or not
  flag = true // assume flag to state n is even or not
}
void accept(int nn)throws IOException
{
  int i;
  n=nn;
  a=new int[n];
  BufferedReader br=new BufferedReader(new
   InputStreamReader(System.in));
  for(i=0;i<n;i++)
  {
  System.out.println("Enter a number");
     a[i]=Integer.parseInt(br.read
     Line());
  }
}
boolean check()
{
 sum = a [ 0 ] + a [n – 1];
 for (int i = 1, j = n – 2; i < (n/2); i + + , j– –)
{
     if (sum ! = (a [i] + a [j]))
       { ans = false;
       break;
   }
}
if (n%2 ! = 0) && (sum ! = (a [n/2] * 2)))
       ans = false;
 if (ans = = true)
{r = (n% 2 = = 0) ? (n/2) : (n/2 + 1);
sum = sum * r; // to store the sum of the sequence
System.out.print ("sum of sequence = "+ sum);
 }
 return ans;
}

}
public static void main(String args[])throws
IOException
{
  Pseudoarithmetic ob= new Pseudoarithmetic();
  ob.accept(5);
  boolean x= ob.check();
  if(x==true)
  {
```

```
    System.out.println("pseudo-arithmetic
                              sequence");
    }
    else
    {
    System.out.println("not a pseudo-arithmetic
                              sequence");
    }
  }
}
```

SECTION–C

Answer 11.

```
import java.io.*;
class Record
{
  String name[];
  int rnk[];
  public Record()
  {
    name=new String[50];
    rnk=new int[50];
  }
  void readvalues()throws IOException
  {
    int i;
    BufferedReader br=new BufferedReader(new
     InputStreamReader(System.in));
    for(i=0;i<50;i++)
    {
    System.out.println("Enter name and rank");
      name[i]=br.readLine();
      rnk[i]=Integer.parseInt(br.readLine());
    }
  }
  void display()
  {
    int i;
    for(i=0;i<50;i++)
    {
System.out.println(name[i]+" "+rnk[i]);
    }
  }
}
class Rank extends Record
{
  int index;
  public Rank()
  {
    super();
    index=0;
  }
  void highest()
  {
    int i;
    for(i=0;i<50;i++)
    {
      if(rnk[i]>rnk[index])
      {
        index=i;
      }
    }
  }
  void display()
  {
    super.display();
  System.out.println("Topper is " + name[index] +
                              "Rank is" + rnk[index]);
  }
}
```

Answer 12.

(a)

```
class Stack
{
  String st[];
  int size,top,ctr;
  public Stack()
  {
    top=-1;
    size=0;
  }
  public Stack(int cap)
  {
    size=cap;
    top=-1;
    st=new String[size];
  }
  void pushname(String n)
  {
    if(top==size-1)
    {
      System.out.println("OVER-
      FLOW");
    }
    else
    {
```

```
        st[++top]=n;
    }
}
String popname()
{
    if(top == -1)
    {
    System.out.println("UNDER-FLOW");
        return "";
    }
    else
    {
        return(st[top--]);
    }
}
void display()
{
    for(ctr=top;ctr>=0;ctr--)
    {
    System.out.println(st[ctr]);
    }
}
}
```

(b) Stack works on a principle of LIFO.

Answer 13.

(a) Method

```
void deletenode (Node start)
{
    if (start = = null)
    {
        System.out.print ("underflow");
    }
    else
    {
    System out.print ("Data Deleted =" + start.info);
    start = start.link;
    }
```

Algorithm

Step 1. It start = = null then
Print "underflow", Goto end

Step 2. If start ! = null then
start = start.link
(move start to its next address)

step 3. End.

(b) Worst case complexity determines the maximum amount of time and resources taken by an algorithm while best case complexity determines the minimum amount of time and resources consumed.

(c) **(i)** B, D, G, H, F, E, C, A

(ii) B, D, G, H

(iii) 4

(iv) A

PHYSICS 2020

Paper 1
(THEORY)
(Maximum Marks : 70)
(Time allowed : Three hours)

*(Candidates are allowed additional 15 minutes for **only** reading the paper. They must NOT start writing during this time.)*

All questions are compulsory.

This question paper is divided into 4 Sections, A, B, C and D as follows :

Section A

Question number 1 is of twelve marks. All parts of this question are compulsory.

Section B

Question numbers 2 to 12 carry 2 marks each with two questions having internal choice.

Section C

Question numbers 13 to 19 carry 3 marks each with two questions having internal choice.

Section D

Question numbers 20 to 22 are long-answer type questions and carry 5 marks each. Each question has an internal choice.

The intended marks for questions are given in brackets [ ].

All working, including rough work, should be done on the same sheet as and adjacent to the rest of the answer. Answers to sub-parts of the same question must be given in one place only. A list of useful physical constants is given at the end of this paper.

A simple scientific calculator without a programmable memory may be used for calculations.

QUESTIONS

SECTION – A

*(Answer **all** questions)*

Question 1.

(A) *Choose the correct alternative (a), (b), (c) or (d) for each of the questions given below :* **[5 × 1]**

(i) *A point charge 'q' is kept at each of the vertices of an equilateral triangle having each side 'a'. Total electrostatic potential energy of the system is :*

(a) $\left(\frac{1}{4\pi\varepsilon_0}\right)\frac{3q^2}{a^2}$ (b) $\left(\frac{1}{4\pi\varepsilon_0}\right)\frac{3q}{a}$

(c) $\left(\frac{1}{4\pi\varepsilon_0}\right)\frac{3q^2}{a}$ (d) $\left(\frac{1}{4\pi\varepsilon_0}\right)\frac{3q}{a^2}$

(ii) ***Curie*** *temperature is the temperature above which:*

(a) *a ferromagnetic substance behaves like a paramagnetic substance*

(b) *a paramagnetic substance behaves like a diamagnetic substance*

(c) *a ferromagnetic substance behaves like a diamagnetic susbtance*

(d) *a paramagnetic substance behaves like a ferromagnetic substance*

(iii) *In an **astronomical telescope of refracting** type :*

(a) *Objective should have small focal length*

(b) *Objective should have large focal length*

(c) *Eyepiece should have large focal length*

(d) *Both objective and eyepiece should have large focal length*

(iv) *In **photoelectric effect** experiment, the slope of the graph of the **stopping potential** versus frequency gives the value of :*

(a) $\frac{h}{e}$ (b) h

(c) $\frac{e}{h}$ (d) $\frac{hc}{e}$

(v) *In a nuclear reactor, cadmium rods are used as :*

(a) *Control rods* (b) *Fuel rods*

(c) *Coolant* (d) *Moderator*

(B) *Answer the following questions briefly and to the point :* **[7 × 1]**

(i) *State **Gauss'** theorem.*

(ii) *A metallic wire having a resistance of 20Ω is bent in order to form a complete circle. Calculate the resistance between any two diametrically opposite points on the circle.*

(iii) *How can a moving coil galvanometer be converted into a voltmeter ?*

(iv) *Write* ***Biot-Savart's law*** *in vector form.*

(v) *What is the* ***phase difference*** *between any two points lying on the same wavefront.*

(vi) *Name the physical principle on the basis of which* ***optical fibres*** *work.*

(vii) *What is* ***Pair production*** *?*

SECTION – B

(Answer ***all*** *questions)*

Question 2. **[2]**

(a) *A uniform copper wire having a cross sectional area of 1 mm^2 carries a current of 5 A. Calculate the* ***drift speed*** *of free electrons in it.*

(Free electron number density of copper = $2 \times 10^{28}/m^3$)

OR

(b) *An electric bulb is rated as 250 V, 750 W. Calculate the :*

(i) *Electric current flowing through it, when it is operated on a 250 V supply.*

(ii) *Resistance of its filament.*

Question 3. **[2]**

Write an expression for ***force per unit length*** *between two long current carrying wires, kept parallel to each other, in vacuum and hence define an* ***ampere****, the SI unit of current.*

Question 4. **[2]**

(i) *Define angle of dip.*

(ii) *State the relation between* ***magnetic susceptibility*** *(χ) and* ***relative permeability*** *(μ_r) of a magnetic substance.*

Question 5. **[2]**

(a) ***Figure 1*** *below shows a metallic rod MN of length l = 80 cm, kept in a uniform magnetic field of flux density B = 0·5 T, on two parallel metallic rails* ***P*** *and* ***Q****. Calculate the emf that will be induced between its two ends, when it is moved towards right with a constant velocity v of 36 km/hr.*

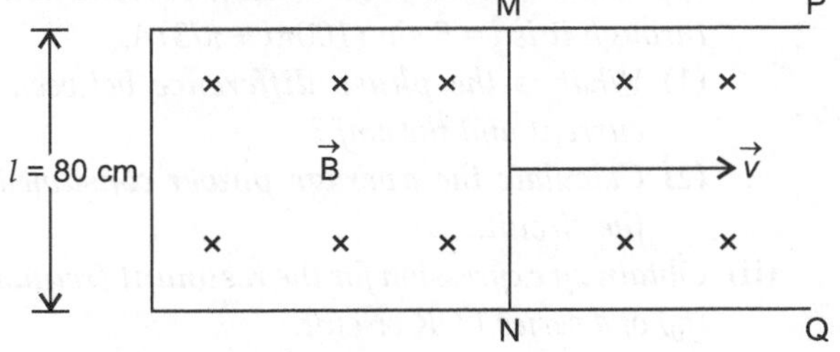

Figure 1

OR

(b) *When current flowing through one coil changes from 0 Amp to 15 Amp in 0·2s, an emf of 750 V is induced in an adjacent coil. Calculate the coefficient of mutual inductance of the two coils.*

Question 6. **[2]**

(i) *State any one use of* ***infrared*** *radiations.*

(ii) *State any one source of* ***ultraviolet*** *radiations.*

Question 7. **[2]**

Where will you keep an object in front of a :

(i) *Convex lens in order to get a* ***virtual*** *and* ***magnified*** *image ?*

(ii) *Concave mirror to get a* ***real*** *and* ***diminished*** *image ?*

Question 8. **[2]**

Draw a labelled graph of angle of deviation (δ) versus angle of incidence (i) for a prism.

Question 9. **[2]**

(i) *State* ***de Broglie*** *hypothesis.*

(ii) *What conclusion can be drawn from* ***Davisson*** *and* ***Germer's*** *experiment ?*

Question 10. **[2]**

Calculate binding energy of oxygen nucleus $\left({}^{16}_{8}O\right)$ from the data given below :

Mass of a proton m_p = 1·007825 u

Mass of a neutron m_n = 1·008665 u

Mass of $\left({}^{16}_{8}O\right) m_o$ = 15·994915 u

Question 11. **[2]**

For a ***radioactive substance****, write the relation between :*

(i) *Half life (T) and disintegration constant (λ).*

(ii) *Mean life (τ) and disintegration constant (λ).*

Question 12. **[2]**

With reference to ***communication systems,*** *what is meant by :*

(i) *modulation ?* **(ii)** *demodulation ?*

SECTION – C

(Answer ***all*** *questions)*

Question 13. **[3]**

Show that intensity of electric field E at a point in broadside on position is given by :

$$E = \left(\frac{1}{4\pi\varepsilon_0}\right)\frac{p}{(r^2 + l^2)^{3/2}}$$

where the terms have their usual meaning.

Question 14. **[3]**

A ***parallel plate capacitor*** *is charged by a battery, which is then disconnected. A dielectric slab having* ***dielectric constant*** *(relative permittivity) K, is now introduced between its two plates in order to occupy the space completely. State, in terms of K, its effect on the following :*

(i) *The capacitance of the capacitor.*

(ii) *The potential difference between its plates.*

(iii) *The energy stored in the capacitor.*

Question 15. **[3]**

(a) *E_1 and E_2 are two batteries having emfs of 3V and 4V and internal resistances of 2 Ω and 1 Ω respectively. They are connected as shown in* ***Figure 2*** *below. Using* ***Kirchhoff's Laws*** *of electrical circuits, calculate the currents I_1 and I_2.*

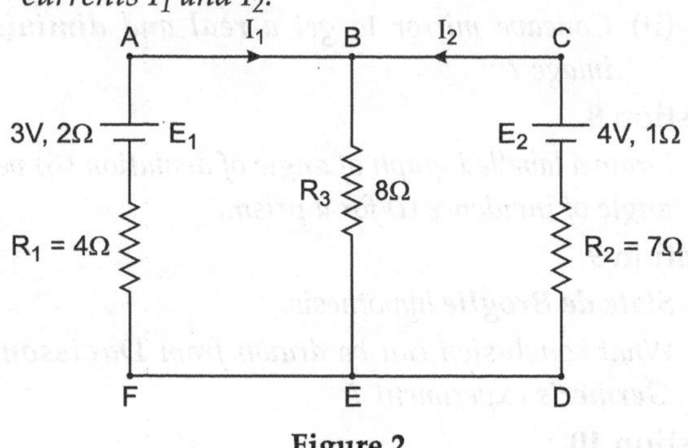

Figure 2

OR

(b) *A* ***potentiometer circuit*** *is shown in* ***Figure 3*** *below. AB is a uniform metallic wire having length of 2 m and resistance of 8 Ω. The batteries E_1 and E_2 have emfs of 4 V and 1·5 V and their internal resistances are 1 Ω and 2 Ω respectively.*

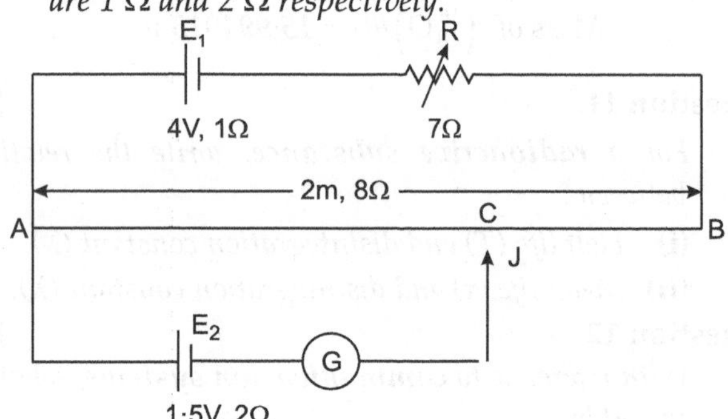

Figure 3

(i) *When the jockey J does not touch the wire AB, calculate :*

(a) *the current flowing through the potentiometer wire AB.*

(b) *the potential gradient across the wire AB.*

(ii) *Now the jockey J is made to touch the wire AB at a point C such that the galvanometer (G) shows no deflection. Calculate the length AC.*

Question 16. **[3]**

For two ***thin lenses*** *kept in* ***contact*** *with each other, show that :*

$$\frac{1}{F} = \frac{1}{f_1} + \frac{1}{f_2}$$

where the terms have their usual meaning.

Question 17. **[3]**

(a) *A* ***compound microscope*** *consists of two convex lenses having focal length of 1·5 cm and 5 cm. When an object is kept at a distance of 1·6 cm from the objective, the final image is virtual and lies at a distance of 25 cm from the eyepiece. Calculate magnifying power of the compound microscope in this set-up.*

OR

(b) *In* ***Young's double slit experiment****, the screen is kept at a distance of 1·2 m from the plane of the slits. The two slits are separated by 5 mm and illuminated with monochromatic light having wavelength 600 nm. Calculate :*

(i) *Fringe width i.e., fringe separation of the interference pattern.*

(ii) *Distance of 10th bright fringe from the centre of the pattern.*

Question 18. **[3]**

Draw the ***energy level diagram*** *of hydrogen atom and show the transitions responsible for :*

(i) *absorption lines of* ***Lyman*** *series.*

(ii) *emission lines of* ***Balmer*** *series.*

Question 19. **[3]**

(i) *State any one difference between energy band diagram of conductors and that of insulators.*

(ii) *Give a relation between α and β for a transistor. (Derivation is* ***not*** *required)*

(iii) *What is the* ***advantage*** *of an LED bulb over the filament electric bulb ?*

SECTION – D

(Answer ***all*** *questions)*

Question 20. **[5]**

(a) (i) *A 400 Ω resistor, a 3H inductor and a 5μF capacitor are connected in series to a 220 V, 50 Hz a.c. source.*

Calculate the :

(1) *Impedance of the circuit.*

(2) *Current flowing through the circuit.*

(ii) *Draw a* ***labelled*** *graph showing the variation of* ***impedance*** *(Z) of a series LCR circuit versus* ***frequency*** *(f) of the a.c. supply.*

OR

(b) (i) *When an alternating emf $e = 310 \sin (100\pi t)$ V is applied to a series LCR circuit, current flowing through it is $i = 5 \sin (100\pi t + \pi/3)$A.*

(1) *What is the* ***phase difference*** *between the current and the emf ?*

(2) *Calculate the* ***average power*** *consumed by the circuit.*

(ii) *Obtain an expression for the* ***resonant frequency*** *(f_0) of a series LCR circuit.*

Question 21. **[5]**

(a) (i) ***Derive*** *an expression for refraction at a single (convex) spherical surface, i.e., a relation between u, v, R, n_1 (rarer medium) and n_2 (denser medium), where the terms have their usual meaning.*

(ii) *Name the* ***phenomenon*** *due to which the Sun appears reddish at sunset.*

OR

(b) (i) *Draw a labelled graph of intensity of diffracted light (I) versus angle (θ) in the **Fraunhofer** diffraction experiment for a single slit diffraction.*

(ii) *State the law of **Malus**.*

(iii) *How will you distinguish **experimentally** between ordinary light and plane polarized light?*

Question 22. **[5]**

(a) (i) *In a semiconductor diode, what is meant by **potential barrier**?*

(ii) *Draw a **labelled** circuit diagram of a **Zener** diode as a **voltage regulator**.*

(iii) *Show with the help of a diagram, how you will obtain an **AND** gate using only **NAND** gates. (Truth table is **not** required).*

OR

(b) (i) *Draw a **labelled** circuit diagram of a transistor acting as a **common emitter amplifier**. What is meant by phase reversal ?*

(ii) *Draw the symbol of a **NAND** gate and write its truth table.*

Useful Constant and Relation :

1 u	=	931 MeV

ANSWERS

SECTION – A

Answer 1.

(A) (i) (c) $U = \frac{1}{4\pi\varepsilon_0}\left[\frac{q_1q_2}{r_{12}} + \frac{q_1q_3}{r_{13}} + \frac{q_2q_3}{r_{23}}\right]$

$$= \frac{1}{4\pi\varepsilon_0}\left[\frac{q^2}{a} + \frac{q^2}{a} + \frac{q^2}{a}\right]$$

$$= \frac{1}{4\pi\varepsilon_0}\cdot\frac{3q^2}{a}$$

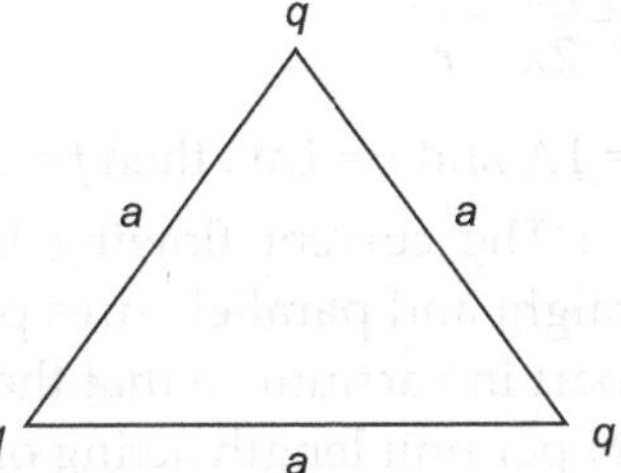

(ii) (a) Above Curie's temperature, a ferro-magnetic substance loses its strong magnetic properties and it becomes paramagnetic.

(iii) (b) Objective is of large focal length. Eyepiece is of short focal length in the case of an astronomical telescope.

(iv) (a) $eV_0 = h\nu$

$$\Rightarrow \quad \frac{V_0}{\nu} = \frac{h}{e}$$

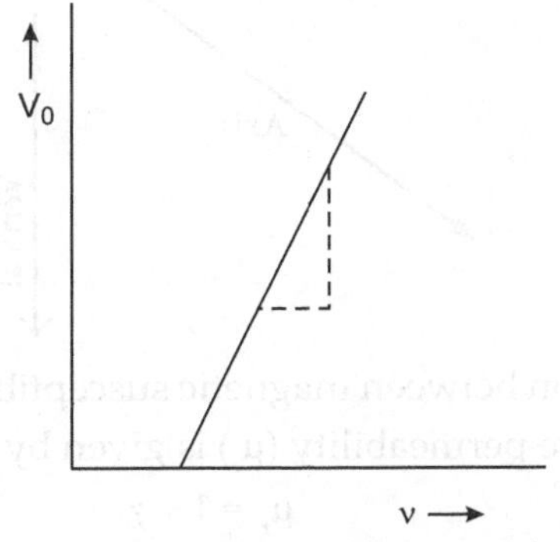

(v) (a) Cadmium rods absorb extra neutrons to control the chain reaction.

(B) (i) Gauss's theorem states that the net electric flux over a closed surface is $\frac{1}{\varepsilon_0}$ times the net electric charge enclosed by the surface.

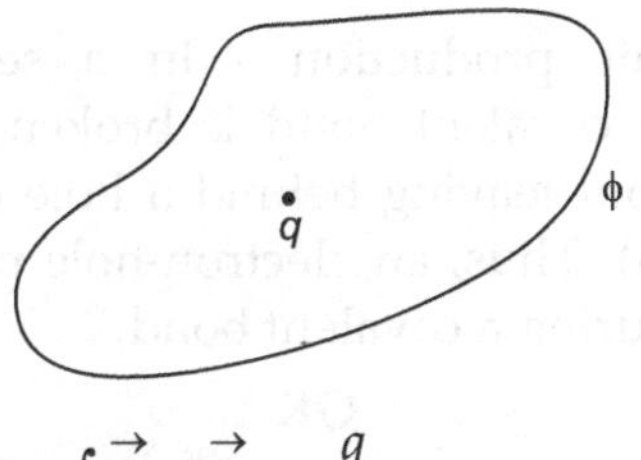

$$\phi = \oint \vec{E}.d\vec{A} = \frac{q}{\varepsilon_0}$$

(ii) Resistance of the wire = 20 Ω. Diameter divides it into two equal parts with resistance 10 Ω each.

∴ Equivalent resistance between diametrically opposite points

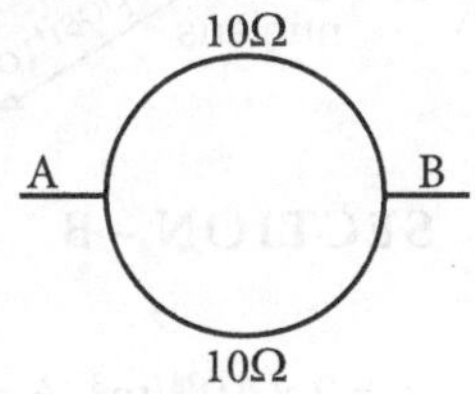

$$\frac{1}{R} = \frac{1}{10} + \frac{1}{10}$$

$$R = \frac{10\times10}{(10+10)} = 5\,\Omega$$ **Ans.**

(iii) Voltmeter is a high resistance device. Therefore, by connecting a suitable high resistance in series with a galvanometer, it can be converted into a voltmeter.

(iv) Biot-Savart's law in vector form.

$$d\vec{B} = \frac{\mu_0}{4\pi}\frac{I(d\vec{l}\times\vec{r})}{r^3}$$

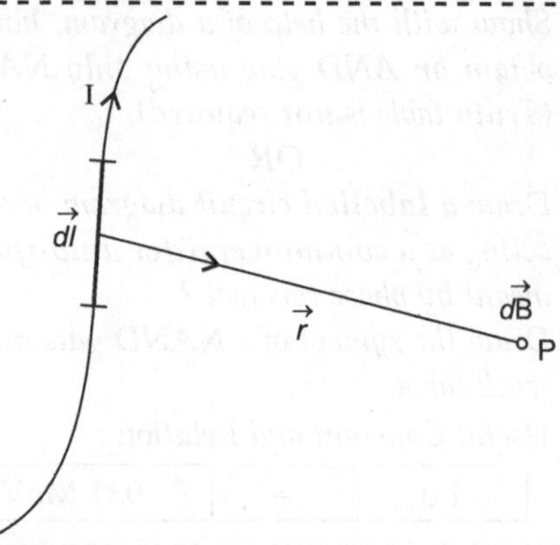

(Magnetic field at a distance r from a current element dl)

(v) Phase difference between any two points on the same wavefront is Zero or have no phase difference.

(Wavefront is the locus of all points at the same phase of vibration)

(vi) Optical fibres work on the principle of Total Internal Reflection.

(vii) Pair production - In a semiconductor, when a covalent bond is broken, the electron comes out leaving behind a hole (deficiency of electron). Thus, an electron-hole pair is formed by rupturing a covalent bond.

OR

Production of an electron – positron pair from a gamma ray (or radiation) photon passes close to an atomic nucleus.

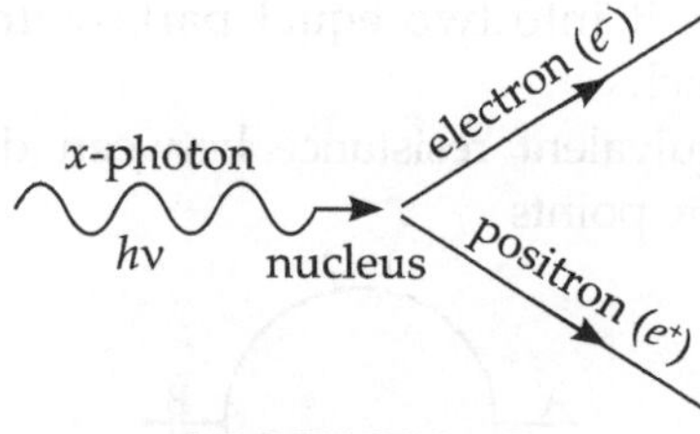

SECTION – B

Answer 2.

(a)

$$n = 2 \times 10^{28}/\text{m}^3, \ A = 1\text{mm}^2 = 10^{-6}\ \text{m}^2$$

$$I = 5\text{A}, \ e = 1.6 \times 10^{-19}\ \text{C}, \ v_d = ?$$

$$I = neAv_d$$

$$v_d = \frac{I}{neA}$$

$$= \frac{5}{2 \times 10^{28} \times 1.6 \times 10^{-19} \times 10^{-6}}$$

$$= 1.5625 \times 10^{-3}\ \text{m/s}$$

$$\approx 1.56\ \text{mm/s}$$ **Ans.**

OR

(b) V = 250 V, P = 750 W, I = ?, R = ?

(i) $P = VI$

$$\Rightarrow I = \frac{P}{V} = \frac{750}{250} = 3\ \text{A}$$

(ii) Also, $P = I^2R$

$$\Rightarrow R = \frac{P}{I^2} = \frac{750}{9} = 83.33\ \Omega$$ **Ans.**

Answer 3.

If two infinitely long straight parallel current carrying conductors are separated by a distance r in vacuum, then the force acting per unit length of each wire is given by,

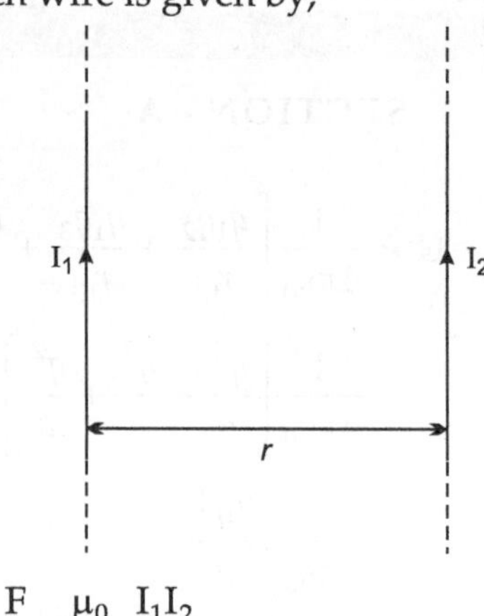

$$f = \frac{F}{L} = \frac{\mu_0}{2\pi} \cdot \frac{I_1 I_2}{r}$$

If $I_1 = I_2 = 1$A and $r = 1$m, then $f = 2 \times 10^{-7}$ N/m

Ampere : The current flowing in each of two long, straight and parallel wires placed exactly 1 metre apart in vacuum so that there is a force of 2×10^{-7} N per unit length acting on the wires.

Answer 4.

(i) Angle of dip : The angle made by the earth's magnetic field with the horizontal in the magnetic meridian is called the angle of dip (magnetic inclination).

OR

It is the angle made by the freely supended magnet with the horizontal.

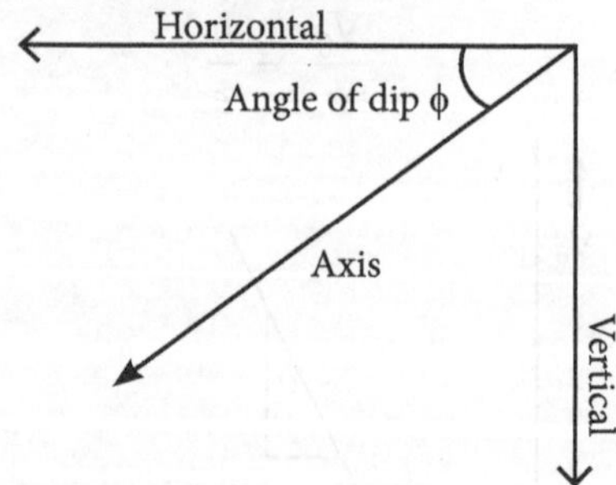

(ii) Relation between magnetic susceptibility (χ) and relative permeability (μ_r) is given by :

$$\mu_r = 1 + \chi$$

Answer 5.

(a) $l = 80$ cm, $B = 0.5$ T

$v = 36$ km/hr $= 10$ m/s

Induced emf :

$e = Bvl$

$= 0.5 \times 10 \times 80 \times 10^{-2}$

$e = 4$ V

OR

(b) $I_1 = 0$ A, $I_2 = 15$ A, $\Delta t = 0.2$ S

$e = 750$ V M = ?

$$|e| = M\frac{\Delta I}{\Delta t}$$

$$\Rightarrow \quad M = \frac{|e|}{(\Delta I / \Delta t)} = \frac{750}{(15-0)/0.2}$$

$= 10$ H

Coefficient of mutual Inductance = 10 H **Ans.**

Answer 6.

(i) IR rays are used in remote control systems (TV remote is an example).

Other uses are as night vision camera, infrared thermometer, drying in sunlight etc.

(ii) Sun emits UV radiations (along with other radiations)

Other sources are mercury lamp, halogen lamp etc.

Answer 7.

(i) In order to get a virtual and magnified image with a convex lens, the object should be placed within the focus or between optical centre and focus (between the principal focus and the lens).

(ii) In order to get a real and diminished image using a concave mirror, the object should be placed beyond the centre of curvature of the mirror.

$[u > 2f]$

Answer 8.

Variation of angle of deviation (δ) with the angle of incidence (*i*) for a prism

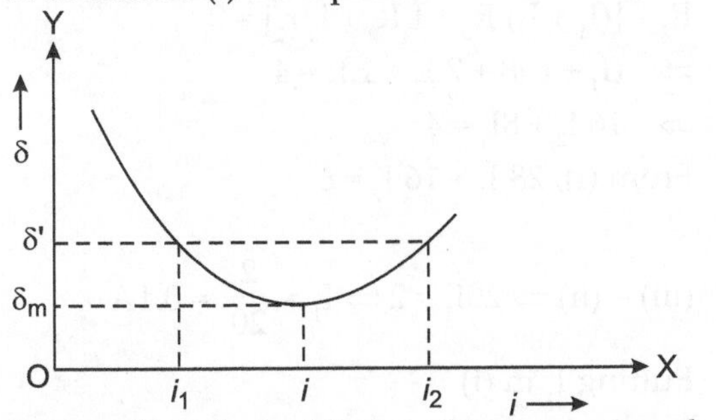

As angle of incidence increases, angle of deviation first decreases and then increases. Same value of δ (shown as δ') for two values of angle of incidence at the first face.

$\delta_m \rightarrow$ angle of minimum deviation for a particular value (*i*) of angle of incidence.

Answer 9.

(i) De-Broglie hypothesis : The waves associated with moving particles are called matter waves. Matter has both particle nature and wave nature. The wave nature of matter is understood from its wavelength (called de-Broglie wavelength) given by :

$$\lambda = \frac{h}{p} \text{ or } \lambda = \frac{h}{mv} \text{ : where } p \text{ is the momentum.}$$

(ii) Davisson-Germer's experiment concluded that electrons exhibited wave nature also. (thereby supporting de-Broglie hypothesis)

Answer 10.

Mass of proton $m_p = 1.007825\ u$

Mass of neutron $m_n = 1.008665\ u$

Mass of $\left({}^{16}_{8}O\right) m_0 = 15.994915\ u$

Binding Energy of ${}^{16}_{8}O$ = ?

Mass defect :

$\Delta m = 8m_p + 8m_n - m_O$

$= (8 \times 1.007825 + 8 \times 1.008665 - 15.994915)u$

$= (8.0626 + 8.06932 - 15.004915)u = 0.137005u$

Binding Energy $= 0.137005\ u$

$= 0.137005 \times 931.5$ MeV. [$\because 1u = 931.5$ MeV]

$= 127.62 \times 1.6 \times 10^{-13}$J [$\because 1\text{ MeV} = 1.6 \times 10^{-13}$ J]

$= 2.042 \times 10^{-11}$ J **Ans.**

[**Note :** Can be obtained as BE = Δmc^2 also, after expressing Δm in kg, $c = 3 \times 10^8$ m/s]

Answer 11.

(i) Half life ($T_{1/2}$) and disintegration constant (λ) are related as

$$T_{1/2} = \frac{0.693}{\lambda}$$

(ii) Mean life (τ) and disintegration constant are related as

$$\tau = \frac{1}{\lambda}$$

Answer 12.

(i) Modulation : Modification of some characteristics (amplitude, frequency etc.) of carrier wave according to the corresponding instantaneous value of the information signal is called modulation. It is used to transmit signals to large distance.

(ii) Demodulation : Reverse process of modulation in which the original information signal is retrieved from the modulated carrier wave.

It is process of separating low frequency modulating wave from modulated wave.

SECTION – C

Answer 13.

Electric field due to dipole (broadside on position)

Consider dipole of length $2l$ and moment $\vec{p}$.

From the figure, resultant electric field intensity at C:

$$\vec{E} = \vec{E}_A + \vec{E}_B$$

$$|\vec{E}_A| = \frac{1}{4\pi\varepsilon_0}\frac{q}{(r^2+l^2)}$$

$$|\vec{E}_B| = \frac{1}{4\pi\varepsilon_0}\frac{q}{(r^2+l^2)}$$

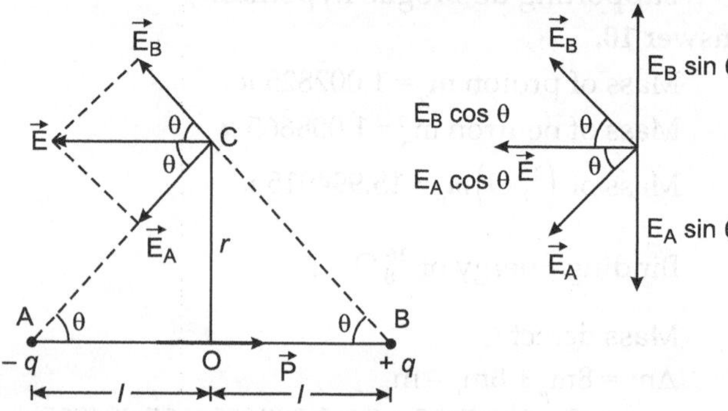

sine components of $\vec{E}_A$ and $\vec{E}_B$ get cancelled each other as $|\vec{E}_A| = |\vec{E}_B|$. The cosine components get added up to give the resultant field.

i.e., $E = E_A \cos\theta + E_B \cos\theta$

$$= \frac{1}{4\pi\varepsilon_0}\cdot\frac{q}{(r^2+l^2)}\cos\theta + \frac{1}{4\pi\varepsilon_0}\cdot\frac{q}{(r^2+l^2)}\cos\theta$$

$$= 2\cdot\frac{1}{4\pi\varepsilon_0}\cdot\frac{q}{(r^2+l^2)}\cos\theta$$

$$= 2\times\frac{1}{4\pi\varepsilon_0}\frac{q}{(r^2+l^2)}\times\frac{l}{(r^2+l^2)^{1/2}}$$

$$= \frac{1}{4\pi\varepsilon_0}\frac{q(2l)}{(r^2+l^2)^{3/2}}$$

$$E = \frac{1}{4\pi\varepsilon_0}\frac{p}{(r^2+l^2)^{3/2}} \text{ as } p = q(2l)$$

The above expression gives the magnitude of the field. The direction of electric field E at C is opposite to the direction of the dipole moment $\vec{p}$.

Answer 14.

We have $C = \frac{\varepsilon_0 A}{d}$, $V = \frac{Q}{C}$ and $U = \frac{Q^2}{2C}$

After inserting the dielectric constant K

(i) Capacitance $C' = \frac{\varepsilon_0 KA}{d} = KC$

$\Rightarrow$ Capacitance increases. It becomes K times the initial value.

(ii) Potential difference between the plates V'

$$= \frac{Q}{C'} = \frac{Q}{KC} = \frac{V}{K}$$

$\Rightarrow$ Potential difference between the plates decreases to $\frac{1}{K}$ times the initial value.

(iii) Energy stored, $U' = \frac{Q^2}{2C'} = \frac{Q^2}{2CK} = \frac{U}{K}$.

$\Rightarrow$ Energy stored decreases to $\frac{1}{K}$ times the initial value.

Answer 15.

(a) $E_1 = 3$ V

$r_1 = 2\ \Omega$

$R_1 = 4\ \Omega$

$E_2 = 4$ V

$r_2 = 1\ \Omega$, $R_2 = 7\ \Omega$, $R_3 = 8\ \Omega$, $I_1 = ?$ $I_2 = ?$

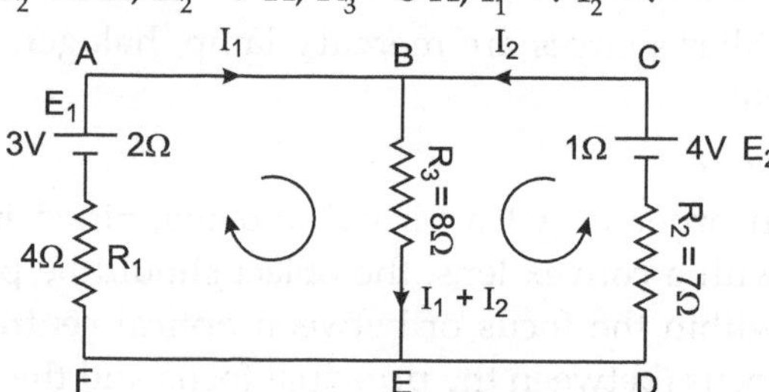

Current through $R_3 = I_1 + I_2$ (Kirchhoff's current rule)

Applying loop rule to loop ABEFA,

$E_1 - [(I_1 + I_2) R_3 + I_1R_1 + I_1r_1] = 0$

$\Rightarrow (I_1 + I_2)8 + 4I_1 + 2I_1 = 3$

$\Rightarrow 14 I_1 + 8I_2 = 3$ ...(i)

Applying loop rule to loop CBEDC

$E_2 - [(I_1 + I_2) R_3 + I_2R_2 + I_2 r_2] = 0$

$\Rightarrow (I_1 + I_2)8 + 7 I_2 + 1.I_2 = 4$

$\Rightarrow 16 I_2 + 8I_1 = 4$ ...(ii)

From (i), $28 I_1 + 16 I_2 = 6$ ...(iii)

(iii) – (ii) $\Rightarrow 20I_1 = 2 \Rightarrow I_1 = \frac{2}{20} = 0.1$A **Ans.**

Putting I_1 in (i)

$I_2 = \frac{3 - 14\times 0.1}{8} = \frac{1.6}{8} = 0.2$A **Ans.**

OR

(b) (i) When jockey J does not touch the wire AB

(a) $E_1 = 4V$

$R = (7 + 8) = 15\ \Omega$

$r = 1\ \Omega$

Current through AB, $I_1 = \frac{E_1}{R + r} = \frac{4}{15 + 1} = \frac{4}{16}$

$= 0.25$ A

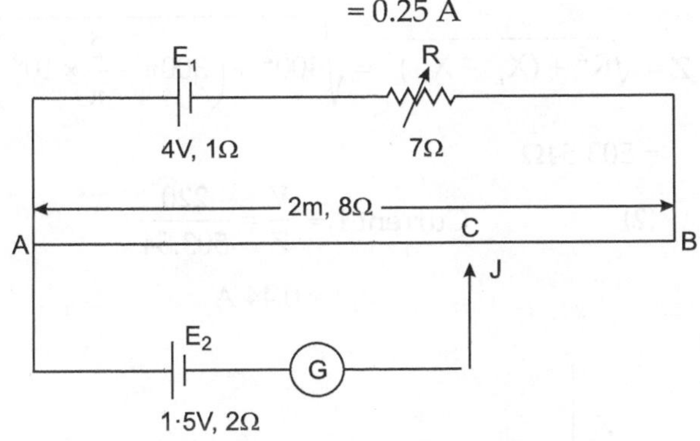

(b) Potential difference across wire AB,

$V_{AB} = I_1 \times 8$

$= 0.25 \times 8 = 2V$

Potential gradient in AB, $k = \frac{V_{AB}}{L} = \frac{2}{2} = 1\ Vm^{-1}$

(ii) Let $AC = l$ metre

$\therefore$ $E_2 = kl$

$\Rightarrow$ $1.5 = 1 \times l$

$\Rightarrow$ $l = 1.5$ m.

Answer 16.

Combined focal length of two thin lenses in contact

Both the Lenses are Convex : Suppose two thin convex lenses L_1 and L_2 of focal lengths f_1 and f_2 are placed in contact in air having a common principal axis. A point-object O is placed on the principal axis at a distance u from the first lens L_1. Its image would be formed by the lens L_1 alone at I_1. distant v' (say) from L_1. Then, from the lens formula, we have

$$\frac{1}{v'} - \frac{1}{u} = \frac{1}{f_1} \qquad ...(i)$$

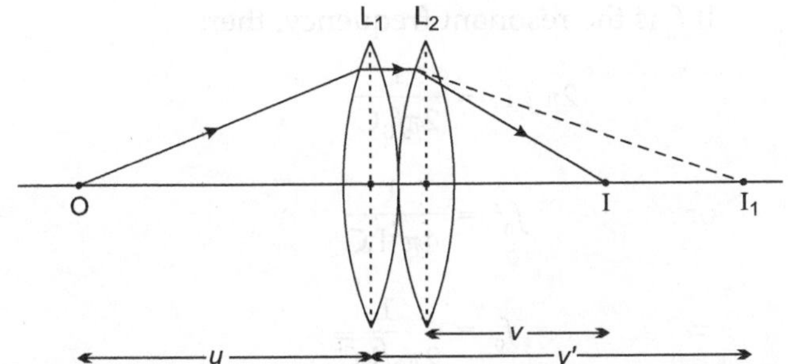

I' serves as a virtual object for the second lens L_2 which forms a final image I at a distance v (say) from it.

Then, we have

$$\frac{1}{v} - \frac{1}{v'} = \frac{1}{f_2}. \qquad ...(ii)$$

Adding eqn. (i) and (ii), we get

$$\frac{1}{v} - \frac{1}{u} = \frac{1}{f_1} + \frac{1}{f_2}. \qquad ...(iii)$$

If we replace these two lenses by a single lens which forms the image of an object distant u from it at a distance v, then the focal length F of this equivalent lens would be given by

$$\frac{1}{v} - \frac{1}{u} = \frac{1}{F} \qquad ...(iv)$$

From eqn. (iii) and (iv), we get

$$\frac{1}{F} = \frac{1}{f_1} + \frac{1}{f_2}. \qquad ...(v)$$

Answer 17.

(a) D = 25 cm

$u_o = -1.6$ cm, $f_o = 1.5$ cm, $f_e = 5$ cm, M = ?

For objective :

$$-\frac{1}{u_o} + \frac{1}{v_o} = \frac{1}{f_o}$$

$$\Rightarrow \quad \frac{1}{v_o} = \frac{1}{f_o} + \frac{1}{u_o}$$

$$\Rightarrow \quad \frac{1}{v_o} = \frac{1}{1.5} + \left(\frac{1}{-1.6}\right)$$

$$\Rightarrow \quad \frac{1}{v_o} = 10\left[\frac{1}{15} - \frac{1}{16}\right]$$

$$= \frac{10}{15 \times 16}$$

$$\Rightarrow \quad v_0 = \frac{16 \times 15}{10}$$

$$= 1.6 \times 15 \text{ cm} = 24 \text{ cm}$$

Now magnifying power

$$|m| = \frac{v_o}{|u_o|}\left(1 + \frac{D}{fe}\right)$$

$$= \frac{1.6 \times 15}{1.6}\left[1 + \frac{25}{5}\right]$$

$$= 15 \times 6 = 90$$

OR

(b) D = 1.2 m, $d = 5 \times 10^{-3}$ m, $\lambda = 600 \times 10^{-9}$ m

(i) $\beta = \frac{D\lambda}{d}$

$$= \frac{1.2 \times 600 \times 10^{-9}}{5 \times 10^{-3}}$$

$= 0.144 \times 10^{-3}$ m

$= 0.144$ mm

(ii) Distance of nth bright fringe from the centre of pattern $x_n = \frac{Dn\lambda}{d}$

Here $n = 10$

$\therefore x_{10} = 10\left(\frac{D\lambda}{d}\right) = 10 \times 0.144$ mm $= 1.44$ mm.

Answer 18.

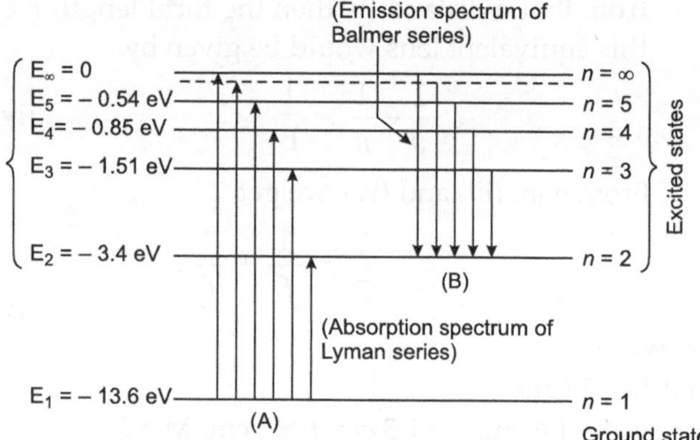

(i) Transitions labelled A shows absorption lines of Lyman series.

(ii) Transitions labelled B show emission lines of Balmer series.

Answer 19.

(i) In conductors, the energy gap between valence band and conduction band is zero mostly, they overlap.

But in insulators, the energy gap is very high so that it is impossible to get the electrons into the conduction band.

(ii) α – current gain in CB configuration

β – current gain in CE configuration

α and β are related as,

$$\alpha = \frac{\beta}{1+\beta}$$

or $$\beta = \frac{\alpha}{1-\alpha}$$

(iii) LED Vs filament bulb : LEDs are energy saving, operating at low voltage, less energy dissipating (heating) and with much longer life compared to ordinary filament lamps.

SECTION – D

Answer 20.

(a) (i) $R = 400\ \Omega$, $L = 3$H, $C = 5\ \mu$F $= 5 \times 10^{-6}$ F

$V = 220$ V, $f = 50$ Hz

(1) Impedance $Z = \sqrt{R^2 + (X_L - X_C)^2}$

$X_L = 2\pi f L$

$= 2\pi \times 50 \times 3$

$= 300\pi\ \Omega$

$$X_C = \frac{1}{2\pi f C} = \frac{1}{2\pi \times 50 \times 5 \times 10^{-6}} = \frac{2}{\pi} \times 10^3\ \Omega$$

$$Z = \sqrt{R^2 + (X_L - X_C)^2} = \sqrt{400^2 + \left(300\pi - \frac{2}{\pi} \times 10^3\right)^2}$$

$= 503.54\Omega$

(2) Current $I = \frac{V}{Z} = \frac{220}{503.54}$

$= 0.44$ A

(ii)

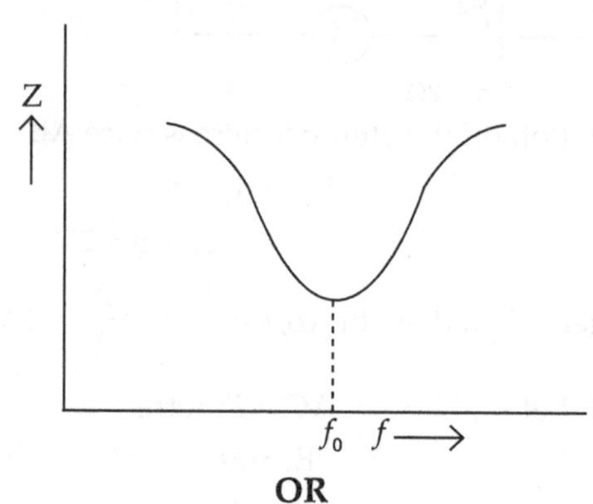

OR

(b) (i) $e = 310 \sin (100\ \pi t)$V

$i = 5 \sin (100\ \pi t + \pi/3)$A

(1) Phase difference between i and e, $\phi = \pi/3$

[current leads by a phase difference $\pi/3$]

(2) Average power consumed.

$$P_{avg} = \frac{1}{2} V_0 I_0 \cos\phi$$

$$= \frac{1}{2} \times 310 \times 5 \times \cos \pi/3$$

$$= \frac{310 \times 5}{4} = 387.5 \text{ W}$$

(ii) For a series LCR circuit, resonance occurs when $X_L = X_C$.

If f_0 is the resonant frequency, then

$$2\pi f_0 L = \frac{1}{2\pi f_0 C}$$

or $$f_0^2 = \frac{1}{4\pi^2 LC}$$

$$\Rightarrow \quad f_0 = \frac{1}{2\pi\sqrt{LC}}$$

Answer 21.

(a) (i) In the figure, XPY is a spherical surface with centre C and radius R, O is the object and I is the image.

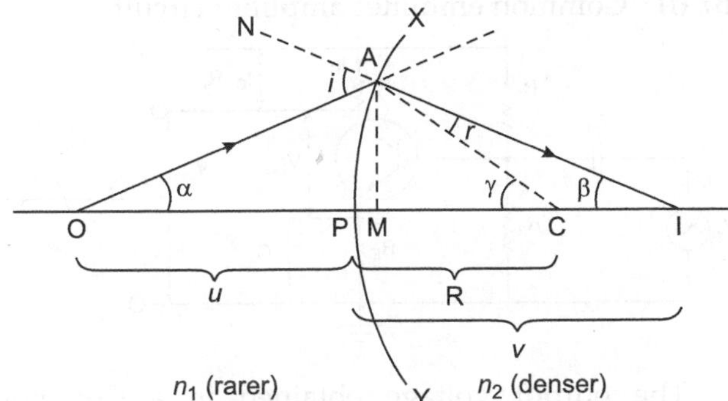

OA and AI are incident and refracted rays.

i and *r* are angles of incidence and refractions. From the figure,

$i = \alpha + \gamma$ and $r = \gamma - \beta$

Now, $\tan \alpha = \frac{MA}{MO}$ or $\alpha = \frac{MA}{MO}$ (When A is very near to P)

Similarly, $\beta = \frac{MA}{MI}$ and $\gamma = \frac{MA}{MC}$

From Snell's law, $\frac{\sin i}{\sin r} = \frac{n_2}{n_1}$

If *i* and *r* are small, $\frac{i}{r} = \frac{n_2}{n_1}$

Also, MO ~ PO, MC ~ PC, MI ~ PI

PO = – *u*, PI = *v* and PC = R

$\therefore \alpha = \frac{PA}{PO} = \frac{PA}{-u}, \beta = \frac{PA}{PI}$ and $\gamma = \frac{PA}{PC} = \frac{PA}{R}$

$i = \alpha + \gamma = PA\left(-\frac{1}{u} + \frac{1}{R}\right)$ and $r = \gamma - \beta = PA\left(\frac{1}{R} - \frac{1}{v}\right)$

Now : $\frac{i}{r} = \frac{n_2}{n_1}$

$\Rightarrow \quad n_1 i = n_2 r$

$\Rightarrow \quad PA\left(-\frac{1}{u} + \frac{1}{R}\right)n_1 = PA\left(\frac{1}{R} - \frac{1}{v}\right)n_2$

$\Rightarrow \quad -\frac{n_1}{u} + \frac{n_1}{R} = \frac{n_2}{R} - \frac{n_2}{v}$

$\Rightarrow \quad \frac{n_2}{v} - \frac{n_1}{u} = \frac{n_2 - n_1}{R}$ **Ans.**

(ii) Sun appears reddish (at sunrise and sunset) because of scattering of light in the atmosphere.

OR

(b) (i) Intensity distribution (w.r.t. Q) for fraunhofer diffraction at single slit.

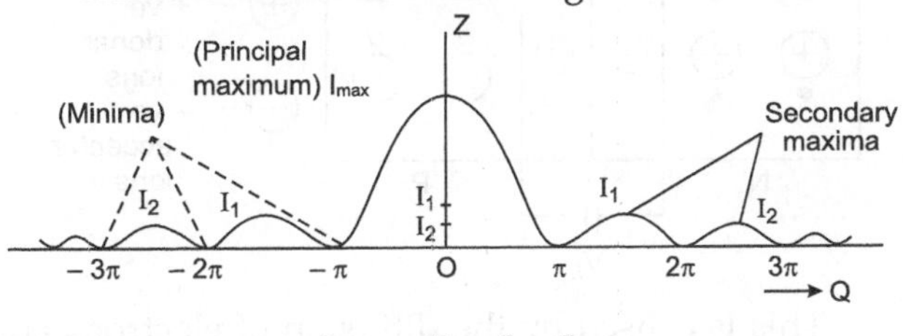

$I_{max} >> I_1 > I_2 > I_3$

(ii) Malus' law

When a completely plane polarised light is incident on an analyser, the intensity of light transmitted by the analyser is directly proportional to the square of the cosine of the angle between the transmission axes of the analyser and the polariser

$I \propto \cos^2 \theta$

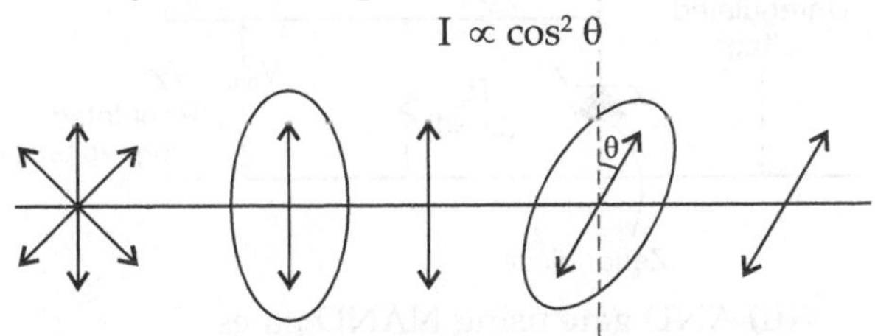

(iii) In order to detect whether the given light is ordinary (unpolarised) light or plane polarised, pass it through polariser. Now, the polariser is rotated keeping the light normal to the plane of the polariser.

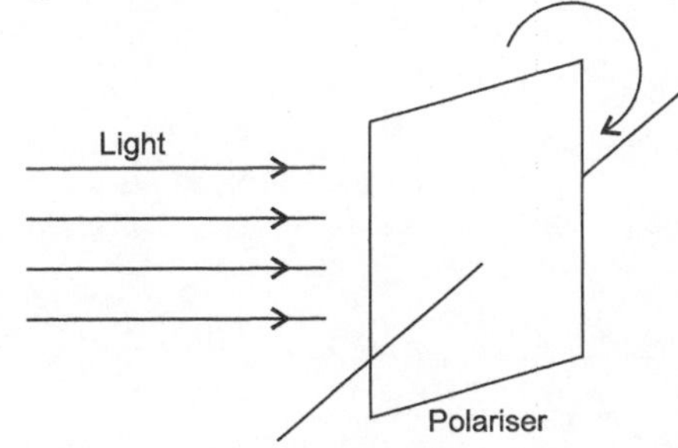

(a) If on complete rotation, there is no change in the intensity of the emerging light, it is unpolarised (ordinary) light.

(b) If the intensity of light changes (gradually varies between maximum and zero twice) in one rotation, then the given light is plane-polarised.

Answer 22.

(a) (i) Potential barrier is the barrier by which the charge carriers (mobile) are stopped by the repelling forces (at the PN junction). It is the potential difference across the depletion region.

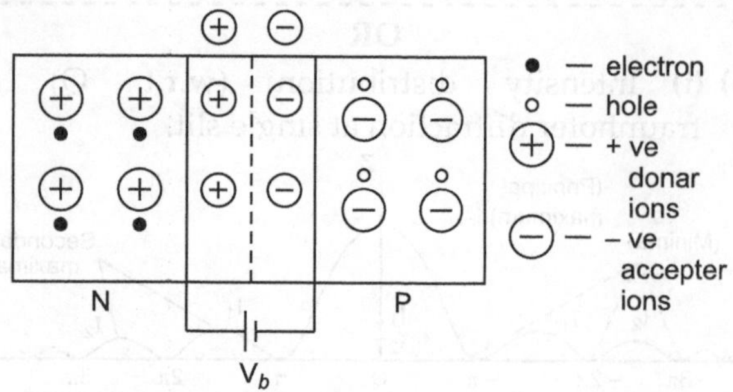

This is caused by the diffusion of electrons and holes across the junction and the subsequent accumulation of immobile charges near the junctions

(ii) Zener diode as voltage regulator

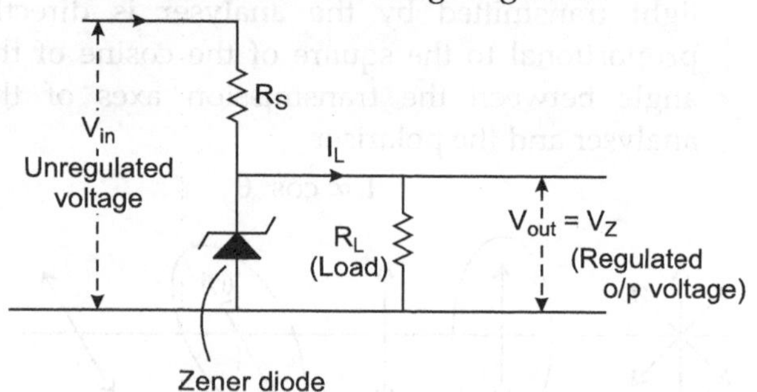

(iii) AND gate using NAND gates

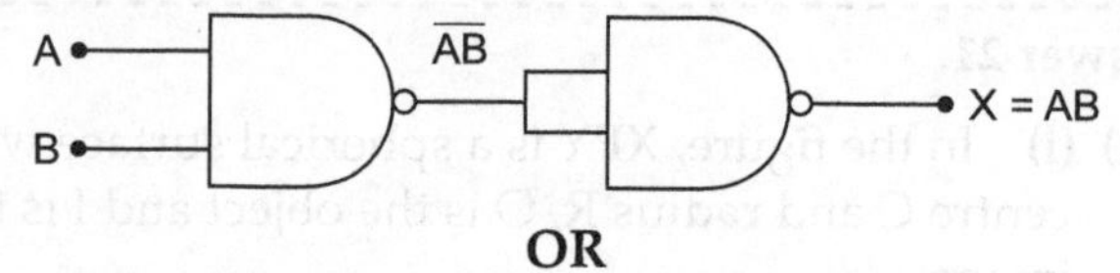

OR

(b) (i) Common emmitter amplifier circuit

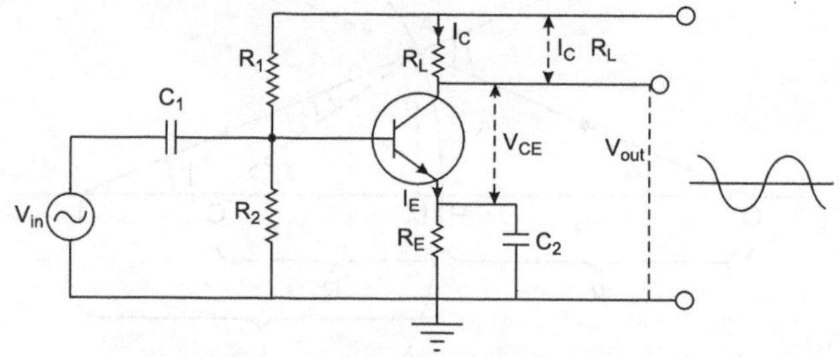

The output voltage obtained in a Common emitter amplifier is 180º out of phase with the input signal voltage. This is called phase reversal.

(ii) NAND gate and truth table

Truth table

A	B	X
0	0	1
0	1	1
1	0	1
1	1	0

A
B
$Y = \overline{A \cdot B}$ X

2 - input NAND gate

PHYSICS 2019

QUESTIONS

SECTION–A

*(Answer **all** questions)*

Question 1.

(A) *Choose the correct alternative (a), (b), (c) or (d) for each of the questions given below:* **[5 × 1]**

(i) *A closed surface in vacuum encloses charges –q and + 3q. The total electric flux emerging out of the surface is:*

(a) *Zero* (b) $2q/\varepsilon_0$

(c) $3q/\varepsilon_0$ (d) $4q/\varepsilon_0$

(ii) *What is the angle of dip at a place where the horizontal component (B_H) and vertical component (B_v) of earth's magnetic field are equal:*

(a) *130°* (b) *60°*

(c) *45°* (d) *90°*

(iii) *A beam of light is incident at the polarizing angle of 35° on a certain glass plate. The refractive index of the glass plate is:*

(a) *sin 35°* (b) *tan 35°*

(c) *tan 55°* (d) *sin 55°*

(iv) *In a gamma ray emission from nucleus:*

(a) *only the number of protons change.*

(b) *the number of protons and neutrons, both change.*

(c) *there is no change in the number of protons and the number of neutrons.*

(d) *only the number of neutrons change*

(v) *The energy associated with light of which of the following colours is minimum:*

(a) *violet* (b) *red*

(c) *green* (d) *yellow*

(B) *Answer the following questions briefly and to the point.* **[7 × 1]**

(i) *Define **equipotential** surface.*

(ii) *Calculate the net emf across A and B shown in Figure 1 below:*

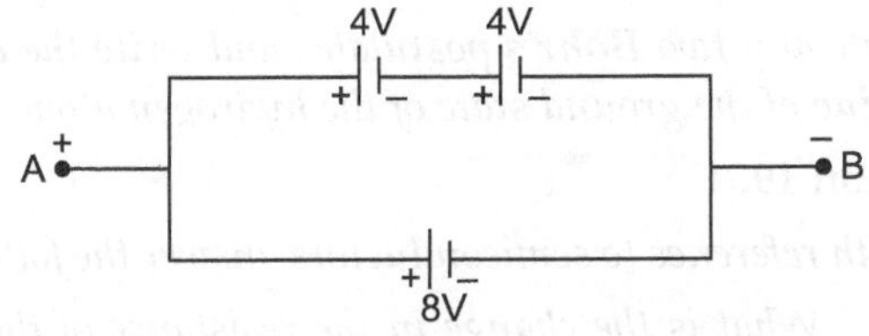

Figure 1

(iii) *Why are the pole pieces of a horseshoe magnet in a **moving coil galvanometer** made cylindrical in shape ?*

(iv) *What is the value of **power factor** for a pure resistor connected to an alternating current source ?*

(v) *What should be the path difference between two waves reaching a point for obtaining constructive interference in **Young's Double Slit** experiment ?*

(vi) *Define **critical angle** for a given medium.*

(vii) *Name the series in the atomic spectra of the hydrogen atom that falls in the ultra violet region.*

SECTION–B

*(Answer **all** questions)*

Question 2. **[2]**

In a potentiometer experiment, the balancing length with a resistance of 2Ω is found to be 100 cm, while that of an unknown resistance is 500 cm. Calculate the value of the unknown resistance.

Question 3. **[2]**

A rectangular loop of area 5 m², has 50 turns and carries a current of 1 A. It is held in a uniform magnetic field of 0·1 T, at an angle of 30°. Calculate the torque experienced by the coil.

Question 4. **[2]**

(a) *An electric current I flows through an infinitely long conductor as shown in **Figure 2(a)** below. Write an **expression** and **direction** for the magnetic field at point P.*

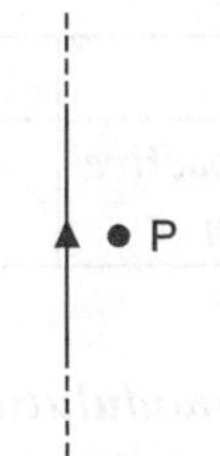

Figure 2(a)

OR

(b) *An electric current I flows through a circular loop as shown in **Figure 2(b)** below. Write an **expression** and **direction** for the magnetic field at the centre of the loop at point P.*

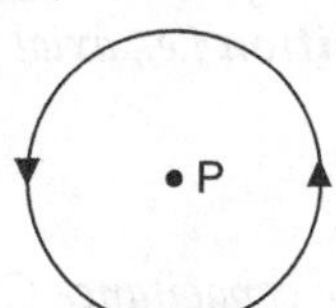

Figure 2(b)

Question 5. [2]

A transformer is used to step up an alternating emf of 200 V to 440 V. If the primary coil has 1000 turns, calculate the number of turns in the secondary coil.

Question 6. [2]

State any two properties of ***microwaves.***

Question 7. [2]

Write any one use for each of the following mirrors :

(a) *Convex*

(b) *Concave*

Question 8. [2]

The deviation produced for violet, yellow and red lights for crown glass are 3.75°, 3.25° and 2.86° respectively. Calculate the dispersive power of the crown glass.

Question 9. [2]

(a) *What is meant by* ***mass defect ?***

(b) *What conclusion is drawn from* ***Rutherford's*** *scattering experiment of α-particles ?*

Question 10. [2]

Define the following with reference to photoelectric effect :

(a) *Threshold frequency (f_0)*

(b) *Stopping potential (V_s)*

Question 11. [2]

(a) *The half-life of radium is 1550 years. Calculate its disintegration constant (λ).*

OR

(b) *Copy and complete the following table for a radioactive element whose half-life is 10 minutes. Assume that you have 30 g of this element at t = 0.*

t (minute)	0	20	30
Amount of radioactive element left in gm	30	...	...

Question 12. [2]

Define ***frequency modulation*** *and state any one advantage of frequency modulation (FM) over amplitude modulation (AM).*

SECTION–C

(Answer ***all*** *questions)*

Question 13. [3]

Obtain an expression for electric potential 'V' at a point in an ***end-on position*** *i.e., axial position of an electric dipole.*

Question 14. [3]

Three capacitors of capacitance $C_1 = 3\mu f$, $C_2 = 6\mu f$ and $C_3 = 10\mu f$, are connected to a 10 V battery as shown in ***Figure 3*** *below :*

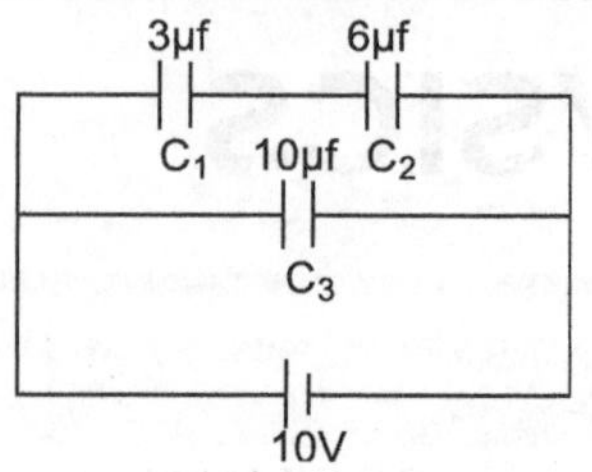

Figure 3

Calculate:

(a) *Equivalent capacitance.*

(b) *Electrostatic potential energy stored in the system.*

Question 15. [3]

(a) *Obtain the balancing condition for the* ***Wheatstone bridge*** *arrangement as shown in* ***Figure 4*** *below :*

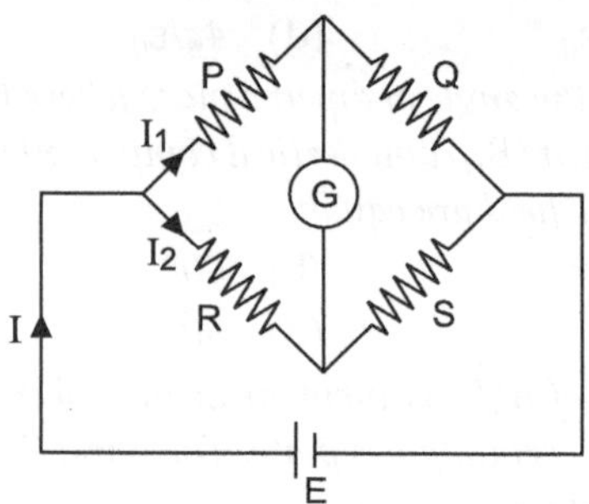

Figure 4

OR

(b) *Draw a labelled circuit diagram of a* ***potentiometer*** *to measure the internal resistance 'r' of a cell. Write the working formula (derivation is not required).*

Question 16. [3]

(a) *A ray of light is incident on a prism whose refractive index is 1·52 at an angle of 40°. If the angle of emergence is 60°, calculate the angle of the prism.*

OR

(b) *Calculate the focal length of a convex lens whose radii of curvature of two surfaces is 10 cm and 15 cm respectively and its refractive index is 1·5.*

Question 17. [3]

Derive the law of reflection using ***Huygen's*** *Wave Theory.*

Question 18. [3]

State any two ***Bohr's*** *postulates and write the* ***energy value*** *of the ground state of the hydrogen atom.*

Question 19. [3]

With reference to semiconductors answer the following:

(i) *What is the change in the resistance of the semiconductor with* ***increase*** *in temperature ?*

(ii) *Name the majority charge carriers in* ***n-type*** *semiconductor .*

(iii) *What is meant by* ***doping ?***

SECTION–D

*(Answer **all** questions)*

Question 20. **[5]**

(a) (i) *An alternating emf of 200 V, 50 Hz is applied to an L–R circuit, having a resistance **R** of 10 Ω and an inductance L of 0·05 H connected in **series**. Calculate :*

(1) *Impedance.*

(2) *Current flowing in the circuit.*

(ii) *Draw a labelled graph showing the variation of inductive reactance (X_L) verses frequency (f).*

OR

(b) (i) *An a.c., source of emf ε = 200 sin ωt is connected to a resistor of 50 Ω.*

Calculate :

(1) *Average current (I_{avg}).*

(2) *Root mean square (**rms**) value of emf.*

(ii) *State any two characteristics of resonance in an LCR series circuit.*

Question 21. **[5]**

(a) *Draw a neat labelled ray diagram showing the formation of an image at the **least distance of distinct vision D** by a **simple microscope.** When the final image is at D, derive an expression for its magnifying power at D.*

OR

(b) *Draw a neat labelled diagram of **Young's** Double Slit experiment. Show that* $\beta = \frac{\lambda D}{d}$ where the terms have their usual meaning (either for bright or dark fringe).

Question 22. **[5]**

(a) (i) *Draw a labelled circuit diagram of a **half wave rectifier** and give its output waveform.*

(ii) *Draw a symbol for **NOR** gate and write its truth table.*

OR

(b) (i) *Draw a neat circuit diagram to study the input and output characteristics of a **common emitter** transistor.*

(ii) *Draw the symbol for AND gate and write its truth table.*

Useful Constants and Relations:

1.	Charge on electron	(e)	1.6×10^{-19} C
2.	Planck's constant	(h)	6.6×10^{-34} Js
3.	Speed of light in vacuum	(c)	3×10^{8} ms^{-1}

ANSWERS

SECTION–A

Answer 1.

(A) (i) (b) $\phi = \frac{q_{enc}}{\varepsilon_0} = \frac{+3q - q}{\varepsilon_0} = \frac{2q}{\varepsilon_0}$ **Ans.**

(ii) (c) Angle of dip, $\tan\delta = \frac{B_V}{B_H}$

$\tan\delta = \frac{B}{B}$, $\delta = 45°$. **Ans.**

(iii) (b) $i_p = 35°$

$\mu = \tan i_p$

$\mu = \tan 35°$ **Ans.**

(iv) (c) *there is no change in the number of protons and the number of neutrons.*

(v) (b) red light

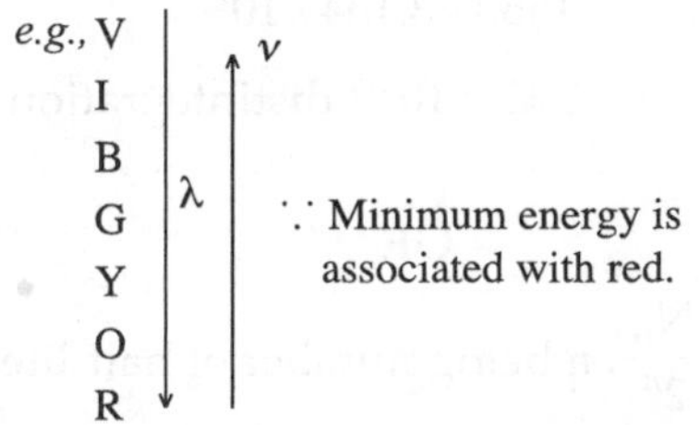

(B) (i) Any surface over which the electric potential is same everywhere is called an equipotential surface.

(ii)

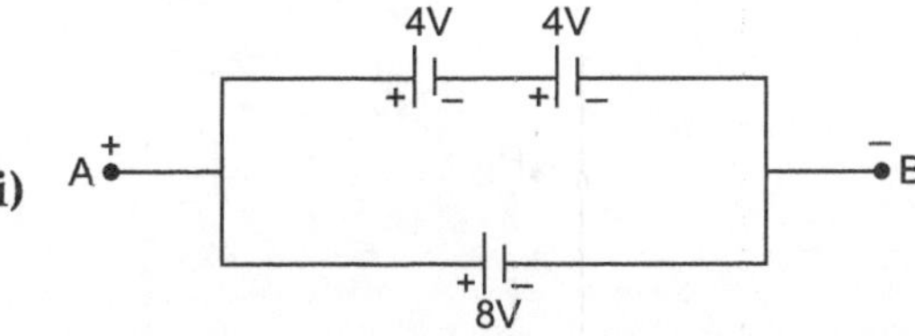

4V and 4V combined in series will have emf of 8V

8V and 8V in parallel will have net emf = 8V.

∴ net emf across A and B = 8 V.

(iii) The pole pieces of a horseshoe magnet in a moving coil galvanometer are made cylindrical in shape to create a uniform radial magnetic field such that in every position, the plane of the coil remains parallel to the lines of force, *i.e.*, in every position, θ = 90°.

(iv) The value of power factor for a pure resistor connected to an alternating source is 1 or cos 0°.

(v) The path difference (x) between two waves reaching a point for constructive interference in Young's double slit experiment is given by,

$x = n\lambda$ Where, n = 0, 1, 2, ... and λ is wavelength.

(vi) The angle of incidence in the denser medium corresponding to which the angle of refraction in the rarer medium is 90° is called critical angle.

(vii) Lyman series.

SECTION–B

Answer 2.

Given : $R = 2\ \Omega$, $l_1 = 100$ cm, $l_2 = 500$ cm

unknown resistance (X) = ?

$$\frac{X}{R} = \frac{l_2}{l_1}$$

$$\Rightarrow \quad \frac{X}{2} = \frac{500}{100}$$

$$\therefore \quad X = 2 \times 5 = 10\ \Omega \qquad \textbf{Ans.}$$

Answer 3.

Given : $N = 50$, $A = 5m^2$, $I = 1A$, $B = 0.1$ T, $\theta = 30°$

Torque experienced by coil (τ) = NIA B sin θ

$$= 50 \times 1 \times 5 \times 0{\cdot}1 \times \sin 30°$$

$$= 50 \times 1 \times 5 \times \frac{1}{10} \times \frac{1}{2} = 12{\cdot}5 \text{ Nm} \qquad \textbf{Ans.}$$

Answer 4.

(a) ∴ By Biot Savart's law

$$\vec{B} = \int d\vec{B} = \frac{\mu_0}{4\pi}\int \frac{I\vec{dl} \times \vec{r}}{r^3}$$

$$\vec{B} = \frac{\mu_0 I}{2\pi r}$$

The magnetic field created by a current carrying wire is radial magnetic field.

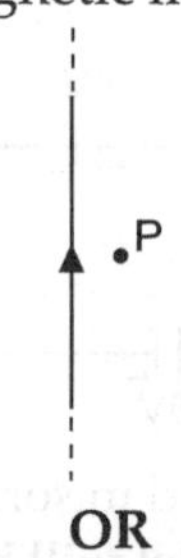

OR

(b) The magnitude of magnetic field at the centre of the loop at point P is:

$$B = \frac{\mu_0 I}{2a}$$

$\mu_0 \rightarrow$ Permeability of free space

I → Current

a → radius of circular loop.

Answer 5.

Given : $E_P = 200$ V, $E_S = 440$ V, $N_P = 1000$, $N_S = ?$

$$\frac{E_S}{E_P} = \frac{N_S}{N_P} \Rightarrow \frac{440}{200} = \frac{N_S}{1000}$$

$$\therefore \quad N_S = \frac{440}{200} \times 1000$$

$$\therefore \quad N_S = 2200 \text{ turns.} \qquad \textbf{Ans.}$$

Answer 6.

1. Microwaves are reflected by metal surfaces.
2. Microwaves have low frequency and high wavelength.

Answer 7.

(a) Convex mirrors are used as reflector in street lights.

(b) Concave mirrors are used as shaving mirrors.

Answer 8.

Given : $\delta_V = 3{\cdot}75°$, $\delta_Y = 3{\cdot}25°$, $\delta_R = 2{\cdot}86°$

$$\text{Dispersive power (w)} = \frac{\delta_V - \delta_R}{\delta_Y}$$

$$= 0{\cdot}2739° \approx 0{\cdot}27° \qquad \textbf{Ans.}$$

Answer 9.

(a) The difference between the sum of the masses of the nucleons constituting a nucleus and the rest mass of the nucleus is called mass defect.

$$\Delta M = [Zm_p + (A - Z)\, m_N - M_{nucleus}]$$

(b) From scattering experiment of α-particles, Rutherford concluded that there is a lot of empty space in an atom.

Answer 10.

(a) The lowest frequency of light which can emit photoelectrons from a metal surface is called the threshold frequency, (f_o) of that material.

(b) The negative potential at which the photoelectric current becomes zero is called 'stopping potential' or 'cut-off potential' (V_S).

Answer 11.

(a) Given, Half-life of radium, T = 1550 years

$$= 1550 \times 3.154 \times 10^7 \text{ second}$$

$$\text{Disintegration constant, } \lambda = \frac{0.6931}{T}$$

$$= \frac{0.6931}{1550 \times 3.154 \times 10^7} \text{s}^{-1}$$

$$= 1.42 \times 10^{-11} \text{ distintegration /second}$$

Ans.

OR

(b) $\because N = \frac{N_0}{2^n}$, n being number of half-life spent.

$t = 20$ min means 2 half-life and $t = 30$ min means 3 half-life.

$$\therefore \quad N_{20\,min} = \frac{N_0}{2^2} = \frac{N_0}{4} \text{ and } N_{30min} = \frac{N_0}{2^3} = \frac{N_0}{8}$$

$$\therefore \quad m_{20\,min} = \frac{m_0}{4} = \frac{30gm}{4} = 7.5\,gm$$

$$\text{and} \quad m_{30\,min} = \frac{m_0}{8} = \frac{30gm}{8} = 3.75\text{ gm}$$

t (minute)	0	20	30
Amount of radioactive element (left in gm).	30	7.5	3.75

Answer 12.

When the frequency of a carrier wave is changed in accordance with the amplitude (intensity) of the modulating wave (signal), it is called frequency modulation.

The advantages of frequency modulation (FM) over amplitude modulation (AM) is that FM is less prone to interference than AM. It has higher efficiency and amplitude is constant unlike AM waves.

SECTION–C

Answer 13.

Electric potential due to a dipole at a point on the end -on position: Figure below shows a dipole consisting of two equal and opposite charges + q and – q separated by a distance $2l$. Let P be a point on the end-on position (on the axis) of the dipole. OP is the axial line of the dipole.

$\longleftarrow l \longrightarrow\longleftarrow l \longrightarrow$

–q O +q P

$\longleftarrow r \longrightarrow$

The potential due to the charge + q at the point P = V_1

$$V_1 = \frac{1}{4\pi\varepsilon_0}\cdot\frac{q}{r-l}$$

The potential due to the charge – q at the point P = V_2

$$V_2 = \frac{1}{4\pi\varepsilon_0}\cdot\frac{(-q)}{(r+l)}$$

Total potential at the point P

$$V = V_1 + V_2$$

$$= \frac{1}{4\pi\varepsilon_0}\cdot\frac{q}{r-l} + \frac{1}{4\pi\varepsilon_0}\cdot\frac{(-q)}{(r+l)}$$

$$= \frac{q}{4\pi\varepsilon_0}\left[\frac{1}{r-l} - \frac{1}{r+l}\right]$$

$$= \frac{q}{4\pi\varepsilon_0}\cdot\left[\frac{r+l-r+l}{(r-l)(r+l)}\right]$$

$$= \frac{1}{4\pi\varepsilon_0}\cdot\frac{2lq}{r^2-l^2}$$

But the dipole moment $p = 2lq$

$$\text{So,} \quad V = \frac{1}{4\pi\varepsilon_0}\cdot\frac{p}{(r^2-l^2)}$$

Suppose $r >> l$. Then l^2 can be neglected compared to r^2. So for a short dipole, the electric potential at a point on the end-on position is,

$$V = \frac{1}{4\pi\varepsilon_0}\cdot\frac{p}{r^2}$$ **Ans.**

Answer 14.

(a) Equivalent capacitance of capacitors 3μF and 6 μF connected in series is

$$C' = \frac{C_1C_2}{C_1+C_2} = \frac{3\times 6}{3+6}\mu F = 2\mu F$$

C′ and (C_3 = 10 μF) are connected in parallel and their equivalent capacitance is

$$C = C' + C_3 = (2+10)\,\mu F = 12\mu F$$

(b) Electrostatic potential energy (U) stored in the system is

$$U = \frac{1}{2}CV^2 = \frac{1}{2}\times 12\mu F \times (10V)^2$$

$$= 600\ \mu J$$

$$= 6 \times 10^{-4}\,J.$$ **Ans.**

Answer 15.

(a) In wheatstone's bridge, four resistances are joined to form a parallelogram. In one diagonal of this parallelogram a galvanometer is attached, and in other diagonal, a cell. The values of resistances are so adjusted that no current flows through the galvanometer. This condition of bridge is called balanced wheatstone's bridge condition. Let P, Q, R and S resistances forms a bridge. Let a current, I flow through the circuit.

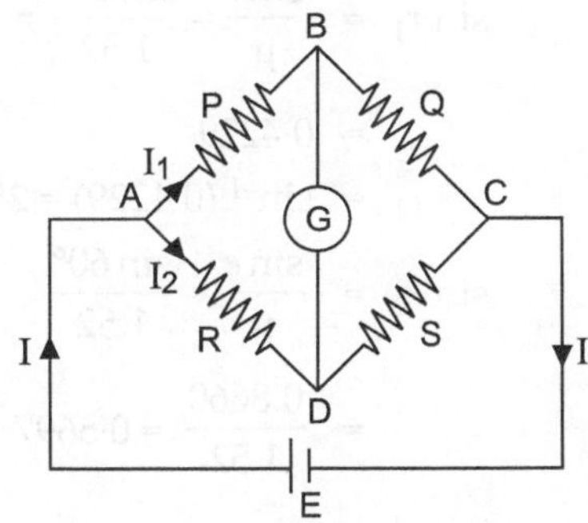

Applying Kirchhoff's second law to closed loop ABDA,

$$I_1P - I_2R = 0 \text{ or } I_1P = I_2R \qquad \ldots(i)$$

Applying Kirchhoff's second law to closed loop BCDB,

$$I_1Q - I_2S = 0 \text{ or } I_1Q = I_2S \quad \dots\text{(ii)}$$

Dividing equation (i) by equation (ii),

$$\frac{I_1P}{I_1Q} = \frac{I_2R}{I_2S} \Rightarrow \frac{P}{Q} = \frac{R}{S}$$

This is the balanced wheatstone bridge condition.

OR

(b) The circuit diagram of a potentiometer to measure the internal resistance 'r' of a cell is given below. The working formula is $r = R\left(\frac{l_1}{l_2} - 1\right)$, l_1 is the distance of null point from point A and l_2 is the distance of null point from A after taking out a suitable resistance R from the resistance box.

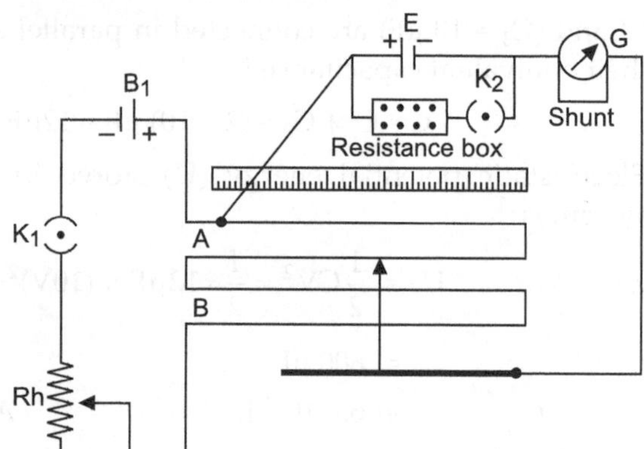

Answer 16.

(a) Given : $\mu = 1{\cdot}52$, $i = 40°$, $e = 60°$

Let r_1 and r_2 be the angles of refraction of two faces

$$\mu = \frac{\sin i}{\sin r_1} = \frac{\sin e}{\sin r_2}$$

$$\Rightarrow \quad \sin r_1 = \frac{\sin i}{\mu} = \frac{\sin 40°}{1.52} = \frac{0.6428}{1.52}$$

$$= 0{\cdot}4229$$

$$\Rightarrow \quad r_1 = \sin^{-1}(0{\cdot}4229) = 25.049°$$

$$\sin r_2 = \frac{\sin e}{\mu} = \frac{\sin 60°}{1.52}$$

$$= \frac{0.8660}{1.52} = 0{\cdot}5697$$

$$\Rightarrow \quad r_2 = \sin^{-1}(0{\cdot}5697) = 34{\cdot}75°$$

Angle of prism A $= r_1 + r_2$

$= 25.049° + 34{\cdot}75°$

$= 59{\cdot}799° = 59{\cdot}7°$ **Ans.**

OR

(b) Given : $R_1 = 10$ cm, $R_2 = -15$ cm, $\mu = 1{\cdot}5$, $f = ?$

$$\frac{1}{f} = (\mu - 1)\left(\frac{1}{R_1} - \frac{1}{R_2}\right)$$

$$\frac{1}{f} = (1{\cdot}5 - 1)\left(\frac{1}{10} + \frac{1}{15}\right)$$

$$\frac{1}{f} = (0{\cdot}5)\left(\frac{15+10}{15 \times 10}\right)$$

$$= (0{\cdot}5)\frac{25}{150}$$

$$\frac{1}{f} = \frac{25 \times 0.5}{150}$$

$$\therefore \quad f = 12 \text{ cm.}$$ **Ans.**

Answer 17.

Let XY be an plane reflecting surface and AMB, the incident plane wavefront. All the particles on AB will be vibrating in phase. Let i be the angle of incidence and r the angle of reflection. In the time the disturbance at A reaches C, the secondary waves from the point B must travel a distance BD equal to AC. With the point B as centre and radius equal to AC a sphere is constructed.

From the point C, we draw tangents CD and CD′.

Then, BD = BD′

In ΔBAC and ΔBDC, BC is common, BD = AC and ∠BAC = ∠BDC = 90°

∴ Δ BAC ≅ Δ BDC

and ∠ABC = ∠BCD

But ∠ABC = i and ∠BCD = r, *i.e.*, $i = r$ or the angle of incidence is equal to the angle of reflection.

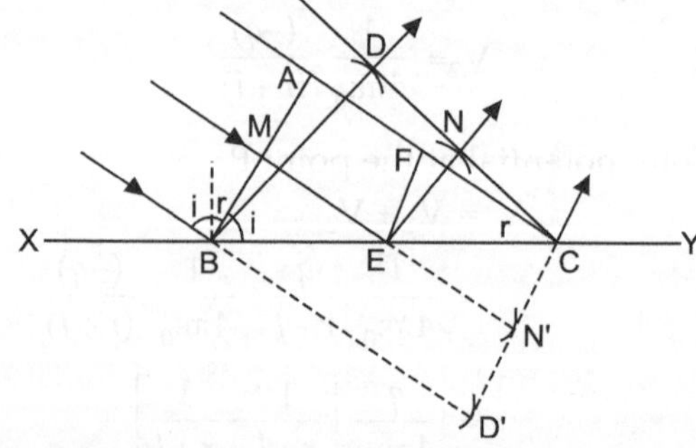

Since AB, AD and XY are in the same plane. Therefore, the incident ray, reflected ray and normal at the point of incidence are all in the same plane. This is the first law of reflection.

Answer 18.

Two Bohr's postulates are:

1. Electrons can revolve only in those orbits in which their angular momentum is an integral multiple of $h/2\pi$, where h is Planck's universal constant.
2. While revolving in stable orbits, the electrons do not radiate energy in spite of their acceleration towards the centre of the orbit. Hence, the atom remains stable and is said to exist in a stationary state.

Answer 19.

(i) The temperature coefficient of resistance of semiconductors is negative, hence their electrical resistance decreases with rise in temperature.

(ii) In an n-type semiconductor the electrons are the majority charge carriers.

(iii) The process of adding trivalent or pentavalent impurity to an intrinsic semiconductor in a controlled manner is called doping.

SECTION–D

Answer 20.

(a) **(i)** Given : $e = 200$ V, L = 0·05 H, R = 10 Ω, f = 50 Hz.

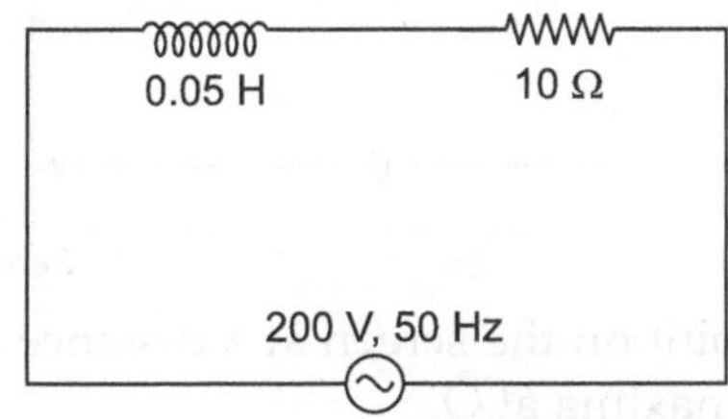

Reactance of coil (X_L) $= 2\pi f L$

$= 2 \times \dfrac{22}{7} \times 50 \times 0{\cdot}05\ \Omega$

$= 15{\cdot}7\ \Omega$

(1) Impedance (Z) $= \sqrt{R^2 + X_L^2}$

$= \sqrt{10^2 + (15.7)^2}$

$= \sqrt{100 + 246.49}$

$= \sqrt{12200} = 18.61\Omega$

(2) Current flowing in the circuit (I_{rms})

$$I_{rms} = \frac{200}{18.61}$$

$= 10{\cdot}74$ A **Ans.**

(ii)

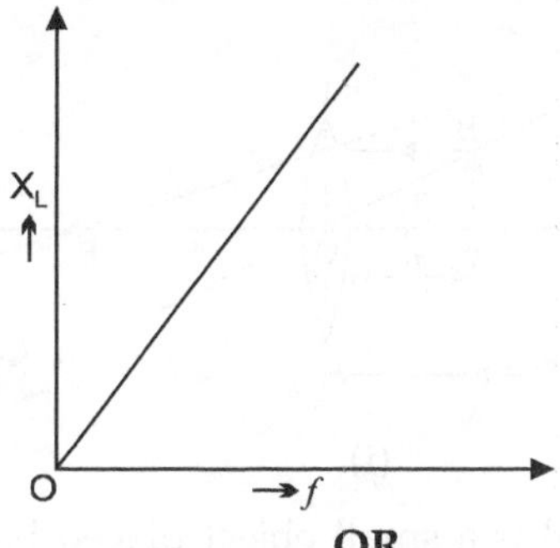

OR

(b) **(i)** Given : $\varepsilon = 200 \sin 314t$

$E_0 = 200$V

$R = 50\ \Omega$

(1)
$$\varepsilon_{av} = \frac{\int_{\omega t=0}^{2\pi} 200 \sin \omega t\, d(\omega t)}{2\pi}$$

$$= \frac{200\left[-\cos (\omega t)_0^{2\pi}\right]}{2\pi}$$

$$= \frac{200[\cos 2\pi - \cos 0]}{2\pi}$$

$$\varepsilon_{av} = 0$$

$$I_{av} = \frac{2E_0}{\pi R} = \frac{2 \times 200}{3.14 \times 50} = 2.55 \text{ A}$$

(2)
$$\varepsilon_{rms} = \frac{E_0}{\sqrt{2}} = \frac{200}{\sqrt{2}}\text{V} = 141.42\text{V}$$

Ans.

(ii) Two characteristics of resonance in series LCR.

1. At resonance impedance of the current is minimum, $Z = R$
2. The net reactance is zero, so the circuit behaves as purely resistive circuit. At resonance frequency, peak current is maximum.

Answer 21.

(a) A microscope is an optical instrument which forms large image of a close and minute object. This image subtends a large visual angle at the eye so that the object looks large. In the simplest form, a simple microscope or magnifying glass, is just a thin, short-focus convex lens carrying a handle. The object to be seen is placed between the lens and its focus and the eye is placed just behind the lens. Then, the eye sees a magnified, erect and virtual image on the same side as the object. The position of the object between the lens and its focus is so adjusted that the image is formed at the least distance of distinct vision (D) from the eye. The image is then seen most distinctly.

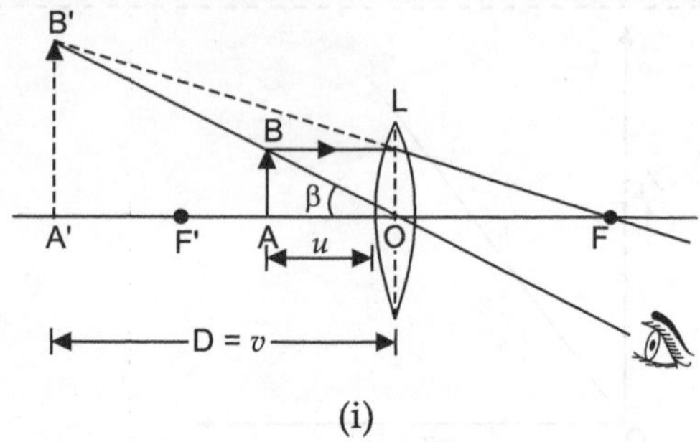

(i)

In fig. (i), AB is a small object placed between a lens L and its first focus F′. Its magnified virtual image A′ B′ is formed at distance D from the lens. Since the eye is just behind the lens, the distance of the image A′ B′ from the eye is also D.

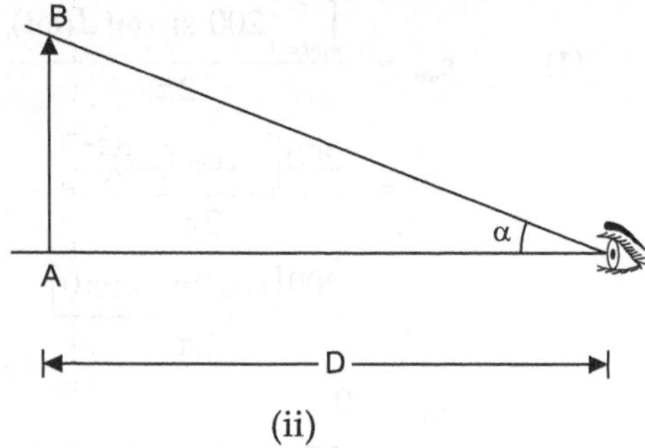

(ii)

Magnifying Power: Let β be the angle subtended by the image A′ B′ at the eye (Fig. (i)] and α the angle subtended by the object AB at the eye when placed at distance D from the eye (Fig. (ii)]. Then, the magnifying power of the simple microscope is given by :

$$M = \frac{\text{Angle subtended by the image at the eye}}{\text{Angle subtended by the object at the eye}}$$

when placed at least distance of distinct vision

$$= \frac{\beta}{\alpha}$$

Since the object AB is small, the angles α and β are also small. Then from Fig. (i) and (ii), we may write :

$$\beta = \tan\beta = AB/OA$$

and $\quad \alpha = \tan\alpha = AB/D$

$$\therefore \quad M = \frac{AB/OA}{AB/D} = \frac{D}{OA}$$

But OA = u (distance of the object from the lens).

$$M = \frac{D}{u}$$

The image A′ B′ is being formed at a distance D. From the lens formula $\frac{1}{v} - \frac{1}{u} = \frac{1}{f'}$ we shall put $v = -D$ and $u = -u$ (with proper sign).

Thus,

$$\frac{1}{-D} - \frac{1}{-u} = \frac{1}{f}$$

or
$$\frac{1}{u} = \frac{1}{D} + \frac{1}{f}$$

or
$$\frac{D}{u} = 1 + \frac{D}{f}$$

putting this value of D/u in equation (i), we get

$$M = 1 + \frac{D}{f}$$ **Ans.**

OR

(b) Let S be a narrow slit illuminated by monochromatic light. S_1 and S_2 are two parallel narrow slits very close to each other and equidistant from S. The light waves from S reach S_1 and S_2 in the same phase and they act as new light sources from which secondary wavelets originate and form interference fringes on a distance screen.

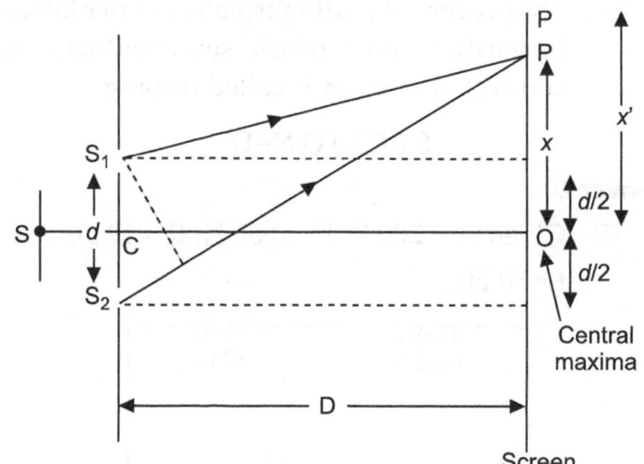

P is a point on the screen at a distance x from the central maxima at O.

Path difference (Δ) of P from S_1 and S_2 = $S_2P - S_1P$

But, $\quad (S_2P)^2 = D^2 + \left(x + \frac{d}{2}\right)^2$

$$(S_1P)^2 = D^2 + \left(x + \frac{d}{2}\right)^2$$

$$(S_2P)^2 - (S_1P)^2 = \left(x + \frac{d}{2}\right)^2 - \left(x - \frac{d}{2}\right)^2$$

or $\quad (S_2P + S_1P)(S_2P - S_1P) = 2dx$

Since, S_1 and S_2 are very close and CO is very large in comparison to S_1S_2, so S_2P, S_1P and D are almost equal to each other.

$$\therefore \quad \text{Path difference } (\Delta) = S_2P - S_1P = \frac{2dx}{2D} \text{ or}$$

$$\Delta = \frac{dx}{D}$$

If P is the centre of nth maxima, then $\frac{dx}{D} = n\lambda$

or, $$x = \frac{Dn\lambda}{d} \quad \ldots(i)$$

Let P′ be the centre of the $(n + 1)^{th}$ maxima at a distance x' from the central maxima

$$\therefore \quad x' = \frac{D(n+1)\lambda}{d} \quad \ldots(ii)$$

The fringe width (β) is given by:

$$\beta = x' - x$$

$$= \frac{D(n+1)\lambda}{d} - \frac{Dn\lambda}{d}$$

or, $$\beta = \frac{D\lambda}{d}$$ **Ans.**

Answer 22.

(a) (i)

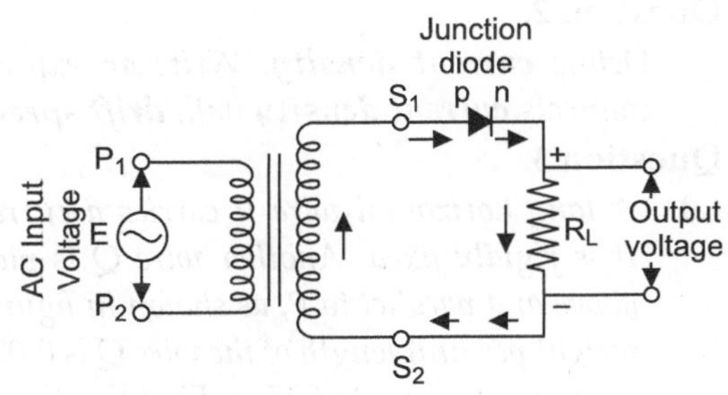

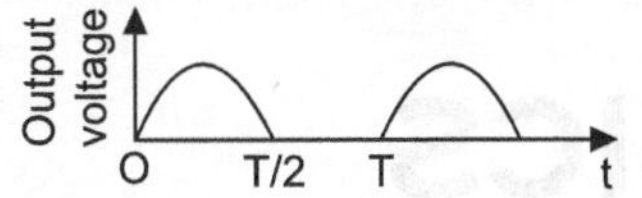

(ii)

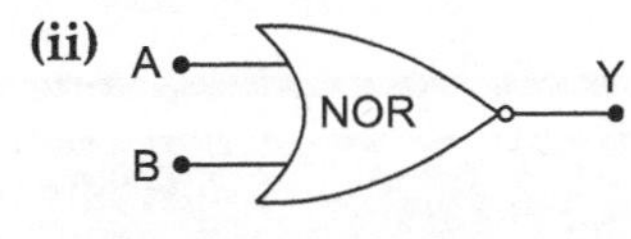

A	B	Y	Y′
0	0	0	1
0	1	1	0
1	0	1	0
1	1	1	0

Truth Table

OR

(b) (i)

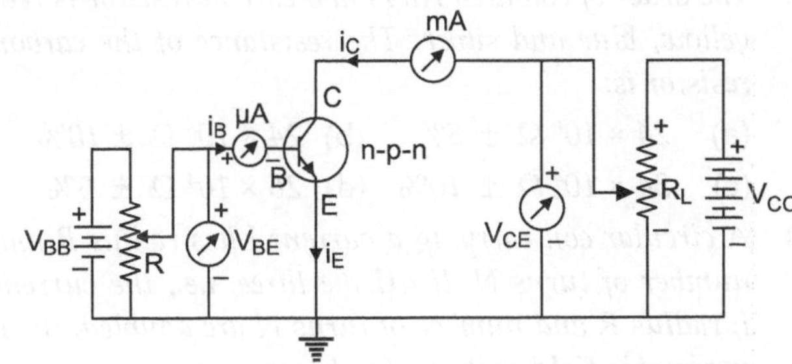

(ii)

A, B → AND → y

Input		Output
A	B	y (= A.B)
0	0	0
0	1	0
1	0	0
1	1	1

Truth Table

PHYSICS 2018

QUESTIONS

SECTION–A

*(Answer **all** questions)*

Question 1.

(A) *Choose the correct alternative (a), (b), (c) or (d) for each of the questions given below:* **[5 × 1]**

(i) *The order of coloured rings in a carbon resistor is red, yellow, blue and silver. The resistance of the carbon resistor is:*

(a) $24 \times 10^6\ \Omega \pm 5\%$ (b) $24 \times 10^6\ \Omega \pm 10\%$

(c) $34 \times 10^4\ \Omega \pm 10\%$ (d) $26 \times 10^4\ \Omega \pm 5\%$

(ii) *A circular coil carrying a current I has radius R and number of turns N. If **all** the three, i.e., the current I, radius R and number of turns N are doubled, then, **magnetic field** at its centre becomes:*

(a) Double (b) Half

(c) Four times (d) One-fourth

(iii) *An object is kept on the principal axis of a **concave mirror** of focal length 10 cm, at a distance of 15 cm from its pole. The image formed by the mirror is:*

(a) Virtual and magnified

(b) Virtual and diminished

(c) Real and magnified

(d) Real and diminished

(iv) ***Einstein's** photoelectric equation is :*

(a) $E_{max} = h\lambda - \phi_0$ (b) $E_{max} = \frac{hc}{\lambda} - \phi_0$

(c) $E_{max} = h\nu + \phi_0$ (d) $E_{max} = \frac{hc}{\lambda} + \phi_0$

(v) *In **Bohr's** model of hydrogen atom, radius of the first orbit of an electron is r_0. Then, radius of the **third** orbit is :*

(a) $\frac{r_0}{9}$ (b) r_0

(c) $3r_0$ (d) $9r_0$

(B) *Answer the following questions briefly and to the point.* **[7 × 1]**

(i) *In a **potentiometer** experiment, balancing length is found to be 120 cm for a cell E_1 of e.m.f. 2 V. What will be the balancing length for another cell E_2 of e.m.f. 1·5 V ? (No other changes are made in the experiment.)*

(ii) *How will you convert a moving coil galvanometer into a **voltmeter** ?*

(iii) *A moving charged particle **q** travelling along the positive X-axis enters a uniform magnetic field B. When will the **force** acting on **q** be **maximum** ?*

(iv) *Why is the core of a transformer **laminated** ?*

(v) *Ordinary (i.e., unpolarised) light is incident on the surface of a transparent material at the **polarising angle**. If it is partly reflected and partly refracted, what is the angle between the reflected and the refracted rays?*

(vi) *Define **coherent** sources of light.*

(vii) *Name a **material** which is used in making **control rods** in a nuclear reactor.*

SECTION–B

*(Answer **all** questions.)*

Question 2. **[2]**

*Define **current density**. Write an expression which connects **current density** with **drift speed**.*

Question 3. **[2]**

(a) *A long horizontal wire P carries a current of 50 A. It is rigidly fixed. Another wire Q is placed directly above and parallel to P, as shown in figure below. The weight per unit length of the wire Q is 0.025 Nm^{-1} and it carries a current of 25 A. Find the distance 'r' of the wire Q from the wire P so that the wire Q remains at rest.*

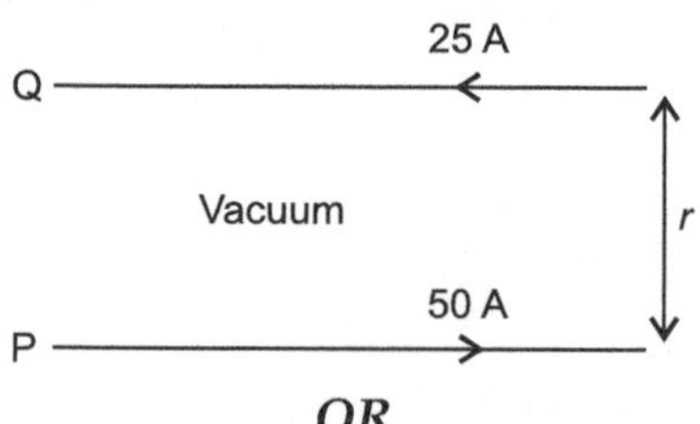

OR

(b) *Calculate **force per unit length** acting on the wire B due to the current flowing in the wire A. (See **figure** below)*

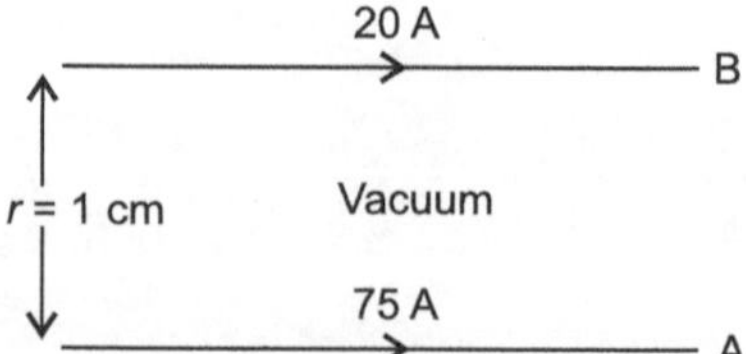

Question 4. **[2]**

(i) *Explain **Curie's law** for a paramagnetic substance.*

(ii) *A rectangular coil having 60 turns and area of 0·4 m^2 is held at right angles to a uniform magnetic field of flux density 5×10^{-5} T. Calculate the **magnetic flux** passing through it.*

Question 5. **[2]**

What is ***motional e.m.f.*** *? State* ***any two*** *factors on which it depends.*

Question 6. **[2]**

(i) *What is the* ***ratio*** *of the speed of gamma rays to that of radio waves in vacuum ?*

(ii) *Name an electromagnetic wave which is used in the* ***radar*** *system used in aircraft navigation.*

Question 7. **[2]**

A biconvex lens made of glass (refractive index 1.5) has two spherical surfaces having radii 20 cm and 30 cm. Calculate its ***focal length.***

Question 8. **[2]**

State ***any two*** *differences between* ***primary*** *rainbow and* ***secondary*** *rainbow.*

Question 9. **[2]**

(i) *State* ***de Broglie*** *hypothesis.*

(ii) *With reference to* ***photoelectric*** *effect, define* ***threshold wavelength****.*

Question 10. **[2]**

Calculate the ***minimum*** *wavelength of the spectral line present in* ***Balmer*** *series of hydrogen.*

Question 11. **[2]**

(a) *What is meant by* ***pair annihilation*** *? Write a* ***balanced*** *equation for the same.*

OR

(b) *What is meant by the terms* ***half-life of a radioactive substance*** *and* ***binding energy of a nucleus*** *?*

Question 12. **[2]**

In a ***communication*** *system, what is meant by* ***modulation*** *? State* ***any two*** *types of modulation.*

SECTION–C

(Answer ***all*** *questions.)*

Question 13. **[3]**

Obtain an expression for intensity of electric field at a point in end on position, i.e., axial position of an electric dipole.

Question 14. **[3]**

Deduce an expression for ***equivalent*** *capacitance C when three capacitors C_1, C_2 and C_3 are connected in* ***parallel.***

Question 15. **[3]**

(a) *ε_1 and ε_2 are two batteries having e.m.f. of 34 V and 10 V respectively and internal resistance of 1 Ω and 2 Ω respectively. They are connected as shown in figure below. Using* ***Kirchhoff's Laws*** *of electrical networks, calculate the currents I_1 and I_2.*

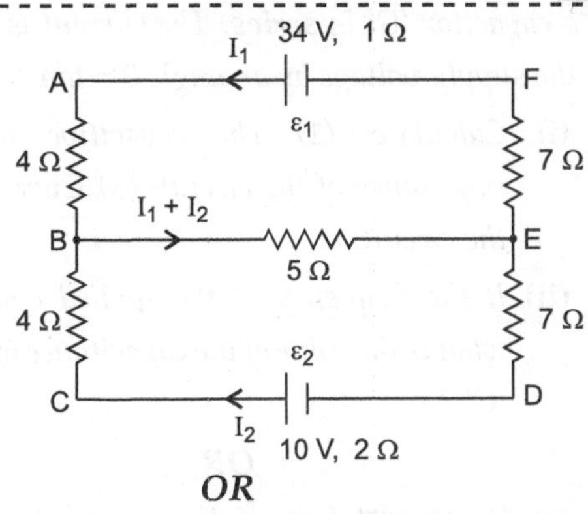

OR

(b) *An electric bulb is marked 200 V, 100 W. Calculate* ***electrical resistance*** *of its filament. If* ***five*** *such bulbs are connected in* ***series*** *to a 200 V supply, how much* ***current*** *will flow through them ?*

Question 16. **[3]**

(a) *For any prism, prove that:*

$$\text{'}n\text{' or } \mu = \frac{\sin\left(\frac{A+\delta_m}{2}\right)}{\sin\left(\frac{A}{2}\right)}$$

where the terms have their usual meaning.

OR

(b) *When two* ***thin*** *lenses are kept in contact, prove that their* ***combined*** *or* ***effective*** *focal length F is given by :*

$$\frac{1}{F} = \frac{1}{f_1} + \frac{1}{f_2}$$

where the terms have their usual meaning.

Question 17. **[3]**

(i) *In* ***Young's*** *double slit experiment, show graphically how intensity of light varies with distance.*

(ii) *In* ***Fraunhofer*** *diffraction, how is the angular width of the central bright fringe affected when slit separation is increased ?*

Question 18. **[3]**

Write one ***balanced*** *equation each to show:*

(i) *Nuclear fission*

(ii) *Nuclear fusion*

(iii) *Emission of β^- (i.e., a negative beta particle)*

Question 19. **[3]**

With reference to semiconductor devices, define a p-type semiconductor and a ***Zener*** *diode. What is the use of a* ***Zener*** *diode ?*

SECTION–D

Answer ***all*** *questions.*

Question 20. **[5]**

(a) *An alternating e.m.f. of 220 V is applied to a circuit containing a resistor R having resistance of 160 Ω and* *a capacitor 'C' in **series**. The current is found to lead the supply voltage by an angle θ = tan⁻¹ (3/4).*

(i) *Calculate: (1) The capacitive reactance, (2) Impedance of the circuit, (3) Current flowing in the circuit.*

(ii) *If the frequency of the applied e.m.f. is 50 Hz, what is the value of the capacitance of the capacitor 'C' ?*

OR

(b) *An AC generator generating an e.m.f. of $\varepsilon = 300 \sin (100\pi t)$ V is connected to a **series** combination of 16 μF capacitor, 1H inductor and 100 Ω resistor. Calculate:*

(i) *Impedance of the circuit at the **given frequency**.*

(ii) ***Resonant** frequency f_o.*

(iii) *Power factor at **resonant** frequency f_o.*

Question 21. **[5]**

(a) *Draw a labelled ray diagram of an image formed by a **refracting telescope** with final image formed at **infinity**. Derive an expression for its magnifying power with the final image at infinity.*

OR

(b) **(i)** *Using **Huygen's** wave theory, derive **Snell's** law of refraction.*

(ii) *With the help of an experiment, state how will you identify whether a given beam of light is polarised or unpolarised.*

Question 22. **[5]**

(a) **(i)** *The forward characteristic curve of a junction diode is shown in figure below:*

Calculate the resistance of the diode at:

(1) *V = 0.5 V* **(2)** *I = 60 mA*

(ii) *Draw **separate** energy band diagrams for conductors, semiconductors and insulators and label each of them.*

OR

(b) **(i)** *The arrangement given below represents a logic gate:*

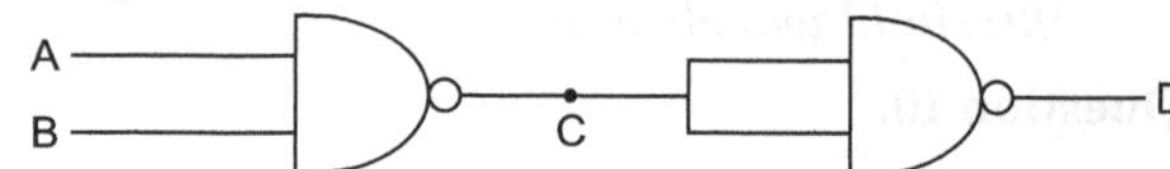

Copy the following truth table in your answer booklet and complete it showing outputs at C and D.

A	*B*	*C*	*D*
0	0		
1	0		
0	1		
1	1		

(ii) *Draw a labelled diagram of a **common emitter** amplifier, showing waveforms of **signal** voltage and **output** voltage.*

ANSWERS

SECTION–A

Answer 1.

(A) **(i)** **(b)** $24 \times 10^6\ \Omega \pm 10\%$

(ii) **(a)** Double

(iii) **(c)** Real and magnified

(iv) **(b)** $E_{max} = \dfrac{hc}{\lambda} - \phi_0$

(v) **(d)** $9\,r_0$

(B) **(i)** Given : $E_1 = 2$ V, $l_1 = 120$ cm, $E_2 = 1{\cdot}5$ V, $l_2 = ?$

We know that

$$\frac{E_1}{E_2} = \frac{l_1}{l_2}$$

$$\Rightarrow \quad \frac{2}{1.5} = \frac{120}{l_2}$$

$$\therefore \quad l_2 = \frac{120 \times 1.5}{2} = 90\,\text{cm} \qquad \textbf{Ans.}$$

(ii) A moving coil galvanometer can be converted into a voltmeter by connecting a very high resistance R in series with the galvanometer and the value of resistance R is given by:

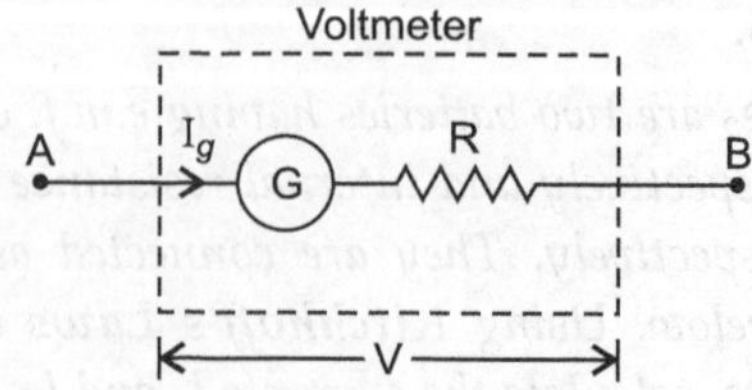

$$R = \frac{V}{I_g} - G$$

where,

I_g = Current in galvanometer.

G = Resistance of galvanometer.

V = Range of voltmeter.

Due to high resistance, galvanometer will not draw much current passing through circuit. The resistance of an ideal voltmeter is infinity.

(iii) The moving charge q will experience maximum force when the uniform magnetic field is perpendicular to the direction of motion of the moving charge.

(iv) The core of a transformer is laminated, (by silicon steels), to minimise the loss of energy due to eddy currents in the iron core. This energy loss is called *iron loss* caused by heating of the core.

(v) At the polarising angle, the angle between the reflected and the refracted rays is 90°.

(vi) Coherent sources are a pair of sources which emit light of same wavelength, same frequency, and equal intensity or amplitude in the same direction. The two light waves must have zero or a constant phase difference.

(vii) In a nuclear reactor, the control rods are made up of cadmium, silver, indium or boron.

SECTION–B

Answer 2.

Current density at a point in a conductor is the ratio of the current at that point in the conductor to the area of cross-section of the conductor at that point, provided area is held perpendicular to the direction of flow of current.

$$\text{Current density} = J = \frac{I}{A}$$

where, I = Electric current

A = Area of cross-section

S. I. unit is A/m².

Expression connecting current density with drift speed

$$J = nev_d,$$

where,

J = Current density.

n = Number of free electrons per unit volume of the conductor.

e = Charge on an electron.

v_d = Drift speed (velocity).

Answer 3.

(a) The magnetic force per unit length on the wire Q due to current in the wire P is:

$$\frac{F}{l} = \frac{\mu_0}{2\pi}\frac{i_1 i_2}{r} \text{ Nm}^{-1}$$

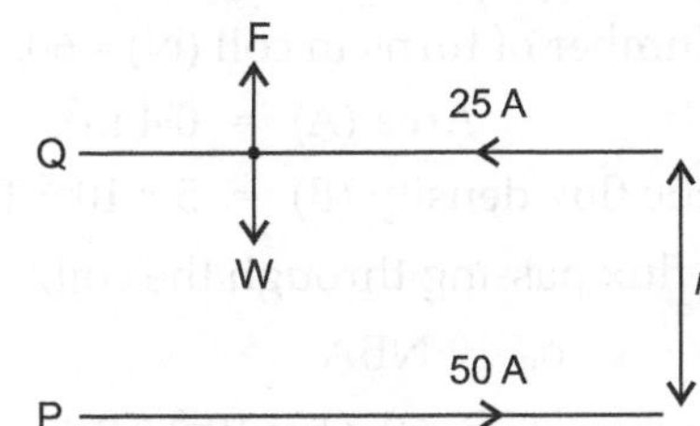

where, r is the distance of wire Q from the wire P.

Since, the currents in P and Q are opposite, Q experiences an upward force and it is balanced by the downward weight of Q.

$$\therefore \quad \frac{F}{l} = \frac{\mu_0}{2\pi}\frac{i_1 i_2}{r} = W$$

where, W is the weight per unit length of the wire Q.

$$\Rightarrow \quad r = \frac{\mu_0}{2\pi}\frac{i_1 i_2}{W}$$

$$= \left(\frac{4\pi \times 10^{-7}}{2\pi}\right) \times \frac{50 \times 25}{0.025}$$

$$= 1 \times 10^{-2} \text{ m} = 1 \text{ cm}$$ **Ans.**

OR

(b) Given : $i_1 = 75$ A, $i_2 = 20$ A, $r = 1$ cm $= 1 \times 10^{-2}$ m

The force per unit length on the wire B due to current flowing in wire A is:

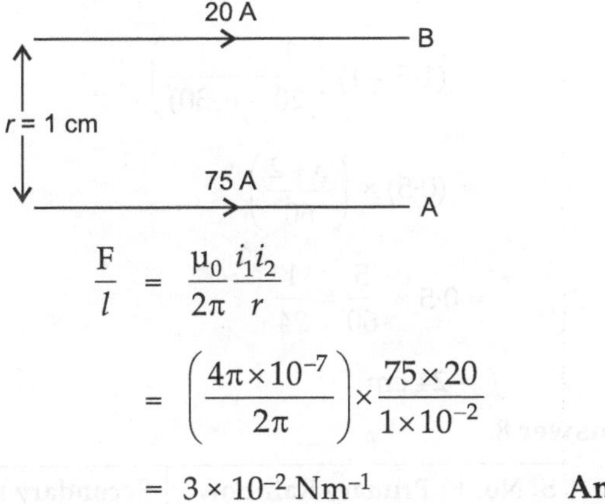

$$\frac{F}{l} = \frac{\mu_0}{2\pi}\frac{i_1 i_2}{r}$$

$$= \left(\frac{4\pi \times 10^{-7}}{2\pi}\right) \times \frac{75 \times 20}{1 \times 10^{-2}}$$

$$= 3 \times 10^{-2} \text{ Nm}^{-1}$$ **Ans.**

Answer 4.

(i) According to Curie's law, the intensity of magnetisation of paramagnetic substances is (i) inversely proportional to the absolute temperature *i.e.,*

$$\chi \propto \frac{1}{T}$$

Here, T = absolute temperature

(ii) and directly proportional to external magnetic field *i.e.,*

$$\chi \propto B$$

Here B = External magneticfield.

(ii) Given : Number of turns in coil (N) = 60,

Area (A) = $0{\cdot}4\ m^2$,

Magnetic flux density (B) = 5×10^{-5} T

Magnetic flux passing through the coil,

$$\phi_B = NBA$$
$$= 60 \times 5 \times 10^{-5} \times 0.4$$
$$= 1{\cdot}2 \times 10^{-3}\ Wb. \qquad \textbf{Ans.}$$

Answer 5.

The e.m.f. induced due to the motion of the conductor in a uniform magnetic field is called the motional e.m.f.

The two factors on which it depends are:

1. length of the conducting rod in the magnetic field.
2. velocity with which it is moving in magnetic field.

Answer 6.

(i) 1 : 1 **(ii)** Microwaves.

Answer 7.

Given : $n = 1{\cdot}5$, $R_1 = 20$ cm, $R_2 = -30$ cm.

The focal length of the bifocal convex lens is given by:

$$\frac{1}{f} = (n-1)\left(\frac{1}{R_1} - \frac{1}{R_2}\right)$$
$$= (1{\cdot}5 - 1)\left(\frac{1}{20} - \frac{1}{(-30)}\right)$$
$$= (0{\cdot}5) \times \left(\frac{3+2}{60}\right)$$
$$= 0{\cdot}5 \times \frac{5}{60} = \frac{1}{24}$$
$$\therefore \quad f = 24\ \text{cm}. \qquad \textbf{Ans.}$$

Answer 8.

S. No.	Primary Rainbow	Secondary Rainbow
1.	The rainbow formed by light rays which have suffered one internal reflection and two refractions is called primary rainbow.	The rainbow formed by light rays which have suffered two internal reflections and two refractions is called secondary rainbow.
2.	The outer edge is red and the inner edge is violet in colour.	The outer edge is violet and the inner edge is red in colour.

Answer 9.

(i) Refer to ISC Paper 2020, Answer 9 (i).

(ii) The maximum wavelength of incident radiation for which photoelectric effect is observed is called threshold wavelength.

Answer 10.

The shortest wavelength of electromagnetic radiation present in Balmer series of hydrogen spectrum :

$$\frac{1}{\lambda} = R\left[\frac{1}{2^2} - \frac{1}{n^2}\right].$$

where, $R = 1{\cdot}097 \times 10^7$ m is Rydberg constant

Here, $n = \infty$ for shortest wavelength,

$$\therefore \quad \frac{1}{\lambda_{min}} = R\left[\frac{1}{2^2} - \frac{1}{\infty^2}\right].$$
$$\Rightarrow \quad \frac{R}{4} = \frac{1.097 \times 10^7}{4}$$
$$\Rightarrow \quad \lambda_{min} = \frac{4}{1.097 \times 10^7}\ \text{m}$$
$$\lambda_{min} = \frac{4 \times 10^{10}}{1.097 \times 10^7}\ \text{Å}$$
$$= 3646\ \text{Å} \qquad \textbf{Ans.}$$

Answer 11.

(a) Whenever an electron and a positron come very close to each other, they are completely annihilated and two γ-photons moving in opposite directions are produced having total energy equal to $2mc^2$. It is represented by the following equation:

$$\underset{\text{(Positron)}}{e^+} + \underset{\text{(Electron)}}{e^-} \longrightarrow \underset{(\gamma\text{-rays})}{h\nu} + \underset{(\gamma\text{-rays})}{h\nu}$$

OR

(b) **Half life:** The half life period of a radioactive substance is defined as the time in which one-half of the atoms of the radioactive substance is disintegrated.

Binding energy: Binding energy is equal to the energy required to split the nucleus into its constituent nucleons, such that the nucleons are at infinite distance apart and there is no interaction between the nucleons. It is the energy equivalent of mass defect. *i.e.,*

$$\text{B.E.} = \text{Mass defect} \times c^2$$
$$= \Delta m \times c^2$$

Answer 12.

The original low frequency message signal cannot be transmitted over long distances. Therefore, it is essential to superimpose the message signal on a high frequency radio wave called the carrier wave.

This process of changing some characteristics like amplitude, frequency or phase of a carrier wave in accordance with the intensity of the modulating signal is called modulation.

The two types of modulation are: (i) Amplitude modulation (AM) and (ii) Frequency modulation (FM).

SECTION-C

Answer 13.

AB is a dipole made up of two charges $-q$ and $+q$ separated by a distance $2l$.

Let E_1 and E_2 be the electric field strengths at point P due to charges, $+q$ and $-q$ respectively.

So, electric intensity at P due to $+q$

$$\vec{E}_1 = \frac{1}{4\pi\varepsilon_0}\cdot\frac{q}{(r-l)^2} \quad \text{(From B to P)}$$

and electric intensity at P due to $-q$

$$\vec{E}_2 = \frac{1}{4\pi\varepsilon_0}\cdot\frac{q}{(r+l)^2} \quad \text{(From P to A)}$$

Since, $E_1 > E_2$, the resultant electric field due to the electric dipole has a magnitude equal to difference of E_1 and E_2 and direction is from B to P.

$$\therefore \quad E = E_1 - E_2$$

$$= \frac{1}{4\pi\varepsilon_0}\cdot\frac{q}{(r-l)^2} - \frac{1}{4\pi\varepsilon_0}\cdot\frac{q}{(r-l)^2}$$

$$= \frac{q}{4\pi\varepsilon_0}\left[\frac{1}{(r-l)^2} - \frac{1}{(r+l)^2}\right]$$

$$E = \frac{q}{4\pi\varepsilon_0}\left[\frac{(r+l)^2-(r-l)^2}{(r^2-l^2)^2}\right]$$

$$= \frac{q}{4\pi\varepsilon_0}\frac{r^2+l^2+2rl-r^2-l^2+2rl}{(r^2-l^2)^2}$$

$$= \frac{q}{4\pi\varepsilon_0}\frac{4rl}{(r^2-l^2)^2}$$

$$= \frac{1}{4\pi\varepsilon_0}\frac{2(q\times 2l)r}{(r^2-l^2)^2}$$

But, $q \times 2l = p$ (dipole moment)

$$\therefore \quad E = \frac{1}{4\pi\varepsilon_0}\frac{2pr}{(r^2-l^2)^2}$$

If $r >> l$, then

$$= \frac{1}{4\pi\varepsilon_0}\cdot\frac{2p}{r^3}$$

$$E = \frac{1}{4\pi\varepsilon_0}\frac{2pr}{r^4}$$

directed from B to P. **Ans.**

Answer 14.

Three capacitors C_1, C_2 and C_3 are connected in parallel between the points A and B across a source of potential difference V. The potential difference between the plates of each capacitor will be same. Let it be V.

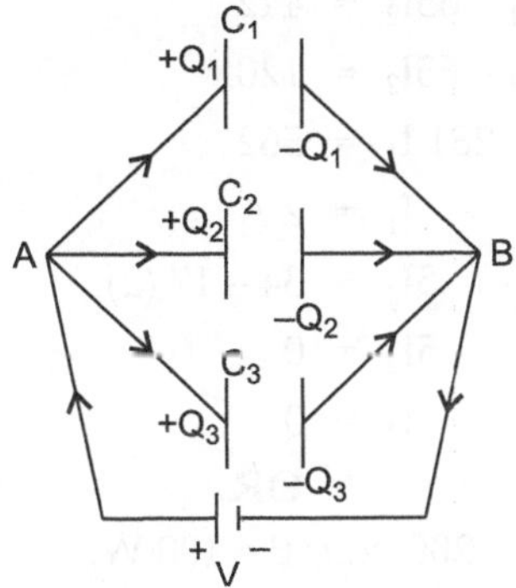

A charge +Q given to the point A by means of an electric source is distributed on all the three capacitors according to their capacitances. If the charges on the capacitors C_1, C_2 and C_3 be Q_1, Q_2 and Q_3 respectively, then

$$Q_1 = C_1V,\ Q_2 = C_2V \text{ and } Q_3 = C_3V$$

Total charge on all the three capacitors is

$$Q = Q_1 + Q_2 + Q_3$$

$$= C_1V + C_2V + C_3V$$

$$= V(C_1 + C_2 + C_3) \quad \ldots\text{(i)}$$

If in place of all the three capacitors, only one capacitor is placed between A and B such that on giving it a charge the potential difference between its plates is V, then

$$Q = CV \quad \ldots\text{(ii)}$$

where, C is the equivalent capacitance of C_1, C_2 and C_3 connected in parallel.

On comparing equations (i) and (ii), we have

$$CV = V(C_1 + C_2 + C_3)$$

or $$C = C_1 + C_2 + C_3$$ **Ans.**

Answer 15.

(a) Applying Kirchhoff's second law in mesh ABEFA,

$$1I_1 + 4I_1 + 5(I_1 + I_2) + 7I_1 = 34$$

or $$17I_1 + 5I_2 = 34 \quad \ldots\text{(i)}$$

Applying Kirchhoff's second law in mesh ACDFA,

$$1I_1 + 4I_1 - 4I_2 - 2I_2 - 7I_2 + 7I_1 = 34 - 10$$

or $12I_1 - 13I_2 = 24$ ...(ii)

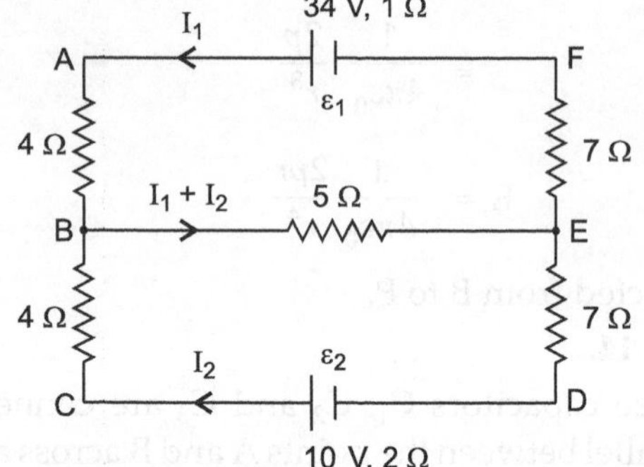

Multiply equation (i) by 13 and equation (ii) by 5, and add both the equations,

$$221I_1 + 65I_2 = 442$$
$$+ 60I_1 - 65I_2 = 120$$
$$281 I_1 = 562$$
$$I_1 = 2 \text{ A}$$
$$5I_2 = 34 - 17(2) \quad \text{[From (i)]}$$
$$5I_2 = 0$$
$$I_2 = 0 \quad \textbf{Ans.}$$

OR

(b) Given : V = 200 volt, P = 100 W.

Resistance of filament of bulb (R)

$$= \frac{V^2}{P}$$
$$= \frac{200 \times 200}{100} = 400\ \Omega$$

Since, 5 bulbs are connected in series

$$R_{eq} = R_1 + R_2 + R_3 + R_4 + R_5$$
$$\therefore \quad R_{eq} = 400 \times 5 = 2000\ \Omega$$

Supply voltage (V) = 200 V

Current through each bulb (I)

$$I = \frac{V}{R_{eg}} = \frac{200}{2000} = 0.1 \text{ A.} \quad \textbf{Ans.}$$

Answer 16.

(a) A ray of light is passing through a triangular prism ABC as shown in the figure.

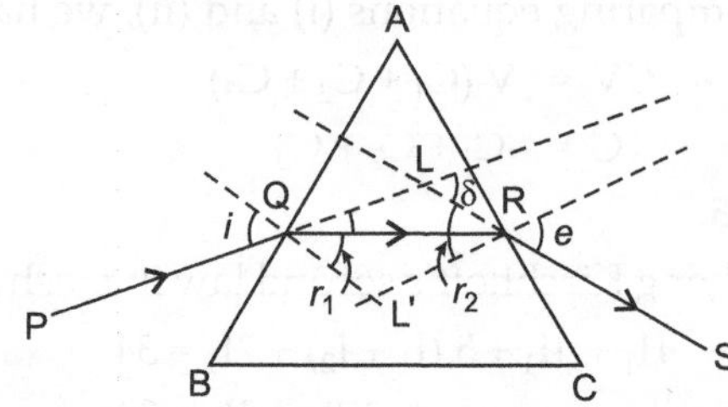

In Δ QLR,

Angle of deviation, δ = ∠ LQR + ∠ LRQ

(Exterior angle property)

$$= (i - r_1) + (e - r_2)$$
$$= (i + e) - (r_1 + r_2) \quad \text{...(i)}$$

In quadrilateral AQL′ R,

$$A + L' = 180° \quad \text{...(ii)}$$

and in Δ QL′R,

$$r_1 + r_2 + L' = 180° \quad \text{...(iii)}$$

From equations (ii) and (iii),

$$r_1 + r_2 = A \quad \text{...(iv)}$$

Putting the value of $(r_1 + r_2)$ in equation (i), we get

$$\delta = i + e - A \quad \text{...(v)}$$

For minimum deviation, $i = e$ and $r_1 = r_2 = r$ and $\delta = \delta_m$

∴ From equations (iv) and (v),

$$\delta_m = 2i - A$$
$$\Rightarrow \quad i = \frac{\delta_m + A}{2}$$
$$\text{and} \quad 2r = A$$
$$\Rightarrow \quad r = \frac{A}{2}$$

Using Snell's law,

Refractive index of glass,

$$n = \frac{\sin i}{\sin r}$$
$$\Rightarrow \quad n = \frac{\sin\left(\frac{\delta_m + A}{2}\right)}{\sin\left(\frac{A}{2}\right)} \quad \textbf{Hence Proved.}$$

OR

(b) Refer to ISC Paper 2020, Answer 16.

Answer 17.

(i)

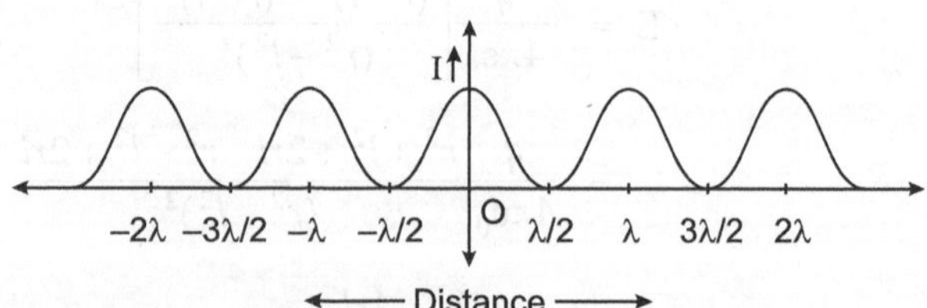

(ii) Angular width of central bright fringe μ

When slit separation is increased, the angular width of central bright fringe decreases.

Answer 18.

(i) Nuclear fission:

$$^{235}_{92}U + ^{1}_{0}n \rightarrow ^{141}_{56}Ba + ^{92}_{36}Kr + 3\,^{1}_{0}n + \text{Energy}$$

(ii) Nuclear fusion:

$${}_1^2\text{H} + {}_1^2\text{H} \rightarrow {}_2^3\text{He} + {}_0^1n + \text{Energy}$$

(iii) Emission of β^{-1}:

$${}_{92}^{239}\text{U} \rightarrow {}_{93}^{239}\text{Np} + {}_{-1}\beta^0 + \bar{v} \text{ (anti-neutrino)}$$

Answer 19.

When a trivalent impurity or III group element like aluminium, boron, gallium, etc., is doped in pure germanium or silicon, then the conductivity of crystal increases due to deficiency of electrons *i.e.,* holes and such a crystal is said to be *p*-type semiconductor. In a *p*-type semiconductor, the charge carriers are holes.

Zener diode is a reverse biased heavily doped *p-n* junction diode. It is operated in the breakdown region. It is designed to operate at the reverse breakdown voltage continuously without being damaged. It is represented by the symbol given below:

Zener diode has a sharp breakdown voltage and it is used as a voltage regulator.

SECTION–D

Answer 20.

(a) Given : $\varepsilon = 220$ V, $R = 160\ \Omega$ and $\theta = \tan^{-1}\left(\frac{3}{4}\right)$

(i) (1) For a RC circuit,

$$\theta = \tan^{-1}\left(\frac{X_C}{R}\right)$$

or $$\tan^{-1}\left(\frac{X_C}{R}\right) = \tan^{-1}\left(\frac{3}{4}\right)$$

$$\therefore \quad \frac{X_C}{R} = \frac{3}{4}$$

$$\Rightarrow \quad X_C = \frac{3}{4} \times R = \frac{3}{4} \times 160\ \Omega$$

$$= 120\ \Omega$$

$\therefore$ Capacitive reactance,

$$X_C = 120\ \Omega$$

(2) Impedance, $Z = \sqrt{R^2 + X_C{}^2}$

$$= \sqrt{(160)^2 + (120)^2}$$

$$= \sqrt{25600 + 14400}$$

$$= \sqrt{40000} = 200\ \Omega$$

(3) Current flowing in the circuit,

$$i = \frac{\varepsilon}{Z}$$

(Z is the effective resistance)

$$i = \frac{220}{200}\text{ A} = 1{\cdot}1\text{ A}$$

(ii) Given : $f = 50$ Hz

We know that

$$X_C = \frac{1}{2\pi f C}$$

$$\Rightarrow \quad C = \frac{1}{2\pi f X_C}$$

$$= \frac{1}{2 \times 3.14 \times 50 \times 120}\text{ F} = \frac{1}{376.80}\text{ F}$$

$$= 26{\cdot}5 \times 10^{-6}\text{ F} = 26{\cdot}5\ \mu\text{F}$$ **Ans.**

OR

(b) Given : $\varepsilon = 300 \sin(100\pi\, t)$ V

$\therefore \quad \varepsilon_0 = 300$ V, $C = 16\ \mu$F, $L = 1$ H, $R = 100\ \Omega$, $\omega = 100\,\pi = 314$ rad s^{-1}.

Inductive reactance

$$(X_L) = \omega L = 314 \times 1 = 314\ \Omega$$

Capacitive reactance

$$(X_C) = \frac{1}{\omega C}$$

$$= \frac{1}{314 \times 16 \times 10^{-6}}\ \Omega = 199\ \Omega$$

(i) Impedance of the circuit

$$(Z) = \sqrt{R^2 + (X_L - X_C)^2}$$

$$= \sqrt{100^2 + (314 - 199)^2} = 152{\cdot}4\ \Omega$$

(ii) Resonant frequency

$$(f_o) = \frac{1}{2\pi\sqrt{LC}}$$

$$= \frac{1}{2 \times 3.14 \times \sqrt{1 \times 16 \times 10^{-6}}}$$

$$= \frac{1}{6.28 \times 4 \times 10^{-3}} = 39{\cdot}80\text{ Hz}$$

(iii) We know that,

$$\text{Power factor } (\cos\phi) = \frac{R}{\sqrt{R^2 + (X_L - X_C)^2}}$$

At resonant frequency,

$$X_L = X_C$$

$$\therefore \quad \cos\phi = \frac{R}{\sqrt{R^2 + 0}} = \frac{R}{R} = 1$$

Answer 21.

(a) A_1B_1 is the image of distant object situated at infinity. If A_1B_1 is at the principal focus of the lens E; the rays will emerge in parallel direction and a virtual magnified image is formed at infinity.

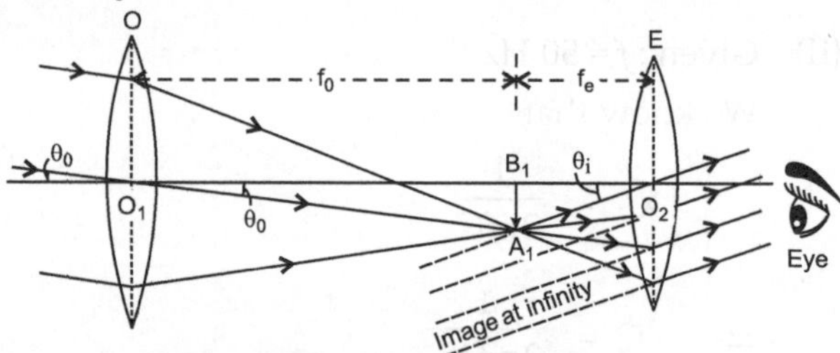

Astronomical refracting telescope when final image is formed at infinity

Magnification in normal adjustment is defined as the ratio of the angle θ_i subtended by the final image at the eye as seen through the telescope to the angle θ_0 subtended by the object at the eye when both the image and object are situated at infinity.

Thus, Magnification, $M = \dfrac{\theta_i}{\theta_0} = \dfrac{\tan\theta_i}{\tan\theta_0}$

($\because \angle\theta_0$ and $\angle\theta_i$ are small)

In $\Delta A_1O_1B_1$, $\tan\theta_0 = \dfrac{A_1B_1}{O_1B_1}$

and in $\Delta A_1O_2B_1$, $\tan\theta_0 = \dfrac{A_1B_1}{O_2B_1}$

$$M = \frac{\tan\theta_i}{\tan\theta_0} = \frac{\dfrac{A_1B_1}{O_2B_1}}{\dfrac{A_1B_1}{O_1B_1}} = \frac{O_1B_1}{O_2B_1}$$

But, $O_1B_1 = f_o$ (focal length of objective)

and $O_2B_1 = f_e$ (focal length of eyepiece)

$\therefore \quad M = \dfrac{f_o}{f_e}$ **Ans.**

OR

(b) (i) Let a plane wavefront ABC be incident from medium 1 on the plane surface AGF. At $t = 0$, the wavefront just touches the surface at A and sends wavelet travelling with a velocity c_1 in medium 1 and with a velocity c_2 in medium 2.

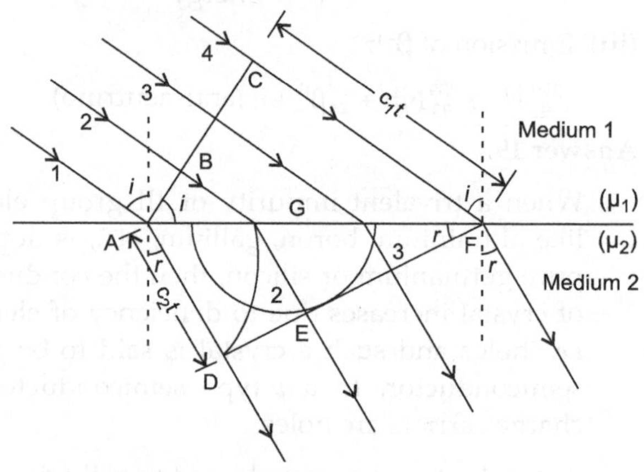

After time interval t, the wave travels the distance CF = c_1t and during which the secondary waves from A travel a distance AD = c_2t. If $\angle$CAF = i and $\angle$AFD = r, where i and r are the angle of incidence and the angle of refraction. Consider ΔACF and ΔADF, then we have

$$\sin i = \sin\angle CAF = \frac{CF}{AF} = \frac{c_1t}{AF}$$

and

$$\sin r = \sin\angle AFD = \frac{AD}{AF} = \frac{c_2t}{AF}$$

$$\therefore \quad \frac{\sin i}{\sin r} = \frac{\dfrac{c_1t}{AF}}{\dfrac{c_2t}{AF}} = \frac{c_1}{c_2}$$

$$= {}_1n_2 = \text{a constant}$$

${}_1n_2$ is the refractive index of the second medium with respect to the first medium and the above equation represent the Snell's law of refraction of light.

(ii) Refer to ISC Paper 2020 Answers 21 b - (iii).

Answer 22.

(a) (i) (1) When V = 0·5 volt, then current (I) = 40 mA = 40 × 10^{-3}A

$\therefore$ Resistance of diode

$$(R_d) = \frac{V}{I} = \frac{0.5}{40\times10^{-3}}$$

$$= 12{\cdot}5\ \Omega$$

(2) When I = 60 mA = 60 × 10^{-3}A, then

V = 0·6 volt

$\therefore$ Resistance of diode

$$(R_d) = \frac{V}{I}$$

$$= \frac{0.6}{60 \times 10^{-3}} = 10\ \Omega$$

(ii) Conductors : In case of conductors, conduction band and valence band overlap each other due to which no forbidden gap is present between them and large number of free electrons are available for electrical conduction. In conductors no energy is required to move the electrons from valence band to conduction band and thus current flows through it, when a p.d. is applied across it. For example: Silver, copper, aluminium etc.

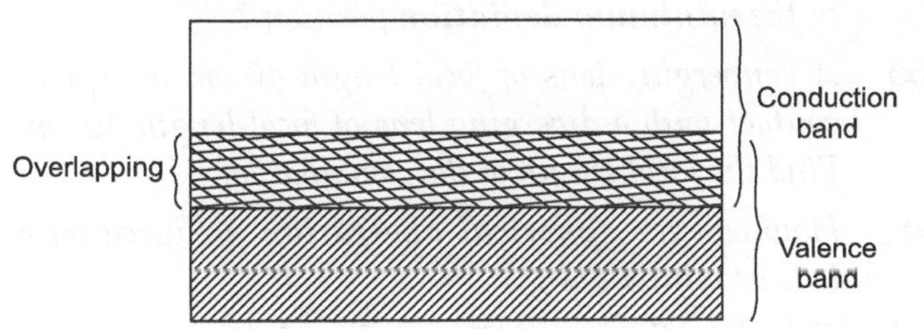

Insulators: In case of insulator, forbidden gap is very large greater than 5 eV. Its valence band is fully filled with the electrons, whereas its conduction band is empty.

The electrons cannot jump from valence band to the conduction band, so no electric current flow through them. For example: Wood, glass, diamond etc.

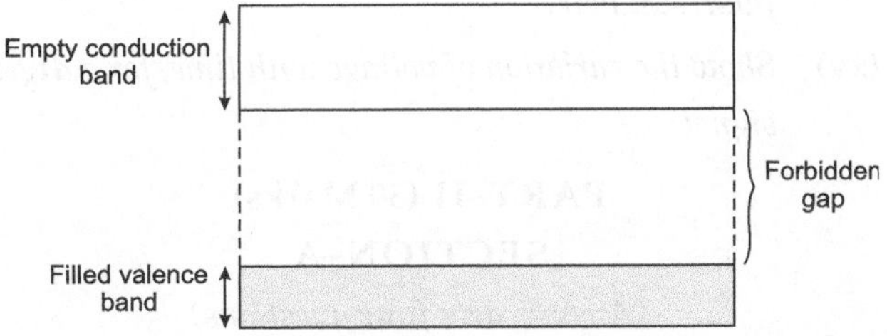

Semiconductor: Semiconductors are the substance whose conductivity lies in between of a good conductor and an insulator. In case of semiconductors, the conduction band is empty and valence band is filled with electrons. Like insulators, forbidden energy gap is not so large in case of semiconductors. The gap is very small. The energy gap is nearly of 1 eV.

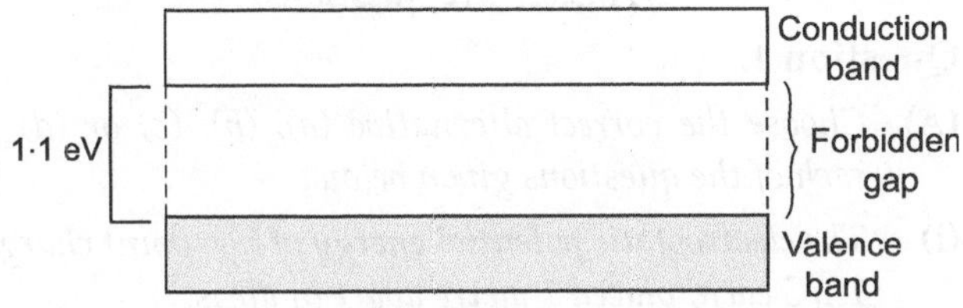

OR

(b) (i) The arrangement acts as AND gate and the truth table is :

A	B	C	D
0	0	1	0
1	0	1	0
0	1	1	0
1	1	0	1

(ii) Circuit diagram of common emitter amplifier is shown below:

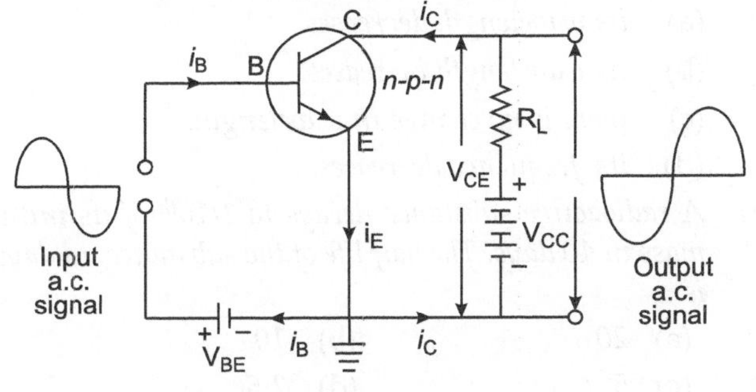

PHYSICS 2017

QUESTIONS

PART–I (20 Marks)

*(Answer **all** questions)*

Question 1.

(A) *Choose the correct alternative (a), (b), (c) or (d) for each of the questions given below:* **[5]**

(i) *The electrostatic potential energy of two point charges, 1 μC each, placed 1 metre apart in air is:*

(a) $9 \times 10^{3} J$ (b) $9 \times 10^{9} J$

(c) $9 \times 10^{-3} J$ (d) $9 \times 10^{-3} eV$

(ii) *A wire of resistance 'R' is cut into 'n' equal parts. These parts are then connected in parallel with each other. The equivalent resistance of the combination is:*

(a) nR (b) R/n

(c) n/R^2 (d) R/n^2

(iii) *Magnetic susceptibility of platinum is 0·0001. Its relative permeability is :*

(a) 1·0000 (b) 0·9999

(c) 1·0001 (d) 0

(iv) *When a light wave travels from air to glass:*

(a) *its wavelength decreases.*

(b) *its wavelength increases.*

(c) *there is no change in wavelength.*

(d) *its frequency decreases.*

(v) *A radioactive substance decays to $1/16^{th}$ of its initial mass in 40 days. The half life of the substance, in days, is:*

(a) 20 (b) 10

(c) 5 (d) 2·5

(B) *Answer **all** questions given below **briefly** and to the point :* **[15]**

(i) ***Maximum*** *torque acting on an electric dipole of moment 3×10^{-29} Cm in a uniform electric field E is 6×10^{-25} Nm. Find E.*

(ii) *What is meant by **drift speed** of free electrons ?*

(iii) *On which conservation principle is **Kirchhoff's second law** of electrical networks based ?*

(iv) *Calculate magnetic flux density of the magnetic field at the centre of a circular coil of 50 turns, having radius of 0·5 m and carrying a current of 5 A.*

(v) *An a.c. generator generates an emf 'ε' where ε = 314 sin (50πt) volt. Calculate the **frequency** of the emf ε.*

(vi) *With what type of source of light are **cylindrical** wavefronts associated ?*

(vii) *How is fringe width of an interference pattern in **Young's double slit experiment** affected if the two slits are brought closer to each other ?*

(viii) *In a **regular** prism, what is the relation between angle of incidence and angle of emergence when it is in the **minimum deviation** position ?*

(ix) *A converging lens of focal length 40 cm is kept in contact with a diverging lens of focal length 30 cm. Find the focal length of the combination.*

(x) *How can the **spherical aberration** produced by a lens be minimised ?***

(xi) *Calculate the **momentum** of a **photon** of energy 6×10^{-19} J.*

(xii) *According to **Bohr**, "Angular momentum of an orbiting electron is quantised". What is meant by this statement?*

(xiii) *Why nuclear fusion reaction is also called **thermo-nuclear** reaction?*

(xiv) *What is the **minimum** energy which a gamma ray photon must possess in order to produce **electron-positron** pair?*

(xv) *Show the variation of voltage with time, for a **digital** signal.*

PART–II (50 Marks)

SECTION–A

*(Answer **any four** questions)*

Question 2.

(a) *Show that **electric potential** at a point P, at a distance 'r' from a fixed point charge Q, is given by* **[4]**

$$V = \left(\frac{1}{4\pi\varepsilon_0}\right)\frac{Q}{r}.$$

(b) *Intensity of electric field at a perpendicular distance of 0·5 m from an infinitely long line charge having linear charge density (λ) is $3{\cdot}6 \times 10^{3}$ Vm^{-1}. Find the value of λ.* **[1]**

Question 3.

(a) *Three capacitors $C_1 = 3\mu F$, $C_2 = 6\mu F$ and $C_3 = 10\mu F$ are connected to a 50 V battery as shown in the figure below:* **[3]**

** Answer is not given due to change in the present syllabus.

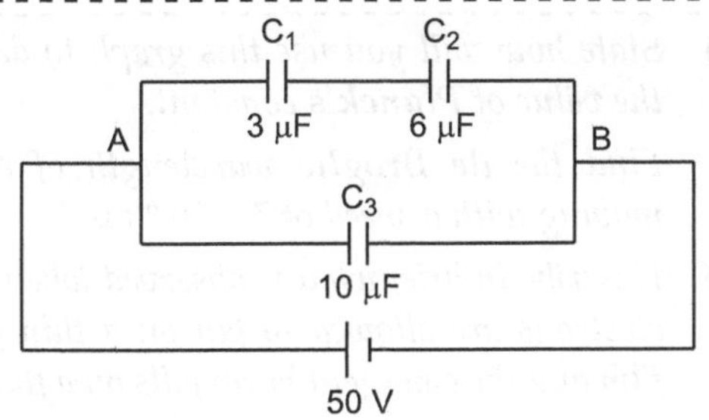

Calculate:

(i) *The equivalent capacitance of the circuit between points A and B.*

(ii) *The charge on C_1.*

(b) *Two resistors $R_1 = 60\ \Omega$ and $R_2 = 90\ \Omega$ are connected in* ***parallel.*** *If electric power consumed by the resistor R_1 is 15 W, calculate the power consumed by the resistor R_2.* **[2]**

Question 4.

(a) *Figure below shows two resistors R_1 and R_2 connected to a battery having an emf of 40 V and negligible internal resistance. A voltmeter having a resistance of 300 Ω is used to measure potential difference across R_1. Find the reading of the voltmeter.* **[3]**

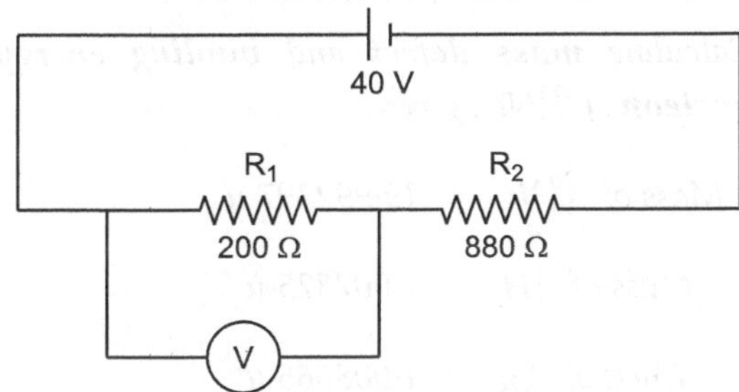

(b) *A moving coil galvanometer has a coil of resistance 59 Ω. It shows a full scale deflection for a current of 50 mA. How will you convert it to an* ***ammeter*** *having a range of 0 to 3 A ?* **[2]**

Question 5.

(a) *In a meter bridge circuit, resistance in the left hand gap is 2 Ω and an unknown resistance X is in the right hand gap as shown in figure below. The null point is found to be 40 cm from the left end of the wire. What resistance should be connected to X so that the new null point is 50 cm from the left end of the wire ?* **[3]**

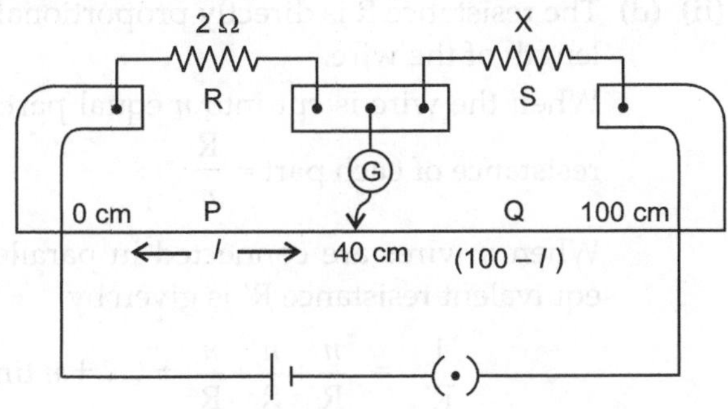

(b) *The horizontal component of earth's magnetic field at a place is $\frac{1}{\sqrt{3}}$ times the vertical component. Determine the* ***angle of dip*** *at that place.* **[2]**

Question 6.

(a) *Using* ***Ampere's circuital law,*** *obtain an expression for the magnetic flux density 'B' at a point 'X' at a perpendicular distance 'r' from a* ***long*** *current carrying conductor. (Statement of the law is not required).* **[3]**

(b) *PQ is a long straight conductor carrying a current of 3 A as shown in figure below. An electron moves with a velocity of $2 \times 10^7\ ms^{-1}$ parallel to it. Find the force acting on the electron.* **[2]**

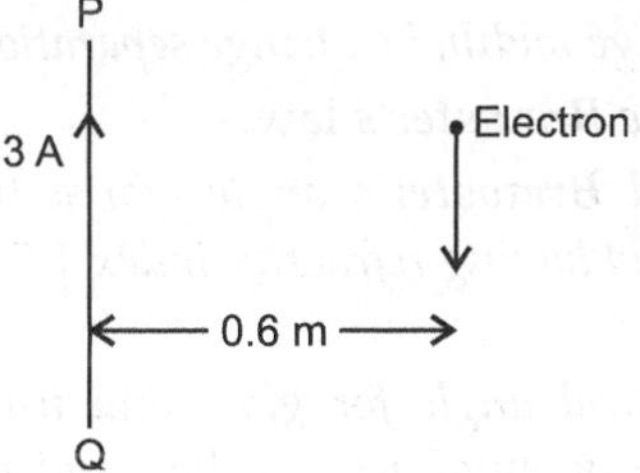

Question 7.

(a) (i) *AB and CD are two parallel conductors kept 1 m apart and connected by a resistance R of 6 Ω, as shown in figure below. They are placed in a magnetic field $B = 3 \times 10^{-2}$ T which is perpendicular to the plane of the conductors and directed into the paper. A wire MN is placed over AB and CD and then made to slide with a velocity $2\ ms^{-1}$. (Neglect the resistance of AB, CD, and MN.)* **[3]**

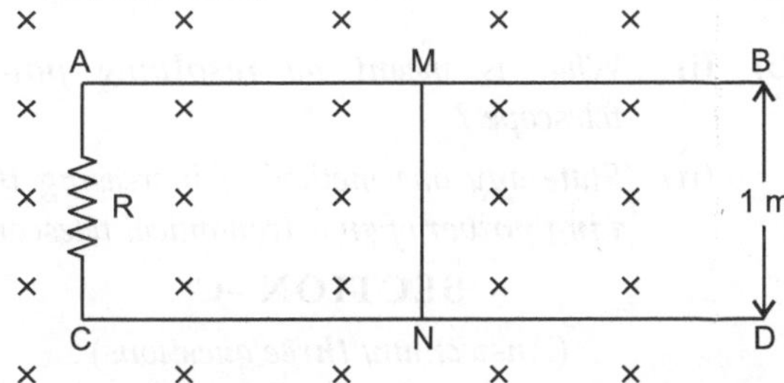

Calculate the induced current flowing through the resistor R.

(ii) *In an* ***ideal*** *transformer, an output of 66 kV is required when an input voltage of 220 V is available. If the* ***primary*** *has 300 turns, how many turns should the* ***secondary*** *have ?*

(b) *In a series LCR circuit, obtain an expression for the* ***resonant frequency.*** **[2]**

SECTION–B

(Answer any ***three*** *questions)*

Question 8.

(a) (i) *State any one property which is* ***common*** *to all* ***electromagnetic*** *waves.* **[3]**

(ii) *Arrange the following electromagnetic waves in increasing order of their frequencies (i.e. begin with the lowest frequency) :*

Visible light, γ rays, X-rays, microwaves, radio waves, infrared radiations and ultraviolet radiations.

(b) (i) *What is meant by* ***diffraction*** *of light ?* **[2]**

(ii) *In* ***Fraunhofer diffraction,*** *what kind of source of light is used and where is it situated ?*

Question 9.

(a) *In* ***Young's double slit experiment*** *using monochromatic light of wavelength 600 nm, 5th bright fringe is at a distance of 0·48 mm from the centre of the pattern. If the screen is at a distance of 80 cm from the plane of the two slits, calculate :* **[3]**

(i) *Distance between the two slits.*

(ii) *Fringe width, i.e., fringe separation.*

(b) (i) *State* ***Brewster's*** law. **[2]**

(ii) *Find* ***Brewster's*** *angle for a transparent liquid having refractive index 1·5.*

Question 10.

(a) *Find critical angle for glass and water pair, given refractive index of glass is 1·62 and that of water is 1·33.* **[2]**

(b) *Starting with an expression for refraction at a single spherical surface, obtain* ***Lens Maker's formula*** **[3]**

Question 11.

(a) *A compound microscope consists of two convex lenses of focal length 2 cm and 5 cm. When an object is kept at a distance of 2·1 cm from the objective, a virtual and magnified image is formed 25 cm from the eye piece. Calculate the* ***magnifying power*** *of the microscope.* **[3]**

(b) (i) *What is meant by* ***resolving power*** *of a telescope ?* **[2]**

(ii) *State any one method of increasing the* ***resolving power*** *of an astronomical telescope.*

SECTION –C

(Answer any ***three*** *questions)*

Question 12.

(a) (i) *Plot a* ***labelled*** *graph of* $|V_S|$ *where* V_S *is* ***stopping potential*** *versus frequency f of the incident radiation.* **[3]**

(ii) *State how will you use this graph to determine the value of* ***Planck's*** *constant.*

(b) (i) *Find the* ***de Broglie*** *wavelength of electrons moving with a speed of* 7×10^6 ms^{-1}. **[2]**

(ii) *Describe in brief what is observed when moving electrons are allowed to fall on a thin graphite film and the emergent beam falls on a fluorescent screen.*

Question 13.

(a) *Draw energy level diagram for hydrogen atom, showing first four energy levels corresponding to n = 1, 2, 3 and 4. Show transitions responsible for :* **[3]**

(i) *Absorption spectrum of* ***Lyman*** *series.*

(ii) *Emission spectrum of* ***Balmer*** *series.*

(b) (i) *Find maximum frequency of X-rays produced by an X-ray tube operating at a tube potential of 66 kV.* **[2]**

(ii) *State any one difference between* ***characteristic*** *X-rays and* ***continuous*** *X-rays.*

Question 14.

(a) *Obtain a relation between* ***half life*** *of a radioactive substance and* ***decay constant*** *(λ).* **[2]**

(b) *Calculate* ***mass defect*** *and* ***binding energy per nucleon*** *of* ${}^{20}_{10}Ne$ *, given* **[3]**

Mass of ${}^{20}_{10}Ne$ = 19·992397 u

Mass of ${}^{1}_{1}H$ = 1·007825 u

Mass of ${}^{1}_{0}n$ = 1·008665 u

Question 15.

(a) *With reference to a semiconductor diode, what is meant by :* **[3]**

(i) *Forward bias,*

(ii) *Reverse bias*

(iii) *Depletion region*

(b) *Draw a diagram to show how* ***NAND*** *gates can be combined to obtain an* ***OR*** *gate.* **[2]**
(Truth table is not required).

ANSWERS

PART–I (20 Marks)

Answer 1.

(A) (i) (c) Electrostatic potential energy,

$$U = \frac{1}{4\pi\varepsilon_0} \cdot \frac{q_1 q_2}{r}$$

$$= 9 \times 10^9 \times \frac{1 \times 10^{-6} \times 1 \times 10^{-6}}{1}$$

$$= 9 \times 10^{-3} \text{ J}$$ **Ans.**

(ii) (d) The resistance R is directly proportional to the length of the wire.

When the wire is cut into n equal parts, then

resistance of each part = $\frac{R}{n}$.

When n wires are connected in parallel then equivalent resistance R′ is given by

$$\frac{1}{R'} = \frac{n}{R} + \frac{n}{R} + \frac{n}{R} + \ldots + n \text{ times}$$

$$\frac{1}{R'} = \frac{n \times n}{R}$$

$$\therefore \quad R' = \frac{R}{n^2}$$ **Ans.**

(iii) **(c)** Relative permeability,

$$\mu_r = 1 + \chi_m$$
$$= 1 + 0{\cdot}0001$$
$$= 1{\cdot}0001$$ **Ans.**

(iv) **(a)** As the light enters from air to the glass, the speed of light decreases but the frequency remains the same.

We know that,

$$c = \lambda\nu$$

where, c = Speed of light
ν = Frequency of light
λ = Wavelength of light

$$\nu = \frac{c}{\lambda}$$

In glass medium, ν remains constant and c decreases. Therefore, λ also decreases to keep the frequency constant. **Ans.**

(v) **(b)** We know that,

$$N = N_0\left(\frac{1}{2}\right)^n$$

Here, N = t = 40 days, T = ?

$$\therefore \quad \frac{N_0}{16} = N_0\left(\frac{1}{2}\right)^{t/T} \qquad [\because n = t/T]$$

$$\frac{N_0}{16} = N_0\left(\frac{1}{2}\right)^{40/T}$$

$$\left(\frac{1}{2}\right)^4 = \left(\frac{1}{2}\right)^{40/T}$$

On comparing the powers

$$\Rightarrow \quad \frac{40}{T} = 4$$

$$\Rightarrow \quad T = \frac{40}{4} = 10 \text{ days.}$$ **Ans.**

(B) **(i)** $\tau = pE \sin\theta$,

For maximum torque $\theta = 90°$, $\sin\theta = 1$, $\tau = PE$

$$\therefore \quad \tau_{max} = pE$$

Here, $\tau_{max} = 6 \times 10^{-25}$ Nm, $p = 3 \times 10^{-29}$ Cm

$$\therefore \quad E = \frac{6\times10^{-25}}{3\times10^{-29}}$$
$$= 2 \times 10^4 \text{ NC}^{-1}$$ **Ans.**

(ii) **Drift velocity:** It is defined as the average velocity gained by the free electrons of a conductor in the opposite direction of the externally applied electric field.

It is given by,

$$\vec{v}_d = \frac{\vec{E}\tau}{m}$$

where, τ is the relaxation time, E is the applied electric field and m is the mass of the electron.

(iii) The Kirchhoff's second law is based on the law of conservation of energy.

(iv) The magnetic flux density of the magnetic field at the centre of a circular coil is given by

$$B = \frac{\mu_0 NI}{2a}$$
$$= \frac{2\pi\times10^{-7}\times50\times5}{2\times0.5}$$
$$= 3{\cdot}14 \times 10^{-4}\,\text{T}$$ **Ans.**

(v) Given : $\varepsilon = 314 \sin(50\pi t)$ volt.

Comparing this equation with $e = e_0 \sin \omega t$, we get

$$\omega = 50\pi$$

We know,

$$f = \frac{\omega}{2\pi}$$
$$= \frac{50\pi}{2\pi}$$
$$= 25 \text{ Hz}$$ **Ans.**

(vi) When the source of light is linear in shape, such as a fine rectangular slit, the wavefront is cylindrical in shape because all the points are equidistant from the linear source lie on a surface of cylinder.

(vii) Fringe width, $\beta = \dfrac{D\lambda}{d}$

If two slits are brought closer to each other, d decreases and $\beta \propto \dfrac{1}{d}$ so fringe width β increases.

(viii) In a regular prism, the deviation is minimum when the angle of incidence is equal to the angle of emergence ($\angle i = \angle e$).

(ix) Given : $f_1 = +40$ cm, $f_2 = -30$ cm

The focal length of the combination is given by

$$\frac{1}{f} = \frac{1}{f_1} + \frac{1}{f_2}$$

$$\frac{1}{f} = \frac{1}{40} - \frac{1}{30}$$

$$f = \frac{-120}{1} = -120 \text{ cm}$$ **Ans.**

(xi) Given: $E = 6 \times 10^{-19}$ J

Momentum of photon,

$$p = E/c$$

$$= \frac{6\times10^{-29}}{3\times10^{8}}$$

$$= 2\times10^{-27} \text{ kg-m/sec}$$ **Ans.**

(xii) The meaning of this statement is that only those circular orbits are permitted in which its angular momentum is an integral multiple of $h/2\pi$, where h is a Planck's constant.

(xiii) Nuclear fusion reaction is called thermo-nuclear reaction because it takes place only at high temperature. To carry nuclear fusion reaction in a bulk material, the temperature of the material has to be raised to 10^6 K, so that the colliding nuclei have enough energy to overcome coulomb's electrostatic force of repulsion. Under this high temperature fusion reaction takes place.

(xiv) The minimum energy X-ray photon possess for pair production

$$= \Delta mc^2$$

$$= 2 \times 9{\cdot}1 \times 10^{-31} \times (3 \times 10^8)^2$$

$$= 1{\cdot}638 \times 10^{-13} \text{ J}$$

$$= \frac{1.638\times10^{-13}}{1.6\times10^{-19}} \text{ eV}$$

$$= 1{\cdot}02 \times 10^6 \text{ eV} = 1{\cdot}02 \text{ MeV}$$

(xv)

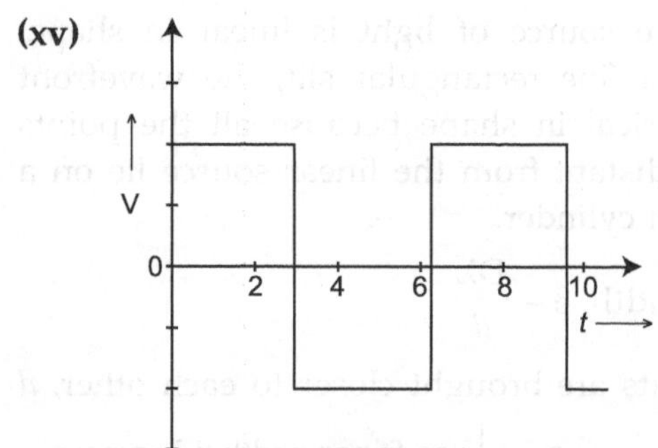

Variation of voltage with time

PART–II (50 Marks)

SECTION–A

Answer 2.

(a) **Electric potential due to a point charge:** Consider a positive point charge Q placed at the origin O. We have to calculate its electric potential at a point P at a distance of r from it as shown in the figure. The electric potential at point P will be equal to the amount of work done in bringing a unit positive charge from infinity to point P.

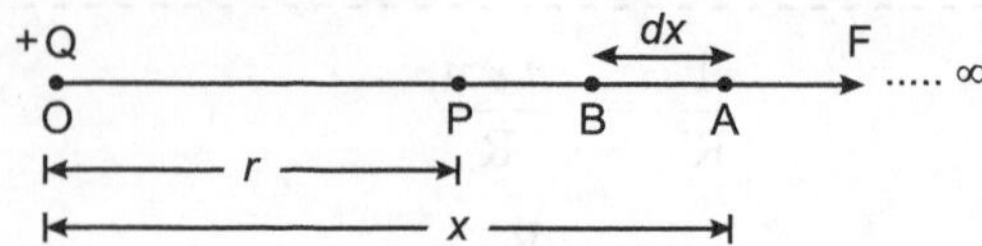

Suppose a test charge q_0 is placed at a point A at a distance of x from O. By Coulomb's law, the electrostatic force acting on charge q_0 is,

$$F = \frac{1}{4\pi\varepsilon_0} \cdot \frac{Qq_0}{x^2}$$

The small work done in moving the test charge q_0 from A to B through small displacement $\vec{dx}$ against the electrostatic force is

$$dW = \vec{F}.\vec{dx} = F\,dx \cos 180° = -F\,dx$$

The total work done in moving the charge q_0 from infinity to the point P is given by

$$W = \int dW = -\int_{\infty}^{r} F\,dx$$

$$= -\int_{\infty}^{r} \frac{1}{4\pi\varepsilon_0} \cdot \frac{Qq_0}{x^2}\,dx$$

$$= -\frac{Qq_0}{4\pi\varepsilon_0} \int_{\infty}^{r} x^{-2}\,dx$$

$$W = -\frac{1}{4\pi\varepsilon_0}\left[-\frac{1}{x}\right]_{\infty}^{r}$$

$$= \frac{Qq_0}{4\pi\varepsilon_0}\left[\frac{1}{r} - \frac{1}{\infty}\right]$$

$$= \frac{1}{4\pi\varepsilon_0} \cdot \frac{Qq_0}{r}$$

The electric potential at P is the work done per unit charge from infinity to the point P.

$$V = \frac{W}{q_0}$$

or $$V = \frac{1}{4\pi\varepsilon_0} \cdot \frac{Q}{r}$$ **Hence Proved.**

(b) Given : $r = 0.5$ m, $E = 3.6 \times 10^3$ Vm^{-1}, $\lambda = ?$

Electric field due to a long line charge having linear charge density (λ) is given by

$$E = \frac{\lambda}{2\pi\varepsilon_0 r}$$

$$\Rightarrow \quad 3.6 \times 10^3 = \frac{\lambda}{2\times3.14\times8.85\times10^{-12}\times0.5}$$

$$\lambda = 2 \times 3.14 \times 8.85 \times 10^{-12} \times 0.5 \times 3{\cdot}6 \times 10^3$$

$$= 1 \times 10^{-7} \text{ C m}^{-1}$$ **Ans.**

Answer 3.

(a) (i)

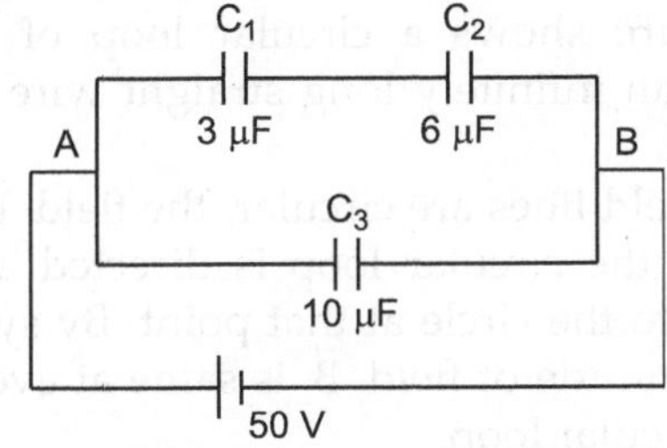

The capacitances C_1 and C_2 are connected in series, so their equivalent capacitance C′ is given by

$$\frac{1}{C'} = \frac{1}{3}+\frac{1}{6}$$

$$\Rightarrow \quad C' = \frac{6}{3} = 2\ \mu F$$

Now, C′ and C_3 are in parallel.

$$\therefore \quad C_{eq} = C' + C_3 = 2 + 10$$

$$= 12\ \mu F$$ **Ans.**

(ii) Since the capacitors C_1 and C_2 are connected in series so, the charge is same.

$$\therefore \quad q = q_1 = q_2 = C'V = 2 \times 50 = 100\ \mu C$$ **Ans.**

(b)

Given : $R_1 = 60\ \Omega$, $R_2 = 90\ \Omega$ and $P_1 = 15$ W.

Since the two resistors are connected in parallel, so the voltage remains constant.

Let the voltage across the resistors be V.

We know,

$$\text{Power, } P = \frac{V^2}{R}$$

For resistor R_1

$$\Rightarrow \quad P_1 = \frac{V^2}{R_1}$$

$$15 = \frac{V^2}{60}$$

$$V^2 = 900,$$

$$V = \sqrt{900}$$

$$V = 30\ V$$

For resistor R_2, V remains same

$$\text{Now,} \quad P_2 = \frac{V^2}{R_2} = \frac{(3)^2}{90}$$

$$= \frac{900}{90} = 10\ W.$$ **Ans.**

Answer 4.

(a)

As R_1 and voltmeter are in parallel, so their equivalent resistance is given by

$$\frac{1}{R'} = \frac{1}{200}+\frac{1}{300}$$

$$\Rightarrow \quad R' = \frac{200\times300}{200+300}$$

$$= \frac{200\times300}{500}$$

$$\Rightarrow \quad R' = 120\ \Omega$$

Now, R′ and R_2 are in series, so their equivalent resistance is given by

$$R = R' + 880$$

$$= 120 + 880 = 1000\ \Omega$$

Since the current in series combination remains the same.

$$\therefore \quad \text{Current, } I = \frac{V}{R}$$

$$= \frac{40}{1000} = \frac{1}{25} A = 0.04\ A$$

∴ Potential drop across the voltmeter

$$V = IR'$$

$$= \frac{1}{25} \times 120 \times 120$$

$$= 4{\cdot}8\ V$$ **Ans.**

(b) Given: $I_g = 50 \times 10^{-3}$ A, $R_g = 59\ \Omega$, $I = 3$ A, $R_S = ?$

$$R_S = \frac{I_g R_g}{I - I_g}$$

$$= \frac{50\times10^{-3}\times59}{3-(50\times10^{-3})}$$

$$= \frac{50\times10^{-3}\times59}{2.95}$$

$$= \frac{2950\times10^{-3}}{2.95}$$

$$= 1000 \times 10^{-3} = 1\ \Omega$$

To convert a galvanometer into an ammeter, connect a 1Ω resistance in parallel to it. **Ans.**

Answer 5.

(a) By Wheatstone bridge principle:

$$\frac{P}{Q} = \frac{R}{S}$$

Case I : When $l = 40$ cm

$$\Rightarrow \quad \frac{l}{100-l} = \frac{2}{X}$$

$$\Rightarrow \quad \frac{40}{100-40} = \frac{2}{X}$$

$$\Rightarrow \quad \frac{40}{60} = \frac{2}{X}$$

$$\Rightarrow \quad 2X = 6$$

$$\therefore \quad X = \frac{6}{2} = 3\,\Omega$$

Case II : When $l' = 50$ cm

$$\frac{P}{Q} = \frac{R}{S}$$

$$\Rightarrow \quad \frac{l'}{100-l'} = \frac{2}{X'}$$

$$\Rightarrow \quad \frac{50}{100-50} = \frac{2}{X'}$$

$$\Rightarrow \quad \frac{50}{50} = \frac{2}{X'}$$

$$\therefore \quad X' = 2\,\Omega$$

$$\because \quad X' < X$$

So resistance need to be decreased and a new resistor should be connected in parallel.

$\therefore$ A resistance X_1 must be connected in parallel to X.

$$\therefore \quad \frac{1}{X'} = \frac{1}{X} + \frac{1}{X_1}$$

$$\Rightarrow \quad \frac{1}{2} = \frac{1}{3} + \frac{1}{X_1}$$

or $$\frac{1}{X_1} = \frac{1}{2} - \frac{1}{3} = \frac{3-2}{6} = \frac{1}{6}$$

$$\Rightarrow \quad X_1 = 6\,\Omega$$

$\therefore$ 6 Ω resistance must be connected in parallel to X to have new null point at 50 cm. **Ans.**

(b) Horizontal component $(B_H) = \frac{1}{\sqrt{3}} \times$ Vertical component (B_V) (Given)

$$\therefore \quad \tan\delta = \frac{B_V}{B_H} = \frac{\sqrt{3}\,B_H}{B_H} = \sqrt{3}$$

$$\Rightarrow \quad \delta = \tan^{-1}\left(\sqrt{3}\right) = 60°$$ **Ans.**

Answer 6.

(a) The figure shows a circular loop of radius r around an infinitely long straight wire carrying current I.

As the field lines are circular, the field $\vec{B}$ at any point of the circular loop is directed along the tangent to the circle at that point. By symmetry, the magnitude of field $\vec{B}$ is same at every point of the circular loop.

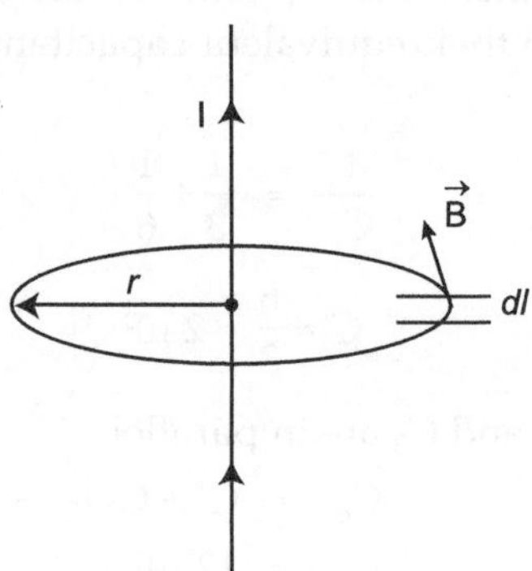

Therefore, $\oint \vec{B}.\vec{dl} = \oint B\,dl\cos 0° = B\oint dl = B.2\pi r$

From Ampere's circuital law,

$$\oint \vec{B}.\vec{dl} = \mu_0 I$$

$$\Rightarrow \quad B\cdot 2\pi r = \mu_0 I$$

$$\therefore \quad B = \frac{\mu_0 I}{2\pi r}$$ **Ans.**

(b) Magnetic field due to current carrying conductor

$$B = \frac{\mu_0 I}{2\pi r}$$

where I = 3A, $r = 0{\cdot}6$ m

$$= \frac{4\pi\times 10^{-7}\times 3}{2\pi\times 0.6} = 10^{-6}\,T$$

$\therefore$ Force acting on electron is given by

$$F = evB$$
$$= 1{\cdot}6\times 10^{-19}\times 2\times 10^{7}\times 10^{-6}$$
$$= 3{\cdot}2\times 10^{-18}\,N.$$ **Ans.**

Answer 7.

(a) (i) Given : $B = 3\times 10^{-2}$ T, $v = 2$ ms^{-1}, $l = 1$ m, $R = 6\,\Omega$, $e = ?$

Induced emf, $e = Blv$
$= 3\times 10^{-2}\times 1\times 2$
$= 6\times 10^{-2}$ V

Induced current, $I = \frac{e}{R}$

$$= \frac{6\times 10^{-2}}{6} = 10^{-2}\,A$$ **Ans.**

(ii) Given : $E_1 = 220$ V, $E_2 = 66 \times 10^3$ V, $N_1 = 300$, $N_2 = ?$

We know that,

$$\frac{E_2}{E_1} = \frac{N_2}{N_1} \quad \text{(transformation ratio)}$$

$$\frac{66 \times 10^3}{220} = \frac{N_2}{300}$$

$$\Rightarrow \quad N_2 = \frac{300 \times 66 \times 10^3}{220}$$

$$= 90 \times 10^3 = 90000.$$ **Ans.**

(b) Resonant frequency of LCR circuit: The current amplitude I_0 for a series LCR circuit is given by

$$I_0 = \frac{E_0}{\sqrt{R^2 + \left(\omega L - \frac{1}{\omega C}\right)^2}}$$

Clearly, I_0 becomes zero both for $\omega \to 0$ and $\omega \to \infty$. The value of I_0 is maximum when

$$\omega L - \frac{1}{\omega C} = 0$$

LCR series circuit is said to be in resonance when inductive reactance is equal to capacitive reactance *i.e.,* $\omega L = \frac{1}{\omega C}$

or $$\omega = \frac{1}{\sqrt{LC}}$$

Then, impendance, $Z = \sqrt{R^2 + \left(\omega L - \frac{1}{\omega C}\right)^2}$

$$= R$$

Clearly, the impedance is minimum. The circuit is purely resistive. The current and voltage are in the same phase and the current in the circuit is maximum. This condition of the LCR circuit is called resonance condition. The frequency at which the current amplitude I_0 attains a peak value is called natural or resonant frequency of the LCR circuit and is denoted by f_r.

Thus, $$\omega_r = 2\pi f_r = \frac{1}{\sqrt{LC}}$$

or $$f_r = \frac{1}{2\pi\sqrt{LC}}$$

SECTION–B

Answer 8.

(a) (i) All electromagnetic waves travel in free space with the same speed *i.e.*, speed of light (*c*).

(ii) Radio waves < Microwaves < Infra-red radiations < Visible light < Ultraviolet radiations < X-rays < γ rays.

(b) (i) The phenomenon of bending of light rays around the corners of an obstacle or aperture and its consequent spreading into the regions of geometrical shadow is called diffraction of light.

(ii) In Fraunhofer diffraction, the source is large (*e.g.,* Sun) and it is placed at infinite distance from the aperture or the obstacle.

Answer 9.

(a) Given : $\lambda = 600$ nm $= 600 \times 10^{-9}$ m, $n = 5$, $y_5 = 0{\cdot}48$ mm $= 0{\cdot}48 \times 10^{-3}$ m, D = 80 cm = 80×10^{-2} m, $d = ?$, $\beta = ?$

(i) We know that,

$$y_n = \frac{nD\lambda}{d}$$

$$y_5 = \frac{5D\lambda}{d}$$

$$\therefore \quad 0{\cdot}48 \times 10^{-3} = \frac{5 \times 80 \times 10^{-2} \times 600 \times 10^{-3}}{d}$$

$$\therefore \quad d = \frac{5 \times 80 \times 10^{-2} \times 600 \times 10^{-9}}{0.48 \times 10^{-3}}$$

$$= 5 \times 10^{-3} \text{ m}$$ **Ans.**

(ii) Fringe width, $\beta = \frac{D\lambda}{d}$

$$= \frac{80 \times 10^{-2} \times 600 \times 10^{-4}}{5 \times 10^{-3}}$$

$$= 9{\cdot}6 \times 10^{-5} \text{ m}$$ **Ans.**

(b) (i) Brewster's law : The angle of incidence at which a beam of unpolarised light falling on a transparent surface is reflected as a beam of completely plane polarised light is called polarising or Brewster's angle. It is denoted by i_p.

Brewster's law states that the tangent of the polarising angle for a given medium is numerically equal to the refractive index of that medium.

i.e., $$\mu = \tan i_p$$

where i_p is the polarising angle and μ is the refractive index of that medium.

(ii) Given : $$\mu = 1{\cdot}5$$

$$\Rightarrow \quad 1{\cdot}5 = \tan i_p$$

or $$i_p = \tan^{-1}(1{\cdot}5)$$

(using trigonometric table)

$$\Rightarrow \quad i_p = 56{\cdot}31°$$ **Ans.**

Answer 10.

(a) Given : $\mu_g = 1{\cdot}62$, $\mu_w = 1{\cdot}33$

We know that,

$$ {}_w\mu_g = \frac{1}{\sin i_C} $$

$$ \therefore \quad \sin i_c = \frac{1}{{}_w\mu_g} = \frac{\mu_w}{\mu_g} $$

$$ \Rightarrow \quad \sin i_c = \frac{1.33}{1.62} $$

$$ \therefore \quad \sin i_c = 0{\cdot}8209 $$

$$ i_c = \sin^{-1}(0{\cdot}8209) = 55{\cdot}18 \approx 55{\cdot}2^\circ $$

(using trigonometric table) **Ans.**

(b) Consider a thin double convex lens of refractive index μ_2 placed in a medium of refractive index μ_1. Here $\mu_1 < \mu_2$. Let B and D be the poles, C_1 and C_2 be the centres of curvature, and R_1 and R_2 be the radii of curvature of the two lens surfaces ABC and ADC respectively.

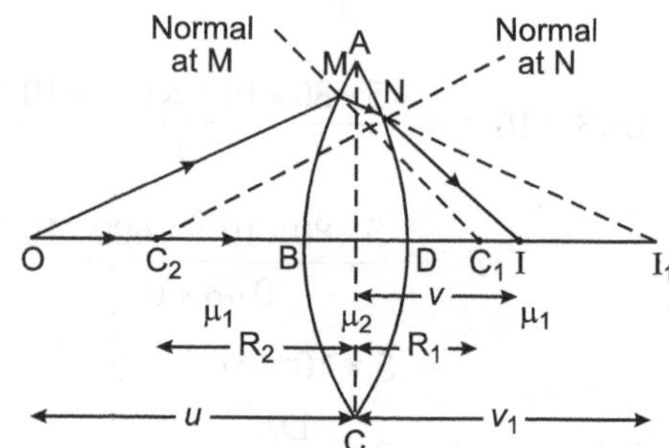

Suppose a point object O is placed on the principal axis in the rarer medium of refractive index μ_1. The ray OM is incident on the first surface ABC. It is refracted along MN, bending towards the normal at this surface. If the lens material were continuous and there were no boundary/secondary surface ADC of the lens, the refracted ray MN would go straight meeting the first refracted at I_1. So, we can treat I_1 as the real image formed by first surface ABC in the medium of refractive index μ_2.

For refraction at surface ABC,

$$ \frac{\mu_2}{v_1} - \frac{\mu_1}{u} = \frac{\mu_2 - \mu_1}{R_1} \quad \ldots(1) $$

But actually the lens material is not continuous. Therefore, the ray MN suffers further refraction at surface ADC and emerges along NI, meeting actually the principal axis at I. Therefore, the relation between the object distance v_1, image distance v and radius of curvature R_2 can be written as :

$$ \frac{\mu_1}{v} - \frac{\mu_2}{v_1} = \frac{\mu_1 - \mu_2}{R_2} \quad \ldots(2) $$

Adding (1) and (2), we get

$$ \frac{\mu_2}{v_1} - \frac{\mu_1}{u} + \frac{\mu_1}{v} - \frac{\mu_2}{v_1} = \frac{\mu_2 - \mu_1}{R_1} - \frac{\mu_1 - \mu_2}{R_2} $$

$$ \frac{\mu_1}{v} - \frac{\mu_1}{u} = (\mu_2 - \mu_1)\left(\frac{1}{R_1} - \frac{1}{R_2}\right) $$

or

$$ \frac{1}{v} - \frac{1}{u} = \left(\frac{\mu_2 - \mu_1}{\mu_1}\right)\left(\frac{1}{R_1} - \frac{1}{R_2}\right) $$

If the object is placed at infinity ($u = \infty$), the image will be formed at the focus *i.e.*, $v = f$

$$ \therefore \quad \frac{1}{f} = \left(\frac{\mu_2 - \mu_1}{\mu_1}\right)\left(\frac{1}{R_1} - \frac{1}{R_2}\right) \quad \ldots(3) $$

$$ \frac{1}{f} = \left(\frac{\mu_2}{\mu_1} - 1\right)\left(\frac{1}{R_1} - \frac{1}{R_2}\right) $$

This is lens maker's formula.

Answer 11.

(a) Given: $f_o = 2$ cm, $f_e = 5$ cm, $u_o = -2{\cdot}1$ cm, $d = 25$ cm

We know that, $\frac{1}{f} = \frac{1}{v_o} - \frac{1}{u_o}$

$$ \Rightarrow \quad \frac{1}{v_o} = \frac{1}{f_o} + \frac{1}{u_o} = \frac{1}{2} + \frac{1}{-2.1} $$

$$ \Rightarrow \quad \frac{1}{v_o} = \frac{-0.1}{4.2} = \frac{1}{42} \text{ cm} $$

$$ \Rightarrow \quad v_o = -42 \text{ cm} $$

$$ \text{Magnification, } m = \frac{v_0}{u_0}\left(1 + \frac{d}{f_e}\right) $$

$$ = \frac{-42}{-2.1}\left(1 + \frac{25}{5}\right) $$

$$ = 20\,(1 + 5) $$

$$ = 20 \times 6 = 120. $$ **Ans.**

(b) (i) Resolving power of a telescope: The resolving power of a telescope is defined as the reciprocal of the smallest angular separation between two close, distant objects whose images are just resolved by the telescope.

$$ \text{R.P. of telescope} = \frac{1}{d\theta} = \frac{D}{1.22\lambda} $$

Here,

$\lambda =$ the wavelength of light

D = the diameter of the objective lens of the telescope

$d\theta =$ the angle subtended by the two close distant objects at the objective lens

(ii) The resolving power of an astronomical telescope can be increased by using the lens of larger diameter of the objective and decreasing the wavelength of light.

SECTION –C

Answer 12.

(a) (i) The graph of stopping potential, $|V_S|$ versus frequency, f of incident radiation is given below:

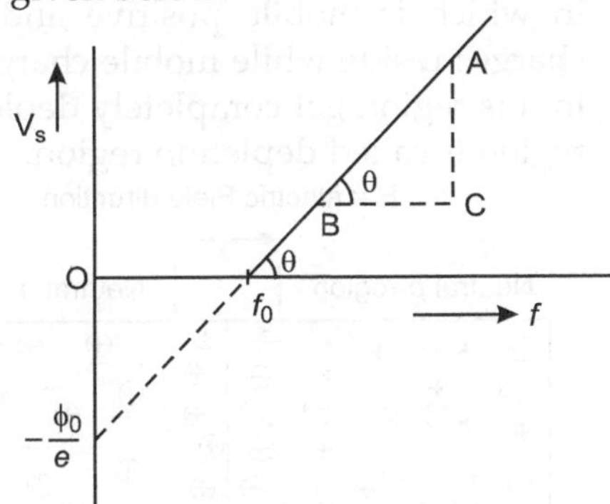

(ii) The value of Planck's constant can be determined from the above graph as

$$\tan\theta = \frac{h}{e}$$

$$h = e\tan\theta$$

$$h = e \times \text{Slope of } (V_s - f) \text{ graph.}$$

(b) (i) de-Broglie wavelength,

$$\lambda = \frac{h}{mv}$$

$$\Rightarrow \quad \lambda = \frac{6.6\times10^{-34}}{9.1\times10^{-31}\times7\times10^{6}}$$

$$= \frac{6.6\times10^{-34}}{63.7\times10^{-25}}$$

$$= 1{\cdot}03 \times 10^{-10}\ \text{m} = 1{\cdot}03\ \text{Å} \simeq 1\ \text{Å} \quad \textbf{Ans.}$$

(ii) Electron diffraction is observed when moving electrons are allowed to fall on a thin graphite film and the emergent beam falls on a fluorescent screen.

Answer 13.

(a) (i)

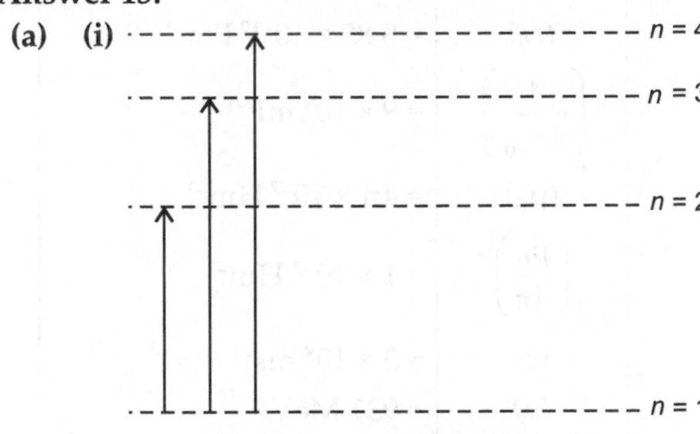

Lyman series

Absorption spectrum of Lyman series

(ii)

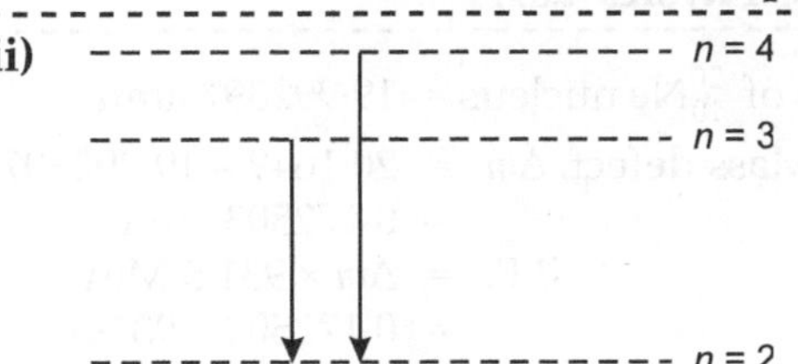

Emission spectrum of Balmer series

(b) (i)

$$\nu_{max} = \frac{eV}{h}$$

where, e charge on electron, $= 1{\cdot}6 \times 10^{-19}$ C, h Planck's constant $= 6{\cdot}62 \times 10^{-34}$ Js, $V = 66 \times 1000$ V.

$$\nu_{max} = \frac{1{\cdot}6\times10^{-19}\times66\times10^{3}}{6{\cdot}6\times10^{-34}}$$

$$= 16 \times 10^{18}\ \text{Hz} = 1{\cdot}6 \times 10^{19}\ \text{Hz}$$

Ans.

(ii) Continuous X-rays consists of radiations of all possible wavelengths with a lower wavelength limit while characteristic X-rays consists of definite wavelength superimposed on the continuous spectrum.

Answer 14.

(a) Let N_0 = Number of radioactive nuclei present in the radioactive sample initially (at $t = 0$)

N = Number of radioactive nuclei left at any instant t

$$\text{At } t = T_{1/2},\ N = \frac{N_0}{2} \quad \ldots\text{(i)}$$

Now, number of atoms present at a time 't' are

$N = N_0 e^{-\lambda t}$, where λ is the radioactive decay constant

$$\therefore \quad \frac{N_0}{2} = N_0 e^{-\lambda T_{1/2}} \quad \text{[from (i)]}$$

$$\text{or} \quad \frac{1}{2} = e^{-\lambda T_{1/2}} \text{ or } e^{\lambda T_{1/2}} = 2$$

Taking natural logarithm on both sides, we get

$$\lambda T_{1/2} \log_e e = \log_e 2$$

$$T_{1/2} = \frac{\log_e 2}{\lambda} = \frac{2.303\log 2}{\lambda} \quad [\because \log_e e = 1]$$

$$= \frac{2.303\times0.3010}{\lambda}$$

$$T_{1/2} = \frac{0.693}{\lambda}$$

(b) The nucleus $^{20}_{10}$Ne contains 10 protons and 10 neutrons.

Mass of 10 protons $= 10 \times 1{\cdot}007825$
$= 10{\cdot}07825$ amu

Mass of 10 neutrons $= 10 \times 1{\cdot}008665$
$= 10{\cdot}08665$ amu

Total mass $= 20{\cdot}1649$ amu

Mass of $^{20}_{10}$Ne nucleus = 19·992397 amu

$\therefore$ Mass defect, Δm = 20·1649 – 19·992397

= 0·172503 amu **Ans.**

B.E. = $\Delta m \times 931{\cdot}5$ MeV

= 0·172503 × 931·5

= 160·68 MeV

$$\text{B.E. per nucleon} = \frac{\text{B.E.}}{\text{Mass number}}$$

$$= \frac{160.68}{20} = 8{\cdot}03 \text{ MeV}$$ **Ans.**

Answer 15.

(a) (i) **Forward Bias :** If the positive terminal of a battery is connected to the *p*-type and the negative terminal to the *n*-type, then the *p-n* junction is said to be forward biased.

In forward biased electrons flow from *n*-region to *p*-region through the junction, which give rise to forward current and offers very low resistance.

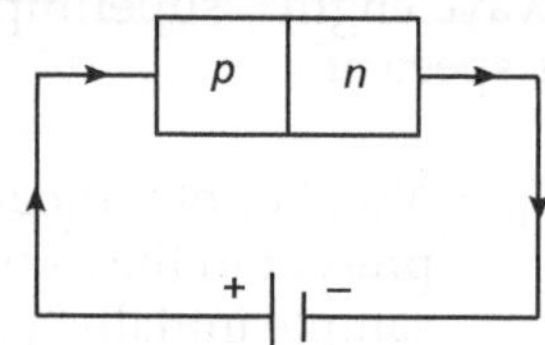

(ii) **Reverse bias :** If the positive terminal of a battery is connected to the *n*-type and negative terminal to the *p*-type then the *p-n* junction is said to be reverse biased.

In reverse bias, the junction offers very high resistance and very small current flows through it.

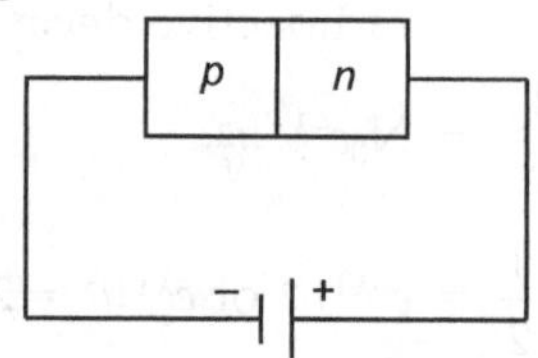

(iii) **Depletion region :** In a *p-n* junction, the diffusion of holes and electrons starts initially. Near the junction, holes recombine in *n*-region to form immobile positive ions. Similarly, electrons recombine in *p*-region to form immobile negative ions. With sufficient accumulation of such immobile ions on both sides, the diffusion stops. So near the junction, there exists a region in which immobile positive and negative charges reside while mobile charge carriers in this region get completely depleted. This region is called depletion region.

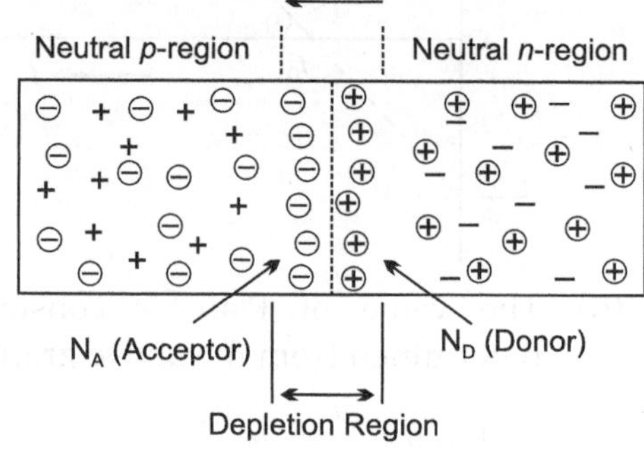

It is a narrow space (region) between *p* and *n* regions which does not contain charge carriers (*i.e.* electrons and holes).

(b) The required figure is shown below:

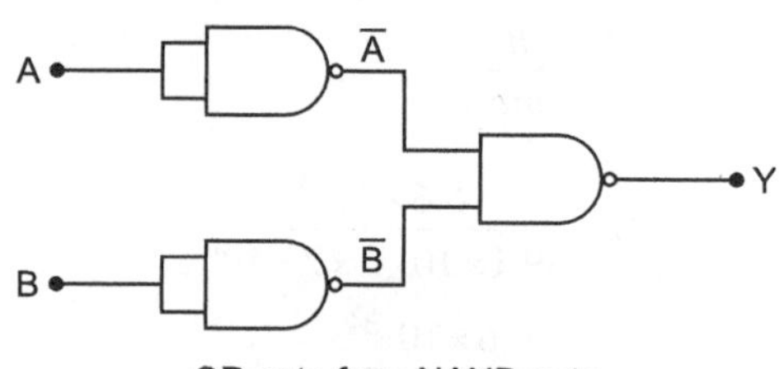

OR gate from NAND gate

Useful Constants and Relations

1.	Charge of a proton	(e)	$= 1{\cdot}6 \times 10^{-19}$ C
2.	Planck's constant	(h)	$= 6{\cdot}6 \times 10^{-34}$ Js
3.	Mass of an electron	(m)	$= 9{\cdot}1 \times 10^{-31}$ kg
4.	Permittivity of vacuum	(ε_0)	$= 8{\cdot}85 \times 10^{-12}$ Fm^{-1}
5.	Constant for Coulomb's law	$\left(\frac{1}{4\pi\varepsilon_0}\right)$	$= 9 \times 10^{9}$ mF^{-1}
6.	Permeability of vacuum	(μ_0)	$= 4\pi \times 10^{-7}$ Hm^{-1}
7.		$\left(\frac{\mu_0}{4\pi}\right)$	$= 1 \times 10^{-7}$ Hm^{-1}
8.	Speed of light in vacuum	(c)	$= 3 \times 10^{8}$ ms^{-1}
9.	Unified atomic mass unit	(u)	= 931 MeV
10.	Electron volt	(1eV)	$= 1{\cdot}6 \times 10^{-19}$ J

PHYSICS 2016

QUESTIONS

PART–I (20 Marks)

*(Answer **all** questions)*

Question 1.

(A) *Choose the correct alternative (a), (b), (c) or (d) for each of the questions given below:* **[5]**

(i) *In figure below a charge Q is fixed. Another charge q is moved along a circular arc MN of radius r around it, from the point M to the point N such that the length of the arc MN = l. The work done in this process is:*

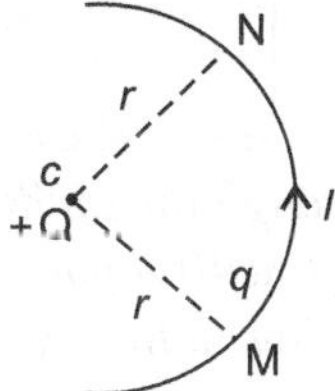

(a) *Zero* (b) $\frac{1}{4\pi\varepsilon_0}\cdot\frac{Qq}{r^2}l$

(c) $\frac{Qq}{2\varepsilon_0 r^2}l$ (d) $\frac{Qq}{2\varepsilon_0 r^2}$

(ii) *A carbon resistor has coloured bands as shown in figure below. The resistance of the resistor is:*

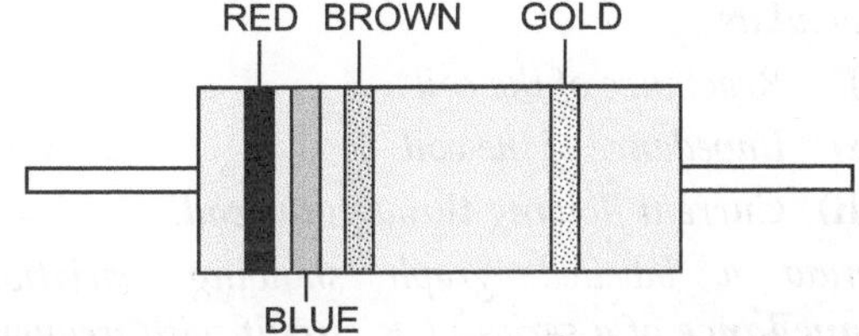

(a) 26 Ω ± 10% (b) 26 Ω ± 5%

(c) 260 Ω ± 5% (d) 260 Ω ± 10%

(iii) *A solenoid L and a resistor R are connected in series to a battery, through a switch. When the switch is put on, current I flowing through it varies with time t as shown in which of the graphs given below :***

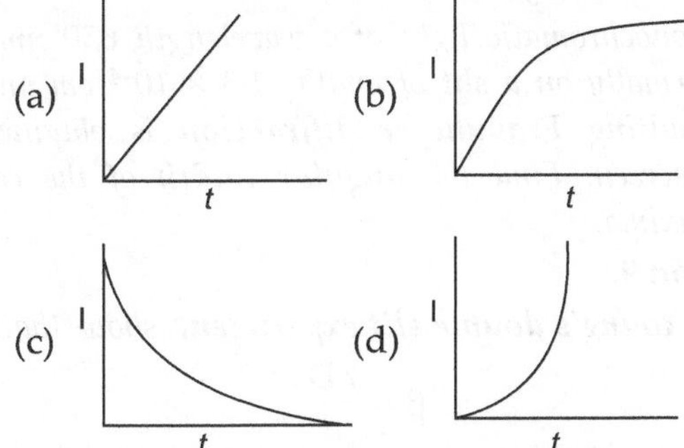

** Answer is not given due to change in the present syllabus.

(iv) *Two thin lenses having optical powers of –10 D and + 6 D are placed in contact with each other. The focal length of the combination is :*

(a) + 0·25 cm (b) – 0·25 cm

(c) + 0·25 m (d) – 0·25 m

(v) *Total energy of an electron in the **ground state** of hydrogen atom is – 13·6 eV. Its total energy, when hydrogen atom is in the **first excited state,** is:*

(a) + 13·6 eV (b) + 3·4 eV

(c) – 3·4 eV (d) – 54·4 eV

(B) *Answer **all** questions given below **briefly** and to the point:* **[15]**

(i) *A charged oil drop weighing $1{\cdot}6 \times 10^{-15}$ N is found to remain suspended in a uniform electric field of intensity 2×10^3 NC^{-1}. Find the **charge** on the drop.*

(ii) *For a metallic conductor, what is the relation between **current density** (J), **conductivity** (σ) and **electric field intensity** E ?*

(iii) *In figure given below, find the value of resistance X for which points A and B are at the same potential:*

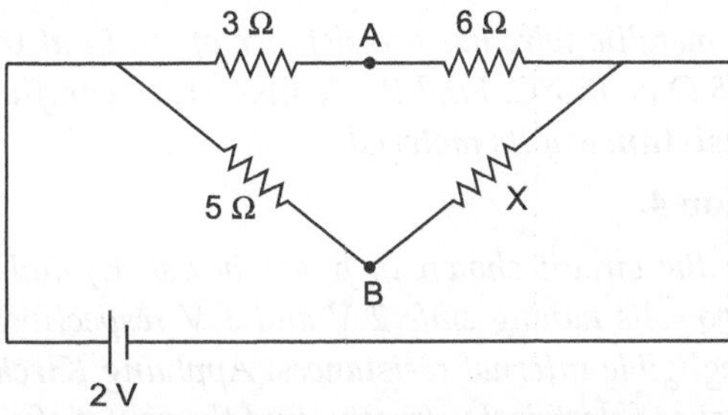

(iv) *Write the expression for the **Lorentz force F** in vector form.*

(v) *A coil has a **self-inductance** of 0·05 Henry. Find magnitude of the emf induced in it when the current flowing through it is changing at the rate 100 As^{-1}.*

(vi) *To which regions of the **electromagnetic spectrum** do the following wavelengths belong:*

(a) 250 nm *(b) 1500 nm*

(vii) *What is the difference between **polarised light** and **unpolarised light ?***

(viii) *Name the **principle** on the basis of which **optical fibres** work.*

(ix) *Calculate **dispersive power** of a transparent material given :*

$n_v = 1{\cdot}56,\ n_r = 1{\cdot}54,\ n_y = 1{\cdot}55$

(x) What is meant by **short-sightedness** ?**

(xi) Two metals A and B have work functions 4 eV and 6 eV respectively. Which metal has lower **threshold wavelength** for photoelectric effect ?

(xii) Calculate **angular momentum** of an electron in the **third** Bohr orbit of hydrogen atom.

(xiii) In a nuclear reactor, what is the function of a **moderator** ?

(xiv) In our nature, where is the **nuclear fusion** reaction taking place continuously ?

(xv) What is the use of a **Zener diode** ?

PART–II (50 Marks)

SECTION–A

*(Answer **any four** questions)*

Question 2.

(a) Two point charges $Q_1 = 400\ \mu C$ and $Q_2 = 100\ \mu C$ are kept fixed, 60 cm apart in vacuum. Find **intensity** of the electric field at **midpoint** of the line joining Q_1 and Q_2. [3]

(b) (i) State **Gauss' law.** [2]

(ii) In an electric dipole, at which point is the **electric potential** zero ?

Question 3.

(a) Obtain an expression for **equivalent capacitance** when three capacitors C_1, C_2 and C_3 are connected in **series.** [3]

(b) A metallic wire has a resistance of 3·0 Ω at 0°C and 4·8 Ω at 150°C. Find the **temperature coefficient of resistance** of its material. [2]

Question 4.

(a) In the circuit shown in figure below, E_1 and E_2 are two cells having emfs 2 V and 3 V respectively, and negligible internal resistances. Applying **Kirchhoff's laws** of electrical networks, find the values of currents I_1 and I_2. [4]

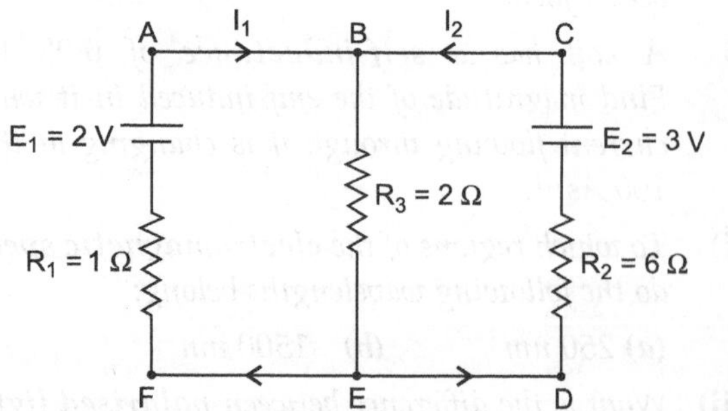

(b) State how a moving coil galvanometer can be converted into an ammeter. [1]

Question 5.

(a) Draw a labelled circuit diagram of a **potentiometer** to measure **internal resistance** of a cell. Write the working formula. (Derivation not required). [3]

(b) (i) Define **Curie temperature.** [2]

(ii) If magnetic susceptibility of a certain magnetic material is 0·0001, find its **relative permeability.**

Question 6.

(a) (i) Two infinitely long current carrying conductors X and Y are kept parallel to each other, 24 cm apart in vacuum. They carry currents of 5 A and 7 A respectively, in the **same** direction, as shown in figure below. Find the position of a neutral point, i.e., a point where resultant magnetic flux density is zero. (Ignore earth's magnetic field). [3]

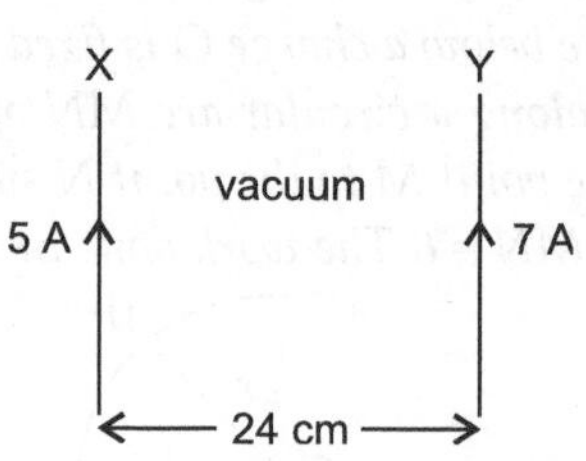

(ii) If current through the conductor Y is **reversed** in direction, will neutral point lie between X and Y, to the left of X or to the right of Y ?

(b) (i) Define **ampere** in terms of force between two current carrying conductors. [2]

(ii) What is an **ideal** transformer ?

Question 7.

(a) A coil having **self-inductance** of 0·7 H and resistance of 165 Ω is connected to an a.c. source of 275 V, 50 Hz. If $\pi = \frac{22}{7}$ [3]

Calculate:

(i) Reactance of the coil

(ii) Impedance of the coil

(iii) Current flowing through the coil.

(b) Draw a labelled graph showing variation of **impedance** of a series LCR circuit with frequency of the a.c. supply. [2]

SECTION–B

*(Answer any **three** questions)*

Question 8.

(a) Derive **Snell's law** of refraction using **Huygen's wave theory.** [3]

(b) Monochromatic light of a wavelength 650 nm falls normally on a slit of width $1 \cdot 3 \times 10^{-4}$ cm and the resulting **Fraunhofer diffraction** is obtained on a screen. Find the **angular width** of the central maxima. [2]

Question 9.

(a) In **Young's double slit experiment,** show that:

$$\beta = \frac{\lambda D}{d}$$

where the terms have their usual meaning. [4]

(b) *A ray of ordinary light is travelling in air. It is incident on air glass pair at a* ***polarising angle*** *of 56°. Find the angle of* ***refraction*** *in glass.* **[1]**

Question 10.

(a) *Find the* ***angle of incidence*** *at which a ray of monochromatic light should be incident on the* ***first surface*** *AB of a* ***regular*** *glass prism ABC so that the emergent ray grazes the adjacent surface AC. (Refractive Index of glass = 1·56)* **[3]**

(b) *State how focal length of a glass lens (Refractive index 1·5) changes when it is completely immersed in :* **[2]**

(i) *Water (Refractive index 1·33)*

(ii) *A liquid (Refractive index 1·65)*

Question 11.

(a) *A convex lens of a focal length 5 cm is used as a* ***simple microscope.*** *Where should an object be placed so that the image formed by it lies at the least distance of distinct vision (D = 25 cm) ?* **[2]**

(b) *Draw a labelled ray diagram showing the formation of an image by a* ***refracting telescope*** *when the final image lies at infinity.* **[3]**

SECTION–C

(Answer any ***three*** *questions)*

Question 12.

(a) *Monochromatic light of wavelength 198 nm is incident on the surface of a metallic cathode whose work function is 2·5 eV. How much potential difference must be applied between the cathode and the anode of a photocell to* ***just stop*** *the photo current from flowing ?* **[3]**

(b) (i) *What is* ***de Broglie*** *hypothesis ?* **[2]**

(ii) *What conclusion can be drawn from* ***Davisson and Germer's*** *experiment ?*

Question 13.

(a) (i) *How are various lines of* ***Lyman series*** *formed? Explain on the basis of* ***Bohr's*** *theory.* **[3]**

(ii) *Calculate the* ***shortest*** *wavelength of electromagnetic radiation present in* ***Balmer series*** *of hydrogen spectrum.*

(b) *State the effect of the following changes on the X-rays emitted by Coolidge X-ray tube :* **[2]**

(i) *High voltage between cathode and anode is increased.*

(ii) *Filament temperature is increased.*

Question 14.

(a) ***Half life*** *of a certain* ***radioactive material*** *is 8 hours.* **[3]**

(i) *Find* ***disintegration constant*** *of this material.*

(ii) *If one starts with 600 g of this substance, how much of it will* ***disintegrate*** *in one day ?*

(b) *Sketch a graph showing the variation of* ***binding energy per nucleon*** *of a nucleus with its* ***mass number.*** **[2]**

Question 15.

(a) *Draw a circuit diagram for the* ***common emitter transistor amplifier.*** *What is meant by* ***phase reversal ?*** **[3]**

(b) *Write the* ***truth table*** *of the following circuit. Name the* ***gate*** *represented by this circuit.* **[2]**

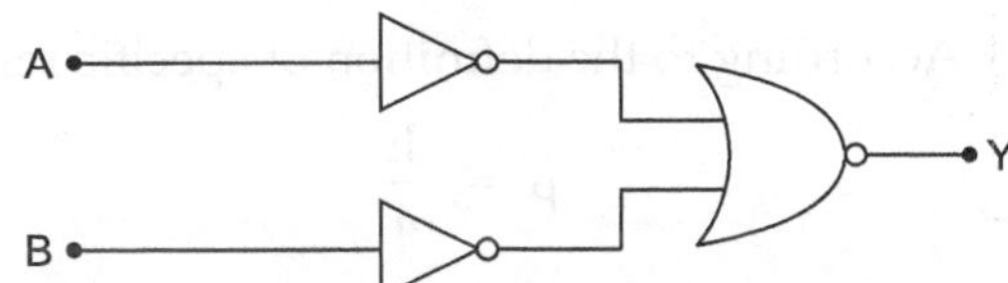

SOLUTIONS

PART–I (20 Marks)

Answer 1.

(A) (i) (a) Since the points M and N are at equal distances from point Q, so these will have same potential.

$$V_M = V_N = \frac{1}{4\pi\varepsilon_0}\frac{q}{r}$$

$$\therefore \quad V_M - V_N = 0$$

The work done in this process

$$W = (V_M - V_N)q = 0.$$

(ii) (c) The resistance of the resistor with Red, Blue, Brown and Gold bands is 260 Ω ± 5%.

(iv) (d) Given, optical powers of lens

$$P_1 = -10\text{ D}$$

and

$$P_2 = +6\text{ D}$$

When these are kept in contact to each other,

Power of combination

$$(P) = P_1 + P_2 = -10 + 6$$

$$P = -4\text{ D}$$

Focal length (in metre)

$$= \frac{1}{P\text{ (in dioptre)}}$$

$$= \frac{1}{4} = -0.25\text{ m} \qquad \textbf{Ans.}$$

(v) **(c)** Total energy in the excited state

$$= \frac{-13.6}{n^2}$$

For first excited state, $n = 2$

$$\therefore \quad \text{Total energy} = \frac{-13.6}{2^2} = \frac{-13.6}{4}$$

$$= -3.4 \text{ eV.}$$ **Ans.**

(B) **(i)** Given : Charge on drop, $q = ?$

Weight, $W = 1.6 \times 10^{-15}$ N

Intensity of electric field, $E = 2\times10^3$ NC^{-1}

When charged oil drop is suspended in uniform electric field then its weight will act in vertically downward direction and force due to uniform electric field will act in upward direction.

$$\therefore \quad qE = mg$$

$$q = \frac{mg}{E}$$

$$= \frac{1.6\times10^{-15}}{2\times10^3}$$

$$= 0.8 \times 10^{-18} \text{ C}$$

or $$q = 8 \times 10^{-19} \text{ C}$$ **Ans.**

(ii) For metallic conductor, if specific resistance is ρ, then conductivity, $\sigma = \frac{1}{\rho}$

According to the definition of specific resistance

$$\rho = \frac{E}{J}$$

where, E is the intensity of electric field at a point inside the conductor and J is the current density.

$$\therefore \quad \sigma = \frac{J}{E}$$

or $$\vec{J} = \sigma\vec{E}$$

(iii) Given circuit can be re-drawn as shown below:

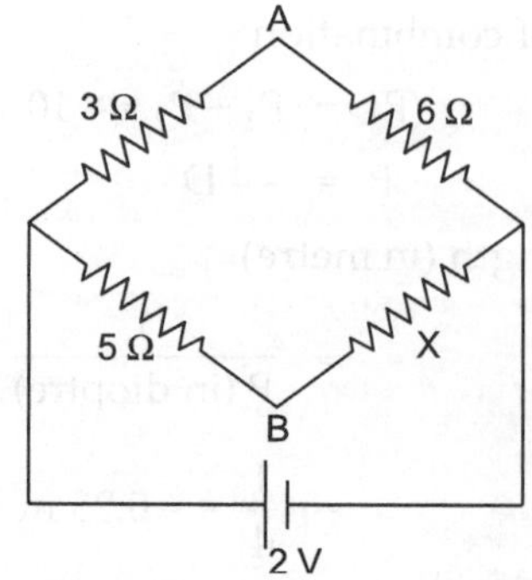

Points A and B are at same potential, so given circuit will behave as a balanced Wheatstone's bridge

$$\therefore \quad \frac{3}{6} = \frac{5}{X}$$

$$\Rightarrow \quad X = 10\,\Omega$$ **Ans.**

(iv) When a particle having charge $+q$, moves with a velocity v, perpendicular to the direction of magnetic field $\vec{B}$, Lorentz force is given by

$$f = qvB$$

The direction of Lorentz force can be calculated on the basis of Fleming's left hand rule. If charged particle is moving in a magnetic field making an angle θ, force F is given by:

$$F = qvB \sin\theta$$

or $$\vec{F} = q(\vec{v}\times\vec{B})$$

(v) Given:

Self-inductance, L = 0.05 Henry

Rate of change of current, $\frac{di}{dt} = 100$ As^{-1}

Magnitude of e.m.f., $e = ?$

We know that,

$$L = \frac{e}{di/dt}$$

$$0.05 = \frac{e}{100}$$

$$e = 5 \text{ V}$$ **Ans.**

(vi) **(a)** Ultraviolet region

(b) Infrared region

(vii) The differences between polarised and unpolarised light are:

1. Polarised light has electric fields oscillating in one direction, whereas, unpolarised light has electric fields oscillating in all directions.
2. Polarised light is coherent in nature whereas unpolarised light is incoherent in nature.
3. Intensity of polarised light depends on the nature of polaroid used whereas, intensity of unpolarised light depends on the nature of source.

(viii) Optical fibres works on the principle of total internal reflection. When light ray strikes at the internal surface of optical fibre cable such that incident angle is greater than critical angle, then incident light ray reflects in the same medium and this phenomenon repeats. In this way, light signal travels from one end of the cable to another end.

(ix) Given:

$n_v = 1.56$
$n_r = 1.54$
$n_y = 1.55$
$\omega = ?$

$$\omega = \frac{n_v - n_r}{n_y - 1} = \frac{1.56 - 1.54}{1.55 - 1} = \frac{0.02}{0.55}$$

$$= \frac{2}{55} = 0.036$$ **Ans.**

(xi) We know that,

$$E = h\nu = \frac{hc}{\lambda}$$

Thus, λ or threshold wavelength is inversely proportional to the energy or the work function. So, metal B which has higher work function of 6 eV will have smaller threshold wavelength.

(xii) We know that,

Angular momentum of an electron in the nth Bohr orbit of hydrogen atom

$$p = \frac{nh}{2\pi}$$

For, $n = 3$

$$p = \frac{3h}{2\pi} = \frac{1.5h}{\pi}$$

$$= 3.15 \times 10^{-34} \text{ Js}$$ **Ans.**

(xiii) The moderator is a material which slows down the speed of the neutrons. In nuclear chain reactions, generally heavy water (D_2O) is used as moderator but some reactors also uses graphite. By controlling the amount of moderator material in the reactor you can maintain the reaction in a controlled state, *i.e.,* the nuclear reaction is sustained, producing heat, but not leading to an uncontrolled explosion.

(xiv) In nature, Sun generates its energy by nuclear fusion of hydrogen nuclei into helium. Fusion of hydrogen atoms at Sun occurs at a solar core at the temperature of 14 million Kelvin.

(xv) **Uses of Zener diode:** Used as voltage references and as shunt regulators to regulate the voltage across small circuits.

PART–II (50 Marks)

SECTION–A

Answer 2.

(a)

400 μC				100 μC
A	E_2	O	E_1	B

$Q_1 = 400\ \mu C = 400 \times 10^{-6}$ C

$Q_2 = 100\ \mu C = 100 \times 10^{-6}$ C

$r = 60$ cm

Consider the point charges kept at point A and point B. If 'O' is the midpoint, where we need to calculate the electric field due to these charges.

At point O, electric field due to point charge kept at A,

$$E_1 = \frac{1}{4\pi\varepsilon_0} \times \frac{q}{r^2}$$

$$= 9 \times 10^9 \times \frac{400 \times 10^{-6}}{(30 \times 10^{-2})^2}$$

At point O, electric field due to point charge kept at B,

$$E_2 = \frac{1}{4\pi\varepsilon_0} \frac{q}{r^2}$$

$$= 9 \times 10^9 \times \frac{100 \times 10^{-6}}{(30 \times 10^{-2})^2}$$

So, resultant electric field $\vec{E} = \vec{E}_1 - \vec{E}_2$

$$\vec{E} = \frac{9 \times 10^9}{(30 \times 10^{-2})^2} \times 10^{-6}(400 - 100)$$

$$= \frac{9 \times 10^9 \times 10^{-6} \times 300}{900 \times 10^{-4}}$$

$$= 3 \times 10^7 \text{ N/C.}$$

(b) (i) **Gauss' law:** According to this law, the total flux linked with a closed surface called Gaussian surface, (the surface need not be a real physical surface, it can also be an hypothetical one) is $\left(\frac{1}{\varepsilon_0}\right)$ times the charge enclosed by the closed surface *i.e.,*

$$\phi = \oint_S \vec{E}.\vec{dA} = \frac{1}{\varepsilon_0}(Q_{enc})$$

(ii) When a point charge is kept at equatorial line of an electric dipole, its electric potential is zero.

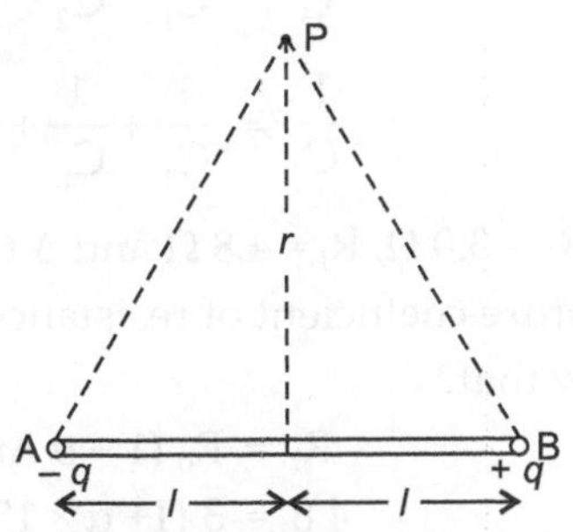

Potential at point P due to charge $+q$,

$$V_1 = \frac{1}{4\pi\varepsilon_0} \frac{q}{BP}$$

Potential at point P due to charge $-q$,

$$V_2 = \frac{1}{4\pi\varepsilon_0}\frac{(-q)}{AP}$$

Resultant electric potential at point P,

$$V = V_1 + V_2$$

$$V = \frac{1}{4\pi\varepsilon_0}\frac{q}{BP} - \frac{1}{4\pi\varepsilon_0}\frac{q}{AP}$$

$$\therefore \quad V = 0 \qquad [\because BP = AP]$$

Answer 3.

(a) Consider three capacitors having capacitance C_1, C_2 and C_3 are connected in series to each other as shown in the figure below.

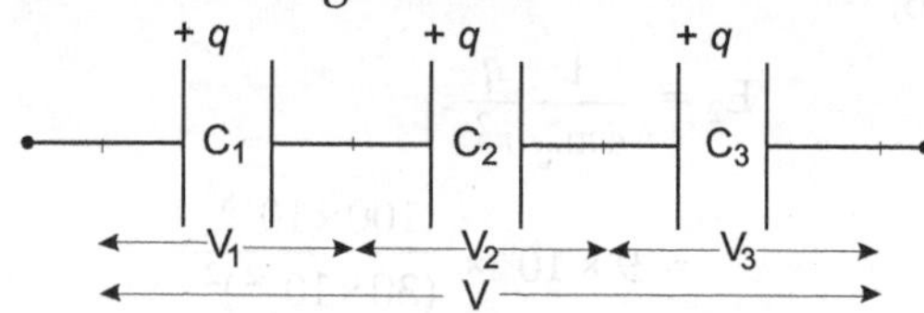

Consider that + q charge is given to first plate of first capacitor. By induction a charge $-q$ will produce at internal surface of second plate of this capacitor and so on.

Consider that potential difference between the plates of capacitor be V_1, V_2 and V_3 respectively, so

$$V_1 = \frac{q}{C_1},\ V_2 = \frac{q}{C_2} \text{ and } V_3 = \frac{q}{C_3}$$

In series, $$V = V_1 + V_2 + V_3$$

$$= \frac{1}{C_1} + \frac{1}{C_2} + \frac{1}{C_3} \qquad ...(1)$$

If equivalent capacitance of combination of capacitors is C then

$$V = \frac{q}{C} \qquad ...(2)$$

By equations (1) and (2), we get

$$\frac{q}{C} = \frac{q}{C_1} + \frac{q}{C_2} + \frac{q}{C_3}$$

$$\frac{1}{C} = \frac{1}{C_1} + \frac{1}{C_2} + \frac{1}{C_3}$$

(b) Given, $R_0 = 3.0\ \Omega$, $R_t = 4.8\ \Omega$ and $\Delta t = 150°C$.

Temperature coefficient of resistance (α) = ?

We know that,

$$R_t = R_0(1 + \alpha\,\Delta t)$$

$$4.8 = 3(1 + \alpha \times 150)$$

$$\therefore \quad \alpha = \frac{4.8 - 3}{3 \times 150}$$

$$= 0.004 \text{ per °C} \qquad \textbf{Ans.}$$

Answer 4.

(a)

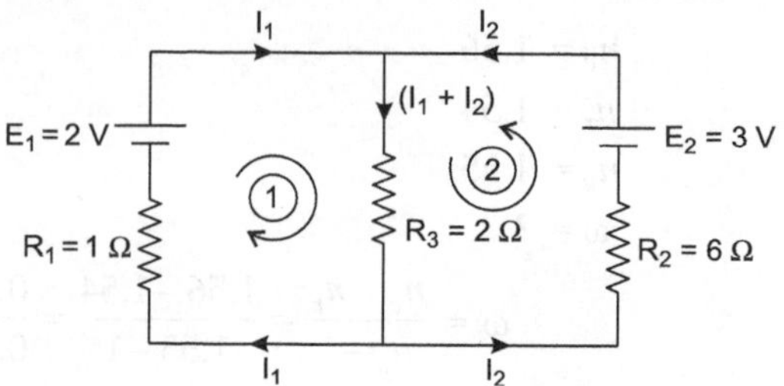

Consider the direction of currents as shown in the figure above.

By applying Kirchhoff's second law in mesh (1), we get

$$R_3(I_1 + I_2) + R_1I_1 - E_1 = 0$$

$$2(I_1 + I_2) + 1 \times I_1 = 2$$

$$\Rightarrow \quad 3I_1 + 2I_2 = 2 \qquad ...(1)$$

By applying Kirhchoff's second law in mesh (2), we get

$$R_3(I_1 + I_2) + R_2I_2 - E_2 = 0$$

$$2(I_1 + I_2) + 6I_2 = 3$$

$$\Rightarrow \quad 2I_1 + 8I_2 = 3 \qquad ...(2)$$

On solving equations (1) and (2), we get

$$I_1 = \frac{1}{2}\text{ A and } I_2 = \frac{1}{4}\text{ A}$$

(b) A galvanometer can be converted into an ammeter by connecting a low resistance (called shunt) in parallel to the galvanometer. The value of shunt is so adjusted that most of the current passes through it. In this way, a galvanometer is converted into ammeter and can measure heavy currents.

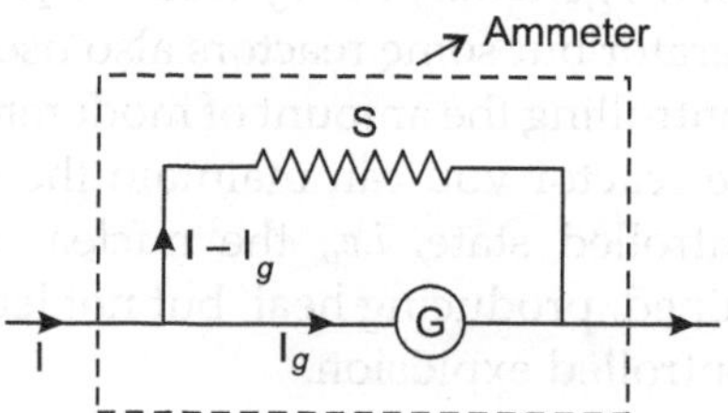

Answer 5.

(a) Refer to ISC Paper 2019 Answer 15 (b).

(b) **(i)** **Curie temperature:** The temperature at which a ferromagnetic material behaves like a paramagnetic material is defined as Curie temperature (T_C).

Or

The minimum temperature at which a ferromagnetic substance is converted into paramagnetic substance is defined as Curie temperature.

This temperature is named after the French physicist Pierre Curie, who in 1985 discovered the laws that relate some magnetic properties to change in temperature.

(ii) Given : Magnetic susceptibility, $\chi_m = 0.0001$

Relative permeability, $\mu_r = ?$

We know that,

$$\mu_r = 1 + \chi_m$$
$$= 1 + 0.0001$$
$$= 1.0001 \quad \textbf{Ans.}$$

Answer 6.

(a) (i) Consider that magnetic field density at point P is zero due to current carrying conductors.

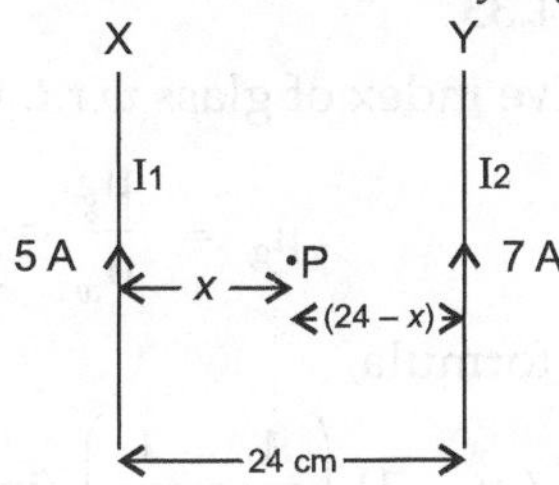

According to right hand palm rule number 1, direction of magnetic field will be perpendicular to the plane of the paper.

1. Downward due to current flowing in conductor X.

2. Upward due to current flowing in conductor Y.

Since at point P, B = 0

$$\therefore \quad B_1 = B_2$$

$$\frac{\mu_0 I_1}{2\pi x} = \frac{\mu_0}{2\pi} \frac{I_2}{(24 - x)}$$

$$\frac{5}{x} = \frac{7}{24 - x}$$

$$5(24 - x) = 7x$$

$$5 \times 24 = 12x$$

$\Rightarrow x = 10$ cm from conductor X.

(ii) It will lie to the left of X (Neutral point always lies near a weaker current).

(b) (i) **Ampere:** One ampere is the current if it is maintained in two straight parallel conductors of infinite length and negligible circular cross-section, placed one metre apart in vacuum, produces a attractive or repulsive force equal to 2×10^{-7} N/m of their length.

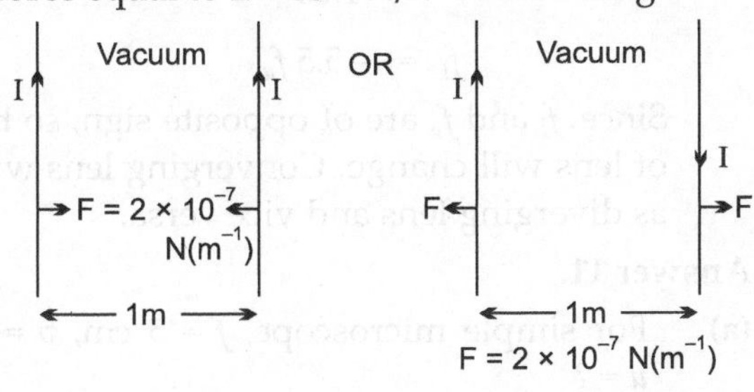

(ii) **Ideal transformer:** A transformer which does not have any energy loss in it, means no core losses, copper losses and any other losses, is called ideal transformer. For ideal transformer $P_{out} = P_{in}$, so its efficiency is $\eta = 100\%$ (But practically, efficiency of transformer lies between 70%–90%). Ideal transformer is a theoretical concept (imaginary transformer), practically it is not possible.

Answer 7.

(a) Given :Self inductance, L = 0.7 H

Resistance, R = 165 Ω

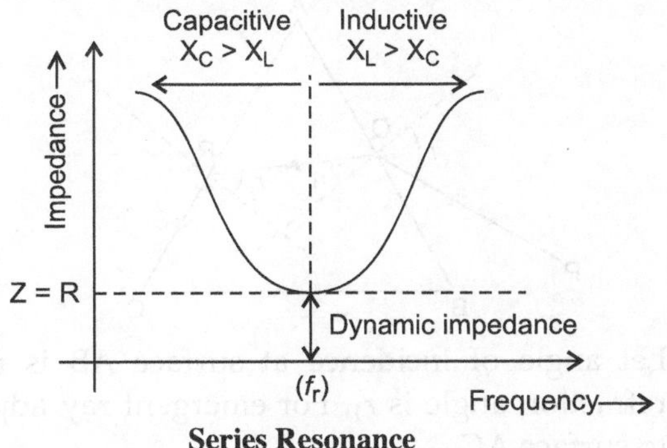

Supply voltage, V = 275 V,

f = 50 Hz

(i) Reactance of coil, $X_L = \omega L$

$$= 2\pi f L \quad [\because \omega = 2\pi f]$$

$$= 2 \times \frac{22}{7} \times 50 \times 0.7$$

$$= 220\ \Omega \quad \textbf{Ans.}$$

(ii) Impedance of the coil, $Z = \sqrt{X_L^2 + R^2}$

$$= \sqrt{220^2 + 165^2}$$

$$= \sqrt{48400 + 27225}$$

$$= 275\ \Omega \quad \textbf{Ans.}$$

(iii) Current flowing through the coil,

$$i = \frac{V}{Z}$$

$$= \frac{275}{275}$$

$$i = 1\text{ A.} \quad \textbf{Ans.}$$

(b) Graph showing variation of impedance of a series LCR circuit with frequency of the a.c. supply:

Series Resonance

SECTION–B

Answer 8.

(a) Refer to ISC Paper 2018, Answer 21 (b).

(b) Given : Wavelength, $\lambda = 650$ nm

$= 650 \times 10^{-9}$ m

Width of slit, $a = 1.3 \times 10^{-4}$ cm

$= 1.3 \times 10^{-6}$ m

The angular width of central maxima = 2θ = ?

$$\sin\theta = \frac{\lambda}{\omega}$$

$$= \frac{650\times10^{-9}}{1.3\times10^{-6}}$$

$$= 0.5$$

$\Rightarrow \quad \theta = 30°$

$\therefore \quad 2\theta = 2 \times 30° = 60°$ **Ans.**

Answer 9.

(a) Refer to ISC Paper 2019 Answer 21 (b).

(b) Given : Polarising angle, $i_p = 56°$

Angle of refraction, r = ?

We know that,

For polarisation of light at air glass surface

$$i_p + r = 90°$$

$$56° + r = 90°$$

$\Rightarrow \quad r = 34°$ **Ans.**

Answer 10.

(a) Given : ${}_an_g = 1.5$

Since emergent ray grazes the adjacent surface AC

$\therefore \quad A = 60°$

$$\sin c = \frac{1}{{}_an_g}$$

$$\sin c = \frac{1}{1.56} = 0.641$$

$$c = \sin^{-1}(0.641)$$

$$= 39.86° \approx 40°$$

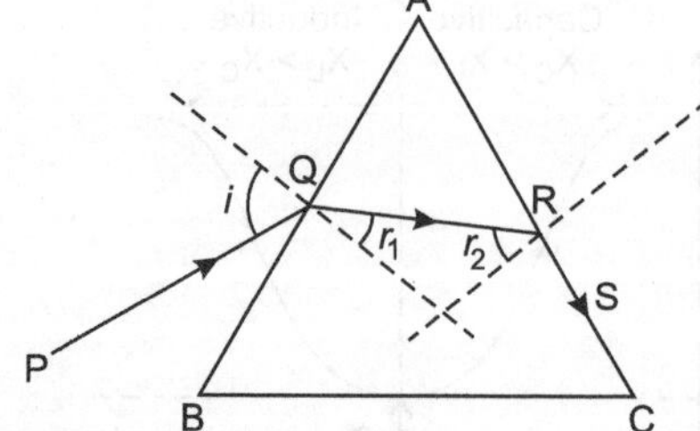

Let angle of incidence at surface AB is i and refraction angle is r_1. For emergent ray adjacent to surface AC,

$$r_2 = c = 40°$$

$\therefore \quad r_1 = A - r_2 = 60° - 40° = 20°$

$${}_an_g = \frac{\sin i}{\sin r_1} = \frac{\sin i}{\sin 20°}$$

$$\sin i = 1.56 \times 0.342 = 0.5335$$

$$i = 32.24°$$ **Ans.**

(b) Given : Refractive index of glass lens, $\mu_g = 1.5$

(i) When it is dipped in water of refractive index, $\mu_w = 1.33$

Refractive index of glass *w.r.t.* water,

$${}_w\mu_g = \frac{\mu_g}{\mu_w} = \frac{1.5}{1.33}$$

By lens formula,

$$\frac{1}{f_a} = ({}_a\mu_g - 1)\left(\frac{1}{R_1} - \frac{1}{R_2}\right) \text{ (in air)} \quad ...(1)$$

$$\frac{1}{f_w} = ({}_w\mu_g - 1)\left(\frac{1}{R_1} - \frac{1}{R_2}\right) \text{ (in water)} \quad ...(2)$$

By equations (1) and (2),

$$\frac{f_w}{f_a} = \frac{{}_a\mu_g - 1}{{}_w\mu_g - 1}$$

$$= \frac{1.5-1}{\frac{1.5}{1.33}-1}$$

$$= \frac{0.5\times1.33}{0.17} = 3.91$$

$\Rightarrow \quad f_w = 3.91 f_a$

It means that when lens is dipped in water its focal length increases.

(ii) For liquid of refractive index 1.65.

$$\frac{1}{f_l} = ({}_l\mu_g - 1)\left(\frac{1}{R_1} - \frac{1}{R_2}\right) \quad ...(3)$$

By equations (1) and (3)

$$\frac{f_l}{f_a} = \frac{({}_a\mu_g - 1)}{({}_l\mu_g - 1)}$$

$$= \frac{1.5-1}{\frac{1.5}{1.65}-1} = \frac{-0.5\times1.65}{0.15}$$

$$f_l = -5.5 f_a$$

Since, f_l and f_a are of opposite sign, so behaviour of lens will change. Converging lens will behave as diverging lens and vice versa.

Answer 11.

(a) For simple microscope, f = 5 cm, v = – 25 cm, u = ?

By lens formula:

$$\frac{1}{f} = \frac{1}{v} - \frac{1}{u}$$

$$\frac{1}{5} = \frac{1}{-25} - \frac{1}{u}$$

$$\Rightarrow \quad -\frac{1}{u} = \frac{1}{25} + \frac{1}{5}$$

$$-\frac{1}{u} = \frac{1+5}{25}$$

$$\Rightarrow \quad u = \frac{25}{6} = -4.17 \text{ cm}$$ **Ans.**

So, object should be placed at a distance of 4.17 cm, in front of the lens.

(b)

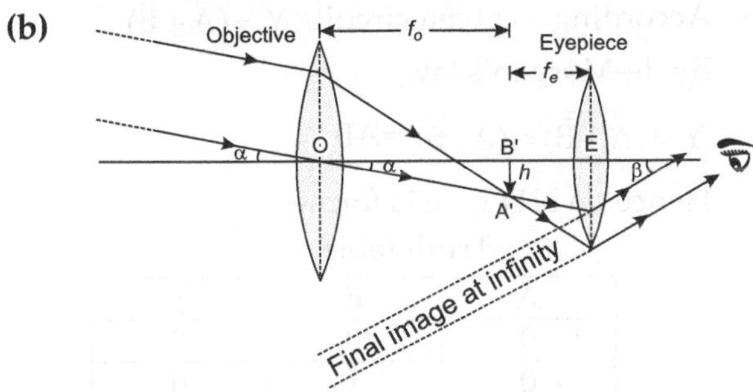

Magnifying power of telescope, $M = \frac{-f_o}{f_e}$

SECTION–C

Answer 12.

(a) Given:

Wavelength of incident light,

$$\lambda = 198 \text{ nm} = 198 \times 10^{-9} \text{ m}$$
$$= 1980 \text{ Å}$$

Work function, ϕ = 2.5 eV

Potential difference, V_0 = ?

We know that,

$$eV_0 = \frac{hc}{\lambda} - \phi_0$$

$$eV_0 = \frac{12375}{\lambda(\text{in Å})} - \phi_0 = \frac{12375}{1980} - 2.5$$

$$= 6.25 - 2.5 = 3.75 \text{ eV}$$

$$1.6 \times 10^{-19} \times V = 3.75 \times 1.6 \times 10^{-19}$$

$$\therefore \quad V = 3.75 \text{ volt.}$$ **Ans.**

(b) Refer to ISC Paper 2020 Answer 9.

Answer 13.

(a) (i) Bohr explained the various lines of spectrums of hydrogen atom theoretically.

Lyman series: Bohr postulated that Lyman series is obtained when an electron jumps to the first orbit ($n_1 = 1$) from any outer orbit (n_2 = 2, 3, 4, ...).

The wave numbers of spectral lines of Lyman series are calculated as:

$$\bar{v} = \frac{1}{\lambda}$$

$$= R\left[\frac{1}{1^2} - \frac{1}{n^2}\right] \text{ where } n = 2, 3, 4......$$

The values of $\bar{v}$ lie in the ultraviolet region of the electromagnetic spectrum.

(ii) For Balmer series, $n_1 = 2$

$$\therefore \quad \frac{1}{\lambda} = R\left[\frac{1}{2^2} - \frac{1}{n^2}\right]$$

Here $n = \infty$ for shortest wavelength

$$\therefore \quad \frac{1}{\lambda} = R\left[\frac{1}{2^2} - \frac{1}{\infty^2}\right] = \frac{R}{4} = \frac{1.097 \times 10^7}{4}$$

$$\Rightarrow \quad \lambda = \frac{4}{1.097 \times 10^7} \text{ m}$$

$$\lambda = 3.646 \times 10^{-7} \text{ m} = 3646 \text{ Å}$$ **Ans.**

(b) (i) When high voltage between cathode and anode is increased, the resultant X-rays' penetration power is increased.

(ii) On increasing the temperature, the number of electrons emitted by the cathode increases and as a result, the intensity of the X-rays increases.

Answer 14.

(a) (i) Given:

Half life of radioactive material, $t_{1/2}$ = 8 hours

Disintegration constant, λ = ?

We know that, For a radioactive material

$$t_{1/2} = \frac{0.6932}{\lambda}$$

$$\Rightarrow \quad \lambda = \frac{0.6932}{8}$$

$$= 0.08665 \text{ per hour}$$ **Ans.**

(ii) Given : N_0 = 600 g

T = 1 day = 24 hours

$T = t_{1/2} \times n$

$24 = 8 \times n$

$\Rightarrow \quad n = 3$

If after n half lives, N gram is left then,

$$N = N_0 \left(\frac{1}{2}\right)^n = 600 \left(\frac{1}{2}\right)^3$$

$$= \frac{600}{8} = 75 \text{ g}$$

$\therefore$ Amount of material disintegrated in one day $= 600 - 75 = 525$ g. **Ans.**

(b)

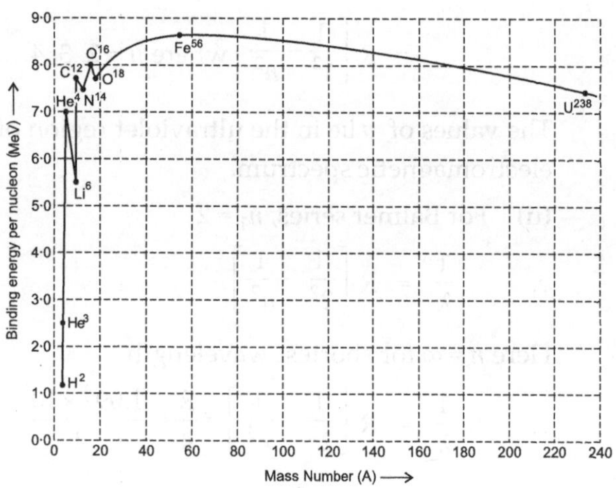

Answer 15.

(a) *p-n-p* transistor as a Common-Emitter Amplifier :

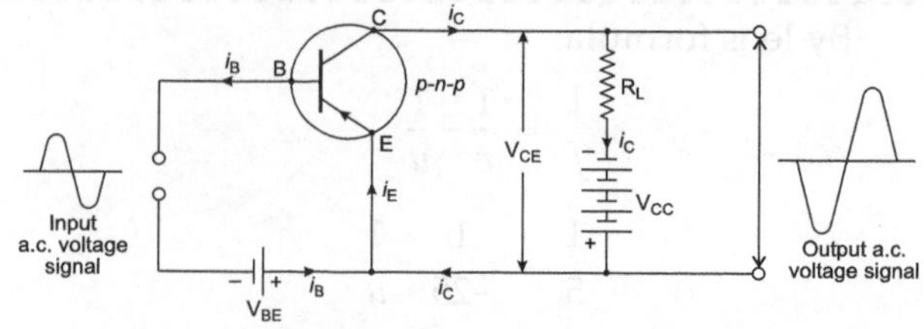

Phase reversal: The phase difference of 180° between the input voltage and output voltage in a common emitter amplifier is known as phase reversal.

(b) The given circuit is

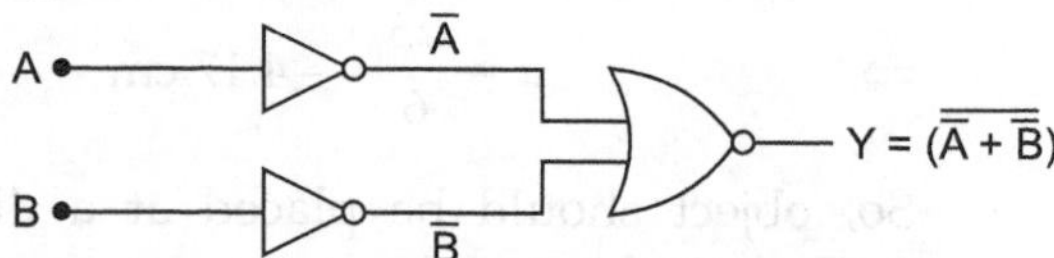

According to given circuit, $Y = (\overline{\overline{A} + \overline{B}})$

By de-Morgan's law,

$$Y = (\overline{\overline{A} + \overline{B}}) = (\overline{\overline{A}}.\overline{\overline{B}}) = AB$$

Hence, AND gate is formed

Truth table

A	B	Y
0	0	0
0	1	0
1	0	0
1	1	1

Useful Constants and Relations:

1.	Charge of a proton	(e)	$= 1.6 \times 10^{-19}$ C
2.	Planck's constant	(h)	$= 6.6 \times 10^{-34}$ Js
3.	Mass of electron	(m)	$= 9.1 \times 10^{-31}$ kg
4.	Permittivity of vacuum	(ε_0)	$= 8.85 \times 10^{-12}$ Fm^{-1}
5.	Constant for Coulomb's law	$\left(\frac{1}{4\pi\varepsilon_0}\right)$	$= 9 \times 10^9$ mF^{-1}
6.	Permeability of vacuum	(μ_0)	$= 4\pi \times 10^{-7}$ Hm^{-1}
7.		$\left(\frac{\mu_0}{4\pi}\right)$	$= 10^{-7}$ Hm^{-1}
8.	Rydberg's constant	(R)	$= 1.097 \times 10^7$ m^{-1}
9.	Speed of light in vacuum	(c)	$= 3 \times 10^8$ ms^{-1}

●●

PHYSICS 2015

QUESTIONS

PART–I (20 Marks)

*(Answer **all** questions)*

Question 1.

(A) *Choose the correct alternative (a), (b), (c) or (d) for each of the questions given below:* **[5]**

(i) *A **short** electric dipole (which consists of two point charges, + q and – q) is placed at the centre O and inside a **large** cube (ABCDEFGH) of length L, as shown in figure below. The electric flux, emanating through the cube is:*

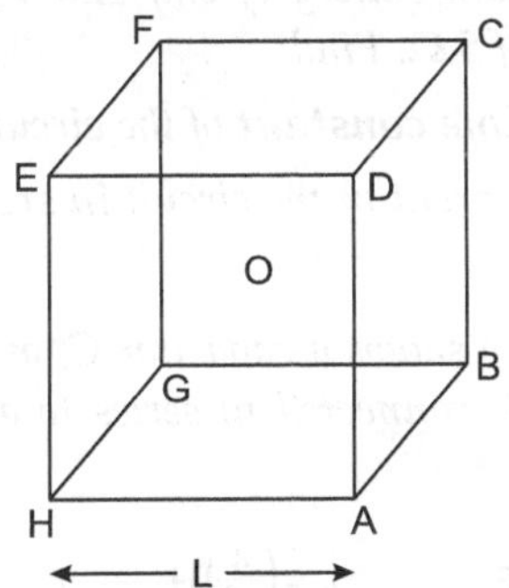

(a) $q/4\pi\varepsilon_0 L$ (b) *Zero*

(c) $q/2\pi\varepsilon_0 L$ (d) $q/3\pi\varepsilon_0 L$

(ii) *The equivalent resistance between points **a** and **f** of the network shown in figure below is:*

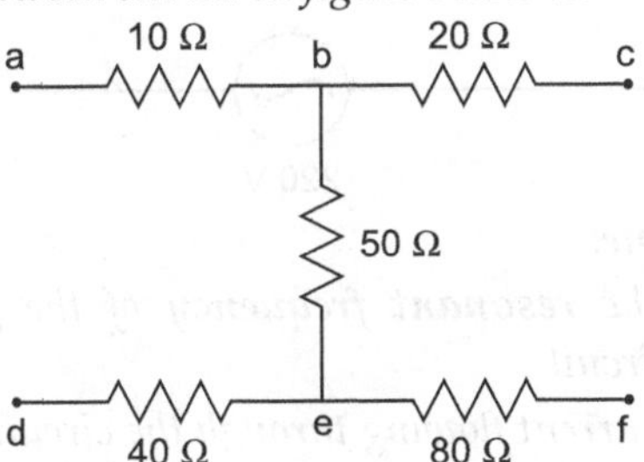

(a) 24 Ω (b) 110 Ω

(c) 140 Ω (d) 200 Ω

(iii) *A moving electron enters a uniform and **perpendicular** magnetic field. Inside the magnetic field, the electron travels along:*

(a) *a straight line* (b) *a parabola*

(c) *a circle* (d) *a hyperbola*

(iv) *A fish which is at a depth of 12 cm in water* $\left(\mu = \frac{4}{3}\right)$ *is viewed by an observer on the bank of a lake. Its **apparent depth** as observed by the observer is:*

(a) *3 cm* (b) *9 cm*

(c) *12 cm* (d) *16 cm*

(v) *If E_p and E_k represent potential energy and kinetic energy respectively, of an orbital electron, then, according to **Bohr's theory**:*

(a) $E_k = -E_p/2$ (b) $E_k = -E_p$

(c) $E_k = -2E_p$ (d) $E_k = 2E_p$

(B) *Answer **all** questions given below **briefly** and to the point:* **[15]**

(i) *What is meant by the term **Quantization of charge?***

(ii) *A resistor **R** is connected to a cell of emf **e** and internal resistance **r**. Potential difference across the resistor **R** is found to be **V**. State the relation between **e, V, R** and **r**.*

(iii) *Three identical cells each of emf 2 V and internal resistance 1 Ω are connected in series to form a battery. The battery is then connected to a parallel combination of two identical resistors, each of resistance 6 Ω. Find the current delivered by the battery.*

(iv) *State how **magnetic susceptibility** is **different** for the three types of magnetic materials, i.e., diamagnetic, paramagnetic and ferromagnetic materials.*

(v) *An emf of 2 V is induced in a coil when current in it is changed from 0 A to 10 A in 0·40 sec. Find the **coefficient of self-inductance** of the coil.*

(vi) *How are electric vector (E), magnetic vector (B) and velocity vector ($\vec{c}$) oriented in an electromagnetic wave ?*

(vii) *State any two methods by which ordinary light can be polarised.*

(viii) *A monochromatic ray of light falls on a **regular prism**. What is the relation between angle of incidence and angle of emergence in the case of **minimum deviation ?***

(ix) *What type of lens is used to correct **long-sightedness ?****

(x) *State any one **advantage** of using a reflecting telescope in place of a refracting telescope.*

(xi) *State **Moseley's law.***

(xii) *Wavelengths of the first lines of the **Lyman series, Paschen series** and **Balmer series**, in hydrogen*

** Answer is not given due to change in the present syllabus.

spectrum are denoted by λ_L, λ_P, and λ_B, respectively. Arrange these wavelengths in increasing order.

(xiii) *What is the **significance** of **binding energy per nucleon** of a nucleus of a radioactive element ?*

(xiv) *Write any one **balanced** equation representing **nuclear fission.***

(xv) *What is the difference between **analog** signal and **digital** signal ?*

PART–II (50 Marks)

SECTION–A

*(Answer **any four** questions)*

Question 2.

(a) *Derive an expression for intensity of electric field at a point in **broadside position** or on an **equatorial line** of an electric dipole.* **[4]**

(b) *Two point charges of 10 C each are kept at a distance of 3 m in vacuum. Calculate their electrostatic potential energy.* **[1]**

Question 3.

(a) *Four capacitors, C_1, C_2, C_3 and C_4 are connected as shown in figure below. Calculate **equivalent** capacitance of the circuit between points X and Y.* **[3]**

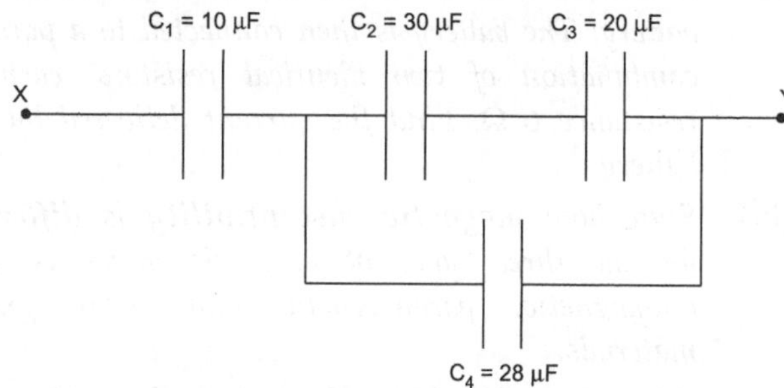

(b) *Draw labelled graphs to show how **electrical resistance** varies with **temperature** for:* **[2]**

(i) *a metallic wire.* **(ii)** *a piece of carbon.*

Question 4.

(a) *Two resistors $R_1 = 400\ \Omega$ and $R_2 = 20\ \Omega$ are connected in **parallel** to a battery. If heating power developed in R_1 is 25 W, find the heating power developed in R_2.* **[2]**

(b) *With the help of a labelled diagram, show that the balancing condition of a **Wheatstone bridge** is:* **[3]**

$$\frac{R_1}{R_2} = \frac{R_3}{R_4},$$ *where the terms have their usual meaning.*

Question 5.

(a) *A 10 m long uniform metallic wire having a resistance of 20 Ω is used as a potentiometer wire. This wire is connected in series with another resistance of 480 Ω and a battery of emf 5 V having negligible internal resistance. If an unknown emf e is balanced across 6 m of the potentiometer wire, calculate:* **[3]**

(i) *the potential gradient across the potentiometer wire.*

(ii) *the value of the unknown emf e.*

(b) **(i)** *Explain the term **hysteresis**.*** **[2]**

(ii) *Name **three elements** of the earth's magnetic field which help in defining **earth's magnetic field** completely.*

Question 6.

(a) *Obtain an expression for magnetic flux density B at the centre of a circular coil of radius R, having N turns and carrying a current I.* **[3]**

(b) *A coil of self inductance 2·5 H and resistance 20 Ω is connected to a battery of emf 120 V having internal resistance of 5 Ω. Find:* **[2]**

(i) *The **time constant** of the circuit.***

(ii) *The current in the circuit in **steady state**.*

Question 7.

(a) *Figure below shows a capacitor C, an inductor L and a resistor R, connected in series to an a.c. supply of 220 V.* **[4]**

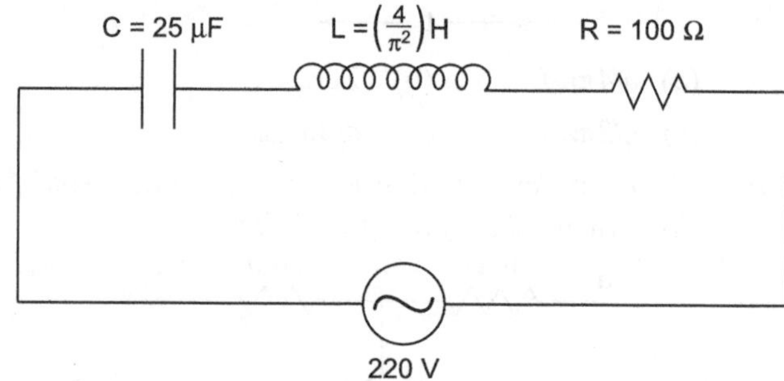

Calculate:

(i) *The **resonant frequency** of the given RLC circuit.*

(ii) *Current flowing through the circuit.*

(iii) *Average power consumed by the circuit.*

(b) *In a series LCR circuit, what is the **phase difference** between V_L and V_C, where V_L is the potential difference across the inductor and V_C is the potential difference across the capacitor ?* **[1]**

SECTION–B

*(Answer any **three** questions)*

Question 8.

(a) *On the basis of **Huygen's wave theory of light**, show that angle of reflection is equal to angle of incidence. You must draw a labelled diagram for this derivation.* **[4]**

(b) *State any one difference between **interference of light** and **diffraction of light**.* **[1]**

** Answer is not given due to change in the present syllabus.

Question 9.

(a) *Laser light of wavelength 630 nm is incident on a pair of slits which are separated by 1·8 mm. If the screen is kept 80 cm away from the two slits, calculate :* **[3]**

(i) *fringe separation i.e., fringe width.*

(ii) *distance of 10th bright fringe from the centre of the interference pattern.*

(b) *Show graphically the intensity distribution in* ***Fraunhofer's single slit diffraction experiment.*** *Label the axes.* **[2]**

Question 10.

(a) *A point object* **O** *is placed at a distance of 15 cm from a convex lens L of focal length 10 cm as shown in figure below. On the other side of the lens, a convex mirror* **M** *is placed such that its distance from the lens is equal to the focal length of the lens. The final image formed by this combination is observed to* ***coincide*** *with the object* **O**. *Find the* ***focal length*** *of the* ***convex mirror.*** **[3]**

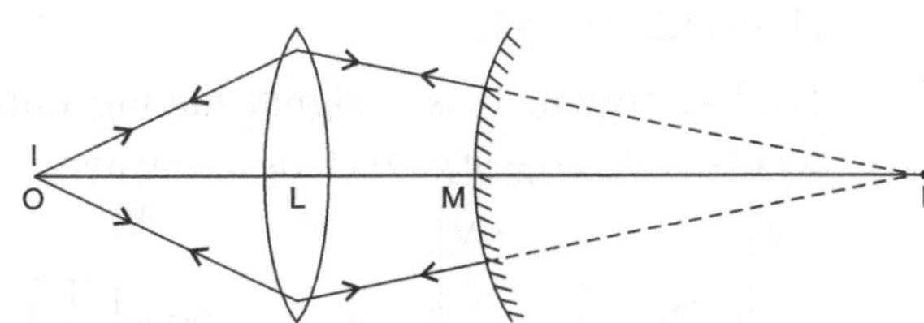

(b) *What is* ***chromatic aberration?*** *How can it be minimized or eliminated ?*** **[2]**

Question 11.

(a) *Draw a labelled ray diagram of an image formed by a* ***compound microscope,*** *when the final image lies at the* ***least distance of distinct vision*** *(D).* **[3]**

(b) *With regard to an* ***astronomical telescope*** *of* ***refracting*** *type, state how you will increase its:* **[2]**

(i) *magnifying power.*

(ii) *resolving power.*

SECTION–C

(Answer any ***three*** *questions)*

Question 12.

(a) *In an experiment of* ***photoelectric effect,*** *the graph of maximum kinetic energy* E_K *of the emitted photoelectrons versus the frequency of the incident light is a straight line* ***AB*** *as shown in figure below:* **[3]**

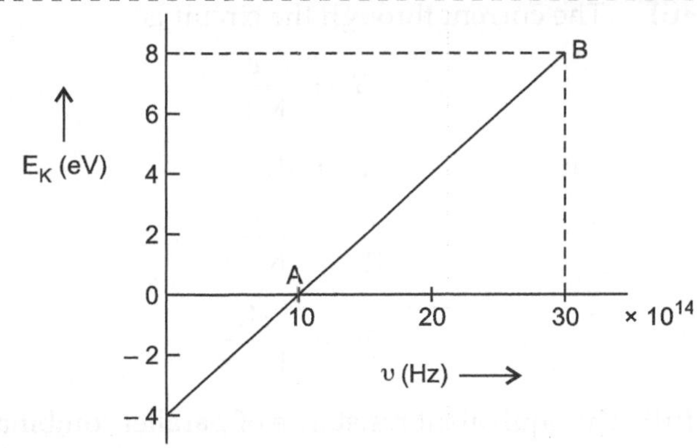

Find:

(i) *Threshold frequency of the metal.*

(ii) *Work function of the metal.*

(iii) *Stopping potential for the photoelectrons emitted by the light of frequency* $\upsilon = 30 \times 10^{14}$ *Hz.*

(b) (i) *State how* ***de-Broglie wavelength*** *(λ) of moving particles varies with their linear momentum (p).* **[2]**

(ii) *State any one phenomenon in which moving particles exhibit wave nature.*

Question 13.

(a) *On the basis of* ***Bohr's*** *theory, derive an expression for the* ***radius*** *of the* $\mathbf{n^{th}}$ *orbit of an electron of hydrogen atom.* **[3]**

(b) *Using the constants given in the end of this paper, find the* ***minimum wavelength*** *of the emitted X rays, when an* ***X-ray tube*** *is operated at 50 kV.* **[2]**

Question 14.

(a) (i) *Define* ***half-life*** *of a radioactive substance.* **[3]**

(ii) *Using the equation* $N = N_0 e^{-\lambda t}$, *obtain the relation between* ***half-life (T)*** *and* ***decay constant*** *(λ) of a radioactive substance.*

(b) *With the help of a suitable* ***example*** *and an* ***equation,*** *explain the term* ***pair production.*** **[2]**

Question 15.

(a) Draw a labelled diagram of a full wave rectifier. Show how output voltage varies with time, if input voltage is a sinusoidal voltage. **[3]**

(b) What is a NAND gate? Write its truth table. **[2]**

ANSWERS

PART–I (20 Marks)

Answer 1.

(A) **(i)** **(b)** Zero **(ii)** **(c)** 140 Ω

(iii) **(c)** a circle **(iv)** **(b)** 9 cm

(v) **(a)** $E_k = -E_p/2$

(B) (i) Quantization of charge is the property by virtue of which all free charges are integral multiple of a basic unit of charge of an electron or proton.

Thus, charge q of a body is given by,

$$q = \pm ne$$

** Answer is not given due to change in the present syllabus.

(ii) The current through the circuit is

$$i = \frac{e}{R+r}$$

But $V = iR$

$$\therefore \quad \frac{V}{R} = \frac{e}{R+r}$$

$$V = \frac{eR}{R+r}$$

(iii) The equivalent resistance of parallel combination of resistors is given by,

$$\frac{1}{R'} = \frac{1}{6}+\frac{1}{6} = \frac{2}{6} = \frac{1}{3}$$

$$R' = 3\,\Omega$$

Total equivalent resistance of the given combination is,

$$R = R' + 3r$$
$$R = 3 + 3\,(1)$$
$$R = 6\,\Omega$$

But $V = IR$

$$\therefore \quad I = \frac{6}{6} = 1\text{ A}$$ **Ans.**

(iv) The magnetic susceptibility is small and negative for diamagnetic material *i.e.*, $\chi_m < 0$, small and positive for paramagnetic material *i.e.*, $\chi_m > 0$, very large and positive for ferromagnetic material *i.e.*, $\chi_m >> 0$.

(v) We have,

$$L = \frac{e}{di/dt}$$

$$= \frac{2}{10} \times 0{\cdot}40$$

$$= 0{\cdot}08\text{ H}$$ **Ans.**

(vi) The electric vector $(\vec{E})$, the magnetic vector $(\vec{B})$ and the velocity vector $(\vec{c})$ are mutually perpendicular to each other.

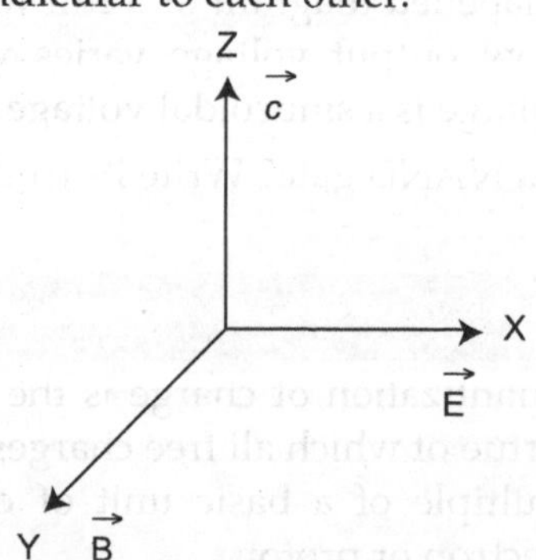

(vii) 1. Polarisation by double refraction.
2. Polarisation by scattering.

(viii) In case of minimum deviation, $\angle i = \angle e$

(x) The advantage is that the image formed is brighter.

(xi) **Moseley's law:** The square root of the frequency of X-rays is directly proportional to the atomic number of the emitting element.

$$\sqrt{\nu} \propto Z$$

where, ν = frequency of X-rays and Z = atomic number.

(xii) $\lambda_L < \lambda_B < \lambda_P$

(xiii) Binding energy per nucleon tells the stability of a nucleus. More the binding energy more is the stability of nucleus.

(xiv) $_{92}U^{235} + {}_0n^1 \rightarrow {}_{92}U^{236} \rightarrow {}_{56}Ba^{141} + {}_{36}Kr^{92} + 3\,{}_0n^1$ + Energy

(xv) **Analog signal:** The voltage continuously varies between 0 and a maximum value (*e.g.*, sinusoidal wave).

Digital signal: It is a signal having only two levels of voltage (*e.g.*, rectangular wave).

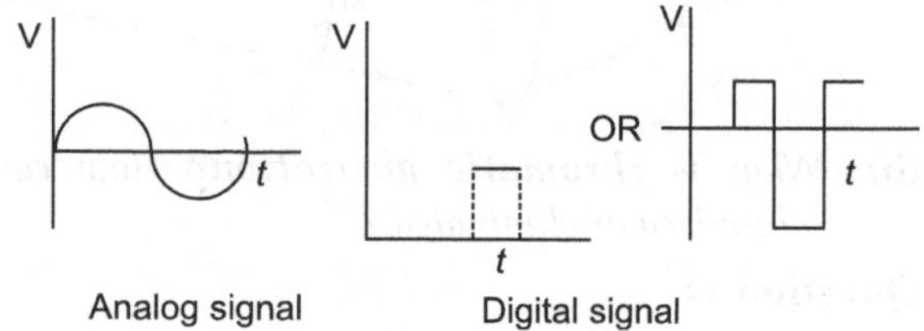

PART–II (50 Marks)

SECTION–A

Answer 2.

(a) Refer to ISC Paper 2020 Answer 13.

(b) We have,

$$U = \frac{1}{4\pi\varepsilon_0}\frac{q_1 q_2}{r}$$

$$= \frac{9\times10^9\times10\times10}{3}$$

$$= 3\times10^{11}\text{ J}$$ **Ans.**

Answer 3.

(a) In the given figure, C_2 and C_3 are connected in series.

$$\therefore \quad \frac{1}{C'} = \frac{1}{30}+\frac{1}{20}$$

$$= \frac{2+3}{60}$$

$$= \frac{5}{60} = \frac{1}{12}$$

$\Rightarrow \quad C' = 12\ \mu F$

Now, C′ and C_4 are connected in parallel.

$C'' = 12 + 28 = 40\ \mu F$

Now, C′ and C_1 are connected in series.

∴ The total equivalent capacitance is given by

$$\frac{1}{C} = \frac{1}{10} + \frac{1}{40}$$

$$= \frac{4+1}{40}$$

$$= \frac{5}{40} = \frac{1}{8}$$

$C = 8\ \mu F$ **Ans.**

(b) (i)

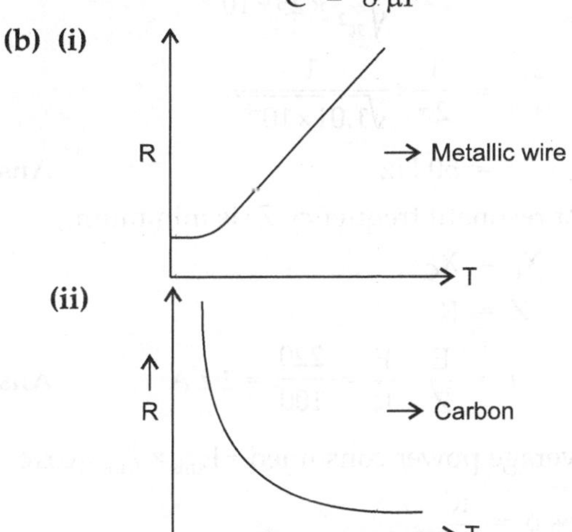

Answer 4.

(a) Given : $R_1 = 400\ \Omega$ and $R_2 = 20\ \Omega$.

Heating power developed in $R_1 = 25$ W

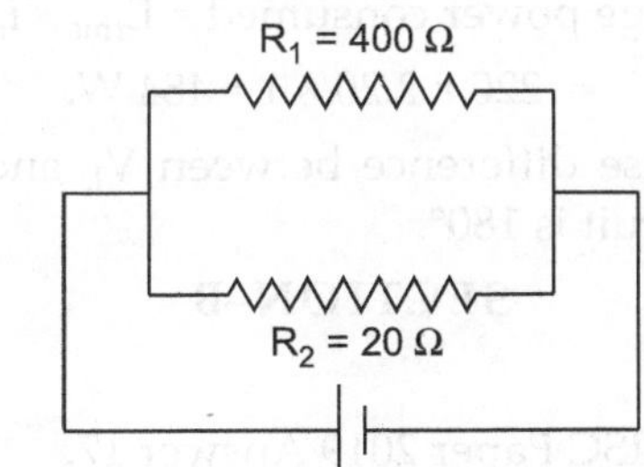

We have,

$P_1 = i_1^2 R_1$

$25 = i_1^2 \times 400$

$i_1 = 0{\cdot}25$ A

$V_1 = i_1 R_1$

$= 0{\cdot}25 \times 400$

$= 100$ V

$V_2 = i_2 R_2$

$100 = i_2 \times 20$

[∵ $V_1 = V_2$ in parallel combination]

$i_2 = 5$ A

$P_2 = i_2^2 R_2$

$= 5^2 \times 20 = 500$ W. **Ans.**

(b) Refer to ISC Paper 2019 Answer 15 (a).

Answer 5.

(a) Given, wire is 10 m long.

Resistance of wire, $R_P = 20\ \Omega$

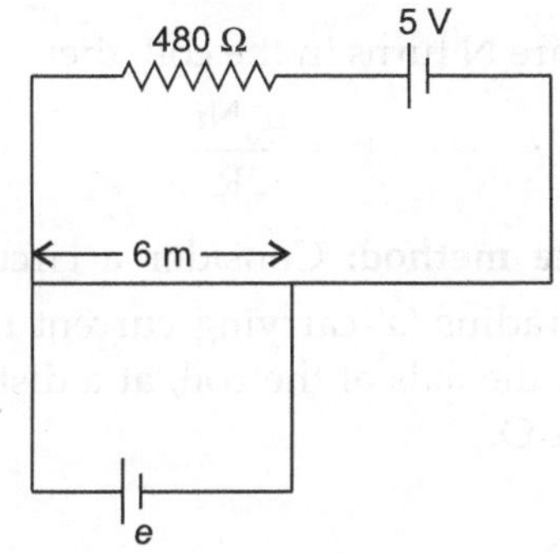

Series resistance $= 480\ \Omega$

$E = 5$ V

Equivalent resistance in series,

$R = 480 + 20 = 500\ \Omega$

We know that, $V = IR$

$5 = I \times 500$

$I = 0{\cdot}01$ A

Potential difference across potentiometer wire is

$V_P = IR_P = 0{\cdot}01 \times 20 = 0{\cdot}2$ V

(i) $K = \dfrac{V_P}{l} = \dfrac{0.2}{10}$

$= 0{\cdot}02$ V/m **Ans.**

(ii) $e = Kl$

$= 0{\cdot}02 \times 6$

$= 0{\cdot}12$ V **Ans.**

(b) (ii) **1.** Angle of declination.

2. Angle of dip.

3. Horizontal component of earth's magnetic field.

Answer 6.

(a) Magnetic field at O due to small element δl,

$$\delta B = \frac{\mu_0}{4\pi} \frac{i \delta l \sin\theta}{R^2}$$

δl R O

But $\theta = 90°$

$$\therefore \quad \delta B = \frac{\mu_0}{4\pi} \frac{i \delta l}{R^2}$$

$\vec{B}$ at O due to the whole circular coil is given by

$$B = \int dB = \int \frac{\mu_0}{4\pi}\frac{idl}{R^2}$$

$$= \frac{\mu_0}{4\pi}\frac{i}{R^2}.2\pi R = \frac{\mu_0 i}{2R}$$

If there are N turns in the coil, then

$$B = \frac{\mu_0 N i}{2R}$$

Alternate method: Consider a circular loop of wire of radius '*a*' carrying current *i*. Let P be a point on the axis of the coil, at a distance *x* from its centre O.

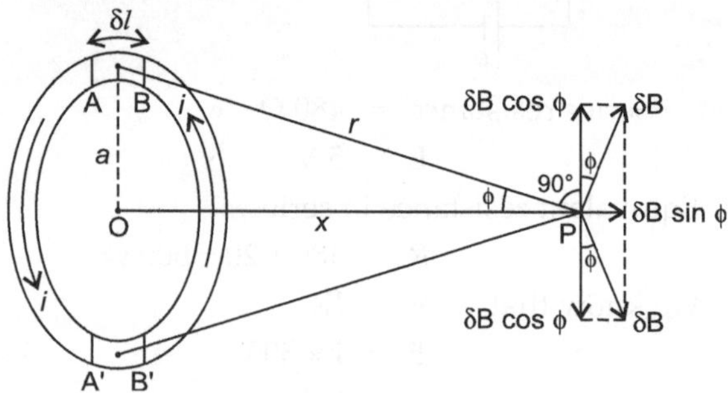

The resultant magnetic field at P is given by

$$B = \int dB \sin\phi$$

But $\sin\phi = \frac{a}{r}$

$$B = \int \frac{\mu_0}{4\pi}\frac{i}{r^2}\times\frac{a}{r}dl = \int \frac{\mu_0}{4\pi}\frac{ia}{r^3}dl$$

Now, since the terms $\mu_0/4\pi$, *i* and *a* are constants and *r* has the same value for all the elements of loop. Thus, the term $\mu_0 ia/4\pi r^3$ can be factored out of integration.

$$\therefore \quad B = \frac{\mu_0}{4\pi}\frac{ia}{r^3}\int dl$$

But $\int dl = 2\pi a$ (circumference of the loop) and $r = (a^2+x^2)^{1/2}$.

$$\therefore \quad B = \frac{\mu_0}{4\pi}\frac{2\pi ia^2}{(a^2+x^2)^{3/2}}$$

$$= \frac{\mu_0 ia^2}{2(a^2+x^2)^{3/2}}$$

If it is a coil of N turns, then each turn will contribute equally to B. Thus,

$$B = \frac{\mu_0 Nia^2}{2(a^2+x^2)^{3/2}}$$

The direction of magnetic field $\vec{B}$ is along the axis of the loop and of the coil.

At the centre of the coil, $x = 0$,

$$\therefore \quad B = \frac{\mu_0 Nia^2}{2(a^2+0)^{3/2}} = \frac{\mu_0 Ni}{2a}$$

(b) (ii) $$i_0 = \frac{E}{R+r} = \frac{120}{20+5}$$

$$= 4{\cdot}8 \text{ A}$$ **Ans.**

Answer 7.

(a) (i) Resonant frequency,

$$f = \frac{1}{2\pi\sqrt{LC}}$$

$$= \frac{1}{2\pi}\times\frac{1}{\sqrt{\frac{4}{\pi^2}\times 25\times 10^{-6}}}$$

$$= \frac{1}{2\pi}\times\frac{1}{\sqrt{1.01\times 10^{-5}}}$$

$$f = 50 \text{ Hz}$$ **Ans.**

(ii) At resonant frequency Z, is minimum.

$\therefore \quad X_L = X_C$

or $\quad Z = R$

$$i = \frac{E}{Z} = \frac{E}{R} = \frac{220}{100} = 2{\cdot}2 \text{ A}$$ **Ans.**

(iii) Average power consumed $= E_{rms} \times i_{rms} \times \cos\phi$

But, $\cos\phi = \frac{R}{Z}$

$$= \frac{R}{100} = \frac{100}{100}$$ [At resonant frequency]

$\therefore$ Average power consumed $= E_{rms} \times i_{rms} \times 1$

$= 220 \times 2.20 \times 1 = 484$ W. **Ans.**

(b) The phase difference between V_L and V_C in an LCR circuit is 180°.

SECTION–B

Answer 8.

(a) Refer to ISC Paper 2019 Answer 17.

(b) Difference between interference of light and diffraction of light is :

Interference: Interference is due to superposition of light waves coming from two coherent sources.

Diffraction: Diffraction is due to interference of secondary wavelets coming from different parts of the same source.

Answer 9.

(a) (i) Given: $\lambda = 630$ nm, $d = 1{\cdot}8$ mm and

D = 80 cm = 0·8 m

Fringe width, $\beta = \frac{D\lambda}{d}$

$$= \frac{0.8 \times 630 \times 10^{-9}}{1.8 \times 10^{-3}}$$

$$= 2{\cdot}8 \times 10^{-4} \text{ m} \qquad \textbf{Ans.}$$

(ii)

$$x_m = \frac{mD\lambda}{d}$$

$$x_{10} = 10 \times 2{\cdot}8 \times 10^{-4}$$

$$= 2.8 \times 10^{-3} \text{ m.} \qquad \textbf{Ans.}$$

(b)

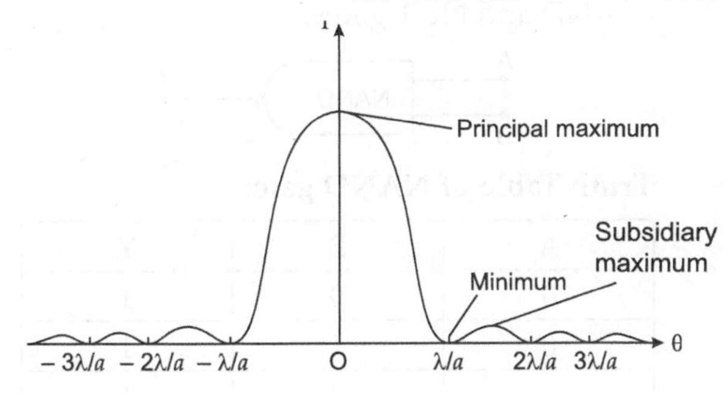

Answer 10.

(a) Given : $f = 10$ cm, $u = -15$ cm

Using lens formula for convex lens:

$$\frac{1}{f} = \frac{1}{v} - \frac{1}{u}$$

$$\frac{1}{10} = \frac{1}{v} + \frac{1}{15}$$

$$\frac{1}{v} = \frac{1}{10} - \frac{1}{15}$$

$$= \frac{3-2}{30} = \frac{1}{30}$$

$\Rightarrow$ $v = 30$ cm

But $2f = R = v - 10$

$2f = 30 - 10$

$2f = 20$

$\therefore$ $f = 10$ cm **Ans.**

Answer 11.

(a)

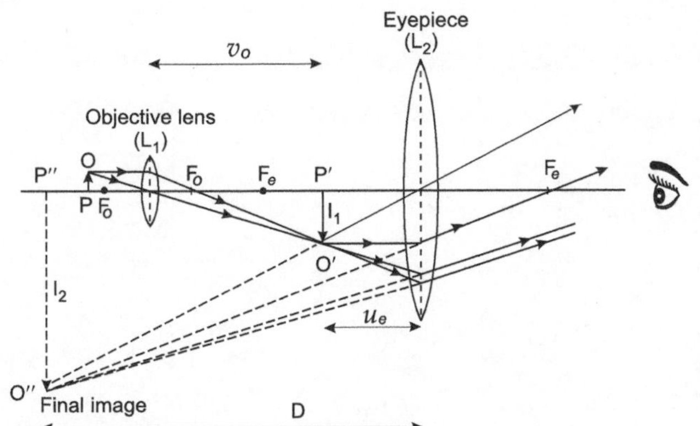

(b) (i) In order to increase the magnifying power of a telescope the focal length f_0 of the objective lens should be large and focal length f_e of the eyepiece should be small.

(ii) Larger the aperture of the objective lens, smaller the limit of resolution or greater the resolving power of the telescope.

SECTION–C

Answer 12.

(a) (i) Threshold frequency = 10×10^{14} Hz.

(ii) Work function = 4eV (–ve intercept on Y-axis)

(iii) Stopping potential = 8 eV from graph.

(b) (i) We know that,

$$\lambda = \frac{h}{p}$$

$$\lambda \ \mu \ \frac{1}{p}$$

From the above equation, it is clear that wavelength is inversely proportional to the momentum.

(ii) Diffraction of electrons.

Answer 13.

(a) Let e, m, v and r be the charge, mass, velocity of the electron and radius of the orbit respectively. The positive charge on the nucleus is Ze, where Z is the atomic number (in case of hydrogen atom, Z = 1). As the centripetal force is provided by the electrostatic force of attraction. We have,

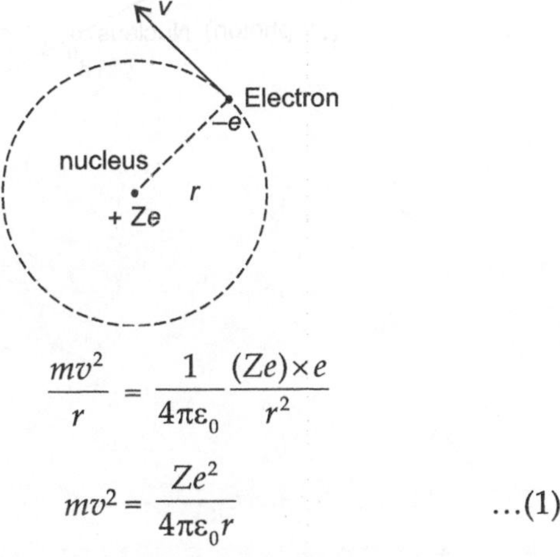

$$\frac{mv^2}{r} = \frac{1}{4\pi\varepsilon_0} \frac{(Ze) \times e}{r^2}$$

$$mv^2 = \frac{Ze^2}{4\pi\varepsilon_0 r} \qquad \ldots(1)$$

From the first postulate, the angular momentum of the electron is

$$mvr = n\frac{h}{2\pi} \qquad \ldots(2)$$

where n (= 1, 2, 3,...) is quantum number. Squaring equation (2) and dividing by equation (1), we get

$$r = n^2 \frac{h^2 \varepsilon_0}{\pi m Z e^2}$$

Since $Z = 1$

$$\therefore \quad r = n^2 \frac{h^2 \varepsilon_0}{\pi m e^2}$$

(b) We have,

$$\lambda_{min} = \frac{hc}{eV}$$

$$= \frac{6.6 \times 10^{-34} \times 3 \times 10^8}{1.6 \times 10^{-19} \times 50 \times 10^3}$$

$$= 2{\cdot}475 \times 10^{-11} \text{m} \qquad \textbf{Ans.}$$

Answer 14.

(a) (i) Half-life: The time-interval in which the mass of a radioactive substance or the number of its atoms is reduced to half of its initial value is called the half life of that substance.

(ii) Refer to ISC Paper 2017, Answer 14 (a).

(b) When an energetic γ-ray photon falls on a heavy substance, it is absorbed by some nucleus of the substance and its energy gives rise to the production of an electron and a positron. This phenomenon is called pair-production and represented by the following equation:

$$h\nu = {}_{-1}\beta^0 + {}_{+1}\beta^0$$

(γ-photon) (electron) (positron)

+Ze
$+1\beta^0$
(γ = photon) Nucleus
$-1\beta^0$

Answer 15.

(a)

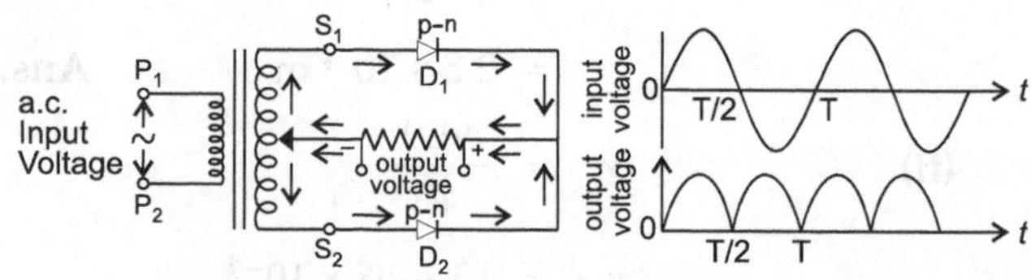

(b) NAND Gate: The NAND gate is a combination of AND and NOT gates.

Truth Table of NAND gate:

A	B	Y
0	0	1
0	1	1
1	0	1
1	1	0

Useful Constants and Relations:

1.	Speed of Light in vacuum	*(c)*	$= 3{\cdot}0 \times 10^8$ ms^{-1}
2.	Charge of a proton	*(e)*	$= 1{\cdot}6 \times 10^{-19}$ C
3.	Planck's constant	*(h)*	$= 6{\cdot}6 \times 10^{-34}$ Js
4.	Constant for Coulomb's law	$\left(\frac{1}{4\pi\varepsilon_0}\right)$	$= 9 \times 10^9$ mF^{-1}
5.	Electron volt	(1eV)	$= 1{\cdot}6 \times 10^{-19}$ J

●●

PHYSICS 2014

QUESTIONS

PART–I (20 Marks)

*(Answer **all** questions)*

Question 1.

(A) *Choose the correct alternative (a), (b), (c) or (d) for each of the questions given below:* **[5]**

(i) *Intensity of electric field at a point at a perpendicular distance 'r' from an infinite line charge, having linear charge density 'λ' is given by:*

(a) $E = \left(\frac{1}{4\pi\varepsilon_0}\right)\frac{\lambda}{r}$ (b) $E = \left(\frac{1}{4\pi\varepsilon_0}\right)\frac{2\lambda}{r}$

(c) $E = \left(\frac{1}{4\pi\varepsilon_0}\right)\frac{\lambda}{r^2}$ (d) $E = \left(\frac{1}{4\pi\varepsilon_0}\right)\frac{2\lambda}{r^2}$

(ii) *If R_1 and R_2 are filament resistances of a 200 W and a 100 W bulb respectively, designed to operate on the same voltage, then :*

(a) $R_1 = R_2$ (b) $R_2 = 2R_1$

(c) $R_2 = 4R_1$ (d) $R_1 = 4R_2$

(iii) *A metallic wire having length of 2 m and weight of 4×10^{-3} N is found to remain at rest in a uniform and transverse magnetic field of 2×10^{-4}T. Current flowing through the wire is:*

(a) *10 A* (b) *5 A*

(c) *2 A* (d) *1 A*

(iv) *When a beam of white light is passed through sodium vapours and then through a spectrometer, spectrum so obtained has two dark lines present in the yellow region. This spectrum is called :*

(a) *band spectrum*

(b) *continuous spectrum*

(c) *absorption spectrum of sodium*

(d) *emission spectrum of sodium*

(v) *If l_3 and l_2 represent angular momenta of an orbiting electron in III and II Bohr orbits*

(a) 3 : 2 (b) 9 : 4

(c) 2 : 3 (d) 4 : 9

(B) *Answer all questions given below briefly and to the point:* **[15]**

(i) *A parallel plate air capacitor has a capacitance of 5 μF. It becomes 50 μF when a dielectric medium occupies the entire space between its two plates. What is the dielectric constant of the medium?*

(ii) *Find the emf of the battery shown in figure below:*

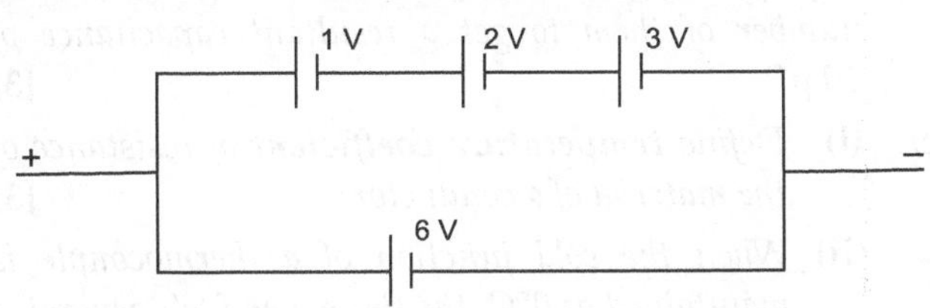

(iii) *Two substances A and B have their relative permeabilities slightly greater and slightly less than 1 respectively. What do you conclude about A and B as far as their magnetic materials are concerned ?*

(iv) *When does a moving charged particle not experience any force while moving through a uniform magnetic field ?*

(v) *What is the turns ratio i.e., **transformer ratio, $n_s : n_p$**, in an ideal transformer which increases a.c. voltage from 220 V to 33000 V ?*

(vi) *What is meant by **coherent sources of light** ?*

(vii) *A ray of light is incident on a transparent medium at polarizing angle. What is the angle between the reflected ray and the refracted ray ?*

(viii) *Name the physical **principle** on which the working of **optical fibres** is based.*

(ix) *What is meant by **shortsightedness** ?***

(x) *How does focal length of a convex lens change with **increase in wavelength** of incident light ?*

(xi) *With reference to photoelectric effect, what is meant by **threshold wavelength** ?*

(xii) *Half life of a certain radioactive element is 3·465 days. Find its disintegration constant.*

(xiii) *Binding energy per nucleon for helium nucleus $\left({}^4_2He\right)$ is 7·0 MeV. Find the value of **mass defect** for helium nucleus.*

(xiv) *Write one balanced reaction representing **nuclear fusion.***

(xv) *Draw the truth table of a NOR gate.*

PART – II (50 Marks)

SECTION–A

*(Answer any **two** questions)*

Question 2.

(a) *An electric dipole of dipole moment $\vec{p}$ is placed in a uniform electric field $\vec{E}$ with its axis inclined to the*

** Answer is not given due to change in the present syllabus.

field. Write an expression for the torque $\vec{\tau}$ experienced by the dipole in **vector** *form. Show* **diagrammatically** *how the dipole should be kept in the electric field so that the* **torque** *acting on it is:*

(i) *maximum* **(ii)** *zero* **[3]**

(b) *You are provided with 8 μF capacitors. Show with the help of a diagram how you will arrange* **minimum** *number of them to get a resultant capacitance of 20 μF.* **[3]**

(c) **(i)** *Define* **temperature coefficient** *of resistance of the material of a conductor.* **[3]**

(ii) *When the cold junction of a thermocouple is maintained at 0°C, the thermo emf 'e', generated by this thermocouple is given by the relation :***

$$e = \left[16.8\theta + \frac{1}{2}(-0.048)\theta^2\right] \times 10^{-6}.$$

where θ is the temperature of the hot junction in °C. Find the **neutral temperature** *of this thermocouple.*

Question 3.

(a) *Draw a* **labelled** *circuit diagram of a* **potentiometer** *to compare emfs of two cells. Write the working formula (Derivation not required).* **[3]**

(b) *How much resistance should be connected to 15Ω resistor shown in the circuit in following figure so that the points M and N are at the same potential:* **[3]**

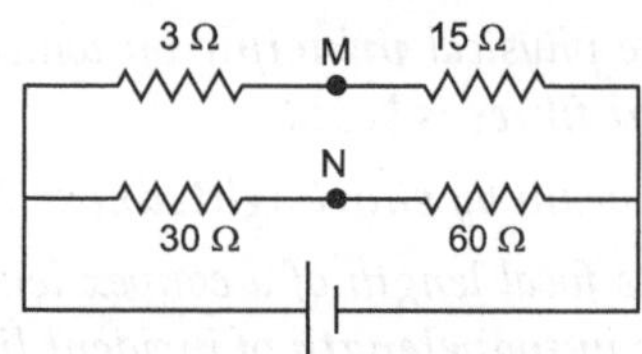

(c) **(i)** *With reference to* **free electron theory of conductivity,** *explain the terms:* **[3]**

(1) Drift speed

(2) Relaxation time

(ii) *What is the* **colour code** *of a carbon resistor having a resistance of 470 Ω and a tolerance of 5% ?*

Question 4.

(a) **(i)** *State* **Tangent Law** *in magnetism.*** **[2]**

(ii) *At a certain temperature, a ferromagnetic material becomes paramagnetic. What is this temperature called ?*

(b) **(i)** *State* **Biot-Savart law.** **[3]**

(ii) *Find magnetic flux density at a point on the axis of a long solenoid having 5000 turns/m when it carrying a current of 2 A.*

(c) *An alternating emf of 110 V is applied to a circuit containing a resistance R of 80 Ω and an inductor L in series. The current is found to lag behind the supply voltage by an angle $\theta = \tan^{-1}(3/4)$. Find the:* **[4]**

(i) *Inductive reactance*

(ii) *Impedance of the circuit*

(iii) *Current flowing in the circuit*

(iv) *If the inductor has a coefficient of self inductance of 0·1 H, what is the frequency of the applied emf ?*

SECTION–B

(Answer any **two** *questions)*

Question 5.

(a) *Name the part of the* **electromagnetic spectrum** *which is:* **[2]**

(i) *Suitable for radar systems used in aircraft navigation.*

(ii) *Produced by bombarding a metal target with high electrons.*

(b) *In* **Young's double slit experiment,** *using monochromatic light, fringes are obtained on a screen placed at some distance from the slits. If the screen is moved by 5×10^{-2} m towards the slits, the change in the fringe width is 3×10^{-5} m. If the distance between the two slits is 10^{-3} m, calculate* **wavelength** *of the light used.* **[3]**

(c) **(i)** *State* **Brewster's law** *of polarization of light.* **[3]**

(ii) *How will you identify* **with the help of an experiment** *whether a given beam of light is of polarised light or of unpolarised light ?*

Question 6.

(a) *A narrow beam of monochromatic light, PQ, is incident normally on one face of an equiangular glass prism of refractive index 1·45. When the prism is immersed in a certain liquid, the ray makes a grazing emergence along the other face (See below figure). Find the* **refractive index** *of this liquid.* **[2]**

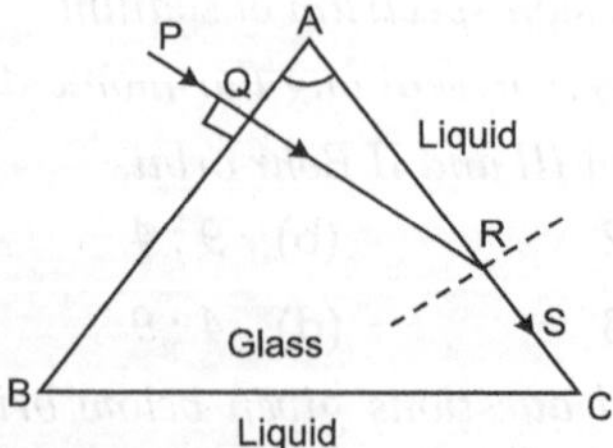

(b) *When two thin lenses of focal lengths f_1 and f_2 are kept coaxially and in contact, prove that their* **combined focal length "f"** *is given by :* **[3]**

$$\frac{1}{f} = \frac{1}{f_1} + \frac{1}{f_2}$$

** Answer is not given due to change in the present syllabus.

(c) *The figure below shows the positions of a point object O, two lenses, a plane mirror and the final image I which coincides with the object. The focal length of the convex lens is 20 cm. Calculate the focal length of the* **concave** *lens.* [3]

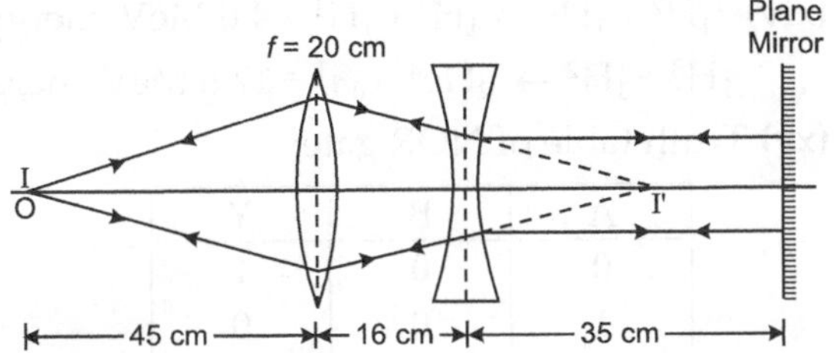

Question 7.

(a) (i) *What is meant by* **dispersive power** *of a transparent material ?* [4]

(ii) *Show that, two thin lenses kept in contact, form an* **achromatic doublet** *if they satisfy the condition :***

$$\frac{\omega}{f}+\frac{\omega'}{f'}=0$$

where the terms have their usual meaning.

(b) (i) *Define* **magnifying power** *of a* **microscope** *in terms of* **visual angle.**

(ii) *What is the advantage of a* **compound microscope** *over a simple microscope ?* [2]

(c) *An astronomical telescope uses two lenses of powers 10 dioptre and 1 dioptre. If the final image of a distant object is formed at infinity, calculate the length of the telescope.* [2]

SECTION–C

(Answer any **two** *questions)*

Question 8.

(a) *Answer the following questions with reference to* **Millikan's oil drop experiment**:** [3]

(i) *What is an atomiser ?*

(ii) *What is the use of an X-ray tube ?*

(iii) *What is the unique property shown by the charge of an oil drop ?*

(b) (i) *Write* **Einstein's photoelectric equation** [3]

(ii) *If the frequency of the incident radiation is increased from 4×10^{15} Hz to 8×10^{15} Hz, by how much will the* **stopping potential** *for a given photo-sensitive surface go up ?*

(c) (i) *What are* **matter waves** *?* [2]

(ii) *Show with the help of a labelled graph how their wavelength (λ) varies with their linear momentum (p).*

Question 9.

(a) *The energy levels of an atom of a certain element are shown in the given figure below. Which one of the transitions A, B, C, D or E will result in the emission of photons of electromagnetic radiation of wavelength 618·75 nm? Support your answer with mathematical calculations.* [3]

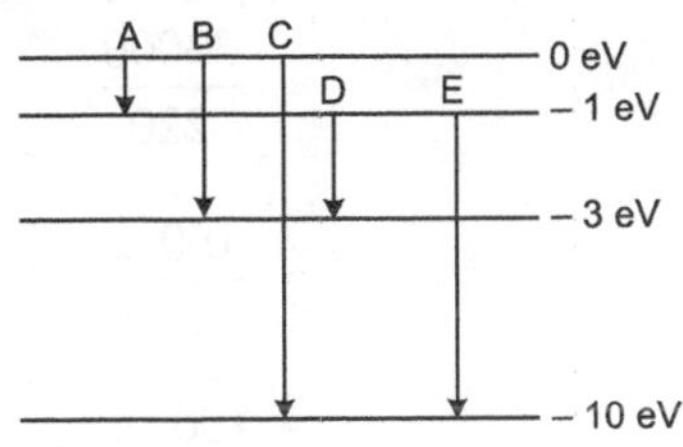

(b) *Voltage applied between cathode and anode of an X-ray tube is 18 kV. Calculate the* **minimum** *wavelength of the X-rays produced.* [2]

(c) *In a* **nuclear reactor**, *what is the function of :* [3]

(i) *The moderator,*

(ii) *The control rods*

(iii) *The coolant.*

Question 10.

(a) (i) *The atomic mass of Uranium $^{238}_{92}U$ is 238·0508 u, while that of Thorium $^{234}_{90}Th$ is 234·0436 u, and that of Helium $^{4}_{2}He$ is 4·0026 u. Alpha decay converts $^{238}_{92}U$ into $^{234}_{90}Th$ as shown below:* [3]

$$^{238}_{92}U \rightarrow {}^{234}_{90}Th + {}^{4}_{2}He + Energy$$

Determine the energy released in this reaction.

(ii) *What is a neutrino ?*

(b) *In semiconductor physics, what is meant by:* [3]

(i) *A rectifier,*

(ii) *An amplifier,*

(iii) *An oscillator.*

(c) *With the help of a diagram, show how you can use several NAND gates to obtain an OR gate.* [2]

SOLUTIONS

PART–I (20 Marks)

Answer 1.

(A) (i) (b) $E=\left(\frac{1}{4\pi\varepsilon_0}\right)\frac{2\lambda}{r}$

(ii) (b) $R_2 = 2R_1$

(iii) (a) 10 A

(iv) (c) absorption spectrum of sodium

(v) (a) 3 : 2

(B) (i) We know that, $C = KC_0$

$50 = K \times 5$

$\Rightarrow K = 10$ **Ans.**

** Answer is not given due to change in the present syllabus.

(ii) Net emf of each row = 6 V.
Since both are in parallel so it will remain 6 V.

(iii) A is paramagnetic material and B is diamagnetic material.

(iv) Charge particle does not experience any force when it is moving parallel to the uniform magnetic field.

(v) Transformer ratio $= \frac{N_s}{N_p}$

$$= \frac{E_s}{E_p}$$

$$= \frac{33000}{220}$$

$$= \frac{150}{1}$$

$= 150 : 1$ **Ans.**

(vi) Two sources are said to be coherent if they emit light wave having sharply defined phase difference that remains constant with time.

(vii) 90°

(viii) Optical fibres works on the principle of total internal reflection of light.

(x) As wavelength increases, focal length of convex lens also increases.

(xi) Threshold wavelength is a wavelength above which photoelectric emission will not take place.

(xii) Given : $T_{1/2}$ = 3·465 days
Disintegration constant,

$$\lambda = \frac{\log_e 2}{T_{1/2}}$$

$$= \frac{0.6931}{3.465}$$

$$= \frac{0.6931}{3.465 \times 24 \times 60 \times 60} \text{ sec.}$$

$\lambda = 2{\cdot}31 \times 10^{-6}$/sec. **Ans.**

(xiii) Given: $\frac{\text{Binding Energy}}{\text{nucleon}} = 7{\cdot}0$ MeV

Mass defect = ?

$$\frac{\text{Binding Energy}}{\text{nucleon}} = \frac{\Delta mc^2}{\text{No. of nucleons}}$$

$$7{\cdot}0 \times 1{\cdot}6 \times 10^{-19} \times 10^6 = \frac{\Delta m \times (3\times10^8)^2}{4}$$

$$\Delta m = \frac{28 \times 1.6 \times 10^{-13}}{9 \times 10^{16}}$$

$= 4{\cdot}97 \times 10^{-29}$ kg.

Ans.

(xiv) $_1H^2 + {_1H^2} \rightarrow {_1H^3} + {_1H^1}$ + 4·0 MeV energy
$_1H^3 + {_1H^2} \rightarrow {_2He^4} + {_0n^1}$ + 17·6 MeV energy

(xv) Truth table of NOR gate:

A	B	Y
0	0	1
1	0	0
0	1	0
1	1	0

PART – II (50 Marks)

SECTION–A

Answer 2.

(a) Expression for torque in the vector form is:

$$\vec{\tau} = \vec{p} \times \vec{E}$$

where $\vec{p}$ is the dipole moment and $\vec{E}$ is the electric field.

(i) If dipole is placed perpendicular to the electric field then torque will be maximum.

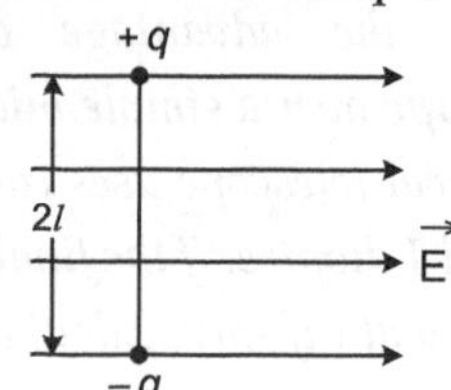

(ii) If dipole is placed parallel to the field then torque = 0

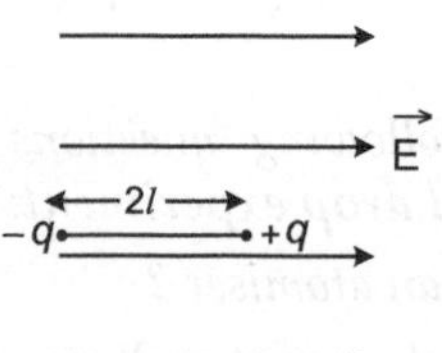

(b) In the arrangement shown, C_3 and C_4 are connected in series.

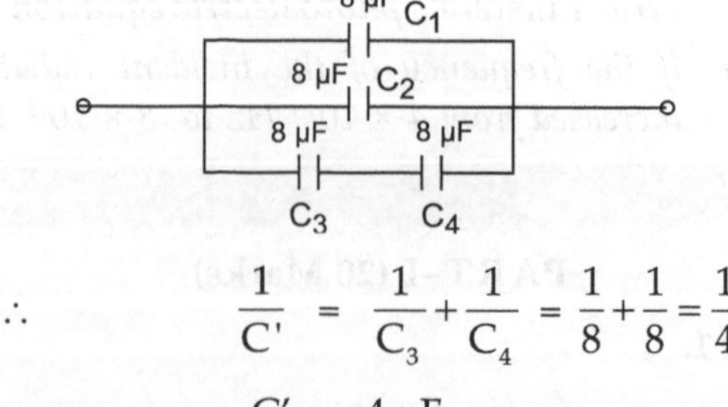

$$\therefore \quad \frac{1}{C'} = \frac{1}{C_3} + \frac{1}{C_4} = \frac{1}{8} + \frac{1}{8} = \frac{1}{4}$$

$C' = 4\ \mu F$

Total equivalent capacitance is given by,

$C = C' + C_1 + C_2 = 4 + 8 + 8$

$C = 20\ \mu F$ **Ans.**

(c) (i) The temperature coefficient of resistance is defined as the increase in resistance per unit original resistance per degree Celsius or Kelvin rise of temperature.

Answer 3.

(a) Refer to ISC Paper 2019 Answer 15 (b).

(b) In balance condition of Wheatstone's bridge,

$$\frac{P}{Q} = \frac{R}{S}$$

$$\frac{3}{X} = \frac{30}{60}$$

$$X = \frac{3\times 60}{30} = 6\ \Omega$$

Since, the resultant of two resistances is less than one of the resistances. Therefore, the two resistances must be connected in parallel.

Let resistance R is connected in parallel with resistance of 15 Ω. Then

$$\frac{1}{X} = \frac{1}{15}+\frac{1}{R}$$

$$\frac{1}{6} = \frac{1}{15}+\frac{1}{R}$$

$$\frac{1}{R} = \frac{1}{6}-\frac{1}{15}$$

$$\frac{1}{R} = \frac{5-2}{30} = \frac{3}{30} = \frac{1}{10}$$

$$R = 10\ \Omega$$

Hence, 10 Ω resistance should be connected in parallel to 15 Ω resistor. **Ans.**

(c) (i) (1) Drift speed: The free electrons in metals move constantly at random with high velocity inside the metal. They frequently collide with the positive ions present within the metal, thus changing their direction of motion. Hence, they have no net motion in any particular direction. When a potential difference is applied across a metal, the free electrons drift with a small constant velocity in the direction opposite to the applied field. This is drift speed.

(2) Relaxation time: It is the average time interval between two successive collisions of a free electron with the positive ions in the metal.

(ii) The colour code of a carbon resistor having a resistance of 470 ± 5% is Yellow, Violet, Brown, Gold.

Answer 4.

(a) (ii) Curie temperature.

(b) (i) Biot-Savart law: The magnetic field at any point due to an element of a conductor carrying current is:

(1) directly proportional to (a) the strength of the current *i*, (b) length of the element *dl*, (c) sine of the angle θ between the element in the direction of current and the line joining the element to the point P, and

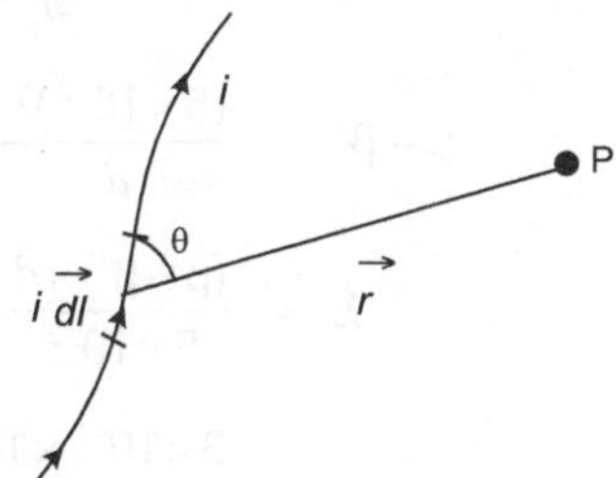

(2) inversely proportional to the square of the distance *r* of the point P from the centre of the element.

i.e.,
$$dB = \frac{\mu_0}{4\pi}\times\frac{Idl\sin\theta}{r^2}$$

(ii) Magnetic flux density,

$$B = \mu_0 ni$$
$$= 4\pi \times 10^{-7} \times 5000 \times 2$$
$$= 0{\cdot}01256\ T$$

(c) (i) Given:

$$R = 80\ \Omega$$
$$E = 110\ V$$
$$\theta = \tan^{-1}(3/4)$$
$$\Rightarrow \tan\theta = 3/4$$
$$\tan\theta = \frac{X_L}{80}$$
$$X_L = \frac{80\times 3}{4} = 60\ \Omega$$ **Ans.**

(ii)
$$Z = \sqrt{R^2+X_L^2}$$
$$= \sqrt{80^2+60^2}$$
$$= 100\ \Omega.$$ **Ans.**

(iii)
$$i = \frac{E}{Z} = \frac{110}{100} = 1{\cdot}1\ A$$ **Ans.**

(iv)
$$L = 0{\cdot}1\ H$$
$$f = ?$$
$$X_L = 2\pi fL = 2\pi f\,L$$
$$60 = 2\pi f\times 0{\cdot}1$$
$$f = \frac{60}{2\pi\times 0.1} = 95{\cdot}54\ Hz.$$ **Ans.**

SECTION-B

Answer 5.

(a) (i) Microwaves or short radiowaves.

(ii) X-rays.

(b) Given: $\Delta\beta = 3 \times 10^{-5}$ m

We know, $\beta = \frac{D\lambda}{d}$

$$\beta' = \frac{(D-(5\times10^{-2})\lambda)}{d}$$

$$\beta - \beta' = \frac{(5\times10^{-2})\lambda}{d}$$

$$\lambda = \frac{(\beta-\beta')\times d}{5\times10^{-2}}$$

or

$$= \frac{3\times10^{-5}\times10^{-3}}{5\times10^{-2}}$$

$$= 6\times10^{-7}\text{ m}$$

$$= 6000\text{ Å}$$ **Ans.**

(c) (i) Brewster's law states that the tangent of angle of polarisation is equal to the refractive index of the medium.

Mathematically, $n = \tan i_p$

where, n = refractive index of the medium and

i_p = polarising angle

(ii) To identify whether a given beam of light is polarised or unpolarised, it is passed through a polaroid and polaroid is rotated about the incident light.

(1) If on rotating the polaroid through one complete rotation, there is no change in the intensity of emergent light in any position, then the given light is unpolarised.

(2) If there is a change in the intensity of emergent light in one complete rotation of the polaroid, the intensity becomes maximum twice and zero twice then the given light is completely plane polarised.

(3) If the intensity of the emergent light changes but never becomes zero, then the given light is partially polarised.

Answer 6.

(a)

$${}_l\mu_g = \frac{1}{\sin C}$$

$$\angle A = 60^\circ$$

$$\angle A + \angle ARP = 90^\circ$$

So, $\angle ARP = 30^\circ$

So, $\angle C = 60^\circ$

$${}_l\mu_g = \frac{1}{\sin 60^\circ} = \frac{2}{\sqrt{3}}$$

$${}_g\mu_l = \sqrt{3}/2 = 0{\cdot}866$$

$${}_a\mu_l = {}_a\mu_g \times {}_g\mu_l$$

$$= 1{\cdot}45 \times 0{\cdot}866$$

$$= 1{\cdot}256$$ **Ans.**

(b) Refer to ISC Paper 2020 Answer 16.

(c) For convex lenses, $\frac{1}{f} = \frac{1}{v} - \frac{1}{u}$

$$\frac{1}{20} = \frac{1}{v} + \frac{1}{45}$$

$$\frac{1}{v} = \frac{1}{20} - \frac{1}{45}$$

$$v = \frac{20\times45}{25} = 36\text{ cm}$$

So 'u' for concave lens = 36 – 16 = 20 cm

For concave lens, $\frac{1}{f} = \frac{1}{v} - \frac{1}{u}$

$$\frac{1}{f} = \frac{1}{\infty} - \frac{1}{20}$$

$$f = -20\text{ cm}$$

So, focal length of concave lens = –20 cm. **Ans.**

Answer 7.

(a) (i) When white light passes through a thin prism, the ratio of the angular dispersion between the violet and the red emergent rays to the deviation suffered by the mean ray (yellow colour) is called dispersive power of an optical medium.

(b) (i) Magnifying power of a microscope is defined as the ratio of visual angle subtended by the image at the eye to the visual angle subtended by the object at the eye when placed at least distance of distinct vision.

i.e., $m = \frac{\beta}{\alpha}$

(ii) In a compound microscope since magnification is produced in two stage magnification is much higher than in a simple microscope.

(c) Given: $P_e = 10$ dioptre and $P_o = 1$ dioptre

We know that,

$$P_e = \frac{1}{f_e}$$

$$10 = \frac{1}{f_e}$$

$$f_e = 0.1\text{ m} = 10\text{ cm}$$

For objective lens, $P_o = \frac{1}{f_o}$

$$f_o = \frac{1}{1} = 1 \text{ m} = 100 \text{ cm}$$

So, $L = f_o + f_e = 100 + 10$

$= 110$ cm **Ans.**

SECTION–C

Answer 8.

(b) (i) Einstein's photoelectric equation is

$$\frac{1}{2} mv^2_{max} = h(\nu - \nu_0)$$

where $h\nu$ = energy of incident photon.

$h\nu_0$ = work function

Also, $\frac{1}{2} mv^2_{max}$ = maximum K.E. of the emitted electron

(ii) Given: $\nu_1 = 4 \times 10^{15}$ Hz, $\nu_2 = 8 \times 10^{15}$ Hz

We know that, K.E. = $h\nu - h\nu_o$

$$K.E_1 = h\nu_1 - h\nu_o$$

$$K.E_2 = h\nu_2 - h\nu_o$$

$$K.E_2 - K.E_1 = h\nu_2 - h\nu_1$$

$$eV_2 - eV_1 = h(\nu_2 - \nu_1)$$

$$e(V_2 - V_1) = 6.6 \times 10^{-34} (8 \times 10^{15} - 4 \times 10^{15})$$

$$e(V_2 - V_1) = 2.64 \times 10^{-18}$$

$$V_2 - V_1 = \frac{2.64 \times 10^{-18}}{1.6 \times 10^{-19}}$$

$V_2 - V_1 = 16{\cdot}5$ volt **Ans.**

(c) (i) The waves associated with moving particles of matter are called de-Broglie waves or matter waves.

(ii) $\lambda = \frac{h}{p}$

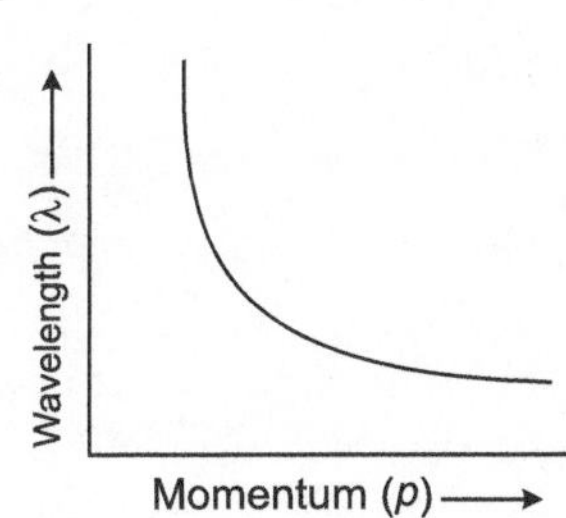

Answer 9.

(a) We know that, $\Delta E = \frac{hc}{\lambda}$

$$= \frac{6.6 \times 10^{-34} \times 3 \times 10^8}{618.75 \times 10^{-9}}$$

$= 3{\cdot}2 \times 10^{-19}$ J

$= 2 \times 1{\cdot}6 \times 10^{-19}$ J

$= 2$ eV

This corresponds to the transition D for which

$$\Delta E = (-1) - (-3)$$

$= 2$ eV **Ans.**

(b) Given: $V = 18 \text{ kV} = 18 \times 10^3$ V

$$\lambda_{min} = \frac{hv}{eV}$$

$$= \frac{6.6 \times 10^{-34} \times 3 \times 10^8}{1.6 \times 10^{-19} \times 18 \times 10^3}$$

$= 6{\cdot}88 \times 10^{-11}$ m. **Ans.**

(c) (i) Moderator: It slows down the energies of thermal neutrons through elastic collisions between its nuclei and the fission neutrons. Heavy water is the best moderator.

(ii) Control rods: These are the rods of cadmium which are used to control the fission rate in the reactor.

(iii) Coolant: The energy released inside the reactor in the form of heat is removed by coolant.

Answer 10.

(a) (i) Atomic mass of Uranium = 238·0508 u

Atomic mass of Thorium = 234·0436 u

Atomic mass of Helium = 4·0026 u.

Mass of reactants = 238·0508 u.

Mass of products = 234·0436 + 4·0026 = 238·0462 u

Δm = 238·0508 − 238·0462 = 0·0046 u

Energy released = 0·0046 × 931

= 4·28 MeV **Ans.**

(ii) Neutrino: These are the particles whose rest mass and charge both are zero but they have energy and momentum.

(b) (i) Rectifier: It is a device which converts an alternating current into a direct current.

(ii) An amplifier is basically a transistor consisting of two *p-n* junctions, one forward biased and the other reverse biased. It is used to amplify a weak signal.

(iii) An oscillator is a device that generates electrical oscillations of constant amplitude and of desired frequency, without any external input.

(c)

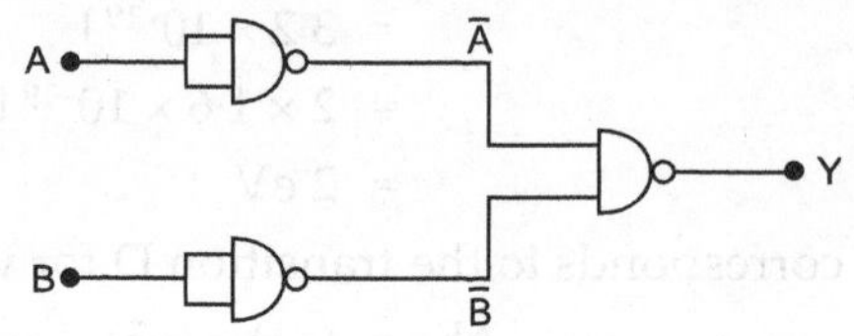

Truth table

A	B	$\overline{A}$	$\overline{B}$	Y
0	0	1	1	0
0	1	1	0	1
1	0	0	1	1
1	1	0	0	1

Useful Constants and Relations

1. Speed of Light in vacuum	(c)	$= 3{\cdot}00 \times 10^8$ m/s
2. Charge of a proton	(e)	$= 1{\cdot}60 \times 10^{-19}$ C
3. Planck's constant	(h)	$= 6{\cdot}6 \times 10^{-34}$ Js
4. Permeability of vacuum	(μ_0)	$= 4\pi \times 10^{-7}$ Hm^{-1}
5. Electron volt	(1 eV)	$= 1{\cdot}6 \times 10^{-19}$ J
6. Unified Atomic Mass Unit	(1 u)	= 931 MeV
7.	(π)	$= 3{\cdot}14$
8.	(ln 2)	$= 0{\cdot}693$

●●

PHYSICS 2013

QUESTIONS

PART – I

*(Answer **all** questions)*

Question 1.

(A) *Choose the correct alternative (a), (b), (c) or (d) for each of the questions given below:* **[5]**

(i) *Relative permittivity of water is 81. If ε_w and ε_0 are permitivities of water and vacuum respective y, then :*

(a) $\varepsilon_0 = 9\varepsilon_w$ (b) $\varepsilon_0 = 81\varepsilon_w$

(c) $\varepsilon_w = 9\varepsilon_0$ (d) $\varepsilon_w = 81\varepsilon_0$

(ii) *Five resistors are connected as shown in adjoining figure .*

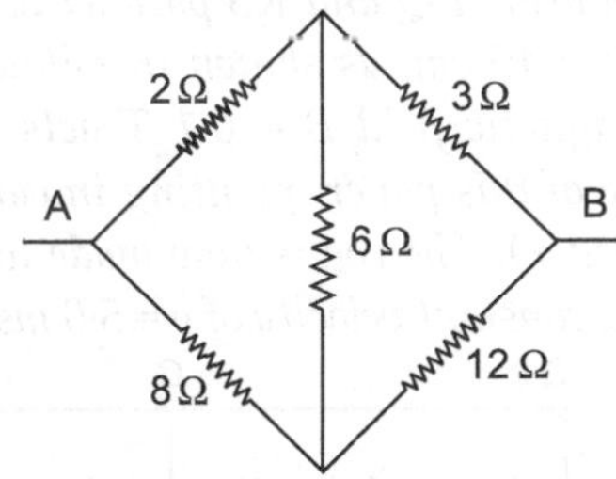

The effective resistance i.e., equivalent resistance between the points A and B is :

(a) 4 Ω (b) 5 Ω

(c) 15 Ω (d) 20 Ω

(iii) *The Biot-Savart's law in vector form is:*

(a) $\overrightarrow{\delta B} = \dfrac{\mu_0}{4\pi}\dfrac{dl(\vec{I} \times \vec{r})}{r^3}$ (b) $\overrightarrow{\delta B} = \dfrac{\mu_0}{4\pi}\dfrac{I(\vec{dl} \times \vec{r})}{r^3}$

(c) $\overrightarrow{\delta B} = \dfrac{\mu_0}{4\pi}\dfrac{I(\vec{r} \times \vec{dl})}{r^3}$ (d) $\overrightarrow{\delta B} = \dfrac{\mu_0}{4\pi}\dfrac{I(\vec{dl} \times \vec{r})}{r^2}$

(iv) *In an astronomical telescope of refracting type:*

(a) *Eyepiece has greater focal length.*

(b) *Objective has greater focal length.*

(c) *Objective and eyepiece have equal focal length.*

(d) *Eyepiece has greater aperture than the objective.*

(v) *The particles which cannot be accelerated by a cyclotron or a Van de Graff generator are :*

(a) *Alpha particles* (b) *Beta particles*

(c) *Neutrons* (d) *Protons*

(B) *Answer all questions given below briefly and to the point:* **[15]**

(i) *A large hollow metallic sphere has a positive charge of 35·4 μC at its centre. Find how much electric flux emanates from the sphere.*

(ii) *A current 'I' flows through a metallic wire of radius 'r' and the free electrons in it drift with a velocity v_d. Calculate the drift velocity of the free electrons through the wire of the same material, having double the radius, when same current flows through it.*

(iii) *Name any one instrument which works on the principle of tangent law in magnetism.***

(iv) *State the SI unit of magnetic dipole moment.*

(v) *Alternating current flowing through a certain electrical device leads over the potential difference across it by 90°. State whether this device is a resistor, capacitor or an inductor.*

(vi) *What is the shape of the wavefront diverging from a point source of light ?*

(vii) *The critical angle for a given transparent medium and air is i_c. A ray of light travelling in air is incident on this transparent medium at an angle of incidence equal to the polarising angle i_p. What is the relation between the two angles i_c and i_p ?*

(viii) *Find the focal length and nature of a lens whose optical power is – 5 D.*

(ix) *What is Modulation ? Explain in brief.*

(x) *What are the dark lines seen in the solar spectrum called ?*

(xi) *What is the relation between wavelength and momentum of moving particles ?*

(xii) *Name the series of lines in the hydrogen spectrum which lies in the ultra-violet region.*

(xiii) *Fill in the blank in the given nuclear reaction:*

$$\_\_\_\_\_\_\_\_ + {}^{27}_{13}Al \rightarrow {}^{25}_{12}Mg + {}^{4}_{2}He$$

(xiv) *Give an example where energy is converted into matter.*

(xv) *To convert a pure semiconductor into n-type semiconductor, what type of impurity is added to it?*

PART – II

SECTION – A

*(Answer any **two** questions)*

Question 2.

(a) (i) *Write an expression (derivation not required) for intensity of electric field in :*

(1) *Axial position.*

(2) *Broad side position of an electric dipole, in terms of its length (2a) dipole moment (p) and distance (r).*

** Answer is not given due to change in the present syllabus.

(ii) *What is the ratio of these two intensities i.e., $E_1 : E_2$, for a short electric dipole ?* [3]

(b) *Three capacitors $C_1 = 6\ \mu F$, $C_2 = 12\ \mu F$ and $C_3 = 20\ \mu F$ are connected to a 100 V battery, as shown in figure below:*

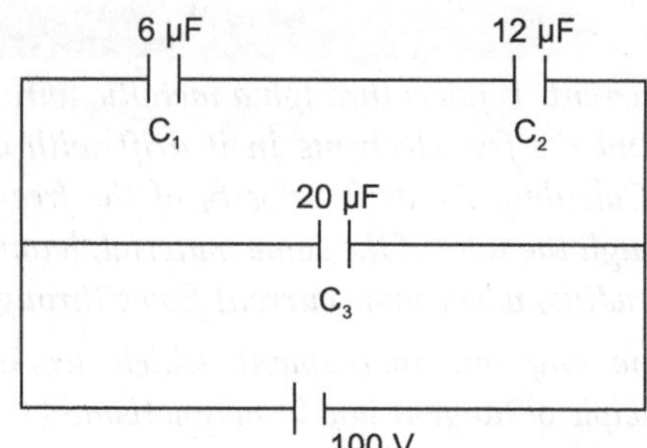

Calculate:

(i) *Charge on each plate of capacitor C_1.*

(ii) *Electrostatic potential energy stored in capacitor C_3.* [3]

(c) *'n' cells, each of emf 'e' and internal resistance 'r' are joined in series to form a row. 'm' such rows are connected in parallel to form a battery of N = mn cells. This battery is connected to an external resistance 'R'.*

(i) *What is the emf of this battery and how much is its internal resistance ?*

(ii) *Show that current 'I' flowing through the external resistance 'R' is given by—*

$$I = \frac{Ne}{mR + nr}$$ [3]

Question 3.

(a) *In the circuit shown below, $E_1 = 17$ V, $E_2 = 21$ V, $R_1 = 2\ \Omega$, $R_2 = 3\ \Omega$ and $R_3 = 5\ \Omega$. Using Kirchhoff's laws, find the currents flowing through the resistors R_1, R_2 and R_3. (Internal resistance of each of the batteries is negligible.)* [4]

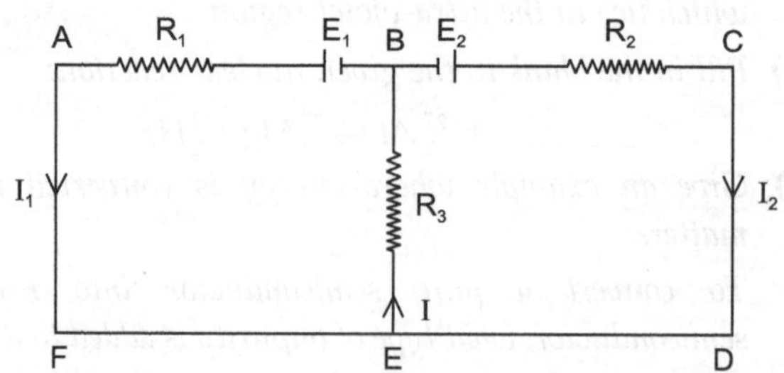

(b) *You are provided with one low resistance R_L and one high resistance R_H and two galvanometers. One galvanometer is to be converted to an ammeter and the other to a voltmeter. Show how you will do this with the help of simple, labelled diagrams.* [2]

(c) (i) *Plot a labelled graph to show variation of thermo-emf 'e' versus temperature difference 'θ' between the two junctions of a thermocouple. Mark 'N' as neutral temperature and 'T' as temperature of inversion.***

(ii) *What is Peltier effect ?*** [3]

Question 4.

(a) *Figure below shows two infinitely long and thin current carrying conductors X and Y kept in vacuum, parallel to each other, at a distance 'a'.* [2]

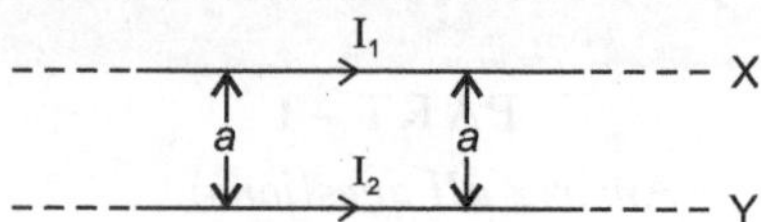

(i) *How much force per unit length acts on the conductor Y due to the current flowing through X? Write your answer in terms of $\left(\frac{\mu_0}{4\pi}\right)$, I_1, I_2 and a.*

(Derivation of formula is not required.)

(ii) *Define ampere, in terms of force between two current carrying conductors.*

(b) *A metallic rod CD rests on a thick metallic wire PQRS with arms PQ and RS parallel to each other, at a distance l = 40 cm, as shown in following figure. A uniform magnetic field B = 0·1 T acts perpendicular to the plane of this paper, pointing inwards (i.e., away from the reader). The rod is now made to slide towards right, with a constant velocity of $v = 5{\cdot}0\ ms^{-1}$.* [3]

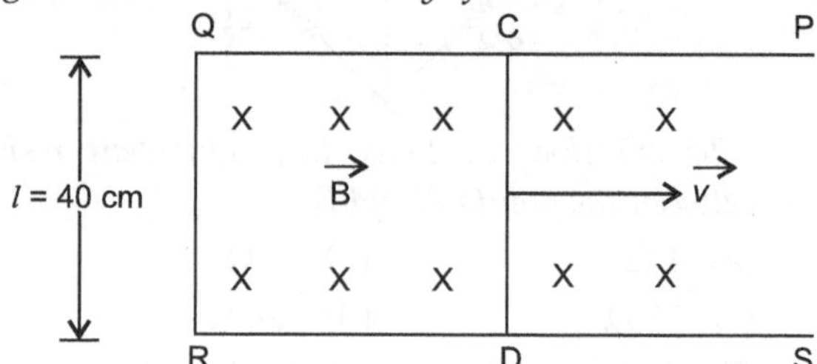

(i) *How much emf is induced between the two ends of the rod CD?*

(ii) *What is the direction in which the induced current flows?*

(c) (i) *Following figure shows a series RCL circuit connected to an ac source which generates an alternating emf of frequency 50 Hz. The readings of the voltmeters V_1 and V_2 are 80 V and 60 V respectively.* [4]

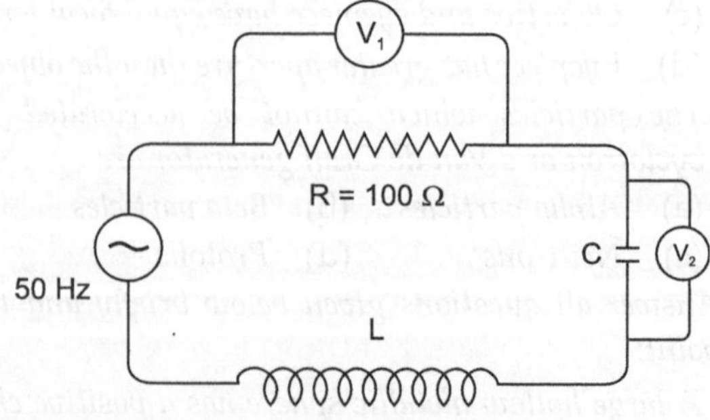

Find :

(1) *the current in the circuit.*

(2) *the capacitance C of the capacitor.*

** Answer is not given due to change in the present syllabus.

(ii) *At resonance, what is the relation between impedance of a series LCR circuit and its resistance R?*

SECTION–B

(Answer any ***two*** *questions)*

Question 5.

(a) (i) *In an electromagnetic wave, how are electric vector $(\vec{E})$, magnetic $(\vec{B})$ vector and velocity of propagation of the wave $(\vec{c})$ oriented ?*

(ii) *How long would gamma radiation take to travel from Sun to Earth, a distance of $1{\cdot}5 \times 10^{11}$ m?* **[2]**

(b) *With the help of a labelled diagram, show that fringe separation β in Young's double slit experiment is given by :* **[4]**

$$\beta = \frac{\lambda D}{d}$$

where the terms have their usual meaning.

(c) (i) *What is the difference between polarised light and unpolarised light based on the direction of electric vector $(\vec{E})$?*

(ii) *What will be the effect on the width of the central bright fringe in the diffraction pattern of a single slit if :*

(1) *Monochromatic light of smaller wavelength is used ?*

(2) *Slit is made narrower ?* **[2]**

Question 6.

(a) *At what angle, a ray of light should be incident on the first face AB of a regular glass prism ABC so that the emergent ray grazes the adjacent face AC?* **[3]**
See figure below. (Refractive Index of glass = 1·6)

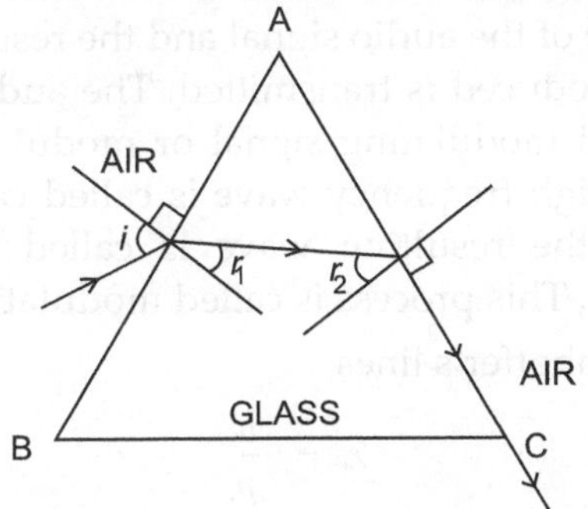

(b) *A convex lens 'L' and a plane mirror 'M' are arranged as shown in figure below. Position of object pin 'O' is adjusted in such a way that the inverted image 'I' formed by the lens mirror combination, coincides with the object pin 'O'. Explain how and when this happens.* **[2]**

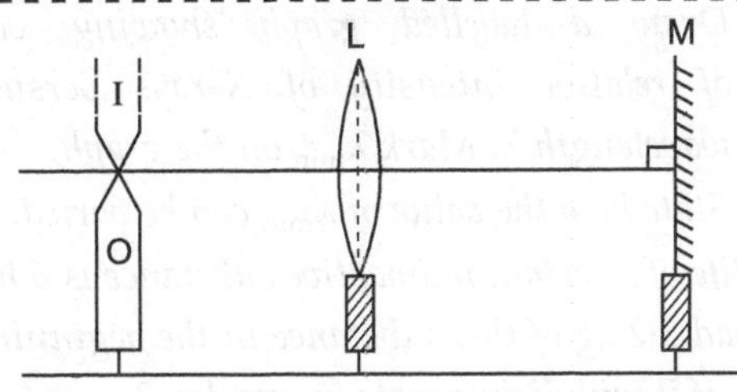

(c) *Starting with an expression for refraction at a single spherical surface, obtain an expression for lens maker's formula.* **[3]**

Question 7.

(a) *Show that the axial chromatic aberration $(f_r - f_v)$ for a convex lens is equal to the product of its mean focal length (f) and dispersive power (ω) of its material i.e. Prove :*** **[3]**

$$f_r - f_v = \omega f.$$

(b) *Draw a labelled diagram of an image formed by a compound microscope, with the image at least distance of distinct vision. Write any one expression for its magnifying power.* **[3]**

(c) *What is meant by long-sightedness ? How can this defect be corrected?*** **[2]**

SECTION–C

(Answer any ***two*** *questions)*

Question 8.

(a) (i) *What is meant by 'Quantization of charge' ?*

(ii) *In Thomson's experiment, prove that the ratio of charge to the mass (e/m) of an electron is given by:*

$$\frac{e}{m} = \frac{1}{2V}\cdot\frac{E^2}{B^2}$$

*where the terms have their usual meaning.*** **[3]**

(b) *In a photo-electric cell, a retarding potential of 0·5 V is required to block the movement of electrons from the cathode when monochromatic light of wavelength 400 nm is incident on its surface. Find the work function of the material of the cathode.* **[3]**

(c) *Name a phenomenon or an experiment which proves:* **[2]**

(i) *Particle nature of electromagnetic radiations.*

(ii) *Wave nature of particles.*

(Description of the phenomenon / experiment is not required).

Question 9.

(a) (i) *State the postulate of Bohr's theory regarding :* **[3]**

(1) *Angular momentum of an electron.*

(2) *Emission of a photon.*

(ii) *Total energy of an electron orbiting around the nucleus of an atom is always negative. What is the significance of this ?*

** Answer is not given due to change in the present syllabus.

(b) (i) *Draw a labelled graph showing variation of relative intensity of X-rays versus their wavelength λ. Mark λ_{min} on the graph.*

(ii) *State how the value of λ_{min} can be varied.* [3]

(c) *Half life of a certain radioactive substance is 6 hours. If you had 3·2 kg of this substance in the beginning, how much of it will disintegrate in one day ?* [2]

Question 10.

(a) (i) *What is the significance of binding energy per nucleon of a nucleus ?*

(ii) *In a certain star, three alpha particles undergo fusion in a single reaction to form ${}^{12}_{6}C$ nucleus. Calculate the energy released in this reaction in MeV.*
Given: m (${}^{4}_{2}He$) = 4·002604 u and m (${}^{12}_{6}C$) =12·000000 u. [3]

(b) *Show by drawing labelled diagrams, the nature of output voltages in case of:* [3]

(i) *A half wave rectifier.*

(ii) *A full wave rectifier.*

(iii) *An Amplifier.*

(In each case, input is an a.c. voltage)
Circuit diagrams of these devices are not required.

(c) *Identify the logic gate whose truth table is given below and draw its symbol:* [2]

A	B	Y
0	0	1
0	1	1
1	0	1
1	1	0

ANSWERS

PART – I

Answer 1.

(A) (i) (d) $\varepsilon_w = 81\varepsilon_0$ (ii) (a) 4 Ω

(iii) (b) $\vec{\delta B} = \frac{\mu_0}{4\pi}\frac{I(\vec{dl}\times\vec{r})}{r^3}$

(iv) (b) Objective has greater focal length.

(v) (c) Neutrons

(B) (i) Electric flux, $\phi = \frac{q}{\varepsilon_0} = \frac{35.4\times10^{-6}}{8.85\times10^{-12}}$

$= 4 \times 10^6$ N m²/C

(ii) We know that, $I = neAv_d$

Drift velocity, $v_d = \frac{I}{neA} = \frac{I}{ne\pi r^2}$ ...(1)

When the radius is doubled,

Drift velocity, $v'_d = \frac{I}{ne\,\pi(2r)^2}$ ...(2)

Now, $\frac{v_d}{v'_d} = \frac{\frac{I}{ne\,\pi r^2}}{\frac{I}{ne\,\pi r^2(4)}}$

[Using equation (1) and (2)]

$\frac{v_d}{v'_d} = 4$

$v'_d = \frac{v_d}{4} = \frac{1}{4}$th of the original

(iv) The S.I. unit of magnetic dipole moment is Am².

(v) Capacitor

(vi) Spherical wavefront

(vii) $n = \tan i_p$

$\tan i_p = \frac{1}{\sin i_c}$

$i_p = \tan^{-1}(\text{cosec } i_c)$

(viii) Given : P = – 5 D

We have,

$P = \frac{1}{F(\text{in m})}$

$F = \frac{1}{P} = \frac{100}{-5} = -20$ cm

This is concave lens. **Ans.**

(ix) **Modulation:** The audio signal is superimposed on a high-frequency wave. Some characteristic of which varies as a function of the instantaneous value of the audio signal and the resulting wave so produced is transmitted. The audio signal is called modulating signal or modulating wave, the high frequency wave is called carrier wave and the resulting wave is called modulated wave. This process is called modulation.

(x) Fraunhoffer's lines

(xi) $\lambda = \frac{h}{p}$

where, λ = wavelength
h = Planck's constant
p = momentum

(xii) Lyman series

(xiii) ${}_1H^2$

(xiv) Pair production

(xv) For getting n-type semiconductor, pentavalent impurity for *e.g.*, antimony, arsenic etc., is added in it.

PART – II
SECTION – A

Answer 2.

(a) (i) (1) $$E_1 = \frac{1}{4\pi\varepsilon_0}\frac{2pr}{(r^2-a^2)^2} \text{ N/C}$$

(2) $$E_2 = \frac{1}{4\pi\varepsilon_0}\frac{p}{(r^2+a^2)^{3/2}} \text{ N/C}$$

(ii) For short electric dipole, a^2 may be neglected.

$$E_1 = 2E_2$$

$$E_1 : E_2 = 2:1$$

(b) (i) In the given figure, C_1 and C_2 are connected in series.

$$\frac{1}{C'} = \frac{1}{6}+\frac{1}{12}$$

$$\frac{1}{C'} = \frac{2+1}{12} = \frac{3}{12} = \frac{1}{4}$$

$$C' = 4\ \mu F$$

Since, the capacitors are connected in series so, same charge flows through each capacitor.

$$\therefore \quad q = C'V$$

$$q = 4\times10^{-6}\times100$$

$$= 400\times10^{-6}$$

$$= 400\ \mu C \qquad \textbf{Ans.}$$

(ii) $$U_E = \frac{1}{2}CV^2$$

$$= \frac{1}{2}\times20\times10^{-6}\times(100)^2$$

$$= 1\times10^{-1} = 0{\cdot}1\ J \qquad \textbf{Ans.}$$

(c)

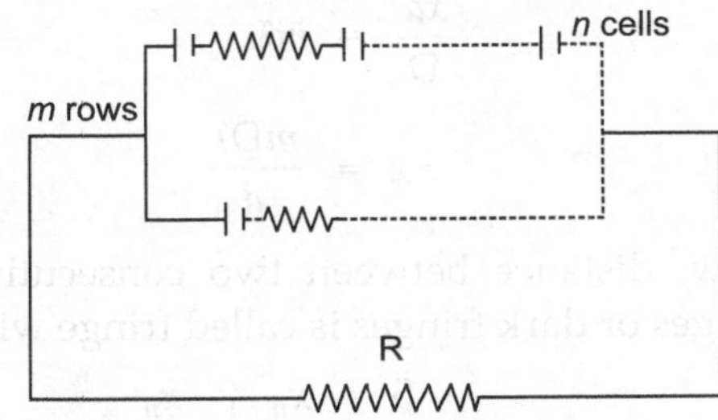

(i) emf $= ne$

Internal resistance = ?

Internal resistance of 1 row $= nr$

$$\frac{1}{r'} = \frac{1}{nr}+\frac{1}{nr}+\ldots\ldots m \text{ times}$$

$$\frac{1}{r'} = \frac{m}{nr}$$

$$r' = \frac{nr}{m} \qquad \textbf{Ans.}$$

(ii) $$I = \frac{ne}{\left(\frac{nr}{m}+R\right)}$$

$$= \frac{ne\times m}{nr+mR} = \frac{nme}{nr+mR}$$

But $nm = N$

So, $$I = \frac{Ne}{nr+mR} \qquad \textbf{Ans.}$$

Answer 3.

(a) Given : $E_1 = 17$ V, $E_2 = 21$ V, $R_1 = 2\ \Omega$, $R_2 = 3\ \Omega$ and $R_3 = 5\ \Omega$.

In loop AFEB,

$$I_1R_1 + (I_1+I_2)R_3 = E_1$$

$$I_1\times2 + (I_1+I_2)5 = 17$$

$$2I_1+5I_1+5I_2 = 17$$

$$7I_1+5I_2 = 17 \qquad \ldots(1)$$

In loop BCDE,

$$I_2R_2 + (I_1+I_2)R_3 = E_2$$

$$I_2\times3 + (I_1+I_2)5 = 21$$

$$5I_1+8I_2 = 21 \qquad \ldots(2)$$

On multiplying equation (1) by 5 and equation (2) by 7, we get

$$35I_1+25I_2 = 85$$

$$35I_1+56I_2 = 147$$

$$- \quad - \qquad -$$

$$-31I_2 = -62$$

$$I_2 = 2\text{ A}$$

$$7I_1+5I_2 = 17 \text{ [from equation (1)]}$$

$$7I_1+10 = 17$$

$$7I_1 = 7$$

$$I_1 = 1\text{ A}$$

$$I = I_1+I_2$$

$$= 1+2 = 3\text{ A} \qquad \textbf{Ans.}$$

(b) Conversion of Galvanometer into Ammeter:

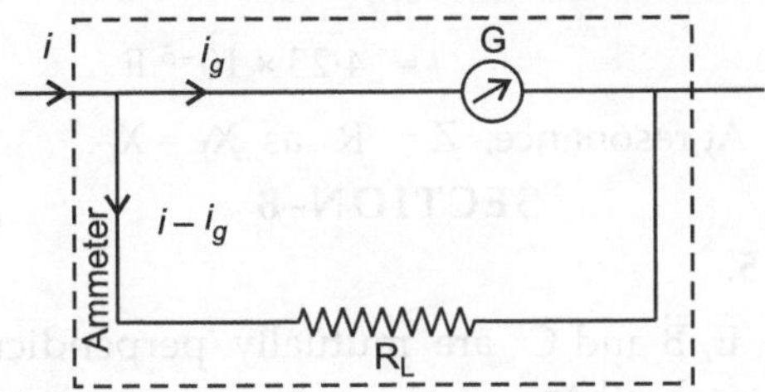

Conversion of Galvanometer into Voltmeter:

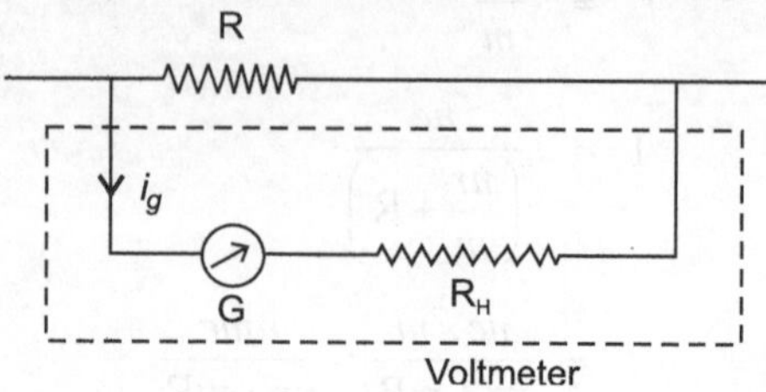

Answer 4.

(a) (i) The force per unit length acts on the conductor Y due to the current flowing through X is,

$$\frac{F}{l} = \frac{\mu_0 2I_1I_2}{4\pi a}$$

(ii) Definition of 1 ampere: 1 ampere is the current which when flowing in each of two infinitely long parallel conductors placed 1 metre apart in vacuum produces between them a force of exactly 2×10^{-7} newton per metre of length.

(b) (i) Induced e.m.f., $e = Bvl = 0{\cdot}1 \times 5 \times 0{\cdot}40 = 0{\cdot}2$ V **Ans.**

(ii) Direction of induced current

$= D \rightarrow C \rightarrow Q \rightarrow R$ (anticlockwise)

(c) (i) (1)

$$f = 50 \text{ Hz}$$
$$I = ?$$
$$V_1 = IR$$
$$80 = I \times 100$$
$$I = \frac{80}{100} = 0{\cdot}8 \text{ A}$$ **Ans.**

(2)

$$V_2 = 60 \text{ V}$$
$$V_2 = I \times X_C$$
$$60 = 0{\cdot}8 \times X_C$$
$$X_C = \frac{60 \times 10}{8}$$
$$= 75\ \Omega$$
$$X_C = \frac{1}{2\pi fC}$$

or

$$C = \frac{1}{2\pi f \times X_C}$$
$$= \frac{1}{2\pi \times 50 \times 75}$$
$$= 4{\cdot}25 \times 10^{-5} \text{ F}$$ **Ans.**

(ii) At resonance, Z = R as $X_L = X_C$.

SECTION–B

Answer 5.

(a) (i) $\vec{E}$, $\vec{B}$ and $\vec{C}$ are mutually perpendicular to each other.

(ii) We know

$$v = \frac{d}{t}$$
$$t = \frac{1.5 \times 10^{11}}{3 \times 10^{8}}$$
$$= 500 \text{ seconds.}$$ **Ans.**

(b) Let 'S' be the monochromatic source of light and S_1, S_2 be the two parallel narrow slits close to each other. According to Huygen's principle, S_1 and S_2 act as new wave source from which secondary wave originate. Let S_1A is the perpendicular drawn on S_2P from S_1. Path difference is given by

$$S_2A = S_2P - S_1P$$

In right angled ΔS_1S_2A, $\sin\theta = \frac{S_2A}{S_1S_2}$

In right angled Δ PCO, $\tan\theta = \frac{OP}{CO}$

If θ is small, then $\frac{S_2A}{S_1S_2} = \frac{OP}{CO}$

$$\frac{S_2A}{d} = \frac{xd}{D}$$

So, Path difference $= S_2A = \frac{xd}{D}$

The intensity of light will be maximum at those places where the path difference between the interfering waves is zero or integral multiple of λ.

So, $\frac{xd}{D} = m\lambda$

or $x = \frac{mD\lambda}{d}$

Now, distance between two consecutive bright fringes or dark fringes is called fringe width (β).

$$\therefore \quad \beta = x_{n+1} - x_n$$
$$= \frac{(n+1)D\lambda}{d} - \frac{nD\lambda}{d}$$
$$\Rightarrow \quad \beta = \frac{D\lambda}{d}$$

(c) (i) 1. In an unpolarised light, the vibrations of the electric vector occur symmetrically in all possible directions in a plane perpendicular to the direction of propagation of light.

2. In polarised light, the vibrations of the electric vector are in a plane perpendicular to the direction of propagation of light and they are confined to a single direction.

(ii) 1. Width of the central bright fringe in the diffraction pattern decreases when light of smaller wavelength is used.

2. Width increases when slit is made narrow.

Answer 6.

(a) Using from figure $\sin r_2 = \frac{1}{n}$

$$\sin r_2 = \frac{1}{1.6} = 0{\cdot}625$$
$$r_2 = 38{\cdot}68°$$
$$r_1 + r_2 = 60°$$
$$r_1 = 60° - 38{\cdot}68°$$
$$r_1 = 21.32°$$
$$n = \frac{\sin i}{\sin r_1}$$
$$\sin i = 1{\cdot}6 \ \sin r_1$$
$$\sin i = 0{\cdot}5817$$
$$i = 35{\cdot}57°$$ **Ans.**

(b) This happens when object is placed at the focus of the convex lens and after refraction from the convex lens the rays becomes parallel and after reflection from plane mirror it converges to focus.

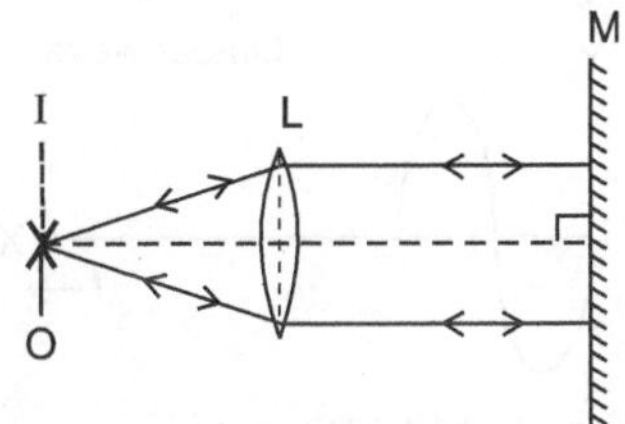

(c) Refer to ISC Paper 2017 Answer 10 (b).

Answer 7.

(b)

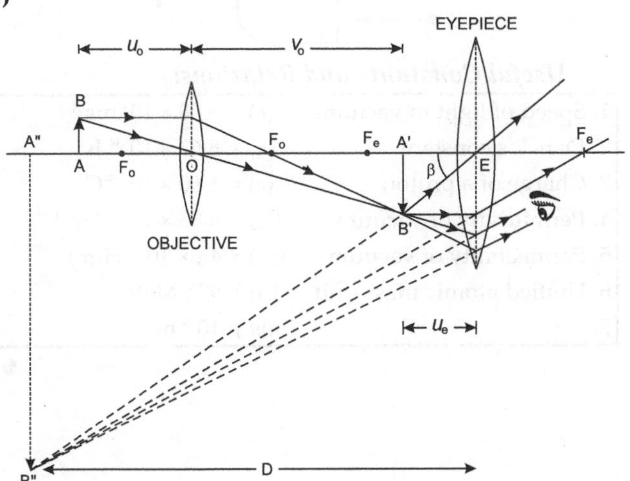

Expression of magnifying power

$$m = -\frac{v_o}{u_o}\left(1 + \frac{D}{f_e}\right)$$

where, v_0 is the image distance from the objective lens

u_o is the object distance from the objective lens

D is the least distance of distinct vision

f_e is the focal length of eye lens.

SECTION–C

Answer 8.

(a) (i) The transfer of electrons from one substance to the other lead to charging of the bodies.

(b) Given: V = 0·5 volt

So, stopping potential

$$V_0 = 0{\cdot}5 \text{ volt}$$
$$\lambda = 400 \text{ nm}$$
$$W = ?$$

We know,

$$E_n = eV_0$$

The photoelectric equation is

$$hv = W + E_n$$
$$\frac{hc}{\lambda} = W + eV_0$$
$$\frac{6.6 \times 10^{-34} \times 3 \times 10^8}{400 \times 10^{-9}} = W + 1{\cdot}6 \times 10^{-19} \times 0{\cdot}5$$
$$W = 4{\cdot}95 \times 10^{-19} - 8 \times 10^{-20}$$
$$= 4{\cdot}15 \times 10^{-19} \text{ J}$$
$$W = 2{\cdot}59 \text{ eV.}$$ **Ans.**

(c) (i) Photoelectric effect.

(ii) Davisson and Germer experiment.

Answer 9.

(a) (i) **(1)** Electrons can revolve only in those orbits in which their angular momentum is an integral multiple of $\frac{h}{2\pi}$, where h is Planck's constant. According to this postulate

$$mvr = \frac{nh}{2\pi}$$

(2) When the atom receives energy from outside, then one of its outer electron leaves its orbit and goes to some higher orbit. This state of atom is called excited state. The electron in the higher orbit stays for 10^{-8} second and then returns back to the lower orbit, while returning back it radiates out energy. The energy of emitted photon is $h\nu$.

$$h\nu = E_2 - E_1$$

(ii) The negative sign shows that the electron is attracted towards the nucleus by electrostatic force and to remove the electron energy must be supplied. The energy needed to remove the electron in the first orbit is + 13·6 eV and this is called ionising energy of hydrogen.

(b) (i) Graph showing variation of relative intensity of X-rays as their wavelength λ is shown below:

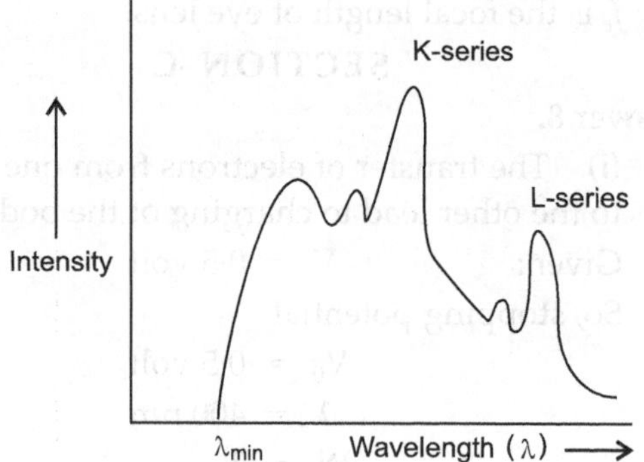

(ii) We know

$$\lambda_{min} = \frac{hc}{eV}$$

λ_{min} can be varied by varying the accelerating potential.

(c) Given: Half life = 6 hrs

Total quantity = 3·2 kg

Total time = one day = 24 hrs.

We have $n = \frac{t}{T} = \frac{24}{6} = 4$ hours

We know, $N = N_0\left(\frac{1}{2}\right)^n = N_0\left(\frac{1}{2}\right)^4$

$= \frac{N_0}{16} = \frac{3.2}{16} = 0{\cdot}2$ kg left

i.e., it will disintegrate = 3·2 – 0·2

= 3 kg in one day. **Ans.**

Answer 10.

(a) (i) **Binding energy per nucleon:** It is the average energy required to remove a nucleon from the nucleus to infinite distance. Higher the average binding energy per nucleon, greater is the stability of the nucleus.

(ii) Mass of 3 alpha particles = 3 × 4·002604 u

= 12·007812 u

Mass of 1 $_6C^{12}$ atom = 12·000000 u

Mass Defect, Δm = 12·007812 – 12·000000

= 0·007812 u

Energy released = 0·007812 × 931·5

= 7·2768 MeV. **Ans.**

(b) (i) A half wave rectifier

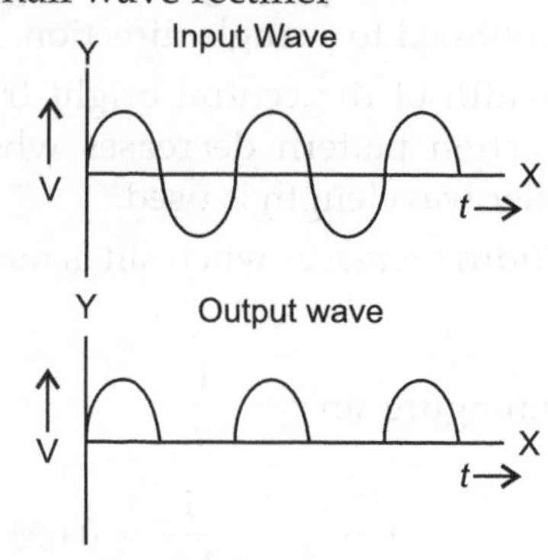

(ii) A full-wave rectifier

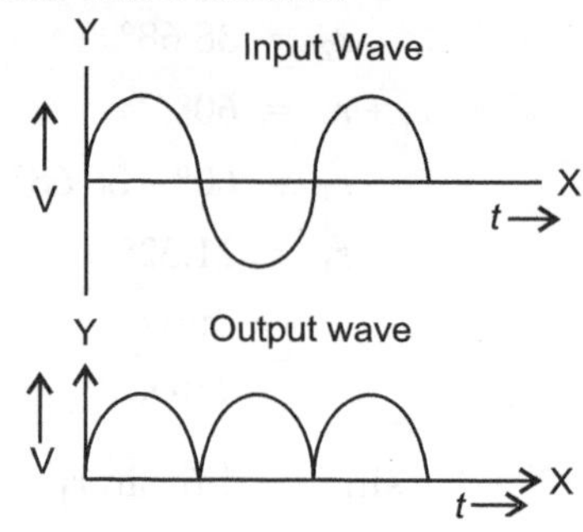

(iii) An amplifier

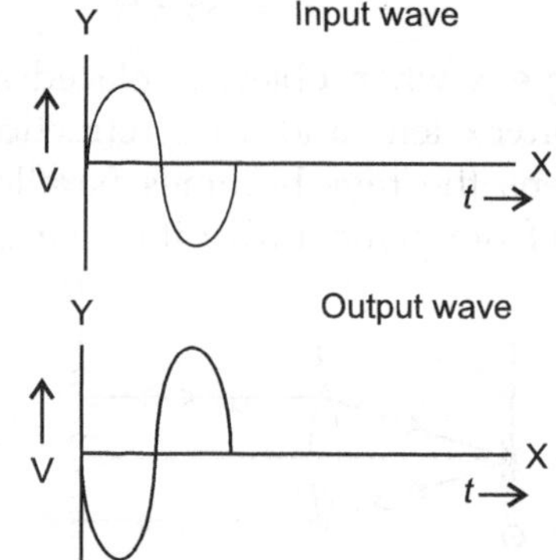

(c) The logic gate is NAND gate.

Symbol :

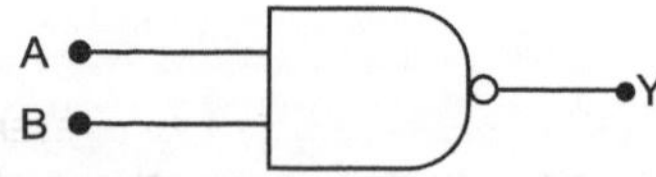

Useful Constants and Relations:

1. Speed of light in vacuum	$(c) = 3{\cdot}00 \times 10^8$ ms^{-1}
2. Planck's constant	$(h) = 6{\cdot}63 \times 10^{-34}$ Js
3. Charge of a proton	$(e) = 1{\cdot}60 \times 10^{-19}$ C
4. Permittivity of vacuum	$(\varepsilon_0) = 8{\cdot}85 \times 10^{-12}$ Fm^{-1}
5. Permeability of vacuum	$(\mu_0) = 4\pi \times 10^{-7}$ Hm^{-1}
6. Unified atomic mass unit	1 u = 931 MeV
7.	1 *nm* = 10^{-9} m

●●

PHYSICS 2012

QUESTIONS

PART–I

*(Answer **all** questions)*

Question 1.

(A) *Choose the correct alternative (a), (b), (c) or (d) for each of the questions given below:* **[5]**

(i) *A body has a positive charge of 8×10^{-19} C. It has :*
(a) *an excess of 5 electrons*
(b) *a deficiency of 5 electrons*
(c) *an excess of 8 electrons*
(d) *a deficiency of 8 electrons*

(ii) *Figure below shows five dc sources (cells). Their emfs are shown in the figure.*

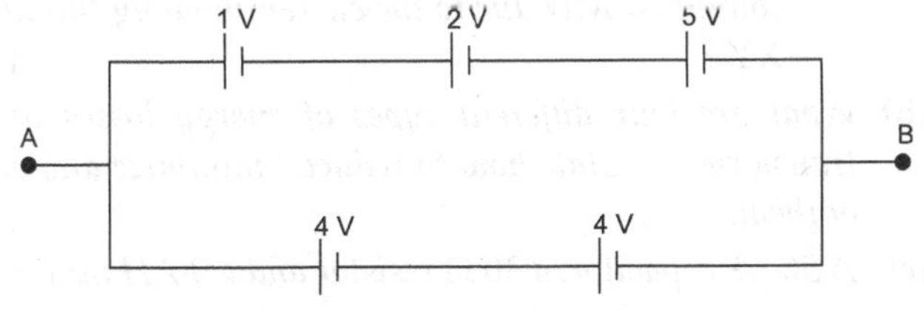

emf of the battery AB is :
(a) *8 V* (b) *16 V*
(c) *4 V* (d) *2 V*

(iii) *Which one of the following graphs below represents variation of reactance 'X_c' of a capacitor with frequency 'f' of an a.c. supply?*

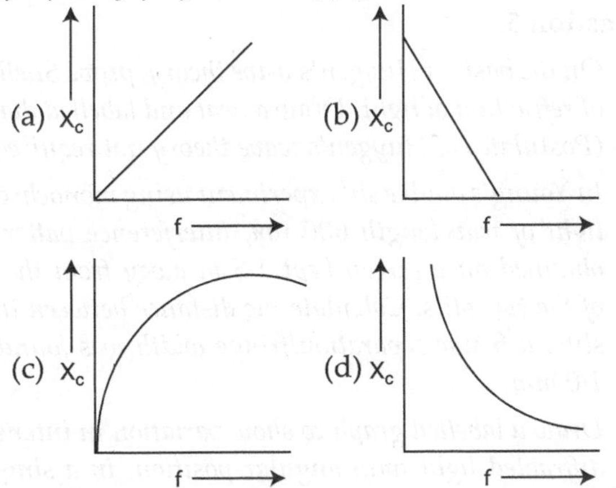

(iv) *White light is passed through sodium vapours contained in a thin walled glass flask and the transmitted light is examined with the help of a spectrometer. The spectrum so obtained is:*
(a) *Absorption spectrum*
(b) *Solar spectrum*
(c) *Band spectrum*
(d) *Continuous spectrum*

(v) *Binding energy of a nucleus is of the order of:*
(a) *electron volt (eV)*
(b) *kilo electron volt (keV)*
(c) *mega electron volt (MeV)*
(d) *a joule (J)*

(B) *Answer all questions briefly and to the point :* **[15]**

(i) *A point charge of 5×10^{-6} C experiences a force of 2×10^{-3} N when kept in a uniform electric field of intensity E. Find E.*

(ii) *Which conservation principle is involved in Kirchhoff's first law of electric circuits ?*

(iii) *Write an expression of magnetic flux density 'B' at a point in end-on position or an axial position of a magnetic dipole. (Derivation not required.)*

(iv) *In a moving coil galvanometer, what is meant by a radial magnetic field ?*

(v) *Variation of alternating current 'I' with time 't' is shown in the graph below.*

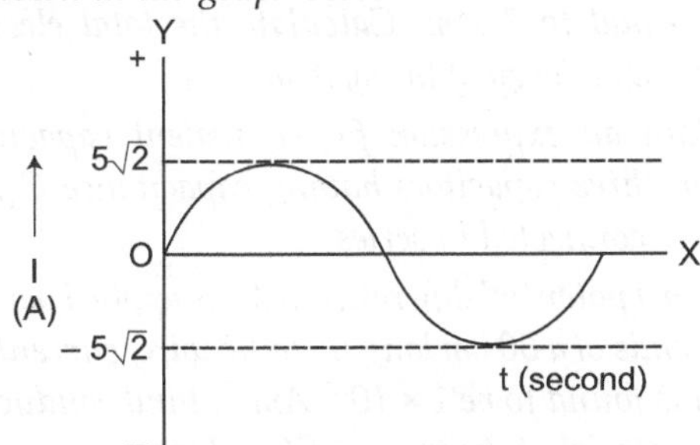

What is the r.m.s. value of this current ?

(vi) *Which electromagnetic radiation has wavelength greater than that of X-rays and smaller than that of visible light ?*

(vii) *How did Fresnel construct a biprism in order to study interference of light ?***

(viii) *State Brewster's law of polarisation of light.*

(ix) *A thin convex lens (L_1) of focal length 80 cm and a thin concave lens (L_2) of focal length f are kept co-axially, 20 cm apart as shown in figure below. When a narrow and parallel beam of light is incident on the convex lens, beam emerging from the concave lens is also a parallel beam. Find f.*

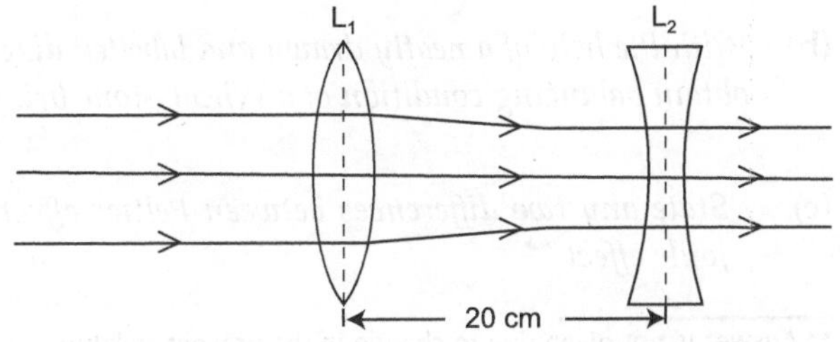

** Answer is not given due to change in the present syllabus.

(x) *What condition must be satisfied by two thin lenses kept in contact so that they form an achromatic doublet, i.e., a combination free from chromatic aberration?* **

(xi) *Threshold frequency of a certain metal for photoelectric emission is 5×10^{14} Hz. Calculate its work function.*

(xii) *What conclusion was drawn by Rutherford based on Geiger-Marsden's experiment on scattering of alpha particles ?*

(xiii) *Write a balanced nuclear reaction showing emission of a β-particles by $^{234}_{90}Th$. (Symbol of daughter nucleus formed in the process is Pa.)*

(xiv) *What is the essential difference between the working of a nuclear reactor and that of a fission bomb ?*

(xv) *State one important use of a Zener diode.*

PART–II

SECTION–A

*(Answer any **two** questions)*

Question 2.

(a) *Three point charges Q_1 = 25 μC, Q_2 = 50 μC and Q_3 = 100 μC, are kept at the corners A, B and C respectively of an equilateral triangle ABC having each side equal to 7·5 m. Calculate the total electrostatic potential energy of the system.* **[4]**

(b) *Obtain an expression for equivalent capacitance C, when three capacitors having capacitance C_1, C_2 and C_3 are connected in series.* **[3]**

(c) *When a potential difference of 3V is applied between the two ends of a 60 cm long metallic wire, current density in it is found to be 1×10^{-7} Am^{-2}. Find conductivity of the material of the wire in SI system.* **[2]**

Question 3.

(a) *In the circuit shown below, E is a battery of emf 6 V and internal resistance 1 Ω. Find the reading of the ammeter A, if it has negligible resistance:* **[4]**

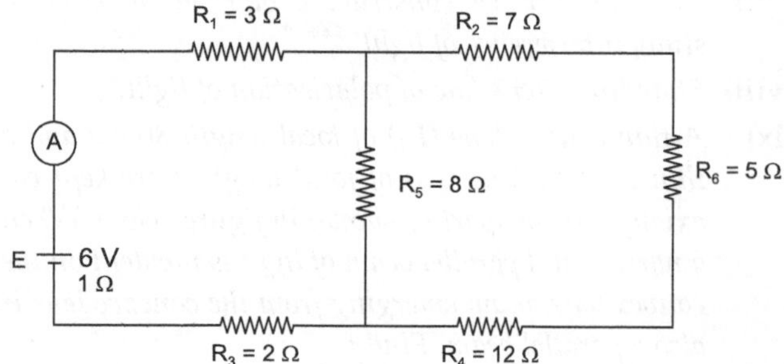

(b) *With the help of a neatly drawn and labelled diagram, obtain balancing condition of a Wheat-stone bridge.* **[3]**

(c) *State any two differences between Peltier effect and Joule effect.* ** **[2]**

** Answer is not given due to change in the present syllabus.

Question 4.

(a) *Figure below shows a point P near a long conductor XY carrying a current I. MN is a short current carrying conductor, kept at the point P, parallel to the conductor XY.*

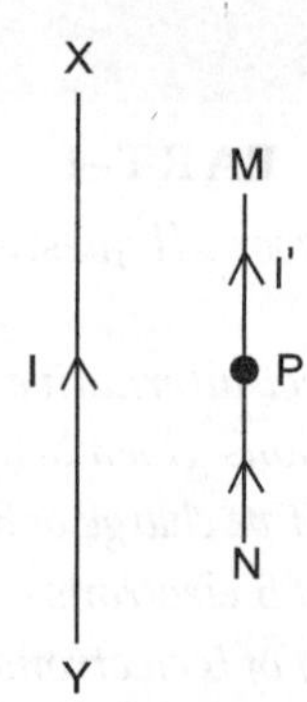

(i) *What is the direction of magnetic flux density 'B' at the point P due to the current flowing through XY ?*

(ii) *What is the direction of the force experienced by the conductor MN due to the current flowing through XY ?* **[3]**

(b) *What are four different types of energy losses in a transformer ? State how to reduce / minimize any one of them.* **[4]**

(c) *A 50 μF capacitor, a 30 Ω resistor and a 0·7 H inductor are connected in series to an a.c. supply which generates an emf 'e' given by e = 300 sin (200 t) volt. Calculate peak value of the current flowing through the circuit.* **[2]**

SECTION–B

*(Answer any **two** questions)*

Question 5.

(a) *On the basis of Huygen's wave theory, prove Snell's law of refraction of light. Draw a neat and labelled diagram. (Postulates of Huygen's wave theory not required)* **[3]**

(b) *In Young's double slit experiment using monochromatic light of wavelength 600 nm, interference pattern was obtained on a screen kept 1·5 m away from the plane of the two slits. Calculate the distance between the two slits, if fringe separation/fringe width was found to be 1·0 mm.* **[3]**

(c) *Draw a labelled graph to show variation in intensity of diffracted light with angular position, in a single slit diffraction experiment.* **[2]**

Question 6.

(a) *You are provided with a narrow and parallel beam of light. State how you will determine experimentally, whether it is a beam of ordinary (unpolarised) light, partially polarised light or completely polarised light.* **[3]**

(b) *For any prism, show that refractive index of its material is given by:* **[3]**

$$n \text{ or } \mu = \frac{\sin\left(\frac{A+\delta_m}{2}\right)}{\sin\left(\frac{A}{2}\right)}$$

where the terms have their usual meaning.

(c) *Figure below shows a parallel beam of monochromatic light incident on a convex spherical surface, radius of curvature R = 30 cm, which separates glass (refractive index = 1·6) from air. Find the position of the image formed due to refraction of light at this single spherical surface.* **[2]**

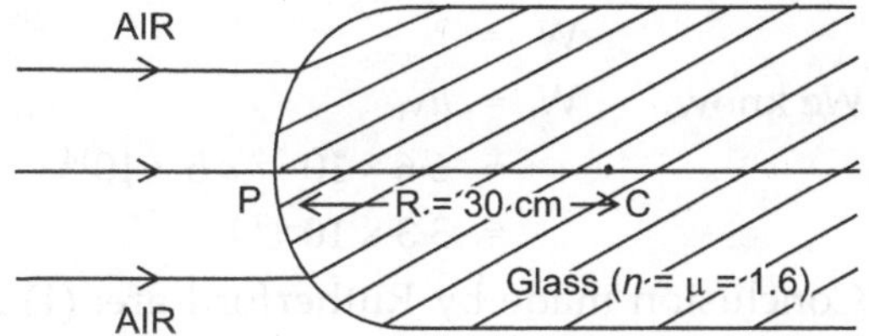

Question 7.

(a) (i) *What is meant by :* **

(1) *Spherical aberration ?*

(2) *Chromatic aberration ?*

(ii) *How can spherical aberration be reduced/ minimized ? Suggest any one method.*** **[3]**

(b) *A compound microscope consists of an objective of focal length 2 cm and an eyepiece of focal length 5 cm. When an object is kept 2·4 cm from the objective, final image formed is virtual and 25 cm from the eyepiece. Determine magnifying power of this compound microscope in this set up i.e., in normal use.* **[3]**

(c) (i) *Define resolving power of a simple astronomical telescope.* **[2]**

(ii) *State one advantage of a reflecting telescope over refracting telescope.*

SECTION—C

*(Answer any **two** questions)*

Question 8.

(a) *Electrons having a velocity $\vec{v}$ of 2×10^6 ms^{-1} pass undeviated through a uniform electric field $\vec{E}$ of intensity 5×10^4 Vm^{-1} and a uniform magnetic field $\vec{B}$·* **[3]**

(i) *Find the magnitude of magnetic flux density B of the magnetic field.*

(ii) *What is the direction of $\vec{B}$ if $\vec{v}$ is towards right and $\vec{E}$ is vertically downwards in the plane of this paper ?*

(b) *Monochromatic light of wavelength 198 nm is incident on the surface of a metal, whose work function is 2·5 eV. Calculate the stopping potential.* **[2]**

(c) *Energy levels of H atom are given by :* **[3]**

$$E_n = -\frac{13.6}{n^2} \text{ eV},$$

where n is principal quantum number. Calculate the wavelength of electromagnetic radiation emitted by hydrogen atom resulting from the transition : n = 2 to n = 1. **[2]**

Question 9.

(a) (i) *What is Compton effect ?*** **[3]**

(ii) *In Coolidge X-ray tube (Modern X-ray tube) how will you vary :*

(1) *intensity of emitted X-rays ?*

(2) *penetrating power of emitted X-rays ?*

(b) *Complete the following table for a radioactive element whose half life is 5 minutes. Assume that you have 32 g of this element at start, i.e., at t = 0.* **[3]**

Time 't' in minutes	0	5	10	15	20	25
Amount of radioactive element left in grams	32					

Now, using this data, plot the "decay curve".

(c) *Calculate the energy released when an electron annihilates a positron.* **[2]**

Question 10.

(a) *Draw a labelled energy band diagram for a solid which is an insulator. What is the main difference between this diagram and that of a semiconductor ?* **[3]**

(b) *Figure below shows the circuit of an electronic device:* **[3]**

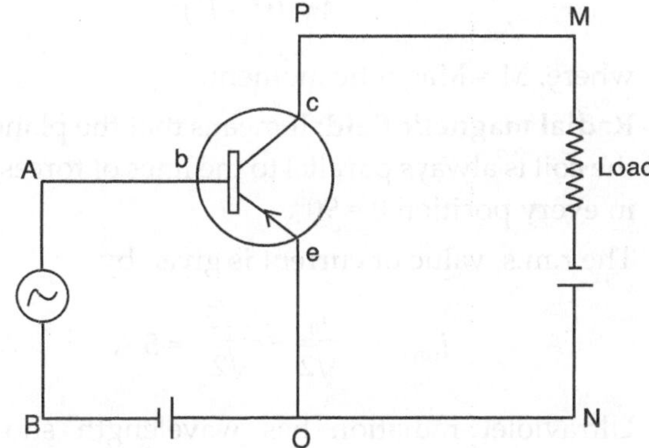

(i) *Which electronic device: a rectifier, an amplifier or an oscillator does the above circuit represent ?*

(ii) *State where the input voltage is applied and where the output voltage is available.*

(iii) *Compare the output voltage of this circuit with its input voltage.*

(c) *Prepare a truth table for the combination of gates shown in figure below:* **[2]**

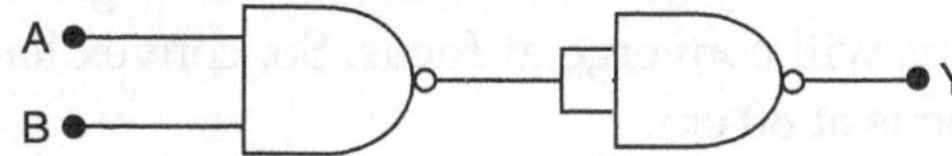

** Answer is not given due to change in the present syllabus.

ANSWERS

PART–I

Answer 1.

(A) (i) (b) a deficiency of 5 electrons

(ii) (a) 8 V

(iii) (d)

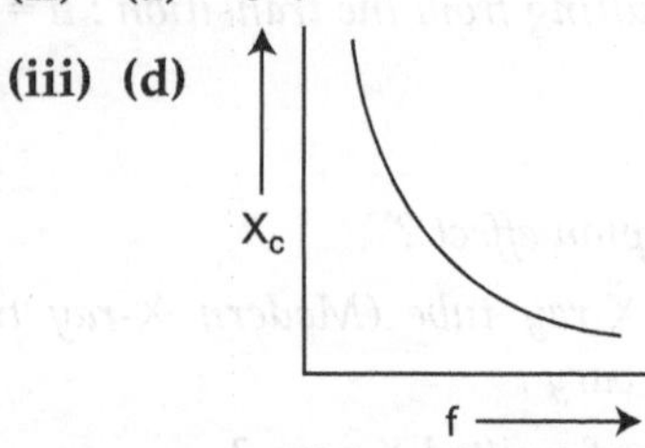

(iv) (a) Absorption spectrum

(v) (c) mega electron volt (MeV)

(B) (i) Given: $q = 5 \times 10^{-6}$ C

$F = 2 \times 10^{-3}$ N

$E = ?$

We know,

$$E = \frac{F}{q}$$

$$= \frac{2\times10^{-3}}{5\times10^{-6}} = 400 \text{ N/C} \quad \textbf{Ans.}$$

(ii) Kirchhoff's first law is based on law of conservation of charge.

(iii) Expression for 'B' in the end-on position is given by

$$B = \frac{\mu_0}{4\pi}\cdot\frac{2Mr}{(r^2-l^2)^2}$$

where, M = Magnetic moment.

(iv) **Radial magnetic field:** It means that the plane of the coil is always parallel to the lines of forces, so in every position θ = 90°.

(v) The r.m.s. value of current is given by

$$i_{rms} = \frac{i_0}{\sqrt{2}} = \frac{5\sqrt{2}}{\sqrt{2}} = 5 \text{ A} \quad \textbf{Ans.}$$

(vi) Ultraviolet radiation has wavelength shorter than visible light and longer than X-rays.

(viii) Brewster's law states that tangent of angle of polarisation is equal to the refractive index of the medium.

i.e., $n = \tan i_p$

where, n is the refractive index of medium and i_p is the angle of polarisation.

(ix) Parallel rays after refraction through convex lens will converge at focus. So, convex lens will focus at 80 cm.

So, object for concave lens will be at 60 cm from it on the right hand side.

We know, $$\frac{1}{f} = \frac{1}{v} - \frac{1}{u}$$

$$\frac{1}{f} = \frac{1}{\infty} - \frac{1}{60}$$

$$\Rightarrow \quad f = -60 \text{ cm.}$$

So, focal length of concave lens is 60 cm. **Ans.**

(xi) Given : $\nu_0 = 5 \times 10^{14}$ Hz.

$W = ?$

We know, $W = h\nu_0$

$= 6{\cdot}6 \times 10^{-34} \times 5 \times 10^{14}$

$= 3{\cdot}3 \times 10^{-19}$ J.

(xii) Conclusion made by Rutherford are: (1) Atom is hollow from inside, (2) The whole of the positive charge and nearly entire mass of atom must be concentrated in an extremely small space at the centre of an atom. This space is called nucleus.

(xiii) ${}_{90}\text{Th}^{234} \rightarrow {}_{91}\text{Pa}^{234} + {}_{-1}\beta^0 + \bar{\nu}$.

(xiv) In nuclear reactor, controlled chain reaction takes place, whereas, in a fission bomb, uncontrolled chain reaction takes place.

(xv) Zener diode works as a voltage regulator or voltage stabilizer.

PART–II

SECTION–A

Answer 2.

(a) We know,

$$U = \frac{1}{4\pi\varepsilon_0}\frac{q_1q_2}{r}$$

So, $$U = \frac{1}{4\pi\varepsilon_0}\left(\frac{q_1q_2}{r_1} + \frac{q_2q_3}{r_2} + \frac{q_3q_1}{r_3}\right)$$

where, $\frac{1}{4\pi\varepsilon_0} = 9 \times 10^9 \text{ Nm}^2\text{ C}^{-2}$

$$= \frac{9\times10^9}{7.5}(25 \times 10^{-6} \times 50 \times 10^{-6} + 50 \times 10^{-6} \times 100 \times 10^{-6} + 100 \times 10^{-6} \times 25 \times 10^{-6})$$

$= 10{\cdot}5$ J. **Ans.**

(b) Let three capacitors C_1, C_2 and C_3 are connected in series. We know charge remains same in series combination.

V_1 $+q-q$ V_2 $+q-q$ V_3 $+q-q$ V_4

C_1 C_2 C_3

$$V_1 = \frac{q}{C_1}$$

$$V_2 = \frac{q}{C_2}$$

$$V_3 = \frac{q}{C_3}$$

$$V = V_1 + V_2 + V_3$$

$$= q\left(\frac{1}{C_1}+\frac{1}{C_2}+\frac{1}{C_3}\right)$$

If the 3 capacitors are replaced by a single capacitor such that on giving it charge q, a capacitance C is obtained then $V = \frac{q}{C}$

$$\therefore \quad \frac{q}{C} = q\left(\frac{1}{C_1}+\frac{1}{C_2}+\frac{1}{C_3}\right)$$

or
$$\frac{1}{C} = \frac{1}{C_1}+\frac{1}{C_2}+\frac{1}{C_3}$$

(c) Given : $V = 3\text{ V}$

$$\text{Length} = 60\text{ cm} = 0{\cdot}60\text{ m}$$

$$J = 1 \times 10^{-7}\text{ A/m}^2$$

$$E = \frac{V}{d}$$

$$= \frac{3}{0.60} = 5\text{ V/m}$$

We know,

$$J = \sigma E$$

$$1 \times 10^{-7} = \sigma \times 5$$

$$\sigma = 2 \times 10^{-8}\ (\Omega\text{ m})^{-1}. \quad \textbf{Ans.}$$

Answer 3.

(a) In the given circuit diagram, R_2, R_6 and R_4 are in series. So,

$$R_1' = 7 + 5 + 12 = 24\ \Omega$$

Now R_1' and R_5 are in parallel. So,

$$\frac{1}{R_2'} = \frac{1}{8}+\frac{1}{24} = \frac{3+1}{24}$$

$$= \frac{4}{24} = \frac{1}{6}$$

$$R_2' = 6\ \Omega$$

Now, R_2', R_1 and R_3 are in series. So,

$$R = R_2' + R_1 + R_3$$

$$= 6 + 3 + 2 = 11\ \Omega$$

We know, $$i = \frac{E}{R+r} = \frac{6}{11+1}$$

$$= \frac{6}{12} = \frac{1}{2}$$

$$i = 0{\cdot}5\text{ A} \quad \textbf{Ans.}$$

(b) The potential at B is equal to potential at D. So, no current flows through the galvanometer. This is called balanced condition of Wheatstone's bridge.

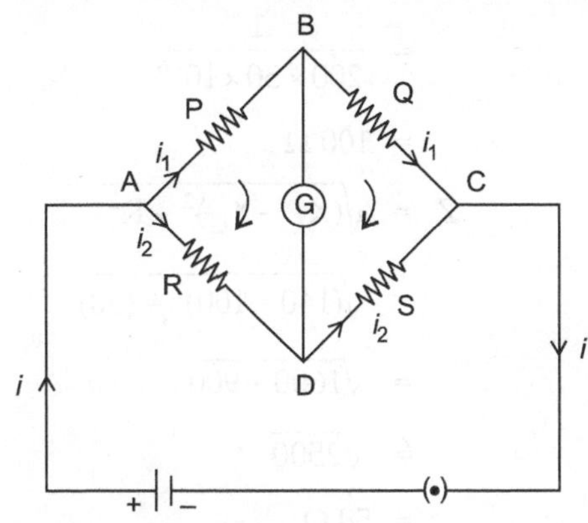

In closed mesh ABDA,

$$-i_1P + i_2R = 0$$

$$i_1P = i_2R \quad \text{...(1)}$$

In closed mesh BCDB,

$$-i_1Q + i_2S = 0$$

$$i_1Q = i_2S \quad \text{...(2)}$$

Divide equation (1) by equation (2), we get

$$\frac{i_1P}{i_1Q} = \frac{i_2R}{i_2S}$$

$$\frac{P}{Q} = \frac{R}{S}$$

Answer 4.

(a) (i) Direction of magnetic flux density 'B' at point P due to current flowing in XY is perpendicular to the plane of paper directed inwards.

(ii) Force experienced by the conductor MN due to the current in XY is towards the conductor XY. *i.e.,* towards LHS (left hand side).

(b) Four types of energy losses in transformer are (1) Copper loss, (2) Eddy current loss, (3) Hysteresis loss, (4) Flux leakage loss.

Copper loss: This loss can be reduced by increasing the cross-sectional area of the conductor or by reducing the winding length.

(c) Given : $C = 50\ \mu F = 50 \times 10^{-6}\text{ F}$

$$R = 30\ \Omega, L = 0{\cdot}7\text{ H}$$

and $$e = 300 \sin (200\,t)\text{ volt.}$$

Comparing the above equation with $e = e_0 \sin \omega t$, we get

$$e_0 = 300\text{ volt},\ i_0 = ?$$

$$\omega = 200\text{ rad s}^{-1} \quad [\because \omega = 2\pi f]$$

Now, $$X_L = \omega L$$

$$= 200 \times 0{\cdot}7 = 140\ \Omega$$

and $X_C = \frac{1}{\omega C}$

$= \frac{1}{200 \times 50 \times 10^{-6}}$

$= 100\ \Omega$

$\therefore \quad Z = \sqrt{(X_L - X_C)^2 + R^2}$

$= \sqrt{(140-100)^2 + (30)^2}$

$= \sqrt{1600+900}$

$= \sqrt{2500}$

$= 50\ \Omega$

Now, peak value of current

$i_0 = \frac{e_0}{Z}$

$= \frac{300}{50} = 6$ A. **Ans.**

SECTION—B

Answer 5.

(a) Refer to ISC Paper 2018 Answer 21 (b) (i).

(b) Given: $\lambda = 600$ nm

$= 600 \times 10^{-9}$ m

$D = 1{\cdot}5$ m

$d = ?$

$\beta = 1$ mm

$= 0{\cdot}001$ m or 1×10^{-3} m

$\beta = \frac{D\lambda}{d}$

$1 \times 10^{-3} = \frac{1.5 \times 600 \times 10^{-9}}{d}$

$d = 9 \times 10^{-4}$ m **Ans.**

(c) Graph showing the variation of intensity of diffracted light with angular positions (θ) is given:

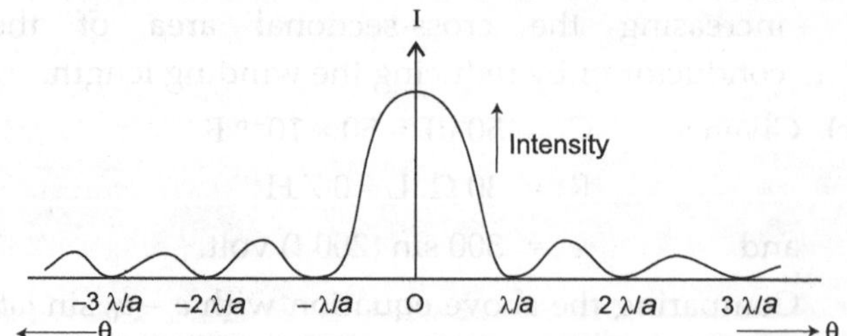

Answer 6.

(a) To detect the given light, it should be passed through a polaroid and is rotated about the incident light: (1) If on rotating the polaroid through one complete rotation, there is no change in the intensity of the emergent light, then the given light is unpolarised, (2) If there is a change in the intensity of emergent light in one complete rotation of the polaroid, the intensity becomes twice maximum and twice zero, then the given light is completely plane-polarised, (3) If the intensity of emergent light changes but never becomes zero, then the given light is partially polarised.

(b) Refer to ISC Paper 2018 Answer 16 (a).

(c) Given : $R = 30$ cm

$n = 1{\cdot}6$

We know, $\frac{n}{v} - \frac{1}{u} = \frac{n-1}{R}$

$\frac{1.6}{v} + \frac{1}{\infty} = \frac{1.6-1}{30}$

$v = \frac{1.6 \times 30}{0.6}$

$v = 80$ cm **Ans.**

Answer 7.

(b) Given: $f_o = 2$ cm, $f_e = 5$ cm

$u_o = -2{\cdot}4$ cm, $v_e = D = 25$ cm, $m = ?$

We know,

$\frac{1}{f_o} = \frac{1}{v_o} - \frac{1}{u_o}$

$\frac{1}{2} = \frac{1}{v_o} + \frac{1}{2.4}$

$\frac{1}{v_o} = \frac{1}{2} - \frac{1}{2.4}$

$\frac{1}{v_o} = \frac{2.4-2}{2 \times 2.4}$

$\therefore \quad v_o = \frac{2 \times 2.4}{0.4}$

$v_o = 12$ cm

$\therefore$ Magnifying power $m = -\frac{v_o}{u_o}\left(1 + \frac{D}{f_e}\right)$

$m = \frac{-12}{-2.4}\left(1 + \frac{25}{5}\right)$

$= \frac{12}{2.4} \times \frac{30}{5}$

$m = 30$ **Ans.**

(c) (i) Resolving power: The ability of the telescope to produce distinctly separate images of two close objects is called resolving power of the telescope.

(ii) One advantage of a reflecting telescope over refracting telescope is that the image formed by reflecting telescope is larger and brighter than the image formed by a refracting telescope.

SECTION—C

Answer 8.

(a) (i) Given :

$$\vec{v} = 2 \times 10^6 \text{ m/s}$$
$$\vec{E} = 5 \times 10^4 \text{ V/m}$$

We know that,

$$\vec{v} = \frac{\vec{E}}{\vec{B}}$$

$$\therefore \quad \vec{B} = \frac{5\times10^4}{2\times10^6}$$
$$= 0.025 \text{ T}$$
$$= 2.5 \times 10^{-2} \text{ T} \quad \textbf{Ans.}$$

(ii) Direction of magnetic field intensity is vertically inside into the plane of paper.

(b) Given :

$$\lambda = 198 \text{ nm}$$
$$= 198 \times 10^{-9} \text{ m}$$
$$W = 2.5 \text{ eV}$$
$$= 2.5 \times 1.6 \times 10^{-19} \text{ J}$$
$$V_0 = ?$$

We know,

$$\frac{hc}{\lambda} = W + E_k$$

$$\frac{6.6\times10^{-34}\times3\times10^8}{198\times10^{-9}} = 2.5 \times 1.6 \times 10^{-19} + E_k$$

$$10 \times 10^{-19} = 4 \times 10^{-19} + E_k$$

$$\therefore \quad E_k = 6 \times 10^{-19} \text{ J}$$

Now,

$$E_k = eV_0$$
$$6 \times 10^{-19} = 1.6 \times 10^{-19} \times V_0$$
$$\therefore \quad V_0 = 3.75 \text{ volt.} \quad \textbf{Ans.}$$

(c) In transition form $E_n = \frac{-13.6}{n^2}$ eV

$$n = 2 \text{ to } n = 1$$
$$\lambda = ?$$

$$\therefore \quad E_2 = \frac{-13.6}{2^2}$$
$$= -3.4 \text{ eV}$$

and

$$E_1 = \frac{-13.6}{1^2}$$
$$= -13.6 \text{ eV}$$

$$\Delta E = E_2 - E_1$$
$$= -3.4 - (-13.6)$$
$$= 10.2 \text{ eV}$$

We know that,

$$\Delta E = \frac{hc}{\lambda}$$

$$\therefore \quad \lambda = \frac{6.6\times10^{-34}\times3\times10^8}{10.2\times1.6\times10^{-19}}$$
$$= 1.213 \times 10^{-7} \text{ m}$$
$$= 1213 \text{ Å}. \quad \textbf{Ans.}$$

Answer 9.

(a) (ii) (1) Intensity: In order to increase the intensity of X-rays, the current through the tube must be increased.

(2) Penetration power: It can be increased by increasing the potential difference applied between the ends of the X-ray tube.

(b)

t	0	5	10	15	20	25
N	32	16	8	4	2	1

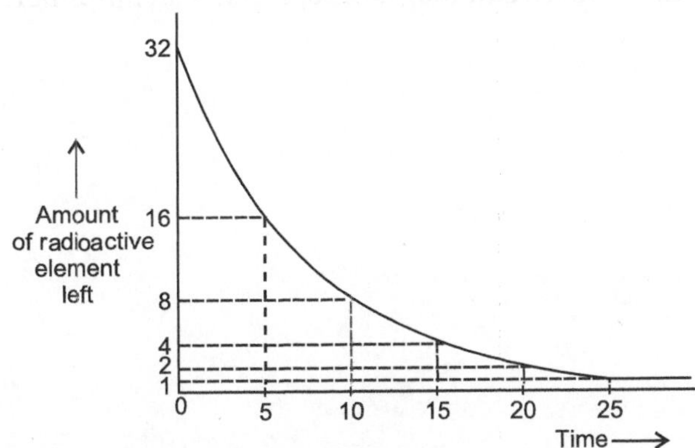

(c) The equation is given by:

$$e^+ + e^- \rightarrow h\nu + h\nu$$

$$\Delta m = 2 \times 9.1 \times 10^{-31}$$
$$= 18.2 \times 10^{-31} \text{ kg}$$
$$\Delta E = \Delta mc^2$$
$$= 18.2 \times 10^{-31} \times (3 \times 10^8)^2$$
$$= 18.2 \times 10^{-31} \times 9 \times 10^{16}$$
$$= 163.8 \times 10^{-15}$$
$$= 1.64 \times 10^{-13} \text{ J}$$
$$= \frac{1.64\times10^{-13}}{1.6\times10^{-19}} \text{ eV}$$
$$= 1.025 \times 10^6 \text{ eV}$$
$$= 1.025 \text{ MeV.} \quad \textbf{Ans.}$$

Answer 10.

(a) Energy band diagram

For an insulator:

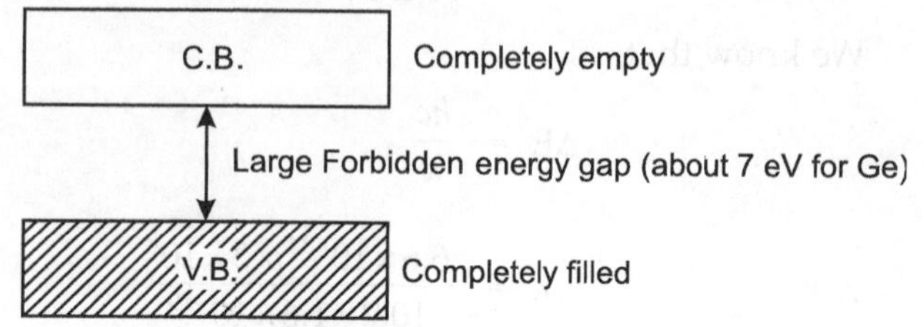

For a Semiconductor :

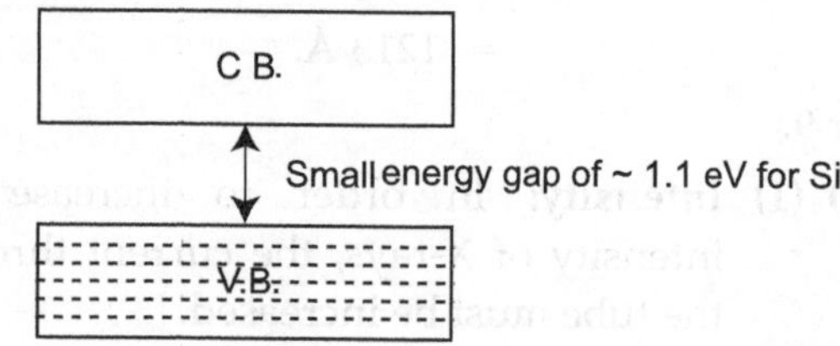

The main difference between the energy band diagram of insulator and that of a semiconductor is that the energy gap between C.B. and V.B. In case of an insulator is about 7 eV whereas in case of a semiconductor it is 1·1 eV in case of Si and 0·7 eV in Ge.

(b) (i) The circuit diagram represents an amplifier.

(ii) Input is applied between AB and output is available across load resistance.

(iii) The output voltage is amplified and is 180° out of phase with its input voltage.

(c) Truth table of the given figure:

A	B	Y_0	Y′
0	0	1	0
0	1	1	0
1	0	1	0
1	1	0	1

where Y′ is the output of the first gate.

Useful Constants and Relations:

1.	Speed of light in vacuum	$(c) = 3{\cdot}0 \times 10^{8}\ ms^{-1}$
2.	Planck's constant	$(h) = 6{\cdot}6 \times 10^{-34}\ Js$
3.	Charge of an electron	$(-e) = -1{\cdot}6 \times 10^{-19}\ C$
4.	Constant for Coulomb's $\left(\frac{1}{4\pi\varepsilon_0}\right)$	$= 9 \times 10^{9}\ mF^{-1}$
5.	Mass of an electron	$(m_e) = 9 \times 10^{-31}\ kg$
6.	1 electron volt	$(1\ eV) = 1{\cdot}6 \times 10^{-19}\ J$
7.	Wavelength	$(1nm) = 10^{-9}\ m$

PHYSICS

2011

QUESTIONS

PART–I

*(Answer **all** questions)*

Question 1.

(A) *Choose the correct alternative (a), (b), (c) or (d) for each of the questions given below :* **[5]**

(i) *Two point charges (+e) and (–e) are kept inside a large metallic cube without touching its sides. Electric flux emerging out of the cube is :*

(a) $\frac{e}{\varepsilon_0}$ (b) $\frac{-e}{\varepsilon_0}$

(c) *Zero* (d) $\frac{2e}{\varepsilon_0}$

(ii) *In current electricity, Ohm's law is obeyed by all:*

(a) *solids* (b) *metals*

(c) *liquids* (d) *gases*

(iii) *When a charged particle is projected perpendicular to a uniform magnetic field, it describes a circular path in which :*

(a) *its speed remains constant.*

(b) *its velocity remains constant.*

(c) *its momentum remains constant.*

(d) *its kinetic energy increases.*

(iv) *Refractive index of a transparent material is :*

(a) *same for all colours.*

(b) *maximum for violet colour.*

(c) *minimum for violet colour.*

(d) *maximum for red colour.*

(v) *Which one of the following graphs in following figure represents variation of de Broglie wavelength (λ) of a particle having linear momentum p :*

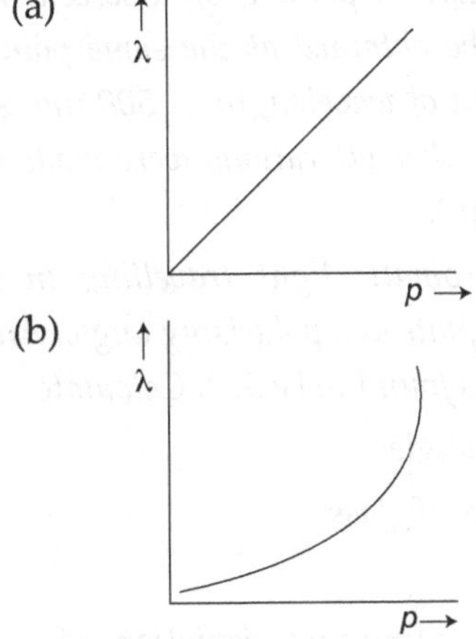

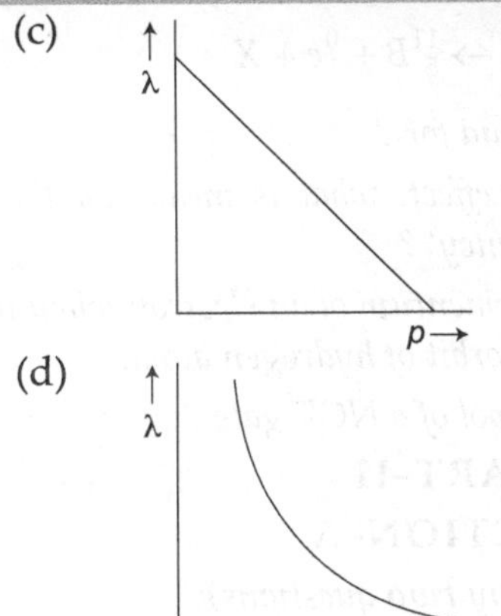

(B) *Answer **all** questions briefly and to the point :***[15]**

(i) *How much work is done in taking an electron around a nucleus in a circular path ?*

(ii) *A 10 m long potentiometer wire carries a steady current. A standard cell of emf 1·018 V is balanced against a length of 254·5 cm of the wire. What is the potential gradient across the potentiometer wire ?*

(iii) *Name any one instrument which works on the principle of Tangent law in magnetism.***

(iv) *An inductor L and a resistor R are connected in series to a battery, through a key/switch. Show graphically, how current decreases with time when the key/switch is opened.***

(v) *An ideal inductor does not consume any power even though both V and I are non zero. Explain in brief.*

(vi) *In Fraunhofer's single slit diffraction experiment, how does semi-angular width θ of the central bright fringe depend on slit width 'a' ?*

(vii) *State one use of a Polaroid.*

(viii) *A ray of light LM incident normally on the surface AC of an isosceles right angled prism ABC (where AB = BC) emerges along PQ, parallel to LM, as shown in given figure.*

What can you say about refractive index μ of the material of the prism ?

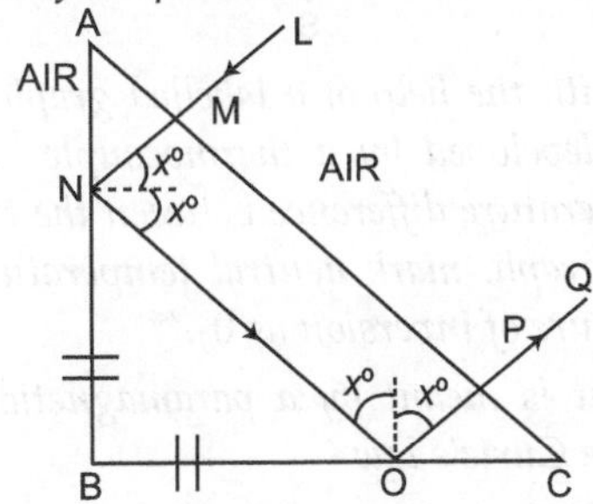

** Answer is not given due to change in the present syllabus.

(ix) *State one condition for obtaining a sustained interference of light.*

(x) *State any one postulate of Huygen's wave theory.*

(xi) *How can an n-type semiconductor be obtained from a pure crystal of germanium ?*

(xii) *In the following nuclear reaction :*

$$^{11}_{6}C \rightarrow ^{11}_{5}B + ^{0}_{1}e + X$$

What does X stand for ?

(xiii) *In photoelectric effect, what is meant by the term 'threshold frequency' ?*

(xiv) *Find angular momentum of an electron when it is in the second Bohr orbit of hydrogen atom.*

(xv) *What is the symbol of a NOT gate ?*

PART–II

SECTION–A

(Answer any **two** *questions)*

Question 2.

(a) *Obtain an expression for intensity of electric field in end on position, i.e., axial position of an electric dipole.* **[4]**

(b) *Three capacitors each of capacitance C are connected in series. Their equivalent capacitance is C_s. The same three capacitors are now connected in parallel. Their equivalent capacitance becomes C_p. Find the ratio $\left(\frac{C_p}{C_s}\right)$.*

(Working must be shown). **[3]**

(c) *A galvanometer with a resistance of 75 Ω produces a full scale deflection with a current of 5 mA. How can this galvanometer be converted into an ammeter which has a range of 0 – 5 A ?* **[2]**

Question 3.

(a) *In the circuit shown in figure below, E_1 and E_2 are batteries having emfs 4·0 V and 3·5 V respectively and internal resistance 1 Ω and 2 Ω respectively. Using Kirchhoff's laws, calculate currents : I_1, I_2 and I_3.* **[4]**

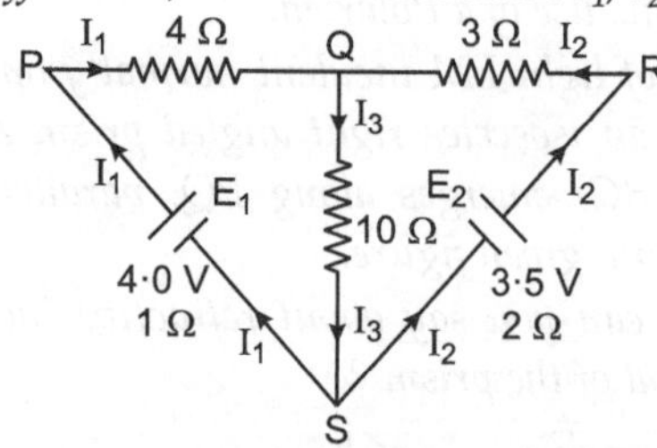

(b) *Show, with the help of a labelled graph, how thermo emf 'e' developed by a thermocouple varies with θ, the temperature difference between the two junctions. On the graph, mark neutral temperature as θ_N and temperature of inversion as θ_1.*** **[2]**

(c) **(i)** *What is meant by a paramagnetic substance ? State Curie's law.*

(ii) *What is the value of magnetic susceptibility of Aluminium if its relative permeability is 1·000022 ?* **[3]**

Question 4.

(a) *Using Ampere's circuital law or Biot-Savart's law, show that magnetic flux density 'B' at a point 'P' at a perpendicular distance 'a' from a long current carrying conductor is given by :*

$$B = \left(\frac{\mu_0}{4\pi}\right)\frac{2I}{a}$$

(Statement of the laws – not required) **[3]**

(b) *A current of 4 A flows in a coil when it is connected to a 12 V dc source. When the same coil is connected to an ac source (12 V, 8 Hz), a current of 2·4 A flows in the coil. Calculate coefficient of self inductance (L) of the coil.* **[4]**

(c) *How much force per unit length acts on a long current carrying conductor X due to a current flowing through another similar conductor Y, kept parallel to it in vacuum?*

Use this equation to define an ampere, the fundamental unit of current. **[2]**

SECTION–B

(Answer any **two** *questions)*

Question 5.

(a) **(i)** *Arrange all the seven types of electro-magnetic radiations in increasing order of their frequencies. (You must begin with a radiation with lowest frequency and end with the one having the highest frequency.)*

(ii) *State how electric vector $\vec{E}$ magnetic vector $\vec{B}$ and velocity vector $\vec{c}$ are oriented in an electromagnetic wave.* **[3]**

(b) *In Young's double slit experiment, using monochromatic light L_1 of wavelength 700 nm, 10th bright fringe was obtained at a certain point P on a screen. Which bright fringe will be obtained at the same point P if monochromatic light of wavelength 500 nm is used in place of L_1 (No other alterations were made in the experimental set up.).* **[3]**

(c) *A certain monochromatic light travelling in air is incident on a glass plate at a polarising angle. Angle of refraction in glass is found to be 32°. Calculate :*

(i) *the polarising angle;*

(ii) *refractive index of glass.* **[2]**

Question 6.

(a) *Calculate angle of minimum deviation (δ_m) for a regular glass prism. (Refractive index of glass = 1·6)* **[3]**

** Answer is not given due to change in the present syllabus.

(b) *Obtain an expression for refraction at a single convex spherical surface, i.e., the relation between μ_1 (rarer medium), μ_2 (denser medium), the object distance u, image distance v and radius of curvature R.* **[3]**

(c) *Where should an object be kept on the principal axis of a convex lens of focal length 20 cm, in order to get an image, which is double the size of the object ?* **[2]**

Question 7.

(a) (i) *What is the use of a spectrometer ?***

(ii) *In a spectrometer, what is the function of :***

(1) *Collimator ?*

(2) *Telescope ?* **[2]**

(b) *Draw a labelled diagram of an image formed by a compound microscope with image at least distance of distinct vision.* **[3]**

(c) *An astronomical telescope consists of two thin convex lenses having focal lengths of 140 cm and 5 cm. The telescope is adjusted to be in normal adjustment.*

(i) *What is the angular magnification, i.e., magnifying power of the telescope in this set up ?*

(ii) *What is the distance between the two lenses equal to ?* **[2]**

Question 8.

(a) *State two important conclusions that can be drawn from Millikan's oil drop experiment to determine the charge of an electron.*** **[3]**

(b) *A monochromatic source of light emits light of wavelength 198 nm. Calculate :*

(i) *energy of each photon;*

(ii) *Momentum of the photon.* **[2]**

(c) (i) *Name a series of lines of hydrogen spectrum which lies in :*

(1) *Visible region*

(2) *Ultraviolet region*

(ii) *Write Bohr's formula to calculate Wavelength (λ) of visible light, emitted by hydrogen, and explain the meaning of each and every symbol used.* **[3]**

Question 9.

(a) *Starting with the law of radioactive disintegration, show that : $N = N_0e^{-\lambda t}$, where the terms have their usual meaning.* **[3]**

(b) *What is meant by pair production ? Explain with the help of an example and a balanced equation.* **[2]**

(c) *An X -ray tube is operated at a tube potential of 40,000 V. Calculate :*

(i) *Kinetic energy of an electron emitted by the filament when it reaches the target/anode.*

(ii) *Wavelengths of all the X-rays emitted by the X-ray tube.* **[3]**

Question 10.

(a) (i) *In the following nuclear reaction, calculate the energy released in MeV :* **[3]**

$$^2_1H + ^2_1H \rightarrow ^3_2He + ^1_0n$$

Given that : *Mass of* = 2·015 u

Mass of e = 3·017 u

Mass of = 1·009 u

(ii) *What is the name of this reaction ?*

(b) *What is meant by the terms :*

(i) *a full wave rectifier ?*

(ii) *an amplifier ?*

(iii) *an oscillator ?* **[3]**

(c) *Using several NAND gates, how can you obtain an AND gate ? Draw a labelled diagram in support of your answer.* **[2]**

** Answer is not given due to change in the present syllabus.

ANSWERS

PART–I

Answer 1.

(A) (i) (c) Zero

(ii) (b) Metals

(iii) (a) Its speed remains constant.

(iv) (b) Maximum for violet colour

(v) (d)

(B) (i) Work done, $W = 0$

(ii) Given : emf, E = 1·018 V, Length, l = 254·5 cm,

Potential gradient, $k = \frac{E}{l} = \frac{1.018}{254.5}$

$= 0.004$ V/cm. **Ans.**

(v) For an ideal inductor the phase angle is 90°.

We know, $P_{av} = E_{rms} \times i_{rms} \times \cos\phi$

$= E_{rms} \times i_{rms} \times \cos 90°$

$= 0$

(vi) We know, semi-angular width, $\theta = \lambda/a$.

where λ is wavelength of light and 'a' is the slit width.

So, as slit width increases the semi-angular width of the central bright fringe decreases.

(vii) Polaroids are used in wind screen of the cars.

(viii) $\angle A = \angle C = 45°$ (isosceles right angled triangle)

$\therefore$ $\angle i = 45°$

For total internal reflection to take place at faces $\sin i > \frac{1}{n}$; where n is the refractive index of prism with respect to air.

$$\sin 45° > \frac{1}{n}$$

$$n > \frac{1}{\sin 45°}$$

$$n > \sqrt{2}$$

Hence, the refractive index of prism is greater than $\sqrt{2}$. **Ans.**

(ix) For sustained interference of light, the two interfering sources must be coherent.

(x) Every particle of the medium situated on the wavefront acts as a new wave source from which the fresh waves originates. These waves are called secondary wavelets.

(xi) 'n' type semiconductor can be obtained from a pure crystal of germanium by adding pentavalent impurity atom such as antimony, phosphorus or arsenic.

(xii) X stands for neutrino.

(xiii) Threshold frequency : The minimum frequency of the incident radiation below which photo-electric emission cannot take place is called threshold frequency.

(xiv) Angular momentum,

$$mvr = \frac{nh}{2\pi} = \frac{2h}{2\pi} = \frac{h}{\pi}$$

$$= \frac{6{\cdot}6 \times 10^{-34}}{3{\cdot}14}$$

$$= 2{\cdot}10 \times 10^{-34} \text{ kg m}^2\text{/s.}$$ **Ans.**

(xv) Symbol of NOT gate :

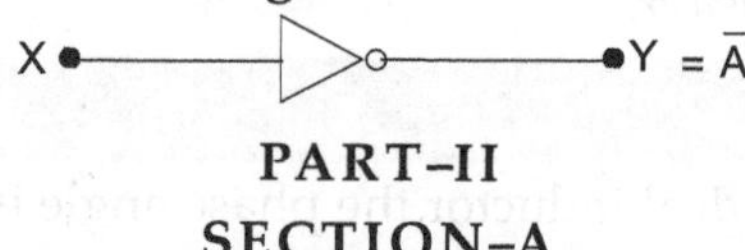

PART–II
SECTION–A

Answer 2.

(a) Let an electric dipole is taken which constitutes $(+q)$ and $(-q)$ charge separated by the distance $2l$.

Let P be a point at a distance r from the centre of the dipole O at which electric field intensity is to be determined. Let E_1 and E_2 be the intensities due to $(+q)$ and $(-q)$ charge respectively.

So, $$E_1 = \frac{1}{4\pi\varepsilon_0}\frac{q}{(r-l)^2};$$

direction along the dipole axis

$$E_2 = \frac{1}{4\pi\varepsilon_0}\frac{q}{(r+l)^2};$$

direction opposite to the dipole axis

Resultant electric field intensity $E = E_1 - E_2$

$$E = \frac{1}{4\pi\varepsilon_0}\frac{q}{(r-l)^2} - \frac{1}{4\pi\varepsilon_0}\frac{q}{(r+l)^2}$$

$$= \frac{1}{4\pi\varepsilon_0} q\left[\frac{1}{(r-l)^2} - \frac{1}{(r+l)^2}\right]$$

$$= \frac{1}{4\pi\varepsilon_0} q\left[\frac{(r+l)^2 - (r-l)^2}{(r-l)^2\,(r+l)^2}\right]$$

$$= \frac{1}{4\pi\varepsilon_0} q\left[\frac{4lr}{(r^2-l^2)^2}\right]$$

$$= \frac{1}{4\pi\varepsilon_0}\frac{2ql \times 2r}{(r^2-l^2)^2}$$

$$E = \frac{1}{4\pi\varepsilon_0}\frac{2pr}{(r^2-l^2)^2} \qquad [\because 2lq = p]$$

If $r >> l$; then l^2 is neglected compared to r^2

$$E = \frac{1}{4\pi\varepsilon_0}\frac{2p}{r^3}$$

(b) Given : Three capacitors each of capacitance C

Series capacitance $= C_s$

Parallel capacitance $= C_p$

$$C_p = C + C + C = 3C \qquad ...(i)$$

and $$\frac{1}{C_s} = \frac{1}{C} + \frac{1}{C} + \frac{1}{C}$$

$$\frac{1}{C_s} = \frac{3}{C}$$

$$C_s = \frac{3}{C} \qquad ...(ii)$$

Now, $$\frac{C_p}{C_s} = \frac{3C}{C/3}$$

[Using equation (i) and (ii)]

$$= 9 : 1$$ **Ans.**

(c) Given :

$$G = 75\,\Omega$$

$$i_g = 5 \text{ mA} = 5 \times 10^{-3} \text{ A}$$

Range = 0 to 5A $\Rightarrow i = 5$ A

We know, $$S = \frac{i_g G}{(i - i_g)}$$

$$= \frac{5 \times 10^{-3} \times 75}{5 - (5 \times 10^{-3})}$$

$$= \frac{375 \times 10^{-3}}{(5 - 0{\cdot}005)}$$

$$= \frac{0{\cdot}375}{4{\cdot}995} = 0{\cdot}075\ \Omega.$$ **Ans.**

A resistor of 0·075 Ω must be connected in parallel to the galvanometer.

Answer 3.

(a) In closed circuit PQSP,

$$1I_1 + 4I_1 + 10I_3 = 4 \quad \ldots(i)$$

Applying Kirchhoff's first law at point S,

$$I_3 = I_1 + I_2$$

Using the value of I_3 in equation (i),

$$I_1 + 4I_1 + 10(I_1 + I_2) = 4$$

$$15I_1 + 10I_2 = 4 \quad \ldots(ii)$$

In closed circuit RQSR,

$$2I_2 + 3I_2 + 10\ (I_1 + I_2) = 3{\cdot}5$$

$$10I_1 + 15I_2 = 3{\cdot}5 \quad \ldots(iii)$$

On solving equations (ii) and (iii), we get

$$I_2 = 0{\cdot}1\text{ A}, I_1 = 0{\cdot}2\text{ A}$$

And $I_3 = I_1 + I_2 = 0{\cdot}3$ A **Ans.**

(c) (i) Paramagnetic substance : It is those substances which when placed in an external magnetic field, are feebly magnetised in the direction of magnetising field.

For example–aluminium, sodium, platinum, etc.

Curie's law : The susceptibility of a paramagnetic substance varies inversely with the absolute temperature of the substance.

$$\chi_m \propto \frac{1}{T}$$

This is known as Curie's law.

(ii) Given : $\mu_r = 1{\cdot}000022$

We know $\mu_r = 1 + \chi_m$

$$1{\cdot}000022 = 1 + \chi_m$$

So, $\chi_m = 0{\cdot}000022$ **Ans.**

Answer 4.

(a) Let XY be a long straight wire carrying current I. Let 'P' is a point at a distance '*a*' from the wire where magnetic field is to be found.

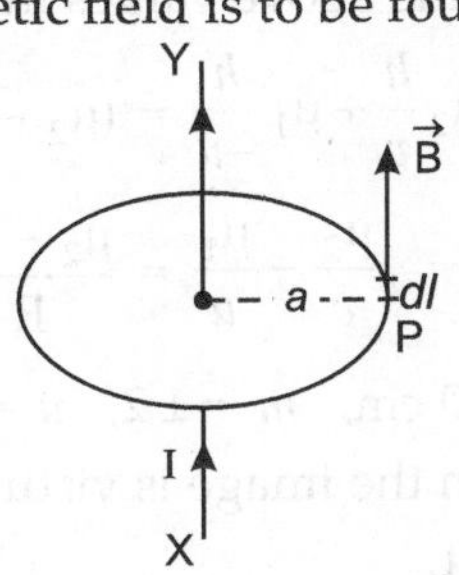

Let us consider a circle element *dl* at P.

$$\oint \vec{B} \cdot \vec{dl} = \oint B \cdot dl$$

$$= B \oint dl$$

Since B is constant around the circle.

$$\oint dl = 2\pi a$$

$$\therefore \quad \oint \vec{B} \cdot \vec{dl} = B \times 2\pi a$$

From Ampere's circuital law,

$$\oint \vec{B} \cdot \vec{dl} = \mu_0 I$$

$$\therefore \quad B \times 2\pi a = \mu_0 I$$

$$\text{or} \quad B = \frac{\mu_0 I}{2\pi a}$$

$$\text{or} \quad B = \left(\frac{\mu_0}{4\pi}\right)\frac{2I}{a}$$

(b) Given : Current of 4 A flow in coil when connected to 12 V d.c. source.

Current of 2·4 A flow in coil when connected to 12 V, 8 Hz a.c. source.

We know,

$$V = I_{dc} R$$

$$12 = 4 \times R$$

$$R = 3\ \Omega$$

When the coil is connected to a.c. source,

$$I_{ac} = \frac{E}{Z}$$

$$2{\cdot}4 = \frac{12}{Z}$$

$$Z = 5\ \Omega$$

But $Z = \sqrt{R^2 + X_L^2}$

$$5 = \sqrt{3^2 + X_L^2}$$

$$\Rightarrow \quad X_L = 4\ \Omega$$

But $X_L = 2\pi f L$

$$\therefore \quad 4 = 2\pi \times 8 \times L$$

$$\Rightarrow \quad L = 0{\cdot}0796\text{ H}$$ **Ans.**

(c) Force per unit length is given by

$$\frac{F}{l} = \frac{\mu_0}{2\pi}\frac{I_1 I_2}{R}$$

where I_1 is the current in X.

I_2 is the current in Y.

R is the distance between them.

Definition of ampere : 1 ampere is the current which when flowing in each of the two infinitely long parallel conductors of negligible circular cross- section, placed 1m apart in vacuum, produces between them a force of exactly $2\pi \times 10^{-7}$ newton per metre of their length.

SECTION–B

Answer 5.

(a) (i) Radio waves, Microwaves, Infrared radiation, Visible light, Ultraviolet radiation, X-rays and Gamma rays.

(ii) The electric vector $(\vec{E})$, magnetic vector $(\vec{B})$ and the velocity vector $(\vec{c})$ are mutually perpendicular to each other.

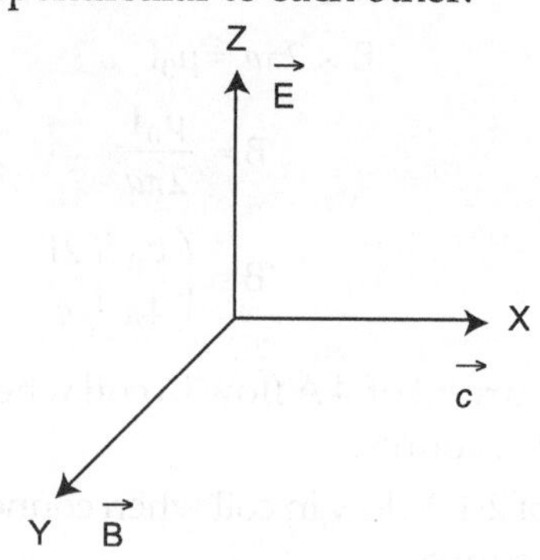

(b) Given : $\lambda_1 = 700$ nm, $n_1 = 10^{th}$ bright fringe

$\lambda_2 = 500$ nm, $n_2 = ?$

We know,

$$n_1\lambda_1 = n_2\lambda_2$$

$$n_2 = \frac{n_1\lambda_1}{\lambda_2}$$

$$= \frac{10 \times 700}{500} = 14$$ **Ans.**

(c) Given : $\angle r = 32°$

(i) $\angle i_p = ?$

$$\angle i_p + \angle r = 90°$$

$$\angle i_p + 32° = 90°$$

$$\angle i_p = 58°$$ **Ans.**

(ii) $n = \tan i_p$

$$n = \tan 58° = 1{\cdot}6$$ **Ans.**

Answer 6.

(a) Given : $A = 60°$, $n = 1{\cdot}6$, $\delta_m = ?$

We know $$n = \frac{\sin\left(\frac{A+\delta_m}{2}\right)}{\sin A/2}$$

$$1.6 = \frac{\sin\left(\frac{60°+\delta_m}{2}\right)}{\sin\left(\frac{60°}{2}\right)}$$

$$\sin\left(\frac{60°+\delta_m}{2}\right) = 1{\cdot}6 \times \frac{1}{2}$$

$$\frac{60°+\delta_m}{2} = \sin^{-1}(0{\cdot}8)$$

$$60° + \delta_m = 106{\cdot}26°$$

$$\therefore \quad \delta_m = 106{\cdot}26° - 60°$$

$$= 46{\cdot}26°$$ **Ans.**

(b) In the figure, SPS′ represents a convex spherical surface of radius of curvature R with its centre of curvature at C. O is a point object and I is its real image formed due to refraction by the surface.

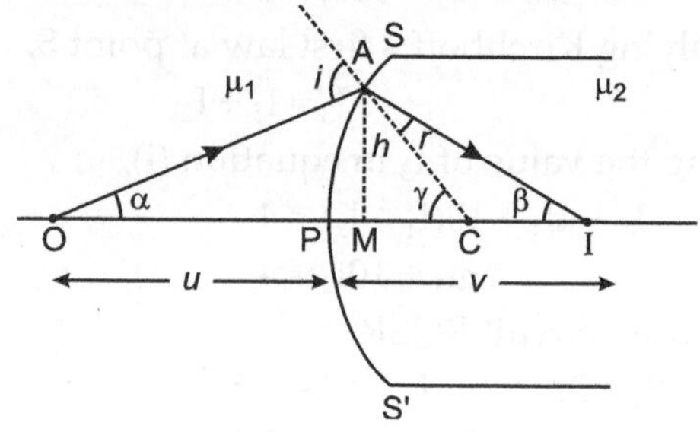

In Δ COA, i is the exterior angle,

So $$i = \alpha + \gamma \quad \text{...(i)}$$

In Δ ACI, γ is the exterior angle,

So $$\gamma = r + \beta$$

$$r = \gamma - \beta \quad \text{...(ii)}$$

By Snell's law,

$$\frac{\sin i}{\sin r} = \frac{\mu_2}{\mu_1}$$

$$\mu_1 \sin i = \mu_2 \sin r.$$

Since aperture is small so $\angle i$ and $\angle r$ will be small.

So,

$\therefore$ $\sin i \cong i$ and $\sin r \cong r$

$$\mu_1 i = \mu_2 r$$

$$\Rightarrow \quad \mu_1(\alpha + \gamma) = \mu_2(\gamma - \beta)$$

[Using (i) and (ii)]

$$\mu_2\beta + \mu_1\alpha = (\mu_2 - \mu_1)\gamma \quad \text{...(iii)}$$

So, $$\alpha = \tan\alpha = \frac{h}{-u};\ \beta = \tan\beta = \frac{h}{v};$$

$$\gamma = \tan\gamma = \frac{h}{R}$$

Substituting the above values in equation (iii),

$$\mu_2\frac{h}{v} + \mu_1\frac{h}{-u} = (\mu_2 - \mu_1)\frac{h}{R}$$

or $$\frac{\mu_2}{v} - \frac{\mu_1}{u} = \frac{\mu_2 - \mu_1}{R}$$

(c) Given : $f = 20$ cm, $m = \pm 2$, $u = ?$

Case I : When the image is virtual, $m = +2$

We know that,

$$m = \frac{f}{u+f}$$

$\Rightarrow$ $$2 = \frac{20}{u+20}$$

$\Rightarrow$ $u + 20 = 10$

$\Rightarrow$ $u = -10$ cm. **Ans.**

Case II : When the image is real, $m = -2$

$\therefore$ $$-2 = \frac{20}{u+20}$$

$\Rightarrow$ $u + 20 = -10$

$\Rightarrow$ $u = -30$ cm. **Ans.**

Answer 7.

(b) Labelled diagram of an image formed by a compound microscope when image is formed at least distance of distinct vision is given below :

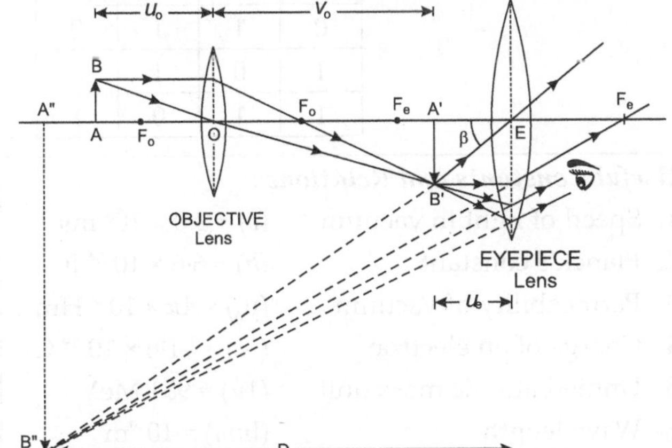

(c) Given : $f_o = 140$ cm, $f_e = 5$ cm.

(i) Angular magnification,

$$m = \frac{-f_o}{f_e}$$

$$= \frac{-140}{5}$$

$$= -28$$ **Ans.**

(ii) Distance between two lenses

$$= f_o + f_e$$

$$= 140 + 5$$

$$= 145 \text{ cm.}$$ **Ans.**

SECTION–C

Answer 8.

(b) Given: $\lambda = 198 \text{ nm} = 198 \times 10^{-9}$ m

(i) Energy of Photon, $E = \dfrac{hc}{\lambda}$

$$= \frac{6{\cdot}6 \times 10^{-34} \times 3 \times 10^{8}}{198 \times 10^{-9}}$$

$$= 1{\cdot}0 \times 10^{-18} \text{ J}$$

(ii) Momentum, $p = ?$

We have

$$\lambda = \frac{h}{p}, \; p = \frac{h}{\lambda}$$

$\because$ $$p = \frac{6{\cdot}6 \times 10^{-34}}{198 \times 10^{-9}}$$

$$= 3{\cdot}33 \times 10^{-27} \text{ kg m/s}$$ **Ans.**

(c) (i) **(1)** Balmer series, **(2)** Lyman series.

(ii) Bohr's formula, $$\frac{1}{\lambda} = Z^2 R \left(\frac{1}{n_1^2} - \frac{1}{n_2^2} \right)$$

For hydrogen $Z = 1$

$\therefore$ $$\frac{1}{\lambda} = R \left(\frac{1}{n_1^2} - \frac{1}{n_2^2} \right)$$

where, n_1 = Lower energy state

n_2 = Higher energy state

λ = Wavelength of light emitted

R = Rydberg's constant

Answer 9.

(a) Let N be the number of atoms present in a radioactive substance at any instant. Then,

$$-\frac{dN}{dt} = \lambda N.$$

Where λ is decay constant.

$$\frac{dN}{N} = -\lambda dt$$

On integrating, we get

$$\log_e N = -\lambda t + C \qquad \text{...(i)}$$

where C is the integration constant. Suppose there were N_0 atoms in the beginning.

Then, $N = N_0$ at $t = 0$

Putting it in equation (i), we get

$$\log_e N_0 = C.$$

Now, the equation (i) becomes

$$\log_e N = -\lambda t + \log_e N_0$$

$$\log_e N - \log_e N_0 = -\lambda t$$

or $$\log_e \frac{N}{N_0} = -\lambda t$$

or $$\frac{N}{N_0} = e^{-\lambda t}$$

$$N = N_0 e^{-\lambda t}$$

(b) Pair production : When an energetic γ-ray photon falls on a heavy substance, it is absorbed by some nucleus of the substance, and its energy give rise to the production to the pair of an electron and a positron. This phenomenon is called pair production.

Equation : $h\nu = {}_{-1}e^0 + {}_{1}e^0$

(γ photon) (Electron) (Positron)

(c) (i) Given : V = 40,000 V

We know, K.E. = $e \times V$

$= 1{\cdot}6 \times 10^{-19} \times 40{,}000$

$= 6{\cdot}4 \times 10^{-15}$ J. **Ans.**

(ii) We know,

$$\lambda_{min} = \frac{hc}{eV}$$

$$= \frac{6{\cdot}6 \times 10^{-34} \times 3 \times 10^{8}}{1{\cdot}6 \times 10^{-19} \times 40{,}000}$$

$$= \frac{1{\cdot}98 \times 10^{-25}}{1{\cdot}6 \times 10^{-19} \times 40{,}000}$$

$= 3{\cdot}09 \times 10^{-11}$

$= 0{\cdot}309$ Å

Range → 0·309 Å to ∞. **Ans.**

Answer 10.

(a) (i) Mass of reactants = 2·015 × 2 = 4·030 u

Mass of products = 3·017 + 1·009

= 4·026 u

Mass defect Δm = 0·004 u

We know, 1u = 931 MeV

So, energy released = 0·004 × 931

= 3·724 MeV **Ans.**

(ii) Fusion reaction.

(b) (i) Full wave rectifier : It converts alternating current into direct current and utilises both halves of each cycle of alternating current. In this, complete conversion of a.c. to d.c. takes place.

(ii) An amplifier : It is device which amplify a signal. A junction diode cannot amplify a signal. A transistor consisting of two *p-n* junctions, one forward biased and the other reverse biased, is used to amplify a weak signal.

(iii) An oscillator : It is a device that generates electrical oscillations of constant amplitude and of desired frequency, without any external input.

(c) To obtain AND gate using NAND gates :

A B Y' Y

Truth Table :

A	B	Y′	Y
0	0	1	0
0	1	1	0
1	0	1	0
1	1	0	1

Useful Constants and Relations :

1. Speed of light in vacuum $(c) = 3{\cdot}0 \times 10^8$ ms^{-1}
2. Planck's constant $(h) = 6{\cdot}6 \times 10^{-34}$ Js
3. Permeability of vacuum $(\mu_0) = 4\pi \times 10^{-7}$ Hm^{-1}
4. Charge of an electron $(-e) = -1{\cdot}6 \times 10^{-19}$ C
5. Unified atomic mass unit $(1u)$ = 931 MeV
6. Wave length $(1nm) = 10^{-9}$m

LL

CHEMISTRY 2020

Paper – 1
(Theory)
(Maximum Marks : 70)
(Time allowed : Three hours)

*(Candidates are allowed additional 15 minutes for **only** reading the paper. They must NOT start writing during this time.)*

All questions are compulsory

Question 1 is of 20 marks having four sub parts, all of which are compulsory.
*Question number 2 to 8 carry 2 marks each, with **two** questions having internal choice.*
*Question numbers 9 to 15 carry 3 marks each, with **two** questions having an internal choice.*
Question number 16 to 18 carry 5 marks each, with an internal choice.
All working including rough work, should be done on the same sheet as, and adjacent to the rest of the answer.
The intended marks for questions of parts of questions are given in brackets [ ].
Balanced equations must be given wherever possible and diagrams where they are helpful.
When solving numerical problems, all essential working must be shown
In working out problems, use the following data :
Gas constant R = 1·987 cal deg^{-1} mol^{-1} = 8·314 JK^{-1} mol^{-1} = 0·0821 dm^3 atm k^{-1} mol^{-1}
1 1 atm = 1 dm^3 atm = 101·3 J. 1 Faraday = 96500 coulombs
Avogadro's number = 6·023 × 10^{23}.

QUESTIONS

Question 1.

(a) *Fill in the blanks by choosing the appropriate word/ words from those given in the brackets :* **[4 × 1]**

(iodoform, volume, mass, haloform, gram equivalent, chloroform, carbylamine. sp^3d^2, high, coke, d^2sp^3, low, gram mole, carbon monoxide)

(i) *Equivalent conductivity is the conducting power of all the ions furnished by one ______ of an electrolyte present in a definite ______ of the solution.*

(ii) *Bleaching powder, on treatment with ethanol or acetone gives ______. This is an example of ______ reaction.*

(iii) *Outer orbital complexes involve ______ hybridization and are spin complexes.*

(iv) *Zinc oxide is reduced by ______ at 1673K to form zinc and ______.*

(b) *Select the correct alternative from the choices given :* **[4 × 1]**

(i) *The packing efficiency of simple cubic structure, body centered cubic structure and face centered cubic structure respectively is:*

(1) *52·4%, 74%, 68%*
(2) *74%, 68%, 52·4%*
(3) *52·4%, 68%, 74%*
(4) *68%, 74%, 52·4%*

(ii) *When acetone is treated with Grignard's reagent, followed by hydrolysis, the product formed is:*

(1) *Secondary alcohol*
(2) *Tertiary alcohol*
(3) *Primary alcohol*
(4) *Aldehyde*

(iii) *Which of the following electrolytes is least effective in causing flocculation of positively charged ferric hydroxide sol?*

(1) $K_3[Fe(CN)_6]$ **(2)** K_2CrO_4
(3) $K_4[Fe(CN)_6]$ **(4)** *KBr*

(iv) *On heating an aliphatic primary amine with chloroform and alcoholic potassium hydroxide, the organic compound formed is an :*

(1) *Alkyl isocyanide* **(2)** *Alkanol*
(3) *Alkanal* **(4)** *Alkyl cyanide*

(c) *Match the following :* **[4 × 1]**

(i) Silicon and phosphorous	*(a) Acetaldehyde*
(ii) Iodoform test	*(b) Xenon hexafluoride*

(iii) *Arrhenius equation* (c) *n-type of semiconductors*

(iv) *Distorted octahedral structure* (d) *Frequency factor*

(d) *Answer the following questions:* **[4 × 2]**

(i) *What is the common name of the polymer obtained by the polymerisation of caprolactam? Is it addition polymer or condensation polymer?*

(ii) *Why Zn^{2+} ions are colourless while Ni^{2+} ions are green and Cu^{2+} ions are blue in colour?*

(iii) *The molar conductivity of NaCl, CH_3COONa and HCl at infinite dilution is 126·45, 91·0 and 426·16 ohm^{-1} cm^2 mol^{-1} respectively. Calculate the molar conductivity (λ_m^∞) for CH_3COOH at infinite dilution.*

(iv) *Identify the compounds A, B, C and D.*

$$C_6H_5COOH \xrightarrow{SOCl_2} A \xrightarrow{NH_3} B \xrightarrow{Br_2/KOH} C \xrightarrow[0-5^\circ C]{NaNO_2+HCl} D$$

Question 2. **[2]**

(a) *An element has atomic weight 93g mol^{-1} and density 11·5 g cm^{-3}. If the edge length of its unit cell is 300 pm, identify the type of unit cell.*
(N_A = 6·023 × 10^{23} mol^{-1})

OR

(b) *Calculate the radius of copper atom. The atomic weight of copper is 63·55 g mol^{-1}. It crystallises in face centered cubic lattice and has density of 8·93 g cm^{-3} at 298K.*
(N_A = 6·023 × 10^{23} mol^{-1})

Question 3.

Complete and balance the following chemical equations : **[2]**

(i) $P_4 + NaOH + H_2O \xrightarrow[\text{Inert atm.}]{\text{heat}}$ ____ + ____

(ii) Cu + dil. $HNO_3 \longrightarrow$ ____ + ____ + ____

Question 4. **[2]**

(i) *Write the chemical equation for the reaction of glucose with bromine water.*

(ii) *Write the zwitter ion structure of glycine.*

Question 5.

(i) *How do antiseptics differ from disinfectants ?* **[2]**

(ii) *Name a substance that can be used as an antiseptic as well as a disinfectant.*

Question 6. **[2]**

An alloy of gold (Au) and cadmium (Cd) crystallises with a cubic structure in which gold atoms occupy the corners and cadmium atoms fit into the face centers. What is the formula of this alloy?

Question 7.

(a) *State reasons for the following:* **[2]**

(i) *Ethylamine is soluble in water whereas aniline is insoluble in water.*

(ii) *Aliphatic amines are stronger bases than aromatic amines.*

OR

(b) *Complete and balance the following equations :*

(i) $C_6H_5NH_2 + CH_3COCl \longrightarrow$ ____ + ____

(ii) $C_2H_5NH_2 + HNO_2 \longrightarrow$ ____ + ____ + ____

Question 8. **[2]**

Draw the structure of xenon tetrafluoride molecule. State the hybridisation of the central atom and the geometry of the molecule.

Question 9. **[3]**

(a) *Calculate the emf and ΔG for the given cell at 25°C:*
$Cr_{(s)}$ / Cr^{3+} (0·1 M) // Fe^{2+} (0·01 M) / $Fe_{(s)}$
Given: $E^\circ_{Cr^{3+}/cr} = -0{\cdot}74$ V, $E^\circ_{Fe^{2+}/Fe} = -0{\cdot}44$ V
(1F = 96500 C, R = 8·314 JK^{-1} mol^{-1})

OR

(b) *Calculate the degree of dissociation (α) of acetic acid, if its molar conductivity ($\wedge_m$) is 39·05 S cm^2 mol^{-1}.*
(Given $\lambda^\circ_{(H^+)}$ = 349·6 S cm^2 mol^{-1} and $\lambda^\circ_{(CH_3COO^-)}$ = 40·95 S cm^2 mol^{-1})

Question 10. **[3]**

Name an important ore of silver. How is silver extracted from its sulphide ore? Give balanced chemical equations involved in the extraction of pure silver.

Question 11. **[3]**

How will you convert the following:

(i) *Chlorobenzene to biphenyl*

(ii) *Propene to 1-bromopropane*

(iii) *Chlorobenzene to aniline*

Question 12. **[3]**

Explain what is observed when:

(i) *A beam of light is passed through a colloidal solution.*

(ii) *An electric current is passed through a colloidal solution.*

(iii) *An electrolyte ($AlCl_3$) is added to a colloidal solution of arsenious sulphide (As_2S_3).*

Question 13. **[3]**

(a) *How will you convert the following: (Give balanced equation)*

(i) *Benzoyl chloride to benzaldehyde.*

(ii) *Methyl chloride to acetic acid.*

(iii) *Acetic acid to methane.*

OR

(b) *A ketone A (C_4H_8O) which undergoes Iodoform reaction gives compound B on reduction. B on heating with conc. H_2SO_4 at 443 K gives a compound C which forms ozonide D. D on hydrolysis with Zn dust gives only E. Identify the compounds A to E. Write the Iodoform reaction with compound A.*

Question 14. **[3]**

A first order reaction is 50% completed in 30 minutes at 300 K and in 10 minutes at 320 K. Calculate the activation energy of the reaction. ($R = 8{\cdot}314\ JK^{-1}\ mol^{-1}$.)

Question 15. **[3]**

Explain the following:

(i) *Transition metals and their compounds generally exhibit a paramagnetic behaviour.*

(ii) *There is an increase in density of elements from titanium (Z = 22) to copper (Z = 29) in the 3d series of transition elements.*

(iii) *$K_2Cr_2O_7$ acts as a powerful oxidising agent in acidic medium.*

Question 16. **[5]**

(a) (i) *The elevation in boiling point when 0·30 g of acetic acid is dissolved in 100 g of benzene is 0·0633°C. Calculate the molecular weight of acetic acid from this data. What conclusion can you draw about the molecular state of the solute in the solution?*

(Given K_b for benzene = $2{\cdot}53\ K\ kg\ mol^{-1}$, at. wt. of C = 12, H = 1, O = 16)

(ii) *Determine the osmotic pressure of a solution prepared by dissolving 0·025 g of K_2SO_4 in 2 litres of water at 25°C, assuming that K_2SO_4 is completely dissociated.*

($R = 0{\cdot}0821$ Lit-atm $K^{-1}\ mol^{-1}$, mol. wt. of $K_2SO_4 = 174\ g\ mol^{-1}$)

OR

(b) (i) *An aqueous solution of a non-volatile solute freezes at 272·4 K, while pure water freezes at 273·0 K. Determine the following :*

(Given $K_f = 1{\cdot}86\ K\ kg\ mol^{-1}$, $K_b = 0{\cdot}512\ K\ kg\ mol^{-1}$ and vapour pressure of water at 298 K = 23·756 mm of Hg)

(1) *The molality of solution*

(2) *Boiling point of solution*

(3) *The lowering of vapour pressure of water at 298 K*

(ii) *A solution containing 1·23g of calcium nitrate in 10g of water, boils at 100·975°C at 760 mm of Hg. Calculate the van't Hoff factor for the salt at this concentration.*

(K_b for water = $0{\cdot}52\ K\ kg\ mol^{-1}$, mol. wt. of calcium nitrate = $164\ g\ mol^{-1}$).

Question 17. **[5]**

(a) (i) *Write the IUPAC names of the following complexes:*

(1) *$[Cu(NH_3)_4]SO_4$*

(2) *$[Co(en)_2Cl_2]$*

(3) *$K_3[Al(C_2O_4)_3]$*

(ii) *With reference to the coordination complex ion $[Fe(H_2O)_6]^{2+}$ answer the following: (at. no. of Fe = 26)*

(1) *Give the IUPAC name of the complex ion.*

(2) *What is the oxidation number of the central metal atom?*

(3) *How many unpaired electrons are there in the complex ion?*

(4) *State the type of hybridisation of the complex ion.*

OR

(b) (i) *Name of the type of isomerism exhibited by the following pairs of compounds :*

(1) *$[Co(ONO)(NH_3)_5]^{2+}$ and $[Co(NO_2)(NH_3)_5]^{2+}$*

(2) *$[Cr(H_2O)_4Cl_2]\ Cl.2H_2O$ and $[Cr(H_2O)_5Cl]\ Cl_2.H_2O$*

(3) *$[Co(NH_3)_6]\ [Cr(CN)_6]$ and $[Cr(NH_3)_6]\ [Co(CN)_6]$*

(ii) *Using the valence bond approach, predict the shape, hybridisation and magnetic behaviour of $[Ni(CO)_4]$. (at. no. of Ni = 28)*

Question 18. **[5]**

(a) (i) *Give balanced chemical equations for the following reactions :*

(1) *Phenol is treated with ice cold alkaline solution of benzene diazonium chloride.*

(2) *Diethyl ether is treated with phosphorous pentachloride.*

(3) *Ethyl alcohol is treated with thionyl chloride.*

(ii) *Give one chemical test each to distinguish between the following pairs of compounds:*

(1) *Ethanol and dimethyl ether*

(2) *Propan-1-ol and propan-2-ol*

OR

(b) (i) *Write chemical equations to illustrate the following name reactions:*

(1) *Williamson's synthesis*

(2) *Esterification reaction*

(3) *Reimer-Tiemann reaction*

(ii) *Identify the compounds A and B in the given reactions :*

(1) $C_2H_5OH \xrightarrow[\Delta]{Cu} A \xrightarrow{dil.NaOH} B$

(2) Phenol (benzene ring with OH) $\xrightarrow{Zn\ dust} A \xrightarrow[[anhy.AlCl_3]]{CH_3COCl} B$

ANSWERS

Answer 1.

(a) **(i)** gram equivalent, volume

(ii) chloroform, haloform

(iii) sp^3d^2, high

(iv) coke, carbon monoxide

(b) **(i)** (3) 52.4%, 68%, 74%

(ii) (2) tertiary alcohol

(iii) (4) KBr

(iv) (1) Alkyl isocyanide

(c) **(i)** silicon and phosphorus – (c) *n*-type semi-conductor

(ii) Iodoform test – (a) Acetaldehyde

(iii) Arrhenius equation–(d) Frequency factor

(iv) Distorted octahedral structure–(b) Xenon hexafluoride

(d) **(i)** Common name of polymer obtained by the polymerization of caprolactum is Nylon-6. It is condensation polymer.

(ii) Electronic configuration : Ni : [Ar] $3d^8\ 4s^2$

Cu : [Ar] $3d^{10}\ 4s^1$

Zn : [Ar] $3d^{10}\ 4s^2$

Ni^{2+} and Cu^{2+} have unpaired electrons in their outer orbitals which can absorb energy and excite to the higher energy levels, and while coming back to the ground state they release energy in visible region to impart colour to the ions showing *d-d* transition. Whereas, the completely filled outer *d*-orbitals in Zn^{2+} ion makes it stable and hence electrons do not show energy exchange.

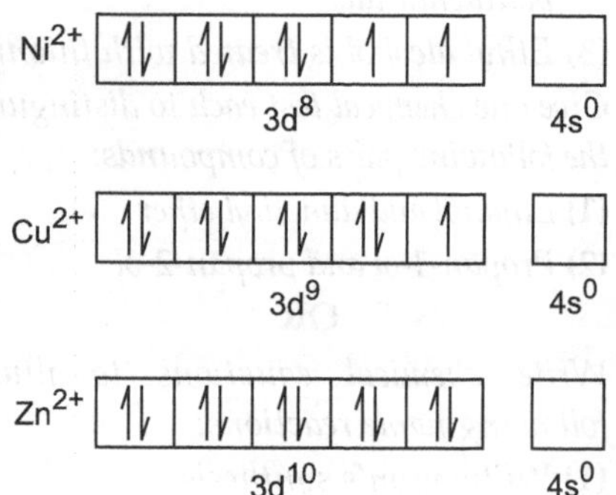

(iii) Molar conductivity of CH_3COOH at infinite dilution can be given by –

$$\lambda^{\infty}_{m}(CH_3COOH) = \lambda^{\infty}_{m}(HCl) + \lambda^{\infty}_{m}(CH_3COONa) - \lambda^{\infty}_{m}(NaCl)$$

$$\lambda^{\infty}_{m}(CH_3COOH) = 426.16 + 91 - 126.45$$

$$= 390.71\ ohm^{-1}\ cm^2\ mol^{-1}$$

Molar conductivity of CH_3COOH at infinite dilution is 390.71 $ohm^{-1}cm^2\ mol^{-1}$.

(iv) The completed sequence is as follows –

COOH (benzene ring) $\xrightarrow{SOCl_2}$ COCl (benzene ring) Benzoyl chloride (A) $\xrightarrow{NH_3}$ $CONH_2$ (benzene ring) Benzamide (B) $\xrightarrow{Br_2\ /\ KOH}$ NH_2 (benzene ring) Aniline (C) $\xrightarrow{NaNH_2\text{ - }HCl,\ 0\text{-}5°C}$ N_2^+Cl (benzene ring) Benzene diazonium chloride (D)

Answer 2.

(a) Given, Atomic weight (M) = 93 g mol^{-1}

Density = 11.5 g cm^{-3}

Edge length (a) = 300 pm = 3×10^{-8} cm

We know that–

$$\text{Density, } d = \frac{(Z \times M)}{N_A \times a^3}$$

Substituting the values –

$$11.5\ g\ cm^{-3} = \frac{(93\,g\,mol^{-1} \times Z)}{6.023 \times 10^{23} \times (3 \times 10^{-8}\,cm)^3}$$

$$\text{Or, } Z = \frac{[11.5\,g\,cm^{-3} \times 6.023 \times 10^{23} \times (3 \times 10^{-8}\,cm)^3]}{93\,g\,mol^{-1}}$$

$$Z = 2.01 \approx 2$$

Hence, the number of atoms present in one unit cell is 2 which indicates a body centered cubic system.

OR

(b) Given, Atomic weight (M) = 63.55 g mol^{-1}

Density = 8.93 g cm^{-3}

Edge length (a) = not given

Face centred cubic system has 4 atoms per unit cell (Z)

We know that –

$$\text{Density, } d = \frac{(Z \times M)}{N_A \times a^3}$$

Substituting the values –

$$8.93\ g\ cm^{-3} = \frac{(63.55\,g\,mol^{-1} \times 4)}{6.023 \times 10^{23} \times a^3}$$

$$\text{Or, } a^3 = \frac{(63.55\,g\,mol^{-1} \times 4)}{6.023 \times 10^{23} \times 8.93\,g\,cm^{-3}}$$

$$= 4.72 \times 10^{-23}\ cm^3$$

$$= 0.472 \times 10^{-24}\ cm^3$$

$$a = 3.6 \times 10^{-8}\ cm$$

Now, radius (r) in fcc system is given by –

$$r = \frac{(\sqrt{2}\,a)}{4}$$

$$= \frac{(1.414 \times 3.6 \times 10^{-8}\,\text{cm})}{4}$$

$= 1.27 \times 10^{-8}$ cm

$= 0.127$ nm

Answer 3.

(i) $P_4 + 3NaOH + 3H_2O \xrightarrow[\text{atmosphere}]{\text{heat and inert}} 3NaH_2PO_2 + PH_3$ (Sodium hypophosphite)

(ii) $3Cu + 8\,HNO_3 \rightarrow 3Cu(NO_3)_2 + 2NO + 4H_2O$ (dil.)

Answer 4.

(i) Glucose reacts with bromine water to give gluconic acid –

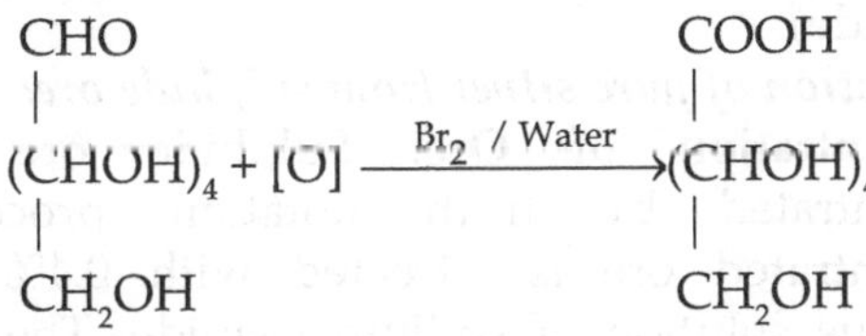

$$\underset{\text{Glucose}}{CHO-(CHOH)_4-CH_2OH} + [O] \xrightarrow{Br_2\ /\ \text{Water}} \underset{\text{Gluconic acid}}{COOH-(CHOH)_4-CH_2OH}$$

(ii) Zwitterionic structure of glycine can be given as follows –

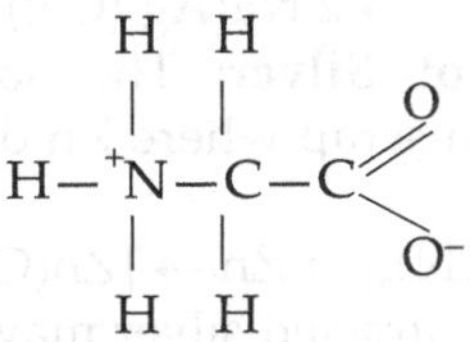

$$H-\overset{+}{N}H_2-CH_2-C(=O)O^-$$

Answer 5.

(i) Antiseptics are the type of substances which hinders or slows down the growth of microbes. These are applied to the living tissues such as wounds, cuts, ulcers and other disease skin surfaces. These are not ingested as antibiotics. Disinfectants are the chemical agents to destroy bacteria. These are applied to inanimate objects such as floors, drainage systems, instruments to sterilise.

(ii) Phenol is a substance which when used in 0.2% concentration works as antiseptic whereas its 1% solution is disinfectant.

Answer 6.

Gold atoms occupy all the corners of a unit cell, So, the number of gold atoms per unit cell $= 8 \times \frac{1}{8} = 1$ atom.

And cadmium atoms fit into the face centres, therefore, the number of cadmium atoms per unit cell $= 6 \times \frac{1}{2} = 3$ atoms

$\therefore$ Ratio of gold (Au) and cadmium (Cd) = 1:3

Hence, formula for the given alloy is $AuCd_3$.

Answer 7.

(a) (i) Ethylamine ($C_2H_5NH_2$) is soluble in water because $-NH_2$ group forms hydrogen bond with water molecule efficiently due to increased electron density on N atom because of (+I) inductive effect of $-C_2H_5$ group. Also, the hydrocarbon part, $-C_2H_5$ is small and do not contribute much towards hydrophobicity.

In Aniline ($C_6H_5NH_2$), the $-NH_2$ group is involved in resonance with the benzene ring with which it is directly attached, hence it does not make efficient hydrogen bonds with water. Secondly, the benzene ring is bulky and hydrophobic to a great extent. Hence, the overall effect is insolubility of aniline in water.

(ii) In aliphatic amines, the alkyl group increase the electron density on N atom due to (+I) inductive effect, hence making the $-NH_2$ group more basic and likely to attract proton.

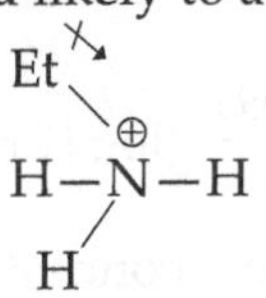

$$Et-\overset{\oplus}{N}H_2-H$$

Whereas, in aromatic amine the lone pair of electrons is involved in the resonance of benzene ring hence unavailable for showing basic character. So, aromatic amines are less basic than aliphatic amines.

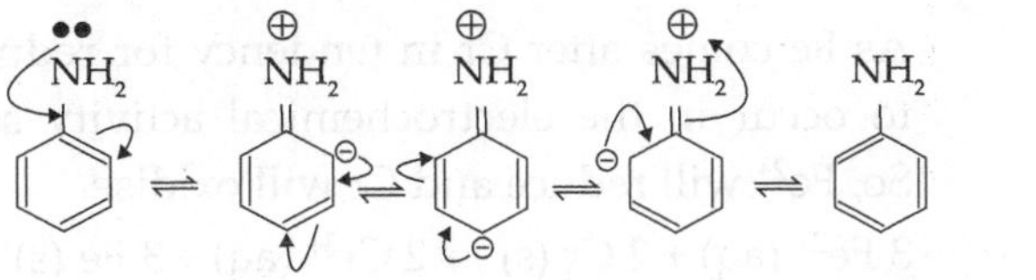

$NH_2 \rightleftharpoons \overset{\oplus}{N}H_2 \rightleftharpoons \overset{\oplus}{N}H_2 \rightleftharpoons \overset{\oplus}{N}H_2 \rightleftharpoons NH_2$

OR

(b) (i) The complete reaction is–

$$\underset{\text{Phenylamine}}{C_6H_5NH_2} + \underset{\text{Acetyl Chloride}}{CH_3COCl} \rightarrow \underset{\text{Acetanilide}}{C_6H_5NHCOCH_3} + \underset{\text{Hydrochloric acid}}{HCl}$$

Aniline + CH_3COCl $\longrightarrow$ Acetanilide + HCl

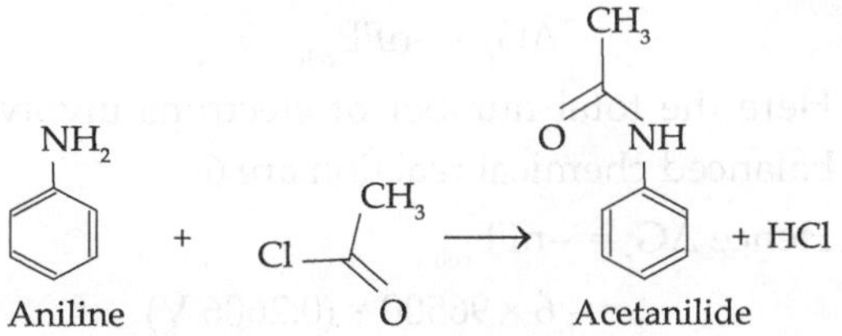

(ii) $\underset{\text{3-methyl pentane}}{CH_3CH_2-NH_2} + \underset{\text{Nitrous acid}}{HONO} \xrightarrow{0.5°C} C_2H_5OH + N_2\uparrow + H_2O$

Answer 8.

Structure of Xenon tetrafluoride molecule –

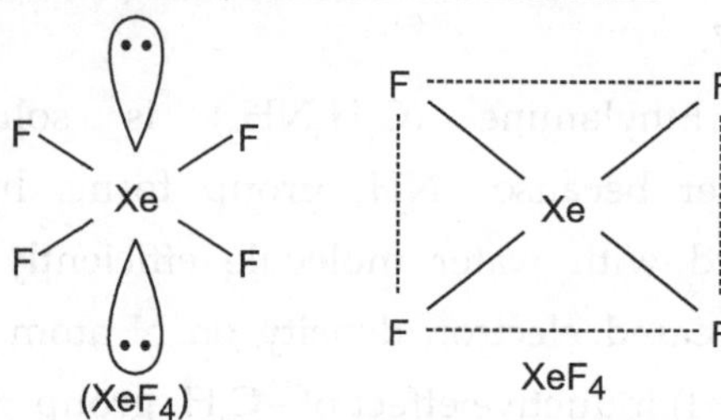

The hybridisation state of central atom i.e. Xe is d^2sp^3.

The geometry is square planar.

Answer 9.

(a) The half-cell reactions can be written as–

$Fe^{2+} + 2e^- \rightarrow Fe$

$E_{el} = -0.44 + \left(\frac{0.0591}{2}\right)(\log [Fe^{2+}])$

Substituting the concentration value and calculating –

$$E_{el} = -0.44 + \left(\frac{0.0591}{2}\right)(\log 0.01)$$
$$= -0.44 - 0.0591$$
$$= -0.4991 \text{ V} \quad ...(i)$$

And

$Cr^{3+} + 3e^- \rightarrow Cr$

$E_{el} = -0.74 + \frac{0.0591}{3}(\log [Cr^{3+}])$

Substituting the concentration value and calculating –

$$E_{el} = -0.74 + \frac{0.0591}{3}(\log 0.1)$$
$$= -0.74 - 0.0197$$
$$= -0.7597 \text{ V} \quad ...(ii)$$

As Fe comes after Cr in tendency for reduction to occur in the electrochemical activity series. So, Fe^{2+} will reduce and Cr will oxidise.

$3\,Fe^{2+}(aq) + 2\,Cr(s) \rightarrow 2\,Cr^{3+}(aq) + 3\,Fe(s)$

So, E_{cell} will be –

$$E_{cell} = -0.4991\text{V} - (-0.7597)\text{ V}$$
$$= 0.2606 \text{ V}$$

ΔG of the reaction is given by –

$$\Delta G_r = -nFE_{cell}$$

Here the total number of electrons involved in balanced chemical reaction are 6.

Hence $\Delta G_r = -nFE_{cell}$

$$= -6 \times 96500 \times (0.2606 \text{ V})$$
$$= -150887.4 \text{ J mol}^{-1}$$
$$= -150.8 \text{ kJ mol}^{-1}$$

OR

(b) Given,

$\lambda^\circ_{(H^+)} = 349.6 \text{ S cm}^2 \text{ mol}^{-1}$

$\lambda^\circ_{(CH_3COO^-)} = 40.9 \text{ S cm}^2 \text{ mol}^{-1}$

Then,

$\lambda^\circ_{CH_3COOH} = \lambda^\circ_{CH_3COO^-} + \lambda^\circ_{H^+}$

$\lambda^\circ_{CH_3COOH} = 349.6 + 40.9 = 390.5 \text{ S cm}^2 \text{ mol}^{-1}$

Degree of dissociation, $a = \frac{\wedge_m}{\wedge^\circ}$

Here, a = Degree of dissociation,

$\wedge_m$ = Molar conductivity

λ° = Molar conductivity at infinite dilution.

$$= \frac{39.05}{390.05} = 0.1$$

Answer 10.

Important ore of silver is argentite (Ag_2S, silver sulphide).

Extraction of pure silver from sulphide ore:

Concentration of Ore: Sulphide ore and is concentrated by froth flotation process. The concentrated ore is treated with 0.4% to 0.7% aqueous solution of sodium cyanide. The argentite ore dissolves in sodium cyanide forming sodium argento cyanide.

$$Ag_2S(s) + 4NaCN(aq) + 2O_2(g) \rightarrow 2\,Na[Ag(CN)_2](aq) + Na_2SO_4(aq)$$

Precipitation of Silver: The solution obtained is treated with Zn scrap where Zn displaces silver from its complex.

$$2[Ag(CN)_2]^- + Zn \rightarrow [Zn(CN)_4]^{2+} + 2Ag$$

Refining: The extracted silver may contain impurities. Hence it is subjected to electrolytic refining to obtain metals of high purity. A block of impure metal is anode while a thin strip of pure silver is cathode. A mixture of $AgNO_3$ solution is electrolyte. On passing current impure silver dissolves and equivalent amount of pure silver is deposited at cathode.

Anode: $Ag \rightarrow Ag^+ + e^-$
(Anode mud)

Cathode: $Ag^+ + e^- \rightarrow Ag$
(Pure metal)

Answer 11.

(i) Chlorobenzene to biphenyl conversion can be achieved by Wurtz reaction–

$$2\,C_6H_5Cl + 2Na \xrightarrow[\Delta]{\text{Dry Ether}} C_6H_5{-}C_6H_5 + 2NaCl$$

Chlorobenzene → Biphenyl

(ii) Propene to 1-bromopropane–

$$CH_3-CH_2=CH_2 \xrightarrow[\text{Ani - Markonikov addition}]{\text{HBr / Peroxide}} CH_3-CH_2-CH_2-Br$$

Propene → 1-Bromopropane

(iii) Chlorobenzene to aniline–

$$\underset{\text{Chloro-benzene}}{C_6H_5Cl} + \underset{\text{Ammonia}}{NH_3} \xrightarrow[200°C,\ 2\ MPa]{Cu_2O} \underset{\text{Aniline}}{C_6H_5NH_2} + \underset{\text{Copper dichloride}}{Cu_2Cl_2} + H_2O$$

Answer 12.

(i) When a beam of light is passed through a colloidal solution, scattering of light by the colloidal particles takes place and the path of light becomes illuminated. This is called Tyndall effect.

(ii) The colloidal particles are charged and carry either a negative or a positive charge. When electric current is passed through the colloidal solution the charged colloidal particles or discharge phase start moving towards the oppositely charged electrode and coagulate after losing their charges on reaching to the electrode.

(iii) When $AlCl_3$ electrolyte is added to colloidal solution of As_2S_3, the positively charged ions of the electrolyte, *i.e.*, Al^{3+} neutralise the charge on colloidal particles and the sol get coagulated.

$$2AlCl_3 + As_2S_3 \rightarrow 2AsCl_3 \downarrow + Al_2S_3$$

Answer 13.

(a) (i) Benzoyl chloride can be converted to benzaldehyde by carrying out Rosenmund's reaction–

$$\underset{\text{Benzoyl chloride}}{C_6H_5COCl} \xrightarrow[Pd/BaSO_4]{\text{Rosenmund's reduction}} \underset{\text{Benzaldehyde}}{C_6H_5CHO} + HCl$$

(ii) Methyl chloride to acetic acid conversion can be done in the following sequence –

$$CH_3Cl + KCN \rightarrow CH_3CN \xrightarrow{H_2O/H^+} \underset{\text{(Acetic acid)}}{CH_3COOH}$$

(iii) Acetic acid can be converted to methane in the following two steps–

1. Ethanoic acid is realed with NaOH to obtain CH_3COONa salt.

$$\underset{\text{Ethanoic acid}}{CH_3COOH} + NaOH \longrightarrow \underset{\text{Sodium ethanoate}}{CH_3COONa} + H_2O$$

2. CH_3COONa is heated strongly with soda lime (NaOH + CaO) to obtain methane and Na_2CO_3.

$$CH_3COONa\ (s) + NaOH\ (s) \longrightarrow \underset{\text{Methane}}{CH_4} + \underset{\text{Sodium carbonate}}{Na_2CO_3}$$

OR

(b) The ketone compound A gives positive iodoform reaction, hence the compound A must contain at least one free methyl group directly attached to carbonyl carbon.

Now, on reduction of ketone A, a secondary alcohol must be obtained, which on dehydration would give an alkene.

As alkene on ozonolyis produces an ozonide compound therefore the compound C should contain one double bond.

As the compound D on hydrolysis produces only one compound, it should be two molecules of acetaldehydes as the starting ketone has 4 C. This implies that, 2 ($CH_3CH=$) group should be present in compound C.

Compound A, which satisfies all the above condition should be But-2-one.

The scheme for the reaction is as follows:

$$\underset{\text{But-2-one [A]}}{CH_3COCH_2CH_3} \xrightarrow{NaOX} CH_3CH_2COONa + CHI_3$$

$$\underset{[A]}{H_3C-\overset{\overset{O}{\|}}{C}-CH_2CH_3} \xrightarrow{\text{Reduction}} \underset{\text{2-Butanol [B]}}{H_3C-\overset{\overset{OH}{|}}{\underset{\underset{H}{|}}{C}}-CH_2CH_3} \xrightarrow{\text{Dehydration}} \underset{\text{2-Butene [C]}}{H_3C-\underset{\underset{H}{|}}{C}=CHCH_3}$$

$$\downarrow O_3 \text{ (Ozonolysis)}$$

$$\underset{\text{2-Butene ozonide [D]}}{H_3C-CH \overset{O}{\diagup\diagdown} CH-CH_3 \ (\text{with } O-O \text{ bridge})}$$

$$\xrightarrow{\text{Zn Dust}} \underset{\text{Ethanal [E]}}{H_3C-\overset{\overset{O}{\|}}{C}-H} + \underset{\text{Ethanal [E]}}{H_3C-\overset{\overset{O}{\|}}{C}-H}$$

Iodoform reaction with compound A (But-2-one) is given as :

$$CH_3COCH_2CH_3 + 3I_2 + 4NaOH \rightarrow CHI_3 + C_2H_5COONa + 3NaI + 3H_2O$$

Answer 14.

By using half life equation, we get,

$$K = \frac{0.693}{t_{1/2}}$$

At 300 K, $K_1 = \frac{0.693}{30\,min} = 0.02315\ min^{-1}$

At 320 K, $K_2 = \frac{0.693}{10\,min} = 0.06935\ min^{-1}$

From the Arrhenius equation

$$\log \frac{K_2}{K_1} = \frac{Ea}{2.303\,R}\left[\frac{1}{T_1} - \frac{1}{T_2}\right]$$

$$Ea = \frac{2.303 \times R \times T_1 \times T_2}{T_2 - T_2} \log\left(\frac{K_2}{K_1}\right)$$

$$\therefore\ Ea = \left(\frac{2.303 \times 8.314 JK^{-1}\ mol^{-1} \times 300K \times 320K}{320K - 300K}\right) \times \log\left(\frac{0.06935}{0.02315}\right) = 43.839\ kJ\ mol^{-1}.$$

Answer 15.

(i) Paramagnetism arises due to the presence of unpaired electrons with each electron having a magnetic moment associated with its spin angular momentum. As transition metals and their *d*-compounds possess vacant unpaired electrons in their valence orbitals, they generally exhibit a paramagnetic behaviour.

(ii) There is an increase in density of elements from titanium (Z = 22) to copper (Z = 29) in the 3d series of transition elements because from titanium to copper, atomic radii decrease, hence volume of the atoms keeps decreasing but the mass increases simultaneously.

$$\text{As density} = \frac{\text{mass}}{\text{volume}}$$

Therefore, density keeps increasing from titanium (Z = 22) to copper (Z = 29).

(iii) Potassium dichromate is powerful oxidising agent in acidic medium. In presence of dilute sulphuric acid, it liberates nascent oxygen, by taking up electrons and getting reduced to Cr^{3+}.

$$\overset{Cr^{6+}}{K_2Cr_2O_7} + 4H_2SO_4 \rightarrow K_2SO_4 + \overset{3+}{Cr_2(SO_4)_3} + 4H_2O + 3[O]$$

Answer 16.

(a) (i) Let solvent (Benzene): Component 1, Solute (acetic acid): Component 2

$W_1 = 100\ g$

$W_2 = 0.30\ g$

Elevation in boiling point–

$\Delta T_b = 0.0633K$

Given, K_b for benzene = 2.53K kg mol^{-1}

Molar mass of acetic acid can be given as–

$$M_2 = \frac{K_b \times W_2 \times 1000}{\Delta T_b \times W_1} = \frac{2.53 \times 0.30 \times 1000}{0.0633 \times 100}$$

$M_2 = 119.90$ kg

Van't Hoff factor (i) is given by–

$$i = \frac{\text{Normal molar mass}}{\text{Abnormal molar mass}} = \frac{60}{119.90}$$

$i = 0.5$

As, $i < 1$, therefore the solute (acetic acid) is associated.

(ii) When K_2SO_4 is dissolved in water, K^+ and SO_4^{2-} ions are produced.

$$K_2SO_4 \rightarrow 2K^+ + SO_4^{2-}$$

Total number of ions produced = 3

∴ Van't Hoff factor, i = 3

Given,

W = 25 mg = 0.025 g

V = 2 L

T = 25°C = (25 + 273) K = 298 K

Also, we know that, R = 0.0821 L atm K^{-1} mol^{-1}

Molar Mass = 174 g mol^{-1}

Applying the following relation for osmotic pressure (π),

$$\pi = i\frac{n}{v}RT$$

$$= i \times \frac{w}{M} \times \frac{1}{v} \times RT$$

$$= 3 \times \frac{0.025}{174} \times \frac{1}{2} \times 0.0821 \times 298$$

$$= 5.27 \times 10^{-3}\ atm$$

OR

(b) (i) (1) $\Delta T_f = 273 - 272.4 = 0.6$ K

Depression of freezing point (ΔT_f) for dilutes solution :

$$\Delta T_f = K_f m$$

Or, $$m = \frac{\Delta T_f}{K_f}$$

Substituting the values –

$$m = \frac{\Delta T_f}{K_f} = \frac{0.6\ K}{1.86\ K\ kg\ mol^{-1}}$$

Molality = 0.32 m

(2) $\Delta T_b = K_b m$

$= 0.512 \text{ K kg mol}^{-1} \times 0.32 \text{ m}$

$\Delta T_b = 0.16 \text{ K}$

$T_s = T_o + \Delta T_b$

$= 373 + 0.16$

$= 373.16 \text{ K}$

(3) As molality of solution is 0.32, it means that 0.32 moles of solute are dissolved in 1 Kg of water (or 1000 g of water, which means 55.55 moles).

So, mole fraction of solute $= \dfrac{0.32}{(55.55 + 0.32)}$

$= 0.0057$

The reduction in vapour pressure of the solvent is given by –

$\Delta p_1 = x_2\, p_1^{\circ}$ (where x_2 is mole fraction of solute and p_1° vapour pressure of water at 298 K)

$= 0.0057 \times 23.756$ mm Hg

$= 0.135$ mm Hg

So, lowering in vapour pressure is 0.135 mm Hg.

(ii) Given : $w = 1.23$ g

$W = 10$ g

$\Delta T_b = 100.975 - 100 = 0.975$

$K_b = 0.52 \text{ K kg Mol}^{-1}$

$$M(\text{observed}) = \frac{1000\, K_b w}{W \Delta T_b}$$

$$= \frac{1000 \times 0.52 \times 1.23}{10 \times 0.975}$$

$M(\text{observed}) = 65.6$

$$\text{Van't Hoff factor } (i) = \frac{M\,(\text{calculated})}{M\,(\text{observed})} = \frac{164}{65.6}$$

$i = 2.5$

Answer 17.

(a) (i) (1) Tetraaminecopper(II) sulphate

(2) Dichloridobis(ethylenediamine)cobalt (II)

(3) Potassium trioxalatoaluminate(III)

(ii) (1) Hexaaquairon(II) ion

(2) Oxidation number of Fe is +2

(3) There are 4 unpaired electrons in the complex ion belonging to Fe^{2+} ion–

$$\left[\overset{x}{Fe}(\overset{0}{H_2O})_6\right]^{2+}$$

Oxidation state of Fe ion = + 2

$x + 0 = 2$

$\Rightarrow \quad x = 2$

Electronic configuration of Fe^{+2} ion = $3d^6\, 4s^0$

As H_2O is a weak field ligand, it will form high spin complex.

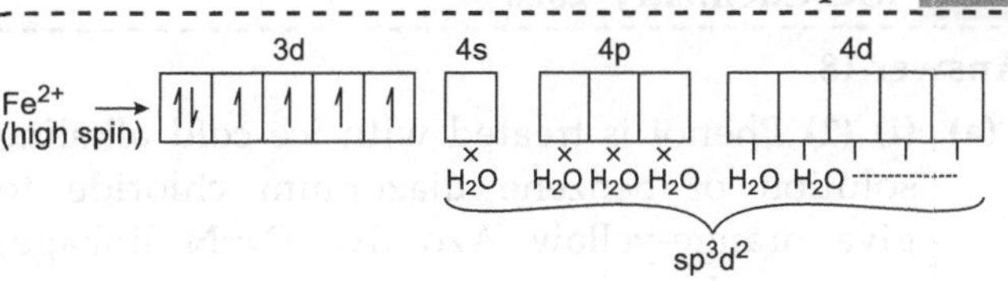

As outer d-orbitals are involved in sp^3d^2 hybridisation it given outer orbital sp^3d^2 hybridisation.

(4) The hybridisation is sp^3d^2.

OR

(b) (i) (1) Linkage isomerism

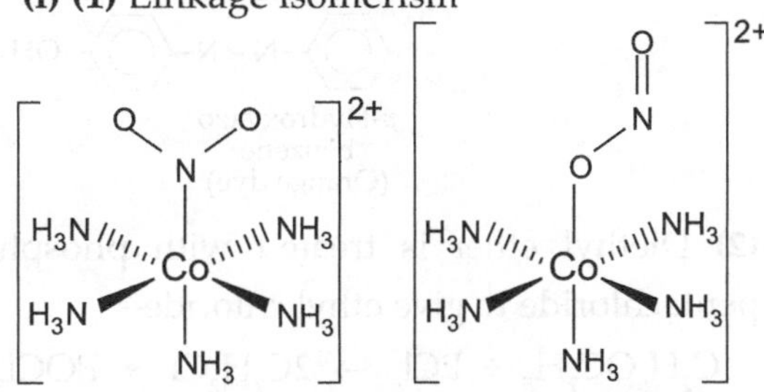

(2) Hydration isomerism occurs when there is an interchange of water molecule and another ligand between the first coordination sphere and the other ligand outside it.

(3) Co-ordination isomerism

(ii) Oxidation state of Ni in $[Ni(CO)_4]$ is 0 as CO is a neutral ligand. Ni has oxidation state of 0 in $[Ni(CO)_4]$ and its configuration is $3d^8\, 4s^2$. As –CO is a strong field ligand, the electrons pair up inside the orbital. Due to which, all the 5 *d*-orbitals gets occupied and the remaining '*s*' and '*p*' orbital gets filled by the incoming electron pairs of the ligand –CO (2 in '*s*' and 6 in '*p*'). One '*s*' orbital and three '*p*' orbitals are filled by –CO as there are 4 CO atoms in the complex. So, the hybridisation is sp^3 and geometry is tetrahedral and as no unpaired electrons are present, it is diamagnetic.

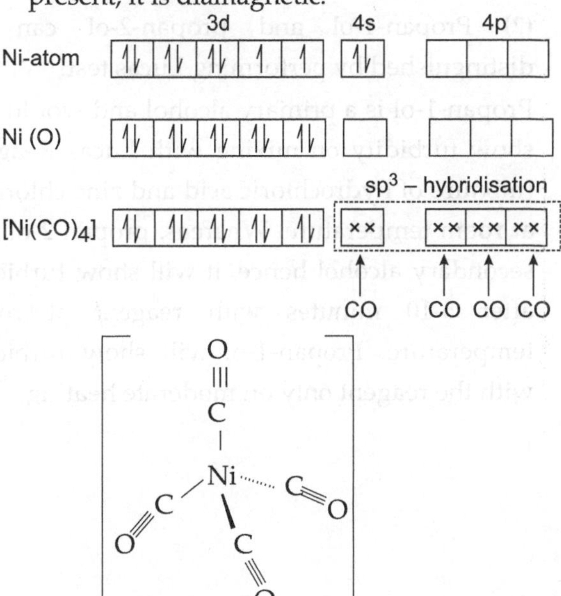

Answer 18.

(a) (i) (1) Phenol is treated with ice cold alkaline solution of benzene diazonium chloride to give orange-yellow Azo dye (N=N linkage) compound.

$$C_6H_5N_2^+Cl^- + C_6H_5OH \xrightarrow[0-5°C]{OH^-(pH-9-10)} C_6H_5-N{=}N-C_6H_4-OH + HCl$$

Benzene diazonium chloride + Phenol → p-Hydroxyazo benzene (Orange dye)

(2) Diethyl ether is treated with phosphorus pentachloride to give ethyl chloride–

$$C_2H_5OC_2H_5 + PCl_5 \rightarrow 2C_2H_5Cl + POCl_3$$

Diethyl ether → Ethyl chloride

(3) Ethyl alcohol is treated with thionyl chloride–

$$C_2H_5OH + SOCl_2 \rightarrow C_2H_5Cl + SO_2 + HCl$$

Ethyl alcohol → Ethyl chloride

(ii) (1) Ethanol and dimethyl ether can be distinguished by reaction with sodium metals and I_2. Ethanol on reaction will liberate H_2 gas forming a yellow Iodoform precipitate while ether doesn't react with sodium metal.

$$2Na + 2C_2H_5OH \longrightarrow 2C_2H_5ONa + H_2$$

Sodium metal, Ethanol, Sodium ethoxide (Yellow ppt.), Hydrogen

$$CH_3OCH_3 + Na \rightarrow \text{No reaction}$$

(2) Propan-1-ol and propan-2-ol can be distinguished by performing Lucas test.

Propan-1-ol is a primary alcohol and would not show turbidity on mixing with Lucas Reagent (mixture of hydrochloric acid and zinc chloride) at room temperature. Whereas, propan-2-ol is a secondary alcohol hence, it will show turbidity after 5-10 minutes with reagent at room temperature. Propan-1-ol will show turbidity with the reagent only on moderate heating.

OR

(b) (i) (1) In Williamson's synthesis, an ether is prepared by the nucleophilic substitution (typically SN_2) of organic halide with an alkoxide ion.

$$R-O^- + R'-X \rightarrow R-O-R' + X^-$$

Alkoxide ion, Halide, Ether

For example :

$$C_6H_5OH + H_3C-I \xrightarrow[H_3C-OH]{H_3C-\overset{\ominus}{O}\overset{\oplus}{Na}} C_6H_5O-CH_3 + NaI$$

Phenol, Methyl Iodide, Anisole, Sodium Iodide

(2) Esterification reaction is the process of combining an organic acid (RCOOH) with an alcohol (ROH) to form an ester (RCOOR) and water in acidic mediums usually. The general reaction is as follows –

$$RC(=O)-OH + R'OH \xrightarrow[heat]{conc.\ H_2SO_4} RC(=O)-O-R' + HOH$$

Acid, Alcohol, Ester, Water

(3) In Reimer-Tiemann reaction a phenol is converted to an o-hydroxy benzaldehyde by reacting phenol with chloroform in presence of a base–

$$C_6H_5OH \xrightarrow[\text{aq. NaOH, } 70°C]{CHCl_3} o\text{-}HOC_6H_4CHO$$

Salicyaldehyde (Main Product)

(ii) (1) $C_2H_5OH \xrightarrow[\Delta]{Cu} CH_3CHO$

Ethyl alcohol → Ethanal (A)

$$CH_3CHO \xrightarrow{\text{dil. NaOH}} HC(=O)-CH_2-CH(OH)-CH_3$$

Ethanal (A) → β-hydroxy acetaldehyde (B)

(2)

$$C_6H_5OH \xrightarrow{\text{Zn dust}} C_6H_6 \xrightarrow[\text{Anhyd. } AlCl_3]{CH_3COCl} C_6H_5COCH_3$$

Phenol → Benzene (A) → Acetyl benzene (B)

●●

CHEMISTRY 2019

QUESTIONS

Question 1.

(a) *Fill in the blanks by choosing the appropriate word/ words from those given in the brackets:* **[4 × 1]**

(more than, primary, cathode, Lucas reagent, two, four, less than, Grignard's reagent, tertiary, anode, zero, equal to, three)

(i) *The elevation of boiling point of 0·5 M K_2SO_4 solution is______ that of 0·5 M urea solution. The elevation of boiling point of 0·5 M KCl solution is______ that of 0·5 M K_2SO_4 solution.*

(ii) *A mixture of conc. HCl and anhydrous $ZnCl_2$ is called ______which shows maximum reactivity with ______ alcohol.*

(iii) *In electrolytic refining the impure metal is made ______ while a thin sheet of pure metal is used as______ .*

(iv) *When the concentration of a reactant of first order reaction is doubled, the rate of reaction becomes ______ times, but for a ______ order reaction, the rate of reaction remains the same.*

(b) *Select the correct alternative from the choices given:* **[4 × 1]**

(i) *The cell reaction is spontaneous or feasible when e.m.f. of the cell is:*

(1) *negative*
(2) *positive*
(3) *zero*
(4) *either positive or negative*

(ii) *Which, among the following polymers, is a polyester?*

(1) *Melamine* (2) *Bakelite*
(3) *Terylene* (4) *Polythene*

(iii) *The correct order of increasing acidic strength of the oxoacids of chlorine is :*

(1) $HClO_3 < HClO_4 < HClO_2 < HClO$
(2) $HClO < HClO_2 < HClO_3 < HClO_4$
(3) $HClO_2 < HClO < HClO_4 < HClO_3$
(4) $HClO_3 < HClO_4 < HClO < HClO_2$

(iv) *A catalyst is a substance which :*

(1) *changes the equilibrium constant of the reaction.*
(2) *increases the equilibrium constant of the reaction.*
(3) *supplies energy to the reaction.*
(4) *shortens the time to reach equilibrium.*

(c) *Match the following:* **[4 × 1]**

(i)	*Diazotisation*	**(a)**	*Anisotropic*
(ii)	*Crystalline solid*	**(b)**	*Reimer-Tiemann reaction*
(iii)	*Phenol*	**(c)**	*Diphenyl*
(iv)	*Fittig reaction*	**(d)**	*Aniline*

(d) *Answer the following questions :* **[4 × 2]**

(i) **(1)** *Which trivalent ion has maximum size in the lanthanoid series i.e., Lanthanum ion (La^{3+}) to Lutetium ion (Lu^{3+}) ? (At. no. of lanthanum = 57 and lutetium = 71)*

(2) *Explain why Cu^{2+} is paramagnetic but Cu^{+} is diamagnetic. (At. No. of Cu = 29)*

(ii) *When a coordination compound $CoCl_3{\cdot}6NH_3$ is mixed with $AgNO_3$, three moles of AgCl are precipitated per mole of the compound. Write the structural formula and IUPAC name of the coordination compound.*

(iii) *Calculate the boiling point of urea solution when 6 g of urea is dissolved in 200 g of water.*

(K_b for water = 0·52 K kg mol^{-1}, boiling point of pure water = 373 K, mol. wt. of urea = 60)

(iv) *Identify the compounds A, B, C and D in the given reaction:*

$$HC \equiv CH \xrightarrow[Hg^{2+}/H_2SO_4]{HO_2} A \xrightarrow[K_2Cr_2O_7+H_2SO_4]{[O]}$$

$$B \xrightarrow{Ca(OH)_2} C \xrightarrow[\text{dry distillation}]{\text{Heat}} D$$

Question 2. **[2]**

(a) *For the reaction A + B → C + D, the initial rate for different reactions and initial concentration of reactants are given below:*

S. No.	Initial Conc. [A] mole L^{-1}	Initial Conc. [B] mole L^{-1}	Initial rate (mole L^{-1} sec^{-1})
1.	1·0	1·0	2×10^{-3}
2.	2·0	1·0	4×10^{-3}
3.	4·0	1·0	8×10^{-3}
4.	1·0	2·0	2×10^{-3}
5.	1·0	4·0	2×10^{-3}

(i) *What is the overall order of reaction ?*

(ii) *Write the rate law equation.*

OR

(b) *25% of a first order reaction is completed in 30 minutes. Calculate the time taken in minutes for the reaction to go to 90% completion.*

Question 3. **[2]**

(i) *Name the type of drug which lowers the body temperature in high fever condition.*

(ii) *What are tranquilizers ? Give one example of a tranquilizer.*

Question 4. **[2]**

Write the balanced chemical equation for each of the following :

(i) *Chlorobenzene treated with ammonia in the presence of Cu_2O at 475 K and 60 atm.*

(ii) *Ethyl chloride treated with alcoholic potassium hydroxide.*

Question 5. **[2]**

(i) *Name the monomer and the type of polymerisation that takes place when PTFE is formed.*

(ii) *Name the monomers of nylon 6, 6.*

Question 6. **[2]**

Name two water soluble vitamins and the diseases caused by their deficiency in the diet of an individual.

Question 7. **[2]**

(a) *How will you obtain the following ? (Give balanced chemical equations):*

(i) *Benzene from phenol*

(ii) *Iodoform from ethanol*

OR

(b) *How will you obtain the following ? (Give balanced chemical equations):*

(i) *Salicylaldehyde from phenol*

(ii) *Propan-2-ol from Grignard's reagent*

Question 8. **[2]**

Show that for a first order reaction the time required to complete 75% of reaction is about 2 times more than that required to complete 50% of the reaction.

Question 9. **[3]**

(a) *When 0·4 g of oxalic acid is dissolved in 40 g of benzene, the freezing point of the solution is lowered by 0·45 K. Calculate the degree of association of acetic acid. Acetic acid forms dimer when dissolved in benzene.*

(K_f for benzene = 5·12 K kg mol^{-1}, at. wt. C = 12, H = 1, O = 16)

OR

(b) *A solution is prepared by dissolving 9·25 g of non-volatile solute in 450 ml of water. It has an osmotic pressure of 350 mm of Hg at 27°C. Assuming the solute is non-electrolyte, determine its molecular mass.* *(R = 0·0821 lit atm K^{-1} mol^{-1})*

Question 10. **[3]**

An element occurs in body centered cubic structure. Its density is 8·0 g/cm^3. If the cell edge is 250 pm, calculate the atomic mass of an atom of this element. (N_A = 6·023 × 10^{23})

Question 11. **[3]**

Describe the role of the following:

(i) *Cryolite in the extraction of aluminium from pure alumina.*

(ii) *NaCN in the extraction of silver from a silver ore.*

(iii) *Coke in the extraction of iron from its oxides.*

Question 12. **[3]**

(i) *Write the IUPAC names of the following :*

(1) *$K_3[Fe(C_2O_4)_3]$* **(2)** *$[Co(NH_3)_5Cl]SO_4$*

(ii) *$[Fe(CN)_6]^{4-}$ is a coordination complex ion.*

(1) *Calculate the oxidation number of iron in the complex.*

(2) *Is the complex ion diamagnetic or paramagnetic?*

(3) *What is the hybridisation state of the central metal atom ?*

(4) *Write the IUPAC name of the complex ion.*

Question 13. **[3]**

(a) *Explain why:*

(i) *Transition elements form alloys.*

(ii) *Zn^{2+} salts are white whereas Cu^{2+} salts are coloured.*

(iii) *Transition metals and their compounds act as catalyst.*

OR

(b) *Complete and balance the following chemical equations:*

(i) *$KMnO_4 + H_2SO_4 + H_2C_2O_4 \rightarrow$ ____ + ____ + ____ + ____*

(ii) *$K_2Cr_2O_7 + H_2SO_4 + KI \rightarrow$ ____ + ____ + ____ + ____*

(iii) *$K_2Cr_2O_7 + H_2SO_4 + FeSO_4 \rightarrow$ ____ + ____ + ____ + ____*

Question 14. **[3]**

Give balanced equations for the following:

(i) *Aniline is treated with bromine water.*

(ii) *Ethylamine is heated with chloroform and alcoholic solution of potassium hydroxide.*

(iii) *Benzene diazonium chloride is treated with ice cold solution of aniline in acidic medium.*

Question 15. **[3]**

Define the following terms with suitable examples:

(i) *Peptization* **(ii)** *Electrophoresis*

(iii) *Dialysis*

Question 16. [5]

(a) (i) *Calculate the mass of silver deposited at cathode when a current of 2 amperes is passed through a solution of $AgNO_3$ for 15 minutes.*

(At. wt. of Ag = 108, 1 F = 96,500 C)

(ii) *Calculate the e.m.f. and ΔG for the cell reaction at 298 K:*

$Mg_{(s)}/Mg^{2+}{}_{(0{\cdot}1\ M)}//Cu^{2+}{}_{(0{\cdot}01M)}/Cu_{(s)}$

Given $E^0_{cell} = 2{\cdot}71\ V$

$1\ F = 96{,}500\ C$

OR

(b) (i) *Define the following terms:*

(1) *Specific conductance*

(2) *Kohlrausch's law*

(ii) *The resistance of a conductivity cell containing 0·001 M KCl solution at 298 K is 1500 ohm. What is the cell constant and molar conductivity of 0·001 M KCl solution, if the conductivity of this solution is $0{\cdot}146 \times 10^{-3}\ ohm^{-1}\ cm^{-1}$ at 298 K ?*

Question 17. [5]

(a) (i) *Explain why:*

(1) *Fluorine has lower electron affinity than chlorine.*

(2) *Red phosphorus is less reactive than white phosphorus.*

(3) *Ozone acts as a powerful oxidising agent.*

(ii) *Draw the structures of the following:*

(1) XeF_6 (2) IF_7

OR

(b) (i) *Explain why:*

(1) *Interhalogen compounds are more reactive than the related elemental halogens.*

(2) *Sulphur exhibits tendency for catenation but oxygen does not.*

(3) *On being slowly passed through water, PH_3 forms bubbles but NH_3 dissolves.*

(ii) *Complete and balance the following reactions :*

(1) $P_4 + H_2SO_4 \rightarrow$ ____ + ____ + ____

(2) $Ag + HNO_3 \rightarrow$ ____ + ____ + ____
(dilute)

Question 18. [5]

(a) (i) *Give balanced equations for the following reactions :*

(1) *Acetaldehyde reacts with hydrogen cyanide.*

(2) *Acetone reacts with phenyl hydrazine.*

(3) *Acetic acid is treated with ethanol and a drop of conc. H_2SO_4.*

(ii) *Give one chemical test each to distinguish between the following pairs of compounds :*

(1) *Acetone and benzaldehyde.*

(2) *Phenol and benzoic acid.*

OR

(b) (i) *Write chemical equations to illustrate the following name reactions:*

(1) *Aldol condensation*

(2) *Cannizzaro's reaction*

(3) *Benzoin condensation*

(ii) *Identify the compounds A and B in the given reactions:*

(1) Benzene $\xrightarrow[AlCl_3\,(anhy)]{CH_3Cl} A \xrightarrow[K_2Cr_2O_7 + H_2S_4]{[O]} B$

(2) $CH_3COCH_3 \xrightarrow[HNO_3\,(conc.)]{[O]} A \xrightarrow{PCl_5}$

ANSWERS

Answer 1.

(a) (i) more than, less than

(ii) Lucas reagent, tertiary

(iii) anode, cathode

(iv) two, zero

(b) (i) (2) Positive

(ii) (3) Terylene

(iii) (2) $HClO < HClO_2 < HClO_3 < HClO_4$

(iv) (4) shortens the time to reach equilibrium

(c) (i) (d) Aniline

(ii) (a) Anisotropic

(iii) (b) Reimer-Tiemann reaction

(iv) (c) Diphenyl

(d) (i) (1) La^{3+} has the maximum size because of lanthanoid contraction.

(2) Cu^{2+} is paramagnetic but Cu^+ is diamagnetic because Cu^{2+} contains unpaired electrons whereas Cu^+ does not have any unpaired electron.

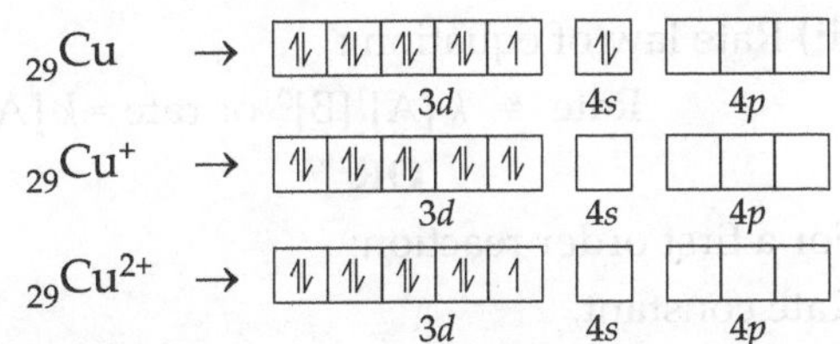

(ii) The complex is $[Co(NH_3)_6]Cl_3$.

The IUPAC name is Hexamminecobalt (III) chloride.

(iii) Moles of urea present in solution

$$= \frac{6\,g}{60\,g} = 0{\cdot}1\ mol$$

Weight of solvent in kg.

$$= \frac{200\,g}{1000\,g} = 0{\cdot}2\ kg.$$

Thus, molality of urea solution

$$= \frac{0.1\,mol}{0.2\,kg} = 0{\cdot}5\ mol\ kg^{-1}.$$

For water, change in boiling point,

$$\Delta T_b = K_b \times m$$
$$= 0{\cdot}52\ K\ kg\ mol^{-1} \times 0{\cdot}5\ mol\ kg^{-1}.$$
$$= 0{\cdot}26\ K$$

Since water boils at 373 K, the elevated boiling point of the urea solution would be 373 + 0·26 K = 373·26 K.

(iv) The compounds A, B, C and D are as shown in the below reaction scheme:

$$\underset{\text{Ethyne}}{HC\equiv CH} \xrightarrow[Hg^{2+},\,H_2SO_2]{HO_2} \underset{\substack{\text{Ethanal}\\(A)}}{CH_3CHO} \xrightarrow[K_2Cr_2O_7 + H_2SO_4]{[O]}$$

$$\underset{\substack{\text{Ethanoic acid}\\(B)}}{CH_3COOH} \xrightarrow{Ca(OH)_2} \underset{\substack{\text{Calcium acetate}\\(C)}}{Ca(CH_3COO)_2}$$

$$\downarrow \text{Heat, Dry distillation}$$

$$\underset{\substack{\text{Acetone}\\(D)}}{CH_3COCH_3} + CaCO_3$$

Answer 2.

(a) On doubling the concentration of reactant A (keeping the concentration of reactant B constant), the rate of reaction is getting doubled which indicates that the reaction is first order with respect to reactant A. Also, on doubling the concentration of reactant B (keeping the concentration of reactant A constant, there is no effect on the reaction rate which indicates that the reaction is zero order with respect to reactant B.

(i) The overall order of the reaction is 1 + 0 = 1, that is first order reaction.

(ii) Rate law of equation :

$$\text{Rate} = k\,[A]^1[B]^0 \text{ or rate} = k\,[A]$$

OR

(b) For a first order reaction:

Rate constant,

$$k = \left(\frac{2.303}{t}\right)\log\frac{[R]^0}{[R]}$$

$$[R] = 100\% - 25\% = 75\%$$

Putting the values given:

$$k = \left(\frac{2.303}{30}\right)\log\frac{100}{75}$$
$$= 0{\cdot}076 \times 0{\cdot}124\ min^{-1}$$
$$= 0{\cdot}00959\ min^{-1}$$
$$= 9{\cdot}59 \times 10^{-3}\ min^{-1}$$

Now, time required for 90% completion,

$$t = ?$$
$$[A]_0 = 100$$
$$[A]_t = 10$$
$$k = 9{\cdot}59 \times 10^{-3}$$

$$9{\cdot}59 \times 10^{-3} = \frac{2.303}{t}\log\frac{100}{10}$$

$$\therefore \quad t = \frac{2.303}{9.59\times 10^{-3}} \times 1 \times 1$$
$$= 0{\cdot}24014 \times 10^3\ min.$$
$$\therefore \quad t = 240{\cdot}14\ min.$$

Answer 3.

(i) Antipyretics lower the body temperature in high fever conditions.

(ii) Tranquilizers are neurologically active drugs which help in reducing conditions of stress and anxiety in a person. They also find use in treatment of mild and severe mental diseases and they induce sleep. Examples: Equanil, veronal, valium, chlordiazepoxide etc.

Answer 4.

(i) The equation can be given as:

$$2\ \underset{\text{Chlorobenzene}}{C_6H_5Cl} + \underset{\text{Ammonia}}{2NH_3} \xrightarrow[Cu_2O]{475\ K,\ 60\ atm} 2\ \underset{\text{Aniline}}{C_6H_5NH_2} + Cu_2Cl_2 + H_2O$$

(ii) The equation can be given as:

$$\underset{\text{Ethylchloride}}{C_2H_5Cl} + KOH\ (alc.) \xrightarrow{\text{Dehydration}} \underset{\text{Ethene}}{CH_2 = CH_2} + KCl + H_2O$$

Answer 5.

(i) PTFE stands for Polytetrafluoroethene or Teflon.

Monomer–tetrafluoroethylene ($F_2C = CF_2$)

Type of polymerisation when PTFE is formed is Addition polymerisation which takes place by free radical polymerisation process.

(ii) Nylon 6, 6 is synthesised by polycondensation of hexamethylenediamine ($NH_2(CH_2)_6NH_2$) and adipic acid ($HOOC-(CH_2)_4-COOH$)

Answer 6.

Name of water soluble vitamin	Disease caused by its deficiency
Vitamin B_1	Beri-beri (retarded growth)
Vitamin C	Scurvy (bleeding gums)

Answer 7.

(a) (i) Benzene can be obtained from phenol by treating it with zinc dust.

$$\text{C}_6\text{H}_5\text{OH} + \text{Zn} \longrightarrow \text{C}_6\text{H}_6 + \text{ZnO}$$

Phenol   Benzene

(ii) Ethanol can be converted into iodoform by treating ethanol with iodine and sodium hydroxide.

$$\underset{\text{Ethanol}}{CH_3CH_2OH} + 4I_2 + 6NaOH \longrightarrow \underset{\text{Iodoform}}{CHI_3} + HCOONa + 5NaI + 5H_2O$$

OR

(b) (i) Salicylaldehyde can be obtained from phenol using the Reimer-Tiemann reaction:

$$\underset{\text{Phenol}}{C_6H_5OH} \xrightarrow[\text{aq. NaOH}]{CHCl_3} \left[C_6H_4(ONa)(CHCl_2) \right] \xrightarrow{NaOH} C_6H_4(ONa)(CHO) \xrightarrow{H^+} \underset{\text{Salicylaldehyde}}{C_6H_4(OH)(CHO)}$$

(ii) Propan-2-ol from Grignard reagent:

$$\underset{\text{Methyl bromide}}{H_3C-Br} \xrightarrow[\text{Dry ether}]{Mg} \underset{\text{Methyl magnesium bromide (Grignard reagent)}}{CH_3MgBr}$$

$$\underset{\text{Ethanal}}{CH_3CHO} + CH_3MgBr \xrightarrow[\text{HOH}]{H^+} \left[H_3C-HC(\overset{-}{O}\overset{+}{Mg}Br)-CH_3 \right] \longrightarrow \underset{\text{Propan-2-ol}}{H_3C-CH(OH)-CH_3} + Mg(OH)Br$$

Answer 8.

As we know, rate constant for a first order reaction is given by,

$$k = \left(\frac{2.303}{t}\right)\log_{10}\frac{[R]^0}{[R]}$$

For completion of 50% reaction, time required is t_{50} and for 75% completion time required is t_{75}. Writing the equations and taking their ratios:

$$k = \frac{2.303}{t}\log_{10}\frac{a}{a-x}$$

or $$t_{75\%} = \frac{2.303}{k}\log\frac{100}{25}$$

or $$t_{75\%} = \frac{2.303}{k}\log \times 0.602$$

$$t_{75\%} = \frac{1.386}{k}$$

$$t_{50\%} = \frac{2.303}{k}\log\frac{100}{50}$$

or $$t_{50\%} = \frac{2.303}{k}\log 2$$

or $$= \frac{2.303}{k} \times 0.3010$$

or $$= \frac{0.693}{k}$$

$$\therefore \quad 2t_{75} = 2 \times t_{50}$$

Hence, it is proved that time required for 75% completion of reaction is 2 times the time required for 50% completion of the reaction.

Answer 9.

(a) Molecular mass (normal) for acetic acid, CH_3COOH is,

$$M_{normal} = 60 \text{ g mol}^{-1}$$

Observed molecular mass of CH_3COOH,

$$M_{observed} = \frac{K_f w_2}{w_1 \Delta T_f}$$

$$= \frac{(5.12 \text{K kg mol}^{-1})(0.4 \times 10^{-3} \text{ kg})}{(0.04 \text{ kg})(0.45 \text{K})}$$

$$= 113{\cdot}77 \times 10^{-3} \text{ kg mol}$$

$$= 113{\cdot}77 \text{ g mol}^{-1}$$

Now, van't Hoff factor,

$$i = \frac{M_{normal}}{M_{observed}}$$

$$= \frac{60 \text{ g mol}^{-1}}{113.77 \text{g mol}^{-1}} \Rightarrow 0.527$$

Let x be the degree of association of CH_3COOH, then,

$$2C_6H_5COOH \leftrightarrow (C_6H_5COOH)_2$$

If x represents the degree of association of the solute then we would have $(1 - x)$ mol of benzoic acid left in unassociated form and correspondingly $x/2$ as associated moles of benzoic acid at equilibrium. Therefore,

The total number of moles after association

$$= 1 - x + \frac{x}{2} = 1 - \frac{x}{2}$$

$$\frac{\text{Normal molar mass}}{\text{Observed molar mass}} = 1 - \frac{x}{2}$$

$$\frac{60 \text{ g mol}^{-1}}{113.77 \text{ g mol}^{-1}} = 1 - \frac{x}{2}$$

$$0.527 = 1 - \frac{x}{2}$$

$$\frac{x}{2} = 0.473 \times 2$$

$$= 0.946 \approx 0.95$$

Hence, degree of association, $x = 0.95$.

or degree of association of acetic acid is 95%.

OR

(b) The osmotic pressure is given by the equation,

$$\pi = \left(\frac{n_2}{V}\right) RT$$

Number of moles,

$$n_2 = \frac{\text{Weight in grams}}{\text{Molecular mass}}$$

So, $$\pi = \left(\frac{w}{M} \times V\right) RT$$

$$350 \text{ mm Hg} = \frac{350}{760} = 0.46 \text{ atm}$$

Putting the given values,

$$\pi = CRT$$

$$0.46 = \frac{n}{V} RT$$

$$0.46 = \frac{n}{450} \times 0.0821 \times 300 \times 1000$$

$$\therefore \quad n = \frac{0.46 \times 450}{0.0821 \times 300 \times 1000}$$

$$= \frac{207}{24630} = 0.0084$$

$$\text{Molar mass} = \frac{\text{Mass}}{\text{Moles}}$$

$$= \frac{9.25}{0.0084} = 1101 \text{ g}$$

Answer 10.

As the lattice is bcc, the number of element atoms per unit cell, Z = 2

Volume would be

$$a^3 = (250 \text{ pm})^3 = (250 \times 10^{-10} \text{ cm})^3$$

$$\text{Density}, d = \frac{Z.M}{V.N_A}$$

Putting the given values,

$$8.0 \text{ g/cm}^3 = \frac{2 \times M}{(6.023 \times 10^{23} \text{ mol}^{-1}) \times (250 \times 10^{-10} \text{ cm})^2}$$

$$M = \left(\frac{8.00 \text{ g cm}^{-3}}{2}\right) \times (6.023 \times 10^{23} \text{ mol}^{-1}) \times (250 \times 10^{-10} \text{ cm})^3$$

$$= 4 \text{ g cm}^{-3} \times 6.023 \times 10^{23} \text{ mol}^{-1} \times 1.56 \times 10^{-23} \text{ cm}^3$$

$$= 37.6 \text{ g mol}^{-1}$$

So, the molecular mass of the element is 37·6 g mol^{-1}.

Answer 11.

(i) Alumina (Al_2O_3) is subjected to electrolytic reduction to give pure aluminium. Alumina is mixed with Cryolite (Na_3AlF_6) which lowers its melting point and also increases the electrical conductivity of alumina, thereby facilitating the electrolytic reduction process.

(ii) Role of NaCN in the extraction of silver (Ag) is to do the leaching of Ag ore in the presence of air (for O_2) where silver ore gets dissolved and from which the silver is obtained later by replacement whereas impurities remain insoluble.

$$4Ag(s) + 8CN^-(aq) + 2H_2O(aq) + O_2(g) \rightarrow 4[Ag(CN)_2]^-(aq) + 4OH^-(aq)$$

$$2[Ag(CN)_2]^-(aq) + Zn(s) \rightarrow 2Ag(s) + [Zn(CN)_4]^{2-}(aq)$$

(iii) Coke is source of heat in blast furnace and acts as a reducing agent for iron ore. Coke is burnt to produce carbon monoxide which later reacts with the iron oxides to reduce them in steps to yield iron. Further it reacts with silica impurities to remove them as soluble impurities.

$$C(s) + \frac{1}{2}O_2 \longrightarrow CO(g)$$

$$3Fe_2O_3 + CO \longrightarrow 2Fe_2O_3 + CO_2$$

$$Fe_3O_4 + 4CO \longrightarrow 3Fe + 4CO_2$$

$$Fe_2O_3 + CO \longrightarrow 2FeO + CO_2$$

Answer 12.

(i) (1) $K_3(Fe(C_2O_4)_3]$ = Potassium trioxalatoferrate(III).

(2) $[Co(NH_3)_5Cl]SO_4$ = Pentaamminechloridocobalt(III) sulphate

(ii) (1) Let the oxidation number of Fe is x.

So, $x + 6(-1) = -4$

$x - 6 = -4$

or $x = +2$

So, oxidation number of Fe is + 2.

(2) As CN^- is strong field ligand there are no unpaired electrons in Fe^{2-} ion, hence it is diamagnetic.

(3) Hybridisation state of central metal atom is d^2sp^3.

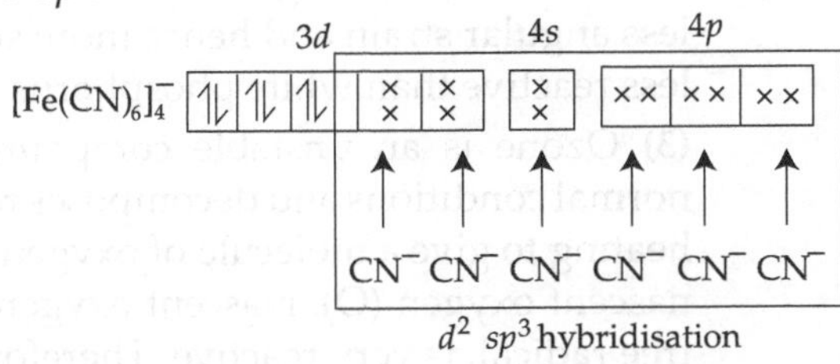

(4) The IUPAC name is Hexacyanoferrate(II).

Answer 13.

(a) **(i)** The tendency of transition metals to form alloys can be explained on the basis of their atomic sizes. Atomic sizes of transition metals are very similar to each other and hence in the crystal lattice atom of the one metal can be easily replaced with atom of the another transition metal.

(ii) Zn^{2+} has $3d^{10}$ configuration and no unpaired electrons, hence it is colourless. Whereas in Cu^{2+} state which has $3d^9$ configuration there is one unpaired electron and it undergoes *d-d* transitions emiting colour so Cu^{2+} salts are blue in colour.

(iii) The transition metals and their compounds are known for their catalytic activity because they have ability to adopt multiple oxidation states they have variable oxidation state giving unstable intermediates and can form complexes.

OR

(b) Complete balanced equations are:

(i) $$2KMnO_4 + 3H_2SO_4 + 5H_2C_2O_4 \longrightarrow K_2SO_4 + 2MnSO_4 + 8H_2O + 10CO_2$$

(ii) $$K_2Cr_2O_7 + 6KI + 7H_2SO_4 \longrightarrow 4K_2SO_4 + Cr_2(SO_4)_3 + 3I_2 + 7H_2O$$

(iii) $$K_2Cr_2O_7 + 6FeSO_4 + 7H_2SO_4 \longrightarrow Cr_2(SO_4)_3 + 3Fe_2(SO_4)_3 + K_2SO_4 + 7H_2O$$

Answer 14.

(i) Aniline reacts with bromine water at room temperature to give 2, 4, 6-tribromoaniline and hydrobromic acid:

$$\underset{\text{Aniline}}{C_6H_5NH_2} + 3Br_2 \xrightarrow{H_2O} \underset{\text{2, 4, 6-Tribromoaniline}}{C_6H_2Br_3NH_2} + \underset{\text{Hydrobromic acid}}{3HBr}$$

(ii) Ethylamine being a primary amine, forms foul-smelling methyl isocyanide on heating with chloroform and alcoholic potassium hydroxide:

$$\underset{\text{Ethyl amine}}{CH_3CH_2NH_2} \xrightarrow[CHCl_3]{Alc.KOH} \underset{\text{Ethyl isocyanide}}{CH_3H_2C-N \equiv C} + 3KCl + 3H_2O$$

(iii) Benzene diazonium chloride is treated with ice cold solution of aniline in acidic medium to obtain an azo dye, *p*-aminoazobenzene:

$$\underset{\text{Benzene diazonium chloride}}{C_6H_5N_2^+Cl^-} + \underset{\text{Aniline}}{C_6H_5NH_2} \xrightarrow[\text{ice cold}]{\text{acid}} \underset{\text{p-aminoazo benzene}}{C_6H_5-N=N-C_6H_4-NH_2} + HCl$$

Answer 15.

(i) Peptisation: The process of conversion of a fresh precipitate into colloidal solution by shaking it with dispersion medium in the presence of electrolyte is called peptisation.

Example : Red coloured solution of $Fe(OH)_3$ is obtained when freshly prepared precipitate is treated with small amount of $FeCl_3$ solution.

(ii) Electrophoresis: The movement of colloidal particles under an applied electric potential is called electrophoresis. Positively charged particles move towards cathode, while negatively charged particles move towards the anode.

Example: Purification of sewage water, proteins, etc.

(iii) Dialysis: It is the process of removing a dissolved substance from a colloidal solution by means of diffusion through a suitable membrane. The molecules or ions diffuse through membrane and pure colloidal solution is left behind.

Example: Removal of sugar from blood in diabetic patients.

Answer 16.

(a) **(i)** Time, $t = 15 \times 60 = 900$ sec;

$$\text{Charge} = \text{Current} \times \text{Time}$$
$$= 2A \times 900s = 1800\ C$$

According to the reaction,

$$Ag^+ (aq) + e^- \rightarrow Ag(s)$$

So, we need 1F or 96500 C to deposit one mole or 108 g of silver, so, by 1800 C, the mass of silver deposited would be:

$$= 108\ g\ mol^{-1} \times \frac{1800\,C}{96500\,C\,mol^{-1}}$$
$$= 2{\cdot}014\ g$$

So, 2·014 g of silver is deposited in 15 minutes.

(ii) Considering the cell reaction as:

$$Cu^{2+}(aq) + Mg(s) \rightarrow Mg^{2+}(aq) + Cu(s)$$

The $E_{(cell)}$ would be given by:

$$E_{(cell)} = E^{\ominus}_{(cell)} - \frac{RT}{nF}\log\frac{[Mg^{2+}]}{[Cu^{2+}]}$$

So, $E_{(cell)}$ will be:

$$E_{(cell)} = 2{\cdot}71\text{ V} - \left(\frac{0.059\text{ V}}{2}\right)\log_{10}\left(\frac{0.1}{0.01}\right)$$

$$= 2{\cdot}68045\text{ V}$$

Now, ΔG of the reaction is given by:

$$\Delta G_r = -nFE_{cell}$$

$$= -2 \times 96500\text{ C mol}^{-1} \times 2{\cdot}68045\text{ V}$$

$$= -517{\cdot}326\text{ kJ mol}^{-1}$$

OR

(b) (i) (1) Specific conductance is the conductance of a conductor whose length is 1 cm and area of cross section is equal to 1 cm^2. It is equal to the reciprocal of specific resistance of the conductor.

(2) The Kohlrausch's law of independent migration of ions states that limiting molar conductivity of an electrolyte at infinite dilution can be represented as the sum of the individual contributions of molar conductivities of the anion and cation of the electrolyte.

(ii) The cell constant is given by:

Cell constant, G = Conductivity × Resistance

Putting the given values,

$$G = 0{\cdot}146 \times 10^{-3}\text{ ohm}^{-1}\text{ cm}^{-1} \times 1500\text{ ohm}$$

$$= 0{\cdot}219\text{ cm}^{-1}$$

Concentration = $0{\cdot}001$ mol L^{-1}

Molar conductivity is given by:

$$\Lambda_m = \frac{1000 \times K}{c}$$

$$= \frac{0.146 \times 10^{-3}\text{ ohm}^{-1}\text{cm}^{-1} \times 1000}{0.001\text{ mol L}^{-1}}$$

$$= 146\text{ cm}^2\text{ ohm}^{-1}\text{ mol}^{-1}$$

Answer 17.

(a) (i) (1) Electron affinity of fluorine is less than that of chlorine because the atomic size of fluorine is very less than chlorine as a results there is a large electronic repulsion between the electrons of fluorine. Hence, in fluorine atom the incoming electron experiences lesser attraction (rather faces repulsion from the existing electrons) towards the nucleus than in chlorine atom.

(2) White phosphorus is discrete P_4 molecule having tetrahedral shape and have very high angular strain in the molecule, as the angles are 60 degrees each. This high angular strain makes white phosphorus unstable and highly reactive. Red phosphorus on the other hand, is a polymeric structure and consists of P_4 tetrahedra linked together, which has much less angular strain and hence more stable and less reactive than white phosphorus.

(3) Ozone is an unstable compound under normal conditions and decomposes readily on heating to give a molecule of oxygen (O_2) and nascent oxygen (O). Nascent oxygen, being a free radical, is very reactive. Therefore, ozone acts as a powerful oxidising agent and it is more reactive than oxygen.

$$\underset{\text{Ozone}}{O_3} \xrightarrow{UV} \underset{\substack{\text{Oxygen}\\\text{molecule}}}{O_2} + \underset{\substack{\text{Oxygen}\\\text{free radical}}}{O^{\bullet}}$$

(ii) (1) Structure of XeF_6

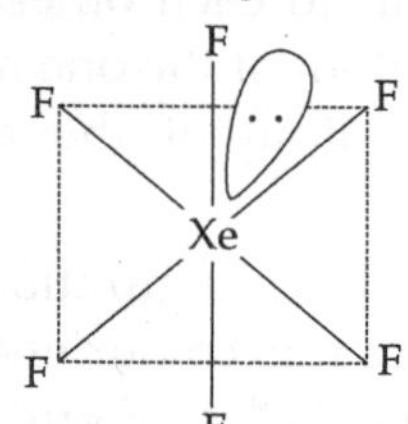

sp^3d^3-hybridisation of Xe

(2) Structure of IF_7

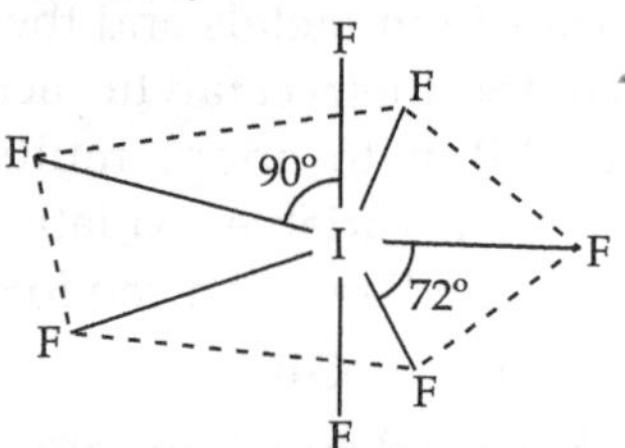

OR

(b) (i) (1) The interhalogen compounds are reactive because the X—X' bond in interhalogen compounds are weaker than X—X bonds in dihalogen molecules, hence easy to break causing more reactivity. Also the X—X' bond in interhalogen compounds are polar due to difference in electronegativities making it easy to cleave the bonds.

(2) Sulphur has greater catenation property than oxygen because S—S bond energy is more than O—O bond energy. This is because, oxygen atom is small in size and the lone pair of electrons in oxygen repel bond pair electrons of O—O to a greater extent than sulphur, in turn making the O—O bond weak. As a result, oxygen exists as a diatomic molecule and sulphur forms polyatomic molecule, that is sulphur shows catenation.

(3) On passing through water, ammonia

(NH_3) molecules form hydrogen bonds with water and hence get dissolved in water, whereas phosphine (PH_3) molecules are not able to form hydrogen bonds with water due to large atomic size of phosphorous.

Hydrogen bonding

(ii) (1) $P_4(s) + 10H_2SO_4 \longrightarrow 4H_3PO_4(aq) + 10SO_2(g) + 4H_2O$

(2) $3Ag(s) + 4HNO_3(dil.) \longrightarrow 3AgNO_3(s) + 2H_2O + NO$

Answer 18.

(a) (i) (1) Acetaldehyde reacts with hydrogen cyanide to give cyanohydrin.

$H_3C(H)C{=}O + H{-}CN \xrightarrow{\text{Step 1}} H{-}C(CH_3)(CN){-}O^- \xrightarrow[H^+]{\text{Step 2}} H{-}C(CH_3)(CN){-}O{-}H$

Cyanohydrin

(2) Acetone reacts with phenyl hydrazine to give corresponding hydrazine.

$C_6H_5NHNH_2 + CH_3COCH_3 \longrightarrow C_6H_5NHN = C(CH_3)_2 + H_2O$

(3) Acetic acid reacts with ethanol in acidic medium to give ethyl acetate.

$$\underset{\text{Acetic acid}}{CH_3COOH} + \underset{\text{Ethanol}}{CH_3CH_2OH} \xrightarrow[\text{Esterification}]{H^+} \underset{\text{Ethyl acetate}}{CH_3COOC_2H_5} + H_2O$$

(ii) (1) Acetone can be distinguished from benzaldehyde by carrying out Iodoform reaction. When acetone is treated with iodine and NaOH/KOH, a yellow precipitate of CHI_3 or Iodoform is obtained, which is not obtained in case of benzaldehyde, which does not give the Iodoform reaction.

Acetone or propanone $\xrightarrow{I_2/KOH}$ CH_3COOK + CHI_3 (yellow ppt) Triiodo-methane or iodoform

(2) Phenol reacts with neutral $FeCl_3$ to give a violet colour complex while benzoic acid does not gives this reaction.

Phenol $\xrightarrow{FeCl_3}$ Violet colour complex

OR

(b) (i) (1) **Aldol condensation** : The aldehydes and ketones having at least one α-hydrogen undergo a reaction in the presence of dilute alkali as catalyst to form β-hydroxy aldehydes (aldol) or β-hydroxy ketones (ketol), respectively. This is known as Aldol reaction.

Ethanal $\xrightleftharpoons{\text{dil. NaOH}}$ 3-Hydroxybutanal

(2) **Cannizzaro reaction:** The aldehydes which do not have an α-hydrogen atom, undergo self-oxidation and reduction disproportionation reaction on heating with concentrated alkali, to give a molecule of carboxylic acid and a molecule of alcohol respectively.

$$\underset{\text{Formaldehyde}}{HCHO + HCHO} + \text{Conc. NaOH} \longrightarrow \underset{\text{Methanol}}{H_3C{-}OH} + \underset{\text{Sodium methanoate}}{HCOONa}$$

(3) **Benzoin condensation:** The Benzoin condensation is a coupling reaction between two aldehydes that allows the preparation of α-hydroxyketones.

$$2\ \underset{\text{Benzaldehyde}}{C_6H_5CHO} \xrightarrow[H_2O/EtOH]{CN^-} \underset{\text{Benzoin}}{C_6H_5COCH(OH)C_6H_5}$$

(ii) (1) The compounds A and B are:

$$\underset{\text{Benzene}}{C_6H_6} \xrightarrow[\text{Friedal crafts reaction}]{CH_3Cl + AlCl_3\ (\text{Anhyd})} \underset{\text{Toluene (A)}}{C_6H_5CH_3} \xrightarrow[K_2Cr_2O_7 + H_2SO_4]{[O]} \underset{\text{Benzoic acid (B)}}{C_6H_5COOH}$$

(2) The compounds A and B are:

$$\underset{\text{Acetone}}{CH_3COCH_3} \xrightarrow[HNO_3]{[O]} HCOOH + \underset{\text{Acetic acid (A)}}{CH_3COOH}$$

$HCOOH \xrightarrow{PCl_5} CO + POCl_3 + 2HCl$

$CH_3COOH \xrightarrow{PCl_5} \underset{\text{Acetyl Chloride (B)}}{CH_3COCl} + POCl_3 + HCl$

●●

CHEMISTRY 2018

QUESTIONS

Question 1. **[20]**

(a) *Fill in the blanks by choosing the appropriate word/words from those given in the brackets:* **[4 × 1]**

(square pyramidal, electrical, 74, 26, sp^3d^2, sp^3d, chemical, 68, 32, tetrahedral, yellow, white, iodoform Lucas)

(i) *A Galvanic cell converts ______ energy into ______ energy.*

(ii) *The percentage of unoccupied spaces in bcc and fcc arrangements are ______ and ______ respectively.*

(iii) *Propan-2-ol on reaction with iodine and sodium hydroxide gives ______ precipitate and the reaction is called ______ test.*

(iv) *The geometry of $XeOF_4$ molecule is ______ and the hybridisation of xenon atom in the molecule is ______.*

(b) *Complete the following statements by selecting the* **correct alternative from** *the choices given:* **[4 × 1]**

(i) *During the course of an S_N1 reaction, the intermediate species formed is:*

(1) *a carbocation*

(2) *a free radical*

(3) *a carbanion*

(4) *an intermediate complex*

(ii) *Purification of aluminium by electrolytic refining is called:*

(1) *Serpeck's process*

(2) *Hoope's process*

(3) *Hall's process*

(4) *Baeyer's process*

(iii) *An aqueous solution of urea freezes at – 0.186°C, K_f for water = 1.86 K kg. mol^{-1}, K_b for water = 0.512 K kg mol^{-1}. The boiling point of urea solution will be:*

(1) *373.065 K* **(2)** *375.186 K*

(3) *373.512 K* **(4)** *373.0512 K*

(iv) *In the dehydration of alcohols to alkenes by heating with concentrated sulphuric acid, the initiation step is:*

(1) *formation of carbocation*

(2) *formation of an ester*

(3) *protonation of alcohol molecule*

(4) *elimination of water*

(c) *Match the following:* **[4 × 1]**

(i)	*Rate constant*	**(a)**	*Dialysis*
(ii)	*Biodegradable polymer*	**(b)**	*Glycine*
(iii)	*Zwitter ion*	**(c)**	*Arrhenius equation*
(iv)	*Purification of colloids*	**(d)**	*PHBV*

(d) *Answer the following questions:* **[4 × 2]**

(i) **(1)** *Why does the density of transition elements increase from Titanium to Copper ?*

(At. No. Ti = 22, Cu = 29)

(2) *Why is zinc not regarded as a transition element ?* *(At. No. Zn = 30)*

(ii) *Identify the compounds A, B, C and D.*

$$CH_3CN \xrightarrow{H_2O/H^+} A \xrightarrow{NH_3} B \xrightarrow{heat} C \xrightarrow{Br_2/KOH} D$$

(iii) *Calculate the osmotic pressure of a solution prepared by dissolving 0.025 g of K_2SO_4 in 2.0 litres of water at 25°C assuming that K_2SO_4 is completely dissociated.*

(Mol. wt. of K_2SO_4 = 174g mol^{-1})

(iv) *What type of isomerism is shown by the following coordination compounds ?*

$[Pt\,Cl_2\,(NH_3)_4]\,Br_2$ and $[Pt\,Br_2\,(NH_3)_4]\,Cl_2$.

Write their IUPAC names.

Question 2. **[2]**

(a) **(i)** *Write the rate law expression for the reaction A + B + C D + E, if the order of reaction is first, second and zero with respect to A, B and C, respectively.*

(ii) *How many times the rate of reaction will increase if the concentration of A, B and C are doubled in the equation in* **(i)** *above ?*

OR

(b) *The rate of reaction becomes four times when the temperature changes from 293 K to 313 K. Calculate the energy of activation (E_a) of the reaction assuming that it does not change with temperature.*

(R = 8.314 J K^{-1} mol^{-1})

Question 3. **[2]**

(a) *How do antiseptics differ from disinfectants ?*

(b) *State the role of the following chemicals in the food industry :*

(i) *Sodium benzoate* **(ii)** *Aspartame*

Question 4. [2]

An aromatic organic compound [A] on heating with NH_3 and Cu_2O at high pressure gives [B]. The compound [B] on treatment with ice cold solution of $NaNO_2$ and HCl gives [C], which on heating with Cu/HCl gives compound [A] again. Identify the compounds [A], [B] and [C]. Write the name of the reaction for the conversion of [B] to [C].

Question 5.

Write the names of the monomers for each of the following polymers: [2]

(a) *Bakelite* **(b)** *Nylon–2–nylon–6*

Question 6. [2]

Name the purine bases and pyrimidine bases present in RNA and DNA.

Question 7. [2]

(a) *How will you obtain the following ? (Give balanced equation.)*

(i) *Picric acid from phenol.*

(ii) *Ethyl chloride from diethyl ether.*

OR

(b) *How will you obtain the following ? (Give balanced equation.)*

(i) *Anisole from phenol*

(ii) *Ethyl acetate from ethanol.*

Question 8. [2]

40% of a first order reaction is completed in 50 minutes. How much time will it take for the completion of 80% of this reaction?

Question 9. [3]

(a) *The freezing point of a solution containing 5.85 g of NaCl in 100 g of water is –3.348°C. Calculate van't Hoff factor 'i' for this solution. What will be the experimental molecular weight of NaCl ? (K_f for water = 1.86 K kg mol^{-1}, at. wt. Na = 23, Cl = 35.5)*

OR

(b) *An aqueous solution containing 12.48 g of barium chloride ($BaCl_2$) in 1000 g of water, boils at 100.0832°C. Calculate the degree of dissociation of barium chloride. (K_b for water = 0.52 K kg mol^{-1}, at. wt. Ba = 137, Cl = 35.5)*

Question 10. [3]

Examine the defective crystal given below and answer the question that follows:

A^+	B^-	A^+	B^-	A^+
B^-		B^-	A^+	B^-
A^+	B^-	A^+		A^+
B^-	A^+	B^-	A^+	B^-

State if the above defect is stoichiometric or non-stoichiometric. How does this defect affect the density of the crystal ? Also, write the term used for this type of defect.

Question 11. [3]

Give reason for each of the following:

(i) *For ferric hydroxide sol the coagulating power of phosphate ion is more than chloride ion.*

(ii) *Medicines are more effective in their colloidal form.*

(iii) *Gelatin is added to icecreams.*

Question 12. [3]

(a) *For the complex ion $[Fe(CN)_6]^{3-}$, state :*

(i) *the type of hybridisation.*

(ii) *the magnetic behaviour.*

(iii) *the oxidation number of the central metal atom.*

(b) *Write the IUPAC name of $[Co(en)_2Cl_2]^+$ ion and draw the structures of its geometrical isomers.*

Question 13. [3]

(a) *Explain why:*

(i) *Mn^{2+} is more stable than Fe^{2+} towards oxidation to +3 state. (At. no. of Mn = 25, Fe = 26)*

(ii) *Transition elements usually form coloured ions.*

(iii) *Zr and Hf exhibit similar properties. (At. no. of Zr = 40, Hf = 72)*

OR

(b) *Complete and balance the following chemical equations:*

(i) $KMnO_4 + KI + H_2SO_4 \rightarrow$ ____ + ____ + ____ + ____

(ii) $K_2Cr_2O_7 + H_2SO_4 + H_2S \rightarrow$ ____ + ____ + ____ + ____

(iii) $KMnO_4 + H_2SO_4 + FeSO_4 \rightarrow$ ____ + ____ + ____ + ____

Question 14. [3]

(a) *Arrange the following in the increasing order of their basic strength:*

$C_2H_5NH_2$, $C_6H_5NH_2$, $(C_2H_5)_2NH$

(b) *Give a balanced chemical equation to convert methyl cyanide to ethyl alcohol.*

(c) *What happens when benzene diazonium chloride reacts with phenol in weak alkaline medium ?*

(Give balanced equation)

Question 15. [3]

Name the sulphide ore of Copper. Describe how pure copper is extracted from this ore.

Question 16. [5]

(a) **(i)** *Calculate the emf and ΔG^0 for the cell reaction at 25°C:*

$$Zn(s)\left|Zn^{2+}_{(aq)}\right|\left|Cd^{2+}_{(aq)}\right|Cd_{(s)}$$

(0.1 M) (0.01 M)

Given $E^0_{Zn^{2+}/Zn} = -0.763$

and $E^0_{Cd^{2+}/Cd} = -0.403V$

(ii) *Define the following terms:*

(1) *Equivalent conductivity*

(2) *Corrosion of metals*

OR

(b) (i) *The specific conductivity of a solution containing 5 g of anhydrous* $BaCl_2$ *(mol. wt. = 208) in 1000* cm^3 *of a solution is found to be 0.0058* $ohm^{-1}\ cm^{-1}$*. Calculate the molar and equivalent conductivity of the solution.*

(ii) *What is an electrochemical series? How is it useful in predicting whether a metal can liberate hydrogen from acid or not ?*

Question 17. **[5]**

(a) (i) *Explain why:*

(1) *Nitrogen does not form pentahalides.*

(2) *Helium is used for filling weather balloons.*

(3) *ICl is more reactive than* I_2*.*

(ii) *Draw the structures of the following:*

(1) $HClO_4$ (2) H_3PO_3

OR

(b) (i) *Explain why:*

(1) *Mercury loses its meniscus in contact with ozone.*

(2) *Halogens are coloured and the colour deepens on moving down in the group from fluorine to iodine.*

(3) *Hydride of sulphur is a gas while hydride of oxygen is a liquid.*

(ii) *Complete and balance the following reactions :*

(1) $NaCl + MnO_2 + H_2SO_4 \rightarrow$ ___ + ___ + ___ + ___

(2) $KMnO_4 + SO_2 + H_2O \rightarrow$ ___ + ___ + ___

Question 18. **[5]**

(a) (i) *Give balanced equations for the following reactions:*

(1) *Benzaldehyde reacts with hydrazine.*

(2) *Acetic acid reacts with phosphorus pentachloride.*

(3) *Acetone reacts with sodium bisulphite.*

(ii) *Give one chemical test each to distinguish between the following pairs of compounds :*

(1) *Ethanol and acetic acid*

(2) *Acetaldehyde and benzaldehyde*

OR

(b) (i) *Write chemical equations to illustrate the following name reactions:*

(1) *Clemmensen's reduction*

(2) *Rosenmund's reduction*

(3) *HVZ reaction*

(ii) *Explain why:*

(1) *Acetaldehyde undergoes aldol condensation, but formaldehyde does not.*

(2) *Acetic acid is a weaker acid as compared to formic acid.*

ANSWERS

Answer 1.

(a) (i) chemical, electrical

(ii) 32, 26

(iii) yellow, iodoform

(iv) square pyramidal, sp^3d^2

(b) (i) (1) a carbocation

(ii) (2) Hoope's process

(iii) (4) 373.0512 K

(iv) (3) protonation of alcohol molecule

(c) (i) Rate constant → Arrhenius equation

(ii) Biodegradable polymer → PHBV

(iii) Zwitter ion → Glycine

(iv) Purification of colloids → Dialysis

(d) (i) (1) The atomic radii decreases from Ti (Z = 22) to copper (Z = 29). On moving from Ti to Cu, the atomic radii decreases due to increase in nuclear charge. The decrease in metallic radius from Ti (Z = 22) to copper (Z = 29) coupled with increase in atomic mass results in a general increase in the density of these elements.

(2) The zinc atom (Z = 30) has completely filled *d*-orbitals ($3d^{10}$) in its ground state as well as in its oxidised state, hence it is not regarded as a transition metal.

(ii) The compound A, B, C and D are:

$$\underset{\text{Acetonitrile}}{CH_3CN} \xrightarrow{H_2O/H^+} \underset{\substack{\text{Acetic acid}\\(A)}}{CH_3COOH} \xrightarrow{NH_3} \underset{\substack{\text{Ammonium acetate}\\(B)}}{CH_3COO^-NH_4^+}$$

$$\underset{\substack{\text{Ammonium acetate}\\(B)}}{CH_3COO^-NH_4^+} \xrightarrow{-H_2O,\ \Delta} \underset{\substack{\text{Acetamide}\\(C)}}{CH_3CONH_2} \xrightarrow{Br_2/KOH} \underset{\substack{\text{Methyl amine}\\(D)}}{CH_3NH_2}$$

(iii)

$$\pi = i\,CRT$$

$$\pi = \frac{n}{V}RT$$

$$= i\frac{w}{mV}RT$$

$$= \frac{3 \times 0.025}{174 \times 2} \times 0.0821 \times 298$$

$$= 5.2728 \times 10^{-3} \text{ atm}$$

(iv) $[Pt\,Cl_2(NH_3)_4]Br_2$ and $[Pt\,Br_2\,(NH_3)_4]Cl_2$ shows ionisation isomerism.

$[PtCl_2(NH_3)_4]Br_2^-$
Dichloridotetraammineplatinum(II) bromide/ Tetraamminedichloridoplatinum (IV)bromide

and $[Pt\,Br_2\,(NH_3)_4]Cl_2^-$
Dibromidotetraammineplatinum(II) chloride/Tetraamminedibromidoplatinum (IV) chloride

Answer 2.

(a) (i) For the given reaction:

$$A + B + C \rightarrow D + E$$

The rate law expression would be:

$$\text{Rate} = k\,[A]\,[B]^2$$

As the order of reaction is zero w.r.t. C, it would not appear in rate law.

(ii) The overall order of reaction is 1 + 2 = 3

Now, when concentration of A is doubled, the reaction rate will get doubled as the rate of reaction depends on concentration of A to the first power. And when concentration of B is doubled the rate of reaction becomes 4 times

As the order of reaction is zero w.r.t. C, any change in its concentration would not affect the rate of reaction. So, in total the rate of reaction increases by 2 × 4 = 8 times when concentration of the reactants are doubled.

OR

(b) Given, $\frac{k_2}{k_1} = 4$

$T_2 = 313;\ T_1 = 293$

Now, substituting the values in equation:

$$\log\frac{k_2}{k_1} = \frac{E_a}{2.303\,R} \times \left[\frac{(T_2 - T_1)}{T_1 T_2}\right]$$

$$\log 4 = \frac{E_a}{2.303 \times 8.314\,JK^{-1}\,mol^{-1}} \times \left[\frac{(313 - 293)}{313 \times 293}\right]$$

$$0.602 = \frac{E_a}{19.1\,JK^{-1}\,mol^{-1}} \times \left[\frac{20}{91709}\right]$$

$$0.602 = \frac{E_a}{19.1\,JK^{-1}\,mol^{-1}} \times (0.000218)$$

$$E_a = \frac{(0.602 \times 19.1)}{0.000218}$$

$$= 52744.03 \text{ J or } 52{\cdot}744 \text{ kJ/mol}$$

Answer 3.

(a) Refer to ISC Paper 2020, Answer 5(i).

(b) In food industry:

(i) Sodium benzoate is used as a preservative.

(ii) Aspartame is used as artificial sweetening agent.

Answer 4.

(a) The compounds [A], [B] and [C] are:

Chlorobenzene (A) $\xrightarrow[Cu_2O]{NH}$ Aniline (B) $\xrightarrow[HCl,\ 0°C]{NaNO_2}$ Benzene diazonium chloride (C) ($NH_2^+Cl^-$) $\xrightarrow{Cu/HCl}$ Chlorobenzene

The reaction for the conversion of B to C is known as Diazotisation reaction.

Answer 5.

(a) Monomers of Bakelite are Phenol and Formaldehyde.

(b) Monomers of nylon–2–nylon–6 are Glycine and 5-aminocaproic acid.

Answer 6.

	DNA	**RNA**
Pyrimidine bases	Cytosine (C), Thymine (T)	Cytosine (C) and Uracil (U).
Purine bases	Adenine (A), Guanine (G)	Adenine (A), Guanine (G)

Answer 7.

(a) (i) Picric acid can be obtained from phenol as follows:

Phenol + 3Conc. HNO_3 $\xrightarrow{\text{Conc. } H_2SO_4}$ 2, 4, 6-Trinitrophenol (Picric acid) + $3H_2O$

(ii) Ethyl chloride from diethyl ether can be obtained by treating it with PCl_5

$$(C_2H_5)_2O + PCl_5 \rightarrow 2C_2H_5Cl + POCl_3$$

OR

(b) (i) Anisole can be obtained from phenol as follows:

$$\underset{\text{Phenol}}{C_6H_5OH} \xrightarrow[-H_2O]{NaOH} \underset{\text{Sodium phenoxide}}{C_6H_5O^-Na^+} \xrightarrow{CH_3Br} \underset{\text{Anisole}}{C_6H_5OCH_3} + NaBr$$

(ii) Ethyl acetate from ethanol can be obtained by esterification:

$$\underset{\text{Ethanol}}{C_2H_5OH} + \underset{\text{Acetic acid}}{CH_3COOH} \xrightarrow[\text{Esterification}]{H^+} \underset{\text{ethyl acetate}}{CH_3COOC_2H_5} + H_2O$$

Answer 8.

For a first order reaction:

Time required is given by,

$$t = \frac{1}{k_1}\log\left[\frac{a}{a-x}\right]$$

where a is the initial concentration.

Hence,

$$\frac{t_{40}}{t_{80}} = \frac{\frac{1}{k_1}\log\left(\frac{100}{100-40}\right)}{\frac{1}{k_1}\log\left[\frac{100}{100-80}\right]}$$

$$\frac{t_{40}}{t_{80}} = \frac{\log\left[\frac{100}{100-40}\right]}{\log\left[\frac{100}{(100-80)}\right]}$$

$$\frac{50}{t_{80}} = \frac{\log\left[\frac{100}{60}\right]}{\log\left[\frac{100}{20}\right]}$$

$$\frac{50}{t_{80}} = \frac{\log(1.66)}{\log(5)}$$

$$\frac{50}{t_{80}} = \frac{0.22}{0.69}$$

$$t_{80} = \frac{0.69 \times 50}{0.22} = 156.8 \text{ min.}$$

Hence, time required for 80% completion would be 156.8 min.

Answer 9.

(a) Depression of freezing point is given by the equation, $\Delta T_f = i\,K_f\,m$

Where, i is the van't Hoff factor, K_f is freezing point constant and m is the molality of solution

Given, $K_f = 1.86 \text{ K kg mol}^{-1}$

$$\Delta T_f = 273.15 \text{ K} - 269.802 \text{ K} = 3.348 \text{ K}$$

$$\text{Molarity, } m = \frac{\left[\frac{5.85}{58.5}\times 1000\right]}{100}$$

$$= 1 \text{ mol kg}^{-1}$$

Putting the values in equation :

$$3.348 \text{ K} = i \times 1.86 \text{ K kg mol}^{-1} \times 1 \text{ mol kg}^{-1}$$

$$i = 1.8$$

$$\text{Now, } i = \frac{\text{Normal molar mass}}{\text{Abnormal molar mass}}$$

So, experimental molecular mass would be

$$\text{Abnormal molar mass} = \frac{\text{Normal molar mass}}{i}$$

$$= \frac{58.5}{1.8} = 32.5 \text{ g mol}^{-1}$$

OR

(b) Elevation of boiling point is given by $\Delta T_b = i\,K_b\,m$

Where, i is the van't Hoff factor, K_b is freezing point constant and m is the molality of solution

Given : $K_b = 0.52 \text{ K kg mol}^{-1}$

$$\Delta T_b = 373.233 \text{ K} - 373.15 \text{ K} = 0.0832 \text{ K}$$

$$\text{Molarity, } m = \frac{\left[\frac{12.48}{207}\right]}{1}$$

$$= 0.06 \text{ mol kg}^{-1}$$

Putting the values in equation :

$$0.083 \text{ K} = i \times 0.52 \text{ K kg mol}^{-1} \times 0.06 \text{ mol kg}^{-1}$$

$$i = \frac{0.083}{0.0312} = 2.67$$

$$\text{Now, } \underset{1-x}{BaCl_2} \rightarrow \underset{x}{Ba^+} + \underset{2x}{2\,Cl^-}$$

(where x is the degree of dissociation of $BaCl_2$)

The total number of moles after dissociation

$$= 1 - x + x + 2x = 1 + 2x$$

$$i = \frac{\text{Normal colligative property}}{\text{Abnormal colligative property}}$$

$$2.67 = \frac{1+2x}{1}$$

$$1 + 2x = (2.67)$$

$$x = \frac{1.67}{2} = 0.835$$

Degree of dissociation is 0.835 of 83·5%.

Answer 10.

In the given crystal equal number of cations and anions are missing, hence this defect is stoichiometric defect. As the ions are missing from the crystal it is a vacancy defect and decreases the density of the crystal. This defect is known as Schottky defect.

Answer 11.

(i) The efficacy of a negative ion to cause coagulation depends upon its valency. Hence phosphate, (PO_4^{2-}) ion has more coagulating power than chloride, Cl^-, for ferric hydroxide, $Fe(OH)_3$ sol. Since according to Hardy-Schulze law phosphate ion has more negative charge as compared to chloride ion, thus the coagulation power of phosphate ion is more than chloride ion.

(ii) Medicines are prepared in colloidal form so that they can be assimilated more effectively in biological systems, this is because colloidal medicines have a large surface area leading to easy assimilation due to their colloidal size.

(iii) The icecream mixture is colloidal solution in nature and gelatin is added as a emulsifier in icecream to stabilise the mixture in order to prevent the crystallisation.

Answer 12.

(a) For the complex $[Fe(CN)_6]^{3-}$

(i) The hybridisation is d^2sp^3.

Electronic configuration of d^2sp^3 hybridised orbitals (as CN^- is a strong ligand) of Fe^{3+}, with six pairs of electrons from six CN^- groups in $[Fe(CN)_6]^{3-}$

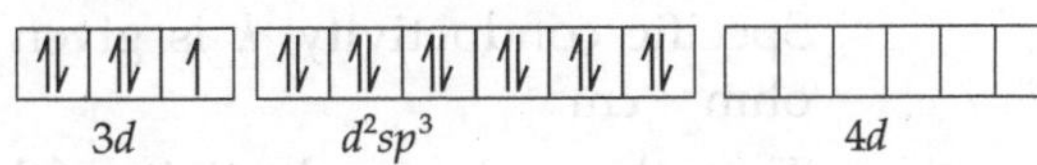

(ii) As there are 5 unpaired electrons, the $[Fe(CN)_6]^{3-}$ compound is paramagnetic.

(iii) Oxidation state of iron is + 3 (Fe^{+3}).

(b) IUPAC name of $[Co(en)_2Cl_2]^+$ is dichloridobis (ethaylenediamine) cobalt(III) ion

Geometrical isomers of $[Co(en)_2Cl_2]^+$

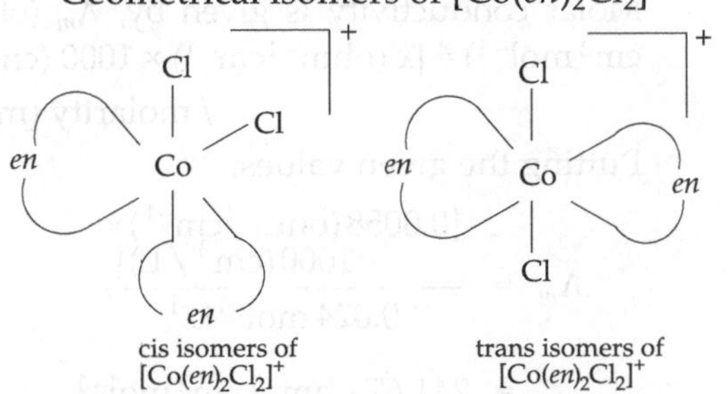

cis isomers of $[Co(en)_2Cl_2]^+$ trans isomers of $[Co(en)_2Cl_2]^+$

Answer 13.

(a) **(i)** Mn^{2+} is more stable than Fe^{2+} towards oxidation to +3 because Mn^{2+} has a d^5 electronic configuration, *i.e.,* extra stable half filled *d* orbital configuration, hence it cannot lose third electron easily whereas Fe^{2+} has a d^6 electronic configuration, which has no such extra stability.

(ii) Formation of coloured ions of transition elements can be attributed to the ability of electrons to make a transition from lower to higher energy *d*-orbitals after absorbing energy from visible light i.e., they undergo *d-d* transitions. The energy of excitation corresponds to the frequency of light absorbed, this frequency lies generally in visible region hence making the ions coloured.

(iii) Zirconium (Zr, Z = 40) belongs to second series of *d*-block transition elements and Hafnium (Hf, Z = 72) belongs to third series of *d*-block transition elements. These show similar chemical and physical properties because they have similar atomic radii (Zr = 159 pm and Hf = 156 pm) due to lanthanoid contraction.

OR

(b) Complete balanced equations are:

(i) $$2KMnO_4 + 10KI + 8H_2SO_4 \rightarrow 2MnSO_4 + 6K_2SO_4 + 8H_2O + 5I_2$$

(ii) $$3H_2S + K_2Cr_2O_7 + 4H_2SO_4 \rightarrow K_2SO_4 + Cr_2(SO_4)_3 + 3S + 7H_2O$$

(iii) $$2KMnO_4 + 10FeSO_4 + 8H_2SO_4 \rightarrow K_2SO_4 + 2MnSO_4 + 5Fe_2(SO_4)_3 + 8H_2O$$

Answer 14.

(a) Increasing order of basic strength would be:

$$C_6H_5NH_2 < C_2H_5NH_2 < (C_2H_5)_2NH$$

(b) Conversion of methyl cyanide to ethyl alcohol:

$$\underset{\text{Methyl cyanide}}{CH_3CN} \xrightarrow[\text{Reduction}]{LiAH_4/\text{ether}} CH_3CH_2NH_2 \xrightarrow{HNO_2} \underset{\text{Ethyl alcohol}}{C_2H_5OH} + N_2 + H_2O$$

(c) Benzene diazonium chloride reacts with phenol in weak alkaline medium to give *p*-hydroxyazobenzene, an orange dye.

$$\underset{\text{Benzene diazonium chloride}}{C_6H_5N_2^+Cl^-} + \underset{\text{Phenol}}{C_6H_5OH} \xrightarrow{NaOH} \underset{\substack{p\text{-Hydroxyazobenzene} \\ \text{(Orange dye)}}}{C_6H_5{-}N{=}N{-}C_6H_4{-}OH} + NaCl + H_2O$$

Answer 15.

Sulphide ore of copper are Copper glance (Cu_2S) and Copper pyrites ($CuFeS_2$). Copper is obtained in pure form from its sulphide ore in the following steps :

1. Concentration of ore through froth floatation method: In this method gangue is removed from the sulphide ore. The finely crushed ore is concentrated by froth-floatation process. The finely crushed ore is suspended in water containing a little amount of pine oil. A blast of air is passed through the suspension. The particles get wetted by the oil and float as a froth which is skimmed. The gangue (unwanted materials) sinks to the bottom, and is then removed.

2. Conversion of sulphide ore to oxide : The concentrated ore is then roasted in a furnace between 500°C and 700°C in the presence of a current of air. Sulphur is oxidised to SO_2 and other impurities are also removed as volatile oxides. The following reaction takes place.

$$2CuFeS_2 + O_2 \longrightarrow Cu_2S + 2FeS + SO_2$$

The roasted ore is the mixed with coke and sand and fed into the blast furnance. (Smelting)

$$2Cu_2S + 3O_2 \longrightarrow 2Cu_2O + 2SO_2$$
$$2FeS + 3O_2 \longrightarrow 2FeO + 2SO_2$$
$$FeO + SiO_2 \longrightarrow FeOSiO_3$$

3. Extraction of copper from cuprous oxide: Auto reduction of cuprous oxide takes place in Bessemer converter.

$$2Cu_2O + Cu_2S \longrightarrow \underset{\text{Blister copper}}{6Cu} + SO_2\uparrow$$

4. Refining: Blister copper is refined by electrolysis. Blocks of blister copper are casted to use as anode and thin sheets of pure copper act as cathode. The cathode plates are coated with graphite in order to remove depositing copper. The electrolyte is copper sulphate ($CuSO_4$) mixed with a little amount of H_2SO_4 to increase the electrical conductivity. Optimum potential difference is 1.3 volt for this electrolytic process. During electrolysis, pure copper (99.99% Cu) is deposited on the cathode plates and impurities which are soluble fall to the bottom of the cell as anode mud or sludge.

$$Cu \longrightarrow Cu^{2+} + 2e^- \quad \text{(at the anode)}$$
$$Cu^{2+} + 2e^- \longrightarrow Cu \quad \text{(at the cathode)}$$

Pure copper cathode
Impure copper anode
Copper (ii) sulphate solution
Anode sludge

Answer 16.

(a) (i) $Zn + Cd^{2+} \longrightarrow Zn^{2+} + Cd$

Here,

$$E_{(cell)}^\circ = E_R^\circ - E_L^\circ$$
$$= -0.403 - (-0.763)$$
$$= 0.36 \text{ V}$$

So, EMF of the cell would be given by:

$$E_{(cell)} = E^0_{cell} - \frac{0.0591}{n} \log \frac{[Zn^{+2}]}{[Cd^{+2}]}$$

$$E_{(cell)} = 0.36 - \frac{0.0591}{2} \log \frac{0.1}{0.01}$$

$$E_{(cell)} = 0.36 - 0.0295 \log 10$$
$$= 0.36 - 0.0294 \times 1$$
$$= 0.3305 \text{ V}$$

Now, $\Delta G^\circ = -nFE^\circ_{(cell)}$

So, putting the values:

$$\Delta G^\circ = -2 \times 96500 \text{ C mol}^{-1} \times 0.36 \text{ V}$$
$$= -69480 \text{ J/mol}$$
$$= -69.48 \text{ kJ/mol.}$$

(ii) **(1) Equivalent conductivity (Λ):** It is the conductivity power of all the ions produced by dissolving one gram equivalent in solution.

(2) Corrosion of metals: The slow and spontaneous process of the conversion of a metal into an undesirable compound (usually oxide) on exposure to atmospheric conditions is called corrosion of metals.

OR

(b) (i) Given, 5 g of $BaCl_2$ is dissolved in 1000 cm^3 (1 L) water.

Specific conductivity, k is given as 0.0058 ohm^{-1} cm^{-1}

Thus, the molar conductivity of the solution can be calculated as:

Molarity of the solution is,

$$M = \frac{\frac{5}{208}}{1L}$$
$$= 0.024 \text{ mol L}^{-1}$$

Molar conductivity is given by, Λ_m (ohm^{-1} cm^2 mol^{-1}) = [k (ohm^{-1}cm^{-1}) × 1000 (cm^3/L)] / molarity (mol/L)

Putting the given values:

$$\Lambda_m = \frac{[0.0058\,(\text{ohm}^{-1}\text{cm}^{-1}) \times 1000\,(\text{cm}^3/\text{L})]}{0.024\,\text{mol}^{-1}\text{L}^{-1}}$$

$$= 241.67 \text{ ohm}^{-1} \text{ cm}^2 \text{ mol}^{-1}$$

Thus, the equivalent conductivity of the solution can be calculated as:

Number of equivalents of $BaCl_2$,

$$n_e = \text{Number of moles} \times \text{Valency}$$

$$= \frac{5}{208} \times 2 = 0.048$$

Normality of solution is given by,

$$N = \frac{n_e}{V \text{ (in cc)}} \times 1000$$

Substituting the given values,

$$N = \frac{0.048}{1000} \times 1000 = 0.048$$

Equivalent conductivity is given by:

$$\Lambda_e = k \times \frac{1000}{N}$$

Substituting the given values:

$$\Lambda_e = k \times \frac{1000}{N}$$

$$= 0.0058 \text{ ohm}^{-1} \text{ cm}^{-1} \frac{1000}{0.048}$$

$$= 120.83 \text{ cm}^2 \text{ ohm}^{-1} \text{ equiv}^{-1}.$$

(ii) **Electrochemical series**: Different metal or metal ion combinations have different values of standard electrode potentials and thus they can be arranged in order of increasing or decreasing values of their standard reduction potentials. The arrangement of various elements in the order of increasing values of standard reduction potentials is called electrochemical series. The most reactive elements are kept at the top and the less reactive at the bottom.

Standard electrode potential for reduction of hydrogen ion (proton) to hydrogen molecule has been assigned as 0.0 V and it comes in the middle of electrochemical series. Now, if the standard electrode potential of an element is negative (less than zero) then its oxidized form is more stable than the reduced form of that element, hence it can liberate hydrogen gas from acids.

Answer 17.

(a) (i) (1) Nitrogen has a configuration of $2s^2$ $2p^3$ and it does not have *d*-orbitals to expand its valency beyond four. Hence, nitrogen does not form pentahalides.

(2) Helium is inert, non-inflammable and light gas hence it is used for filling weather balloons.

(3) ICl is more reactive than I_2 because the I—Cl bond is weaker than the I—I bond as there is a difference in radius of I and Cl atoms hence electron cloud is not equally shared and thus bond becomes weak.

(ii) (1) Structure of $HClO_4$

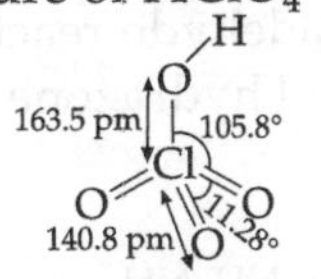

163.5 pm, 105.8°, 140.8 pm, 112.8°

(2) Structure of H_3PO_3

150 pm, 155 pm, 139 pm

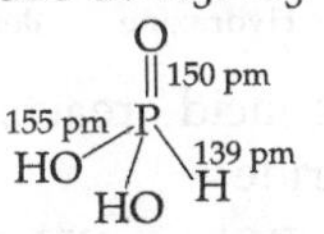

OR

(b) (i) (1) Mercury loses its meniscus in contact with ozone due to formation of mercurous oxide, which dissolves in mercury. It starts sticking to the glass and loses meniscus. This is also known as tailing of mercury and is a laboratory test for ozone gas.

$$2Hg + O_3 \longrightarrow Hg_2O + O_2$$

(2) Halogens are coloured due to absorption of radiations in visible region which results in excitation of outer electrons of higher energy level. The smaller size halogens absorb high energy radiation as the electrons need more energy to get excited, thus, the colour emitted is of low frequency. Going down the group, higher frequency radiations are absorbed hence colour deepens going down the group. Fluorine being small absorbs violet colour and shows the complementary colour yellow whereas iodine absorbs yellow colour and shows complementary colour violet.

(3) H_2S that is hydride of sulphur is a gas because of absence of hydrogen bonding within molecules which is present in H_2O, *i.e.*, hydride of oxygen. It is a simple covalent compound hence occurs as a gas. There is a great electronegativity difference between oxygen and hydrogen atoms hence a partial negative charge is developed on oxygen and a partial positive charge is developed on hydrogen atom. Due to the small size of oxygen atom H_2O forms intermolecular H-bond and gets associated hence occurs as liquid.

(ii) (1) $2NaCl + MnO_2 + 2H_2SO_4 \longrightarrow MnSO_4 + 2Na_2SO_4 + Cl_2 + 2H_2O$

OR

$2NaCl + MnO_2 + 3H_2SO_4 \longrightarrow 2NaHSO_4 + MnSO_4 + Cl_2 + 2H_2O$

(2) $2KMnO_4 + 5SO_2 + 2H_2O \longrightarrow 2MnSO_4 + K_2SO_4 + 2H_2SO_4$

Answer 18.

(a) (i) (1) Benzaldehyde reacts with hydrazine to give phenyl hydrazone

$C_6H_5CHO + NH_2NH_2 \longrightarrow C_6H_5CH{=}N{-}NH_2 + H_2O$

Benzaldehyde Hydrazine Benzaldehyde hydrazone

(2) Acetic acid reacts with phosphorus pentachloride

$CH_3COOH + PCl_5 \rightarrow CH_3COCl + POCl_3 + HCl$

(3) Acetone reacts with sodium bisulphite to give crystalline bisulphate addition product.

$(CH_3)_2C{=}O + NaHSO_3 \longrightarrow (CH_3)_2C(OH)SO_3Na$

(ii) (1) Ethanol and acetic acid can be distinguished by adding a solution of sodium bicarbonate at room temperature. Acetic acid will give effervescence (bubbles) due to evolution of CO_2, whereas ethanol would not give any test.

$CH_3COOH + NaHCO_3(aq) \longrightarrow CH_3COONa + CO_2(g) + H_2O(l)$

(2) Acetaldehyde gives positive iodoform test, whereas benzaldehyde would not give the test.

$[CH_3CHO] \xrightarrow[I_2+NaOH]{NaOI} CH_3I + HCOO^-Na^+$

Acetaldehyde Iodoform

Also, acetaldehyde and benzaldehyde can be distinguished with the help of Fehling's reagent. Acetaldehyde is oxidised to ethanoate ion and a red precipitate of cuprous oxide is obtained in the process. Whereas benzaldehyde does not react with Fehling's reagent.

$C_2H_5CHO + 2Cu^{2+} + 5OH^- \longrightarrow CH_3COO^- + Cu_2O + 3H_2O$

Acetaldehyde

OR

(b) (i) (1) **Clemmensen's reduction:** The Clemmensen's reduction is an organic reaction used to reduce an aldehyde or ketone to an alkane using amalgamated zinc and hydrochloric acid. The substrate must be unreactive to the strongly acidic conditions of the Clemmensen reduction.

$R_1COR_2 \xrightarrow[HCl]{Zn(Hg)} R_1CH_2R_2 + H_2O$

$CH_3COCH_3 + 4[H] \xrightarrow{Zn\text{-}Hg/HCl} CH_3CH_2CH_3 + H_2O$

(2) **Rosenmund's reduction:** Acyl chloride or acid chloride on hydrogenation in presence of catalyst Palladium on barium sulphate ($BaSO_4$) yields corresponding aldehydes. This reaction is called Rosenmund's reduction and is used for the preparation of aldehydes.

$RCOCl \xrightarrow[Pd\text{-}BaSO_4]{H} RCHO + HCl$

R = Alkyl, Aryl group

(3) **HVZ reaction :** This is also known as Hell Volhard-Zelinsky reaction Carboxylic acids having an α-hydrogen are halogenated at the α-position on treatment with chlorine or bromine in the presence of small amount of red phosphorus to give α-halocarboxylic acids. The reaction is known as Hell-Volhard-Zelinsky (HVZ) reaction.

$RCH_2COOH \xrightarrow[(ii)\ H_2O]{(i)\ X_2/\text{Red phosphorus}} R{-}CH(X){-}COOH \quad X = Cl, Br$

α-Halocarboxylic acid

(ii) (1) In aldol condensation, such aldehydes and ketones which have at least one α-hydrogen undergo the reaction in presence of dilute alkali as catalyst to form β-hydroxy aldehydes (aldol) or β-hydroxy ketones (ketol), respectively. As benzaldehyde does not have an α-hydrogen, it does not undergo the aldol condensation reaction.

$2CH_3CHO \xrightleftharpoons{dil.\ NaOH} CH_3CH(OH)CH_2CHO$

Acetaldehyde 3-Hydroxybutanal

(2) Acetic acid is a weak acid than formic acid. The methyl group attached to carbonyl carbon imparts a + I inductive effect, that enhances electron availability and thus reduces the chances of removal of proton from acetic acid and also destabilises the carboxylate anion by increasing the negative charge developed on it after loss of proton.

●●

CHEMISTRY 2017

QUESTIONS

PART–I (20 Marks)

*Answer **all** questions.*

Question 1.

(a) *Fill in the blanks by choosing the appropriate word/ words from those given in the brackets:* **[5]**

(iodoform, acetaldehyde, positive, greater, acidic, acetone, disaccharide, negative, increases, glucose, decreases, chloroform, polysaccharide, lactose, lesser, basic, cationic hydrolysis, anionic hydrolysis)

(i) *Calcium acetate on heating gives_____which gives_______on heating with iodine and sodium hydroxide solution.*

(ii) *On dilution of a solution, its specific conductance_______while its equivalent conductance _______ .*

(iii) *Sucrose is a_________and yield upon hydrolysis, a mixture of______ and fructose.*

(iv) *More_____the standard reduction potential of a substance, the _____ is its ability to displace hydrogen from acids.*

(v) *An aqueous solution of CH_3COONa is _____ due to _______ .*

(b) *Complete the following statements by selecting the **correct alternative from** the choices given:* **[5]**

(i) *In a face centered cubic lattice, atom (A) occupies the corner positions and atom (B) occupies the face centre positions. If one atom of (B) is missing from one of the face centered points, the formula of the compound is :*

(1) A_2B_5 **(2)** A_2B_3

(3) AB_2 **(4)** A_2B

(ii) *The half life period of a first order reaction is 20 minutes. The time required for the concentration of the reactant to change from 0·16 M to 0·02 M is:*

(1) *80 minutes* **(2)** *60 minutes*

(3) *40 minutes* **(4)** *20 minutes*

(iii) *For a spontaneous reaction $\Delta G°$ and $E°$ cell will be respectively:*

(1) *– ve and + ve* **(2)** *+ ve and – ve*

(3) *+ ve and + ve* **(4)** *– ve and – ve*

(iv) *The conjugate acid of HPO_4^{2-} is :*

(1) H_3PO_3 **(2)** H_3PO_4

(3) $H_2PO_4^-$ **(4)** PO_4^{3-}

(v) *The polymer formed by the condensation of hexamethylenediamine and adipic acid is :*

(1) *Teflon* **(2)** *Bakelite*

(3) *Dacron* **(4)** *Nylon-66*

(c) *Answer the following questions:* **[5]**

(i) *Why the freezing point depression (ΔT_f) of 0·4 M NaCl solution is nearly twice than that of 0·4 M glucose solution ?*

(ii) *Identify the order of reaction from each of the following units of rate constant (k):*

(a) *mol L^{-1} sec^{-1}* **(b)** *mol^{-1} L sec^{-1}*

(iii) *Specific conductivity of 0·20 M solution of KCl at 298 K is 0·025 S cm^{-1}. Calculate its molar conductivity.*

(iv) *Name the order of reaction which proceeds with a uniform rate throughout.*

(v) *What are the products formed when phenol and nitrobenzene are treated separately with a mixture of concentrated sulphuric acid and concentrated nitric acid?*

(d) *Match the following:* **[5]**

(i)	*Diazotisation*	**(a)**	*Bakelite*
(ii)	*Argentite*	**(b)**	*Nernst equation*
(iii)	*Thermosetting plastics*	**(c)**	*Aniline*
(iv)	*Electrochemical cell*	**(d)**	*Ethylenediamine*
(v)	*Bidentate ligand*	**(e)**	*Froth floatation process*

PART–II (50 Marks)

SECTION–A

*Answer any **two** questions.*

Question 2.

(a) (i) *Determine the freezing point of a solution containing 0·625g of glucose ($C_6H_{12}O_6$) dissolved in 102·8 g of water.* **[2]**

(Freezing point of water = 273 K , K_f for water = 1·87 K kg mol^{-1}, at. wt. C = 12, H = 1, O = 16).

(ii) *A 0·15 M aqueous solution of KCl exerts an osmotic pressure of 6·8 atm at 310 K. Calculate the degree of dissociation of KCl. (R = 0·0821 Lit. atm K^{-1} mol^{-1})* **[2]**

(iii) *A solution containing 8·44g of sucrose in 100 g of water has a vapour pressure 4·56 mm of Hg at 273 K. If the vapour pressure of pure water* *is 4·58 mm of Hg at the same temperature, calculate the molecular weight of sucrose.* **[1]**

(b) (i) *When ammonium chloride and ammonium hydroxide are added to a solution containing both Al^{3+} and Ca^{2+} ions, which ion is precipitated first and why ?*** **[2]**

(ii) *A solution of potassium chloride has no effect on litmus whereas, a solution of zinc chloride turns the blue litmus red. Give a reason.*** **[2]**

(c) *How many sodium ions and chloride ions are present in a unit cell of sodium chloride crystal ?* **[1]**

Question 3.

(a) (i) *Lead sulphide has face centered cubic crystal structure. If the edge length of the unit cell of lead sulphide is 495 pm, calculate the density of the crystal.*

(At. wt. Pb = 207, S = 32) **[1]**

(ii) *For the reaction:*

$2H_2 + 2NO \rightleftharpoons 2H_2O + N_2$, *the following rate data was obtained:* **[3]**

S. No.	[NO] mol L^{-1}	[H_2] mol L^{-1}	Rate : mol L^{-1} sec^{-1}
1.	0·40	0·40	$4·6 \times 10^{-3}$
2.	0·80	0·40	$18·4 \times 10^{-3}$
3.	0·40	0·80	$9·2 \times 10^{-3}$

Calculate the following:

(1) *The overall order of reaction.*

(2) *The rate law.*

(3) *The value of rate constant (k).*

(b) (i) *The following electrochemical cell is set up at 298 K:* **[2]**

$Zn \mid Zn^{2+} (aq) (1M) \mid\mid Cu^{2+} (aq) (1M) \mid Cu$

Given $\rightarrow E^\circ_{Zn^{2+}/Zn} = -0·761$ V,

$E^\circ_{Cu^{2+}/Cu} = +0·339$ V

(1) *Write the cell reaction.*

(2) *Calculate the emf and free energy change at 298 K.*

(ii) *Answer the following :*** **[2]**

(1) *What is the effect of temperature on ionic product of water (Kw)?*

(2) *What happenes to the ionic product of water (Kw) if some acid is added to it ?*

(c) *Frenkel defect does not change the density of the ionic crystal whereas, Schottky defect lowers the density of ionic crystal. Give a reason.* **[2]**

Question 4.

(a) (i) *Name the law or principle to which the following observations conform :* **[3]**

(1) *When water is added to a 1·0 M aqueous solution of acetic acid, the number of hydrogen ion (H^+) increases.***

(2) *When 9650 coulombs of electricity is passed through a solution of copper sulphate, 3·175 g of copper is deposited on the cathode.*

(At. wt. of Cu = 63·5)

(3) *When ammonium chloride is added to a solution of ammonium hydroxide, the concentration of hydroxyl ions decreases.***

(ii) *What is the difference between the order of a reaction and its molecularity?* **[2]**

(b) (i) *Explain why high pressure is required in the manufacture of sulphur trioxide by contact process. State the law or principle used.*** **[2]**

(ii) *Calculate the equilibrium constant (Kc) for the formation of NH_3 in the following reactions :*** **[1]**

$$N_2(g) + 3H_2(g) \rightleftharpoons 2NH_3(g)$$

At equilibrium, the concentration of NH_3, H_2 and N_2 are $1·2 \times 10^{-2}$, $3·0 \times 10^{-2}$ and $1·5 \times 10^{-2}$ respectively.

(c) *Explain the following :*** **[2]**

(i) *Hydrolysis of ester (ethyl acetate) begins slowly but becomes fast after sometime.*

(ii) *The pH value of acetic acid increases on addition of a few drops of sodium acetate.*

SECTION–B

Answer any ***two*** *questions.*

Question 5.

(a) *Write the formula of the following compounds :* **[2]**

(i) *Potassium trioxalatoaluminate(III).*

(ii) *Hexaaquairon(II) sulphate.*

(b) *Name the types of isomerism shown by the following pairs of compounds :* **[1]**

(i) $[Cu(NH_3)_4][Pt(Cl_4)]$ and $[Pt(NH_3)_4][CuCl_4]$

(ii) $[Co(Pn)_2Cl_2]^+$ and $[Co(tn)_2Cl_2]^+$

(c) *For the coordination complex ion $[Co(NH_3)_6]^{3+}$.* **[2]**

(i) *Give the IUPAC name of the complex ion.*

(ii) *What is the oxidation number of cobalt in the complex ion ?*

(iii) *State the type of hybridisation of the complex ion.*

(iv) *State the magnetic behaviour of the complex ion.*

Question 6.

(a) *Give balanced equations for the following reactions:* **[3]**

(i) *Potassium permanganate is heated with concentrated hydrochloric acid.*

(ii) *Lead sulphide is heated with hydrogen peroxide.*

(iii) *Ozone is treated with potassium iodide solution.*

** Answer is not given due to change in the present syllabus.

(b) *Discuss the theory involved in the manufacture of sulphuric acid by contact process.*** **[2]**

Question 7.

(a) (i) *What are the types of hybridisation of iodine interhalogen compounds IF_3, IF_5 and IF_7 respectively ?* **[3]**

(ii) *Draw the structure of xenon hexafluoride (XeF_6) molecule and state the hybridisation of the central atom.*

(b) *Give the balanced equations for the conversion of argentite (Ag_2S) to metallic silver.* **[2]**

SECTION–C

*Answer any **two** questions.*

Question 8.

(a) *How can the following conversions be brought about:*

(i) *Acetaldehyde to propan-2-ol.* **[1]**

(ii) *Nitrobenzene to p-aminoazobenzene.* **[1]**

(iii) *Acetic acid to methylamine.* **[2]**

(iv) *Aniline to benzene.* **[1]**

(b) (i) *How will you distinguish between primary, secondary and tertiary amines by Hinsberg's test?* **[1]**

(ii) *Why do alcohols possess higher boiling points as compared to those of corresponding alkanes?* **[1]**

(c) *Identify the compounds **A, B** and **C**:* **[3]**

(i) $C_6H_5COOH \xrightarrow{PCl_5} A \xrightarrow{H_2-Pd/BaSO_4} B \xrightarrow[\text{distilled}]{\text{KCN alc}} C$

(ii) $H-C\equiv C-H \xrightarrow[\text{dil.}H_2SO_4+HgSO_4]{H_2O} A \xrightarrow[\text{[Ni]}]{H_2} B \xrightarrow[\text{conc.}H_2SO_4]{140°C} C$

Question 9.

(a) *Give balanced equations for the following name reactions:* **[3]**

(i) *Friedel-Crafts reaction (alkylation)*

(ii) *Williamson's synthesis.*

(iii) *Aldol condensation.*

(b) *Give chemical test to distinguish:* **[3]**

(i) *Ethyl alcohol and sec-propyl alcohol.*

(ii) *Acetaldehyde and acetic acid.*

(c) (i) *Deficiency of which vitamin causes the following diseases:* **[4]**

(1) *Scurvy* **(2)** *Night blindness.*

(ii) *Write two differences between globular and fibrous proteins.*

Question 10.

(a) *An aliphatic unsaturated hydrocarbon (A) when treated with $HgSO_4/H_2SO_4$ yields a compound (B) having molecular formula C_3H_6O. (B) on oxidation with concentrated HNO_3 gives two compounds (C) and (D). Compound (C) when treated with PCl_5 gives compound (E). (E) when reacts with ethanol gives a sweet smelling liquid (F). Compound (F) is also formed when (C) reacts with ethanol in the presence of concentrated H_2SO_4.* **[4]**

(i) *Identify the compound A, B, C, D, E and F.*

(ii) *Give the chemical equation for the reaction of (C) with chlorine in the presence of red phosphorus and name the reaction.*

(b) *Answer the following:* **[3]**

(i) *What is the common name of the polymer obtained by the polymerisation of caprolactum? Is it addition polymer or condensation polymer?*

(ii) *Name the two organic compounds which have the same molecular formula C_2H_6O. Will they react with PCl_5 ? If they react, what are the products formed?*

(c) *Give balanced equations for the following reactions :* **[3]**

(i) *Methyl magnesium bromide with ethyl alcohol.*

(ii) *Acetic anhydride with phosphorus penta-chloride.*

(iii) *Acetaldehyde with hydroxylamine.*

ANSWERS

PART–I

Answer 1.

(a) (i) acetone, iodoform

(ii) decreases, increases

(iii) diasaccharide, glucose

(iv) negative, greater

(v) basic, anionic hydrolysis

(b) (i) (1) A_2B_5 **(ii)** (2) 60 minutes

(iii) (1) – ve and + ve **(iv)** (3) $H_2PO^-_4$

(v) (4) Nylon-66

(c) (i) Depression in freezing point (ΔT_f) is directly proportional to Van't Hoff factor (i)

$$\Delta T_f \propto i$$

In case of NaCl $i = 2$

In case of glucose, $i = 1$

Therefore, freezing point depression (ΔT_f) is nearly twice than that of glucose at same concentration (0·4 M).

(ii) **(a)** Zero order reaction.

(b) Second order reaction.

** Answer is not given due to change in the present syllabus.

(iii) Specific conductivity (k) $= 0{\cdot}025\ S\ cm^{-1}$

Concentration (C) $= 0{\cdot}20$ M

Molar conductivity (Λ_m) = ?

$$\Lambda_m = \frac{k \times 1000}{C}$$

$$= \frac{0.025 \times 1000}{0.20}$$

$$\Lambda_m = 125\ S\ cm^2\ mol^{-1}$$

(iv) Zero order reaction proceeds with a uniform rate throughout.

(v)

Phenol $\xrightarrow[\text{Conc. } HNO_3]{\text{Conc. } H_2SO_4}$ 2, 4, 6-trinitrophenol (picric acid) $+ H_2O$

Nitrobenzene $\xrightarrow[\text{Conc. } HNO_3]{\text{Conc. } H_2SO_4}$ m-Dinitrobenzene $+ H_2O$

(d) (i) Diazotisation → Aniline

(ii) Argentite → Froth floatation process

(iii) Thermosetting plastics → Bakelite

(iv) Electrochemical cell → Nernst equation

(v) Bidentate ligand → Ethylenediamine

PART–II
SECTION–A

Answer 2.

(a) (i) Wt. of solute (W_B) $= 0{\cdot}625$ g

Wt. of solvent (W_A) $= 102{\cdot}8$ g

Molar mass of solute (M_B) = 180

Freezing point of water $= T_1 = 273$ K

Freezing point of solution $= T_2$

Depression in freezing point

$$= \Delta T_f = T_2 - T_1$$

K_f for water $= 1{\cdot}87\ K\ kg\ mol^{-1}$

$$\Delta T_f = T_2 - T_1$$

$$= \frac{K_f \times W_B \times 1000}{W_A \times M_B}$$

$$T_2 - 273 = \frac{1.87 \times 0.625 \times 1000}{102.8 \times 180}$$

$$= 0.0631\ K$$

$$T_2 = 273 - 0.0631$$

$$= 272.936\ K$$

(ii) Concentration of solution (C) $= 0{\cdot}15$ M

Osmotic pressure (π) $= 6{\cdot}8$ atm

Temperature (T) = 310 K

$R = 0{\cdot}0821\ Lit\ atm\ K^{-1}\ mol^{-1}$

Van't Hoff factor (i) = ?

$$\pi = i\,.\,CRT$$

$$i = \frac{\pi}{CRT}$$

$$= \frac{6.8}{0.15 \times 0.0821 \times 310}$$

$$i = 1{\cdot}78$$

Degree of dissociation,

$$\alpha = \frac{i-1}{m-1}$$

Observed molar mass = 2

$$\alpha = \frac{i-1}{2-1}$$

$$= 1.78 - 1 = 0.78$$

Degree of dissociation (α) = 0.78 or 78%

(iii) Wt. of sucrose (W_B) = 8.44 g

Wt. of water (W_A) = 100 g

Vapour pressure of solution (P_s) = 4.56 mm of Hg

Vapour pressure of pure water (P_o) = 4.58 mm of Hg

Molar mass of water (M_A) = 18

Molar mass of sucrose (M_B) = ?

$$M_B = \frac{W_B \times M_A}{W_A \times \left(\frac{P_o - P_s}{P_o}\right)}$$

$$= \frac{8.44 \times 18}{100 \times \left(\frac{4.58 - 4.56}{4.58}\right)}$$

$$= \frac{8.44 \times 18}{100 \times 4.36 \times 10^{-3}}$$

$$M_B = 347{\cdot}90\ g\ mol^{-1}$$

(c) Sodium chloride has face centred cubic structure. Therefore, the number of sodium ions and chloride ions present in a unit cell of sodium chloride crystal is 4.

Answer 3.

(a) (i) Edge length of unit cell

(a) = 495 pm

$= 495 \times 10^{-10}$ cm

Formula mass of lead sulphide (PbS),

M = 239·3 g/mol

No. of atoms in fcc unit cell

(Z) = 4

N_A, Avogadro number = $6{\cdot}023 \times 10^{23}$

Density of crystal

$$(d) = \frac{Z \times M}{a^3 \times N_A}$$

$$= \frac{4 \times 239.3}{(495 \times 10^{-10})} \times 6{\cdot}023 \times 10^{23}$$

$$d = 13{\cdot}1 \text{ g/cm}^3$$

(ii) Initial rate expression can be written as

$$\text{Rate} = k\,[NO]^a\,[H_2]^b$$

Substituting the values of concentration of NO and H_2 from experimental data 1 and 2

$$(\text{Rate})_1 = k\,(0{\cdot}40)^a\,(0{\cdot}40)^b \quad \ldots(i)$$

$$(\text{Rate})_2 = k\,(0{\cdot}80)^a\,(0{\cdot}40)^b \quad \ldots(ii)$$

Dividing equation (ii) by equation (i)

$$\frac{(\text{Rate})_2}{(\text{Rate})_1} = \frac{k\,(0.80)^a\,(0.40)^b}{k\,(0.40)^a\,(0.40)^b}$$

$$= \frac{18.4 \times 10^{-3}}{4.6 \times 10^{-3}}$$

$$(2)^a = 4$$

$$a = 2$$

Thus, the order of reaction with respect to NO is **2**.

On comparing experiment data (1) and (3)

$$(\text{Rate})_3 = k\,(0{\cdot}40)^a\,(0{\cdot}80)^b \quad \ldots(iii)$$

$$(\text{Rate})_1 = k\,(0{\cdot}40)^a\,(0{\cdot}40)^b \quad \ldots(iv)$$

Dividing equation (iii) by equation (iv)

$$\frac{(\text{Rate})_3}{(\text{Rate})_1} = \frac{k\,(0.40)^a\,(0.80)^b}{k\,(0.40)^a\,(0.40)^b}$$

$$= \frac{9.2 \times 10^{-3}}{4.6 \times 10^{-3}}$$

$$(2)^b = 2$$

$$b = 1$$

Thus, order of reaction with respect to H_2 is 1.

(1) Overall order of the reaction = 2 + 1 = 3

(2) The rate law for the reaction

$$\text{Rate} = k\,[NO]^2\,[H_2]$$

(3) Rate constant can be calculated by substituting the values of rate of [NO] and $[H_2]$ for any experiment

$$k = \frac{\text{Rate}}{[NO]^2[H_2]}$$

$$= \frac{4.6 \times 10^{-3}}{(0.40)^2\,(0.40)}$$

$$= \frac{4.6 \times 10^{-3}}{0.064}$$

$$= 71{\cdot}875 \times 10^{-3}$$

Thus, the value of rate constant

$$k = 71.875 \times 10^{-3} \text{ mol}^{-2}\,\text{L}^2\text{s}^{-1}$$

(b) **(i) (1)** Cell reactions:

At anode $Zn \rightarrow Zn^{2+} + 2e^-$ $E° = 0{\cdot}761V$

At cathode $Cu^{2+} + 2e^- \rightarrow Cu$ $E° = +0{\cdot}339V$

Overall Cell Reaction:

$Zn(s) + Cu^{2+}(aq) \rightarrow Zn^{2+}(aq) + Cu(s)$

(2) EMF (E°) = E° (cathode) – E° (anode)

= 0.339 – (– 0.761)

E° = 1.1 V

Free energy change

$$\Delta G° = -nFE°$$

$$= -2 \times 96500 \times 1{\cdot}1$$

$$= -212300 \text{ J}$$

$$\Delta G° = -212.3 \text{ kJ}$$

(c) Frenkel defect arises due to the presence of holes in the cationic lattice site and cation occupies an interstitial position. Hence density remains the same, whereas Schottky defect arises when same number of cations and anions are missing from their normal site and a pair of holes are formed hence density decreases.

Answer 4.

(a) **(2)** Faraday's 1st law of electrolysis

(ii) Difference between order of reaction and molecularity:

	Order	Molecularity
1.	Order is sum of powers of concentration terms in the rate law expression.	Molecularity is the number of reacting species undergoing simultaneous collision in the simple reaction.
2.	Order of reaction is determined experimentally.	Molecularity is a theoretical concept.
3.	Order of reaction can be zero, negative or a fraction.	Molecularity of the reaction is always a whole number other than zero.
4.	Order of the reaction is given by the slowest step.	Molecularity of the slowest step is same as the order of reaction.

SECTION–B

Answer 5.

(a) **(i)** $K_3[Al(C_2O_4)_3]$

(ii) $[Fe(H_2O)_6]SO_4$

(b) **(i)** Co-ordination isomerism

(ii) Ligand isomerism

(c) **(i)** Hexaaminecobalt(III) ion

(ii) + 3

(iii) d^2sp^3 hybridisation

(iv) Diamagnetic

Answer 6.

(a) (i) $2KMnO_4 + 16HCl \xrightarrow{heat} 2MnCl_2 + 5Cl_2 + 2KCl + 8H_2O$

(ii) $PbS + 4H_2O_2 \xrightarrow{\Delta} PbSO_4 + 4H_2O$

(iii) $O_3 + 2KI + H_2O \longrightarrow I_2 + 2KOH + O_2$

Answer 7.

(a) (i)

$IF_3 \longrightarrow sp^3d$

$IF_5 \longrightarrow sp^3d^2$

$IF_7 \longrightarrow sp^3d^3$

(ii)

F, F, F, Xe, F, F, F

sp^3d^3-hybridisation of Xe

(b) $Ag_2S + 4NaCN \rightleftharpoons 2Na[Ag(CN)_2] + Na_2S$

Argentite — Sodium argentocyanide

$2Na[Ag(CN)_2] + Zn \rightarrow Na[Zn(CN)_4] + 2Ag\downarrow$

Sodium tetracyano zincate(II) — Silver

SECTION–C

Answer 8.

(a) (i) $CH_3CHO + CH_3MgBr \longrightarrow$

Acetaldehyde — Methyl magnesium Bromide (Grignard's reagent)

$H_3C-C(CH_3)(H_3)-OMgBr \xrightarrow{H^+/H_2O}$

$H_3C-C(CH_3)(H)-OH + Mg(OH)Br$

Propan-2ol

(ii)

Nitrobenzene ($C_6H_5NO_2$) $+ 6[H] \xrightarrow{Sn/HCl}$ Aniline ($C_6H_5NH_2$) $\xrightarrow[\text{ice-cold}]{\text{Acid}}$ Benzene diazonium chloride ($C_6H_5N^+\equiv N$)

$NH_2 \equiv Cl$

$C_6H_5-N=N-N(H)-C_6H_5$

↓ Rearrangement

$C_6H_5-N=N-C_6H_4-NH_2$

p-Aminoazobenzene

(iii)

$CH_3-\overset{O}{\overset{||}{C}}-OH \xrightarrow[\text{Conc.}H_2SO_4]{NH_3} CH_3COONH_4 \xrightarrow{\Delta}$

Acetic acid — [Schmidt reaction]

$CH_3CONH_2 \xrightarrow{Br_2/KOH} CH_3NH_2 + CO_2 + N_2$

Methyl amine

(iv)

Aniline ($C_6H_5NH_2$) $\xrightarrow[273\ K]{NaNO_2,\ HCl}$ Benzene diazonium chloride ($C_6H_5N_2^+Cl^-$) $\xrightarrow[H_2O]{H_3PO_2}$ Benzene

(b) (i) **Hinsberg's test for amines:** In this test, the amine is heated with benzene sulphonyl chloride ($C_6H_5SO_2Cl$) in the presence of excess of alkali.

Different amines gives different observations:

1. Primary amines give clear solution which on acidification yields an insoluble material.
2. Secondary amines give an insoluble substance which remains unaffected on addition of acid.
3. Tertiary amines do not react and remains insoluble in alkali and can be dissolved in acids.

(ii) Alcohols possess higher boiling points as compared to those of corresponding alkanes due to the presence of hydrogen-bonding in them whereas in alkanes, the only inter-molecular forces are Vander Waals dispersion forces.

(c) (i)

$C_6H_5COOH \xrightarrow{PCl_5} C_6H_5COCl \xrightarrow[\text{S, quinoline}]{H_2,Pd\text{-}BaSO_4}$

Benzoic acid — Benzoyl chloride (A)

$C_6H_5CHO \xrightarrow[\text{distill water}]{\text{alc.KCN}} C_6H_5-CH\begin{cases}OH\\COC_6H_5\end{cases}$

Benzaldehyde (B) — (C)

(ii) $H-C\equiv C-H \xrightarrow[\text{dil.}H_2SO_4 + HgSO_4]{H_2O} H_3C-CHO \xrightarrow[{[Ni]}]{H_2}$

Ethene — Acetaldehyde (A)

H_3C-CH_2-OH

Ethyl alchol (B)

↓ Conc. H_2SO_4, 140°

$CH_3CH_2-OCH_2CH_3$

Diethyl ether (C)

Answer 9.

(a) **(i) Friedel-Crafts reaction (alkylation) :** Benzene and other aromatic compounds react with alkyl halides in the presence of anhydrous $AlCl_3$ to form alkyl benzene.

$$C_6H_6 \text{ (Benzene)} + CH_3Cl \text{ (Methyl Chloride)} \xrightarrow{\text{anhy. } AlCl_3} C_6H_5CH_3 \text{ (Toluene)} + HCl$$

(ii) Refer to ISC Paper 2020, Answer 18 (b) (i) (1)

(iii) Refer to ISC Paper 2019, Answer 18 (b) (i) (1)

(b) (i) Ethyl alcohol and sec-propyl alcohol can be distinguished by Lucas Test.

Ethyl alcohol is a primary alcohol, hence won't show turbidity on mixing with Lucas Reagent ($HCl + ZnCl_2$) at room temperature. Whereas, sec-propyl alcohol is a secondary alcohol. Hence, it will show turbidity after 5 - 10 minutes with reagent at room temperature.

(ii) Acetaldehyde gives iodoform test while acetic acid does not gives iodoform test.

1. $$\underset{\text{Acetaldehyde}}{CH_3CHO} + NaOI \xrightarrow{NaOH,\ I_2} \underset{\text{Yellow ppt (Iodoform)}}{CHI_3} + HCOONa + 2NaOH$$

2. $$\underset{\text{Acetic acid}}{CH_3COOH} \xrightarrow{NaOH,\ I_2} \text{No reaction}$$

(c) (i) (1) Vitamin C

(2) Vitamin A

(ii) Difference between globular and fibrous proteins

	Fibrous Proteins	Globular Proteins
1.	These types of proteins consist of linear thread like molecules which tend to lie side by side to form fibres. **Example**—Keratin, fibroin, myosin etc.	In this type of proteins, the molecules are folded together into compact units forming spheroidal shapes. **Example**—Albumin, insulin etc.
2.	These proteins are insoluble in water.	These proteins are soluble in water.
3.	They are stable to moderate changes in pH and temperature.	They are very sensitive to changes in pH, temperature.

Answer 10.

(a) (i) (A) $CH_3-C\equiv CH$ (Propyne)

(B) $CH_3-\overset{O}{\overset{\|}{C}}-CH_3$ (Acetone)

(C) CH_3-COOH (Acetic acid)

(D) $HCOOH$ (Formic acid)

(E) CH_3COCl (Acetylchloride)

(F) $CH_3COOC_2H_5$ (Ethyl acetate)

(ii) $$\underset{\text{(Acetic acid) (C)}}{CH_3COOH} \xrightarrow{Cl_2,P} \underset{(1)}{CH_2(Cl)COOH} \xrightarrow{Cl_2,P}$$

$$\underset{(2)}{Cl_2CHCOOH} \xrightarrow{Cl_2,P} \underset{(3)}{Cl_3C-COOH}$$

1. ⇒ monochloro acetic acid

2. ⇒ dichloroacetic acid

3. ⇒ trichloroacetic acid

This reaction is known as **Hell Volhard Zelinsky reaction.**

(b) (i) Polymer obtained from the polymerisation of caprolactum is known as **Nylon-6.**

$$\text{Caprolactum} \xrightarrow{533-543\ K} \underset{\text{Amino Caproic acid}}{H_2N-(CH_2)_5-COOH} \xrightarrow{\Delta} \underset{\text{Nylon-6}}{\left[NH-(CH_2)_5-\overset{O}{\overset{\|}{C}}\right]}$$

This is a condensation polymer.

(ii) $\underset{\text{Ethyl alcohol}}{CH_3CH_2OH}$ and $\underset{\text{Dimethyl ether}}{CH_3-O-CH_3}$

Yes, both of the compounds react with PCl_5.

1. $$\underset{\text{Ethanol}}{CH_3CH_2OH} \xrightarrow{PCl_5} \underset{\text{Ethyl chloride}}{C_2H_5Cl} + POCl_3 + HCl$$

2. $$\underset{\text{Dimethyl ether}}{CH_3-O-CH_3} \xrightarrow{PCl_5} \underset{\text{Methyl chloride}}{2CH_3Cl} + PCl_3$$

(c) (i) $$\underset{\text{Methyl magnesiumbromide}}{CH_3MgBr} + \underset{\text{Ethanol}}{CH_3CH_2OH} \longrightarrow \underset{\text{Methane}}{CH_4} + Mg\begin{smallmatrix}Br \\ OC_2H_5\end{smallmatrix}$$

(ii) $$\underset{\text{Acetic anhydride}}{(CH_3CO)_2O} + PCl_5 \longrightarrow \underset{\text{Acetyl chloride}}{2CH_3COCl} + POCl_3$$

(iii) $$\underset{\text{Acetaldehyde}}{H_3C(H)C=O} + \underset{\text{Hydroxylamine}}{NH_2OH} \xrightarrow{H^+} \underset{\text{Oxime}}{H_3C(H)C=NOH} + H_2O$$

●●

CHEMISTRY 2016

QUESTIONS

PART–I (20 Marks)

*Answer **all** questions.*

Question 1.

(a) *Fill in the blanks by choosing the appropriate word/words from those given in the brackets:* **[5]**

(Henry's, aldol condensation, absence, do not, ohm, Raoult's, increases, common ion effect, easily, three, solubility product, ohm^{-1}, two, four, ohm^{-1} cm^2, Cannizzaro, ohm^{-1} cm^{-1}, zero, decreases, presence)

(i) *Ideal solutions obey_________law and they _________form azeotropic mixtures.*

(ii) *Benzaldehyde undergoes_________reaction due to_________of α-hydrogen atom.*

(iii) *The solubility of silver chloride_________ in the presence of sodium chloride because of_________.***

(iv) *The unit of conductance is_______and that of specific conductance is_______.*

(v) *When the concentration of a reactant of first order reaction is doubled, the rate becomes _______times, but for _____ order reaction, the rate remains same.*

(b) *Complete the following statements by selecting the* **correct alternative from** *the choices given:* **[5]**

(i) *Electrochemical equivalent is the amount of substance which gets deposited from its solution on passing electrical charge equal to:*

(1) *96,500 coulomb* **(2)** *1 coulomb*

(3) *60 Coulomb* **(4)** *965 coulomb*

(ii) *The complex ion $[Ni(CN)_4]^{2-}$ is:*

(1) *Square planar and diamagnetic*

(2) *Tetrahedral and paramagnetic*

(3) *Square planar and paramagnetic*

(4) *Tetrahedral and diamagnetic*

(iii) *Wohler's synthesis is used for the preparation of :***

(1) *Glycine* **(2)** *Amino acids*

(3) *Urea* **(4)** *Proteins*

(iv) *When SO_2 gas is passed through acidified $K_2Cr_2O_7$ solution, the colour of the solution changes to:*

(1) *Red* **(2)** *Black*

(3) *Orange* **(4)** *Green*

(v) *In the equation $CH_3COOH + Cl_2 \xrightarrow[-HCl]{Red\ P} A$, the compound A is:*

(1) *CH_3CH_2Cl* **(2)** *$ClCH_2COOH$*

(3) *CH_3Cl* **(4)** *CH_3COCl*

(c) *Answer the following questions:* **[5]**

(i) *What is the order of reaction whose rate constant has the same unit as the rate of reaction?*

(ii) *What is the pH value of a solution whose hydroxyl ion concentration is 1×10^{-2} M ?***

(iii) *Calculate the number of coulombs required to deposit 5·4 g of Al when the electrode reaction is:*

$$Al^{3+} + 3e^- \rightarrow Al$$

[Atomic weight of Al = 27 g/mol]

(iv) *Write the reaction to prepare acetaldehyde from hydrogen gas and an acid chloride.*

(v) *The edge length of unit cell of a body centered cubic (bcc) crystal is 352 pm. Calculate the radius of the atom.*

(d) *Match the following:* **[5]**

(i)	*Weak electrolyte***	**(a)**	*pH of a solution***
(ii)	*Colour in crystals*	**(b)**	*Iodoform*
(iii)	*Acetone*	**(c)**	*Tollen's reagent*
(iv)	*Sorensen***	**(d)**	*Ostwald dilution law***
(v)	*Ammonical silver nitrate*	**(e)**	*F-centre*

PART–II (50 Marks)

*Answer **six** questions choosing **two from Section A**, **two from Section B** and **two from Section C.***

SECTION–A

*Answer any **two** questions.*

Question 2.

(a) (i) *A 10% aqueous solution of cane sugar (mol. wt. 342) is isotonic with 1.754% aqueous solution of urea. Find the molecular mass of urea.* **[2]**

(ii) *The molecular weight of an organic compound is 58 g mol^{-1}. What will be the boiling point of a solution containing 48 gram of the solute in 1200 gram of water ?* **[2]**

[K_b for water = 0.513 °C kg $mole^{-1}$; Boiling point of water = 100°C.]

(iii) *What will be the value of van't Hoff factor (i) of benzoic acid if it dimerises in aqueous solution?*

** Answer is not given due to change in the present syllabus.

How will the experimental molecular weight vary as compared to the normal molecular weight ? **[1]**

(b) (i) *Determine the pH value of 0·001 M acetic acid solution if it is 2% ionised at this concentration. How can the degree of dissociation of this acetic acid solution be increased ?* ** **[2]**

(ii) *The solubility product of $PbCl_2$ at 298 K is 1.7×10^{-5}. Calculate the solubility of $PbCl_2$ in g / lit at 298 K.* ** **[2]**

[Atomic weights : Pb = 207 and Cl = 35·5]

(c) *Graphite is anisotropic with respect to conduction of electric current. Explain.* **[1]**

Question 3.

(a) (i) *In a body centred and face centred arrangement of atoms of an element, what will be the number of atoms present in respective unit cells? Justify your answer with calculation.* **[2]**

(ii) *A compound AB has a simple cubic structure and has molecular mass 99. Its density is 3·4 g cm^{-3}. What will be the edge length of the unit cell ?* **[2]**

(b) (i) *For the reaction:* ** **[2]**

$2NO_{(g)} \rightleftharpoons N_{2(g)} + O_{2(g)};\ \Delta H = -\text{heat}$

$K_e = 2{\cdot}5 \times 10^2$ at 298 K

What will happen to the concentration of N_2 if:

(1) *Temperature is decreased to 273 K.*

(2) *Pressure is reduced.*

(ii) *In a first order reaction, 10% of the reactant is consumed in 25 minutes. Calculate :* **[2]**

(1) *The half-life period of the reaction.*

(2) *The time required for completing 87·5% of the reaction.*

(c) *Water acts as Bronsted acid as well as a Bronsted base. Give one example each to illustrate this statement.* ** **[2]**

Question 4.

(a) (i) *Consider the following cell reaction at 298 K:* **[3]**

$2Ag^+ + Cd \rightarrow 2\,Ag + Cd^{2+}$

The standard reduction potentials (E°) for Ag^+/Ag and Cd^{2+}/Cd are 0·80 V and – 0·40 V respectively:

(1) *Write the cell representation.*

(2) *What will be the emf of the cell if the concentration of Cd^{2+} is 0·1 M and that of Ag^+ is 0·2 M ?*

(3) *Will the cell work spontaneously for the condition given in (2) above ?*

(ii) *What is a buffer solution ? How is it prepared? Explain the buffer action of a basic buffer with a suitable example.* ** **[2]**

(b) *Explain the following:* **[2]**

(i) *When NaCl is added to $AgNO_3$ solution, a white precipitate is formed.*

(ii) *An aqueous solution of ammonium chloride is acidic in nature.* **

(c) *A 0.05 M NH_4OH solution offers the resistance of 50 ohm to a conductivity cell at 298 K. If the cell constant is 0·50 cm^{-1} and molar conductance of NH_4OH at infinite dilution is 471·4 $ohm^{-1}\ cm^2\ mol^{-1}$, calculate:* **[3]**

(i) *Specific conductance*

(ii) *Molar conductance*

(iii) *Degree of dissociation*

SECTION–B

*Answer any **two** questions.*

Question 5.

(a) *Write the IUPAC names of the following:* **[2]**

(i) $[Co(NH_3)_4SO_4]\ NO_3$

(ii) $K[Pt\ (NH_3)\ Cl_3]$

(b) *What type of isomerism is exhibited by the following pairs of compounds:* **[1]**

(i) $[PtCl_2\ (NH_3)_4]\ Br_2$ and $[PtBr_2\ (NH_3)_4]\ Cl_2$

(ii) $[Cr(SCN)\ (H_2O)_5]^{2+}$ and $[Cr(NCS)\ (H_2O)_5]^{2+}$

(c) *How does $K_2\ [PtCl_4]$ get ionized when dissolved in water? Will it form precipitate when $AgNO_3$ solution is added to it ? Give a reason for your answer.* **[2]**

Question 6.

(a) *Give balanced equations for the following reactions:* **[3]**

(i) *Silver nitrate is added to dilute solution of sodium thiosulphate.*

(ii) *Potassium dichromate is treated with acidified ferrous sulphate solution.*

(iii) *Phosphorus reacts with conc. sulphuric acid.*

(b) *How will you obtain pure potassium permanganate ($KMnO_4$) crystals from its ore, pyrolusite ? Give the steps involved and the reactions.* **[2]**

Question 7.

(a) (i) *Sulphur dioxide acts as an oxidizing agent as well as a reducing agent. Give one reaction each to show its oxidizing nature and its reducing nature.* **[3]**

(ii) *Explain why an aqueous solution of potassium hexacyanoferrate(II) does not give the test for ferrous ion.*

(b) *What is meant by Lanthanide contraction? Write the general electronic configuration of inner transition elements.* **[2]**

**Answer is not given due to change in the present syllabus.

SECTION–C

*Answer any **two** questions.*

Question 8.

(a) *How can the following conversions be brought about :*

(i) *Acetaldehyde to acetaldehyde phenyl hydrazone.* **[1]**

(ii) *Benzoic acid to aniline.* **[1]**

(iii) *Methyl chloride to acetone.* **[2]**

(iv) *Benzene to benzene diazonium chloride.* **[1]**

(b) (i) *Glycerol (propane 1, 2, 3 triol) is more viscous than ethylene glycol (ethane 1, 2, diol). Explain.* **[1]**

(ii) *How can urea be detected by Biuret test ?*** **[1]**

(c) *Identify the compounds* **A, B** *and* **C:** **[3]**

(i) $C_2H_5OH \xrightarrow{PCl_5} A \xrightarrow{KCN} B \xrightarrow{H_3O^+} C_2H_5COOH \xrightarrow[\Delta]{NH_3} C$

(ii) $C_6H_5COOH \xrightarrow{SOCl_2} A \xrightarrow{NH_3} B \xrightarrow{Br_2/KOH} C$

Question 9.

(a) *Give balanced equations for the following name reactions:* **[3]**

(i) *Benzoin condensation*

(ii) *Wurtz-Fittig reaction*

(iii) *Carbylamine reaction*

(b) *Give chemical test to distinguish:* **[3]**

(i) *Formaldehyde and acetaldehyde*

(ii) *Dimethyl ether and ethyl alcohol*

(c) (i) *Write the structure of three ethers with molecular formula $C_4H_{10}O$.* **[4]**

(ii) *Starting with Grignard's reagent, how will you prepare propanoic acid?*

Question 10.

(a) *An organic compound A has the molecular formula C_7H_6O. When A is treated with NaOH followed by acid hydrolysis, it gives two products B and C. When B is oxidized, it gives A, when A and C are each treated separately with PCl_5, they give two different products D and E.* **[3]**

(i) *Identify A, B, C, D and E.*

(ii) *Give the chemical reaction when A is treated with NaOH and name the reaction.*

(b) *Answer the following:* **[4]**

(i) *What do you observe when glucose solution is heated with Tollen's reagent ?*

(ii) *Name the monomers and the type of polymerisation in each of the following polymers:*

(1) Terylene

(2) Polyvinyl chloride

(c) *Give balanced equations for the following reactions :* **[3]**

(i) *Ethylamine with nitrous acid.*

(ii) *Diethyl ether with phosphorus pentachloride.*

(iii) *Aniline with acetyl chloride.*

ANSWERS

PART–I

Answer 1.

(a) (i) Raoult's, do not

(ii) Cannizzaro, absence

(iv) ohm^{-1}, $ohm^{-1}\ cm^{-1}$

(v) two, zero

(b) (i) (2) 1 coulomb

(ii) (1) Square planar and diamagnetic

(iv) (4) Green

(v) (2) $ClCH_2COOH$

(c) (i) Zero Order Reaction

(iii) $Al^{3+} + 3e^- \rightarrow Al$

For 1 mol (27g) of Al requires

$$= 3 \times 96500\ C$$

$$1\ g\ \text{of Al requires} = \frac{3 \times 96500}{27}$$

$$5{\cdot}4\ g\ \text{of Al requires} = \frac{3 \times 96500}{27} \times 5{\cdot}4$$

$$= 57900\ C$$

(iv) $\underset{\text{Acetyl chloride}}{CH_3-\overset{O}{\overset{\|}{C}}-Cl} + \underset{\text{Hydrogen gas}}{H_2} \xrightarrow[\text{Xylene}]{Pd/BaSO_4} \underset{\text{Aldehyde}}{CH_3-\overset{O}{\overset{\|}{C}}-H} + HCl$

(v) Bcc crystal

$$a = 352\ pm,$$

$$r = ?$$

$$\frac{\sqrt{3}}{4} a = r$$

$$\frac{1.732}{4} \times 352 = 1{\cdot}732 \times 88\ pm$$

$$r = 152{\cdot}42\ pm$$

(d) (ii) Colour in crystals → F-centre

(iii) Acetone → Iodoform

(v) Ammonical silver nitrate → Tollen's reagent

PART–II

SECTION–A

Answer 2.

(a) (i) Van't Hoff factor $(i) = 1$

$$\pi_{canesugar} = \pi_{urea}$$

$$C_1RT = C_2RT$$

$$C_1 = C_2$$

$$\left[\frac{10}{342}\right] = \left[\frac{1.754}{x}\right]$$

$$x = \frac{1.754 \times 342}{10} \quad x = 59{\cdot}9868 \text{ g}$$

Molecular mass of urea is 59·9868 g.

(ii) Molecular weight = 58 gm/mol.

Boiling point of solution = ?

48 gm solute in 1200 gm H_2O

K_b for water = 0·513°C

Boiling point of water = 100°C (T_b)

$$\Delta T_b = \frac{1000 \times K_b \times w}{M \times W}$$

$$\Delta T_b = \frac{0.513 \times 100 \times 48}{1200 \times 58}$$

$$\Delta T_b = 0{\cdot}353°C$$

As we know that,

$$\Delta T_b = T - T_b$$

$$0{\cdot}353°C = T - 100°C$$

$$100{\cdot}353°C = T$$

∴ B.P. of the solution is 100·353°C.

(iii) $i = \dfrac{\text{Particles after association}}{\text{Particles before association}}$

$$= \frac{1}{2} = 0{\cdot}5$$

The experimental molecular weight of benzoic acid is double of the normal molecular weight because number of particles reduces to almost one half due to the formation of dimer.

(c) Graphite is a hexagonal crystal, that is carbon atoms in it are arranged in layers parallel to the plane. Thus it exists in the form of layer structure. The distance between adjacent layers is 3·4 Å. Carbon atoms are arranged in different layers and each atom is covalently bonded to three of its neighbouring atom in the same layer. The electrical conductivity is more parallel to the layer whereas the electrical conductivity is less perpendicular to the layer.

Answer 3.

(a) (i) Body centred cubic: Number of atoms at the corners per unit cell

$$= \frac{1}{8} \times 8 = 1$$

Number of atoms present at the centre of the cube = 1

∴ Total number of atoms in bcc arrangement

$$= 1 + 1 = 2$$

Face centred cubic : Number of atoms present at the corners per unit cell

$$= \frac{1}{8} \times 8 = 1$$

Number of atoms present at the faces per unit cell

$$= \frac{1}{2} \times 6 = 3$$

∴ Total number of atoms in fcc arrangement

$$= 1 + 3 = 4$$

(ii) Simple cubic structure molecular mass = 99

$$\rho = 3{\cdot}4 \text{ g/cm}^3$$

Edge length = ?

$$\rho = \frac{Z \times M}{a^3 \times N_A} \text{ g/cm}^3$$

$$3{\cdot}4 = \frac{1 \times 99}{a^3 \times 6.023 \times 10^{23}}$$

$$a^3 = \frac{99}{3.4 \times 6.023 \times 10^{23}}$$

$$a = 3{\cdot}64 \times 10^{-8} \text{ cm}$$

(b) (ii) t = 10% reactant → 25 minute

(1) $$t_{1/2} = \frac{0.693}{K}$$

$$K = \frac{2.303}{t} \log \frac{100}{90}$$

$$K = \frac{2.303}{25} \log \frac{100}{90}$$

$$K = 0{\cdot}004215$$

$$t_{1/2} = \frac{0.693}{K} = \frac{0.693}{0.004215}$$

$$t_{1/2} = 165 \text{ minutes.}$$

(2) $$t = \frac{2.303}{K} \log \frac{100}{12.5}$$

$$t = \frac{2.303}{0.0042} \log \frac{100}{12.5}$$

$$t = 495{\cdot}14 \text{ minutes.}$$

Answer 4.

(a) (i) (1) $Cd\left|Cd^{2+}_{(0.1M)}\right|\left|2Ag^{2+}_{(0.2M)}\right|2Ag$

(2) $E°_{(red)\,cell} = E°_{(red)\,cathode} - E°_{(red)\,anode}$

$$= 1{\cdot}2$$

$$E_{cell} = E^\circ_{(red)cell} - \frac{0.059}{n}\log\left[\frac{M}{M^{n+}}\right]$$

$$[n = 2]$$

$$= 1{\cdot}2 - \frac{0.059}{n}\log\frac{[2Ag][Cd^{2+}]}{[Cd][Ag^+]}$$

$$= 1{\cdot}2 - \frac{0.059}{2}\log\left[\frac{0.1}{0.04}\right]$$

$$E_{cell} = 1{\cdot}1882177 \text{ volt}$$

(3) Since E_{cell} is +ve therefore, cell reaction will be spontaneous.

$$\Delta G^\circ = -nFE_{cell} = -229{\cdot}32 \text{ kJ}$$

Since E^0 is positive, ΔG will be negative so the cell will work spontaneously.

(b) (i) In the silver nitrate test there was an anion present in the solution, that will cause silver cations to precipitate silver usually as halides. Here, it precipitates as chlorides,

$$AgNO_3 + NaCl \rightarrow \underset{\text{white precipitate}}{AgCl} + \underset{\text{aq.}}{NaNO_3}$$

(c) (i) Specific conductance :

$$k = \frac{1}{\rho} = \frac{1}{R}\left(\frac{l}{A}\right)$$

$$R = 50 \text{ ohm}$$

$$\frac{l}{a} = \text{cell constant} = 0{\cdot}50 \text{ cm}^{-1}$$

$$k = \frac{1}{50} \times 0{\cdot}50$$

$$k = 0{\cdot}01 \text{ ohm}^{-1} \text{ cm}^{-1}.$$

(ii)

$$\Lambda_m = \frac{k \times 1000}{M}$$

$$k = 0{\cdot}01 \text{ ohm}^{-1} \text{ cm}^{-1}$$

$$\Lambda_m = \frac{0.01 \times 1000}{0.05}$$

$$\Lambda_m = 200 \text{ ohm}^{-1} \text{ cm}^2/\text{mol}$$

(iii)

$$\alpha = \frac{\Lambda_m}{\Lambda_m^\infty}$$

$$\alpha = \frac{200}{471.4} = 0.4242$$

SECTION–B

Answer 5.

(a) (i) Tetraaminesulphatocobalt(III) nitrate

(ii) Potassiumaminetrichloroplatinate(II)

(b) (i) Ionization isomerism

(ii) Linkage isomerism

(c) When $K_2[PtCl_4]$ is dissolved in water then it ionizes as follows:

$$K_2PtCl_4 \longrightarrow 2K^+ + [PtCl_4]^{2-}$$

and when $AgNO_3$ is added to $K_2[PtCl_4]$ then it will not form the white precipate with $AgNO_3$ solution since Cl^- ions are not free instead of it grey precipitate is obtained.

$$K_2[PtCl_4] + 2AgNO_3 \longrightarrow 2KCl\,(aq) + \underset{\text{Grey precipitate}}{PtCl_2\,(AgNO_3)_2\,(s)}$$

Answer 6.

(a) (i) $2AgNO_3 + Na_2S_2O_3 \longrightarrow \underset{\text{White ppt.}}{Ag_2S_2O_3} + 2NaNO_3$

$$Ag_2S_2O_3 + H_2O \longrightarrow \underset{\text{Black}}{Ag_2S} + H_2SO_4$$

(ii) $K_2Cr_2O_7 + 7H_2SO_4 + 6FeSO_4 \longrightarrow K_2SO_4 + Cr_2(SO_4)_3 + 3Fe_2(SO_4)_3 + 7H_2O$

(iii) $P_4 + 10H_2SO_4 \longrightarrow 4H_3PO_4 + 10SO_2 + 4H_2O$

(b) Potassium permanganate is prepared from mineral pyrosulphite (MnO_2)

(1) Conversion of Pyrosulphite ore (MnO_2) to Potassium Manganate K_2MnO_4 :

$$MnO_2 + 4KOH + O_2 \xrightarrow{\text{Heat}} 2K_2MnO_4 + 2H_2O$$

(2) Oxidation of Potassium Manganate (K_2MnO_4) to potassium permanganate ($KMnO_4$)

$$2K_2MnO_4 + Cl_2 \longrightarrow 2\,KCl + 2KMnO_4$$

Answer 7.

(a) (i) The structural formula of sulphur dioxide is SO_2, and its a covalent planar molecule which is V-shaped with an angle of 120 degrees.

Sulphur dioxide as oxidising agent :

Claus Process:

$$\underset{\text{sulphur dioxide}}{SO_2} + 2H_2S \longrightarrow \underset{\text{elemental sulphur}}{3S} + \underset{\text{water}}{2H_2O}$$

Sulphur dioxide as reducing agent:

$$SO_2 + Cl_2 \longrightarrow \underset{\text{sulphuryl chloride}}{SO_2Cl_2}$$

(ii) Potassium hexacyanoferrate (II) does not give ferrous ion test, because it is coordination compound and coordination compounds do not ionise in aqueous solution, thus does not give ferrous ion test.

$$K_4[Fe(CN)_6] \rightleftharpoons 4K^+ + [Fe(CN)_6]^{4-}$$

(b) The steady decrease in the atomic and ionic radii of lanthanide elements with increasing atomic number due to increase in nuclear charge is called Lanthanide contraction.

General electronic configuration of Inner Transition elements :

$$ns^2 (n-1) d^{0-1} (n-2) f^{1-14}$$

SECTION–C

Answer 8.

(a) Conversions :

(i) Acetaldehyde to Acetaldehyde phenyl hydrazone

$$\underset{\text{acetaldehyde}}{CH_3CHO} \xrightarrow[KrCr_2O_7/H_2SO_4]{[O]} \underset{\text{acetic (acid)}}{CH_3COOH}$$

$$\xrightarrow[(CH_3COO)_2Ca]{Ca(OH)_2} \underset{\text{acetaldehyde phenyl hydrazone}}{CH_3-CH{=}NNHC_6H_5}$$

(ii) Benzoic acid to aniline

$$C_6H_5COOH \xrightarrow{+NH_3} C_6H_5COONH_4 \xrightarrow{+H_2O}$$

$$C_6H_5CONH_2 \xrightarrow[\text{Hoffmann Bromide Reaction}]{Br_2/KOH} \underset{\text{(aniline)}}{C_6H_5NH_2}$$

(iii) Methyl chloride ⟶ Acetone

$$\underset{\text{(methyl choride)}}{CH_3Cl + HOH} \xrightarrow{-HCl} CH_3OH \xrightarrow{I_2,P}$$

$$CH_3I \xrightarrow{KCN} CH_3CN \xrightarrow[KMnO_4]{[O]}$$

$$CH_3COOH \xrightarrow{Ca(OH)_2} (CH_3COO)_2Ca \xrightarrow{distil} CH_3COCH_3$$

(iv) Benzene to benzene diazonium chloride

$$\underset{\text{(Benzene)}}{C_6H_6} \xrightarrow[\text{Conc. } H_2SO_4 \; 50°C]{\text{Conc. } HNO_3} \underset{\text{Nitrobenzene}}{C_6H_5NO_2} \xrightarrow[Sn + HCl]{6[H]} \underset{\text{Aniline}}{C_6H_5NH_2}$$

$$\xrightarrow[0\text{-}5°C]{NaNO_2 + HCl} \underset{\text{(Benzene diazonium chloride)}}{C_6H_5NH^{+}{\equiv}NCl^{-}} + 2H_2O$$

(b) (i) Viscosity depends on the amount of hydrogen bonding between molecules and presence of more —OH groups, glycerol involved in intramolecular hydrogen bonding within the molecule, has 3OH groups per molecule whereas ethylene glycol has 2OH groups per molecule. Thus glycerol is more viscous than ethylene glycol.

(c) (i)

$$\underset{\text{Ethanol}}{C_2H_5OH} \xrightarrow{PCl_5} \underset{\text{Ethyl chloride}}{C_2H_5Cl} \xrightarrow{KCN} \underset{\text{Ethyl cyanide}}{C_2H_5CN}$$

$$\xrightarrow[\Delta]{H_3O^+} \underset{\text{Ethanoic acid}}{C_2H_5COOH} \xrightarrow{NH_3} \underset{\text{Acetanamide}}{C_2H_5CONH_2}$$

(c) (ii)

$$C_6H_5COOH \xrightarrow{SOCl_2} \underset{\text{(Benzoylchloride) (A)}}{C_6H_5COCl} \xrightarrow{NH_3} \underset{\text{(Benezamide) (B)}}{C_6H_5CONH_2}$$

$$\xrightarrow{Br/Kon} \underset{\text{(Aniline)}}{C_6H_5NH_2}$$

Answer 9.

(a) (i) Refer to ISC Paper 2019, Answer 18 (b) (i) (3).

(ii) Wurtz-Fittig Reaction:

$$\underset{\text{Chlorobenzene}}{C_6H_5Cl} + \underset{\text{Sodium}}{2Na} + \underset{\text{Methyl chloride}}{ClCH_3} \xrightarrow{\text{dry ether}} \underset{\text{(Tolene)}}{C_6H_5CH_3} + 2NaCl$$

(iii) Carbylamine Reaction:

$$\underset{\text{Aniline}}{C_6H_5NH_2} + CHCl_3 + \underset{\text{(Alc.)}}{3KOH} \xrightarrow{\Delta}$$

$$\underset{\text{Phenyl isocyanide}}{C_6H_5NC} + 3KCl + H_2O$$

(b) (i) Formaldehyde and Acetaldehyde

1. With freshly prepared sodium nitroprusside solution and excess of dil. NaOH acetaldehyde produces a deep blue colour.
2. Formaldehyde does not gives the above test.

(ii) Dimethyl ether and ethyl alcohol

1. Ethyl alcohol gives Iodoform test, a pale yellow precipitate of iodoform is formed.

$$\underset{\text{Ethyl alcohol}}{CH_3CH_2OH} + 4I_2 + 6NaOH \longrightarrow$$

$$\underset{\text{Iodoform}}{CHI_3} + HCOONa + 5NaI + 5H_2O$$

2. Dimethyl ether does not gives iodoform test.

(c) (i) structural formula of three ethers with molecular formula → $[C_4H_{10}O]$

1. $H_5C_2-O-C_2H_5$
   Dimethyl ether
2. $H_3C-O-C_3H_7$
   Methyl propyl ether
3. $CH_3-O-CH(CH_3)-CH_3$

(ii)

$$\underset{\text{Ethyl magnesium bromide (Grignard's Reagent)}}{CH_3CH_2MgBr} \xrightarrow{HCHO} \underset{\text{Addition product}}{[CH_3CH_2CH_2OMgBr]}$$

$\xrightarrow{H^+/H_2O} CH_3CH_2CH_2OH \xrightarrow{[O]}$
1-propanol

$CH_3CH_2CHO \xrightarrow{[O]} CH_3CH_2COOH + MgBrOH$
Propanoic acid

Answer 10.

(a) (i) (A) = C_6H_5CHO (Benzaldehyde)

(B) = $C_6H_5CH_2OH$ (Benzyl alcohol)

(C) = C_6H_5COOH (Benzoic Acid)

(D) = $C_6H_5CHCl_2$ (Benzal Chloride)

(E) = C_6H_5COCl (Benzoyl Chloride)

(ii) **Cannizzaro Reaction:**

$2C_6H_5CHO + NaOH \xrightarrow{50\% NaOH}$
Benzaldehyde

$C_6H_5CH_2OH + C_6H_5COOH$
Benzyl alcohol, Sodium benzoate

(b) (i)

$$\underset{\text{Glucose}}{CHO-(CHOH)_4-CH_2OH} + \underset{\text{Tollen s reagent}}{Ag_2O} \xrightarrow{\text{water}} \underset{\text{Gluconic acid}}{COOH-(CHOH)_4-CH_2OH} + \underset{\text{Silver mirror}}{2Ag\downarrow}$$

(ii)

Polymer	Monomer	Type of Polymerisation
(1) Terylene	Ethylene Glycol and Terephthalic acid	Condensation polymerisation
(2) Polyvinyl chloride	Vinyl chloride	Addition Polymerisation

(c) **Balanced Chemical Reaction:**

(i) Ethylamine with nitrous acid

$$\underset{\text{ethylamine}}{CH_3CH_2NH_2} + \underset{\text{nitrous acid}}{HONO} \longrightarrow \underset{\text{ethanol}}{CH_3CH_2OH} + H_2O + N_2$$

(ii) Diethyl ether with phosphorus pentachloride

$$C_2H_5OC_2H_5 + PCl_5 \longrightarrow \underset{\text{ethyl chloride}}{2C_2H_5Cl} + \underset{\text{phosphoryl chloride}}{POCl_3}$$

(iii) $CH_3COCl + HNHC_6H_5 \longrightarrow \underset{\text{Acetanilide}}{CH_3CONHC_6H_5} + HCl$

●●

CHEMISTRY 2015

QUESTIONS

PART–I (20 Marks)

*Answer **all** questions.*

Question 1.

(a) *Fill in the blanks by choosing the appropriate word/ words from those given in the brackets:* **[5]**

(increases, decreases, positive, efficient, 68, non-efficient, no α-hydrogen, α-hydrogen, negative, Rosenmund's, greater, Cannizzaro, 74, common-ion effect, lesser, buffer action, diamagnetic, paramagnetic)

(i) *The more________is standard reduction potential of a metal, the________is its ability to displace hydrogen from acids.*

(ii) *Both ccp and hcp are________packings and occupy about________% of the available space.*

(iii) *Solubility of silver chloride________in the presence of sodium chloride because of ________.***

(iv) *Benzaldehyde undergoes________in reaction on treatment with concentrated sodium hydroxide because it has________atom.*

(v) *The transition metals show________character because of the presence of unpaired electrons and Cu^+ is________because its electronic configuration is $[Ar]3d^{10}$.*

(b) *Complete the following statements by selecting the **correct alternative from** the choices given:* **[5]**

(i) *The molal freezing point constant of water is 1·86 K kg mol^{-1}. Therefore, the freezing point of 0·1M NaCl solution in water is expected to be:*

(1) – 1·86°C **(2)** – 0·372°C

(3) – 0·186°C **(4)** + 0·372°C

(ii) *For a first order reaction the rate constant for decomposition of N_2O_5 is 6×10^{-4} sec^{-1}. The half-life period for the decomposition in seconds is :*

(1) 11·55 **(2)** 115·5

(3) 1155 **(4)** 1·155.

(iii) *When acetaldehyde is treated with Grignard reagent, followed by hydrolysis the product formed is:*

(1) Primary alcohol

(2) Secondary alcohol

(3) Carboxylic acid

(4) Tertiary alcohol.

(iv) *The geometry of XeF_6 molecule and the hybridization of Xe atom in the molecule is:*

(1) Distorted octahedral and sp^3d^3

(2) Square planar and sp^3d^2

(3) Pyramidal and sp^3

(4) Octahedral and sp^3d^3.

(v) *In the complexes $[Fe(CN)_6]^{3-}$ and $[Pt(en)(H_2O)_2(NO_2)(Cl)]^{2+}$ the respective oxidation numbers of central metal atoms are :*

(1) + 3 and + 4 **(2)** + 6 and + 4

(3) + 6 and + 3 **(4)** + 3 and + 3.

(c) *Answer the following questions:* **[5]**

(i) *What is the effect of temperature on the ionic product of water? How will it change the pH value of a neutral solution?***

(ii) *How many hours does it take to reduce 3 moles of Fe^{3+} to Fe^{2+} with 2·0 A current intensity ?*

(iii) *How is urea prepared by Wohler synthesis?***

(iv) *Two liquids A and B form type II non ideal solution which shows a minimum in its temperature-mole fraction plot (T-χ diagram). Can the two liquids be completely separated by fractional distillation ?*

(v) *The aqueous solution of sodium acetate is basic. Explain.***

(d) *Match the following :* **[5]**

(i) Disaccharide	**(a)** Lucas reagent
(ii) Carbylamine	**(b)** Condensation polymer
(iii) Dacron	**(c)** Obnoxious smell
(iv) Low spin complex, d^2sp^3	**(d)** Sucrose
(v) Anhydrous $ZnCl_2$ + conc. HCl	**(e)** Hexaammine-cobalt(III)ion

PART–II (50 Marks)

*Answer **six** questions choosing **two** from **Section A**, **two** from **Section B** and **two** from **Section C.***

SECTION–A

*Answer any **two** questions.*

Question 2.

(a) (i) *A solution containing 0·5g of KCl dissolved in 100g of water and freezes at – 0·24°C. Calculate the degree of dissociation of the salt. (K_f for water = 1·86°C) Atomic weights [K = 39, Cl = 35·5]* **[3]**

** Answer is not given due to change in the present syllabus.

(ii) *If 1·71 g of sugar (molar mass = 342) are dissolved in 500 ml of an aqueous solution at 300 K, what will be its osmotic pressure ?* [1]

(iii) *0·70 g of an organic compound when dissolved in 32 g of acetone produces an elevation of 0·25°C in the boiling point. Calculate the molecular mass of organic compound. (K_b for acetone = 1·72K mol^{-1})* [1]

(b) (i) *What is the difference between order of a reaction and the molecularity of a reaction ?* [2]

(ii) *A substance decomposes by following first order kinetics. If 50% of the compound is decomposed in 120 minutes, how long will it take for 90% of the compound to decompose ?* [2]

(c) *Name the crystal structure of the copper metal.* [1]

Question 3.

(a) (i) *Chromium metal crystallises with a body centered cubic lattice. The edge length of the unit cell is found to be 287 pm. Calculate the atomic radius. What would be the density of chromium in g/cm³ ?* [2]

(atomic mass of Cr = 52·99)

(ii) *Why does sodium chloride on heating with sodium vapours acquire yellow colour ?* [1]

(iii) *The equilibrium constant for the reaction :***

$N_{2(g)} + 3H_{2\,(g)} \rightleftharpoons 2NH_{3(g)}$ *at 715 K, is* $6{\cdot}0 \times 10^{-2}$.

If, in a particular reaction, there are 0·25 mol L^{-1} *of* H_2 *and 0·06 mol* L^{-1} *of* NH_3 *present, calculate the concentration of* N_2 *at equilibrium.* [1]

(iv) *Calculate the concentration of* OH^- *ions in solution when* $[H^+] = 6{\cdot}2 \times 10^{-2}$ *mol* L^{-1}.** [1]

(v) *State the Le-Chatelier's principle.*** [1]

(b) *For a crystal of sodium chloride, state :* [2]

(i) *The type of lattice in which it crystallizes.*

(ii) *The coordination number of each sodium ion and chloride ion in the crystal lattice.*

(iii) *The number of sodium ions and chloride ions present in a unit cell of sodium chloride.*

(iv) *The structural arrangement of the sodium chloride crystal.*

(c) *Consider the following reaction :*** [2]

$$N_2O_{4(g)} + Heat \rightleftharpoons 2NO_{2(g)}$$

How is the composition of equilibrium mixture affected by :

(i) *A change in temperature*

(ii) *A change in pressure*

(iii) *A change in concentration of* N_2O_4

(iv) *The removal of* NO_2 *from the reaction mixture*

Question 4.

(a) *The specific conductance of a 0·01 M solution of acetic acid at 298K is* $1{\cdot}65 \times 10^{-4}$ ohm^{-1} cm^{-1}. *The molar conductance at infinite dilution for* H^+ *ion and* CH_3COO^- *ion are 349·1* ohm^{-1} cm^2mol^{-1} *and 40·9* ohm^{-1} cm^2mol^{-1} *respectively.* [3]

Calculate :

(i) *Molar conductance of the solution.*

(ii) *Degree of dissociation of* CH_3COOH.

(iii) *Dissociation constant for acetic acid.*

(b) (i) *Calculate the e.m.f. of the following cell reaction at 298K :* [2]

$$Mg_{(s)} + Cu^{2+}\ (0.0001\ M) \times Mg^{2+}\ (0.001M) + Cu_{(s)}$$

The standard potential (E°) of the cell is 2.71 V.

(ii) *The solubility product* (K_{sp}) *of* $BaSO_4$ *is* 1.5×10^{-9}. *Calculate the solubility of barium sulphate in pure water and in 0.1 M* $BaCl_2$.** [2]

(c) *Explain the following :***

(i) *When* NH_4Cl *and* NH_4OH *are added to a solution containing both,* Fe^{3+} *and* Ca^{2+} *ions, which ion is precipitated first and why ?* [2]

(ii) *Dissociation of* H_2S *is suppressed in acidic medium.* [1]

SECTION–B

*Answer any **two** questions.*

Question 5.

(a) *Write the IUPAC names of the following coordination compounds :* [1]

(i) $[Cr(NH_3)_4(H_2O)_2]Cl_3$

(ii) $[PtCl_2(NH_3)_4]\ [PtCl_4]$

(b) *State the hybridization and magnetic property of* $[Fe(CN)_6]^{3-}$ *ion according to the valence bond theory.* [1]

(c) (i) *What type of isomers are* $[Co(NH_3)_5Br]SO_4$ *and* $[Co(NH_3)_5SO_4]Br$ *? Give a chemical test to distinguish between them.* [2]

(ii) *Write the structures of optical isomers of the complex ion* $[Co(en)_2Cl_2]^+$ [1]

Question 6.

(a) *Give balanced chemical equations for the following reactions :* [3]

(i) *Fluorine is passed through cold, dilute NaOH solution.*

(ii) *Hydrogen peroxide is treated with acidified* $KMnO_4$ *solution.*

(iii) *Sulphuric acid is treated with hydrogen sulphide.*

(b) *Draw the structure of xenon tetrafluoride molecule and state the hybridization of the central atom and the geometry of the molecule.* [2]

** Answer is not given due to change in the present syllabus.

Question 7.

(a) *Name the important ore of silver. Write all the steps and reactions involved in the Cyanide process for the extraction of silver from its ore.* **[3]**

(b) *Explain the following :* **[2]**

(i) *Why do transition metal ions possess a great tendency to form complexes ?*

(ii) *The paramagnetic character in 3d-transition series elements increases upto Mn and then decreases.*

SECTION–C

Answer any ***two*** *questions.*

Question 8.

(a) *How can the following conversions be brought about :*

(i) *Glycerol to formic acid* **[1]**

(ii) *Chlorobenzene to phenol* **[1]**

(iii) *Diethyl ether to ethanol* **[1]**

(iv) *Phenol to aniline.* **[2]**

(b) (i) *How is iodoform prepared from ethanol ? Give balanced equation.* **[1]**

(ii) *What will be the product formed when chlorobenzene is heated with sodium metal in the presence of dry ether ?* **[1]**

(c) *Identify the compounds* ***A, B, C, D, E and F*** *:* **[3]**

$$CH_3COCH_3 \xrightarrow[(O)]{Conc.HNO_3} A \xrightarrow{SOCl_2} B$$

$$\xrightarrow{NH_3} C$$

$$\xrightarrow{LiAlH_4} D \xrightarrow{HNO_2} E \xrightarrow{CH_3COCl} F$$

Question 9.

(a) *Give balanced equations for the following name reactions :* **[3]**

(i) *Reimer-Tiemann reaction*

(ii) *Rosenmund reaction*

(iii) *Hoffmann's degradation reaction*

(b) *Give one chemical test to distinguish between the following pairs of compounds :*

(i) *Ethylamine and diethylamine*

(ii) *Acetaldehyde and benzaldehyde* **[3]**

(c) (i) *Arrange the following compounds in the ascending order of their basic strength and give reasons for your answer :* **[2]**

Methylamine, Aniline, Ethylamine, Diethyl ether

(ii) *Name the monomers and the type of polymerization in each of the following polymers :* **[2]**

(1) *Polyester* **(2)** *Bakelite*

Question 10.

(a) *An organic compound A with molecular formula C_2H_7N on reaction with nitrous acid gives a compound B. B on controlled oxidation gives compound C. C reduces Tollen's reagent to give silver mirror D and B reacts with D in the presence of concentrated sulphuric acid to give sweet smelling compound E. Identify A, B, C, D and E. Give the reaction of C with ammonia.* **[3]**

(b) *Give balanced equations for the following reactions :* **[4]**

(i) *How will you convert ethyl amine to methyl amine ?*

(ii) *What is the effect of denaturation on the structure of proteins ?*

(iii) *Name the nitrogen base residues present in DNA.*

(c) *Give balanced equations for the following reactions :* **[3]**

(i) *Aniline is treated with nitrous acid and HCl at low temperature.*

(ii) *Acetyl chloride is treated with ethyl alcohol.*

(iii) *Formaldehyde is treated with ammonia.*

ANSWERS

PART–I

Answer 1.

(a) (i) negative, greater

(ii) efficient, 74

(iv) Cannizzaro, no α-hydrogen

(v) paramagnetic, diamagnetic

(b) (i) (2) – 0·372°C

(ii) (3) $t_{1/2} = \dfrac{0.693}{K}$

$$= \frac{0.693}{6\times10^{-4}}$$

$$= \frac{693\times10^3}{6\times10^{-4}} = 1155 \text{ sec}$$

(iii) (2) Secondary alcohol

(iv) (1) Distorted octahedral and sp^3d^3,

(v) (1) +3 and +4

(c) (ii) Given, Current = 2·0 ampere

$$3Fe^{3+} + 3e^- \longrightarrow 3Fe^{2+}$$

Charge required = 3 Faraday

= 3 × 96500 coulomb

= 289500 coulomb

We know, Charge = Current × Time

** Answer is not given due to change in the present syllabus.

or $\quad \text{Time} = \dfrac{289500}{2}$

$= 144750 \text{ sec} = 40{\cdot}20 \text{ hrs.}$

(iv) Two liquids A and B form type II non-ideal solution which shows a minimum in its temperature-mole fraction plot can not be completely separated by fractional distillation because they form a constant boiling azeotropic mixture. Thus, upon fractional distillation either A or B is obtained pure along with composite composition.

(d)	**(i)** Disaccharide	**(d)**	Sucrose
	(ii) Carbylamine	**(c)**	Obnoxious smell
	(iii) Dacron	**(b)**	Condensation poloymer
	(iv) Low spin complex, d^2sp^3	**(e)**	hexaamine-cobalt(III)
	(v) Anhydrous $ZnCl_2$ + Conc. HCl	**(a)**	Lucas reagent

PART–II

SECTION–A

Answer 2.

(a) (i) $\Delta T_f = 0 - (-0.24° \text{ C}) = 0.24° \text{ C}$

$$M_2 = \frac{1000\, K_f W_2}{\Delta T_f W_1}$$

$$= \frac{1000 \times 1.86 \times 0.5}{0.24 \times 100}$$

$= 38.75 \text{ g mol}^{-1}$

Theoretical molar mass of KCl = 39 + 35.5 = 74.5

$$i = \frac{\text{Calculated mol mass}}{\text{Theoretical mol mass}}$$

$$= \frac{74.5}{38.75} = 1.92$$

$KCl = K^+ + Cl^-$

1 mole $\quad 0 \quad 0$

$1 - \alpha \quad \alpha \quad \alpha$

Total number of moles after dissociation = $1 + \alpha$

$$i = \frac{1+\alpha}{1}$$

$i = 1 - \alpha = 1.92 - 1$

$\Rightarrow 0.92$

Percentage dissociation = 92%

(ii) Given : Wt. of solute $(w) = 1{\cdot}71\text{g}$

Molar mass of solute $(m) = 342$

Volume of solution (V) = 500 ml = 0·5 L

Temperature (T) = 300 K

Gas constant (R) = 0·0821 Lt atm K^{-1} mol^{-1}

$$\text{Osmotic pressure } (\pi) = \frac{wRT}{mV}$$

$$= \frac{1{\cdot}71 \times 0{\cdot}0821 \times 300}{342 \times 0{\cdot}5}$$

$= 0{\cdot}25 \text{ atm}$

Thus, Osmotic pressure of the solution is 0·25 atm.

(iii) Given : Wt. of the solute $(w) = 0{\cdot}70$ g

Wt. of the solvent (W) = 32 g

Elevation of boiling point $(\Delta T_b) = 0{\cdot}25°C$

$K_b = 1{\cdot}72 \text{ K kg mol}^{-1}$

The relation of elevation of boiling point and molecular mass is :

$$m = \frac{1000 \times K_b \times w}{W \times \Delta T_b}$$

$$= \frac{1000 \times 1{\cdot}72 \times 0{\cdot}70}{32 \times 0{\cdot}25}$$

$= 150{\cdot}5$ g

Thus, the molecular mass of the organic compound is 150·5g.

(b) (i) Refer to ISC Paper 2017, Answer 4 (a) (ii)

(ii) Given : Half life $(T_{1/2})$ = 120 minutes

For first order reaction, half life $(T_{1/2})$

$$= \frac{0{\cdot}693}{\text{Rate constant (K)}}$$

Here, $\quad K = \dfrac{0{\cdot}693}{120} = 0{\cdot}00578 \text{ min}^{-1}$

Let initial concentration a mol and time t in minutes required for 90% decomposition of the compound.

Now, $\quad t = \dfrac{2{\cdot}303}{K} \log \dfrac{a}{a-x}$

$$= \frac{2{\cdot}303}{0{\cdot}00578} \log 10 = 399 \text{ minutes}$$

Thus, 398 minutes are required for 90% decomposition of the compound.

(c) The crystal structure of copper metal is face centered cubic (fcc) or cubic close packing (ccp).

Answer 3.

(a) (i) For body centred cubic cell atomic radius,

$$r = \frac{\sqrt{3} \times \text{edge length}}{4}$$

$$= \frac{\sqrt{3} \times 287}{4} = 124{\cdot}27 \text{ pm.}$$

Thus, atomic radius of chromium metal is 124·27 pm.

Density of the unit cell of the solid

$$= \frac{\text{Mass of unit cell}}{\text{Volume of unit cell}}$$

$$= \frac{Z \times \text{atomic mass}}{N_A \times a^3}$$

For body centered cubic lattice, the number of atoms in unit cell (Z) is

$$= 1 + \left(\frac{1}{8} \times 8\right) = 1 + 1 = 2$$

The volume of the cubic lattice

= (Edge length)^3

= $(287 \times 10^{-10})^3$

∴ Density of the unit cell is

$$= \frac{2 \times 52{\cdot}99}{6{\cdot}023 \times 10^{23} \times (287)^3 \times 10^{-30}} \text{ gm/cm}^3$$

= 7·44 g/cm^3

Thus, the density of chromium is 7·44 g/cm^3.

(ii) Sodium chloride on heating with sodium vapours acquires yellow colour because sodium chloride crystal undergoes metal excess defect with sodium vapours when heated. The chloride ions diffuses to the surface of the crystals and combine with Na atoms which get ionized to Na^+ ions by losing electrons. These electrons get trapped in anion vacancies and act as F-centres which impart colour to the crystal.

(b) (i) For a crystal of sodium, the type of lattice in which it crystallises is face centered cubic (fcc).

(ii) The coordination number of each sodium ion and chloride ion in the crystal lattice is 6 : 6.

(iii) The number of sodium ions and chloride ions present in a unit cell is 4 each.

(iv) The structural arrangement of the sodium chloride crystal is cubic close packing with Cl^- ions at corners and face centres and Na^+ ions at edge centres and body centre with respect to Na^+ ion is octahedral.

Answer 4.

(a) (i) Molar conductance,

$$\Lambda_m = \frac{K \times 1000}{C}$$

$$= \frac{1{\cdot}65 \times 10^{-4} \times 1000}{0{\cdot}01}$$

= 16·5 ohm^{-1} cm^2 mol^{-1}

(ii) Degree of dissociation,

$$\alpha = \frac{\Lambda_m}{\Lambda_m^{\infty}}$$

Λ_m = 16·5 ohm^{-1} cm^2 mol^{-1}

$\Lambda_{(CH_3COOH)} = \lambda^{\infty}[H^+] + \lambda^{\infty}[CH_3COOH^-]$

= 349·1 + 40·9

= 390 ohm^{-1} cm^2 mol^{-1}

$$\alpha = \frac{16{\cdot}5}{390}$$

= 0·0423

(iii) $CH_3COOH \rightleftharpoons CH_3COO^- + H^+$

	CH_3COOH	CH_3COO^-	H^+
Initial	0·01 M	0	0
At equilibrium	0·01 (1 – α)	0·01 α	0·01 α

$$\text{Dissociation constant } K_d = \frac{[CH_3COO^-][H^+]}{[CH_3COO^-]}$$

$$= \frac{0{\cdot}01\alpha \times 0{\cdot}01\alpha}{0{\cdot}01(1-\alpha)}$$

$$= \frac{0{\cdot}01\alpha^2}{1-\alpha} = 0{\cdot}01\alpha^2$$

[Taking 1 – α = 1, as α is negligible as compared to 1]

= 0·01 × (0·0423)^2

K_d = 1·86 × 10^{-5}

Hence, the dissociation constant (K_d) for acetic acid is 1·86 × 10^{-5}.

(b) (i) $Mg_{(s)} + Cu^{2+}$ (0·0001M) × Mg^{2+} (0·001 M) + Cu

By Nernst equation the EMF of the cell at 298 K is :

$$E = E^\circ - \frac{0{\cdot}0591}{2} \log \frac{[Mg^{2+}]}{[Cu^{2+}]}$$

$$= 2{\cdot}71 - \frac{0{\cdot}0591}{2} \log \frac{0{\cdot}001}{0{\cdot}0001}$$

= 2·71 – 0·02955 = 2·68 V

Thus, the EMF of the cell is 2·68V

SECTION–B

Answer 5.

(a) (i) tetraamminediaquachromium(III) chloride

(ii) tetraamminedichloroplatinum(IV) tetrachloroplatinate(II).

(b) Hybridization of $[Fe(CN)_6]^{3-}$: d^2sp^3

Magnetic property : Due to presence of one unpaired electrons in 3*d* shell, $[Fe(CN)_6]^{3-}$ ion is paramagnetic in nature.

(c) (i) $[Co(NH_3)_5Br]SO_4$ and

I

$[Co(NH_3)_5SO_4]Br$ are Ionization isomer.

II

When the two isomers are treated with slightly acidic $BaCl_2$ solution

$[Co(NH_3)_5Br]SO_4$ produces white ppt. ($BaSO_4$) but $[Co(NH_3)_5SO_4]Br$ does not.

(ii) Optical Isomers of $[Co(en)_2Cl_2]^+$

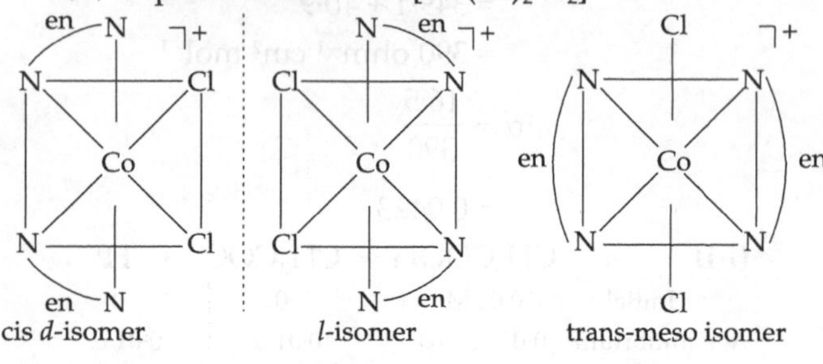

Answer 6.

(a) (i) $2F_2 + 2NaOH \rightarrow OF_2 + 2NaF + H_2O$

Fluorine, Sodium hydroxide (cold and dil.), Oxygen difluoride, Sodium fluoride, Water

(ii) $2KMnO_4 + 5H_2O_2 + 3H_2SO_4 \rightarrow$

Potassium permangnate (Violet), Hydrogen Peroxide, Sulphuric acid

$2MnSO_4 + 8H_2O + 5O_2 + K_2SO_4$

Manganese sulphate (colourless)

(iii) $H_2SO_4 + H_2S \rightarrow 2H_2O + SO_2\uparrow + S$

Sulphuric acid, Hydrogen sulphide, Water, Sulphur dioxide, Sulphur

(b) Structure of Xenon tetrafluoride :

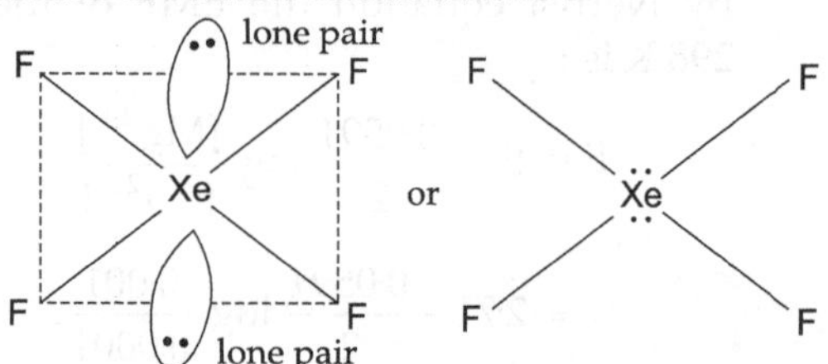

Hybridization of the central atom : sp^3d^2

Geometry of XeF_4 molecule : Square planar

Answer 7.

(a) Refer to ISC Paper 2020, Answer 10.

(b) (i) The transition metal ions possess a great tendency to form complexes because of their small size, high ionic charge and availability of partially filled *d*-orbitals for bond formation. These partially filled *d*-orbitals can easily accommodate ligands electrons and consequently transition metals ions form complexes. Examples $[Fe(CN_6)]^{3-}$, $Ni(CO)_4$ etc.

(ii) In the 3*d* transition series as we move from Sc (21) to Mn (25), the number of unpaired electrons increases and hence paramagnetic character increases. After Mn, the pairing of electrons in the *d*-orbital starts and the number of unpaired electrons decreases and hence, paramagnetic character decreases.

SECTION–C

Answer 8.

(a) (i)

$$\underset{\text{Glycerol}}{\begin{matrix}CH_2OH\\|\\CHOH\\|\\CH_2OH\end{matrix}} + \underset{\text{Oxalic acid}}{\begin{matrix}COOH\\|\\COOH\end{matrix}} \xrightarrow[\Delta]{110°C}$$

$$\underset{\text{Glycerol mono-oxalate}}{\begin{matrix}CH_2O\cdot OC\cdot COOH\\|\\CHOH\\|\\CH_2OH\end{matrix}} \xrightarrow[383K]{-CO_2} \begin{matrix}CH_2O\cdot OCH\\|\\CHOH\\|\\CH_2OH\end{matrix}$$

$$\xrightarrow{+H_2O} \begin{matrix}CH_2OH\\|\\CHOH\\|\\CH_2OH\end{matrix} + \underset{\text{Formic acid}}{HCOOH}$$

(ii)

$$\underset{\text{Chloro benzene}}{C_6H_5Cl} + NaOH \xrightarrow[\substack{320atm\\-HCl}]{\substack{Cu\text{-catalyst}\\623K}} C_6H_5O^-Na^+ \xrightarrow[-NaCl]{HCl} \underset{\text{Phenol}}{C_6H_5OH}$$

(iii) $\underset{\text{Diethyl ether}}{C_2H_5OC_2H_5} + H_2O \xrightarrow[\text{Super heated steam}]{} \underset{\text{Ethanol}}{2C_2H_5OH}$

(iv) $\underset{\text{(Phenol)}}{C_6H_5OH} \xrightarrow[(\Delta)]{\text{Zn dust}} \underset{\text{(Benzene)}}{C_6H_6} \xrightarrow[(Conc.H_2SO_4)\ 60°]{Conc.HNO_3}$

$\underset{\text{(Nitrobenzene)}}{C_6H_5NO_2} \xrightarrow[(6[H])]{Sn/HC} \underset{\text{(Aniline)}}{C_6H_5NH_2}$

(b) (i) When ethanol is warmed (60°C) with iodine in the presence of an alkali (NaOH or Na_2CO_3), iodoform is obtained

$$\underset{\text{Ethanol}}{C_2H_5OH} + 4I_2 + 2Na_2CO_3 \longrightarrow \underset{\text{Iodoform}}{CHI_3} + HCOONa + CO_2 + 5NaI + 5H_2O$$

(ii) The product formed will be biphenyl $[C_6H_5-C_6H_5]$ and sodium chloride (NaCl)

Wurtz Fittig reaction.

(c) $\underset{\text{(Acetone)}}{CH_3COCH_3} \xrightarrow[(O)]{\text{conc. }HNO_3} \underset{\substack{\text{(Acetic acid)}\\(A)}}{CH_3COOH} \xrightarrow{SOCl_2}$

$$\underset{\substack{\text{(Acetylchloride)}\\ \text{(B)}}}{CH_3COCl} \xrightarrow{NH_3} \underset{\substack{\text{(Acetamide)}\\ \text{(C)}}}{CH_3CONH_2} \xrightarrow{LiAlH_4}$$

$$\underset{\substack{\text{(Ethylamine)}\\ \text{(D)}}}{CH_3CH_2NH_2} \xrightarrow{NHO_2} \underset{\substack{\text{(Ethyl Alcohol)}\\ \text{(E)}}}{CH_3CH_2OH} \xrightarrow{CH_3COCl}$$

$$\underset{\substack{\text{(Ethyl acetane)}\\ \text{(F)}}}{CH_3COOC_2H_5}$$

Answer 9.

(a) (i) Phenol + $CHCl_3 + 3NaOH \xrightarrow{\Delta}$ Salicylaldehyde (o-hydroxyl benezaldehyde) + $3NaCl + 2H_2O$

(ii) Rosenmund's reduction : It is a method for preparation of aldehyde by catalytic hydrogenation of acid chlorides. During reaction, H_2 is passed over acid chloride in presence of Pd-$BaSO_4$ partially poisoned by sulphur or quinoline. For example.

$$\underset{\text{acetyl chloride}}{CH_3-\overset{\overset{O}{||}}{C}-Cl} + H_2 \xrightarrow[S]{Pd-BaSO_4} \underset{\text{acetaldehyde}}{CH_3-\overset{\overset{O}{||}}{C}-H} + HCl$$

(iii) $$\underset{\text{Acetamide}}{CH_3CONH_2} + Br_2 + 4KOH \xrightarrow{\text{heat}} \underset{\text{Methylamine}}{CH_3NH_2} + K_2CO_3 + 2KBr + 2H_2O$$

(b) (i) These can be distinguished by the carbylamine test. Ethylamine is primary (1°) amine therefore, it gives carbylamine test *i.e.*, when heated with an alcoholic solution of KOH and $CHCl_3$, it gives an unpleasant foul smell of isocyanide which is easily detectable.

$$\underset{\text{Ethylamine}}{C_2H_5NH_2} + \underset{\text{Trichloromethane}}{CHCl_3} + \text{alc. } KOH \longrightarrow \underset{\text{Ethyl isocyanide}}{C_2H_5NC} + 3KCl + 3H_2O$$

On the other hand, diethylamine is a secondary (2°) amine and hence does not give this test.

(ii) When acetaldehyde is heated with Fehling's solution, red precipitate of cuprous oxide is formed.

$$\underset{\text{Acetaldehyde}}{CH_3CHO} + \underset{\substack{\downarrow\\ \text{Fehlings solution}}}{2CuO} \longrightarrow \underset{\text{(Red ppt.)}}{Cu_2O(s)\downarrow} + \underset{\text{Acetic Acid}}{CH_3COOH}$$

On the other hand, benzaldehyde does not give red precipitate with Fehling's solution.

(c) (i) $CH_3-\ddot{N}H_2$ (I); Aniline, $C_6H_5\ddot{N}H_2$ (II); $C_2H_5\ddot{N}H_2$ (III)

I: + I effect increases availability of lone pair on Nitrogen

II: + M effect decreases lone pair availability on Nitrogen

Basicity order :

Increasing order of basic strength

Diethylether < Aniline < methylamine < ethylamine

Reason : This is due to the + I effect of the alkyl group and aromatic amines are weaker bases than aliphatic amines because the one pair of electron is in the conjugation in aromatic amines and are not donated easily due to the steric effects of alkyl groups.

(ii) (1) Monomer :

$HOOC-C_6H_4-COOH$ — benzene-1, 4-dicarboxylic acid (terephthalic acid)

$HO-CH_2-CH_2-OH$ — ethane-1, 2-diol (ethylene glycol)

Type of polymerisation : Condensation Polymerisation

(2) Phenol (C_6H_5OH) and Formaldehyde ($H-\overset{\overset{O}{||}}{C}-H$)

Type of polymerisation : Condensation Polymerisation

Answer 10.

(a)

$$\underset{(C_2H_7N)}{A} \xrightarrow{HNO_2} B + D \xrightarrow{[H^+]} E$$

$B \xrightarrow{[O]} C$; $C \xrightarrow{\text{Tollens Reagent}} D$

A : $C_2H_5NH_2$ (ethylamine)

B : C_2H_5OH (ethanol)

C : CH_3CHO (ethanal)

D : CH_3COOH (ethanoic acid)

E : $CH_3COOC_2H_5$ (ethyl acetate)

$$\underset{\text{Acetaldehyde (C)}}{CH_3C(H)=O} + \underset{\text{Ammonia}}{NH_3} \longrightarrow \underset{\text{Acetaldehyde ammonia}}{CH_3-C(OH)(H)-NH_2}$$

(b) (i) $\underset{\text{(Ethylamine)}}{C_2H_5NH_2} \xrightarrow{HONO} C_2H_5OH$

$$\xrightarrow[{[O]}]{K_2Cr_2O_7/H_2SO_4} CH_3CHO \xrightarrow{[O]}$$

$$CH_3COOH \xrightarrow{NH_3} CH_3COONH_4 \xrightarrow{\Delta}$$

$$CH_3CONH_2 \xrightarrow{B_2/KOH} \underset{\text{(Methylamine)}}{CH_3NH_2}$$

(ii) Proteins are the long-chain polymer of amino acids. Thus, upon denaturation amino acids are produced. During denaturation, secondary and tertiary structures of proteins are destroyed but primary structures remain as such and again the globular proteins are converted into fibrous proteins and their biological activity is lost.

$$\underset{\text{Protein}}{H_2N-CH(R)-\overset{O}{\overset{\|}{C}}NH-CH(R')-COOH} \xrightarrow{\Delta} \underset{\text{Amino acid}}{H_2N-CH(R)COOH}$$

(iii) Nitrogen base residues present in DNA are :

(a) Adenine (b) Guanine

(c) Thymine (d) Cytosine

(c) (i) $\underset{\text{Aniline}}{C_6H_5NH_2} \xrightarrow[0-5°C]{HNO_2 + \text{conc. } HCl} \underset{\text{Benzene diazonium chloride}}{C_6H_5N_2^+Cl^-}$

(ii) $\underset{\text{Acetylchloride}}{CH_3COCl} + C_2H_5OH \longrightarrow \underset{\text{Ethylacetate}}{CH_3COOC_2H_5} + HCl$

(iii) $\underset{\text{Formaldehyde}}{6HCHO} + 4NH_3 \longrightarrow \underset{\text{Urotropine}}{(CH_2)_6N_4} + 6H_2O$

●●

CHEMISTRY 2014

QUESTIONS

PART–I (20 Marks)

*Answer **all** questions.*

Question 1.

(a) *Fill in the blanks by choosing the appropriate word/ words from those given in the brackets :* **[5]**

(increases, formic acid, decreases, less, zero, small, paired, atoms, unpaired, ions, pentagonal bipyramidal, electrical, more, ethylamine, molecules, propanoic acid, methylamine, chemical)

(i) *An electrochemical cell converts ______ energy to ______ energy.*

(ii) *The crystal of graphite is made up of ______ while that of sodium chloride is made up of ______ .*

(iii) *Ethyl isocyanide, on hydrolysis with dilute sulphuric acid, gives ______ and ______ .*

(iv) *The molar conductance of a solution ______ with dilution, while its specific conductance ______ with dilution.*

(v) *The van't Hoff factor of acetic acid solution is ______ than one and the value of normal colligative property is ______ than the observed colligative property of this solution.*

(b) *Complete the following statements by selecting the* ***correct alternative*** *from the choices given :* **[5]**

(i) *Of the following terms used for denoting concentration of a solution, the one which does not get affected by temperature is :*

(1) *Molarity* **(2)** *Molality*
(3) *Normality* **(4)** *Formality*

(ii) *The solubility of calcium hydroxide is s mol litre^{-1}. The solubility product under the same condition will be :***

(1) $4s^3$ **(2)** $2s^3$
(3) $2s^2$ **(4)** s^3

(iii) *A current liberates 0·50 g of hydrogen in 2 hours. The weight of copper (At. wt. = 63·5) deposited at the same time by the same current through copper sulphate solution is :*

(1) *63·5 g* **(2)** *31·8 g*
(3) *15·9 g* **(4)** *15·5 g*

(iv) *Natural rubber is a :*

(1) *Polyester* **(2)** *Polyamide*
(3) *Polyisoprene* **(4)** *Polysaccharide*

(v) *Among the following halogens, the one which does not form an oxyacid is :*

(1) *Fluorine* **(2)** *Chlorine*
(3) *Bromine* **(4)** *Iodine*

(c) *Answer the following questions :* **[5]**

(i) *What is the [OH⁻] concentration of an acid whose pH is 5 at 25°C ?***

(ii) *What happens when a nickel rod is dipped into a copper sulphate solution ? Justify your answer.*

$$\left[E^0_{Ni^{+2}/Ni} = -0.25\text{ V and } E^0_{Cu^{+2}/Cu} = +0.34\text{ V}\right]$$

(iii) *Write the equation for the preparation of acetanilide from aniline.***

(iv) *Define Raoult's law for the elevation of boiling point of a solution.*

(v) *An ionic compound is made up of A cations and B anions. If A cations are present at the alternate corners and B anion is present on the body of the diagonal, what is the formula of the ionic compound ?*

(d) *Match the following :* **[5]**

(i)	*Molal depression constant*	**(a)**	*Infinite dilution*
(ii)	*Acetaldehyde*	**(b)**	*mol l^{-1} sec^{-1}*
(iii)	*Rate of reaction*	**(c)**	*Iodoform*
(iv)	*Optical activity*	**(d)**	*K kg mol^{-1}*
(v)	*Kohlrausch's law*	**(e)**	*Lactic acid*

PART–II (50 Marks)

*Answer **six** questions choosing **two from Section A, two from Section B and two from Section C.***

SECTION–A

*Answer any **two** questions.*

Question 2.

(a) (i) *What will be the vapour pressure of a solution containing 5 moles of sucrose ($C_{12}H_{22}O_{11}$) in 1 kg of water, if the vapour pressure of pure water is 4·57 mm of Hg ? [C = 12, H = 1, O = 16]* **[3]**

(ii) *A 2 molal solution of sodium chloride in water causes an elevation in the boiling point of water by 1·88 K. What is the value of van't Hoff factor ? What does it signify ?*
[K_b = 0·52 K kg mol^{-1}] **[2]**

(b) (i) *Write the mathematical expression relating the variation of rate constant of a reaction with temperature.* **[4]**

(ii) *How can you graphically find the activation energy of the reaction from the above expression ?*

** Answer is not given due to change in the present syllabus.

(iii) *The slope of the line in the graph of log k(k = rate constant) versus is – 5841. Calculate the activation energy of the reaction.*

(c) *Define Frenkel defect in solid crystal.* [1]

Question 3.

(a) *Explain giving reasons why :* [4]

(i) *Ionic solids conduct electricity in molten state, but not in solid state.*

(ii) *Solution of sodium chloride has no effect on litmus, but a solution of zinc chloride turns blue litmus red.***

(b) *In a crystal of diamond :* [2]

(i) *How many carbon atoms are present per unit cell ?*

(ii) *What type of lattice does diamond crystallize in ?*

(iii) *How many carbon atoms surround each carbon atom ?*

(iv) *How are they arranged ?*

(c) (i) *What is standard hydrogen electrode ?* [1]

(ii) *0·05 M NaOH solution offered a resistance of 31·6 ohm in a conductivity cell at 298 K. If the cell constant of the cell is 0·367 cm^{-1}, calculate the molar conductivity of the NaOH solution.* [3]

Question 4.

(a) (i) *K_c for the reaction $SO_{2(g)} \frac{1}{2} + O_{2(g)} \rightleftharpoons SO_{3(g)}$ is 61·7 at 60°C. What is its unit? Calculate K_p for the reaction and write its unit.*** [3]

(ii) *What happens to the equilibrium in a reversible reaction if a catalyst is added to it ?*** [1]

(b) *State the effect of the following on the reaction $2SO_{2(g)} + O_{2(g)} \rightleftharpoons 2SO_{3(g)} + 189{\cdot}4$ kJ at equilibrium :*** [2]

(i) *Temperature is increased.*

(ii) *Concentration of SO_2 is increased.*

(iii) *Pressure is decreased.*

(iv) *Helium is added at constant pressure.*

(c) (i) *0·3605 g of a metal is deposited on the electrode by passing 1·2 amperes of current for 15 minutes through its salt solution. The atomic weight of the metal is 96. What is the valency of the metal ?* [3]

(ii) *Explain why phenolphthalein is used as an indicator in acid-base titration.*** [1]

SECTION–B

*Answer any **two** questions.*

Question 5.

(a) *Write the formula of the following compounds :* [2]

(i) *Triamminetriaquachromium (III) chloride.*

(ii) *Potassiumhexacyanoferrate (III).*

(b) *Name the types of isomerism shown by the following pairs of compounds :* [2]

(i) *$[CoCl(H_2O)(NH_3)_4Cl_2$ and $[CoCl_2(NH_3)_4]Cl.H_2O$*

(ii) *$[Pt(NH_3)_4][PtCl_6]$ and $[Pt (NH_3)_4Cl_2][PtCl_4]$*

(c) *For the complex ion of $[Co(NH_3)_6]^{3+}$:* [1]

(i) *State the hybridisation of the complex.*

(ii) *State the magnetic nature of the complex.*

Question 6.

(a) *Write balanced chemical equations for the following reactions :* [3]

(i) *Ozone and lead sulphide.*

(ii) *Chlorine is passed through hot concentrated NaOH solution.*

(iii) *Sulphuric acid is treated with phosphorus.*

(b) *Give reasons for the following :* [2]

(i) *Zn^{+2} salts are white but Cu^{2+} salts are blue in colour.*

(ii) *Fluorine gives only one oxide but chlorine gives a series of oxides.*

Question 7.

(a) *How is potassium dichromate prepared from a sample of chromite ore ? Give balanced equations for the chemical reactions inolved.* [3]

(b) *For the molecule IF_7 :*

(i) *Draw the structure of the molecule.* [2]

(ii) *State the hybridisation of the central atom.*

(iii) *State the geometry of the molecule.*

SECTION–C

*Answer any **two** questions.*

Question 8.

(a) *How can the following conversions be brought about :*

(i) *Acetic acid to methyl cyanide.* [2]

(ii) *Acetaldehyde to formaldehyde.* [3]

(iii) *Nitrobenzene to 2, 4, 6- tribromoaniline.* [2]

(b) *Identify the reagents A, B, C, D, E and F required for the following conversion :* [3]

$$C_6H_5NO_2 \xrightarrow{A} C_6H_5NH_2 \xrightarrow{B} C_6H_5N_2^+Cl^- \xrightarrow{C}$$

$$C_6H_5Cl \xrightarrow{D} C_6H_5OH \xrightarrow{H_2SO_4} E + F$$

Question 9.

(a) *The deficiency of which vitamin will cause the following diseases :* [2]

(i) *Scurvy* (ii) *Haemorrhages*

(b) *Give one chemical test to distinguish between the following pairs of compounds :* [3]

(i) *Ethanol and 2 propanol.*

(ii) *Aniline and ethylamine.*

(c) *Write the structures of all enantiomers possible for lactic acid.*** [1]

** Answer is not given due to change in the present syllabus.

(d) *Give balanced equations for the following reactions :* **[4]**

(i) *Acetaldehyde is heated with hydroiodic acid in the presence of red phosphorous.*

(ii) *Calcium acetate is subjected to dry distillation.*

(iii) *Sodium ethoxide is treated with ethyl bromide.*

(iv) *Benzaldehyde is treated with sodium bisulphite.*

Question 10.

(a) *An organic compound A with molecular formula C_7H_8 on oxidation by chromylchloride in the presence of CCl_4 gives a compound B which gives positive tollen's test. The compound B on treatment with NaOH followed by acid hydrolysis gives two products C and D. C on oxidation gives B which on further oxidation gives D. The compound D on distillation with sodalime gives a hydrocarbon E. Below 60°C, concentrated nitric acid reacts with E in the presence of concentrated sulphuric acid forming a compound F. Identify the compounds A, B, C, D, E and F.* **[3]**

(b) *Give balanced equations for the following name reactions :* **[3]**

(i) *Clemmensen's reduction.*

(ii) *Kolbe's electrolytic reaction.*

(iii) *Balz-Schiemann's reaction.*

(c) (i) *What do you observe when glucose is treated with bromine water ?* **[2]**

(ii) *What is isoelectric point ?*

(d) *Answer the following :*** **[2]**

(i) *What is biuret test ?*

(ii) *Write balanced equation for the formation of biuret.*

ANSWERS

PART–I

Answer 1.

(a) (i) chemical, electrical
(ii) atoms, ions
(iii) ethylamine, formic acid (or vice versa)
(iv) increases, decreases
(v) more, less (or vice versa)

(b) (i) (2) Molality (iii) (3) 15·9 g.
(iv) (3) Polyisoprene (v) (1) Fluorine.

(c) (ii) Red copper metal starts precipitating.

This is because $E^\circ_{Ni^{2+}/Ni}$ is less than $E^\circ_{Cu^{2+}/Cu}$.
So Ni displaces Cu^{+2} ions from the solution.

$$Ni + Cu^{+2}_{(aq)} \longrightarrow Ni^{+2}_{(aq)} + \underset{\text{metal}}{Cu}$$

(iv) According to Raoults law : 'Elevation of boiling point of a solution is directly proportional to the lowering in vapour pressure caused by the number of particles of solute present in the solution'.

(v) Formula of ionic compound is $A^{2+}B^-$ or AB_2.

(d) (i) Molal depression constant (a) K kg mol^{-1}
(ii) Acetaldehyde (c) Iodoform
(iii) Rate of reaction (b) mol l^{-1} sec^{-1}
(iv) Optical activity (e) Lactic acid
(v) Kohlrausch's law (a) Infinite dilution

PART–II
SECTION–A

Answer 2.

(a) (i) According to Raoults law :

$$P_{solution} = P^{\infty}_{solvent} \times X_{solvent}$$

$$= 4{\cdot}57 \times \frac{\frac{1000}{18}}{\frac{1000}{18}+5}$$

$$= 4{\cdot}57 \times \frac{55{\cdot}55}{55{\cdot}55+5}$$

$$= 4{\cdot}192 \text{ mm of Hg.}$$

(ii) Elevation in boiling point ΔT_b is a colligative property and

$$\Delta T_b = i \times K_b \times \text{Molality}$$

Given : $\Delta T_b = 1{\cdot}88$ K.

$K_b = 0{\cdot}52$ K kg Mol^{-1}

Molality = 2

$$\therefore \quad i = \frac{1{\cdot}88}{0{\cdot}52 \times 2}$$

$$i = 1{\cdot}8076 \approx 2$$

Value of '*i*' in this case signifies that when 1 mole of NaCl is dissolved in water it dissociates and furnishes 1·8076 moles of particles in the solution.

(b) (i) Mathematical expression is Arrhenius equation :

$$k = Ae^{-Ea/RT}$$

where k is rate constant of the given reaction
A is another constant called frequency factor.
E_a is activation energy
R is gas constant and
T is temperature in Kelvin.

(ii) Finding activation energy graphically :
Log form of Arrhenius equation is

$$\log_e k = \log_e A - \frac{E_a}{RT}$$

** Answer is not given due to change in the present syllabus.

or $\log_{10} k = \log_{10} A - \dfrac{E_a}{2 \cdot 303RT}$

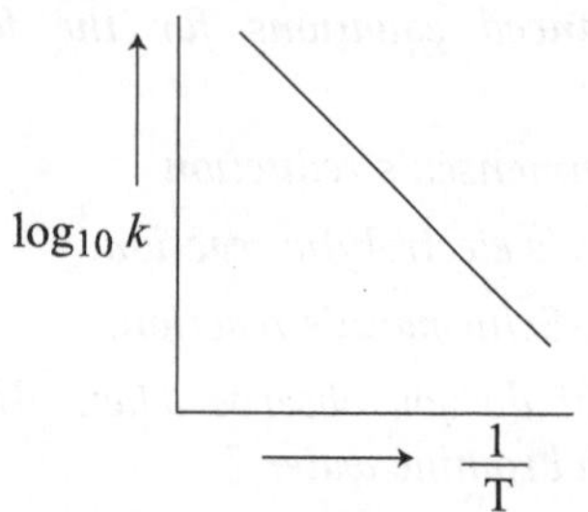

It is a straight line equation ($y = mx + c$) when different values of rate constant k are calculated at different temperatures for a given reaction and if a graph is plotted between $\log_{10} k$ against $\dfrac{1}{T}$ then slope of the graph line should be equal to $-\dfrac{E_a}{2 \cdot 303R}$

Using $\dfrac{dy}{dx} = -\dfrac{E_a}{2 \cdot 303R}$

value of E_a can be calculated

(iii) $\text{Slope} = -\dfrac{E_a}{2 \cdot 303R}$

or $-E_a = -5841 \times 2 \cdot 303 \times 8 \cdot 314$

$E_a = 111838 \cdot 45 \text{ J mol}^{-1}$

$= 111 \cdot 838 \text{ kJ mol}^{-1}$

(c) Frenkel defect : This defect arises in a crystal when an ion, generally cation, leaves its lattice site and occupies an interstitial site so that electrical neutrality as well as stoichiometry is maintained.

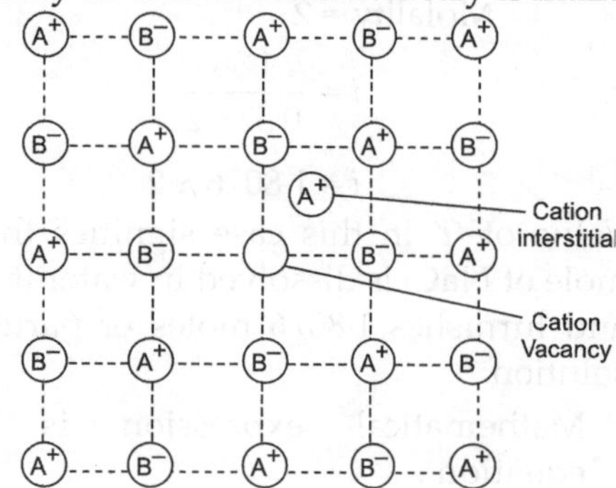

Frenkel defect in crystal [$A^+ B^-$]

Answer 3.

(a) (i) To conduct electricity ions of a substance should be free to move to oppositely charged electrodes. In solid state of a substance (ionic compound) ions are not free to move but in molten (liquid) state they are free to move.

(b) (i) Eight carbon atoms are present per unit cell.

(ii) Diamond crystallizes in a network type of lattice with fcc arrangement of carbon atoms.

(iii) Four carbon atoms surround each carbon atom.

(iv) Four carbon atoms are tetrahedrally arranged around each central carbon atom

Tetrahedral arrangement

(c) (i) Standard hydrogen electrode : It is a reference electrode against which the electrode potentials of all electrodes are measured.

When hydrogen gas at 1 atm pressure is adsorbed over platinum electrode dipped in 1M HCl at 25°C, it is standard hydrogen electrode and its potential is, $E^\circ = \pm 0$ volt.

A standard hydrogen electrode can be represented as : Pt, $H_2(g)$ (1 atm)/H^+ (1 mol) at 25°C

(ii) Given : Molarity of NaOH solution = 0·05 m

Resistance = 31·6 ohm

Cell constant = $0 \cdot 367 \text{ cm}^{-1}$

Now, Molar conductivity

$= \dfrac{\text{Specific conductivity} \times 1000}{\text{Molarity}}$

$= \dfrac{\text{Cell constant}}{\text{Resistance}} \times \dfrac{1000}{\text{Molarity}}$

$= \dfrac{0 \cdot 367 \times 1000}{31 \cdot 6 \times 0 \cdot 05}$

$\Lambda_m = 232 \cdot 3 \text{ ohm}^{-1} \text{ cm}^2 \text{ mol}^{-1}$

Answer 4.

(c) (i) Given : weight 'w' of the metal = 0·3605 g

Current = 1·2 amp.

Time = 15 min = 15 × 60 sec.

Atomic weight of metal = 96

According to Faraday's first law of electrolysis :

Electrochemical equivalent

$Z = \dfrac{w}{I \times t}$

$= \dfrac{0 \cdot 3605}{1 \cdot 2 \times 15 \times 60} = 3 \cdot 34 \times 10^{-4} \text{ g c}^{-1}$

$\text{Valency} = \dfrac{\text{Atomic wt.}}{\text{Equivalent wt.}}$

$= \dfrac{96}{3 \cdot 34 \times 10^{-4} \times 96500} = 2 \cdot 98 \approx 3$

($\because$ Equivalent wt = Z × F)

$\therefore$ Valency of metal is 3.

SECTION–B

Answer 5.

(a) (i) $[Cr(NH_3)_3(H_2O)_3]Cl_3$

(ii) $K_3[Fe(CN)_6]$

(b) (i) Hydrate isomerism

(ii) Coordination isomerism.

(c) (i) Complex has d^2sp^3 hybridisation.

(ii) Complex is diamagnetic in nature.

Answer 6.

(a) (i) $\underset{\text{Lead sulphide}}{PbS} + \underset{\text{Ozone}}{4O_3} \longrightarrow \underset{\text{Lead sulphate}}{PbSO_4} + 4O_2$

(ii) $\underset{\text{Hot conc.}}{6NaOH + 3Cl_2} \longrightarrow \underset{\text{Sodium chlorate}}{NaClO_3} + 5NaCl + 3H_2O$

(iii) $P_4 + 10H_2SO_4 \longrightarrow \underset{\text{Phosphoric acid}}{2H_3PO_4} + 10SO_2 + 4H_2O$

(b) (i) Zinc and copper are transition metals and their ions are generally coloured due to *d-d* transition of unpaired electron.

But in case of Zn^{2+} 3*d* is completely filled and no unpaired electrons are available for transition. So light fallin on these ions gets completely transmitted and their salts appear white. However ion Cu^{2+} 3*d* has 9 electrons. The unpaired electron absorbs particular wavelengths in visible region of light and the transmitted light shows the complementary.

(ii) Fluorine has only two orbits. In 2nd orbit it has s^2p^5 configuration with no vacant '*d*' orbitals. So it cannot form more than one oxide *i.e.*, OF_2.

Chlorine, on the other hand, has 3 orbits in its atoms and 3*d* is lying vacant in it. Therefore, it can show variable covalency and gives a series of oxides. Like Cl_2O, ClO_2 etc.

Cl atom in ground state

3*s* 3*p* 3*d*

Cl atom in excited state

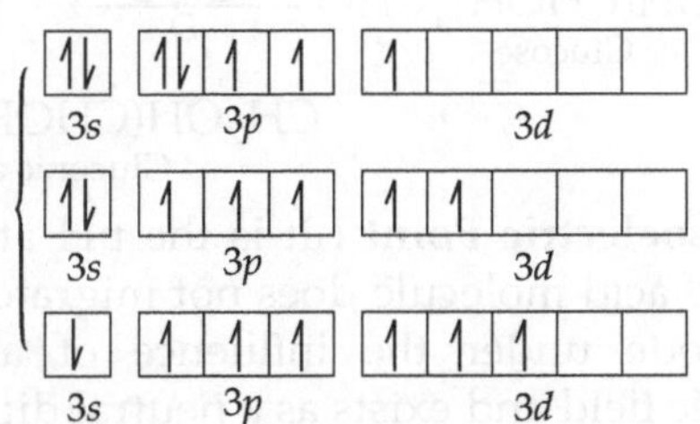

Answer 7.

(a) Preparation of potassium dichromate from chromite ore involves following steps :

1. Conversion of chromite ore to sodium chromate by roasting finely powdered chromite ore with sodium carbonate and quick lime in excess of air.

$$\underset{\text{Chromite ore}}{4FeCr_2O_4} + \underset{\text{Sodium carbonate}}{8Na_2CO_3} + 7O_2 \xrightarrow{CaO + air} \underset{\text{Ferric oxide}}{2Fe_2O_3} + \underset{\text{Sodium chromate}}{8Na_2CrO_4} + 8CO_2$$

2. Conversion of sodium chromate to sodium dichromate by the action of conc. H_2SO_4

$$2Na_2CrO_4 + \underset{\text{conc.}}{H_2SO_4} \longrightarrow \underset{\text{Sodium dichromate}}{Na_2Cr_2O_7} + Na_2SO_4 + H_2O$$

3. Conversion of sodium dichromate to potassium dichromate by treating it with KCl solution.

$$Na_2Cr_2O_7 + 2KCl \longrightarrow \underset{\text{Potassium dichromate}}{K_2Cr_2O_7} + 2NaCl$$

From resulting solution, potassium dichromate is crystallised.

(b) (i)

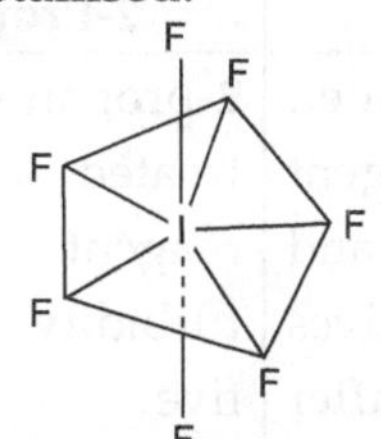

Structure of IF_7

(ii) Hybridisation of I is sp^3d^3.

(iii) Geometry is pentagonal bipyramidal.

SECTION–C

Answer 8.

(a) (i) $\underset{\text{Acetic acid}}{CH_3COOH} + NH_3 \rightarrow \underset{\text{Ammonium acetate}}{CH_3COONH_4} \xrightarrow[-H_2O]{\Delta}$

$$\underset{\text{Acetamide}}{CH_3CONH_2} \xrightarrow[\Delta]{P_2O_3} H_2O + \underset{\text{Methyl Cyanide}}{CH_3CN}$$

(ii) $\underset{\text{Acetaladehyde}}{CH_3CHO} \xrightarrow{(O)} \underset{\text{Acetic acid}}{CH_3COOH} \xrightarrow[\Delta(-H_2O)]{+NH_3}$

$$\underset{\text{Acetamide}}{CH_3CONH_2} \xrightarrow{Br_2/KOH} \underset{\text{Methylamine}}{CH_3NH_2}$$

$$\xrightarrow{NaNO_2 + HCl} \underset{\text{Methyl alcohol}}{CH_3OH} \xrightarrow{(O)} \underset{\text{Formaldehyde}}{HCHO}$$

(iii) $C_6H_5NO_2$ (Nitrobenzene) $+ 6H \xrightarrow[-2H_2O]{Sn/HCl}$ $C_6H_5NH_2$ (Aniline) $\xrightarrow[-3HBr]{3Br_2}$ 2, 4, 6, tribromoaniline

(b) A ⟶ Sn + conc. HCl.
B ⟶ $NaNO_2$ + conc. HCl (0°C – 5° C)
C ⟶ $CuCl_2$ + HCl
D ⟶ NaOH/HCl

E = (benzene ring with OH and SO_3H ortho) 2-hydroxy benzene sulphonic acid.

F = (benzene ring with OH and SO_3H para) 4-hydroxy benzene sulphonic acid.

Answer 9.

(a) (i) Vitamin C (ii) Vitamin K.

(b) (i)

Ethanol	2-Propanol
Ethanol when treated with Lucas reagent (anhy. $ZnCl_2$ and conc. HCl) gives turbidity after heating.	2-propanol when treated with Lucas reagent gives turbidity within five minutes without heating.

(ii) Aniline and ethylamine can be distinguished using azodye test.

Aniline	Ethylamine
Aniline reacts with HNO_2 ($NaNO_2$ + dil. HCl) at 0 – 5°C to give benzene diazonium chloride, which on further heating with β-naphthol gives red dye.	Ethylamine gives a brisk effervescence due to evolution of N_2 gas under similar conditions.

(d) (i)

$$\underset{\text{Acetaldehyde}}{CH_3CHO} + 4HI \xrightarrow[\Delta]{\text{red P}} \underset{\text{Ethane}}{CH_3{\cdot}CH_3} + H_2O + 2I_2$$

(ii) $$\underset{\text{Calcium acetate}}{(CH_3COO)_2Ca} \xrightarrow{\Delta} \underset{\text{Acetone}}{CH_3COCH_3} + CaCO_3$$

(iii) $$\underset{\text{Sodium ethoxide}}{C_2H_5ONa} + \underset{\text{Ethyl bromide}}{C_2H_5Br} \longrightarrow \underset{\text{Diethyl ether}}{C_2H_5OC_2H_5} + NaBr$$

(iv) $$\underset{\text{Benzaldehyde}}{C_6H_5CHO} + \underset{\text{Sodium bisulphite}}{NaHSO_3} \longrightarrow \underset{\text{Benzaldehyde sodium bisulphite}}{C_6H_5-CH(OH)-SO_3Na}$$

Answer 10.

(a) A → $C_6H_5CH_3$, Toluene B → C_6H_5CHO, Benzaldehyde
C → $C_6H_5CH_2OH$ Benzyl alcohol D → C_6H_5COOH, Benzoic acid
E → C_6H_6, Benzene F → $C_6H_5NO_2$ Nitrobenzene

(b) (i) **Clemmensen's reduction** : It is a reaction in which reduction of Aldehyde to alkane takes place using Zn/Hg and HCl.

$$\underset{\text{Acetaldehyde}}{CH_3CHO} + 4H \xrightarrow[+HCl]{Zn/Hg} \underset{\text{Ethane}}{CH_3-CH_3} + H_2O$$

(ii) **Kolbe's electrolytic reaction** : In this reaction aqueous solution of sodium salt of carboxylic acid is electrolyzed and alkane is formed at anode.

$$CH_3COONa_{(aqu.)} \rightleftharpoons CH_3COO^- + Na^+$$
$$2H_2O \longrightarrow 2OH^- + 2H^+$$

At anode :

$$2CH_3COO^- - e^- \longrightarrow 2CH_3CO \xrightarrow{-CO_2} \underset{\text{(Ethane)}}{CH_3-CH_3}$$

At cathode : $2H_2O + 2e^- \longrightarrow 2OH^- + H_2$

(iii) **Balz-Schiemann reaction** : In this reaction, aniline is first converted to benzene diazonium fluoroborate which on heating gives fluorobenzene.

$$\underset{\text{Aniline}}{C_6H_5NH_2} \xrightarrow[273-278k]{NaNO_2 / HBF_4} \underset{\text{Benzene-diazonium Fluoroborate}}{C_6H_5N_2BF_4} \xrightarrow{\Delta} \underset{\text{Fluorobenzene}}{C_6H_5F} + N_2 + BF_3$$

(c) (i) When glucose is treated with bromine water, Bromine water gets decolourised.

$$\underset{\text{Glucose}}{CH_2OH(CHOH)_4CHO} \xrightarrow[D]{Br_2/H_2O} \underset{\text{Gluconic acid}}{CH_2OH(CHOH)_4COOH}$$

(ii) **Isoelectric Point** : It is the pH at which an amino acid molecule does not migrates to either electrode under the influence of an applied electric field and exists as a neutral dipolar ion.

••

CHEMISTRY 2013

QUESTIONS

PART–I (20 Marks)

*Answer **all** questions.*

Question 1.

(a) *Fill in the blanks by choosing the appropriate word/ words from those given in the brackets :* **[5]**

(zero, first, second, increased, decreased, anode, cathode, active, inactive, potassium cyanide, internal, external, dependent, independent, red, benzoic acid, benzoin, common ion effect, salt hydrolysis, alkali, potassium hydroxide)

(i) *In a galvanic cell, electrons flow from _____ to ______ through the connecting wires.*

(ii) *Racemic mixtures are optically _______ because of ________ compensation.*

(iii) *Half life period of a ______ order reaction is _______ of the concentration of the reactant.*

(iv) *Benzaldehyde when treated with an alcoholic solution of ______ forms _______.*

(v) *Solubility of calcium oxalate is _________ in the presence of ammonium oxalate because of ___________.* **

(b) *Complete the following statements by selecting the **correct alternative** from the choices given :* **[5]**

(i) *The compound which is optically active is :*

(1) *1-butanol*

(2) *2-butanol*

(3) *1-propanol*

(4) *2-methyl-1-propanol*

(ii) *The salt which will not hydrolyse in aqueous solution is :*

(1) *Copper sulphate*

(2) *Sodium sulphate*

(3) *Potassium cyanide*

(4) *Sodium carbonate*

(iii) *Copper has the face centred cubic structure. The coordination number of each ion is :*

(1) *4* **(2)** *12*

(3) *14* **(4)** *8*

(iv) *For the reaction $2SO_2 + O_2 \rightleftharpoons 2SO_3$, the unit of equilibrium constant is :*

(1) $L\ mol^{-1}$ **(2)** $J\ mol^{-1}$

(3) $mol\ L^{-1}$ **(4)** $[L\ mol^{-1}]^2$

(v) *The deficiency of vitamin D causes :*

(1) *Rickets* **(2)** *Gout*

(3) *Scurvy* **(4)** *Night blindness*

(c) *Answer the following questions :* **[5]**

(i) *Two metallic elements A and B have the following standard oxidation potentials : $A = 0{\cdot}40V$, $B = -0{\cdot}80V$. What would you expect if element A was added to an aqueous salt solution of element B ? Give a reason for your answer.*

(ii) *Two moles of NH_3 are introduced into one litre flask in which it dissociates at high temperature as follows :* **

$2NH_3(g) \rightleftharpoons N_2(g) + 3H_2(g)$. Determine Ke, if at equilibrium 1 mole of NH_3 remains.

(iii) *Give balanced equation for the preparation of salicylaldehyde from phenol.*

(iv) *If the half life period for a first order reaction is 69·3 seconds, what is the value of its rate constant ?*

(v) *Define cryoscopic constant.*

(d) *Match the following :* **[5]**

(i)	*Colligative property*	**(a)**	*Polysaccharide*
(ii)	*Nicol prism*	**(b)**	*Osmotic pressure*
(iii)	*Activation energy*	**(c)**	*Aldol condensation*
(iv)	*Starch*	**(d)**	*Polarimeter*
(v)	*Acetaldehyde*	**(e)**	*Arrhenius equation*

PART–II (50 Marks)

*Answer **six** questions choosing **two** from **Section A**, **two** from **Section B** and **two** from **Section C.***

SECTION–A

*Answer any **two** questions.*

Question 2.

(a) **(i)** *Ethylene glycol is used as an antifreeze agent. Calculate the amount of ethylene glycol to be added to 4 kg of water to prevent it from freezing at – 6°C. (K_f for $H_2O = 1{\cdot}85\ K\ mole^{-1}\ kg$)* **[3]**

(ii) *The freezing point of a solution containing 0·3 gms of acetic acid in 30 gms of benzene is lowered by 0·45 K. Calculate the van't Hoff factor.*

(At. wt. of C = 12, H = 1, O = 16, K_f for benzene = $5{\cdot}12\ K\ kg\ mole^{-1}$). **[2]**

** Answer is not given due to change in the present syllabus.

(b) *Name the law or principle confirmed by the following observations :*

(i) *When water is added to 0·01 M aqueous solution of acetic acid the number of hydrogen ions increases.*** **[1]**

(ii) *When 96500 coulombs of electricity is passed through acidulated water, 5·6 litres of oxygen at S.T.P. is liberated at the anode.* **[1]**

(c) *Arrange Ag, Cr and Hg metals in the increasing order of reducing power. Given :* **[1]**

$$E^\circ_{Ag^+/Ag} = +0{\cdot}80\ V$$

$$E^\circ_{Cr^{+3}/Cr} = -0{\cdot}74\ V$$

$$E^\circ_{Hg^{+2}/Hg} = +0{\cdot}79\ V$$

(d) *In a first order reaction, 10% of the reactant is consumed in 25 minutes. Calculate :* **[2]**

(i) *The half life of the reaction.*

(ii) *The time required for completing 17% of the reaction.*

Question 3.

(a) *Explain giving reasons why (Give equations in support of your answer) :***

(i) *A solution of NH_4Cl and NH_4OH acts as a buffer.* **[2]**

(ii) *Cu is precipitated as CuS while Zn is not precipitated when H_2S is passed through an acidic solution of $Cu(NO_3)_2$ and $Zn(NO_3)_2$ respectively.* **[2]**

(b) **(i)** *What is Schottky defect in a solid ?* **[1]**

(ii) *A bcc element (atomic mass 65) has a cell edge of 420 pm. Calculate its density in gms/cm³.* **[3]**

(c) *The rate of the reaction $H_2 + I_2 \rightleftharpoons 2HI$ is given by :***

$$Rate = 1{\cdot}7 \times 10^{-19}\ [H_2][I_2]\ at\ 25°C$$

The rate of decomposition of gaseous HI to H_2 and I_2 is given by :

$$Rate = 2{\cdot}4 \times 10^{-21}\ [HI]^2\ at\ 25°C.$$

Calculate the equilibrium constant for the formation of HI from H_2 and I_2 at 25°C. **[2]**

Question 4.

(a) **(i)** *Give Lewis' definition for acids and bases.*** **[1]**

(ii) *The solubility of Ag_2CrO_4 at 25°C is $8{\cdot}0 \times 10^{-5}$ moles/litre. Calculate its solubility product.*** **[1]**

(b) **(i)** *Define molar conductance of a solution. State its unit. How is it related to the specific conductance of a solution ?* **[2]**

(ii) *Calculate the value of E_{cell} at 298 K for the following cell :* **[3]**

$$Al/Al^{3+}\ (0{\cdot}01M)\ //\ Sn^{2+}\ (0{\cdot}015M)\ /\ Sn$$

$$E^\circ_{Al^{3+}/Al} = -1{\cdot}66\ volt$$

and

$$E^\circ_{Sn^{2+}/Sn} = -0{\cdot}14\ volt$$

(c) **(i)** *Calculate the degree of hydrolysis of 0·2 (M) sodium acetate solution.* **[1]**

*(Hydrolysis constant of sodium acetate = $5{\cdot}6 \times 10^{-10}$ and ionic product of $H_2O = 10^{-14}$ at 25°C)***

(ii) *Explain why high pressure is used in the manufacture of ammonia by Haber's process. State the law or principle used.* **[2]**

SECTION–B

*Answer any **two** questions.*

Question 5.

(a) *Give the IUPAC names of the following coordination compounds :* **[2]**

(i) *$K_2[Zn(OH)_4]$*

(ii) *$[Co(NH_3)_5(CO_3)]Cl$.*

(b) *For the complex ion $[Fe(CN)_6]^{3-}$ state :* **[1]**

(i) *The geometry of the ion.*

(ii) *The magnetic property of the ion.*

(c) *What type of structural isomers are : $[Co(NH_3)_5Br]SO_4$ and $[Co(NH_3)_5SO_4]Br$? Give a chemical test to distinguish the isomers.* **[2]**

Question 6.

(a) *For the molecule XeF_2 :* **[2]**

(i) *Draw the structure of the molecule indicating the lone pairs.*

(ii) *State the hybridisation of the central atom.*

(iii) *State the geometry of the molecule.*

(b) *Give balanced chemical equations for the following reactions :* **[3]**

(i) *Fluorine treated with dilute sodium hydroxide solution.*

(ii) *Hydrogen sulphide treated with concentrated sulphuric acid.*

(iii) *Potassium iodide treated with acidified potassium permanganate solution.*

Question 7.

(a) *In the extraction of zinc from zinc blende :* **[2]**

(i) *Give an equation to show how zinc oxide is converted to zinc.*

(ii) *How is impure zinc finally electrorefined ?*

(b) *Explain why :*

(i) *Transition elements form coloured compounds.*

[3]

** Answer is not given due to change in the present syllabus.

(ii) *Interhalogen compounds are more reactive than their constituent elements.*

(iii) *Cu^{+} is diamagnetic but Cu^{2+} is paramagnetic. (Z = 29)*

SECTION–C

*Answer any **two** questions.*

Question 8.

(a) *How can the following conversions be brought about :*

(i) *Nitro benzene to benzene diazonium chloride.* **[2]**

(ii) *Propanoic acid to ethylamine.* **[3]**

(iii) *Benzoic acid to benzaldehyde.* **[2]**

(b) *Identify the compounds A, B, C, D, E and F :* **[3]**

$$HC \equiv CH \xrightarrow[Hg^{2+}]{dil.H_2SO_4} [A] \xrightarrow{[O]} [B] \xrightarrow{SOCl_2} [C]$$

$$[A] \xrightarrow{NH_2OH} [E]$$

$$[C] \xrightarrow[\Delta]{CH_3COONa} [D]$$

$$[C] \xrightarrow[\Delta]{C_6H_5OH} [F]$$

Question 9.

(a) (i) *Acetamide is heated with bromine and sodium hydroxide solution.* **[4]**

(ii) *Benzaldehyde is treated with 50% sodium hydroxide solution.*

(b) *Give one chemical test to distinguish between the following pairs of compound :*

(i) *Acetone and phenol.* **[4]**

(ii) *Formic acid and acetic acid.*

(c) (i) *Name the type of isomerism exhibited by the following pairs of compound :* **

(1) *$(C_2H_5)_2NH$ and $CH_3–NH–C_3H_7$*

(2) *1-butanol and 2 methyl-1-propanol.*

(ii) *Name the type of isomerism that the compound with molecular formula $C_3H_6O_2$ exhibits. Represent the isomers.* ** **[2]**

Question 10.

(a) *Write balanced chemical equations for the following reactions :* **[4]**

(i) *Oxalic acid is treated with acidified potassium permanganate solution.*

(ii) *Benzoic acid is treated with a mixture of concentrated nitric acid and concentrated sulphuric acid.*

(iii) *Methyl magnesium iodide is treated with carbon dioxide and the product hydrolyzed in acidic medium.*

(iv) *Ethylacetate is treated with ammonia.* **

(b) *An organic compound [A] having molecular formula C_2H_7N on treatment with nitrous acid gives a compound [B] having molecular formula C_2H_6O. [B] on treatment with an organic com-pound [C] gives a carboxylic acid [D] and a sweet smelling compound [E]. Oxidation of [B] with acidified potassium dichromate also gives [D].* **[4]**

(i) *Identify [A], [B], [C], [D] and [E].*

(ii) *Write balanced chemical equation of [D] with chlorine in the presence of red phosphorus and name the reaction.*

(c) *Acetamide is amphoteric in nature. Give two equations to support this statement.* ** **[2]**

ANSWERS

PART–I

Answer 1.

(a) (i) anode, cathode

(ii) inactive, external

(iii) first, independent

(iv) potassium cyanide, benzoin

(b) (i) (2) 2-butanol

(ii) (2) Sodium sulphate

(iii) (2) 12

(iv) (1) L mol^{-1}

(v) (1) Rickets

(c) (i) Metal A will displace metal B from its salt solution because oxidation potential of A is higher than that of B.

(iii)

$$\underset{\text{Phenol}}{C_6H_5OH} + \underset{\text{Chloroform}}{CHCl_3} + 3NaOH \rightarrow \underset{\text{Salicyaldehyde}}{C_6H_4(OH)CHO} + 3NaCl + 2H_2O$$

(iv) If half life of first order reaction is T then its rate constant,

$$K = \frac{0{\cdot}693}{T} = \frac{0{\cdot}693}{69{\cdot}3} = 0{\cdot}01 \text{ sec}^{-1}$$

(v) **Cryoscopic constant :** It is the depression of freezing point of the solvent produced on dissolving one mole of a substance in 1000 g of solvent. It is also called molal depression constant.

** Answer is not given due to change in the present syllabus.

(d) (i) Colligative property (b) Osmotic pressure
(ii) Nicol prism (d) Polarimeter
(iii) Activation energy (e) Arrhenius equation
(iv) Starch (a) Polysaccharide
(v) Acetaldehyde (c) Aldol condensation

PART–I
SECTION–A

Answer 2.

(a) (i) Depression in freezing point

$$\Delta T_f = \frac{1000 \times K_f \times \omega}{W \times M}$$

In the given case

$W = 4\ kg = 4000\ g.$

$K_f = 1{\cdot}85$

$\Delta T_f = 0 - (-6°) = 6°C$

Mol. wt. of ethylene glycol $CH_2OH-CH_2OH = 62$

∴ Amount of ethylene glycol required,

$$\omega = \frac{M \times W \times \Delta T_f}{1000 \times K_f}$$

$$= \frac{62 \times 4000 \times 6}{1000 \times 1{\cdot}85}$$

$= 804{\cdot}32$ g.

(ii) If i is the van't Hoff factor then depression in freezing point

$$\Delta T_f = \frac{i \times 1000 \times K_f \times \omega}{W \times M}$$

Given $\Delta T_f = 0{\cdot}45$ K,

$\omega = 0{\cdot}3$ g

$W = 30$ g

M of acetic acid, $CH_3COOH = 60$

$K_f = 5{\cdot}12$

$$\therefore \quad i = \frac{\Delta T_f \times W \times M}{1000 \times K_f \times \omega}$$

$$= \frac{0{\cdot}45 \times 30 \times 60}{1000 \times 5{\cdot}12 \times 0{\cdot}3}$$

$i = 0{\cdot}5273$

van't Hoff factor is 0·5273.

(b) (ii) Faraday's first law of electrolysis.

(c) Ag < Hg < Cr

(d) (i) For first order reaction,

$$K = \frac{2{\cdot}303}{t} \log \frac{a}{a-x}$$

$$= \frac{2{\cdot}303}{25} \log \frac{100}{100-10}$$

$= 0{\cdot}0042145$

$$\therefore \quad \text{Half life,} \quad T = \frac{0{\cdot}693}{K}$$

$$= \frac{0{\cdot}693}{0{\cdot}0042145}$$

$T = 164{\cdot}43$ minutes.

(ii) Again, for the given reaction

$$K = \frac{2{\cdot}303}{t} \log \frac{a}{a-x}$$

$$t = \frac{2{\cdot}303}{0{\cdot}0042145} \log \frac{100}{100-17}$$

$t = 44{\cdot}21$ minutes.

∴ Time required for completing 17% of the reaction is 44·21 minutes.

Answer 3.

(b) (i) **Schottky defect :** It is a stoichiometric defect in which some of the lattice points in a crystal are occupied such that cation-anion ratio remains the same and electrical neutrality is maintained.

⊕ ⊖ ⊕ ⊖ ⊕ ⊖
⊖ ○ ⊖ ⊕ ⊖ ⊕
⊕ ⊖ ⊕ ⊖ ⊕ ⊖ Vacancies
⊖ ⊕ ⊖ ⊕ ○ ⊕
⊕ ⊖ ⊕ ⊖ ⊕ ⊖

Schottky defect in ionic crystal

(ii) Density, $\rho = \dfrac{Z \times M}{a^3 \times N_A}$

Given, $Z = 2$ (for bcc arrangement)

$M = 65$

$a = 420\ pm = 420 \times 10^{-10}$ cm

$N_A = 6{\cdot}023 \times 10^{23}$

$$\therefore \quad \rho = \frac{2 \times 65}{(420 \times 10^{-10})^3 \times 6{\cdot}023 \times 10^{23}}$$

$$= \frac{2 \times 65 \times 10^4}{(42)^3 \times 6{\cdot}023}$$

$= 2{\cdot}91\ g\ cm^{-3}$

Density of given element is $2{\cdot}91\ g\ cm^{-3}$.

Answer 4.

(b) (i) **Molar conductance :** It is the conductance of all ions furnished by one mole of an electrolyte present in a definite volume of the solution.

Its unit is $ohm^{-1}\ cm^2\ mol^{-1}$. Molar conductance Λ_m is related to specific conductance k by the following equation

$$\Lambda_m = \frac{k \times 1000}{M}, \text{ where}$$

k is Λ_{sp} and M is molar concentration of the solution.

(ii) Given cell is :

$Al / Al^{3+} (0{\cdot}01\ M) \,||\, Sn^{2+} (0{\cdot}015\ M)/Sn$

Given, $E^\circ_{Al^{3+}/Al} = -1{\cdot}66V$,

$E^\circ_{Sn^{2+}/Sn} = -0{\cdot}14V$

$$\therefore \quad E^\circ_{cell} = E^\circ_R - E^\circ_L$$

$$= -0{\cdot}14 - (-1{\cdot}66)$$

$$= -0{\cdot}14 + 1{\cdot}66$$

$$= 1{\cdot}52V$$

and net cell reaction is

$2Al + 3Sn^{2+} \longrightarrow 2Al^{3+} + 3Sn$

so net cell reaction involves transfer of 6 mole of electrons

$\therefore \quad n = 6$

According to Nernst equation, at 298 K

$$E_{cell} = E^\circ_{cell} - \frac{0{\cdot}059}{n} \log \frac{[Al^{3+}]^2}{[Sn^{2+}]^3}$$

$$= 1{\cdot}52 - \frac{0{\cdot}059}{6} \log \frac{[0{\cdot}01]^2}{[0{\cdot}015]^3}$$

$$= 1{\cdot}52 - 0{\cdot}01447$$

$$E_{cell} = 1{\cdot}50553V.$$

SECTION–B

Answer 5.

(a) (i) Potassiumtetrahydroxozincate (II)

(ii) Pentaamminecarbonatocobalt (III) chloride

(b) (i) Octahedral geometry

(ii) Ion is paramagnetic.

(c) $[Co(NH_3)_5 Br] SO_4$ and $[Co(NH_3)_5SO_4] Br$ are ionisation isomers.

Aqueous solution of $[Co(NH_3)_5 Br] SO_4$ will give a white ppt. with $BaCl_2$ solution. The ppt. is insoluble in conc. HNO_3 confirming presence of free SO_4^{2-} ions. Aqueous solution of $[Co(NH_3)_5SO_4]$ Br will not give this test.

$[Co(NH_3)_5Br] SO_4 \rightleftharpoons [Co(NH_3)_5Br]^{2+} + SO_4^{2-}$

$BaCl_2 + SO_4^{2-} \rightarrow BaSO_4$ (white ppt.) $+ 2Cl^-$

$[Co(NH_3)_5SO_4]Br \rightleftharpoons [Co(NH_3)_5SO_4]^+ + Br^-$

$BaCl_2 + Br^- \rightarrow$ No ppt

Answer 6.

(a) (i)

(Structure of XeF_2: F — Xe — F with three lone pairs on Xe; lp labels, } → Lone pair of electrons (lp))

(ii) Central atom Xe shows sp^3d hybridisation.

(iii) Molecule is linear.

(b) (i) $2F_2 + 2NaOH \longrightarrow 2NaF + H_2O + OF_2$

(ii) $H_2S + \underset{conc.}{H_2SO_4} \longrightarrow S + SO_2 + 2H_2O$

(iii) $2KMnO_4 + 8H_2SO_4 + 10KI \longrightarrow 6K_2SO_4 + 2MnSO_4 + 8H_2O + 5I_2$

Answer 7.

(a) (i) $\underset{\text{Zinc oxide}}{ZnO} + \underset{\text{coke}}{C} \xrightarrow[1673K]{\text{heated}} \underset{\text{zinc}}{Zn} + CO \uparrow$

(ii) *Electro refining of zinc* :

Electrolyte used : $ZnSO_4$ solution containing a little dilute H_2SO_4

Anode : Block of impure zinc

Cathode : Thin sheet of pure zinc.

On passing current pure zinc is obtained at cathode.

Anode reaction : $\underset{\text{impure}}{Zn} \rightarrow Zn^{2+}_{(aq)} + 2e^-$ (Anode Mod)

Cathode reaction : $Zn^{2+}_{(aq)} + 2e^- \rightarrow \underset{\text{(pure metal)}}{Zn}$

(b) (i) Refer to ISC Paper 2018, Answer 13 (a) (ii).

(ii) Refer to ISC Paper 2019, Answer 17 (b) (1).

(iii) Refer to ISC Paper 2019, Answer 1 (d) (i) (2).

SECTION–C

Answer 8.

(a) (i)

$\underset{\text{Nitrobenzene}}{C_6H_5NO_2} + 6H \xrightarrow{Sn + HCl} C_6H_5NH_2 + 2H_2O$

$C_6H_5NH_2 \xrightarrow[5°C]{+ HNO_2 + HCl} \underset{\text{benzene diazonium chlorid}}{C_6H_5N_2Cl} + 2H_2O$

(ii) $\underset{\text{Propanoic acid}}{CH_3CH_2COOH} \xrightarrow[-SO_2, -HCl]{SOCl_2} \underset{\text{Propanoyl chloride}}{CH_3CH_2COCl}$

$\xrightarrow[-NH_4Cl]{NH_3 \text{ (excess)}} \underset{\text{Propionamide}}{CH_3CH_2CONH_2} \xrightarrow{Br_2/KOH} \underset{\text{Ethylamine}}{CH_3CH_2NH_2}$

(iii)

$$C_6H_5COOH + SOCl_2 \xrightarrow[-HCl]{-SO_2} C_6H_5COCl$$

Benzoic acid

$$\xrightarrow[Pd/BaSO_4]{+H_2,\ -HCl} C_6H_5CHO$$

Benzaldehyde

(b) A is CH_3CHO → Acetaldehyde
B is CH_3COOH → Acetic acid
C is CH_3COCl → Acetyl chloride
D is $CH_3COOCOCH_3$ → Acetic anhydride
E is $CH_3CH=NOH$ → Acetaldoxime
F is $C_6H_5OCOCH_3$ → Phenyl acetate

Answer 9.

(a) (i) This reaction is called Hoffmann's bromamide degradation reaction.

$$\underset{\text{Acetamide}}{CH_3CONH_2} + Br_2 + 4NaOH \xrightarrow{\Delta} \underset{\text{Methanamine}}{CH_3NH_2} + 2NaBr + Na_2CO_3 + 2H_2O$$

(ii) $$\underset{\text{benzaldehyde}}{2C_6H_5CHO} + \underset{\text{50\% sodium hydroxide}}{NaOH} \longrightarrow C_6H_5COONa + C_6H_5CH_2OH$$

This is Cannizzaro's reaction.

(b) (i) When acetone is heated with iodine solution and aqueous NaOH in a water bath, yellow crystal of iodoform are obtained having a characteristic odour.

$$\underset{\text{Acetone}}{CH_3COCH_3} + 3I_2 + 4NaOH \longrightarrow \underset{\text{Iodoform}}{CHI_3 \downarrow} + CH_3COONa + 3NaI + 3H_2O$$

Phenol does not give this test.

(ii) Formic acid gives a silver mirror with Tollen's reagent.

$$\underset{\text{formic acid}}{HCOOH} + Ag_2O \longrightarrow H_2O + CO_2 + \underset{\text{silver mirror}}{2Ag (\downarrow)}$$

Acetic acid does not give this test.

Answer 10.

(a) (i) $5(COOH)_2 + 2KMnO_4 + 3H_2SO_4 \longrightarrow K_2SO_4 + 2MnSO_4 + 10CO_2 + 8H_2O$

(ii)

$$\underset{\text{Benzoic acid}}{C_6H_5COOH} + \underset{\text{conc.}}{HNO_3} \xrightarrow{\text{conc. } H_2SO_4} \underset{\text{Meta nitro benzoic acid}}{m\text{-}NO_2C_6H_4COOH} + H_2O$$

(iii) $$O=C=O + CH_3MgI \longrightarrow \left[CH_3-\overset{O}{\overset{\|}{C}}-OMgI \right] \xrightarrow[\text{acid}]{+HOH} CH_3COOH + Mg(OH)I$$

(b) (i) [A] → $CH_3CH_2NH_2$ or Ethylamine
[B] → CH_3CH_2OH or Ethanol
[C] → $(CH_3CO)_2O$ or Ethanoic anhydride
[D] → CH_3COOH or Ethanoic acid.
[E] → $CH_3COOC_2H_5$ or Ethyl ethanoate.

(ii) $$CH_3COOH + Cl_2 \xrightarrow{\text{Red P}} CH_2ClCOOH + HCl$$

Hell Volhard Zelinsky reaction or HVZ reaction.

(c) Basic : $CH_3CONH_2 + HCl \rightarrow CH_3CONH_2.HCl$ or $[CH_3CONH_3]^+Cl^-$

Acidic : $CH_3CONH_2 + HgO \rightarrow (CH_3CONH)_2Hg + H_2O$

●●

CHEMISTRY 2012

QUESTIONS

PART–I

*Answer **all** questions.*

Question 1.

(a) *Fill in the blanks by choosing the appropriate word/ words from those given in the brackets :* **[5]**

(Raoult's, Arrhenius, lateral, sodium, magnesium, negative, positive, non-ideal, ideal, iron, copper, van't Hoff, s, p, ethanol, ethanoic acid, methanoic acid, methanol, propanoic acid)

(i) *For a spontaneous change to take place, the ΔS of the system should be______ and ΔG of the system should be___________.***

(ii) *Hydrolysis of methyl propanoate gives _______ and_______.*

(iii) *Solutions which strictly obey_________ law are called________ solutions.*

(iv) *π bonds are formed by the______ overlap of _________ orbitals.***

(v) *Zinc can displace_______ from $CuSO_4$ solution, but cannot displace ______ from $MgSO_4$ solution.*

(b) *Complete the following statements by selecting the **correct alternative** from the choices given :* **[5]**

(i) *The quantity of electricity required to deposit 1·15 g of sodium from molten NaCl (Na = 23, Cl = 35·5) is :*

(1) *1 F* **(2)** *0·5 F*

(3) *0·05 F* **(4)** *1·5 F*

(ii) *When acetic acid is reacted with calcium hydroxide and the product is distilled dry, the compound formed is :*

(1) *Calcium acetate* **(2)** *Acetone*

(3) *Acetaldehyde* **(4)** *Acetic anhydride*

(iii) *The $[OH^-]$ concentration of a weak base is given by :***

(1) ck_b **(2)** $\sqrt{ck_b}$

(3) $\sqrt{k_b / c}$ **(4)** $\sqrt{k_b}$

(iv) *In a plot of log k vs 1/T, the slope is :*

(1) $-E_a/2{\cdot}303$ **(2)** $E_a/2{\cdot}303\ R$

(3) $E_a/2{\cdot}303$ **(4)** $-E_a/2{\cdot}303\ R$

(v) *Among the following coordination compounds, the one giving a white precipitate with $BaCl_2$ solution is :*

(1) $[Cr(H_2O)_5Br]SO_4$

(2) $[Cr(H_2O)_5SCN]$

(3) $[Co(NH_3)_5SO_4]Br$

(4) $[Pt(NH_3)_6]Cl_4$

(c) *Answer the following questions :* **[5]**

(i) *A solution X is prepared by dissolving three moles of glucose in one litre of water and a solution Y is prepared by dissolving 1·5 moles of sodium chloride in one litre of water. Will the osmotic pressure of X be higher, lower or equal to that of Y ? Give a reason for your answer.*

(ii) *Give one example (equation) of a homogeneously catalysed reaction and name the catalyst.*

(iii) *Write the formula of the product formed when formaldehyde reacts with ammonia and name the product.*

(iv) *If the ionization (dissociation) constant of acetic acid is k_a , what will be the pH of a solution containing equal concentrations of acetic acid and sodium acetate ?***

(v) *What is the electronic configuration of chromium atom (Z = 24) ? Give a reason for your answer.*

(d) *Match the following :* **[5]**

(i)	*Nernst equation*	**(a)**	*Water*
(ii)	*Lactic acid***	**(b)**	*Constant volume*
(iii)	*Amphiprotic solvent*	**(c)**	*Ammonia*
(iv)	*Lewis base*	**(d)**	*Optical isomers***
(v)	*Isochoric process*	**(e)**	*Electrochemical cells.*

PART–II

*Answer **six** questions choosing **two** from **Section A**, **two** from **Section B** and **two** from **Section C.***

SECTION–A

*Answer any **two** questions.*

Question 2.

(a) **(i)** *A solution of urea in water has a boiling point of 100·18°C. Calculate the freezing point of the solution. (K_f for water is 1·86 K kg mol^{-1} and K_b for water is 0·512 K kg mol^{-1}).* **[2]**

(ii) *A solution of lactose containing 8·45 g of lactose in 100 g of water has a vapour pressure of*

** Answer is not given due to change in the present syllabus.

4·559 mm of Hg at 0°C. If the vapour pressure of pure water is 4·579 mm of Hg, calculate the molecular weight of lactose. **[3]**

(b) (i) *The molecular weight of H_2S is more than that of H_2O, but H_2S is a gas and H_2O is a liquid. Explain.* **[2]**

(ii) *When potassium cyanide reacts with water, will the resulting solution be acidic, alkaline or neutral ? Justify your answer.*** **[2]**

(c) *What is the hybridization of the carbon atom in ethyne molecule ? What is the H—C—H bond angle ?*** **[1]**

Question 3.

(a) (i) *State the second law of thermodynamics in terms of the entropy of the universe.*** **[1]**

(ii) *Calculate the maximum work that can be obtained from the given electro-chemical cell constructed with two metals M and N.* **[3]**

$$[E^{\circ}_{M^{+2}/M} = -0{\cdot}76\ V, E^{\circ}_{N^{+2}/N} = +0{\cdot}34\ V]$$

The cell reaction is $M + N^{+2} \rightarrow M^{+2} + N$

(b) (i) *To precipitate group III cations NH_4Cl should be added to the solution before the addition of ammonium hydroxide. Explain why.*** **[2]**

(ii) *A study of chemical kinetics of the reaction A + B → products, gave the following data at 25°C :* **[2]**

Experiment	[A]	[B]	$\frac{d[Products]}{dt}$
1	1·0	0·15	$4{\cdot}20 \times 10^{-6}$
2	2·0	0·15	$8{\cdot}40 \times 10^{-6}$
3	1·0	0·20	$5{\cdot}60 \times 10^{-6}$

Find :

(1) *The order of reaction with respect to A.*

(2) *The order of reaction with respect to B.*

(3) *The rate law.*

(c) (i) *What are F-centres in an ionic crystal ?* **[1]**

(ii) *Why are crystals having F-centres paramagnetic ?* **[1]**

Question 4.

(a) (i) *The central atom of methane and water is in the same state of hybridization, but the shapes of the two molecules are different. Discuss.*** **[2]**

(ii) *The conductivity of 0·2 M KCl solution is 3×10^{-2} ohm^{-1} cm^{-1}. Calculate its molar conductance.* **[2]**

(b) (i) *Draw the valence shell molecular orbital diagram of oxygen molecule and predict its magnetic nature.*** **[2]**

(ii) *Calculate the solubility of lead chloride in water, if its solubility product is $1{\cdot}7 \times 10^{-5}$.** (Pb = 206, Cl = 35·5)* **[2]**

(c) *For a crystal of diamond, state :* **[2]**

(i) *The hybridisation of the carbon atom.*

(ii) *The coordination number of each carbon atom.*

(iii) *The type of lattice in which it crystallizes.*

(iv) *The number of carbon atoms present per unit cell.*

SECTION–B

*Answer any **two** questions.*

Question 5.

(a) *Write the formulae of the following coordination compounds :* **[2]**

(i) *potassiumtetracyanonickel(0).*

(ii) *triamminetrinitrocobalt(III).*

(b) *$[CoF_6]^{3-}$ is a coordination complex ion.* **[2]**

(i) *What is the oxidation number of cobalt in the complex ?*

(ii) *How many unpaired electrons are there in the complex ?*

(iii) *State the magnetic behaviour of the complex.*

(iv) *Give the I.U.P.A.C. name of the complex.*

(c) *Draw the structural isomer of $[Co(NH_3)_5NO_2]Cl_2$ and name the type of isomerism.*

Question 6.

(a) *Give the equations for the conversion of argentite (Ag_2S) to metallic silver.* **[2]**

(b) *Give balanced equations for the following reactions :* **[3]**

(i) *Acidifed potassium permanganate and oxalic acid.*

(ii) *Ozone and mercury.*

(iii) *Action of heat on a mixture of sodium chloride and concentrated sulphuric acid.*

Question 7.

(a) *Explain why transition metals form complex compounds.* **[2]**

(b) (i) *What is the hybridization of the chlorine atom in ClF_3 molecule ?* **[2]**

(ii) *Draw the structure of the molecule and state its geometry.*

(c) *Name the inert gases used for :* **[1]**

(i) *Filling sodium vapour lamps.*

(ii) *Obtaining light of different colours in neon signs.*

** Answer is not given due to change in the present syllabus.

SECTION–C

*Answer any **two** questions.*

Question 8.

(a) *How can the following conversions be brought about :*

(i) *Ethanol to methylamine.* **[3]**

(ii) *Benzene to phenol.* **[2]**

(b) *Complete the following reactions and name the reactions :* **[3]**

(i) $\underline{\qquad\qquad} + 3I_2 + 4KOH \rightarrow CHI_3 + CH_3COOK + 3KI + 3H_2O$

(ii) $C_3H_7NH_2 + CHCl_3 + 3KOH\ (alc.) \xrightarrow{\Delta} \underline{\qquad\qquad} + 3KCl + 3H_2O$

(c) *Name the type of polymerisation (addition or condensation) and name the monomers in each of the following polymers:* **[2]**

(i) *Protein.*

(ii) *Polyethylene.*

Question 9.

(a) **(i)** *What type of isomers are glucose and fructose?* **[1]**

(ii) *Name the functional group common to both glucose and fructose.*

(b) **(i)** *Identify the products A, B and C :* **[3]**

CH_3 (toluene) $\xrightarrow{\text{Alk. } KMnO_4}$ A $\xrightarrow[\text{(soda lime)}]{\Delta,\ NaOH,\ CaO}$ B

A $\xrightarrow{Cl_2,\ FeCl_3}$ C

(ii) *Identify the reagents X, Y and Z.*

$$C_2H_5Cl \xrightarrow{X} C_2H_5CN \xrightarrow{Y} C_2H_5CH_2NH_2 \xrightarrow{Z} C_2H_5CH_2NHCOCH_3$$

(c) *Give balanced equations for the following reactions:* **[3]**

(i) *Benzaldehyde and hydroxylamine.*

(ii) *Benzoic acid and phosphorus pentachloride.*

(iii) *1-butanol and hydrogen chloride.*

(d) *Give one good chemical test to distinguish between the following pairs of compounds :* **[3]**

(i) *Methanal and ethanal.*

(ii) *Urea and benzoic acid.***

Question 10.

(a) *An aliphatic hydrocarbon A on treatment with sulphuric acid in the presence of $HgSO_4$ yields a liuqid B with molecular formula C_2H_4O. B on oxidation with acidified potassium dichromate yields C which gives effervescence with sodium bicarbonate. C when treated with $SoCl_2$ gives D. When D reacts with ethanol it gives a sweet smelling liquid E. E is also formed when C reacts with ethanol in the presence of conc. H_2SO_4.* **[4]**

(i) *Identify A, B, C, D and E.*

(ii) *Draw the structure of the isomer of compound B.*

(iii) *Write the balanced equation of the conversion of A to B.*

(b) **(i)** *The compound C_6H_{12} shows optical isomerism. Draw the structural formula of the compound and name it.*** **[1]**

(ii) *Name any three types of isomerism that the compound with molecular formula C_4H_7Cl can give rise to. Also represent the structures of the compound relevant to these isomers.*** **[3]**

(c) *Give equations to show what happens when a mixture of potassium cyanate and ammonium sulphate is strongly heated. Name the reaction.*** **[2]**

ANSWERS

PART–I

Answer 1.

(a) **(ii)** methanol, propanoic acid,

(iii) Raoult's, ideal,

(v) copper, magnesium.

(b) **(i)** (3) 0·05 F

(ii) (2) Acetone

(iv) (4) $- E_a/2{\cdot}303\ R$

(v) (1) $[Cr(H_2O)_5Br]SO_4$

(c) **(i)** Osmotic pressure of X and Y will be equal.

Reason : This is because osmotic pressure is a colligative property. Its value depends upon the number of particles of solute present in the solution. In the given case 3 moles of glucose give 3 moles of molecules (particles) and 1·5 moles of NaCl, due to complete dissociation it provides 1·5 moles of Na^+ ions and 1·5 moles of Cl^- ions. Total 3 moles of ions (particles).

** Answer is not given due to change in the present syllabus.

$$NaCl \xrightleftharpoons{water} Na^+ + Cl^-$$

1·5 moles → 1·5 moles + 1·5 moles = 3 moles.

(ii) Hydrolysis of ethyl acetate to give ethanol and acetic acid is catalysed by dil H_2SO_4.

$$\underset{\text{ethyl acetate}}{CH_3COO{\cdot}C_2H_5} + \underset{\text{Water}}{HOH} \xrightarrow[\text{dil.}H_2SO_4]{\text{Catalyst}} \underset{\text{acetic acid}}{CH_3COOH} + \underset{\text{ethanol}}{C_2H_5OH}$$

(iii) Compound formed is urotropine or hexamethylene tetramine and its formula is $(CH_2)_6N_4$

(v) Electronic configuration of Cr atom is :

$$1s^2, 2s^2p^6, 3s^2\,p^6d^5, 4s^1$$

This is because an electronic configuration with half-filled and completely filled orbitals is more stable. $_{24}Cr$ has $3d$ orbitals half-filled which is more stable configuration than normally expected *i.e.*, $3d^4\,4s^2$.

(d) (i) Nernst equation — **(e)** Electrochemical cells

(iii) Amphiprotic solvent — **(a)** Water

(iv) Lewis base — **(c)** Ammonia

(v) Isochoric process — **(b)** Constant volume

PART–II

SECTION–A

Answer 2.

(a) (i) Given : Boiling point of solution = 100·18°C

$$K_b = 0{\cdot}512 \text{ K kg mol}^{-1}$$

$$K_f = 1{\cdot}86 \text{ K kg mol}^{-1}$$

$$\Delta T_b = 100{\cdot}18° - 100° = 0{\cdot}18°C$$

$$\Delta T_b = K_b \times \text{Molality}$$

$$\therefore \quad \text{Molality} = 0{\cdot}18/0{\cdot}512$$

Now $\Delta T_{freezing} = K_f \times \text{Molality}$

$$= 1{\cdot}86 \times \frac{0{\cdot}18}{0{\cdot}512}$$

$$= 0{\cdot}6539$$

∴ Freezing point of solution = 0 – 0·6539°C

= – 0·6539°C.

(ii) Given :Weight of lactose w = 8·45 g.

Weight of water W = 100 g

Vapour pressure of pure water ($p°$)

= 4·579 mm of Hg

Vapour pressure of solution (p_s)

= 4·559 mm of Hg.

Now, according to Raoult's law.

$$\frac{p° - p^s}{p°} = \frac{w \times M}{m \times W}$$

$$\frac{4{\cdot}579 - 4{\cdot}559}{4{\cdot}579} = \frac{8{\cdot}45 \times 18}{m \times 100}$$

$$m = \frac{8{\cdot}45 \times 18 \times 4{\cdot}579}{0{\cdot}02 \times 100}$$

$$= 348{\cdot}23 \approx 348$$

Molecular mass of lactose is 348.

(b) (i) This is because, H_2O has intermolecular hydrogen bond between its molecule (H—O—H......H—O—H) while H_2S has weak Vander Waals forces between its molecule. Hence, the molecules of H_2O are strongly packed than H_2S, thus water is a liquid at room temperature while H_2S is a gas.

Answer 3.

(a) (ii) According to given cell reaction

$$E°_{\text{anode }(M^{+2}/M)} = -0{\cdot}76 \text{ V}$$

$$E°_{\text{Cathode }(N^{+2}/N)} = +0{\cdot}34 \text{ V}$$

$$n = 2$$

Now, $w_{\text{electrical}} = -\,n\,FE°_{\text{cell}}$

$$= -2 \times 96500 \times [0{\cdot}34 - (-0{\cdot}76)]$$

$$= -2 \times 96500 \times 1{\cdot}1$$

$$= -212{\cdot}3 \text{ K joule.}$$

(b) (ii) 1.

$$\frac{r_2}{r_1} = \left(\frac{c_2}{c_1}\right)^n$$

$$\frac{8{\cdot}40 \times 10^{-6}}{4{\cdot}20 \times 10^{-6}} = \left(\frac{2}{1}\right)^n$$

$$2 = (2)^n$$

$$n = 1$$

Order of reaction w.r.t. A is 1.

2.

$$\frac{r_2}{r_1} = \left(\frac{c_2}{c_1}\right)^n$$

$$\frac{0{\cdot}20}{0{\cdot}15} = \left(\frac{5{\cdot}60 \times 10^{-6}}{4{\cdot}20 \times 10^{-6}}\right)^n$$

$$\log\frac{r_2}{r_1} = n \log\frac{c_2}{c_1}$$

$$\log \frac{0.20}{0.15} = n \log \left(\frac{5.60\times10^{-6}}{4.20\times10^{-6}}\right)$$

$$\log (0.20 - 0.15) = n(5.60 \times 10^{-6} - 4.20 \times 10^{-6})$$

$$0.125 = n \times 0.125$$

$$n = 1$$

Order of r, w.r.t. B is 1

3. Rate law, $k = [A]^1 [B]^1$

(c) (i) **F-centres** : In ionic compounds having non-stoichiometric defects aninoic vacancies are occupied by electrons. These are called F-centres. These electrons remain delocalised and are responsible for imparting colours to crystals.

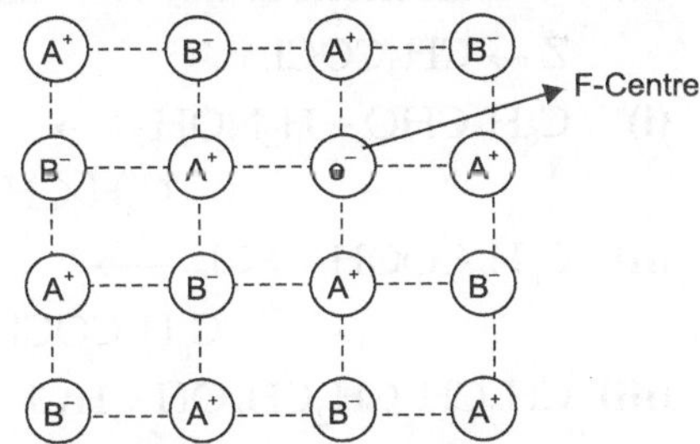

Ionic Crystal with F-Centre

(ii) Free, unpaired electrons occupying anionic vacancies undergo spin in a particular direction. This produces a magnetic moment and ionic crystals with F-centres, therefore, shows paramagnetic behaviour.

Answer 4.

(a) (ii) Given :

Conductivity $= 3 \times 10^{-2}$ ohm^{-1} cm^{-1}

Molarity = 0·2 M

Now, $\Lambda_{molar} = \frac{1000\times\text{Conductivity}}{\text{Molarity}}$

$= \frac{10000\times3\times10^{-2}}{0.2}$

$= 150$ ohm cm^2 mol^{-1}

(c) (i) sp^3 hybridization

(ii) Coordination number is 4.

(iii) Type of lattice : Network type with fcc arrangement of C atoms.

(iv) Number of particles per unit cell is 4.

SECTION–B

Answer 5.

(a) (i) $K_4[Ni(CN)_4]$

(ii) $[Co(NH_3)_3(NO_2)_3]$

(b) (i) $[CoF_6]^{3-}$

$x + (6 \times -1) = -3$

$\therefore$ Oxidation number of Co, 'x' $= +3$

(ii) 4 unpaired electrons.

(iii) Complex ion is paramagnetic.

(iv) Hexafluorocobaltate(III) ion.

(c) Structural isomer $\rightarrow [Co(NH_3)_5Cl]\ Cl.NO_2$.

Type of isomerism $\rightarrow$ Ionisation isomerism.

Answer 6.

(a) $\underset{\text{Argentite}}{Ag_2S} + 4NaCN \rightleftharpoons 2Na[Ag(CN)_2] + Na_2S$

$Na_2S + 2O_2 \rightleftharpoons Na_2SO_4$

$2Na[Ag(CN)_2] + Zn \rightarrow Na_2[Zn(CN)_4] + \underset{\text{Silver metal}}{2Ag\downarrow}$

(b) (i) $2KMnO_4 + 3H_2SO_4 + 5(COOH)_2 \longrightarrow K_2SO_4 + 2MnSO_4 + 8H_2O + 10CO_2$

(ii) $O_3 + 2Hg \longrightarrow Hg_2O + O_2$

(iii) $NaCl + \underset{\text{conc.}}{H_2SO_4} \xrightarrow{\Delta < 200°C} NaHSO_4 + HCl\uparrow$

Answer 7.

(a) Transition metals are d-block elements with valence shell configuration $ns^2(n-1)\ d^{1-10}$. These metals and their ions easily coordinate to a number of negative ions or neutral molecules having lone pairs of electrons to form complex. This is due to :

1. Small size and high nuclear charge of these metals.
2. Availability of vacant d-orbitals of suitable energy to accommodate lone pairs of electrons donated by the ligands.

(b) (i) The hybridization of Cl atom in ClF_3 is sp^3d.

Cl atom [↿⇂] [↿⇂|↿⇂|↿] [| | | |]
3s 3p 3d

Cl atom in ClF_3 with sp^3d hybridisation [↿⇂|↿⇂|↿|↿|↿] [| | |]
sp^3d hybrid 3d

(ii) ClF_3 molecule structure.

It is a 'T' shaped molecule.

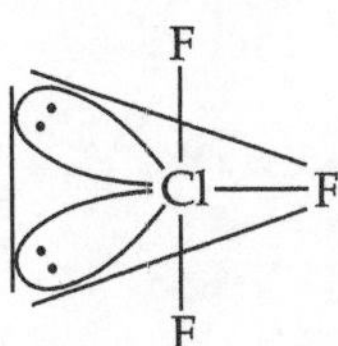

ClF_3 molecule

(c) (i) Neon gas.

(ii) Argon mixed with neon gas.

SECTION–C

Answer 8.

(a) (i) Ethanol to methylamine.

$$\underset{\text{Ethanol}}{C_2H_5OH} \xrightarrow[-H_2O]{+[O]} \underset{\text{Acetic acid}}{CH_3COOH} \xrightarrow{+NH_3}$$

$$\underset{\text{Ammonium acetate}}{CH_3COONH_4} \xrightarrow[-H_2O]{\Delta} \underset{\text{Acetamide}}{CH_3CONH_2}$$

$$\xrightarrow{Br_2,\ NaOH} \underset{\text{Methylamine}}{CH_3NH_2}$$

(ii) Benzene to phenol.

$$\underset{\text{Benzene}}{C_6H_6} + \underset{\text{conc.}}{H_2SO_4} \xrightarrow{-H_2O} \underset{\text{Benzene sulphonic acid}}{C_6H_5SO_3H} \xrightarrow[-H_2O]{NaOH}$$

$$\underset{\text{Sodium salt of benzene}}{C_6H_5SO_3Na} \xrightarrow{\text{heated with fused NaOH}} \underset{\text{Sodium phenoxide}}{C_6H_5ONa} \xrightarrow[-NaCl]{+HCl} \underset{\text{Phenol}}{C_6H_5OH}$$

(b) **(i) Iodoform reaction**

$$CH_3COCH_3 + 3I_2 + 4KOH \longrightarrow CHI_3 + CH_3COOK + 3KI + 3H_2O$$

(ii) Carbylamine reaction

$$C_3H_7NH_2 + CHCl_3 + 3KOH\ (\text{alc.}) \xrightarrow{\Delta} C_3H_7NC + 3KCl + 3H_2O$$

(c) (i) Condensation polymerisation.
monomer → amino acids.

(ii) Addition polymerization
monomer → ethylene

Answer 9.

(a) (i) Functional isomers.

(ii) Alcoholic group.

(b) (i) A → Benzoic acid, B → Benzene, C → 3, 5 dichloro benzoic acid.

$$\underset{\text{Toluene}}{C_6H_5CH_3} \xrightarrow{\text{al } KMnO_4} \underset{\substack{\text{Benzoic acid}\\ [A]}}{C_6H_5COOH} \xrightarrow[\Delta]{NaOH,\ CaO} \underset{\substack{\text{Benzene}\\ [A]}}{C_6H_6}$$

$$C_6H_5COOH \xrightarrow{Cl_2,\ FeCl_3} \underset{\substack{\text{3, 5 dichloro benzoic acid}\\ [C]}}{C_6H_3Cl_2COOH}$$

(ii) X → alcoholic KCN, Y → $LiAlH_4$, Z → CH_3COCl.

(c) (i) $C_6H_5CHO + H_2NOH \longrightarrow C_6H_5CH{=}NOH + H_2O$

(ii) $C_6H_5COOH + PCl_5 \longrightarrow C_6H_5COCl + POCl_3 + HCl$

(iii) $CH_3CH_2CH_2CH_2OH + HCl \longrightarrow CH_3CH_2CH_2CH_2Cl + H_2O$

(d) (i) Ethanol, on warming with iodine and NaOH solution gives yellow crystalline precipitate of iodoform while, methanol does not gives this test.

Answer 10.

(a) (i) A → Acetylene (C_2H_2)
B → Acetaldehyde (CH_3CHO)
C → Acetic acid (CH_3COOH)
D → Acetyl chloride (CH_3COCl)
E → Ethyl acetate ($CH_3COOC_2H_5$)

(ii) Isomer of acetaldehyde.

```
H  H
|  |
C = C
|  |
O  OH
```

(iii) $\underset{[A]}{CH\equiv CH} + H_2O \xrightarrow[HgSO_4]{H_2SO_4\ +} \underset{[B]}{CH_3CHO}$

●●

CHEMISTRY 2011

QUESTIONS

PART–I

*Answer **all** questions.*

Question 1.

(a) *Fill in the blanks by choosing the appropriate word/words from those given in the brackets:* **[5]**

(concentrated sulphuric acid, methylamine, ethylamine, sp^2, 2-propanol, s^{-1}, sp^3, glycol, Cannizzaro's reaction, pyramidal, mol l^{-1} s^{-1}, Hofmann's degradation, glycerol, concentrated nitric acid, square planar, octahedral, concen-trated hydrochloric acid)

(i) *Nitrogen atom in ammonia undergoes _______ hybridization and the geometry of the molecule is_______.*

(ii) *For a first order reaction, the unit of rate is _______ and that of rate constant is _______.*

(iii) *When acetamide is treated with bromine and caustic soda, it gives _________ as the main product and the reaction is called _______.***

(iv) *_______ is an example of trihydric alcohol and _______ is an example of dihydric alcohol.*

(v) *Aqua regia is a mixture of __________ and __________ in the ratio of 3 : 1.*

(b) *Complete the following statements by selecting the **correct alternative** from the choices given :* **[5]**

(i) *Out of the following solutions, the one having the highest boiling point will be :*

(1) *0·1 M NaCl* **(2)** *0·1 M $BaCl_2$*

(3) *0·1 M KNO_3* **(4)** *0·1 M $K_4[Fe(CN)_6)]$*

(ii) *75% of a first order reaction was completed in 32 minutes. When was 50% of the reaction completed ?*

(1) *24 minutes* **(2)** *16 minutes*

(3) *8 minutes* **(4)** *4 minutes*

(iii) *When zinc granule is dipped into copper sulphate solution, copper is precipitated because :*

(1) *Both, copper and zinc have a positive reduction potential.*

(2) *Reduction potential of copper is higher than that of zinc.*

(3) *Reduction potential of zinc is higher than that of copper.*

(4) *Both, zinc and copper have a negative reduction potential.*

(iv) *Among the following compounds, the one showing geometric isomerism is :***

(1) *2-chloro propane*

(2) *2-bromo-2-chlorobutane*

(3) *1, 2 dichloro ethene*

(4) *Glycine*

(v) *Of the following compounds, the one which is a Lewis acid is :***

(1) *PCl_3* **(2)** *$AlCl_3$*

(3) *NCl_3* **(4)** *$AsCl_3$*

(c) *Answer the following questions :* **[5]**

(i) *0·1 M urea solution shows less depression in freezing point than 0·1 M $MgCl_2$ solution. Explain.*

(ii) *What is the pH of a solution whose hydroxyl ion concentration is 10^{-2} M ?***

(iii) *If neutral litmus solution is added to sodium acetate solution, what will you observe and why ?***

(iv) *State why the boiling point of HF is very high.*

(v) *Define piezoelectricity and give one use of piezoelectric crystals.*

(d) *Match the following :* **[5]**

(i)	*Biuret***	**(a)**	*DNA*
(ii)	*Urotropine*	**(b)**	*Amines*
(iii)	*Purine*	**(c)**	*Urea***
(iv)	*Frasch process*	**(d)**	*Formaldehyde*
(v)	*Hinsberg's reagent*	**(e)**	*Sulphur*

PART–II

*Answer **six** questions choosing **two** from **Section A**, **two** from **Section B** and **two** from **Section C**.*

SECTION–A

*Answer any **two** questions.*

Question 2.

(a) (i) *46 gms of ethyl alcohol is dissolved in 18 gms of water. Calculate the mole fraction of ethyl alcohol. (At. wt of C = 12, O = 16, H = 1).* **[1½]**

(ii) *The osmotic pressure of 0·01 molar solution of an electrolyte is found to be 0·65 atm at 27°C. Calculate the van't Hoff factor. What conclusion can you draw about the molecular state of the solute in the solution ?* **[2½]**

______

** Answer is not given due to change in the present syllabus.

(b) (i) *State Faraday's first law of electrolysis.* [1]

(ii) *How many electrons will flow when a current of 5 amperes is passed through a solution for 200 seconds ?* [1]

(c) *Give reasons for the following :***

(i) *A reaction/process will be spontaneous when it is exothermic and randomness is increasing.* [2]

(ii) *The number of hydronium ions increases when one litre of water is added to 1M acetic acid.* [2]

Question 3.

(a) (i) *What are semiconductors ? What is the effect of increasing temperature on the conductivity of a semiconductor ?* [1½]

(ii) *A compound AB has a cubic structure and molecular mass 99. Its density is 3·4 g cm^{-3}. What is the length of the edge of the unit cell?* [3]

(b) (i) *What is the maximum work that can be obtained by the isothermal expansion of one mole of an ideal gas at 273 K from 2·24 dm^3 to 22·4 dm^3 ?*** [2½]

(ii) *State the geometry of PCl_5 molecule. Draw its structure.* [1]

(c) *Give two differences between a sigma bond and a pi bond.*** [2]

Question 4.

(a) (i) *What is meant by common ion effect ?*** [1]

(ii) *Give the conjugate acid and the conjugate base for NH_3.*** [1]

(b) *Consider the reaction $2Ag^+ + Cd \rightarrow 2Ag + Cd^{2+}$. The standard reduction potentials of Ag^+/Ag and Cd^{2+}/Cd are + 0·80 volt and – 0·40 volt, respectively.* [4]

(i) *Give the cell representation.*

(ii) *What is the standard cell emf, E° ?*

(iii) *What will be the emf of the cell if concentration of Cd^{2+} is 0·1 M and Ag^+ is 0·2 M ?*

(iv) *Will the cell work spontaneously for the condition given in (iii) above ?*

(c) (i) *What is meant by promoter ? Give an example.* [1]

(ii) *The solubility product of $BaSO_4$ is $1{\cdot}5 \times 10^{-9}$. Find out its solubility in pure water.*** [1]

(iii) *What is the dissociation constant of 0·1 M solution of a weak acid HA which is 4·5% ionized at 20°C ?*** [2]

SECTION–B

*Answer any **two** questions.*

Question 5.

(a) *Give the IUPAC names for the following :* [2]

(i) $Na_3[AlF_6]$ (ii) $[Co(NH_3)_6]Cl_3$

(b) *For the complex ion of $[Fe(CN)_6]^{3-}$:* [3]

(i) *Show the hybridization diagrammatically.*

(ii) *Is it an inner orbital complex or an outer orbital complex ?*

(iii) *State its magnetic property.*

Question 6.

(a) *Give balanced chemical equations for the following :*

(i) *Chlorine gas is passed through cold, dilute NaOH.* [3]

(ii) *Sulphur dioxide gas is passed through NaOH solution.*

(iii) *Zinc is added to sodium argentocyanide solution.*

(b) *Iron is ferromagnetic in nature. Explain why.* [2]

Question 7.

(a) *State the common oxidation state of :* [1]

(i) *Lanthanides* (ii) *Actinides*

(b) *In a given transition series, there is no significant change in the atomic radii of elements with increase in atomic number. Explain why.* [2]

(c) *Give reactions and the conditions required for preparation of the following compounds :* [2]

(i) XeF_6 (ii) $XeOF_4$

SECTION–C

*Answer any **two** questions.*

Question 8.

(a) *Carry out the following conversions :* [3]

(i) *Methyl chloride to acetic acid.*

(ii) *Benzene to benzoic acid.*

(iii) *Ethanol to acetone.*

(b) *Deficiency of what vitamins will cause the following diseases :* [2]

(i) *Night blindness.*

(ii) *Scurvy.*

(c) *Give balanced equations for the following :* [5]

(i) *Glycerol is heated with oxalic acid at 110°C (383 K).*

(ii) *Acetamide is heated with sodium hydroxide.***

(iii) *Acetone reacts with hydrogen in the presence of heated copper.*

Question 9.

(a) *Identify A to F :* [3]

$$A \xrightarrow{LiAlH_4} C_2H_5OH \xrightarrow{PBr_3} B \xrightarrow{KCN} C \xrightarrow{D}$$

$$C_3H_7NH_2 \xrightarrow{HNO_2} E \xrightarrow[K_2Cr_2O_7H^+]{[O]} F$$

(b) *Give one good chemical test to distinguish between the following pairs of organic compounds :* [3]

(i) *Benzaldehyde and acetone.*

(ii) *Methylamine and dimethylamine.*

** Answer is not given due to change in the present syllabus.

(c) *Draw the isomers of 2-hydroxy propionic acid.* ** [1]

(d) *Give an example (equation) for each of the following name reactions :* [3]

(i) *Aldol condensation.*

(ii) *Reimer-Tiemann reaction.*

(iii) *Rosenmund's reduction.*

Question 10.

(a) *An organic compound A has the molecular formula of C_7H_6O. When A is treated with NaOH followed by acid hydrolysis, it gives two products, B and C. When B is oxidised, it gives A. When A and C are each treated separately with PCl_5, they give two different organic products D and E.* [4]

(i) *Identify A to E.*

(ii) *Give the chemical reaction when A is treated with NaOH and name the reaction.*

(b) *Draw a pair of isomers for each of the following and name the type of isomerism :*** [4]

(i) C_4H_{10} (ii) $C_2H_2Cl_2$

(iii) CH_3COCH_3 (iv) $C_4H_{10}O$

(c) *What are polyamides ? Give one example of a polyamide and name its monomers.* [2]

ANSWERS

PART–I

Answer 1.

(a) (i) sp^3, pyramidal.

(ii) mol l^{-1} s^{-1}, s^{-1}.

(iii) methylamine, Hofmann's degradation.

(iv) Glycerol, glycol.

(v) Concentrated hydrochloric acid, concentrated nitric acid.

(b) (i) (4) 0·1 M $K_4[Fe(CN)_6]$

(ii) (2) 16 minutes

(iii) (2) Reduction potential of copper is higher than that of zinc.

(c) (i) Depression in freezing point is a colligative property which depends upon number of particles present in the solution.

$MgCl_2$ when dissolved in water gives three particles per molecules,

($MgCl_2 \rightleftharpoons Mg^{2+} + 2Cl^-$) while urea remains undissociated.

(iv) Besides the van der Waals forces, HF can form intermolecular hydrogen bonds. This is due to the high electronegativity of F. This means more energy is needed to break the hydrogen bond as well as the van der Waals forces in HF and hence higher boiling point.

(v) **Piezoelectricity** : Piezoelectricity is the process of using crystals (like quartz) to convert mechanical energy into electrical energy, or vice versa. Piezoelectric devices include Piezoelectric igniters (utane lighters, gas grills, gas stoves), quartz watches, microphone sensors, etc.

(d) (ii) Urotropine (b) Formaldehyde

(iii) Purine (c) DNA

(iv) Frasch process (d) Sulphur

(v) Hinsberg's reagent (e) Amines.

PART–II

SECTION–A

Answer 2.

(a) (i)

$$X_{alc} = \frac{n_{alc}}{n_{alc} + n_{water}}$$

$$= \frac{46/46}{(46/46)+(18/18)}$$

$$= \frac{1}{2} = 0{\cdot}5.$$

(ii) Given : $= 0{\cdot}65$ atm; $c = 0{\cdot}01$ M

$R = 0{\cdot}0821$ l atm K^{-1} Mol^{-1},

$T = 273 + 27 = 300$ K

Now, $\pi = i \times CRT$

$$0{\cdot}65 = i \times 0{\cdot}01 \times 300 \times 0{\cdot}0821$$

$$i = \frac{0{\cdot}65}{0{\cdot}01 \times 300 \times 0{\cdot}0821}$$

$$= 2{\cdot}639.$$

van't Hoff factor $i = 2{\cdot}639$. It indicates that the electrolyte undergoes dissociation in given solution.

(b) (i) **Faraday's first law of electrolysis** : The law states that the amount of a substance decomposed as a result of electrolysis is directly proportional to the quantity of electricity passed.

$$m = Z \times Q$$

or $m = ZIt$.

(ii) 1 F = charge on 1 mole of e^-

or 96500 coulombs = charge on $6{\cdot}023 \times 10^{23}$ electrons.

$\therefore$ 5 × 200 coulombs

$$= \frac{6{\cdot}023 \times 10^{23} \times 1000}{96500} \text{ electrons charge}$$

$= 6{\cdot}24 \times 10^{21}$ electrons

$= 6{\cdot}24 \times 10^{21}$ electrons will flow.

Answer 3.

(a) (i) **Semiconductors :** These are the crystals which allow only partial conduction of current through them. Their conductivities are intermediate between those of conductors and insulators. Their conductivity increases with increase in temperature.

(ii) Given : Density, $\rho = 3{\cdot}4$ g cm^{-3}, M. mass = 99, Z = 1 (cubic structure), $N_A = 6{\cdot}023 \times 10^{23}$

Now, $$\rho = \frac{Z \times M.Mass}{a^3 \times N_A}$$

$$3{\cdot}4 = \frac{1 \times 99}{a^3 \times 6{\cdot}023 \times 10^{23}}$$

$$a = \sqrt[3]{99 / 3{\cdot}4 \times 6{\cdot}023 \times 10^{23}}$$

Edge length, $a = 3{\cdot}64 \times 10^{-8}$ cm.

(b) (ii) PCl_5 in neutral medium has trigonal bipyramidal geometry.

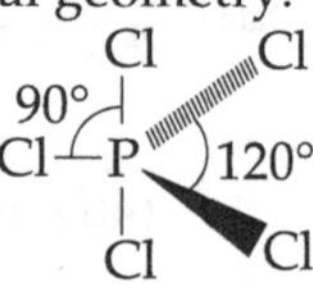

Answer 4.

(b) (i) Cd | Cd^{2+} | | Ag^{1+} | Ag

(ii) Given : $E^\circ_{Ag^+/Ag} = +0{\cdot}80$ V,

$E^\circ_{Cd^{2+}/Cd} = -0{\cdot}40$V.

Now, $$E^\circ_{cell} = E^\circ_{cathode} - E^\circ_{anode}$$

$$= 0{\cdot}80 - (-0{\cdot}40) = 1{\cdot}20 \text{ V.}$$

(iii) $$E_{cell} = E^\circ_{cell} - \frac{0{\cdot}0591}{n} \log \frac{[Product]}{[Reactant]}$$

$$= 1{\cdot}20 - \frac{0{\cdot}0591}{2} \log \frac{[0{\cdot}1]}{[0{\cdot}2]^2}$$

$$= 1{\cdot}20 - \frac{0{\cdot}0591}{2} \times {\cdot}3979$$

$$= 1{\cdot}20 - 0{\cdot}01175 = 1{\cdot}188 \text{ V}$$

(iv) Cell will work spontaneously because E_{cell} is a positive (+ ve) value.

(c) (i) **Promoters :** These are substances which when added to substrate in very small amounts along with the catalyst, enhance the effectiveness of the catalyst. These are not catalysts themselves. Example : In Haber's process iron powder acts as catalyst while Al_2O_3 is a promoter.

SECTION–B

Answer 5.

(a) (i) Sodium hexafluroaluminate(III).

(ii) Hexaamminecobalt(III) chloride.

(b) (i) In $[Fe(CN)_6]^{3-}$ oxidation state of Fe is + 3

Fe atom in (Z = 26) ground state

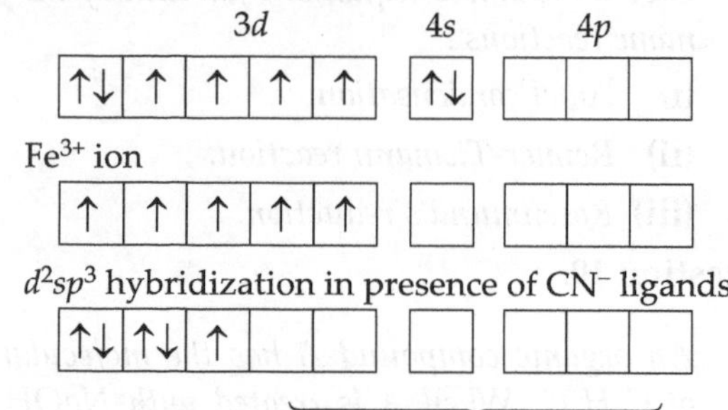

d^2sp^3 hybridization in presence of CN^- ligands

d^2sp^3 hybrid orbitals

(ii) It is an inner orbital complex.

(ii) The ion is paramagnetic due to the presence of an unpaired electron.

Answer 6.

(a) (i) $$2NaOH + Cl_2 \longrightarrow NaCl + NaClO + H_2O$$

(ii) $$2NaOH + SO_2 \longrightarrow Na_2SO_3 + H_2O$$

(iii) $$Zn + 2Na[Ag(CN)_2] \longrightarrow Na_2[Zn(CN)_4] + 2Ag \downarrow$$

(b) Ferromagnetic substances have unpaired electrons and are attracted strongly in a magnetic field.

Fe_{26} has $4s^2 3d^6$ valence shell configuration with four unpaired electrons in $3d$. So it is ferromagnetic when all these four unpaired electrons get aligned in the same direction in a given magnetic field.

Fe_{26} atom's valence shell

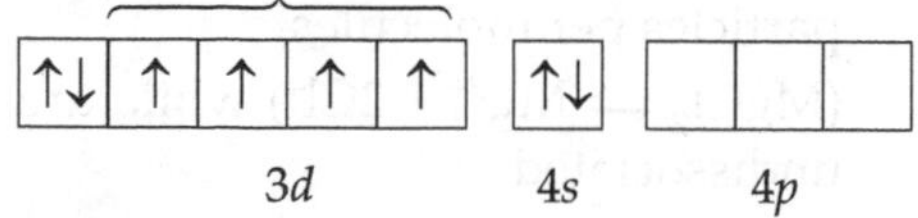

Answer 7.

(a) (i) Oxidation state of lanthanides is + 3. Some lanthanides show + 2, + 4 oxidation state as well.

(ii) Oxidation state of actinides is + 3 but many actinides show + 2, + 4 + 5, + 6, + 7 oxidation states also.

(b) In a transition series the electrons are filled in inner d orbitals so there is no increase in the number of shells but there is an increase in the nuclear charge and an increase in the repulsion between inner orbital electrons also. The two opposite forces *i.e.,* increased attraction by the nucleus and increased repulsion between electrons counter balance each other, as a result there is no significant change in atomic radii of transition element.

(c) (i) $Xe + 3F_2 \xrightarrow[300°C,\ 50\ atm.]{1:20\ ratio,\ Ni\ vessel} XeF_6$

(ii) $XeF_6 + H_2O \xrightarrow[with\ water]{Partial\ hydrolysis} XeOF_4 + 2HF$

SECTION–C

Answer 8.

(a) (i) Methyl chloride to acetic acid.

$$\underset{\text{Methyl chloride}}{CH_3Cl} \xrightarrow[-KCl]{+KCN} CH_3CN \xrightarrow{+2H_2O}$$

$$CH_3COONH_4 \xrightarrow[-NH_4Cl]{+HCl} \underset{\text{Acetic acid}}{CH_3COOH}$$

(ii) Benzene to benzoic acid

$$C_6H_6 + CH_3Cl \xrightarrow[-HCl]{\text{anhydrous } AlCl_3} C_6H_5CH_3$$

$$2C_6H_5CH_3 + 3O_2 \xrightarrow[(CH_3COO)_2Mn]{442-573K} 2C_6H_5COOH + 2H_2O$$

(iii) Ethanol to acetone

$$\underset{\text{Ethanol}}{2C_2H_5OH} \xrightarrow[-2H_2O]{+2O} 2CH_3COOH \xrightarrow[-2H_2O]{+Ca(OH)_2}$$

$$(CH_3COO)_2Ca \xrightarrow{\Delta} \underset{\text{Acetone}}{CH_3COCH_3} + CaCO_3$$

(b) (i) Vitamin A (ii) Vitamin C

(c) (i)

$$\underset{\text{Glycerol}}{\begin{matrix} CH_2OH \\ | \\ CHOH \\ | \\ CH_2OH \end{matrix}} + \underset{\text{Oxalic acid}}{\begin{matrix} COOH \\ | \\ COOH \end{matrix}} \xrightarrow{110°C} \begin{matrix} CH_2O.OC.COOH \\ | \\ CHOH \\ | \\ CH_2OH \end{matrix}$$

$$\xrightarrow[\text{(Water of crystallization) } 110°C]{+H_2O} \underset{\text{Glycerol}}{\begin{matrix} CH_2OH \\ | \\ CHOH \\ | \\ CH_2OH \end{matrix}} + \underset{\text{Formic acid}}{HCOOH} + CO_2$$

(iii) $\underset{\text{Acetone}}{CH_3COCH_3} + H_2 \xrightarrow[\Delta]{Cu} CH_3CHOHCH_3$

Propanol-2

Answer 9.

(a) $A \rightarrow CH_3CHO$, $B \rightarrow C_2H_5Br$, $C \rightarrow C_2H_5CN$, $D \rightarrow LiAlH_4$, $E \rightarrow C_3H_7OH$, $F \rightarrow C_2H_5COOH$.

(b) (i) Benzaldehyde gives silver mirror when heated with Tollen's reagent. Acetone does not give this test.

(ii) Methylamine gives offensive smell of methyl isocyanide when heated with chloroform and alcoholic KOH dimethylamine does not give this test.

(d) (i)

$$\underset{\text{Acetaldehyde}}{2CH_3CHO} \xrightarrow[\text{dil NaOH}]{\Delta \text{ with}} \underset{\text{aldol}}{CH_3-\overset{\overset{OH}{|}}{CH}-CH_2-CHO}$$

(ii)

$$\underset{\text{Phenol}}{C_6H_5OH} + CHCH_3 + 3NaOH \longrightarrow \underset{\text{Salicylaldehyde}}{C_6H_4(OH)CHO} + 3NaCl + 2H_2O$$

$$\underset{\text{Ethanoyl chloride}}{CH_3COCl} + H_2 \xrightarrow[\text{quinolene}]{Pd\ BaSO_4} \underset{\text{Ethanal}}{CH_3CHO} + CHl$$

Answer 10.

(a) (i) $A \rightarrow C_6H_5CHO$, $B \rightarrow C_6H_5CH_2OH$, $C \rightarrow C_6H_5COOH$
$D \rightarrow C_6H_5CHCl_2$, $E \rightarrow C_6H_5COCl$.

(ii) $2C_6H_5CHO + NaOH \rightarrow C_6H_5COONa + C_6H_5CH_2OH$

This is Cannizzaro's reaction.

(c) Polyamides are condensation polymers formed from acids and amines.

Nylon 6, 6 is a polyamide. Monomer units present in it are hexamethylene diamine and adipic acid.

●●

BIOLOGY

2020

Paper – 1
(Theory)
(Maximum Marks : 70)
(Time allowed : Three hours)

*(Candidates are allowed additional 15 minutes for **only** reading the paper.*
They must NOT start writing during this time.)
*This paper comprises **TWO PARTS** – Part I and Part II.*
*Answer **all** questions.*
*Part I consists of **one** question of 20 marks having six subparts.*
Part II consists of Sections A, B and C.
*Section A consists of **seven** questions of **two** marks each.*
*Section B consists of **seven** questions of **three** marks each, and*
*Section C consists of **three** questions of **five** marks each.*
Internal choices have been provided in two questions in Section A, two questions in Section B and in all three questions of Section C.
The intended marks for questions or parts of questions are given in brackets [ ].

QUESTIONS

PART I (20 Marks)

*(Answer **all** questions)*

Question 1.

(a) *Answer the following questions briefly and to the point:* **[8 × 1]**

(i) *How many chromosomes are present in male gamete of a rat?*

(ii) *Why is haemophilia uncommon in females?*

(iii) *Name the diseaseresistant variety of cowpea developed by plant breeding technique.*

(iv) *Define Brood parasitism.*

(v) *Name the vegetative propagule of Bryophyllum.*

(vi) *Which geological era was dominated by reptiles?*

(vii) *Define polygenic inheritance.*

(viii) *What is Dobson unit?*

(b) *Each of the following sub-parts, (i) to (iv) has four choices. Choose the best option in each case :* **[4 × 1]**

(i) *If 10 individuals in a laboratory population of 100 mice die during the period of one year, the death rate in this population will be:*

(1) *110* **(2)** *0.01*
(3) *0.1* **(4)** *90*

(ii) *The flowers which open their petals to expose their reproductive parts to allow pollination are called:*

(1) *Cleistogamous* **(2)** *Geitonogamous*
(3) *Chasmogamous* **(4)** *Autogamous*

(iii) *Which of the following is paired incorrectly:*

(1) *Cyclosporin A - Trichoderma polysporum*
(2) *Streptokinase - Saccharomyces cerevisiae*
(3) *Swiss Cheese - Propionibacterium*
(4) *Penicillin - Penicillium*

(iv) *The pathogen which causes Syphilis:*

(1) *Neisseria* **(2)** *Chlamydia*
(3) *Treponema* **(4)** *Papilloma virus*

(c) *Give one significant contribution of each of the following scientists:* **[4 × 1]**

(i) *F. Griffith*

(ii) *P. Ehrlich*

(d) *Expand the following:* **[2 × 1]**

(i) *IUI* **(ii)** *ADA*

(e) *Define the following:* **[2 × 1]**

(i) *Biopiracy* **(ii)** *Aneuploidy*

(f) *Give the reason for each of the following:* **[2 × 1]**

(i) *Cyanobacteria increase the productivity in paddy fields.*

(ii) *The shape of the pyramid of biomass in an aquatic ecosystem is inverted.*

PART II

SECTION–A (14 Marks)

*(Answer **all** questions)*

Question 2.

Give one significant difference between the following:

(i) *Linkage and crossing over*

(ii) *Transition and transversion* **[2]**

Question 3.

Study the graph given below and answer the questions that follow: **[2]**

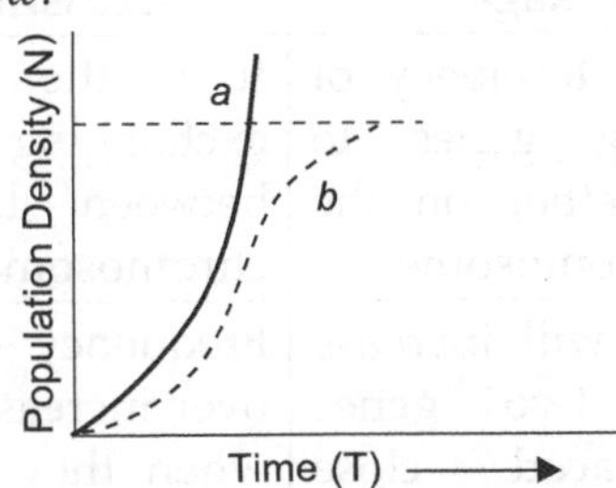

(i) *In the absence of predators which one of the two curves would appropriately depict the prey population? Give a reason.*

(ii) *Time has been shown on X-axis and there is a parallel dotted line shown above. Explain the significance of this dotted line.*

Question 4.

What is biogas? Name any two main constituents of biogas. **[2]**

Question 5.

Explain two characteristics of cancer cells. **[2]**

Question 6. **[2]**

(a) *Draw a labelled diagram of a germinating pollen grain with at least four labellings.*

OR

(b) *Draw a labelled diagram of a mature human ovum with at least four labellings.*

Question 7.

What is outbreeding? How is it useful in animal breeding? **[2]**

Question 8.

(a) *What is biomagnification? Write two effects of biomagnification.* **[2]**

OR

(b) *Write a short note on the contribution of Ahmed Khan of Bangalore.*

SECTION–B (21 Marks)

*(Answer **all** questions)*

Question 9.

Give three adaptations in organisms by which they avoid predation. **[3]**

Question 10.

(a) *Define decomposition. Explain main steps involved in the process of decomposition.* **[3]**

OR

(b) *Write three causes and three effects of cultural eutrophication.*

Question 11. **[3]**

(i) *Write two differences between homologous organs and analogous organs.*

(ii) *Give one example of homologous organs and one example of analogous organs found in plants.*

Question 12.

Describe the process of double fertilisation in angiosperms. What is its significance? **[3]**

Question 13.

What is a bioreactor? Explain important features of a Stirred tank bioreactor. **[3]**

Question 14.

Give three significant differences between asexual reproduction and sexual reproduction. **[3]**

Question 15. **[3]**

(a) *Explain the process of spermatogenesis in humans.*

OR

(b) *Give an account of hormonal control of oogenesis.*

SECTION–C (15 Marks)

*(Answer **all** questions)*

Question 16.

(a) *Explain the various steps involved in Recombinant DNA technology.* **[5]**

OR

(b) *Explain the steps involved in the production of human insulin by Recombinant DNA technology.*

Question 17.

(a) (i) *Give an account of Meselson and Stahl's experiment.* **[5]**

(ii) *What is the significance of Meselson and Stahl's experiment?*

OR

(b) (i) *Describe the Oparin Haldane Theory of origin of life.*

(ii) *The tadpole larva of amphibians resembles fishes. How does this observation support evolution?*

Question 18.

(a) *Draw a labelled diagram to show the life cycle of Plasmodium.* **[5]**

OR

(b) *Draw a labelled diagram to show replication of HIV in human cells.*

ANSWERS

PART I (20 Marks)

Answer 1.

(a) (i) 21 chromosomes .

(ii) Haemophilia is due to the presence of a recessive sex-linked gene h, carried by the X chromosome. A female becomes haemophilic only when both X chromosomes carry the gene (X^hX^h). A female having only one allele for haemophilia (XX^h) appears normal. Thus, the possibility of a human female becoming haemophilic is extremely rare because she has to be homozygous recessive for the trait.

(iii) Pusa Komal variety of Cowpea.

(iv) Brood parasitism, also called social parasitism, is the unique population interaction in which the parasitic bird deposits its eggs in the nests or broods of another individual (host) and let them incubate. For example- a cuckoo lays its eggs in the nest of a crow.

(v) Adventitious buds in the notches of a leaf.

(vi) Mesozoic Era.

(vii) Polygenic inheritance, also known as quantitative inheritance, refers to a single inherited phenotypic trait that is controlled by two or more different genes.

(viii) Dobson unit is used for measuring the thickness of ozone in a column of air from ground to top of the atmosphere. One Dobson Unit is the equivalent of 2.69×10^{16} molecules of ozone per square cm.

(b) (i) (3) 0·1

(ii) (3) Chasmogamous

(iii) (2) Streptokinase - *Saccharomyces cerevisiae*

(iv) (3) *Treponema*

(c) (i) Frederick Griffith illustrated the process of transformation in bacteria *Streptococcus pneumonia.*

(ii) Ecologist Paul Ehrlich gave the rivet popper hypothesis to help in understanding the contribution of species richness.

(d) (i) Intra Uterine insemination

(ii) Adenosine Deaminase

(e) (i) The term biopiracy also known as biological theft is illegal or unethical exploitation of the natural occurring biochemical or genetic resources of a country without providing prior information or compensation to the country.

(ii) Aneuploidy refers to the chromosomal variation due to a loss or a gain of one or more chromosomes deviating from normal genome number for that species, due to non-disjunction of the homologous chromosome.

(f) (i) Cyanobacteria is a nitrogen fixing bacteria and thus help in the management of the nutrient in the soil and maintain the soil structure, increasing the productivity of paddy crops.

(ii) In an aquatic ecosystem, the biomass of the fish (top consumer) is greater than the producers (phytoplankton).

PART II
SECTION–A (14 Marks)

Answer 2.

(i)

Linkage	Crossing Over
It is the tendency of inheriting genes to stay together on the same chromosome	It is the process of exchanging genes between homologous chromosomes
Linkage will increase if the two genes are placed close to each other on a chromosome.	Frequency of crossing over increases in genes when they are placed far from each other.
More parental lines are produced by linkage.	More recombinants are produced by crossing over
Linkage decreases with age.	Crossing over increases with age.
It helps to maintain the parental traits in offspring	It leads to the variation in parental traits

(ii)

Transition	Transversion
In transition, a purine base is substituted with a purine base or a pyrimidine base is substituted with a pyrimidine base. (C ↔ T or A ↔ G)	In transversion, a purine base is substituted with a pyrimidine or a pyrimidine with a purine.

Answer 3.

(i) The curve *'a'* would depict the prey population as there would be no competition for food and shelter resources in the absence of predators, leading to their exponential growth with time.

(ii) The dotted line in the above graph represents the carrying capacity (K) of the population. It represents the size of a population that the environment can hold by providing necessary resources. It signifies the limit of habitat, i.e. limited resources in a given habitat to support growth up to a certain level beyond which no further growth can take place.

Answer 4.

Biogas is the mixture of gases produced by the anaerobic breakdown of organic matter such as agricultural waste, manure, municipal waste, plant material, sewage, green waste or food waste by bacteria.

The two main constituents of biogas are methane and carbon dioxide.

Answer 5.

Characteristics of cancer cells:

1. Cancer cells show uncontrolled mitotic divisions which result in unorganised growth.
2. Cancer cells lose their property of differentiation and proliferation.
3. Cancer cells also fail in Apoptosis or programmed cell death.
4. Cancer cells acquire the capacity to draw out their supply of blood and blood vessels (Angiogenesis).

Answer 6.

(a)

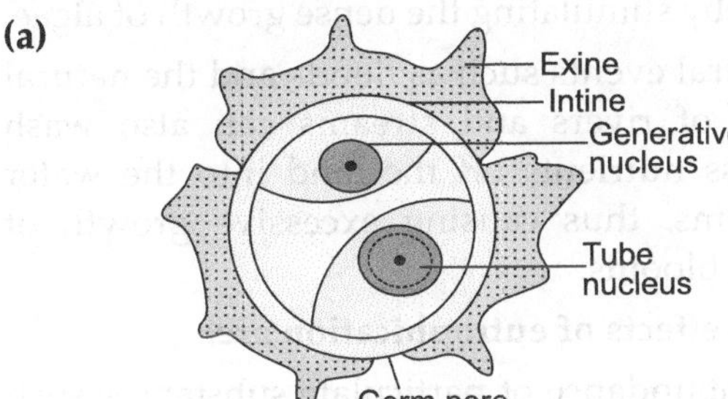

OR

(b) **Mammalian Oocyte**

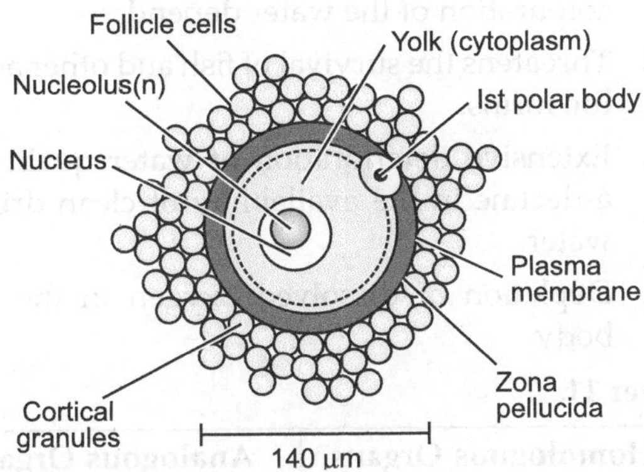

Answer 7.

Out-breeding is the breeding of the unrelated animals, which may be between individuals of the same breed but having no common ancestors for 4-6 generations (out-crossing) or between different breeds (cross-breeding) or different species (interspecific hybridisation).

Significance of outbreeding in animals:

1. Outcrossing is the best breeding method for animals that are below average in productivity in milk production, growth rate in beef cattle, etc.
2. A single outcross often helps to overcome inbreeding depression.
3. Cross-breeding allows the desirable qualities of two different breeds to be combined. The progeny hybrid animals may themselves be used for commercial production.

Answer 8.

(a) Biomagnification refers to the accumulation of non-biodegradable toxic substances in the food chain.

Following are the effects of biomagnification on living organisms and the environment:

1. Biomagnification makes humans more prone to cancer, kidney problems, liver failure, birth defects, respiratory disorders, and heart diseases.
2. The pesticides and chemicals such as DDT and mercury released into the lakes and rivers are ingested by the aquatic organisms. These get accumulated in their body tissues and are transferred to other organisms that feed on them.

OR

(b) Ahmed Khan, a plastic sack manufacturer in Bengaluru was successful in finding a problem of plastic waste accumulation :

1. In collaboration with R. V. College of Engineering and the Bangalore City Corporation, he developed polyblend, a powdered structure manufactured from waste plastic which when mixed with bitumen (an important ingredient in constructing roads), increased the water repelling property of roads that increased the average lifetime of roads three times.
2. The raw material for creating polyblend is any plastic film waste. So, against the price of ₹ 0.40 per kg that rag pickers had been getting for plastic waste, Khan now offers ₹ 6. So, he was successful in finding a problem with plastic waste accumulation.

SECTION–B (21 Marks)

Answer 9.

Prey animals must constantly be on the lookout for their predators. To survive, they must escape from or defend themselves against those animals trying to eat them. Thus, they developed various adaptations to avoid predation.

1. **Camouflage:** Camouflage is an adaptation in which a species evolves to resemble its background. It is the technique used by many animals of resembling and blending in with their surroundings. It is a kind of colouring, body shape, and/or behaviour which animals use to confuse, distract, startle, or hide. For example, many species of frogs and insects camouflage in their surroundings and escape their predators.

2. **Highly developed senses:** Prey animals often rely on escape as their form of protection. Therefore, they must have a well-developed sense of sight, hearing, and smell so they can detect the presence of a predator in enough time to escape.
3. **Warning Signals:** Warning signals are used to scare away predators and to warn other animals. Some animals are poisonous, taste bad, or have a painful bite or sting. Many of these animals are brightly coloured with bold patterns. These colours serve as a warning signal to predators. For example, a skunk is associated with its bad smell, when threatened.
4. **Defensive Weapons:** Prey animals have adaptations to fight or avoid their predators. These weapons could be physical like thorns, wings, hooves, teeth, claws, antlers, etc or chemicals, like snakes have venom.
5. **Behavioural Adaptations:** Some animals trick their predators into thinking they are hurt or dangerous. Others will congregate into large groups or have highly adapted forms of hiding from their attackers.

Answer 10.

(a) Decomposition is the process that involves the breakdown of complex organic matter or biomass from the body of dead plants and animals with the help of decomposers into inorganic raw materials such as carbon dioxide, water and other nutrients.

The various steps involved in decomposition are as follows:

1. **Fragmentation:** It is the first step in the process of decomposition. It involves the breakdown of detritus into smaller pieces by the action of detritivores such as earthworms.
2. **Leaching:** The fragmented particles may contain a lot of water-soluble nutrients which are inorganic in nature. These nutrients get dissolved in the water and seep into the soil and get precipitated in the process of leaching.
3. **Catabolism:** Once the complex material is broken down into smaller particles and the inorganic nutrients are removed, it is time to convert the detritus into simpler inorganic compounds. This process is carried out by various fungal and bacterial enzymes by the process of catabolism.
4. **Humification:** The next step is humification which leads to the formation of a dark-coloured colloidal substance called humus, which acts as a reservoir of nutrients for plants.
5. **Mineralisation:** The humus is further degraded by the action of microbes, which finally leads to the release of inorganic nutrients into the soil. This process of releasing inorganic nutrients from the humus is known as mineralisation.

OR

(b) Various causes for eutrophication include:

1. Eutrophication is predominantly caused by the use of phosphorus and nitrate-rich fertilizers.
2. Direct sewage discharge and industrial wastes result in high amounts of chemical nutrients, thereby stimulating the dense growth of algae.
3. Natural events such as floods and the natural flow of rivers and streams can also wash excess nutrients off the land into the water systems, thus causing excessive growth of algal blooms.

Adverse effects of eutrophication are:

1. The abundance of particulate substances such as phytoplankton, zooplankton, bacteria, fungi and debris on which the turbidity and colouration of the water depend.
2. Threatens the survival of fish and other aquatic life forms.
3. Extensive deterioration of water quality and a decline in the availability of clean drinking water.
4. Depletion of dissolved oxygen in the water body.

Answer 11.

(i)

Homologous Organs	Analogous Organs
Those organs which are similar in their structure, morphology and anatomy but different in their functions are known as homologous organs.	Those organs which are different in their structure, morphology and anatomy but similar in their functions are called analogous organs.
Inherited from common ancestors.	They are inherited from different ancestors.
They have a similar developmental pattern.	Dissimilar developmental pattern.

(ii) **Homologous organs:** Leaves of Pitcher plant and Cactus are homologous in their functions.

Analogous organs: Thorns of Bougainvillea and tendrils of Cucurbita are both modifications of a stem arising from axillary bud but perform different functions like protection and climbing, respectively.

Answer 12.

Double fertilisation in angiosperms involves a fusion of a female gametophyte with two male gametes. The haploid sperm-egg combines to form a diploid zygote. The other sperm nucleus fuses with the other two haploid polar nuclei of the megagametophyte to form a triploid nucleus, which develops into an endosperm and is called the primary endosperm nucleus. This entire phenomenon of fertilisation that involves the fusion of the egg and one male gamete and the fusion of the other male gamete with the secondary or the polar nuclei is called double fertilisation.

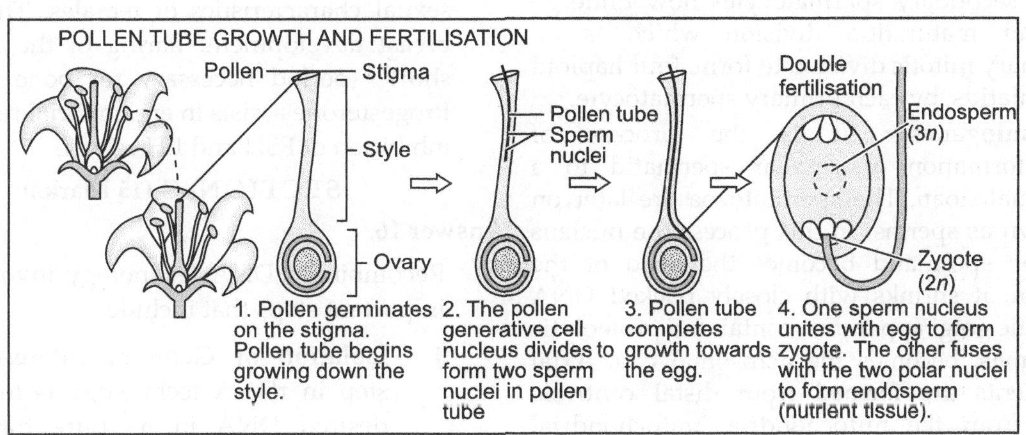

Significance: Double fertilisation leads to the development of triploid endosperm which provides sufficient food material to the developing healthy seeds. It introduces genetic variations in the species. It restores the diploid condition by fusion of haploid male and female gametes

Answer 13.

A bioreactor is a vessel or a fermenter like an apparatus that is used to grow organisms such as bacteria or fungi under controlled environmental conditions e.g. pH, temperature, dissolved oxygen, etc. It can be used for the production of compounds of pharmaceutical importance, antibodies or vaccines.

A stirred tank reactor is cylindrical has a curved base that assists in the mixing of the reactor contents. The stirrer facilitates even mixing and oxygen availability throughout the bioreactor. Alternatively, air can be bubbled through the reactor. The function of the agitation system is to provide good mixing and thus increase mass transfer rates through the bulk liquid. The agitation system consists of the agitator and the baffles that are used to break the liquid flow to increase turbulence and mixing efficiency.

The components of a bioreactor are:

1. An agitator system
2. An oxygen delivery system
3. A foam control system
4. Temperature control system
5. pH control system
6. Sampling ports to withdraw small volumes of culture periodically.

Answer 14.

Asexual Reproduction	Sexual Reproduction
Single parent is involved.	Two parents are involved.
There is no fusion of gametes.	Fusion of male and female gametes occurs.
Somatic cells act as reproductive units	Sex cells act as reproductive units.
Genetic makeup of offspring is exactly similar to the parent	Offspring are genetically different from parents
Meiotic divisions may or may not present.	Meiotic divisions are present
Rate of production of offspring is rapid	Rate of production of offspring is slow.

Answer 15.

(a) The process of formation of sperms is called spermatogenesis. It occurs in the male gonads-testes. Testes are made up of many seminiferous tubules lined by germinal epithelium. Cells of this layer divide to form spermatozoa in the following four steps:

1. **Multiplication Phase:** At sexual maturity, the undifferentiated primordial germ cells divide several times by mitosis to produce a large number of spermatogonia. These spermatogonia move towards the lumen of seminiferous tubules and enter the growth phase.
2. **Growth Phase:** Each spermatogonium actively grows to a larger spermatocyte by obtaining

nourishment from the nursing cells. They are now called primary spermatocytes.

3. **Maturation Phase:** Each primary spermatocyte undergoes two successive divisions, called maturation divisions. The first maturation division is reductional or meiotic. Hence, the primary spermatocyte divides into two haploid daughter cells called secondary spermatocytes. Both secondary spermatocytes now undergo second maturation division which is an ordinary mitotic division to form, four haploid spermatids, by each primary spermatocyte.
4. **Spermiogenesis:** It is the process of transformation of circular spermatid to a spermatozoan. The spermatozoa are later on known as sperms. In this process the nucleus of the spermatid becomes the head of the sperm, it shrinks with closely packed DNA and the golgi apparatus, containing proteolytic enzymes, becomes the acrosomal cap. Axial filaments are formed from distal centriole and from the mitochondria, mitochondrial spiral is formed. Also the loss of cytoplasm takes place and only a sheath remains around mitochondria.

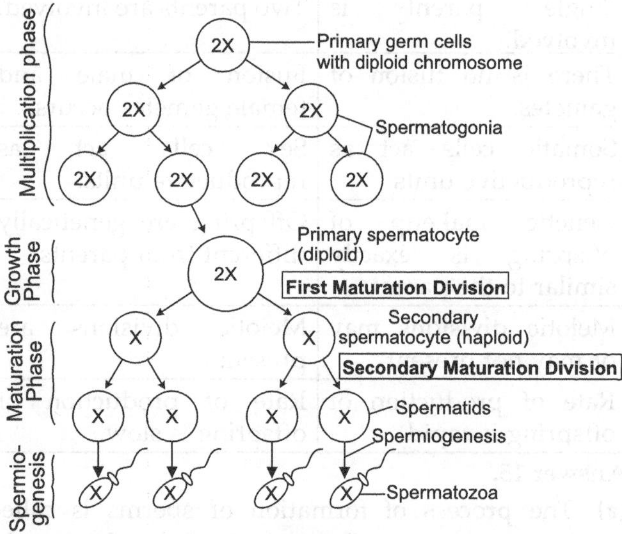

OR

(b) In females, the ovarian cycle is regulated by hormones secreted by the hypothalamus, pituitary, and ovaries. The hypothalamic hormone GnRH (gonadotropin-releasing hormone) causes the release of FSH (follicle-stimulating hormone) and LH (luteinizing hormone) from the anterior pituitary. Simultaneously, estrogen and progesterone are released from the developing follicles present in ovaries. FSH stimulates the development of ova, which further develop in structures called follicles. Follicle cells in turn produce inhibin, which inhibits FSH production. LH also plays a role in the development of ova, as well as in the induction of ovulation and stimulation of estradiol and progesterone production by the ovaries. Estradiol and progesterone are steroid hormones that prepare the body for pregnancy. Estradiol is the reproductive hormone in females that assists in endometrial regrowth, ovulation, and calcium absorption; it is also responsible for the secondary sexual characteristics of females. These include breast development, flaring of the hips, and a shorter period necessary for bone maturation. Progesterone assists in endometrial regrowth and inhibition of FSH and LH release.

SECTION–C (15 Marks)

Answer 16.

(a) Recombinant DNA technology involves several important steps that include:

1. **Isolation of Gene of Interest:** The first step in rDNA technology is to isolate the desired DNA in its pure form i.e. free from other macromolecules. Since DNA exists within the cell membrane along with other macromolecules such as RNA, polysaccharides, proteins, and lipids, it must be separated and purified which involves enzymes such as lysozymes, cellulase, chitinase, ribonuclease, proteases etc.

2. **Restriction Enzyme Digestion:** Once DNA is purified; restriction enzymes are used to isolate the gene of interest. Restriction enzymes act as molecular scissors that cut DNA at specific locations. This step involves the incubation of the purified DNA with the selected restriction enzyme, at conditions optimal for that specific enzyme. The vector DNA is also processed using the same procedure.

3. **Amplification Using PCR:** Polymerase Chain Reaction or PCR is a method of making multiple copies of a DNA sequence using the enzyme – DNA polymerase in vitro. It helps to amplify a single copy or a few copies of DNA into thousands to millions of copies. The cut fragments of DNA can be amplified using PCR and then ligated with the cut vector.

4. **Ligation of DNA Molecules:** The purified DNA and the vector of interest are cut with the same restriction enzyme and ligated using the enzyme 'DNA ligase'. The resulting DNA molecule is a hybrid of two

DNA molecules – the interest molecule and the vector. Hence, this new hybrid DNA molecule is also called a recombinant DNA molecule.

5. **Insertion of Recombinant DNA into Host:** In this step, the recombinant DNA is introduced into recipient host cells mostly, a bacterial cell. This process is called transformation. Hosts are treated to make them competent to accept new DNA. The processes used may be thermal shock, Ca^{2+} ion treatment, electroporation etc.

6. **Isolation of Recombinant Cells:** The transformation process generates a mixed population of transformed and non-transformed cells. The selection process involves filtering the transformed host cells only. Gel electrophoresis is often used to separate pieces of DNA which in turn makes it possible to determine which plasmids took up the gene of interest. Several other methods involve growing the cells in antibiotic (e.g. tetracycline) agar plates.

OR

(b) The first major product of genetic engineering was human insulin. It can be produced using recombinant DNA technology by the following steps:

1. The gene responsible for producing human insulin protein is isolated.
2. Then, a circular piece of DNA called a plasmid is removed from a bacterial cell. Special enzymes called restriction enzymes are used to cut the plasmid ring.
3. With the open plasmid ring, the gene for insulin is inserted into the plasmid ring and the ring is closed.
4. The bacterial DNA now contains the human insulin gene and is inserted into a bacteria.
5. Many plasmids with the insulin gene are inserted into many bacterial cells. When the bacterial cells reproduce by dividing, the human insulin gene is also reproduced in the newly created cells.
6. Human insulin protein molecules produced by bacteria are gathered and purified. By culturing the genetically engineered bacteria, limitless supplies of insulin may be produced.

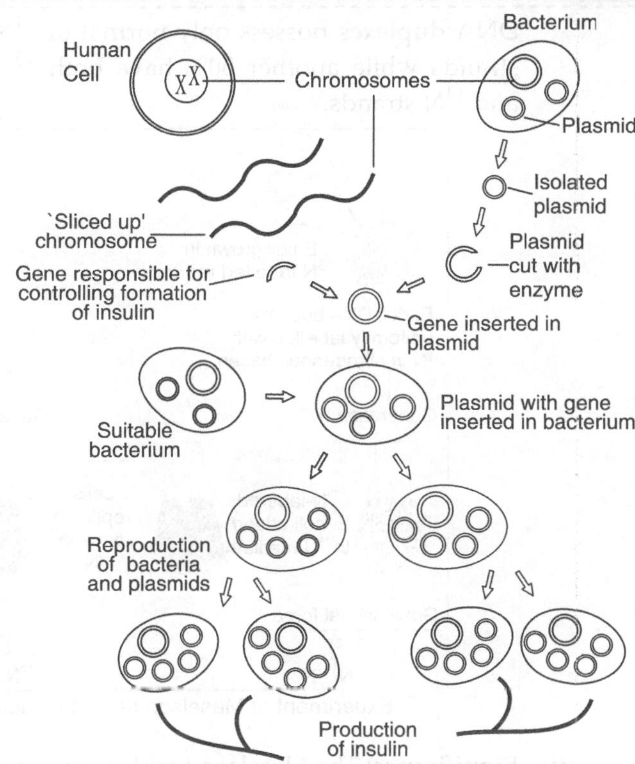

Answer 17.

(a) (i) Semiconservative replication of DNA was proved by the work of Mathew Meselson and Franklin Stahl (1958). They grew *Escherichia* coli for many generations in a medium having a heavy isotope of nitrogen ^{15}N, till the bacterial DNA became completely labelled with heavy isotope. The labelled bacteria were then shifted to a fresh medium having normal or ^{14}N nitrogen. Samples were taken for each generation and the DNA was tested for the heavy isotope of nitrogen through density gradient centrifugation using caesium chloride. They found that the DNA of the first generation was hybrid or intermediate between ^{15}N and ^{14}N.

The second generation of bacteria contained two types of DNA, 50% light and 50% intermediate. This observation is possible only if the two strands of DNA duplex separate at the time of replication and act as a template for the synthesis of new complementary strands of DNA having normal or ^{14}N. This will produce two DNA duplexes with one old strand (^{15}N) and one new strand (^{14}N).

During the formation of second-generation, ^{15}N and ^{14}N strands of DNA duplex separate to function as templates so that 50% of new

DNA duplexes possess only normal or ^{14}N strands while another 50% have both ^{15}N and ^{14}N strands.

In this way at each replication, one strand of parent DNA is conserved in the daughter while the second is freshly synthesised.

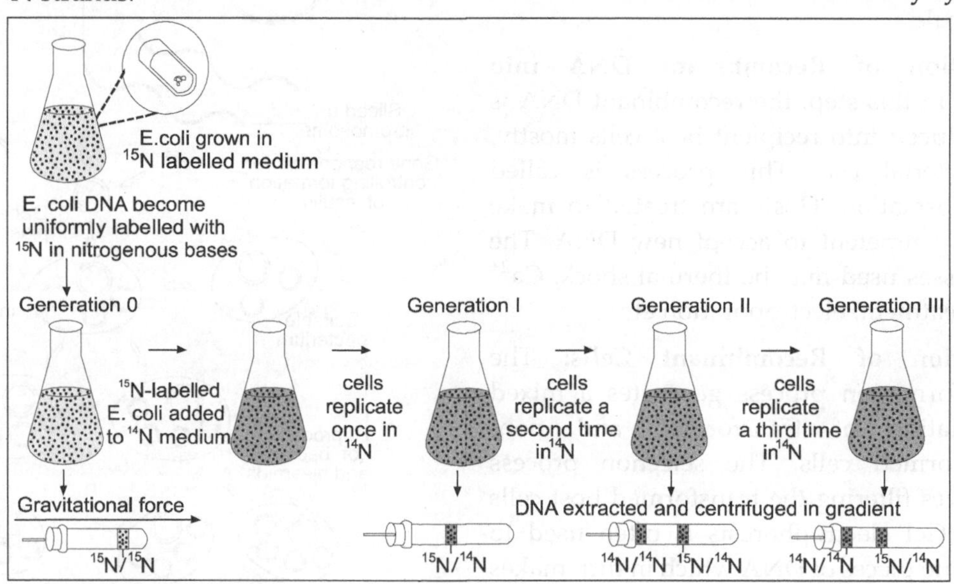

Experiment of Meselson and Stahl demonstrating semi-conservative replication

(ii) **Significance:** The Meselson-Stahl experiment supported Watson and Crick's hypothesis that DNA replication was semi-conservative Each of the two new molecules has one strand of new material and one strand of old material.

This enabled researcher to explain how DNA replicates thereby provides a physical basis for the genetic phenomenon of heredity and diseases, and how organisms preserved and transferred genetic information of DNA to their offspring.

Based on the result, Meselson and Stahl made three conclusions:

1. Each DNA molecule divided evenly between the two subunits of DNA, and that the subunits stayed intact throughout the observed replication cycles
2. If a parent passes on one subunit of DNA to its offspring, then half of the parental DNA is conserved in the offspring DNA, and half of the parental DNA is not.
3. For every parental DNA molecule, two new molecules were made. Therefore, the amount of DNA after each replication increased by a factor of two.

OR

(b) (i) Oparin-Haldane theory is also called a chemical theory or naturalistic theory. They believed that the origin of life was a gradual process of changes thought to have occurred about 4000-3600 million years ago. According to Oparin Haldane theory of evolution, life originated on early earth through physicochemical processes of atoms combining to form molecules, molecules react to produce inorganic and organic compounds. Organic compounds interacted to produce different forms of macromolecules which organised to form the first living system or cells. Thus, according to the theory 'life' originated on earth spontaneously from non-living matter. First inorganic compounds and then organic compounds were formed in accordance with ever-changing environmental conditions. This is referred to as chemical evolution which cannot occur under present environmental conditions upon the earth. The conditions suitable for the origin of life existed only upon primitive earth.

1. Free atoms ↓	(Hydrogen, Carbon, Nitrogen)
2. Inorganic molecules ↓	(H_2, H_2O, CH_4, NH_3, CO_2, HCN)
3. Simple organic molecules ↓	(Sugars, Fatty acids, Glycerol, Amino Acids, Nitrogen bases *i.e.*, Purines, Pyrimidines)
4. Complex organic molecules ↓	(Polysaccharides, Proteins, Lipids, Nucleotides, Nucleic acids)
5. Coacervates ↓	(Cluster of complex organic molecules bounded by fatty acid membrane, division by budding)
6. Eobionts ↓	(Nucleic acid controlled coacervates)
7. Prokaryotes ↓	(Without organised nucleus)
8. Eukaryotes	(With well organised nucleus)

(ii) Haeckel formulated his theory as "Ontogeny recapitulates phylogeny", also known as the recapitulation theory. Ontogeny is the growth (size change) and development (structure change) of an individual organism; phylogeny is the evolutionary history of a species. Haeckel claimed that the development of advanced species passes through stages represented by adult organisms of more primitive species.

The resemblance of tadpole larva of amphibians with fishes shows that the amphibians might have evolved from fishes. For example, cold blooded animals, respiration by gills early in life, laying eggs in water, etc.

However, with due course of their evolution, amphibians divert from fishes with more adaptation towards land habitat, such as presence of lungs, 3-chamberd heart etc. Thus, amphibians are evolved from fishes.

Answer 18.

(a)

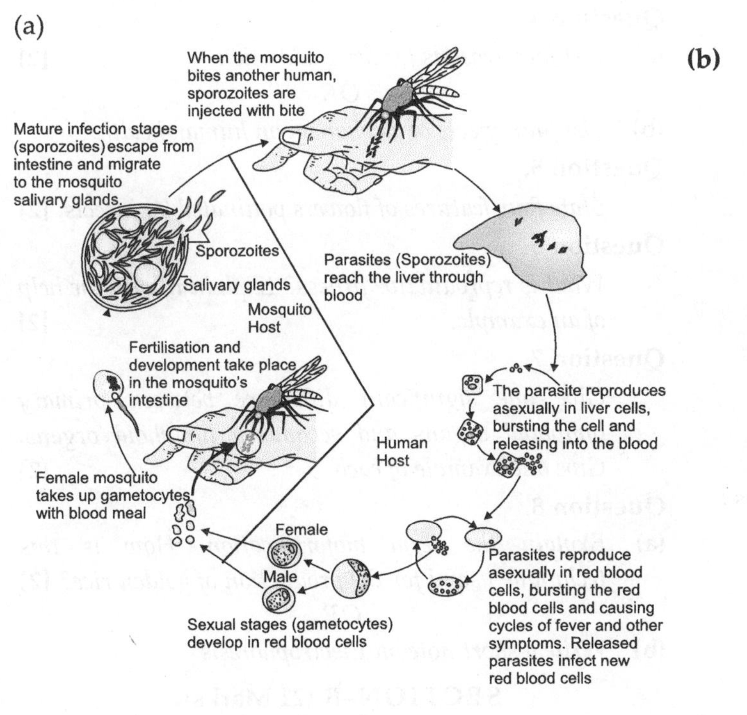

OR

(b)

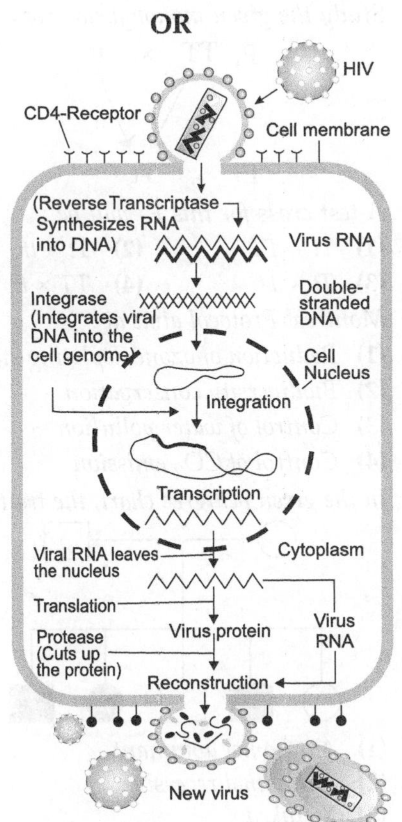

BIOLOGY 2019

QUESTIONS

PART I (20 Marks)

*(Answer **all** questions)*

Question 1.

(a) *Answer the following questions briefly and to the point:* **[8 × 1]**

(i) *Name the antibody which is most effective in allergies.*

(ii) *What is the function of GEAC?*

(iii) *What is a clone?*

(iv) *What do detritus food chains begin with?*

(v) *Give the full form of EFB.*

(vi) *How many chromosomes are present in meiocytes of a fruit fly?*

(vii) *Name the common ancestor of apes and man.*

(viii) *Give the scientific term used for the preservation of germplasm at a very low temperature.*

(b) *Each of the following sub-parts, (i) to (iv) has four choices. Choose the best option in each case :* **[4 × 1]**

(i) *Eyelids in human foetus separate in :*

(1) *14 weeks* **(2)** *16 weeks*

(3) *24 weeks* **(4)** *40 weeks*

(ii) *Study the given monohybrid cross :*

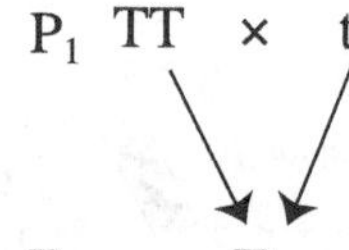

A test cross for this F_1 will be :

(1) *Tt × TT* **(2)** *Tt × tt*

(3) *Tt × Tt* **(4)** *TT × tt*

(iii) *Montreal Protocol aims at :*

(1) *Reduction of ozone depleting substances*

(2) *Biodiversity conservation*

(3) *Control of water pollution*

(4) *Control of CO_2 emission*

(iv) *In the given pedigree chart, the trait shown is :*

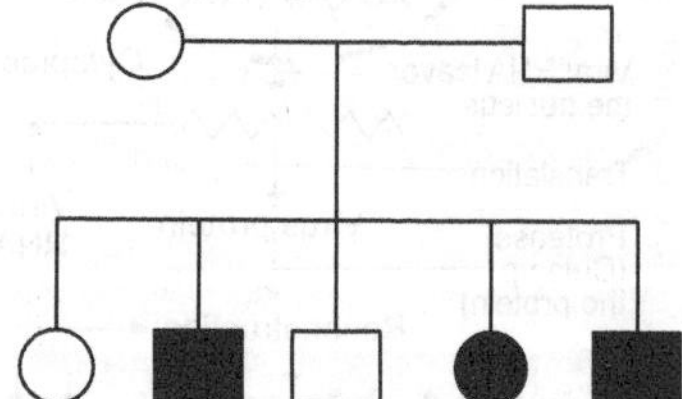

(1) *Autosomal dominant*

(2) *Autosomal recessive*

(3) *X-linked*

(4) *Y-linked*

(c) *Give one significant contribution of each of the following scientists :* **[4 × 1]**

(1) *Wallace* **(2)** *R. Mishra*

(3) *G. Gamow* **(4)** *Sanger*

(d) *Define the following :* **[2 × 1]**

(i) *Carrying capacity*

(ii) *Homologous chromosomes*

(e) *Give a reason for each of the following :* **[2 × 1]**

(i) *Bagging is essential in artificial hybridisation.*

(ii) *Climax stage is achieved quickly in secondary succession as compared to primary succession.*

PART II

SECTION–A (14 Marks)

*(Answer **all** questions)*

Question 2.

Enumerate any four essential features of good and effective poultry farm management practices. **[2]**

Question 3.

What is a single cell protein? How is it significant for human welfare? **[2]**

Question 4.

(a) *List four reasons for drug addiction.* **[2]**

OR

(b) *List four effects of alcoholism on human health.*

Question 5.

State four features of flowers pollinated by insects. **[2]**

Question 6.

What is reproductive fitness? Explain it with the help of an example. **[2]**

Question 7.

Give one significant difference between primary lymphoid organs and secondary lymphoid organs. Give one example of each. **[2]**

Question 8.

(a) *Explain the term biofortification. How is this technique useful for the production of golden rice?* **[2]**

OR

(b) *Write a short note on Electrophoresis.*

SECTION–B (21 Marks)

*(Answer **all** questions)*

Question 9.

Explain the evolution of long neck of giraffe according to Charles Darwin. **[3]**

Question 10.

(a) *Draw a labelled diagram of the T.S. of a mature anther.* **[3]**

OR

(b) *Draw a labelled diagram of the internal structure of human ovary.*

Question 11.

Describe the structure of a nucleosome with the help of a well-labelled diagram. **[3]**

Question 12.

(a) *Explain the Rivet Popper hypothesis.* **[3]**

OR

(b) *Define :*

(1) Standing crop

(2) Stenothermal organisms

(3) Niche

Question 13.

Give the biological names of the following : **[3]**

(i) *The mould from which penicillin is obtained.*

(ii) *Baker's yeast.*

(iii) *The microbe used to control insect larvae growing on cotton.*

(iv) *The microbe used to produce Swiss cheese.*

(v) *The fungus that is being developed as a bio-control agent.*

(vi) *A symbiotic nitrogen fixing bacterium found in root nodules.*

Question 14.

Explain the different types of endosperms in angiosperms. **[3]**

Question 15.

A homozygous pea plant with round seed coat and yellow cotyledons is crossed with another homozygous pea plant having wrinkled seed coat and green cotyledons. **[3]**

(i) *Give the type of gametes produced by plants of F_1-generation.*

(ii) *Give the dihybrid phenotypic ratio with the corresponding phenotypes.*

(iii) *State the Mendel's principle involved in this cross.*

SECTION–C (15 Marks)

*(Answer **all** questions)*

Question 16.

(a) *Describe the physico-chemical events that take place during fertilisation in humans.* **[5]**

OR

(b) (i) *Define and give the role of amniocentesis.*

(ii) *Name the causative agent and give any one symptom of Gonorrhoea.*

(iii) *What is the significance of dispersal of seeds? Give any two points.*

(iv) *What are seasonal breeders? Give an example.*

(v) *How is the chromosome number maintained in sexually reproducing organisms?*

Question 17.

(a) (i) *What are restriction endonucleases? Give the rules of their nomenclature.* **[5]**

(ii) *Explain the mechanism of action of restriction endonucleases that makes them suitable for genetic engineering.*

OR

(b) (i) *Explain what are the desirable characteristics of an ideal cloning vector used in rDNA technology.*

(ii) *Describe two vectorless methods of gene transfer used in rDNA technology.*

Question 18.

(a) *Give a graphic representation of carbon cycle in nature.* **[5]**

OR

(b) *Give a graphic representation of phosphorus cycle in nature.*

ANSWERS

PART I (20 Marks)

Answer 1.

(a) (i) Ig E.

(ii) To appraise activities involving large scale use of hazardous microorganisms and recombinants in research and industrial production from the environmental angle.

(iii) A clone is a group of identical cells that share a common ancestry, *i.e.,* they are derived from the same cell. They are morphologically and genetically similar individuals.

(iv) Dead and decomposed organic matter.

(v) The European Federation of Biotechnology.

(vi) 8 or 4 pairs

(vii) Dryopithecus or Ramapithecus

(viii) Cryopreservation.

(b) (i) (3) 24 weeks

(ii) (2) Tt × tt

(iii) (1) Reduction of ozone depleting substances

(iv) (2) Autosomal recessive

(c) (1) **Alfred Russel Wallace** discovered the concept of evolution by natural selection and divided the earth into biogeographical regions.

(2) **R. Mishra** is the Father of Ecology in India. His research laid the foundations for understanding of tropical communities and their succession, environmental responses of plant populations and productivity and nutrient cycling in tropical forest and grassland ecosystems.

(3) **George Gamow**, a physicist suggested that the genetic code should be made up of three nucleotides. He stated that since there are only four bases and if they have to code for 20 amino acids, the code should constitute a combination of bases.

(4) **Frederick Sanger** made several important contributions to genetics and molecular biology for DNA and protein sequencing. He was awarded a Nobel Prize in Chemistry for his work on the structure of proteins, especially that of insulin.

(d) (i) The carrying capacity is the maximum population size of the species that the environment can sustain indefinitely, given the food, habitat, water and other necessities available in the environment without the environmental degradation.

(ii) A pair of corresponding chromosomes of same shape, size and sequence of genes, one from each parent.

(e) (i) Bagging is essential in artificial hybridisation in order to prevent contamination from foreign, self and unwanted pollen grains.

(ii) Due to the presence of nutrients and the spores that are left behind, climax stage is achieved quickly in secondary succession as compared to primary succession.

PART II

SECTION–A (14 Marks)

Answer 2.

Following poultry farm management practices are effective :

1. **Farm Management :** Poultry farming involves both egg layers and broilers, for high yield.
2. **Shelter :** Poultry birds are needed to be maintained in a strictly hygienic manner with proper housing. A good and clean environment with proper ventilation should be provided.
3. **Food :** Broilers are raised for meat. They are fed with protein and fat rich food. Vitamin supplementary helps them for good growth rate. Egg layers are fed with diet rich in vitamins A and K which help in more egg production.
4. **Disease Management :** Poultry farming should have a disease-free condition. Regular and proper cleaning, sanitation, spraying of pesticides and disinfectants should be done. Vaccinations help to prevent the chance of occurrence and spreading of infections.

Answer 3.

Single cell protein (SCP) refers to dried microbial cells or protein extracted from pure microbial cell culture. It refers to the whole cells of bacteria, yeast, filamentous fungi or algae. SCP is not pure protein. It also contains carbohydrates, lipids, nucleic acids, mineral salts and vitamins. The composition depends upon the organism and the substrate on which it grows.

Significance for human welfare :

1. They can be used as a protein supplement for both humans or animals. Microorganisms like algae, fungi, yeast and bacteria have very high protein content in their biomass.
2. It is cost effective and easy to digest.
3. Reduces pressure on agriculture.
4. The fat content is low and there is no pathogenicity and toxicity present in SCP.

Answer 4.

(a) Following could be the reason for drug addiction:

1. **Stress :** To escape financial, career, family or social pressure, person abuse the drugs. Depression, frustration and other emotional turbulence contribute to the reason for drug addiction.
2. **Curiosity :** Drugs advertisements in news-papers, literature and on radio make a person curious to have personal experience of the taste of a particular drug.
3. **Peer pressure :** Constant description by friends about the "good feeling" creates a temptation.
4. **To overcome frustrations and depressions :** The desire to get solace or relief from personal problems initiates the use of drugs.

OR

(b) Following could be the effect of Alcoholism :

1. Excessive alcohol consumption leads to organ failure or prevent them from properly balancing blood sugar levels.
2. A person suffering from alcohol abuse commonly, becomes dull, anti-social, depressed, tired, aggressive, etc. leading to low self esteem.
3. Dilated pupils and mood swings, loss of sleep (insomnia), lowers immunity.
4. Continous alcohol usage ultimately leads to failure of vital organs like liver and kidneys.

Answer 5.

Characteristics of flowers pollinated by insects are :

1. These flowers are often attractive, with bright petals and are fragrant to attract the insect.
2. They often have small, compact and sticky stigmas to allow the insect to perch on it.
3. Pollen grains are sticky and spiny so that they can stick easily to insect's body parts.
4. Many of the insect-pollinated flowers also secrete nectar which attracts bees, butterflies or other similar insects to the flowers.

Answer 6.

Reproductive fitness is the individuals reproductive success equal to average contribution to gene pool. For example : Ascaris lays 2 crore eggs.

Answer 7.

Primary lymphoid organs	Secondary lymphoid organs
Contains either B cells or T cells.	Contains both T and B cells.
Antigen cannot enter. For example : Thymus, Bone marrow.	Antigen can enter. For example : Spleen, tonsils.

Answer 8.

(a) Biofortification is the enrichment of crops with nutrients of higher levels of vitamins, minerals, proteins and fat content. Golden rice is an example of GM crop. The grains of golden rice are yellow as it contains β-carotene which is precursor of vitamin A. It is produced to control night blindness and to overcome vitamin A deficiency.

OR

(b) When electric field is applied across two electrodes that are totally submerged in a colloidal solution, the particles (colloidal) tend to move towards the electrode depending on their electric charge. This movement of particles under the effect of electric field is known as electrophoresis. When extracted DNA is subjected to electro-phoresis following points are observed :

1. DNA fragments are separated by agarose gel electrophoresis.
2. DNA fragments being negatively charged move towards positive electrode.
3. DNA Fragments are separated according to size/ charge ratio.
4. Separated fragments, when stained with ethidium bromide and exposed to UV rays, they form orange coloured bands.
5. The bands formed are extracted by elution or southern blotting.

SECTION–B (21 Marks)

Answer 9.

Darwinian Theory of Evolution explains that evolution is the result of natural selection and natural selection is based on the inherited characters of organisms. Adaptive ability of organisms is the one which helps organisms in evolution through natural selection. Initially, there were two varieties, short necked and long necked giraffe. Giraffe's long neck is the result of their adaptation to their nature since the ground vegetation disappeared and is replaced by high trees. The elongated necks are the outcome of their attempt to feed leaves on tall trees. Long neck could feed easily and they survived whereas short necked could not feed easily and they could not survive. This character then passed on from generation to generation leading to evolution.

Answer 10.

(a)

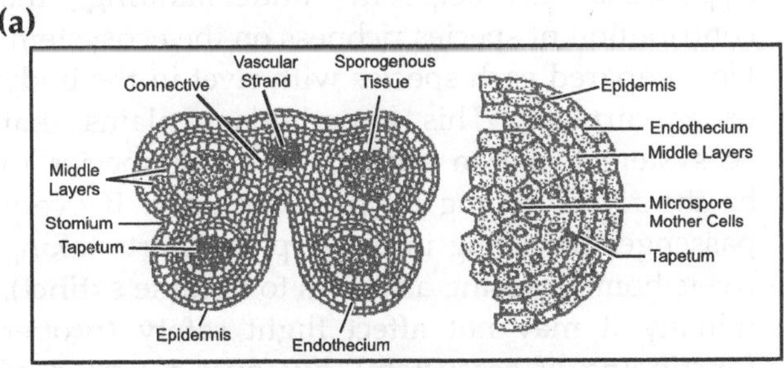

Fig. TS of mature anther

OR

(b)

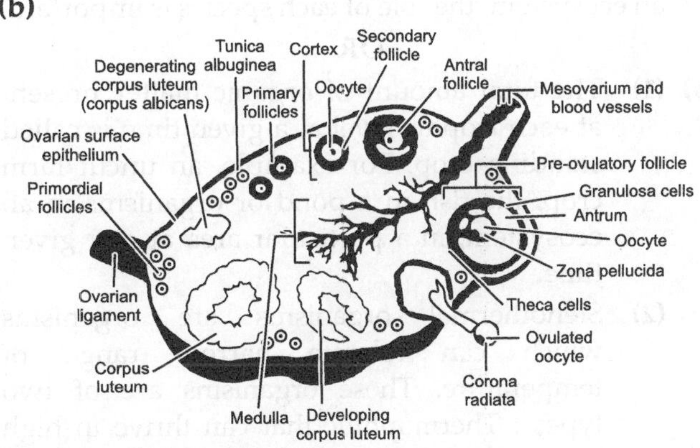

Fig. Internal structure of human ovary

Answer 11.

A nucleosome is the basic structural unit of DNA packaging in eukaryotes. The negatively charged DNA is wrapped around positively charged histone octamer to form nucleosome. Each nucleosome has a core histone octamer and a linker DNA. The core particle consists of 8 histones proteins which occur in pairs : H_2A, H_2B, H_3 and H_4. About 200 bases of DNA are involved with each nucleosome. DNA winds around the core (1¾ or 1.65 turns involving approximately 146 base pairs) and two successive nucleosomes are joined together by linker DNA H_1 histone (approximately 80 bp) which holds the entire complex. There are repeating units in chromatin fibre like beads on string.

Structure of a nucleosome :

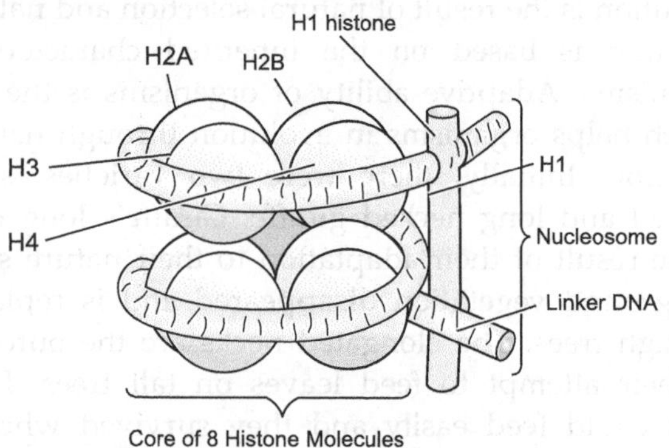

Fig. A nucleosome

Answer 12.

(a) Ecologist Paul Ehrlich gave rivet popper hypothesis to help in understanding the contribution of species richness on the ecosystem. He compared each species with rivet in the body of an airplane. This hypothesis explains that ecosystem to be an airplane and the species to be the rivets joining all parts together. If every passenger travelling in the airplane starts taking rivets home (causing a species to become extinct), initially it may not affect flight safety (proper functioning of ecosystem), but over a period of time the plane becomes weak and dangerous (species become endangered and then extinct). In an ecosystem, the role of each species is important.

OR

(b) (1) The total amount of organic matter present at each trophic level at a given time is called standing crop. For example, an uncut farm crop, the fish in a pond or organisms in an ecosystem, in a particular area at any given time.

(2) Stenothermal organisms are organisms which can tolerate narrow range of temperature. These organisms are of two types : Thermophilic that can thrive in high temperatures only. Examples: coral reefs and some reptiles and Cryophilic that can only survive in low temperature regions. Examples : arctic crustaceans, salmons and seals.

(3) A niche is an organism's functional role in its ecosystem. It is the specific area where an organism inhabits.

Answer 13.

(i) *Penicillium notatum*

(ii) *Saccharomyces cerevisiae*

(iii) *Bacillus thuringiensis*

(iv) *Propionibacterium sharmanii*

(v) *Trichoderma polysporum or Aspergillus niger*

(vi) *Rhizobium leguminosarum*

Answer 14.

The different types of endosperms found in angiosperms are :

1. **Nuclear endosperms** : The endosperm nucleus gives rise to a number of free nuclei which stay in the peripheral layer. A large central vacuole appears in the embryo sac. Cell wall formation takes place from the periphery towards the centre leading to the formation of cellular endosperm. Examples : Maize, Wheat, Rice, Sunflower, coconut, etc.
2. **Cellular Endosperm :** Division of the primary endosperm nucleus is immediately followed by cell wall formation so that the endosperm is cellular from the beginning. Examples : Balsam, *Datura, Petunia, Villarsia,* etc.
3. **Helobial Endosperm** : A partition wall develops between the two nuclei resulting in the first division of the endosperm. Endosperm development is intermediate between the free nuclear and cellular type. A partition wall develops between the two nuclei resulting in the first division of the endosperm. Examples : *Asphodelus, Echium,* etc.

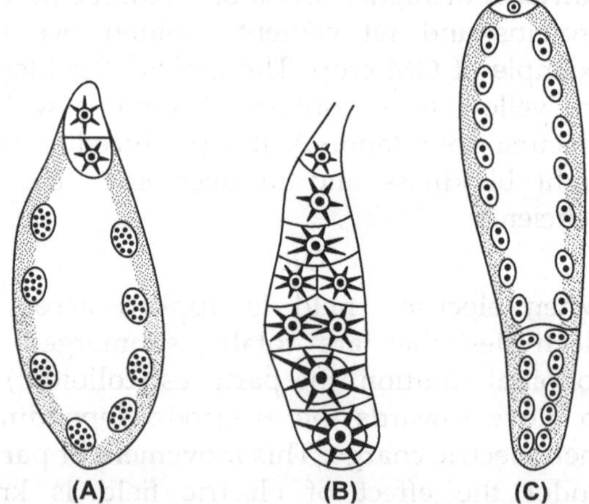

A. Nuclear type B. Cellular type C. Helobial type

Answer 15.

(i) F_1 generation will be yellow cotyledon and round seed.

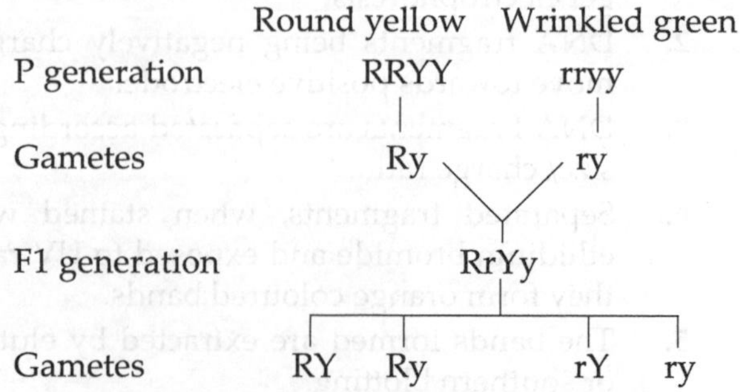

(ii)

Gametes	RY	Ry	rY	ry
RY	RRYY	RRYy	RrYY	RrYy
	round yellow	round yellow	round yellow	round yellow
Ry	RRYy	RRyy	RrYy	Rryy
	round yellow	round green	round yellow	round green
rY	RrYY	RrYy	rrYY	rrYy
	round yellow	round yellow	wrinkled yellow	wrinkled yellow
ry	RrYy	Rryy	rrYy	Rryy
	round yellow	round green	wrinkled yellow	wrinkled green

9 round yellow : 3 round green : 3 wrinkled yellow : 1 wrinkled green.

(iii) As all the F_1 plants were with yellow cotyledon and round seed, so allele Y for yellow cotyledon is dominant over allele y for green cotyledon and allele R for round seed is dominant over allele r for wrinkled seed.

The appearance of all the four possible phenotypic combinations in F_1 in the ratio 9 : 3 : 3 : 1 is possible if the two pairs of characters are believed to behave independent of each other. Each pair of contrasting characters bear no permanent association with particular other character.

The law of Independent Assortment States :

When the two parents differ from each other in two or more pairs of contrasting characters or factors, then the assortment of alleles of one character is independent of assortment of alleles of other characters. Each member of an allelic pair may combine randomly with either of another pair during the formation of gametes.

SECTION–C (15 Marks)

Answer 16.

(a) Fertilisation is defined as a process of union of a haploid male gamete or sperm with a haploid female gamete or ovum to form a diploid cell, the zygote.

Process of Fertilisation : During copulation (coitus), male discharges the semen (coagulated form) into the vagina of the female, where the sperms reach the ampulla partly by the movement of their tails and partly by the action of uterus by peristalsis. The sperms present in the semen travel a long way from vagina through the uterus to the fallopian tube.

The sperm can survive in the female's reproductive tract for 1 to 3 days and it can fertilise the ovum in 12 to 24 hours following ovulation.

Events of Fertilisation :

1. **Activation of sperm and ovum :** Sperm secrete the chemical hyaluronidase and surface protein called antifertilisin (composed of acidic amino acid). The ovum secretes a chemical named fertilisin which mixes with the water to form egg water which attracts the sperms of its own species. Thus, the liquefaction of coagulated semen occurs in this way. The fertilisin of an egg interacts with the anti fertilisin of sperm of the same species. This attraction between fertilisin and anti-fertilisin makes the sperms stick to the egg surface.
2. **Capacitation of sperm :** Capacitation is the process of acquiring the capacity to fertilize the egg by the sperm. Freshly ejaculated sperms are unable to fertilise the ovum. Thus, they undergo a series of changes collectively known as capacitation. Capacitation occurs while sperm reside in the female reproductive tract for a period of time. It is associated with removal of adherent seminal plasma proteins, reorganisation of plasma membrane lipids and proteins. It also seems to involve an influx of extracellular calcium, increase in cyclic AMP and decrease in intracellular pH.
3. **Penetration of sperm :** The ovum is surrounded by three membranes such as corona radiata, zona pellucida and the vitelline membranes. At first the sperm passes through corona radiata to reach zona pellucida. There, it releases the enzyme hyaluronidase or sperm lysin from its acrosome. This enzyme dissolves zona pellucida as a result of which the sperm reaches the plasma membrane of the egg. The above changes on the sperm head are called ***acrosome reaction***. Acrosin digests zona pellucida and calcium ions are needed for this.
4. **Cortical Reaction :** Prior to fertilisation, the egg is in a quiescent state, arrested in metaphase of the second meiotic division. Upon binding of a sperm, the egg rapidly undergoes a number of metabolic and physical changes that collectively are called *egg activation*. Prominent effects include a rise in the intracellular concentration of calcium, completion of the second meiotic division and the so-called cortical reaction.

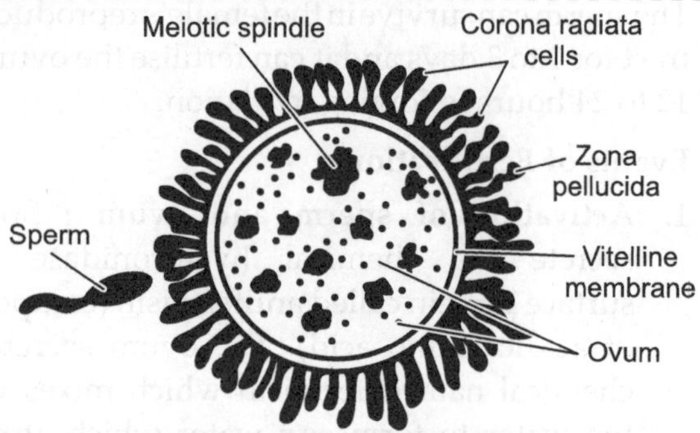

Fig. : Penetration of sperm

5. **Zonal Reaction :** Once one sperm has entered the egg (ovum) the vitelline membrane thickens and is converted into fertilisation membrane. This membrane is rigid and never allows other sperms to pass through this membrane. The egg cytoplasm extends around the head of sperm to form a fertilisation cone that engults the sperm so that the head and the mid piece enter the ovum. The formation of fertilisation membrane around the egg blocks polyspermy and allows only single sperm to enter.
6. **Amphimixis :** A sperm consists of three parts: head, middle piece and tail. Shortly before or after entering the egg, the sperm loses its tail. After the sperm enters the egg, the membrane of head and middle piece dissolves, liberating nucleus, centrosome and mitochondria. The sperm nucleus enlarges to form the male pronuclear and the nucleus of the ovum becomes female pronuclear.

   The male pronuclear moves inwards and then change its direction to meet the egg nucleus. The chromosomes (haploid set) of the sperm and the chromosomes (haploid set) of the egg or ovum are set free by the breakdown of their nuclear envelops.

   Mixing up of the chromosomes of a sperm and an ovum resulting in a diploid zygote nucleus is known as amphimixis or karyogamy. The fertilised egg is now called the zygote.

OR

(b) (i) Amniocentesis is a medical procedure in which a small sample of amniotic fluid is drawn out of the amniotic sac through a needle inserted in the abdomen. The fluid is then analysed for parental diagnosis of genetic defects or the sex of the foetus.

(ii) Gonorrhoea is an sexually transmitted disease which can affect both men and women. Gonorrhoea is caused by a bacterium *Neisseria gonorrhoeae*. The symptoms are itching in genital region, fluid discharge, pain and swelling, sterility, burning sensation while urinating, abnormal discharge of liquids from the opening of the penis, vaginal bleeding between periods etc.

(iii) Dispersal of seeds away from the main plant by number of methods is very important. They may include one or a combination of gravity, wind, water or animals. Seed dispersal has several advantages for the plant such as :

1. Protection from predation and avoids intraspecific competition, leading to rapid evolution.
2. Distribution of seeds avoids the problem of overcrowding and provide suitable conditions for germination of seeds.

(iv) Seasonal breeders are animal species that successfully mate only during certain times of the year. They become reproductively active only during favourable seasons. These times of year allow for the optimisation of survival of young ones due to factors such as ambient temperature, food and water availability and changes in the predation behaviours of other species. For examples : tiger, cat, pig.

(v) In sexually reproducing organisms, the gametes undergo meiosis and therefore, each gamete contains only half set of chromosomes. During gametogenesis the chromosome number is reduced to half by meiosis and during fertilisation, the diploid number is restored in the zygote.

Answer 17.

(a) (i) Restriction Endonuclease is a type of restriction enzyme that cuts within the DNA at a specific site. This enzyme acts as an important tool in genetic engineering for cutting DNA sequences from a specific site. They play an important role in gene manipulation. The enzymes recognise a specific six-base pair sequence known as the recognition sequence or palindromic sequence and cut the sequence at a specific site.

Smith and Nathans proposed the rules for the Nomenclature of restriction endonucleases :

1. The enzyme names should begin with a three-letter acronym in which the first letter was the first letter of the genus from which the enzyme was isolated and the next two letters were the first two letters of the species name.
2. Fourth alphabet is the name of the strain from which the enzyme is isolated and

the roman number is the order in which the enzyme was isolated. Example : EcoRI.

Derivation of the EcoRI name		
Abbreviation	**Meaning**	**Description**
E	*Escherichia*	genus
co	*coli*	specific species
R	RY13	strain
I	First identified	order of identification in the bacterium

(ii) Restriction endonuclease enzymes recognise a specific palindromic base sequence in the DNA and produce a double-stranded cut in the DNA at the sugar-phosphate backbone of both strands. This digestion produces "sticky" ends. For *e.g.*,

G|AATTC
CTTAA|G

Recognition sequences in DNA differ for each restriction enzyme, producing differences in the length, sequence and strand orientation (5′ end or 3′ end) of a sticky-end or overhang. When cut by the same restriction enzyme, the resultant DNA fragments have the same kind of 'sticky-ends' and these can be joined together using DNA ligases. The foreign DNA and vector DNA are cut by same restriction enzymes.

Action of Restriction enzyme

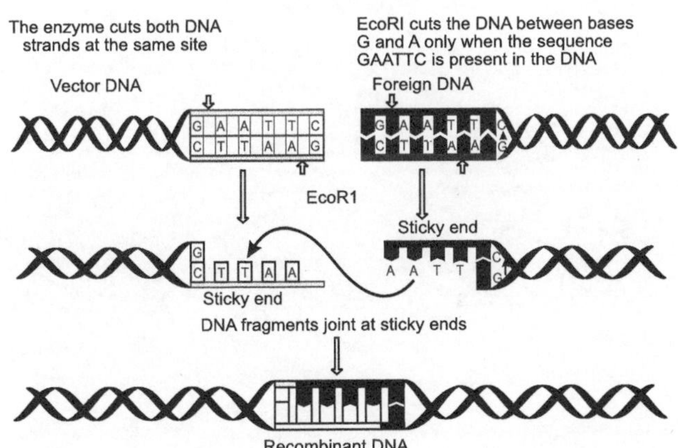

OR

(b) (i) A cloning vector is a small piece of DNA, taken from a virus, a plasmid, or the cell of a higher organism, that can be stably maintained in an organism and into which a foreign DNA fragment can be inserted for cloning purposes. The vector therefore contains features that allows the convenient insertion or removal of a DNA fragment to or from vector.

An ideal cloning vector must have following characteristics :

1. It must have an Origin of Replication (ori). It is the sequence from which replication starts and the foreign gene linked to it will clone itself and it also controls the copy number of the linked DNA.
2. It must be small in size less than 10 kB.
3. It must be self-replicating inside host cell.
4. It must possess restriction site for Restriction Endonuclease enzymes.
5. It must have the cloning sites for the ligation of the alien DNA which is carried out at the recognition site.
6. Introduction of donor DNA fragment must not interfere with replication property of the vector.
7. It must possess some marker genes for later identification of recombinant cell. These will help in identifying and eliminating non-transformants and selectively permitting the growth of transformants.

(ii) The term direct or vector less transfer of DNA is used when the foreign DNA is directly introduced into the genome.

1. **Electroporation** : Electroporation involves the use of high field strength, electrical impulses to reversibly permeabilise the cell membranes for the uptake of DNA. This technique can be used for the delivery of DNA into intact plant cells and protoplasts. Electroporation has been successfully used for the production of transgenic plants of many cereals *e.g.*, rice, wheat, maize.
2. **Particle Bombardment (Biolistics) :** Particle (or micro projectile) bombardment is the most effective method for gene transfer. This method is versatile due to the fact that it can be successfully used for the DNA transfer in mammalian cells and microorganisms. In this the gold or tungsten beads coated with DNA are bombarded with high velocity on the host tissue. The foreign gene gets incorporated in host DNA.
3. Other methods include microinjection where a micropipette is used to inject the DNA into the nucleus. Transformation of bacteria can also be done by making the bacteria more competent to take up the foreign gene by placing them in cold calcium chloride or by heat stock. Liposome mediated transfer can also be used.

Answer 18.

(a)

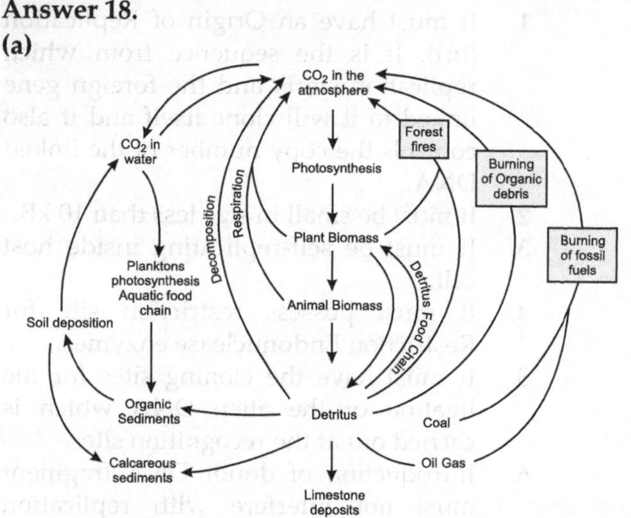

Fig. : Graphic representation of carbon cycle

OR

(b)

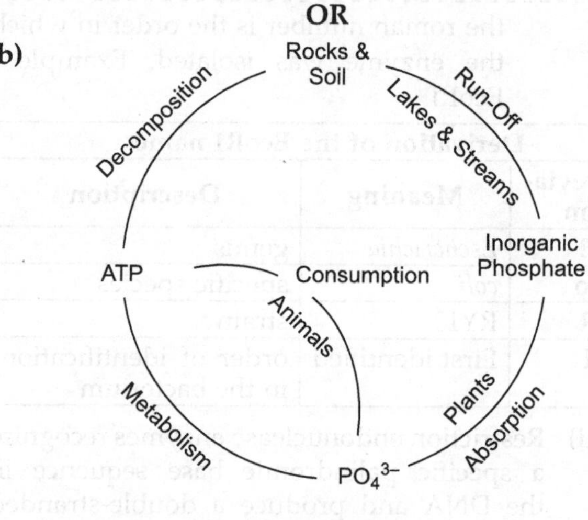

Fig. : Graphic representation of the phosphorus cycle

●●

BIOLOGY 2018

QUESTIONS

PART I (20 Marks)

*(Answer **all** questions)*

Question 1.

(a) *Answer the following questions briefly and to the point:* **[8 × 1]**

(i) *Give a significant point of difference between Oestrous and Menstrual cycle.*

(ii) *Give the biological name of the organism causing typhoid.*

(iii) *If the haploid number of chromosomes in a plant species is 20, how many chromosomes will be present in the cells of the shoot tip ?*

(iv) *Name a plant which flowers every twelve years.*

(v) *Name the diagnostic test for AIDS.*

(vi) *Name the terminal stage of ageing in the life cycle of plants.*

(vii) *Which organisms constitute the last trophic level?*

(viii) *What is emasculation ?*

(b) *Each of the following questions has four choices. Choose the best option in each case :* **[4 × 1]**

(i) *Length of the DNA with 23 base pairs is:*

(A) *78·4 Å* **(B)** *78·2 Å*

(C) *78 Å* **(D)** *74·8 Å*

(ii) *Opium is obtained from :*

(A) *Papaver somniferum*

(B) *Cannabis sativa*

(C) *Erythroxylum coca*

(D) *Datura metel*

(iii) *According to Abiogenesis, life originated from :*

(A) *Non-living matter*

(B) *Pre-existing life*

(C) *Oxygen*

(D) *Extra-terrestrial matter*

(iv) *The largest unit in which gene flow is possible is:*

(A) *Organism* **(B)** *Population*

(C) *Species* **(D)** *Genes*

(c) *Give one significant contribution of each of the following scientists :* **[4 × 1]**

(A) *P. Maheshwari* **(B)** *E. Wilson*

(C) *M. S. Swaminathan* **(D)** *H. Boyer*

(d) *Define the following :* **[2 × 1]**

(i) *Biopatent* **(ii)** *Parthenocarpy*

(e) *Give a reason for each of the following :* **[2 × 1]**

(i) *Pollen grains of wind pollinated flowers are produced in large quantities.*

(ii) *Equilibrium of a forest ecosystem can be disturbed by uncontrolled hunting of big predators.*

PART II

SECTION–A (14 Marks)

*(Answer **all** questions)*

Question 2.

(a) *A woman with blood group O married a man with blood group AB. Show the possible blood groups of the progeny. List the alleles involved in this inheritance.* **[2]**

OR

(b) *If the mother is a carrier of colour blindness and the father is normal, show the possible genotype and phenotype of the offspring of the next generation, with the help of a punnett square.*

Question 3.

Define life span. Give the life span of an elephant. **[2]**

Question 4.

Give two characteristic features of each of the following:

(a) *Ramapithecus*

(b) *Cro-Magnon man* **[2]**

Question 5.

(a) *List any four effects of global warming.* **[2]**

OR

(b) *State any four measures to control noise pollution.*

Question 6.

Define BOD. What is its significance in an aquatic ecosystem ? **[2]**

Question 7.

Give one significant difference between each of the following pairs :

(a) *Humoral immunity and cell mediated immunity.*

(b) *Benign tumour and malignant tumour.* **[2]**

Question 8.

Give four causes of infertility in males. **[2]**

SECTION–B (21 Marks)

*(Answer **all** questions)*

Question 9.

(a) *Draw a labelled diagram of L.S. of human testis.* **[3]**

OR

(b) *Draw a labelled diagram of the mature embryo sac of angiosperms.*

Question 10.

Explain gene therapy, with reference to treatment of SCID. **[3]**

Question 11.

Study the table given below. Do not copy the table, but write the answers in the correct order. **[3]**

Scientific Name	Commercial Product	Use
(a)	Streptokinase	(b)..............
Monascus purpureus	(c)..............	(d)..............
(e)..............	Lactic acid	(f)..............

Question 12.

Explain industrial melanism. **[3]**

Question 13.

Describe the tissue culture technique in plants. **[3]**

Question 14.

Define the following :

(a) *Spermiogenesis* **(b)** *Reproductive health*

(c) *Amenorrhea* **[3]**

Question 15.

(a) *Define the following :*

(i) *Hotspots* **(ii)** *Ramsar Sites*

(iii) *Red data book* **[3]**

OR

(b) *Define the following :*

(i) *Biodiversity*

(ii) *Eutrophication*

(iii) *PAR*

SECTION–C (15 Marks)

*(Answer **all** questions)*

Question 16.

(a) *Describe post transcriptional processing of RNA in eukaryotes.* **[5]**

OR

(b) *Describe Avery, McLeod and McCarty's experiment. State its significance.*

Question 17.

(a) *Write a short note on Chipko Movement.* **[5]**

OR

(b) *Write a short note on Joint Forest Management.*

Question 18.

(a) *What does PCR stand for ? Describe the different steps of PCR.* **[5]**

OR

(b) *Given an account of the Blue-White method of selection of recombinants.*

ANSWERS

PART I (20 Marks)

Answer 1.

(a) (i)

Oestrous cycle	Menstrual cycle
The cycle occurs in non-primates.	The cycle occurs in primates.
The cycle consists of short period of oestrous (heat) followed by non-oestrous period.	The cycle consists of three phases : follicular phase, ovulation and luteal phase.

(ii) Typhoid is a bacterial infection caused by the bacteria *Salmonella typhi*.

(iii) Shoot tip cells are diploid, so they will have 2n *i.e.,* 40 chromosomes will be present in the cells of the shoot tip.

(iv) Kurinji or Neelakurinji (*Strobilanthes kunthiana*) is a shrub that blossoms only once in 12 years.

(v) ELISA, which stands for Enzyme-Linked Immune Sorbent Assay, is used to detect the presence of AIDS.

(vi) Senescence refers to the process or condition of growing old in the life cycle of plants.

(vii) Tertiary consumers *i.e.,* carnivores or apex predators constitute the last trophic level.

(viii) Emasculation is the removal of anthers from bisexual flower before maturing for cross fertilisation.

(b) (i) (D) 74·8 Å

(ii) (A) *Papaver somniferum*

(iii) (A) Non-living matter

(iv) (C) Species or (B) Population

(c) (A) **Panchanan Maheshwari** was an Indian botanist known for his invention of the technique of test-tube fertilisation of angiosperms or plant tissue culture.

(B) **Edward O. Wilson** coined the term biodiversity.

(C) **M. S. Swaminathan** was an Indian geneticist who is known for developing short duration high-yielding varieties of rice. He brought Green Revolution in India.

(D) **Herbert W. Boyer, Stanley N. Cohen,** and **Paul Berg** together are known for their work on the first artificial Recombinant-DNA.

They also discovered the restriction enzyme endonuclease.

(d) (i) **Biopatents :** Patents granted in the field of biology are known as biopatents. It is a legal right given to the owner for a biological discovery which allows the patent holder to exclude others from making, using or selling it for a limited period of time.

(ii) Parthenocarpy refers to development of fruit without fertilisation. The fruit produced resembles a normally produced fruit but is seedless. For example, varieties of pineapple, banana, cucumber, grape, orange.

(e) (i) Pollen grains of wind pollinated flowers are produced in large quantities because lot of them get wasted during the process of pollination. Hence, to increase the chance of pollination they are produced in large quantities.

(ii) Hunting creates an imbalance in the natural elements of the environment. If a predator, like a wolf or mountain lion is hunted down, their prey, deer will increase in number. They inturn will feed on vegetation. In the long term, their uncontrollable number will lead to shortage of vegetation, which will affect other herbivores. This will increase interspecific and intraspecific competition and ultimately disturb the food chains.

PART II

SECTION–A (14 Marks)

Answer 2.

(a)

	♀		♂
Blood Group	O	X	AB
Genotype	I°I°		I^AI^B
	↓		⋀
Gametes	I°		I^A I^B

♂ / ♀	I^A	I^B
I°	$I^AI°$ (A)	$I^BI°$ (B)

Therefore, possible blood group of progeny is Blood group A and Blood Group B.

Alleles involved are : I^O, I^A, I^B.

OR

(b) Carrier female × Normal male

X^CX X XY

Normal female but carrier/Normal male	X^C	Y	Normal female-1
X^C	X^CX^C	X^CY	Normal male-1
	Normal female	Normal male	Carrier female-1
X^C	X^CX^C	X^CY	Colour blind male-1
	Carrier female	Colour blind male	1 : 1 : 1 : 1 : 1

Answer 3.

Life span is the time period from birth to natural death of an organism which measures the length of time an individual person lives. It refers to the longest period over which the life of any organism or species may extend. Elephants are known for their majestic size and resilience, but also for their long lives. Elephants can live up to 20-90 years in the wild.

Answer 4.

(a) The fossil evidence clearly indicates that Ramapithecus, lived in Africa and Asia (about 10-15 million years ago) were the forerunners of Homonids. They had Chimpanzee-like feet, flexible wrists and small canines. Short face, small brain, cranial capacity of around 650 cc and thickly enameled teeth are its characteristic feature.

(b) Following distinctive features are regarded as common feature of the Cro-Magnon race : Long headed skull, flattened forehead, the slightly projecting brow ridges and the occipital bone projects backward. The cranial capacity is large, about 1650-1660 cc. Although the skull is relatively long and narrow, the face appears quite short and wide. They had errect posture and were bipedal.

Answer 5.

(a) The various consequences of Global warming are :

1. Global warming has resulted in melting of glaciers and ice caps and leading to odd climatic conditions.
2. Extinction of a large number of plant and animal species.
3. Decrease in food production.
4. Rise in mean sea-level.

OR

(b) Measures to control noise pollution :

1. Using silencers to control noise from automobiles, ducts, exhausts etc.

2. Minimum use of loudspeakers and amplifiers especially near silence zones.
3. Proper lubrication and better maintainence will reduce noise from the machine.
4. Plantation of coniferous trees on high-ways or green muflers.

Answer 6.

Biochemical oxygen demand (BOD) refers to the amount of dissolved oxygen needed by aerobic organisms required to breakdown organic material present in a given water sample at certain temperature over a specific period of time. It is commonly expressed in milligrams of oxygen consumed per litre of sample during 5 days of incubation at 20°C and is often used as a *"surrogate parameter of the degree of organic pollution of water"*.

BOD is an important water quality parameter because it provides an index to assess the effect of discharged waste water on the environment. The higher the BOD value, the greater the amount of organic matter available for oxygen consuming bacteria.

Answer 7.

(a)

Humoral Immunity	Cell mediated Immunity
Mainly involves B-lymphocytes which produce the antibodies.	Consists of T-lymphocytes.
Generation and maturation occurs in bone marrow.	Originates in bone marrow and complete development occurs in thymus.
Defends body against extracellular microbes and their toxins.	Defends body against intracellular microbes, viruses, parasites (leishmania), bacteria and tumour cells.
Antibodies are produced.	Different cells are produced like Helper T-cell, Killer T-cells etc.

(b)

Benign tumour	Malignant tumour
Confined to affected organ.	Spreads to other organs of body.
No latent stage.	latent stage is present.
No metastasis.	Shows metastasis.
Causes little damage to the body.	Causes extensive damage to the body.

Answer 8.

Causes of infertility in males are :

1. Y chromosome deletion and other genetic problem leads to infertility in males.
2. Blockage in the vas deferens, acidity in urethra causes infertility.
3. Deformities in the shape and size of the sperm.
4. Age, senility and injury to reproductive part are some other causes of infertility.

SECTION–B (21 Marks)

Answer 9.

(a)

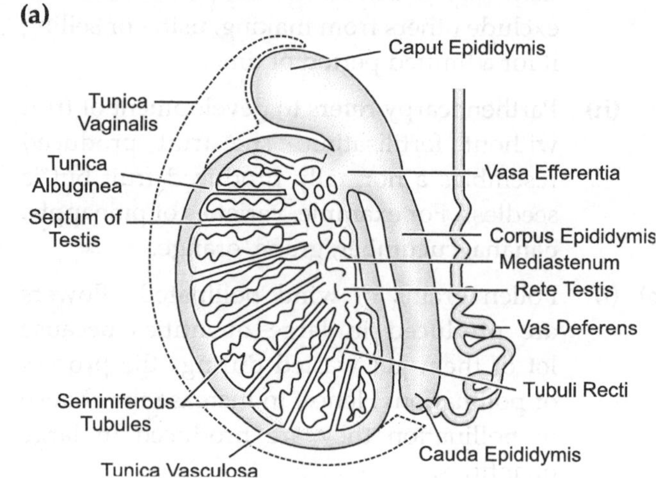

OR

(b)

Chalazal end
Antipodals
Embryo sac
Polar nuclei
Central cell
Egg
Synergids
Filiform apparatus
Micropylar end

Answer 10.

Gene therapy is an experimental technique in which genes are introduced into individual cells or tissues, to cure some hereditary diseases in which a defective allele of a gene is replaced with normal one. Deficiency of Adenosine Deaminase causes SCID. The technique of gene therapy was initiated in 1972.

Severe Combined Immune Deficiency (SCID) is a potentially fatal primary immuno deficiency in which there is combined absence of T-lymphocyte and B-lymphocyte function. There are at least 13 different genetic defects that can cause SCID. This condition is generally considered to be the most serious of the primary immune deficiencies. Cells of the bone marrow of the patient are cultured outside the body. Isolated cells are exposed to retrovirus in cell culture and take up ADA gene. Transgenic cells are then reimplanted into patients bone marrow/blood. This process has to be done repeatedly.

Another type of SCID is caused by mutations in a gene that encodes an enzyme called adenosine deaminase (ADA). ADA is essential for the metabolic function of a variety of body cells but especially T-cells. The absence of this enzyme leads to an accumulation of toxic metabolic by-products within lymphocytes that cause the cells to die. For patients with SCID due to ADA deficiency, replacement therapy with a modified form of the enzyme (from a cow, called PEG-ADA) has been used with some success.

Answer 11.

(a) *Streptococcus.*

(b) to dissolve blood clots.

(c) Statins.

(d) Blood cholestrol lowering agent

(e) Lactic acid bacteria (LAB).

(f) Curd production.

Answer 12.

Prior to industrialisation in Europe, white or dull gray peppered moth, *Biston betolaria* was dominant species than the black peppered moth, *Biston carbonaria*. This was due to the presence of lichen covered trees, which provide camouflage to the former moth. However, with the onset of industrialisation and increased pollution, lichen dissappeared and trees eventually became dark due to deposition of soot and smoke. This dark background exposed the white coloured moth and gave advantage to the dark moth from the predators. This lead to increased number of dark moth in the area. Thus, dark moth were selected by nature (natural selection) and light ones failed to survive due to selective feeding by the predator birds.

Answer 13.

Plant tissue culture is an advanced technique of plant breeding that developed around the 1950s. Plant tissue culture makes use of parts of a plant to generate multiple copies of the plant in a very short duration. The technique is based on the property of totipotency of plant cell which means that any cell from any part of the plant can be used to generate a whole new plant.

Types of Plant tissue culture :

1. Seed Culture	2. Embryo Culture
3. Callus Culture	4. Organ Culture
5. Protoplast Culture	6. Anther Culture

An example of use of plant tissue culture in the lab :

Step 1: A new gene of interest is introduced into tissue of a plant. Thus altering its genotype.

Step 2 : The transformed plant cells are then isolated and placed into sterile nutrient agar.

Step 3 : Each cell grows into a cluster of cells called a callus.

Step 4 : Calluses are separated and a plant hormone called cytokinin is added which stimulates growth of small leaf shoots.

Step 5 : Shoots are separated and the addition auxin stimulates the growth of roots.

Step 6 : New plants are transferred to soil and grown to observe how the alteration of its genotype has affected its phenotype under extreme environmental conditions such as high salt, increased CO_2 and drought.

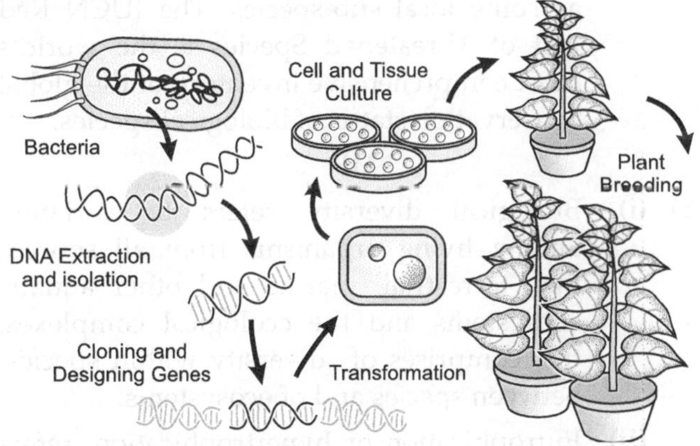

Tissue culture is used to develop thousands of genetically identical plants from one single parent plant known as soma clones, and this process is known as micro-propagation. The method offers an advantage over other methods as it can be used to develop disease free plants from disease-rode plants by using their meristems (apical and axillary) as explants.

Answer 14.

(a) Spermiogenesis is the final stage of spermatogenesis, which includes the maturation of spermatids into mature, motile spermatozoa. The transformation of spermatid into sperms takes place in sertoli cells.

(b) Reproductive health is a state of complete physical, mental and social well-being in all matters relating to the reproductive system. It also refers to the ability of the individual to produce healthy offspring.

(c) Lactational amenorrhea is the temporary postnatal infertility that occurs when the female is amenorrheic (not menstruating) and fully breast-feeding. Breastfeeding prevents the resumption of normal ovarian cycles by disrupting the pattern of pulsatile release of GnRH from the hypothalamus and hence LH from the pituitary.

Answer 15.

(a) (i) A biodiversity hotspot is a biogeographical region with significant levels of biodiversity richness that is threatened with extinction.

(ii) Ramsar site is a wetland site designated of international importance under the Ramsar Convention. Ramsar sites are designated because they meet the criteria for identifying Wetlands of International importance.

(iii) Red Data Book is a state document established for documenting rare and endangered species of animals, plants and fungi, as well as some local sub-species. The IUCN Red List of Threatened Species is the world's most comprehensive inventory of the global conservation status of biological species.

OR

(b) (i) Biological diversity refers to variation among living organisms from all sources like, terrestrial, marine and other aquatic ecosystems and the ecological complexes; this comprises of diversity within species, between species and of ecosystems.

(ii) Eutrophication or hypertrophication, refers to enrichment of a water body with an excess amount of inorganic nutrients. This process induces growth of plants and algae and, due to the biomass load, result in oxygen depletion of the water body. For example algal bloom.

(iii) Photosynthetically Active Radiation (PAR) is the amount of light available for photosynthesis with wavelengths between 400 and 700 nm.

SECTION–C (15 Marks)

Answer 16.

(a) Initially, DNA transcribed *m*RNA is called primary transcript or pre-*m*RNA. It is inactive and non-functional and hence undergoes post-transcriptional process to change into mature and functional *m*RNA.

Post-transcriptional modification of *nn*RNA includes :

1. **Splicing** : In molecular biology, splicing refers to editing of the precursor messenger RNA (pre-mRNA) transcript into a mature messenger RNA (mRNA) wherein introns are removed and exons are joined together *i.e.,* they are ligated. Splicing occurs within the nucleus after transcription.

pre-*m*RNA	Exon 1	Introns	Exon 2

↓ Splicing

*m*RNA	Exon 1	Exon 2

2. **Capping** : While the pre-mRNA is synthesised, a 7-methylguanosine cap is added to the 5′ end of the growing transcript by a 5′-to-5′ phosphate linkage. In molecular biology, the five-prime cap (5′ cap) is a specially altered nucleotide on the 5′ end of some primary transcripts, this process is known as mRNA capping. This is a highly regulated and vital process for forming stable and mature messenger RNA able to undergo translation during protein synthesis. The cap protects the nascent mRNA from degradation. Moreover, initiation factors involved in protein synthesis recognise the cap to help initiate translation by ribosomes.

3. **Tailing** : An enzyme called poly (A) polymerase (PAP) is a part of the protein complex that cleaves the pre-mRNA and it immediately adds a string of approximately 200 A nucleotides, called the poly (A) tail, to the 3′ end of the just-cleaved pre-mRNA. The poly (A) tail protects the mRNA from degradation, aids in the export of the mature mRNA to the cytoplasm, and is involved in binding proteins involved in initiating translation.

OR

(b) In 1944, Oswald Avery, Colin MacLeod and Maclyn Mc Carty, gave an experimental demonstration, that DNA is the substance that causes bacterial transformation as identified Griffith's experiment.

This was done to determine if rough bacteria could be transformed into smooth bacteria, hence passing along the genetic information causing the transformation. By isolating and purifying this chemical component, they could deduce if it had characteristics of a protein or DNA molecule.

Griffith's transforming principle includes, that hereditary material can be extracted from one organism and introudced into another.

1. They prepared cultures containing the heat-killed **S** strain of *Pneumonococci* and then removed lipids and carbohydrates from the solution.
2. Next they treated the solutions with different digestive enzymes DNAase, RNAase and protease to destroy the targeted compound.
3. Finally, they introduced living **R** strain cells to the culture to see which cultures would develop transformed S strain bacteria.

   They observed that in DNAase-treated culture, S strain bacteria failed to grow (*i.e.*, no DNA – no transformation). In the other two cultures, R-strain transformed into S-strain bacteria. This indicated that DNA was the genetic component that was being transferred between cells.

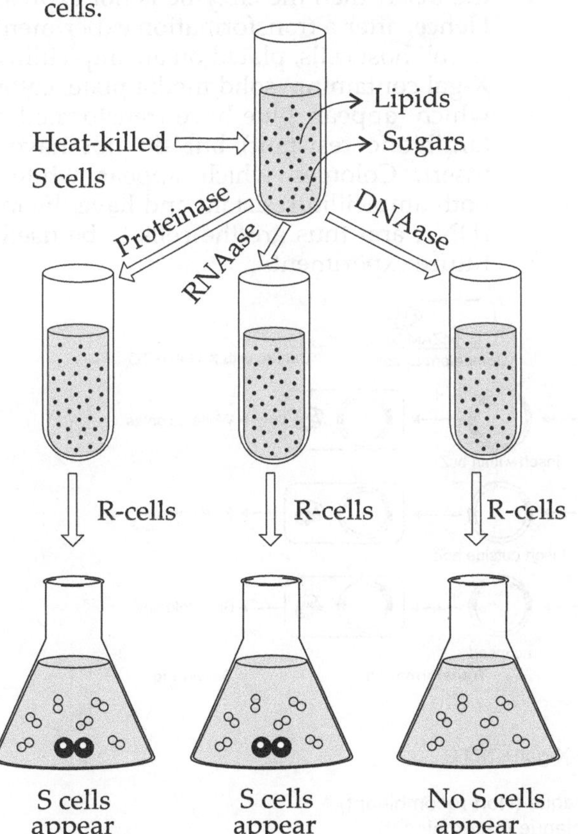

Answer 17.

(a) 1. The Chipko movement or Chipko Andolan was a forest conservation movement. The name of the movement comes from the word 'embrace', as the villagers hugged the trees, and prevented the contractors from felling them.

2. The movement dates back to 1730 AD, when 363 Bishnois of Khejarli village of Rajasthan, sacrificed their lives to save Khejri trees.
3. A large group of them from 84 villages led by a lady called Amrita Devi laid down their lives in an effort to protect the trees. All over the world it created a precedent for starting of non-violent protest in India.
4. This inspired to slow down the rapid deforestation, expose vested interests, increase ecological awareness, and demonstrate the viability of people power.
5. Above all, it stirred up the existing civil society in India, which began to address the issues of tribal and marginalised people. The success achieved by this protest led to similar protests in other parts of the country.
6. From their origins as a spontaneous protest against logging abuses in Uttar Pradesh in the Himalayas, supporters of the Chipko movement, mainly village women, have successfully banned the felling of trees in a number of regions and influenced natural resource policy in India. This became a turning point in the history of eco-development.

OR

(b) 1. The Indian Forest Policy of 1988 (MoEF, 1988) and the subsequent government resolution on participatory forest management (MoEF, 1990) emphasize the need for people's participation in natural forest management. Joint Forest Management (JFM) officially refers to partnerships in forest movement involving both the state forest departments and local communities.

2. The policies and objectives are detailed in the Indian comprehensive National Forest Policy of 1988 and the Joint Forest Management Guidelines of 1990 of the Government of India.
3. Although schemes vary from state to state and are known by different names in different Indian languages, usually a village committee known as the Forest Protection Committee (FPC) and the Forest Department enter into a JFM agreement.
4. Under joint forest management (JFM), village communities are entrusted with the protection and management of nearby forests. The areas concerned are usually degraded or even deforested areas.
5. Villagers agree to assist in the safeguarding of forest resources through protection from fire, grazing, and illegal harvesting in exchange for which they receive non-timber forest

products and a share of the revenue from the sale of timber products.

6. The communities are required to organise forest protection committees, village forest committees, village forest conservation and development societies, etc. Each of these bodies has an executive committee that manages its day-to-day affairs.

Answer 18.

(a) PCR (Polymerase Chain Reaction) method for synthesis of new DNA strands was developed by Kary Mullis in the 1980s. It is the technique of producing multiple copies of gene of interest or gene amplification, in vitro, using primers and DNA polymerase.

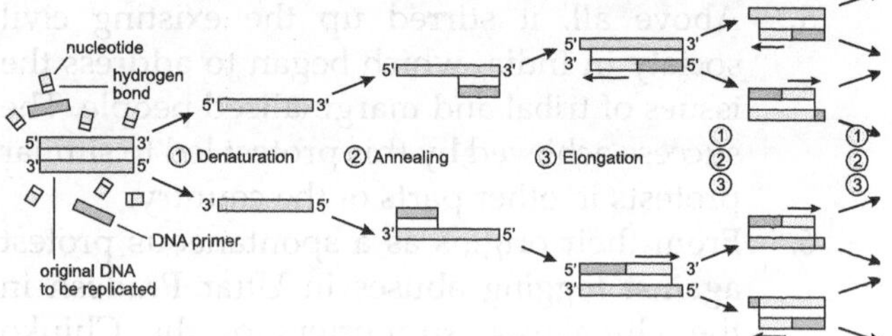

There are three main steps in the process of PCR Denaturation, Annealing and Extending.

1. **Denaturation :** The double stranded template of DNA is heated at a temperature of 94-96°C to separate them into two strands where each strand serves as a template for DNA synthesis.

2. **Annealing :** During this step the temperature is cooled to 40-60°C and the primers are annealed to single stranded DNA.

(iii) **Extending :** This step is final step where temperature is increased to 72°C and the DNA is elongated with the help of heat stable enzyme Taq DNA polymerase, *d*NTPs and Mg^{++} ions.

The process is repeated 20-40 times to get the desired amount of DNA (amplicons). The newly synthesized DNA is subjected to same procedure to double the DNA content from the previous cycle.

OR

(b) Blue-white method of selection is a method of screening for the presence of recombinant plasmid. This method is based upon the insertional inactivation of the lac Z gene present on the vector (*e.g.*, PUC19). This gene expresses the enzyme β-galactosidase whose activity can cleave a colourless substrate called X-gal into blue coloured product. Isopropyl β-D-1-thiogalactopyranoside (IPTG) is used along with X-gal for blue-white screening. IPTG is a non-metabolizable analog of galactose that induces the expression of lac Z gene. IPTG act as a inducer for β-galactosidase. If the lac Z gene is inactivated due to the presence of the insert then the enzyme is not expressed. Hence, after a transformation experiment on E. coli host cells, plated on an ampicillin and X-gal containing solid media plate, colonies which appear blue have transformed cells (antibiotic resistant) but do not have the insert. Colonies which appear white are both ampicillin resistant and have the insert rDNA and thus are the cells to be used for future experiment.

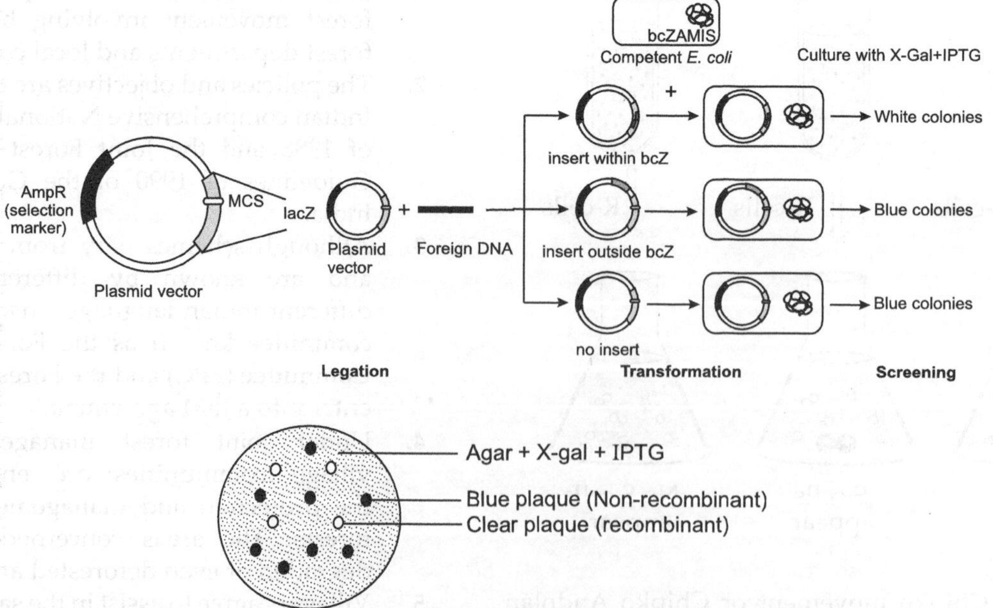

Blue-White Method for Selection

BIOLOGY 2017

QUESTIONS

PART I (20 Marks)

*(Answer **all** questions)*

Question 1.

(a) *Give a brief answer for each of the following :* [4]

(i) *Why do green plants start evolving CO_2 instead of O_2, at high temperatures ?***

(ii) *Define Apomixis.*

(iii) *What is a Recon ?***

(iv) *Why are the spores of Bacillus thuringiensis used as bioinsecticide ?*

(b) *Each of the following question(s)/statement(s) has four suggested answers. Choose the correct option in each case.* [4]

(i) *Initiation codon of protein synthesis in Eukaryotes is :*

(A) *GUA* (B) *GGA*
(C) *CCA* (D) *AUG*

(ii) *Type of interaction where an individual sacrifices its own welfare (life) for the benefit of another animal of its own species is :*

(A) *Altruism* (B) *Scavenging*
(C) *Protocooperation* (D) *Commensalism*

(iii) *Wings of insect and birds are examples of :*

(A) *Analogous* (B) *Homologous*
(C) *Vestigial* (D) *Atavism*

(iv) *The pressure of the cell contents on the cell wall is known as :*

(A) *Wall pressure* (B) *Osmotic pressure*
(C) *Turgor pressure* (D) *Diffusion pressure*

(c) *Give scientific terms for each of the following :* [4]

(i) *An act of expelling the full term foetus from mother's uterus at the end of gestation.*

(ii) *Entry of pollen tube into an ovule through integuments.*

(iii) *An alternative form of the single gene which influences the same character and produces different expressions in different individuals of a species.*

(iv) *The study of human population covering all aspects and parameters.*

(d) *Expand the following abbreviations :* [4]

(i) *MTP* (ii) *IVF*
(iii) *HIV* (iv) *DPD***

(e) *Name the scientists who have contributed to the following :* [4]

(i) *Discovered fossil of Cro-Magnon man.*

(ii) *Classified active and passive absorption of water by roots.*

(iii) *Reported Haemophilia*

(iv) *Discovered double fertilisation.*

PART II (50 Marks)

SECTION–A

*(Answer any **two** questions)*

Question 2.

(a) *Differentiate between Apes and Man with respect to the following characteristics :* [3]

(i) *Posture* (ii) *Brow ridges*
(iii) *Cranial capacity*

(b) *Define protobionts.* [1]

(c) *What is cognogeny ?*** [1]

Question 3.

(a) *Explain any three molecular (genetic) evidences in favour of organic evolution.* [3]

(b) *Define biogenesis.* [1]

(c) *Define fossils.* [1]

Question 4.

(a) *List any three drawbacks of Darwinism.* [3]

(b) *State Hardy Weinberg's principle.* [1]

(c) *Differentiate between Directional natural selection and Disruptive natural selection.* [1]

SECTION–B

*(Answer any **two** questions)*

Question 5.

(a) *Give four anatomical differences between a dicot leaf and monocot leaf.*** [4]

(b) *Briefly describe the secretory phase of the menstrual cycle.* [4]

(c) *Define :*

(i) *Menarch*

(ii) *Actinomorphic symmetry*** [2]

Question 6.

(a) *Give a graphic representation of the Hatch Slack or C_4 cycle.*** [4]

(b) *Give two significant differences between :***

(i) *Transpiration and Guttation.*

(ii) *Chlorophyll 'a' and Chlorophyll 'b'* [4]

(c) *Define the following :*

(i) *Amniocentesis* (ii) *Polyembryony* [2]

** Answer is not given due to change in the present syllabus.

Question 7.

(a) *Describe K^+ transport stomatal mechanism*** **[4]**

(b) *Draw a neat labelled diagram of L.S. of anatropous ovule.* **[4]**

(c) *Differentiate between the following :*

(i) *Spermatogenesis and oogenesis.*

(ii) *Apocarpous ovary and syncarpous ovary.*** **[2]**

SECTION–C

*(Answer any **two** questions)*

Question 8.

(a) *Explain Pleiotropy with reference to phenylketonuria.* **[4]**

(b) *Explain the mechanism of transcription in a prokaryotic cell.* **[4]**

(c) *Explain Rh factor incompatibility during pregnancy.*** **[2]**

Question 9.

(a) *Discuss the various In-situ and Ex-situ strategies for conservation of biodiversity.* **[4]**

(b) *List any four applications of tissue culture.* **[4]**

(c) *Mention the causative agent and preventive measures for each of the following :*

(i) *Gonorrhoea*

(ii) *Pneumonia* **[2]**

Question 10.

(a) *Name the components of lac operon and discuss their role.* **[4]**

(b) *Give the significance of transgenic animals.* **[4]**

(c) *Give one significant difference between* **[2]**

(i) *Electroporation and Gene Gun.*

(ii) *ECG and EEG.***

ANSWERS

PART I (20 Marks)

Answer 1.

(a) (ii) The term Apomixis has been derived from two Greek word *"Apo"* meaning away from and *"mixis"* which refers to an act of mixing or mingling. It refers to replacement of the normal sexual reproduction by asexual reproduction, without fertilisation.

(iv) *Bacillus thuringiensis* is used as a biological pesticide because of its specific toxicity against target pests in their larval stage. During sporulation, Bt strains produce crystalline proteins (proteinaceous inclusions), called δ endotoxin, which have insecticidal action.

(b) (i) (D) AUG

(ii) (A) Altruism

(iii) (A) Analogous

(iv) (C) Turgor pressure

(c) (i) Parturition **(ii)** Mesogamy

(iii) Multiple Allele **(iv)** Demography

(d) (i) Medical Termination of Pregnancy

(ii) In Vitro Fertilisation

(iii) Human Immuno Deficiency Virus

(e) (i) Mac Gregor **(ii)** Kramer

(iii) John Otto **(iv)** G. Nawaschin

PART II (50 Marks)

SECTION–A

Answer 2.

(a) (i) Posture : Man has upright stance and bipedal movement whereas apes are actually quadrupeds on the ground and brachiators (arm swingers) in trees with semi-erect posture.

(ii) Brow Ridges : In man brow ridges are inconspicuous whereas in apes they are prominent.

(iii) Cranial Capacity : In man the cranial capacity is of 1350-1600 cm^3 whereas in apes it is 390-650 cm^3.

(b) Protobionts are considered to be the evolutionary precursors of prokaryotic cells which resemble very simple cells. They are thought to have originated as an array of abiotically produced microspheres of diverse organic and inorganic compounds enclosed by lipid membranes.

Answer 3.

(a) The evidences in favour of organic evolution are as follows :

1. **Evidence from paleontology and metabolic process :** Most of the biochemical reactions are found similar from bacteria to humans. For *e.g.*, Kreb cycle, Electron transport system, Glycolytic pathway etc. Energy released is stored in ATP molecules during oxidation of glucose and nitrogenous wastes in all living organism is produced in the form of ammonia.

2. **Evidences from biochemical findings :** Every living organism shows a degree of similarity in DNA base sequences and amino acids. Genes are composed of triplet code of nucleobases, which is universal in nature. About 99% of gene sequences in organism are similar. Homology is found in amino

** Answer is not given due to change in the present syllabus.

acid sequence of enzymes and hormones like cytochrome C and Haemoglobin. Similarities in chromosome structure, number and structure of actin tubulin protein also support organic evolution.

3. **Evidences from cytology :** Cell is the structural and functional unit of all living organisms. Every organism consist of cells, with similar set of structural and functional properties. Certain cell organelles like Mitochondria, ER, Golgi bodies are similar in all the eukaryotes.

(b) Biogenesis : Biogenesis refers to production of life from pre-existing organisms or matter. Biogenesis is observed at all patterns of life including like when bacteria divides, production of seeds from plants etc.

(c) Fossils : Fossils are the remains or impression of prehistoric animals or plants, naturally preserved in the layers of earth's crust, in petrified form.

Answer 4.

(a) 1. Darwin was not able to explain that characters are transmitted from generation to generation.

2. Sometimes survival of the fittest would not be possible because variations selected by nature may be harmful to the life of organism and ignored the mutations.

3. He considered individual (not population) as unit of evolution and hence unable to explain the connecting link.

(b) According to Hardy-weinberg law, in a large randomly mating population, the allelic frequencies of sexually reproducing organisms will remain constant.

$$(p + q) = 1$$

or $$p^2 + 2pq + q^2 = 1$$

(c)

Directional natural selection	Disruptive natural selection
The directional selection describes the change that occurs when a single phenotype is favoured, causing the allele frequency to continuously shift in one direction.	Disruptive natural selection describes the changes in population genetics in which extreme values for a trait are favoured over intermediate values.

SECTION–B

Answer 5.

(b) During the secretory phase, progesterone hormone is produced by the ovaries. Progesterone and estrogen are secreted by the corpus luteum that develops from the Graafian follicle. Graafian follicle ruptures to release the ovum due to LH surge and uterine contractions decreases.

Progesterone stimulates the further build-up of the cells and vascularisation in the endometrium of the uterus and stimulates the glands in the uterus to secrete uterine milk that maintain the endometrium and prevents it from breaking down. In case ovum is not fertilised, corpus luteum degerates to corpus albicans, leading to fall in progesterone level. Therefore, this phase of the menstrual cycle is called the secretory phase (Days 14-28 of the menstrual cycle). Decline of progesterone level starts menstrual phase.

(c) (i) **Menarche :** It refers to the time when menstruation first begins in young females. During the menarche period, menstruation may be irregular and unpredictable, it is also known as female puberty.

Answer 6.

(c) (i) **Amniocentesis :** Amniocentesis or amniotic fluid test (AFT) is a prenatal test used in prenatal diagnosis of chromosomal abnormalities and fetal infections, and for sex determination. In this small amount of amniotic fluid is removed from the sac surrounding the foetus for testing. The amniotic fluid (less than one ounce) is taken with a fine needle which is inserted into the uterus through the abdomen, under ultrasound guidance. The fluid is then tested in the laboratory for analysis.

(ii) **Polyembryony :** Polyembryony refers to the phenomenon of two or more embryos developing from a single fertilised egg. It occurs in both animals as well as in plants.

Answer 7.

(b)

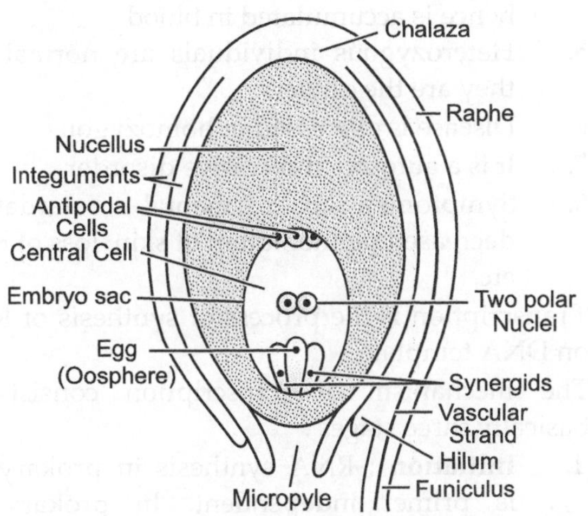

Structure of a anatropous ovule

(c) (i)

Spermatogenesis	Oogenesis
It is the process of formation of Spermatozoa from Spermatogonia inside the testes of male.	It is the process of formation of Ova from Germinal cells in ovary.
Starts at puberty.	Starts before birth

SECTION–C

Answer 8.

(a) **Pleiotropy** occurs when one gene influences two or more unrelated phenotypic effects.

A mutation in a pleiotropic gene may have an effect on several traits simultaneously due to the gene coding for a product used by a myriad of cells or different targets that have the same signaling function.

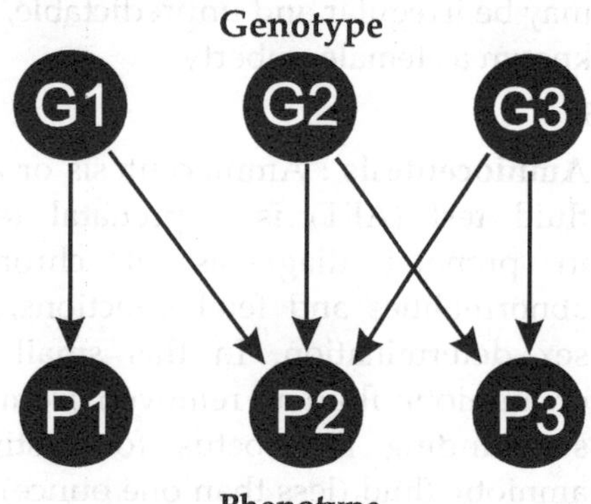

Phenylketonuria :

1. Enzyme deficiency is due to abnormal autosomal recessive gene on chromosome 12.
2. It is inborn metabolic disorder.
3. Phenylketonuria is a condition caused due to deficiency of phenylalanine hydroxylase enzyme .
4. When this enzyme is missing, the body is unable to break down phenylalanine and hence is accumulated in blood.
5. Heterozygous individuals are normal but they are the carriers.
6. Disease is expressed in homozygous.
7. It is a autosomal recessive disorder.
8. Symptoms are : mental retardation, decreased pigmentation of skin, loss of hairs etc.

(b) Transcription is the process of synthesis of RNA on DNA template.

The mechanism of transcription consists of basically three stages :

1. **Initiation :** RNA synthesis in prokaryotes is primer independent. In prokaryotes, transcription begins with binding of RNA polymerase to promoter region in DNA. At this stage DNA is unwouned and forms open complex. RNA polymerase, together with sigma factor binds to promotor DNA. RNA polymerase creates transcription bubble, which separates the two strands of DNA helix, by breaking the H-bonds between complementary strands.
2. **Elongation :** After initiation RNA polymerase releases its sigma factor and forms tertiary elongation complex which consists of RNA polymerase, template DNA and nascent RNA chain. The RNA polymerase synthesizes RNA in 5′ to 3′ direction. The region of unwounded DNA called the 'Transcription bubble' appears to move along the DNA with the polymerase. The polymerase seems to unwind DNA in front of the transcription bubble and rewinds DNA at its rear.
3. **Termination :** When polymerase reaches the termination signal it leaves the DNA, the hydrogen bonds of the RNA-DNA helix break, freeing the newly synthesized RNA strand. Termination in prokaryotes is of two types Rho-independent and Rho-dependent termination.

   In **Rho-dependent termination,** Rho factor binds to RNA polymerase, replacing initiation factor, sigma factor. At the transcription bubble, Rho pulls the RNA transcript and the template DNA strand apart, releasing the RNA molecule and ending transcription.

   Rho-independent termination depends on specific sequences in the DNA template strand. As the RNA polymerase approaches the end of the gene being transcribed, it attach to a region rich in C and G nucleotides. The RNA transcribed from this region folds back on itself, and the complementary C and G nucleotides bind together. The result is a stable hairpin structure that halts the polymerase action.

   In a termination, the hairpin is followed by a stretch of U nucleotides in the RNA, which matches up with A nucleotides in the template DNA. The complementary U-A region of the RNA transcript forms only a weak interaction with the template DNA. This, coupled with the stalled polymerase, produces enough instability for the enzyme to fall off and liberate the new RNA transcript.

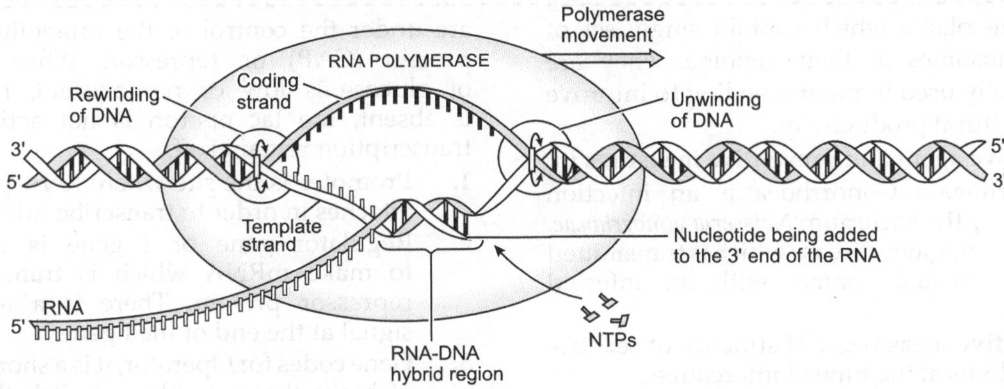

Answer 9.

(a) Conservation refers to protection, preservation, management, or restoration of wildlife and natural resources such as forests and water.

Conservation can be broadly divided into two main types :

1. In-Situ, **2.** Ex-Situ

Biodiversity Conservation

IN-SITU	EX-SITU
1. Protected Areas	1. Seed Banks, Gene Banks
– National Parks	2. Long Term Captive Breeding
– Sanctuaries	3. Animal Translocations
– Biosphere Reserves	4. Tissue Culture Banks
2. Sacred Forests and Lakes	5. Cryopreservation of Gametes and Embryos
	6. Botanical Gardens
	7. Zoological Gardens or Zoos.

1. **In-Situ Conservation Methods :** In-situ conservation refers to conservation of species in their natural habitats and is one of the most appropriate ways of conserving biodiversity. It is also known as on-site conservation. Various natural habitats are national parks, wildlife sanctuaries and biosphere reserves. A national park is a reserved area which is strictly reserved for the betterment of the wildlife and where activities like forestry, grazing on cultivation are not permitted.
2. **Ex-Situ Conservation Methods :** Ex-situ conservation refers to the preservation of components of biological diversity outside their natural habitats. It is also known as off-site conservation. This involves conservation of genetic resources, as well as wild and cultivated species. Some of these include :
   (i) Gene banks, *e.g.*, seed banks, sperm and ova banks, field banks.
   (ii) In vitro plant tissue and microbial culture collections.
   (iii) Captive breeding of animals and artificial propagation of plants.
   (iv) Collecting living organisms for zoos, aquaria, and botanical gardens for research and public awareness.

Ex-situ conservation is an important measure against extinction. These measures also have a valuable role to play in recovery programmes for endangered species. It also helps in maintaining domesticated plants which cannot survive in nature unaided.

Ex-situ conservation plays a central role in public education and awareness raising by bringing members of the public into contact with plants and animals.

(b) Applications of plant tissue culture :

1. **Micropropagation :** Micropropagation involves culturing of whole plant through tissue culture. Hundreds of plantlets can be developed from a single piece of callus. So this technique is quite economic.
2. **Somaclonal variation :** Plants produced by tissue culture show similarity in their characters but few show variations. They are known as somaclonal variants. The regenerated variants were tested for their resistance to herbicide resistance, heavy metals, temperature and drought.
3. **Virus eradication :** Plant viruses can be removed by apical meristem culture. The apical meristem culture is the only way to obtain a clone of virus free plant which can be multiplied vegetatively under control conditions that would protect them from the chance of reinfection.

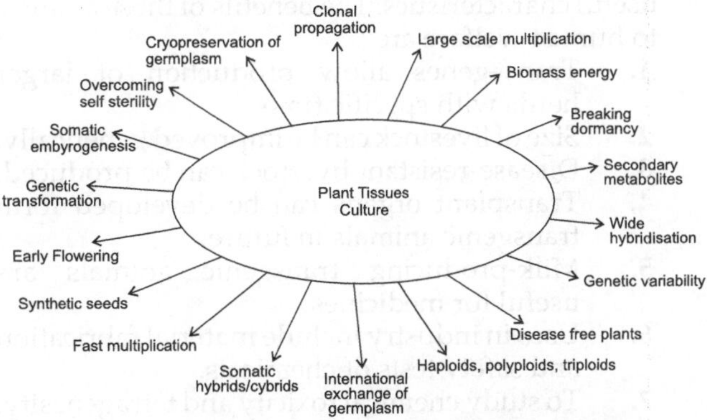

4. **Production of androgenic haploids :** Anthers are produced from anther culture, ovule culture or pollen culture. Haploids

are the plants which contain single set of chromosomes in their genome. They are generally used in plant breeding to improve agricultural productivity.

(c) The causative agent and preventive measures in:

(i) Gonorrhoea : Gonorrhoea is an infection caused by the bacterium *Neisseria gonorrhoeae.* It is an contagious disease which is transmitted through sexual contact with an infected person.

Preventive measures : Abstinence of sex, use of condoms at the time of intercourse.

(ii) Pneumonia : Pneumonia is an infection of lungs which can be caused by bacteria (*Streptococcus pneumonia*), virus or fungi.

Prevention : Maintain personal and public hygiene. Avoid sharing glasses, utensils, food or water with infected persons.

Answer 10.

(a) The lac operon encodes for the genes needed to utilise lactose as an energy source. The lac operon contains an operator, promoter, and structural genes. These genes are transcribed together and are under the control of the catabolite activator protein (CAP) or repressor. When the level of glucose is low or non-existent, but lactose is absent, the lac operon is not activated and transcription remains off.

1. Promoter is the site where RNA polymerase attaches in order to transcribe mRNA.
2. Regulator gene or I gene is transcribed to make mRNA which is translated to a repressor protein. There is a termination signal at the end of the I gene.
3. Gene codes for Operator; it is a short sequence of bases that acts like a switch that can be recognised by repressor protein.
4. Lac Z, lac Y and lac A are all structural genes (genes that code for polypeptides). **Z** codes for β-galactosidase; which hydrolyse lactose into glucose and galactose **Y** codes for lactose permease, a protein that functions to actively bring lactose from outside to cell to the inside, even against a concentration gradient. **A** codes for transacetylase, an enzyme that is also needed to breakdown many sugars related to lactose.

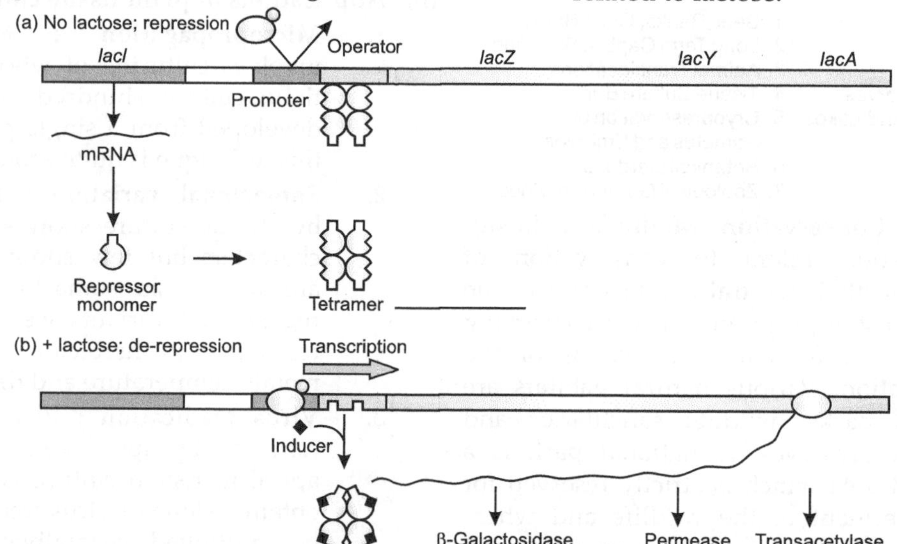

(b) Transgenic animals are genetically modified organisms which carry foreign genes with some useful characteristics. The benefits of these animals to human welfare are :

1. Trans–genes allow production of larger herds with specific traits.
2. Size of livestock can be improved genetically.
3. Disease-resistant livestock can be produced.
4. Transplant organs can be developed form transgenic animals in future.
5. Milk-producing transgenic animals are useful for medicines.
6. Uses in industry include material fabrication and safety tests of chemicals.
7. To study chemical toxicity and tetragenesity.
8. Helps in the improvement of quality of human race.
9. To produce industrial or consumer products (fibres for multiple uses).
10. To enhance the animals interactions with humans.

(c) (i)

Electroporation	Gene Gun
Electroporation is a method that uses short pulses of high voltage to carry DNA across the cell membrane. This shock is thought to cause temporary formation of pores in the cell membrane, allowing DNA molecules to pass through.	In this technique, DNA is coated onto gold particles/platinum/ tungsten and loaded into a device which generates a force to achieve penetration of the DNA into the cells.
Can be used only for naked protoplast.	Can be used for cells with cell wall.

●●

BIOLOGY

2016

QUESTIONS

PART I (20 Marks)

*(Answer **all** questions)*

Question 1.

(a) *Give a brief answer for each of the following :* **[4]**

(i) *What is central dogma ?*

(ii) *Define cryopreservation.*

(iii) *What is symbiosis ?*

(iv) *Explain the term perianth.***

(b) *Each of the following question(s)/statement(s) has four suggested answers. Choose the correct option in each case.* **[4]**

(i) *The curve showing the amount of light absorbed at each wavelength is:***

(A) Action spectrum

(B) Absorption spectrum

(C) Quantum yield

(D) Quantum requirement

(ii) *After fertilisation, the integuments of an ovule develop into :*

(A) Seed *(B) Seed coat*

(C) Fruit *(D) Fruit wall*

(iii) *Meselson and Stahl's Experiment proved:*

(A) Transduction

(B) Transformation

(C) DNA is the genetic material

(D) Disruptive DNA replication

(iv) *The act of expelling the full term foetus from the uterus is termed as :*

(A) Gestation *(B) Implantation*

(C) Parturition *(D) Capacitation*

(c) *Give scientific terms for each of the following :* **[4]**

(i) *The smallest unit of DNA which can mutate.*

(ii) *Type of water absorption by roots where metabolic energy is required.***

(iii) *Statistical study of human population.*

(iv) *Multiple effects of a gene on the phenotype of an organism.*

(d) *Expand the following abbreviations :* **[4]**

(i) *rDT* **(ii)** *BAC*

(iii) *SSBP* **(iv)** *IUCD*

(e) *Name the scientists who have contributed to the following :* **[4]**

(i) *Reverse transcription*

(ii) *Photorespiration***

(iii) *Principle of limiting factors***

(iv) *Photolysis of water***

PART II (50 Marks)

SECTION–A

*(Answer any **two** questions)*

Question 2.

(a) *Mention three features of the Neanderthal Man.* **[3]**

(b) *Differentiate between connecting link and missing link.* **[1]**

(c) *What is adaptive radiation ?* **[1]**

Question 3.

(a) *Give an account of Lederberg's replica plating experiment to show the genetic basis of evolution.* **[3]**

(b) *Define phylogeny.* **[1]**

(c) *What is Founder's effect ?* **[1]**

Question 4.

(a) *With reference to the levels of organisation, differentiate between living organisms and non-living objects.* **[3]**

(b) *Mention one cause for variation in nature.* **[1]**

(c) *What is the difference between the teeth of apes and the teeth of man ?* **[1]**

SECTION–B

*(Answer any **two** questions)*

Question 5.

(a) *Give a graphic representation of the C_3 cycle.*** **[4]**

(b) *Discuss the role of cambium in secondary growth of dicot stems.*** **[4]**

(c) *State two advantages of vegetative propagation.* **[2]**

Question 6.

(a) *Explain the role of hormones during the menstrual cycle.* **[4]**

(b) *Give four adaptations shown by flowers pollinated by wind.* **[4]**

(c) *Give two differences between heart wood and sap wood.*** **[2]**

Question 7.

(a) *Explain chemiosmotic hypothesis for ATP synthesis.*** **[4]**

(b) *Draw a neat labelled diagram of the vertical section of a monocot leaf.*** **[4]**

(c) *Mention any two functions of the human placenta.* **[2]**

** Answer is not given due to change in the present syllabus.

SECTION–C

(Answer any two questions)

Question 8.

(a) *Explain the process of sex determination in honey bees.* **[4]**

(b) *Define complete linkage. Give an example of a cross, showing complete linkage.* **[4]**

(c) *Write a short note on Multiple Ovulation Embryo Transfer Technology.* **[2]**

Question 9.

(a) *Given an account of artificial chromosomes in transfer of genetic material.* **[4]**

(b) *Mention any four methods involved in the treatment of cancer.* **[4]**

(c) *What is RNA interference ? Give any one application of RNA interference.* **[2]**

Question 10.

(a) *What is Integrated Pest Management ?* **[4]**

(b) *Explain the structure of a typical antibody molecule.* **[4]**

(c) *Why are biofertilizers preferred over chemical fertilisers ?* **[2]**

ANSWERS

PART I (20 Marks)

Answer 1.

(a) (i) The 'Central Dogma' is the process by which the instructions in DNA are converted into a functional product. The central dogma of biology describes the unilateral flow of information from the DNA to the mRNA to the ribosomes.

(ii) Cryopreservation is a process where cells, tissues or germplasm are preserved by cooling at low to sub-zero temperatures (such as –196°C).

(iii) Symbiosis is the interaction between two different organisms living in close physical association, typically to the advantage of both. Example : *Rhizobium* in root nodules of legumes, lichens, etc.

(b) (ii) Seed coats

(iii) Disruptive DNA replication

(iv) Parturition

(c) (i) Muton

(iii) Demography

(iv) Pleiotrophy

(d) (i) Recombinant DNA Technology

(ii) Bacterial Artificial Chromosome

(iii) Single Stranded Binding Protein

(iv) Intra-Uterine Contraceptive Device

(e) (i) Howard Temin and David Baltimore

PART II (50 Marks)

SECTION–A

Answer 2.

(a) Features :

1. Lower jaw was strong with strong jaw muscles with no chin.
2. Cranial capacity was 1300-1600 cc.
3. Less neotenised skull than modern human and prominent eyebrow ridges.

(b)

Connecting Links	Missing Links
They are living organisms having the characters of two different groups. *e.g.*, Protopterus, Peripatus.	They are extinct organisms. They had the characters of two different groups of organisms. *e.g.*, Seymouria, Archaeopteryx.

(c) Due to environmental pressure when one species develops new features to adapt, and eventually evolve into separate species, then it is called adaptive radiation.

Answer 3.

(a) The Lederberg's Replica plating experiment : In 1952, Esther and Joshua Lederberg performed an experiment that helped to show that many mutations are random, not directed. Bacteria is grown into isolated colonies on plates and these colonies can be reproduced from an original plate to new plates by "stamping" the original plate with a cloth and then stamping empty plates with the same cloth. Bacteria from each colony are picked up on the cloth and then deposited on the new plates by the cloth. Colonies which could grow were resistant to antibiotic.

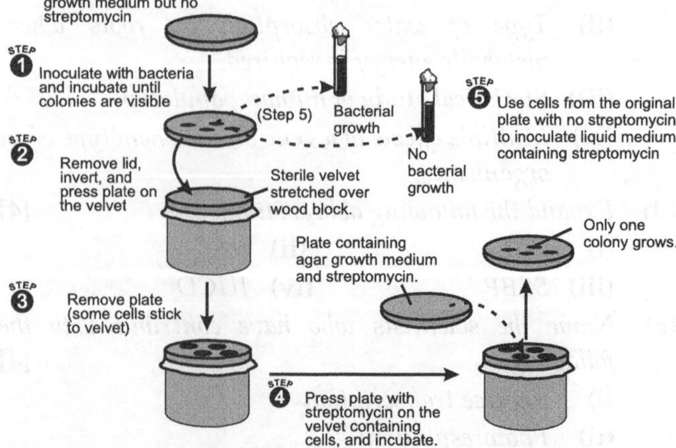

Fig. : Esther and Joshua Lederberg Experiment

The hypothesis for the experiment is that the antibiotic resistant strains of bacteria which survived the application of antibiotics, had the resistance before their exposure to the antibiotics, not as a result of the exposure.

(b) Phylogeny is the sequence of events involved in the evolution of species.

(c) The Founder's effect is the loss of genetic variation that occurs when a new population is established by a very small number of individuals from a larger population and so can cause a change in the allelic frequency leading to genetic drift.

Answer 4.

(a)

Living organisms	Non-living objects
Formed of organic molecules.	Formed of inorganic molecules.
Level of organisation is present (*i.e.*, cell to organism).	Level of organisation is absent in non-living object.
Living organism possess nutrition, adaptation, reproduction, etc.	Biological processes are not present in non-living object.

(b) Mating patterns due to migration is the cause of variation in nature. Crossing over and gene flow also cause variation in nature.

(c)

Apes	Man
Large canines.	Small canines.
Diastema present.	Diastema absent.

SECTION–B

Answer 5.

(c) Advantages :

1. The offspring are genetically identical and therefore advantageous traits can be preserved. It is a rapid method.
2. Only one parent is required which eliminates the need for special mechanisms such as pollination, etc. Thus it is less expensive.

Answer 6.

(a) Role of hormones in the menstrual cycle :

1. **FSH (Follicle Stimulating Hormone) :**
   (i) This hormone is secreted by anterior pituitary, it stimulates the growth of selected ovarian follicles and maturation of the primary oocyte.
   (ii) It also stimulates the follicular cells of Graafian follicle to secrete a female sex hormone named Oestradiol (Oestrogen).
2. **Role of Luteinising Hormone (LH) :** It induces the mature Graafian follicle to burst and eject its egg into fallopian tube and stimulates the formation of corpus luteum. It's surge causes ovulation.
3. **Role of Progesterone :**
   (i) It maintains the hypertrophy of endometrial lining in the uterus and fallopian tube.
   (ii) It is required for proper implantation of foetus in the wall of uterus.
   (iii) It stimulates the endometrial glands to secrete a nutrient fluid for foetus.
   (iv) It inhibits the release of LH and hence inhibits its production.
4. GnRH released by hypothalamus regulates the release of FSH and LH from anterior pituitary.
5. Oestrogen from ovary repairs the endometrial lining. High oestrogen stimulates LH and cuts off FSH.

(b) Adaptations :

1. Flowers are colourless, inconspicuous and without fragrance and versatile anther is present.
2. No nectar is required to attract insects.
3. The stigma is feathery or net–like to catch pollen grains.
4. Pollen is produced in large quantities.

Answer 7.

(c) 1. The placenta acts to provide oxygen and nutrients to the foetus, whilst removing carbon dioxide and other waste products. It stores fat, glycogen and iron.

2. It acts as an endocrine tissue and produce several hormones like human chorionic gonadotropin (HCG), progesterone, estrogen, etc.

SECTION–C

Answer 8.

(a) It is a type of sex determination in which the male is haploid while the female is diploid. Haplodiploidy occurs in some insects like bees, ants and wasps. Male insects are haploid because they develop parthenogenetically from unfertilised eggs. The phenomenon is called arrhenotoky. Meiosis does not occur during the formation of sperms.

Females grow from fertilised eggs and are hence diploid. Queen bee picks up all the sperms from the drone during nuptial flight and stores the same in her seminal vesicle. Formation of worker bees (diploid females) and drones (haploid males) depends upon the brood cells visited by the queen. While visiting the smaller brood cells, the queen emits sperms from its seminal receptacle after laying the eggs.

Males are normally fertile haploids due to development from unfertilised eggs. Occasionally diploid infertile males are also produced from heterozygous females through fertilisation.

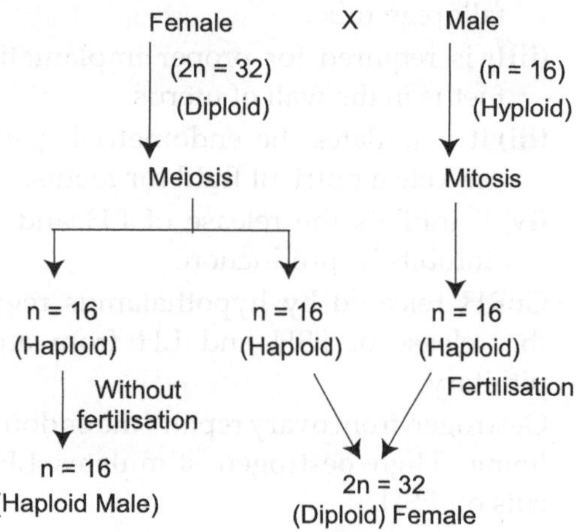

Haplo-diploid sex determination system

(b) Complete linkage : Linkage of genes on a chromosome which is not altered and is inherited as such from generation to generation without any cross over is called complete linkage. In this type of linkage, genes are closely associated and tend to remain together. They are transmitted together in their parental combination. Example, Genes for body colour and types of wings in *Drosophila*.

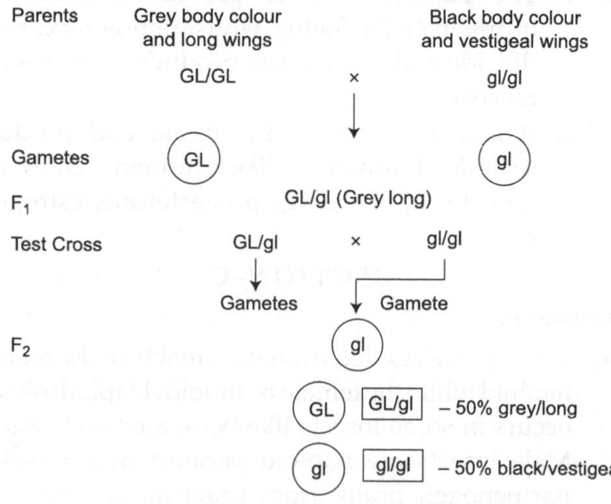

Fig. Complete Linkage

Only parental combinations are formed in F_2 generation and no appearance of any new combinations. 100% Parental combinations indicated that the gene for grey body colour is completely linked with long wings.

(c) Multiple Ovulation Embryo Transfer (MOET) Technology provides chances of successful production of hybrids. In this method, cow is administered with FSH like hormone, to induce follicular maturation and super ovulation. In each cycle, 4-10 eggs mature simultaneously. The cow is either mated with an elite bull or artificially inseminated. The blastocysts at 8-32 cell stage are recovered non-surgically and transferred to surrogate mothers, who may not have superior quality who carries it to full form. The genetic mother is available for another round of super ovulation. Ova is fertilised by artificial insemination. This technology is successfully used for cattle, sheep, rabbits, buffaloes etc. High milk yielding breeds of female and high quality meat yielding bull (lean meat with less lipids) have been bred successfully to increase herd size in a short period and to improve the breed of cattles. By deep freezing it is possible to preserve 7 day old foetus for years.

Answer 9.

(a) Artificial chromosomes are synthetic chromosomes consisting of fragments of DNA integrated into a host chromosome. These artificial chromosomes are introduced into host cells to propagate and can be used to transfect other cells, introducing new DNA. Because these chromosomes are more useful in cloning larger fragments of DNA, it is easier and quicker to clone and transform genes. Artificial chromosomes vectors also make it easier to store through bacterial cells rather than mammalian cells.

There are two types of artificial chromosomes : Bacterial Artificial Chromosome (BAC). They are constructed from bacterial genomes and Yeast Artificial Chromosomes (YAC) are constructed from yeast genome. The construction of BAC is based on F-plasmid and it can accept DNA inserts of the range of 300-350 Kb. It is more stable than Yeast Artificial Chromosome. Although yeast artificial chromosomes can contain more base pairs (over one million) than bacterial artificial chromosomes, bacterial artificial chromosomes and are more common than yeast artificial chromosomes because they are more stable, making them easier to work with a smaller risk of rearrangement due to the circular shape of a plasmid.

YAC can accept large fragments of foreign DNA (upto 1 million base), it has centromeric and telomeric regions, ori site and selectable markers along with autonomously replication sequences. Both BAC and YAC are used as vectors for HGP (Human genome Project).

(b) The four methods involved in the treatment of cancer:

1. **Surgery** : Cancers in organs like the breast, prostate and testicles can be cured by the

removal of those parts through surgery. This is the oldest known treatment option, which is effective if metastasis has not taken place and if cancer is localised to a small portion of the body. It cannot be employed for leukaemias.

2. **Radiation therapy :** This is the most commonly used therapeutic option against cancer. During this therapy, high-energy X-rays and high-energy gamma-rays emitted from radium are focused specifically on the cancerous cells. These high energy radiations damage the molecules making up the cancerous cells. This treatment can independently shrink the tumor size or obliterate the cancer cells.
3. **Chemotherapy :** This therapy involves the use of chemicals which can inhibit the active cell division process, so that the process of apoptosis (cell death) gets stimulated within the cancer cells. This treatment is generally chosen as an option in case of cancers which have spread to different parts of the body as medicines and can travel to different parts of the body.
4. **Immunotherapy :** This novel therapeutic option aims to boost up the immune system so that it can attack the tumours. For example, in Radioimmunotherapy, radioisotope linked monoclonal antibodies are used for the treatment of cancer. A significant success has been achieved in the treatment of breast cancer through immunotherapy.

(c) RNA interference is silencing a specific mRNA by introducing a gene that produces the sense and antisense strand of mRNA so that a double stranded mRNA is produced which cannot be translated into the specific protein, inhibiting gene expression. This technique is used to create nematode resistant plants, delay in ripening of fruits, to know gene functions, production of vaccines etc.

Answer 10.

(a) Integrated pest management (IPM), also known as Integrated Pest Control (IPC) is a broad-based approach that integrates practices for economic control of pests. IPM aims to suppress pest populations below the Economic Injury Level (EIL).

IPM can be defined as an ecosystem-based strategy that focuses on long-term prevention of pests or their damage through a combination of techniques such as biological control, habitat manipulation, modification of cultural practices, and use of resistant varieties. Pesticides are used only after monitoring indicates they are needed according to established guidelines, and treatments are made with the goal of removing only the target organism. Pest control materials are selected and applied in a manner that minimizes risks to human health, beneficial to non-target organisms, and the environment.

The functions of IPM are :

1. IPM focuses on long-term prevention of pests or their damage by managing the ecosystem.
2. In IPM, monitoring and correct pest identification help us to decide whether management is needed.
3. IPM programmes combine management approaches for greater effectiveness.
4. The most effective, long-term way to manage pests is by using a combination of methods that work better together than using separately.

Approaches for managing pests are often grouped in the following categories :

1. **Biological control :** Biological control is the use of *natural enemies*—predators, parasites, pathogens and competitors—to control pests and their damage.
2. **Cultural controls :** Cultural control are practices that reduce pest establishment, reproduction, dispersal and survival. For example, changing irrigation practices can reduce pest problems, since too much water can increase root disease and weeds.
3. **Mechanical and physical controls :** Mechanical and physical controls kill a pest directly, block pests out, or make the environment unsuitable for it. Traps for rodents are examples of mechanical control.
4. **Chemical control :** Chemical control is the use of pesticides. In IPM, pesticides are used only when needed and in combination with other approaches for more effective, long-term control.

Benefits of IPM :

1. Economic in nature.
2. Environmental friendly.

(b) **Structure of antibody :** Antibodies are glycoproteins which are highly specific to specific antigens. They are also known as immunoglobulins (Igs). Antibodies are produced by plasma cells which in turn are formed by B-lymphocytes. The plasma cells produce about 2000 molecules of antibodies.

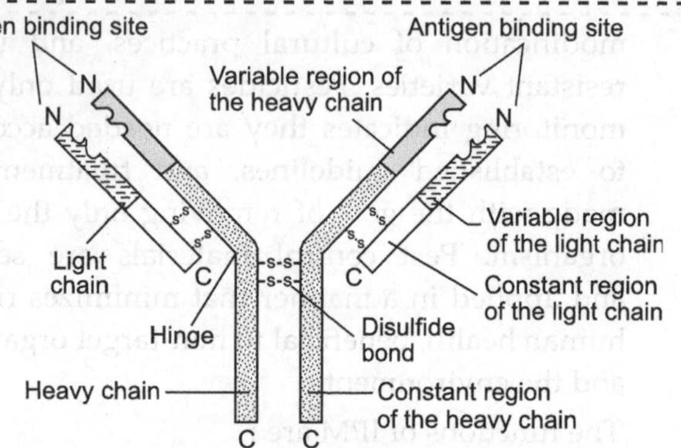

Fig. Structure of antibody

Antibody is a 'Y' shaped structure. It consists of four polypeptide chains, two heavy or H chains and two light or L chains. The four polypeptide chains are held together by disulphide bonds (–s–s–) to form a 'Y' shaped structure. The region holding arms and stem of antibody is termed as hinge. Each chain of the antibody includes two distinct regions, the variable region and the constant region. Variable regions constitute the antigen-binding site. (paratope). This part of antibody recognises and binds to the specific antigen forming an antigen-antibody complex. Since most antibodies carry two antigen binding sites, they are said to be bivalent.

(c) Biofertilisers are organisms that enrich the nutrient quality of the soil such as Azotobacter while chemical fertilisers degrade the soil fertility if excessively used and cause land pollution also.

Biofertilisers are preferred over chemical fertilisers because :

1. They do not pollute environment.
2. They are cheap and easy to produce.

BIOLOGY

2015

QUESTIONS

PART I (20 Marks)

*(Answer **all** questions)*

Question 1.

(a) *Give a brief answer for each of the following:* **[4]**

(i) *What is heterosis ?*

(ii) *Why is non-cyclic photo phosphorylation considered as a non-cyclic pathway ?***

(iii) *Define test cross.*

(iv) *What are introns ?*

(b) *Each of the following question(s)/statements(s) has four suggested answers. Choose the correct option in each case.* **[4]**

(i) *Triple Fusion involves :*

(A) *Fusion of one male gamete with female gamete.*

(B) *Fusion of tube nucleus with generative nucleus.*

(C) *Fusion of two polar nuclei.*

(D) *Fusion of second male gamete with two polar nuclei.*

(ii) *An EEG represents spontaneous electrical activity of the :***

(A) *Kidney* (B) *Spinal cord*

(C) *Heart* (D) *Brain*

(iii) *The genotype of a person with Turner's syndrome will be :*

(A) *44 + XXY* (B) *44 + XYY*

(C) *44 + XO* (D) *44 + XXYY*

(iv) *Transcription is the transfer of genetic code from a DNA molecule to :*

(A) *RNA molecule*

(B) *Second DNA molecule*

(C) *Ribosomal sub unit*

(D) *Sequence of amino acids in a protein molecule.*

(c) *Give a scientific term for each of the following:* **[4]**

(i) *The first formed category of photosynthetic organisms.***

(ii) *The surgical removal of a section of fallopian tube.*

(iii) *An animal behaviour which benefits others but is of no advantage to itself.*

(iv) *The hydrostatic pressure developed inside the cell on the cell wall due to endosmosis.***

(d) *Expand the following abbreviations :* **[4]**

(i) *STD* (ii) *NADP***

(iii) *MRI*** (iv) *DDT*

(e) *Name the scientists who are associated with the following :* **[4]**

(i) *Discovered the fossil of Australopithecus*

(ii) *Microspheres*

(iii) *Coined the term Diffusion Pressure Deficit***

(iv) *Invented the CT scan***

PART II (50 Marks)

SECTION–A

*(Answer any **two** questions)*

Question 2.

(a) *Give any three characters that have developed during human evolution.* **[3]**

(b) *Explain the term chemogeny.* **[1]**

(c) *Give any two distinctive features of Dryopithecus.* **[1]**

Question 3.

(a) *Explain the evolution of giraffe's neck according to Lamarck's theory of evolution.* **[3]**

(b) *Give two chromosomal similarities between man and apes.* **[1]**

(c) *Name any two temporary embryonic structures in vertebrates which provide evidence for evolution.* **[1]**

Question 4.

(a) *Persons suffering from sickle cell anaemia are at an advantage in Malaria infected areas. Explain.* **[3]**

(b) *Define the term gene flow.* **[1]**

(c) *What are analogous organs ? Describe with one example from the plant kingdom.* **[1]**

SECTION–B

*(Answer any **two** questions)*

Question 5.

(a) *With the help of diagrams, name and describe the different types of placentation seen in angiosperms.*** **[4]**

(b) *Give four points of anatomical differences between a monocot stem and a dicot stem.*** **[4]**

(c) *Define the following terms :*** **[2]**

(i) *Racemose inflorescence*

(ii) *Osmotic pressure*

Question 6.

(a) *Draw a diagram of the internal structure of the human ovary.* **[4]**

** Answer is not given due to change in the present syllabus.

(b) *Define the term water potential. What are its components ? Explain.*** [4]

(c) *Give definition and importance of :* [2]

(i) *Imbibition*** (ii) *Parturition.*

Question 7.

(a) *Give four adaptations in flowers pollinated by insects.* [4]

(b) *Describe the mass flow hypothesis for translocation of organic solutes (food) in plants.*** [4]

(c) *Write a brief note on the causes of infertility.* [2]

SECTION–C

*(Answer any **two** questions)*

Question 8.

(a) *Give any four reasons for Mendel's success.* [4]

(b) *Briefly describe the technique employed in DNA fingerprinting.* [4]

(c) *Give any two features of Genetic Code.* [2]

Question 9.

(a) *Explain the mechanism of action of T cells to antigens.* [4]

(b) *Explain how insulin can be produced using recombinant DNA technology.* [4]

(c) *What is pisciculture ? Give one advantage.* [2]

Question 10.

(a) *Name the causative organism and preventive measures for each of the following :* [4]

(i) *Swine flu* (ii) *Typhoid*

(iii) *Filariasis* (iv) *Syphilis*

(b) *State four causes and four consequences of population growth.* [4]

(c) *Differentiate between :* [2]

(i) *Cannabinoids and Barbiturates*

(ii) *Biotic potential and Carrying capacity.*

ANSWERS

PART I (20 Marks)

Answer 1.

(a) (i) Heterosis is defined as the superiority of the hybrid over the average performance of the parents of the hybrid. It is a phenotypic phenomenon.

(iii) When F_1 hybrid is crossed back only with homozygous recessive parent, it is called test cross. (Tt × tt)

(iv) The regions of gene that do not form part of mRNA and are removed during RNA processing during the mRNA synthesis are said to be introns.

(b) (i) (D) Fusion of second male gamete with two polar nuclei.

(iii) (C) 44 + XO

(iv) (A) RNA molecule

(c) (ii) Tubectomy

(iii) Commensalism

(d) (i) **STD :** Sexually Transmitted Disease

(iv) **DDT :** Dichloro Diphenyl Trichloro-ethane

(e) (i) Raymond Dart (ii) Dr. Sidney Fox

PART II (50 Marks)

SECTION–A

Answer 2.

(a) The characters that have developed during human evolution are :

1. Enlargement and rounding of cranium.
2. Narrowing and elevation of nose.
3. Development of curves in vertebral column for erect posture.
4. Ability to learn.
5. Low fertility rate.

(b) The term chemogeny refers to the chemical evolution where complex organic compounds are synthesised in following steps :

1. Synthesis of simple compound.
2. Origin of organic compounds.
3. Formation of complex organic molecule.
4. Formation of nucleoprotein.

(c) **The features of Dryopithecus are :**

1. Semi-erect body posture.
2. Brow ridges not present.

Answer 3.

(a) Lamarck had proposed "Inheritance of acquired character." His classic example involved the neck of giraffes have evolved from deer like ancestors. Earlier giraffes had the short neck and forelimbs and they grazed on grass. He observed that giraffe's neck became longer through the generations as the giraffe stretches its neck to reach the leaves on the tall trees because the ground vegetation disappeared and is replaced by high trees. This resulted in gradual elongation of neck and forelimbs. Its offspring would inherit the longer neck and continued stretching would make it longer still over several generations. This theory is referred to as 'law of use and disuse'

(b) 1. DNA content and banding pattern of chromosome 3 and 6 are similar.

2. Similar type of coding sequences are present.
3. Similarly in DNA sequences and amount.

** Answer is not given due to change in the present syllabus.

4. Close similarities in chromosome number (apes = 24 pairs, humans = 23 pairs).

(c) Presence of notochord and gill clefts are the temporary embryonic structures in vertebrates which provide evidence for evolution.

Answer 4.

(a) Sickle cell anaemia is a genetic disorder caused by a variant form of haemoglobin in red blood cells reducing the oxygen carrying capacity. Individuals homozygous (Hb^SHb^S) for this disorder die at an early stage. In heterozygotes (Hb^AHb^S) also, the red blood cells with variant form of haemoglobin become sickle shaped and were unable to bind oxygen efficiently. Since sickle-shaped RBCs effectively kill the malaria parasite, the individual heterozygous for this variant are able to cope with malaria infections much better than normal person. Natural selection has allowed this condition to remain in Africa.

(b) Gene flow is the transfer of alleles or genes from one population to another, as a result of interbreeding between members of the two populations thereby changing the composition of the gene pool of the receiving population.

(c) The organs which have the same function and are superficially alike but are quite different in the fundamental structure and embryonic origin are called analogous organs.

E.g., Sweet potato is a root tuber and potato is a stem tuber, but both store food and are grown underground.

SECTION–B

*(Answer any **two** questions)*

Answer 6.

(a) Refer to ISC Paper 2019, Answer 10. (b)

(c) (ii) Parturition : It is the act of expelling the full-term foetus from the mother's uterus at the end of gestation period.

Answer 7.

(a) Refer to ISC Paper 2019, Answer 5.

(c) Refer to ISC Paper 2018, Answer 8.

SECTION–C

Answer 8.

(a) Reasons for Mendel's success are :

1. Mendel concentrated on the results of one trait at a time.
2. When the behaviour of one trait was established, only then he considered two characters together.
3. He conducted a large number of crosses (2000-3000) to eliminate the chance factor and to obtain a valid and accurate explanation.
4. Most of all, he actually counted the number of offspring of each category and maintained accurate records for each generation in each experiment.
5. *Pisum sativum* is with seven homologous pairs of chromosomes. The unit factors for the seven characters identified are located on different homologous pairs in this plant.

(b) Steps involved in DNA fingerprinting :

1. **Isolating the desired DNA :** It can be done either chemically, by using a detergent to wash the extra material from the DNA, or mechanically, by applying a large amount of pressure in order to "squeeze out" the DNA.
2. **Cutting the DNA into several pieces of different size :** This is done using one or more restriction enzymes.
3. **Sorting the DNA pieces by size :** The process by which the size separation is done is called gel electrophoresis. The DNA is poured into a gel, such as agarose, and an electrical charge is applied to the gel. The different-sized pieces of DNA will be separated by size, with the smaller pieces towards the bottom and the larger pieces towards the top.
4. **Denaturation :** Denaturing the DNA fragments, so that all of the DNA is rendered single-stranded. This can be done either by heating or chemically treating the DNA in the gel using alkali.

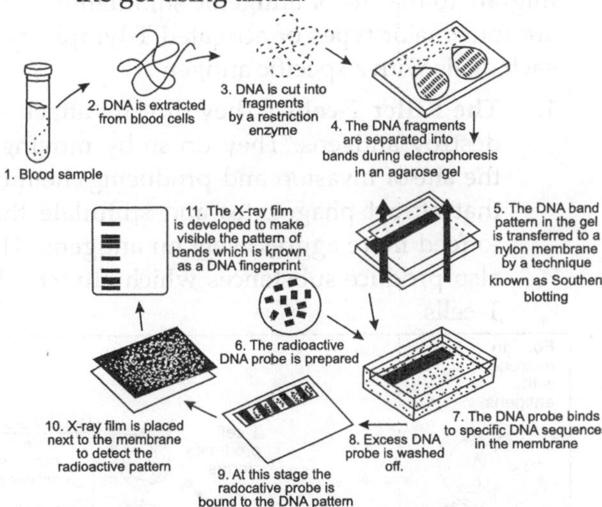

5. **Blotting the DNA. :** The gel with the size-fractionated DNA is applied to a sheet of nitrocellulose paper or nylon membrane, and then baked to permanently attach the DNA to the sheet. The Southern Blot is now ready to be analysed.

In order to analyse a Southern Blot, a radioactive genetic probe is used in a *hybridisation reaction* with the DNA. If an X-ray is taken of the Southern Blot after a radioactive probe has been allowed to bound with the denatured DNA on the paper, only the areas where the radioactive probe binds will show themselves on the film *(autoradiography)*. This allows researchers to identify, in a particular person's DNA, the occurrence and frequency of the particular genetic pattern contained in the probe.

(c) Features of Genetic code are :

1. It is non-overlapping or non-ambiguous. Each nucleotide is a part of only one codon and is read only once during translation and is degenerative.
2. It is almost universal. In nearly all organisms, most codons have the same amino acid meaning and are always triplet.

Answer 9.

(a) T-cells respond to antigens by producing a clone of T-cells. Each T-cell recognises a specfic antigen. Therefore, the body contains separate T-cells for every antigen that it encounters. When a specific antigen enters the tissues, the T-lymphocytes which are programmed to recognise that particular antigen return to the specific T-lymphocyte domains of the lymphoid tissues where they transform into lymphoblasts. Lymphoblasts then divide mitotically to produce activated T-lymphocytes which enter the circulation and migrate to the site of antigenic stimulation. There are four major types of activated T-lymphocytes, each deals with a specific antigen :

1. **The Killer T-cells :** They directly attack and destroy antigens. They do so by moving to the site of invasion and producing chemicals that attract phagocytes and stimulate them to feed more aggressively on antigens. They also produce substances which attract other T-cells.

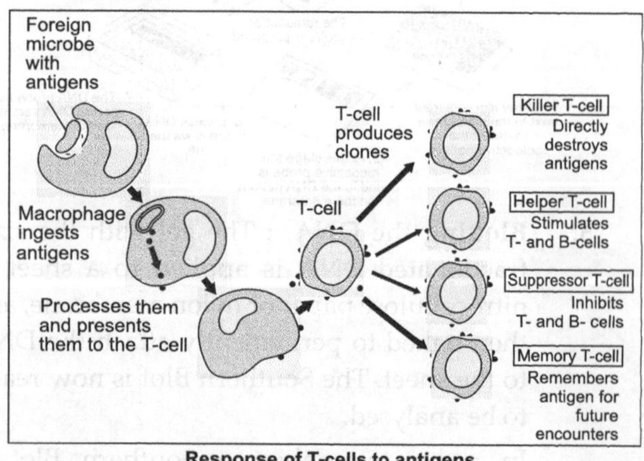

Response of T-cells to antigens

2. **Helper T-cells :** They act to stimulate antibody production by B-cells.
3. **Supressor T-cells :** They suppress the total immune reaction keeping it or to prevent it from attacking the body's own cells.
4. **Memory T-cells :** They have the ability to produce more effector cells and they remember antigen for future encounters.

(b) Refer to ISC Paper 2020, Answer 16 (b).

(c) Pisciculture is the branch of animal husbandry that deals with rational deliberate culturing of fish or fishes to marketable size in a controlled water body. It is beneficial in providing disease free fish at a fast pace.

Answer 10.

(a) **(i) Swine flu :**

Causative organism : Swine influenza virus (H_1N_1).

Preventive measures : (1) H_1N_1 vaccination. **(2)** Use the Paracetamol and anti-viral drug.

(ii) Typhoid :

Causative agent : *Salmonella typhi.*

Preventive measures : (1) Vaccination **(2)** Proper sanitary conditions to avoid spread of bacteria **(3)** Use of boiled water, protection of eatables from flies **(4)** Administration of drugs.

(iii) Filariasis :

Causative organism : *Wuchereria bancrofti.*

Preventive measure : (1) Vector control **(2)** Chemotherapy.

(iv) Syphilis :

Causative organism : *Treponema/allidum.*

Prevention measures : (1) Hygienic conditions **(2)** To avoid having multiple sex partners **(3)** Use of antibiotics like tetracycline, erythromycin.

(b) **Causes :**

1. In older days, many deaths occurred due to diseases. But the development in medical sciences has controlled the diseases and brought about decrease in the death rate.
2. Development in agriculture, better techniques to grow more and better quality food.
3. Food available at all the times due to better storage conditions. So people do not die of starvation when food is not available.
4. Awareness against Lack of education, desire for male child and religious beliefs are some other factors.

Consequences :

1. Poverty—If the family is large, it will become poor with limited income and also will lead to unemployment.
2. Unhygienic living conditions due to more people living in a small area. This would increase diseases in people.
3. Housing problem will arise as more houses will be required for the increasing population but they are not built at such a high rate and shortage of drinking water will be there.
4. Pollution will be an added problem of high population. Large population will cause environmental pollution.

(c) (i) 1. **Cannabinoids** : These are a class of diverse chemical compounds that repress neurotransmitter release in the brain and are obtained from cannabis sativa.

2. **Barbiturates** : These are a type of depressant drug that causes relaxation and sleepiness and are obtained from derivatives of barbituric acid.

(ii) 1. **Biotic potential** : It is defined as inherent capacity of organism to produce their offspring's under natural favourable environmental conditions.

2. **Carrying capacity** : It is defined as the capacity of environment or an ecosystem to support and sustain a level of population in a given environment under ideal conditions.

●●

BIOLOGY 2014

QUESTIONS

PART I (20 Marks)

*(Answer **all** questions)*

Question 1.

(a) *Mention one significant difference between each of the following :* **[5]**

(i) *Parenchyma and Sclerenchyma.***

(ii) *Epistasis and dominance.*

(iii) *Hormones of ovulatory phase and hormones of luteal phase.*

(iv) *Symplastic movement and apoplastic movement.***

(v) *Phenotype and Genotype.*

(b) *Give reasons for the following :* **[5]**

(i) *Testes descend into the scrotum before birth.*

(ii) *Secondary growth does not occur in monocot stems.***

(iii) *Nitrogenous fertilisers are not applied in fields where leguminous crops grow.*

(iv) *Genetic code is 'universal'.*

(v) *At higher temperatures, green plants start evolving CO_2 instead of O_2.***

(c) *Each of the following questions/statements has four suggested answers. Rewrite the correct answer in each case :* **[5]**

(i) *Typhoid is classified as a :*

(A) Viral disease *(B) Genetic disorder*

(C) Bacterial disease *(D) Protozoan disease*

(ii) *Bt cotton is resistant to :*

(A) Insects *(B) Herbicides*

(C) Salt *(D) Drought*

(iii) *Roots and shoots lengthen through the activity of :***

(A) Apical meristem

(B) Vascular Cambium

(C) Lateral meristem

(D) Cork Cambium

(iv) *An antiviral protein released from infected and dying cells is :*

(A) Antigen *(B) Antibody*

(C) Antiserum *(D) Interferon*

(v) *Opening and closing of stomata is due to:***

(A) Ca^{2+} *(B) Na^+*

(C) K^+ *(D) Cl^-*

(d) *State the best known contribution of :* **[3]**

(i) *Alec Jeffery*

(ii) *P.K. Sethi***

(iii) *Hugo de Vries*

(e) *Expand the following :* **[2]**

(i) *SCID* **(ii)** *ZIFT*

PART II (50 Marks)

SECTION–A

*(Answer any **two** questions)*

Question 2.

(a) *Describe the Miller and Urey experiment on the origin of life.* **[3]**

(b) *Define the following :* **[2]**

(i) *Frame shift mutations*

(ii) *Genetic drift.*

Question 3.

(a) *Name and define the three types of natural selection.* **[3]**

(b) *State the following :* **[2]**

(i) *Hardy Weinberg's principle*

(ii) *Theory of recapitulation*

Question 4.

(a) *Mention the important features of the Neanderthal man.* **[3]**

(b) *What are homologous organs ? How do they help in providing evidence for organic evolution ?* **[2]**

SECTION–B

*(Answer any **two** questions)*

Question 5.

(a) *Describe the different types of vascular bundles.*** **[4]**

(b) *Give three anatomical differences between a monocot root and a dicot root.*** **[3]**

(c) *Explain the effect of light and temperature on photosynthesis.*** **[3]**

Question 6.

(a) *Explain the transpiration pull theory for ascent of sap.*** **[4]**

(b) *Explain the process of spermatogenesis in humans.* **[3]**

(c) *Define the following :* **[3]**

(i) *Placentation* **(ii)** *Parthenocarpy*

(iii) *Diffusion***

** Answer is not given due to change in the present syllabus.

Question 7.

(a) *Why are xylem and phloem classified as complex tissues ? Describe the structure of phloem.*** [4]

(b) *Describe the ultra-structure of chloroplast.*** [3]

(c) *State three functions of the placenta.* [3]

SECTION–C

*(Answer any **two** questions)*

Question 8.

(a) *Describe the experiment performed by Griffith. What conclusions did he infer from his observations ?* [4]

(b) *What is artificial insemination ? Mention two ways in which it is useful in breeding of dairy animals.* [3]

(c) *What is single cell protein ? Give its source and significance.* [3]

Question 9.

(a) *How did Hershey and Chase prove that DNA is the genetic material ?* [4]

(b) *Give one main application of each of the following:*** [3]

(i) *MRI* (ii) *Ultrasound* (iii) *ECG*

(c) *Explain the role of stem cells in medical treatment.* [3]

Question 10.

(a) *Write short notes on :* [4]

(i) *Multiple Alleles*

(ii) *Artificial measures to control population.*

(b) *What complications will arise if the blood of an Rh positive person is transfused to an Rh negative person and vice versa ?* [3]

(c) *State any three goals of the Human Genome Project.* [3]

ANSWERS

PART I (20 Marks)

Answer 1.

(a)

(ii)

Epistasis	Dominance
In it, two pairs of genes are involved out of which one gene inhibits or masks the expression of another non-allelic gene.	Out of a pair of alelomorphic characters, the one which appears in F_1 generation, is called dominant and the phenomenon is called dominance.

(iii)

Hormones of Ovulatory phase	Hormones of Luteal phase
Hormones (Estrogen and Luteinizing hormone) bring about ovulation and cause a graafian follicle to develop into a corpus luteum which produces progesterone.	Hormone (progesterone) stimulates the uterine lining development before implantation of a fertilised egg.

(v)

Phenotype	Genotype
These are the external observable characters, controlled by genes.	It is the genetic constitution of an organism.

(b) (i) Sperm development requires a temperature which is a few degrees less than the normal body temperature. Scrotum has almost no fat insulation, so it keeps the testes at a cooler temperature. Also contraction or relaxation of muscles moves the scrotum close to or far from testes and heat of the body according to the environmental temperatures.

(iii) In the root nodules of leguminous plants, nitrogen fixing bacteria, like Rhizobium is present. It converts atmospheric nitrogen into soluble nitrogenous compound, hence enriching the soil with nitrogen.

(iv) Genetic code is universal because it consists of only four nitrogenous bases-adenine, guanine, thymine and cytosine. These four bases in different combinations form all the various types of proteins, formed by genetic coding and is present in all living organisms.

(c) (i) (C) Bacterial disease.

(ii) (A) Insects.

(iv) (D) Interferon.

(d) (i) DNA finger printing.

(iii) Gave the term "mutation".

(e) (i) Severe Combined Immuno Deficiency.

(ii) Zygote Intrafallopian Transfer.

PART II (50 Marks)

SECTION–A

Answer 2.

(a) Oparin-Haldane's concept of biochemical basis of origin of life was put to a test by Stanley Miller and Harold C. Urey (1953) in laboratory by creating the probable conditions of primitive earth in the laboratory. They designed their apparatus of glass tubes and flasks as shown in the figure and created an atmosphere containing hydrogen, ammonia, methane and water vapour in one chamber of the apparatus and allowed

** Answer is not given due to change in the present syllabus.

condensed liquids to accumulate in another chamber. Energy was supplied by heating the liquid containing chamber by electric sparks from two tungsten electrodes in the gaseous chamber. The experiment was run continuously for a week and then they analysed the chemical composition of the liquid inside the apparatus. They found that the liquid contained a large number of complex organic compounds including some amino acids such as glycine, alanine and aspartic acid. Some more complex organic compounds were collected in vitro after a few days *e.g.* purines, pyrimidine, sugar, glycerol, formic acid, short chain tatty acids and acetic acid.

However, from the result of their experiment, they suggested that the electrical discharges produced during lightening in the primitive atmosphere of earth containing hydrogen, ammonia, nitrogen and water vapour might have resulted in the formation of amino acids and other essential building blocks (sugars, nucleotides etc.) of living organisms, these could thus have formed life on the primitive earth. Thus, the experiment of Miller and Urey provides support for the biochemical concept of origin of life of Oparin and Haldane.

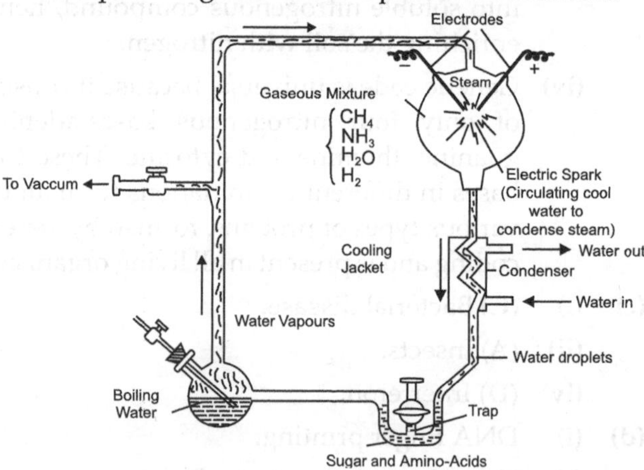

(b) (i) A frameshift mutation is a genetic mutation caused by a deletion or insertion of one or more nucleotide pairs in a DNA sequence at one or more places that shifts the way the sequence is read.

(ii) Genetic (gene frequency) or allele frequency drift refers to the change in a type of genes in a small population due to a random occurrence.

Answer 3.

(a) Three types of natural selection are :

1. **Stabilising or Balancing Selection** : It leads to the elimination of organisms having over specialised characters and maintains homogenous population which is genetically constant. It favours the average or normal phenotypes while eliminates the individual with extreme expression *e.g.,* sickle cell anaemia in human beings. Also the phenotypic features of organisms coincide with optimal environmental conditions.
2. **Directional or Progressive Selection** : In this selection, the population changes towards one particular direction along with change in environment. As environment is undergoing continuous change, the organisms having acquired new characters survive, and others are eliminated *e.g.,* Industrial melanism.
3. **Disruptive or Diversifying Selection** : It is a type of natural selection which favours extreme expression of certain traits to increase variance in a population. It breaks a homogenous population into many adapted forms population is represented by both extreme phenotypes and results in balanced polymorphism. *e.g.* Three types of snails in sea. It simultaneously favours the individuals at both extremes of the distribution curve and eliminates the intermediate types.

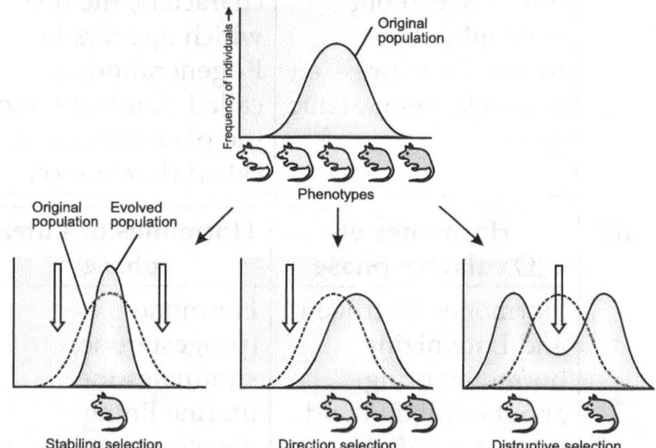

(b) (i) The Hardy-Weinberg principle states that allele and genotype frequencies in a population will remain constant from generation to generation in the absence of other evolutionary influences.

$$p^2 + 2pq + q^2 = 1$$

or

$$(p + q)^2 = 1$$

(ii) It states that ontogeny recapitulates phylogeny. It means that embryos, in their development repeat the evolutionary history of their ancestors in a short, abbreviated form.

Answer 4.

(a) A typical Neanderthal man had :

1. Height less than 150 cm.

2. Heavily built with strong and outwardly curved thigh bones.
3. Cranial capacity of 1300-1600 cc.
4. Brow ridges prominent.
5. Skull thick-boned, depressed and protruded behind.
6. Forehead was low and sloping.
7. Devised and used flint tools.
8. Lived in caves and used animal skin to cover themselves.
9. Buried their dead with ceremonies.
10. Used stone tools and weapons.

(b) Organs having different functions but similar embryogenic origin internal structure, basic plan, development and having similar relationship with adjacent organs are called homologous organs.

They indicate close relationship between their possessors *e.g.*, the forelimbs of human, wing of bird, leg of horse and flipper of a seal are all apparently different from one another in structure and function, yet they are all built on the same pentadactyl plan, having the same number of bones, muscles, blood vessels and nerves arranged in the same pattern and developed similarly. Evidences indicate that the organisms having homologous structures must have arisen from common ancestors through successive generations. Thus, they help in providing evidence for organic evolution.

SECTION–B

Answer 6.

(b) Refer to ISC Paper 2020, Answer 15 (a).

(c) (i) **Placentation** : Arrangement of ovules within the ovary.

(ii) **Parthenocarpy** : Phenomenon of development of fruits without the act of fertilisation.

Answer 7.

(c) In mammals, placenta performs the following functions :

1. It helps in the nutrition of embryo as the nutrients like amino acids, mono-sugars, vitamins etc., diffuse from maternal blood into foetal blood through placenta.
2. It helps in respiration of the embryo as O_2 of the maternal blood and CO_2 of foetal blood diffuse through placenta.
3. It also helps in excretion of the nitrogenous wastes of foetal blood like urea which diffuse into maternal blood through placenta.
4. It also acts as an endocrine gland as it secretes certain hormones like estrogen, relaxin, progesterone and Human Chorionic Gonadotropin (HCG).
5. Helps in the transport of antibodies from maternal blood to the foetus.

SECTION–C

Answer 8.

(a) The bacterium *Diplococcus pneumoniae* causes pneumonia in humans. Frederick Griffith observed two strains of this bacteria. One strain called the smooth type (S) has polysaccharide forming a large capsule around the cell. The colony of such cells has a glistening appearance. The other strain bacterial cells do not have the polysaccharide layer and the colony formed by these cells has an irregular appearance and is called the rough type (R). The S-strain is virulent while the R-strain is non-virulent.

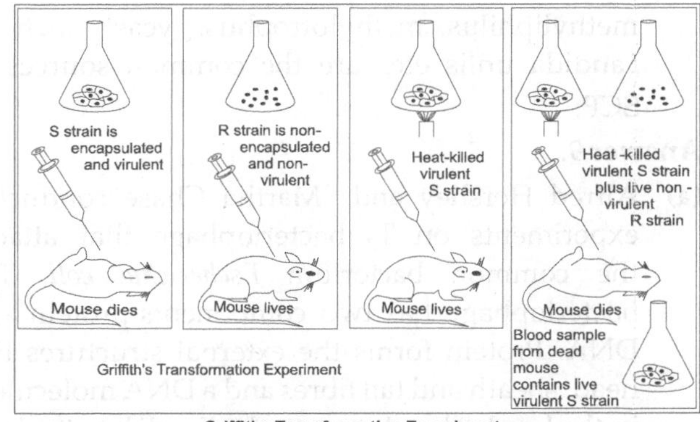

Griffiths Transformation Experiment

In his experiments Griffith injected mice in the lab with live R-type of bacteria. They did not suffer from the disease. When he injected 'S'-type of bacteria, the mice died from the disease. However when heat killed S-type of bacteria were injected into the mice, they did not suffer from pneumonia. Then he injected the mice with a mixture of living R-type (non-virulent) with heat-killed S-type (virulent) bacteria, the mice suffered from the disease and died. Griffith observed that in the blood of dead mice, both R & S type of bacteria were present. He thus concluded that heat killed smooth type bacteria caused a transformation of the living rough type bacteria. This was described as Griffith effect or bacterial transformation. This experiment suggested that DNA and not proteins is the genetic material.

(b) Artificial insemination is done to get better variety of animals. In this method semen from the desired type of animal (*e.g.*, bull) is collected and preserved by chemical methods or freezing. This preserved semen is then injected into the genital tract of the chosen cow during its maximum fertility period. Normal reproductive process then occurs and the progeny thus obtained is a hybrid of desired characters.

Usefulness in breeding animals :

1. It is economical as semen from a desired animal *e.g.*, bull can be transported to far away places while transporting the animal is not easy.
2. Semen of healthy males can be used to inseminate a number of females.
3. Spread of diseases can be controlled.
4. Helps to improve a particular breed and is easy to store.
5. It reduces the cost of transporation and can be used at convenient place and time.

(c) Refer to ISC Paper 2019, Answer 3.

Sources—*Chlorella* (algae), *Rhodopseudomonas capsulata* (bacteria), *Trichoderma* (fungi). Cyano bacteria such as spirulina, bacteria such as methyliphilus methylotrophus, yeast such as candida utilis etc., are the common sources of SCP.

Answer 9.

(a) Alfred Hershey and Martha Chase conducted experiments on T_2 bacteriophage that attacks the common bacterium *Escherichia coli.* The bacteriophage has two components-protein and DNA. Protein forms the external structures like head, sheath and tail fibres and a DNA molecule is in the head. The phage attacks *E. coli* by attaching its tail fibres to the bacterial wall and injecting its genetic material into the bacterial cell to produce new phages.

Hershey and Chase labelled the DNA and protein components of the phage separately with specific radioactive tracers and then followed these components through the life cycle of the phage. They developed two strains of the virus, one with labelled protein and other with labelled DNA. Almost all proteins contain sulphur which is not found in DNA while all DNA molecules contain phosphorus which is not found in proteins. The T_2 phages grown in the presence of radioactive sulphur (^{35}S) had labelled proteins and T_2 phages grown in presence of radioactive phosphorus (^{32}P) had labelled DNA. ^{35}S is incorporated in the protein coat whereas ^{32}P is incorporated in the DNA of T_2 phage.

After developing these strains, Hershey and Chase allowed each strain to infect the bacteria. Soon after infection the bacterial cells were gently agitated in a blender loosen the adhering phage particles. It was observed that only radioactive ^{32}P was found in the bacterial cells and ^{35}S was present only in the surrounding medium and not inside the bacterial cells. When they studied the viral progeny for radioactivity, it was found that it had only ^{32}P and no ^{35}S. The phages grown in ^{35}S did not transmit their radioactivity to their daughter phages whereas the phages P^{32} pass their radioactivity to the daughter phage particle.

The result clearly shows that only DNA is the genetic material and protein coat does not even enter the bacterial cells and as such not protein but DNA is the genetic material, which was passed on from one generation of T_2 to the next.

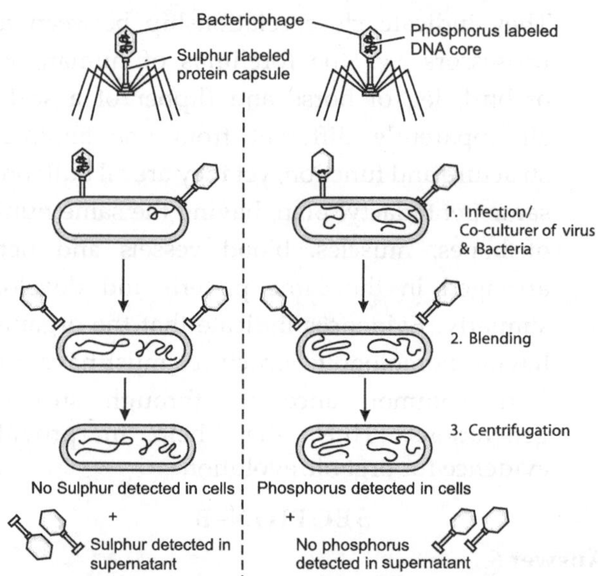

(c) Stem cells are found in all organisms and are capable of dividing to form new cells which can be differentiated into various types of specialised cells.

Role of stem cells in medical treatment :

1. These cells have been used for treatment of various cancers like leukemia and lymphoma.
2. It may also be used for treatment of severe autoimmune diseases like multiple sclerosis.
3. Embryonic stem cells, directed to differentiate into specific cell types, offer the possibility of a renewable source of replacement cells and tissues to treat diseases like Alzheimer's disease, spinal cord injury, stroke, burns, diabetes, arthritis and heart diseases.

Answer 10.

(a) (i) Multiple Alleles : Most genes occur in two alternative forms, both controlling the same character and occupying the same locus in homologous chromosomes. These different forms of the same gene are called alleles. Some genes may occur in more than two allelic forms and they are collectively called multiple alleles. A set of such multiple alleles may contain 3 to 20 or even more members which occupy the same locus in homologous chromosomes. In such a set of multiple alleles one member is always dominant and one recessive to all others. E.g., Human blood groups.

(ii) Artificial Measures to Control Population :

1. **Barrier Contraception :** Most popular barriers are the condom and the diaphragm that keep the sperms out or prevent them from going very far into women's reproductive system. E.g., Condom diaphragms etc.
2. **Intra Uterine Device (IUD) :** It is a small device that is placed in the uterus to prevent pregnancy. Once the IUD is in place, it can provide birth control for 5 to 10 years, depending on type. IUD increase phagocytes of sperms within the uterus, like copper T.
3. **Contraceptive Pills :** Birth control pills, which contain hormones, prevent a woman from getting pregnant as it prevents ovulation.
4. **Chemical methods :** Spermicidal creams, jellies, foams can also be used.
5. **Surgical methods :** Surgical methods like tubectomy, vasectomy.

(b) The Rh-factor or Rh-antigen was first reported in Rhesus monkeys RBCs by Landsteiner. Later it was found in most of human population. 85-99% of population, depending upon the race, have the Rh-factor, hence are Rh +ve. There is no antibody against Rh-antigen in human body. The Rh-antigen is produced due to a dominant gene, hence Rh + ve individuals are presented as RR or Rr with Rh–ve as rr. Rh antigens induce formation of Rh antibodies in recipients blood.

Rh –ve blood can be given safely to a Rh +ve individual. But when Rh +ve blood is transfused into Rh –ve person, then during first transfusion, there is no complexity arising because of absence of Rh-antibody in the recipient's blood; but this transfusion induces the synthesis of antibodies in recipient's blood. In case of subsequent transfusion, more antibodies are formed which destroy the RBCs in the Rh+ve blood and agglutinate the blood.

(c) The Human Genome Project (HGP) is an international research program that is to spell out the complete genetic inheritance of human beings and selected experimental animals. The HGP's goal is to map the entire genome by determining the sequence of nucleotides in the DNA of each of the 22 + X and Y chromosome and make them accessible for further biological study.

Goals of Human Genome Project (HGP) are :

1. To sequence the entire genome that includes above 3 million base pairs.
2. To obtain a physical map of human genome by cloning DNA into YACs and BACs and cosmids.
3. To identify thousands of genetic markers and map them in the genome.
4. To develop technological advances in genetic methodologies such as gene cloning sequence/improve tools for data analysis.
5. To address the ethical, legal and social implications of the results obtained from the project.

●●

BIOLOGY 2013

QUESTIONS

PART I (20 Marks)

*(Answer **all** questions)*

Question 1.

(a) *Give one significant difference between each of the following :* **[5]**

(i) *Implantation and Parturition.*

(ii) *Active absorption and passive absorption.***

(iii) *Haemodialysis and Peritoneal dialysis.***

(iv) *Simple fruit and Aggregate fruit.*

(v) *Auricles and Ventricles.***

(b) *Explain what would happen if :*** **[5]**

(i) *Excess fertilisers are added to soil.*

(ii) *Blood clots in the coronary artery.*

(iii) *Beta cells in the islets of langerhans are damaged.*

(iv) *Silicon emulsion is applied over the surface of leaves.*

(v) *Magnesium element is deficient in the soil.*

(c) *Each of the following questions/statements have four suggested answers. Rewrite the correct answer in each case :*** **[3]**

(i) *The cell division in the tunica region of shoot apex is :*

(A) Periclinal *(B) Horizontal*
(C) Anticlinal *(D) Radial*

(ii) *The dark coloured dead wood present in the central region of old trees is:*

(A) Spring wood *(B) Heart wood*
(C) Sap wood *(D) Cambium*

(iii) *Dwarfism accompanied with mental retardation is due to hyposecretion of :*

(A) Growth hormone
(B) Thyroxine hormone
(C) Parathormone
(D) Adrenalin hormone

(iv) *Oxygen is released in photosynthesis by :*

(A) Photophosphorylation
(B) Photolysis of water
(C) Photorespiration
(D) Photons

(v) *The spinal nerve is :*

(A) A mixed nerve *(B) A sensory nerve*
(C) A motor nerve *(D) A cranial nerve*

(vi) *The cells of the areolar tissue which produce heparin are :*

(A) Fibrocytes *(B) Mast cells*
(C) Macrophages *(D) Chondriocytes*

(d) *Mention the most significant function of the following :*** **[3]**

(i) *Semi circular canals*

(ii) *Bowman's capsule*

(iii) *Parenchyma*

(iv) *Leg haemoglobin*

(v) *Guard cells*

(vi) *Alveoli*

(e) *State the best known contribution of :*** **[2]**

(i) *Munch* **(ii)** *Ronald Ross*
(iii) *Marshall Hall* **(iv)** *Huxley*

(f) *Expand the following :* **[2]**

(i) *ABA*** **(ii)** *FSH*
(iii) *AIDS* **(iv)** *DDT*

PART II (50 Marks)

SECTION–A

*(Answer any **three** questions)*

Question 2.

(a) *Give four differences between root apex and shoot apex.*** **[4]**

(b) *Explain the development of the different types of endosperms in angiosperms.* **[3]**

(c) *Explain briefly :*** **[3]**

(i) *Capillary water*

(ii) *Osmosis*

(iii) *Aeroponics*

Question 3.

(a) *Explain the C_4 cycle of photosynthesis.*** **[4]**

(b) *State three advantages and three disadvantages of vegetative reproduction.* **[3]**

(c) *Mention one role and one deficiency symptom of the following elements in plant nutrition :*** **[3]**

(i) *Phosphorus* **(ii)** *Iron*

(iii) *Chlorine*

Question 4.**

(a) *What are tropic hormones ? Describe the feedback control of tropic hormones with an example.* **[4]**

(b) *Explain the conduction of nerve impulse through a nerve fibre.* **[3]**

(c) *Draw a labelled diagram of the T.S. of bone.* **[3]**

** Answer is not given due to change in the present syllabus.

Question 5.

(a) *Explain the role of pancreas in digestion of various food materials.*** [4]

(b) *Briefly describe the stages in clotting of blood.*** [3]

(c) *Define :* [3]

(i) *Reparative regeneration***

(ii) *Capacitation*

(iii) *Menarche.*

Question 6.

(a) *State four differences between transpiration and guttation.*** [4]

(b) *Give an account of the secretory phase of menstrual cycle.* [3]

(c) *Define :*** [3]

(i) *Radial vascular bundle*

(ii) *Rigor mortis*

(iii) *Root pressure*

SECTION–B

*(Answer any **two** questions)*

Question 7.

(a) *Differentiate between apes and man with respect to the following characteristics :* [4]

(i) *Posture* (ii) *Cranium*

(iii) *Brow ridges* (iv) *Locomotion*

(b) *Define :* [3]

(i) *Vestigial organs* (ii) *Variations*

(iii) *Neo-Darwinism*

(c) *Give three differences between Natural Selection and Artificial Selection.* [3]

Question 8.

(a) *State four characteristics of the Cro magnon man.* [4]

(b) *Explain the basic postulates of Darwinism.* [3]

(c) *Archaeopteryx is a connecting link between reptiles and birds. Justify the statement by giving two characteristics of each group.* [3]

Question 9.

(a) *Persons suffering from G6PD deficiency are resistant to malaria. Explain.* [4]

(b) *Define :* [3]

(i) *Genetic Erosion* (ii) *Bioinsecticides*

(iii) *Antigen* (iv) *Psychosis*

(c) *Define Biofortification.* [3]

Question 10.

(a) *List the activities of Community Health Services.*** [4]

(b) *Give three early diagnostic symptoms of cancer.* [3]

(c) *Define :* [3]

(i) *Carrying capacity* (ii) *Implant*

(iii) *Carcinoma.*

ANSWERS

PART I (20 Marks)

Answer 1.

(a)	(i)	Implantation	Parturition
		It is the fixing of blastocyst in the endometrium of uterus.	It is the act of expelling full term foetus at the end of gestation from the uterus.
	(iv)	Simple fruit	Aggregate fruit
		These are formed from a single carpel or more but fused carpels of a flower.	It develops from several ovaries of free carpels of a flower.

(f) (ii) Follicle Stimulating Hormone

(iii) Acquired Immuno Deficiency Syndrome

(iv) Dichloro diphenyl trichloroethane

PART II (50 Marks)

SECTION–A

Answer 2.

(b) Formation of endosperm is initiated by mitotic division of primary endosperm nucleus, which is triploid and is formed by triple fusion of polar nuclei to secondary nucleus (2n) and with male gamete (n). Endosperm is formed before the division of zygote. In angiosperms, endosperm development are of three types :

1. **Nuclear type :** It is the most common type of endosperm formation. In this type the primary endosperm nucleus gives rise to a number of free nuclei which remain as peripheral layer and a large central vacuole appears in the embryo sac. Cell wall formation takes place later leading to the formation of cellular endosperm *e.g.*, Coconut, Areca.

2. **Cellular type :** In it, first and subsequent divisions of primary endosperm nucleus are accompanied by wall formation, hence endosperm is cellular from the very beginning *e.g., Petunia, Datura.*

3. **Helobial type :** It is an intermediate type between the nuclear and cellular types. The first division of endosperm nucleus is followed by wall formation, but the subsequent divisions are free nuclear. The chamber towards the micropylar end of embryo sac is much larger. A large number of nuclei are formed in it by free nuclear

** Answer is not given due to change in the present syllabus.

divisions while the nucleus of chalazal chamber divides to form a fewer, free nuclei or may not divide at all *e.g.*, Drimya.

Answer 3.

(b) Advantages of Vegetative Propagation :

1. Plants produced by vegetative reproduction are exactly similar to parental type. Thus the best varieties of potatoes, oranges, apple etc., can be preserved by vegetative reproduction without loss of their fine quality. In this way pure lines of desired types can be selected and propagated all over the world.
2. This is a quick method and thus, by this means, flowers and fruits can be produced early.
3. This is more certain method of producing new plants than that of seed reproduction.
4. It is more economical for the plant because no energy is wasted in producing large and showy petals, nectar, fruits, seeds etc.

Disadvantages of Vegetative Propagation :

1. Due to overcrowding of large number of plants near the parent plants, there is severe competition between the members of the same species. Thus many plants become weak and stunted.
2. There may be degeneration of species due to lack of sexual stimulus.
3. Due to lack of genetic variability in vegetative reproductive structure there may not be evolution of new species.

Answer 5.

(c) (ii) Before fertilising the ovum, the sperm prepares itself for the act of fertilisation and undergoes many changes. These changes in a sperm just prior to fertilisation are called capacitation.

(iii) Menarche is the onset of menstruation in a young female at the age of about 12-13 years. It signals the possibility of fertility.

Answer 6.

(b) The secretory phase of the menstrual cycle extends from 12-14 days. After ovulation, LH stimulation completes the development of the graafian follicle into a structure called the corpus luteum. Progesterone is then released by corpus luteum and together with estrogen produced by the ovaries, produces changes in the secondary sex organs to prepare the body for pregnancy. Since these changes are due to the progesterone, this phase is also known as progesterone phase or luteal phase. Secretory phase is associated with following changes in uterus and ovary :

1. **Changes in Uterus :** In secretory phase, thickness of endometrium progressively increases. The arteries of the endometrium become coiled and glycogen content of the endometrial epithelium increases. Towards the end of the secretory phase, the endometrium is thick, soft and richly supplied with blood. These changes are the obvious preparation for providing a suitable environment for fertilised ovum. The uterine glands become active and are filled with secretions and uterine movements are reduced so that uterus may not contract. However in the absence of pregnancy, these measures are aborted.
2. **Changes in Ovary :** If ovum is not fertilised, progesterone levels decrease as the corpus luteum undergoes degeneration and the thickened endometrium sheds in the form of a menstrual flow and once again initiating the menstrual cycle.

SECTION–B

Answer 7.

(a)

Features	Apes	Man
(i) Posture	Semi erect.	Fully erect.
(ii) Cranium	Flattened with cranial capacity under 650 cm^3.	Rounded with cranial capacity of about 1450 cm^3.
(iii) Brow ridges	Prominent.	Inconspicuous.
(iv) Locomotion	Walk semi-erect on the outer edges of the feet and knuckles of hands, soles do not lie flat on the ground.	Walk fully erect on the soles of feet which lie flat on the ground while walking.

(b) (i) Organs which are non-functional in the possessor but were functional in their ancestors and in related animals are called vestigial organs. It is an evidence of organic evolution. *e.g.*, nictitating membrane of human eye, vermiform appendix.

(ii) Any difference between cells, individual organism or group of organisms of any species caused either by genetic differences or by the effect of environmental factors is known as variation. Genetic drift, mutation, recombination, gene flow, migration, selection are some sources of variation.

(iii) According to Neo-Darwinism, adaptations are caused due to many forces, natural selection being one of them and that characters are not inherited as such but there are character determiners which control their development. It explains the reason for variation.

(c)

Natural Selection	Artificial Selection
1. Nature selects the best or most favourable adaptation.	Man selects the characteristics it wants in an organism.
2. It causes great diversity in nature.	It has led to evolution of a few economically important plants and animals only.
3. It takes hundreds of years for a new species to emerge.	It leads to the formation of new species within a few years.

Answer 8.

(a) Cromagnon man or *Homo sapiens fossils* lived about 50,000 years ago and became extinct about 20,000 years ago. Its characteristic features are :

1. Height was about 180 cms with sturdy body.
2. Cranial capacity was about 1660 cm^3.
3. Face was perfectly orthognathous with a narrow, and elevated nose, arched and broad forehead, moderate brow ridges.
4. Jaws were strong with man-like dentition and a well developed chin.
5. He was a cave dweller, hunter and direct ancestor of the modern man.
6. He knew the art of painting and carving. He made tools and carved ornaments from ivory, made finely chipped stone arrows and spearheads.
7. Used animal skin to sew crude garments.
8. Used fire.
9. Had some religious beliefs and elaborate burial customs.
10. He neither practiced agriculture nor animal domestication. **(any four)**

(b) The basic postulates of Darwinism are :

1. **Enormous Power of Fertility :** Every living organism has power to reproduce rapidly. Hence there is a danger of population overrunning the food and other facilities. As a result there are chances of elimination of a number of individuals because the available resources are less and that has resulted in competition among organisms where every individual tries to be better than the other. Darwin suggested that population of each species remains more or less constant because offspring perish in good number before themselves becoming capable of reproducing.
2. **Struggle for Existence :** Over population resulted in struggle which Darwin called the struggle for existence, which are of three types:
   (i) Intra-specific struggle : The struggle within the species.
   (ii) Inter-specific struggle : The struggle between different groups of animals *e.g.* carnivorous animals eat the lower animals.
   (iii) Environment struggle : Struggle with the adverse environmental conditions, such as extreme temperature, lack of sunlight, nutrients etc.
3. **Variations and Heredity :** During competition every individual tries to become better and shows variation. But all the variations are not significant from evolutionary point of view. Some may be useful and some harmful. The useful variations are inherited in progeny of those organisms in which they arose and therefore, the progeny have better chances of survival.
4. **Survival of the Fittest or Natural Selection :** During struggle for existence only those individuals which are better adapted to the environment and show favourable variations are selected. The less fit and unfit individuals are rejected by nature. Darwin stated that the variations are sorted out naturally. Therefore, the organisms which sort out such variations are said to be fittest.
5. **Origin of Species :** According to Darwin, new variations appear in every generation and get inherited to the new generation. As a result, the offspring are distinct from their ancestors. This results in the formation of new species. Gradual modification of the existing species leading to the formation of new species is known as speciation.

(c) Reptile-like characteristics of Archaeopteryx :

1. Tail was long with free caudal vertebrae.
2. Teeth present in jaws.
3. A weak and keel-less sternum.

Bird-like characters of Archaeopteryx :

1. Presence of feathers on the body.
2. Fore-limbs modified into wings.
3. Four toes in the foot adapted for perching.

Due to these characteristics of archaeopteryx it is considered as a connecting link between reptiles and birds.

Answer 9.

(a) G6PD is the enzyme glucose-6-phosphate dehydrogenase which decomposes H_2O_2 formed during metabolism. The deficiency of this enzyme is an inborn error of metabolism controlled by specific gene. When persons suffering from malaria are given *Primaquine* it causes haemolysis because H_2O_2 produced during metabolism is not decomposed due to the deficiency of G6PD enzyme. This H_2O_2 haemolyses the blood. The haemoglobin of haemolysed blood collects on cell membrane and RBCs are deformed. In such distorted RBCs, malarial parasite fails to survive and multiply. It means persons suffering from G6PD deficiency are resistant to malaria. Such persons are therefore, favoured by natural selection in area infested with malarial parasite. For this reason G6PD deficiency is found in 50% Negroes living in areas with malarial infection.

(b) (i) Genetic erosion refers to the process in which a plant or animal species faces a gradual or drastic loss of its unique gene pool and genetic adversity due to urban expansion, deforestion etc.

(ii) Bioinsecticides are living organisms like other insects, bacteria etc., which themselves or their metabolic products are used to control insect pests.

(iii) A substance that is capable of stimulating an immune response, leading to acquired immunity is called antigen.

(iv) Psychosis is a kind of mental illness in which the person has no idea about his condition, loses contact with reality and cannot function in society.

(c) Biofortification is a breeding process of specific crops such as rice, wheat, beans, and other cereals and legumes, with the aim of enhancing both their mineral density and bioavailability. According to WHO, 2 billion people worldwide are suffering from anaemia and is mainly due to deficiency of iron in their diet. It is relatively new intervention to improve human nutrition worldwide, with a special emphasis on the populations of poor and developing countries. Basically, it is a process of breeding staple crops to the higher level of essential nutrients either through selective breeding or genetic modifications.

The main objectives of biofortification are :

1. Vitamin content.
2. Oil content and quality.
3. Protein content and quality.
4. Micronutrient and mineral quality.

Methods of Biofortification involve :

1. **Selective breeding :** This is the conventional method which requires crops which have naturally occurring high nutritious value to be crossbred with high-yielding varieties.
2. **Genetic modification :** Altering the genetic makeup of a crop by introducing foreign genes from the wild crop of same species or other species that code for the increased production of certain nutrients or disease resistance could make the host crop rich in nutrients and increase its quality. One of the examples is golden rice which has been enriched with beta-carotene, a precursor of Vitamin A.

Some of the Biofortification projects include :

1. Hybrid maize developed with twice the amount of amino acids lysine and tryptophan compared with existing maize.
2. Wheat variety Atlas 66 having high protein content has been used as donor for improving cultivated wheat.

Answer 10.

(b) Early diagnostic symptoms of cancer are :

1. Troublesome and persistent cough.
2. Unaccustomed indigestion, constipation or diarrhoea.
3. Abdominal pain and pain or difficulty in urination.

(c) (i) Far a given region, carrying capacity is the maximum number of individuals of a given species that an area's resources can sustain indefinitely without significantly depleting or degrading those resources.

(ii) A contraceptive implant is a flexible plastic rod that is placed under the skin of the upper arm. It releases a low, steady dose of a progesterone hormone to thicken cervical mucus and thin the lining of the uterus (endometrium), suppressing ovulation.

(iii) Carcinomas are tumours made up of mainly epithelial cells that cover or line the body organs, e.g., breast cancer.

●●

BIOLOGY 2012

QUESTIONS

PART I

*(Answer **all** questions)*

Question 1.

(a) *Give one difference between each of the following :* **[5]**

(i) *Ventricular systole and Ventricular diastole.***

(ii) *Sertoli cells and Spermatids.*

(iii) *Dwarfism and Cretinism.***

(iv) *Antibodies and Interferons.*

(v) *Glucocorticoids and Mineralocorticoids.***

(b) *Give reasons for the following :*** **[5]**

(i) *Nerve impulse travels in one direction.*

(ii) *Jamming of wooden doors and windows takes place during rainy season.*

(iii) *A cut plant wilts fast even if its cut end is dipped in water.*

(iv) *Urine excreted during summer months is hypertonic.*

(v) *A person has difficulty in focusing on nearer objects, as the age increases.***

(c) *Give a scientific term for each of the following :* **[3]**

(i) *A single isolated contraction of muscle fibre.***

(ii) *Inhibition of lateral bud growth by terminal bud.***

(iii) *Specialised structure through which guttation occurs.***

(iv) *Development of embryo from the egg without the process of fertilisation.*

(v) *Process of splitting of water molecules during photosynthesis.***

(vi) *Passing out of urine.***

(d) *Mention the most significant function of each of the following :* **[3]**

(i) *Tapetum cells*

(ii) *Serotonin***

(iii) *Lenticels***

(iv) *Cerebrospinal fluid***

(v) *Islets of Langerhans***

(vi) *Bundle sheath***

(e) *State the most significant contribution of the following scientists :* **[2]**

(i) *Hans Berger*** **(ii)** *Dixon and Jolly***

(iii) *J. B. Lamarck* **(iv)** *William Harvey***

** Answer is not given due to change in the present syllabus.

(f) *Expand the following :*** **[2]**

(i) *OP* **(ii)** *RuBP*

(iii) *IBA* **(iv)** *PEP*

PART II

SECTION–A

*(Answer any **three** questions)*

Question 2.

(a) *Give any four anatomical differences between monocotyledonous and dicotyledonous leaf.*** **[4]**

(b) *Explain the phases of growth in meristem of plants.*** **[3]**

(c) *Draw a neat labelled diagram of a matured anatropous ovule before fertilisation.* **[3]**

Question 3.

(a) *Give an account of activity of cambium in the secondary growth of the stem.*** **[4]**

(b) *Write three differences between C_3 and C_4 cycles.*** **[3]**

(c) *Mention two advantages each of the following :* **[3]**

(i) *Hydroponics***

(ii) *Turgidity to Plants***

(iii) *Cross pollination*

Question 4.

(a) *Explain the movement of water cell to cell across the root from the soil to the xylem.*** **[4]**

(b) *Draw a labelled diagram of T. S. of hyaline cartilage. Write a brief note on its functions.*** **[3]**

(c) *What is the full form of ADH ? How does ADH control osmoregulation in human kidney ?*** **[3]**

Question 5.

(a) *Explain the process of oogenesis in humans.* **[4]**

(b) *State three differences between red muscle fibre and white muscle fibre.*** **[3]**

(c) *Mention a cause and symptom of each of the following :*** **[3]**

(i) *Emphysema* **(ii)** *Renal calculi*

(iii) *Diarrhoea.*

Question 6.**

(a) *Describe the structure of an artery and a vein. Explain how their structure helps in their functioning.* **[4]**

(b) *Write three differences between short day and long day plants.* **[3]**

(c) *Name the 2nd, 3rd and 8th cranial nerves in man and write a function of each.* **[3]**

SECTION–B

*(Answer any **two** questions)*

Question 7.

(a) *Explain mass selection and pure line selection. How is pure line selection a better method for crop improvement ?* **[4]**

(b) *Write short notes on :* **[3]**

(i) *Atavism* **(ii)** *Protoplast fusion*

(iii) *Rh factor*

(c) *Mention the chromosomal abnormalities concerned with the following conditions in humans :* **[3]**

(i) *Turner's Syndrome*

(ii) *Klinefelter's Syndrome*

(iii) *Down's Syndrome.*

Question 8.

(a) *How does the human body protect itself from infections ?* **[4]**

(b) *Write short notes on the following :* **[3]**

(i) *Biomedical Engineering.***

(ii) *Stem cells.*

(iii) *Cryopreservation.*

(c) *Give an account of Darwin's finches.* **[3]**

Question 9.

(a) *Explain the convergent and divergent evolution with suitable examples.* **[4]**

(b) *What is manure ? How does green manure differ from biofertilisers ?* **[3]**

(c) *What is IPM ? Give an example of bio-insecticides and bioherbicides and how do they help in pest control.* **[3]**

Question 10.

(a) *What is mental illness ? Explain any three methods of treatment of mental illness.*** **[4]**

(b) *What are Koch's postulates ? Why are they not applicable to viruses ?*** **[3]**

(c) *Name the causative agent and the main symptom of each of the following diseases :* **[3]**

(i) *Filariasis*

(ii) *Rabies***

(iii) *Chicken pox***

ANSWERS

PART I

Answer 1.

(a)

(ii)	Sertoli cells	Spermatids
	These cells provide mechanical support and protection to the developing germ cells of testis and also help in their nutrition and maturation.	These are the matured, haploid, primary spermatocytes which will form the spermatozoan.
(iv)	**Antibodies**	**Interferons**
	These are produced by the B-cells and defend the body against infection.	These are toxins produced by virus infected cells and protect the body cells against viruses.

(c) **(iv)** Parthenogenesis

(d) **(i)** It serves as a nutritive tissue for pollen mother cells and pollen grains.

(e) **(iii)** Proposed the 'Theory of Inheritance of Acquired characters'.

PART II

SECTION–A

Answer 2.

(c) Refer to ISC Paper 2017, Answer 7 (b)

** Answer is not given due to change in the present syllabus.

Answer 3.

(c) **(iii) Cross-pollination :**

1. New varieties of plants can be produced.
2. Offspring are much healthier and better adapted to changed environment.

Answer 5.

(a) The phenomenon of formation of haploid ova (egg) from the diploid egg mother cells of ovary (female gonad) is called oogenesis. It occurs in three phases *i.e.,* Multiplication, Growth and Maturation.

Process of oogenesis includes the following steps :

1. **Multiplication phase :** Some cells of the germinal epithelium of ovary function as primary germ cells and are called oogonia, which multiply by repeated mitotic division and become primary oocytes when cell division ceases.
2. **Growth phase :** This growth phase of oogenesis is very prolonged and the increase in size of primary oocyte is very considerable. It is a slow process and may take a long period. In human female, the oogonia are already formed at the time of birth and continue growth only after attaining sexual maturity. It contains diploid number of chromosomes.
3. **Maturation phase :** Now the primary oocyte undergoes first meiotic division and gives rise to two daughter cells, larger one called

secondary oocyte and smaller one as first polar body. The secondary oocyte divides again and gives rise to haploid ootid or the ovum and the second polar body. Size of egg depends upon the amount of reserve food material. Thus one ovum is formed from one primary oocyte as a result of oogenesis. Oogenesis is also controlled by FSH hormone.

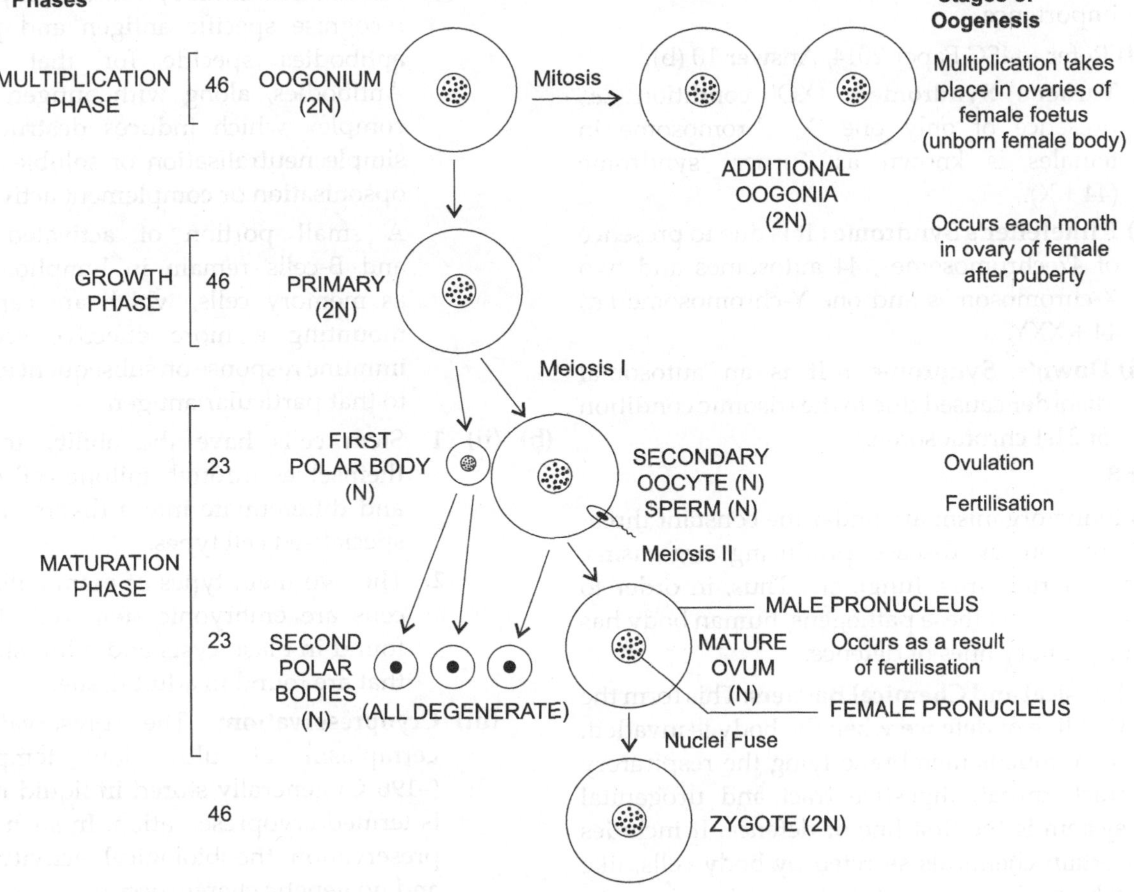

Stages in oogenesis

SECTION–B

Answer 7.

(a) Mass Selection : Mass selection is the most common and ancient method of crop improvement. Desirable individual plants are selected solely on the basis of their phenotype and the resulting seeds are pooled and the progenies are grown in mass. The progeny is individually tested. Selection is accomplished by eliminating the undesirable ones and saving the best.

Pure Line Selection : A desirable plant is selected from a genetically mixed population. It is then selfed during subsequent generations and a true breeding species with desirable characters is picked out and multiplied on a large scale. A pure line is thus a collection of true breeding plants resulting from self pollination of a single homozygous individual. Thus considerable homozygosity can be achieved after ten generations of inbreeding.

Pure line selection is a better method for crop improvement because :

1. The variety developed is genetically more durable.
2. Fewer plants are selected and tested.

(b) (i) Atavism : It is the reappearance or refunctioning of certain ancestral characters after several generations, which have either completely disappeared or greatly reduced *e.g.*, a baby born with a short tail.

Atavism or reversion provides a strong evidence of organic evolution. These characters were present and functional in the ancestors but in due course of evolution they were lost as they were not important in the changed organisms.

(ii) Protoplast Fusion : The protoplasts from different varieties of species can be made to fuse with the help of certain compounds such as PEG-polyethylene glycol or by electro fusion. This provides a chance for an extended variety of hybrid plants between non related species and genera. Hybrids of wheat and rye, and turnip and cabbage, between species of Brassica, Nictotiana, Petunia and Solanum have been produced by protoplast fusion. In 1978, tomato and potato protoplasts were fused. The resulting hybrid is called pomato. It is of great scientific interest and agronomic importance.

(iii) Refer to ISC Paper 2014, Answer 10 (b).

(c) (i) **Turner's Syndrome** : 'XO' condition *i.e.*, presence of only one 'X' chromosome in females is known as Turners syndrome (44 + X).

(ii) **Klinefelter's Syndrome** : It is due to presence of 47 chromosomes, 44 autosomes and two X-chromosomes and one Y-chromosome *i.e.*, 44 + XXY.

(iii) **Down's Syndrome** : It is an autosomal disorder caused due to the trisomic condition of 21st chromosome.

Answer 8.

(a) All living organism are under the constant threat of invasion by disease producing organisms, like bacteria, virus, fungi, etc. Thus, in order to combat against these pathogens, human body has three primary lines of defence.

1. **Physical and Chemical Barriers:** This form the first line of defence when the body is invaded. Skin, mucus membrane lying the respiratory tract, throat, digestive tract and urogenital system is the first line of defence. It includes certain chemicals secreted by body cells, like sebum, lysozyme (tears), gastric juice, etc. which inhibits the growth of pathogens.
2. **Nonspecific Resistance (Innate Immunity):** When the first line of defence fails against pathogen attack, body's second line of defence comes to light. It destroys invaders in a generalised way without targeting specific individuals. Action of phagocytic cells (e.g. macrophages and monocytes), inflammation, fever, release of pyrogens and interferon forms the nonspecific immunity.
3. **Specific Resistance (Acquired Immunity):** This includes the third line of defence. This system relies on antigens, which are specific substances found in foreign microbes. They serve as the stimulus to produce an immune response. Lymphocytes are the functional units of the immune system. They express the activity in two ways:

(i) **Cell-mediated immune system** : When an antigen is detected by a macrophage, it alerts T-cells to become activated. The T-cells secrete interleukin-2, which causes the proliferation of certain cytotoxic T cells, which cause direct destruction of foreign antigens. Helper –T cells stimulate B cells, to produce antibodies.

(ii) **Humoral immune system** : B-lymphocytes recognise specific antigen and produce antibodies specific for that antigen. Antibodies, along with antigen from a complex which induces destruction by simple neutralisation of soluble antigen, opsonisation or complement activation.

A small portion of activated T-cells and B-cells remain in lymphoid tissue as memory cells, which are capable of mounting a more effective secondary immune response on subsequent exposure to that particular antigen.

(b) (ii) 1. Stem cells have the ability to renew themselves through mitotic cell division and differentiate into a diverse range of specialised cell types.

2. The two main types of mammalian stem cells are embryonic stem cells that are found in blastocysts and adult stem cells that are found in adult tissue.

(iii) **Cryopreservation:** The preservation of germplasm at ultra low temperature (–196°C) generally stored in liquid nitrogen is termed cryopreservation. In such kind of preservation the biological activity ceases and no genetic change occurs.

(c) Darwin studied the climate conditions, fauna and flora of Galapagos island, during his voyage around the world. These islands are situated nearly 600 miles away of the West coast of South America and have variable environmental conditions. Darwin called these islands as a living laboratory of evolution. He noticed nearly 20 related varieties of small birds in these islands which differed mainly in the shape and size of their beak and in the colour of their plumage. These birds are now called Darwin's finches. Finches are, in fact, a group of small birds found in these islands. The general bird fauna of these islands shows affinity with the American species but the Darwin's finches differ considerably from American finches. However, a related species of Darwin's finches was reported from South American mainland. Therefore, Darwin concluded that the American mainland species was the original one from which different forms migrated to the different islands of Galapagos and adapted to the different environmental conditions of these islands. These adapted forms are considered to be the new species. Thus, Darwin's finches provide a beautiful example of adaptive radiation supporting the concept of evolution.

Answer 9.

(a) Convergent evolution is the process by which unrelated or distantly related organisms evolve similar body forms, coloration, organs and adaptations. It is also called parallel evolution. *e.g.,* Some Marsupials of Australia resemble placental mammals that live in similar habitats on other continents. Australia separated from other continents more than 50 million years ago. Marsupials arrived in Australia before it separated and evolved in isolation even before the placental mammals. Natural selection has favoured changes that made the two groups more similar. Their phenotypes have converged.

Divergent evolution is also referred to as adaptive radiation. In it the original form of organisms migrated to and adapted to different environments in different areas. They gradually changed and formed new species. Darwin's finches, homologous organs are good examples of divergent evolution *e.g.,* the fore limbs of mammals, seal of whale, wings of bat, leg of horse, human hand, though different superficially and perform different functions but have almost same structure.

(b) Manures are partially decayed organic wastes added to the soil to increase crop productivity. They supply all essential elements required by crop plants. They also improve the physical conditions of soil by preventing erosion and leaching and increase the water holding capacity of the soil.

Green Manure is the principal supplementary means of adding organic matter to the soil. It is a quick growing crop, and ploughed under to incorporate into the soil. The green manure crop supplies organic matter as well as additional nitrogen particularly if it is a legume crop which has the ability to fix atmospheric nitrogen with the help of its root nodule bacteria. The green manure crops also protect erosion and leaching. For example, leguminous plants are used as green manuring.

Biofertilisers are nutrient inputs of biological origin for plant growth. It is the use of microorganisms to increase the availability of nutrients like nitrogen and phosphorus to crops. Nitrogen fixing bacteria are examples of biofertilisers.

(c) Integrated Pest Management (IPM) : IPM involves several pest-control methods which promotes biological control methods instead of the use of chemical pesticides, thus, saving environment from agrochemical problems. It involves following few control methods :

1. Soil conditioning, crop rotation, proper sanitation practices.
2. Eliminating pest by starvation.
3. Mixing crop with specific weed on which insect may feed instead of crop.

Bioinsecticides : These are living organisms usually insects, which help to control other insectpests *e.g.,* lady bugs control aphids in crops.

Bioherbicides : These are insects or micro-organisms which are used to control weeds *e.g.,* the fungus–*Phytophthora palmivora* checks the growth of milkweed vines in citrus orchards. These organisms feed selectively or produce diseases in weeds and thus eliminate them.

Answer 10.

(c)

Disease	Causative Agent	Symptoms
(i) Filariasis	*Wuchereria bancrofti*	Abnormal, huge enlargement of limbs, scrotum etc., legs may become so swollen that they resemble those of an elephant.

●●

BIOLOGY 2011

QUESTIONS

PART I

*(Answer **all** questions)*

Question 1.

(a) *Mention one significant difference between each of the following :*** **[5]**

(i) *Growth and Development*

(ii) *Muscle twitch and Tetanus.*

(iii) *Heartwood and Sapwood*

(iv) *Leghaemoglobin and Hemoglobin*

(v) *Collateral vascular bundle and Concentric vascular bundle.*

(b) *Give reasons for the following :* **[5]**

(i) *Adrenaline is referred to as emergency hormone.***

(ii) *Despite availability of plenty of water, leaves of certain plants wilt during the day and recover in the evening.***

(iii) *Hybrid seeds should be raised every year.*

(iv) *The wing of a bat is said to be homologous to the wing of a bird and analogous to the wing of an insect.*

(v) *Symptoms of deficiency of certain nutrients appear in the old leaves first.***

(c) *Give a scientific term for each of the following :* **[3]**

(i) *The development of more than one embryo in a seed.*

(ii) *The process of growing old.***

(iii) *Method of inducing early flowering in plants by pre-treatment of their seeds at lower temperatures.***

(iv) *Determination of the age of a tree by counting the number of annual rings.***

(v) *The type of growth in which the volume of the body increases without the increase in the number of body cells.***

(vi) *A condition when the muscles deteriorate and the person becomes an invalid.***

(d) *Mention the most significant function of each of the following :* **[3]**

(i) *Fovea centralis***

(ii) *Lymphocytes*

(iii) *Bundle of His***

(iv) *Calyptra***

(v) *Bulliform cells***

(vi) *Quiescent centre***

(e) *State the best known contribution of the following scientists :* **[2]**

(i) *Ernest Hackel* **(ii)** *Carl Landsteiner***

(iii) *Robert Koch*** **(iv)** *Hensen***

(f) *Expand the following :*** **[2]**

(i) *G-6PD* **(ii)** *DPD*

(iii) *MRI* **(iv)** *SAN*

PART II

SECTION–A

*(Answer any **three** questions)*

Question 2.

(a) *What are guard cells ? Explain their role in regulating transpiration.*** **[4]**

(b) *Explain tunica corpus theory of origin of shoot apex.*** **[3]**

(c) *Give one function and one deficiency symptom of each of the following in plants :*** **[3]**

(i) Magnesium (ii) Calcium (iii) Molybdenum

Question 3.

(a) *Describe the development of female gametophyte in Angiosperms.* **[4]**

(b) *Explain the mass flow hypothesis of transport of food.*** **[3]**

(c) *Differentiate between cyclic and non-cyclic photophosphorylation.*** **[3]**

Question 4.

(a) *Explain in detail the digestion of carbohydrates, as the food passes through the alimentary canal.*** **[4]**

(b) *Describe step by step what happens in the different phases of the cardiac cycle in human beings.*** **[3]**

(c) *Write the effects of cytokinins on plants.*** **[3]**

Question 5.

(a) *Write about the chemical changes which occur during contraction of skeletal muscles.*** **[4]**

(b) *Draw a neat labelled diagram of the L.S. of a kidney.*** **[4]**

(c) *What is chloride shift ?*** **[2]**

Question 6.

(a) *Explain how the human ear helps in hearing.*** **[4]**

(b) *Briefly describe the events that occur during the proliferative phase of the menstrual cycle.* **[3]**

** Answer is not given due to change in the present syllabus.

(c) *Mention the site of secretion and function of the following :*** [3]

(i) Glucocorticoids (ii) Calcitonin (iii) Glucagon

SECTION–B

*(Answer any **two** questions)*

Question 7.

(a) *Explain the evolution of the long neck of giraffe according to Darwin and Lamarck.* [4]

(b) *Explain briefly :* [4]

(i) *Environmental resistance*

(ii) *Albinism*

(iii) *Plant introduction*

(iv) *Paleontology*

(c) *Write two uses of each of the following :*** [2]

(i) *Emblica officinalis*

(ii) *Adhatoda vasica*

Question 8.

(a) *Give four applications of tissue culture in crop improvement.* [4]

(b) *What do you understand by the term population growth ? Give three ways of discouraging population growth.* [4]

(c) *Define :* [2]

(i) Coacervates (ii) Gene bank

Question 9.

(a) *Explain the role of Rh factor in blood incompatibility.* [4]

(b) *State the main morphological changes that occurred in the ancestors of modern man.* [3]

(c) *Describe briefly the functions of the following :*** [3]

(i) CT Scan (ii) External prosthesis (iii) Pacemaker

Question 10.

(a) *Explain the role of a genetic counsellor.* [4]

(b) *Write the causative agent and the main symptoms of the following diseases :* [4]

*(i) Poliomyelitis*** *(ii) Typhoid*

*(iii) Tuberculosis*** *(iv) Cholera***

(c) *State two similarities between the chromosomes of man and apes.* [2]

ANSWERS

PART I

Answer 1.

(a) **(iii)** Hybrid seeds must be produced every year to obtain maximum advantage of heterosis. When hybrid seeds are cross pollinated, they lose their hybrid vigour for next generation. Hence, they should produced every year to maintain the superiority of the traits.

(iv) Wing of bat is homologous to the wing of a bird as both are different in function but have the same embryonic origin and development. Bat's wing is analogous to insect's wing as both perform the same function but have different origin.

(c) (i) Polyembryony

(d) (ii) They produce antibodies to destroy the pathogens which enter the body.

(e) (i) Gave the 'recapitulation theory' or the 'biogenetic law'.

PART II

SECTION–A

Answer 3.

(a) Ovules are present in the ovary, attached to the placenta in the ovary wall. It contains the female gametophyte.

Development of female gametophyte: A hypodermal cell of the nucleus of ovule differentiates into archesporial cell. It may directly act as megaspore mother cell and undergo meiosis and form four haploid megaspores. Three of them degenerate while the remaining one becomes the mother cell of the female gametophyte. It grows in size and forms an embryo sac. The megaspore

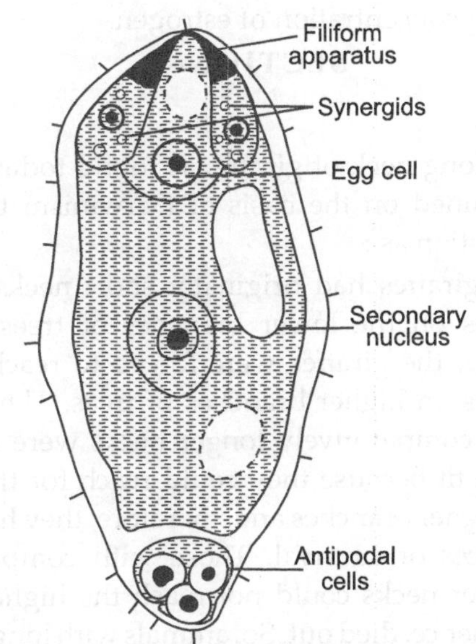

nucleus divides mitotically and forms eight nuclei, which arrange themselves in a particular manner in the embryo sac. Three nuclei towards the micropylar end of ovule organise into an egg apparatus. Three nuclei at the chalazal end form the antipodal cells and the remaining two

** Answer is not given due to change in the present syllabus.

nuclei, called polar nuclei migrate to the centre of the embryo sac. They later fuse to form a single, diploid secondary nucleus. The large, central cell of the egg apparatus is the egg or the female gamete. It is partially surrounded by two lateral synergid cells. The mature gametophyte or embryo sac thus has two polar nuclei, three antipodal cells, one egg and two synergids.

Answer 6.

(b) Proliferative phase (or Follicular phase) begins with the end of menstruation. In this phase, growth and rapid division of the cells of walls of uterus, fallopian tube etc., occurs. At the start of the cycle, hypothalamus signals the pituitary gland to release follicle stimulating hormone (FSH).

Changes in the Ovary : FSH stimulates the ovaries for the production and maturation of follicles. Each follicle contain an immature egg, which changes into mature Graafian follicle, and in turn release oestrogen.

Changes in Uterus : Oestrogen released from mature follicle stimulates the endometrium and its glands start growing, the epithelial cells divide and grow rapidly. The growth of endometrial stroma occurs and become highly vascularised. The endometrium gradually becomes 3-5mm thick. The myometrial contractions become stronger and secretion of the glands at the cervix of uterus increases.

All these changes in the uterus are due to the rising concentration of estrogen.

SECTION–B

Answer 7.

(a) The long neck of giraffe, as found today, can be explained on the basis of Darwinism theory of evolution as :

The giraffes had originally short necks. As the leaves on the lower branches of trees became scarce, the giraffes were forced to reach for the leaves on higher branches of trees. The animals with comparatively longer necks were certainly more fit because they could reach for the leaves on higher branches and, therefore, they had better chances of survival. Those with comparatively shorter necks could not reach the higher leaves and hence, died out. So, animals with longer necks were selected by the nature, they fed comfortably and reproduced. Thus, with the passage of time and generation after generation, the present day giraffes evolved.

According to Lamarck's theory, the giraffe's neck became longer through generations as the giraffe stretched its neck to reach the leaves on tall trees. The information about the stretching of neck was transmitted in some way to the gametes. Thus, the long neck became a permanent feature in the inheritance of acquired characters.

(b) (i) **Environmental Resistance:** The sum total of inhibitory environmental factors, both biotic and abiotic, which regulate population size and do not permit unlimited growth of population is called environmental resistance. Because of environmental resistance, the population is unable to reach full biotic potential.

(ii) **Albinism :** The individuals suffering from albinism are called albino. They lack melanin pigment in their skin, hair, iris of eye etc. Such persons are susceptible to bright sun rays and develop eye disorders and skin diseases. Albinism is caused due to recessive autosomal gene mutation. Hence these individuals lack the enzyme tyrosinase which is essential for the synthesis of melanin. The gene for albinism is a recessive gene (a) which can be expressed only in homozygous (aa) condition. A person with its dominant allele (A) will be normal.

(iii) **Plant introduction :** The process of introducing new plants from their growing place into a new locality with different climate is called plant introduction. The new crop varieties may be introduced in the form of seeds or cuttings. It is a method of rapid crop improvement and requires minimum effort and cost e.g., Groundnut is introduced in India from Philippines, Cinchona from Peru and Papaya from West-Indies.

(iv) **Paleontology :** The study of fossils is called paleontology and it reveals the existence of life in past and illustrates the course of evolution of plants and animals. The sedimentary rocks with fossils can be compared with the leaves of a book with evolutionary history written in the language of fossils. The fossils provides direct evidence of evolution, e.g., *Archaeopteryx lithographica* is a connecting link between reptiles and birds. Its fossil was obtained by Andreas Wagner in 1861 from limestone rocks of Solenhofen in Bavaria. It possess teeth in jaws, fingers with claws, long tail with free caudal vertebrae sternum without keel like reptiles and feathers, wings, beak and V-shaped furcula like birds. Hence it is a strong evidence of reptilian origin of birds.

Answer 8.

(a) Application of tissue culture :

1. Tissue culture provides rapid multiplication under favourable conditions. A small plant bearing five to six leaves is produced within a few weeks which under normal conditions may need several months.
2. By tissue culture techniques, large number of offsprings or plantlets can be obtained every year.
3. The progeny of these plants can be obtained in millions, which multiply with difficulty by conventional methods.
4. Cloning can be done throughout the year in a very small space under controlled conditions.
5. Offsprings of sterile plants can be obtained or rare hybrids of extra ordinary characters e.g., oil palm.
6. Tissue culture ensure the production of disease–free plants. In the virus infected plants, only the meristematic cells of the apical dome are devoid of virus. Therefore, by using shoot apical meristem culture, virus free plants can be raised.

(b) Population growth refers to the total increase in the number of organisms occupying a certain area. **The various methods to discourage population growth are :**

1. The marriage age should be increased. It will decrease reproductive span.
2. People should be educated about the negatives of overpopulation by Government agencies like radio, television, newspaper, magazines, posters, etc.
3. Family planning methods :
   (i) Use of oral contraceptive by women and encouragin the use of condoms.
   (ii) Use of vaginal diaphragms.
   (iii) Use of intra-uterine contraceptive devices (IUCD) like loop or copper-T.
   (iv) Surgical techniques of birth control like tubectomy in females & vasectomy in males.

(c) **(i) Coacervates** : They are the Aggregates or clusters of large, complex organic molecules, capable of growth and replication, which are supposed to have given rise to cell like structures.

(ii) Gene bank : Gene bank is a place where valuable plant material, likely to become irretrievably lost in the wild or in cultivation, is preserved in a viable condition.

Answer 9.

(a) Refer to ISC Paper 2014, Ans. 10 (b).

(b) The characteristics that have been developed in human beings during evolution are :

1. Enlargement and rounding of cranium.
2. Narrowing and elevation of nose.
3. Formation of chin.
4. Reduction in body hair.
5. Increase in the size of the brain and intelligence in man.
6. Development of curves in vertebral column for erect posture.
7. Man's legs are longer than his arms.
8. Able to walk on two legs.
9. Flattening of face.

Answer 10.

(a) Roles of genetic counselor are :

(i) Genetic counsellors help to identify families at possible risk of genetic condition by gathering and analysing family history and inheritance patterns and calculating chances of recurrence.

(ii) They provide information about genetic testing and related procedures.

(iii) They trained to present complex and difficult-to-comprehend information about genetic risks, testing, and diagnosis to families and patients.

(iv) Genetic counsellors can help families understand the significance of genetic conditions in relation to cultural, personal, and familial contexts.

(b) (ii) Typhoid : Causative agent : *Salmonella typhi.*

Symptoms : 1. Cough.

2. Prolonged fever, often rising on the fifth day and after.

3. Abdominal pain and loss of appetite.

4. Rose-coloured spots on the chest in the second week of fever.

5. Constipation, may be followed by diarrhoea.

(c) The similarities between the chromosomes of man and apes are:

1. The banding pattern of individual human chromosomes is very similar and some times identical to the banding pattern of chromosomes of apes.

2. The banding pattern of chromosomes no. 6 and 3 of man and apes is very similar.

●●

ENVIRONMENTAL SCIENCE

2020

(Maximum Marks : 80)

(Time allowed : Three hours)

*(Candidates are allowed additional 15 minutes for **only** reading the paper. They must NOT start writing during this time).*

*Answer **Question 1** (compulsory) from **Part I** and **five** questions from **Part II**, choosing **at least one** question from each of the Sections **A, B** and **C**.*

The intended marks for questions or parts of questions are given in brackets [ ].

QUESTIONS

PART–I (20 Marks)

*Answer **all** questions.*

Question 1.

(i) *Define deep ecology. How is the concept of deep ecology different from shallow ecology?* **[2]**

(ii) *Mention any two factors influencing the growth of population.* **[2]**

(iii) *Give any four characteristics of 'r' strategy.* **[2]**

(iv) *Give any two differences between a national park and a sanctuary.* **[2]**

(v) *Differentiate between primary pollutants and secondary pollutants. Give one example each.* **[2]**

(vi) *Give any two socio-economic benefits of cottage industries.* **[2]**

(vii) *What is food adulteration ? Name two of the common food adulterants.* **[2]**

(viii) *Give the full forms of :* **[2]**

(a) *MFN* **(b)** *EPR*

(ix) *Classify the natural resources based on its distribution, giving examples.* **[2]**

(x) *What is IPR ? State any two categories of IPR.* **[2]**

PART–II (50 Marks)

*Answer **five** questions in all, choosing at least **one** question from each of the Sections **A, B** and **C**.*

SECTION–A

Question 2.

(a) *What is sustainable development ? Explain any five basic principles of sustainable development.* **[6]**

(b) *Give a biographical sketch of Wendell Berry with special reference to his work on environment.* **[4]**

Question 3.

(a) *What are soil indicators ? Name any five types of soil indicators and give one significance of each one of them.* **[6]**

(b) *What is meant by demographic transition ? Explain the first three stages of demographic transition.* **[4]**

Question 4.

(a) *Discuss any six functions of the State Pollution Control Board.* **[6]**

(b) *Explain any four threats to the ecosystem.* **[4]**

SECTION–B

Question 5.

(a) *Define sustainable agriculture. Explain any five elements of sustainable agriculture.* **[6]**

(b) *Write a short note on the privatisation of water in Cochabamba.* **[4]**

Question 6.

(a) *What is meant by new urbanism ? Discuss any five goals of smart growth in new urbanism.* **[6]**

(b) *Discuss any two social and two ecological impacts of green revolution.* **[4]**

Question 7.

(a) *What is food preservation and processing ? Why do we need it? Discuss any three methods of food preservation and processing.* **[6]**

(b) *What is meant by food additives ? Discuss any three health hazards caused by food additives.* **[4]**

SECTION–C

Question 8.

(a) *What is meant by natural capital ? What is the importance of preserving it ? State the methods of regenerating natural capital.* **[6]**

(b) *Give four advantages and four disadvantages of GNP.* **[4]**

Question 9.

(a) *Discuss the ecological values of the rain forests of Amazonia.* **[6]**

(b) *What is Physical Accounting System ? What is its aim ? State two advantages and two disadvantages of Physical Accounting System.* **[4]**

Question 10.

(a) *What is globalization ? Give any five advantages of globalisation.* **[6]**

(b) *Explain any four impacts of hazardous waste.* **[4]**

ANSWERS

PART–I

Answer 1.

(i) Deep ecology is a modern ecological thought that recognises the value of every living organism irrespective of its utility to man. Concept of deep ecology is different from shallow ecology because shallow ecology believes in conservation of the environment only for human need.

Deep ecologists believe that the world does not exist as a resource to be freely exploited by humans. The shallow ecologists do wish to save the world but only for man's own need.

(ii) Some factors influencing the growth of population are :

1. Availability of family planning and birth control methods.
2. Access to health care or medical technology.
3. Government policies and social norms.
4. Cost of education brings up children to provide an incentive to reduce family size.
5. Economic growth and development.
6. Immigration levels or strictness for stagnancy in the population. *(Any two)*

(iii) The characteristics of r- strategies are :

1. Small size of organism.
2. Energy used to make each individual is low.
3. Many offspring are produced.
4. Early maturity.
5. Short life expectancy.
6. Each individual reproduces only once.
7. Little or no parental care.
8. Most offspring die before reaching reproductive age. *(Any four)*

(iv) *(Any two)*

S. No.	National Park	Sanctuary
1.	Habitat centric.	Species centric.
2.	Ownership of land not allowed.	Private ownership of land not allowed.
3.	No human habitation allowed.	Restricted human habitation is possible.
4.	Well defined boundaries.	Boundary not well defined.

(v)

Primary pollutants	Secondary pollutants
Emit directly from a source to the atmosphere. Example: Ash, smoke, dust, oxides of carbon, sulphur, particulate matter, fluoride, benzene, etc.	Formed as a result of chemical or physical interactions between the primary pollutants and the other atmospheric components. Example : Photochemical oxidants (NO_2, SO_3, O_3), smog, secondary particulate matter, Ozone, acid rain, PAN, etc.

(vi) Two socio-economic benefits of cottage industries are as follow :

1. They not only constitute major manufacturing activities in themselves but are also often the only available means to provide additional employment and raise the level of living for both rural and urban populations.

2. It plays a prominent role in overall economic development of the country and contributes significantly to India's Gross Domestic Products and export earnings.

(vii) When some poor quality, inferior or harmful substances are deliberately added or mixed with food that spoils the quality and nature of food items. It is considered as food adulteration.

The two common food adulterants are :

1. Chalk powder in sugar
2. Starch in milk

(viii) (a) MFN – Most Favoured Nation

(b) EPR – Extended Producer Responsibility

(ix) The natural resources based on their distribution are :

1. **Ubiquitous resources :** They are present everywhere we live. Example : Air, land and water.

2. **Localised resources :** They are natural resources found at certain places. Example : Copper, coal, petroleum etc.

(x) Intellectual Property Rights (IPR) is the rights of ownership of intangible intellectual properties like creative work, idea or invention.

1. Patent
2. Trademark
3. Copyright
4. Trade secret
5. Geographical indicator
6. Industrial design

Answer 2.

(a) Sustainable development is defined as the development which meets with the basic needs of human by using resources without compromising the ability of future generations to meet their own needs in a way that allows for them to renew and exist for others.

The basic principles of sustainable development are :

1. The principle of sustainable management of resources: Utmost care and effort must be made towards sustainability. We need to live within the carrying capacity of an ecosystem.

2. The holistic approach: All things are interrelated. Any change would lead to an environmental disbalance, if not dealt in a judicious manner.

3. The principle of social justice: Fundamental human rights must be guaranteed to all. Equal opportunities to acquire desired knowledge and skill should be available to everyone.

4. The principle of integration: Local, regional and national activities must be coordinated such that the implementation of any policies, plans are socially and economically viable to all.

5. The principle of utilizing local resources: All efforts need to be made in making use of all resources available locally, so as to maintain the uniqueness of a particular region.

6. Guaranteed fundamental human right for each and every individual: Every individual must have equal access to educational and development skills so that each can hold a respectable place in the society.

7. The principle of public participation: Adequate access to information affecting social or economic life and environment, information on decision making processes must be provided to all. Public participation on decision making must be strengthened.

8. The principle of social responsibility: To enable sustainable development and make higher quality of life possible. Unsustainable patterns of production and consumption has to be stopped.

(b) Wendell Erdman Berry is an American novelist, poet, essayist, environmental activist, cultural critic, and farmer. He is a prolific writer of both fiction and non-fiction and is a member of the Fellowship of Southern Writers and, a recipient of The National Humanities Medal. He has also significantly contributed towards environment.

Some of his prominent work towards environment are :

1. In 1979, Berry engaged in non-violent civil disobedience against the construction of a nuclear power plant at Marble Hill, Indiana.

2. In 2009, Berry joined over 2,000 others in non-violently blocking the gates to a coal-fired power plant in Washington, D.C. and participated in a rally in Louisville on how to manage coal ash.

3. Berry combined with along with several other non-profit organizations and rural electric co-operations to file petition against and protest the construction of a coal-burning power plant in Clark County, Kentucky. As a result the construction of the plant was cancelled in 2011.

4. He believed that good life encompasses sustainable agriculture, appropriate technology, and connection to place, healthy rural communities, good food, and the miracle of life, fidelity, frugality and reverence. He has written more than 30 books. He spoke for sustainable agriculture and also came out strongly against the death sentence.

Answer 3.

(a) The indicators which provide physical, chemical, biological properties, processes that are measured to monitor change in soil which disturb the performance of ecosystem functioning in the soil of interest are called soil indicators. Soil indicators help us in monitoring soil quality. They tell us about the health of the soil and how suitable it would be to grow any particular type of crop.

The various types of indicators along with their significances are:

1. Visual indicators : Exposure of subsoil, change in soil color shows that quality of soil is changing.

2. Physical indicators : Primarily reflect limitations to root growth, seedling emergence, infiltration, or movement of water within the soil profile.

3. Chemical indicators : Include measurements of pH, salinity, organic matter, phosphorus concentrations.

4. Biological indicators : Measurements of micro and macro-organisms, their activity, or byproducts which play an important role in the formation and stability of soil aggregates.

5. Microbial population includes bacteria, fungi, actinomycetes and algae. Soil should contain larger proportion of beneficial organisms in order to be healthy.

(b) A model used to represent the movement of high birth to low birth and high death rates to low death rates as a country develops from a pre-industrial to an industrialised economic system. The concept is used to explain how population growth and economic development of a country are connected. The concept of demographic transition has four stages including the pre-industrial stage, the transition stage, the industrial stage, and the post-industrial stage.

Stage 1 : Birth rates and death rates are both high and about same. In this stage, population growth rate is very low.

Stage 2 : The gap between birth and death rate curves increases as population growth rate is much higher than death rate.

Stage 3 : Birth rate goes down and curves of both birth and death rate come very close population growth is very low.

Stage 4 : Birth rate curve goes down more than death rate curve. Net population of the region decreases. *(Any three)*

Answer 4.

(a) **Some functions of the State Pollution Control Board are :**

1. To advise the State Government regarding water pollution control or location of industries.

2. Planning comprehensive programme for prevention, control and abatement of pollution of streams and wells.

3. To inspect trade effluents and waste water treatment plants.

4. To evolve economical and reliable methods of disposal, treatment and reuse of waste water in agriculture.

5. Making, revoking or varying any order for preservation or control of discharge of waste into streams and wells for disposal of effluents.

6. To conduct and encourage investigations and research related aspects of water pollution.

7. To collaborate with the Central Board for training personnel for handling water pollution programmes and organising related mass education programmes.

8. Laying down the standards of treatment of sewage and trade effluents to be discharged into any stream. *(Any six)*

(b) **The major threats to the ecosystem are :**

1. **Environmental pollution** : Air, land, water and soil destroy the health of ecosystem which result in release of noxious substances and destructive chemicals into the environment and disturb the ecological balance. The implication of pollution is inference with natural cycles like oxygen cycle, nitrogen cycle etc.

2. **Over-exploitation of natural resources:** Activities like over hunting, over fishing, over mining and excessive logging have led to reduction in community structures, population distributions and species breeding.

3. **Eutrophication** : The excessive concentration of chemical nutrients in water bodies to an extent encourages the dense growth of plant life and algae blooms. Its effects are depletion of oxygen, extensive deterioration of water quality, poisoning of seafood which results in endangered aquatic life forms.

4. **Climate change and global warming** : Changes in climates and global temperatures directly affect the abiotic factors essential for sustaining the biotic elements. The present rate of rising global temperatures is destroying and altering the coral reefs, mountain regions, water cycles, which are vital ecosystem resources.

5. **Invasive species** : Species often compete for food with the native species and can also alter the habitat. This gradually destroys ecosystems and leads to the extinction of species. *(Any four)*

Answer 5.

(a) Sustainable agriculture is farming in sustainable ways, which means meeting society's present food and textile needs and preserving the environment by conserving water and energy and by limiting the use of pesticides and fertilizers for future generation to meet their needs. These practices are intended to protect the environment, expand the Earth's natural resource base, and maintain and improve soil fertility.

Five elements of sustainable agriculture are :

1. **Mixed farming** : A practice where farmers manage crops and livestock are managed together.

2. **Mixed cropping** : Growing of two or more crops simultaneously on the same piece of land.

3. **Crop rotation** : A process in which different types of crops are grown alternatively in the same soil or field.

4. **Soil management** : Managing and maintaining soil fertility by testing soil and recommended doses of nitrogen, phosphorus, potassium and other nutrients to meet the needs of crops.

5. **Use of biofertilisers** : The organisms which increase the nutrient availability to crop plants either directly or through soil enrichment.

(b) Case study of water privatization of water in Cochabamba :

The Cochabamba Water War was a series of protests that took place in Cochabamba, Bolivia's fourth largest city, between December 1999 and April 2000 in response to the privatization of the city's municipal water supply company SEMAPA. The wave of demonstrations and police violence were described as a public uprising against water prices.

The tensions erupted when a new firm, Aguas del Tunari – a joint venture involving Bechtel – was required to invest in construction of a long-envisioned dam (a priority of Mayor Manfred Reyes Villa) - so they had drastically raised water rates. Protests, largely organised through the Coordinadora in Defense of Water and Life, a community coalition, erupted in January, February, and April 2000, culminating in tens of thousands marching downtown and battling police.

The World Bank had suggested the privatisation to repair 13 and maintain the water work which reduce its efficiency and was supplying water at 40% of its capacity.

Initiatives for mitigation: Middle class people protested. The main beneficiaries of the subsidy, farmers, teachers, police joining hands under the banner of Morales, Campichino movement.

In retaliation to police firing, the protesters became wild and paralysed the country by nationwide road blockades.

Result: The controversial amendment was scrapped, and the agreement was revoked. Water supply continues to function at 40% capacity. Economically weaker sections have to make do with contaminated water from water tankers. Cochabamba's once rich green fertile valley has been reduced to arid dust bowl.

Answer 6.

(a) New Urbanism is an urban design movement which promotes environmentally friendly habits by creating walkable neighbourhoods containing a wide range of housing and job types. It is a concept where the planners attempt to integrate the best features of both, urban and rural living. It arose in the United States in the early 1980s, and has gradually influenced many aspects of real estate development, urban planning, and municipal land-use strategies. New urbanism attempts to address the ills associated with urban sprawl and post-Second World War suburban development.

Smart Growth promotes a shift in the conventional development patterns, and reaches out across disciplines. It also advocates compact, transit-oriented, walkable, bicycle-friendly land use, including neighbourhood schools, complete streets, and mixed-use development with a range of housing choices. The term "smart growth" is particularly used in North America.

Smart growth values long-range, regional considerations of sustainability over a short-term focus. Its sustainable development goals are :

1. To create a positive self-image for the community.
2. To do away with substandard housing.
3. To solve all problems concerned with urbanisation such as housing, transportation, pollution, etc.
4. To improve communication.
5. To make cities liveable.
6. Regional planning for open space.
7. Balanced development of job and housing.
8. Strategies to reduce traffic congestion.
9. Increase affordable housing.
10. Safe streets, green building.
11. Communities designed for pedestrian and transit.
12. Universally accessible public spaces.
13. Appropriate technology and heritage conservation.

(b) Some social impacts of Green revolution are as follows :

1. **Regional inequalities:** The new agricultural strategy has restricted to only a few regions due to that only some areas got benefits.

2. **Change in attitudes:** The change in the attitudes of farmers in areas where new agricultural strategy has being practiced. Increase in productivity has raised the status of agriculture from a low-level subsistence activity to a money-making activity

3. Inter-personal inequalities: The new technology requires a huge amount of investment which can be afforded by rich farmers so, they were much benefitted. This inequality was increased in rural India.

4. Question of labour absorption: The adoption of new technology had reduced labour absorption in agriculture. The growth of output was also slow to generate adequate employment opportunities.

Some economical impacts of Green revolution are as follows :

1. Increase in agricultural production: The first major direct effect of the green revolution has been the sharp increase in agricultural production. As a result of new agricultural strategy, food grains output increased substantially. Due to the use of HYV seeds, it disturbs the nitrogen cycle.

2. Impact on employment: Capital-intensive techniques of the new agricultural strategy are supposed to be also at the same time labor-intensive which is characterized by frequent application of water, fertilizers, insecticides, double cropping, larger volumes of transportation, marketing and food-processing.

3. Increase in productivity : The productivity of agriculture, as measured in terms of yield per hectare, has increased.

4. Excessive use of the Fertilizers and Pesticides: Due to the excessive use of the Fertilizers and Pesticides it results in the eutrophication which in turn results in the salinization of the soil and makes the plants vulnerable to diseases. Due to the use of the pesticides Biomagnification has increased which kills the useful organisms and the toxicity of the soil increases. *(Any two)*

Answer 7.

(a) Food preservation and processing is a technique of transformation of raw ingredients into food by means of physical and chemical means and packaging to prevent food poisoning, food spoilage, infections, microbial contamination in the food and take clean, butched components and use them to produce attractive and marketable food products. Its advantages surplus produce may be conserved and reduces the work at home.

Food preservation is to prevent and remove microbial contamination and kill pathogens, to inhibit the growth of microorganisms like bacteria or fungi to the food and slows down the oxidation of fats which causes rancidity.

There are some methods for food preservation and processing such as :

1. Pickling is done to preserve meat, vegetables by action of acetic acid which prevent microbial growth.

2. Salting prevents microbial growth by sprinkling salt on fish, meat.

3. Pasteurization is the process of heating milk below its boiling point to kill bacteria and make safe milk and sterilization is to remove microbes from food by boiling at 100°C.

4. Other traditional methods are smoking, salting, keeping the food in excess amount of sugar and salt, etc. There is no single method of food preservation that provides protection against all hazards for an unlimited period of time.

(b) Any of various chemical substances added to foods to produce specific desirable effects like to keep food fresh or enhance their colour, flavour or texture are known as food additives. Additives such as salts, spices foaming agents, artificial sweeteners to preserve foods and make them more palatable.

Some health hazards caused by food additives are :

1. Anamine causes lesions in liver and kidneys, allergy, respiratory disorder, insomnia.

2. Congo red – Stomach pain, mental retardation, brain, and nerve damage.

3. Lead chromate – Anaemia, abortion, paralysis, mental retardation and brain damage.

4. Malachite green – Tumour of lungs, breast, ovary and liver. Abnormalities of skin and eyes.

5. Orange I – Diaorrhea. *(Any three)*

Answer 8.

(a) Natural capital can be defined as stock of natural resources like geology, soil, air, water and all living things. It includes both renewable and non-renewable natural resources. It provide us with ecosystem goods and services which are essential for our survival on earth.

It is important to preserve natural capital because it provides fundamental life support functions to mankind. Two of the components of natural capital (clean water, and fertile soil) underpin our economy and society, thus it is important to preserve natural capital.

Methods of regenerating the natural capital are :

1. Natural resources should be used in an appropriate manner since their depletion will lead to an increase in the market prices which will have not only the financial implications for the country but also environmental implications as services provided by ecosystems are damaged and unable to function effectively which in turn causes flow on effects. For example, as greenhouse gas emissions increase and areas responsible for carbon sequencing decreases, global temperatures rises, weather pattern changes, sea levels increases, terrestrial and aquatic ecosystems re-adjust and land usability pattern changes.

2. Protection and appropriate pricing of environmental resources should essentially taken care of by economic theories and practices since the natural income is monetary income derived from natural capital. If economic and societal development is allowed to grow uncheck, stocks of natural capital will continue to decline, resulting in problems for natural life support systems, increased market prices and a decrease in the quality of human life.

3. New easy and economical alternatives should also be used for the services or products provided by natural capital in order to avoid large dependency on the natural resources.

4. People should be made aware about the minimum levels or time spans required for resources to replenish or renew lack of knowledge regarding the interaction and dependences between resources and their true values, usefulness or necessity, poor management of trans-boundary resources and inequalities between developed and developing nations is one of the major cause of the decline in natural capital thus it should be taken care both at the community and individual level.

5. Sustainability of the resources should be focused for the long term benefits of the resources along with the other sustainability concepts and tools, including ecological footprints, environmental accounting and eco-effectively.

(b) **Four advantages of GNP are :**

1. Helps to achieve economic sustainable development and indicates the economic condition of the country.

2. Emphasize that growth should take place without polluting the environment.

3. Equitable distribution of benefits of its development. Possible to compare the wealth of one country with that of another.

4. Promotes economic welfare for a long period of time. It is easy to calculate.

Four disadvantages of GNP are :

1. It does not actually measure GNP of a nation as GNP for a nation may increase after a year with increase in poverty and decrease in median income. Unpaid economic activities are ignored.

2. It does not take into account a number of factors to calculate total wellbeing of a nation like environmental factors, malnutrition rate, health of population etc. Also does not take into account locally produced and consumed goods.

3. Environmental impact not taken into consideration.

4. Comparison of incomes of two countries may be distorted by changing exchange rate.

5. Does not include sale of used goods.

Answer 9.

(a) Amazons are the largest rainforests and are also known as 'lungs of ecosystem'. About 30 percent of world's species resides in amazon rainforests. Ecological importance of rainforests of Amazonia are :

1. Amazon rainforest is of incalculable biological and ecological value. It's home to about 390 billion trees and more than 16,000 plant species and millions of animal species.

2. Among the animals that live in the rainforest are some of the world's rarest and most colorful birds.They are the home of various species of animals and plants.

3. The 390 billion trees across the Amazon rainforest locks up massive amounts of carbon in their leaves, branches and trunks.

4. Through transpiration, the Amazon rainforest is responsible for creating 50-75 percent of its own precipitation.

5. Regulation of chemical composition of air, thus in turn regulates the climate and water.

6. Maintains the nutrients in the atmosphere and prevents the flooding. Helps in the formation of the soil.

(b) Physical Accounting System refers to the natural resources and environmental accounting of stocks and changes in stocks in physical (non-monetary) units, e.g. weight, area, or number. Qualitative measures, expressed in terms of quality classes, types of uses or ecosystem characteristics, may supplement quantitative measures.

Aim : The basic principle underlining this method is that material flow balances in physical unit of material, water and energy flow within a defined system boundary. Thus, in such systems, physical measures are defined to suit the environmental policy.

Advantages :

1. Physical accounting is more accurate evaluation of natural resources such as groundwater water and forest cover have no market price.

2. Physical accounting is more long term as the changes in volume, number and quantity can be compared over generations and not to be influenced by short period market fluctuations.

Disadvantages :

1. Physical accounting is more difficult to interpret as it does not have any single dimension as monetary accounting (rupees, dollars, etc.).

2. The qualitative aspects of natural resources do not get reflected readily.

3. Accurate accounting for qualitative differences requires more and more disaggregation which makes interpretation cumbersome.

Answer 10.

(a) Globalisation is the process of interaction and integration among people, companies and governments worldwide in order to develop the global economy. With major technological advances, globalisation benefits society as a whole. Five advantages of globalisation are :

1. As globalisation increases, more and more companies are setting up businesses in other countries. This in turn increases the employment opportunities and allows the interchange of cultures.

2. As a result of globalisation, people have access to the best quality of goods and services throughout the world. Companies have to strive to provide better quality goods and services to the consumer and the consumer has the liberty of choosing whichever product he thinks is best suited for his needs. Competition keeps prices relatively low, inflation is likely to occur.

3. With the increase in globalisation, it has become easier for people to move across borders to different parts of the world to acquire better education. To meet the increasing demands, there is an increase in the production giving options to the manufacturers as well.

4. Attracts foreign capital as well as foreign updated technology, improving the quality of production. Thus, allows the better communication and understanding between nations.

5. Helps boost the average growth rate of the country's economy through improving in allocative efficacy and efficiency of the resources. Developing countries benefit as there is a flow of money, decrease in currency difference.

(b) Four impacts of hazardous wastes are :

1. The main danger is water pollution. The chemicals that pour into our waterways make streams, rivers, lakes and aquifers unsafe to use for drinking or agricultural purposes. Ground water pollution can lead to lead poisoning.

2. Emissions pour into the air, these chemicals destroy the ozone, which then warms the Earth leading to global warming and greenhouse effect. The polluted air has an adverse effect on human health and causes diseases like respiratory disorders.

3. Use of synthetic pesticides and fertilizers decrease the fertility of soil on over use. Also degrades the performance of nutrients and disturbs the natural cycle like oxygen cycle, phosphorus cycle, nitrogen fixation in roots. Toxic wastes in the soil can denude the landscape.

4. Biomedical wastes and radioactive wastes are dangerous to health as they do not dispose off easily and are also responsible for diseases like AIDS, cancer, asthama etc. Needles, medicinal wastes and radioactive materials present nearby domestic fields may cause diseases.

5. The non-degradable toxic waste may enter the food chain and through biomagnification affect both, animals and human beings.

●●

ENVIRONMENTAL SCIENCE 2019

QUESTIONS

PART–I (20 Marks)

*Answer **all** questions.*

Question 1.

(i) *State the basic principles of stewardship of land.* **[2]**

(ii) *Given any two impacts of genetic erosion on ecosystem.* **[2]**

(iii) *Define carrying capacity. State any two factors that determine the carrying capacity of an ecosystem.* **[2]**

(iv) *Give any two scarce resources which are diverted to make luxury goods.* **[2]**

(v) *List any two benefits of local market to the villagers.* **[2]**

(vi) *Distinguish traditional agriculture from modern agriculture.* **[2]**

(vii) *List any two ecological impacts of the Transnational Companies.* **[2]**

(viii) *Mention any two chemical indicators of soil quality.* **[2]**

(ix) *Give two objectives of Basel Convention.* **[2]**

(x) *Give the full forms of :* **[2]**

(a) *IPR* (b) *GNP*

PART–II (50 Marks)

*Answer any **five** questions in all, choosing at least **one** question from each of the Sections A, B and C.*

SECTION–A

Question 2.

(a) *What is WWF ? List the goals and strategies put forth by WWF.* **[6]**

(b) *What is social forestry ? Mention any four types of social forestry. How does it differ from Joint Forestry Management ?* **[4]**

Question 3.

(a) *What are life strategies ? State their types. Mention any four characteristics of each of the types stated by you.* **[6]**

(b) *Discuss four indicators of water quality and give one significance of each of them.* **[4]**

Question 4.

(a) *What are the different techniques of monitoring air quality of a place ? Discuss briefly any two of the methods.* **[6]**

(b) *'Women empowerment can bring about a major change in human population.' Discuss.* **[4]**

SECTION–B

Question 5.

(a) *What are the pull and push factors of urbanisation ? Explain any four negative impacts of urbanisation.* **[6]**

(b) *Discuss urban development planning and traffic management methods used in Curitiba.* **[4]**

Question 6.

(a) *Discuss the different types of irrigation systems. Explain any four advantages and any four disadvantages of each type.* **[6]**

(b) *What is a dry toilet ? How does it work ?* **[4]**

Question 7.

(a) *Discuss the problems related to food security in the third world countries. Give any two solutions to tackle these problems.* **[6]**

(b) *Mention any eight main features of the Food Security Law 2013.* **[4]**

SECTION–C

Question 8.

(a) *What is meant by defensive expenditure ? Explain its purpose and importance.* **[6]**

(b) *What is ozone layer ? How does it get depleted? State any two steps taken to control ozone layer depletion.* **[4]**

Question 9.

(a) *Differentiate between free trade and protectionism. State any two advantages and any two disadvantages of each.* **[6]**

(b) *State the two types of externalities. Give two example of each type.* **[4]**

Question 10.

(a) *What is meant by Cost Benefit Analysis ? Mention any three advantages and two disadvantages of Cost Benefit Analysis.* **[6]**

(b) *Discuss the case study of aquaculture in India to understand the impact of free trade.* **[4]**

ANSWERS

PART–I

Answer 1.

(i) The principles of stewardship are :

1. Maximising efficiency and striving to reduce the consumption of renewable and non-renewable resources; long-term optimisation versus short-term maximisation of production.

2. Sustaining and encouraging natural biological diversity and complexity; maintaining natural areas and functions on the land (*i.e.*, wildlife habitat conservation).

(ii) Genetic erosion has following impacts on the ecosystem :

1. **Loss of biodiversity :** Loss of genetic diversity in terms of individual genes or recombinant genes ultimately leads to loss of biodiversity.

2. **Disturbance in food chain and food web :** When endangered species die before attaining their reproductive age, there are high chances of those species to get extinct. Disappearance of species from the ecosystem will result in disturbance in food chain and food web.

(iii) Carrying capacity describes the maximum number of individuals or species a specific environment's resources can sustain for an indefinite period of time without degrading it. Factors which affect carrying capacity are :

1. **Overpopulation :** Overpopulation leads to overconsumption of resources. Therefore, the resources will get exhausted within a short duration.

2. **Natural disaster :** Earthquake, volcano, hurricanes, floods and droughts destroy not only the area but also flora and fauna of that place. In addition, the quality of air, land and water is also affected due to which carrying capacity is reduced.

(iv) Two scarce products are :

1. **Ivory :** It is used to make jewellery, piano, etc.

2. **Sandalwood :** It is used to make decor items and furnitures.

(v) Local markets play an important role in improving agricultural marketing. They can :

1. Provide a location at which farmers can meet with traders.

2. Offer vegetables, fruits, flowers and grains that are directly brought to the market from the field.

3. Make prices more competitive.

4. Make marketing a more pleasurable activity and provide a focal point for rural activities.

(Any two)

(vi) Differences between traditional and modern agriculture are as follows :

S. No.	Traditional Agriculture	Modern Agriculture
1.	Uses manure and compost.	Uses chemical fertilizers.
2.	More farm labours and no machines.	Less farm labours and more machines.
3.	Depended on seasonal rainfall.	Well occupied irrigation system and so not dependent on seasonal rainfall.
4.	Animals are used to plough the field.	Tractors used for ploughing the field.
5.	Production is less but the crops are of high quality.	Production is more but the crops are of low quality.

(vii) Transnational companies impact the environment in following ways :

1. A significant amount of the green house emissions are generated by them, in which they are responsible for global warming.

2. Furthermore, they are also the users of ozone-depleting chlorofluorocarbons (CFCs) and they are producers of CFCs.

3. Transnational corporations are responsible for the pollution of land, air, wetlands, water and the oceans.

4. Through their mining activities and commercial logging, there is a contribution to deforestation by the TNC. As a result of such mining and logging negative effects such as flooding, loss of topsoil rapid run-off of rain has occurred.

(Any two)

(viii) Chemical indicators of soil quality are soil pH, cation exchange capacity, salinity, nutrient cycling rate, electrical conductivity.

(ix) Objectives of basel convection are :

1. To reduce hazardous waste generation and to the promote environmentally sound management of hazardous wastes.

2. To restrict trans-boundary movements of hazardous wastes except where it is perceived to be in accordance with the principles of environmentally sound management.

3. To implement regulatory system applying to cases where trans-boundary movements are permissible.

(Any two)

(x) (a) **IPR** – Intellectual Property Rights.

(b) **GNP** – Gross National Product.

PART–II
SECTION–A

Answer 2.

(a) WWF is World Wildlife Fund. It is a leading organisation that works for the conservation of wildlife and endangered species.

Goals of WWF are to stop the degradation of the planet's natural environment and to build a future in which humans can live in harmony with nature, by :

1. Conserving the world's biological diversity.

2. Ensuring that the use of renewable natural resources is sustainable.

3. Promoting the reduction of pollution and wasteful consumption.

Strategies of WWF are :

1. **Forest**–WWFs are increasing protected areas and bringing more forests under improved management, halting deforestation, particularly in deforestation fronts and restoring degraded forest landscapes.

2. **Oceans**–By highlighting the importance of healthy oceans, WWF is working with communities, governments, businesses including the finance industry to end support for damaging activities and invest in creating a healthy marine environment. They are protecting and restoring critical coastal environments like coral reefs and mangroves. WWF is supporting coastal communities particularly in developing countries to manage their marine resources, and using their influence with industry, governments, retailers and consumers to make a real difference.

3. **Wildlife**–WWF is tackling illegal trade and over-exploitation by strengthening regulations and making sure they are properly enforced. They are also influencing the markets and consumer choices that drive demand for wildlife products.

4. **Food**–WWF is promoting sustainable agriculture. It is also creating awareness of food loss and food wastage.

5. **Climate and energy**–WWF promotes installation and usage of alternative and renewable sources of energy like solar energy, wind energy, tidal energy, etc. It also takes measures on human factors that induces climate change.

6. **Freshwater**–WWF runs freshwater conservation projects in more than 50 countries from protecting iconic species like sturgeon and river dolphins, to restoring wetlands and keeping rivers free flowing. They also constantly push for better protection for freshwater habitats at a national and international level.

(b) Social forestry is the management and protection of forest and afforestation of barren and deforested lands with the purpose of helping environmental, social and rural development.

Types of social forestry :

1. Farm forestry is the management of trees for a specific purpose within a farming context. Typically these are timber plantations on private land.

2. Agro forestry is the combination of agriculture and tree growing in order to produce both agricultural products and tree products on a commercial basis. The purpose of this scheme is to gain positive interactions between the two systems at both the paddock level and the enterprise level.

3. Planting of trees on the sides of roads, canals and railways, along with planting on wastelands is known as 'extension' forestry. Extension forestry helps in increasing the boundaries of forests. Under this project, there has been creation of forests on the village common lands, government wastelands and panchayat lands.

4. Community forestry is a village-level forestry activity, decided collectively and implemented on communal land, where local populations participate in the planning, establishing, managing and harvesting of forest crops, and so receive a major proportion of the socio-economic and ecological benefits from the forest.

Answer 3.

(a) Life strategy is a representation of the pattern of survival and reproduction events typical for a member of the species. It also represents an "optimization" of trade-offs between growth, survival, and reproduction. All living things need energy and nutrients to grow, maintain their bodies, and reproduce. In nature, these resources are in limited supply, and there is often competition for access to them (*e.g.*, to sunlight and minerals for plants or food sources for animals). Thus, each organism will have non-infinite resources to divide among activities like growth, body maintenance, and reproduction. Organisms with inherited traits that cause them to distribute their resources in a more effective way will tend to leave more offspring than organisms lacking these traits, causing the traits to increase in the population over generations by natural selection. The life strategies can be of *r*-type or *k*-type.

***r*-selected organisms :**

1. They usually mature rapidly and have an early age of first reproduction.

2. They have a relatively short lifespan.

3. They have a large number of offspring at a time, and few reproductive events, or are semelparous.

4. These organisms have a high mortality rate and a low offspring survival rate.

5. They need minimal parental care. ***(Any four)***

***k*-selected organisms :**

1. They usually mature more slowly and have a later age of first reproduction.

2. They have a longer lifespan.

3. They have few offspring at a time and more reproductive events spread out over a longer span of time.

4. They have a low mortality rate and a high offspring survival rate.

5. They need more parental care. ***(Any four)***

(b) Indicators of water quality are :

1. **pH** : pH measures the concentration of hydrogen ions in water on a scale of 0 to 14 where

7 is neutral and anything above 7 is alkaline and anything below 7 is acidic. The pH of most natural water is between 6 and 8.5. Levels below 4.5 and above 9.5 are usually lethal to aquatic organisms. pH affects the solubility of organic compounds, metals, and salts. In highly acidic waters, certain minerals can dissolve and release metals and other chemical substances into the water.

2. **Levels of dissolved oxygen :** Dissolved oxygen is essential for a healthy aquatic eco-system. Fish and aquatic animals need the oxygen dissolved in the water to survive. Drastic alterations in the usual levels in dissolved oxygen can cause great amounts of damage to environmental ecosystems.

3. **Levels of metals :** A number of metals, such as copper, manganese, and zinc are essential to biochemical processes that sustain life. However, high concentrations of these and other metals in water can be toxic to animals and humans if they are ingested, or if they are found in animals that are then consumed by humans. Dissolved metals are generally more toxic than metals bound in complexes with other molecules.

4. **Temperature :** Aquatic organisms are dependent on certain temperature ranges for optimal health. Temperature affects many other parameters in water, including dissolved oxygen, types of plants and animals present and the susceptibility of organisms to parasites, pollution and disease. Causes of temperature change in water include weather conditions, shade and discharges from urban sources or groundwater inflows.

Answer 4.

(a) Different techniques of monitoring air quality are – Ambient monitoring, satellite monitoring, emission monitoring, mobile monitoring and on-road monitoring.

1. **Mobile monitoring :** A van equipped with instruments is run for a day or a week or more, to collect ambient monitoring data for all the criteria pollutants. The flexibility of the station to relocate between sites is what makes it mobile, but the measurements taken from the unit are considered ambient. The data is used for understanding pollution exposure during commute; specially to understand the acute health impacts of being exposed to augmented pollution levels on the roads.

2. **Satellite monitoring :** Satellite monitoring is a multiple step procedure that involves estimating pollution levels using a modelling exercise based on multiple assumptions. The assumptions include on-ground measurements from ambient air monitoring stations, and results of a global chemical transport model that itself is based on estimated emission inventories. Hence for accurate estimates from satellite monitoring, data from on-ground monitoring stations are crucial.

(b) 1. The decision to limit family largely depends on a couple's cultural background and education. The idea of limiting family size is counteracted by many cultural attitudes. Such attitudes will change if people are educated.

2. People should be made aware as to how they themselves and whole human race will be affected with such an enormous growing population. People should be made aware of how the quality of life will be much better with less population.

3. Women have legal rights and freedom relating to childbirth and her reproductive health. In most of the male dominating society, women are not aware about obtainability of such right or not empowered to claim their freedom.

4. Some communities and cultures prohibit use of contraception and/or abortion. Child marriage and forced pregnancy are some of the evil practice which are still going on in many of the backward communities. These evil practices hamper the eminence of next generation of human population.

5. Access to education helps people to understand the importance and need of leaving behind such evil practices. It helps people to respect each other and fosters a positive growth. It is not just enough if we reduce the population but, the quality of life and livelihood is also vital.

SECTION–B

Answer 5.

(a) The reasons that people migrate are determined by push and pull factors, which are forces that either induce people to move to a new location, or oblige them to leave old residences. Push factors are conditions that can force people to leave their homes and are related to the country from which a person migrates. Push factors include non-availability of enough livelihood opportunities, poverty, rapid population growth that surpasses available resources. Pull factors are exactly the opposite of push factors. They are the reasons why people move to a particular area.

Rural-Urban migration has detrimental effects on village as well as city economy, culture and quality of life. Over urbanisation has caused unbalanced distribution of population, which in turn is leading to various problems like :

1. Shortage of resources and space in urban areas.
2. Decline in quality of environment and life in urban areas.
3. Unemployment.
4. Lack of motivation and interest in rural areas to practice agriculture.

5. Lack of interest and motivation to practice indigenous and exotic occupations like pottery, carving, sculpture making, etc., which are representatives of our culture and tradition.

(Any four)

(b) 1. In the 1970s the city turned away from conventional large housing complexes, preferring to break up large projects into smaller and better integrated ones, using the already existing infrastructure. Even the smallest complexes have schools, day-care centres and health centres included.

2. Funding is made available for self-build projects, for the purchase of urbanised lots, for urbanisation and regulation of irregularly occupied areas and for partnerships with the private sector where building advantages are traded for lots and housing units.

3. In order to overcome the lack of resources to meet housing needs, the Municipal Housing Fund was created in 1990. This provides funds for low-income housing from private developers through levies charged on new development permissions.

4. Public transport is used by more than 1,600,000 passengers each day and accounts for over 70 per cent of daily journeys. As a result, Curitiba has one of the lowest levels of air pollution in Brazil, few traffic jams and a low household expenditure on transport.

5. Concentric circles of local bus routes connect to five major radial routes. On these routes articulated buses move in their own lanes, each carrying 270 passengers. They go as fast as an underground system but have been provided for less than two per cent of the capital construction cost of an underground system.

6. The buses are clean and fares are cheap. No bus is allowed to operate on the roads once it is more than ten years old.

Answer 6.

(a) **Different types of Irrigation Systems are :**

1. Overhead Irrigation : These are like sprinklers.

Advantages of Overhead Irrigation :

1. Overhead irrigation rains down to cleanse and hydrate gardens and lawns.
2. It can be easily installed and moved.
3. Versatile sprinklers often include programmable features to minimise labour.
4. Overhead systems benefit any terrain, no matter how uneven or rough, and easily adjust to suit different areas.

Disadvantages of Overhead Irrigation :

1. Increased risk of plant disease due to wet foliage.
2. Winds often blow overhead sprays off target.
3. Wasting misdirected water and increasing evaporation.
4. Runoff due to over-watering and nearby concrete add to water loss.

2. **Surface Irrigation :** Surface methods include flooding and furrow irrigation, as well as basins around trees and shrubs.

Advantages of Surface Irrigation :

1. Low costs.
2. Minimal labour.
3. Less instrumentation required.
4. Water stays in the root zone, and foliage stays dry.

Disadvantages of Surface Irrigation :

1. Potential over-watering and wasteful runoff.
2. If soil lacks proper sloping or does not absorb readily, water cannot move through the garden.
3. Standing water damages plants.
4. Standing water may also reduce the yields for edible crops.

3. **Drip Irrigation :** It is the slow and controlled watering systems, which deliver water directly to plants rather than areas.

Advantages of Drip Irrigation :

1. Drip irrigation keeps soil evenly moist, yet full of oxygen.
2. With balance instead of fluctuations, roots flourish.
3. Less wastage of water.
4. Improves seed germination.

Disadvantages of Drip irrigation : They are limited, but drip systems must be carefully managed and monitored.

1. Too little water results in poor root development and plant stress.
2. Sensitivity to clogging.
3. Moisture distribution problem.
4. Salinity hazards.

(b) A *dry toilet* is a *toilet* that operates without flush water, unlike a flush *toilet*. Dry toilets are also called as Composting toilets. They use the natural processes of decomposition and evaporation to recycle human waste. Waste entering the toilets is over 90% water, which is evaporated and carried back to the atmosphere through the vent system. The small amount of remaining solid material is converted to useful fertilizing soil by natural decomposition. A composting toilet must perform three completely separate processes :

1. Compost the waste and toilet paper quickly and without odour.
2. Ensure that the finished compost is safe and easy to handle.
3. Evaporate the liquid.

The correct balance between oxygen, moisture, heat and organic material is needed to ensure a rich environment for the aerobic bacteria that transform the waste into fertilizing soil. This ensures odour-free operation and complete decomposition of waste. When human waste is properly composted, the end product does not contain any pathogens or viruses (these are destroyed by bacterial breakdown). This nutrient-rich fertilizer can then be used on plants or around the base of trees, as part of the natural cycling of nutrients, reducing the need for commercial fertilizers and preserving local water quality.

Answer 7.

(a) Food security, as defined by the United Nations Committee on World Food Security, means that all people, at all times, have physical, social, and economic access to sufficient, safe, and nutritious food that meets their food preferences and dietary needs for an active and healthy life.

Problems related to food security are :

1. The impact on children, mothers and elderly are very evident as seen in malnutrition and hunger related deaths.
2. Reduction in the life expectancy and quality of life.
3. Children succumb to hunger within short period as they cannot stand long period of starvation and they die even before the arrival of emergency assistance.
4. Adults living in poverty are at greater risk for a number of health issues, such as diabetes, heart disease and stroke.

Solutions to the problems related to global food security are :

1. Support domestic food production to stabilise and guarantee fair prices to farmers and consumers by re-establishing floor prices and publicly owned national grain reserves.
2. The major solution is to control food wastage. Foremost, precedence should be centred on balancing food production with demand to reduce the problem of food wastage.
3. Implementation of best farming techniques to increase the food yield.
4. Provide subsidies or discount on food for people from low income.
5. Bio-technologically fortify the food products with enriching nutrients. *(Any two)*

(b) Salient features of food security law 2013 are :

1. **Coverage and entitlement under Targeted Public Distribution System (TPDS) :** Up to 75% of the rural population and 50% of the urban population will be covered under TPDS, with uniform entitlement of 5 kg per person per month.
2. **State-wise coverage :** Corresponding to the all India coverage of 75% and 50% in the rural and urban areas, State-wise coverage will be determined by the Central Government.
3. **Subsidised prices under TPDS and their revision :** Food grains under TPDS will be made available at subsidised prices of ₹ 3/2/1 per kg for rice, wheat and coarse grains for a period of three years from the date of commencement of the Act. Thereafter prices will be suitably linked to Minimum Support Price (MSP).
4. **Identification of Households :** Within the coverage under TPDS determined for each State, the work of identification of eligible households is to be done by States.
5. **Nutritional Support to women and children :** Pregnant women and lactating mothers and children in the age group of 6 months to 14 years will be entitled to meals as per prescribed nutritional norms under Integrated Child Development Services (ICDS) and Mid-Day Meal (MDM) schemes. Higher nutritional norms have been prescribed for malnourished children up to 6 years of age.
6. **Maternity Benefit :** Pregnant women and lactating mothers will also be entitled to receive maternity benefit of not less than ₹ 6,000.
7. **Women Empowerment :** Eldest woman of the household of age 18 years or above to be the head of the household for the purpose of issuing of ration cards.
8. **Grievance Redressal Mechanism :** Grievance redressal mechanism at the District and State levels. States will have the flexibility to use the existing machinery or set up separate mechanism.
9. **Food Security Allowance :** Provision for food security allowance to entitled beneficiaries in case of non-supply of entitled food grains or meals.
10. **Penalty :** Provision for penalty on public servant or authority, to be imposed by the State Food Commission, in case of failure to comply with the relief recommended by the District Grievance Redressal Officer. *(Any eight)*

SECTION–C

Answer 8.

(a) In environmental accounting, defensive expenditures are expenditures that seek to minimise potential damage to oneself. An expenditure can be considered 'defensive' when it is sustained to avoid or to reduce the effect of a negative externality originating from production or consumption activities or to contain or compensate for the damage derived from this externality. Defensive expenditure study can be done to measure the following :

1. Expenditures sustained to avoid the production of externalities (for example, the adoption of "clean" techniques and processes).

2. Expenditures sustained to restore the "environmental functions" damaged by the economic activities (for example, the improvement and restoration of damaged areas).

3. Expenditures aimed to avoid damages (for example, by moving people to less polluted places) or to protect them (for example, by the installation of double glazing to protect against noise pollution). This refers to damages arising when the externality is neither avoided nor opposed.

4. Expenditures sustained to compensate the people for the harm caused to them (the example often cited in the literature is that of medical expenses for an illness related to the degradation of the environment).

Importance of defensive expenditure : The information contained in the account of the defensive expenditures is extremely relevant, both for economic as well as environmental analysis. This information will enable researchers to quantify a very significant aggregate, which can be used for analysing the interrelationship between the economy and the environment.

(b) Ozone (O_3) is a molecule formed by three atoms of oxygen. While O_2, which we normally refer to as oxygen, is essential for all aerobic forms of life. Ozone at the higher levels of the atmosphere is a product of UV radiation acting on oxygen (O_2) molecule. The higher energy UV radiations split apart some molecular oxygen (O_2) into free oxygen (O) atoms. It is caused by the release of chlorofluorocarbons (CFCs) and other Ozone-Depleting Substances (ODS), which were used widely as refrigerants, insulating foams, and solvents. When CFCs reach the stratosphere, the ultraviolet radiation from the sun causes them to break apart and release chlorine atoms which react with ozone, starting chemical cycles of ozone destruction that deplete the ozone layer.

We can protect ozone layer by :

1. Avoid the consumption of gases dangerous to the ozone layer, due to their content or manufacturing process. Some of the most dangerous gases are CFCs (chlorofluorocarbons), halogenated hydrocarbon, methyl bromide and nitrous oxide.

2. We should maintain air conditioners, as their malfunctions cause CFCs to escape into the atmosphere.

3. CFCs should be replaced with hydro-chlorofluorocarbons (HCFCs), which deplete stratospheric ozone, but to a much lesser extent than CFCs. *(Any two)*

Answer 9.

(a) Free trade is whereby trade transaction is allowed to continue freely without any hindrance from the government in form of tariffs and restrictive quotas. Governments often impose restriction on imports this, is usually done with the aim of blocking foreign companies from taking control of local markets as well as companies. On the other hand protectionism is whereby government restricts trade through various ways therefore making it difficult for foreign countries to trade freely. Protectionism basically protects local businesses as well as workers by putting restrictions on international or foreign businesses.

Advantages of free trade :

1. Local companies also receive access to the latest technologies from their multinational partners. As local economies grow, so do job opportunities. Multinational companies provide job training to local employees.

2. Many governments subsidise local industry segments. After the trade agreement removes subsidies, those funds can be put to better use.

Disadvantages of free trade :

1. As development moves into isolated areas, indigenous cultures can be destroyed. Local peoples are uprooted. Many suffer disease and death when their resources are polluted.

2. Many developing countries do not have laws to protect patents, inventions, and new processes. The laws they do have are not always strictly enforced.

Advantages of protectionism :

1. Protectionism keeps the domestic economy rolling. Since there is a decrease in imports, domestic firms have less competition, and so are able to continue. The domestic economy will also be strengthened because unemployment will be down due to the domestic firms and they will be able to produce and sell more goods with a lot less difficulty, giving firms less reason to decrease its costs by decreasing its workforce.

2. Protectionism makes domestic firms less competitive in the export market, as import barriers raise domestic prices through higher costs for mediocare inputs this means that export products also become more expensive and decrease in market share against the international competition.

Disadvantages of protectionism :

1. Consumers pay more with protectionism. Without a system of competitive pricing, domestic companies are free to raise their prices without raising the quality of their goods. When a business has no competition then the consumer is left without options.

2. Consumers will have limited choices of goods as a result of limitations on foreign goods.

(b) Externalities are defined as the positive or negative consequences of economic activities on unrelated third parties. Economic activities that have positive effects on unrelated third parties are considered positive externalities. Examples of positive externalities are :

1. The CSR initiatives by a company like offering free treatment to people or free education to children on charity.

2. Tree plantation by a firm, society benefits from the enhanced environment, but no payment is made.

3. Maintenance of public parks and green spaces by a firm for free. *(Any two)*

Economic activities that have negative effects on unrelated third parties are considered negative externalities. Examples of negative externalities are :

1. The working of a factory may cause environmental pollution (air, water), but no penalty is charged.

2. The use of fossil fuels like coal, petroleum etc in factories leads to depletion of these non-renewable resources, which has major ecological impact.

3. The spread of MNCs in developing countries may cause the non-competitive local industries to shut down - spreading further unemployment. But no penalties are imposed on such MNCs. *(Any two)*

Answer 10.

(a) Cost benefit analysis is defined as a process analysis by which we decide whether an idea related to our business is going in right way or not. It is a calculation of all costs involved in a process or in a business deal and total possible profit obtained from deal or business proposal.

Advantages of cost benefit analysis are :

1. By the help of cost benefit analysis we can simplify our complex business decisions. By the help of this analysis we can measure the estimate of profit for a project.

2. Cost benefit analysis provides us the objectives related to our business or our projects.

3. Cost benefit analysis also helps us to set a goal related to our projects because all the needs and benefits related to it can be easily studied in starting.

Disadvantages of cost benefit analysis are :

1. The method based on collecting data is generally biased due to any accidental condition.

2. Previously calculated and analysed amount for subjectivity involvement may be biased because of change in customer and employee satisfaction.

3. All the calculation based on present value is same but future value may be changed. *(Any two)*

(b) 1. Free trade is meant to eliminate unfair barrier to global commerce and raise the economy in developing as well as developed nations alike. India produced 10.8 million tonnes of fish in 2016, the third highest in the world, after China and Indonesia.

2. India's share of global fish production was around 6%, according to the FAO (The Food and Agricultural Organisation). In aquaculture, India is second only to China, with a 7% share in global production. Implementation of free trade on aquaculture permits more harvesting and exporting of fishes to other countries. This could have negative impacts as well. For example– Approximately 200 million people live along the Bay of Bengal's coasts and of these a major proportion are partially or wholly dependent on its fisheries. But the fisheries of the Bay of Bengal have been under pressure for decades and are now severely depleted.

3. Many once-abundant species have all but disappeared. Particularly badly affected are the species at the top of the food chain. The bay was once feared by sailors for its man-eating sharks; they are now rare in these waters. Other apex predators like grouper, croaker and rays have also been badly hit.

4. Catches now consist mainly of species like sardines, which are at the bottom of the marine food web. In the 1960s, western aid agencies encouraged the growth of trawling in India, so that fishermen could profit from the demand for prawns in foreign markets. This led to a "pink gold rush", in which prawns were trawled with fine mesh nets that were dragged along the sea floor.

5. But along with hauls of "pink gold" these nets also scooped up whole seafloor ecosystems as well as vulnerable species like turtles, dolphins, sea snakes, rays and sharks. Furthermore, environmental consequences like emission of green houses gases, disturbed bio-geochemical cycles, pollution, species extinction are likely to occur.

6. Absence of stringent rules and regulation of export-import also encourages illegal and unaccounted supply of goods to the foreign countries. For example– In 2015 Sri Lankan authorities claimed to have spotted approximately 40,000 Indian trawlers in Sri Lanka's territorial waters. Seventy trawlers were seized and many fishermen were arrested.

••

ENVIRONMENTAL SCIENCE

2018

PART–I (20 MARKS)

*Answer **all** questions.*

Question 1.

(i) *Mention two disadvantages of macro irrigation.* **[2]**

(ii) *Who coined the term deep ecology and when ?* **[2]**

(iii) *Differentiate between insitu and exsitu methods of wildlife conservation.* **[2]**

(iv) *With reference to urban environment, mention any two problems and their solutions.* **[2]**

(v) *Differentiate between immigration and emigration.* **[2]**

(vi) *Mention any two advantages of Physical Accounting.* **[2]**

(vii) *Write the full form of each of the following :* **[2]**

(a) *CNG* (b) *GATT*

(viii) *Classify natural resources based on their origin and distribution.* **[2]**

(ix) *Explain the terms :* **[2]**

(a) *Industrial design* (b) *Patents*

(x) *Mention any four factors regulating population size of an area.* **[2]**

PART–II (50 Marks)

*Answer any **five** questions in all, choosing at least **one** question from each of the Sections A, B and C.*

SECTION-A

Question 2.

(a) *Write a short note on Gary Snyder.* **[6]**

(b) *What is meant by a survivorship curve ? Explain the three shapes of survivorship curves.* **[4]**

Question 3.

(a) *What is Joint Forestry Management ? State its objectives.* **[6]**

(b) *Write a short note on Earth First.* **[4]**

Question 4.

(a) *What is NAMP ? Discuss any five objectives of this programme.* **[6]**

(b) *Discuss the basic Gandhian approach for rural development. How would you contrast Gandhian approach with the present growth paradigm ?* **[4]**

SECTION-B

Question 5.

(a) *Discuss the adoption of sustainable management of traffic in Bogota.* **[6]**

(b) *Discuss the successes and failures of Project Tiger. Support your answer giving reasons.* **[4]**

Question 6.

(a) *Define population. Explain any three factors that affect the birth rate and any two factors that affect the death rate in human population.* **[6]**

(b) *"Deforestation is a major problem in Amazonian rain forests." Discuss any four solutions to control this problem.* **[4]**

Question 7.

(a) *What is WTO ? Discuss any two principles and any three functions of WTO.* **[6]**

(b) *Explain the different zones of the biosphere reserves.* **[4]**

SECTION-C

Question 8.

(a) *Explain any six ecological impacts of Green Revolution.* **[6]**

(b) *What is meant by IPM ? Discuss any three ways by which IPM is implemented.* **[4]**

Question 9.

(a) *Describe the physical indicators of soil quality.* **[6]**

(b) *Write a short note on Bamako Convention.* **[4]**

Question 10.

(a) *What is meant by natural capital ? Why do we need to preserve it ? Discuss four ways to regenerate natural capital.* **[6]**

(b) *Explain any four steps to reduce vehicular pollution.* **[4]**

ANSWERS

PART–I

Answer 1.

(i) Two disadvantages of macro irrigation are :

1. Macro irrigation may lead to water scarcity as this system involves uneven distribution of water. It makes the soil unproductive as harmful underground salts come to the surface level due to water logging. Land can also become marshy there.

2. Macro irrigation can contribute to soil erosion. Many diseases are caused due to spread of mosquitoes, worms and insects on account of stationary water in canals.

(ii) The term "deep ecology" was coined by the Norwegian philosopher **Arne Naess** in 1973.

(iii)

In-situ method of conservation	Ex-situ method of conservation
This is the conservation, maintenance and recovery of species in their natural habitat. For example, wild life sanctuaries, national parks.	This is the conservation, maintenance and recovery of species in a manmade habitat. For example, zoo, aquarium.
In-situ conservation is more dynamic.	Ex-situ conservation is more static.

(iv) **Problems : 1.** Due to urbanization, there is environmental degradation specially in the quality of water, air and noise.

2. Urbanization is a major concern for management researchers because it results in decline in the quality of living for urban inhabitants.

Remedies : 1. To develop social mechanisms which will assist to reduce inequality and make sure the basics like— health, sanitation, education to reach those who have been underprivileged of the same.

2. To fulfill all the demands for jobs, shelter, water, roads, transport and other urban infrastructure is overwhelming. So, as to raise the standard of living.

(v)

Immigration	Emigration
Immigration is the movement of people into a destination country of which they are not natives or where they do not possess citizenship in order to settle or reside there, or to take-up employment as a migrant worker or temporarily as a foreign worker.	Emigration is the act of leaving one's resident country with the intent to settle elsewhere.

(vi) **The advantages of physical accounting are as follows :**

1. It helps to measure the impact of various processes and operation on the air, water, soil, our health, safety and society.

2. It provides a database for corrective action and future plans.

(vii) **1. CNG** – Compressed Natural Gas.

2. GATT – General Agreement on Tariffs and Trade.

(viii) On the basis of origin, the natural resources can be classified as :

1. Biotic resources : Resources obtained from living and organic matter like fossil fuels, food, wood, leather, etc.

2. Abiotic resources : Resources which are obtained from non-living materials such as ores, metal, sand, soil, oxygen, air, land, water etc.

On the basis of distribution, the natural resources are classified as :

1. Ubiquitous resource : The resources which are available everywhere on the earth. *e.g.*, air and water.

2. Localised resource : The resources which are available at selected locations on the earth *e.g.*, mines of various minerals, indigenous plants and animals, types of soil, etc.

(ix) Industrial design is the work of creating and developing plans and process that optimises the function, value and appearance of the product produced by that particular industry. The designing is made by considering the mutual benefit of both the consumer and producer.

A patent is a set of exclusive rights granted by a sovereign state or inter-governmental organization to an inventor or assignee for a limited period of time in exchange for detailed public disclosure of an invention.

(x) The factors that regulate the population size of an area are : food, temperature, predator, oxygen availability, light availability, presence of competitors and parasites.

PART–II

SECTION-A

Answer 2.

(a) Gary Snyder is a poet, environmentalist, Zen Buddhist and educator. He was born in San Francisco, California, on May 8, 1930. In 1947, Snyder received a scholarship to Reed College. There, he studied Asian culture; in addition, he delved into the relationship between indigenous people and their land. He graduated in 1951, and then spent a year studying anthropological linguistics at Indiana University. In 1955, Snyder read his poem "Berry Feast" at San Francisco's Six Gallery. Snyder's writing focuses on environmental concerns and Zen Buddhism. He is an environmental activist who is known for his simple, clear style, as well as his first person descriptions of his experiences in the natural world. In 1975, his collection Turtle Island was awarded the Pulitzer Prize for Poetry. He also published the poetry collections Riprap (1959) and Myths and Texts (1960), which documented his personal experiences.

(b) A survivorship curve is a graph showing the number or proportion of individuals surviving to each age for a given species or group.

1. **Type I or convex curves** : These are characterized by high age-specific survival probability in early and middle life which further continues as a rapid decline in survival in later life.

2. **Type II or diagonal curves** : These are a transition between Types I and III, where mortality rate or survival probability is regardless of age.

3. **Type III or concave curves** : These curves have the greatest mortality and low rates of death. This type of curve is characteristic of species that produce a large number of offspring.

Answer 3.

(a) Joint Forestry Management is a practice of developing partnerships between the forest user groups and the Forest Department for forest protection and growth. The objectives of Joint Forestry Management are :

1. To boost local economy by creating new posts and jobs.

2. To fosters a sense of responsibility, participation and unity.

3. To improve awareness and knowledge about the uses and sustainable management of forest resources, thus ensuring a long-term benefit to the local people.

4. To enhance aesthetic values of the area.

5. To combat threats to forests like forest fire, over-exploitation of resources, etc.

(b) Earth First is an environmental group focused on the protection of wildlife. Earth First was formed in 1980 in South-western United States. *"No compromise in defence of Mother Earth"* is the slogan of this organisation. Earth first uses direct action to oppose, stop and eventually reverse the forces that are responsible for the destruction of the Earth and its inhabitants. Many NGOs, private sectors, citizens or residents are a part of these activities. The famous non-violent campaign "Redwood Summer" was launched by the Earth's first in response to Louisiana Pacific Lumber Mill's decision to double the rate of logging. By 1987, Earth First became primarily associated with direct action to prevent logging, building of dams, and other forms of development which may cause destruction of wildlife habitats or the despoliation of wild places. This attracted the support and participation from many countries like United States, United Kingdom, Canada, Australia, Netherlands, Belgium, Philippines, Czech Republic, India, Mexico, France, Germany, New Zealand, Poland, Nigeria, Slovakia, Ireland, Italy and Spain.

Answer 4.

(a) National Air Quality Monitoring Programme (NAMP) is a nation-wide programme of ambient air quality monitoring that is launched by central state pollution control board. Under NAMP air pollutants like Sulphur dioxide, Oxides of Nitrogen, Respirable Suspended Particulate Matter (SPM) and Fine Particulate Matter are identified for regular monitoring at all the locations. The network of NAMP consists of six hundred and eighty-three operating stations covering three hundred cities or towns in twenty-six states and four Union Territories of the country. The objectives of NAMP are as follows :

1. To determine status and trends of ambient air quality.

2. To ascertain whether the prescribed ambient air quality standards are violated.

3. To identify non-attainment cities.

4. To obtain the knowledge and understanding necessary for developing preventive and corrective measures.

5. To understand the natural cleansing process undergoing in the environment through dispersion, dry deposition, precipitation, etc.

(b) The village is the basic unit of the Gandhian ideal social order. The Gandhian approach attaches supreme importance to moral values and gives primacy to moral values over material conditions. Following are the requirements of rural development :

1. **Decentralization** : Gandhi firmly believes that village republics can be built only through decentralisation of social and political power.

2. **Self-sufficient** : The village should be self-sufficient as far as its basic needs like food, clothing and other commodities are concerned. The village also has to maintain trade relationship with other villages to import and export commodities.

3. **Trusteeship** : Gandhiji emphasized the principle of trusteeship in social and economic affairs. He firmly believed that all social property should be held in trust.

4. **Employment opportunities** : Gandhiji emphasized on generating maximum employment opportunities in villages to uplift the standard of living of villagers.

5. **Industrialization** : Gandhiji maintained that industrialization replaces manpower and

adds to unemployment. In a country like India, industrialisation also forces labourers to migrate to urban area.

The present growth paradigm contrasts the Gandhian ideas. Gandhian approach focussed a lot on holistic development of villages, but as a result of technology and economic progress, urbanisation is being induced. All villages are not empowered to be self-sufficient as far as its basic commodities are concerned. Many villagers are migrating to urban areas in search of jobs, better facilities or education. Such a migration is resulting in loss of indigenous arts, occupation and agricultural activities.

SECTION-B

Answer 5.

(a) In Bogota more than 300 km of bicycle lanes, stretching from the slum areas and suburbs into the capital's centre, have been built. The bicycle lanes are an ongoing project under concurrent development. Since the construction of the lanes, bicycle use has increased by 5 times in the city. It is estimated that there are between 300,000 and 400,000 trips made daily in Bogota by bicycle. A large portion of these rides are in the southern, poorer areas.

The public transport system of Bogota has also been improved. There is no metro in the city, but instead TransMilenio, an affordable and rapid bus transit system, has been implemented. Trans Milenio includes numerous elevated stations in the centre of a main avenue. Users pay at the station by smart card and await the arrival of the bus, whose doors open at the same time as the sliding glass doors of the station. The buses have dedicated lanes and on the station extra lanes allows express buses to pass through without stopping.

In early 2006, commuters completed 10,50,000 trips a day because of restriction on private cars in the inner city during rush hour, the TransMilenio buses are running three times as fast as a typical New York bus, which equals 28 km an hour. In Bogota vehicle traffic is reduced by 40% through a measure that restricts vehicles, according to the last number in their license plate, from travelling during peak hours in the entire urban area. In addition car free Sundays (Ciclovia) and the laying down of thousands of parking spots have made the roads more pedestrian-friendly. On car-free Sundays, the city is now using public streets as a large open park and there are plenty free exercises for families to attend. Bogota has become a healthier and safer city with greater social integration and cheap, sustainable transport.

(b) Project Tiger is a tiger conservation programme launched on 1 April, 1973 by the Government of India. The main objective of this project was to ensure protection of Bengal tigers in their natural habitats. The project aimed to combat various factors that led to decline in the population of tigers like poaching, loss of habitat, increasing human population and depletion of prey. The tiger reserves covering a total area of 37, 761 sq km was established as a habitat and environment for the tigers. The tiger population was seen as the heritage of our country. The project authorities took measures to mitigate human intervention in the areas of tiger reserves. The project has core-buffer strategy with buffer area having no human activity. The livestock grazing in these tiger reserves has been curbed thereby reducing the competition for food. This resulted into increased number of prey species and this increased food opportunities have resulted in tiger population increase. However, there were certain shortcomings of this project. In 2006, the Indian government passed a Forest Rights Act which recognizes the rights of some forest dwelling communities in forest areas. This gives an opportunity for tiger poaching. The project tiger committee merely formed an advisory group which lacked staff at the grassroot level. The implementation of the rules, regulations and plans were not up to the mark. The local communities and the forest dwellers were not involved in this project.

Answer 6.

(a) Population is defined as the group of organisms belonging to a particular species residing in a geographically localized area and having the ability to interbreed. The population includes all individuals that live in that certain area.

The factors that affect the birth rate are :

1. **Number of people in the reproductive age and marriage :** Population of a country widely depends on this factor. If the number of people of age group 18 – 35 years are more, then this implies that this population is in the reproductive age. Many cultures and religions encourage this young age group to get married and start a family.

2. **Availability and affordability of birth control tools :** Many people are ignorant about the ill effects of population explosion. In some cases, they might be unaware of the birth control practices and tools or they are not able to afford these tools. In such cases, they give birth to offsprings and increase the population.

3. **Religious beliefs and cultural norms** : In some countries, religious beliefs and cultural factors favour large families and strongly oppose abortion and some forms of birth control.

The factors affecting death rate are :

1. **Availability of medical facilities** : With the technological advancement in the field of medicine and health, many lives taking diseases are curable. The life expectancy of human beings can be increased. Vaccines, immunization and other medical treatments have saved thousands of life.

2. **Occupation:** Some occupations demand high intensity work, which creates physical and mental stress. Such stress wares out the body and invites various diseases. Continuous exposure to medicines and stress full life may reduce the life expectancy.

(b) **The measures to combat deforestation are :**

1. **Reforestation** : Reforestation is the practice of replenishing existing forests that have been deplete through deforestation. Reforestation is a very important procedure in order to save our planet.

2. **Energy plantation** : Energy plantation refers to growing certain species of trees and shrubs which are harvestable in a comparably shorter time and are specifically meant for fuel. The fuel wood may be used either directly in wood burning stoves and boilers or processed into methanol, ethanol and producer gas.

3. **Forest harvesting of non-timber forest products** : Non-Timber Forest Products (NTFPs) are preferred over timber products because, it does not involve logging of trees. (NTFPs) are biological resources other than timber that can be harvested from forest for subsistence for trade. Therefore, sustained use of NTFPs is the key to sustainable forest management.

4. **Exploring alternative resources for livelihood** : We depend on forest resources especially trees for various day to day needs. Renewable sources of energy like solar, hydropower, geothermal, wind energy, etc., are clean sources of energy. Using renewable sources of energy not only reduces the need to cut trees for food, but also contributes in controlling pollution and other ill-effects on environment.

Answer 7.

(a) World Trade Organization is the expansion of WTO. The World Trade Organization (WTO) deals with the global rules of trade between nations. The World Trade Organization (WTO) is the only global international organization dealing with the rules of trade between nations. At its heart are the WTO agreements, negotiated and signed by the bulk of the world's trading nations and ratified in their parliaments. The goal is to help producers of goods and services, exporters and importers conduct their business. It is an organization for trade opening. It is a forum for governments to negotiate trade agreements. It is a place for them to settle trade disputes.

Principles of WTO

1. **Non-discrimination** : Non-discrimination is a fundamental principle of the multilateral trading system and is recognized in the Preamble of the WTO Agreement as a key instrument to achieve the objectives of the WTO. Under the WTO agreements, countries cannot normally discriminate between their trading partners.

Fundamentally, WTO is built on a foundation of non-discrimination which has two major components : the Most Favoured Nation rule and the National Treatment Policy.

2. **Transparency** : WTO members are required to publish their trade regulations and to make sure any decisions affecting trade are notified to other WTO members and to the WTO Secretariat itself. The WTO also has a Trade Policy Review Mechanism (TPRM) which periodically reviews the laws and regulations of a WTO member country.

Functions of WTO

1. To provide the institutional framework for the administration of the substantive code which encompasses a spectrum of norms governing the conduct of member countries in the arena of global trade.

2. To provide an integrated structure of the administration, thus, to facilitate the implementation, administration and fulfillment of the objectives of the WTO Agreement and other Multilateral Trade Agreements.

3. To ensure the optimum use of world resources.

(b) There are three zones of biosphere namely Core zone, Buffer zone and Transition zone.

1. **Core Zone** : This is the zone where the human activity is not allowed. This area is legally protected and undisturbed ecosystem.

2. **Buffer zone** : The second zone is the buffer zone. This zone supports limited human activities like research, education, etc .

3. **Transition zone** : Transition zone is the outermost area of biosphere reserve. Human

activities like settlements, cropping, recreation and forestry are carried out without disturbing the environment. Reserve management and local people help to maintain the harmony of the biosphere.

SECTION-C

Answer 8.

(a) Ecological impacts of green revolution are :

1. **Salinization and Waterlogging :** The High Yielding Varieties of wheat and rice require several watering. Many a times the farmers over-irrigate their crops.

2. **Fallowing :** Extension in the network of canals and drilling of tube-wells and pumping sets have enabled the farmers to harvest two cereals (rice and wheat) and one fodder or cash crop (millet or cotton) in the same field in one agricultural year.

3. **Noise pollution :** After the adoption of HYV, the agriculture has become so intensive that farmers are generally harvesting three crops in a year.

4. Soil erosion and leaching of top cover of soil is a universal phenomenon. Although it may be observed in all types of geo-climatic conditions, its intensity is more serious in the arid, semi-arid, mountainous and undulating regions.

5. HYV as discussed above, demand a lot of moisture through irrigation at regular intervals. The energy requirements in the form of chemical fertilizers are also high.

6. The valuable resource of underground water has, thus, become increasingly vulnerable after the adoption of the HYV of wheat and rice.

(b) IPM includes all the pest control techniques by integrating appropriate measures that discourage the development of pest populations and keep pesticides and other interventions to levels that are economically justified and reduce or minimize risks to human health and the environment. A careful evaluation of each pest infestation is the first step of IPM. The life cycle of the pest, nature of the damage caused by the pest, predators, effects of weather and other such factors are considered before a control plan is implemented. For example :

1. Weeds like Queen Anne's lace (Daucus carota) or can dwell **beneficial insects** such as lacewing or lady beetle larvae that help keep aphid and other pest populations at tolerable levels.

2. Clover growing in a lawn may be viewed as an unwanted weed, but as a legume it is synthesizing nitrogen for the soil and the flowers are providing nectar to honey bees and other **pollinators**. Therefore, tolerance for some weeds may be part of an IPM plan.

3. A woodpecker makes holes in the trunk of a peach tree, but at the same time it feeds on the insect **larvae** that may cause more damage to the tree. IPM is an integral component of crop management which looks at all aspects necessary to grow a crop. The control options are usually a combination of physical, biological and chemical aspects to manage crops pests and diseases.

Answer 9.

(a) Soil quality indicators are used to evaluate how well the soil functions. Physical indicators provide information about soil hydrologic characteristics, such as water entry and retention, that influences availability to plants. Following are the physical indicators of soil quality :

1. Aggregate stability refers to the ability of soil aggregates to resist disintegration when disruptive forces associated with tillage and water or wind erosion are applied.

2. Available water capacity is the maximum amount of available water a soil can provide. It is an indicator of a soil's ability to retain water and make it sufficiently available for plant use.

3. Bulk density reflects the soil's ability to function for structural support, water and solute movement and soil aeration.

4. Infiltration is an indicator of the soil's ability to allow water movement into and through the soil profile. Soil temporarily stores water, making it available for root uptake, plant growth and habitat for soil organisms.

5. Slaking indicates the stability of soil aggregates, resistance to erosion and suggests how well soil can maintain its structure to provide water and air for plants and soil biota when it is rapidly wetted.

6. Soil pores exist between and within aggregates and are occupied by water and air. Macropores are large soil pores, usually between aggregates. Macropores drain freely by gravity and allow easy movement of water and air. They provide habitat for soil organisms and plant roots can grow into them.

(b) The Bamako Convention is a treaty of African nations prohibiting the import of any hazardous waste. The convention was negotiated by twelve nations of the Organisation of African Unity at Bamako, Mali in January, 1991, which came into force in 1998. Radioactive waste was also covered under hazardous waste category. The treaty intended to protect human health and the environment from dangers posed by hazardous wastes by reducing their generation to a minimum in terms of quantity or hazard potential. The motivation for the Bamako convention arose from the following factors :

1. The failure of the Basel Convention to prohibit trade of hazardous waste to less developed countries.

2. The realization that many developed nations were exporting toxic wastes to Africa.

3. To minimize and control transboundary movements of hazardous wastes within the African continent.

4. To prohibit all ocean and inland water dumping or incineration of hazardous wastes.

5. To ensure that disposal of wastes is conducted in an environmentally sound manner.

Answer 10.

(a) Natural capital can be defined as the world's stocks of natural assets which include geology, soil, air, water and all living things. It is from this natural capital that humans derive a wide range of ecosystem services, which make human life possible. The ecosystem services include the food, the water and the plant materials we use for fuel, building materials and medicines. The other less visible ecosystem services includes the climate regulation, natural flood defences, the billions of tonnes of fuel and natural gas stored by peatlands, the pollination of crops by insects, etc. The cultural ecosystem services such as the inspiration we take from wildlife and the natural environment and aesthetic value also adds to the reasons as to why we should preserve the natural capital.

Following are the ways to regenerate natural capital :

1. **Sustainable development :** Sustainable development is development that meets the needs of the present without compromising the ability of future generations to meet their own needs. The agenda of sustainable development is holistic and universal. The concept of sustainability links the concern for the carrying capacity of natural systems with the social, political and economic challenges faced by humanity. For sustainable development to be achieved, it is crucial to harmonise three core elements : economic growth, social inclusion and environmental protection.

2. **Controlling pollution :** Appropriate measures should be taken to control the pollution of air, soil and water. Pollution of air, water and soil degrades and hampers the growth of flora and fauna on the earth and thus the functioning of ecosystem.

3. **Human population control :** Increase in population implies an increase in demand for resources like food, water, metals, fuel and other materials. These materials are directly or indirectly extracted from the environment. Such a speedy and huge scale of resource extraction impacts the ecosystem by fostering imbalance in nature and pollution.

(b) Following measures can be implemented to curb air pollution from vehicles :

1. **Catalytic converters :** It is an exhaust emission control device that converts toxic gases and pollutants to less toxic pollutants by catalysing a redox reaction. Installation of catalytic converters can reduce the release of pollutants into the air.

2. Alternative sources of energy like electricity, solar, LPG etc., should be used as fuels to run the vehicles. These sources of energy are less polluting.

3. Actions have to be taken to control the rate of traffic growth. Public awareness and participation in traffic management is a must. Every citizen is equally responsible and accountable for traffic management along with the government.

4. Using public transports or car-pooling whenever possible can considerably help in reducing pollution. Engines of vehicles can be switched off while waiting at the traffic signal. Vehicles should be properly maintained.